AUTO TECHNOLOGY
Theory & Service

Second Edition

Saverio G. Bono
William J. deKryger

 Delmar Publishers Inc.®

NOTICE TO THE READER

Publisher and author do not warrant or guarantee any of the products described herein or perform any independent analysis in connection with any of the product information contained herein. Publisher and author do not assume, and expressly disclaim, any obligation to obtain and include information other than that provided to them by the manufacturer.

The reader is expressly warned to consider and adopt all safety precautions that might be indicated by the activities described herein and to avoid all potential hazards. By following the instructions contained herein, the reader willingly assumes all risks in connection with such instructions.

The publisher and author make no representations or warranties of any kind, including but not limited to, the warranties of fitness for particular purpose or merchantability, nor are any such representations implied with respect to the material set forth herein, and the publisher and author take no responsibility with respect to such material. The publisher and author shall not be liable for any special, consequential or exemplary damages resulting, in whole or in part, from the readers' use of, or reliance upon, this material.

Cover photos: Engine analyzer courtesy Bear Automotive Service Equipment Company
Auto wire-frame model courtesy Chrysler Motors

Delmar Staff
Administrative Editor: Joan Gill
Project Editor: Marlene McHugh Pratt
Production Coordinator: Sandra Woods
Design Coordinator: Susan C. Mathews

For information, address Delmar Publishers Inc.,
2 Computer Drive West, Box 15-015
Albany, New York 12212

COPYRIGHT © 1990
BY DELMAR PUBLISHERS INC.

All rights reserved. Certain portions of this work copyright 1986. No part of this work covered by the copyright hereon may be reproduced or used in any form or by any means--graphic, electronic, or mechanical, including photocopying, recording, taping, or information storage and retrieval systems--without written permission of the publisher.

Printed in the United States of America
Published simultaneously in Canada
By Nelson Canada
A Division of The Thomson Corporation

10 9 8 7 6 5 4 3 2 1

Library of Congress Cataloging-in-Publication Data
Bono, Saverio G.
    Automotive technology: theory and service. -- 2nd ed. / Saverio G. Bono, William J. deKryger.
      p. cm.
    Rev. ed. of: Auto mechanics / William J. deKryger, Robert T. Kovacik, Saverio G. Bono. c1986.
    ISBN 0-8273-3811-2
    1. Automobiles. 2. Automobiles--Maintenance and repair.
I. DeKryger, William J. II. DeKryger, William J. Auto mechanics.
III. Title.
TL205.B63 1990
629.28'722--dc20   89-39637
                                                                                 CIP

Table of Contents

Preface   xi

PART I: Career Opportunities And the Automobile

1. Careers and Opportunities   2

1.1 The Automotive Industry   2
1.2 Increasing Needs for Service   2
1.3 Automotive Service Businesses   3
1.4 Automotive Technician Certification   4
1.5 Training for a Career in Automotive Service   7
1.6 Going to Work   7

2. The Automobile   10

2.1 Basic Parts of the Automobile   10
2.2 Engine   10
2.3 Drivetrain   14
2.4 Chassis   18
2.5 Body   22

3. Using Service Manuals and Shop Forms   25

3.1 Service Manuals   25
3.2 Using Manufacturer's Service Manuals   26
3.3 Forms and Record-Keeping   27

4. Safety Around the Automobile   33

4.1 Attitude   33
4.2 Good Safety Habits   33
4.3 Safety While Working   33

5. Using Fasteners   43

5.1 Measurement Systems   43
5.2 Threaded Fasteners   44
5.3 Washers   46
5.4 Threaded Diameters and Designations   46
5.5 Determining Thread Pitch   47
5.6 Grade Markings   47
5.7 Torque Requirements for Threaded Fasteners   47
5.8 Chemical Thread Compounds   48
5.9 Nonthreaded Fasteners   48

6. Using Hand Tools   52

6.1 Tool Measurements   52
6.2 Wrenches   52
6.3 Screwdrivers   54
6.4 Pliers   55
6.5 Hammers and Mallets   56
6.6 Chisels and Punches   56
6.7 Files   56
6.8 Taps   57
6.9 Dies   57
6.10 Reamer   57
6.11 Screw Extractor   57
6.12 Pry Bars   57

7. Using Power Tools   60

7.1 Electrical Tools   60
7.2 Pneumatic Tools   61
7.3 Hydraulic Tools   62
7.4 Cleaning Equipment   63

8. Using Measuring Tools   66

8.1 Mechanical Measuring Tools   66
8.2 Fluid Measuring Tools   69
8.3 Electrical Testers and Diagnostic Equipment   71

9. Troubleshooting   75

9.1 Automotive Diagnostics   75
9.2 Troubleshooting   75
9.3 Troubleshooting Charts   75
9.4 The Test Drive   75
9.5 Pinpointing a Problem   77

PART II: The Automotive Engine

10. Piston Engine Fundamentals   81

10.1 Energy, Force, Work, and Power   81
10.2 Converting Heat Energy to Motion   82

- 10.3 Piston Engine Parts 83
- 10.4 The Four-Stroke Cycle 86
- 10.5 Multiple-Cylinder Engines 87
- 10.6 Engine Classifications 90

11. Engine Measurements and Performance 94
- 11.1 Engine Size 94
- 11.2 Compression Ratio 95
- 11.3 Horsepower and Torque 96
- 11.4 Horsepower Losses 97
- 11.5 Friction 97
- 11.6 Measuring Horsepower 98
- 11.7 Engine Efficiency 98
- 11.8 Overall Vehicle Efficiency 99

12. Engine Construction 102
- 12.1 Cylinder Block 102
- 12.2 Crankshaft 103
- 12.3 Engine Bearings 106
- 12.4 Gaskets and Seals 109

13. Pistons, Rings, and Connecting Rods 112
- 13.1 Pistons 112
- 13.2 Piston Construction 115
- 13.3 Rings 116
- 13.4 Piston Pins 116
- 13.5 Connecting Rods 116

14. Cylinder Head and Valve Train 120
- 14.1 Cylinder Head Design 120
- 14.2 Overhead Camshaft Valve Train 123
- 14.3 Overhead Camshaft Drive Mechanism 123
- 14.4 Pushrod Valve Trains 125
- 14.5 Valves 125
- 14.6 Valve Guides 126
- 14.7 Lifters 127
- 14.8 Pushrod 128
- 14.9 Rocker Arm 128
- 14.10 Camshaft 128

PART III: Automotive Engine Systems

15. The Lubrication System 134
- 15.1 Lubrication 134
- 15.2 Functions of Engine Oil 134
- 15.3 Oil Viscosity and Service Ratings 134
- 15.4 Natural and Synthetic Engine Oils 136
- 15.5 Oil Consumption 136
- 15.6 Engine Lubrication System 137

16. Lubrication System Service 143
- 16.1 Oil Deterioration 143
- 16.2 Routine Lubrication System Service 144
- 16.3 Oil and Filter Change 144
- 16.4 Leaks From Drain Plug 145
- 16.5 Replacing Valve Cover Gasket 145
- 16.6 Replace Oil Pressure Sender 147
- 16.7 Lubrication System Service With Oil Pan Removed 147

17. The Cooling System 150
- 17.1 Heat Transfer 150
- 17.2 Parts of a Liquid Cooling System 151
- 17.3 Coolant 151
- 17.4 Radiator 152
- 17.5 Water Pump 152
- 17.6 Fan 153
- 17.7 Water Jacket 155
- 17.8 Thermostat 155
- 17.9 Temperature Warning System 157
- 17.10 Radiator Pressure Cap 157
- 17.11 Coolant Recovery System 159
- 17.12 Heater System 159
- 17.13 Oil Cooler 160

18. Cooling System Service 163
- 18.1 Cooling System Preventive Maintenance 163
- 18.2 Check/Add Coolant 163
- 18.3 Check and/or Replace Drive Belts 164
- 18.4 Check for and/or Repair Leaks 165
- 18.5 Check Coolant Mixture 170
- 18.6 Flush Cooling System 170
- 18.7 Replace Thermostat 171
- 18.8 Replace Water Pump 171
- 18.9 Overheating Problems 171

19. Fuel System Fundamentals 174
- 19.1 Automotive Fuels 174
- 19.2 Gasoline 174
- 19.3 Basic Fuel Delivery System 177

19.4	Heated Intake Manifold 182	23.4	EFI Air Measurement 239
19.5	Diesel Fuel 185	23.5	Diesel Mechanical Fuel Injection 241
19.6	Diesel Fuel System 185		
19.7	Turbocharging 186		
19.8	Supercharging 187		

24. Fuel Injection Service — 245

- 24.1 Preliminary Checks 245
- 24.2 Fuel Injection Preventive Maintenance 245
- 24.3 Troubleshooting Fuel Injection Systems 248
- 24.4 Checking for Fuel Delivery 249
- 24.5 Fuel Injection System Repairs 251
- 24.6 Diesel Fuel System Service 254

20. Fuel System Service — 192

- 20.1 Checking for Fuel 192
- 20.2 Fuel System Preventive Maintenance 193
- 20.3 Checking for Leaks 193
- 20.4 Repairing Leaks 194
- 20.5 Replacing Fuel and Air Filters 196
- 20.6 Testing a Fuel Pump 197
- 20.7 Replacing the Fuel Pump 198
- 20.8 Diesel Fuel Filters and Water Separators 199
- 20.9 Turbocharger Service 200
- 20.10 Supercharger Service 201

25. The Exhaust System — 259

- 25.1 Basic Exhaust System 259

26. Exhaust System Service — 267

- 26.1 Exhaust System Problems 267
- 26.2 Exhaust System Preventive Maintenance 268
- 26.3 Muffler and Connector Pipe Replacement 271
- 26.4 Catalytic Converter Problems 271

21. The Carburetor — 204

- 21.1 Carburetion 204
- 21.2 Control of Air Flow 204
- 21.3 Metering 204
- 21.4 Atomization 206
- 21.5 Vaporization 206
- 21.6 Carburetor Vacuum Operation 206
- 21.7 Basic Carburetor Circuits 206
- 21.8 Multiple-Venturi Carburetors 211
- 21.9 Additional Carburetor Equipment 212
- 21.10 Feedback Carburetor 217
- 21.11 The Future of the Carburetor 217

PART IV: Automobile Electrical and Electronic Systems

27. Electrical and Electronic Fundamentals — 276

- 27.1 Basic Automotive Electrical Circuits 276
- 27.2 Matter and Atoms 277
- 27.3 Direction of Electrical Flow 280
- 27.4 DC and AC 281
- 27.5 Electrical Measurements 281
- 27.6 Electrical Circuits 283
- 27.7 Magnetism 283
- 27.8 Magnetic Induction 285
- 27.9 Electronics and Semiconductors 286

22. Carburetor Service — 220

- 22.1 Carburetor Preventive Maintenance 220
- 22.2 Carburetor Problems 221
- 22.3 Driveability Problems 225
- 22.4 Carburetor Adjustments 228
- 22.5 Carburetor Overhaul 230

28. Batteries — 290

- 28.1 Battery Function 290
- 28.2 Electrochemical Action 290
- 28.3 Battery Chemistry 292
- 28.4 Battery Ratings 295
- 28.5 Maintenance-Free Batteries 295
- 28.6 Physical Size 296

23. Fuel Injection — 232

- 23.1 Advantages of Fuel Injection 232
- 23.2 Gasoline Electronic Fuel Injection 232
- 23.3 Common Fuel Injection Systems 237

29. Battery Service — 298

- 29.1 Jump-Starting Procedures 298
- 29.2 Battery Preventive Maintenance 299

29.3	Visual Inspection 299		34.8	Cylinder Numbering and Firing Order 360	
29.4	Instrument Testing 300		34.9	Detonation Control 360	
29.5	Battery Removal and Cleaning 302		34.10	Distributorless Ignition 362	
29.6	Battery Replacement 304		34.11	Diesel Ignition Timing 362	
29.7	Battery Chargers 304				
29.8	"Magic" Battery Chemicals 305		**35.**	**Ignition System Service**	**365**
			35.1	Checking for Spark 365	
30.	**The Starting System**	**307**	35.2	Ignition System Service 366	
30.1	Starting an Engine 307		35.3	Visual Inspection 366	
30.2	Electric Motor Operation 307		35.4	Troubleshooting With Instruments 370	
30.3	Starter Motor 308		35.5	Testing Electronic Ignition Control Units 373	
30.4	Starting System 309		35.6	Spark Plug Replacement 373	
30.5	Typical Starter Motors 311		35.7	Replacing Breaker Points 375	
30.6	Starter Motor Wiring 312		35.8	Setting Ignition Timing 378	
30.7	Starter Motor Circuits 312		35.9	Distributor Removal and Replacement 379	
31.	**Starting System Service**	**316**			
31.1	Starter Preventive Maintenance 316		**36.**	**Electrical Devices**	**382**
31.2	Starting System Troubleshooting 317		36.1	Wiring Circuits 382	
31.3	On-Car Service 318		36.2	Circuit Protection 384	
31.4	Starter Solenoid and Relay Testing 319		36.3	Relays 385	
31.5	Starter Removal 323		36.4	Warning Devices 387	
31.6	Starter Inspection 324		36.5	Instruments 388	
31.7	Starter Repair 324		36.6	Lights 391	
			36.7	Windshield Wiper and Washer 393	
32.	**The Charging System**	**328**	36.8	Horn 394	
32.1	Charging System Function 328		36.9	Comfort Accessories 394	
32.2	Alternator 328		36.10	Audio Systems 395	
32.3	Alternator Regulation 333		36.11	Anti-Theft System 395	
32.4	Charging Indicator 335				
			37.	**Electrical Service**	**398**
33.	**Charging System Service**	**338**	37.1	Basic Electrical Checks 399	
33.1	Charging System Preventive Maintenance 338		37.2	Electrical System Service 399	
33.2	Charging System Electrical Tests 340		37.3	Electrical Troubleshooting 399	
33.3	Fusible Link Service 343		37.4	Electrical Service 402	
33.4	Alternator Off-Car Service 346				
			38.	**Electronic Devices**	**407**
34.	**The Ignition System**	**350**	38.1	Automotive Electronics 407	
34.1	Purpose of the Ignition System 350		38.2	Control Systems 408	
34.2	Ignition System Parts 350		38.3	Self Diagnostics 412	
34.3	Ignition Coil 350		38.4	Other Uses of Electronic Controls 414	
34.4	Distributor 351				
34.5	Primary Circuit 352		**39.**	**Electronic Device Service**	**416**
34.6	Secondary Circuit 356		39.1	Indicating Devices 417	
34.7	Combustion and Engine Power Output 358		39.2	Self Diagnostics 417	

39.3   Open-Loop and Closed-Loop Operation   419
39.4   Sensors   421
39.5   Control Modules   422
39.6   Mechanical Output Devices   422

PART V: Emissions Systems

40.   Emission Control Fundamentals   428

40.1   Air Pollution   428
40.2   Automotive Emissions   430
40.3   Government Regulation and Air Pollution   432
40.4   Government Regulation and Fuel Mileage   435

41.   Emission Control Systems   438

41.1   Factors That Effect Emissions   438
41.2   Engine Controls   438
41.3   Vacuum-Operated Controls   439
41.4   Vacuum-Operated Control System   440
41.5   Electronically Operated Control Systems   440
41.6   Emission Control Equipment   440
41.7   Evaporative Emission Controls   440
41.8   Positive Crankcase Ventilation System   443
41.9   Fuel System Modifications   443
41.10   Combustion Control   445
41.11   Engine and Intake Manifold Modifications   445
41.12   Ignition Timing Modifications   447
41.13   Exhaust Gas Recirculation   448
41.14   Air Injection   450
41.15   Catalytic Converter   451

42.   Emission Control Systems Service   455

42.1   Emission Control Devices and Tampering   455
42.2   Emission Controls Preventive Maintenance   456
42.3   Vacuum, Vapor, and Fuel Lines   458
42.4   Vacuum Control Systems   458
42.5   Thermostatically Controlled Air Cleaner   459
42.6   Vapor Recovery System   460
42.7   PCV System   461
42.8   Air Injection System   462
42.9   EGR System   463
42.10   Catalytic Converter   464

PART VI: Automotive Engine Service

43.   Engine Problem Diagnosis   469

43.1   Diagnostic Procedures   469
43.2   Vacuum Tests   469
43.3   Compression Tests   470
43.4   Cylinder Leakage Test   472
43.5   Engine Oil Pressure Test   473
43.6   Locating Engine Noises   474

44.   Cylinder Head Service   476

44.1   On-Car Cylinder Head Service   476
44.2   Inspecting the Cylinder Head   477
44.3   Inspecting the Valve Train   477
44.4   Valve Clearance Adjustment   478
44.5   Removing Valve Springs   478
44.6   Checking for Valve Spring Damage   481
44.7   Measuring Valve Stem Clearance   481
44.8   Measuring Camshaft Lobe Lift   481
44.9   Off-Car Cylinder Head Service   482
44.10   Cylinder Head Removal   482
44.11   Cleaning and Visual Inspection   484
44.12   Checking for Physical Damage   484
44.13   Measuring Valves and Valve Guides   485
44.14   Machining and Repair Operations   485
44.15   Cylinder Head Reassembly and Installation   487

45.   Piston, Ring, and Rod Service   490

45.1   Engine Removal   490
45.2   Preparing for Piston Removal   490
45.3   Removing Pistons and Connecting Rods   492
45.4   Removing Pistons From Connecting Rods   493
45.5   Piston Service   493
45.6   Connecting Rod Bearing and Journal Service   495
45.7   Connecting Rod Bearing Inspection   496
45.8   Installing Pistons on Connecting Rods   496
45.9   Installing Piston Rings   496

46.   Cylinder Block and Crankshaft Service   498

46.1   Pre-Disassembly Measurements   498
46.2   Cylinder Block Service   499

- 46.3 Cylinder Block Cleaning and Inspection 499
- 46.4 Cylinder Measurement 500
- 46.5 Bearing Surface Measurement 500
- 46.6 Cylinder Boring and Refinishing 503
- 46.7 Crankshaft and Camshaft Journal Measurement 503
- 46.8 Cylinder Block Reassembly 504
- 46.9 Engine Reinstallation 508
- 46.10 Engine Break-In Procedure 509

PART VII: Automotive Drivetrains

47. Drivelines 512
- 47.1 Driveline Design 512
- 47.2 The Effects of Torque 513
- 47.3 Torque-Tube Drive 514
- 47.4 Hotchkiss Drive 514
- 47.5 Driveshaft 515
- 47.6 Center Support Bearing 516
- 47.7 Universal Joint 516

48. Driveline Service 519
- 48.1 Diagnosing Driveline Problems 521
- 48.2 Driveline Service 521
- 48.3 Driveline Balancing 523

49. Manual Transmissions and Transaxles 525
- 49.1 Transmission Design 525
- 49.2 Manual Transmission Construction 528
- 49.3 Four-Speed Transmission Power Flow 531
- 49.4 Five-Speed Transmission 534
- 49.5 Other Transmission Parts 534
- 49.6 Transaxle Design 534
- 49.7 Transaxle Drivetrain 535

50. Manual Transmission and Transaxle Service 541
- 50.1 Diagnosing Transmission and Transaxle Problems 543
- 50.2 Driveline and Driving Axle Removal 544
- 50.3 Remove Transmission Parts and Linkage 545
- 50.4 Transmission and Transaxle Removal 546

51. The Clutch 554
- 51.1 Clutch Design 554
- 51.2 Clutch Variations 557
- 51.3 Clutch Linkage 557

52. Clutch Service 560
- 52.1 Diagnosing Clutch Problems 561
- 52.2 Clutch Service 563

53. Automatic Transmissions And Transaxles 567
- 53.1 Conventional Automatic Transmission Design 567
- 53.2 Fluid Coupling 567
- 53.3 Torque Converter 568
- 53.4 Planetary Gearset 569
- 53.5 Apply Devices 571
- 53.6 Hydraulic Principles 572
- 53.7 Hydraulic Controls 574
- 53.8 Computer Controlled Automatic Transmissions 578
- 53.9 Transmission Oil Cooling 578
- 53.10 Conventional Automatic Transaxle Design 578
- 53.11 Conventional Automatic Transaxle Operation 579
- 53.12 Continuously Variable Transmission (CVT) 581

54. Automatic Transmission and Transaxle Service 586
- 54.1 Automatic Transmission and Transaxle Diagnosis 589
- 54.2 Testing 590
- 54.3 Adjustment and Light Service 591
- 54.4 Automatic Transmission and Transaxle Overhaul 593
- 54.5 Continuously Variable Transmission Diagnosis and Service 594

55. Differentials, Driving Axles, and 4WD 597
- 55.1 Differential Design 597
- 55.2 Differential Construction 600
- 55.3 Limited-Slip Differential 600
- 55.4 Driving Axles 602
- 55.5 Four-Wheel Drive (4WD) 603

56. Differential, Driving Axle, and 4WD Service 609
- 56.1 Diagnosing Differential Problems 610

56.2	Differential Service 611	
56.3	Driving Axle Service 619	
56.4	Transfer Case Service 621	

PART VIII: The Automotive Chassis

57. The Suspension System   625

- 57.1 Suspension Functions 625
- 57.2 Springs 625
- 57.3 Shock Absorbers 628
- 57.4 Front Suspension 629
- 57.5 Rear Suspension 633
- 57.6 Passive, Semi-Active, and Active Suspension 638

58. Suspension System Service   644

- 58.1 Diagnosing Suspension System Problems 646
- 58.2 Suspension System Service 649

59. Steering and Wheel Alignment   658

- 59.1 Steering System Operation 658
- 59.2 Manual Steering 660
- 59.3 Power Steering 660
- 59.4 Steering Column 666
- 59.5 Wheel Alignment 667

60. Steering and Wheel Alignment Service   671

- 60.1 Diagnosing Steering Problems 672
- 60.2 Steering System Service 674
- 60.3 Wheel Alignment 676

61. Tires and Wheels   681

- 61.1 Tire Function 681
- 61.2 Tire Construction 681
- 61.3 Temporary Use Tires 685
- 61.4 Retreaded Tires 685
- 61.5 Wheels 685

62. Tire and Wheel Service   688

- 62.1 Diagnosing Tire Problems 691
- 62.2 Tire and Wheel Service 692

63. The Brake System   696

- 63.1 Brake System Design 696
- 63.2 Brake System Construction 697
- 63.3 Disc Brakes 699
- 63.4 Drum Brakes 700
- 63.5 Disc and Drum Brake Combinations 702
- 63.6 Power Brakes 703
- 63.7 Anti-Lock Brake System (ABS) 704
- 63.8 Parking Brake Operation 706

64. Brake System Service   709

- 64.1 Brake System Problem Diagnosis 711
- 64.2 Bleeding Brakes 714
- 64.3 Master Cylinder Rebuild 716
- 64.4 Disc Brake Service 716
- 64.5 Drum Brake Service 719
- 64.6 Power Brake Service 721
- 64.7 Anti-Lock Brake System (ABS) Service 721

PART IX: Auxiliary Systems

65. Heating and Air Conditioning Systems   725

- 65.1 Heating System 725
- 65.2 Air Conditioning System 727

66. Heating and Air Conditioning Service   732

- 66.1 Diagnosing Heater Problems 733
- 66.2 Heater Service 733
- 66.3 Diagnosing Air Conditioner Problems 733
- 66.4 Air Conditioner Maintenance 735
- 66.5 Air Conditioner Service 737

67. Safety Systems   742

- 67.1 Safety Devices 742
- 67.2 Occupant Protection 744

68. Safety Systems Service   748

- 68.1 Diagnosing Occupant Protection System Problems 748
- 68.2 Occupant Protection Systems Service 748

Glossary   753

Index   782

Preface

Automotive Technology: Theory and Service is the cornerstone of a comprehensive learning program. This textbook is designed for the secondary school student who is considering a career in the automotive service industry.

The automobile industry—manufacturing, sales, and service—accounts for almost 20 percent of all employment in the United States. Within this huge industry, career opportunities are expanding most rapidly in the area of automobile service and repair. Automotive technology is one of the fastest-growing employment fields in the nation.

For these reasons, *Automotive Technology: Theory and Service* provides the student with a comprehensive overview of the automobile service and repair career field. In addition, the textbook provides valuable insights into the skills and attitudes that lead to job and career success.

The vocational content of the textbook is complemented by comprehensive technical treatment of the service and repair tasks performed by auto technicians. The majority of the content is aimed specifically at building job-qualifying skills among students. The total effect is a blend of the theory learned in the classroom with the practical skills acquired in the automotive shop. This blend of knowledge and skill acquisition provides a complete learning experience.

ORGANIZATION

This textbook is organized into nine parts. Each part concentrates upon a major area of knowledge and/or skill involved in automotive service and repair. Within each part are units that represent natural learning experiences for students. These units are divided further into numbered topics. The numbering system provides a convenient and useful system of cross referencing throughout the textbook. The student is guided swiftly to referenced material through the use of unit numbers in the titles of major discussion topics. Further enhancing this ease of reference is the inclusion of numbered topics in the Table of Contents.

Major systems and components of the automobile are presented in a two-unit format. This arrangement permits a natural and meaningful transition from classroom instruction to hands-on shop experience.

One major exception is made to this format in the interest of a logical learning progression. Part II consists entirely of knowledge units introducing the student to the basic principles of automotive engine design and operation. This information is necessary for a clear understanding of the automotive systems that follow. However, at this stage, the student is not prepared to undertake major engine service procedures. Thus, the service units pertaining to major engine repairs are positioned later in the text, in Part VI.

Each unit is presented in a readable and understandable format. At the beginning of each unit is a Unit Preview, which provides a brief overview of the subject matter. Immediately following the preview is a listing of major achievement goals, entitled Learning Objectives. These elements provide a focus for the study and practice activities in the unit. These introductory elements also provide both student and instructor with references by which learning performance can be rated.

Throughout this textbook, a strong emphasis is placed on safe working habits. In keeping with this emphasis, skill units include a thorough discussion of Safety Precautions pertaining to the tasks in those units.

Unit Highlights at the conclusion of each unit provide a brief summary of the major topics. Important words and phrases defined in the unit are listed under the heading Terms. Review Questions are presented in a variety of multiple-choice styles. This format resembles that of the technician certification tests administered by the National Institute for Automotive Service Excellence (ASE, formerly NIASE). Essay Questions are provided to test students understanding of text material and concepts or service procedures presented in each unit. Finally, Supplemental Activities are suggested to enrich and reinforce the knowledge or skills emphasized in the unit.

FEATURES

Illustrations are positioned in relationship to the text content to which they apply. All illustrations are cited in the text to provide solid learning reinforcement.

Safety is emphasized throughout the text. An entire unit (Unit 4, "Safety Around the Automobile") is devoted to a thorough discussion of safe working habits. Particular attention is paid to proper use of tools and power equipment. Development of a safety-minded attitude is stressed.

Safety is emphasized further by the inclusion of special cautions for all shop procedures in which potential hazards exist. Each of these cautions, printed in boldface type for emphasis, begins with the words SAFETY CAUTION.

Other boldface notations reinforce other important concepts. Notes introduced by the word CAUTION!!! call for special care to avoid damage to a part during a specific procedure. Notes introduced by the word NOTE call attention to an important concept or to a helpful suggestion in performing a task. A special TROUBLESHOOTING AND MAINTENANCE heading, found in many service units, calls attention to general maintenance techniques and possible hazards associated with a particular service/repair procedure.

A Shop Manual complements this textbook. The shop manual contains individual Job Sheet assignments and written Tests. The Job Sheets are step-by-step procedures designed to verify and reinforce the student's shop practice experiences. The Tests are presented in a format similar to that used by ASE in its technician certification tests.

ACKNOWLEDGMENTS

The authors wish to thank Russell J. Mukai, auto technician instructor at North Hollywood High School, Los Angeles Unified School District, for his valuable contributions as a reviewer and technical consultant.

Photography for this textbook was done by James L. Camp and Robert Koch, except where otherwise credited.

The authors also wish to thank the following companies for supplying illustrations and photographs used throughout this book: Chevrolet Motor Division, Pontiac Motor Division, Oldsmobile Division, Buick Motor Division, and Cadillac Motor Division, General Motors Corporation; Delco-Morraine and Hydra-Matic, Divisions of General Motors Corporation; Chrysler Corporation; Ford Motor Company; American Motors Corporation; American Honda Motor Company, Inc.; American Isuzu, Inc.; Mazda (North America), Inc.; Mercedes-Benz of North America, Inc.; Mitsubishi Motor Sales of America, Inc.; Nissan Motor Corporation in U.S.A.; Volkswagen of America, Inc.; and Volvo of America, Inc.

Other companies that supplied illustrations and/or photographs are: Champion Spark Plug Company; Snap-On Tools Corporation; Gould, Inc.; Moog Automotive, Inc.; Robert Bosch Corporation; BFGoodrich; Sun Electric Corporation; Inland Manufacturing Company; Kleer-Flo Company; L.S. Starrett Co.; Storm Vulcan; Ammco Tools, Inc.; Sioux Tools, Inc.; Bear Auto Service Equipment Co.; The Black & Decker Mfg. Co.; Fel-Pro, Inc.; Hastings Manufacturing Co.; Dana Corporation; Stanadyne Diesel Systems Group; and TRW, Inc.

PART I: Career Opportunities And the Automobile

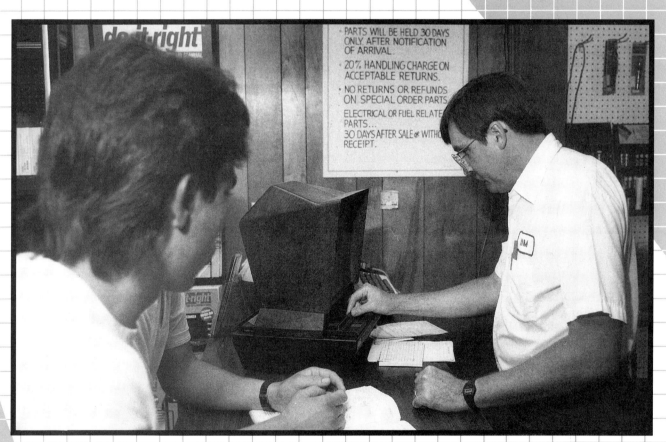

With proper training, experience, and motivation, automotive technicians can choose from several career paths. The first part of this book provides an overview of the automotive servicing business. Information on using service manuals, shop forms, and tools is presented. Unit 4 is devoted to the safety precautions necessary to prevent injuries when working around automobiles. The technician in this photo is reading part numbers on microfilm.

1 Careers and Opportunities

UNIT PREVIEW

The automotive industry is one of the nation's largest provider of jobs and career opportunities. A large number of all workers in the United States are associated, directly or indirectly, with the automotive industry. This book deals with automotive mechanics, one of the many job opportunities in the automotive field. This unit discusses the types of jobs available in automotive service and repair.

LEARNING OBJECTIVES

When you have completed your assignments and exercises in this unit, you should be able to:

- Explain why the need for automotive service is increasing.
- Identify major types of automotive service businesses.
- Identify jobs available in automotive service.
- Describe the basic duties of general and specialty technicians.
- Describe characteristics of a good employee.

1.1 THE AUTOMOTIVE INDUSTRY

Automotive mechanics is one of many different occupations included in the automotive industry. Automobiles are manufactured, sold, and used by people. Thus, the automotive industry can be divided into three basic areas of job and career opportunities:

- Manufacturing
- Sales
- Service.

Manufacturing. Included in the manufacturing process are design, engineering, production of parts, and assembly of automobiles. Automobile manufacturing companies employ people in many different skill areas such as metal working processes, plastics, electronics, paint and finish technology, tire manufacturing, etc.

Sales. Automobiles in the United States are sold through independent dealerships, which contract to sell the products of manufacturers. Some automobile dealers carry more than one product line.

Service. This textbook deals with the *servicing* of automobiles. Servicing includes *maintenance* and *repair*. It is the only one of these three areas that applies to the full life of every automobile. To perform properly, an automobile must be maintained on a regular basis. Maintenance is the care and upkeep of mechanical and other parts of an automobile, usually performed according to a schedule. Regular maintenance includes procedures such as oil and filter changes, lubrication, replacement of belts and hoses, and tune-ups **[Figure 1-1]**.

Repair is the replacement or fixing of parts that wear out, break, or malfunction. Some repairs—servicing of brakes or replacement of bolt-on parts, such as shock absorbers or alternators—are considered minor work. Other repairs, such as engine or transmission overhauls, are considered major work.

1.2 INCREASING NEEDS FOR SERVICE

The need for automotive service is ongoing. Most routine service intervals have been extended on modern automobiles. However, regularly scheduled maintenance is as important as ever. In addition, several factors contribute to the increasing need for trained and certified automotive technicians. These factors are:

- Increasing use of computers and electronics
- Increased average age of automobiles
- Emissions and fuel economy requirements.

Figure 1-1. Regular maintenance procedures are routine parts of a technician's job.

Increasing Use of Computers and Electronics

Modern automobiles include several computers and electronic devices that control vehicle operations. For example, on new vehicles, computers may be used to control fuel flow, engine operation, automatic transmission shifting, anti-skid braking, vehicle height and cornering, shock absorber settings, and many other operations. Computerized and electronic systems are lightweight and compact and contribute to the operation, comfort, and safety of modern automobiles.

Servicing these highly technical systems and parts requires special knowledge and new diagnostic equipment **[Figure 1-2]**. Computers are not maintainable or repairable. Instead, computers that fail are removed and replaced with new units.

Increased Average Age of Automobiles

At one time, it was common practice for motorists to replace their automobiles on an average of every three years. However, increased prices of new automobiles and higher interest rates encourage car owners to keep their vehicles longer. As a vehicle ages and accumulates mileage, the more maintenance and repairs it requires.

Emissions and Fuel Economy Requirements

Many states have laws that require periodic inspections of passenger vehicles. In these cases, vehicles must pass exhaust emissions and/or safety inspections before registration can be renewed. Vehicles that fail such inspections must be brought up to standards at the owners' expense.

Federal regulations governing emissions control and minimum fuel mileage standards require that all major systems on the automobile function properly for a specified number of years and miles. Preventive maintenance and repairs are necessary to keep these systems operating at maximum efficiency.

1.3 AUTOMOTIVE SERVICE BUSINESSES

Automotive technicians are employed in a variety of automotive businesses. Some of these businesses offer a full range of automotive maintenance and repair services. Others offer specialized services, such as electrical or brake repairs. From a career standpoint, two important facts stand out. First, automotive service businesses are large in number. Second, these businesses are needed in every community in the country.

The majority of automotive service businesses fall into one of the following categories:

- New-car dealerships
- Independent garages
- Specialty shops
- Service stations
- Fleet garages
- Auto supply and accessory stores, and department stores.

New-Car Dealerships

Many automotive service functions are performed at dealerships **[Figure 1-3]**. Dealer service departments usually coordinate any type of service activity. These activities may include preparing new and used automobiles for delivery to customers, servicing customers' automobiles, and body repairs. Dealerships also serve as manufacturers' representatives in the performance of *warranty* repairs. A warranty is a manufacturer's guarantee that certain parts of an automobile will perform as designed for a designated period of time. A warranty repair is the repair or replacement of a defective part at the manufacturer's cost.

Another important function in most dealerships is the parts department. A parts department keeps commonly used parts available for installation by the service department and for sale to the public. A well-run parts department is a profitable part of a dealership.

Figure 1-2. Diagnostic equipment is used by technicians to help locate automotive problems.

Figure 1-3. New-car dealer service centers usually offer all services required for the automobiles they sell.

Independent Garages

The *independent garage* is the primary source for automotive servicing in many communities **[Figure 1-4]**. An independent garage typically offers customers a complete line of services. Maintenance services include oil and filter changes, chassis lubrication, tune-ups, and brake inspections. Most independent garages also perform major and minor mechanical repairs. Many independent garages overhaul engines and transmissions.

Specialty Shops

The automotive *specialty shop* usually concentrates on one type of automotive service, or a limited number of services. This specialization allows the shop owner or manager to maintain close control of operations and expenditures. Examples of specialty shops are tire, muffler, tune-up, air conditioning, and brake repair facilities **[Figure 1-5]**.

Service Stations

A *service station* is a gasoline station that offers automotive servicing. One or more *service bays*, or work areas, may be used for maintenance and repair services. Some service stations perform only simple maintenance services and sell and install tires and batteries. Others perform tune-ups, brake repairs, and other services. Still others may offer major repairs, such as engine overhauls **[Figure 1-6]**.

In many communities, the availability of a certified technician can greatly increase the volume of business done by a service station.

Fleet Garages

Many businesses, government agencies, and other organizations have *fleets* of automobiles. A fleet may consist of as few as two or three automobiles, or it may number in the thousands. Small fleets usually are serviced by independent shops. Large fleets often are serviced in a garage owned by the organization that operates the fleet. Garages for major fleets usually have complete maintenance and repair capabilities **[Figure 1-7]**.

Auto Supply and Accessory Stores, And Department Stores

Auto supply and accessory stores find it profitable to sell and install certain products. These products include tires, shock absorbers, mufflers, and batteries. Most department store chains offer automotive services at some, if not all, of their locations **[Figure 1-8]**.

1.4 AUTOMOTIVE TECHNICIAN CERTIFICATION

Jobs in automotive service have a variety of names. The following discussions apply to basic work functions.

Figure 1-5. A specialty shop offers limited services, usually specializing in one phase of automobile repair and service.

Figure 1-6. Many service stations provide service as well as fuel for automobiles. Some service stations specialize in simple maintenance. Others perform major repairs.

Figure 1-4. An independent garage can perform many services for different brands of automobiles.

Figure 1-7. Government and private industries have fleets of automobiles that are serviced regularly in fleet garages.

Figure 1-8. Major department store chains have service centers. These centers usually install parts and make repairs with parts that are sold through the stores.

Master Automobile Technician

The National Institute for Automotive Service Excellence (formerly NIASE, now referred to as ASE) offers tests and certifications for technicians **[Figure 1-9]**. Many states recognize ASE certifications and/or base their own certifications on ASE standards.

ASE offers tests and certifications in eight specialty areas of automobile repair. A technician who passes all required tests and meets experience requirements is certified as a *Master Automotive Technician*. These eight certification areas are:

- Engine Repair
- Engine Performance.
- Suspension and Steering
- Brakes
- Automatic Transmission/Transaxle

Figure 1-9. Many automotive shops use only ASE certified technicians.

- Manual Drive Train and Axles
- Electrical Systems
- Heating and Air Conditioning

A master automobile technician must be skilled in all areas of automobile maintenance and repair. In many automotive service businesses, technicians are divided into two general categories:

- Line technician—light work
- Line technician—heavy work.

Line technician—light work. The *line technician—light work* performs nonspecialized minor services **[Figure 1-10]**. These services include oil and filter changes, lubrication, accessory installation, tune-ups, and replacement of belts, hoses, and lights. Also included in this category is the inspection of new automobiles prior to delivery to customers, which includes brakes, air conditioning and heating.

Line technician—heavy work. The *line technician—heavy work* performs nonspecialized services, including disassembly, repair, and replacement of

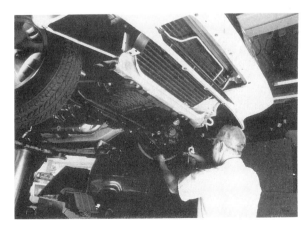

Figure 1-10. A line technician may perform only light-work duties, such as chassis lubrication.

parts, and reassembly of major parts. These services include engine, transmission, and differential overhauls [Figure 1-11]. The heavy-work line technician must be competent in all areas of automotive service.

Automotive Specialty Technician

The *automotive specialty technician* concentrates on servicing a single area, such as electrical systems, brakes, or transmissions [Figure 1-12]. These specialties require advanced and continuing training in a particular field.

ASE also offers certification tests in other automotive specialties. These include Auto Body Repair and Heavy-Duty Truck tests and certifications.

Service Writer

The person who greets customers at a service center is the *service writer* or *service advisor* [Figure 1-13]. A service writer needs a good knowledge of automobiles, a friendly attitude, and the ability to interact with people efffectively. Customers discuss their automotive problems and needs with the service writer. The service writer then makes a preliminary diagnosis and prepares a cost estimate for the customer.

Shop Foreman

A service department with a number of technicians usually has one or more shop foremen. The *shop foreman* supervises the work of several technicians. The foreman must be able to work effectively with people and must have thorough knowledge of automotive mechanics. The shop foreman schedules work for technicians and helps to solve mechanical problems. The foreman also may check the completed work to be sure that the customer's automobile has been serviced properly.

Service Manager

All maintenance and repair jobs are controlled by the *service manager*. The service manager must be a good technician and have some business knowledge. The service manager is responsible for the profit earned or lost by the department, as well as for customer satisfaction.

Parts Department

The sale of automotive parts requires a working knowledge of the automobile and servicing. Service departments require parts to perform most repair jobs. A *parts department* maintains a stock of often-used replacement parts [Figure 1-14]. If a needed part is not in stock, the parts department worker must know where to obtain it quickly.

Figure 1-11. A line technician who does heavy work repairs larger, more complicated automotive assemblies.

Figure 1-12. A specialty technician repairs only one part of an automobile, such as the brake system.

Figure 1-13. A service writer greets the customer and makes a preliminary diagnosis of a problem for the technician.

CAREERS AND OPPORTUNITIES

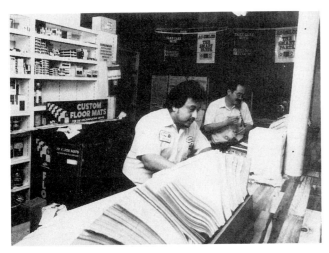

Figure 1-14. Parts departments keep stock of often-used parts. These parts are available to a technician when a replacement is necessary to complete a repair.

A *parts manager* runs the parts department. The parts manager is responsible for ordering, stocking, and selling parts and accessories.

1.5 TRAINING FOR A CAREER IN AUTOMOTIVE SERVICE

The course in which you are using this text probably is the first—and most important—step in your training. In this course, you will learn the basics of how automobiles operate and how they are serviced.

To move ahead in the automotive field, you should continue your education in one or more specialized areas **[Figure 1-15]**. You also may be able to obtain valuable training through work in an apprenticeship program. For information on such programs in your community or area, consult with your school counselor or placement office.

Courses in automotive servicing are available in high schools, community colleges, vocational schools, and the armed forces. Some automotive manufacturers also furnish training supplies and programs, either through public schools or through their own regional training centers. You also may be able to work in a service department while attending school. Many dealers and garages offer apprentice programs.

1.6 GOING TO WORK

Becoming a valuable worker requires more than learning job skills. When you get a job, you will enter into a business transaction with your employer. A business transaction is an exchange of goods or services that have value. When you become an employee, you sell your time, skills, and effort. Your employer pays you money for these resources.

Both parties in a business transaction have responsibilities. The obligations of an employer to a worker include:

- Instruction and supervision. You should be told what is expected of you on the job. There should be a supervisor who can observe your work and tell you if it is satisfactory **[Figure 1-16]**.
- A clean, safe place to work. An employer should provide adequate and safe facilities for you to work in. These facilities include fire extinguishers and first-aid kits.
- Wages. You should know how much you are to be paid before accepting a job. Your employer should pay you on designated paydays.
- Fringe benefits. When you are hired, you should be advised about any benefits, in addition to wages, that you can expect. Fringe benefits usually include paid vacations and employer contributions to health insurance and retirement plans.

Figure 1-15. Automotive classes help students learn more about servicing all systems in an automobile.

Figure 1-16. Supervisors rely on their experience to help young technicians learn needed skills.

- Opportunity and fair treatment. Opportunity means that you are given a chance to succeed and possibly to advance in a company. Fair treatment means that all employees are treated equally, without prejudice or favoritism.

On the other side of this business transaction, employees have responsibilities to their employers. Your obligations as a worker will include:

- Regular attendance and punctuality. A good employee is reliable and arrives at work on time. Businesses cannot operate successfully unless their workers are punctual and on the job regularly.
- Following directions. As an employee, you are part of a team. Doing things "your way" may not serve the best interests of the business.
- Responsibility. Be willing to answer for your work-related obligations and your conduct.
- Productivity. Remember, you are paid for your time as well as your skills and effort. You have a duty to use time on the job as effectively as possible.
- Attitude. With a positive attitude, you increase your value beyond your skills and time. One reason for this is that your attitude will have a positive effect on other employees and customers.
- Loyalty. Being loyal means that you act in the best interests of your employer, both on and off the job.
- Honesty. Customers deserve, and rightfully expect, honesty from automotive technicians. You will be doing a great service to your profession and place of employment if you are honest with customers.

Getting Along at Work

In addition to your obligations to an employer, you also will have certain responsibilities toward your fellow workers. You will be a member of a team. Teamwork means cooperation with, and caring about, other workers. An important strength of a valuable employee is the ability to work in harmony with fellow employees. You also should strive for harmonious relations with your supervisors and with those you supervise.

Pride in Your Work

Doing your best on the job comes naturally if you take pride in your work. Doing a job right the first time is the pathway to profits. The more pride you take in your work, the more likely a customer is to be satisfied. This satisfaction works to everyone's benefit.

Pride in your work means more than just doing the work correctly and well. Also important is your attitude toward customer satisfaction. For example, stained or soiled upholstery will not please a customer, no matter how well you perform a mechanical repair [Figure 1-17]. Treat other people's property as if it were your own.

Figure 1-17. Many service shops use paper floormats and other protective devices to prevent soiling customers' automobiles.

UNIT HIGHLIGHTS

- Automotive mechanics can be the beginning of a career in servicing—the maintenance and repair of automobiles.
- The automotive industry consists of three main areas of job and career opportunities: manufacturing, sales, and service.
- Manufacturing starts with ideas and raw materials. It concludes with a finished product.
- Automotive service is performed by many different types of businesses.
- Automotive service will always be in demand to keep automobiles operating.
- A variety of automotive servicing jobs are available.
- An employer and employees have obligations and responsibilities to one another.

TERMS

servicing	maintenance
repair	fleet
fleet garage	parts manager
service writer	shop foreman
service manager	independent garage

specialty shop
service bay
automotive specialty technician
Master Automotive Technician
warranty
service station
parts department
line technician—heavy work
line technician—light work

REVIEW QUESTIONS

DIRECTIONS: The following questions are similar to those used on automotive technician certification tests. Answer each question by circling the letter of the correct choice.

1. Which of the following statements about large automotive service facilities is true?
   I. A shop foreman supervises the work of several technicians in the shop.
   II. The service manager is in charge of all the technicians in the shop.
   A. I only
   B. II only
   C. Both I and II
   D. Neither I nor II

2. Technician A says that computers and electronics in many modern cars control the operation of automatic transmissions and fuel systems.
   Technician B says that the average age of automobiles is 12 years.
   Who is correct?
   A. A only
   B. B only
   C. Both A and B
   D. Neither A nor B

3. A technician who passes all eight ASE automotive certification tests is certified as a
   A. Line technician—light work.
   B. Line technician—heavy work.
   C. Master Automotive Technician.
   D. Automotive Technician.

Questions 4 and 5 are not like the first three questions. Each has the word EXCEPT. For each question, look for the choice that does *not* apply. Read each question carefully before you choose your answer.

4. All of the following businesses engage in most types of automotive servicing EXCEPT
   A. dealership.
   B. independent garage.
   C. specialty shop.
   D. fleet garage.

5. Major automotive job and career areas include all of the following EXCEPT
   A. manufacturing.
   B. automotive computer maintenance and repair.
   C. sales.
   D. service.

ESSAY QUESTIONS

1. Describe the three main sources of automotive-industry jobs.
2. What conditions create an increasing need for automotive service?
3. Describe ASE certification for automotive technicians.

SUPPLEMENTAL ACTIVITIES

1. Describe the major stages of automobile manufacturing.
2. Identify the duties of a service writer.
3. Describe the activities of a fleet garage.
4. Identify the differences between light-work and heavy-work line technicians.
5. Describe the operation of an auto dealership parts department.
6. Discuss the responsibilities employees have to their employers.

2 The Automobile

UNIT PREVIEW

An automobile is made up of approximately 15,000 separate parts. These parts are grouped together into systems. For example, many separate parts are assembled to form the engine. When the engine operates, it furnishes turning force, which is transferred to driving wheels through a series of parts called a drivetrain. This unit discusses the major mechanical parts that make the automobile work.

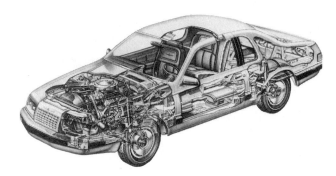

Figure 2-1. An automobile consists of thousands of parts.
FORD MOTOR COMPANY

LEARNING OBJECTIVES

When you have completed your assignments and exercises in this unit, you should be able to:

❏ Describe the basic parts of a reciprocating piston engine.
❏ Describe how fuel is drawn in and burned in an automotive engine.
❏ Identify the differences between a transmission and a transaxle.
❏ Describe the major types of automotive body construction.
❏ Identify disc and drum brakes.

2.1 BASIC PARTS OF THE AUTOMOBILE

The automobile uses thousands of different parts, as illustrated in **Figure 2-1.** However, all of these parts can be divided into four basic assemblies:

- Engine
- Drivetrain
- Chassis
- Body.

2.2 ENGINE

An *engine* is a machine that converts energy into mechanical motion and force. The most common form of engine in passenger vehicles is the internal-combustion piston engine. An *internal-combustion engine* is one in which combustion, or burning, of air and fuel takes place inside the engine.

Automobile engines produce mechanical motion in the form of rotation. This motion is transmitted through a series of parts to turn the driving wheels and move the vehicle.

Piston Engine

The most widely used passenger-car engine is the *reciprocating piston engine.* In this type of engine, a piston moves up and down in a cylinder **[Figure 2-2].** The major parts of a reciprocating piston engine are:

- Cylinder block
- Cylinder head
- Valve train
- Piston
- Connecting rod
- Crankshaft
- Manifolds.

The following descriptions apply to gasoline engines. Diesel engines share the same major components.

Cylinder block. The *cylinder block*, or *engine block*, **[Figure 2-3]** forms the lower part of the engine. Parts within the block move to create mechanical energy from pressure created by the burning of fuel. Large round holes called *cylinders* are formed in the block. Each cylinder produces mechanical energy from the chemical energy of burning fuel. The engine uses this energy to create *torque,* or turning force, that moves the vehicle. The block is the largest single engine component. Most other engine parts are attached to or operate within the block. It is sometimes considered the foundation of the complete engine. The block contain internal passageways to carry lubricant and coolant to other parts of the engine.

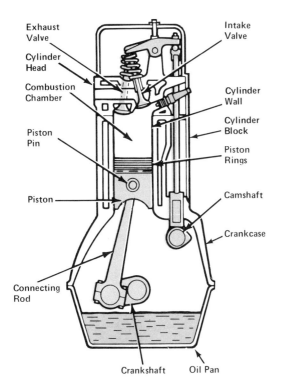

Figure 2-2. The basic parts of a reciprocating piston engine.
CHRYSLER CORPORATION

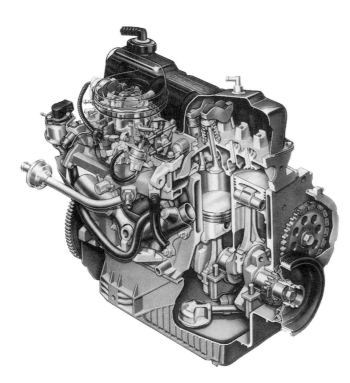

Figure 2-3. The piston movement acts to draw in an air-fuel mixture and push out exhaust gases.
FORD MOTOR COMPANY

Cylinder head. The *cylinder head,* or upper part of the engine, is bolted to the top of the cylinder block and seals the top end of the cylinders. Hollow passageways within the cylinder head are provided for an air-fuel mixture to enter and for burned gases to exit. Coolant also circulates through the head to carry away excess heat.

Valve train. A *valve train* is a series of parts used to open and close the intake and exhaust ports. A *valve* is a movable part that opens and closes a passageway. Valve movement is controlled by a *camshaft,* the metal shaft onto which *cams* have been cast or machined. Cams are projections that cause other parts to move **[refer to Figure 2-2]**.

Piston. A *piston* is a machined, round part that moves up and down within a cylinder. The top of the piston forms the bottom of the *combustion chamber,* the area where fuel is burned. In a *four-stroke cycle* engine the piston moves downward on an intake stroke to draw in an air-fuel mixture. Then the piston is moved upward on a compression stroke to squeeze and compress the mixture tightly. When combustion occurs, the piston is pushed down by the pressure of expanding, heated gases to create a power stroke. Finally, the piston is moved upward in the cylinder again on an exhaust stroke to force out burned gases.

Connecting rod. A *connecting rod* transfers force when the piston moves upward or downward. The top of the connecting rod is connected to the piston. The bottom of the connecting rod fits around a machined area on the crankshaft.

Crankshaft. The *crankshaft* is a shaft located at the bottom of the engine block. A shaft is a cylindrical part used to turn, or to be turned by, other parts. The crankshaft has offset, or extended, portions where the connecting rods are attached. The distance that the piston moves, up and down, in the cylinder depends upon the crank pin offset.

As each connecting rod is pushed down by a piston, the motion is changed from a reciprocating, or up-and-down, movement to a rotating movement by the crankshaft.

Manifolds. To direct gases into and out of the engine, hollow *manifolds* are used **[Figure 2-4]**. A manifold is a part with several tubular sections through which fluids (gases in this case) can pass. The *intake manifold* delivers air and fuel to the cylinder head intake ports. The *exhaust manifold* carries exhaust gases away from the cylinder head exhaust ports.

Automotive engines usually have four, six, or eight cylinders. An engine also has a corresponding number of pistons and connecting rods, all of which turn a single crankshaft.

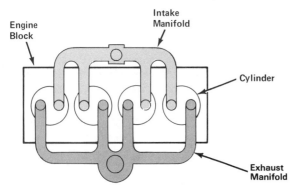

Figure 2-4. A manifold directs gases in and out of combustion chambers.

Piston Engine Systems

Basic systems required for operation of a gasoline-fueled piston engine are:

- Lubrication
- Cooling
- Fuel
- Exhaust
- Ignition.

Lubrication system. All moving parts in the engine require *lubrication*. Lubrication in the engine is the application of oil to moving parts to reduce friction. *Friction* is the resistance to motion that occurs when two objects rub against one another. Friction between mechanical parts causes heat and wear. A *lubrication system* is used in engines to distribute engine oil. A typical lubrication system consists of an oil pan, an oil pump, and oil galleries [Figure 2-5].

The engine oil is stored in an *oil pan* that is bolted to the bottom of the engine block. Oil is drawn from the oil pan by an *oil pump*. *Oil galleries*, or small passageways, direct oil to the moving parts of the engine. Extra lubrication is provided by the rotating crankshaft, which delivers oil to the connecting rods and indirectly to the cylinder walls.

Cooling system. The engine produces force by burning a fuel mixture, so engine parts become very hot. To maintain proper temperatures, an engine *cooling system* is provided [Figure 2-6]. Most engines are *water-cooled*, or cooled by a liquid. Some engines are *air-cooled*.

A water-cooled system uses *coolant*. Coolant is a fluid that contains special chemicals mixed with water. Coolant flows through passages in the engine and through a *radiator*. The radiator accepts hot coolant from the engine and lowers its temperature. Air flowing around and through a radiator takes heat from the coolant. The lower-temperature coolant is returned to the engine through a pump.

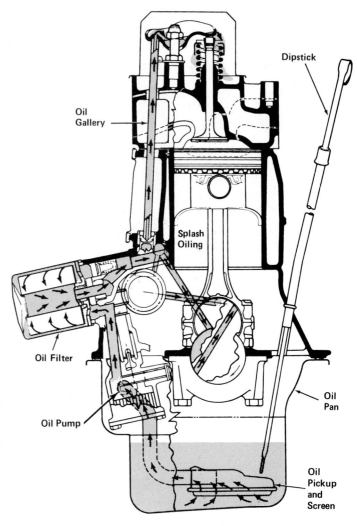

Figure 2-5. A lubrication system distributes oil throughout the engine. BUICK MOTOR DIVISION—GMC

Fuel system. For fuel to burn in an automotive engine, it must be mixed with air. This air-fuel mixture is introduced into the cylinder at the cylinder head. A *carburetor* [Figure 2-7] or *fuel injection system* (discussed in Unit 23) provides an air-fuel mixture to be burned in the cylinders.

Exhaust system. As the piston returns to its uppermost position on the exhaust stroke, it forces the burned air-fuel mixture, or exhaust gases, out of the engine. The exhaust then passes through the *exhaust system*, a series of parts shown in **Figure 2-8**.

Ignition system. A *spark plug* is installed in the cylinder head and extends into the combustion chamber. The spark plug has a gap formed by two electrodes. In most cases, a *distributor* provides high-voltage electrical current through cables to each spark plug in an engine at the correct time. When electrical current jumps the spark plug gap, a

The Automobile 13

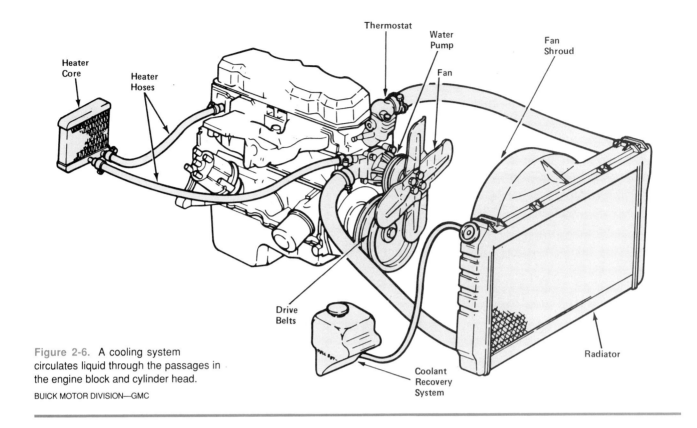

Figure 2-6. A cooling system circulates liquid through the passages in the engine block and cylinder head.
BUICK MOTOR DIVISION—GMC

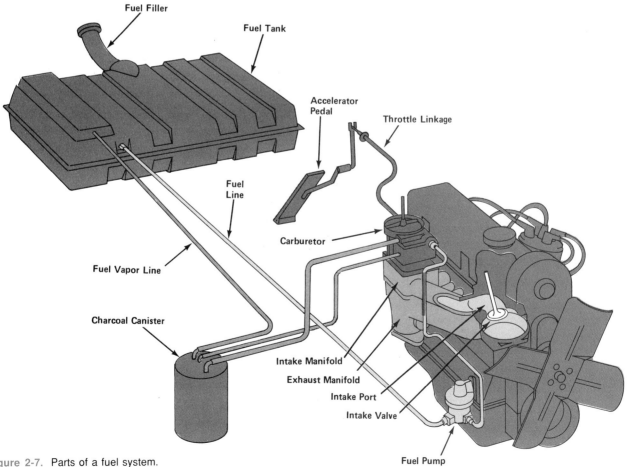

Figure 2-7. Parts of a fuel system.

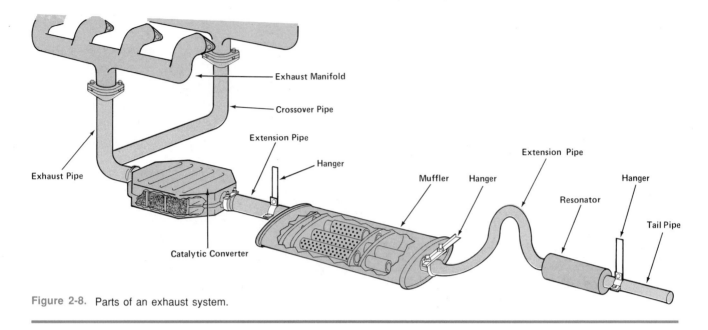

Figure 2-8. Parts of an exhaust system.

spark is created to ignite the air-fuel mixture. The spark plug fires when the piston is nearing the top of its compression stroke **[Figure 2-9]**.

The ignition system is a subsystem, or portion, of the vehicle's electrical system. Other electrical subsystems include the starting system, charging system, and accessory circuits. The sources of electricity in the electrical system are the *battery* and the *alternator*. The *starting system* uses battery power to turn the crankshaft and start the engine. The alternator is driven by the engine and provides electrical power while the engine is running. In addition to the alternator, the *charging system* has a *voltage regulator*. The voltage regulator controls the electrical output from the alternator to the battery. Accessory circuits deliver electricity to power the lights, safety systems, and accessories, such as radios, computers, and other electrically operated equipment.

2.3 DRIVETRAIN

The engine creates torque, or turning force, to turn the driving wheels of a vehicle. The *drivetrain* includes the parts and assemblies that transmit torque from the engine to the driving wheels. The term *power train* describes the combination of *both* the engine and the drivetrain.

Most vehicles transmit engine torque to either rear wheels only, front wheels only, or all four wheels. A *rear-wheel drive (RWD)* drivetrain with a manual transmission is shown in **Figure 2-10**. A drivetrain includes:

- Clutch
- Transmission or Transaxle
- Driveline

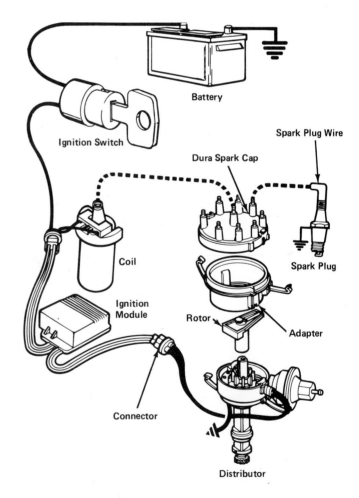

Figure 2-9. Parts of an ignition system.

The Automobile

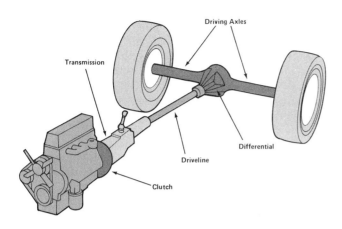

Figure 2-10. A drivetrain for a conventional rear-wheel drive (RWD) automobile.

- Differential
- Driving axles.

Figure 2-11 shows a *front-wheel drive (FWD)* drivetrain with a transaxle. In a FWD vehicle, the drivetrain transmits engine torque only to the front wheels.

On *four-wheel drive (4WD)* vehicles, additional mechanisms transmit torque to two more wheels so that all four wheels are driven by the engine **[Figure 2-12]**.

Most FWD automobiles have transaxles with the engines positioned over the driving wheels. A transaxle combines the functions of the transmission and differential. A transaxle drivetrain does not have a driveline **[Figure 2-13]**. Instead it has two driving axles.

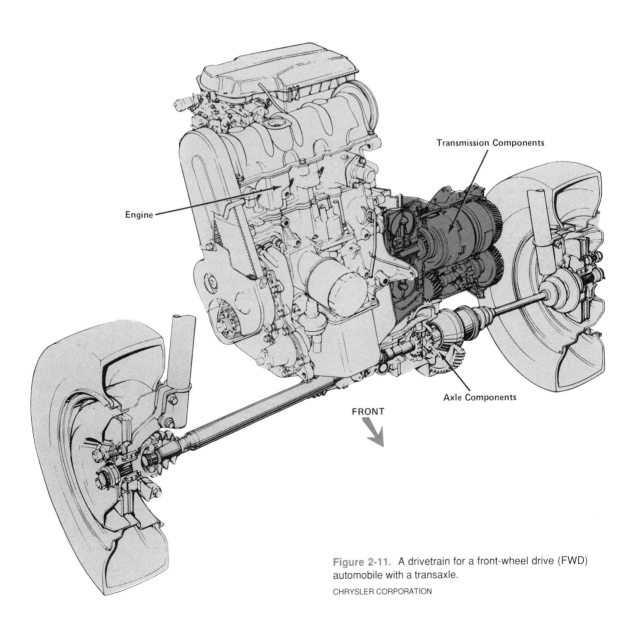

Figure 2-11. A drivetrain for a front-wheel drive (FWD) automobile with a transaxle.

CHRYSLER CORPORATION

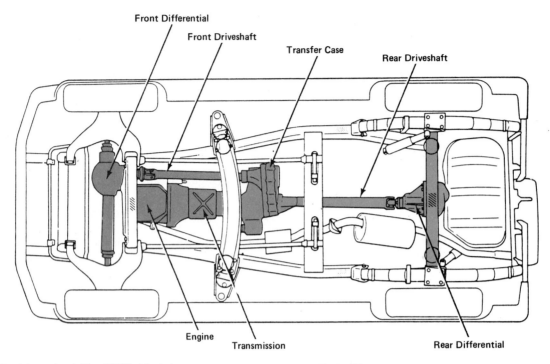

Figure 2-12. A four-wheel drive (4WD) drivetrain. MITSUBISHI MOTOR SALES OF AMERICA, INC.

Clutch

A *clutch* is a mechanical coupling assembly that connects or disconnects the flow of engine torque into a manual transmission or transaxle. The clutch is located within a housing between the rear of the engine and the front of the manual transmission or transaxle [Figure 2-14].

When the driver depresses the clutch pedal, engine torque is disconnected from the transmission and the rest of the drivetrain. When the clutch pedal is released, engine torque flows into the transmission and drivetrain to turn the driving wheels. A clutch allows the transmission to be shifted easily.

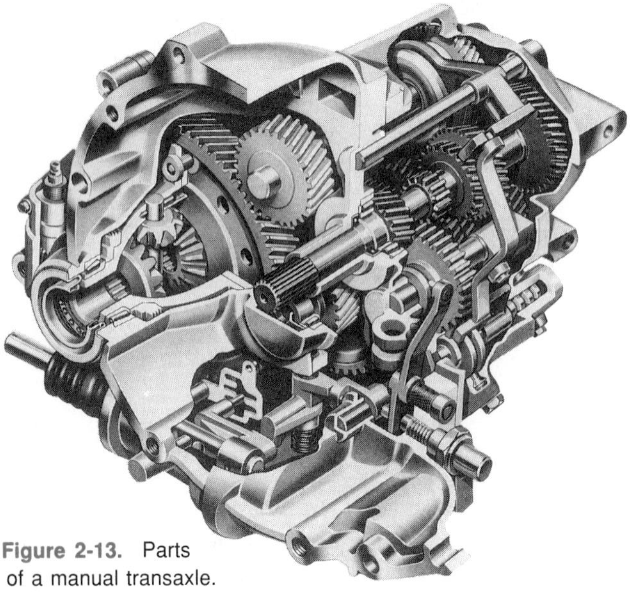

Transmission and Transaxle

Transmissions and *transaxles* include sets of gears. Through mechanical principles, the gears increase either torque or turning speed. Thus, the engine's torque output can be increased or decreased to match the load on the vehicle. When gears increase torque, rotational speed is reduced. When gears decrease torque, rotational speed is increased.

For example, a driver uses "low" gear (First) to increase engine torque and move the weight of the vehicle from rest. Then, for faster road speeds, the driver upshifts into "higher" forward gears (Second, Third, Fourth, and Fifth) that provide less torque but greater road speeds. When the load on the engine increases as the vehicle travels up a steep grade, the driver downshifts to a lower gear. Transmissions and transaxles also include gearsets to reverse the rotation of engine torque so that the vehicle can move backwards.

Manual transmission or transaxle. To shift gears easily in a *manual transmission* or *transaxle* [refer to Figures 2-13 and 2-14], the driver depresses the clutch pedal. This action interrupts the flow of torque from the engine to the transmission. Then the driver moves a shift lever by hand to select and engage the required set of gears. Finally, the driver releases the clutch pedal to apply engine torque to the transmission, drivetrain, and driving wheels.

Figure 2-13. Parts of a manual transaxle. FORD MOTOR COMPANY

The Automobile

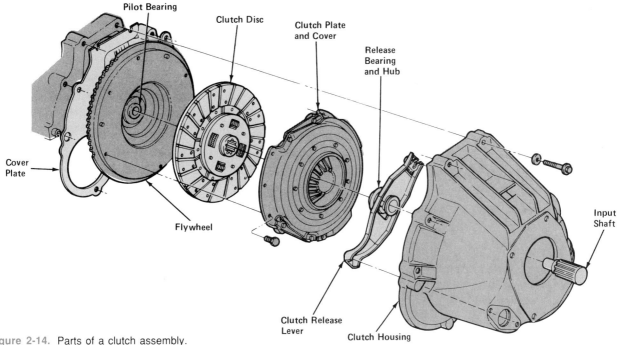

Figure 2-14. Parts of a clutch assembly.
FORD MOTOR COMPANY

Automatic transmission or transaxle. An *automatic transmission* or *transaxle* shifts automatically through forward gears. The driver selects a gear range by moving a selector lever **[Figure 2-15]**. The automatic transmission upshifts and downshifts automatically as necessary.

In place of a clutch, a *torque converter* couples engine torque to automatic transmissions and transaxles. A torque converter uses the force of flowing fluid instead of a mechanical coupling to transfer engine torque.

With a torque converter, the driver does not have to depress a pedal to couple or uncouple engine torque from an automatic transmission. The torque converter accepts turning force constantly from the engine. However, at low engine speeds, the engine produces and the torque converter transfers very little torque. As a result, there is not enough torque to move the vehicle efficiently.

When the driver depresses the accelerator and engine speed increases, more torque is developed and transmitted through the torque converter to the automatic transmission or transaxle gear mechanism. This torque turns the transmission gears, drivetrain parts, and driving wheels.

Driveline

A *driveline* transfers torque from a rear-wheel drive (RWD) transmission toward the driving wheels **[refer to Figure 2-10]**. The driveline includes both a

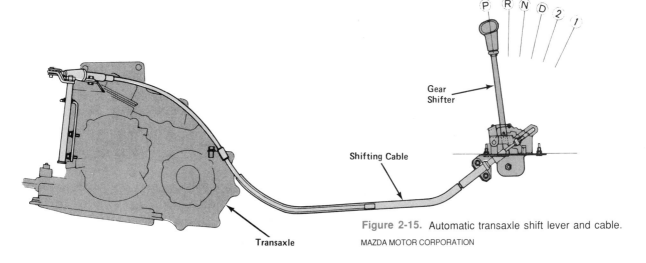

Figure 2-15. Automatic transaxle shift lever and cable.
MAZDA MOTOR CORPORATION

metal *driveshaft* and flexible joints called *universal joints.* The universal joints in the driveline allow it to transmit torque smoothly as the vehicle travels over bumps and dips in the road. Drivelines also are used in four-wheel drive (4WD) vehicles to transfer torque from a *transfer case* to a differential **[refer to Figure 2-12]**. A transfer case is a mechanism used in 4WD vehicles to apply torque to two additional wheels.

Differential

In a RWD or 4WD vehicle, the driveline transfers torque to a separate *differential* unit mounted between the two driving wheels **[Figure 2-16]**. The differential is a gear mechanism that allows the driving wheels to turn at different speeds when cornering. This action reduces tire wear on the driving wheel tires.

Front-wheel drive vehicles also need a differential mechanism. Transaxles include both a transmission and a differential unit in the same housing.

Driving Axles

From a differential, torque flows to metal *driving axles.* The driving axles turn the driving wheels **[Figure 2-17]**.

2.4 CHASSIS

The *chassis* is the supporting structure of a vehicle. The power train, body, and all other parts are mounted on the chassis **[Figure 2-18]**. Systems and parts associated with the chassis include:

* Suspension
* Steering
* Brakes
* Tires and wheels.

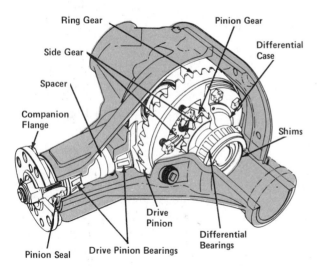

Figure 2-16. Conventional differential used in rear-wheel drive (RWD) and four-wheel drive (4WD) vehicles. FORD MOTOR COMPANY

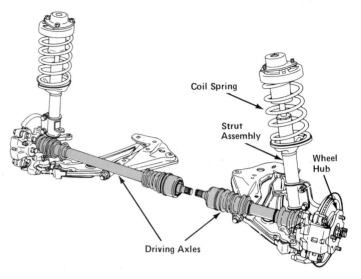

Figure 2-17. Driving axles for a front-wheel drive (FWD) automobile with transaxle. NISSAN MOTOR CORPORATION

Figure 2-18. The body of an automobile is supported by the chassis. FORD MOTOR COMPANY

Two types of chassis commonly used in modern vehicles are: frame and unitized body.

Frame

A *frame* is a large supporting assembly that fits below the floor and stretches the length of an automobile. The frame is made of thick, steel members formed into shapes such as those shown in **Figure 2-19**. The frame supports the engine, transmission, body, and most other assemblies of the automobile. The body is discussed in Topic 2.5. Frames currently are used primarily on large automobiles, trucks, and buses.

Unitized Body

The frame and body are combined into a single unit in many modern automobiles. This design is called a *unitized body.* Together, all of these parts of a unitized body create a structure that supports and

The Automobile

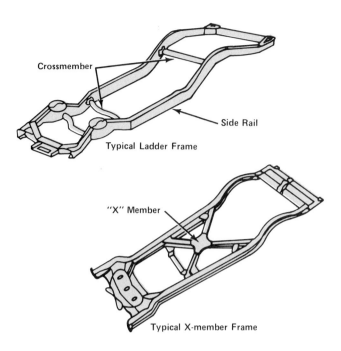

Figure 2-19. Two types of full automobile frames.

Figure 2-20. Body panels can be attached to a space frame.

holds the vehicle together. The strongest part of a unitized body is the *underbody.* Metal in the underbody is thinner than metal in frames, but underbody sections are overlapped, welded, and shaped to provide strength.

Some automobiles use a partial frame, or *stub frame*, for extra support in the engine and/or rear suspension areas.

A *space frame* includes tubes or pressed sheet-metal braces that are welded into a strong, light, web-like structure. The engine, drivetrain, and body panels can be attached to a space frame [**Figure 2-20**].

Suspension

The chassis is supported by the suspension. The suspension is connected to the frame or underbody and holds the wheels in proper alignment. The main parts of the suspension are the springs and shock absorbers. The ride and handling characteristics of an automobile depend upon the *suspension.*

Springs cushion the ride for both passengers and automobile parts. Much of the up-and-down movement from rough road surfaces is absorbed by the springs. The most common types of springs are coil springs and leaf springs. **Figure 2-21** shows different types of springs.

The action of springs is controlled by *shock absorbers.* Shock absorbers limit spring movement and resist the continuing flexing of springs as wheels encounter uneven surfaces [**Figure 2-22**].

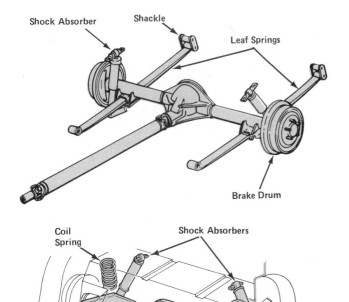

Figure 2-21. Two different types of springs are used on many automobiles.
CHEVROLET MOTOR DIVISION—GMC

Steering

To turn an automobile, a *steering system* is built into the front end of the automobile. When the driver turns the steering wheel, a shaft from the steering column turns a steering gear. The steering gear moves *tie rods* that connect to the front wheels. The tie rods move the front wheels to turn the vehicle left or right. **Figure 2-23** shows two common steering systems.

Some recent passenger automobiles also steer the rear wheels slightly for better cornering and ease in parking. This system is known as *four-wheel steering (4WS)* and is described in Unit 59.

Brakes

When the driver depresses the brake pedal, fluid pressure causes parts to move and create friction, or rubbing resistance, that slows and stops the vehicle.

Two types of brakes are used on automobiles: *disc brakes* and *drum brakes.* Most passenger vehicles have disc brakes at the front wheels and drum brakes at the rear wheels. Older vehicles had drum brakes at all four wheels, and luxury and sports cars may have disc brakes at all four wheels.

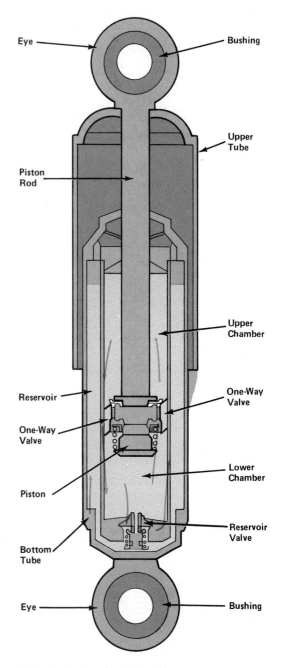

Figure 2-22. Parts of a shock absorber.

Disc brakes operate in much the same way as hand brakes on a bicycle. Disc brake parts include a *rotor, caliper unit,* and *friction pads* **[Figure 2-24]**. When the brake pedal is depressed, the U-shaped caliper forces the pads against both sides of the rotor that rotates with a wheel. The pad materials rub against the rotor to create friction that slows and stops the wheel.

Drum brakes include a *drum, wheel cylinder,* and *brake shoes* and *linings* **[Figure 2-25]**. When the brakes are applied, the brake shoes are forced against the inside of the hollow brake drum that rotates with a wheel. The linings on the brake shoes create friction to slow the wheel.

The Automobile

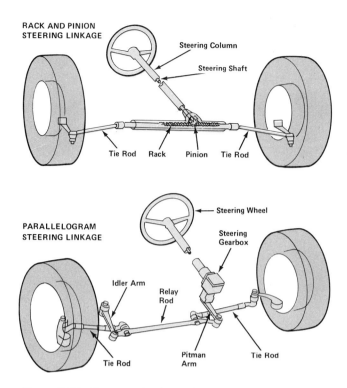

Figure 2-23. Common steering systems.

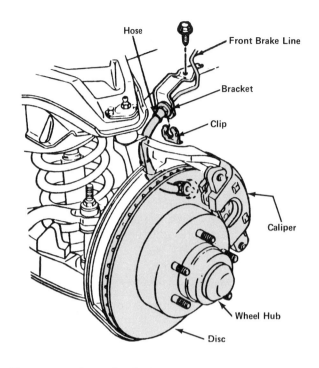

Figure 2-24. Parts of a disc brake assembly.
BUICK MOTOR DIVISION—GMC

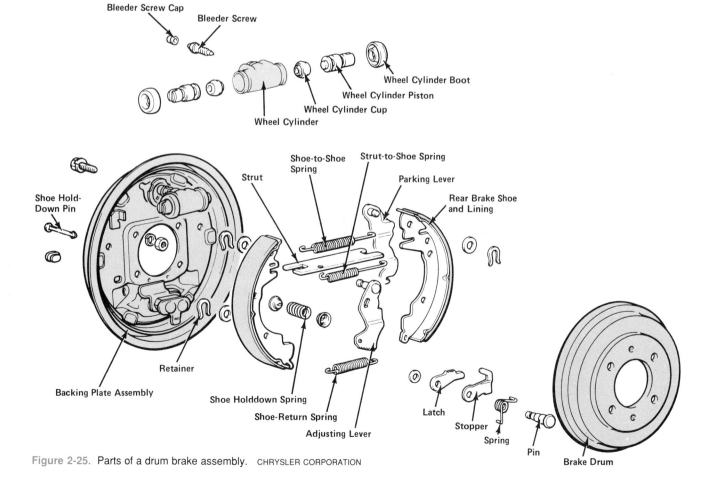

Figure 2-25. Parts of a drum brake assembly. CHRYSLER CORPORATION

Tires and Wheels

The only contact a vehicle has with the road are its tires. Tires are made of rubber and other materials and are filled with air to cushion the ride of an automobile. The tires are mounted on metal wheels and turn on axles or spindles. Wheels and tires come in many different sizes. Their sizes must be matched to one another and to the automobile.

2.5 BODY

The *body* holds, carries, and protects passengers and cargo. The exterior panels of the body form a protective shell and give the automobile its shape. Interior body parts provide comfort for passengers and space for cargo.

Body Size and Styles

Automobiles and light-duty trucks come in many sizes. The smallest are called mini-compacts. Other sizes are subcompact, compact, intermediate, and full-size. Light-duty trucks include pickup, van, and utility vehicle designs in a variety of sizes. Common automobile body styles are shown in **Figure 2-26.**

Aerodynamics

Aerodynamics is the science that studies the ability of a body to move through air. An automotive body that moves through air easily provides greater fuel economy and efficiency **[Figure 2-27].**

Body Materials

Most automotive body parts and panels are made from thin steel sheets stamped and welded into specific shapes. Plastic and fiberglass materials also are used for body parts.

Figure 2-27. Airflow is measured as it passes over an automobile in a wind-tunnel test.
FORD MOTOR COMPANY

Figure 2-26. Common automotive body styles.

UNIT HIGHLIGHTS

- Almost all automobiles use internal-combustion, reciprocating piston engines.
- The basic parts of an automotive engine include the block, cylinder head, pistons, connecting rods, and crankshaft.
- An automotive engine burns a mixture of air and fuel.
- An electrical spark is created to ignite the air-fuel mixture.
- A rear-wheel drive (RWD) vehicle's drivetrain can include a clutch, transmission, driveline, differential, and driving axles.
- Front-wheel drive (FWD) vehicles have a transaxle that includes a transmission and a differential in one housing that turn driving axles.
- The automobile body may have a separate frame or may be of the unitized-body type.

TERMS

- engine
- cylinder block
- cylinder head
- valve
- four-stroke cycle
- crankshaft
- lubrication system
- oil pump
- cooling system
- air-cooled
- radiator
- fuel injection system
- combustion chamber
- distributor
- alternator
- power train
- clutch
- transmission
- automatic transmission
- driveline
- universal joint
- differential
- chassis
- unitized body
- space frame
- springs
- steering system
- drum brake
- brake shoe
- brake lining
- caliper unit
- body
- transfer case
- torque
- manifold
- exhaust manifold
- reciprocating piston engine
- front-wheel drive (FWD)
- underbody
- four-wheel steering (4WS)
- charging system
- internal combustion engine
- cylinder
- valve train
- piston
- connecting rod
- lubrication
- oil pan
- oil gallery
- water-cooled
- coolant
- carburetor
- exhaust system
- spark plug
- battery
- drivetrain
- transaxle
- manual transmission
- manual transaxle
- automatic transaxle
- driveshaft
- cam
- driving axle
- frame
- stub frame
- suspension
- shock absorber
- disc brake
- brake drum
- rotor
- friction pad
- aerodynamics
- torque converter
- camshaft
- intake manifold
- starting system
- voltage regulator
- rear-wheel drive (RWD)
- four-wheel drive (4WD)
- tie rod
- wheel cylinder

REVIEW QUESTIONS

DIRECTIONS: The following questions are similar to those used on automotive technician certification tests. On a separate sheet of paper, write the letter of the correct choice.

1. Which of the following statements is correct?
   I. A clutch is used with a manual transaxle.
   II. A transmission is part of a transaxle.
   A. I only
   B. II only
   C. Both I and II
   D. Neither I nor II

2. Technician A says that the differential allows the driving wheels to turn at different speeds during cornering.
   Technician B says power passes from the differential to the driving wheels through the driving axles.
   Who is correct?
   A. A only
   B. B only
   C. Both A and B
   D. Neither A nor B

3. Which of the following parts are found in a braking system?
   A. Friction pads
   B. Gears
   C. Manifold
   D. Distributor

Questions 4 and 5 are not like those above. Each has the word EXCEPT. For each question, look for the choice that does *not* apply. Read each question carefully before you choose your answer.

4. All of the following describe a common automotive engine EXCEPT
   A. reciprocating piston engine.
   B. external combustion.
   C. internal combustion.
   D. burns air-fuel mixture.

5. All of the following are part of the suspension EXCEPT
   A. coil springs.
   B. calipers.
   C. leaf springs.
   D. shock absorbers.

ESSAY QUESTIONS

1. What are the four basic assemblies that make up an automobile?
2. Describe the major parts of a gasoline piston engine.
3. What types of drivetrains are used on automobiles?

SUPPLEMENTAL ACTIVITIES

1. Identify and describe the basic parts of an internal-combustion piston engine.
2. Look at the underside of an automobile and determine whether the driving wheels are at the front or at the rear.
3. Identify the individual parts of a drivetrain.
4. Locate and identify the parts of front and rear brakes used on a vehicle.
5. Refer to Figure 2-23 and identify the type of steering used on a vehicle selected by your instructor.

3 Using Service Manuals And Shop Forms

UNIT PREVIEW

Service manuals are books that describe automotive servicing and repair. Automotive manufacturers and specialized publishing companies furnish many types of manuals. In repair businesses, shop forms are used for work orders, inventory control, and as records of work performed.

LEARNING OBJECTIVES

When you have completed your assignments and exercises in this unit, you should be able to:

- ❏ Find specific information in service manuals and service bulletins.
- ❏ Demonstrate how to use a flat rate manual to determine the time and costs of a repair.
- ❏ Find specific information in specifications tables.
- ❏ Describe information disclosure laws and explain how they affect servicing procedures.
- ❏ Demonstrate how to write a repair order.

3.1 SERVICE MANUALS

A *service manual* is a book that explains how to service and repair an automobile. Some service manuals include exact, step-by-step instructions on how to diagnose and repair parts and systems on a specific automobile. Other service publications give only the highlights on a repair job.

Technicians use different types of service publications. The most common are:

- Manufacturer's service manuals
- Manufacturer's service bulletins
- General and specialty repair manuals
- Flat rate manuals.

Manufacturer's Service Manuals

Automotive manufacturers put out a series of service manuals each year that explain service procedures related to the vehicles of that model year.

Manufacturer's service manuals are updated each year. New manuals are issued whenever new automobiles are introduced. One or more service manuals is produced by the manufacturer for each automotive model **[Figure 3-1]**.

Figure 3-1. Manufacturers publish service manuals with complete servicing information for each model they produce.

Manufacturer's service manuals provide a technician with step-by-step procedures on all servicing aspects of a vehicle, plus *specifications.* Specifications are measurements for tightening, fitting, and testing various parts of an automobile. These manuals also describe special tools and special servicing procedures needed.

Manufacturer's service manuals explain repair procedures for experienced technicians in dealerships. The writers of the manuals assume that the technicians are experienced in all types of auto repairs. The proper use of manufacturer's service manuals is discussed in greater detail in Topic 3.2.

Manufacturer's Service Bulletins

A *manufacturer's service bulletin* notifies technicians of changes or corrections in service procedures or specifications **[Figure 3-2]**. The service bulletins usually are one- or two-page descriptions of a change or addition to information in a service manual.

General Repair Manuals

General repair manuals and *specialty repair manuals* are printed by publishing companies rather than automotive manufacturers. General repair manuals have some service procedures and specifications for several automotive models and years. Information in general repair manuals is condensed, or shortened, and more general in nature. This allows the coverage of more subjects in less space. One volume of a general repair manual may contain information from 20 or more manufacturer's service manuals. The widely used general repair manuals are Chilton, Mitchell, and Motor **[Figure 3-3]**.

Figure 3-2. Regular service bulletins are issued by automobile manufacturers to provide the latest service procedures or changes.

In addition to manuals, information on servicing automobiles is published in many automotive service magazines. Tool manufacturers also have servicing publications describing the use of their tools or equipment.

Flat Rate Manuals

A *flat rate manual* lists the cost of parts and labor times for automotive repair jobs. A flat rate manual, shown in **Figure 3-4,** is used to provide a written estimate of repairs for the customer. In addition, service managers and shop foremen base technicians' pay on the times listed for specific jobs.

3.2 USING MANUFACTURER'S SERVICE MANUALS

To use service manuals correctly, the technician must be able to find specific information quickly and accurately. There are two basic steps in using service manuals: Choosing the correct manual and locating information in it.

Choose the Correct Manual

A variety of manuals are available in most shops. Choose the correct manual based on the make, model, year, and type of work to be performed on an automobile.

Figure 3-3. General repair manuals offer service procedures for a wide variety of automobiles.

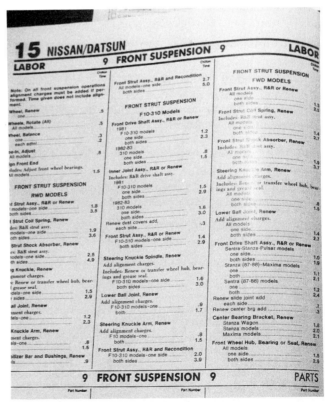

Figure 3-4. A flat-rate manual gives information on the cost of parts and labor for specific service procedures.

USING SERVICE MANUALS AND SHOP FORMS 27

Locate Desired Information

To locate the necessary information, use the table of contents and look up the proper section or sections. Check out the service procedures and specifications tables in those sections.

Using the table of contents. The table of contents is located at the front of the service manual. A table of contents is shown in **Figure 3-5.**

The table of contents gives a general description of service procedures for all parts of an automobile. A section listing, beside the table of contents, gives a section number where service procedures can be found.

Section Guides. The first section page has a guide, or more detailed table of contents **[Figure 3-6]**. Use this to locate the correct page for information that you need.

Information within sections. Each section contains step-by-step repair procedures for one specific automotive assembly **[Figure 3-7]**. The information includes special tools, safety procedures, and safety precautions.

Specifications tables. Torque and other specification tables usually are listed at the end of each section. Tighten fasteners to the torque specifications listed **[Figure 3-8]**. Improper torquing, or making a part too tight or too loose, can cause that part to fail or to become disconnected.

3.3 FORMS AND RECORD-KEEPING

As in any other business, written records must be kept neatly and correctly. Commonly used forms include parts orders, dispatch sheets, and repair orders.

Dispatch Sheet

A work schedule, or *dispatch sheet,* keeps track of appointments. The work schedule at a garage will be posted for use by the service manager. The service manager uses the dispatch sheet to schedule customers' service appointments.

Repair Order

After an appointment is made, the service manager, or an assistant, may fill out a *repair order,* or *work order.* A repair order includes spaces for information about the customer and the vehicle, services performed, parts used, and billing **[Figure 3-9]**.

Many states have *information disclosure* laws to protect consumers from repair fraud. The garage must furnish written job estimates to the customer **[Figure 3-10]**. Additional approval must be received if the

TABLE OF CONTENTS	

GENERAL INFORMATION
  A. General Information
  B. Maintenance and Lubrication
	0.

CLIMATE CONTROL
  C. Electronic Climate Control
  D1. R-4 Compressor
  D2. DA-6 Compressor
	1.

BODY, FRAME AND BUMPERS
  A. Frame and Body Mounts
  B. Bumpers
  C. Chassis Sheet Metal
	2.

STEERING, SUSPENSION, WHEELS AND TIRES
  3. Diagnosis
  A. Wheel Alignment
  B. Steering
     B1. Power Rack and Pinion
     B3. Power Steering Pump
     B4. Steering Wheel and Column
  C. Front Suspension
  D. Rear Suspension
  E. Wheels and Tires
	3.

DRIVE AXLES	4.
BRAKES	5.

ENGINE
DIESEL — 4.3 Liter V-6, Code T
  6. Diagnosis
  A. Mechanical
  B. Cooling
  C. Fuel Injection
  D. Electrical
  E. Emission Controls
  F. Exhaust
  H. Vacuum Pump
GASOLINE — HT4100 V-8, Code 8
  6. Diagnosis
  A. Mechanical
  B. Cooling
  C. Fuel System and DFI
  D. Electrical
  E. Emission Controls
  F. Exhaust
  H. Vacuum Pump
	6.

TRANSAXLE
  A. Maintenance and On-Car Service
     440 — T4 Transaxle
	7.

ELECTRICAL
  A. Electrical Troubleshooting
  B. Lighting Systems
  C. Instrument Panel and Gages
  D. Computer System Diagnosis
     D1. ECM Trouble Code Diagnosis
     D2. BCM Trouble Code Diagnosis
	8.

ACCESSORIES	9.
INDEX	10.

Figure 3-5. The table of contents at the front of a service manual tells a technician where to look for service procedures.

CADILLAC MOTOR CAR DIVISION—GMC

440-T4 DIAGNOSIS AND ON-CAR SERVICE
CONTENTS

GENERAL DESCRIPTION	
MAINTENANCE AND	
ADJUSTMENTS	7A-1
Checking Fluid Level	7A-1
Changing Fluid and Filter	7A-2
T.V. Cable Adjustment	7A-3
Manual Linkage	7A-3
Shift Control Cable	7A-3
Park/Neutral and Back-Up Lamp Switch	7A-4
Shift Indicator	7A-4
DIAGNOSIS	
Preliminary Checking Procedure	7A-5
Road Test Procedure	7A-6
Fluid Pressure Test Procedure	7A-7
Fluid Leak Diagnosis	7A-7
Vacuum Modulator Diagnosis	7A-7
Clutch Plate Diagnosis	7A-8
Computer Diagnostics	7A-9
Torque Converter	7A-10
General Description	7A-10
Converter Stator Operation	7A-11
Converter Inspection	7A-11
Converter Vibration Test Procedure	7A-11
Converter Flushing Procedure	7A-12
Torque Converter Replacement	7A-13
Viscous Converter Clutch	7A-13
GENERAL SERVICE	
PROCEDURES	7A-15
Engine Coolant in Transmission	7A-15
Oil Cooler and Cooler Line Flushing	7A-15
Case Porosity Repair Procedure	7A-17
ON CAR SERVICE	
Transaxle Filler Tube and/or Seal	7A-17
T.V. Cable and/or Seal	7A-17
Converter to Flexplate Bolts	7A-18
Speed Sensor, Driven Gear and Seal	7A-18
Governor, Cover and Seal	7A-18
Pan and Gasket	7A-18
Filter and Lip Seal Ring	7A-18
Scavenger Oil Scoop	7A-18
Governor Pipe Assembly	7A-21
Accumulators, 1-2 and 3-4	7A-21
Vacuum Modulator and 0-Ring	7A-21
Reverse Servo Assembly	7A-21
1-2 Servo Assembly	7A-21
Cooler Lines and/or Fittings	7A-21
Shift Control Cable	7A-22
Transaxle Mount	7A-23
Case Side Cover Pan	7A-25
Valve Body	7A-26
Thermal Element	7A-27
Pressure Switch, 3rd and 4th/Thermistor Switch	7A-27
Solenoid, Check Ball and Wiring Harness	7A-27
440-T4 Transaxle Assembly	7A-28

Figure 3-6. A section guide contains specific information about servicing procedures. CADILLAC MOTOR CAR DIVISION—GMC

---

cost will be higher than the original written estimate. Replaced parts of reasonable size and weight must be returned to the customer when they request.

When the customer arrives for the scheduled appointment, a service writer notes customer complaints or labor instructions on the repair order. These instructions tell the technician what problems to diagnose and/or what repairs to perform.

The repair order and the automobile are dispatched, or sent, to the technician assigned to do the servicing. The technician follows instructions on the work order to begin service.

Ordering Parts

Vehicle dealerships are required to maintain an adequate inventory of commonly used parts. Parts managers use parts order forms to order additional stocks and also to make special orders of seldom-used parts. An independent repair shop usually will not stock large inventories of parts. Instead, it will rely on nearby parts stores and on dealership parts departments.

Parts requisition. Any new parts that are needed are written in the *parts requisition* section of the repair order by the technician [Figure 3-10]. The technician takes the repair order to the parts department to get the parts. As new parts are issued, the counter attendant writes parts prices in the parts requisition.

Labor charges. The technician turns in the repair order when work on the automobile is completed. The service manager, or service writer, adds the labor charges.

Billing. All labor charges and parts charges are totaled in the billing department. Sales tax and any extra charges are added and totaled. The customer pays the charges in the billing department.

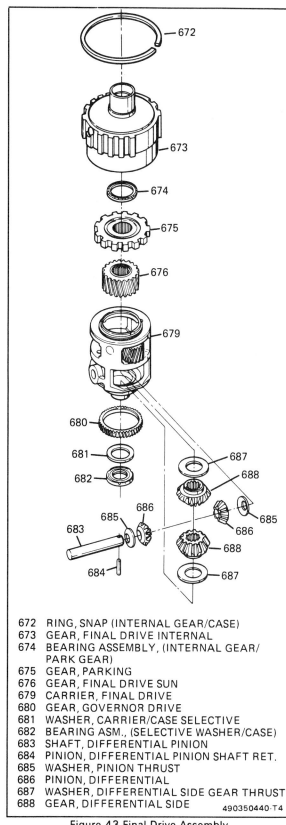

672 RING, SNAP (INTERNAL GEAR/CASE)
673 GEAR, FINAL DRIVE INTERNAL
674 BEARING ASSEMBLY, (INTERNAL GEAR/PARK GEAR)
675 GEAR, PARKING
676 GEAR, FINAL DRIVE SUN
679 CARRIER, FINAL DRIVE
680 GEAR, GOVERNOR DRIVE
681 WASHER, CARRIER/CASE SELECTIVE
682 BEARING ASM., (SELECTIVE WASHER/CASE)
683 SHAFT, DIFFERENTIAL PINION
684 PINION, DIFFERENTIAL PINION SHAFT RET.
685 WASHER, PINION THRUST
686 PINION, DIFFERENTIAL
687 WASHER, DIFFERENTIAL SIDE GEAR THRUST
688 GEAR, DIFFERENTIAL SIDE

Figure 43 Final Drive Assembly

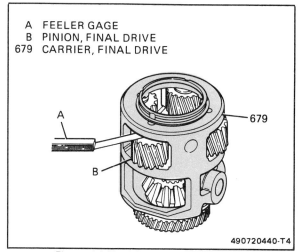

Figure 44 Final Drive Pinion End Play

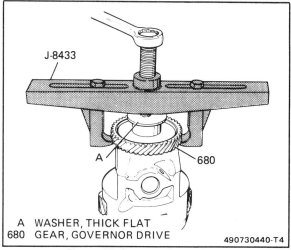

Figure 45 Governor Drive Gear

Install or Connect

- Drive gear - tap into position with a soft mallet

Inspect

- Pinions (688) and side gears (686) for damage

Pinion Replacement Procedure

Disassemble (Figure 46)

- Retaining pin (684) with a pin punch
- Pinion shaft (683)
- Pinions (688), side gears (686), and washers (685 and 687)

Inspect

- Washers (685 and 687) and carrier (679) for damage.

Assemble (Figure 43)

Figure 3-7. Step-by-step service procedures give exact descriptions of the work that needs to be done. CADILLAC MOTOR CAR DIVISION—GMC

DESCRIPTION OF USAGE	THREAD SIZE	ASM. TORQUE
CONNECTOR COOLER FITTING	1/4 - 18	41.0 N·m (30 lb.-ft.)
MODULATOR TO CASE	M8 X 1.25 X 20.0	27.0 N·m (20 lb.-ft.)
PUMP COVER TO CHANNEL PLATE	M6 X 1.0 X 94.0	14.0 N·m (10 lb.-ft.)
PUMP COVER TO PUMP BODY	M8 X 1.25 X 20.0	27.0 N·m (20 lb.-ft.)
PUMP COVER TO PUMP BODY (TORX. HD.)	M8 X 1.25 X 20.0	27.0 N·m (20 lb.-ft.)
GOVERNOR CONTROL BODY TO COVER (TORX. HD.)	M6 X 1.0 X 22.0	14.0 N·m (10 lb.-ft.)
PIPE PLUG	1/8 - 27	14.0 N·m (10 lb.-ft.)
CASE TO DRIVE SPROCKET SUPPORT	M8 X 1.25 X 23.5	27.0 N·m (20 lb.-ft.)
MANIFOLD TO VALVE BODY	M6 X 1.0 X 35.0	14.0 N·m (10 lb.-ft.)
GOVERNOR TO CASE	M8 X 1.25 X 25.0	27.0 N·m (20 lb.-ft.)
PRESSURE SWITCH	1/8 - 27	14.0 N·m (10 lb.-ft.)
SOLENOID TO VALVE BODY	M6 X 1.0 X 14.0	14.0 N·m (10 lb.-ft.)
MANUAL DETENT SPRING TO VALVE BODY	M6 X 1.0 X 16.0	14.0 N·m (10 lb.-ft.)
CASE SIDE COVER TO CHANNEL PLATE (NUT)	M6 X 1.0 X 20.0	14.0 N·m (10 lb.-ft.)
PUMP COVER TO VALVE BODY	M6 X 1.0 X 45.0	14.0 N·m (10 lb.-ft.)
PUMP COVER TO CHANNEL PLATE	M6 X 1.0 X 85.0	14.0 N·m (10 lb.-ft.)
VALVE BODY TO CASE (TORX. HD.)	M8 X 1.25 X 90.0	27.0 N·m (20 lb.-ft.)
VALVE BODY TO CASE	M8 X 1.25 X 70.0	27.0 N·m (20 lb.-ft.)
PUMP BODY TO CASE	M8 X 1.25 X 95.0	27.0 N·m (20 lb.-ft.)
VALVE BODY TO CHANNEL PLATE	M6 X 1.0 X 35.0	14.0 N·m (10 lb.-ft.)
VALVE BODY TO CHANNEL PLATE (TORX. HD.)	M6 X 1.0 X 60.0	14.0 N·m (10 lb.-ft.)
CHANNEL PLATE TO CASE (TORX. HD.)	M8 X 1.25 X 30.0	27.0 N·m (20 lb.-ft.)
CHANNEL PLATE TO DRIVEN SPROCKET SUPPORT (TORX. HD.)	M8 X 1.25 X 45.0	27.0 N·m (20 lb.-ft.)
VALVE BODY TO CASE	M8 X 1.25 X 70.0	27.0 N·m (20 lb.-ft.)
CHANNEL PLATE TO CASE (TORX. HD.)	M8 X 1.25 X 45.0	27.0 N·m (20 lb.-ft.)
VALVE BODY TO DRIVEN SPROCKET SUPPORT (TORX. HD.)	M8 X 1.25 X 90.0	27.0 N·m (20 lb.-ft.)
SIDE COVER TO CASE	M8 X 1.25 X 16.0	13.0 N·m (10 lb.-ft.)
ACCUMULATOR COVER TO CASE	M8 X 1.25 X 30.0	27.0 N·m (20 lb.-ft.)
OIL SCOOP TO CASE	M8 X 1.25 X 30.0	13.0 N·m (10 lb.-ft.)
RETAINER GOVERNOR CONTROL BODY	M8 X 1.25 X 30.0	27.0 N·m (20 lb.-ft.)
TRANSMISSION OIL PAN TO CASE	M8 X 1.25 X 16.0	13.0 N·m (10 lb.-ft.)
MANUAL SHAFT TO INSIDE DETENT LEVER (NUT)	M10 X 1.5	34.0 N·m (25 lb.-ft.)

Figure 3-8. Specifications tables give measurements and tightening requirements that must be used during service.
CADILLAC MOTOR CAR DIVISION—GMC

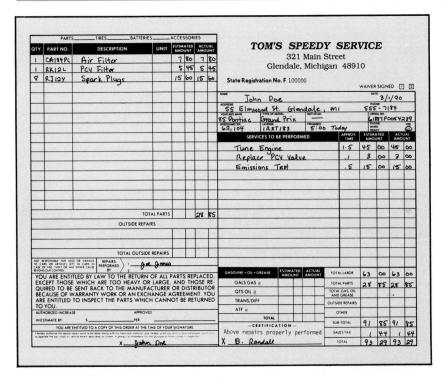

Figure 3-9. A repair order is filled out with information about service needed, parts replacement, and the cost of repairs.

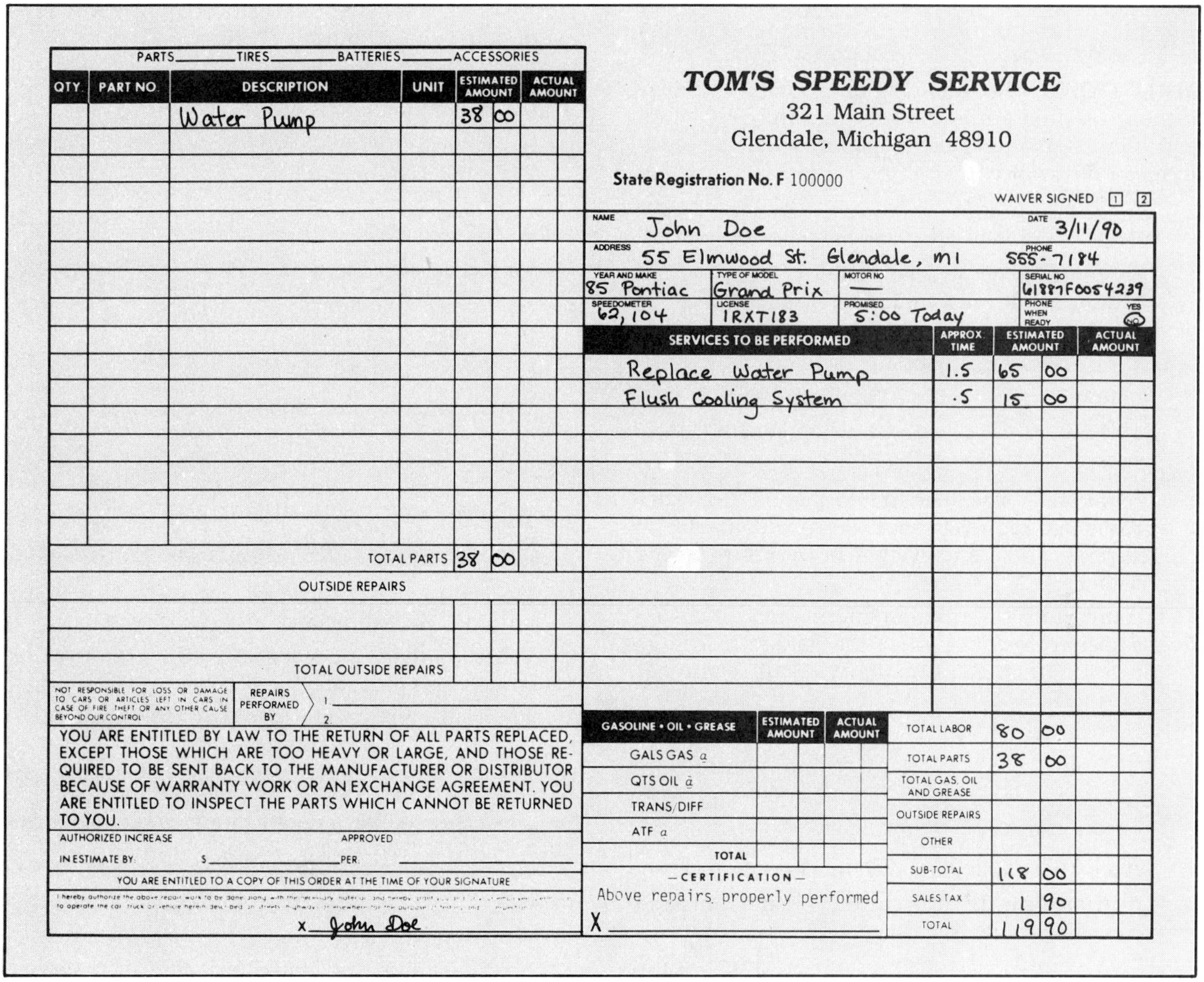

PARTS____TIRES____BATTERIES____ACCESSORIES

QTY.	PART NO.	DESCRIPTION	UNIT	ESTIMATED AMOUNT	ACTUAL AMOUNT
		Water Pump		38 00	
			TOTAL PARTS	38 00	

OUTSIDE REPAIRS

TOTAL OUTSIDE REPAIRS

NOT RESPONSIBLE FOR LOSS OR DAMAGE TO CARS OR ARTICLES LEFT IN CARS IN CASE OF FIRE, THEFT OR ANY OTHER CAUSE BEYOND OUR CONTROL

REPAIRS PERFORMED BY 1 ____ 2 ____

YOU ARE ENTITLED BY LAW TO THE RETURN OF ALL PARTS REPLACED, EXCEPT THOSE WHICH ARE TOO HEAVY OR LARGE, AND THOSE REQUIRED TO BE SENT BACK TO THE MANUFACTURER OR DISTRIBUTOR BECAUSE OF WARRANTY WORK OR AN EXCHANGE AGREEMENT. YOU ARE ENTITLED TO INSPECT THE PARTS WHICH CANNOT BE RETURNED TO YOU.

AUTHORIZED INCREASE IN ESTIMATE BY ____ $ ____ APPROVED PER ____

YOU ARE ENTITLED TO A COPY OF THIS ORDER AT THE TIME OF YOUR SIGNATURE

X John Doe

TOM'S SPEEDY SERVICE
321 Main Street
Glendale, Michigan 48910

State Registration No. F 100000

WAIVER SIGNED 1 2

NAME John Doe — DATE 3/11/90
ADDRESS 55 Elmwood St. Glendale, MI — PHONE 555-7184
YEAR AND MAKE 85 Pontiac — TYPE OF MODEL Grand Prix — MOTOR NO — — SERIAL NO 61887F0054239
SPEEDOMETER 62,104 — LICENSE 1RXT183 — PROMISED 5:00 Today — PHONE WHEN READY YES (NO)

SERVICES TO BE PERFORMED	APPROX. TIME	ESTIMATED AMOUNT	ACTUAL AMOUNT
Replace Water Pump	1.5	65 00	
Flush Cooling System	.5	15 00	

GASOLINE • OIL • GREASE	ESTIMATED AMOUNT	ACTUAL AMOUNT
GALS GAS @		
QTS OIL @		
TRANS/DIFF		
ATF @		
TOTAL		

— CERTIFICATION —
Above repairs properly performed
X ____

TOTAL LABOR	80 00
TOTAL PARTS	38 00
TOTAL GAS, OIL AND GREASE	
OUTSIDE REPAIRS	
OTHER	
SUB-TOTAL	118 00
SALES TAX	1 90
TOTAL	119 90

Figure 3-10. A job estimate often is required before any service work can be performed.

UNIT HIGHLIGHTS

- Service manuals are a major source of information for technicians.
- Manufacturer's service manuals have more specific service information than general repair manuals.
- Service manuals generally are divided into the following sections: Table of contents, automotive parts systems, and specifications tables.
- A repair order includes spaces for customer complaints, directions to technicians, parts requisitions, labor charges, and billing totals.

TERMS

service manual
specifications
general repair manual
specialty repair manual
flat rate
flat rate manual
dispatch sheet
repair order
information disclosure
work order
parts requisition
billing
manufacturer's service manual
manufacturer's service bulletin

REVIEW QUESTIONS

DIRECTIONS: The following questions are similar to those used on automotive technician certification tests. On a separate sheet of paper, write the letter of the correct choice.

1. Which of the following statements is correct?
   I. Step-by-step procedures for repairing an automobile are found in service manuals.
   II. Specifications tables list correct measurements for repair jobs.
   A. I only
   B. II only
   C. Both I and II
   D. Neither I nor II

2. Technician A says that an information disclosure provision requires that worn-out parts of reasonable size and weight be returned to the customer.
   Technician B says that an information disclosure provision does not allow the garage to charge more than the estimate without customer approval.
   Who is correct?
   A. A only
   B. B only
   C. Both A and B
   D. Neither A nor B

3. Which of the following statements is correct?
   I. A repair order lists customer complaints and services to be performed.
   II. A completed repair order includes billing information.
   A. I only
   B. II only
   C. Both I and II
   D. Neither I nor II

Questions 4 and 5 are not like those above. Each has the word EXCEPT. For each question, look for the choice that does *not* apply. Read each question carefully before you choose your answer.

4. All of the following information sources would be of help to an automotive technician EXCEPT
   A. a sales brochure.
   B. a manufacturer's service bulletin.
   C. a general repair manual.
   D. a flat rate manual.

5. Forms must be filled out to complete all of the following procedures EXCEPT
   A. taking inventory.
   B. making appointments.
   C. giving a repair estimate.
   D. moving the automobile.

ESSAY QUESTIONS

1. Describe the types of manuals used by automotive technicians.
2. What are the basic steps in locating information in a service manual?
3. What information does a repair order include?

SUPPLEMENTAL ACTIVITIES

1. Select a service manual and locate the section for repairing engines.
2. Find a torque specifications table in a service manual.
3. Look in a flat rate manual and report how many hours and tenths of hours are required to replace a clutch on a specified automobile.
4. At $40 an hour, figure out the flat rate labor charge on a repair job selected by your instructor.
5. Fill out all sections of a repair order and total the charges correctly.

4 Safety Around The Automobile

UNIT PREVIEW

Safe working habits prevent accidents. Automotive technicians must be aware of personal safety, shop safety, and emergency procedures. No one outgrows the need for safety, regardless of experience. Shop safety is everyone's responsibility.

LEARNING OBJECTIVES

When you have completed your assignments and exercises in this unit, you should be able to:

- ❏ Demonstrate how to use hand tools and power tools safely.
- ❏ Demonstrate how to use compressed air safely.
- ❏ Safely use jacks, hoists, and jackstands.
- ❏ Identify and describe the different kinds of fire extinguishers.

4.1 ATTITUDE

Carelessness and a poor working attitude can cause injury to yourself or to someone else. Only proper conduct and a good working attitude can prevent accidents. Always use proper safety procedures and avoid shortcuts.

4.2 GOOD SAFETY HABITS

Always work in an area with proper ventilation. Running engines emit exhaust fumes that contain *carbon monoxide,* a poisonous gas. Also, fuel leakage builds up a dangerous vapor in the air, which can be ignited by the slightest spark.

Keep the floor in your area clean and free of grease and oil. Grease and oil spots are slippery and cause accidents.

Keep your work area floor clear of tools and parts. Someone could trip and injure themselves.

4.3 SAFETY WHILE WORKING

With a little common sense, you can protect yourself from dangers and avoid most problems when you work around automobiles. ***Don't take chances.*** Avoid unsafe situations and horseplay. Follow all safety recommendations and give your complete attention to the job you are doing.

To learn about specific work hazards on your job, ask your instructor or refer to the manufacturer's service manual *before you begin work.*

As a general guide for your continuing safety and health, read about and follow the following safety practices whenever you work on and around automobiles.

- Work on vehicles only when you are feeling fit and well. Never work on cars if you are tired, ill, or have taken drugs that cause drowsiness, or that impair mental or physical functioning.
- Whenever possible, work with helpers or partners to help you with work difficulties and safety precautions.
- Perform all repairs and maintenance according to the vehicle manufacturer's recommendations in the service manual.
- Make sure you have the correct tools, parts, supplies, and materials before you begin work. Do not substitute or try to improvise with makeshift tools, parts, or supplies.
- Always wear eye protection when you work on vehicles. *Safety glasses, face shields,* and *goggles* prevent eye injuries, infections, and possible blindness **[Figure 4-1]**.
- Know the location of your shop's first aid kit and of water fountains to flush out eyes and to clean minor wounds.
- Whenever possible, avoid working on a running engine or one which is still hot. Serious burns

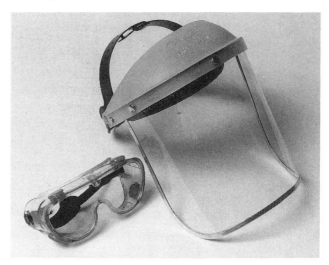

Figure 4-1. Safety glasses and face shields are worn to protect eyes and face.

result from touching hot exhaust system, cooling system, or other engine parts. If possible, allow the engine to cool down for several hours before beginning work.

- If you must work on a running engine, stay clear of all moving parts, including belts, pulleys, and fan blades. A spinning fan may be impossible to see. Serious injury or loss of limbs can result if you are caught in moving machinery.
- Remove all watches, rings, neck chains or bracelets, and all other jewelry from your body before you start work. These items can become caught in moving parts or short out electrical circuits and cause serious injury.
- Do not wear loose clothing that can become caught in moving parts. Tuck in shirttails and roll up long sleeves or wear short-sleeved shirts or blouses when working on a running engine.
- If you have long hair, use rubber bands to tie it in a tight ponytail. Pin the ponytail to the top of your head and wear a cap or hair net.
- Work in well-ventilated areas. Never run the engine in a closed garage or other enclosed area. Carbon monoxide gas from the exhaust is poisonous.
- Work in well-lighted areas and use a safety droplight with a wire cage around the bulb. If a bare bulb breaks, the hot filament can ignite fuel or other flammable vapors and cause a fire.
- Position the vehicle on a flat and level surface before beginning work. Avoid working on a vehicle parked on an incline.
- Place the transmission in PARK (automatic transmissions) or in a gear (manual transmissions) and set the parking brake securely.
- If you are going to raise one end of a vehicle, place wheel chocks both in front and in back of the wheels that remain on the ground. This precaution will prevent the vehicle from rolling.
- Raise and support the vehicle properly and safely according to the recommendations in the manufacturer's service manual.
- Always use pairs of equal-height *safety stands*, or *jackstands*, to support a vehicle safely off the ground. Never place any part of your body under a vehicle that is supported only by a jack.
- Always disconnect the grounded (usually the Negative) battery terminal when you work on the electrical system. This precaution will prevent sparks, fires, and damage to electrical parts.
- Know the location of fire extinguishers that will extinguish Classes A, B, and C fires **[Figure 4-2]**.
- Do not use a heater with an open flame to heat the work area. Use an electric heater placed as far away as possible from the vehicle and from any gasoline or other flammable liquids.
- Don't smoke near vehicles or engines. Sparks or flames from cigarettes and other smoking materials can ignite fuel vapors, flammable liquids, and hydrogen gas produced by the battery.
- Collect oil, fuel, brake fluid, and other liquids only in approved safety containers **[Figure 4-3]**. Dispose of liquids in accordance with all local, state, and federal environmental and safety standards.
- Avoid spilling or storing fuel or other flammable liquids near any source of open flame or sparks, including gas water heaters and electrical switches. Wipe up all spills immediately. Store soaked rags and flammable liquids in approved metal safety containers with tight-fitting lids to prevent fires **[Figure 4-3]**.

Knowing Shop Layout

A *shop layout* shows the location of work areas and equipment in a particular shop. A typical automotive repair shop is illustrated in **Figure 4-4**. A layout, or drawing, of the same shop is shown in **Figure 4-5**.

In case of emergency, you should know the locations of safety equipment and the telephone in your shop. Locate the work areas, workbenches, lifts, and equipment. Some shops have painted lines on the floor to mark off work areas and danger areas.

Location of Safety Devices

Know where all exits are located. Be sure that you can locate—and know how to use—each fire extinguisher in the shop. Telephones should have emergency numbers listed beside them. Know the location of first-aid kits and how to use them properly.

Avoiding Hazards in the Shop Area

Be aware of potential hazards. The automotive service area is no place for horseplay. To avoid hazards, always use tools, flammable and toxic liquids, and other automotive supplies safely and treat them with respect. If you observe hazards or unsafe conditions, report them to your instructor immediately.

Hand tools. *Hand tools* are tools that are operated by hand, such as screwdrivers, wrenches, and hammers. Keep hand tools in good condition. Tools that slip can cause cuts and bruises. If a tool falls into a moving part, it may fly out and cause serious personal injury.

Use the proper tool. Using the wrong tool can damage parts, the tool itself, or cause injury. Do *not* use broken or bent tools.

SAFETY AROUND THE AUTOMOBILE

FIRES

A — CLASS A FIRES ORDINARY COMBUSTIBLE MATERIALS SUCH AS WOOD, PAPER, TEXTILES AND SO FORTH. REQUIRES...COOLING-QUENCHING

B — CLASS B FIRES FLAMMABLE LIQUIDS, GREASES, GASOLINE, OILS, PAINTS AND SO FORTH. REQUIRES...BLANKETING OR SMOTHERING

C — CLASS C FIRES ELECTRICAL EQUIPMENT, MOTORS, SWITCHES AND SO FORTH. REQUIRES...A NONCONDUCTING AGENT

EXTINGUISHERS

TYPE	USE		OPERATION
FOAM — SOLUTION OF ALUMINUM SULPHATE AND BICARBONATE OF SODA	OK FOR	A B	*FOAM*: DON'T PLAY STREAM INTO THE BURNING LIQUID. ALLOW FOAM TO FALL LIGHTLY ON FIRE
	NOT FOR	C	
CARBON DIOXIDE — CARBON DIOXIDE GAS UNDER PRESSURE	NOT FOR	A	*CARBON DIOXIDE*: DIRECT DISCHARGE AS CLOSE TO FIRE AS POSSIBLE. FIRST AT EDGE OF FLAMES AND GRADUALLY FORWARD AND UPWARD
	OK FOR	B C	
DRY CHEMICAL	MULTI-PURPOSE TYPE — OK FOR A B C	ORDINARY BC TYPE — NOT FOR A, OK FOR B C	*DRY CHEMICAL*: DIRECT STREAM AT BASE OF FLAMES. USE RAPID LEFT-TO-RIGHT MOTION TOWARD FLAMES
SODA-ACID — BICARBONATE OF SODA SOLUTION AND SULPHURIC ACID	OK FOR	A	*SODA-ACID*: DIRECT STREAM AT BASE OF FLAME
	NOT FOR	B C	

Figure 4-2. Fire extinguisher reference chart. FORD MOTOR COMPANY

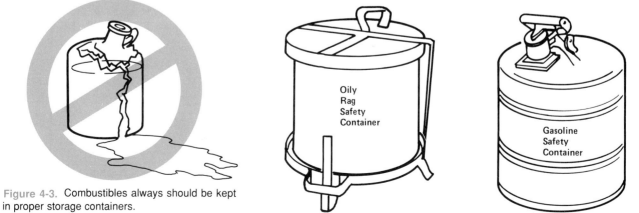

Figure 4-3. Combustibles always should be kept in proper storage containers.

Oily Rag Safety Container

Gasoline Safety Container

Figure 4-4. A typical shop layout.

Be careful when using sharp or pointed tools that can slip and cause injury. If a tool, such as a chisel, is supposed to be sharp, make sure it is sharp. Dull tools that slip also can cause injury.

Power tools. *Power tools* are operated with an outside source of power, such as electricity, compressed air, or hydraulic pressure. Always wear safety glasses when using power tools. Because power tools exert great force, serious injury can result from carelessness.

Do *not* use a power tool without permission from your instructor. Be sure you know how to operate the tool properly before using it. Read the operating instructions carefully.

When working with larger power tools—like bench or floor equipment—check machines for signs of damage and proper adjustments. Place all *safety guards* in position [Figure 4-6]. A safety guard is a protective cover over a moving part that prevents injury. Wear safety glasses or a face shield. Make sure everyone is, and all tools are, clear of moving parts before starting the machine. Always keep hands and clothing away from moving parts.

Never leave a power tool unattended when it is running. If you leave, turn off the machine. Anyone passing a running, unattended machine can be hurt seriously.

If a machine does not operate properly, turn it off and notify your instructor immediately.

After using a power tool, turn it off and wait until it has completely stopped. Disconnect the power source of the tool, then clean, oil, and adjust the tool as necessary.

Do *not* enter a work area where a power tool is being used. Obey the rules outlined by your instructor.

Compressed air. *Compressed air* is air that is under pressure and directed through hoses to perform work. Compressed air is used to inflate tires, spray paint, and drive tools. Compressed air equipment can be dangerous when not used properly.

When using compressed air, always wear safety glasses or a face shield. Particles of dirt and pieces of metal blown under high pressure can penetrate your skin or get into your eyes.

Before using a compressed air system, check all hose connections. Always hold the *air nozzle*, or air control device, securely when starting or shutting off compressed air. A loose nozzle under pressure can whip suddenly and can cause serious injury.

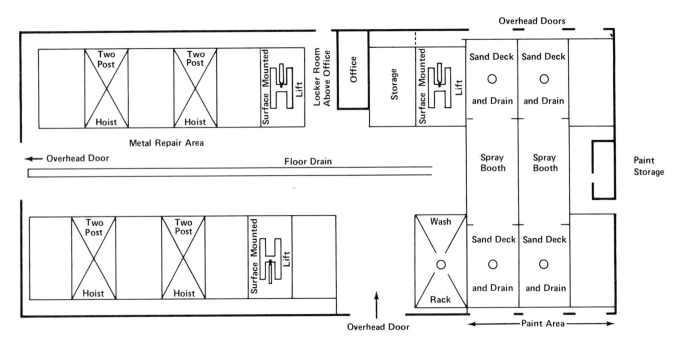

Figure 4-5. Floorplan of a shop layout.

SAFETY AROUND THE AUTOMOBILE

Figure 4-6. Safety guards on a grinder help prevent pieces from flying out and causing injury.

Do *not* point an air nozzle at anyone. Do *not* blow dirt from your clothes or hair. Do *not* use compressed air to clean the floor or a workbench. The high pressure of the compressed air will blow particles everywhere, possibly causing an injury.

Hydraulic jacks and hoists. An automobile is raised off the ground by a *hydraulic jack* or *hoist*. A jack is a portable tool that is moved under the automobile to raise it off the ground [**Figure 4-7**]. A hoist is mounted permanently in a work area [**Figure 4-8**]. An automobile must be moved onto the hoist.

Extreme caution is necessary when using jacks and hoists. Most automobiles weigh more than a ton. If a vehicle falls, it will seriously injure or kill anyone beneath it.

Before you use a jack or hoist, you must understand thoroughly how it works. Get permission from the instructor to use the jack or hoist.

The doors, hood, and trunk lid of the automobile should be closed before raising the vehicle. These parts could be damaged when the vehicle is lifted. All passengers should be out of the vehicle before it is raised. *Never* jack up a vehicle when someone is underneath it.

Never use a jack or hoist that is not working properly. When the vehicle is raised, connect the locking device, if a locking device is part of the system. If you are using a floor jack, lower the car onto a properly positioned jackstand under the vehicle before you get underneath. Ask your instructor for permission before you get underneath the vehicle. Wear safety glasses or a face shield whenever you work under an automobile.

Jackstands. *Jackstands,* also called *safety stands,* are important safety devices used with a jack or hoist.

Figure 4-7. A floor jack is a portable tool used to raise an automobile.

Jackstands are variable-height supports that hold a vehicle safely and securely off the ground. Jackstands are placed under a sturdy chassis member, such as the frame or axle housing [**Figure 4-9**].

Chain hoists and cranes. Heavy parts of the automobile, such as engines, are removed with *chain hoists* [**Figure 4-10**] or *cranes*. Another term for chain hoist is *chain fall*. Cranes sometimes are called "cherry pickers." To prevent serious injury, chain hoists and cranes must be attached properly to the parts being lifted and the load must be balanced.

Get permission from your instructor before you attach a hoist or crane to an automobile part. Attach the lifting chain or cable to the system that is to be removed. Have your instructor check the attachment.

Place the chain hoist or crane directly over the assembly that is to be removed. Make sure the chain or cable is secure before lifting.

Figure 4-8. Hoists are large, permanently installed devices that raise an entire automobile for a technician to work under.

Figure 4-9. Jackstands, or safety stands, are safety supports used to support an automobile for service.

Figure 4-10. A chain hoist, mounted to an overhead pulley, is used to remove engines and other heavy assemblies.

Exhaust gases from running engines. Many service procedures require that you run the engine in the shop. To do this, the exhaust should be directed outside of the shop. Many shops have an exhaust ventilation system [Figure 4-11]. Some shops may require that the automobile be started and run outside, or with the shop doors fully opened.

Before you start the vehicle, get permission from the instructor. Block the wheels to prevent the vehicle from moving. Place the transmission lever in PARK for automatic transmissions or in neutral for manual transmissions. Set the emergency brake. Do *not* stand in front of or behind the automobile or let anyone else stand there when you start the car.

Painting equipment. *Spray painting* is done with compressed air. Paint is flammable and toxic. Painting should be done in a spray booth with a ventilation system to remove flammable and toxic fumes. The painter must wear a respirator, or filtering mask, and safety glasses to protect against the paint spray.

Cleaning equipment. The cleaning of parts is necessary in all repair shops. Parts may be cleaned either by steam or in a solvent, *never* with gasoline. A *steam cleaner* uses hot water vapor, or steam, to melt and flush dirt off parts [Figure 4-12]. When you use a steam-cleaner, wear safety glasses or a face shield and gloves to protect against burns.

Solvent is a chemical that dissolves dirt and grease from parts. Some solvents are harmful chemicals that can damage clothing, skin, and eyes. When you use these solvents, wear safety glasses or a face shield, gloves, and overalls or a shop coat [Figure 4-13].

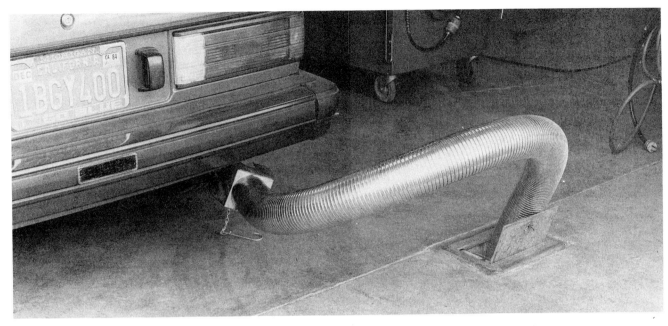

Figure 4-11. An exhaust-gas ventilation system draws exhaust fumes from an automobile tailpipe and directs it outside.

Figure 4-12. A steam cleaner removes heavy dirt from many parts of an automobile.

Figure 4-13. Parts are cleaned in a solvent tank. Safety glasses, a shop apron, gloves, and long sleeves prevent burns while cleaning in a solvent tank.

Equipment Defects and Misuse

Be on the lookout for defective and misused shop equipment. Make sure each piece of equipment has the proper safety guard. Guards should be located over all fans, belts, pulleys, and grinders.

Electric cords should be in good condition, not be worn or frayed. Defective electrical cords and connections may cause fire or could even electrocute someone. Remove cords from sockets by pulling on the plug, not the cord. Position electrical cords out of the way and store them properly after you have finished work.

Cylinders of compressed gas, such as the kind used for welding, must *not* be stored near heat. All cylinders must be used in a ventilated area. Secure them to a solid support with a chain or cable. Cylinders should *not* be used to support other equipment or parts. Improper storage can cause cylinders to explode.

Do *not* use faulty welding equipment. Gas welding must be confined to a protected area with no combustibles, oil, or grease nearby to cause an explosion. Electric welding also must be done in a protected place or in a special booth. Welders must wear proper safety equipment. Fire extinguishers must be handy and well-marked. Never coil electrical welding cables during use.

Fire Prevention and Fire Extinguishers

Gasoline, cleaning solvents, and paints are only some of the combustibles found in automotive service areas. These combustibles can catch fire or cause an explosion easily if safety measures are not followed.

The most important sign that any technician should heed is the "NO SMOKING" sign. All "NO SMOKING" signs should be large and displayed properly. No one should smoke, or have an open flame, in a service area, including customers.

All combustibles must be kept in fireproof containers designed for storing such liquids [**refer to Figure 4-3**]. These storage containers must prevent leaking and evaporation. Improperly stored combustibles can be ignited by a spark as small as from a light switch. Never store combustibles in glass containers. Glass containers can fall, break, and explode.

Wipe up spilled gasoline immediately. Put gasoline-soaked rags outside to dry, away from flames and in a "NO SMOKING" area. All oil and paint rags must be kept in closed, fireproof containers.

When you work on a leaking fuel system, catch leaking gasoline in a container or with rags. Fix the leak as quickly as possible. Do *not* allow metal tools to touch the battery while you work on the fuel system, because this may cause sparks and an explosion.

Keep the shop doors open or the ventilation system operating to prevent the buildup of harmful fumes.

All shop areas should have fire extinguishers. Know the location and the specific purpose of each fire extinguisher. Only class B and C fire extinguishers can be used to extinguish gasoline or electrical fires. Also, know how to use fire extinguishers on all types of fires [**refer to Figure 4-2**]. The sooner a fire is reached, the easier it is to control.

Do *not* play with fire extinguishers. Besides causing someone to slip or cause eye damage, empty extinguishers are useless in a real emergency.

Safety Rules in the Shop Area

All shops have safety rules. Follow the rules at all times [**Figure 4-14**]. Before you begin work, learn all of the safety rules. These rules are designed to protect you and your job.

Emergency Procedures

If an accident occurs or someone is hurt, notify your instructor immediately. Do *not* give first aid or move an injured person if you don't know what you are doing. Trying to help may do more harm than good.

Operating Cars Safely

Do *not* move a car in the shop or make a road test without your instructor's permission.

Special precautions. When moving a car in the work area, first test the brakes. Then, buckle up the safety belt. Use extreme care when driving a car in the shop. Make sure no one is under another car, that the way is clear, and that there are no tools or parts under the car.

When making a road test, first check the brakes. Fasten your safety belt tightly. A road test is a drive to identify a specific problem or verify the repair of a problem. Observe all traffic laws. Drive only as far as is necessary to check the automobile. *Never* make jackrabbit starts, turn corners too quickly, or drive faster than conditions allow. A careless road test can damage the vehicle severely or cause an accident and injuries.

Figure 4-14. Always follow a shop's safety rules.
CLIFF CREAGER

SAFETY AROUND THE AUTOMOBILE

UNIT HIGHLIGHTS

- Injuries can occur if you become careless and sloppy.
- Personal safety includes wearing the proper clothes, wearing the proper safety devices, and using the proper tools.
- A shop layout shows the location of work areas and equipment.
- The improper use of hand and power tools can cause hazards in the shop area.
- Compressed air is dangerous when not used properly.
- Extreme caution is necessary when using jacks and hoists because an automobile may weigh a ton or more.
- To run an engine in the shop, the exhaust must be directed outside of the shop.
- Each piece of equipment should be inspected for defects and misuse both before and after use.
- Combustibles must be stored in a special area and handled carefully to prevent fires and explosions.
- Know the location and use of all fire extinguishers in the shop area.

TERMS

carbon monoxide
face shield
hand tools
safety guard
air nozzle
hydraulic hoist
chain hoist
spray painting
solvent
goggles

safety glasses
shop layout
power tools
compressed air
hydraulic jack
jackstand
crane
steam cleaner
safety stand

REVIEW QUESTIONS

DIRECTIONS: The following questions are similar to those used on automotive technician certification tests. Answer each question by circling the letter of the correct choice.

1. Which of the following statements is correct?
   I. Long hair and jewelry can become caught in moving parts.
   II. It's alright to get under a car only supported by a jack if you'll only be there a few minutes.
   A. I only
   B. II only
   C. Both I and II
   D. Neither I nor II

2. Technician A says that screwdrivers, wrenches, and hammers are typical hand tools.
   Technician B says that a power tool can be left running if the technician puts up a "power on" sign.
   Who is correct?
   A. A only
   B. B only
   C. Both A and B
   D. Neither A nor B

3. You should store flammable liquids in
   A. a metal safety container with a tight-fitting lid.
   B. a glass bottle.
   C. a plastic bottle.
   D. an uncovered drain pan.

4. Which of the following statements is correct?
   I. Carbon dioxide fire extinguishers are recommended to extinguish Class A fires.
   II. Dry chemical, multi-purpose type fire extinguishers are recommended to extinguish types A, B, and C fires.
   A. I only
   B. II only
   C. Both I and II
   D. Neither I nor II

Question 5 is not like those above. It has the word EXCEPT. For question 5, look for the choice that does *not* apply. Read the entire question carefully before you choose your answer.

5. All of the following statements are true EXCEPT
   A. compressed air is used to drive tools.
   B. a jack is a portable tool that raises an automobile.
   C. a hoist is mounted permanently in a work area.
   D. jackstands are used to keep jacks aligned properly.

ESSAY QUESTIONS

1. Why is safety important?
2. Describe how an automotive technician dresses properly for work in the shop area.
3. Which types of fire extinguishers should be used for different types of fires?

SUPPLEMENTAL ACTIVITIES

1. Make a sketch of the school shop layout. Indicate any possible safety hazards. Note the locations of all exits, fire extinguishers, and water fountains.
2. Demonstrate proper and safe use of a power tool chosen by your instructor.
3. Study the procedure for raising an automobile on the school hoist. Demonstrate the correct and safe way to raise and support an automobile.
4. Choose a fire extinguisher that is used in your work area. Study the instructions carefully. Describe what type of fire the extinguisher is designed to control. Explain how to control a fire with that extinguisher.

5 Using Fasteners

UNIT PREVIEW

Fasteners attach and hold parts together. Most automotive assemblies are held together with fasteners. This unit discusses the different types of fasteners and how they are used.

LEARNING OBJECTIVES

When you have completed your assignments and exercises in this unit, you should be able to:

- ❏ Describe different kinds of screws and bolts.
- ❏ Identify the kinds of nuts found on automobiles.
- ❏ Describe the differences between the conventional system and the metric system in measuring threaded diameters, thread pitch, and interpreting designations.
- ❏ Describe nonthreaded fasteners.
- ❏ Measure conventional and metric screws, bolts, and nuts and correctly identify their sizes and thread pitches.

5.1 MEASUREMENT SYSTEMS

For automotive systems to operate properly, the parts must fit together securely. Fasteners, parts, and the tools needed to work on them are made to specific sizes, or measurements. Automotive manufacturers use the two most common measurement systems—the *conventional system* and the *metric system.*

The Conventional System

The measurement system found most frequently in the United States is the conventional system. Measurements used in this system require knowing many different combinations of numbers [**Figure 5-1**]. For example, one foot contains 12 inches. Rulers and other measuring instruments divide each inch into 16, 32, 64, or 100 equal parts.

Sizes must be known for fasteners, tools, and parts. These sizes are stated in inches or parts of inches. The parts of inches are expressed either as fractions or in decimals.

The Metric System

The metric system is also called the international system, because it is used in most countries. The metric system is easier to use. There is no need to memorize that 12 inches make a foot, three feet make a yard, and so on. Just remember that each metric unit of measurement is 10 times the size of the previous unit [**Figure 5-2**]. For example, 10 millimeters (10 mm) equal 1 centimeter (1 cm). A meter (39.4 inches) equals 1,000 mm or 100 cm.

Decimal fractions always are used in the metric system. Units smaller than 1 meter are listed as 0.01 for centimeter and 0.001 for millimeter. Each smaller metric unit of measurement is 1/10 the size of the previous unit. Some metric fasteners on American cars can be identified by their blue color.

Conventional/Metric Conversions

Converting from conventional to metric, or from metric to conventional is common in automotive servicing. Conversion tables or booklets usually are available in shops. A typical automotive *conventional/metric conversion* chart is shown in **Figure 5-3**.

UNITS OF LENGTH MEASUREMENT

12 Inches = 1 Foot
3 Feet = 1 Yard

UNITS OF WEIGHT MEASUREMENT

16 Drams = 1 Ounce
16 Ounces = 1 Pound
2000 Pounds = 1 Ton

Figure 5-1. Conventional length and weight measurements.

Number of Meters	Prefix	Symbol
1,000	Kilometer	km
100	Hectometer	hm
10	Decameter	dam
1	Meter	
0.1	Decimeter	dm
0.01	Centimeter	cm
0.001	Millimeter	mm
0.000 001	Micrometer	μm

Figure 5-2. Metric length measurements.

5.2 THREADED FASTENERS

A thread is a spiral groove on the outside or inside of a fastener. An external thread fits into a matching internal thread within another fastener or within a part. Fasteners hold parts and assemblies together securely when torqued, or tightened, properly.

Threaded fasteners commonly used in automobiles include:

- Screws
- Bolts
- Nuts
- Studs.

	multiply	by	for equiv. no. of:
ACCELERATION	Foot/sec^2	0.304 8	metre/sec^2 (m/s^2)
	Inch/sec^2	0.025 4	metre/sec^2
TORQUE	Pound-inch	0.112 98	newton-metres (N·m)
	Pound-foot	1.355 8	newton-metres
POWER	horsepower	0.746	kilowatts (kw)
PRESSURE or STRESS	inches of water	0.2488	kilopascals (kPa)
	pounds/sq. in.	6.895	kilopascals (kPa)
ENERGY or WORK	BTU	1 055.	joules (J)
	foot-pound	1.355 8	joules (J)
	kilowatt-hour	3 600 000. or 3.6 x 10^6	joules (J=one W's)
LIGHT	foot candle	10.76	lumens/metre^2 (lm/m^2)
FUEL PERFORMANCE	miles/gal	0.425 1	kilometres/litre (km/l)
	gal/mile	2.352 7	litres/kilometre (l/km)
VELOCITY	miles/hour	1.609 3	kilometres/hr. (km/h)
LENGTH	inch	25.4	millimetres (mm)
	foot	0.304 8	metres (m)
	yard	0.914 4	metres (m)
	mile	1.609	kilometres (km)
AREA	inch^2	645.2	millimetres^2 (mm^2)
		6.45	centimetres^2 (cm^2)
	foot^2	0.092 9	metres^2 (m^2)
	yard^2	0.836 1	metres^2
VOLUME	inch^3	16387.	mm^3
	inch^3	16.387	cm^3
	quart	0.946 4	litres (l)
	gallon	3.785 4	litres
	yard^3	0.764 6	metres^3 (m^3)
MASS	pound	0.453 6	kilograms (kg)
	ton	907.18	kilograms (kg)
	ton	0.907 18	tonne (t)
FORCE	kilogram	9.807	newtons (N)
	ounce	0.278 0	newtons
	pound	4.448	newtons
TEMPERATURE	degree fahrenheit	0.556 (°F −32)	degree Celsius (°C)

Figure 5-3. Conventional/metric conversions. FORD MOTOR COMPANY

Screws

A *screw* is a small-diameter fastener that threads into a threaded hole. A screw has threads on the outside, or *external threads*. A screw holds two parts together.

Some screws have hexagonal heads that can be turned with a wrench that fits around the head. An example of this type is a *capscrew* [**Figure 5-4**].

A *machine screw* [**Figure 5-5**] has a slotted head. A screwdriver blade fits into the screw slot.

A *Phillips screw* has a hollow, cross-shaped depression in its head. A Phillips screwdriver has a matching tip to turn the screw.

An *Allen head screw* [**Figure 5-6**] has a hollow hexagonal depression in its head. Special hex-shaped metal bars called Allen wrenches are used to turn Allen head screws.

A *sheet metal screw* is used to join two pieces of sheet metal. Sheet metal is thin metal, such as the metal used for car bodies. The sheet metal screw is tapered and pointed at the end [**Figure 5-7**]. This allows it to be started easily in a hole in the sheet metal. Sheet metal threads are wide so that the screw will cut and hold while it is being turned.

Self-tapping screws have points that cut their own threads into metal. A self-tapping screw is hardened to aid in cutting threads. The external thread makes internal threads as the screw is tightened into place [**Figure 5-8**].

Bolts and Nuts

A *bolt* is a larger fastener with external threads. Bolts usually have hexagonal heads [**Figure 5-9**], but may have Allen, Torx, square heads, or other shapes of heads.

One form of Torx bolt [**Figure 5-10**] has a special, modified hexagonal depression in it, similar to an Allen head. Another form of Torx bolt has a raised portion with the Torx shape. Torx fasteners are used extensively on newer vehicles.

Bolts are secured with *nuts.* Nuts are fasteners with internal threads. Nuts are turned and tightened on the threaded shaft of a bolt. Most automotive nuts have hexagonal heads [**Figure 5-11**]. Square-head nuts are used occasionally, as on top-post battery clamps.

Self-locking nuts typically include plastic inserts [**Figure 5-12**]. These plastic inserts resist movement and prevent the nut from loosening and unscrewing due to vibration. Self-locking nuts are used where extra locking action is needed. As the self-locking nut is tightened, the bolt is squeezed solidly into the plastic material.

A slotted nut, or *castellated nut,* is used when it is important for the nut to remain in one position without turning. A castellated nut has slots and

Figure 5-4. Capscrew.

Figure 5-5. Machine screw.

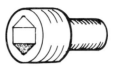
Figure 5-6. Allen head screw.

Figure 5-7. Sheet metal screw.

Figure 5-8. Self-tapping screw.

Figure 5-9. Bolt.

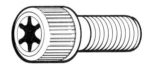
Figure 5-10. Torx bolt.

Figure 5-11. Hex nut.

Figure 5-12. Self-locking nut.

Figure 5-13. Castellated nut and cotter pin.

holes in it [**Figure 5-13**]. A *cotter pin* consists of a folded, flattened steel wire. The pin is inserted through holes in parts and slots in the castellated nut, and the ends of the cotter pin are spread to keep parts from moving or loosening.

Speed Nuts

A *speed nut,* shown in **Figure 5-14,** is pressed over the threads of a bolt or stud. A speed nut is used to replace conventional nuts in some applications.

Figure 5-14. Speed nut.

Studs

A *stud* is a fastener with external threads on each end [**Figure 5-15**]. To hold parts together, a stud first is tightened into a threaded hole in a part. A second part fits over the stud. Finally, a nut fits over the exposed end of the stud and is tightened.

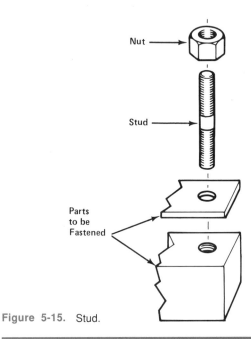

Figure 5-15. Stud.

5.3 WASHERS

Washers are used between a screw, bolt, or nut and the part being tightened. *Flat washers* and *lockwashers* are used frequently in automobiles.

Flat washers spread out the load of a tightened nut over a larger area [**Figure 5-16**]. This spreading-out action prevents damage to a part surface.

Lockwashers have sharp edges that dig into fasteners and parts to prevent them from moving. Lockwashers, shown in **Figure 5-17,** are placed under a nut or screw head.

5.4 THREADED DIAMETERS AND DESIGNATIONS

Once you have determined which type of threaded fastener is required, you must find one that fits properly. Fasteners come in many sizes in both conventional and metric measurements.

Figure 5-16. Flat washer.

Figure 5-17. Lock washers.

Conventional System

The bolt shown in **Figure 5-18** has a conventional size designation of 1/2-20 NF x 1. This designation tells the technician that the diameter of the bolt is 1/2 inch. There are 20 threads per inch. The bolt has a *national fine thread (NF).* A fine thread is one that has narrow spaces between threads. The length of the bolt, from the base of the head, is 1 inch.

The figure of 20 threads per inch determines whether the bolt will fit with the threads of a part or nut. The NF also determines whether the threads will match.

Bolt, screw, or stud size is the outside thread diameter. Nut size is the inside thread diameter.

A *national coarse thread (NC)* designation is used, too. A coarse thread has more space between threads than a fine thread. Fine and coarse threads are based on what is called the Unified System (nonmetric).

A length of one inch means that the threaded portion of the shaft (below the head) is one inch long. The bolt must be long enough to pass through

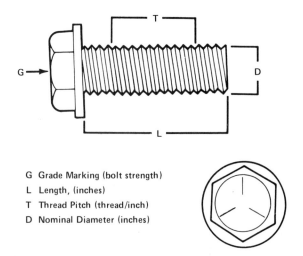

G Grade Marking (bolt strength)
L Length, (inches)
T Thread Pitch (thread/inch)
D Nominal Diameter (inches)

Figure 5-18. Conventional system bolt measurement.
FORD MOTOR COMPANY

the part and screw completely into another part or nut to hold securely.

Metric System

The bolt shown in **Figure 5-19** has a metric size designation of M 6.0 x 1. This designation tells the technician that the bolt has metric (M) threads. The diameter of this bolt is 6.0 millimeters. The *thread pitch* is indicated by the numeral 1, meaning that there is 1mm between each thread. A bolt with a higher pitch number has a coarser thread. In the metric system, no NC or NF thread designations are used.

If the metric designation is followed by a number for length, the length is given in metric units, usually millimeters. This is the length of the threaded shaft, below the bolt head.

5.5 DETERMINING THREAD PITCH

Many manufacturers use both conventional and metric system fasteners on the same automobile, as well as fine and coarse thread sizes. Because different sizes of fasteners are *not* interchangeable, the technician must be able to determine exactly what is needed.

Use a *thread gauge,* either conventional or metric, to determine thread pitch **[Figure 5-20]**. Find the blade that matches and fits into the threads and read the pitch designation on the blade.

5.6 GRADE MARKINGS

Fasteners are graded for strength. You must know how to interpret the *grade markings* on fasteners. Always use a fastener of similar quality when a replacement is necessary. A lower grade bolt or nut probably will break and fail. As in thread designations, there are grade markings for both conventional and metric systems.

Figure 5-20. Thread gauge.

Conventional systems for bolt and nut grade markings are shown in **Figure 5-21**. No marks means the fastener is of the lowest quality. Eight marks indicate the highest quality, strongest bolt. Grade markings for nuts are indicated with dots. No dots means lowest quality, while six dots indicate highest strength.

Metric grade markings for bolts and nuts are shown in **Figure 5-22**. Grade markings for bolts are in numerals from 4.8 to 10.9. No numerals means lowest strength, while 10.9 means highest strength. Grade markings for nuts also are in numerals. No numerals means lowest quality, while 10 means highest quality.

5.7 TORQUE REQUIREMENTS FOR THREADED FASTENERS

If fasteners are tightened too tight or too loose, a failure of parts or of the fastener can result. To determine how much a fastener must be tightened, *torque specifications* are furnished by all manufacturers. Torque specifications are given in pound-inches (lb.-in.), pound-feet (lb.-ft.), newton-meters (Nm), or Kilogram-meters (KgM).

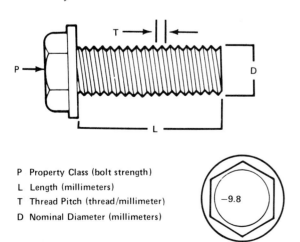

P Property Class (bolt strength)
L Length (millimeters)
T Thread Pitch (thread/millimeter)
D Nominal Diameter (millimeters)

Figure 5-19. Metric bolt measurement.
FORD MOTOR COMPANY

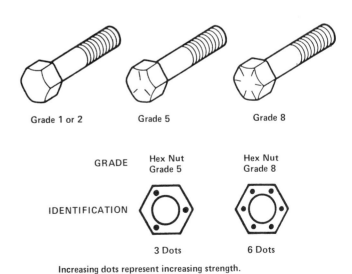

Figure 5-21. Conventional system grade markings for bolts and nuts. FORD MOTOR COMPANY

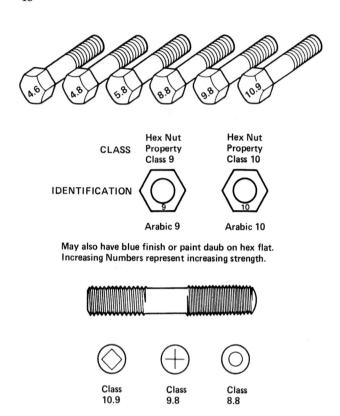

Figure 5-22. Metric grade markings for bolts, nuts, and studs. FORD MOTOR COMPANY

Figure 5-23. Thread sealants. SAVERIO BONO

5.8 CHEMICAL THREAD COMPOUNDS

Under some conditions, special chemical compounds may be needed to help fasteners do their job. Special sealing compounds are applied to the threads of fasteners to keep them from unscrewing and/or to prevent damage to threads.

Thread-locking Compounds and Sealants

Thread-locking compounds are used where vibration would cause the fastener to unscrew. *Thread sealants* are used in locations where the threads of a bolt or other fastener protrudes into an area where liquids, such as coolant, could corrode and damage the fastener **[Figure 5-23]**.

Anti-Seize Compound

Anti-seize compound **[Figure 5-24]** prevents the metals of the fastener and of the part into which it screws from reacting with one another and seizing, or becoming stuck together. Anti-seize compound is often used on spark plug threads that screw into aluminum cylinder heads to prevent damage when they are removed.

5.9 NONTHREADED FASTENERS

Several other types of fasteners are used to hold automotive parts together. Unlike bolts and screws,

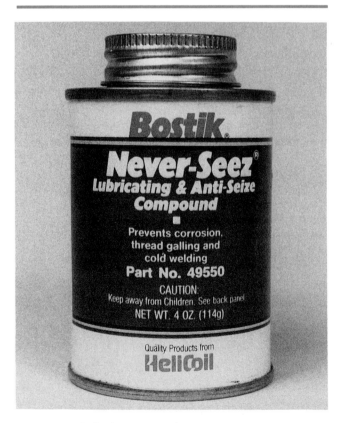

Figure 5-24. Anti-seize compound. SAVERIO BONO

Using Fasteners

these fasteners do not have threads to lock them in place. Common *nonthreaded fasteners* used on automobiles are:

- Dowel pins
- Keys
- Rivets
- Retaining rings.

Dowel Pins

Dowel pins align parts. Dowel pins are solid or hollow cylinders that fit into holes in the parts that are connected. Dowels may be straight, tapered, or have slits (called a "roll pin") **[Figure 5-25]**.

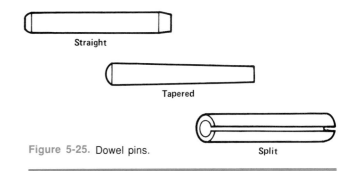

Figure 5-25. Dowel pins.

Keys

Figure 5-26 shows a *key* fastener. A key is a small piece of metal that positions and holds two parts so that they rotate together. Keys are positioned in slots, or *keyways,* in the connecting parts. Keys are used to connect gears and pulleys to shafts.

Rivets

A solid *rivet* is a soft metal pin or dowel with an expanded head on one end. Holes are drilled in two pieces of sheet metal or other materials. The rivet is inserted and the cylindrical end is expanded to hold the materials together securely. **Figure 5-27** shows common types of solid metal rivets.

Tubular *expanding rivets* **[Figure 5-28]** are used commonly in automotive work. A special tool pulls a metal shank within the hollow rivet to expand the shank, or tubular portion. The shank breaks off and is removed. The expanded tubular rivet remains in place to hold parts together securely. Often this type of rivet is called a pop rivet.

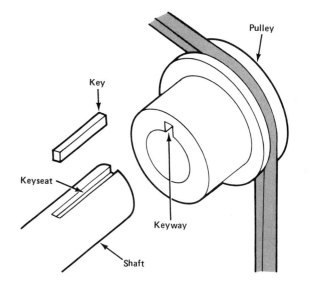

Figure 5-26. Key and keyway.

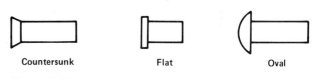

Figure 5-27. Rivets.

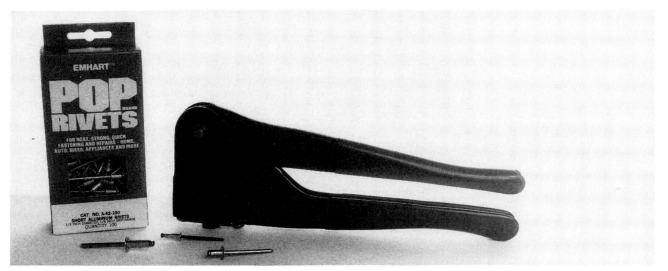

Figure 5-28. Expanding rivet and riveting tool. SAVERIO BONO

Retaining Rings

A *snap ring*, or *retaining ring*, fits into a groove on the inside or outside of a shaft or cylinder [**Figure 5-29**]. Retaining rings are used in transmissions to hold gears on shafts. A retaining ring, like a spring, is made to keep a specific shape. If a retaining ring is temporarily pushed out of shape, it will spring back to its original position.

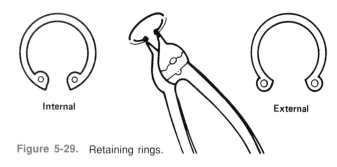

Figure 5-29. Retaining rings.

UNIT HIGHLIGHTS

- Measurements are made using the conventional system and the metric system.
- Inches are commonly divided into 16ths, 32nds, and 64ths for automotive work.
- Threaded fasteners include all screws, bolts, nuts, and washers.
- Bolts are classified by thread diameter, thread pitch, and length of the shaft.
- Threaded pitch is the distance between threads and is measured with a thread gauge.
- Grade markings indicate the strength of a bolt or nut.
- Nonthreaded fasteners include dowel pins, keys, rivets, and retaining rings.

TERMS

conventional system	metric system
threaded fasteners	external threads
capscrew	machine screw
Allen head screw	sheet metal screw
self-tapping screw	bolt
nut	self-locking nut
castellated nut	cotter pin
speed nut	stud
washer	flat washer
lockwasher	Phillips screw
thread pitch	thread gauge
grade marking	torque specification
anti-seize compound	nonthreaded fasteners
dowel pin	key
keyway	rivet
snap ring	retaining ring
expanding rivet	screw
national fine thread (NF)	thread-locking compound
national coarse thread (NC)	conventional/metric conversion
thread sealant	

REVIEW QUESTIONS

DIRECTIONS: The following questions are similar to those used on automotive technician certification tests. On a separate sheet of paper, circle the letter of the correct choice.

1. Technician A says that machine screws and Allen head screws are the same except for size.
   Technician B says that bolts and nuts can both be castellated.
   Who is correct?
   A. A only
   B. B only
   C. Both A and B
   D. Neither A nor B

2. The bolt designation 1/2-20 NF x 1 indicates that this bolt
   A. is a metric fastener.
   B. has a strength rating of 20.
   C. has fine threads.
   D. has an Allen head.

3. Which of the following statements is correct?
   I. Grade markings found on bolts and nuts are indicated with dots and letters.
   II. Torque specifications determine how far a bolt can be turned.
   A. I only
   B. II only
   C. Both I and II
   D. Neither I nor II

4. Technician A says that dowel pins must be hollow to be called fasteners.
   Technician B says that a key fastener is used to connect gears and a shaft.
   Who is correct?
   A. A only
   B. B only
   C. Both A and B
   D. Neither A nor B

Question 5 is not like the first four questions. It has the word EXCEPT. For question 5, look for the choice that does *not* apply. Read the entire question carefully before you choose your answer.

5. All of the following statements about washers are correct EXCEPT
   A. they are placed between a bolt head and a nut.
   B. lock washers are flat, with smooth edges.
   C. they can help prevent damage by spreading out tightening loads.
   D. they can help to hold bolts and nuts securely.

ESSAY QUESTIONS

1. What two major systems are used for measurements of fasteners and automotive parts, and what units are commonly used in automotive work?
2. How are bolts and nuts described?
3. Why is the manufacturer's specification for torque important?

SUPPLEMENTAL ACTIVITIES

1. Identify and describe the differences between capscrews, machine screws, Allen head screws, and sheet metal screws.
2. Determine whether bolts and nuts furnished by your instructor are conventional or metric size.
3. Install a castellated nut and cotter pin on a bolt.
4. Measure and determine the threaded diameters, strength designations, and thread pitches of bolts furnished by your instructor.
5. Determine the strength grade of one conventional and one metric set of bolts and nuts.

6 Using Hand Tools

UNIT PREVIEW

Common hand tools include wrenches, sockets, screwdrivers, pliers, hammers, and files. Special tools also may be required for some jobs. As a technician, you will need to know how to use hand tools and how to choose the right tool for the job. This unit provides an overview of hand tools and their uses.

LEARNING OBJECTIVES

When you have completed your assignments and exercises in this unit, you should be able to:

❏ Describe the different types of automotive wrenches.
❏ Describe the different types of handles and accessories that can be used with a socket wrench.
❏ Identify different types of pliers.
❏ Describe how the cutting teeth on a file are identified.

6.1 TOOL MEASUREMENTS

The first step in choosing the right tool is to select the correct size. The size of a wrench opening and the bolt or nut that it fits is indicated by stamped or engraved numbers [Figure 6-1]. The opening of the wrench must fit snugly on the fastener, not too loose or too tight, to prevent damage to the fastener and to the wrench. The tool measurement will be either in the conventional or metric system.

Common sizes for tools with conventional measurements include: 1/4, 5/16, 3/8, 7/16, 1/2, 9/16, 5/8, 11/16, 3/4, 13/16, 7/8, 15/16, 1 inch, and larger. Tools smaller than 1/4 inch are made for precision work. Commonly used metric sizes include 6, 7, 8, 9, 10, 11, 12, 13, 14, 17, 19 mm, and larger. Smaller sizes, including half sizes (for example, 4.5 mm) are made for precision work.

6.2 WRENCHES

A *wrench* is the proper tool to loosen or tighten a bolt or nut. Pliers should *not* be used on bolts or nuts or else the corners of the fastener will become rounded and impossible to grip. To work properly, wrenches must be the proper size and correct shape. Openings formed on one or both ends of a wrench fit around a fastener. The angle of the wrench handle to the opening is important in tight working conditions [Figure 6-2]. The technician must be able to grip and move the tool. Commonly used wrenches include:

- Open-end
- Adjustable open-end
- Box-end
- Combination
- Socket
- Special wrenches.

Open-End

A selection of *open-end wrenches* is shown in [Figure 6-3]. Each wrench has two different-sized openings

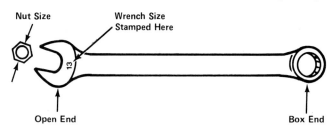

Figure 6-1. Wrench size measurements.

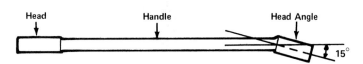

Figure 6-2. Parts of a wrench.

Figure 6-3. Open-end wrenches.

to fit a different size fastener. The open ends grip two sides of a fastener so that it can be turned for loosening or tightening. The handle is often placed at an angle to the opening to allow the wrench to be used in two different positions. Open-end wrenches are useful for quickly turning fasteners that are not extremely tight. For extremely tight fasteners, use a box-end wrench or socket wrench, discussed below.

A variation of the open-end wrench is a "flare-nut" or "tubing" wrench, that grips a hex on five sides to prevent slippage. The open sixth side can be used on metal compression fittings, such as those used on fuel lines.

Adjustable Open-End

An *adjustable open-end wrench* [Figure 6-4] has a fixed and a moveable jaw that form an opening. To use an adjustable open-end wrench, adjust the jaws tightly around the fastener. Although these wrenches are useful, they are not a substitute for a set of open-end wrenches. Their thick jaws and long handles make them hard to use in tight areas.

Box-End

The heads on a *box-end wrench* are circular and shaped to fit around nuts and bolts. The box-end, or box, wrenches shown in **Figure 6-5** have different sized heads on each end. The head openings have either six or 12 grooves that grip hexagonal fasteners. Because of head design, a technician can apply more force to the box-end wrench without the wrench slipping.

Combination

When an open end and a box end are combined on the same wrench, it is known as a *combination wrench* [Figure 6-6]. Both ends of the wrench have the same head sizes. The box end usually is at an angle to the wrench handle to allow room for fingers to grip the wrench [refer to Figure 6-2].

Socket

A *socket wrench* is a circular tool that fits over a fastener and surrounds it. Six to 12 grooves, or points, inside the socket grip the corners of the fastener. The socket has a squared hole, called a drive opening, where the socket is attached to the drive handle [Figure 6-7].

The handle may have a *ratchet* built in to operate the drive. A ratchet is a mechanism inside the handle that allows drive action in either turning direction. A lever near the drive end can be positioned to switch the ratchet's operating direction to either clockwise or counterclockwise. The drive is the square peg which is connected to the socket.

Figure 6-4. Adjustable wrenches.

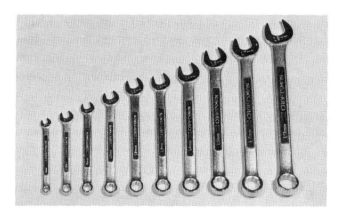

Figure 6-6. Combination wrenches.

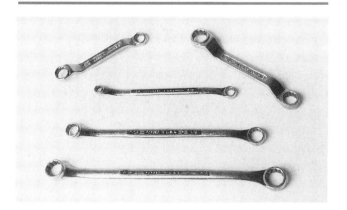

Figure 6-5. Box-end wrenches.

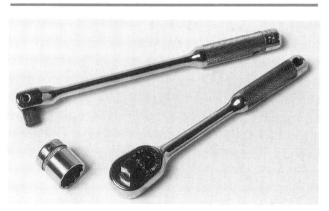

Figure 6-7. Socket, ratchet handle, and breaker bar.

Drive sizes. *Drive size* refers to the measurement of the square drive peg that fits a corresponding hole in the socket. A 1/4-inch drive is light-duty, 3/8-inch drive is medium-duty, and 1/2-inch drive is heavy-duty. To turn large sockets with sufficient torque, larger drive sizes are used. Commonly-used socket wrench drives can be as large as 1 inch.

Handles and accessories. A variety of handles can be used with a socket wrench. Handles and handle accessories, shown in **Figure 6-8,** include:

- Speed handle
- Breaker bar
- T-handle
- Extension
- Flexible joints and accessories.

Specialized handles allow the technician to use socket wrenches in places where another wrench might not work.

A *speed handle* is a long crank with a handle that turns freely to remove loosened fasteners. A *breaker bar* is long and hinged to allow greater turning force and turning angle. A *T-handle* is flexible in tight areas and gives the technician more turning force when the handle is extended.

Extensions allow the drive handle to be positioned to clear obstructions and be turned easily. Extensions of different lengths connect between the drive and the socket.

A flexible *universal joint* permits you to work with a socket and/or extension at an angle up to 45 degrees. Special screwdriver- and Allen-head socket drivers also are available. Drive-size adapters also can be used to drive sockets with a handle of a different drive size.

Torque. If a bolt or nut holding a part must be torqued to specifications, a *torque wrench* is required. A torque wrench indicates how much torque, or turning force, is applied.

A torque wrench consists of a long, flexible bar and a socket drive. The proper socket is placed onto the drive and the fastener starts to be tightened. A beam torque wrench shows the torque reading on an indicator on the handle **[Figure 6-9]**. A micrometer type torque wrench is adjusted to proper torque and "clicks" when that torque is reached.

Special Wrenches

Some fasteners require special tools because of head design. Allen and Torx wrenches are tools used on fasteners with special head designs **[Figure 6-10]**.

6.3 SCREWDRIVERS

A *screwdriver* is a shaft, or shank, of metal with a handle at one end and a blade at the other end **[Figure 6-11]**. The shaft is imbedded in a plastic, metal, or wood handle. The blade fits the head of a screw for turning. A screwdriver blade that is too large or too small will damage the fastener and the screwdriver. Always choose the proper size of screwdriver to match the screw head.

Slotted Screwdrivers

The screwdriver that many people call "common" actually is named the *slotted screwdriver.* A slotted screwdriver has a flat blade.

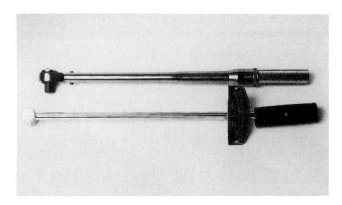

Figure 6-9. Torque wrenches.

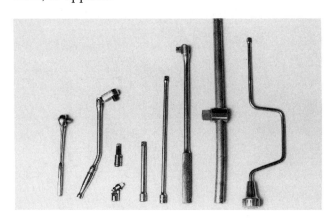

Figure 6-8. Socket wrench accessories.

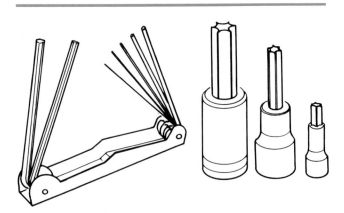

Figure 6-10. Allen wrenches, left, and Torx drivers, right.

Using Hand Tools

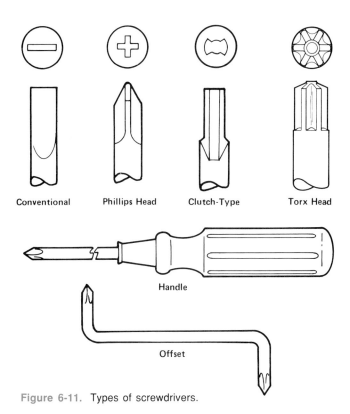

Figure 6-11. Types of screwdrivers.

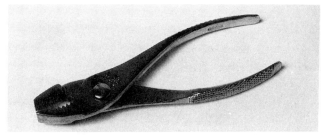

Figure 6-12. Slip-joint adjustable pliers.

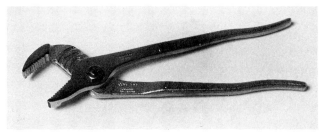

Figure 6-13. Arc-joint pliers.

Special Screwdrivers

Three commonly-used special screwdrivers are the *Phillips*, *Torx*, and *clutch-head screwdrivers*. Another special tool is the *offset* screwdriver, which is used in tight areas.

6.4 PLIERS

Pliers have jaws to hold, bend, rotate, or cut parts or fasteners. Pliers should *not* be used for loosening or tightening bolts or nuts. Common pliers used by technicians are:

- Slip-joint adjustable
- Arc-joint
- Locking
- Longnose
- Diagonal cutting
- Special-purpose.

Slip-Joint Adjustable

Slip joint, or "combination," pliers are shown in **Figure 6-12**. A slot on one jaw allows the slip joint to move and adjust the space between the two jaws.

Arc-Joint

A variation of slip-joint adjustables are *arc-joint pliers*. Grooves, or channels, allow a wider range of adjustment without slipping **[Figure 6-13]**.

Locking

Parts and fasteners are held in position by using *locking pliers.* The jaws of locking pliers are adjusted by turning a screw in the handle. The adjusted jaws can be locked into position by pulling in on the lower handle **[Figure 6-14]**. To release the pliers, push on the lever above the lower handle.

Longnose

The jaws of *longnose pliers* are long and narrow to fit in tight working spaces **[Figure 6-15]**. A variation of these pliers, which have very small and pointed jaws, are known as *needlenose* pliers.

Diagonal Cutting

Some pliers, called *diagonal cutting pliers*, or *cutters*, are used to cut electrical connections, cotter pins, and similar thin metal. Jaws on diagonal cutting pliers have hardened cutting edges **[Figure 6-16]**.

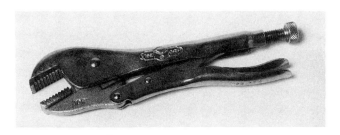

Figure 6-14. Locking pliers.

Figure 6-15. Needlenose pliers.

Figure 6-16. Diagonal pliers.

Special-Purpose

Many special pliers are manufactured for special purposes. From left to right in **Figure 6-17** are brake-spring pliers, lock ring pliers, snap-ring pliers, and battery connection pliers.

6.5 HAMMERS AND MALLETS

Hammers have hardened steel heads and are used to drive punches and chisels. The head of a soft-faced mallet is made from brass, plastic, leather, or rubber to prevent damage to soft or machined surfaces. From top to bottom in **Figure 6-18** are a rubber *mallet*, brass mallet, and ball peen hammer.

6.6 CHISELS AND PUNCHES

Chisels and *punches* are used to separate or align parts during servicing **[Figure 6-19]**. Chisels are used to cut heads off rivets, cut metal pins, or to loosen fasteners that cannot be removed in a normal manner. Because chisels are struck with hammers and may occasionally shatter or splinter, eye protection must be worn when using a hammer and chisel.

A pin punch is used to drive pins in or out. An aligning punch is used to align holes in two parts that will be connected. A center punch is used to provide a small depression to start a drill bit accurately. From left to right in **Figure 6-19** are a large chisel, long pin punch, short pin punch, centerpunch, and small chisel.

6.7 FILES

Files are used to remove metal and smooth surfaces. A file is a cutting tool with rows of hardened teeth. A file has a pointed end, or *tang*, that fits into a handle **[Figure 6-20]**. Never use a file without a handle.

The teeth of a file may be single-cut (run in one direction) or double-cut (cross each other). The distance between the cutting teeth determines how the file works. A file with teeth that are wide apart will remove metal quickly, but leave a rough surface. A file with fine teeth is used for polishing or smoothing **[Figure 6-21]**. Commonly used files can be flat, triangular, square, round, or tapered like a knife blade.

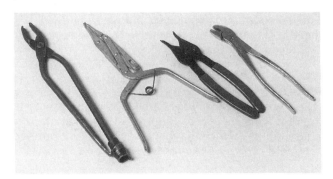
Figure 6-17. Special-purpose pliers, brake spring, locking spring, and battery pliers.

Figure 6-19. Chisels and punches.

Figure 6-18. Hammer and mallets.

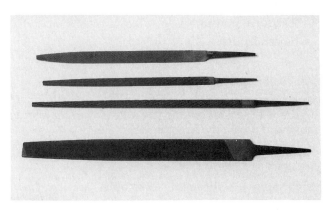
Figure 6-20. Files.

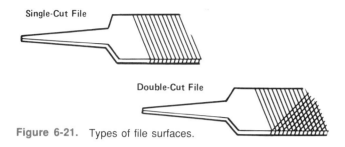

Figure 6-21. Types of file surfaces.

6.8 TAPS

A *tap* is a tool that cuts new internal threads [**Figure 6-22**]. The end of the tap is square to fit into a special tap wrench. To cut threads, the tap is held at 90 degrees and slowly turned with the tap wrench into a drilled hole in a metal part.

6.9 DIES

A *die* is a tool that cuts new external threads [**Figure 6-23**]. Dies are either hexagonal or round to fit into special wrenches called die stocks. To cut threads, the die is held at 90 degrees and slowly turned with the stock onto a metal rod.

6.10 REAMER

Some holes in metal parts must be sized more precisely than can be done with a drill. A *reamer* has sharp, tapered cutting edges designed to remove only small amounts of metal at a time from a drilled hole [**Figure 6-24**]. Reamers are turned either with a wrench or by hand.

6.11 SCREW EXTRACTOR

Broken bolts or studs often are removed with pliers if part of a broken fastener protrudes above the surface. If a broken part of a bolt is inside a hole below the surface, a *screw extractor* is used [**Figure 6-25**].

A hole of the proper size is drilled into the broken shaft of the bolt or stud. The screw extractor is driven into the drilled hole. Then the screw extractor is turned counterclockwise with a wrench to loosen and remove the broken shaft.

6.12 PRY BARS

To position or disassemble some automotive systems, a *pry bar* is sometimes needed. A pry bar is a long, thick steel lever [**Figure 6-26**].

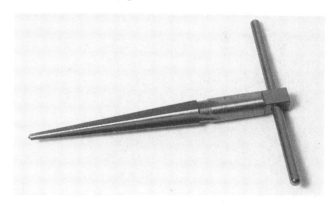

Figure 6-24. Reamer.

Figure 6-25. Screw extractor.

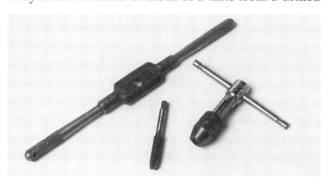

Figure 6-22. Tap (center) and tap wrenches.

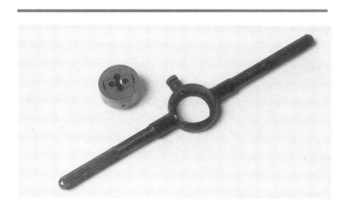

Figure 6-23. Die and die stock.

Figure 6-26. Using a pry bar to position a heavy assembly.

UNIT HIGHLIGHTS

- Wrenches are classified as: open-end, adjustable open-end, box-end, combination, socket, torque, and special.
- Screwdrivers have blades that fit into different-shaped fastener heads.
- Pliers have jaws that hold, bend, rotate, or cut parts or fasteners.
- Chisels and punches remove fasteners with the aid of hammers or mallets.
- Files remove, smooth, and polish metal, depending on the cutting teeth of the tools.
- Taps cut internal threads, dies cut external threads.
- Special tools such as reamers and screw extractors are useful in auto work.

TERMS

wrench
box-end wrench
socket wrench
speed handle
T-handle
universal-joint
screwdriver
Phillips screwdriver
clutch-head screwdriver
offset screwdriver
locking pliers
hammer
chisel
file
die
reamer
ratchet
adjustable open-end wrench
needlenose pliers
open-end wrench
combination wrench
drive size
breaker bar
extension
torque wrench
slotted screwdriver
Torx screwdriver
slip-joint adjustable pliers
arc-joint pliers
longnose pliers
mallet
punch
tap
pry bar
screw extractor
tang
diagonal cutting pliers

REVIEW QUESTIONS

DIRECTIONS: The following questions are similar to those used on automotive technician certification tests. On a separate sheet of paper, write the letter of the correct choice.

1. Technician A says that metric fasteners are used on some cars.
   Technician B says that conventional fasteners are used on some cars.
   Who is correct?
   A. A only
   B. B only
   C. Both A and B
   D. Neither A nor B

2. Technician A says that the length of a wrench handle is the drive size.
   Technician B says that a flexible joint allows a socket and a drive to rotate at different speeds.
   Who is correct?
   A. A only
   B. B only
   C. Both A and B
   D. Neither A nor B

3. Which of the following statements is correct?
   I. Chisels and punches must *not* be struck with a hammer or mallet because they may shatter.
   II. A double-cut file has cutting teeth that crisscross.
   A. I only
   B. II only
   C. Both I and II
   D. Neither I nor II

Questions 4 and 5 are not like those above. Each has the word EXCEPT. For each question, look for the choice that does *not* apply. Read each question carefully before you choose your answer.

4. All of the following wrenches are used frequently in automotive mechanics EXCEPT
   A. combination.
   B. allen.
   C. torx.
   D. monkey.

5. All of the following types of pliers can be adjusted to fit a part or fastener EXCEPT
   A. arc-joint.
   B. locking.
   C. slip-joint.
   D. diagonal cutting.

ESSAY QUESTIONS

1. What types of wrenches are commonly used in automotive work?
2. Describe pliers commonly used in automotive work and what they might be used for.
3. Why are safety precautions necessary when using a hammer and chisel?

SUPPLEMENTAL ACTIVITIES

1. Correctly identify tools laid out by your instructor.
2. Make a list of tools that are in the shop inventory.
3. Demonstrate how to operate a socket wrench with a ratchet handle.
4. Correctly identify slotted, Allen, Phillips, Torx, and clutch-head screwdrivers.
5. Identify special tools laid out by your instructor.
6. Identify the size of wrenches needed for fasteners laid out by your instructor.

7 Using Power Tools

UNIT PREVIEW

Power tools make a technician's job easier, because they operate faster and with more force than hand tools. Power is furnished by either electricity, compressed air, or hydraulic fluid under pressure. The force and speed of power tools requires that special safety precautions be observed.

LEARNING OBJECTIVES

When you have completed your assignments and exercises in this unit, you should be able to:

❏ Use electrical power tools safely and correctly.
❏ Identify the parts of a portable electric drill.
❏ Use pneumatic power tools safely and correctly.
❏ Use hydraulic power tools safely and correctly.
❏ Describe different types of parts cleaning machines and how they operate.

SAFETY PRECAUTIONS

Carelessness or mishandling of power tools can cause serious injury. Be sure you know how to operate the tool properly before using it. Read all instructions carefully. Do not use a power tool without obtaining permission from your instructor. And *always* wear safety glasses when using a power tool.

When you work with large power tools, such as bench or floor equipment, check the machines for damage before you begin work. Also check for proper adjustments. Place all safety guards in position. Wear safety glasses or a face shield. Make sure the work area is clear of bystanders and parts before starting a machine. Keep hands and clothing away from moving parts.

Never leave a power tool unattended while it is running. If a machine does not operate properly, turn it off and notify your instructor.

When you finish using a power tool, turn it off. Wait until the tool has stopped completely and then disconnect it from its power source. Clean, oil, and adjust it, if necessary.

Do not enter a work area where a power tool is being used. Stay away from power tools that are operating.

Obey all safety rules outlined by your instructor.

7.1 ELECTRICAL TOOLS

An electrical cord connects electrical tools to an outlet, which is a source of electrical current. Current flows through the cord and powers a motor or other device inside the tool. An on/off switch controls the operation of the tool.

Droplight

Adequate light is necessary when working under and around automobiles. A *droplight* [Figure 7-1] can be hand held or attached to parts with a hook. The droplight handle is insulated to protect against electric shocks. A plastic or metal cage surrounds the bulb to prevent breakage. Some droplights also include electrical sockets for portable power tools. The light has a long cord for freedom of movement in the shop area. Many shops now use fluorescent work lights, which create less heat for additional safety.

Figure 7-1. A droplight is used when working under an automobile and in dark areas.

Electric Impact Wrench

An *electric impact wrench* has an electric motor that turns a socket drive. The tool provides sharp bursts of power that help to loosen tight nuts or bolts. A trigger, or switch, on the handle starts or stops the wrench. Another switch controls the direction of rotation [Figure 7-2]. Only special, hardened impact sockets, extensions, and flex joints are used with impact tools.

Portable Electric Drill Handle

A *portable electric drill* can be moved around easily for hand use. An electric motor provides the power and is operated by a trigger on the drill handle. An electric drill has an adjustable *chuck* to hold bits securely [Figure 7-3].

Using Power Tools

Figure 7-2. Electric impact wrench.

Figure 7-4. Electric grinder.

Figure 7-3. Portable electric drill.

Figure 7-5. Electric sander and polisher.

An electric drill turns a *drill bit* to bore holes. A drill bit is a hardened steel shaft that has a sharp point and a spiral groove and cuts holes and removes material when turned by a drill handle.

Drill handle sizes are measured when the chuck is opened to its widest setting. Common automotive drill sizes are 1/4, 3/8, and 1/2 inch.

Grinders

A *grinder* is a motor with a rotating shaft on which a grinding wheel or a wire buffing wheel can be mounted [Figure 7-4]. A grinding wheel is solid and is used to remove metal and to sharpen tools or parts. A wire wheel, made from separate strands of wire, is used to clean rust and corrosion from tools or parts.

Sanders and Polishers

One electric-powered tool can be used as a *sander* and as a *polisher* [Figure 7-5]. This tool has interchangeable discs for a variety of jobs. A sanding disc smooths and removes metal. A polishing disc polishes and waxes an automobile's finish.

7.2 PNEUMATIC TOOLS

Pneumatic tools are powered by compressed air pressure. Air is compressed by an electric motor and a compressor that draws air into a tank and increases its pressure.

Flexible hoses attach the air supply to a tool through mating couplers. A finger-operated valve on a tool allows air pressure to flow through the tool to spin a turbine wheel or power other mechanisms.

Compressed air can blow particles into eyes and harm other body parts. Always wear eye protection and direct the air *away* from other people. Make sure that all connections are secure and that the hoses are in good condition and not leaking.

Blowgun

A *blowgun* [Figure 7-6] provides a flow of air when the valve is opened. Blowguns are used for blowing debris off parts during cleaning.

Air Impact Wrench

An *air impact wrench* [Figure 7-7] provides great torque and speed for turning bolts and nuts. A reversing switch controls the direction of rotation.

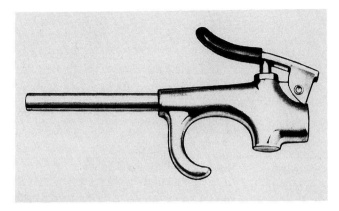

Figure 7-6. Compressed-air blowgun.
SNAP-ON TOOL CORPORATION

Figure 7-7. Air impact wrench.
SNAP-ON TOOL CORPORATION

Most air impact wrenches also include a valve that controls the amount of torque applied. Uses of this tool include removing wheel lugs and turning tight fasteners on large assemblies. Only special, hardened impact sockets, extensions, and flex joints should be used with impact tools.

Air Ratchet

An *air ratchet,* shown in **Figure 7-8,** uses air to turn the ratchet drive. It is used to spin smaller nuts and bolts on and off quickly.

Air Chisel

The *air chisel* is a hammering tool often used to cut off bodywork or light metal parts, such as exhaust system parts **[Figure 7-9].**

Figure 7-8. Air ratchet. SNAP-ON TOOL CORPORATION

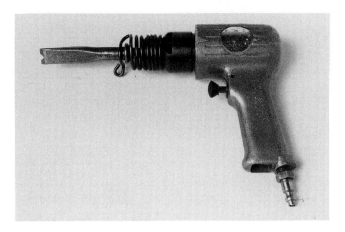

Figure 7-9. Air chisel.

Air Drill

For safety around flammable liquids, an *air drill* is powered by compressed air **[Figure 7-10].** A pneumatic tool eliminates the safety hazard of sparks from an electric motor.

7.3 HYDRAULIC TOOLS

Hydraulic tools operate on oil or liquid pressure. Oil is pumped into a system, where it puts pressure on pistons that move to do work. Hydraulically operated tools can provide great pushing or pulling force.

Floor Jack

A *floor jack* has wheels that allow it to be positioned easily **[refer to Figure 4-7].** It is used to raise the weight of an automobile. To raise the *saddle,* or part of the jack that contacts the automobile, turn the handle clockwise to close the control valve. Then move the handle up and down in a pumping motion to create hydraulic pressure that raises the saddle. To lower the jack, *slowly* turn the handle counterclockwise to release hydraulic pressure. *Always use jackstands to support a vehicle safely after you raise it with a jack.*

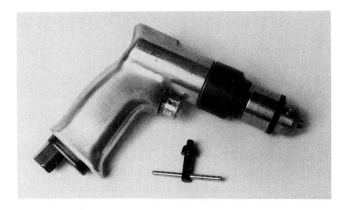

Figure 7-10. Air drill.

Using Power Tools

Hoist

A lift, or *hoist,* raises the entire automobile. Most modern hoists are double-post designs **[refer to Figure 4-8]**, although some single-post models remain in use. A control lever is placed at a convenient location near the hoist. The lever controls compressed air that creates hydraulic pressure to move the posts of the hoist upward. *Always set the safety locks after the vehicle has been raised.* When the safety locks are released and the pressure is lowered, the hoist posts move downward. Lower the hoist to the floor when it is not being used.

Engine Lift

A crane, or *engine lift,* is used to remove or install an automotive engine. The engine lift operates using a hydraulic cylinder that pushes up on the long support arm. The lift is on wheels, so it can be rolled anywhere in the shop **[Figure 7-11]**. The lift control can raise or lower heavy assemblies slowly and safely to aid in positioning.

Hydraulic Press

A *hydraulic press* **[Figure 7-12]** forces, or presses, tightly fitted parts together or apart. The table portion of the press, where parts are positioned, must withstand the tons of force applied. A handle at the side of the hydraulic press is pumped to increase pressure. A valve releases pressure.

7.4 CLEANING EQUIPMENT

Cleaning parts is necessary for careful inspection of defects. Several types of cleaning equipment are found in the shop.

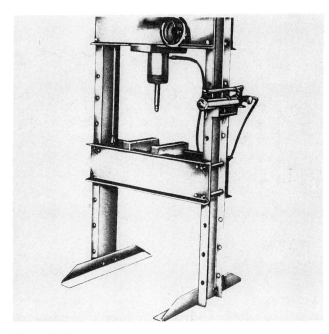

Figure 7-12. Hydraulic press. SNAP-ON TOOL CORPORATION

High Pressure and Steam Cleaners

High pressure and heat are effective methods for cleaning away automotive grease and grime. A *high-pressure spray cleaner* **[Figure 7-13]** forces compressed air and water through a nozzle to clean parts. A *steam cleaner* is used to heat water and create steam. Steam is mixed with a detergent solution and sprayed on the parts to wash away dirt and grease.

Solvent Cleaner

A *solvent cleaner* is a tank that holds and circulates non-caustic cleaning solvent **[Figure 7-14]**. Cleaning solvent thins grease, oil, and dirt buildups so they can be washed away. A solvent cleaner usually has strainers or screens to keep small parts from becoming lost. The solvent cleaner has an electric motor that pumps cleaning solvent through a flexible tube to

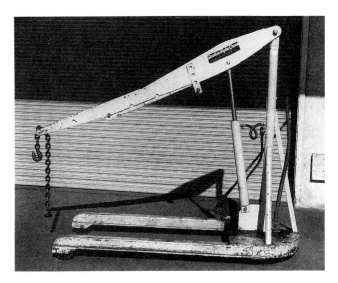

Figure 7-11. A crane, or engine lift.

Figure 7-13. High-pressure spray cleaner.

Figure 7-14. Solvent cleaner. KLEER-FLOW COMPANY

flush off deposits. Solvent returns to the tank and is filtered to remove dirt and grease that has accumulated.

Cold Tank

Certain metals, such as aluminum and brass, must be cleaned in a *cold tank*. A cold tank is used to soak parts for long periods of time **[Figure 7-15]**. As its name implies, a cold tank is not heated. Face shields, protective clothing, and rubber gloves must be worn when using the harsh chemicals in a cold tank.

Figure 7-15. Cold tank. KLEER-FLOW COMPANY

Hot Tank

Large iron and steel parts, such as engine blocks, are cleaned in a *hot tank* **[Figure 7-16]**. A hot tank is heated and contains highly caustic chemicals. Some metals, such as aluminum, will dissolve in a hot-tank solution. Wear face shields, rubber gloves, and protective clothing when working around this equipment.

Glass Bead Blaster

For final cleaning, a *glass bead blaster* sprays small beads of glass at a part **[Figure 7-17]**. The glass beads abrade, or wear off, remaining small amounts of dirt or residue.

Figure 7-16. Hot tank.
KLEER-FLOW COMPANY

Figure 7-17. Glass bead blaster.
INLAND RADIATOR EQUIPMENT, INC.

USING POWER TOOLS

UNIT HIGHLIGHTS

- Electrical power tools include drills, impact wrenches, grinders, sanders, and polishers.
- Pneumatic tools include blowguns, air impact wrenches and ratchets, air chisels and drills.
- Hydraulic tools use fluid pressures to create powerful pushing or pulling forces for jacks, hoists, and hydraulic presses.
- Cleaning tools include high pressure and steam cleaners, solvent tanks, and bead blasters.

TERMS

droplight	electric impact wrench
portable electric drill	chuck
drill bit	grinder
sander	polisher
pneumatic tools	blowgun
air impact wrench	air ratchet
air chisel	air drill
hydraulic tools	floor jack
hoist	engine lift
hydraulic press	steam cleaner
solvent cleaner	cold tank
hot tank	glass bead blaster
saddle	high-pressure spray cleaner

REVIEW QUESTIONS

DIRECTIONS: The following questions are similar to those used on automotive technician certification tests. Answer each question by circling the letter of the correct choice.

1. Which of the following statements is correct?
   I. A grinder is an electric power tool.
   II. An impact wrench is a hydraulic power tool.
   A. I only   C. Both I and II
   B. II only   D. Neither I nor II

2. Technician A says that it is safe to leave a power tool running if you will only be away for a moment.
   Technician B says that it is safe to wear long sleeves when working with electrical power tools if you are careful.
   Who is correct?
   A. A only   C. Both A and B
   B. B only   D. Neither A nor B

3. Which of the following would best remove carbon deposits safely from aluminum pistons?
   A. Steam cleaner   C. Bead blaster
   B. Hot tank   D. Grinder or polisher

4. Which of the following statements is correct?
   I. Check power machinery for damage before beginning work.
   II. Avoid horseplay when working with power machinery.
   A. I only   C. Both I and II
   B. II only   D. Neither I nor II

Question 5 is not like those above. It has the word EXCEPT. For question 5, look for the choice that does *not* apply. Read the entire question carefully before you choose your answer.

5. The dangers of pneumatic tools include all of the following EXCEPT
   A. compressed air blowing particles into your eyes.
   B. sparks from motors when working around flammable liquids.
   C. powerful turning forces.
   D. moving parts and turning shafts.

ESSAY QUESTIONS

1. Describe common electrical power tools and their uses.
2. What pneumatic and hydraulic power tools are commonly used in automotive work?
3. What safety hazards are present when using power tools, and what safety precautions are needed?

SUPPLEMENTAL ACTIVITIES

1. Demonstrate the safe use of power tools chosen by your instructor.
2. Demonstrate the safe use of a hydraulic jack and a hoist.
3. Safely use cleaning equipment selected by your instructor to clean parts.

8 Using Measuring Tools

UNIT PREVIEW

Precise measurements are vital to proper automotive performance and reliability. A technician must be able to measure accurately many critical parts in an automobile to determine size and wear. This unit provides an overview of common automotive measuring tools and how they are used.

LEARNING OBJECTIVES

When you have completed your assignments and exercises in this unit, you should be able to:

- ❑ Describe the types of measuring tools used in automotive work.
- ❑ Use feeler gauges and wire gauges properly.
- ❑ Use an outside or inside micrometer accurately.
- ❑ Use a dial indicator to measure movement in a part or assembly.
- ❑ Use a cooling system tester and coolant hydrometer.
- ❑ Use a voltmeter, ammeter, and battery load tester.
- ❑ Hook up a diagnostic "scope" properly and observe firing traces.

8.1 MECHANICAL MEASURING TOOLS

A tool that must be operated by hand and adjusted to take readings is called a *mechanical measuring tool*. Many different mechanical measuring tools are used by a technician.

Rulers and Straightedges

A *ruler* is a straight strip of stiff or flexible material marked with divisions for measuring lengths. Rulers for the conventional system of measurements commonly have divisions every 1/8, 1/16, 1/32, or 1/64 of an inch **[Figure 8-1]**. The smallest common division on most metric rulers is 1 millimeter (1 mm) **[Figure 8-2]**.

A *straightedge* is an accurately machined, thick metal bar whose edges are straight and parallel. Straightedges do not have ruler markings. When a straightedge is placed across a warped part, there will be a gap, or clearance, between the straightedge

Figure 8-1. Conventional ruler.

Figure 8-2. Metric ruler.

and the part. The clearance is measured with feeler gauges, discussed below, to determine how warped the part is.

Feeler Gauge

Flat metal strips or blades are used to measure the space between two surfaces. These strips, called *feeler gauges*, usually come in sets **[Figure 8-3]**. The correct measurement is indicated when a blade slides through the space being measured with a light drag, or slight rubbing action. Feeler gauges come in conventional and metric sizes. Thicknesses are etched or stamped on each strip.

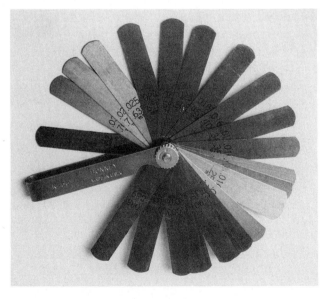

Figure 8-3. Feeler gauge set. L.S. STARRETT COMPANY

Using Measuring Tools

Wire Gauge

A round *wire gauge* is used as a feeler gauge on some automotive parts. A typical wire gauge is shown in **Figure 8-4**. Wire gauges are used to measure spark plug gaps.

Outside Micrometer

Measurements of the outside size of parts are made with an *outside micrometer* [**Figure 8-5**]. Outside micrometers have either conventional or metric measurements. Parts of a micrometer, often called a "mike," are:

- Frame
- Anvil
- Spindle
- Sleeve
- Lock Nut
- Thimble
- Ratchet.

Measurements are taken between the *anvil* and the *spindle*. Open the micrometer gap by turning the *thimble* counterclockwise. Place the part to be measured between the anvil and spindle. Turn the ratchet handle clockwise with your fingertips to move the spindle outward. When the spindle and anvil are snugged against the part, the ratchet will "click." Take your measurement by noting the alignment of the markings on the *sleeve* and on the thimble.

Conventional system measurement. To read the measurement, add the sleeve and thimble numbers together. **Figure 8-6** shows how the two measurement scales appear on the micrometer. Each scale has divisions that indicate numbers. Measurements for an outside micrometer with the conventional system are read by using the following procedure.

The outside micrometer is a very accurate tool. Each complete turn of the thimble, for example, moves the spindle only 0.025 in. This distance can be read on the sleeve as the thimble moves in or out with the rotating spindle. The sleeve has a scale that begins with 0 and moves up to 1, 2, etc. Between those numbers are three slashes, which provide four divisions between numbers. Each division indicates one rotation of the thimble, or a 0.025-in. movement by the spindle. Thus, four divisions indicate four turns, or 0.100 in. between the numbers on the sleeve.

An even more accurate measurement is made when reading the scale on the beveled end of the thimble. The edge of the thimble facing the sleeve is beveled, or tapered. The bevel scale has 25 divisions. Each division indicates 1/25th of a turn. If a complete turn is 0.025 in., then 1/25th of a turn is 0.001 in.

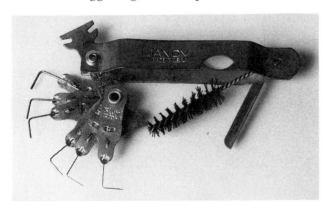

Figure 8-4. Wire gauge.

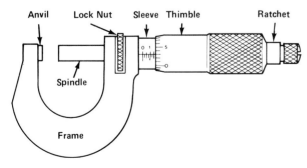

Figure 8-5. Outside micrometer.

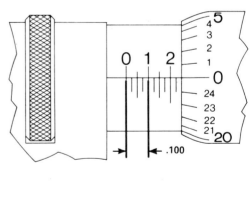

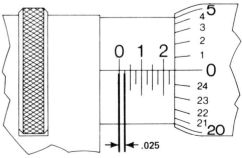

Figure 8-6. Conventional micrometer measurement readings.
CLEVITE

Add the readings of both scales to get a measurement. In **Figure 8-6,** from the sleeve, take the highest numbered division showing and add the unnumbered division. Sleeve readings of 2 (numbered division) and 0.050 (two unnumbered divisions) add up to 0.250. Add to this total the division on the bevel, 0.000. In **Figure 8-6,** the total reading indicated is 0.250 inch.

Metric system measurement. On a metric micrometer, the sleeve has a scale that is divided into two divisions separated by a line **[Figure 8-7]**. The lower division indicates half, or 0.50, of a millimeter and is not numbered. The upper division indicates 1 mm, and the scale is numbered every 5 mm. One complete rotation of the thimble indicates 0.50 mm. Two complete rotations indicate 1 mm.

The bevel scale has 50 divisions. Each division indicates 1/50th of a turn. If a complete turn is 0.50 mm, then 1/50th of a turn is 0.01 mm.

Take a measurement by reading the numbers on the spindle and bevel scales. In **Figure 8-7,** start at the upper-division sleeve scale for a reading of 10. Add the bottom scale division of 0.50. Now add the bevel scale reading of 0. In **Figure 8-7,** the total is 10.500 mm.

Inside Micrometer

Measurements of large hole sizes, or distances between parallel surfaces, are taken with an *inside micrometer.* An inside micrometer can be used to measure the diameter of a cylinder or bearing bore. Adjustments for different sizes are made by attaching spacing collars and rods to the micrometer. Conventional and metric measurements are taken in the same way as with outside micrometers. The use of an inside micrometer can be difficult. Care must be taken to avoid tilting the micrometer in the bore when taking a measurement.

Plastigage

A disposable material is used to measure close clearances between some parts. This material is a plastic thread, called *Plastigage,* which flattens between parts that are clamped together **[Figure 8-8]**. Measurements are taken when the parts are separated. Plastigage comes in an envelope with measurements printed on the outside. The width of the flattened Plastigage is compared to the width of the markings on the envelope. In **Figure 8-8,** the measurement indicated is 0.003 inch.

Small Hole Gauge

Holes too small for an inside micrometer are measured with a *small hole gauge,* sometimes called a *split-ball gauge.* A small hole gauge has a split ball at the end of the handle. The split-ball end is placed in the hole, and the handle is turned. The turning handle expands the split ball until it touches the walls of the hole. The gauge is removed, and the ball is measured with an outside micrometer.

Telescoping Gauge

A *telescoping gauge* measures the inside diameter of a hole. Telescoping gauges are spring-loaded to make them expand to hole sizes when they are inserted. Telescoping gauges do not contain graduated markings. Instead, an outside micrometer is used to take the measurement.

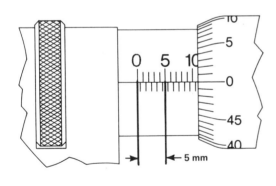
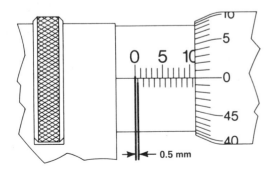

Figure 8-7. Metric micrometer readings.
CLEVITE

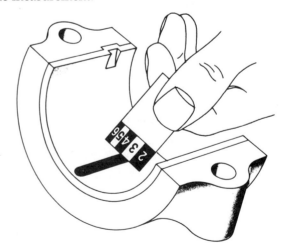

Figure 8-8. Checking a clearance with Plastigage.
CLEVITE

Using Measuring Tools

Vernier Calipers

Vernier calipers often are used instead of micrometers to measure inside and outside diameters. Vernier calipers have jaws that adjust against the outer edges or inner surfaces of a part **[Figure 8-9]**. Adjustments are made by moving the jaws closer together or farther apart.

Dial Indicator

Automotive parts are measured for back-and-forth movement, or *play,* and *runout,* or side-to-side movement, with a *dial indicator.* A dial indicator is a measuring tool with a *plunger* that acts on a gauge **[Figure 8-10]**. The plunger touches the part that is being measured and transfers any movement from that part to the gauge. The pointer on the gauge indicates how much play or runout there is. A magnetic base often is used when holding a dial indicator in position for measurements. Dial indicator measurements are in conventional (thousandths of an inch) or metric (hundredths of a millimeter) units.

Drive Belt Tension Gauge

A *drive belt* is a belt that operates equipment in the engine compartment. A drive belt must have the proper tension, or tightness, for it to operate properly. Tension may be measured with a *drive belt tension gauge* **[Figure 8-11]**. A tension guide determines whether the belt must be tightened or loosened.

8.2 FLUID MEASURING TOOLS

Tools in this category are used to check pressures and other measurements of air and liquids.

Tire Pressure Gauge

Automotive tires are inflated with compressed air. To check inflation, a *tire pressure gauge* is used. Air pressure is measured in pounds per square inch (psi) or kilopascals (kPa). Both dial- and pencil-type gauges are available **[Figure 8-12]**.

Figure 8-9. Vernier calipers.

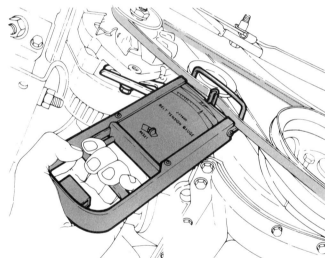

Figure 8-11. Drive-belt tension gauge.
AMERICAN MOTORS CORPORATION

Figure 8-10. Dial indicator. L.S. STARRETT COMPANY

Figure 8-12. Tire pressure gauges.

Compression Tester

Each cylinder develops pressure when the piston rises and compresses the air-fuel mixture. A *compression tester,* shown in **Figure 8-13,** is inserted into a spark plug hole, and the engine is cranked with the starter motor. The technician notes how the pressure rises and the final reading. Readings from all cylinders are taken and compared to manufacturer's specifications. Compression readings are indicated in pounds per square inch (psi) or kilopascals (kPa).

Vacuum and Pressure Gauge

Some engine operations are checked by using a *vacuum and pressure gauge.* A vacuum reading is obtained by connecting the gauge to the intake manifold. A pressure reading is taken when the gauge is connected to the fuel system to measure fuel pump output pressure. Vacuum measurements are recorded in inches of mercury (Hg). Pressure measurements are recorded in pounds per square inch (psi).

Cylinder Leakage Tester

A *cylinder leakage tester* is used to check piston and valve condition **[Figure 8-14]**. This tester directs air into a closed/sealed cylinder and measures the amount of leakage taking place. This measurement is expressed as a percentage, as 10 percent leakage, 15 percent leakage, and so on.

Cooling System Tester

A *cooling system tester,* shown in **Figure 8-15,** is installed on the radiator neck after the radiator cap is removed. To check for leaks, the tester is pumped until the system reaches operating pressure and the gauge is watched for evidence of pressure loss. The tester also is used to check pressure cap condition.

Vacuum Diaphragm Tester

A *vacuum diaphragm tester* consists of a pump, a gauge, and a rubber hose **[Figure 8-16]**. Diaphragms use engine vacuum to produce mechanical motion and control different engine functions. If the diaphragm will not hold vacuum, it is defective.

Battery Hydrometer

The density of battery acid, or *electrolyte,* is tested with a *battery hydrometer.* The battery hydrometer has a suction tube, a rubber bulb that creates suction, and a clear reservoir **[Figure 8-17]**. The reservoir also houses a float or floating balls to record measurements. The bulb is squeezed, inserted into the battery, and released to draw electrolyte into the reservoir. The electrolyte raises the float or balls.

SAFETY CAUTION

Always wear eye protection when working around batteries. Electrolyte is an acid. It will eat through clothes or paint and can cause eye and skin damage.

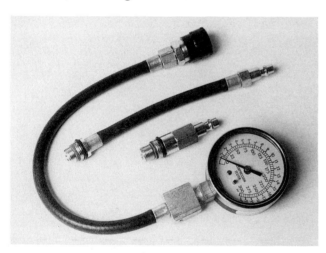

Figure 8-13. Compression tester.

Figure 8-14. Cylinder leakage tester.

Figure 8-15. Cooling system tester. CLIFF CREAGER

USING MEASURING TOOLS

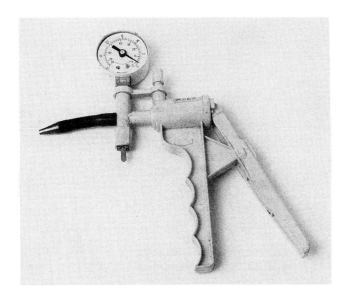

Figure 8-16. Vacuum diaphragm tester.

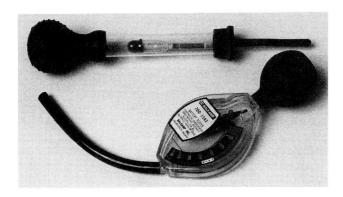

Figure 8-17. Battery hydrometers.

Coolant Hydrometer

Antifreeze is added to water to create *coolant*. As discussed in Unit 17, a mixture of antifreeze and water will provide protection against cooling system freezing, boiling, and corrosion. To determine the concentration of the coolant mixture, a *coolant hydrometer* [Figure 8-18] is used. The hydrometer provides a reading that indicates the freezing and/or boiling points of the coolant. This information helps the technician to determine the concentration of antifreeze in the mixture.

8.3 ELECTRICAL TESTERS AND DIAGNOSTIC EQUIPMENT

To determine the condition of electrical parts and systems and find defects, several types of electrical testers are used. In addition, sophisticated computer-type equipment is used to diagnose electrical and electronic malfunctions. These devices also are used to determine the amount of pollution produced by a vehicle during emissions inspections.

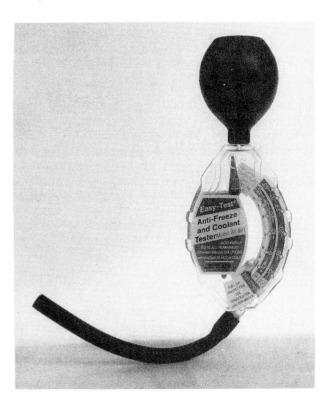

Figure 8-18. Coolant hydrometer. SAVERIO BONO

Circuit Tester

To check electrical flow in wiring, a *circuit tester* is used. A circuit tester is a tapered metal probe attached to a handle with a light bulb inside [Figure 8-19]. A connecting wire is attached to an electrical source or ground. The bulb lights when electrical current flows through the tester. This tester also may be called a *continuity tester* or a *test light*.

Voltmeter

A *voltmeter* is used to check the amount of voltage in an electrical circuit. A voltmeter consists of an indicator, which gives voltage measurements, and connecting leads [Figure 8-20].

Ammeter

Electric current flow in an electrical circuit is measured with an *ammeter* [Figure 8-21]. An ammeter

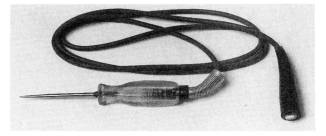

Figure 8-19. Continuity tester.

Figure 8-20. Battery voltage tester.

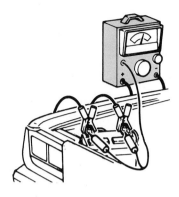

Figure 8-22. Battery tester. AMERICAN MOTORS CORPORATION

Figure 8-21. An ammeter usually is part of a unit that also contains a voltage tester.

Figure 8-23. Dwell-tachometer.

The dwell meter also is used to test operation of electronically controlled carburetors. Ignition systems are discussed in Unit 35.

has a pointer that indicates the amount of electrical current flowing in a circuit. Electrical flow is discussed in Unit 27.

Battery Tester

The battery must be able to provide and maintain a certain level of electrical current. To check electrical performance, a *battery tester* is connected to the battery [Figure 8-22]. The tester draws current from the battery and an ammeter and voltmeter on the tester indicate the battery condition.

Dwell-Tachometer

The tachometer portion of a *dwell-tachometer* tester is used to check and adjust engine speed [Figure 8-23]. The tachometer is connected to the ignition system and reads engine revolutions per minute (rpm).

The dwell meter measures *dwell,* or how long the contact points are closed in older ignition systems.

Oscilloscope

Ignition system and engine operation measurements are displayed on an *oscilloscope* screen, [Figure 8-24] similar to the screen on a television set. Oscilloscope leads are connected to wires in the ignition system. Signals from the wires are displayed on the oscilloscope screen. The traces, or luminous lines, on the screen indicate the operating condition of the systems being measured. Oscilloscopes also can be used to check alternator and other circuit voltages and waveforms.

Exhaust Gas Analyzer

Automotive emissions are checked with an *exhaust gas analyzer.* An exhaust gas analyzer is a machine that takes readings from a probe [Figure 8-25]. The probe is placed inside the exhaust pipe to draw exhaust gases into the analyzer. This tester is used when adjusting or troubleshooting the fuel and emissions systems.

Using Measuring Tools

Computerized Diagnostic Tester

Some service centers use a larger machine that can furnish the technician with a printout of an automobile's condition. This machine is called a *computerized diagnostic tester* [Figure 8-26]. Wires are connected to electrical parts of the automobile. Inputs are directed to a computer. The computer controls both the oscilloscope screen and a printer.

Many instrument companies produce compact, hand-held small computers for diagnosis. Plug-in cartridges enable the same tester to check different manufacturer's computer systems. Another variation are testers based on personal computers, or "PCs," with test programs stored on magnetic disks.

Figure 8-26. Computerized diagnostic tester.
SUN ELECTRIC CORPORATION

Figure 8-24. Oscilloscope.
CLIFF CREAGER

Figure 8-25. Exhaust gas analyzer.

UNIT HIGHLIGHTS

- Feeler gauges and wire gauges are used to measure the distance between two surfaces.
- Outside micrometers are made for both coventional and metric measurements.
- Measurements on the sleeve and thimble scales of the outside micrometer are added to obtain the proper measurement.
- An inside micrometer measures hole sizes or the distance between parallel surfaces.
- A small hole gauge, telescoping gauge, and vernier calipers are inside measuring tools for holes.
- A dial indicator measures play and runout.
- Fluid measuring tools are used with gases and liquids.
- Hydrometers indicate coolant and electrolyte densities.
- An oscilloscope can be used to diagnose ignition and other electrical system problems.

TERMS

straightedge
feeler gauge
outside micrometer
spindle
sleeve

ruler
wire gauge
anvil
thimble
inside mircrometer

Plastigage	small hole gauge
telescoping gauge	vernier calipers
play	runout
dial indicator	plunger
drive belt	drive belt tension gauge
tire pressure gauge	compression tester
cooling system tester	electrolyte
battery hydrometer	coolant hydrometer
circuit tester	voltmeter
battery tester	ammeter
dwell-tachometer	dwell
oscilloscope	exhaust gas analyzer
coolant	cylinder leakage tester
computerized diagnostic tester	mechanical measuring tool
vacuum and pressure gauge	vacuum diaphragm tester
split-ball gauge	

REVIEW QUESTIONS

DIRECTIONS: The following questions are similar to those used on automotive technician certification tests. Answer each question by circling the letter of the correct choice.

1. Technician A says that a ruler and a straightedge are exactly the same, except that one is metric.
   Technician B says that a feeler gauge and a wire gauge measure the space between two surfaces.
   Who is correct?
   A. A only
   B. B only
   C. Both A and B
   D. Neither A nor B

2. Which of the following are used to take accurate inside measurements?
   A. Ruler
   B. Vernier calipers
   C. Straightedge
   D. Cylinder leakage tester

3. Which of the following statements is correct?
   I. Fluid-measuring tools are used to check vacuum and pressure.
   II. A hydrometer checks the liquid level in a battery or radiator.
   A. I only
   B. II only
   C. Both I and II
   D. Neither I nor II

4. Technician A says that a dwell meter is used to check how fast the engine is running.
   Technician B says that a tachometer is used to check how long ignition contact points remain closed.
   Who is correct?
   A. A only
   B. B only
   C. Both A and B
   D. Neither A nor B

Question 5 is not like the first four questions. It has the word EXCEPT. For question 5, look for the choice that does *not* apply. Read the entire question carefully before you choose your answer.

5. All of the following statements about outside micrometers are correct EXCEPT
   A. both conventional and metric micrometers are available.
   B. reading an outside micrometer requires adding sleeve and thimble numbers.
   C. measurements are taken between the anvil and the thimble.
   D. each rotation of a conventional micrometer thimble moves the spindle 0.025.

ESSAY QUESTIONS

1. Briefly describe mechanical measuring tools that are used by automotive technicians.
2. Briefly describe commonly used fluid measuring tools for automotive work and their uses.
3. Briefly describe electrical measuring tools used in automotive troubleshooting.

SUPPLEMENTAL ACTIVITIES

1. Demonstrate how to use feeler gauges.
2. Use a micrometer to measure a part your instructor has furnished.
3. Demonstrate how to use an inside micrometer.
4. Use a dial indicator to measure movement on a part or assembly selected by your instructor.
5. Use a cooling system tester and coolant hydrometer on a vehicle selected by your instructor.
6. Check the battery condition of a vehicle selected by your instructor and give ammeter and voltmeter readings.
7. Demonstrate how to correctly hook up an oscilloscope and observe ignition system traces.
8. Describe the safety cautions that must be observed during battery electrolyte testing.

9 Troubleshooting

UNIT PREVIEW

Automotive technicians follow specific procedures when diagnosing automotive problems. The use of diagnostic procedures is called troubleshooting. Troubleshooting charts in service manuals help to pinpoint problem areas and decrease diagnostic time for a technician. Troubleshooting charts also direct a technician in conducting necessary tests and inspections.

LEARNING OBJECTIVES

When you have completed your assignments and exercises in this unit, you should be able to:

❏ Use a troubleshooting chart to trace the possible causes of a problem.
❏ Describe test driving procedures from a service manual.
❏ Make a visual inspection for automotive problems.
❏ Troubleshoot a problem chosen by your instructor.

9.1 AUTOMOTIVE DIAGNOSTICS

In automotive terms, *diagnostics* refers to the general process of checking automotive parts and systems to find what is right and what is wrong. A lot of time, effort, and customer goodwill can be wasted by doing unnecessary repairs based on a guess. You can save time, effort, and possible lost pay by diagnosing automotive problems correctly and doing only necessary repairs or adjustments.

9.2 TROUBLESHOOTING

The more specific process of locating and identifying the causes of problems is called *troubleshooting.* Troubleshooting is a step-by-step process of elimination. The technician makes checks to eliminate certain parts or functions as the cause of a problem. Troubleshooting progresses from simple checks and inspections to more involved test procedures.

Service manuals furnish a technician with procedures, guides, and charts for troubleshooting automotive problems.

Defining a Problem

A first step is to speak with the driver. When asking questions about the complaint, concentrate on four areas:

- *What happens?* Examples: the engine stalls; the transmission shifts harshly; an exhaust smell is noticeable inside the car.
- *When does the problem occur?* Examples: when the car is cold; when the car is fully warmed up; when the car is turning.
- *Where does the problem occur?* Examples: up or down hills; on flat roads; on highways; around corners.
- *How long has the problem been noticeable?* Examples: gradually over a period of months; for a week; it just began today.

However, do *not* rely on the driver's information alone. Make your own diagnosis. Although information from the driver can be helpful, the technician is responsible for correctly diagnosing the problem and performing only necessary repairs.

9.3 TROUBLESHOOTING CHARTS

Repair manuals have *troubleshooting charts* for many automotive systems. Troubleshooting charts enable a technician to locate causes of problems quickly. **Figure 9-1** shows a typical troubleshooting chart for a brake system. First, find the condition in the left column. Then move to the center column for possible causes. The right column lists repairs to be made to correct the conditions noted.

9.4 THE TEST DRIVE

A *test drive* is a form of troubleshooting that requires specific driving procedures and safety precautions by the technician. Many repair manuals outline different driving procedures that should be followed when checking for specific problems. **Figure 9-2** shows test driving procedures used in checking a driveline. Follow the chart boxes downward to trace specific causes of problems.

Driving safety is always important. Before you begin, buckle up the seat belt and test the brakes. Avoid jackrabbit starts, hard turning, or sudden

GENERAL BRAKE SYSTEM DIAGNOSIS GUIDE

CONDITION	POSSIBLE SOURCE	RESOLUTION

• Brakes grab or lock-up when applied.	• Tires worn or incorrect pressure.
	• Grease or fluid on linings — damaged linings.
	• Improper size or type of linings.
	• Other brake system components:
	  • Bolts for caliper attachment loose or missing.
	  • Worn, damaged or dry wheel bearings.
	  • Improperly adjusted parking brake.
		• Inflate tires to correct pressure. Replace tires with worn tread.
		• Inspect, service or replace.
		• Replace with correct brake in axle sets.
		• Inspect, service or replace as required.

• Brake warning light on.	• Hydraulic system. • Shorted light circuit. • Parking brake not returned. • Brake warning switch.	• See Master Cylinder Diagnosis Guide. • Correct short in warning circuit. • See "Parking brake will not release or fully return" below. • Replace.
• Intermittent loss of pedal.	• Loose wheel bearing.	• Adjust as required. • Perform steps under "Excessive pedal travel or pedal travel goes to floor".
• Rough engine idle or stall, brakes applied — power brakes only.	• Vacuum leak in neutral switch. • Vacuum booster.	• Check lines for leaks. Service or replace as required. • Check vacuum booster for internal leaks. Replace if required.
• Parking brake control will not latch (manual release).	• Kinked or binding release cable. • Control assembly.	• Inspect, service or replace. • Inspect, service or replace.
• Parking brake control will not latch (automatic release).	• Vacuum leak. • Vacuum switch. • Control assembly.	• Service as required. • Test. Replace if necessary. • Service or replace.
• Parking brake will not release or fully return (manual release).	• Cable disconnected. • Control assembly binding. • Parking brake linkage binding. • Rear brakes.	• Connect cable or replace. • Service or replace. • Service or replace. • Check rear brakes shoe retracting springs and parking brake levers.
• Parking brake will not release or fully return (automatic release).	• Vacuum line leakage or improper connections. • Neutral switch. • Control assembly.	• Inspect and service. • Adjust or replace. • Service or replace.

Figure 9-1. A troubleshooting chart in a service manual. FORD MOTOR COMPANY

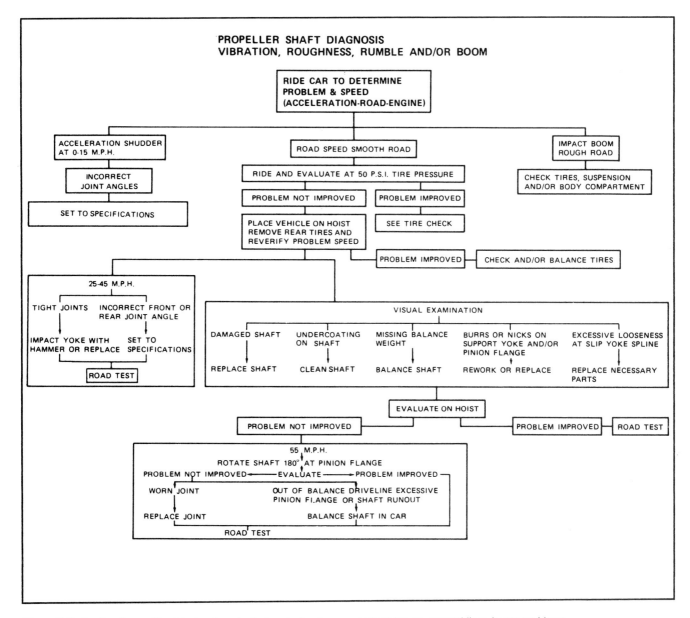

Figure 9-2. Another form of troubleshooting chart gives instructions on test driving an automobile to locate problems.
CHEVROLET MOTOR DIVISION—GMC

braking. Improper driving can make a problem worse or create a new problem. Carelessness during the test drive can result in an accident. Drive only in a manner that is required to diagnose the problem.

Consult the proper troubleshooting chart before the test drive. Test drive procedures change when different parts of the automobile are checked. Different brands of automobiles may have different test driving procedures. Procedures may call for low speeds, hard acceleration, or sudden gear changes. Follow the troubleshooting charts carefully. Using the wrong driving procedures may lead to an improper diagnosis. Try to take the test drive in an area where traffic is light. The fewer the distractions the better.

9.5 PINPOINTING A PROBLEM

Following a test drive, and after checking the troubleshooting chart, the next step is to pinpoint the problem. By now, the technician should have a general idea of where to look. To locate the problem, two types of inspections can be made:

- Visual inspection
- Measurement inspection.

Visual Inspection

Make a *visual inspection* by looking around the automobile in the area where the likely problem exists. Look for leaks and loose or broken parts.

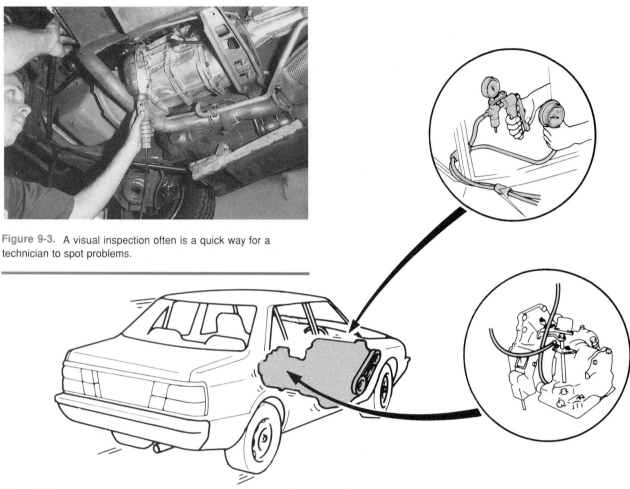

Figure 9-3. A visual inspection often is a quick way for a technician to spot problems.

Figure 9-4. A measurement inspection, using gauges and special tools, often is necessary to diagnose a problem.
MAZDA MOTOR CORPORATION

Shake the parts to check for looseness, and listen for noises [Figure 9-3]. Many automotive problems can be seen if a technician simply examines parts carefully.

Measurement Inspection

If visual inspection does not expose a problem, a *measurement inspection* may be needed. A measurement inspection may include one or more specific procedures. Check out the suspected problem area by using the proper measuring equipment for the job [Figure 9-4]. Specific tests are used, depending on which parts or conditions of the automobile are to be measured.

Rechecking Problem Areas

Once a problem area has been found and repaired, recheck the automobile. Two or more problems could be responsible for a single complaint. Check the operation of the automobile to be sure it is satisfactory. If a problem persists, repeat all of the necessary troubleshooting procedures until a correct diagnosis is made.

UNIT HIGHLIGHTS

- Troubleshooting is investigating and analyzing an automotive problem.
- Troubleshooting charts speed up a technician's work by identifying the condition and locating the possible source of the problem.
- A test drive requires specific driving procedures to troubleshoot an automobile.
- Locate problems by visual and/or measurement inspection.

TERMS

diagnostics	troubleshooting
troubleshooting charts	test drive
measurement inspection	visual inspection

REVIEW QUESTIONS

DIRECTIONS: The following questions are similar to those used on automotive technician certification tests. On a separate sheet of paper, write the letter of the correct choice.

1. Which of the following types of information is found in a troubleshooting chart?
   A. Questions to ask the driver
   B. How long the problem has existed
   C. Conditions, possible causes, and necessary repairs
   D. The proper lubricants to use

2. Which of the following statements is correct?
   I. Test drive procedures can change when different parts of the automobile are being checked.
   II. Test drive procedures can change when different makes of automobiles are being checked.
   A. I only
   B. II only
   C. Both I and II
   D. Neither I nor II

3. Technician A says that the same safety procedures are used regardless of a test drive diagnosis.
   Technician B says that finding the problem is more important than safety during a test drive.
   Who is correct?
   A. A only
   B. B only
   C. Both A and B
   D. Neither A nor B

Questions 4 and 5 are not like those above. Each has the word EXCEPT. For each question, look for the choice that does *not* apply. Read each question carefully before you choose your answer.

4. All of the following statements about troubleshooting are correct EXCEPT
   A. the procedure requires investigating and analyzing a problem.
   B. the most accurate source of information about problems always is the driver.
   C. service manuals provide guides for troubleshooting.
   D. it is necessary to know when the problem occurs.

5. All of the following statements about troubleshooting are correct EXCEPT
   A. specific information about problem conditions helps the technician.
   B. troubleshooting charts help the technician to find solutions quickly.
   D. measurement tests never indicate the cause of problems.
   C. a test drive can be helpful in pinpointing problem areas.

ESSAY QUESTIONS

1. What are the differences between diagnostics and troubleshooting?
2. Why is it important to get as detailed a description of car problems as possible before beginning troubleshooting?
3. Describe the steps that can be involved in troubleshooting.

SUPPLEMENTAL ACTIVITIES

1. Locate a troubleshooting guide in a shop service manual.
2. Identify a possible problem based on an explanation furnished by your instructor.
3. Select a troubleshooting chart and explain the various test driving procedures that must be followed.
4. Make a visual inspection and locate a problem in an automobile system selected by your instructor.
5. Perform a measurement inspection on a part furnished by your instructor.

PART II: The Automotive Engine

Career opportunities in engine service include troubleshooting, and engine repair and rebuilding. This part of the book provides detailed knowledge on piston engine fundamentals, engine measurements and performance, and engine construction. In addition, piston, ring, connecting rod, cylinder head and valve train construction and operation are covered. Here, a technician is putting a new piston ring on a piston.

10 Piston Engine Fundamentals

UNIT PREVIEW

An automotive engine changes heat energy into mechanical energy that moves a vehicle. Modern automotive engines are powered by the pressure created by burning fuel inside the engine.

The operating principles of the engine can be divided into four basic functions. The engine must bring in fuel and air, heat and confine the fuel and air to a small area, burn the fuel to create pressure, and expel the burned gases.

Automotive engines consist of multiple power-producing units, or cylinders, and can be classified according to shape, size, fuel used, and many other characteristics.

The most common type of engine used for automobiles has been in use for over 100 years. Although other engine types have been proposed or used, the piston engine designed by Nikolaus Otto in 1876 remains the most practical and popular design for automobiles.

LEARNING OBJECTIVES

When you have completed your assignments and exercises in this unit, you should be able to:

❏ Define and explain the terms, energy, force, work, and power.
❏ Describe how an automobile engine produces power.
❏ Identify and describe the functions of the internal parts of an automotive engine.
❏ Identify and describe the operating functions of common automotive engines.
❏ Classify engines according to specific characteristics.

10.1 ENERGY, FORCE, WORK, AND POWER

To understand how an automobile engine works, you must understand some basic terms that describe physical processes. For example, think about what happens if your car stalls in traffic. If you are strong enough by yourself, or get enough people to help you, you can push the car off to the side of the road. The ability to move heavy objects is called *energy*.

When you push the car, you are applying *force* to it. Force applied to an object causes motion. If you have to push the car very far, you will have done a good bit of *work*. Work is the result of applying force to an object and causing it to move. The more force applied and the more motion caused, the more work done.

Power is the rate at which work is done. *Rate* means how fast you are able to do the work of pushing.

If you have more energy, you can apply more force and do more work. The faster the work is done, the more power you are producing and using. These concepts are summarized in **Figure 10-1**.

An automobile engine uses the energy present in its fuel to apply force and move the vehicle. The engine produces *torque*, a twisting force that turns the driving wheels. As the engine applies force and moves the vehicle, work is performed. The rate at which this work is done, often called engine power, is measured in *horsepower (hp)*, *watts (w)*, or *kilowatts (kw)*.

Energy can take many forms. Human beings and animals store energy chemically in their muscles. A

ENERGY = The Ability to do Work
WORK = Force that Causes Motion
POWER = How Fast Work is Done

Force Is Applied To Push The Car

Figure 10-1. The concepts of power.

tightly wound spring from an alarm clock stores *mechanical energy.* A fire extinguisher uses a combination of *chemical energy* and *physical energy* to spray a non-burning substance on a fire. When a gas stove is lit, *heat energy* is present for cooking.

10.2 CONVERTING HEAT ENERGY TO MOTION

Thermodynamics is the scientific study of how mechanical energy and heat energy are related. An automobile *engine* is a machine, or mechanical device, that changes heat energy into mechanical energy. Heat energy is produced by the rapid burning of gasoline. This heat energy is converted into mechanical energy to turn the wheels that move a vehicle.

External Combustion Piston Engine

Two types of engines commonly have been used to convert heat energy to motion. In one type, fuel is burned outside of the engine. These engines are called *external combustion* engines. External combustion simply means "outside burning."

An old-fashioned steam locomotive is an example of an external combustion engine. The burning of fuel takes place outside of the engine areas that apply the force to move the locomotive. Fuel (wood, coal, or oil) is burned in a firebox under a boiler filled with water. The heat turns the water into steam. The steam pushes against a tight-fitting plug, called a *piston,* inside a hollow area called a *cylinder* [**Figure 10-2**].

The steam exerts great *pressure* against the piston. Pressure is the amount of force applied to a given area. The force pushes the piston through the cylinder. The piston is connected by a rod to the driving wheels of the locomotive. As the piston is moved in the cylinder by the steam pressure, the driving wheels of the locomotive begin to turn. This action is illustrated in **Figure 10-3**.

Internal Combustion Piston Engine

Another type of engine for converting heat energy to mechanical energy is called an *internal combustion* engine. The typical automotive engine is an internal combustion engine. Internal combustion means "inside burning." Fuel (gasoline or diesel) is burned inside a cylinder in the engine, above the piston. The burning produces heat, which in turn produces pressure. The pressure forces the piston down the cylinder [**Figure 10-4**].

The piston is connected by a rod to a *crankshaft.* The crankshaft and connecting rod convert the downward push of the piston into a spinning, or rotary motion. The rotary motion of an engine's crankshaft is connected through the drivetrain to turn the driving wheels of the vehicle. **Figure 10-5** shows a cutaway view of an automotive piston engine.

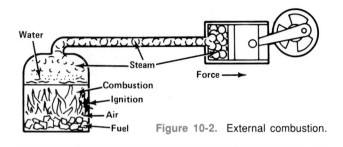

Figure 10-2. External combustion.

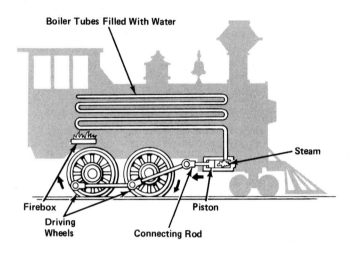

Figure 10-3. Steam pressure moves a piston that rotates the driving wheels.

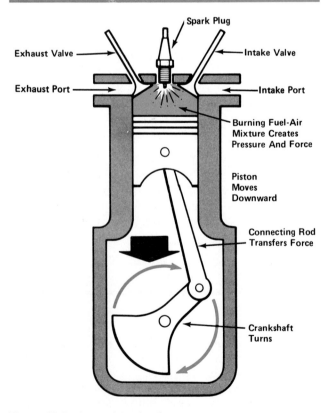

Figure 10-4. Internal combustion.

Figure 10-5. Cutaway view of an internal combustion engine.

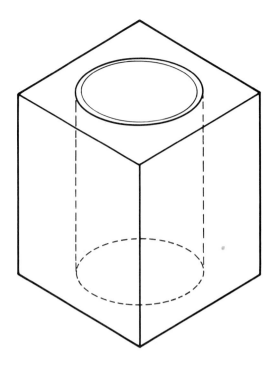

Figure 10-6. A cylinder is formed by boring a large hole in a solid piece of metal. The drilled, solid metal block is called the cylinder block.

10.3 PISTON ENGINE PARTS

An automotive internal combustion engine is made up of many parts. The major parts of an engine include:

- Cylinder block
- Pistons and rings
- Crankshaft
- Connecting rods
- Bearings
- Cylinder head and valves
- Valve train and camshaft
- Flywheel
- Fuel system
- Exhaust system.

Cylinder Block

Imagine starting to build an internal combustion piston engine from scratch. First you would need a large chunk, or block, of metal **[Figure 10-6]**. This part is called the *cylinder block.* To form the cylinder, you could drill, or bore, a large hole, called the *bore.*

Piston, Rings, Crankshaft, Connecting Rod

A piece of metal can then be formed into a cylindrical piston shape. To make the piston fit tightly in the cylinder, springy metal *piston rings* are inserted around the piston. These form a tight seal for the high-pressure gases that power the engine. A metal *connecting rod,* connects the piston to the crankshaft. To convert the downward motion of the piston into a rotary motion, a crankshaft is attached to the bottom of the block. The motion of the connecting rods and crankshaft is similar to the motion of pedaling a bicycle. Your legs go up and down, and the bicycle "crank" converts that motion into a turning motion. **Figure 10-7** illustrates these parts.

Bearings

The connecting rod must be able to spin around the crankshaft. Also, the crankshaft must be able to turn freely in the engine block. To permit these motions, soft metal parts called *insert bearings* are used. These bearings also protect the connecting rod and crankshaft from rubbing against each other and wearing out. **Figure 10-8** shows connecting rod bearing inserts. Many bearings are used in the engine to support and protect rotating parts and to allow them to turn freely.

Cylinder Head and Valves

To close off the end of the cylinder above the piston, a thick piece of metal is attached. Because it is at the top of the engine, it is called the *cylinder head.* A sealing *gasket,* called a *head gasket,* seals the separation between the head and the cylinder block.

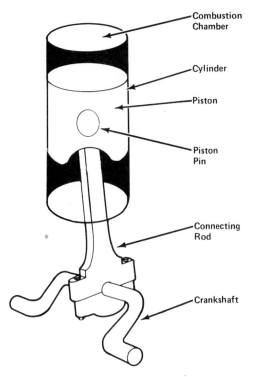

Figure 10-7. Up-and-down movement of a piston is converted into rotary motion at the crankshaft.

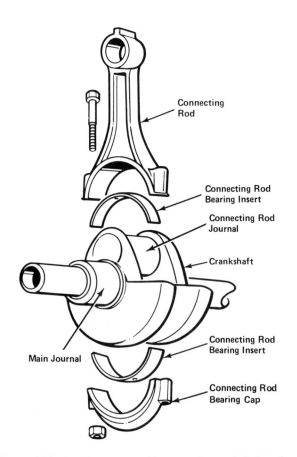

Figure 10-8. Bearings are used between the crankshaft and connecting rods.

The cylinder head has hollow passages that are molded, or cast, into the metal. Separate hollow passages, called *intake* and *exhaust ports,* allow air and gasoline vapor to flow into the cylinder, and burned gases to flow out. To close the ports during the burning of the gasoline vapor, tapered parts called *valves* are fitted. A valve acts like a cork in the neck of a bottle. When the valve is closed, the cylinder is sealed. When the valve is opened, gases can flow around the valve into or out of the cylinder. Also in the cylinder head is a threaded hole for a *spark plug.* The spark plug acts like the fuse in a cannon. When a high-voltage spark jumps the gap at the end of the spark plug, the gasoline vapor is ignited and burned rapidly. These parts are illustrated in **Figure 10-9.**

Valve Train and Camshaft

The valves must form a tight seal to contain the force of combustion. Large springs are attached to the stems of the valves to pull the valves shut. The parts that operate to push the valves open are known as the *valve train.* To open the valves, a rotating metal shaft with projections, called a *camshaft,* is used. A *cam,* or *cam lobe,* is a projection on the camshaft, with a shape similar to an egg. The cam lobe pushes the valve downward to open the port. The camshaft can be mounted over the cylinder head, as shown in **Figure 10-10.** Such an arrangement is known as an *overhead camshaft (OHC).*

Another way of opening the valves is to position the camshaft inside a hollow area of the block. Intermediate parts, shown in **Figure 10-11,** then are used to transfer the pushing motion to the valve stem. This arrangement is known as a *pushrod* valve train.

The camshaft may be driven, or turned, by a gear arrangement, a chain, or a reinforced, toothed rubber belt. Camshaft drive mechanisms are shown in **Figure 10-12.**

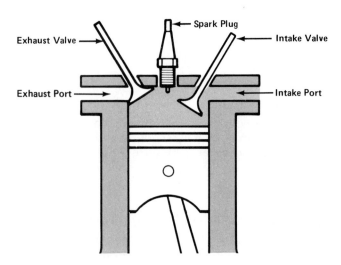

Figure 10-9. A cylinder head and valve assembly.

Piston Engine Fundamentals

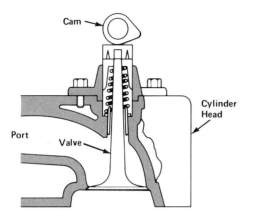

Figure 10-10. An overhead camshaft assembly.

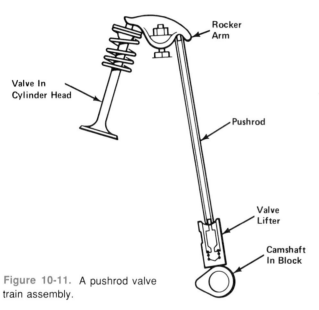

Figure 10-11. A pushrod valve train assembly.

Flywheel

A heavy round metal wheel, called a *flywheel*, is attached to the crankshaft at the rear of the engine [Figure 10-13]. Once the flywheel starts spinning, its momentum helps the engine to run smoothly.

Basic Fuel System

To supply gasoline vapor to the basic engine, a fuel system is needed. A *carburetor* mixes gasoline and air to produce a fine mist. The carburetor is connected to the *intake port* of the cylinder head by a hollow metal part called an *intake manifold*. Figure 10-14 shows a cutaway view of the carburetor, intake manifold, and intake valves. Recent engines use a *fuel injection* system instead of a carburetor to provide fuel (see Unit 23).

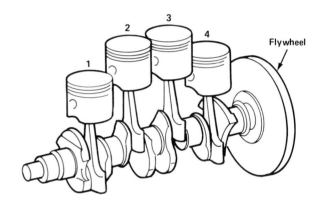

Figure 10-13. A flywheel rotates with the crankshaft to help the engine run smoothly.

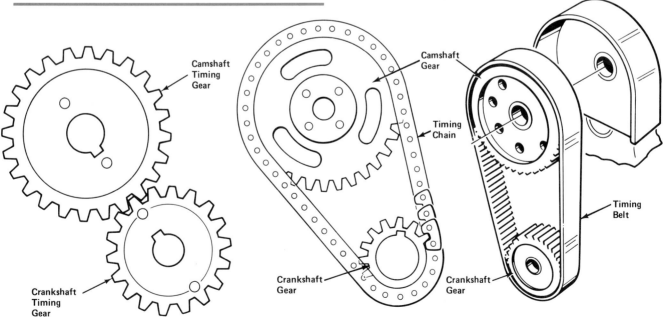

Figure 10-12. Camshafts can be operated by gears, chains, or special belts.

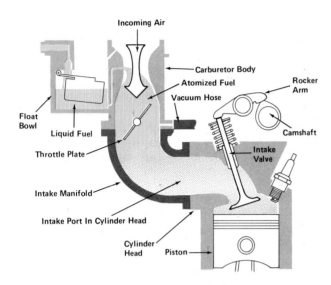

Figure 10-14. In many engines, the air-fuel mixture is introduced into the cylinders through a carburetor, intake manifold, and intake valves.

Basic Exhaust System

After the air-fuel mixture is burned, the gases must be removed from the cylinder in preparation for a fresh *intake charge*, or mixture of air and vaporized fuel. The *exhaust valve* is pushed open by an exhaust cam to allow the burned gases to be forced out of the cylinder. A hollow, tubular part, called an *exhaust manifold*, is bolted to the cylinder head at the exhaust port, as shown in **Figure 10-15**. Connecting metal tubes lead to the muffler, shown in **Figure 10-16**, which reduces the noise from the combustion within the engine.

10.4 THE FOUR-STROKE CYCLE

The basic engine must draw in an air-fuel mixture, compress it, burn it, and expel the gases. This sequence of events makes up what is known as the *four-stroke cycle*. A *stroke* is the movement of the piston in the cylinder, from one end to the other. A cycle is a sequence that is repeated. The four-stroke cycle, illustrated in **Figure 10-17**, includes the following strokes:

- Intake
- Compression
- Power or ignition
- Exhaust.

Follow the illustrations in **Figure 10-17** as you read the following discussion.

Intake

The crankshaft and camshaft are both turning as the intake stroke begins. As the crankshaft turns, the connecting rod pulls the piston downward. At the same time, the intake valve leading to the intake manifold and carburetor is opened by the camshaft.

Because the piston rings seal the piston tightly in the cylinder, a low-pressure area is formed as the piston moves down. The atmospheric pressure of the outside air forces air to flow down through the carburetor. The carburetor discharges a fine mist of gasoline into the incoming air. This intake charge fills the cylinder as the piston reaches the bottom of the intake stroke. When the piston is all the way down, it is at *bottom dead center (BDC)*.

Compression

As the piston nears the bottom of its intake stroke and starts back up, the camshaft rotation allows the intake valve to be pulled closed by the valve spring. Both intake and exhaust valves now are closed. The air-fuel mixture is trapped as the piston continues to rise. The rising piston compresses, or squeezes the mixture. Compressing the mixture heats it for better burning, and confines it to a very small area for more intense combustion.

The *compression ratio* is a set of numbers that expresses how much the mixture is compressed. [**Figure 10-18**]. Typical compression ratios for gasoline engines range from as low as approximately 7 to 1 to as high as 12 to 1. When the piston reaches the very top of its travel, it is at *top dead center (TDC)*. The volume of space through which the piston travels from BDC to TDC is called the *displacement* of the cylinder. In a multiple-cylinder engine, the total displacement is the sum of all the individual cylinders' displacements.

Power or Ignition

At a point near the top of the compression stroke, a high-voltage current is sent to the spark plug. The current jumps a gap at the end of the spark plug to form a spark. The spark ignites the tightly compressed air-fuel mixture, causing it to burn rapidly. This burning takes place within the cylinder head in a small area above the piston, called the ***combustion***

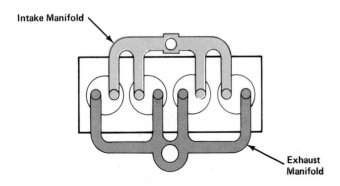

Figure 10-15. Exhaust gases are routed outward through an exhaust manifold.

PISTON ENGINE FUNDAMENTALS

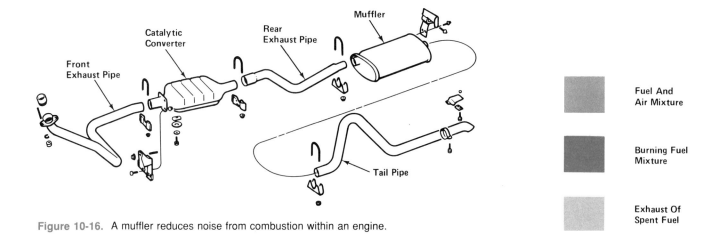

Figure 10-16. A muffler reduces noise from combustion within an engine.

chamber. The combustion causes heat, and the expanding mixture creates pressure in the cylinder. The pressure pushes the piston downward. The connecting rod pushes against the crankshaft and forces it to turn.

Exhaust

Near the bottom of the power stroke, the camshaft pushes the exhaust valve open. As the piston reaches BDC, it reverses direction and travels up on the exhaust stroke and pushes the burned gases out through the exhaust port.

At this point, the piston is again near the top of the cylinder. Quickly, the rotation of the camshaft allows the exhaust valve to close and opens the intake valve, and the piston begins an intake stroke immediately.

At highway speeds, the four-stroke cycle repeats itself in each cylinder thousands of times every minute.

10.5 MULTIPLE-CYLINDER ENGINES

Each cylinder is a power-producing unit. To provide the force needed to move a heavy automobile, engines have several cylinders. These cylinders can be arranged differently within the cylinder block. **Figure 10-19** shows the most common cylinder arrangements, or *configurations,* for passenger car engines.

Piston Motion

In a multiple-cylinder engine, the crankshaft is made so that the movements of the pistons are staggered. That is, in an engine with an even number of cylinders, half of the pistons move upward as the others move downward. This arrangement helps to balance the forces created by the engine so that it runs smoothly.

In a four-cylinder engine, two pistons will be moving upward at the same time that two are

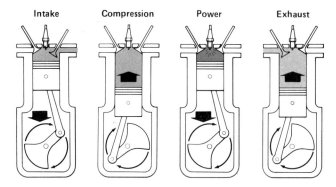

Figure 10-17. The four-stroke cycle.

moving downward. One of the two moving upward will be on its compression stroke, and the other will be on its exhaust stroke. One of the two pistons moving downward will be on its intake stroke, and the other will be on its power stroke.

Cylinder Numbering

On the end of the crankshaft, opposite the flywheel, a *power pulley* is attached. Belts are stretched between the power pulley grooves and the pulleys of engine accessories. When the engine operates, the belts force the accessory device pulleys to turn **[Figure 10-20]**.

The engine cylinders are numbered from the power pulley end toward the flywheel end. The cylinders in an in-line engine are numbered one through four, five, or six, from the power pulley rearward.

V-type and flat engines may have different numbering systems. **Figure 10-21** illustrates some cylinder numbering arrangements.

Firing Order

The order in which the power strokes occur within the cylinders usually is not the same as the cylinder numbering. See Unit 12 for an explanation of how

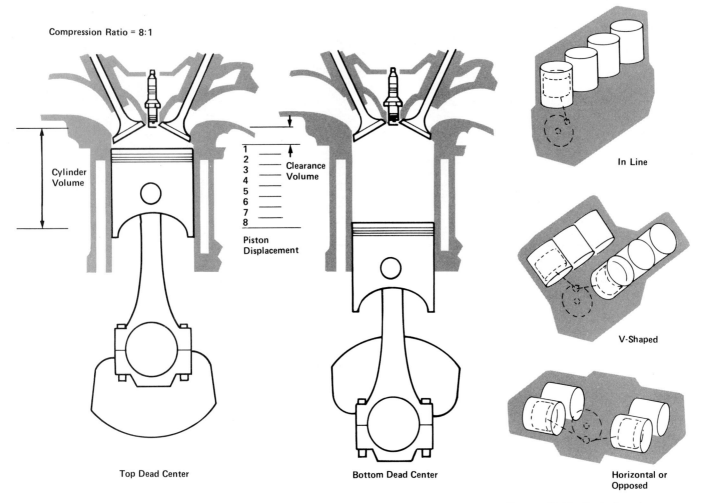

Figure 10-18. The compression ratio indicates how much the air-fuel mixture is squeezed on the compression stroke.

Figure 10-19. Typical engine cylinder configurations.

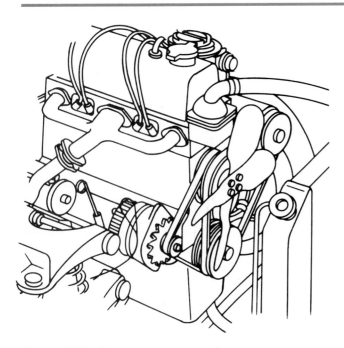

Figure 10-20. Engine accessory operation.

firing order is determined. Examples of cylinder numbering and firing order arrangements are shown in **Figure 10-22.**

Engine Speed and Power Overlap

When referring to engine speed, the rotational speed of the crankshaft is specified in *revolutions per minute (rpm).* At idle, engine speed typically is between 600 to 900 rpm, according to the manufacturer's specifications. On the highway at 55 mph or more, most passenger vehicle engines run at between 2,200 and 3,000 rpm.

Two revolutions of the engine crankshaft are necessary for the four-stroke cycle to be completed in any single cylinder. The more cylinders an engine has, the more power strokes, or *power pulses,* the crankshaft receives per revolution. In engines with more than four cylinders, the power pulses overlap each other to provide a smoother application of force to the crankshaft. Thus, the more cylinders, the smoother an engine will run.

PISTON ENGINE FUNDAMENTALS

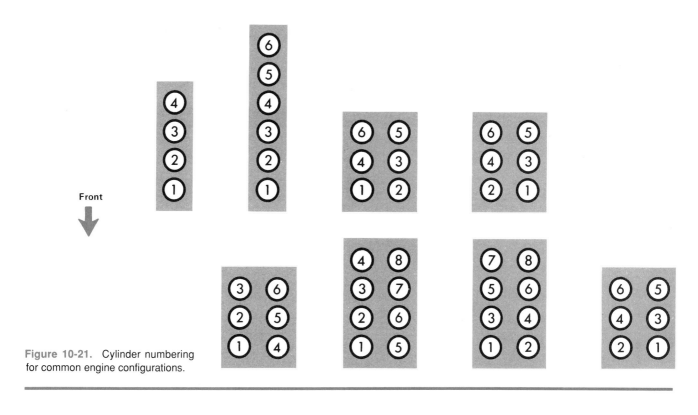

Figure 10-21. Cylinder numbering for common engine configurations.

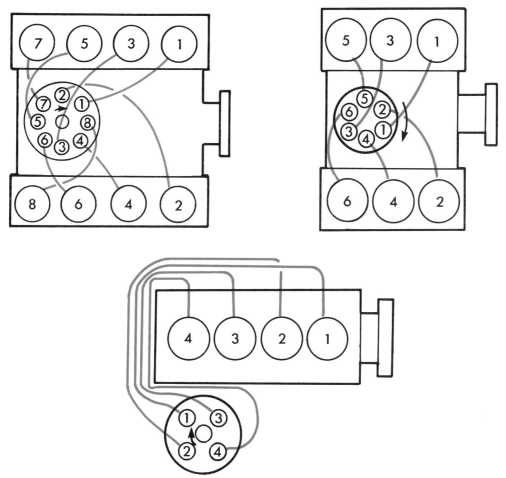

Figure 10-22. Cylinder numbering and firing order arrangements.

10.6 ENGINE CLASSIFICATIONS

Automobile engines can be classified according to several features. These features include:

- Number of cylinders
- Cylinder arrangement
- Number of valves
- Valve arrangement
- Valve train
- Combustion chamber shape
- Fuel used
- Ignition system
- Cooling system
- Lubrication system
- Strokes per cycle
- Reciprocating vs. Rotary
- Use.

Number of Cylinders

Automobile engines can be classified by total number of cylinders. Current automotive engines include 3-, 4-, 5-, 6-, 8-, and 12-cylinder models. Prototype 10-cylinder engines also have been built as models for production.

Cylinder Arrangement

Engines can be classified according to cylinder arrangement. In other words, an engine can be of the inline, V, or flat type. Other, more complicated arrangements also have been used.

Number of Valves

Until the mid-1980s, most automotive engines had only two valves per cylinder. However, recent production engines include four valves per cylinder (two intake and two exhaust) for better "breathing," or gas flow.

These engines may be referred to as having a "4-valve" head (four valves per cylinder). Or, they may be known as "16 valve engines" (total number of valves, for a 4-cylinder) or "24-valve engines" (for a 6 cylinder), and so on. Refer to **Figure 10-5** for a cutaway view of a 4-valve per cylinder, 16-valve four-cylinder engine.

Valve Arrangement

Although the valve-in-head design is now universal for automotive engines, other valve arrangements are possible. One or both of the valves can be located within the block, instead of in the head. Most automobile engines built prior to 1955 had such an arrangement, as do small engines, such as those used for lawnmowers.

Valve Train

As mentioned above, engine valve trains can be either pushrod type or overhead camshaft (OHC) type. Two separate camshafts can be used for exhaust and intake valves. These *dual over-head camshafts (DOHC)* are used in high-performance cars.

Combustion Chamber Shape

Engines can be classified according to the shape of the combustion chamber used. **Figure 10-23** shows two of the most commonly used variations in shape for automobile engines.

Fuel Used

Fuels other than gasoline can be used to provide the heat energy to drive an automotive engine. Practical fuel alternatives include propane, diesel fuel, *methanol* (poisonous "wood alcohol"), and *ethanol* (grain alocohol, such as that used in liquor).

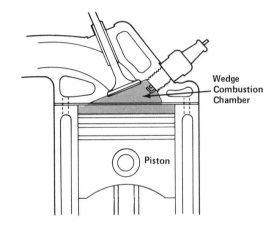

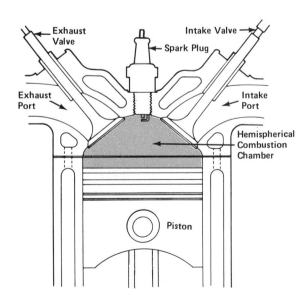

Figure 10-23. Common combustion chamber configurations.

Piston Engine Fundamentals

Ignition System

The air-fuel mixture can be ignited by an electrical spark, as explained above. Such an arrangement is known as *spark ignition*. However, *diesel engines* have no spark plugs. A diesel engine's compression ratio is much higher–between 20:1 and 23:1–than that of a gasoline engine. The greater heat generated by this higher compression is sufficient to ignite the fuel for the power stroke. Thus, a diesel engine also is known as a *compression ignition* engine.

Cooling System

Heat is generated by the combustion of the air-fuel mixture within the combustion chamber. This heat is sufficient to melt and destroy some engine parts, such as aluminum pistons. Some form of cooling system is needed to dissipate, or draw away, this excess heat.

Air cooling. Engines can be cooled by shaping the outside metal of the engine block and head into *cooling fins* and forcing air over them. *Air cooling* has been used for motorcycle and automotive engines **[Figure 10-24]**.

Liquid cooling. A more efficient way to cool the engine is to form hollow passages in the block and head, and circulate cool liquid through them. The heat is conducted into the liquid. *Liquid cooling* allows the engine to run at a more consistent temperature in hot and cold weather. A liquid cooling system is shown in **Figure 10-25**.

Lubrication System

In addition to cooling requirements, all moving parts of the engine must be lubricated. An *oil pan* is fitted to the bottom of the engine to hold a supply of motor oil. Modern automotive engines are lubricated by oil flowing from an *oil pump* under pressure to moving parts **[Figure 10-26]**. This system is known as a *pressurized lubrication system*. In most engines, the lubricant also helps to cool the engine.

As the crankshaft rotates, it splashes oil on the cylinder walls below the pistons. This activity is known as *splash lubrication*. Modern engines use a form of splash lubrication, called *throw-off*. Oil is allowed to "leak" from crankshaft main bearing lubrication points. The oil is thrown onto the cylinder walls as the crankshaft rotates.

Strokes per Cycle

Some engines operate on a *two-stroke cycle*. This design is used for smaller motorcycles, chain saws, some lawnmowers, and similar applications. In the past, small automobiles also were powered by two-stroke engines. Each time the piston in a two-stroke engine reaches TDC, the spark plug fires a compressed mixture. Prototypes of advanced-design, two-stroke engines have been discussed as possible future automotive engines.

Reciprocating vs. Rotary

Most automobile engines are reciprocating engines. That is, the pistons travel back and forth in the cylinders. A small number of automobiles are powered by *rotary engines*. A rotary engine uses a curved, triangular rotor to perform the functions of the four-stroke cycle. A rotary engine four-stroke cycle is illustrated in **Figure 10-27**. In recent years, only the Mazda RX-7 sports car has been produced with a rotary engine.

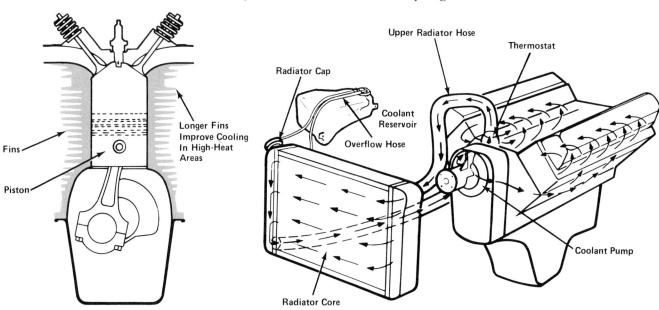

Figure 10-24. Air cooling system.

Figure 10-25. Liquid cooling system.

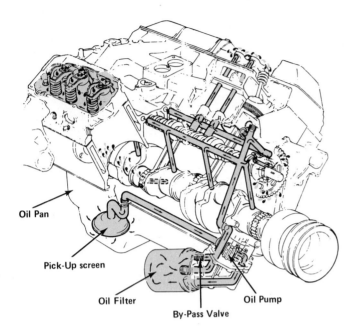

Figure 10-26. Lubrication system.

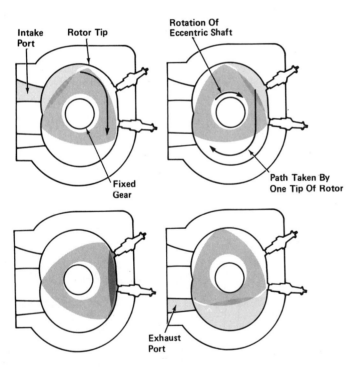

Figure 10-27. Rotary engine cycle.

Use

Engines can also be classified by use. Examples include automobile, truck, and motorcycle engines, stationary engines (for agricultural water pumps and electrical generators), airplane engines, and so on.

UNIT HIGHLIGHTS

- The terms energy, force, work, and power have specific meanings.
- An internal-combustion engine creates force to move a vehicle by burning fuel.
- The basic parts of an internal-combustion piston engine are the cylinder block, piston and rings, crankshaft, connecting rod, cylinder head, valves, camshaft, valve train, flywheel, fuel system and exhaust system.
- The internal-combustion piston engine remains the most popular and durable automotive powerplant.
- Engines can be classified by several characteristics of construction, operation, and use.

TERMS

energy	force
work	power
torque	horsepower (hp)
watt (w)	mechanical energy
chemical energy	physical energy
heat energy	thermodynamics
engine	external combustion
piston	cylinder
pressure	internal combustion
crankshaft	cylinder block
piston ring	connecting rod
insert bearing	cylinder head
valve	spark plug
valve train	camshaft
cam	pushrod
flywheel	carburetor
intake port	intake manifold
intake charge	exhaust valve
exhaust manifold	four-stroke cycle
stroke	bottom dead center (BDC)
top dead center (TDC)	compression ratio
displacement	combustion chamber
configuration	power pulley
power pulse	methanol
ethanol	spark ignition
diesel engine	compression ignition
cooling fins	air cooling
liquid cooling	oil pan
oil pump	splash lubrication
throw-off lubrication	two-stroke cycle

rate
kilowatt (kw)
cam lobe
head gasket
revolutions per minute (rpm)
overhead camshaft (OHC)
fuel injection
gasket
rotary engine
bore
exhaust port
pressurized lubrication system
dual overhead camshaft (DOHC)

REVIEW QUESTIONS

DIRECTIONS: The following questions are similar to those used on automotive technician certification tests. On a separate sheet of paper, write the letter of the correct choice.

1. An internal-combustion gasoline piston engine converts
   A. gasoline to torque.
   B. air-fuel mixture to force.
   C. heat energy to mechanical energy.
   D. horsepower to watts.

2. On which strokes of the four-stroke cycle are both valves closed?
   A. Intake and compression
   B. Power and exhaust
   C. Compression and power
   D. Intake and exhaust

3. Which of the following statements is correct?
   I. Valves and piston rings form a tight seal to contain the force produced by combustion that push the piston down.
   II. The head gasket forms a seal between the cylinder head and the block.
   A. I only
   B. II only
   C. Both I and II
   D. Neither I nor II

4. Technician A says rpm stands for revolutions per minute.
   Technician B says rpm stands for reciprocating piston movement.
   Who is correct?
   A. A only
   B. B only
   C. Both A and B
   D. Neither A nor B

5. How many crankshaft revolutions are necessary to complete all four strokes of the four-stroke cycle?
   A. 4
   B. 2
   C. 1
   D. 4, 5, 6, 8, or 12, depending on the number of cylinders

ESSAY QUESTIONS

1. Define the terms energy, force, work, and power in your own words.
2. What are the basic parts of an internal-combustion piston engine and their functions?
3. Describe the four-stroke cycle.

SUPPLEMENTAL ACTIVITIES

1. Identify parts removed from an engine selected by your instructor.
2. Classify the engines in your family's or friends' cars by as many characteristics as you can identify.
3. If available, inspect a shop cutaway engine to locate the basic parts of an engine discussed in Topic 10.3
4. Identify the engine parts that act together to change the reciprocating piston motion to rotary motion.

11 Engine Measurements And Performance

UNIT PREVIEW

Engine size is a measure of the total volume displaced by the movement of the pistons. Displacement can be calculated from engine measurements.

Displacement and compression ratio affect the force and power output of an engine. Engine power is the rate at which the engine can do the work of producing torque.

Engine power is lost due to friction, restrictions in the intake and exhaust systems, power used to drive engine accessories, and heat losses.

Overall vehicle efficiency is limited by heat and frictional losses and the resistance to movement through the air.

LEARNING OBJECTIVES

When you have completed your assignments and exercises in this unit, you should be able to:

❏ Make measurements and determine an engine's displacement.
❏ Calculate the compression ratio of an engine.
❏ Explain the relationships of different types of horsepower ratings.
❏ Describe how engine power can be lost.
❏ Calculate the efficiency of an engine.
❏ Identify and describe factors that affect vehicle efficiency.

11.1 ENGINE SIZE

Displacement, discussed in Unit 10, is the volume through which a piston moves from BDC to TDC on a stroke. Total engine displacement is the sum of all the individual cylinder displacements during one revolution of the crankshaft.

When referring to engine size, the quantity that is measured is displacement, not the physical dimensions of an engine. A "big" engine is one with a great deal of displacement. A "smaller" engine is one with less displacement.

Cylinder volume in conventional units is expressed in *cubic inches*. One cubic inch is the volume contained in a cube that measures one inch on each side.

In *International Standard of Units (SI)*, or metric, measurements, volume is measured in *cubic centimeters (cc)*. A cubic centimeter is the volume contained in a cube that measures one centimeter, or 10 millimeters, on each side. 1000 cc = 1 liter = 61.0 cubic inches.

Bore and Stroke

A simplified formula for engine displacement is:

$$ED = 0.785 \times BD^2 \times LS \times NC$$

Where:
ED = engine displacement
BD = bore diameter
LS = length of stroke
NC = number of cylinders

The stroke can be determined by turning the crankshaft and measuring the piston's travel in the cylinder. Stroke also can be determined by measuring the crankshaft. The individual crankshaft throws, where the connecting rods are attached, are offset from the centerline of the main journals. The stroke is equal to the amount of offset from the crankshaft centerline multiplied by two.

For example, say a V-6 engine has a bore of 3.800 inches (9.65 cm) and a stroke of 3.400 inches (8.64 cm). The displacement of this engine is:

$$ED = 0.785 \times 3.800^2 \times 3.400 \times 6$$
$$= 231.24 \text{ cubic inches}$$

OR

$$ED = 0.785 \times 96.5^2 \times 86.4 \times 6$$
$$= 3789.56 \text{ cc}$$
$$= 3.8 \text{ liters}$$

Displacement

The larger the cylinders or the longer the stroke, the greater an engine's displacement will be. Large cylinders and long strokes mean that large volumes of air-fuel mixture will be drawn into the cylinders. The more fuel that can be burned, the greater the chemical energy that can be turned into mechanical energy. Thus, in general, the larger an engine's displacement, the more torque produced.

More energy is required to move a larger mass. Larger, heavier vehicles are provided with large-displacement engines. Large-displacement engines produce more torque than smaller-displacement engines, and they also consume more fuel.

Smaller, lighter vehicles can be adequately powered by lower-displacement engines that use less fuel.

Power, as discussed in Unit 10, is the rate at which work is done. In the case of an engine, power is the rate at which torque is produced. The faster an engine produces torque, the more powerful it is. In addition to displacement, another factor that influences engine power is the compression ratio.

11.2 COMPRESSION RATIO

The more the air-fuel mixture becomes squeezed on the compression stroke, the greater the heat produced by compression. In addition, the pressure, or force per unit area, applied to the piston top is greater. These factors combine to produce more force on the power stroke.

The *compression ratio* is a comparison between the volume above the piston at BDC and the volume above the piston at TDC (clearance volume), as shown in **Figure 11-1**. The volume of the combustion chamber in the cylinder head also must be considered when determining compression ratio.

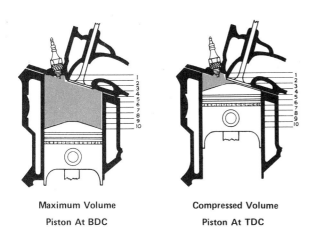

Figure 11-1. Compression ratio. BLACK & DECKER

Imagine a cylinder containing 98 cubic inches with the piston at BDC and 2 cubic inches in the cylinder head. The total volume above the piston at BDC is 98 + 2 = 100 cubic inches. Now, consider the situation when the piston reaches TDC.

There are 8 cubic inches left in the cylinder and the same 2 cubic inches in the cylinder head. The total volume above the piston at TDC is 8 + 2 = 10 cubic inches. If the volume is 100 cubic inches at BDC and 10 cubic inches at TDC, the compression ratio is 100 ÷ 10 = 10.0:1. The formula for determining compression ratio is:

CR = (VBDC + VCH) ÷ (VTDC + VCH)

Where:
CR = compression ratio
VBDC = volume in cylinder above piston at BDC
VCH = volume in cylinder head
VTDC = volume in cylinder above piston at TDC

The higher the compression ratio, the more power an engine theoretically can produce. Also, as the compression ratio increases, the heat produced as the piston rises on the compression stroke also increases. Gasoline with a low *octane rating* may explode rather than burn. This can cause serious engine damage, as explained in Unit 20. The higher a gasoline's octane rating, the less likely it is to explode.

Ideally, ignition should occur so that maximum pressure is produced slightly after the piston reaches TDC on the power stroke. Such normal combustion is illustrated in **Figure 11-2**.

The compression ratio of an engine must be suited to the type of fuels available. In other words, as compression ratio increases, the octane rating of the gasoline also should be increased to prevent abnormal combustion.

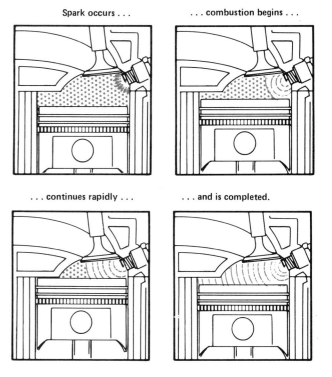

Figure 11-2. Normal combustion. CHAMPION SPARK PLUGS

11.3 HORSEPOWER AND TORQUE

Torque is a turning or twisting force. The engine crankshaft applies torque that is transmitted through the drivetrain to turn the driving wheels of the vehicle. In an engine, *horsepower* is the rate at which torque is produced.

Horsepower

In the late 1700s in England, James Watt formed a company to sell his newly-invented steam engine to coal mine operators. At that time, horses were used in the mines to lift heavy buckets of coal. In order to show the relative power of his steam engine, James Watt had to find out how powerful a horse was. Through experiments, it was found that the average horse could lift a 330-pound bucket of coal 100 feet in one minute, as shown in **Figure 11-3**. The formula for determining work is:

Work = force x distance

Thus, the work a horse can do equals 330 lb x 100 feet or 33,000 *foot-pounds (ft-lb)* per minute. A foot-pound is a unit of work. It is equivalent to the force necessary to move 1 pound a distance of 1 foot.

In SI metric units, work is measured in *joules (J)*, after the scientist James P. Joule. A Joule is equal to one *Newton-meter (Nm)* of work per second. A Newton-meter is the force required to move 9.8 kilograms (approximately 21.6 pounds) 1 meter in 1 second. The Newton-meter is named after Sir Isaac Newton.

Torque

Engines produce power by turning a crankshaft in a circular motion. To convert terms of force applied in a straight line to force applied in a circular motion, the formula is:

Torque = force x radius

A 10-pound force applied to a wrench 1 foot long will produce 10 *pounds-feet (lb-ft)* of torque. Imagine that the 1-foot-long wrench is connected to a shaft. If 1 pound of force is applied to the end of the wrench, 1 pound-foot of torque is produced. Ten pounds of force applied to a wrench 2 feet long will produce 20 pounds-feet of torque **[Figure 11-4]**.

Torque is measured in pounds-feet. To standardize and prevent confusion, this text specifies torque in *pounds-feet* and work in *foot-pounds.* In both cases, force causes motion through a certain distance. In SI units, torque is specified in Newton-meters (Nm).

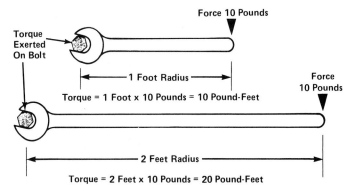

Figure 11-4. Force and torque.

The Relationship of Horsepower to Torque

If the torque output of an engine at a given speed (rpm) is known, horsepower can be determined by this formula:

HP = (torque x rpm) ÷ 5,252

Where:

HP = horsepower

rpm = revolutions per minute

An engine produces different amounts of torque based on the rotational speed of the crankshaft and other factors. A mathematical representation, or graph, of the relationship between horsepower and torque in one engine is shown in **Figure 11-5**.

This graph shows that torque drops off above about 1,700 rpm. Brake horsepower increases steadily until about 3,500 rpm, and then drops. The third line on the graph indicates horsepower needed to overcome the resistance to movement of engine internal parts against each other. This resistance is known as *friction.*

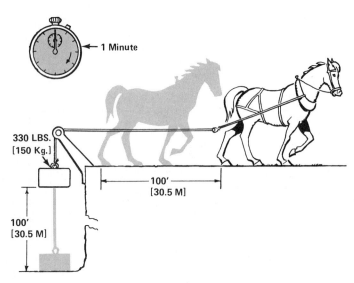

Figure 11-3. Work and horsepower.

ENGINE MEASUREMENTS AND PERFORMANCE

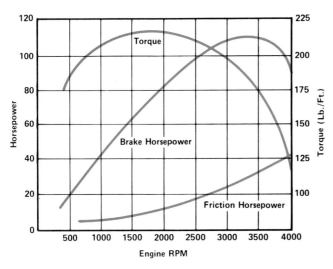

Figure 11-5. Horsepower and torque.

11.4 HORSEPOWER LOSSES

The usable power produced by an engine is known as *brake horsepower (bhp)*. To measure brake horsepower, a braking, or stopping, device is used to put a load on an engine. The more stopping force the engine can resist, the more powerful it is.

The power that an engine could theoretically produce from combustion pressure inside the cylinders is known as *indicated horsepower (ihp)*. A graph of the pressure in a cylinder during the four-stroke cycle is shown in **Figure 11-6.**

Brake horsepower is less than indicated horsepower because some of an engine's power is used by friction between internal engine parts. The power necessary to overcome friction within an engine is known as *frictional horsepower (fhp)*. The relationship between brake horsepower, indicated horsepower, and frictional horsepower is:

$$bhp = ihp - fhp$$

Where:
bhp = brake horsepower
ihp = indicated horsepower
fhp = frictional horsepower

11.5 FRICTION

Different types of friction may be present between parts of an engine, including dry friction, greasy friction, and viscous friction.

Dry Friction

Dry friction occurs when objects directly contact each other, as metal parts rubbing against other metal parts. Dry friction causes rapid wear and heat that rapidly ruins metal parts. The friction between a belt and a pulley is dry friction.

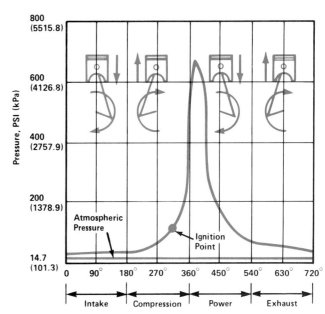

Figure 11-6. Pressure in a cylinder during the four-stroke cycle.

Greasy Friction

Greasy friction occurs when a thin film of grease or oil is present between surfaces. Within an engine, a thin film of oil remains on bearing surfaces after the engine is shut off, as shown in **Figure 11-7.** Greasy friction occurs before an engine starts, just as the crankshaft and other parts begin to turn.

After starting, oil is pumped between metal surfaces by the lubrication system to eliminate dry and greasy friction. Until the oil is warmed up and pumped everywhere within an engine, greasy friction does not provide adequate lubrication. Thus, parts will wear rapidly.

Viscous Friction

Viscous friction is the resistance to motion between layers of liquid, such as oil. *Viscosity* is the property of liquids that causes them to resist flowing. A highly viscous oil is thick and heavy.

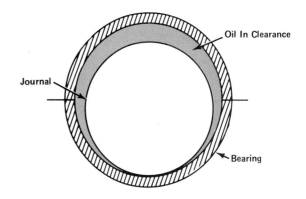

Figure 11-7. A thin film of oil remains on bearing surfaces after the engine has been turned off.

The engine lubrication system pumps oil into the clearance between bearing surfaces. Within the space of the clearance, microscopic layers of oil adhere, or stick, to each other as the shaft spins. This causes a wedge-shaped film of oil to build up and support the weight of the shaft **[Figure 11-8]**.

With proper lubrication to prevent dry and greasy friction, bearing surfaces and other rubbing parts should last a long time.

11.6 MEASURING HORSEPOWER

A *dynamometer* is a device for measuring horsepower. A *brake dynamometer* applies a braking, or stopping, force to an engine. The braking force can be developed in two ways. The most common way is to force a turbine, or propeller, to turn inside a container filled with liquid **[Figure 11-9]**. An electrical generator across which a variable load is connected can also be turned to provide braking force.

For testing purposes, an engine can be removed from the vehicle and tested separately. In addition, all engine accessories, such as the waterpump, alternator, and so on, can be powered from outside sources. The rating from such a test would be an engine's *gross horsepower*. The rating with the accessories connected to the engine would be less, and is known as *net horsepower*. Power used to operate an engine accessory is known as a *parasitic loss* of horsepower. A *chassis dynamometer* measures the power output at a vehicle's driving wheels.

Horsepower output is affected by atmospheric conditions, such as the barometric pressure of the air, humidity, and air temperature. The readings taken with a dynamometer are known as *observed horsepower*. Observed horsepower is corrected to a standard known as *corrected horsepower* by reference to tables listing atmospheric conditions.

The conventional unit of corrected horsepower is in the *Society of Automotive Engineers (SAE) horsepower*.

In SI units, power is measured in *watts (w)* or *kilowatts (kw)*, named after James Watt. This unit also is used to measure electricity. One horsepower is equal to 746 watts, or 0.746 kw. A kilowatt is 1,000 watts.

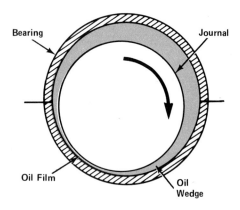

Figure 11-8. A wedged-shaped film of oil supports the weight of a shaft in a bearing.

11.7 ENGINE EFFICIENCY

Efficiency is a measure of how well a device can convert energy to work. To find the efficiency of a given device, the output is divided by the input, when both are stated in the same units, or terms.

Mechanical Efficiency

Mechanical efficiency is a comparison of how much horsepower an engine produces at the flywheel, compared to the theoretical horsepower the engine could produce from the pressures generated in the combustion chamber. The formula for determining mechanical efficiency is:

$$\text{Mechanical efficiency} = \text{bhp} \div \text{ihp}$$

Where:
bhp = brake horsepower
ihp = indicated horsepower

Volumetric Efficiency

An engine can be thought of as a pump that draws gases in and forces them out. *Volumetric efficiency* is a comparison of how much fuel and air are actually drawn into an engine compared to how much could be drawn in.

At low speeds and when the throttle plate is fully open, atmospheric pressure has time to fill

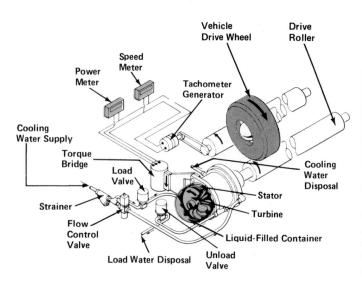

Figure 11-9. Parts of a brake dynamometer.

the cylinders fully. At higher speeds and when the throttle is closed, less air-fuel mixture is drawn in. Smaller amounts of air-fuel mixture in the cylinder mean that less pressure is developed during combustion. The formula for determining volumetric efficiency is:

> Volumetric efficiency = aov ÷ aip
>
> Where:
> aov = actual air output volume
> aip = maximum possible air input volume

Pumping losses are drops in volumetric efficiency due to restrictions in the intake air passages. Throttle plates, valves, and cylinder head passages all may be restrictive.

A diesel engine has no throttle plate in the air intake. Instead, different amounts of fuel are introduced into the cylinder to regulate the power output. Thus, a diesel engine has fewer pumping losses.

Pumping losses can be lessened by forcing air through the intake system with a *turbocharger* or *supercharger*. These devices pump air through the intake system at greater than atmospheric pressure and thus produces a higher volumetric efficiency. Turbochargers and superchargers are discussed in Unit 20.

Thermal Efficiency

Thermal efficiency is a comparison of the energy present in the fuel used and the energy output of the engine. The customary unit of heat energy is the *British thermal unit (Btu)*. One horsepower is equivalent to 42.4 Btu per minute. Gasoline has approximately 110,000 Btu per gallon. For a gasoline engine, the formula used to determine thermal efficiency is:

> TE = (bhp × 42.4 Bpmin) ÷ (110,000 Bpg × gpmin)
>
> Where:
> TE = thermal efficiency
> bhp = brake horsepower
> Bpmin = Btu per minute
> Btu = British thermal unit
> Bpg = Btu per gallon
> gpmin = gallons used per minute

Thermal efficiency for a typical gasoline engine is less than 25 percent. Diesel fuel contains more heat energy than gasoline, and a diesel engine has a somewhat better thermal efficiency. Part of this efficiency is due to higher compression ratios of between 20:1 and 23:1. As a result, vehicles with diesel engines produce better fuel mileage.

However, diesel engines have disadvantages as well. Passenger vehicle diesel engines have less horsepower and weigh more than gasoline engines of the same size (displacement). As a result, most passenger vehicles with diesel engines have slow acceleration and low top speeds.

Thermodynamics

Thermodynamics is the scientific study of the mechanical action or relations of heat. The first law, or principle, of thermodynamics is this: Energy can be neither created nor destroyed. In other words, energy exists or doesn't exist. However, energy may be present in different forms, such as heat energy, chemical energy, mechanical energy, and so on. Energy also can be changed from one form to another.

The second law of thermodynamics states that heat cannot be converted completely to another form of energy, for example mechanical energy. Heat energy, such as that present during combustion, can never be converted completely into mechanical energy, or work. Some energy always is lost, no matter how efficient the engine or device used to make the conversion.

11.8 OVERALL VEHICLE EFFICIENCY

Current gasoline piston engines waste about two-thirds of the heat energy present in gasoline. To prevent overheating, the cooling system must carry away a third of the heat produced during combustion. Another third of the heat energy is lost in hot exhaust gases.

Of the energy left, five percent is lost to internal engine friction. Another 10 percent is lost to friction in the drivetrain components. By the time the power reaches the driving wheels at the end of the drivetrain, about 19 percent is left.

However, *rolling resistance,* the friction between the tires and the road, uses some of this remaining power.

Air resistance, the result of the vehicle moving through the air, also takes its toll. Air resistance increases as speed increases. The two factors that contribute to air resistance are the vehicle's *coefficient of drag,* or C_d, and its *frontal area*. Drag is air resistance. The lower the coefficient of drag, the less air resistance the vehicle has. Frontal area is the area against which air resistance is present as the vehicle travels forward.

Since the early 1980s, vehicle manufacturers have been paying increasing attention to *streamlining,* or reducing air resistance.

As the diagram in **Figure 11-10** shows, only about 15 percent of the energy produced from the burning of gasoline finally goes toward moving the vehicle.

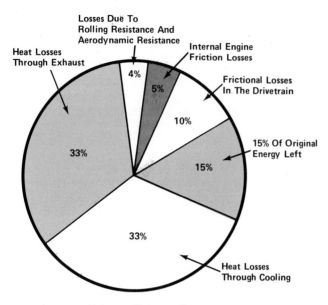

Figure 11-10. Volume efficiency diagram.

UNIT HIGHLIGHTS

- Engine size is a measure of displacement and can be measured in conventional units (cubic inches) or SI units (cubic centimeters).
- Displacement can be calculated by using measurements of the cylinder bore and stroke in a mathematical formula.
- Larger engine displacements and increased compression ratios create more torque and horsepower.
- Horsepower is the rate at which an engine produces torque.
- Three types of friction may be present within an engine: dry friction, greasy friction, and viscous friction.
- Engine efficiency is limited by internal friction, pumping losses, parasitic losses, and heat losses.
- Overall vehicle efficiency is limited by heat losses, frictional losses, and aerodynamic drag.

TERMS

cubic inch
horsepower
joule (J)
foot-pound (ft-lb)
brake horsepower (bhp)
dry friction
viscous friction
dynamometer
gross horsepower
cubic centimeter (cc)
customary unit
Newton-meter (Nm)
pound-foot (lb-ft)
friction
greasy friction
viscosity
brake dynamometer
net horsepower
parasitic loss
observed horsepower
turbocharger
kilowatt (kw)
volumetric efficiency
supercharger
thermodynamics
air resistance
frontal area
compression ratio
International System of Units (SI)
frictional horsepower (fhp)
octane rating
chassis dynamometer
corrected horsepower
watt (w)
mechanical efficiency
pumping loss
thermal efficiency
rolling resistance
coefficient of drag (C_d)
streamlining
British thermal unit (Btu)
indicated horsepower (ihp)
Society of Automotive Engineers (SAE)

REVIEW QUESTIONS

DIRECTIONS: The following questions are similar to those used on automotive technician certification tests. On a separate sheet of paper, write the letter of the correct choice.

1. Which of the following statements is correct?
   I. 1 liter = 1000 cc = 61.0 cubic inches.
   II. 1500 cc = 91.5 cubic inches.
   A. I only
   B. II only
   C. Both I and II
   D. Neither I nor II

2. The volume in the cylinder above a piston at BDC is 76 cubic inches, and at TDC it is 6 cubic inches. The volume in the cylinder head is 4 cubic inches. The compression ratio is:
   A. 12.66 to 1
   B. 7.60 to 1
   C. 8.00 to 1
   D. None of the above.

3. Technician A says that work is measured in pounds-feet (lb-ft).
   Technician B says that torque is measured in foot-pounds (ft-lb).
   Who is correct?
   A. A only
   B. B only
   C. Both A and B
   D. Neither A nor B

4. What quantity is found by using the equation bhp = ihp - fhp?
   A. Volumetric efficiency
   B. Thermal efficiency
   C. Brake horsepower
   D. Net horsepower

Question 5 is not like the first four questions. It has the word EXCEPT. For question 5, look for the choice that does *not* apply. Read the entire question carefully before you choose your answer.

5. All of the following are correct EXCEPT
   A. work = force x distance.
   B. torque = force x radius.
   C. horsepower = (torque x rpm) ÷ 5,252.
   D. mechanical efficiency = ihp ÷ bhp.

ESSAY QUESTIONS

1. Explain how to calculate the displacement of an engine.
2. What factors increase engine torque and horsepower output?
3. What factors decrease engine horsepower output?

SUPPLEMENTAL ACTIVITIES

1. Make a list of engine displacement figures for engines in vehicles owned by relatives or friends. Correctly convert the displacement figures from conventional units (cubic inches) to SI units (cubic centimeters), or vice versa.
2. Measure the bore and stroke of a shop engine and correctly calculate the displacement of the engine.
3. Use equipment to measure the volume of a cylinder head and of a cylinder when the piston is at TDC and BDC. Calculate the compression ratio.
4. Discussion topic: What effect does enlarging a cylinder bore (making the bore bigger) have on the compression ratio of an engine?
5. Visit several local service stations. Find the octane rating stickers on the pumps, and make a list of the octane ratings for different grades and brands of gasoline.

12 Engine Construction

UNIT PREVIEW

The cylinder block is the main engine casting. The block forms the foundation for the entire engine, with other engine components being attached at various points. Manufacturing processes used to form the block include casting and machining operations.

Precision insert bearings support the crankshaft and allow it to turn freely. The crankshaft changes the reciprocating motion of the pistons into rotary motion that powers the driving wheels.

LEARNING OBJECTIVES

When you have completed your assignments and exercises in this unit, you should be able to:

- ❏ Describe the processes used to manufacture cylinder blocks and crankshafts.
- ❏ Explain how crankshaft design determines engine firing order.
- ❏ Identify the main parts of the crankshaft.
- ❏ Describe the characteristics of engine bearings.
- ❏ Identify and describe the parts used to keep liquids and gases from leaking or becoming contaminated.

12.1 CYLINDER BLOCK

The cylinder block for a liquid-cooled engine includes the cylinder bores and the crankshaft attachment area, known as the *crankcase*. The cylinder block also has machined surfaces where other parts are mounted. These include the cylinder heads, oil and water pumps, motor mounts, transmission, and, in some engines, the camshaft. Many other smaller components also are attached to the cylinder block.

Cylinder Block Design

The number of cylinders and their configuration determines the basic shape of the engine. **Figure 12-1** shows the two most common engine block shapes, the inline and the V-type.

In an inline-type cylinder block, the cylinders are usually formed so that they are more or less perpendicular (at a right angle, 90 degrees) to the horizontal. The cylinders also may be at an angle, or *slant* to the horizontal.

In a V-type block the angle between the two cylinder areas, or *banks* of cylinders, can vary. The angle between the cylinder banks on a V-8 engine is 90 degrees. The angle between the cylinder banks on a V-6 engine is normally 60 degrees. However, V-6 engines derived from V-8 engines (by eliminating the two rear cylinders) may have an angle of 90 degrees. The flat-type engine can be thought of as a V-type that has been spread apart to form approximately a 180-degree angle.

The casting process. The basic cylinder block is formed by a process called *casting*. A mold with the overall shape of the engine is made. Molds for cylinder blocks are made of special sand that is treated and packed into a form. Molten, or melted, metal is poured into the mold. As the metal cools, it takes on the shape of the mold.

To form hollow areas, parts called *cores* are inserted into the mold. A core is simply a solid sand plug. The molten metal flows around the cores. When the cores are removed, the core area remains hollow. **Figure 12-2** shows the type of cores used in casting cylinder blocks.

Cylinder blocks can be cast from iron or aluminum. In today's smaller cars, aluminum frequently is used to reduce vehicle weight for better fuel economy.

Machining operations. *Sand cast* surfaces are fairly rough, like the surface of the sand mold. However, sealing surfaces must be very smooth. Any nicks, scratches, or roughness will allow liquids or gases to penetrate and leak past the seal area.

To make the cylinder walls smooth, the cylinders and all other block sealing surfaces are *machined*. Machining means that a power machine is used to cut or grind a surface to the desired size, shape, or finish.

In addition to machining, holes are drilled and tapped for attaching bolts and studs.

After casting, the sand cores are removed through large holes in the side of the block. These holes are machined and fitted with soft metal plugs, called *core plugs* or "freeze plugs."

Inserts and liners. Because aluminum is not as strong as cast iron, the piston rings (made of cast iron, sometimes chrome plated) would quickly wear into plain aluminum cylinders and cause deep gouges.

ENGINE CONSTRUCTION

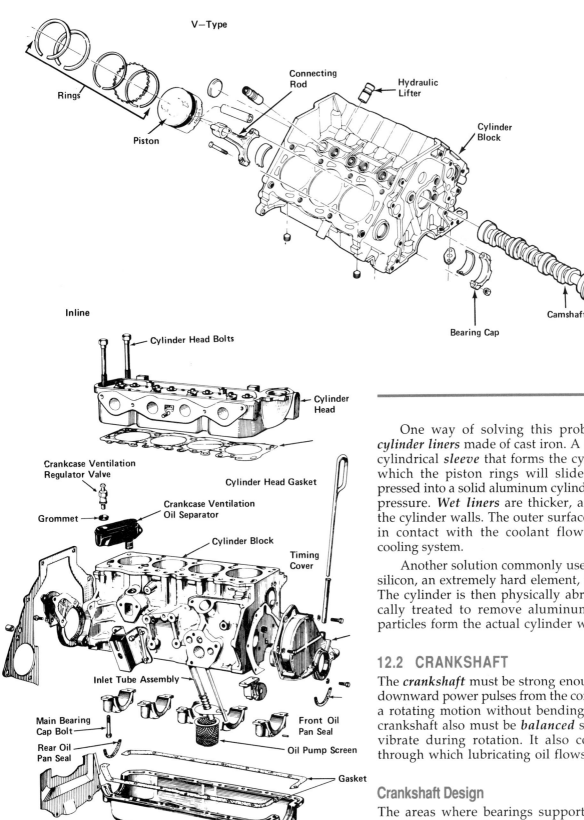

Figure 12-1. Engine block designs. FORD MOTOR COMPANY

One way of solving this problem is to insert *cylinder liners* made of cast iron. A cylinder liner is a cylindrical *sleeve* that forms the cylinder surface on which the piston rings will slide. *Dry liners* are pressed into a solid aluminum cylinder opening under pressure. *Wet liners* are thicker, and actually form the cylinder walls. The outer surface of a wet liner is in contact with the coolant flowing through the cooling system.

Another solution commonly used today is to add silicon, an extremely hard element, to the aluminum. The cylinder is then physically abraded and chemically treated to remove aluminum so that silicon particles form the actual cylinder wall surface.

12.2 CRANKSHAFT

The *crankshaft* must be strong enough to change the downward power pulses from the connecting rods into a rotating motion without bending or twisting. The crankshaft also must be *balanced* so that it does not vibrate during rotation. It also contains passages through which lubricating oil flows.

Crankshaft Design

The areas where bearings support and protect the spinning crankshaft are called *journals*, as shown in **Figure 12-3**. *Rod journals* are located on the *throws*, or offset areas of the crankshaft where the connecting rods are attached. The *main journals* are the areas where the crankshaft is attached to the crankcase at the bottom of the block.

Figure 12-2. Engine block casting cores.

The number of cylinders determines the number of throws on the crankshaft. However, in V-8 engines and some V-6 engines, two connecting rods share a single throw, as shown in **Figure 12-4**.

Even-firing and Uneven-firing Engines

The angle of the throws determines the firing order of the engine. To spread the power pulses out evenly, V-8 engines have a 90-degree offset between throws. Inline 6-cylinder engines have 120 degrees between throws. Inline 4-cylinder engines have 180 degrees between throws. These are known as *even firing* engines and run smoothly. The number of degrees of crankshaft revolution between power pulses is evenly divisible into 720 degrees. The 720 degrees represents the two full revolutions of the crankshaft necessary for the four-stroke cycle. Crankshaft designs are shown in **Figure 12-5**.

V-6 engines derived from V-8s are *uneven-firing* engines. In these engines, two connecting rods share a single *crankpin*, or attaching area. This makes it impossible to evenly space the power pulses from the two cylinders. Consequently, the engine idles roughly and unevenly.

To solve this problem, the crankpins can be individually offset, or *splayed*, on the throw, as shown in **Figure 12-6**. This splaying makes it possible to apply the power pulses evenly to the crankshaft.

Crankshaft Construction

Crankshafts can be cast, *forged*, or machined from a solid bar, or *billet*. Forging is a process of hammering a hot piece of metal into shape. Forged crankshafts usually are manufactured for high-performance applications.

During crankshaft manufacture, *counterweights* are formed to balance the offset throws. In addition, fine balancing is done by drilling out areas of the counterweights and throws or by adding small weights.

Oil passages to supply lubricant to the connecting-rod and main-journal bearings also are drilled during manufacture **[Figure 12-7]**.

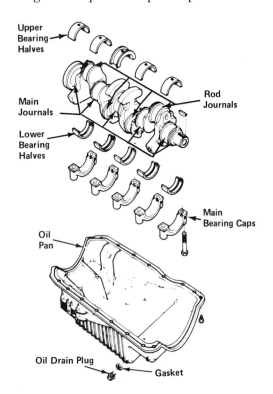

Figure 12-3. Crankshaft bearings are located at journals to support and protect the spinning crankshaft. FORD MOTOR COMPANY

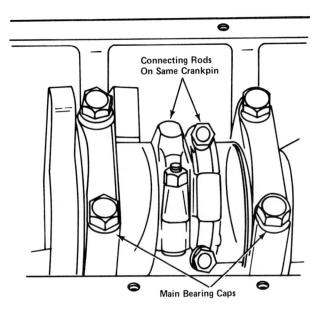

Figure 12-4. Two connecting rods can share a single throw, or offset portions of a crankshaft.

Engine Construction

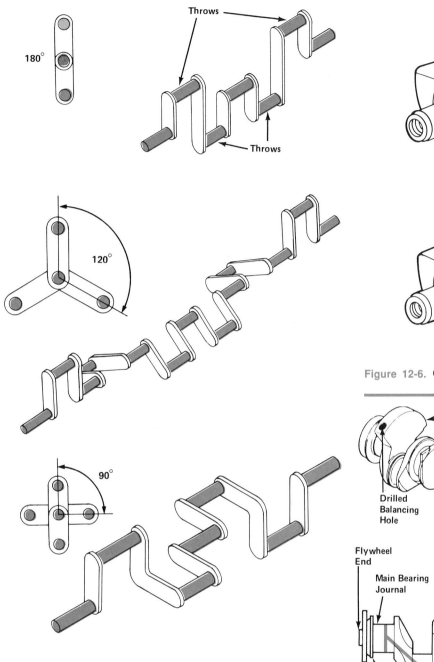

Figure 12-5. Crankshaft designs.

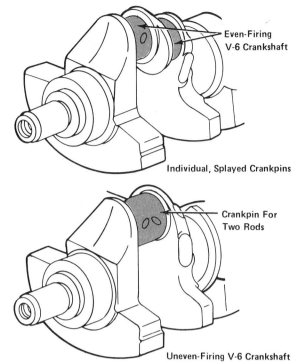

Figure 12-6. Crankpin offset.

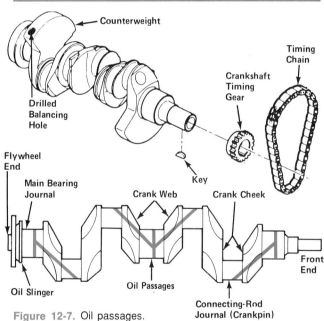

Figure 12-7. Oil passages.

The crankshaft absorbs a great deal of force during engine operation. Therefore, care must be taken so that *stress*, or applied force, does not crack or bend the crankshaft. Stress is most likely to cause metal fatigue and cracks at sharp edges. To relieve stress, the edges of the drilled crankshaft oil holes are *chamfered*, or beveled. In addition, the machined crankpin areas next to the counterweights are rounded by machining a small curved area, or *fillet*, as shown in **Figure 12-8.** Many engines, especially high-performance gas and diesel engines, have rolled fillets.

Attaching the Flywheel and Pulley

The rear part of the crankshaft is drilled and tapped for flywheel attachment. Near the front of the crankshaft, a *timing gear*, sprocket, or pulley is attached to drive the camshaft or camshafts. A *damper pulley* or power pulley, is attached to the outside front end of the crankshaft. In many cases, the damper or *harmonic balancer,* is a separate unit pressed onto the front of the crankshaft. The pulley

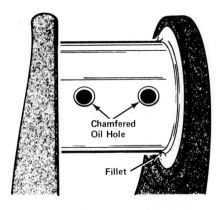

Figure 12-8. Fillets are machined in crankpin areas next to counterweights.

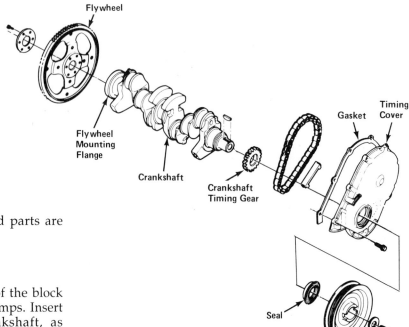

Figure 12-9. Crankshaft and flywheel assembly.

is then bolted on. These attachments and parts are shown in **Figure 12-9**.

Attaching the Crankshaft to the Block

The crankshaft is secured to the bottom of the block with *bearing caps,* or machined metal clamps. Insert bearings are placed arround the crankshaft, as shown in **Figure 12-10**.

The number of main bearings can vary. The more main bearings, the less likely the crankshaft is to *whip,* or flex, during rotation. A well-supported crankshaft has one more main bearing than the number of throws.

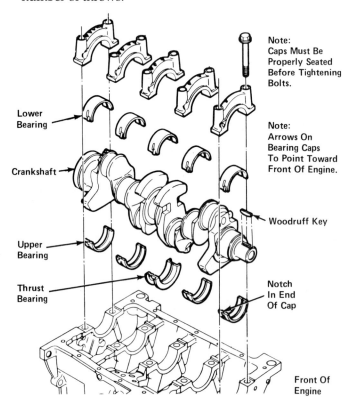

Figure 12-10. Bearing caps hold the connecting rods to the crankshaft.

12.3 ENGINE BEARINGS

One of the most important factors in engine life is the durability of engine bearings. If these bearings fail, the entire engine will be ruined. The locations of the engine bearings, *bushings,* and sleeves are shown in **Figure 12-11**.

Bearing Design

Friction is the resistance between two rubbing or sliding materials. Excess friction eventually will wear and ruin moving parts. To reduce friction, two types of bearings can be used: *anti-friction* ball and roller bearings, and friction, or insert bearings. Crankshaft bearings are *friction bearings* and also are known as *precision insert bearings,* shown in **Figure 12-12**.

The *alloy,* or combination of metals, used for bearings may include babbitt material (lead or tin with small amounts of copper or antimony). Other metals that may be included are copper-lead alloys, and aluminum with lead, copper, and/or tin. In addition, babbitt metal is sometimes used as an overplate, or outer coat, over other materials for easy bearing break-in.

Bearing Clearance and Lubrication

Under normal conditions when the engine is running, spinning shafts should ride on a microscopic wedge of oil, as shown in **Figure 12-13**. The *clearance,* or

ENGINE CONSTRUCTION

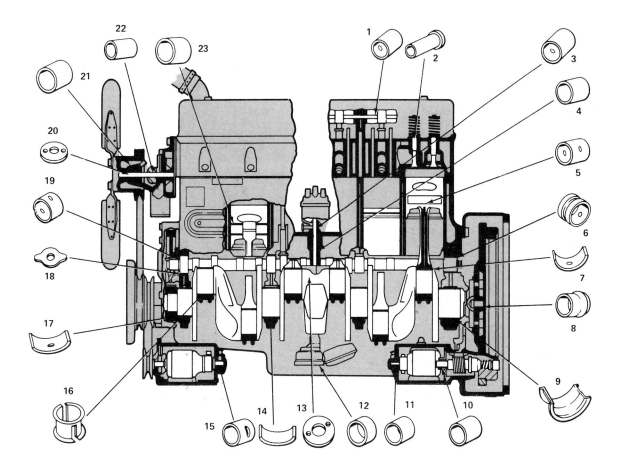

1 Rocker Arm Bushing	9 Flanged Main Bearing	17 Front Main Bearing
2 Valve Guide Bushing	10 Starting Motor Bushing, Drive End	18 Camshaft Thrust Plate
3 Distributor Bushing, Upper	11 Starting Motor Bushing, Commutator End	19 Camshaft Bushing
4 Distributor Bushing, Lower	12 Oil Pump Bushing	20 Fan Thrust Plate
5 Piston Pin Bushing	13 Distributor Thrust Plate	21 Water Pump Bushing, Front
6 Camshaft Bushing	14 Intermediate Main Bearing	22 Water Pump Bushing, Rear
7 Connecting Rod Bearing	15 Alternator Bushing	23 Piston Pin Bushing
8 Clutch Pilot Bushing	16 Connecting Rod Bearing, Floating Type	

Figure 12-11. Engine bearing, bushing, and sleeve locations.

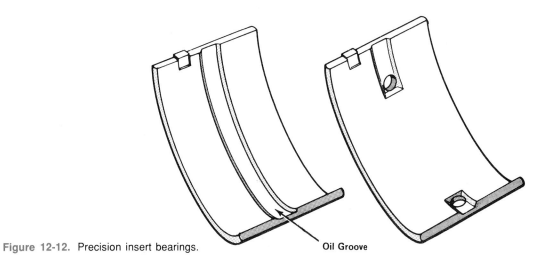

Figure 12-12. Precision insert bearings.

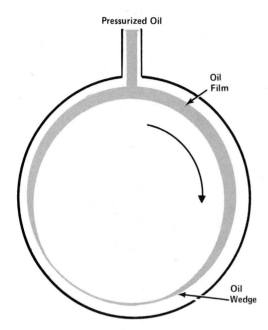

Figure 12-13. Rod bearing clearance and lubrication.

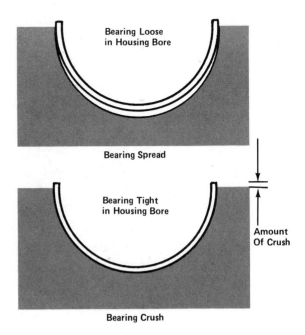

Figure 12-14. Bearing spread and crush.

distance, between the bearing material and the crankshaft usually ranges from 0.0005 to 0.0025 in. (0.01 to 0.06 mm). As the clearance increases with wear, the oil flow will increase, causing a drop in oil pressure. Under these conditions, the shaft may rub against the bearing surface and wear even faster.

Oil for crankshaft bearing lubrication flows through a long gallery in the cylinder block. Each main bearing has its own oil supply passageway from this gallery. Passageways drilled in the crankshaft carry oil from the main-bearing journals to the rod journals.

Excessive wear in one bearing will lower oil pressure to every bearing after it. Frequently, bearings will fail in progression. Bearings closest to the oil pump get adequate lubrication, but those farther away become starved for oil. Bearing failure creates metal particles that get into the oil and increase wear even further.

Bearing Retaining Devices

The diameter of a precision insert bearing is slightly larger than the opening into which it fits. This *bearing spread* allows the bearing to snap into place and be held by spring tension. This makes engine reassembly much easier. In addition, the bearing halves project slightly above the openings. When the connecting rod or main bearing caps are tightened, these projections are forced together. This *bearing crush* holds the bearing in place. Bearing spread and crush are illustrated in **Figure 12-14.**

Two other commonly used retaining devices, the locking *tang* and the *dowel,* are illustrated in **Figure 12-15.** A tang is a projecting piece in the bearing that fits into a matching opening in the crankshaft. A

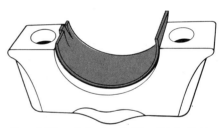

Spread Allows A Bearing To "Snap" Into Place.

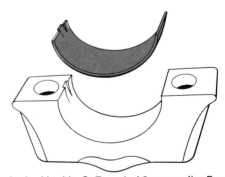

Bearing Locking Lip, Or Tang, And Corresponding Recess In Cap.

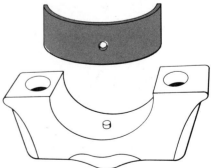

Use Of A Dowel To Retain The Bearing Half.

Figure 12-15. Main bearing retaining devices.

dowel is a cylindrical part on the crankshaft that fits in a matching drilled opening in the bearing. The purpose of tangs and dowels is to prevent the bearing from spinning in its bore.

Thrust Bearing

To prevent front-to-back motion of the crankshaft, one of the main bearings has *flanges,* or projections, on its edges, as shown in **Figure 12-16.** If the crankshaft moves, the sides of the throws next to the crankpin are stopped by the flanges.

12.4 GASKETS AND SEALS

A *gasket* forms a seal by being compressed between two parts. Gaskets can be made of soft materials, such as cork, rubber, paper, asbestos, or combinations of these materials. Gaskets also can be made of soft metals, such as brass, copper, aluminum, or *malleable* (soft) steel sheet metal. Gaskets are placed between stationary parts where liquid or gases could leak in or out. An engine overhaul gasket set is shown in **Figure 12-17.** Most gaskets are designed to be used only once. A new gasket is required when the parts are reassembled.

Head Gasket

To seal and contain the pressures of combustion within the engine, a ***head gasket*** is placed between the cylinder head and the block, as shown in **Figure 12-18.** The head gasket also seals oil passages between the block and the head. Also, the gasket controls the flow of coolant between the block and head.

Seals

Gaskets around a rotating part would quickly wear out and leak. Instead, *seals* are used for these dynamic, or moving applications. Like gaskets, seals should never be reused. Always install new seals when reassembling parts.

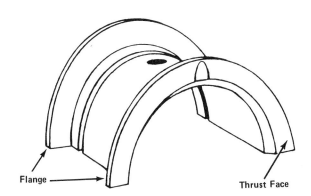

Figure 12-16. A thrust bearing has flanges on its edges to prevent front-to-back crankshaft motion.

Figure 12-17. Engine gasket set.

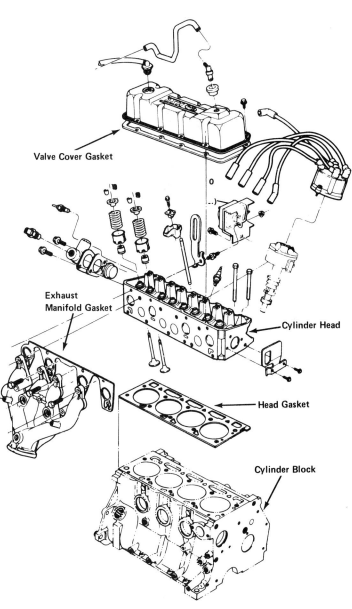

Figure 12-18. Head gasket installation.

Lip-type seal. A *lip-type seal* consists of three parts, as shown in **Figure 12-19**. These parts are a metal or plastic casing, a rubber sealing element, and a *garter spring*. The garter spring helps to hold the seal against a turning shaft.

Ring seals. A *ring seal* can be used between stationary parts and around rotating shafts. A ring seal can be round, square, or half-round in section, as shown in **Figure 12-20**. Ring seals can be made from rubber or hollow metal tubing.

Packing. *Packing* material, made from braided fabric, is often used as a rear main bearing seal.

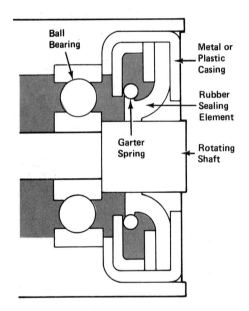

Figure 12-19. Lip-type seal installation.

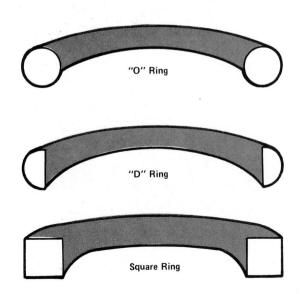

Figure 12-20. Ring seals.

UNIT HIGHLIGHTS

- Common cylinder block designs include the inline-type, V-type, or flat-type.
- Casting and machining operations are used to manufacture cylinder blocks.
- The design of the crankshaft determines the firing order of the engine and how smoothly it will run.
- Rod journals and main journals are machined areas for bearing support and attachment.
- Gaskets and seals are used to keep liquids and gases from leaking or becoming contaminated.

TERMS

crankcase	timing gear
slant	damper pulley
bank	harmonic balancer
casting	bearing cap
core	whip
sand cast	bushing
cylinder liner	anti-friction bearing
sleeve	friction bearing
dry liner	precision insert bearings
wet liner	alloy
balance	clearance
journal	bearing spread
rod journal	bearing crush
throw	tang
core plug	crankshaft
main journal	dowel
even firing	flange
uneven firing	gasket
crankpin	malleable
splay	head gasket
forge	seal
billet	packing
counterweight	lip-type seal
stress	garter spring
chamfer	ring seal
machining	fillet

ENGINE CONSTRUCTION

REVIEW QUESTIONS

DIRECTIONS: The following questions are similar to those used on automotive technician certification tests. On a separate sheet of paper, write the letter of the correct choice.

1. Technician A says that sand cores are used to form the cylinders during block manufacture.

   Technician B says that cylinders are machined during block manufacture.

   Who is correct?
   - A. A only
   - B. B only
   - C. Both A and B
   - D. Neither A nor B

2. Technician A says that the crankshaft is supported by bearings on the throws.

   Technician B says that drilled crankshaft oil holes are filleted and that machined areas next to crankpins are chamfered to relieve stress.

   Who is correct?
   - A. A only
   - B. B only
   - C. Both A and B
   - D. Neither A nor B

3. Which of the following statements is correct?
   - I. The power pulses of an uneven-firing engine are equally spaced.
   - II. Splayed crankpins are used to even out the application of power pulses to V-8 engine crankshafts.
   - A. I only
   - B. II only
   - C. Both I and II
   - D. Neither I nor II.

4. Technician A says that soft metal gaskets may be reused if they are not badly damaged.

   Technician B says that new gaskets and seals always should be used when reassembling parts.

   Who is correct?
   - A. A only
   - B. B only
   - C. Both A and B
   - D. Neither A nor B

Question 5 is not like questions one through four. It has the word EXCEPT. For question 5, look for the choice that does *not* apply. Read the entire question carefully before you choose your answer.

5. All of the following are alloys used in bearings EXCEPT
   - A. babbitt metal.
   - B. copper-lead alloys.
   - C. stainless steel.
   - D. aluminum with lead.

ESSAY QUESTIONS

1. What procedures are necessary to manufacture cast metal parts?
2. Describe crankshaft construction and features.
3. What are the features of precision insert bearings?

SUPPLEMENTAL ACTIVITIES

1. Explain why some parts of castings must be machined to a smooth surface.
2. Identify the parts of a shop crankshaft, and determine the number of cylinders and design of engine it is from.
3. Describe and demonstrate how precision insert bearings can be held in place.
4. Name the materials you can identify in gaskets selected by your instructor.

13 Pistons, Rings, And Connecting Rods

UNIT PREVIEW

The harsh operating conditions within the engine require light yet strong pistons and connecting rods. Manufacturing processes play an important role in determining how much heat and stress these parts can withstand. This unit provides an overview of piston, ring, and connecting rod designs and their functions.

LEARNING OBJECTIVES

When you have completed your assignments and exercises in this unit, you should be able to:

❑ Describe how pistons, rings, and connecting rods are designed to solve operating problems.
❑ Identify and describe the major types of pistons, rings, and connecting rods.
❑ Measure and identify different types of pistons.

13.1 PISTONS

The piston must be able to withstand the heat and pressures of combustion and of transferring force to the connecting rod. The piston also must change direction at rates varying from about 10 times to hundreds of times each second. At the same time, the piston must remain tightly sealed within the cylinder.

To meet these requirements, the piston must be both light and strong. It also must be constructed so that heat, pressure, and the forces of movement are controlled and directed properly. Modern pistons are cast or forged from high-strength aluminum alloys.

Piston Shape

The side view of a piston shown in **Figure 13-1** illustrates some commonly used shapes for piston heads, or tops. The piston head may be flat, concave, dome, or recessed, depending on the combustion chamber shape and the desired compression ratio.

The cutaway shape of the piston *skirt*, or lower part, allows the piston to clear the rotating crankshaft counterweights, as shown in **Figure 13-2**.

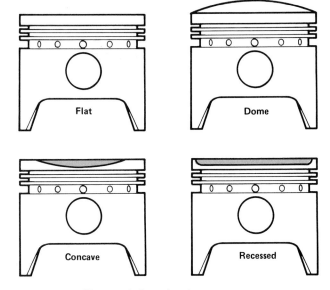

Figure 13-1. Shapes of piston heads.

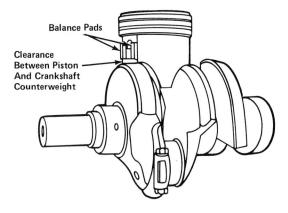

Figure 13-2. Piston skirts have a cutaway design for clearance of crankshaft counterweights. CHEVROLET MOTORS DIVISION—GMC

Piston Parts

Figure 13-3 illustrates the major external parts of a piston, and **Figure 13-4** shows a cross-sectional view of its internal construction. The thicker reinforcing area around the hole is known as a *boss*.

A *piston pin* attaches the piston to the connecting rod. **Figure 13-5** shows a piston, connecting rod, and piston pin assembly.

112

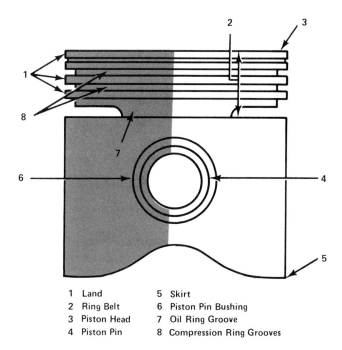

1 Land
2 Ring Belt
3 Piston Head
4 Piston Pin
5 Skirt
6 Piston Pin Bushing
7 Oil Ring Groove
8 Compression Ring Grooves

Figure 13-3. External piston construction.

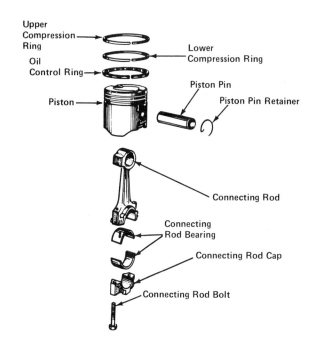

Figure 13-5. Piston, connecting rod, and piston pin assembly.
FORD MOTOR COMPANY

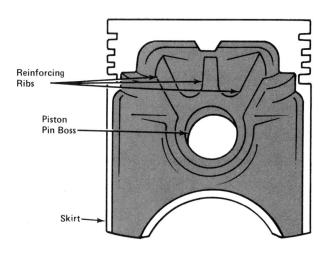

Figure 13-4. Cutaway view of piston construction.

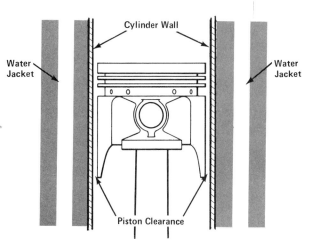

Figure 13-6. Piston clearance is the space between piston and cylinder wall.

Piston Clearance

The piston is slightly smaller in diameter than the cylinder *bore*, or hole diameter. The distance between the piston and cylinder wall, usually about 0.003 inch (0.08mm), is called *piston clearance*. Under normal circumstances this clearance [**Figure 13-6**] is filled with a thin film of oil.

If the pistons are badly worn, there will be too much piston clearance. As the power stroke begins, the piston tilts sharply against the cylinder wall and makes noise, known as *piston slap*. Besides the annoying noise, piston slap can damage the piston, rings, and cylinder wall. Piston slap can result from using pistons of too small a size for the cylinder bore.

When an engine is rebuilt, the cylinders are rebored to a larger size and fitted with oversized pistons of the proper size.

Controlling Piston Expansion

Because automotive pistons are made of aluminum, they heat and expand more quickly than cast iron cylinder walls. If piston clearance is too small, the piston will *seize*, or become stuck, in the cylinder.

The thicker area where the piston pin attaches to the piston expands less than the thinner piston skirt area. To control piston expansion, different methods can be used, including:

- Steel struts and belts
- Cam grinding
- Barrel shape.

Steel struts and belts. Steel expands less, and at a slower rate, than aluminum. Thus, steel "belts" or *struts* can be inserted inside the piston when it is cast **[Figure 13-7]**. A belt is a circular band inside the piston. A strut is a flat structural piece. Both help to resist expansion.

Cam grinding. The most common way of controlling piston expansion is to machine the piston so that it is not perfectly round. *Cam grinding* produces an oval-shaped piston, as shown in **Figure 13-8**. As the piston heats, it expands parallel to the piston pin and thus becomes round when hot.

Barrel shape. Some pistons are ground so the lower portion of the skirts is closer to the cylinder wall than the upper part **[Figure 13-9]**. Since the upper part of the piston gets hotter, it expands until the entire skirt is straight when hot.

Thrust Forces and Piston-Pin Offset

As the piston reaches the top of the compression stroke and begins the power stroke, two events occur. First, heated, expanding gases exert force against the piston head. Second, the connecting rod angle changes as the connecting rod goes over top dead center (TDC). These two events cause the piston to tilt and press against the cylinder wall. The force exerted by the piston against the cylinder wall is known as *side thrust*. The side of the piston that exerts the most force is known as the *major thrust face*. The opposite side of the piston tilts in the opposite direction and applies less force. This side is known as the *minor thrust face*, as shown in **Figure 13-10.**

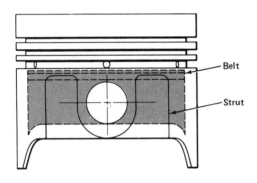

Figure 13-7. Piston struts and belts help to control piston expansion.

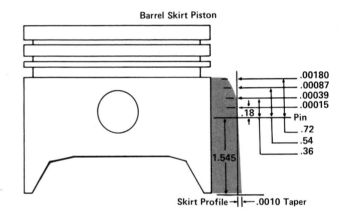

Figure 13-9. The skirt portion of a piston is made wider because it expands less than the upper portion of the piston.

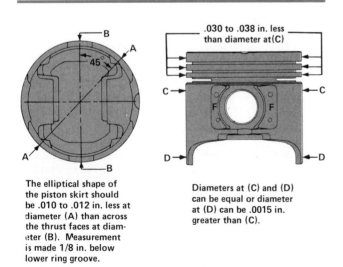

Figure 13-8. Pistons often are cam ground to an oval configuration. Pistons become round when heated.

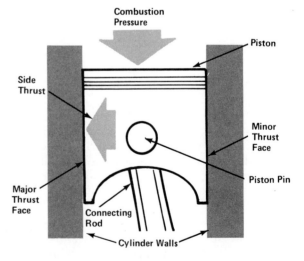

Figure 13-10. Combustion chamber pressure on the top of a piston causes it to thrust, or tilt, against the cylinder wall.

If the piston pin is mounted exactly in the center of the piston, side thrust can cause piston slap. However, mounting the piston pin slightly off center can reduce side thrust. When the piston pin is *offset*, or moved, more force is exerted against the minor thrust face. As the piston and connecting rod move over TDC, the piston rocks onto its major thrust side, thus eliminating piston slap **[Figure 13-11]**.

Ring-Groove Strengthening

The top groove and piston ring are subjected to very high pressure and heat from combustion. In addition, relatively little lubrication is present at the top of the cylinder. These factors cause the uppermost ring to wear more rapidly than the other rings.

To protect this groove, an insert of cast iron, iron alloy, or stamped steel is designed into cast pistons **[Figure 13-12]**. Many diesel engines use pistons of this type. Forged aluminum piston grooves can be *metal sprayed*, or sprayed with a thin coat of molten metal that hardens to protect the grooves.

13.2 PISTON CONSTRUCTION

The piston must be able to change direction quickly at the top and bottom of the strokes. Such a change in direction is easier for a light object than a heavy one. Aluminum is used for pistons because it is a relatively light material and conducts heat well. The heat is conducted away from the piston through the plugs into the cylinder walls. Piston manufacturing processes include:

- Casting or forging
- Machining and finishing.

Casting or Forging

Pistons can be cast or forged. Casting is cheaper and quicker, and allows the insertion of separate parts. Forging produces stronger, denser parts but is more expensive. Heat is conducted better through dense materials. **Figure 13-13** shows the differences in heat concentrations between cast and forged pistons. The cooler a piston operates, the less likely it is to be damaged by heat.

Machining and Finishing

After forging or casting, the pistons are machined to proper size and finish. The piston skirt under the piston pin area includes thicker portions, called *balance pads*. These balance pads can be ground down to equalize the weight of each piston. This helps to equalize the force transmitted by each piston during engine operation.

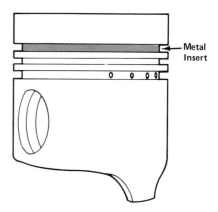

Figure 13-12. The top piston ring groove usually has a special metal insert to resist wear.

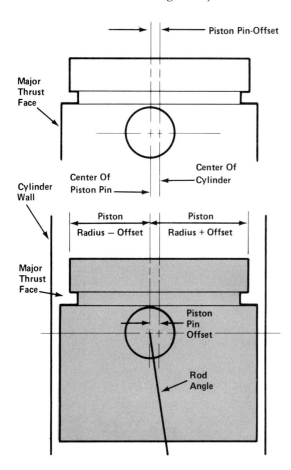

Figure 13-11. Piston pins are offset to help prevent piston slap.

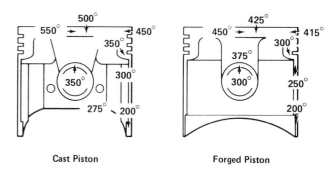

Figure 13-13. Heat concentrations in pistons differ, depending upon materials and manufacturing techniques used.

To prevent *scuffing,* or rubbing, the surface of the piston skirt may be machined smooth and tin plated. In other cases, the aluminum piston skirt is lightly grooved to help carry oil for lubrication.

The diameter of the piston *lands,* or areas between the ring grooves, is about 0.020 inch (0.5 mm) smaller than the piston skirt.

A groove, called a *heat dam,* is often cut into the piston above the top groove. This groove helps to minimize heat transfer to the ring area. The groove cut below the oil ring groove also may be called a heat dam groove.

Grinding, as mentioned above, also is a part of piston manufacture. A special machine, called a *cam grinder,* is used to machine a round cast or forged piston into the oval cam or barrel shape.

In addition, the piston pin holes are machined. In some cases bronze bushings are pressed into the openings to protect the softer aluminum from excessive wear.

13.3 RINGS

Piston ring grooves are machined deeply enough so the rings can be compressed and not protrude past the piston skirt. The upper rings on a piston that hold the pressure generated during combustion are called *compression rings*. The lower ring that scrapes excess oil off the lower cylinder walls is called an *oil control ring*.

Compression Rings

A piston ring is not a complete circle. There is a break, or *ring gap,* in the ring to allow it to be stretched for installation around the piston. The ring gap is shown in **Figure 13-14**. After the piston is installed in the cylinder, the gap is reduced to about three or four thousandths of an inch per inch of bore size. Thus, a 3-inch diameter piston would have ring gaps of from 0.009 to 0.012 thousandths of an inch.

The compression ring is made of cast iron. Compression rings sometimes are plated with harder metals, such as chromium, molybdenum, or aluminum oxides for longer wear. Synthetic plastic resins containing fluorine (Dupont Teflon) also have been used for ring coatings. Because of the great pressures within a cylinder (up to 1,200 psi), two rings are used. Part of the pressure of combustion gases "blows by" the top ring through the gap. The gaps on the lower rings are *staggered,* or purposely misaligned. Most of the pressure that gets by the first compression ring is trapped behind the second ring. Compression rings often are manufactured to provide a special ring profile **[Figure 13-15]**.

Oil Control Rings

The lower oil control ring is made from grooved cast iron or separate pieces of steel sheet metal. Its function is to scrape excessive oil off the lower cylinder walls. Holes cut in the oil control ring piston groove allow the oil to drain back through the piston into the oil pan. Expander-type rings include an inner spring-like expander that provides force to keep the oil control ring snugly against the cylinder walls.

Many different types of piston rings are made for different engines. **Figure 13-15** shows a selection of the most popular types of modern piston rings.

13.4 PISTON PINS

The *piston pin,* also called a *wrist pin,* is inserted through the holes in the piston and connecting rod. The pin may be free to pivot in both the connecting rod and piston, and is held in by retaining rings in the ends of the piston pin bore. This type of attachment is known as a *free-floating piston pin.* However, it also may be fixed to either the connecting rod or to the piston. Such an attachment is known as a *semi-floating piston pin.* This type of attachment does not require spring retainers to hold the pin in place.

Spring *retainers,* or clips, shown in **Figure 13-16,** are used to hold free-floating piston pins from moving against the cylinder walls.

13.5 CONNECTING RODS

The connecting rod must be strong enough to transfer the power impulses to the crankshaft, yet as light as possible. Connecting rods for production engines are made of cast iron or forged steel. The long outer portion of the rod is partially hollowed out. In section, the shape resembles the capital letter I. Balance pads on the rods allow material to be ground off to equalize rod weights.

The top end of the rod (the small end) is connected to the piston by the piston pin. With free-floating piston pins, a bushing is pressed into the small end of the rod. The bushing helps to reduce friction and wear. The lower part of the rod that connects around the crankpin is known as the *big end.* A *rod cap* holds the rod and bearing inserts around

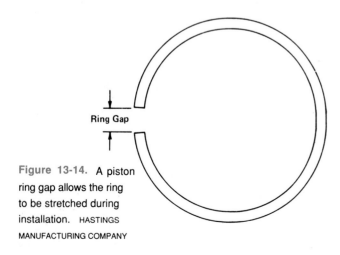

Figure 13-14. A piston ring gap allows the ring to be stretched during installation. HASTINGS MANUFACTURING COMPANY

Pistons, Rings, and Connecting Rods

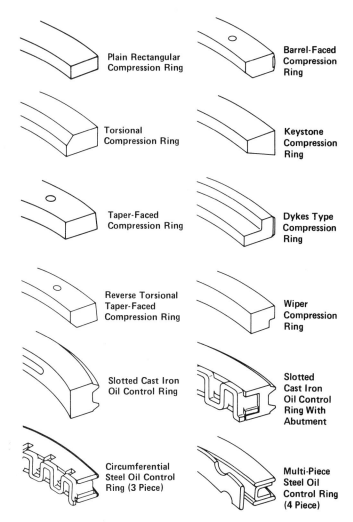

Figure 13-15. Types of compression and oil-control piston rings. HASTINGS MANUFACTURING COMPANY

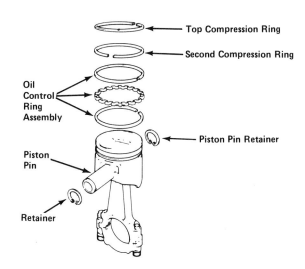

Figure 13-16. Free-floating piston pins are kept from striking cylinder walls by spring retainers. CHEVROLET MOTOR DIVISION—GMC

the crankpin, as shown in **Figure 13-17**. The big end hole is machined after the bearing cap is bolted on. Therefore, each bearing cap can be used only with its original mating connecting rod—and only facing in the original position.

V-6 engines often have offset connecting rods, as shown in **Figure 13-18**. The offset is necessary to properly align the connecting rod on the limited space available on a V-6 crankpin.

Lubricating oil is forced through the block galleries to the crankshaft and into the main and connecting rod bearings. Small holes, called *oil spurt holes*, sometimes are drilled into the big end or bearing cap. The spurt holes allow oil to flow through and flush out contaminants. Oil spurt holes also can be positioned to squirt oil at the underside of pistons to lubricate piston pins as well as the camshaft and lifters (see Unit 15). This also helps cool the pistons [**Figure 13-19**]. Some connecting rods have holes drilled from the big end through the center part to the small end. Oil flows through the connecting rod to the piston pin.

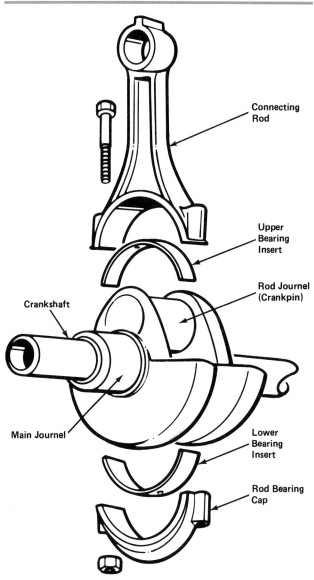

Figure 13-17. Parts of a connecting rod.

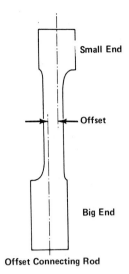

Figure 13-18. An offset connecting rod often is used in V-6 engines.

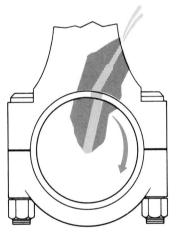

Figure 13-19. An oil spurt hole in a connecting rod.
FEDERAL-MOGUL

UNIT HIGHLIGHTS

- A piston must be strong enough to withstand the forces of combustion but light enough to be able to change directions quickly.
- The clearance between the piston and the cylinder wall is approximately 0.003 in. (0.08 mm).
- Pistons are made of aluminum and are either cast or forged.
- Two types of rings are used on pistons: compression rings and oil-control rings.
- The piston pin can be of either free-floating or semi-floating design.
- Production engine connecting rods are cast iron or forged steel.

TERMS

skirt	piston pin
boss	bore
piston clearance	piston slap
seize	struts
cam grinding	piston side thrust
major thrust face	minor thrust face
offset	metal spray
balance pad	scuff
land	heat dam
cam grinder	compression ring
oil control ring	ring gap
stagger	free-floating piston pin
semi-floating piston pin	retainer
wrist pin	big end
rod cap	oil spurt hole

REVIEW QUESTIONS

DIRECTIONS: The following questions are similar to those used on automotive technician certification tests. On a separate piece of paper, write the letter of the correct choice.

1. Technician A says that wear can cause piston slap.
   Technician B says too small a piston can cause piston slap.
   Who is correct?
   A. A only
   B. B only
   C. Both A and B
   D. Neither A nor B

2. Which of the following statements is correct?
   I. Piston expansion can be controlled by cam-grinding the piston.
   II. Piston expansion can be controlled by casting belts or struts into the piston.
   A. I only
   B. II only
   C. Both I and II
   D. Neither I nor II

3. Why is piston pin offset used?
   A. To control piston expansion
   B. To lessen piston slap
   C. To direct the power impulses to the crankshaft evenly
   D. To hold the piston pin in the connecting rod

4. Technician A says that the gaps on piston rings are staggered to reduce heat transfer to the piston skirt.
   Technician B says that the gaps on piston rings are staggered to reduce piston slap.
   Who is correct?
   A. A only
   B. B only
   C. Both A and B
   D. Neither A nor B

Question 5 is not like the first four questions. It has the word EXCEPT. For question 5, look for the choice that does *not* apply. Read the entire question carefully before you choose your answer.

5. Parts of a piston include all of the following EXCEPT
   A. piston rings.
   B. pin boss.
   C. skirt.
   D. lands.

ESSAY QUESTIONS

1. Describe different piston designs and their purposes.
2. Identify and describe different types of piston rings.
3. Describe the design of connecting rods.

SUPPLEMENTAL ACTIVITIES

1. Examine and measure pistons selected by your instructor for taper, shape, and construction features.
2. Measure piston pin offset in pistons selected by your instructor.
3. Describe how piston pin offset can improve the operation of the engine.
4. Identify free-floating and semi-floating piston pins.

14 Cylinder Head And Valve Train

UNIT PREVIEW

Cylinder heads are cast and machined in much the same way as cylinder blocks. The design of the cylinder head combustion chamber affects the efficiency, power output, and production of exhaust emissions in an engine.

Although overhead camshaft designs are theoretically more efficient, pushrod-type designs are still in use. The valve train can include parts that operate mechanically and/or hydraulically.

The shape of the cam lobes on a camshaft determines when the valves open and close, and how long they remain open. The camshaft usually drives the distributor, and in the latest model engines, may drive oil, fuel, and water pumps.

LEARNING OBJECTIVES

When you have completed your assignments and exercises in this unit, you should be able to:

- ❏ Identify the processes used to manufacture cylinder heads.
- ❏ Describe how overhead camshaft and pushrod type valve trains operate the valves.
- ❏ Describe the advantages of overhead camshaft and pushrod type valve trains.
- ❏ Describe how valves are fitted into the cylinder head.
- ❏ Describe the mechanisms that can be used to make sure that the valves close fully.
- ❏ Identify and describe the functions and operation of the camshaft.

14.1 CYLINDER HEAD DESIGN

The cylinder head forms the top of the combustion chamber. Combustion chamber design can vary with the intended purpose of the engine. Cylinder head designs for passenger cars must combine reasonable fuel economy characteristics with efficient and nearly complete burning of fuel. Two types of cylinder heads commonly used in passenger vehicle engines are:

- Wedge combustion chamber
- Hemispherical combustion chamber.

These shapes are shown in **Figure 14-1.** In a wedge head design, two valves are parallel to the centerline of the engine. The two valves in a hemi-head design are on opposite sides of the combustion chamber. For a discussion of four-valve cylinder heads, see Unit 41.

Cylinder Head Materials

Cylinder heads are cast from iron, iron alloy, or aluminum alloy. Aluminum is both light in weight and transfers heat more efficiently than iron or iron alloy. Efficient heat transfer allows higher compression ratios to be used for greater engine efficiency and power.

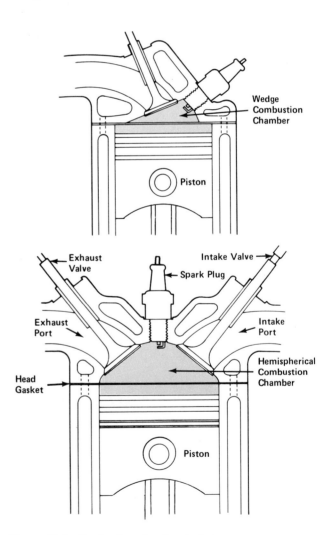

Figure 14-1. Combustion chamber designs.

Cylinder Head and Valve Train

Cylinder Head Construction

Cylinder heads are manufactured in much the same way as cylinder blocks. A typical cylinder head is shown in **Figure 14-2**.

Casting. A casting mold is made, and sand cores are inserted to form hollow areas within the cylinder head. Molten metal is poured in and allowed to cool. The cores are broken out and removed, and the cylinder head is cleaned of any remaining sand. Another casting method involves molds of expanded-bead polystyrene. A cylinder head made from expanded-bead polystyrene is packed in a sand mold. When molten aluminum is poured onto it, the foam is burned away. The aluminum takes the place of the foam and solidifies. No sand is used in this process, called the "lost foam process."

Machining operations. Sealing surfaces are machined smooth. Holes are drilled and tapped for attaching bolts or studs. In sand-cast heads, the large holes, through which sand was removed from the cores, are machined. They then are fitted with soft metal plugs, called "core plugs."

The valve seat areas are machined from the metal of the cylinder head and hard metal *valve seat inserts* are pressed into the machined holes. Valve seat inserts are circular metal rings with a shape that matches the shape of the valve. These inserts are used in aluminum cylinder heads to provide a sealing surface for valve seating.

In most cast-iron cylinder heads, the valve seats are machined directly into the cylinder head. The seats then are induction hardened to produce a strong, tough surface for seating the valves.

The *valve guide* area is machined from the metal of the cylinder head or holes are drilled for pressed-in guides. Valve guides are the hollow cylindrical parts in which the stem of the valve moves. Aluminum cylinder heads have pressed in guides.

Attaching the cylinder head to the cylinder block. The cylinder head is held to the cylinder block with bolts or studs and nuts **[Figure 14-3]**. During engine assembly and service, the bolts must be tightened and loosened in a specific order shown in the manufacturer's service manual.

Intake and Exhaust Passages

The intake and exhaust passages and ports are made to specific sizes. These sizes are a compromise to provide good driveability at both low and high engine speeds. In general, smaller passages will allow more torque at low engine speeds and larger passages will produce greater horsepower at high engine speeds.

Each intake and exhaust passage may be formed separately in the head; or the intake or exhaust passages for adjacent cylinders may be *siamesed,* or have a common, thin wall between the ports.

Cylinder head design in modern engines is of two basic types. One type is called *counter flow* and has all intake and exhaust ports on one side, cast either separately or in siamese fashion. The other type is called a *cross-flow head.* In this design, the intake ports are on one side of the head and the exhaust ports on the other. The cross-flow design allows for straighter passageways and improved volumetric efficiency. A disadvantage lies in the location of the exhaust system away from the intake passsageways. The result is difficulty in preheating the incoming air-fuel mixture. These configurations are illustrated in **Figure 14-4**.

Coolant and Oil Passages

Coolant and oil flow from the block through the head gasket into the head. To allow the removal of

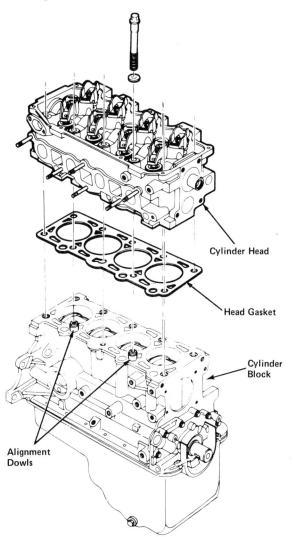

Figure 14-2. A typical cylinder head. FORD MOTOR COMPANY

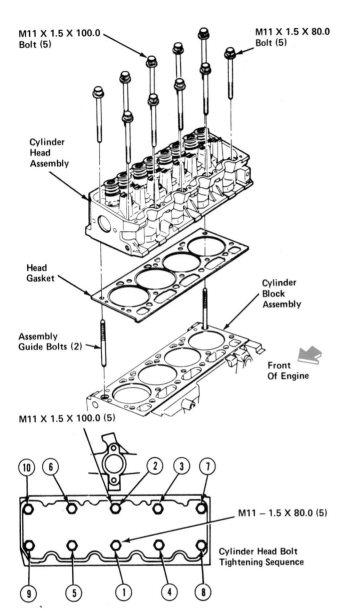

Figure 14-3. Bolts or studs hold a cylinder head to the cylinder block. FORD MOTOR COMPANY

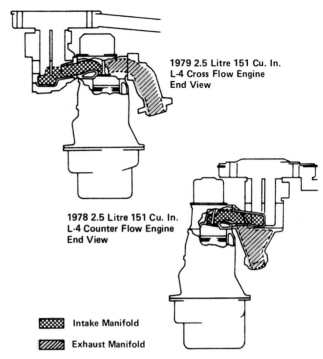

Figure 14-4. Cylinder head intake and exhaust passage designs.
PONTIAC MOTOR DIVISION—GMC

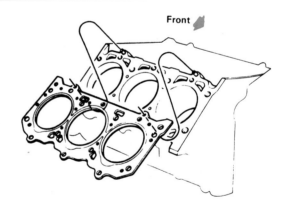

Figure 14-5. Coolant and oil flow are directed from the engine block, through a head gasket, and into the cylinder head.
CHEVROLET MOTOR DIVISION—GMC

casting cores, coolant and oil passages are sometimes made larger than desirable. In this case, smaller holes in the head gasket can be used to restrict the flow of liquids to acceptable levels **[Figure 14-5]**.

Overhead Valve and Overhead Camshaft Designs

As explained in Unit 10, the camshaft can be located above the head or in the block. For overhead camshafts, camshaft bearing supports and cap attachment points must be cast and machined as part of the head **[Figure 14-6]**.

Emission Control Considerations

Emission control is the name given to techniques that reduce the level of harmful gases in the engine exhaust. Wedge-head and hemi-head designs cause the air-fuel mixture to be burned differently.

Wedge-head design. The spark plug in a wedge-head is placed near the center of the *turbulence* to ignite the mixture **[refer to Figure 14-1]**. The narrow area farthest from the spark plug is known as the *quench area.* The mixture in the quench area is cooled by its contact with the surface of the piston and head. This contact quenches, or puts out, the *flame front,* or edge of the burning mixture. A small part of the fuel in this quench area remains unburned when the exhaust gases are forced out. This undesirable

CYLINDER HEAD AND VALVE TRAIN

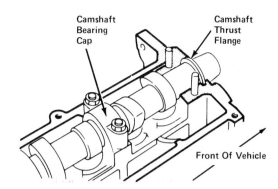

Figure 14-6. A cylinder head with an overhead camshaft is manufactured with bearing supports and cap attachment points. FORD MOTOR COMPANY

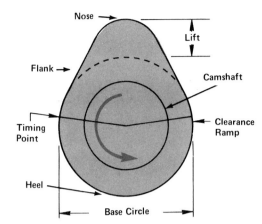

Figure 14-7. Parts of a cam lobe.

characteristic has been reduced in the hemispherical combustion chamber.

Hemispherical-head design. The spark plug in a hemi-head is located in the center of the combustion chamber **[refer to Figure 14-1]**. Smaller quench areas result in more complete burning of fuel. However, there is less turbulence. Consequently, the air-fuel mixture is not mixed as well as in a wedge chamber.

14.2 OVERHEAD CAMSHAFT VALVE TRAIN

Overhead camshaft (OHC) designs can use a *single overhead camshaft (SOHC)* with intake and exhaust lobes on the same shaft. Another design has two separate camshafts, one each for intake and exhaust. Such an arrangement is known as a *dual overhead camshaft (DOHC)* design.

Overhead camshaft designs can use one of two methods to move the valves:

- Cam followers
- Rocker arms.

Cam Followers

The camshaft *lobes*, or individual cam shapes, shown in **Figure 14-7**, rotate as the camshaft is turned. A *lifter* is a part that changes the rotary motion of the cam lobe into a *reciprocating*, or back-and-forth, motion.

As the *nose*, or high spot, of the cam lobe rotates into position over the valve, the hollow *cam follower* moves downward and opens the valve. A replaceable *valve adjustment shim* is used under the follower **Figure 14-8]**. Cam followers also are known as bucket lifters.

Rocker Arms

Another way of changing the rotary motion of the cam lobes into a reciprocating motion is to use *rocker arms*. The rocker arm pivots on a supporting *rocker*

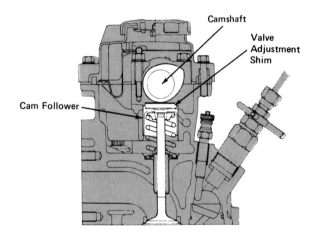

Figure 14-8. The camshaft pushes on a cam follower to move a valve assembly. FORD MOTOR COMPANY

shaft or on an independent pivot point. One end of the rocker is held upward by the valve spring pressure. The other end rides against a cam lobe. As the nose of the cam pushes up against the rocker, the rocker pivots down and opens the valve. A *valve adjusting screw* and *locknut* may be part of the rocker assembly, as shown in **Figure 14-9**.

14.3 OVERHEAD CAMSHAFT DRIVE MECHANISM

Two mechanisms are used in passenger vehicles to drive overhead camshafts:

- Toothed belt and pulleys
- Chain and sprockets.

The crankshaft makes two full revolutions to complete the 4-stroke cycle in any cylinder. However, each valve opens only once during the entire cycle. Thus, the camshaft must complete only one revolution for every two revolutions of the crankshaft. The camshaft must turn at one-half crankshaft speed. In addition, the motion of the camshaft must be coordinated to the action of the crankshaft.

To accomplish the speed reduction and the coordination of motion, a form of *gear drive* can be used.

Gears are toothed wheels that *mesh*, or fit together, as they rotate. The meshing gear teeth are all the same size. However, the sizes of the gears and the total number of teeth on each gear can differ. If one gear has twice as many teeth as another, it will rotate only half as many revolutions as the smaller gear.

This principle is used to rotate all camshafts at the proper speed. A sprocket or toothed pulley is attached to the front of the camshaft. A gear, called a *timing gear*, with half as many teeth is attached to the front of the crankshaft. As the crankshaft turns, the camshaft turns at half crankshaft speed [Figure 14-10].

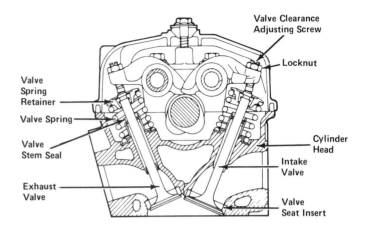

Figure 14-9. Parts of a valve assembly that uses rocker arms.
CHRYSLER CORPORATION

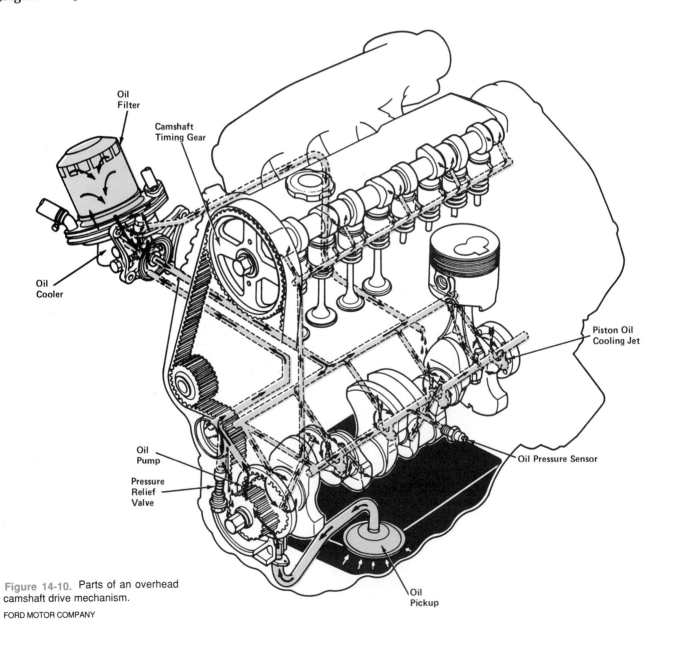

Figure 14-10. Parts of an overhead camshaft drive mechanism.
FORD MOTOR COMPANY

The turning camshaft also can be used to drive other assemblies. The camshaft usually drives the ignition distributor, either directly or through a gear. In addition, it can also drive oil, fuel, and/or water pumps.

Some overhead camshafts run in precision insert bearings similar to those used for the crankshaft. Other camshafts use fully circular *camshaft bearings*, or bushings.

14.4 PUSHROD VALVE TRAINS

Remember, power is the rate at which work is done. To produce more horsepower, today's smaller engines must run faster. This extra speed can cause problems.

The intermediate parts for a pushrod valve train, discussed in Topic 14.7, are larger and heavier than those for an overhead camshaft design. More weight means more momentum and the possibility of *valve float* at high engine speeds. Valve float occurs when the valve springs are unable to close the valves at the proper time in the four-stroke cycle.

Fewer and lighter parts in the valve train cause less valve float. Thus, overhead cam designs are better for smaller engines that run faster. However, pushrod valve trains have some servicing advantages. For example, the cylinder head can be removed without disturbing the camshaft drive and timing mechanism on a pushrod engine.

14.5 VALVES

The valves in an automobile engine must be able to function reliably under harsh conditions for many miles. During this time, the valves may open and close several hundred million times.

Valve Design

Suppose you drive for one hour at 55 mph (88 km/h). During this time, each of the valves in a 4-cylinder engine open and close approximately 90,000 to 100,000 times.

The exhaust valves will be exposed to combustion temperatures of 3,800 to 4,500 degrees F (2,093 to 2,482 degrees C). The valves will be red hot during operation, yet they must retain their hardness and shape.

Each valve weighs about five ounces, but it must seal against pressures of up to 1,200 psi (8,274 kPa).

Valve Construction

The parts of a typical automotive *poppet valve* are shown in **Figure 14-11**. A poppet valve is one that moves up and down to open and close.

Intake valves usually are larger, to assure a good flow of air-fuel mixture into the cylinder under atmospheric pressure. Exhaust valves can be smaller because the hot exhaust gases are under pressure as they are forced to leave the cylinder.

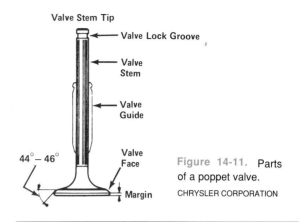

Figure 14-11. Parts of a poppet valve. CHRYSLER CORPORATION

Exhaust valves are made of stainless steel or high-strength steel alloys. Some exhaust valves are made by welding a **Stellite** head to a steel stem. Stellite is an extremely hard alloy of cobalt, chromium, and tungsten metals.

High-performance exhaust valves sometimes are filled with metallic sodium. The sodium liquifies at high temperatures to conduct heat away from the valve head. Valve heads may be formed to be either elastic (flexible) or rigid, as shown in **Figure 14-12**.

Valve Seats and Sealing

The valve seat is the area in the cylinder head against which the face of the valve forms a seal.

The valve face and seat may be machined to the same angle, usually 45 or 30 degrees. Alternately, the valve face and seat may be machined to different angles, usually a variance of about one degree. This variance forms an *interference angle* between the face and seat for better sealing, as shown in **Figure 14-13**.

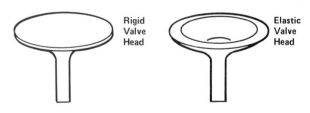

Figure 14-12. Valve head designs.

Integral valve seat. An *integral valve seat* is machined from the material of a cast-iron cylinder head. Integral valve seat areas are *induction hardened*, or heat treated, to make them more resistant to wear.

Valve seat insert. A *valve seat insert* is a circular, hardened steel or Stellite ring pressed into a machined space in the head. Inserts must be used in aluminum cylinder heads.

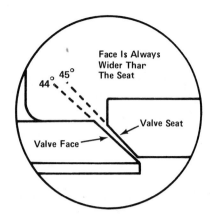

Figure 14-13. A valve face and valve seat may be machined at different angles, called an interference angle.

Valve Spring Assembly

A valve spring assembly, illustrated in **Figures 14-14 and 14-15,** can include the following parts:

- Valve spring seat
- Valve stem seal
- Single or double valve spring
- Spring damper
- Valve spring retainer
- Valve locks
- Valve rotator.

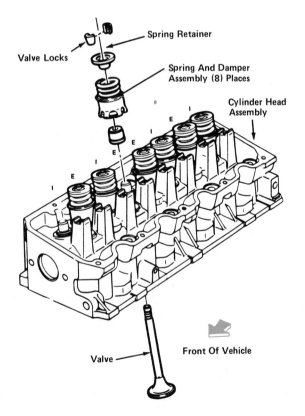

Figure 14-14. Parts of a valve spring assembly.
FORD MOTOR COMPANY

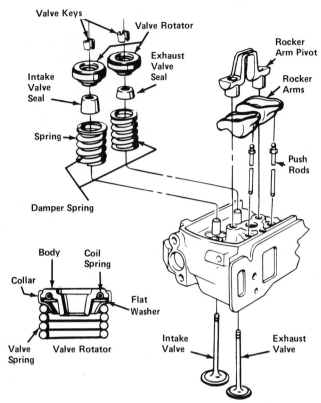

Figure 14-15. A valve rotator rotates a valve in a complete circle to help reduce heat buildup and remove deposits.
CHEVROLET MOTOR DIVISION—GMC

Some valve spring assemblies use a single spring. If necessary, double valve springs or spring dampers can be used to control *valve spring surge,* or vibration.

Valve springs tend to wind and unwind as they are compressed and released. This causes the valve face to turn in relation to the seat, helping to even out wear, heat buildup, and remove deposits. Some engines have specially made *valve rotators* in the valve spring retainer to rotate the valve in a complete circle **[Figure 14-15].**

14.6 VALVE GUIDES

Integral valve guides are machined holes in the metal of the cylinder head. *Replaceable valve guides,* shown in **Figure 14-16,** are bushings made of cast iron, steel, or bronze.

The guide supports and positions the valve stem in relation to the valve seat. In addition, heat is transferred from the valve stem to the guide, and from the guide to the cooling system. The guide is located near the hollow coolant passages within the cylinder head.

There must be a very small clearance between the valve stem and the inside diameter of the valve guide, usually between 0.0006 and 0.003 inch (6 ten-thousandths to 3 thousandths) (0.02 to 0.08 mm). This clearance is necessary for lubrication. However,

CYLINDER HEAD AND VALVE TRAIN

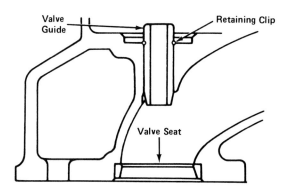

Figure 14-16. A replaceable valve guide is a bushing that supports and positions a valve stem. FORD MOTOR COMPANY

excessive clearance around the intake valve stem will cause oil to be drawn into the combustion chamber and burned during the intake stroke.

14.7 LIFTERS

The valve train must be able to open the valves. To accomplish this, either solid or hydraulic lifters may be used. Hydraulic lifters have largely replaced the older solid lifters on V-type engines, and are used on newer four-cylinder engines. Solid lifters, however, continue to be used. Advantages of hydraulic lifters are that they do not require routine adjustment and they operate more quietly. Hydraulic lifters also compensate automatically for changes in temperature and valve train wear.

Solid Lifter

Valve clearance, or *valve lash*, must be maintained between the valve tip and the valve train when the cam is not applying pressure to open the valve. *Solid lifters*, or mechanical lifters, are made from a single piece of solid metal.

One way to adjust valve clearance is with an adjusting screw and locknut built into the rocker arm, as shown in **Figure 14-17**. Valve clearance adjustments must be made at regular intervals, as specified by the vehicle manufacturer. On cam follower designs, thinner or thicker valve shims are installed **[refer to Figure 14-8]**.

Hydraulic Lifter

Another way of opening the valve is to use *hydraulic lifters*. Most pushrod valve trains use hydraulic lifters. Some manufacturers install hydraulic lifters in overhead cam engines. A hydraulic lifter, shown in **Figure 14-18**, contains a movable plunger, springs, and other parts.

Pressurized oil from the engine lubricating system fills the inner part of the lifter whenever the engine is operating. As the cam lobe rotates to push the lifter, a *check valve* within the lifter traps the

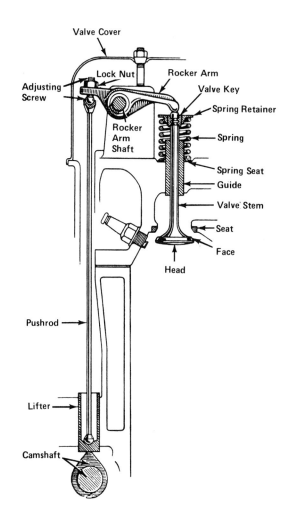

Figure 14-17. Parts of a valve train with a solid valve lifter. BLACK & DECKER

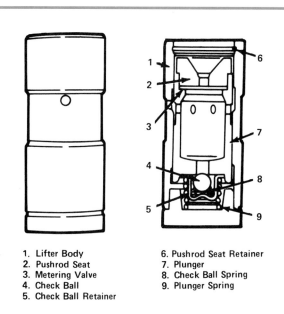

1. Lifter Body
2. Pushrod Seat
3. Metering Valve
4. Check Ball
5. Check Ball Retainer
6. Pushrod Seat Retainer
7. Plunger
8. Check Ball Spring
9. Plunger Spring

Figure 14-18. Parts of a hydraulic valve lifter. CHEVROLET MOTOR DIVISION—GMC

oil inside. Because liquids cannot be compressed, the hydraulic lifter acts like a solid lifter, and transfers force to open the valve.

As the cam continues to rotate, the valve spring closes the valve. Force from the valve spring pushes against the rocker arm, pushrod, and hydraulic lifter. This force causes the lifter to follow the cam lobe. As the camshaft continues to turn, the lifter reaches the *base circle,* or "flat" part of the cam lobe. At this point, the valve train is in a "relaxed" state. Any oil that has leaked or seeped out during the high-pressure portion of the operational cycle will be replaced. Replacement occurs as engine oil pressure again opens the check valve, filling the lifter body to take up any looseness. Hydraulic lifter operation is shown in **Figure 14-19.**

Solid lifters typically cause a "tapping" noise as the valve clearance is opened and closed. Hydraulic lifters operate silently and require only an initial adjustment during engine assembly.

Hydraulic lash adjuster. Several recent vehicles use a *hydraulic lash adjuster,* similar to a hydraulic lifter, in overhead camshaft valve trains. Such an arrangement is shown in **Figure 14-20.**

14.8 PUSHROD

Pushrods transfer the up and down movement of the lifter to the rocker arm. Pushrods usually are hollow for lightness. In many engines, lubricating oil is pumped up the hollow pushrod for upper valve train lubrication **[Figure 14-21].**

14.9 ROCKER ARM

Rocker arms can use leverage to increase the movement transmitted by the valve train. The *rocker arm ratio* is calculated by taking two measurements from the center, or pivot point, of the rocker. One measurement is the distance to the valve tip. The other measurement is the distance to the pushrod tip

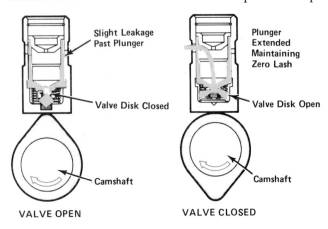

Figure 14-19. Hydraulic valve lifter operation.
FORD MOTOR COMPANY

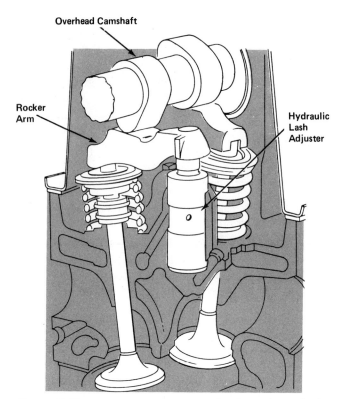

Figure 14-20. A hydraulic lash adjuster often is used to maintain valve assembly clearances. FORD MOTOR COMPANY

[Figure 14-22]. An optimum rocker arm ratio reduces lifter movement while providing adequate valve opening.

Adjustable Rocker Arms

Rocker arms for solid-lifter valve trains can be cast or forged and usually are mounted on rocker shafts. Adjusting screws and locknuts are added for valve clearance adjustments.

Nonadjustable Rocker Arms

Stamped, or pressed, sheet metal or cast aluminum rocker arms typically are used with hydraulic valve lifters. The rocker arm is mounted on a *pivot stud,* or pivot point. During engine assembly, the rocker arm is positioned to center the hydraulic lifter plunger in its bore **[Figure 14-23].**

Roller Rocker Arms

Needle bearing rollers are used in rocker arms to decrease valve train friction, and improve engine torque and horsepower output. Recent passenger cars with high-performance engines use roller rocker arms and dual overhead camshafts **[Figure 14-24].**

14.10 CAMSHAFT

Each cylinder in a typical engine has an intake cam lobe and an exhaust cam lobe to operate the valves.

Cylinder Head and Valve Train

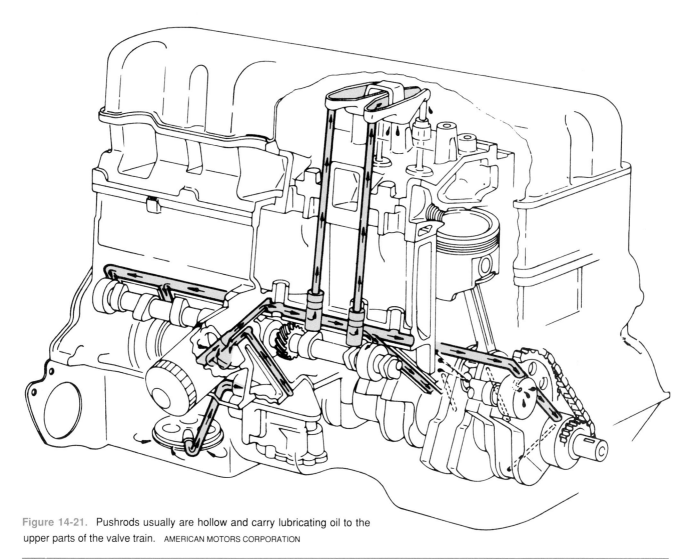

Figure 14-21. Pushrods usually are hollow and carry lubricating oil to the upper parts of the valve train. AMERICAN MOTORS CORPORATION

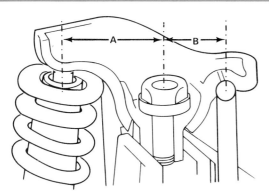

Figure 14-22. Rocker arm ratios are determined by making measurements from the pivot point to the valve tip and pushrod tip.

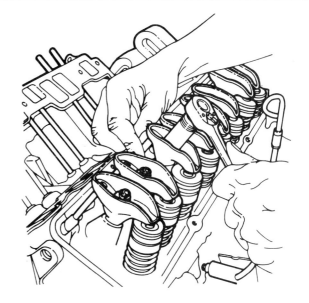

Figure 14-23. Non-adjustable rocker arms are positioned to center the hydraulic lifter plunger during assembly. CHEVROLET MOTOR DIVISION—GMC

Each lobe performs three functions. It opens a valve at the proper time. It allows the valve to remain open for a sufficient period. Finally, it allows the valve to close at the proper time.

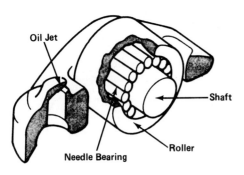

Figure 14-24. Roller rocker arm.
MITSUBISHI MOTOR SALES OF AMERICA, INC.

Valve Timing and Duration

Figure 14-25 shows a typical valve timing diagram. Refer to the figure as you read the following discussion. The diagram shows when the intake and exhaust valves open and close during the intake, compression, power, and exhaust strokes. The valve timing events are stated in relation to the rotation of the crankshaft:

1. Intake valve opens, 20 degrees BTDC (before top dead center)
2. Intake valve closes, 45 degrees ABDC (after bottom dead center)
3. Spark plug fires, 10 degrees BTDC
4. Exhaust valve opens, 45 degrees BBDC (before bottom dead center)
5. Exhaust valve closes, 20 degrees ATDC (after top dead center).

Intake valve opening can be determined by adding together the parts of the valve timing spiral, moving clockwise:

20 degrees BTDC —> TDC =	20 degrees
TDC —> BDC =	180 degrees
BDC —> 45 degrees ABDC =	45 degrees
Total intake valve duration =	245 degrees

Exhaust valve duration is determined in the same manner:

45 degrees BBDC —> BDC =	45 degrees
BDC —> TDC =	180 degrees
TDC —> 20 degrees ATDC =	20 degrees
Total exhaust valve duration =	245 degrees

Valve overlap can be determined by noting how far before TDC the intake valve opens and how far after TDC the exhaust valve closes:

Intake opens at 20 degrees BTDC =	20 degrees
Exhaust closes at 20 degrees ATDC =	20 degrees
Total valve overlap =	40 degrees

Because of its weight and momentum, a pushrod valve train does not follow the contours of the camshaft exactly. More valve overlap becomes necessary to fully exhaust burnt gases and fully fill the cylinder with fresh air-fuel mixture. The less valve overlap, the more smoothly and efficiently an engine will idle. Because overhead camshaft valve trains follow the cam contours more exactly, less valve overlap is needed. This condition provides a smooth idle and efficient high-speed operation.

Camshaft Construction

The camshaft is cast or machined from hardenable iron alloy or steel. The cam lobes are ground to the proper shape and position in relation to one another. The bearing surfaces also are ground smooth, and the distributor drive gear is machined into the shaft. The cam lobes are then flame or induction hardened. Pushrod valve train camshafts also may have an *eccentric*, or off-center, circular lobe to operate the fuel pump **[Figure 14-26]**. Recent designs also include hollow, tubular shafts for light weight with cam lobes welded to the shaft.

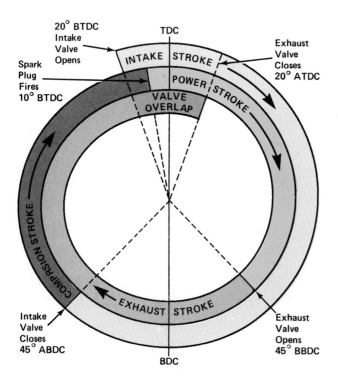

Figure 14-25. Valve timing and duration diagram.
CHEVROLET MOTOR DIVISION—GMC

CYLINDER HEAD AND VALVE TRAIN

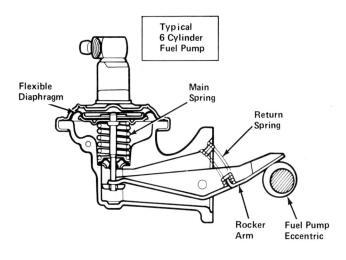

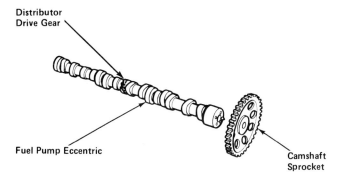

Figure 14-26. Parts of a camshaft assembly.
CHRYSLER CORPORATION

UNIT HIGHLIGHTS

- Cylinder heads are manufactured using the same processes used to make cylinder blocks.
- Wedge-head and hemispherical-head designs affect engine emissions differently.
- Overhead camshaft designs are theoretically more efficient than pushrod designs for small, fast-running engines.
- Complete valve closing can be assured by maintaining a specified valve clearance with solid lifters, or by using hydraulic lifters.
- The shape of the camshaft lobe determines when the valves will open and close, and how much valve overlap will occur.

TERMS

valve seat insert
siamese
valve guide
emission control
quench area
turbulence
lifter
nose
rocker arm
valve adjusting screw
gear drive
camshaft bearing
poppet valve
interference angle
induction hardening
valve rotator
replaceable valve guide
valve lash
hydraulic lifter
reciprocating
rocker arm ratio
valve overlap
base circle
cross flow cylinder head
single overhead camshaft (SOHC)
valve stem insert

flame front
lobe
cam follower
valve adjustment shim
rocker shaft
locknut
mesh
valve float
Stellite
integral valve seat
valve spring surge
integral valve guide
valve clearance
solid lifter
check valve
hydraulic lash adjuster
pivot stud
eccentric
counter flow
timing gear
dual overhead camshaft (DOHC)

REVIEW QUESTIONS

DIRECTIONS: The following questions are similar to those used on automotive technician certification tests. On a separate sheet of paper, write the letter of the correct choice.

1. Technician A says that aluminum is used for cylinder heads because it is light in weight.
   Technician B says that aluminum is used for cylinder heads because it transfers heat efficiently.
   Who is correct?

   A. A only   C. Both A and B
   B. B only   D. Neither A nor B

2. Which of the following statements is correct?
   I. The more quench area in the combustion chamber, the more unburned fuel in the exhaust gases.
   II. The more quench area in the combustion chamber, the more efficiently the engine will run.

   A. I only   C. Both I and II
   B. II only   D. Neither I nor II

3. Which of the following is used for cam-follower type valve adjustments?
   A. Replaceable valve shims
   B. Roller rocker arms
   C. Non-adjustable rocker arms
   D. Adjustable rocker arms

4. Technician A says that all valve trains must have valve clearance to allow the valves to be opened.

   Technician B says that hydraulic lifters or lash adjusters maintain a small valve clearance automatically.

   Who is correct?
   A. A only
   B. B only
   C. Both A and B
   D. Neither A nor B

Question 5 is not like those above. It has the word EXCEPT. For question 5, look for the choice that does *not* apply. Read the entire question carefully before you choose your answer.

5. All of the following are usually found in a pushrod-type valve train EXCEPT
   A. lifter.
   B. pushrod.
   C. rocker arm.
   D. cam follower.

ESSAY QUESTIONS

1. What differences are there between wedge-head and hemispherical-head combustion chambers?
2. Describe the advantages and disadvantages of over-head camshaft and pushrod engine designs.
3. What methods are used to make sure that valves open and close completely?

SUPPLEMENTAL ACTIVITIES

1. Examine a cylinder head. Identify and describe the manufacturing techniques and design that were used to make the cylinder head.
2. Examine a cylinder head and corresponding piston and correctly identify the quench areas.
3. Describe how a hydraulic valve lifter works.
4. Using a dial indicator and a degree wheel, make a diagram of the opening and closing of the valves on an engine.
5. Explain what valve overlap is, and how it can affect engine performance.

PART III: Automotive Engine Systems

Common maintenance and repair work for all vehicles include lubrication, cooling, fuel, and exhaust system service. Beginning with this part of the book, the units follow a two-part format. First, a knowledge unit details the construction and operation of a single system. The service unit that follows provides specific information on troubleshooting, preventive maintenance, and repairs. In this photo, a technician is cleaning a radiator.

15 The Lubrication System

UNIT PREVIEW

The engine lubrication system provides lubricating oil for all parts that move and/or contact other parts within the engine.

To provide proper lubrication, engine oil must be able to perform many tasks under difficult conditions. To help meet these requirements, petroleum-based engine oils require many additives.

The lubrication system consists of several parts and components that function together to provide adequate lubrication to internal engine parts.

LEARNING OBJECTIVES

When you have completed your assignments and exercises in this unit, you should be able to:

❏ Identify and describe the functions engine oil must perform within an engine.
❏ Identify and describe what additives are necessary for petroleum-based engine oils.
❏ Identify the SAE viscosity and API service ratings for gasoline and diesel engine oils.
❏ Explain how engine oil can be lost or consumed.
❏ Identify and describe the parts of an engine lubrication system.

15.1 LUBRICATION

The word *lubrication* comes from a Latin verb that means "to make slippery." *Lubricants* are materials, such as oil and grease, that reduce friction and make surfaces slippery.

A lubrication system consists of parts and components that circulate lubricant to parts that otherwise would become worn through friction. An engine lubrication system provides engine oil to all bearing surfaces and between all moving parts within the engine.

The life expectancy of major engine parts depends on proper lubrication. These major parts include bearings, crankshaft and camshaft journals, camshaft lobes, pistons and piston rings, and cylinder walls. If the engine lubrication system fails, friction can cause deep gouges and extreme heat. *Seizing*, which occurs when metal parts become hot enough to melt and stick to each other, can damage the engine beyond repair.

15.2 FUNCTIONS OF ENGINE OIL

The oil within the engine lubrication system must perform several functions, including:

- Reducing friction
- Absorbing bearing shock loads during the power stroke
- Forming a seal between the piston rings and cylinder walls
- Cooling and cleaning internal engine parts.

15.3 OIL VISCOSITY AND SERVICE RATINGS

Two groups, the *Society of Automotive Engineers (SAE)* and the *American Petroleum Institute (API)* have established standards for engine oil.

SAE Viscosity Rating

The *viscosity* ratings of engine oils are established through tests. A specified quantity of oil at a specific temperature is passed through a calibrated hole. The time required for the oil to pass through this hole establishes its viscosity rating number. For example, oil may be tested at 212 degrees F (100 degrees C) and assigned a single number rating. An example of this is SAE 30, as shown in **Figure 15-1**.

Figure 15-1. A single-viscosity engine oil maintains one viscosity rating at operating temperatures.

The Lubrication System

Single-viscosity oils commonly used for automobiles include SAE 10, 20, 30, 40, and 50. The higher the viscosity index number, the thicker the oil remains when heated. Oils with higher viscosity numbers can protect the engine better under conditions of extreme heat, such as high-speed driving and desert conditions. Oils with lower viscosity ratings are used more often when cold temperature starting is a concern.

Multi-viscosity oils have been chemically modified to stabilize the rate of viscosity change [**Figure 15-2**]. Thus, these oils have low viscosity when cold, yet sufficient viscosity when hot to provide adequate engine protection. A multi-viscosity oil can have a rating such as SAE 20W-50.

The "W" indicates winter, or cold weather, test conditions. "20W" indicates that the oil has been checked for viscosity at 0 degrees F (-18 degrees C) and has the viscosity of 20. The "50" indicates that the oil also has been checked for viscosity at 212 degrees F (100 degrees C) and found to have a viscosity equivalent to SAE 50. Such a multi-viscosity oil can flow easily at low temperatures to allow the engine to start and run well under cold conditions. The same oil also will protect engine parts from wear even in extremely hot conditions.

The choice of a proper viscosity engine oil depends on the vehicle manufacturer's recommendations and the outside temperatures to be encountered, as shown in **Figure 15-3**.

API Service Rating

Engine oil also is rated for its ability to perform under a variety of engine conditions. The API ratings for increasingly severe conditions in spark-ignition engines are SA, SB, SC, SD, AE, SF, and SG. An oil graded API SG meets the highest current standards and can be used in all automotive engines. The 1989 and later model engines must use oils with the SG ratings. In the future, oils graded SH, SI, or higher may become available.

Oils graded SA and SB are ***nondetergent oils*** and are totally unsuitable for automotive engines.

The other API ratings indicate the year in which each was required to gain manufacturers' warranty approval:

SC: 1964

SD: 1968

SE: 1972

SF: 1980

SG: 1989

The API ratings for increasingly severe conditions in compression-ignition (diesel) engines are CA, CB, CC, CD, and CE. **Refer to Figures 15-1** and **15-2**. An oil with the API service classification CE meets the highest current standards for automotive diesel engines. As with the S-scale for spark-ignited engines, the C-scale is open-ended. In the future, CF, CG, and higher-rated oils may become available.

CAUTION!!!

Refer to the vehicle owner's manual, manufacturer's service manual, or service bulletins to determine the API oil service ratings necessary for specific engines. Using the wrong type of oil, especially in diesel engines, can cause engine damage and void the manufacturer's new-vehicle warranty.

Figure 15-2. A multi-viscosity engine oil changes viscosity when it is cold and when it is hot.

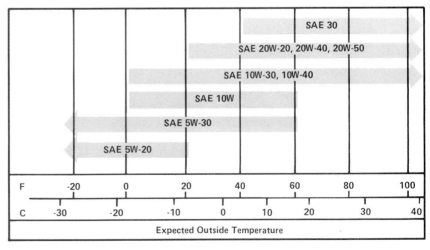

Figure 15-3. Engine oil recommendation chart.

15.4 NATURAL AND SYNTHETIC ENGINE OILS

The majority of engine oils are refined from petroleum, or crude oil, pumped from underground deposits. Refined petroleum oils, by themselves, cannot adequately lubricate modern engines. Chemicals, called additives, are blended with the oil to improve performance.

Engine Oil Additives

Several chemical additives are used to improve the lubricating ability of petroleum engine oils. These additives include:

Detergent/dispersants keep particles of carbon and other contaminants suspended within the oil.

Viscosity index improvers allow the oil to maintain an adequately thick film as the oil is heated.

Pour point depressants allow cold oil to remain thin enough to flow to the engine parts during starting.

Extreme pressure additives help prevent the oil from being squeezed out of oil clearances when heavy loads are applied.

Anti-wear additives chemically coat bearings and other moving parts. This helps prevent wear even if the parts touch or rub slightly under heavy loads.

Corrosion inhibitors help prevent the formation of acids in the oil that can damage metal parts in the engine.

Oxidation inhibitors help prevent oil *oxidation*, the chemical breakdown of oil.

Foam inhibitors reduce the formation of foam, or oil mixed with air, which can occur when the crankshaft churns the oil during normal operation. Foamy oil cannot support bearing loads and oxidizes more easily.

Oil rated for API service SG contains all the additives necessary for proper lubrication of spark-ignited automotive engines.

Synthetic Oils

Man-made, or *synthetic*, lubricating oils are created chemically. These oils reduce friction more than petroleum oils. However, some synthetic oils may have less film strength and resistance to *scuffing*, or rubbing due to greasy friction, than petroleum oils. Synthetic oils that meet the API Service Rating SG are safe to use in modern engines.

Synthetic oils cost three to four times as much as high-quality petroleum oils. Manufacturers of such oils claim the higher price is offset by extended oil change intervals of 15,000 to 25,000 miles or more. However, vehicle manufacturers will not honor new-vehicle warranties when oil and filter changes are performed at such long intervals.

Special Oils for Turbocharged Engines

In mid-1984, the Valvoline Oil Company introduced an oil for the special problems of turbochargers, discussed in Unit 20. This oil, Turbo V, has additional detergent/dispersants and oxidation inhibitors. Other oil companies produce similar products. Synthetic oils seem to lend themselves particularly well to turbocharger applications. They are highly resistant to heat and chemical breakdown.

15.5 OIL CONSUMPTION

Years ago, new-vehicle oil consumption of a quart in 600 miles was considered normal. Current vehicles rarely use more than one quart of oil in 1,000 or more miles, and some vehicles use as little as one quart of oil in 3,000 miles. Oil can be lost through leakage from the engine. Oil also can be consumed if there is excess clearance in the engine and oil can enter the combustion chamber.

Leakage

Oil can be lost by leakage through dried, cracked, or seeping gaskets. The most frequent source of engine oil leaks on older cars is through the valve cover gasket. This source has been minimized in recent years through the use of silicone sealers instead of gaskets. Other common leakage sources include:

- Oil pan gasket
- Crankshaft seals
- Oil pressure sending unit
- Timing cover gasket and/or seal
- Fuel pump gasket
- Distributor drive O-ring gasket
- Engine casting porosity.

Burning

Oil can be consumed by being burned in the combustion chamber. Worn cylinders and piston rings can cause *oil pumping*. The vacuum above the piston during the intake stroke can draw oil up, past worn rings, into the combustion chamber. In addition, worn intake valve guides can allow oil to be sucked down the valve stem, into the combustion chamber. During combustion, the oil is burned.

It is not unusual for small amounts of oil vapors and blowby gases to be burned in engines that are in good condition. The vapors and gases are drawn through the *positive crankcase ventilation (PCV) system* into the intake manifold. A PCV system keeps harmful vapors and gases from being released into the atmosphere, as discussed in Units 41-43.

In a worn engine, blowby gases carry large amounts of oil vapor into the PCV system to be burned. In some cases, this overload or a clogged PCV system can cause oil to be blown into the air filter. A clogged PCV system also can cause pressure inside

The Lubrication System

the engine that can rupture gaskets and seals and cause oil leaks.

15.6 ENGINE LUBRICATION SYSTEM

An engine lubrication system, shown in **Figure 15-4**, consists of several parts, including:

- Oil pan
- Pump
- Filter
- Oil distribution system
- Pressure indicator
- Oil level indicator.

Oil Pan

The *oil pan*, bolted to the engine crankcase area, serves as a reservoir for oil. A *drainplug* is provided to allow the engine oil to drain out during oil changes. An oil pump and pickup, shown in **Figure 15-5**, are mounted on the crankcase inside the oil pan.

Oil Pump

An *oil pump*, shown in **Figure 15-6**, consists of the following parts:

- Pickup and screen
- Housing and cover
- Gears or rotors
- Oil pressure relief valve.

Pickup and screen. The oil pickup is a hollow, flat cup mounted at the end of a tube. Mounted over the cup is a screen to prevent large particles from entering the pump and damaging the gears and housing. The pickup tube leads to the intake, or low pressure, side of the pump.

Housing and cover. The housing encloses the gears or rotors and contains a drive shaft for one of the gears, as shown in **Figure 15-7**.

Gears or rotors. Two gears or rotors are meshed in the housing. One of the gears is driven, either by a shaft from the distributor drive gear, or by an OHC cam belt. At the point where the oil pump gears unmesh, or separate, a low pressure area is created. Atmospheric pressure forces the oil up through the pickup to the low pressure area. The oil is then carried around the outside of the gear, in the chambers formed between the teeth, housing, and side plates **[Figure 15-8]**. As the oil reaches the outlet port, it is forced out, under pressure, as the teeth mesh again. Thus, oil is forced out of the pump under pressure to the oil filter.

Oil pressure relief valve. To prevent excess pressure from developing, a relief valve is used. A spring holds a check valve or ball on a seat in the oil outlet. When oil pressure develops more force against the check valve than the spring, the valve is pushed open. This action uncovers a relief passage that allows oil flow back to the pump inlet. In action, the relief valve acts as a controlled leak and allows only enough oil out to maintain a controlled pressure.

Oil Filter

The *oil filter* is located in the lubrication system between the oil pump and the engine parts that require lubrication. Its purpose is to remove harmful particles from the oil. The filter usually is screwed directly onto the side of the cylinder block or mounted on an adapter **[Figure 15-9]**. Some newer General Motors vehicles have the oil filter mounted inside the oil pan under a large cover which must be removed for servicing.

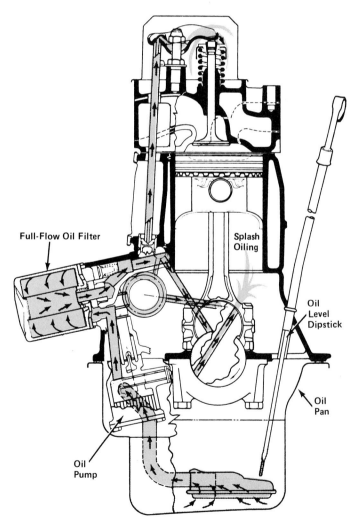

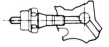

Figure 15-4. Engine lubrication system.
BUICK MOTOR DIVISION—GMC

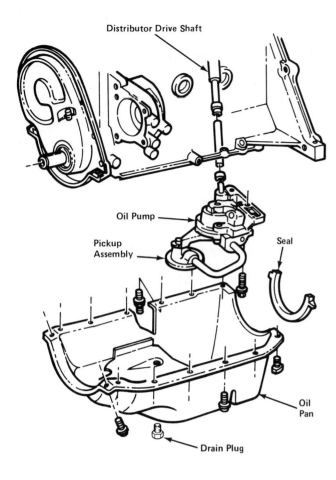

Figure 15-5. Oil pan and oil pump mountings.
BUICK MOTOR DIVISION—GMC

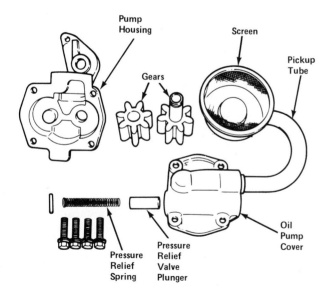

Figure 15-6. Parts of an oil pump.
BUICK MOTOR DIVISION—GMC

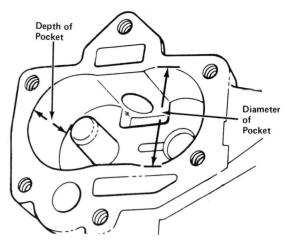

Figure 15-7. Oil pump housing.
BUICK MOTOR DIVISION—GMC

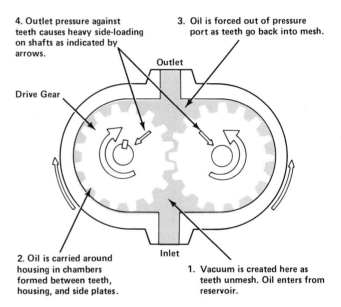

Figure 15-8. Oil pump gear operation.

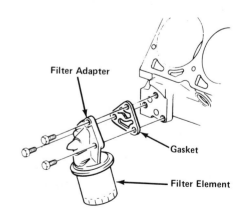

Figure 15-9. An oil filter usually is mounted on an adapter that bolts to the engine block.
BUICK MOTOR DIVISION—GMC

The Lubrication System

The oil filter usually consists of a pleated element made of special filtering paper inside a metal housing **[Figure 15-10]**.

Oil filtering systems used in current vehicles are of the *full-flow* type with bypass systems. Full-flow means that all of the oil normally passes through the filter before it can flow through the galleries to the moving parts of the engine.

However, if the filter becomes excessively clogged with contaminant particles, the oil pressure could drop and cause oil starvation to moving parts. To prevent this situation, a *bypass valve* is used to redirect unfiltered oil to the engine. Dirty oil is preferable to no oil at all.

The oil bypass valve may be located in the filter mount on the engine, as shown in **Figure 15-11,** or within the filter.

Oil Distribution System

The drilled passages through which the oil passes to bearing surfaces and moving parts are called *oil galleries.*

Splash lubrication throws oil from the crankshaft to the cylinder walls, piston pins, timing chain and/or gears, and other parts, as shown in **Figure 15-12.**

Oil spurt holes in the connecting rods spray oil onto the lower cylinder walls, camshaft, and lifters. In pushrod engines, drain holes also allow oil to flow from the upper valve train parts downward onto lifters and camshaft lobes.

Pressure Indicator

Low oil pressure can cause lack of lubrication and rapid engine damage. A warning system to alert the driver to low oil pressure is part of the engine lubrication system.

Warning lamp. A warning system has a pressure sensor screwed into an oil gallery, wiring, and a lamp on the dash **[Figure 15-13]**.

Oil pressure moves a *diaphragm,* or flexible metal partition, within the sensor to interrupt the

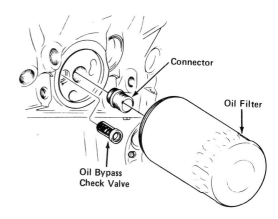

Figure 15-11. A bypass valve redirects engine oil flow if the oil filter becomes clogged. BUICK MOTOR DIVISION–GMC

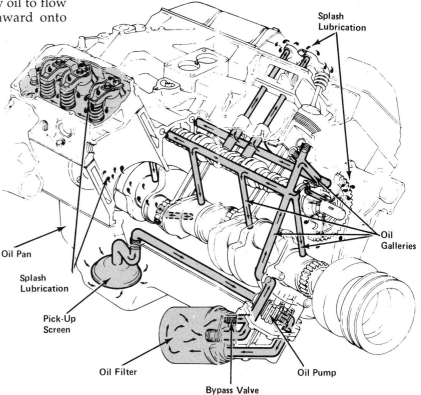

Figure 15-10. Parts of an oil filter.
CHRYSLER CORPORATION

Figure 15-12. Engine lubrication system and splash lubrication.
AMERICAN MOTORS CORPORATION

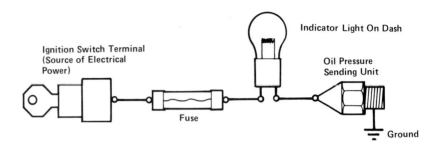

Figure 15-13. Oil pressure indicator circuit.

electrical circuit to the lamp [**Figure 15-14**]. When oil pressure falls below the level necessary for safe operation, the sensor completes the electrical path to ground. The warning lamp on the dashboard or instrument panel lights to warn the driver.

Electrical oil pressure gauge. Some vehicles are equipped with an oil pressure gauge, which registers the pressure developed within the lubrication system. A sensor, shown in **Figure 15-15**, is screwed into an oil gallery. Oil pressure moves a diaphragm connected to a variable *resistor*. A resistor lowers the amount of voltage and current passing through an electrical circuit. A gauge, or instrument, reacts to the amount of current passing through the circuit, and moves a needle over a scale to indicate the oil pressure.

Other types of oil pressure gauges operate mechanically instead of electrically. Oil travels up through a tube to the back of the gauge. A springy, flexible, hollow tube, called a **Bourdon tube**, uncoils as the pressure increases. A needle connected to the tube moves over a scale to indicate oil pressure.

Oil Level Indicator

A lack of oil will cause engine parts to become starved for oil. Friction and heat buildup will rapidly damage or destroy moving parts.

Dipstick. The simplest type of liquid level indicator is a *dipstick*, shown in **Figure 15-16**. The dipstick extends downward into the lubricant supply.

Before checking a dipstick, shut off the engine and wait for a few minutes. This delay allows oil to drain back to the oil pan from the upper areas of the engine. Wipe the dipstick to remove the oil splashed on it when the crankshaft was rotating. The dipstick is marked to indicate a level at which oil should be added and a level beyond which no oil should be added. Too much oil in the pan can be whipped into foam by the crankshaft, which results in engine damage or leaks.

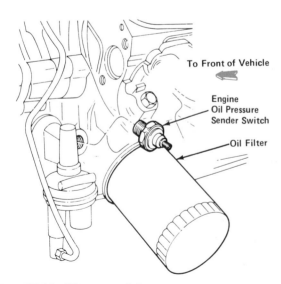

Figure 15-14. Oil pressure light sensor.
FORD MOTOR COMPANY

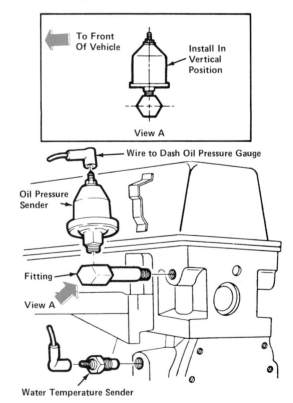

Figure 15-15. Oil pressure gauge sensor.
FORD MOTOR COMPANY

The Lubrication System

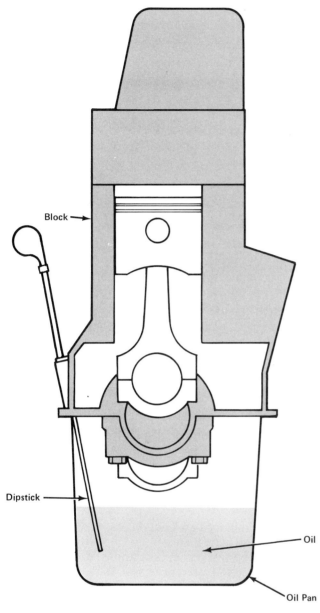

Figure 15-16. Engine oil dipstick.

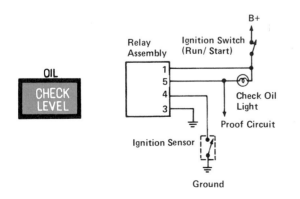

Figure 15-17. Electronic engine oil level indicator.

Electronic oil level indicator. Some vehicles built since 1980 have an electronic oil level indicator. A warning light will be activated if the amount of oil in the pan falls below a safe level. A circuit for such a warning system is shown in **Figure 15-17**.

UNIT HIGHLIGHTS

- The engine lubrication system provides oil to all surfaces within the engine that require lubrication.
- Engine oil performs several functions including reducing friction, absorbing shock loads, forming a seal, and cooling and cleaning internal engine parts.
- Engine oils have an SAE viscosity rating and an API service rating.
- Petroleum-based engine oils require several chemical additives to help the oil provide adequate lubrication.
- Oil can be lost through leakage or by being burned in the combustion chamber.
- An engine lubrication system consists of an oil pan, oil pump, oil filter, oil distribution system, pressure indicator, and oil level indicator.

TERMS

lubrication	lubricant
seize	oil pan
oil filter	detergent-dispersant
nondetergent oil	extreme pressure
additive	viscosity index improver
pour point depressant	anti-wear additive
corrosion inhibitor	oxidation inhibitor
oxidation	synthetic
foam inhibitor	viscosity
single-viscosity oil	multi-viscosity oil
scuffing	oil pumping
drainplug	oil pump
bypass valve	oil gallery
splash lubrication	diaphragm
resistor	Bourdon tube
full-flow oil filtering system	American Petroleum Institute (API)
Society of Automotive Engineeers (SAE)	dipstick
positive crankcase ventilation (PCV) system	

REVIEW QUESTIONS

DIRECTIONS: The following questions are similar to those used on automotive technician certification tests. On a separate sheet of paper, write the letter of the correct choice.

1. Technician A says an engine oil rated for API service SG is suitable for all modern gasoline automotive engines.

   Technician B says that an engine rated for API service CC is suitable for all modern automotive diesel engines.

   Who is correct?
   A. A only
   B. B only
   C. Both A and B
   D. Neither A nor B

2. Which of the following statements is correct?
   I. Oil can be lost through leakage.
   II. Oil can be consumed by being burned in the combustion chamber.
   A. I only
   B. II only
   C. Both I and II
   D. Neither I nor II

3. An oil pressure relief valve is used to
   A. prevent excess pressure as engine speed increases.
   B. prevent oil from lubricating nonmoving parts.
   C. allow a full-flow oil filtering system to bypass a clogged filter.
   D. allow an oil pressure warning system to operate correctly.

Questions 4 and 5 are not like those above. Each has the word EXCEPT. For each question, look for the choice that does *not* apply. Read each question carefully before you choose your answer.

4. The functions of engine oil include all of the following EXCEPT
   A. reducing friction and forming a seal.
   B. preventing excess oil pressure.
   C. absorbing shock loads.
   D. cooling and cleaning engine parts.

5. Methods of distributing oil within the engine include all of the following EXCEPT
   A. galleries.
   B. splash lubrication.
   C. oil spurt holes and drain holes.
   D. positive crankcase ventilation (PCV) system.

ESSAY QUESTIONS

1. What do the SAE and API ratings for engine oil indicate?
2. Describe the functions of engine oil.
3. What are the parts and functions of a lubrication system?

SUPPLEMENTAL ACTIVITIES

1. Make a survey of the API service grades of oil for sale at an auto parts store. Make a list of the names and prices of each oil available. Report to your class on price differences between different SAE viscosities and different API service-rated oils.

2. Talk to relatives and neighbors, and check their vehicle owner's manuals for oil viscosity and oil change interval recommendations. Report to your class on any differences you find. Note differences among 4-, 6-, and 8-cylinder engines. Also note differences among types of vehicles (foreign and domestic, passenger vehicles, pickup trucks, vans, and so on).

3. Talk to a parts counter person at an auto parts store. Find out how many different oil filters are made for domestic and foreign vehicles. Report to your class how many different types are kept in stock.

4. Examine a vehicle chosen by your instructor. Locate all external physical parts of the lubrication system and check the oil level.

16 Lubrication System Service

UNIT PREVIEW

Regular oil and filter changes are necessary to remove byproducts of normal engine operation and ensure proper engine lubrication.

In addition to scheduled maintenance, lubrication system service is necessary from time to time. This service includes stopping leaks from oil pan drain openings, replacing valve cover gaskets, and replacing oil pressure sender units.

Lubrication system service during engine overhaul includes checking and/or replacing the oil pump and replacing the pan gasket.

LEARNING OBJECTIVES

When you have completed your assignments and exercises in this unit, you should be able to:

- ❏ Identify the problems that make regular oil and filter changes necessary.
- ❏ Explain why the oil filter should be changed at every oil change.
- ❏ Perform an oil and filter change.
- ❏ Replace valve cover gaskets.
- ❏ Replace an oil-pressure sender unit.
- ❏ Identify and describe lubrication-system service performed during engine overhaul.

SAFETY PRECAUTIONS

Wear eye protection at all times in the shop area.

Oil and filter changes require that the vehicle be safely raised and supported. If you are using safety stands or ramps to support the vehicle, also block all wheels remaining on the ground. This will prevent the vehicle from moving forward or backward.

Use caution around hot surfaces such as exhaust manifolds, pipes, and mufflers. Hot oil that drains from the oil pan or oil filter can cause burns and skin irritation.

Always refer to the vehicle manufacturer's service manual for specific procedures.

16.1 OIL DETERIORATION

Although you may have heard that oil itself never "wears out," other problems occur that make frequent and regular oil changes necessary.

Oxidation

Oxidation is the chemical combination of oil with oxygen, which causes oil to deteriorate, or break down chemically. Oil oxidation is increased by the high temperatures common in modern engines. Oxidation causes the formation of carbon and varnish, which increase friction and can restrict (reduce the size of) oil passages and clog the oil pump pickup screen. Synthetic oils typically produce less varnish at high temperatures than petroleum oils.

Acid Formation

Air is drawn through the engine when it is running. The air circulates through oil return passages and above the oil in the crankcase.

Air contains moisture, or water vapor. Thus, water is formed as a byproduct of the combustion of a *hydrocarbon* fuel, such as gasoline or diesel fuel. The mineral element *sulfur* is present in most lubricating oil and gasoline. When water combines with sulfur, *sulfuric acid,* a powerfully corrosive substance, is formed. Other acids also can be formed when water vapor combines with other mineral elements in oil and motor fuels.

As the oil circulates, this acid is carried to all moving parts of the engine such as bearings, crankshaft and camshaft journals and lobes, cylinder walls, and so on. Acids can rapidly corrode and etch the surfaces of these parts.

When a vehicle is driven for long distances or in hot weather, heat evaporates most water vapor from the oil. This lessens acid formation. However, the oil in vehicles driven mainly for short distances or in extremely cold conditions may never get hot enough to evaporate all the moisture. Thus, moisture can remain in the oil to form more and stronger acids.

Another consideration, important in cold climates, is *dilution*. When a vehicle is driven in colder winter temperatures, the carburetor choke is used extensively during engine startup. Short-trip driving under these conditions may never allow the engine to fully warm up. Thus, the choke stays open most of the time. This can lead to oil dilution from the gasoline in excessively rich air-fuel mixtures. Under these conditions, the oil must be changed more often.

Sludge Formation

Enough moisture may be present to form drops of water in the crankcase. This water can be whipped by the spinning crankshaft into the oil. The combination of water and oil produces a frothy substance known as *water sludge.*

Particles of carbon and other contaminants can combine with acid to produce a thick, soft, tar-like sludge. Sludge accumulates inside the engine and upper valve train parts. In addition, sludge clogs oil screens and oil passages and cuts off oil circulation. To minimize the harmful effects of acid and sludge formation, oil and filter changes must be done regularly.

16.2 ROUTINE LUBRICATION SYSTEM SERVICE

Routine lubrication system service includes changing the oil and oil filter at regular intervals. How often the oil should be changed is determined by the type of driving done and the climate.

Oil Change Intervals

Vehicles driven 600 miles or less per month should have an oil and filter change every three months. Vehicles driven longer distances should have oil and filter changes as recommended by the vehicle manufacturer.

Smaller, 4-cylinder engines run hotter and at higher rpm, and contain less oil, than larger V-6 and V-8 engines. Frequent oil and filter changes for four-cylinder engines can help to prolong engine life and reliability.

Refer to the manufacturer's service manual for recommended oil and filter change intervals. However, remember that any maintenance performed more frequently than the recommended minimum intervals cannot cause harm. More frequent oil and oil filter changes will lengthen engine life.

Oil Filter

On most engines, draining the engine oil does not drain the oil filter. Up to a quart of dirty oil may remain in the filter. If an oil change is done but the oil filter is not changed, clean oil is pumped to the filter and mixed with dirty oil as soon as the engine starts.

Oil filters should be changed at every oil change to ensure maximum engine life and service.

16.3 OIL AND FILTER CHANGE

Before beginning this service, run the engine until the oil is fully warmed up. This allows the oil to drain freely and removes a maximum amount of contaminants. When the engine has warmed up, shut it off and follow these steps:

1. Raise the vehicle and support it safely.
2. Position an oil catch pan under the drainplug.
3. Use a wrench to loosen the drainplug a few turns.
4. When the plug is loose, press inward while unscrewing the plug.
5. When the plug comes out, pull it out of the way quickly to allow the hot oil to drain, as shown in **Figure 16-1.**
6. While the oil is draining, clean all dirt and contaminants from the drainplug, especially around the threads. Examine the plug and pan for stripped or damaged threads. If a gasket is used, check it for damage. Repair of stripped drainplug and/or oil pan threads is discussed in Topic 16.4. Allow the oil to drain for at least five minutes.
7. Install a new gasket on the drainplug, if one is used, and reinstall the plug to the manufacturer's specified torque.

NOTE: In most cases, the same catch pan cannot catch both pan oil and filter oil at the same time. Replace the drainplug before repositioning the catchpan under the filter.

8. Position the catch pan beneath the oil filter. Filter removal is illustrated in **Figure 16-2.** Be sure you have the correct replacement filter. Several types of oil filter wrenches are made, as shown in **Figure 16-3.** Confined working spaces on some vehicles require the use of one specific type of wrench rather than another.
9. Be sure that the old oil filter gasket comes off with the old filter.

CAUTION!!!

The rubber mounting gasket on the new filter must be in the same location relative to the mounting threads as the gasket on the old filter.

10. Install the new filter and tighten or torque it as recommended in the manufacturer's service manual or service bulletins. On most engines, tighten it three-quarters to one full turn after the gasket contacts the base.
11. Remove the oil filler cap. Add the correct amount of engine oil. Replace the oil filler cap.
12. Start the engine and let it idle slowly until the oil pressure light goes out or the gauge indicates proper pressure.
13. Check for leaks around the drainplug and oil filter mount and retighten if necessary.

LUBRICATION SYSTEM SERVICE 145

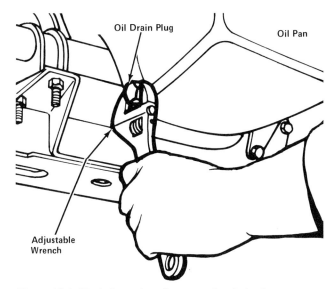

Figure 16-1. To drain engine oil, remove the drain plug.
CHRYSLER CORPORATION

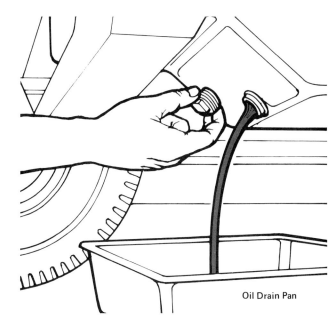

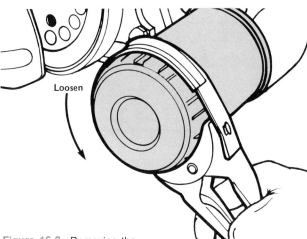

Figure 16-2. Removing the oil filter with an oil filter wrench.
CHRYSLER CORPORATION

Figure 16-3. Oil filter wrenches differ in design.

14. Stop the engine and check the engine oil level with the dipstick.

16.4 LEAKS FROM DRAIN PLUG

Damaged drain plug and/or oil pan drain hole threads that cause leaks can be repaired. Replacement metal drain plugs, shown in **Figure 16-4,** are available from auto parts stores. The replacement plugs have a tap-like end that cuts new threads into the oil pan drain hole.

To install a replacement metal plug, align the plug perpendicular to the drain hole surface. Start the tapered end of the plug in by hand. Continue tightening with a wrench until the plug has threaded itself in and compressed the gasket.

Expandable rubber drain plugs, shown in **Figure 16-5,** are also available to stop leaks from badly damaged oil pan drain hole threads. A special tool, supplied with the plug, is used to stretch the plug for installation. When the tool is removed, the plug contracts and tightly seals the drain plug hole.

16.5 REPLACING VALVE COVER GASKET

Before replacing a valve cover gasket, remove belts, hoses, wires, brackets, and engine accessories blocking access to the cover. Make drawings of how each part is attached to aid in reassembly. Then proceed as follows:

1. Remove the valve cover attaching bolts as shown in **Figure 16-6.** Start loosening the bolts near the center of the cover and proceed outward in a crisscross pattern, as shown in **Figure 16-7.**

2. If the cover is stuck, tap it gently from side to side with a soft-faced mallet until it loosens. Remove the cover. Use scrapers to remove any

Figure 16-4. Replacement drain plugs are often self-tapping.

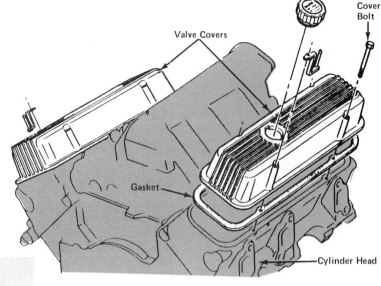

Figure 16-6. Valve cover assembly. BUICK MOTOR DIVISION—GMC

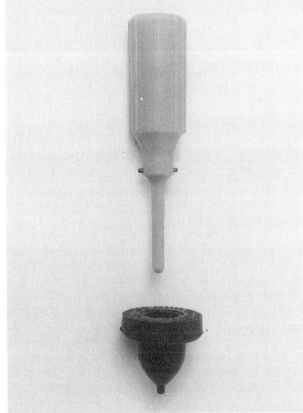

Figure 16-5. Expandable rubber drain plug.

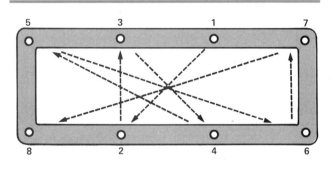

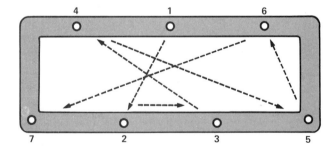

Figure 16-7. Crisscross bolt tightening and loosening patterns for valve covers.

remains of the gasket on the cover. Do not bend the edge flanges of the cover during cleaning.

3. Check the cover for bent flanges by placing it on a flat surface. A block of wood and a hammer can be used to flatten bent flanges on pressed sheet-metal covers. Bent, cracked, or badly damaged cast-aluminum covers cannot be straightened and must be replaced.

4. Check the bolt holes for *dimples.* A dimple is a rounded depression in a thin piece of metal caused by overtightened attachment bolts. Dimples can be flattened by using the ball peen end of a hammer. Support the hollow side of the dimple against a piece of wood held in a vise. Place the ball peen of one hammer against the projecting side of the dimple. Hit the face of the first hammer with a soft-faced hammer. Do not use excess force. Hit the dimple only hard enough to make it level with the edge flange. Recheck the flange against a flat surface.

LUBRICATION SYSTEM SERVICE

SAFETY CAUTION
Wear eye protection at all times when hammering.

5. Use shop towels to plug oil drain holes in the cylinder head. Wrap shop towels around the upper valve train of the cylinder head. Use scrapers to remove any traces of gasket material from the valve cover sealing surfaces. Brush or blow off any traces of crumbled gasket.
6. Remove all towels. Refer to the vehicle manufacturer's service manual for recommended sealants. Apply *room temperature vulcanizing (RTV)* rubber sealant as recommended. **Figure 16-8** shows a sample pattern for RTV sealant on a specific valve cover.

CAUTION!!!
Gasket residue can contaminate the lubrication system. Make sure all traces of gasket material have been removed before attempting to replace the valve cover and gasket. To prevent contamination of the oxygen sensor used in emissions control systems, use only the low-volatility RTV sealants recommended by the vehicle manufacturer.

7. If RTV is not used, press the new gasket into place around the cover. Replace the cover on the cylinder head. Start the bolts in by hand. Tighten the bolts gently to the manufacturer's recommended torque in a crisscross pattern, from the center outward **[refer to Figure 16-7]**.

16.6 REPLACE OIL PRESSURE SENDER

Oil pressure sender switches may develop leaks, or be electrically defective, and will allow lubricating oil to be pumped out of the engine rapidly. To replace a sender, remove the wire connector from the sender, as shown in **Figure 16-9**.

Use a special eight-point oil sender socket to loosen lamp-type warning system senders. Unscrew and remove the sender.

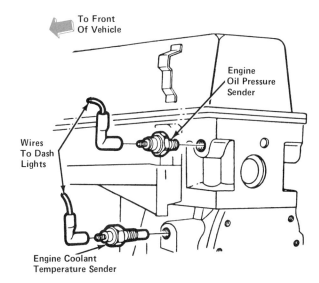

Figure 16-9. Oil pressure sender replacement.
FORD MOTOR COMPANY

Thread the new sender in by hand. Tighten with the special socket to the manufacturer's recommended torque. Replace the wire connector.

16.7 LUBRICATION SYSTEM SERVICE WITH OIL PAN REMOVED

When an engine is disassembled for repairs, the oil pump should be checked for wear. Always install a new oil pump when rebuilding an engine. All gaskets, including the oil pan gasket, must be replaced with new gaskets during reassembly procedures.

Oil Pump

If an engine is being disassembled, the oil pump gears or rotors can be checked for wear and clearance, as shown in **Figure 16-10**. Measure the clearance between the housing and the gear teeth and between the cover plate and the gear teeth. Refer to the

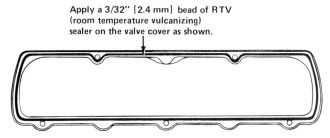

Figure 16-8. Applying sealant on a valve cover.
BUICK MOTOR DIVISION—GMC

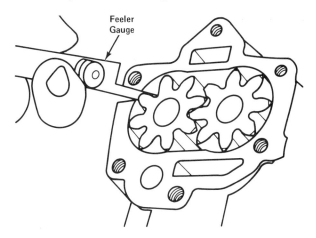

Figure 16-10. Measuring oil pump gear clearances.
BUICK MOTOR DIVISION—GMC

manufacturer's service manual for the proper clearances. If there is any doubt about the condition of the pump, install a new oil pump.

Oil Pan Gasket

Procedures for oil pan gasket replacement are similar to those for valve cover gasket replacement. All traces of the old gasket must be removed from the oil pan and cylinder block sealing surfaces. Bent flanges and dimples on pressed sheet-metal pans should be straightened before reinstallation. Bent, cracked, or badly damaged cast-aluminum oil pans cannot be straightened and must be replaced.

SAFETY CAUTION

Most engines must be raised and supported to allow removal of the oil pan. Be sure this is done carefully to prevent damage to external engine components and connections and personal injury.

If recommended by the manufacturer, sealant materials must be applied to the pan and/or cylinder block surfaces, as shown in **Figure 16-11**. Oil pan gaskets may be made in a single piece or in several pieces, as shown in **Figure 16-12**.

Reinstall the bolts and tighten them in the manufacturer's specified sequence to the proper torque.

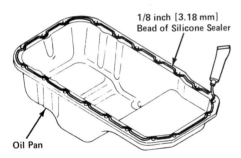

Figure 16-11. Applying sealant on an oil pan.
FORD MOTOR COMPANY

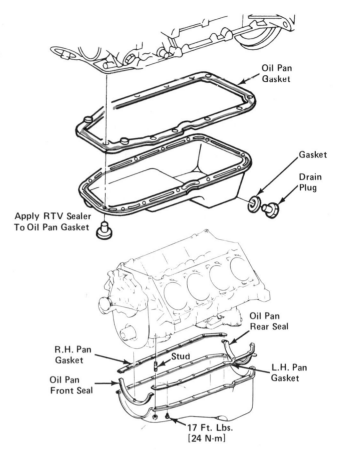

Figure 16-12. Oil pan gaskets. BUICK MOTOR DIVISION—GMC

UNIT HIGHLIGHTS

- Problems caused by oxidation, acid formation, oil dilution, and sludge require that oil and filter changes be done regularly.
- Vehicles driven mainly for short distances or in cold weather require more frequent oil and filter changes.
- Oil filters should be changed every time engine oil is changed.
- Leaks from oil drain plug openings can be repaired.
- When valve cover gaskets are replaced, no gasket residue must be allowed to enter the engine.
- Bent and dimpled flanges on pressed sheet metal valve covers and oil pans can be repaired.
- Damaged cast-aluminum covers and oil pans must be replaced.

- The oil pressure sender may cause an oil leak and require replacement.
- During engine overhaul, oil pumps must be checked or replaced.
- Gaskets should never be reused.

TERMS

oxidation
hydrocarbon
water sludge
room temperature vulcanizing (RTV)
sulfur
sulfuric acid
dimple
dilution

REVIEW QUESTIONS

DIRECTIONS: The following questions are similar to those used on automotive technician certification tests. On a separate sheet of paper, write the letter of the correct choice.

1. Technician A says vehicles driven longer distances never need oil or filter changes.
   Technician B says vehicles driven short distances and in cold weather need frequent oil and filter changes.
   Who is correct?
   A. A only
   B. B only
   C. Both A and B
   D. Neither A nor B

2. How should the oil filter be tightened?
   A. As tight as possible by hand.
   B. With an oil filter wrench.
   C. With a torque wrench.
   D. To the manufacturer's service manual or service bulletin specifications.

3. Which of the following statements is correct?
   I. Bent flanges on pressed sheet metal covers and oil pans can be straightened.
   II. Bent flanges on cast-aluminum covers and oil pans can be straightened.
   A. I only
   B. II only
   C. Both I and II
   D. Neither I nor II

Questions 4 and 5 are not like those above. Each has the word EXCEPT. For each question, look for the choice that does *not* apply. Read each question carefully before you choose your answer.

4. All of the following problems make oil changes necessary EXCEPT
   A. oil "wearing out" with use and age.
   B. oil oxidizing, especially at high temperatures.
   C. acids forming in the oil when moisture and chemical elements combine.
   D. sludge forming in the oil from moisture and contaminant particles.

5. All of the following conditions could cause faulty indications in a lamp-type oil-pressure warning system EXCEPT
   A. a loose or disconnected electrical connector.
   B. a faulty sender.
   C. a burned-out bulb in the dash.
   D. an oil pickup screen clogged with sludge.

ESSAY QUESTIONS

1. What conditions make frequent and regular oil and oil filter changes necessary?
2. Describe lubrication system maintenance and services.
3. Why is it important to use only manufacturer-recommended RTV sealants?

SUPPLEMENTAL ACTIVITIES

1. Perform an oil and filter change on a vehicle.
2. Remove and replace a valve cover gasket on a vehicle. If the cover is made of pressed sheet metal, correct any warping and/or dimples. Report to the class on the deposits found inside the cover and on upper valve train parts. Describe the amount and nature of the deposits.
3. Remove and replace an oil pressure sender unit.
4. Measure clearances on a shop oil pump. Refer to the manufacturer's service manual and determine if the pump requires replacement.

17 The Cooling System

UNIT PREVIEW

The heat generated during combustion is great enough to melt or damage engine parts. The cooling system must be able to remove approximately one third of the heat developed during combustion to prevent engine damage.

The parts of the cooling system work together to circulate air and liquid that transfer heat away from the engine. The pressurized cooling system of a modern vehicle allows the engine to operate at a stable temperature and provide heat for passenger comfort.

LEARNING OBJECTIVES

When you have completed your assignments and exercises in this unit, you should be able to:

❏ Describe how heat is transferred away from the combustion chamber.
❏ Identify and describe the parts of a vehicle's cooling system.
❏ Explain why an antifreeze/water coolant mixture must be used in the cooling system.
❏ Describe how the radiator pressure cap helps to prevent boiling.

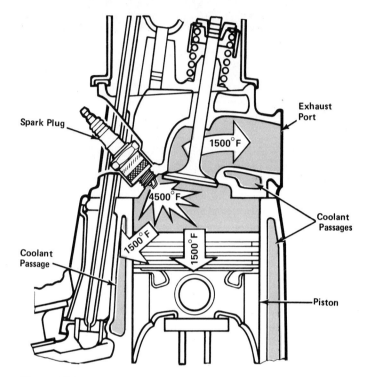

Figure 17-1. Maximum combustion chamber temperatures.
CHRYSLER CORPORATION

17.1 HEAT TRANSFER

Combustion chamber temperatures can reach 4,500 degrees F (2,468 degrees C) **[Figure 17-1]**. The average combustion chamber temperature is approximately 2,000 degrees F (1,080 degrees C). This is more than enough to melt aluminum pistons, warp the cylinder block and head surfaces, and ruin engine oil. An engine cooling system must transfer enough heat away from the engine to prevent damage. Heat energy in the automobile is transferred in three ways:

* Conduction
* Convection
* Radiation.

Conduction

When something hot contacts something cooler, heat is transferred to the cooler object. This process is known as *conduction.* For example, when hot gases in the combustion chamber touch the cooler cylinder walls, the walls absorb the heat from the gases and become hot. This heat then is conducted to the coolant which is in direct contact with the cylinder walls.

Convection

The saying, "heat rises," describes *convection.* Convection operates wherever fluids (gases or liquids) and gravity are present. Heated atoms or molecules of fluid are less dense, and thus lighter, than cooler atoms or molecules. Therefore, the heated portion of the fluid rises upward, and the cooler portion sinks downward.

For example, in a room the temperature at the ceiling is warmer than the temperature near the floor.

Radiation

The warmth of sunlight felt on your skin is a result of heat *radiation* from the sun. Heat energy can travel from one location to another through empty space.

The Cooling System

Within an engine cooling system, all three forms of heat transfer take place.

17.2 PARTS OF A LIQUID COOLING SYSTEM

A basic cooling system, shown in **Figure 17-2,** consists of several parts that function together to assure a stable engine operating temperature. These parts include:

- Coolant
- Radiator
- Water pump
- Fan
- Water Jacket
- Thermostat
- Temperature warning system
- Radiator pressure cap
- Coolant recovery system
- Heater system
- Automatic transmission oil cooler.

17.3 COOLANT

Coolant is a solution of antifreeze and water. Antifreeze is a liquid compound called *ethylene glycol* which is mixed with anti-corrosion chemicals. Water alone will promote rust, corrosion, and *electrolysis* in the cooling system.

Electrolysis is a chemical and electrical decomposition process that occurs when two dissimilar metals are joined in the presence of moisture. Minerals and metals in the cooling system can cause water to become slightly acid. An electrical current begins to flow that can actually cause metals in the cooling system and engine, such as brass, copper, and aluminum, to be corroded away. A fresh, proper mixture of antifreeze and water will prevent acid buildup and corrosion. Although ethylene glycol does not evaporate easily, the anti-corrosion chemicals in antifreeze are used up in approximately one year. Antifreeze also contains a lubricant to help lubricate the water pump seal.

Within the engine, a hollow space filled with coolant, called a *water jacket,* surrounds the cylinders and combustion chambers. Heat is conducted into the coolant by the heated metal.

Another function of antifreeze is to prevent freezing. When water freezes, it expands. Such expansion can exert enough pressure to crack the water jackets in engine blocks and cylinder heads, and to rupture radiators.

If the coolant boils, liquid no longer is in contact with the heated cylinder walls and head passages.

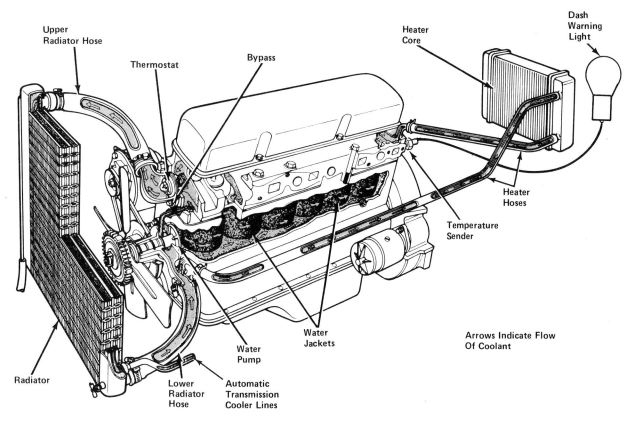

Figure 17-2. Parts of a cooling system. CHRYSLER CORPORATION

Heat conduction from the metal of the cylinder walls into the coolant stops, and excess heat builds up inside the cylinders. This can cause the pistons to scuff and seize to the cylinder walls and warp block and cylinder head surfaces. A half-and-half mixture of antifreeze and water has a boiling temperature of 230 degrees F or approximately 18 degrees (8.5 degrees C) higher than plain water. This provides added protection from boiling. The radiator pressure cap will also affect the boiling point as explained in Topic 17.10.

In summary, a high-quality antifreeze is important for several reasons: It prevents cooling system freeze-up, rust, corrosion, and acid buildup. It raises the boiling point of the coolant. Finally, it provides blended lubricants for water pump seal lubrication.

17.4 RADIATOR

The heated coolant is forced under pressure through an outlet in the cylinder head and into the upper radiator hose. The coolant then flows into the *radiator*. A radiator is made of metals, such as brass, copper, or aluminum, that conduct heat well. Hot coolant passes through and heats the hollow tubes [**Figure 17-3**]. Small, thin cooling fins made of copper or aluminum are in contact with the tubes to increase the surface area exposed to air.

The radiator tubes may be arranged in a vertical (downflow) or horizontal (crossflow) pattern.

Crossflow radiators allow coolant to travel a longer distance, thus giving up more heat, than downflow radiators of the same height. In addition, convection that aids cooling can occur more readily in a crossflow radiator. The flat, horizontal shape of a crossflow radiator also can fit more easily into the front of vehicles with steeply sloped hoods [**Figure 17-4**].

Heat is conducted from the hot tubes and fins into air passing through the radiator core. The air carries the heat away from the vehicle. The cooled coolant leaving the radiator outlet travels through the lower radiator hose to the water pump, where it is again pumped through the cooling system.

Most radiators have a drain located near the bottom, as shown in **Figure 17-5,** for flushing purposes. On radiators without drains, the lower radiator hose must be removed.

Some of the radiators on newer automobiles have end tanks that are made from a composite material which is similar to plastic.

17.5 WATER PUMP

A *water pump*, or coolant pump, shown in **Figure 17-6**, is operated by a belt. The drive belt may be a V-type belt, a serpentine belt, or on OHC engines, it can be the belt that operates the camshaft [**Figure 17-7**].

Small, fan-like blades on the *impeller*, or rotor, draw liquid into the center of the pump. The blades then force the liquid outward through centrifugal force as the shaft spins. The faster the engine turns,

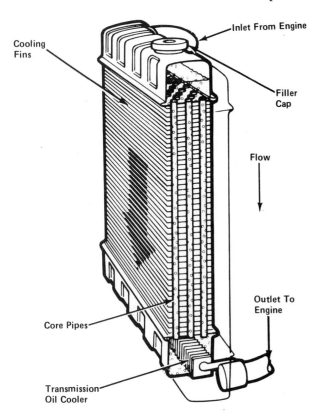

Figure 17-3. Parts of a downflow radiator.
CHRYSLER CORPORATION

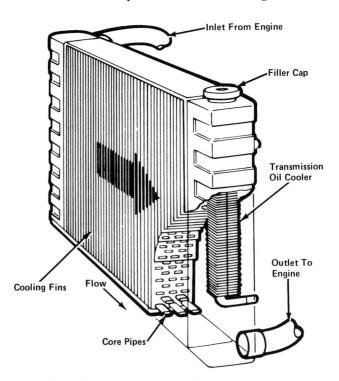

Figure 17-4. Parts of a crossflow radiator.
CHRYSLER CORPORATION

THE COOLING SYSTEM   153

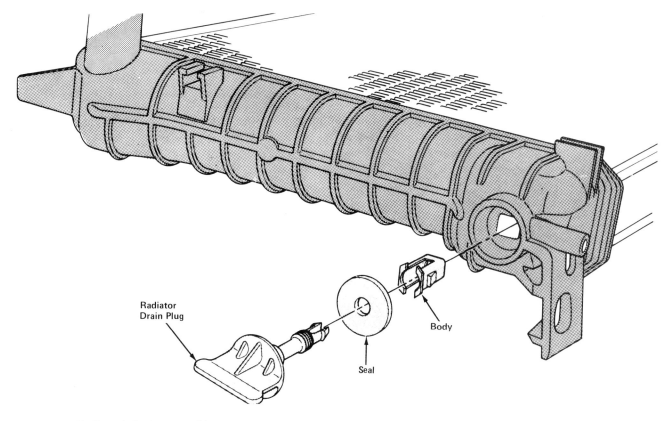

Figure 17-5. Radiator drain plug assembly. FORD MOTOR COMPANY

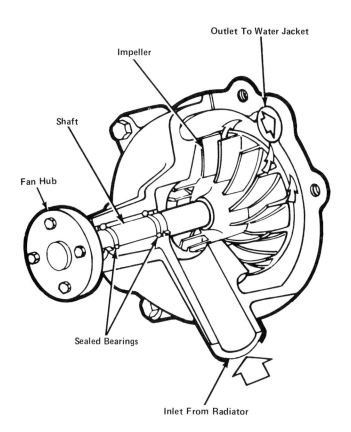

Figure 17-6. Parts of a water pump. CHRYSLER CORPORATION

the faster the pump circulates coolant. Seals prevent the coolant from contacting the pump bearings.

17.6 FAN

Cooling fans can be driven by electricity or engine power.

Electric Fan

Engines mounted *transversely*, or crossways, in front-wheel drive vehicles have electrically driven fans **[Figure 17-8]**.

An electrical switch turns the fan on and off in response to coolant temperature. The coolant switch can be located in the radiator or hoses, or screwed into the coolant passage in the cylinder head **[Figure 17-9]**.

Belt-Driven Fan

Engines mounted *longitudinally*, or front-to-back, generally use a fan mounted on the water pump shaft. The fan helps to draw air through the radiator at low road speeds or when the vehicle is stopped and idling. The drive belt turns both the water pump and fan, as shown in **Figure 17-10**. Two types of fan blades can be used: rigid or flex-blade **[Figure 17-11]**.

PART III: AUTOMOTIVE ENGINE SYSTEMS

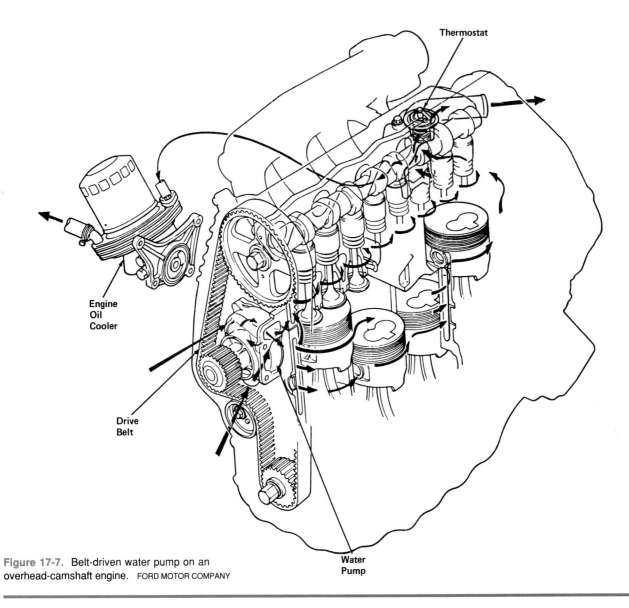

Figure 17-7. Belt-driven water pump on an overhead-camshaft engine. FORD MOTOR COMPANY

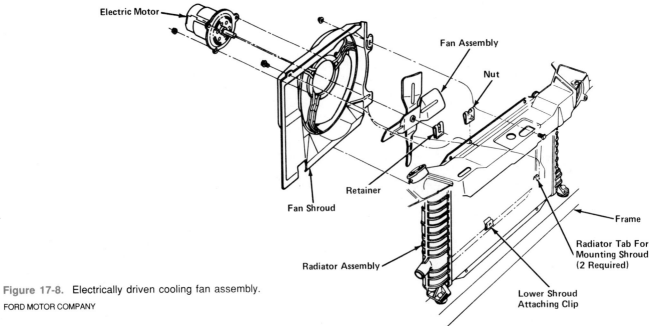

Figure 17-8. Electrically driven cooling fan assembly. FORD MOTOR COMPANY

The Cooling System

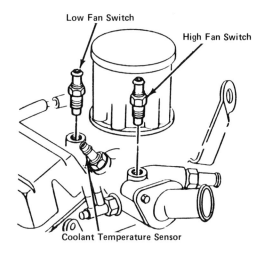

Figure 17-9. Engine fan and temperature switches.
BUICK MOTOR DIVISION—GMC

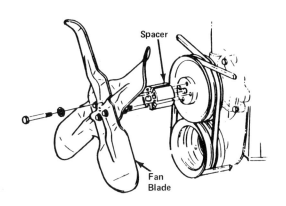

Figure 17-10. A drive belt turns both the water pump and cooling fan. BUICK MOTOR DIVISION—GMC

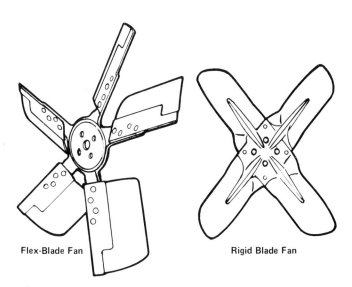

Figure 17-11. Two types of blades are used on cooling fans.
CHRYSLER CORPORATION

A rigid fan tends to make more noise and use more energy than a flexible fan. Flexible fan blades straighten at high engine speeds. This straightening action moves less air, uses less energy, and makes less noise. At faster road speeds, sufficient airflow is provided by the motion of the vehicle through the air.

Viscous Drive Fan Clutch

If the fan turns at full speed at all times, energy is wasted and fuel economy suffers. A *viscous drive fan clutch,* shown in **Figure 17-12,** is a mechanism that changes fan speed in response to engine temperature. When an engine is cold, a special silicone oil transfers less torque, which allows clutch slippage and turns the fan slower. As the engine heats up, the heated oil transfers more torque, allows less clutch slippage and turns the fan faster.

Such a fan clutch can be operated by a *thermostatic coil.* A thermostatic coil allows more or less silicone oil into the working chamber in response to engine temperatures **[Figure 17-13]**.

A thermostatic coil also is known as a *bimetallic spring.* This coil or spring is made from two different metals attached back to back and wound into a coil. Different metals expand and contract at different rates, causing the coil to wind or unwind in response to engine temperatures.

17.7 WATER JACKET

Hollow passages in the engine block and cylinder head surround the areas closest to the cylinders and combustion chambers. Coolant flow through the block and head can be a *series, parallel,* or *series-parallel* flow **[Figure 17-14]**. Included in the water jacket exterior block surface are soft plugs and a block drain plug **[Figure 17-15]**.

17.8 THERMOSTAT

The cooling system *thermostat* is a heat-operated valve that opens and closes in response to the heat of coolant to maintain a stable temperature. On nearly

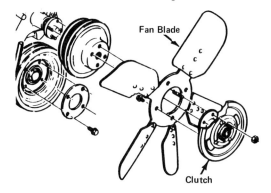

Figure 17-12. Viscous drive fan clutch assembly.
BUICK MOTOR DIVISION—GMC

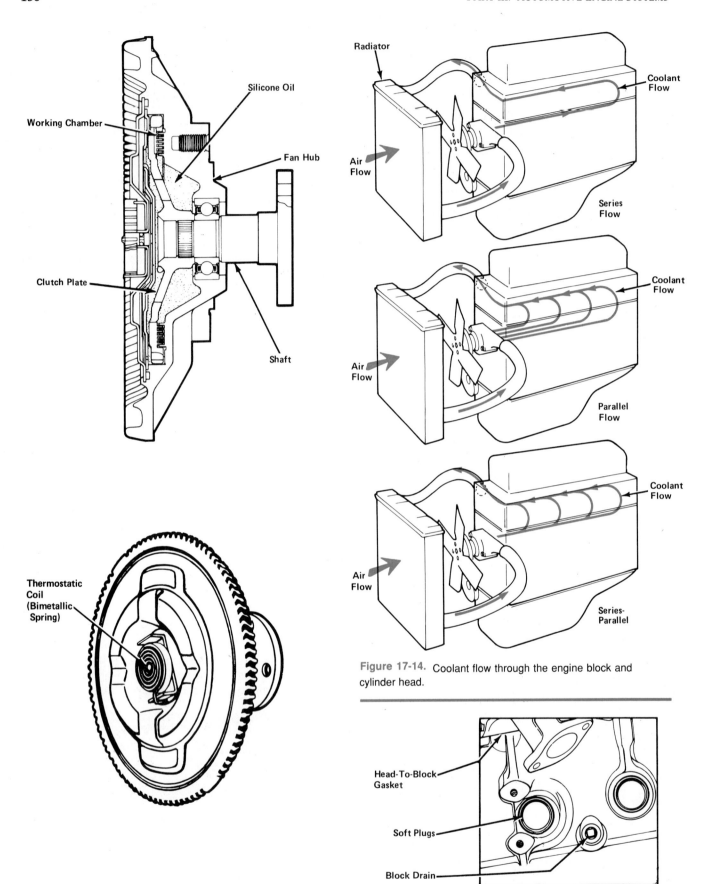

Figure 17-13. Parts of a fluid coupling fan drive. CHRYSLER CORPORATION

Figure 17-14. Coolant flow through the engine block and cylinder head.

Figure 17-15. Soft plugs and a drain plug are located in the water jacket. CHRYSLER CORPORATION

THE COOLING SYSTEM

all vehicles, the thermostat is located under the water outlet from the cylinder head [Figure 17-16].

A spring holds the valve closed when the coolant is cold. As the coolant heats, a special plastic, wax-like substance within the temperature sensing bulb of the thermostat expands. This expansion forces the valve to open [Figure 17-17].

When the coolant is cold, the thermostat remains shut. Coolant circulates within the engine block and cylinder head and through a *coolant bypass* to the water pump inlet [Figure 17-18].

As the coolant heats, the thermostat begins to open and coolant flows through the upper radiator hose to the radiator, where it is cooled. In operation, the thermostat remains open in proportion to the amount of heat that must be transferred to the radiator. If the coolant flowing to the radiator is hotter than the thermostat temperature rating, the thermostat will remain fully open.

Vehicles built since the late 1970s are designed to have a 190-195 degrees F (88-91 degrees C) thermostat to heat the engine quickly to a high temperature. This increases engine efficiency and reduces exhaust emissions. If the coolant does not reach this temperature, computer controls that affect engine operation and emissions may not operate properly (see Units 23 and 41).

17.9 TEMPERATURE WARNING SYSTEM

To alert the driver of an overheating condition, a temperature gauge and/or a lamp warning system is provided by the manufacturer. A temperature sensor is screwed into a threaded hole in the water jacket [Figure 17-19].

Two types of senders are used. One lights a warning lamp. The other type operates a temperature gauge [Figure 17-20]. Heated coolant causes the first type of sender to complete an electrical circuit to light a warning lamp. The other type of sender changes resistance to cause a gauge needle to move over a scale.

17.10 RADIATOR PRESSURE CAP

The boiling temperature of pure water is 212 degrees F (100 degrees C) at sea level. At sea level, which is the same all over the world, the weight of the atmosphere exerts a pressure of 14.7 psi (101.35 kPa).

At higher elevations, less atmospheric pressure is exerted and the boiling temperature of water decreases. At elevations below sea level (such as Death Valley in California), added atmospheric pressure causes the boiling temperature to rise.

A spring above the relief valve of a radiator pressure cap exerts internal pressure sealing the cooling system to a pressure of approximately 15 psi [Figure 17-21]. Each additional pound (0.45 kg) of pressure above atmospheric pressure increases the

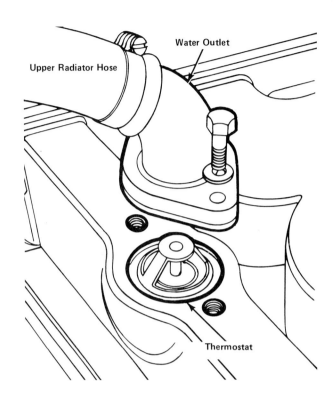

Figure 17-16. Thermostat location. CHRYSLER CORPORATION

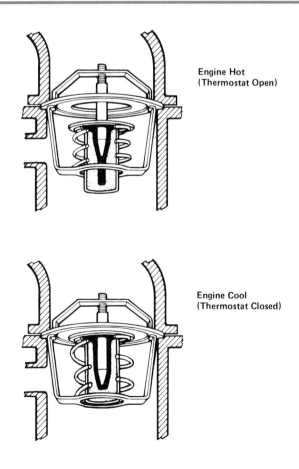

Figure 17-17. Thermostat operation. CHRYSLER CORPORATION

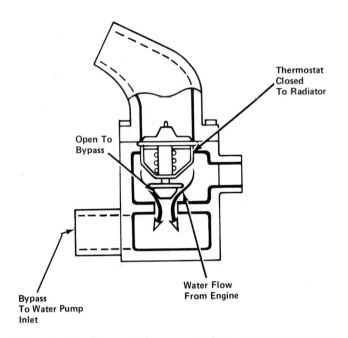

Figure 17-18. Thermostat bypass operation. CHRYSLER CORPORATION

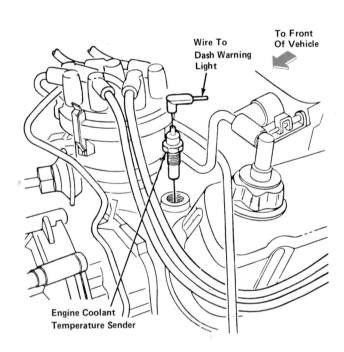

Figure 17-19. Coolant temperature sender. FORD MOTOR COMPANY

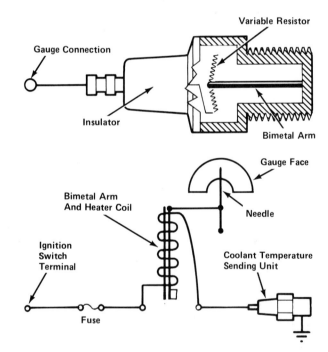

Figure 17-20. Temperature warning light and gauge circuits.

boiling temperature of water approximately 3 degrees F (1.4 degrees C). Thus, a 15-lb. (6.8 kg) pressure cap raises the boiling point of water to 257 degrees F (111 degrees C). This increase occurs only in a sealed and pressurized cooling system. A proper coolant mixture will further increase this by 9 degrees F (4.25 degrees C), to 266 degrees F (134 degrees C). Some 1982 and later vehicles have 18-lb radiator pressure caps.

If the coolant overheats, pressure builds against the relief valve spring and eventually lifts the pressure seal. Hot, expanding coolant flows out of the radiator neck through an overflow tube into a coolant recovery system.

As the engine cools, the coolant contracts, forming a low-pressure area within the radiator and cooling system. A vacuum vent valve allows coolant to be drawn in from a reservoir to fill the space. This

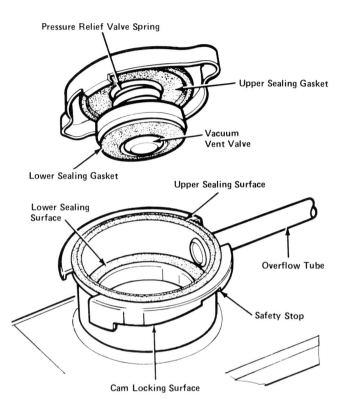

Figure 17-21. Parts of a radiator pressure cap assembly. CHRYSLER CORPORATION

prevents the thin walls of the radiator from being crushed by atmospheric pressure [Figure 17-22].

17.11 COOLANT RECOVERY SYSTEM

A *coolant recovery system* is illustrated in **Figure 17-23.** The system consists of a special pressure cap with an upper sealing gasket, an overflow tube, and a recovery container.

Coolant recovery systems are standard equipment in every new passenger vehicle. On older vehicles without such systems, it is normal to see an air space below the radiator neck. This allows for coolant expansion. If coolant expands above this level, it is forced to flow through the overflow tube and is lost.

In a coolant recovery system, as the coolant is heated and expands, it flows into the recovery container. When the engine cools, coolant flows from the recovery container back through the overflow hose and into the radiator, keeping the system full.

In this manner, coolant loss is eliminated and excess air is kept out of the system. Air contains oxygen, which combines with metals in the cooling system to form rust and corrosion.

17.12 HEATER SYSTEM

A hot-liquid heater system, shown in **Figure 17-24,** is part of the cooling system. When a heat control is

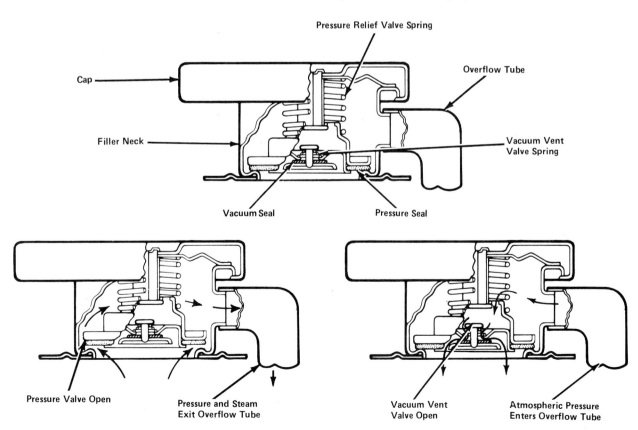

Figure 17-22. Radiator pressure cap operation. CHRYSLER CORPORATION

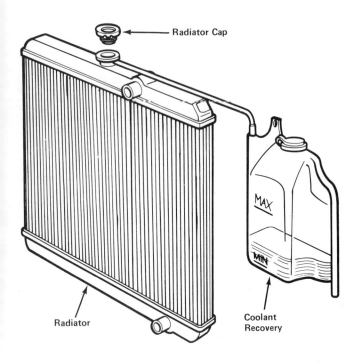

moved, a valve opens and heated coolant flows through *heater hoses* to a *heater core*, or small radiator. Air is directed or blown over the hot heater core inside a plenum, or air box, under the dash or in the engine compartment. The heated air is blown into the passenger compartment. Movable doors can be controlled to blend cool air with heated air for more or less heat and to direct air to defrost the windshield **[Figure 17-25]**.

17.13 OIL COOLER

Radiators for vehicles with automatic transmissions have a sealed *heat exchanger*, or form of radiator, located in the coolant outlet tank **[Figure 17-26]**. Metal or rubber hoses carry hot automatic transmission fluid to the heat exchanger. The coolant passing over the sealed heat exchanger and cools the fluid, which then is returned to the transmission.

Some vehicles with diesel engines use a similar arrangement to cool the engine oil **[Figure 17-27]**.

Figure 17-23. Coolant recovery system. CHRYSLER CORPORATION

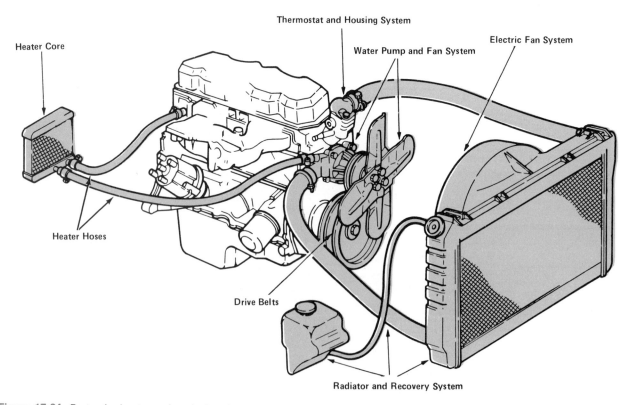

Figure 17-24. Parts of a heater and coolant system.
BUICK MOTOR DIVISION—GMC

THE COOLING SYSTEM 161

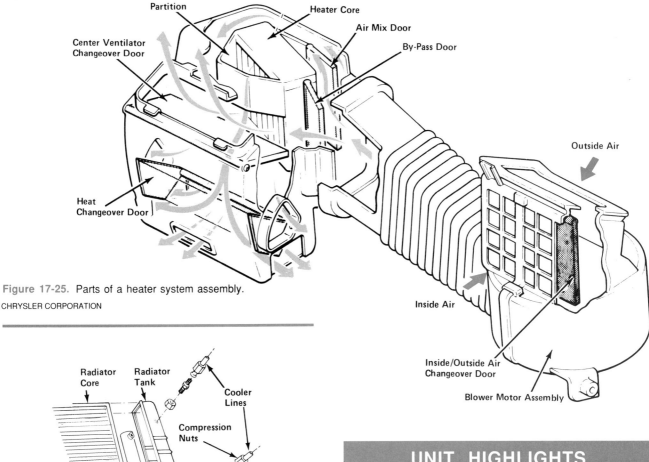

Figure 17-25. Parts of a heater system assembly.
CHRYSLER CORPORATION

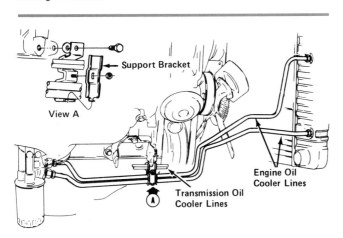

Figure 17-26. Parts of an automatic transmission oil cooler at the engine radiator.

Figure 17-27. Parts of a diesel engine oil cooler.
BUICK MOTOR DIVISION—GMC

UNIT HIGHLIGHTS

- Heat transfer within a vehicle's cooling system takes place through conduction, convection, and radiation.
- Coolant is a mixture of antifreeze and water that helps to prevent freezing, boiling, rust, corrosion, and electrolysis. The cooling system should be flushed and the coolant changed yearly.
- Parts of a cooling system include: coolant, radiator, water pump, drive belt, fan, water jacket, thermostat, and temperature warning system. Other parts are: radiator pressure cap, coolant recovery system, connecting hoses, heater core, and oil cooler.
- The radiator pressure cap helps prevent the coolant from boiling. If the coolant boils, heat conduction from the interior of the engine cannot occur.

TERMS

conduction	convection
heater hose	coolant
electrolysis	water jacket
heat exchanger	radiator
water pump	impeller

transversely	longitudinally
viscous drive fan clutch	thermostatic coil
series	parallel
series-parallel	thermostat
coolant bypass	heater core
ethylene glycol	bimetallic spring
radiation	coolant recovery system

REVIEW QUESTIONS

DIRECTIONS: The following questions are similar to those used on automotive technician certification tests. On a separate sheet of paper, write the letter of the correct choice.

1. Which methods of heat transfer occur within an engine?
   A. Conduction
   B. Convection
   C. Radiation
   D. All of the above

2. Technician A says that antifreeze is used in coolant to prevent freezing and boiling.
   Technician B says that antifreeze is used in coolant to prevent rust, corrosion, and electrolysis.
   Who is correct?
   A. A only
   B. B only
   C. Both A and B
   D. Neither A nor B

3. To warn a driver of an overheating condition, a temperature sensor is placed in the
   A. water jacket.
   B. thermostat.
   C. coolant recovery system.
   D. coolant overflow assembly.

4. Which of the following statements is correct?
   I. The radiator pressure cap helps to prevent the coolant from boiling.
   II. A leak in any part of the cooling system will release pressure and can cause the coolant to boil.
   A. I only
   B. II only
   C. Both I and II
   D. Neither I nor II

Question 5 is not like the first four questions. It has the word EXCEPT. For question 5, look for the choice that does *not* apply. Read the entire question carefully before you choose your answer.

5. All of the following could cause insufficient heat from a heater EXCEPT
   A. a clogged heater core or heater hose.
   B. a jammed heater control valve.
   C. a thermostat stuck in the closed position.
   D. a thermostat stuck in the open position.

ESSAY QUESTIONS

1. What are the functions of the parts of a liquid cooling system?
2. What effect would removing the thermostat have on a modern engine?
3. How does a coolant recovery system work?

SUPPLEMENTAL ACTIVITIES

1. On a vehicle chosen by your instructor, locate all externally visible parts of the cooling system. Locate the water pump, fan, radiator and heater hoses, automatic transmission cooler lines, radiator drain, soft plugs, and block drain.

2. Examine vehicles owned by your relatives and friends. Report to your class what the pressure cap markings are and how many have coolant recovery systems.

3. Visit several auto parts stores and make a list of the ingredients listed on bottles of different brands of antifreeze. Report to your class what differences you find.

4. Ask your instructor for permission to use a voltmeter for the following experiment. Cut an orange, lemon, or other citrus fruit in half. Push a penny in one part and a dime in another part of the fruit. Set the voltmeter to the lowest DC voltage scale and connect the leads. Reverse the leads if the meter reads backwards. How much voltage do you observe? What does this experiment have to do with cooling system maintenance?

5. Discussion Topic: Before the 1950s, radiator pressure caps were rated from three to seven pounds, on average. Why do you think the pressures were so low?

18 Cooling System Service

UNIT PREVIEW

Routine cooling system maintenance consists of checking, repairing, and/or replacing cooling system parts. These parts include drive belts, hoses, gaskets, soft plugs, thermostats, and water pumps.

The coolant should be checked regularly and be flushed and replaced approximately once a year to prevent rust, corrosion, and electrolysis.

In addition to cooling system faults, overheating can be caused by factors outside of the cooling system itself.

LEARNING OBJECTIVES

When you have completed your assignments and exercises in this unit, you should be able to:

❏ Locate, identify, and repair common cooling system leaks.
❏ Check coolant mixture strength.
❏ Detect cooling system leaks, including head gasket leaks.
❏ Replace drive belts, a thermostat, and a water pump.
❏ Identify and describe possible causes of overheating.

SAFETY PRECAUTIONS

Wear eye protection when working around heated cooling systems. Hot coolant can cause burns, blisters, and eye damage.

Before attempting to open a radiator cap, touch the metal of the radiator carefully to determine how hot the coolant is. Opening the radiator cap releases the pressure that prevents hot coolant from boiling. Boiling coolant can gush and spurt out of the radiator neck and cause injury. If the radiator is uncomfortably hot to the touch, do not attempt to open the radiator cap until the system cools.

Another way to check for pressure is to squeeze the upper radiator hose. If it feels hot and tight, the system is pressurized.

Remove jewelry from hands, wrists, and around the neck. Roll up long sleeves tightly or change into a short-sleeved shirt or blouse when working around the fan and/or belts. Remember that the fan cannot be seen when it spins.

CAUTION!!!

Electrically driven fans can start at any time, including after the engine is shut off. Disconnect the battery negative terminal or electrical fan power connections before attempting removal procedures.

The main ingredient in antifreeze, ethylene glycol, is poisonous if swallowed.

18.1 COOLING SYSTEM PREVENTIVE MAINTENANCE

Routine preventive maintenance for the cooling system consists of the following procedures:

• Check and/or add coolant to proper levels in radiator and coolant recovery container
• Check and/or adjust/replace drive belts
• Check and/or repair leaks
• Check coolant mixture strength
• Flush cooling system
• Replace thermostat
• Replace water pump.

18.2 CHECK/ADD COOLANT

First, look in the coolant recovery container. The liquid level in the container should be between the MINIMUM mark and the MAXIMUM mark (approximately half full).

Before you attempt to open the pressure cap, you must determine if it is safe to do so. An overheated cooling system can spray hot coolant and steam.

SAFETY CAUTION

Severe burns and scalding can result if overheated coolant sprays from the radiator neck. Wait until the upper tank of the radiator feels cool to the touch, or the top radiator hose feels unpressurized, before you open the pressure cap.

You must check the coolant level in the recovery bottle and in the radiator neck. The coolant level in the radiator should be at the top of the neck. A faulty upper gasket on a radiator cap can allow coolant to be forced into the recovery container but not be drawn back to the radiator neck.

Fill the recovery container and radiator neck with coolant of the required concentration. In most cases, a 50 percent concentration is suitable.

18.3 CHECK AND/OR REPLACE DRIVE BELTS

Most drive belts, often called *V-belts,* are so named because their cross-sectional shape resembles the letter V. A drive belt transfers force through friction between the *sides* of the belt and the pulley flanges **[Figure 18-1]**. Late-model vehicles may use a single serpentine belt with multiple drive ribs.

Belt Condition

To transfer adequate force, drive belts must be in good condition and properly tightened. To check belt condition, twist the belt and look at the inside surface for defects as illustrated in **Figure 18-2**.

Belt Tension

A V-belt does not require extreme tightening to transmit turning force. However, an excessively loose belt will slip, cause a squealing noise, and eventually become *glazed*, or smooth. The friction surfaces of a glazed belt are too slick to transfer force properly.

Belts can be checked for proper tension with a tension gauge, or by pressing on the belt with your fingers, as shown in **Figure 18-3**.

Belts can be replaced or belt tension adjusted by moving an idler pulley toward or away from the power pulley. In most cases, the idler pulley is the alternator drive pulley. A bottom pivot bolt and top adjustment bolt on the alternator unit must be loosened. The alternator must be pushed inward or outward **[Figure 18-4]**. Any other belts in front of the belt to be replaced must be removed first.

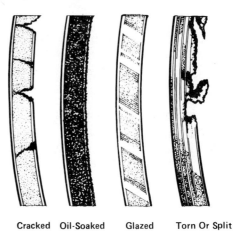

Figure 18-2. Drive belt defects. CHRYSLER CORPORATION

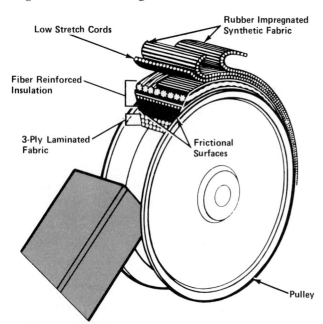

Figure 18-1. Parts of a drive belt. CHRYSLER CORPORATION

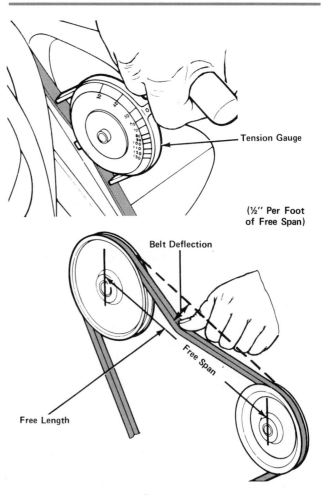

Figure 18-3. Checking drive belt tension. CHRYSLER CORPORATION

COOLING SYSTEM SERVICE 165

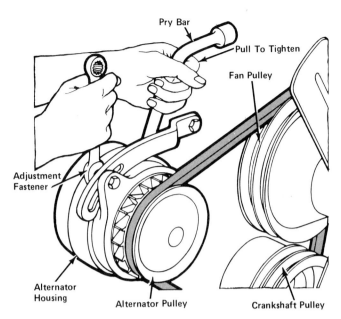

Figure 18-4. Drive belt replacement. CHRYSLER CORPORATION

A loosened belt to be replaced must be worked off from the pulleys and from around the fan. A new belt is replaced in the proper pulley grooves. The pivoting idler unit then is pulled away from the engine to tighten the belt. When the belt tension is correct, the adjustment bolt and pivot bolt are tightened. All other belts are replaced in reverse order of removal.

After approximately 200 miles of use, new belts will stretch. They should be checked and retightened if necessary.

18.4 CHECK FOR AND/OR REPAIR LEAKS

Loss of pressure and/or leaks can occur at many locations within the cooling system. Some leaks occur when the system is cold and parts are contracted to form gaps. Other leaks occur when the system is hot and under pressure. The following areas are the most common sources of leaks:

- Hoses
- Gaskets
- Radiator
- Heater core and heater control valve
- Water pump
- Soft plugs (core plugs).

Pressure Checks

Even small leaks will allow pressure to be lost from the cooling system. Loss of pressure will cause the coolant to boil at a lower temperature. A *cooling system pressure tester,* shown in **Figure 18-5,** checks cooling system pressure-holding capacity. A cooling system pressure tester consists of an air pump and a

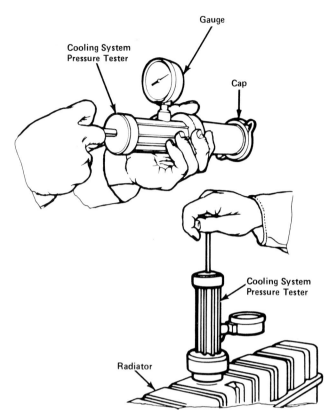

Figure 18-5. Pressure testing a cooling system.
CHRYSLER CORPORATION

gauge. Pressure is created by pumping the handle. If leakage occurs, the gauge pressure reading will drop.

The pressure tester can be used to check both the entire cooling system and the radiator cap. If the pressure reading drops when connected to the radiator, look for external leaks. Check hoses, soft plugs, the water pump, and so forth. An external leak usually will leave a puddle on the floor. If the radiator cap will not hold its rated pressure, replace the cap.

If no external leaks are found, check for internal leaks. Remove the oil dipstick and check for water contamination—which shows up as a milky color.

Head gasket leakage, explained below, may be detected by connecting the pressure tester to the radiator neck, applying low pressure, and running the engine at slow speeds. Excessive pressure buildup indicates that combustion chamber gases are entering the cooling system.

Hoses

Hoses are made of layers of rubber and cloth. Rubber is subject to deterioration from heat, vibration, and oil and fuel vapors. Hoses are manufactured as either molded or flexible types **[Figure 18-6]**.

Check hoses for leakage, hardening, swelling, or *chafing,* as shown in **Figure 18-7.**

166　　PART III: AUTOMOTIVE ENGINE SYSTEMS

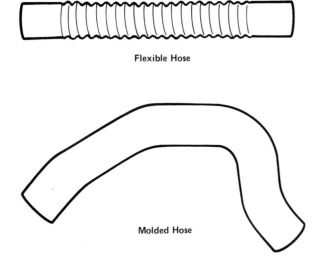

Figure 18-6. Cooling system hoses. CHRYSLER CORPORATION

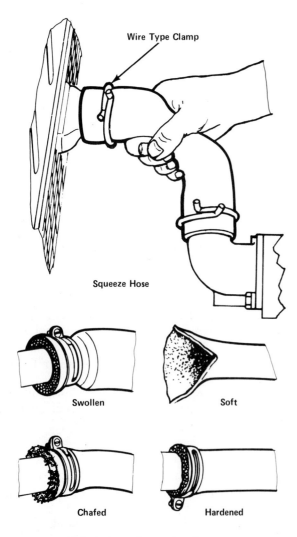

Figure 18-7. Defects in cooling system hoses.
CHRYSLER CORPORATION

Several types of hose clamps are used, as shown in **Figure 18-8**. The *worm-drive clamp* can be reused and tightened without loss of holding power, unlike the other types shown. When replacing hoses, a good policy is to replace wire or banded clamps with worm-drive clamps. Hose replacement is illustrated in **Figure 18-9**.

CAUTION!!!

Use care when removing hoses, especially radiator hoses, to avoid damaging the hose sealing surface. Do not use a screwdriver or excessive force on radiators with molded "plastic" tanks. They are quite fragile and can be damaged easily.

Gaskets

Gaskets that can cause cooling system leaks and problems include the following:

- Radiator pressure cap
- Water outlet (thermostat gasket)
- Water pump gasket
- Intake manifold gasket
- Cylinder head gasket.

Radiator pressure cap. Inspect the radiator pressure cap rubber seals and vacuum relief valve as shown in **Figure 18-10**. If the seals are cracked, stiff, or brittle, or the vacuum valve is sticky or broken, replace the cap. Check the radiator neck for bent, broken, or otherwise damaged flanges. If not badly damaged, radiator necks can be repaired.

Other gaskets. If thermostat, water pump, and/or cylinder head gaskets are leaking, the units themselves are replaced or reconditioned before replacement.

Thermostat and water pump gasket replacement are explained below. Head gasket replacement is covered in Unit 44, Cylinder Head Service.

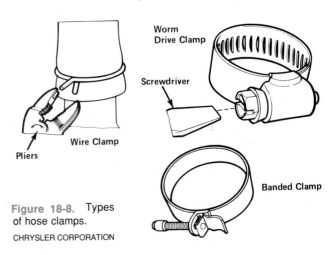

Figure 18-8. Types of hose clamps.
CHRYSLER CORPORATION

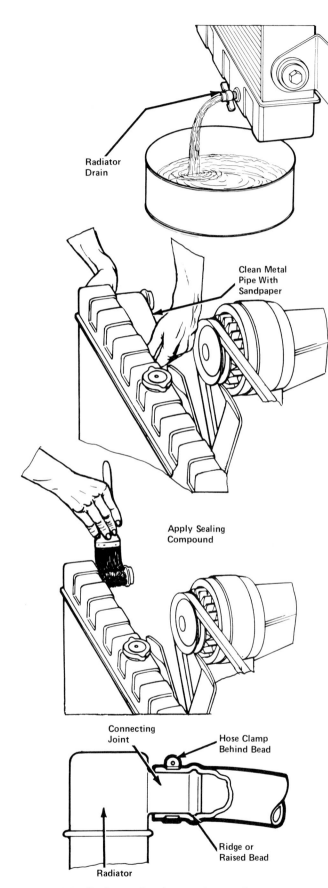

Figure 18-9. Cooling system hose replacement.
CHRYSLER CORPORATION

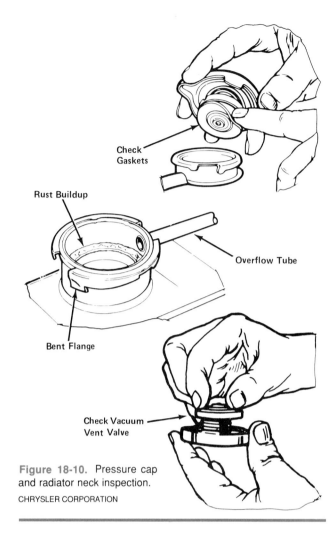

Figure 18-10. Pressure cap and radiator neck inspection.
CHRYSLER CORPORATION

Leakage around the thermostat and water pump gaskets can be seen during a cooling system pressure test.

Cylinder head gasket. Indications of a blown head gasket include:

- Coolant on the engine oil dipstick
- Oily deposits inside the radiator neck
- Coolant on the sparkplug firing tips after the engine has cooled overnight.

Suspected head gasket leaks can be verified by use of a *combustion leak detector,* shown in **Figure 18-11.** Enough coolant is drained from the radiator to leave at least a one-inch space above the fluid. Then, the engine is started. Special test fluid is poured into the tester. The bulb then is squeezed to draw gases from above the coolant through the test fluid. If combustion chamber gases are present, indicating a blown head gasket, the fluid will change color.

Another method for detecting head gasket leakage is to use the probe of an *exhaust gas analyzer* over the radiator neck to detect combustion

Figure 18-11. Combustion leak detector for checking head gasket leaks.

chamber gases. An exhaust gas analyzer is an electronic diagnostic instrument used to determine amounts of harmful gases in engine exhaust fumes. This is discussed in Unit 43.

Radiator

Most radiator leak repairs require the removal of the radiator from the vehicle. The coolant must be drained and all hoses and oil cooler lines disconnected. Bolts holding the radiator are loosened and removed.

The radiator inlet and outlet tanks and core are joined together at the *seams*, or lines of contact, where leaks can occur. In the past, most radiators were made of metals such as brass and copper that could be *soldered*, or joined, together to prevent leaks, replace tanks, or replace radiator cores.

Soldering is a process of using heat, a cleaning substance called *flux*, and *solder* to join parts. Solder is an alloy, or mixture, of lead and tin. When cleaned metal parts are heated sufficiently, applied solder will melt and join the parts together.

However, the aluminum radiators used on many modern automobiles cannot be soldered. In addition, plastic inlet and outlet tanks are used on newer vehicles with aluminum cores. On such radiators, metal tabs on the core are *crimped*, or tightly bent, to join the tanks to the core. A rubber O-ring seal prevents leaks, as shown in **Figure 18-12.**

Figure 18-13 shows one type of crimping operation used to join the core to a radiator tank.

If the metal is sound and sufficiently thick, radiator core tube leaks can be repaired. Radiator cores with badly corroded tubes must be replaced.

Brass or copper tubes can be soldered. Aluminum cores can be patched with an *epoxy resin* plastic material. Epoxy resin is a two-part material that hardens when the two parts are mixed. An epoxy repair is illustrated in **Figure 18-14.**

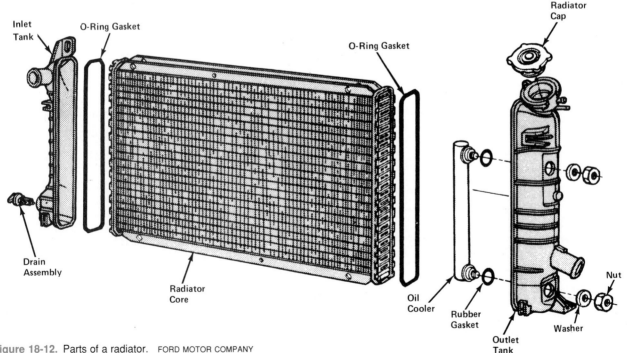

Figure 18-12. Parts of a radiator. FORD MOTOR COMPANY

COOLING SYSTEM SERVICE

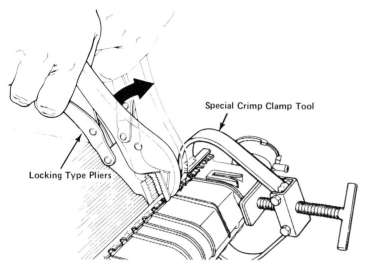

Figure 18-13. Crimping operation to join radiator core to radiator tank. FORD MOTOR COMPANY

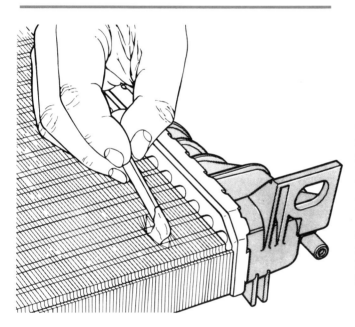

Figure 18-14. Applying epoxy repair material to seal a radiator core. FORD MOTOR COMPANY

Oil coolers in the outlet tank can be replaced or repaired when the tank is removed from the core.

Heater Core and Heater Control Valve

Heater core repairs are performed in the same way as radiator repairs. Leaks from heater cores usually show up inside the vehicle, on the front passenger floor space under the dash.

Heater Control Valve.
Stuck heater control valves often can be lubricated with light oil and manually moved to allow them to operate freely. Removing a leaking control valve simply involves loosening clamps, removing hoses, and unfastening the old valve. Replacement is essentially the reverse of this procedure. A typical heater control valve is shown in **Figure 18-15.**

Water Pump

The water pump shaft turns on anti-friction ball bearings that are lubricated by a sealed lubricant supply. If the seals fail, coolant will wash the lubricant from the bearings and ruin them. In addition, coolant can leak from around the front of the shaft and from the *weep hole.* The weep hole is the ventilation hole located at the bottom of the shaft housing.

On a running engine, ruined, noisy water pump bearings can be heard through a technician's stethoscope or rubber tubing.

SAFETY CAUTION

Whenever you work near a running engine, keep hands and clothing away from the moving fan, belts, and pulleys. Do not allow the stethoscope or rubber tubing to be caught by moving parts.

There is another test that can be performed with the engine off and the fan belt removed. This involves grasping the fan and attempting to move it in and out and up and down **[Figure 18-16].** More than 1/8 inch (3.2 mm) of movement at the ends of the blades indicates worn bearings that require water pump replacement.

Special tools are available to loosen bolts that attach fan blades and viscous drive units with the radiator in place. Water pump removal with the radiator in place requires that the belt, fan, and spacer or viscous drive unit be removed first.

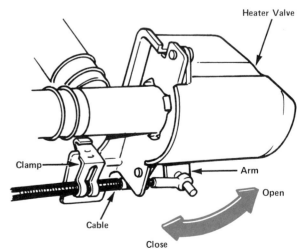

Figure 18-15. Heater control valve assembly.
AMERICAN HONDA MOTOR CO., INC.

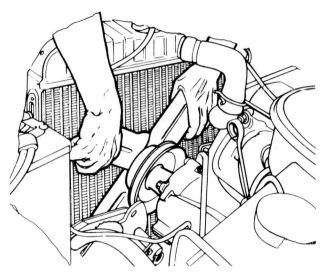

Figure 18-16. Checking water pump bearings.
CHRYSLER CORPORATION

CAUTION!!!

The radiator core surface is very delicate and can easily be damaged. When working at the front of the engine near the radiator, place a piece of cardboard over the radiator surface to protect the core.

Soft Plugs

Excess water in coolant mixture can cause soft plugs, also known as core plugs, in the block and/or cylinder head to rust or corrode through. This will cause leaks, as shown in **Figure 18-17.**

Before you attempt to replace soft plugs, drain the cooling system. Gaining access to soft plugs may require removing engine accessories and/or intake or exhaust manifolds. In difficult cases, it may be necessary to remove the entire engine or transmission.

Rusted or corroded soft plugs can be removed by driving a punch through the metal plug. The punch then is used as a lever to remove the old plugs.

Special tools and drivers are available to help the technician drive in soft plugs in difficult locations.

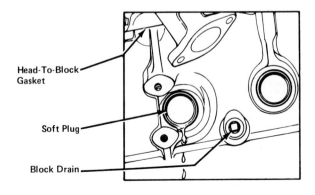

Figure 18-17. Engine block soft plugs.   CHRYSLER CORPORATION

18.5   CHECK COOLANT MIXTURE

Coolant mixtures can be checked with a *coolant hydrometer,* shown in **Figure 18-18.** A coolant hydrometer measures the freezing point of the mixture of antifreeze and water. For example, a half-and-half mixture has a freezing point of -34 degrees F (-51 degrees C). A higher freezing temperature would indicate more than 50 percent water, because pure water has a freezing point of +32 degrees F (0 degrees C).

Different percentages of antifreeze can be mixed with water to prevent freezing at varying temperatures. Refer to the charts printed on antifreeze containers. However, never use more than 70 percent antifreeze in a coolant mixture. In severely cold temperatures, pure antifreeze will gel, or thicken, and will not flow through the system. Too much antifreeze also can cause overheating in hot weather.

To change the proportion of antifreeze to water in the coolant mixture, liquid must be drained from the system so that additional antifreeze or water can be added. Coolant mixtures should be routinely flushed out and replaced with fresh coolant.

18.6   FLUSH COOLING SYSTEM

Before flushing is done, chemical cleaners can be added to the cooling system to help dissolve rust and scale deposits. Two types of cleaners are commonly used.

Heavy-duty cooling system cleaners consist of powdered phosphoric acid. The thermostat must be removed to prevent damage before heavy-duty cleaners are used. After use, a chemical neutralizer (baking soda) is put into the system to neutralize any remaining phosphoric acid before flushing.

SAFETY CAUTION

Be careful not to inhale fumes when pouring the powdered cleaner into the radiator. If powder gets on skin, wash the area immediately. Wear safety glasses.

Liquid cooling system cleaners consist of a milder solution of phosphoric acid dissolved in water and mixed with detergent. Liquid cleaners can be used without removing the thermostat. The flushing procedure removes the cleaning solution.

Reverse flushing is a procedure of forcing clean liquid backwards through the cooling system. This carries away rust, scale, corrosion, and other contaminants. A *flushing gun* that operates on compressed air is used to force clean water and air through the system **[Figure 18-19].**

18.7   REPLACE THERMOSTAT

The sole purpose of the thermostat is to control the minimum operating temperature of the engine. It is

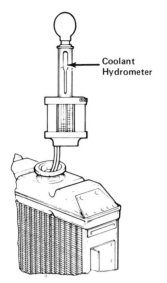

Figure 18-18. Checking coolant mixture with a coolant hydrometer. CHRYSLER CORPORATION

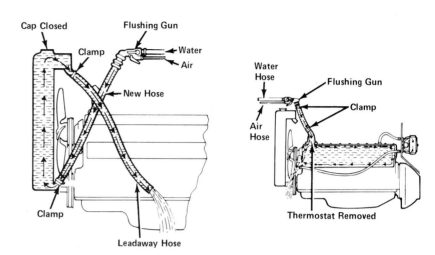

Figure 18-19. Reverse flushing a cooling system. CHRYSLER CORPORATION

usually located on the cylinder head or on the intake manifold for V-type engines.

A defective thermostat may cause engine overheating. However, an engine which fails to reach normal operating temperature always indicates a defective thermostat. The thermostat must be removed to be tested for proper operation. When replacing a thermostat, care must be taken to be sure the correct end is facing the radiator, and that it is properly seated in the machined recess before the thermostat housing is installed [**Figure 18-20**]. RTV sealer can be used for sealing the housing.

18.8 REPLACE WATER PUMP

The cooling system must be drained before water pump removal. Vehicles with belt-driven fans have components that must be removed to permit access to the water pump. These components include belts, the fan and spacer or viscous drive clutch, and other engine accessories. In many cases, the radiator and *shroud,* shown in **Figure 18-21,** must be removed to gain access to the water pump. A shroud is a hollow duct that helps to direct air toward the fan for better cooling.

The water pump is attached to the cylinder block as shown in **Figure 18-22.** Loosen and remove the bolts in a crisscross pattern from the center outward.

Insert a rag into the block opening and scrape off any remains of the old gasket. Apply the vehicle manufacturer's recommended sealant to the new gasket and sealing surfaces. If the bolt threads extend into the water jacket, coat them with sealant. Insert the bolts by hand, then tighten in a crisscross pattern, from the center outward, to the manufacturer's recommended torque.

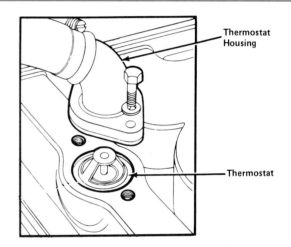

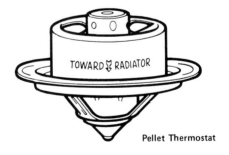

Figure 18-20. Thermostat installation. CHRYSLER CORPORATION

18.9 OVERHEATING PROBLEMS

Overheating can be caused by many factors. Typical causes include:

1. Cooling system clogged by rust, corrosion, or scale deposits
2. Loose or slipping fan belt

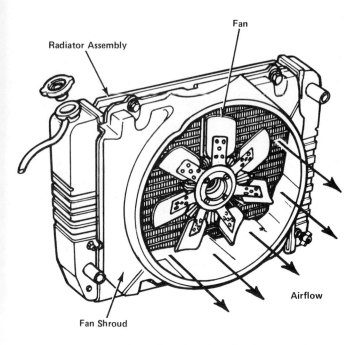

Figure 18-21. Radiator, shroud, and cooling fan assembly.
CHRYSLER CORPORATION

3. Improper operation of a viscous fan drive clutch
4. Pressure loss through leaks (including defective radiator cap)
5. Thermostat installed incorrectly
6. Airflow through radiator blocked
7. Incorrect mixture of coolant and water
8. Faulty water pump (loose impeller)
9. Collapsed radiator hoses
10. Heavy loads imposed by using the air conditioner, carrying excess weight, or towing trailers
11. Incorrect ignition timing and/or carburetor adjustment
12. Engine vacuum leaks
13. Automatic transmission overheating
14. Dragging brakes.

Only the first nine causes involve the cooling system itself. The remaining causes involve other systems or parts of the automobile. The automobile technician must be prepared to look past the obvious causes in diagnosing overheating problems.

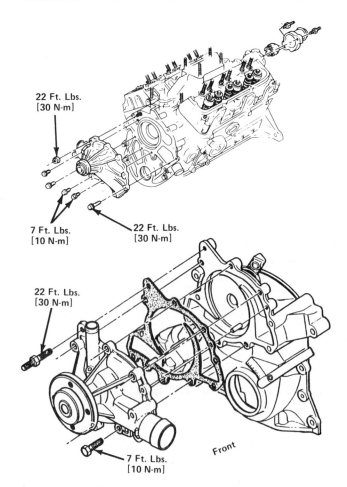

Figure 18-22. Water pump mounting assembly.
BUICK MOTOR DIVISION—GMC

UNIT HIGHLIGHTS

- Routine cooling system maintenance includes checking and/or replacement of drive belts, and checking and/or repairing leaks. Other routine maintenance procedures include: determining coolant mixture strength, replacing thermostats, replacing the antifreeze, and flushing the cooling system.
- Leaks can be detected by visual inspection or by using a cooling system pressure tester.
- Leaking thermostat, water pump, and cylinder head gaskets usually indicate more than the need to replace the gaskets. Replacement or reconditioning of the affected units also may be required.
- Overheating can be caused by many problems, some of which are not directly related to the cooling system.

TERMS

glazed	chafing
worm-drive clamp	combustion leak detector
exhaust gas analyzer	seam
solder	crimp

epoxy resin
coolant hydrometer
flushing gun
flux
cooling system pressure tester
heavy duty cooling system cleaner
weep hole
reverse flushing
shroud
V-belt
liquid cooling system cleaners

REVIEW QUESTIONS

Directions: The following questions are similar to those used on automotive technician certification tests. On a separate sheet of paper, write the letter of the correct choice.

1. Technician A says drive belts can become glazed if they are installed too loosely.
   Technician B says drive belts can become split if they are installed too tightly.
   Who is correct?
   A. A only
   B. B only
   C. Both A and B
   D. Neither A nor B

2. Which of the following statements is correct?
   I. When tested, a radiator pressure cap should open at its indicated limit and release pressure.
   II. When tested, a radiator pressure cap should hold pressure up to its indicated limit.
   A. I only
   B. II only
   C. Both I and II
   D. Neither I nor II

3. Which of the following could cause overheating?
   A. Automatic transmission problems
   B. Use of radial tires
   C. Manual transmission problems
   D. Weak battery

Questions 4 and 5 are not like those above. Each has the word EXCEPT. For each question, look for the choice that does *not* apply. Read each question carefully before you choose your answer.

4. All of the following can be used to detect a blown head gasket EXCEPT
   A. visual inspection and/or cooling system pressure tester.
   B. combustion leak detector or exhaust gas analyzer.
   C. cylinder compression or leakdown test.
   D. oil pressure tester.

5. All of the following indicate water pump seal and/or bearing problems EXCEPT
   A. excessive movement of water pump shaft.
   B. leakage from a water pump weep hole.
   C. excessive noise from water pump bearings when the engine is running.
   D. free spinning of a viscous fan drive clutch when warm.

ESSAY QUESTIONS

1. What can test instruments indicate about cooling system operating conditions?
2. Describe the steps necessary to replace a drive belt.
3. What methods are available to repair a radiator?

SUPPLEMENTAL ACTIVITIES

1. Inspect a vehicle's cooling system and use a pressure tester to diagnose possible problems. Report to your class what problems were found during inspection.
2. Check drive belts on a vehicle for proper tension. Remove and replace all belts and tighten them properly.
3. Perform visual inspection on a vehicle to detect head gasket problems. With the equipment available in your shop, make tests to verify whether the vehicle has a blown or leaky head gasket.
4. Clean and flush the cooling system and replace the thermostat and radiator hoses on a vehicle.
5. Replace the water pump on a vehicle selected by your instructor.

19 Fuel System Fundamentals

UNIT PREVIEW

To provide proper performance, automotive fuels are carefully formulated and blended. Gasoline and diesel fuel are the most widely used petroleum-based engine fuels.

A basic fuel system includes a fuel tank, fuel pump, filters, and either a carburetor or fuel injection system to meter fuel properly. Turbochargers and superchargers can be used to pressurize the intake manifold to produce a denser intake charge and more powerful combustion pressures. These devices are being used increasingly to produce more engine horsepower and torque output.

LEARNING OBJECTIVES

When you have completed your assignments and exercises in this unit, you should be able to:

- Describe the characteristics of gasoline and diesel engine fuels.
- Describe the causes of abnormal combustion in gasoline and diesel engines.
- Visually identify and describe the parts of a basic fuel system on a vehicle.
- Identify the differences between different types of gasoline fuel injection and diesel fuel injection systems.
- Identify and describe the main parts of turbochargers and superchargers and explain the functions of each.

19.1 AUTOMOTIVE FUELS

Fuels used in automotive engines include petroleum fuels such as gasoline, diesel fuel, propane, and butane. Alcohol-type fuels, such as methanol and ethanol, also can be used. In addition, approximately 10 percent alcohol can be mixed with 90 percent gasoline to produce a *gasohol* blend.

CAUTION!!!

Some fuel system components designed for use exclusively with gasoline may not be compatible with alcohol or gasohol fuels. These include rubber, plastic, and some metal parts. Use of alcohol or gasohol fuels may damage or destroy such parts, and cause fuel system failure. Follow the vehicle manufacturer's recommendations in the service manual and bulletins.

19.2 GASOLINE

Gasoline is the most widely used engine fuel. Gasoline, like other petroleum fuels, is a compound of hydrogen and carbon atoms, known as **hydrocarbon (HC)** molecules. To perform well as an engine fuel, gasoline must have several properties, including:

- Correct volatility and resistance to ice formation
- Resistance to abnormal combustion
- Chemical stability
- Low sulfur content.

Volatility

A gasoline engine runs on a vapor composed of gasoline and air. Gasoline becomes *atomized* and mixed with air molecules in the carburetor and/or intake manifold to form this vapor. Atomize means to reduce to tiny particles or a fine spray. A liquid that evaporates easily is said to be *volatile.* Liquids, including gasoline, become more volatile as they are heated.

Petroleum refineries anticipate the air temperatures to be encountered and blend gasolines for the correct volatility for engine performance. In cold weather, gasoline must have greater volatility to vaporize more easily at low temperatures for easy starting and running. In warm temperatures, too volatile a gasoline blend produces combustion and fuel delivery problems.

Gasoline must be delivered in a liquid form to the parts of the fuel system that atomize it. If the fuel vaporizes before it reaches the carburetor or fuel injection system, *vapor lock* will stop the engine. Vapor lock is the formation of bubbles of vaporized fuel, blocking the flow of liquid fuel in the fuel system.

Gasoline also will vaporize more easily at lower atmospheric pressures, such as at high altitudes. Gasoline refineries produce different blends of gasoline for different geographical and temperature regions and altitudes to prevent as many problems as possible. Blends are changed several times a year, depending on the geographic location.

De-icers. Water vapor from the air condenses on the inside of the gasoline tank. Water eventually will settle at the bottom of the fuel tank, fuel lines, fuel pump, and carburetor or fuel injection pump.

FUEL SYSTEM FUNDAMENTALS

When the temperature drops low enough, ice crystals can form, blocking the flow of gasoline, especially where atomization takes place. To prevent this problem, chemical *de-icers* are added to gasoline to prevent ice particles from forming and adhering to cold metal surfaces.

Resistance to Abnormal Combustion

Normal combustion begins with a spark from a spark plug. The combustion continues evenly until nearly all of the fuel in the combustion chamber is burned. Combustion is discussed in Unit 10.

Detonation. Increased heat and pressure are created as the flame front from combustion moves across the combustion chamber from the spark plug. This heat and pressure compress the remaining gases and cause them to explode, or *detonate,* after the spark plug fires **[Figure 19-1]**.

For maximum power output, maximum combustion pressure should be produced just after the piston reaches top dead center. Detonation interferes with this controlled combustion process. Detonation causes excess heat and pressure on the piston crown, or top, at the wrong time.

When detonation occurs, a knocking or hammering sound is heard. However, detonation that occurs at high speeds generally cannot be heard because of engine and road noise.

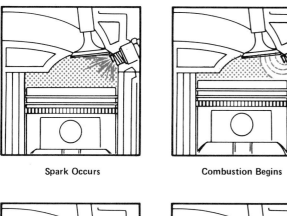

Spark Occurs

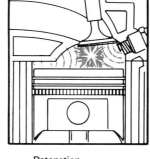
Combustion Begins

Combustion Continues
(Spontaneous Ignition)

Detonation
(Flame Fronts Collide)

Figure 19-1. Combustion chamber detonation.
CHAMPION SPARK PLUG COMPANY

Severe detonation can lead to holes melted and blown through the top of the piston crown. Connecting rod bearings, piston rings, and other parts can be damaged severely. Even mild detonation decreases fuel economy because the combustion pressures build and peak erratically.

Automotive engineers design engines so that under normal conditions detonation is prevented while engine efficiency is maintained. Key elements include shapes of combustion chambers and pistons, compression ratios, intake charge flow patterns, ignition timing, and fuel specifications. Such designs promote *swirl,* a circular or whirlpool-shaped flow that helps fuel mix with air more completely, and promote *turbulence.* Turbulence is an irregular pattern of flow. Swirl and turbulence are used to promote better vaporization and combustion and reduce detonation **[Figure 19-2]**. An engine's tendency to detonate is increased by the following factors:

- High compression ratios
- Ignition spark that occurs too soon
- Lean air-fuel mixtures
- Low humidity
- Overheating
- Low octane fuel.

Octane rating. *Octane rating* is a measure of an engine fuel's ability to resist ignition through heat and pressure (anti-knock quality). *Iso-octane* is a hydrocarbon fuel that is assigned an octane rating of 100. Gasoline octane ratings are related to this standard. The higher the octane rating, the more the gasoline matches (or in the case of aviation gasoline for airplanes, exceeds) iso-octane's resistance to ignition through heat and pressure.

Two methods are used to determine octane ratings, the Research method and the Motor method. These two methods use a standardized variable-compression, single-cylinder, fuel research engine. The Research method, which produces the Research Octane Number (RON), operates at relatively low speed and low inlet air temperatures. The Motor method, which produces the Motor Octane Number (MON), operates at higher speeds and higher air inlet temperatures. The Research method rating numbers usually are 4-6 points higher than those for the Motor method.

The octane ratings on Environmental Protection Agency (EPA) required stickers on gasoline pumps reflect an average of the two methods:

$$(R + M) \div 2$$

Where:
R = research
M = motor

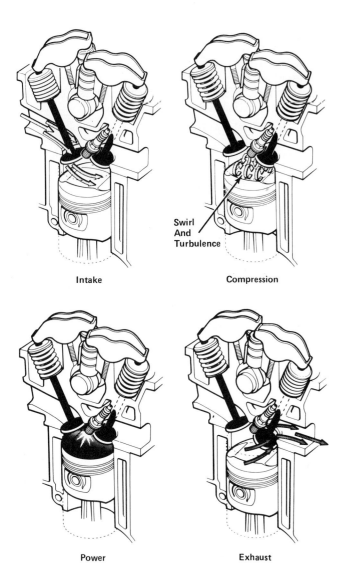

Figure 19-2. Combustion chambers are designed to encourage swirl and turbulence and reduce detonation. FORD MOTOR COMPANY

Preignition. Deposits in the combustion chamber can cause *preignition*, or "ping" or "knock." Preignition occurs when a glowing deposit or metal part becomes heated enough to begin ignition *before* a spark plug fires **[Figure 19-3]**. Preignition can be caused by any of the following:

- Low-octane fuel used in an engine with a high compression ratio
- Hot carbon deposits in the combustion chamber from burned gasoline or oil
- An overheated sharp metal edge in the combustion chamber.

Preignition can produce holes in piston crowns and lead to the same types of damage caused by detonation. The symptoms and results of preignition and detonation are very similar and often are confused.

As an engine is used, deposits form on the piston crown and within the combustion chamber. Deposits on older engines that used leaded gasoline were relatively soft and would flake off beyond a certain depth. In addition, lead in gasoline served as a "lubricant" to cushion the impact of valve faces against valve seats. Newer vehicles, using unleaded gasoline, must employ harder materials such as Stellite and stainless steel for valve heads and seats. Cast iron heads use induction-hardened valve seats. Deposits in engines that use unleaded gasoline are much harder and do not flake off as easily.

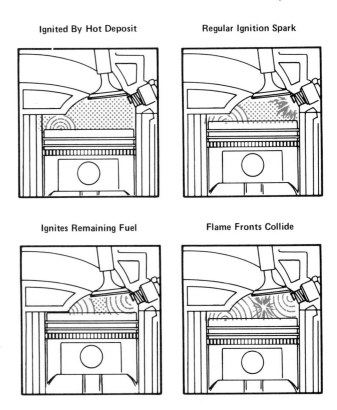

Figure 19-3. Combustion chamber preignition.
CHAMPION SPARK PLUG COMPANY

Chemicals can be added to gasoline to increase octane rating. Beginning in 1915, *tetraethyl lead* was added to gasoline. Such automotive *leaded gasolines* in the late 1960s achieved octane ratings as high as 102, measured by the Research method.

Because of health problems associated with leaded fuel, "regular" and "premium" leaded gasolines will be phased out in the near future.

Beginning in the mid-1970s, engines and emission (smog) control systems were redesigned to run on lower-octane unleaded fuel. Different methods of refining and different chemical additives produce unleaded gasoline with average octane ratings of 87 to 92.

If no detonation problems occur, higher-octane fuel will neither cause the engine to run better nor improve fuel economy.

FUEL SYSTEM FUNDAMENTALS

Combustion chamber deposit buildup of any kind causes an increase in compression ratio. The more deposits in the combustion chamber, the higher octane fuel the engine requires to prevent preignition and detonation.

Dieseling. *Dieseling*, or *run-on*, in gasoline engines is caused by excessive heat within the combustion chamber and/or excessive idle speed. Dieseling occurs when combustion chamber parts or deposits are hot enough to provide ignition after spark ignition stops. Cooling the combustion chamber, slowing the idle speed, and using a higher-octane gasoline all can help prevent dieseling.

Chemical Stability

Gasoline molecules tend to degrade, or break down, and form heavy, gummy deposits. These deposits can coat valve stems, valve lifters, carburetor parts, and piston rings. Improved chemical processes have increased gasoline's chemical stability and reduced its tendency to form deposits.

Low Sulfur Content

As discussed in Units 10 and 17, sulfur in combination with water forms sulfuric acid. The amount of sulfur in gasoline and lubricating oil depends upon the geographic areas from which petroleum comes and the refining processes used. Excessive sulfur also can reduce the effectiveness of additives used to raise the octane rating.

Fuel Contamination

Excessive fuel contamination by water, rust, debris, or chemicals can cause many fuel system problems. Fuel injection system pumps and injectors can become clogged or corroded by contamination. In severe cases, the fuel tank and lines must be emptied and cleaned to eliminate the sources of contamination.

19.3 BASIC FUEL DELIVERY SYSTEM

A basic fuel system, illustrated in **Figure 19-4,** consists of the following components:

- Fuel tank
- Connecting lines
- Fuel pump
- Fuel filter
- Air filter
- Fuel atomization and vaporization systems (carburetor or fuel injection).

Fuel Tank

The fuel tank is made either of pressed sheet metal with welded seams or of a reinforced plastic material. A filler neck with a cap allows fuel to be added, as shown in **Figure 19-5.**

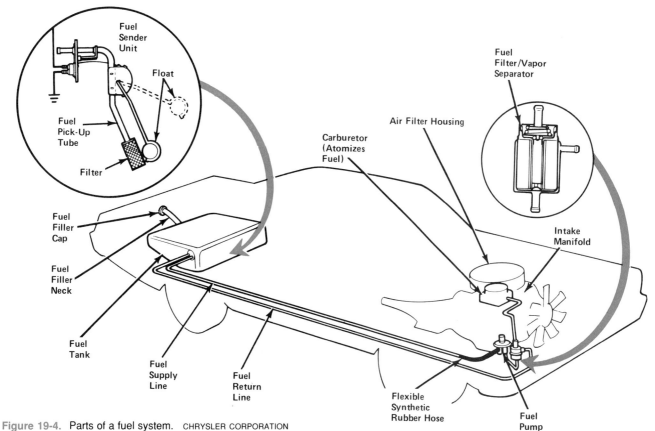

Figure 19-4. Parts of a fuel system. CHRYSLER CORPORATION

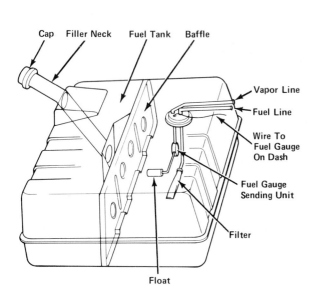

Figure 19-5. Parts of a fuel tank.

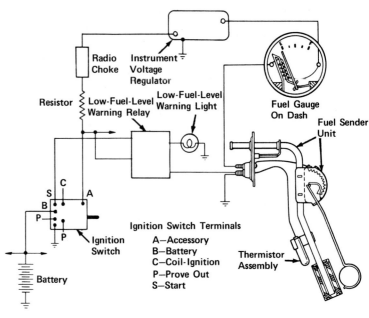

Figure 19-6. Fuel gauge system.

An electrical fuel gauge sender is located in the tank and operates in basically the same way as an oil-pressure or water-temperature gauge sender. A diagram of such a unit is shown in **Figure 19-6.** Many of the newer fuel injection systems have an electric fuel pump inside the fuel tank.

Connecting Lines

Small-diameter metal tubing and synthetic rubber hoses connect the fuel tank to the fuel pump. Synthetic rubber hoses are used where vibration might crack or bend metal lines.

Fuel Pump

The fuel metering and atomization system is located higher in the vehicle than the fuel tank. A mechanical or electrical pump is used to draw the fuel from the tank and deliver it to the carburetor or fuel injection system.

Mechanical pump. A *mechanical fuel pump* is driven by engine power, usually from the camshaft. The power is transmitted either through an eccentric lobe or a lobe, pushrod, and rocker arm arrangement **[Figure 19-7].**

Mechanical pumps produce suction by moving a synthetic rubber diaphragm against a spring inside a closed housing. This action produces a low-pressure area and draws fuel into the housing through a one-way valve. As the camshaft turns, the spring expands. This forces the diaphragm in the opposite direction, pushing the fuel out of the pump through another one-way valve **[refer to Figure 19-7].**

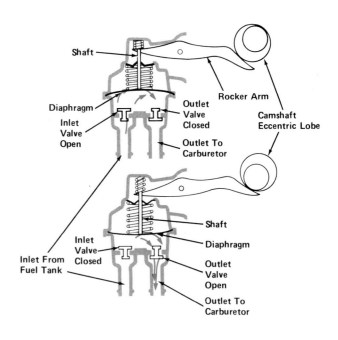

Figure 19-7. Mechanical fuel pump assembly.

Mechanical pumps in the engine compartment are subject to heat. In addition, during fuel intake, a low pressure area is created in the fuel line. Both conditions can lead to vapor lock. Many modern mechanical fuel pumps include a vapor separator and vapor return line to the fuel tank. These components reduce the tendency toward vapor lock. Mechanical pumps are used extensively for carburetor systems. These pumps produce relatively low pressure outputs of from 3 to 5 psi.

FUEL SYSTEM FUNDAMENTALS

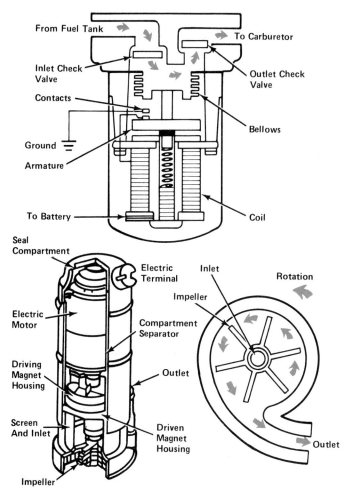

Figure 19-8. Parts of an electrical fuel pump. FORD MOTOR COMPANY

Electrical pump. There are two types of *electrical fuel pumps*, shown in **Figure 19-8**. One type is an *impeller pump* that operates like a water pump. The second type operates like a mechanical fuel pump, except that a magnetic field is used to produce the force to compress a spring and move a synthetic rubber diaphragm.

Electrical pumps often are mounted at or in the fuel tank to protect them from heat. In addition, impeller-type pumps keep the fuel line pressurized at all times to reduce vapor formation that can lead to vapor lock. The pumps are activated by the ignition switch.

Electrical fuel-injection system pumps are high-pressure pumps, producing from approximately 40 to 80 psi (see Unit 23).

Fuel Filter

Moisture and contaminants can collect in the fuel tank. Particles of rust from the tank can clog fuel lines and small passages in the fuel metering and atomization system. A screen is located in the fuel tank over the fuel outlet. One or more *fuel filters* in the system trap sediment, rust particles, and other contaminants that leave the tank.

Fuel filters can be located anywhere between the fuel in the tank and the carburetor or fuel-injection system intake. The filters shown in **Figure 19-9** are typical for carburetor-equipped vehicles. Fuel-injected vehicles have larger, metal canister filters with threaded end fittings. Use the filter recommended in the manufacturer's service manual.

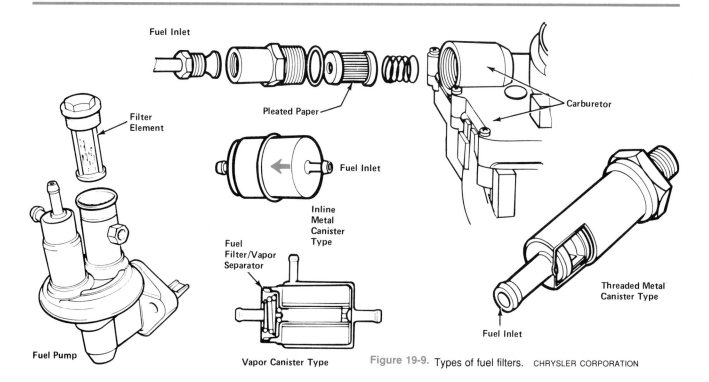

Figure 19-9. Types of fuel filters. CHRYSLER CORPORATION

Air Filter

Air contains dust and dirt particles and other contaminants. To prevent contaminants from entering the engine with the air-fuel mixture, an *air filter* is mounted in a housing [Figure 19-10]. Air drawn in on the intake stroke passes through the housing and filter to a carburetor or fuel-injection system. The resulting air-fuel mixture then passes through an intake manifold into the cylinders.

Fuel Atomization and Vaporization Systems

The fuel system should provide the correct air-fuel mixture to operate the engine efficiently at varying speeds and loads. This mixture is expressed as a number of parts of air to one part of gasoline. At sea level, the chemically correct mixture is 14.7 parts of air to 1 part of gasoline, measured by weight. This is expressed by the ratio 14.7:1. A chemically correct air-fuel mixture is called *stoichiometric.* At higher elevations, air contains less oxygen. Thus, more parts of air are required for complete combustion.

Air-fuel ratios with higher concentrations of air, such as 16:1, 18:1, or 20:1, are called *lean mixtures.* Air-fuel ratios with fewer parts of air to gasoline, such as 12:1, 10:1, or 8:1, are called *rich mixtures.* Beyond these lean and rich limits, the engine will not run properly.

During warmup and on heavy acceleration, the engine requires a mixture as rich as 11.5:1. After warmup and during low-load cruising, mixtures as lean as 18:1 can be used. However, the most power for the amount of fuel consumed occurs at an air-fuel ratio of 14.7:1 (at sea level).

Carburetor

A *carburetor* is a vacuum-operated device that supplies a fine spray of gasoline into the incoming air stream. The amount of gasoline drawn in by vacuum is proportional to the amount of air drawn through the *barrel* of the carburetor. The barrel is the throat opening or *throttle bore.* The amount of air that passes through the carburetor can be controlled by operating a movable valve called a *throttle plate* in an opening at the bottom of the carburetor bore.

A low-pressure area can be created by narrowing the opening at a point within the bore [Figure 19-11]. This point is called a *venturi*. To maintain a constant volume of flow, air must speed up as it passes through this venturi, or narrow restriction.

A tube end placed within the venturi will have suction, or vacuum, applied to it. If the opposite end is in a supply of gasoline, the gasoline is drawn through the tube, into the venturi [Figure 19-12].

The end of the tube is designed and placed so that the fuel is drawn out in a finely atomized spray. Some carburetors have multiple venturis, known as *booster venturis,* one within the other, to further increase the speed of air flow [refer to Figure 19-12].

After the fuel is atomized into the intake manifold, three factors combine to vaporize the fuel. These factors are: the flow of the mixture, low pressure in the manifold, and heat within the manifold [Figure 19-13].

The basic carburetor described above would function correctly only at high engine speeds. Many additional fuel passages, parts, and linkages are required. Together, they provide the range of correct air-fuel mixtures necessary at different engine speeds, loads, and temperatures. These parts and their operation are discussed in Unit 22.

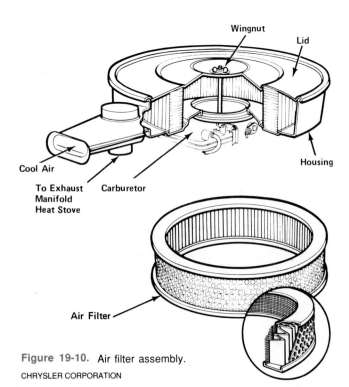

Figure 19-10. Air filter assembly.

CHRYSLER CORPORATION

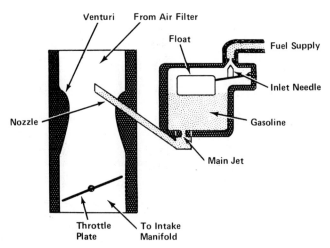

Figure 19-11. Basic carburetor assembly.

CHEVROLET MOTOR DIVISION—GMC

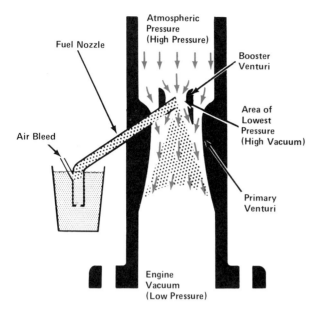

Figure 19-12. Carburetor venturi principle.
CHEVROLET MOTOR DIVISION—GMC

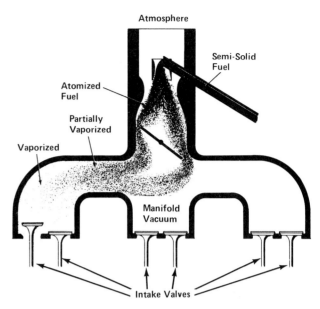

Figure 19-13. Fuel atomization and vaporization.
CHEVROLET MOTOR DIVISION—GMC

Electronic Fuel Injection Systems

Electronic fuel injection also provides a range of air-fuel mixtures. This range of mixtures allows the engine to run properly and efficiently at widely varying speeds, loads, and temperatures. Fuel injection systems are discussed in Unit 23.

Electronic fuel injection (EFI) systems spray fuel under pressure through *injectors*, or small nozzles, into the intake manifold. The amount of fuel sprayed is regulated by an electronic computer that receives signals from many sensors on the engine. The computer reacts rapidly to sensor input signals.

A list of the *input* and *output* electrical signals for a throttle-body EFI system is shown in **Figure 19-14**. Input signals are electrical currents from temperature and other sensors that indicate information, or facts, such as coolant temperature. Output signals are electrical currents to units that can modify engine operation, such as turning on an electrical cooling fan.

EFI systems operate much more rapidly and efficiently than carburetors. Such systems can provide better fuel economy, performance, and less harmful exhaust gases.

Two general forms of fuel injection are currently in use, *single-point* and *multi-point* fuel injection.

Single-point fuel injection. Single-point fuel injection sprays fuel at a device, a *throttle body*, that contains a throttle plate like a carburetor **[Figure 19-15]**. Another name for such a system is *throttle body injection (TBI)*. One or two injectors may be used in a single throttle body. In addition, multiple throttle bodies may be used.

A simplified diagram of TBI operation is shown in **Figure 19-16**. Fuel is atomized and mixed within the intake manifold in the same way as in a carbureted fuel system.

Multi-point fuel injection. The terms *port fuel injection, multi-point fuel injection, and multi-port fuel injection* are alternate terms for the same system. A separate injector is provided for each intake port for better fuel distribution.

In carbureted or TBI fuel systems, the amount of fuel distributed to the cylinders through the intake manifold can vary **[Figure 19-17]**. A manifold's design and shape, and its distance from the carburetor or throttle body, influence how equally fuel is distributed.

A port fuel injection system has separate, individual injectors for each cylinder. Each injector is positioned in an intake manifold runner. The injector is located just in front of the cylinder head intake port, ahead of the intake valve. A metal line called a *fuel rail* supplies fuel to each injector. This design supplies each cylinder with an equal mixture of air and fuel.

Electrical signals to the injectors cause them to open to spray fuel. The longer or more often the electrical signal opens the injectors, the more fuel is sprayed. When the electrical signal is shut off, springs close the injector.

The spraying of the fuel from all injectors can occur independently of the opening of the intake valves. Such a system is known as *continuous fuel injection*. In a continuous fuel injection system, the amount of air allowed through the throttle valve determines the air-fuel ratio.

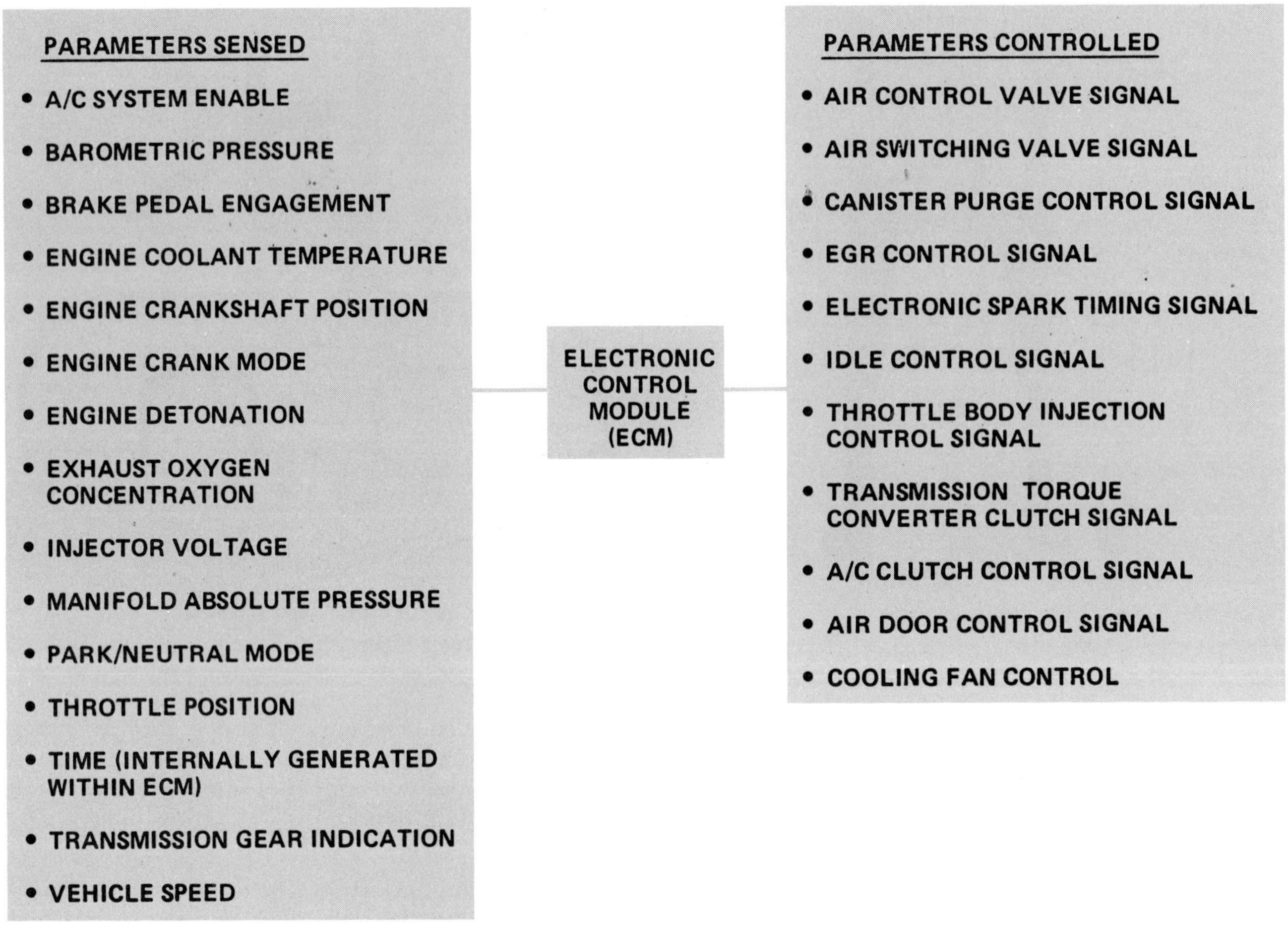

PARAMETERS SENSED	PARAMETERS CONTROLLED
• A/C SYSTEM ENABLE	• AIR CONTROL VALVE SIGNAL
• BAROMETRIC PRESSURE	• AIR SWITCHING VALVE SIGNAL
• BRAKE PEDAL ENGAGEMENT	• CANISTER PURGE CONTROL SIGNAL
• ENGINE COOLANT TEMPERATURE	• EGR CONTROL SIGNAL
• ENGINE CRANKSHAFT POSITION	• ELECTRONIC SPARK TIMING SIGNAL
• ENGINE CRANK MODE	• IDLE CONTROL SIGNAL
• ENGINE DETONATION	• THROTTLE BODY INJECTION CONTROL SIGNAL
• EXHAUST OXYGEN CONCENTRATION	• TRANSMISSION TORQUE CONVERTER CLUTCH SIGNAL
• INJECTOR VOLTAGE	• A/C CLUTCH CONTROL SIGNAL
• MANIFOLD ABSOLUTE PRESSURE	• AIR DOOR CONTROL SIGNAL
• PARK/NEUTRAL MODE	• COOLING FAN CONTROL
• THROTTLE POSITION	
• TIME (INTERNALLY GENERATED WITHIN ECM)	
• TRANSMISSION GEAR INDICATION	
• VEHICLE SPEED	

These parameters are inputs and outputs of the ELECTRONIC CONTROL MODULE (ECM).

Figure 19-14. Input and output signals for an electronic fuel injection system. BUICK MOTOR DIVISION—GMC

Fuel injectors also may be grouped so that half of them spray at one time and half at another time. Such a system is known as *non-sequential fuel injection*, or *pulsed fuel* injection [Figure 19-18]. Some fuel remains in the intake manifold until a cylinder begins its intake stroke.

Alternately, fuel can be sprayed from individual injectors, in sequence, just before their respective intake valves open. This system is known as *sequential fuel injection (SFI)*. A block diagram of an SFI system is shown in Figure 19-19.

19.4 HEATED INTAKE MANIFOLD

To help vaporize fuel, heat from the exhaust manifold or the cooling system can be routed under or near the intake manifold [Figure 19-20].

Exhaust Manifold Valve

A valve can be used in the exhaust manifold to increase heat within the intake manifold. This valve, sometimes called a *heat riser valve*, is similar to a throttle plate in a carburetor [Figure 19-21]. The valve can be operated by a thermostatic spring and/or a vacuum motor.

When the engine is cold and idling slowly, a thermostatically controlled vacuum motor closes the flap almost completely. Exhaust gases pass out of the exhaust manifold more slowly and cause the engine and intake manifold to heat up more quickly. The result is better fuel vaporization. This system is known as *early fuel evaporation (EFE)* on General Motors vehicles.

Hot coolant also can be routed through the intake manifold to accomplish the same purpose.

Electrical Manifold Heater

Another way to heat the intake manifold to increase gasoline vaporization is to use an electrical heating grid or element [Figure 19-22]. Electrical manifold heaters are used with carburetors and TBI fuel-injection systems.

A computer electronic control unit (ECU) receives coolant temperature and engine rpm signals from sensors. Based on a program, the ECU operates a relay to switch electrical current ON or OFF to the heating grid.

When the ECU switches the relay ON, electrical current warms the heater grid. Heat from the

FUEL SYSTEM FUNDAMENTALS

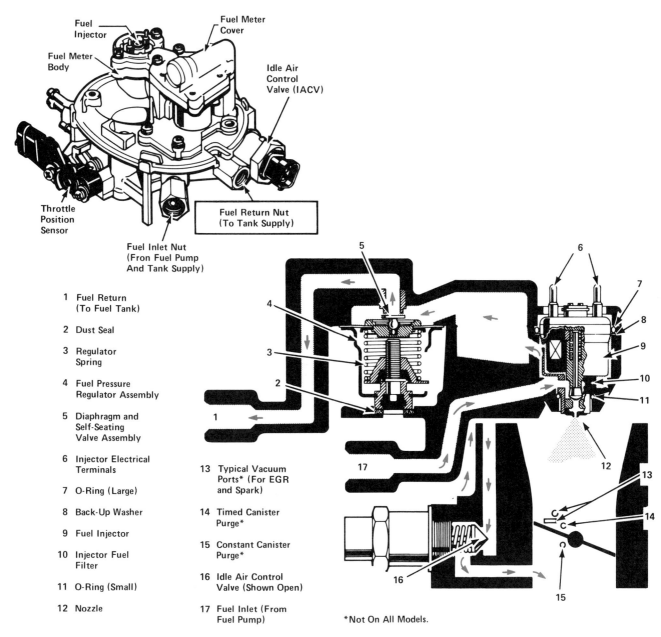

Figure 19-15. Throttle body fuel injection system. BUICK MOTOR DIVISION—GMC

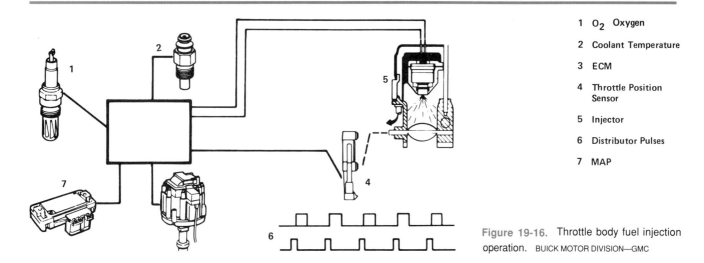

Figure 19-16. Throttle body fuel injection operation. BUICK MOTOR DIVISION—GMC

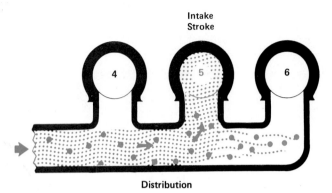

Figure 19-17. Fuel distribution in the intake manifold.
CHEVROLET MOTOR DIVISION—GMC

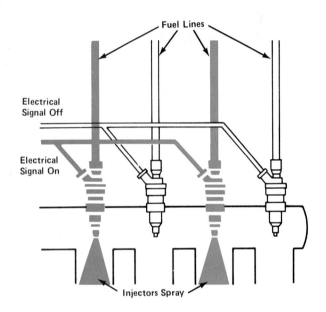

Figure 19-18. Non-sequential fuel injection operation.
ROBERT BOSCH CORPORATION

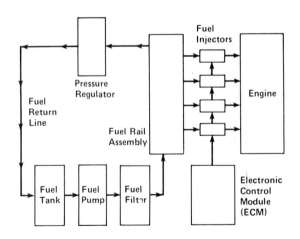

Figure 19-19. Sequential fuel injection system.
BUICK MOTOR DIVISION—GMC

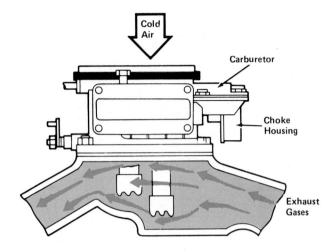

Figure 19-20. Heat from the exhaust manifold is routed near the intake manifold.

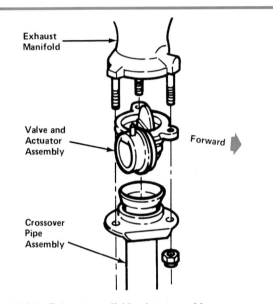

Figure 19-21. Exhaust manifold valve assembly.
CHEVROLET MOTOR DIVISION—GMC

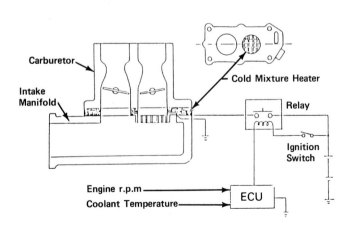

Figure 19-22. Electrical manifold heater.
MITSUBISHI MOTOR SALES OF AMERICA, INC.

FUEL SYSTEM FUNDAMENTALS

grid warms the surrounding air and metal of the intake manifold for better fuel vaporization.

19.5 DIESEL FUEL

Diesel fuel, unlike gasoline, is composed of heavier, more viscous hydrocarbon molecules. Because of this composition, diesel fuel contains more potential heat energy than gasoline.

Volatility Rating

Diesel fuel is graded for volatility as Number 1 (more volatile, for colder temperatures) or Number 2 (less volatile, for more moderate temperatures). Most passenger car diesel engines use Number 2 diesel fuel.

Cetane Rating

A diesel engine depends on the heat of compression to ignite the fuel. A *cetane rating* measures the ease with which diesel fuel will ignite under the heat and pressure of compression.

The cetane rating of diesel fuel has a direct bearing on engine operation in cold weather. A high cetane rating is desirable in cold climates.

A higher cetane rating means the fuel will ignite easily. Diesel fuel with too low a cetane rating has a long ignition delay period, followed by rapid combustion. This causes *diesel knocking*, a condition similar to detonation in a gasoline engine.

19.6 DIESEL FUEL SYSTEM

All diesel injection systems are multi-point injection systems that inject fuel into each cylinder. A schematic diagram of a diesel fuel-injection system is shown in **Figure 19-23**.

In a diesel engine, only air is compressed by the piston. Diesel fuel is sprayed directly into the cylinder under high pressure through a *mechanical fuel injection nozzle* **[Figure 19-24]**. A spring holds the nozzle closed. Diesel fuel is pumped into the nozzle under high pressure. This forces the spring to open and the nozzle to spray fuel.

Diesels have no throttle plate in the air induction system. The speed of the engine is controlled by the amount of diesel fuel injected and the time at which it is injected.

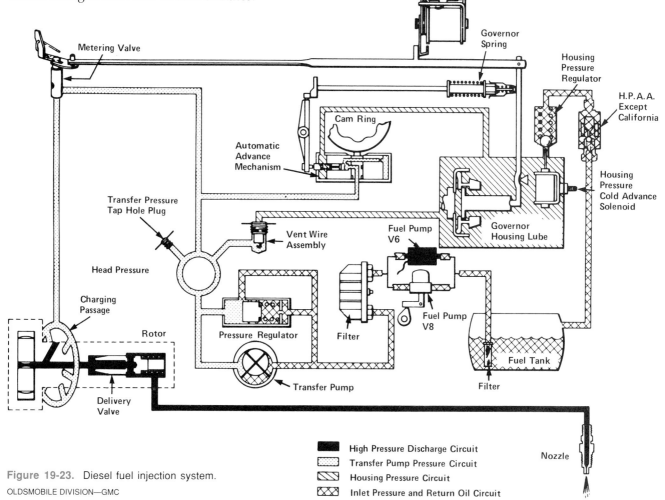

Figure 19-23. Diesel fuel injection system.
OLDSMOBILE DIVISION—GMC

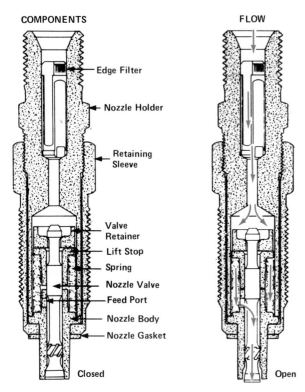

Figure 19-24. Diesel mechanical fuel injection nozzle.
OLDSMOBILE DIVISION—GMC

19.7 TURBOCHARGING

In the past, several methods were used to gain more engine performance:

- Increased engine displacement
- Increased compression ratio
- Higher-lift and longer-duration camshafts
- Rich air-fuel mixtures.

Because of the need for reasonable fuel economy and lower emissions, most of these methods no longer can be used. To produce horsepower and torque more efficiently from today's smaller gasoline and diesel engines, *turbocharging* is used increasingly. Turbocharging is a method of *supercharging*. Supercharging is forcing more air-fuel mixture into a cylinder than can be drawn in by atmospheric pressure alone. A type of compressor, or pump, forces air and fuel under pressure into the intake manifold.

Turbochargers on gasoline engines can be used with carburetors or with fuel injection systems. When used with a carburetor, a turbocharger can either force or draw air through the carburetor. Technicians call such systems "blow-through" and "suck-through" designs, respectively. An example of a turbocharger that draws air through a carburetor is shown in **Figure 19-25**.

The denser mixture produced by turbocharging creates greater heat, pressure, and force during combustion. This greater force enables a smaller displacement engine to produce more torque and horsepower.

A *turbocharger* uses the wasted heat energy of the exhaust gases flowing from the engine to drive a compressor. A schematic view of a turbocharging system is shown in **Figure 19-26**.

The turbocharger unit consists of two separate impellers, or *turbines*, mounted on a shaft within a housing **[Figure 19-27]**. As the hot exhaust gases are forced into the turbine wheel, they give up their heat and kinetic energy. This heat and energy cause the turbine, turbine shaft, and compressor to spin at very high speeds, approximately 100,000 rpm. The spinning compressor draws in the air-fuel mixture, compressing and pressurizing it into the intake manifold.

To prevent excessive pressure from building up, a *wastegate*, or pressure relief valve, is part of the turbocharger system **[refer to Figure 19-27]**. The wastegate operates like a pressure relief valve in a lubrication system. Pressures in turbocharged passenger car intake manifolds generally are limited to 15 psi (103 kPa) or less above atmospheric pressure.

Turbocharging has some drawbacks. The exhaust turbine creates a restriction within the exhaust manifold. Excessive *backpressure* from such a restriction can reduce engine efficiency.

Also, as the intake air is compressed, it is heated, making it less dense (less oxygen). This loss of density reduces combustion efficiency. An *intercooler*, a type of radiator heat exchanger, can be mounted in the airstream to cool the intake air charge. This restores most of the lost efficiency. **Figure 19-28** shows an intercooler used with a turbocharger.

Most turbochargers do not begin to function well until engine rpm builds enough to force exhaust gases out rapidly. This delay in function is called *turbo lag*. At low rpm, the engine power output is comparable to a nonturbocharged, or *normally aspirated* engine.

The turbocharger shaft, spinning at speeds of up to 100,000 rpm, is lubricated by oil from the engine lubrication system. Clean oil is critical to the life of the turbocharger. Special oils formulated for use with turbochargers can be used to help prevent damage to turbocharger and engine bearings.

In the late 1970s and early 1980s, special precautions were necessary to prevent damage to turbocharger bearings. For example, instruction manuals warned drivers to warm up turbocharged engines gently and fully before hard acceleration. In addition, the manual instructed drivers to idle fully-heated turbocharged engines for five minutes before shutting them off.

These measures were necessary to warm and cool the lubricating oil and turbocharger bearings. If lubricating oil heats rapidly to high temperatures,

FUEL SYSTEM FUNDAMENTALS

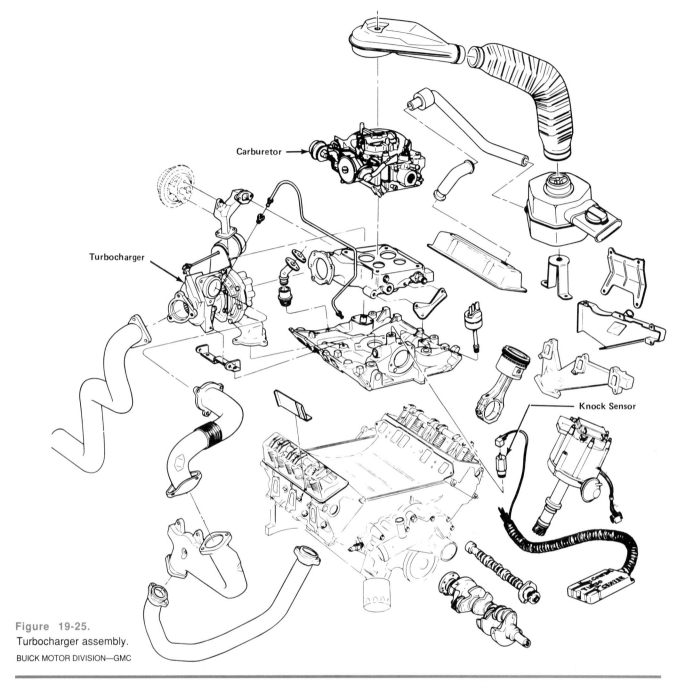

Figure 19-25.
Turbocharger assembly.
BUICK MOTOR DIVISION—GMC

coking can occur. Coking is the formation of gum, varnish, and carbon caused by excessively heated oil.

Recent turbocharger designs incorporate a hollow water jacket in the housing around the turbocharger exhaust-end bearing. Engine coolant flows through the water jacket to lower the temperature of the turbocharger bearings and lubricating oil. This design reduces oil coking and extends the life of the turbocharger bearings.

19.8 SUPERCHARGING

As mentioned above, turbocharging has some disadvantages. For example, in most turbocharger applications on small-displacement engines, the engine must be turning at high speeds to operate the turbocharger efficiently. These vehicles typically have little torque when starting from a stop unless the engine is "revved up" to produce a good flow of exhaust gases.

In addition, turbocharger turbine wheels, housings, and bearings must be made from expensive, heat-resistant materials.

To overcome these problems and provide a pressurized intake system at lower cost, some recent vehicles use a *supercharger* in place of a turbocharger. A supercharger is a compressor driven by the engine crankshaft. Recent vehicles use a supercharger driven by a belt on the engine crankshaft,

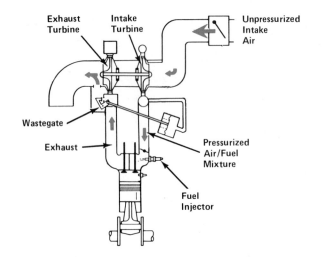

Figure 19-26. Turbocharger air-fuel and exhaust flow. FORD MOTOR COMPANY

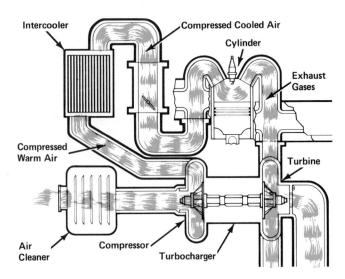

Figure 19-28. Turbocharger installation with intercooler.

some with an intercooler [Figure 19-29]. A magnetic clutch operates to turn the internal compressor parts, much as an air-conditioning system compressor.

Rootes-type Supercharger

A popular design for the supercharger compressor is a Rootes-type compressor. In a *Rootes-type supercharger*, two figure-8 shaped rotors turn to compress air. Gears attached to the end of the rotor shafts keep the rotors synchronized [Figure 19-30].

As the rotors turn, air is drawn in, compressed against the inner walls of the supercharger housing, then discharged through the outlet to the intake manifold [Figure 19-31]. The pressurized intake charge creates additional engine torque and horsepower, as with a turbocharger.

Other types of compressors also can be used as superchargers. Centrifugal, or rotary, compressors have an impeller that resembles a turbocharger impeller. As the shaft turns, air is compressed and forced through the intake manifold.

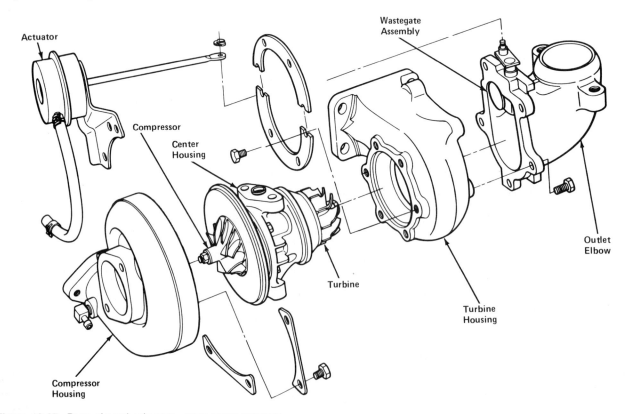

Figure 19-27. Parts of a turbocharger. FORD MOTOR COMPANY

Fuel System Fundamentals

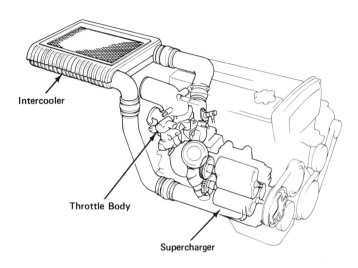

Figure 19-29. Belt-driven supercharger with intercooler.

However, superchargers also have disadvantages. Because a supercharger draws energy directly from the crankshaft, engines with superchargers suffer *parasitic losses* of horsepower. An extreme example of this was the Mercedes-Benz W196 racing car engine of the 1930s. The 3-liter (183 cubic-inch) engine produced more than 640 horsepower on methanol fuel with a supercharger driven by gears off the front of the engine crankshaft. However, approximately 120 horsepower was required just to turn the supercharger!

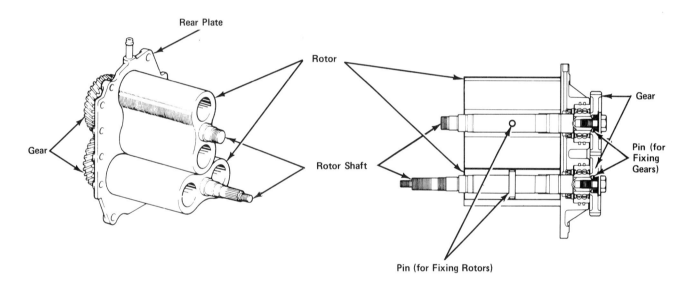

Figure 19-30. Rootes-type supercharger compressor rotors.

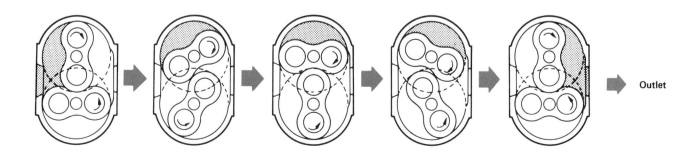

Figure 19-31. Rootes-type compressor operation.

UNIT HIGHLIGHTS

- Gasoline and diesel fuel are the most common hydrocarbon engine fuels.
- Gasoline must vaporize properly, resist the formation of ice crystals, resist abnormal combustion, and have low sulfur content.
- Octane rating is a way of measuring gasoline's resistance to ignition through applied heat and pressure.
- Detonation is an uncontrolled explosion of remaining air-fuel mixture *after* the spark plug fires. Preignition is the ignition of the air-fuel mixture *before* the spark plug fires.
- A basic fuel system consists of a fuel tank, connecting lines, fuel pump, fuel filter, air filter, and fuel atomization system.
- A fuel atomization and vaporization system can be either a carburetor or a fuel injection system.
- Fuel is vaporized in the intake manifold by swirl, turbulence, heat, and low pressure, or vacuum.
- Diesel fuel contains more heat energy per unit volume than gasoline and vaporizes less easily.
- Cetane rating is a way of measuring diesel fuel's ease of ignition through applied heat and pressure.
- Turbocharging is a way of using an engine's exhaust gases to drive a compressor to pressurize the air-fuel mixture into the intake manifold.
- Belt-driven superchargers are used on some recent vehicles to provide a pressurized intake system at lower cost than a turbocharger.

TERMS

gasoline
hydrocarbon (HC)
volatile
de-icer
swirl
octane rating
tetraethyl lead
preignition
fuel-vapor separator
parasitic losses
impeller pump
air filter
stoichiometric
rich mixture
throttle bore

gasohol
atomize
vapor lock
detonate
turbulence
iso-octane
leaded gasoline
dieseling
booster venturi
mechanical fuel pump
fuel filter
electrical fuel pump
lean mixture
carburetor
barrel

throttle plate
input signal
throttle body
fuel rail
injector
cetane rating
coking
supercharger
wastegate
intercooler
Rootes-type supercharger
electronic fuel injection (EFI)
continuous fuel injection
sequential fuel injection (SFI)
normally aspirated

venturi
output signal
single-point fuel injection
multi-point fuel injection
heat riser valve
diesel knocking
turbocharger
turbine
backpressure
turbo lag
mechanical fuel injection nozzle
throttle body injection (TBI)
non-sequential fuel injector
early fuel evaporation (EFE)

REVIEW QUESTIONS

DIRECTIONS: The following questions are similar to those used on automotive technician certification tests. On a separate sheet of paper, write the letter of the correct choice.

1. Technician A says that multi-point fuel injection can be used on a gasoline engine.
   Technician B says that a single-point fuel injection system sprays fuel at a throttle body.
   Who is correct?
   A. A only
   B. B only
   C. Both A and B
   D. Neither A nor B

2. Which of the following gasoline air-fuel ratios theoretically would be better for engine performance at altitudes above sea level?
   A. 13.7:1
   B. 14.7:1
   C. 15.7:1
   D. 10:1

3. Which of the following statements is correct?
   I. A turbocharged small-displacement engine can produce as much horsepower as a normally aspirated larger-displacement engine.
   II. A supercharged small-displacement engine can produce as much torque as a normally aspirated larger-displacement engine.
   A. I only
   B. II only
   C. Both I and II
   D. Neither I nor II

Questions 4 and 5 are not like the first three questions. Each has the word EXCEPT. For each question, look for the choice that does *not* apply. Read each question carefully before you choose your answer.

4. All of the following will promote detonation EXCEPT
   A. high compression ratio.
   B. lean fuel mixture.
   C. ignition spark occurring too soon.
   D. high humidity in the air.

5. All of the following statements about a venturi are correct EXCEPT
   A. it is also known as a throttle bore.
   B. it is a restriction within a throttle bore.
   C. it causes the airflow through a throttle bore to speed up.
   D. it creates a low-pressure area to draw fuel up a tube immersed in a supply of gasoline.

ESSAY QUESTIONS

1. Describe the characteristics of a useful engine fuel.
2. Describe a basic fuel system.
3. Why might multi-point (port) fuel injection be more efficient than single-point (throttle-body) fuel injection?

SUPPLEMENTAL ACTIVITIES

1. Examine a vehicle and identify all visible fuel system components. Determine what type of fuel and fuel atomization system the vehicle uses.

2. Examine all flexible rubber fuel and vapor lines on a vehicle chosen by your instructor. Can you locate any leaks or cracks? Make a simple drawing of the fuel and vapor lines and report on their location and condition to your class.

3. Survey your relatives and friends. How many have vehicles with carburetors? How many have vehicles with fuel injection? How many have diesel engines? How many have vehicles with turbocharged engines? How many have vehicles with superchanged engines? Report the results of the survey to your class.

4. Visit new-car dealerships and obtain advertising booklets on vehicles with turbocharged, supercharged, and normally aspirated engines. Make a table of horsepower-to-displacement ratios by dividing the number of rated horsepower by the number of cubic inches or liters of displacement for each engine. Include EPA estimated mileage figures for all engines. Report the results of the survey to your class.

5. Discussion topic: Would turbocharging and supercharging be more or less effective at higher altitudes? Why or why not?

20 Fuel System Service

UNIT PREVIEW

Fuel system maintenance helps to ensure reliable and efficient engine operation at all speeds, loads, and temperatures. Basic maintenance of gasoline engines consists of filter replacement, checking for and repairing fuel leaks, and cleaning exterior carburetor linkages.

Diesel fuel systems are quite different and contain many specialized parts and components. However, many of the basic maintenance procedures are much the same.

Fuel system service procedures may require the replacement of individual parts or entire components.

Before turbochargers are removed for service or replacement, basic engine performance factors must be checked. A visual inspection of the turbocharger also must be performed.

LEARNING OBJECTIVES

When you have completed your assignments and exercises in this unit, you should be able to:

❏ Check for the presence of fuel.
❏ Test for fuel contamination
❏ Identify and describe sources of fuel leaks.
❏ Replace damaged metal and synthetic rubber fuel lines.
❏ Perform basic tests on gasoline fuel pumps.
❏ Replace both mechanical and electrical gasoline fuel pumps.
❏ Identify and describe specific differences between gasoline and diesel filtering systems.

SAFETY PRECAUTIONS

Perform fuel system work only in well-ventilated areas. Familiarize yourself with the location and operation of shop fire extinguishers before beginning such work.

Working around fuel systems can be hazardous. One gallon of gasoline, totally vaporized, has the explosive power of 14 sticks of dynamite—enough to destroy a large building and injure many people. Diesel fuel, although not as volatile, is flammable and can cause serious fires.

Be aware of sources of ignition for gasoline vapors. These include:

- Lighted cigarettes
- Ignition sparks
- Electrical switches
- Electric drills and other tools
- Air compressor motors
- Heat from exhaust pipes
- Flames from gas or oil heaters
- Welding equipment
- Drop lights.

Wear eye protection when working with motor fuels or using compressed air to clean parts. Fuel explosions and compressed air can hurl objects, liquids, and particles into the eyes. Liquid fuel can cause serious eye and skin burns.

Protective gloves should be worn to avoid prolonged skin contact with motor fuels and cleaning solvents.

Fuel should be drained into and stored only in approved metal containers with caps that can be tightly closed. Glass or plastic containers should not be used. They can break or leak and cause a fire and/or explosion.

Before disconnecting fuel lines, prepare proper metal containers to catch fuel and use pinch-off clamps on flexible fuel lines to prevent uncontrolled fuel flow.

Fuel-soaked rags should be placed in tightly closed, approved, fireproof metal containers. Spontaneous combustion of oily or fuel-soaked rags can cause fires and explosions.

20.1 CHECKING FOR FUEL

Before assuming any fuel system problem, check for fuel. Determine if fuel is present in the tank and at the carburetor or fuel injection system. Also check for the presence of ignition spark.

To check for fuel in a carburetor, first be sure the engine is not running. Open the *choke plate,* shown in **Figure 20-1.** Shine a flashlight down the carburetor bore and pump the accelerator linkage quickly two or

FUEL SYSTEM SERVICE

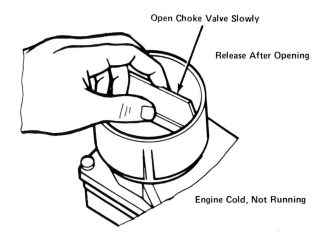

Figure 20-1. Checking for fuel in the carburetor by opening the choke plate. CHRYSLER CORPORATION

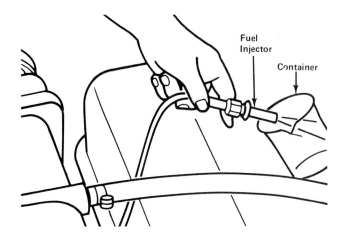

Figure 20-2. Checking for fuel in an injection system by removing an injector and placing it in a container to catch fuel spray.

three times. Look, listen, and/or smell for a strong squirt of fuel each time.

Checking for fuel in a fuel injection system can generally be done at a fuel rail connection or an individual injector. Refer to the manufacturer's service manual for detailed procedures. Disable the ignition system and have a metal container ready to catch fuel pumped through the rail or injector **[Figure 20-2]**.

Test for Fuel Contamination

The most common fuel contaminants are water, rust particles from the interior of the tank, and debris in the fuel.

SAFETY CAUTION

Keep a fire extinguisher capable of extinguishing class B flammable materials fully charged and readily available.

To check for contamination, locate the drain plug on the bottom of the fuel tank. Loosen the plug just enough so that fuel drips into a large metal can.

Allow approximately eight ounces of fuel to flow into the can, then tighten the drain plug securely and wipe off any excess fuel from the tank bottom. Pour the drained fuel into a clear glass container and visually inspect for rust particles and debris. Allow the fuel to stand for half an hour. Water in the fuel will settle to the bottom with a visible separation line.

If the fuel is contaminated, the tank must be drained completely. Then the fuel supply line must be removed from the carburetor or fuel rail. Fresh, uncontaminated fuel is used to flush the system completely to the carburetor or fuel rail. The fuel filter must be replaced and the carburetor or fuel injectors also must be disassembled and/or cleaned.

20.2 FUEL SYSTEM PREVENTIVE MAINTENANCE

Preventive maintenance for basic fuel system components consists of the following items:

- Checking for and/or repairing leaks
- Inspecting and/or replacing filters.

20.3 CHECKING FOR LEAKS

Leaks can occur at any point in the fuel system. This includes everything from the filler cap and filler neck of the fuel tank forward to the intake manifold. Liquid fuel, especially gasoline, can cause fires that could result in personal injury and damage to the vehicle. Check for leaks at the following points, shown in **Figure 20-3**:

- Filler cap
- Filler neck connection to tank
- Fuel tank seams
- Fuel and vapor line connections
- Metal and synthetic rubber fuel and vapor lines
- Fuel filter
- Fuel pump
- Carburetor
- Fuel injection pump
- Fuel injection lines and injectors
- Carburetor gaskets or fuel injection pump seals
- Intake manifold gasket.

Make a thorough visual inspection. Intermittent leaks often leave stains on parts. Some leaks occur only when the system is pressurized. Run your hand gently over lines and connections. Wetness can indicate fuel leakage. Smell the wetness to determine if it is fuel.

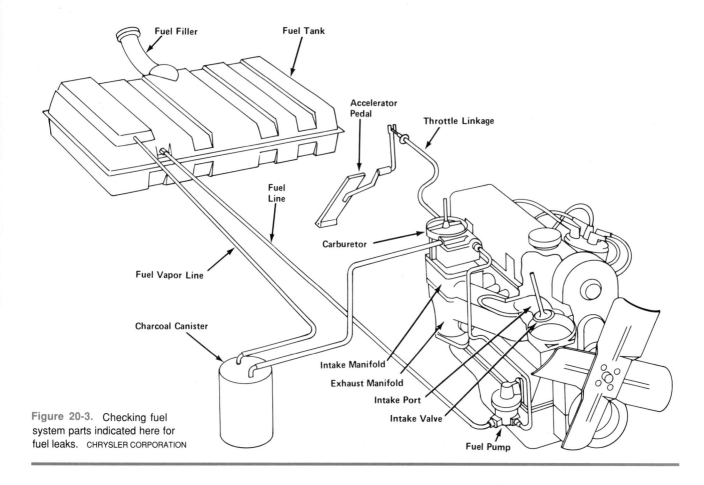

Figure 20-3. Checking fuel system parts indicated here for fuel leaks. CHRYSLER CORPORATION

SAFETY CAUTION

Be extremely careful when checking for leaks when the engine is running. Hot exhaust system components can cause fuel to vaporize and/or explode. Ignition sparks can ignite fuel vapors. Keep a fire extinguisher capable of extinguishing class B flammable materials fully charged and readily available.

Start the engine and let it run slowly at idle. Check for leaks between the fuel pump and the carburetor or fuel injection system.

SAFETY CAUTION

Stay away from moving fans, belts, and pulleys when the engine is running. Roll up long sleeves, remove all jewelry, and tie long hair back when working on a running engine.

20.4 REPAIRING LEAKS

Most leaks are repaired by replacing gaskets, fuel lines, and fuel line fittings. If a part without a replaceable gasket leaks, the entire unit must be replaced. After repairs are done, check fuel system parts for leaks again.

Fuel Tank and Related Parts

If a fuel tank cap is defective, replace it. Rubber sleeves or couplings connecting fuel tank filler necks to tanks can be replaced separately. Metal fuel tanks can be repaired by using special safety precautions and welding techniques. Epoxy compounds can be used to patch small leaks.

SAFETY CAUTION

Do not attempt any fuel tank repair requiring heat, welding, or hammering. Welding of metal fuel tanks is a specialized technique. Fuel vapors, even in an empty and thoroughly washed tank, can explode violently if the tank is heated or crushed.

Fuel and Vapor Lines

Leaks caused by loose clamps on synthetic rubber lines in good condition can be repaired by tightening or replacing the clamp. Fuel or vapor lines that are cracked or leaking should be replaced immediately.

SAFETY CAUTION

The fuel tank should be drained before attempting to replace synthetic rubber or metal fuel lines. Store the

FUEL SYSTEM SERVICE

fuel in a clean, tightly sealed, approved metal container away from any sources of ignition.

Fuel line replacement. Metal lines can become closed, abraded (worn through rubbing), or rusted through. A special type of *flare,* or end shape, is used on metal fuel lines and brake lines. This is called a double lap flare. Preparation of a new metal fuel line is illustrated in **Figure 20-4.**

Short sections (12 inches or less) of damaged tubing can be replaced with hose. Cut out the defective section and replace it with approved fuel line hose. Overlap each end approximately two inches to allow for proper clamping and sealing.

Sections longer than 12 inches should be replaced with steel tubing. The steel tubing should be connected at each end with approved fuel line hose and clamps. Be sure to properly secure any new line or hose to prevent damage from vibration. Any damaged or leaking fuel line hose must be replaced with a new hose.

Leakage From Fuel System Parts

If a part such as a fuel pump or fuel filter is leaking, the entire unit must be replaced. Leakage from carburetor sections can be repaired during carburetor overhaul. Leakage from fuel injection system components may be repairable, or may require replacement.

Gaskets

Leaks around fuel pump, carburetor, or other fuel system gaskets generally require partial disassembly or removal of the unit. In some cases, sealants can be used to stop leakage. In other cases, entire gaskets must be replaced.

CAUTION!!!

Use only sealants compatible with the fuel used. Refer to the manufacturer's service manual for recommended procedures and sealants. In some cases, sealants are not to be used on gaskets.

If a molded plastic or rubber fuel system part is cracked or broken, the entire part should be replaced.

Carburetor and Intake Manifold Vacuum Leaks

A vacuum leak into the intake manifold will allow excess air to be drawn into the intake manifold. The excess air will produce an excessively lean air-fuel mixture in the cylinders.

A lean mixture can cause rough running, poor driveability, stalling, and *lean misfire.* Lean misfire occurs when one or more cylinders misfire because the air-fuel mixture is too lean to ignite properly. Lean

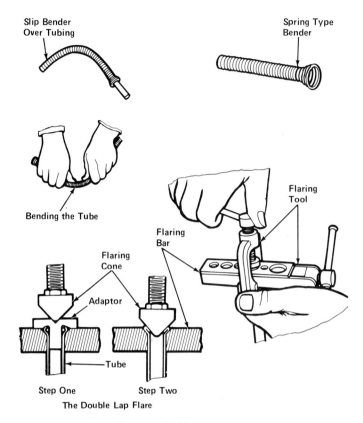

Figure 20-4. Preparing a new fuel line. CHRYSLER CORPORATION

misfire causes large amounts of unburned fuel to pass through the engine and increases exhaust pollution.

A simple method to check for vacuum leaks around the base of the carburetor and at the intake manifold gasket is to use a can of carburetor spray cleaner. Check the manufacturer's service manual and bulletins for approved spray cleaners that will not damage oxygen sensors.

Spray cleaner chemical can be drawn in through a vacuum leak to enrich the air-fuel mixture. If a lean air-fuel mixture is enriched, engine rpm will increase slightly.

SAFETY CAUTION

Carburetor spray cleaner may ignite and cause a fire. Have a fully charged fire extinguisher capable of extinguishing classes B and C fires ready for use during this test.

To perform the test, connect a low-rpm tachometer to the engine. Start the engine and let it warm up to normal operating temperature. Remove the air cleaner housing. With the engine at normal curb idle, spray carburetor spray cleaner briefly at different locations around the base of the carburetor. Then spray around the intake manifold gasket surfaces. A momentary increase in engine rpm indicates a vacuum leak.

If vacuum leaks are found, remove the carburetor and/or intake manifold. Replace gaskets as necessary. Replace all parts, then recheck for vacuum leaks.

CAUTION!!!

Some fuel injection components may be harmed by carburetor spray cleaner. Refer to the manufacturer's service manual for recommended testing procedures for fuel injection units.

Replacement of carburetor, fuel injection system, or intake manifold gaskets requires component removal. Refer to the manufacturer's service manual for correct removal and replacement procedures.

20.5 REPLACING FUEL AND AIR FILTERS

Fuel filters may be installed in a fuel line or inside a fuel system component. PCV filters, or "crankcase breather filters," always should be checked and/or replaced when air filters are replaced.

In-line Fuel Filter

Fuel filters may be installed between sections of synthetic rubber fuel lines. This type of filter is removed by loosening the clamps and pulling the filter from the lines. Replace the rubber fuel line sections and reinstall the new filter with the arrow pointing *toward* the carburetor or fuel-injection system **[Figure 20-5]**.

Screw-on Fuel Filter

Fuel filters are removed from screw-on housings by loosening the housing cover and removing the filter. New gaskets generally are supplied with the filter. Tighten the housing cover as recommended by the vehicle manufacturer **[Figure 20-6]**.

Fuel Filter in Carburetor Housing

Replacement of fuel filters within carburetor housings requires special care. An open-end and a *flare-nut wrench* must be used together to prevent the metal fuel line from becoming bent, twisted, or crimped. Hold the open-end wrench tightly as the flare-nut wrench is turned to loosen the fitting **[Figure 20-7]**.

Unscrew the fitting. Then, unscrew and remove the large housing nut. Note the position of the filter and spring as it is removed and replace the new filter in the same position.

Replace Air and PCV Filter

Replacing most air filters requires removing the top of an air cleaner housing. Wipe the inner surfaces of

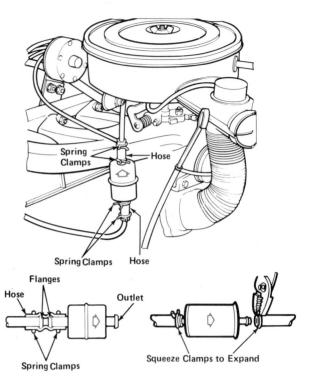

Figure 20-5. Inline fuel filter replacement.
CHRYSLER CORPORATION

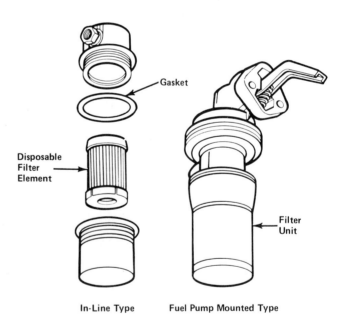

Figure 20-6. Screw-on fuel filter assembly.
CHRYSLER CORPORATION

the housing with a lightly oiled rag to remove dust and other contaminants.

Pleated-paper air cleaner elements cannot be effectively cleaned. Replace the element if thick dust is evident on the surface of the paper or if it has been in service for more than one year or 12,000 miles.

FUEL SYSTEM SERVICE

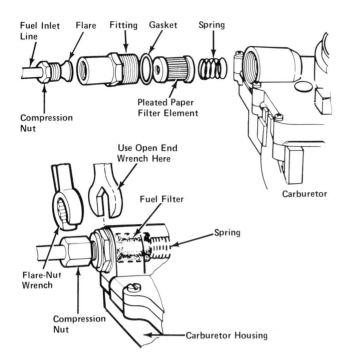

Figure 20-7. Removing parts of a carburetor fuel filter.
CHRYSLER CORPORATION

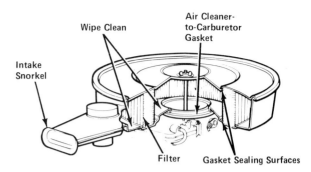

Figure 20-8. Air cleaner element replacement.
CHRYSLER CORPORATION

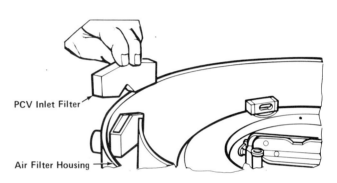

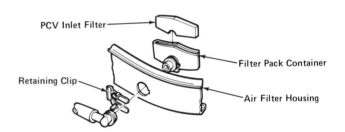

Figure 20-9. Replacing PCV filter. CHRYSLER CORPORATION

Check that the air cleaner-to-carburetor gasket, shown in **Figure 20-8,** is in good condition. Replace the gasket if torn or broken.

Also check and/or replace the PCV filter, shown in **Figure 20-9,** if fitted. PCV systems are explained in Unit 42.

20.6 TESTING A FUEL PUMP

Three tests may be made to determine if a fuel pump is defective and needs to be replaced:

* Pressure test
* Output volume and aeration test
* Inlet vacuum test.

SAFETY CAUTION

Have a properly charged fire extinguisher capable of extinguishing class B fires readily available during fuel pump testing. Diesel injection fuel pump testing requires special high-pressure test instruments and procedures. Do not attempt diesel fuel injection pump testing without the manufacturer's specific instructions and test equipment.

Pressure Test

A pressure test is conducted by connecting a pressure/vacuum gauge to a T-fitting in the fuel line, as shown in **Figure 20-10.** Refer to the manufacturer's specifications for correct pressure output and test procedures.

If the pump has a vapor return to the fuel tank, disconnect and plug the line tightly. After your pressure reading has been recorded, note the time necessary for the pressure to dissipate. A serviceable diaphragm-type pump should hold pressure for at least 30 seconds. This check does not apply to electrical impeller-type pumps.

Output Volume and Aeration Test

The following test will determine whether a diaphragm-type pump can provide a sufficient volume of gasoline and is not leaking. Disconnect the fuel line from the carburetor, and disable the ignition system. Connect a length of hose to the end of the line. Direct the hose into a large metal container away from the engine **[Figure 20-11].**

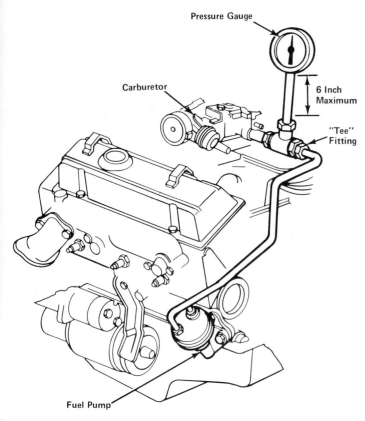

Figure 20-10. Fuel pump pressure test. CHRYSLER CORPORATION

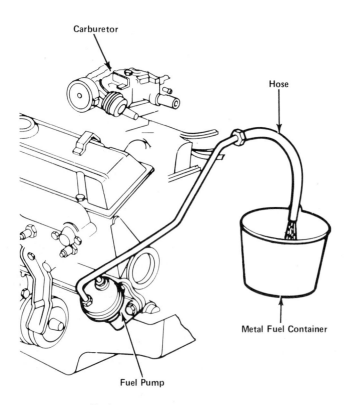

Figure 20-11. Fuel pump output test. CHRYSLER CORPORATION

Start the engine and let it idle. Hold the hose down in the bucket. Check for *aeration,* the presence of air bubbles, as the fuel covers the end of the hose. Let the engine run for 15 seconds. Aeration and/or less than 8 ounces (0.25 liter) of fuel produced usually indicates a leaking fuel pump diaphragm, or a cracked fuel inlet hose. Check manufacturer's specifications for exact volumes.

Similar problems in an electrical pump can be caused by a number of other factors, including:

- Dirty or corroded electrical ground or power connections
- Weak battery
- Faulty fuel pump relay (switch) unit
- Faulty safety relay unit.

Refer to the manufacturer's service manual to check and repair these problems.

Kinked or blocked fuel lines, especially in cold weather, may also cause insufficient pump output from electrical or mechanical pumps.

Inlet Vacuum Test

To check inlet vacuum, remove the fuel supply line from the fuel pump. Connect a vacuum gauge to the inlet connection. Crank the engine and compare the reading to manufacturer's specifications. If the pressure, volume, and vacuum tests meet specifications, the fuel line may be kinked, blocked, plugged, or cracked.

Checking for fuel line blockage. To check for blockage, remove the fuel supply line from the fuel pump and apply low-pressure compressed air (3 to 5 psi) to the line **[Figure 20-12]**. Be sure to remove the filler cap.

SAFETY CAUTION

To prevent fuel from pouring down the line, this test is best performed when the tank is one-fourth or less full. Raise the front of the vehicle so that the fuel lines tilt toward the tank.

Have a helper listen for bubbles at the filler neck. If ice formation is suspected, rags soaked in boiling water or heat lamps can be used to heat the fuel lines. This will melt the ice.

SAFETY CAUTION

Never use a torch of any kind to attempt to melt ice in fuel lines. Immediate explosions and fire can occur.

20.7 REPLACING THE FUEL PUMP

Modern fuel pumps are not repairable. They must be replaced if faulty.

FUEL SYSTEM SERVICE

SAFETY CAUTION

Before starting, disconnect the battery ground terminal to prevent electrical sparks that could ignite fuel vapors during fuel pump replacement.

Mechanical Fuel Pump

To replace a mechanical fuel pump, first loosen and remove the fuel lines, shown in **Figure 20-13.** Then, loosen the mounting bolts and gently remove the pump. Clean the gasket mounting surface and install a new gasket and pump. If the position of the camshaft eccentric prevents pump replacement, turn the engine slightly by hand to reposition the eccentric.

Electrical In-Tank Pump

Electrical fuel pumps mounted within the fuel tank must be removed from the tank, as shown in **Figure 20-14.**

Clean the mounting surface and install a new gasket and pump. Double check all connections and allow fuel vapors to dissipate. Clean and reconnect the electrical connections. Reconnect the battery ground terminal when finished.

20.8 DIESEL FUEL FILTERS AND WATER SEPARATORS

Fuel injection systems can be damaged severely by small contaminant particles and water that might pass through a carbureted gasoline fuel system.

Diesel fuel injection systems are particularly subject to water and particle contamination. Electronic systems that warn of water in the fuel can be part of a diesel fuel tank sender unit.

Modern diesel fuel filter combination units, shown in **Figure 20-15,** can include several elements:

- Water separator
- Water sensor sender
- Water drain
- Fuel heater
- Hand primer pump
- Filter change indicator sender.

Diesel fuel systems have a number of fuel conditioning components, which differ among manufacturers. However, all systems have fuel filters, which must be replaced routinely according to prescribed maintenance schedules.

Many systems have water sensors located either in the fuel tank or in a separate water separator. When the water sensor light is lit, the system must have the water drained as soon as possible.

Another feature common to diesels is a fuel heater. Warming the fuel prevents *clouding* (formation of small wax particles) and plugging of the fuel filter during cold-temperature operation.

A fuel filter change signal, available on some systems, warns of an excessive pressure drop, which indicates filter plugging.

All these features are included in the unit shown in **Figure 20-15.**

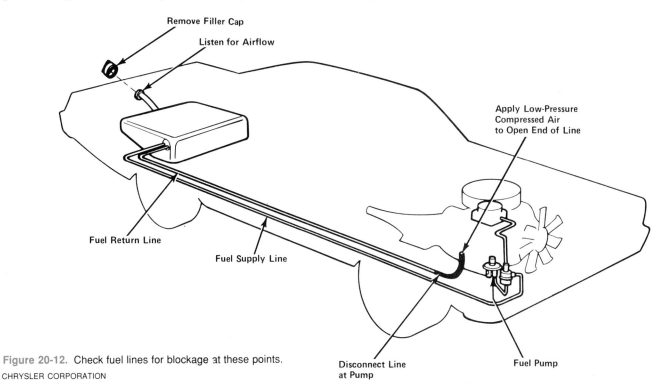

Figure 20-12. Check fuel lines for blockage at these points.
CHRYSLER CORPORATION

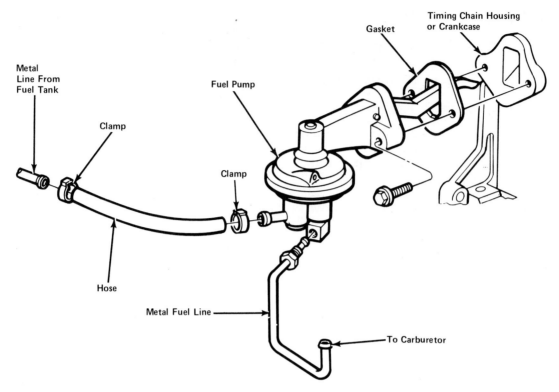

Figure 20-13. Mechanical fuel pump mounting. CHRYSLER CORPORATION

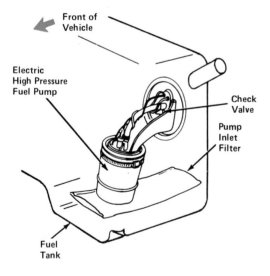

Figure 20-14. Electrical fuel pump mounting.
FORD MOTOR COMPANY

20.9 TURBOCHARGER SERVICE

Most passenger car turbocharger service needs are caused by lubrication problems. Oil *coking*, or the formation of gum, varnish, and carbon through heat, can damage or destroy bearings and shafts. Excess clearance on such parts then can cause rubbing damage to the compressor wheels and housings.

CAUTION!!!

Care must be exercised whenever a basic engine bearing or internal part is damaged or changed on a turbocharged engine. The oil and oil filter should be changed to remove possible contaminants. In addition, the turbocharger should be flushed with clean engine oil to remove contaminants before reuse. API Service SG and synthetic oils will help prevent oil coking and resultant damage.

A turbocharger may not be the cause of an engine running problem. Refer to Unit 15 for basic engine troubleshooting procedures.

To inspect for turbocharger damage, remove intake and exhaust tubing and use a light to inspect the compressor blades **[Figure 20-16]**.

Check that air ducts and clamps are properly connected and tight. Rotate the compressor wheel by hand and check for binding or rubbing. Lift both the compressor and turbine wheels and check the shaft and bearings for excessive clearance play. Refer to the manufacturer's service manual for specific checking procedures.

Some manufacturers require that turbochargers be replaced as complete units. Other manufacturers recommend field servicing for minor damage or wear. Refer to the vehicle manufacturer's service manuals for correct procedures.

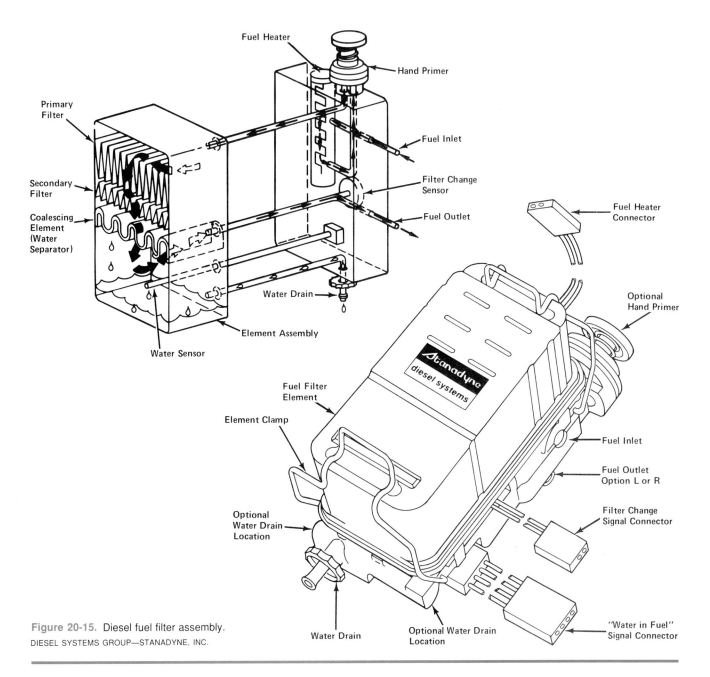

Figure 20-15. Diesel fuel filter assembly.
DIESEL SYSTEMS GROUP—STANADYNE, INC.

20.10 SUPERCHARGER SERVICE

Check supercharger engine mounting brackets for tightness. Retighten to the manufacturer's torque specifications if required. Check, tighten, or replace worn drive belts. Check lubricant reservoirs and add lubricant if necessary. Also check air ducts and clamps for proper connections and tightness.

UNIT HIGHLIGHTS

- Before assuming that a fuel system part must be defective, check for the presence of fuel in the tank. Also check for fuel at the fuel metering and atomization system.
- Basic fuel system maintenance consists of checking for leaks and replacing filters.
- Most leaks are repaired by replacing gaskets, fuel lines, and fuel line fittings. If a part without a replaceable gasket leaks, the entire unit must be replaced.
- Diesel fuel system filter combination units can include fuel heaters, water separators and drains, warning light systems, and hand-operated primer pumps.
- Basic fuel pump tests include pressure, output volume, and inlet vacuum tests.
- Diesel fuel injection pumps require special testing procedures and test equipment.

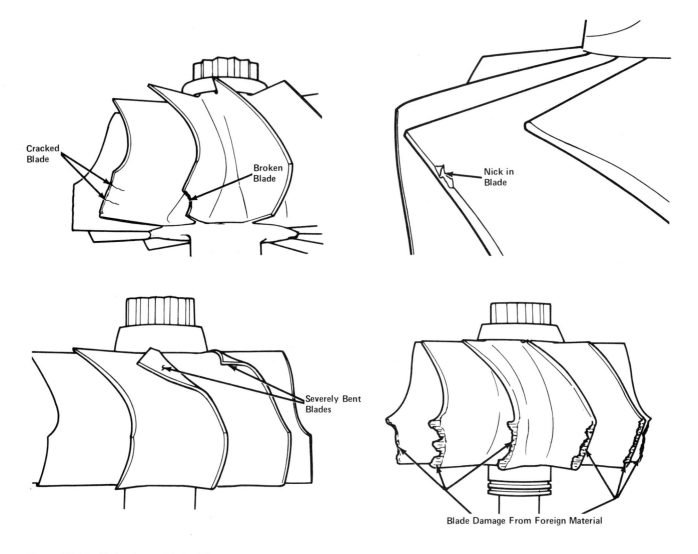

Figure 20-16. Turbocharger blade defects. BUICK MOTOR DIVISION—GMC

- The main cause of turbocharger damage is related to lubrication problems. Turbocharger compressor vanes can be inspected visually for damage.
- Supercharger mounting bolts, drive belts, lubricant supplies, and hose connections are checked during routine maintenance.

TERMS

flare
flare-nut wrench
coking
choke plate
clouding
aeration
lean misfire

REVIEW QUESTIONS

DIRECTIONS: The following questions are similar to those used on automotive technician certification tests. On a separate sheet of paper, write the letter of the correct choice.

1. Which of the following statements is correct?
   I. Basic fuel system checks include looking for the presence of fuel and fuel leaks.
   II. Basic fuel system service includes replacing fuel and air filters.
   A. I only
   B. II only
   C. Both I and II
   D. Neither I nor II

2. Technician A says that ice crystals can block a gasoline fuel line in cold weather.

   Technician B says that diesel fuel wax particles can block a diesel fuel filter in cold weather.

   Who is correct?
   A. A only
   B. B only
   C. Both A and B
   D. Neither A nor B

3. When replacing a mechanical fuel pump, the pump does not seem to fit close enough to the engine to fasten the bolts. What is the most likely cause?
   A. Wrong fuel pump
   B. Fuel pump arm jammed inside pump
   C. Fuel pump arm too long
   D. None of the above

Questions 4 and 5 are not like those above. Each has the word EXCEPT. For each question, look for the choice that does *not* apply. Read each question carefully before you choose your answer.

4. All of the following should be done when conducting a gasoline fuel pump output volume test EXCEPT
   A. disconnecting the battery ground cable.
   B. disabling the ignition system.
   C. using a metal container to catch fuel.
   D. having a class B fire extinguisher ready.

5. Turbocharger inspection includes checking for all of the following EXCEPT
   A. compressor blade damage.
   B. turbine wheel binding or rubbing.
   C. turbocharger rotational speed.
   D. excessive bearing clearance.

ESSAY QUESTIONS

1. Describe how to check for fuel contamination.
2. What fuel pump checks can be done to locate problems?
3. Describe how to check turbochargers for damage.

SUPPLEMENTAL ACTIVITIES

1. Locate all fire extinguishers in the shop area.
2. Safely perform a check for fuel on a carbureted fuel system.
3. Replace fuel, air, and PCV filters on a vehicle chosen by your instructor.
4. Safely perform fuel pump pressure and output volume tests on a vehicle chosen by your instructor. Report the results of the tests to your class.
5. Safely remove and replace both a mechanical and an electrical fuel pump on vehicles chosen by your instructor.
6. If available, perform an inspection on a turbocharger or supercharger according to the manufacturer's instructions. Report on any defects found to your class.

21 The Carburetor

UNIT PREVIEW

An engine should operate efficiently and smoothly, with low exhaust emissions, under all loads, speeds, and temperatures. A carburetor is one way of providing a proper mixture of air and gasoline for different operating conditions.

The carburetor provides a precisely metered amount of gasoline to be atomized into the incoming air stream. Separate circuits within the carburetor provide a varying air-fuel ratio for good driveability under many different conditions.

Only electronically controlled feedback carburetors are used on recent passenger vehicles.

LEARNING OBJECTIVES

When you have completed your assignments and exercises in this unit, you should be able to:

- ❏ Describe the basic processes involved in carburetion.
- ❏ Identify and describe the six basic carburetor circuits.
- ❏ Visually locate and identify carburetor parts and accessories.
- ❏ Describe different types of carburetors and their uses.

21.1 CARBURETION

Carburetion means enriching a gas (usually air) by combining it with a carbon-containing compound (a hydrocarbon fuel, usually gasoline). Four general stages, or steps, are involved in carburetion:

- Control of air flow
- Metering
- Atomization
- Vaporization.

21.2 CONTROL OF AIR FLOW

The amount of air that flows through a carburetor is controlled by the throttle valve, or throttle plate. The throttle valve is located at the bottom of the carburetor [Figure 21-1].

The throttle plate is linked to the accelerator pedal within the passenger compartment by metal

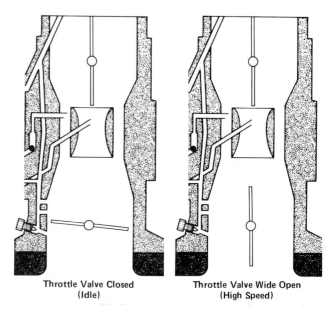

Throttle Valve Closed (Idle)   Throttle Valve Wide Open (High Speed)

Figure 21-1. A throttle valve controls airflow into the carburetor.

rods or a cable mechanism. The throttle plate can be opened slightly to produce a uniform and correct idle speed. This adjustment is made with an electrical and/or mechanical device on the carburetor.

21.3 METERING

Metering means measuring. In the case of carburetion, fuel is metered into the airstream passing through the bore, or venturi, of the carburetor. The amount of fuel is varied according to the amount of air passing through the carburetor. Other factors also influence the amount of fuel metered into the air. These factors include engine temperature, load and speed requirements, and the content of exhaust fumes. The amount of fuel metered into the airflow is controlled by *jets.* Jets are precisely sized, calibrated holes in a hollow passage for fuel or air. Three basic types of jets are used:

- Fixed restriction
- Variable restriction
- Air-bleed jet.

Fixed Restriction

Jets may be formed by drilling in the metal at the end of a fuel passage within the carburetor. However, in most cases, jets are small metal fittings with holes that are screwed or pressed into the fuel passages [Figure 21-2].

The Carburetor

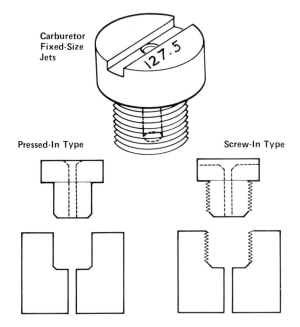

Figure 21-2. Jets control fuel flow in a carburetor.

Such a jet can control the amount of fuel delivered to the carburetor venturi, as shown in **Figure 21-3**.

Variable Restriction

In some carburetors, a tapered or stepped needle moves within the jet *orifice,* or hole. This allows different amounts of fuel to flow through the hole. The larger the step or needle diameter positioned within the hole, the less fuel can flow through the jet **[Figure 21-4]**. Needle movement can be controlled by mechanical linkage, vacuum, or the movement of an electrical solenoid or stepping motor. Fixed and variable restrictions are often used in the main metering circuit of the carburetor.

Some carburetors use a screw with a needle-shaped end to perform the same function. The needle end of the screw fits into a conical (cone-shaped) passage. Turning the screw inward or outward

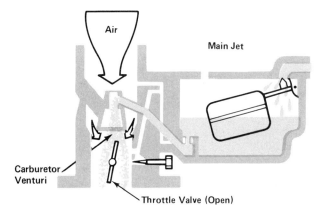

Figure 21-3. Main metering circuit.

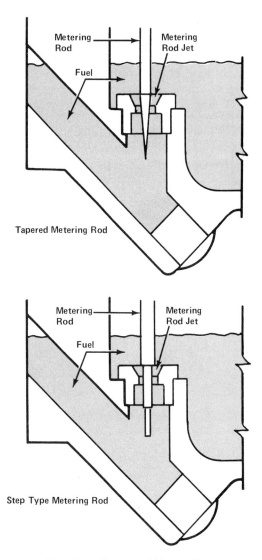

Figure 21-4. Metering rods move within a carburetor to vary fuel flow.

decreases or increases the fuel flow in proportion to the space left open. Such an arrangement is known as a *needle valve.* Most *idle mixture control screws* are needle valves **[Figure 21-5]**. On many carburetors built since the late 1970s, the mixture control screws have been capped or sealed. This is done to prevent readjustment, which might increase harmful exhaust gas emissions.

Air-Bleed Jet

Because a basic carburetor provides fuel in proportion to air flow, the mixture could become overrich at higher engine rpms. This condition would result in increased exhaust emissions and poor fuel economy. To assist in fuel ratio control and help atomize the fuel, special jets called *air-bleed jets* can be used **[Figure 21-6]**.

As fuel is consumed more quickly at high engine speeds, the fuel level drops, uncovering a perforated

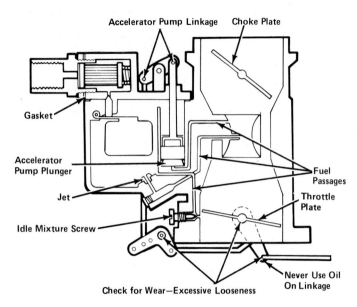

Figure 21-5. Parts of a carburetor. CHRYSLER CORPORATION

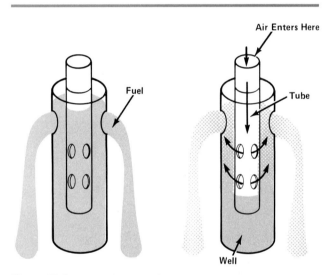

Figure 21-6. Air-bleed jets lean the air-fuel mixture at high rpm.

tube. Air enters this tube from the top. The air mixes progressively with the fuel passing out the sides of the jet as the fuel level drops. This action provides a leaner, more correct mixture at low-load high cruising speeds. Air-bleed jets are also known as *emulsion tubes.* Since the mixture is discharged as an emulsion, atomization takes place readily.

To make the air-fuel mixture leaner or richer, additional air or fuel passages can be opened or closed. These methods are discussed in the following topics.

21.4 ATOMIZATION

The metered air-fuel emulsion is drawn into the air stream in the form of tiny droplets. Despite the name of this process, *atomization,* fuel particles are not reduced to the size of atoms. Very small droplets of fuel are drawn out of passages, called *discharge ports.* From these ports, the droplets enter the air stream that flows into the intake manifold.

21.5 VAPORIZATION

Because an atomized droplet is small, its surface area is in contact with a relatively large amount of surrounding air. In addition, the venturi is a low-pressure area. These factors—emulsification, vaporization, and low pressure—combine to create a fine mist of fuel below the venturi in the bore.

Vaporization takes place below the venturi, in the intake manifold, and within the cylinder. As discussed in Unit 20, swirl, turbulence, and heat within the intake manifold and cylinder also help to vaporize fuel.

21.6 CARBURETOR VACUUM OPERATION

The vacuum developed within the intake manifold varies according to throttle position. As the throttle is opened suddenly, vacuum drops. Without a mechanism to compensate for this problem, the mixture would become too lean.

Conversely, when the throttle is closed suddenly after it has been open, a high vacuum exists within the intake manifold. This high vacuum, along with a sudden loss of air, causes excess fuel to be drawn into the cylinders, causing an overly rich mixture. In addition, exhaust gases may be drawn into the cylinder during valve overlap by the high intake manifold vacuum. This excess amount of exhaust gases takes the place of oxygen in the intake charge, causing poor combustion.

To compensate for these and other operating conditions, several *circuits,* or systems of mechanisms, are included within the carburetor. These circuits provide the proper air-fuel mixture for widely varying engine speed, load, and temperature conditions.

21.7 BASIC CARBURETOR CIRCUITS

The basic circuits within a carburetor are:

- Float circuit
- Idle circuit
- Accelerator pump circuit
- Main metering circuit
- Enrichment circuit
- Choke circuit.

Float Circuit

A constant level of gasoline within the carburetor is needed to supply the engine during various operating conditions. This fuel level is maintained within the *float bowl,* a miniature fuel reservoir within the carburetor.

A lightweight part called a *float* rests on the surface of the fuel, as shown in **Figure 21-7**. A needle valve controls the flow of fuel into the float bowl.

As fuel is consumed, the fuel level drops and the float descends. This action opens the needle valve farther. Fuel from the fuel pump enters the float bowl and raises the float, decreasing the needle valve opening and the flow of fuel.

During normal vehicle operation, the needle valve remains open just enough to maintain a constant level of fuel within the float bowl **[Figure 21-8]**.

Idle Circuit

Idle speed is the speed at which the engine runs without the accelerator pedal depressed. During idling, only a small gap exists between the almost-closed throttle plate and the carburetor bore. The flow of air through the venturi is not sufficient to draw fuel into the carburetor bore.

The pistons are still moving on their intake strokes, attempting to draw in an intake charge. However, the supply of air is restricted, or throttled, so a high vacuum exists below the throttle plate.

An opening called the *idle discharge port* is positioned just below the edge of the throttle plate **[Figure 21-9]**. The high vacuum draws fuel from the idle discharge port to allow the engine to run at idle.

Because of the relatively small amount of air, air-fuel mixture at idle is rich, approximately 12:1.

Low-speed operation. The amount of fuel flowing through the port is not sufficient for the vehicle to run smoothly at speeds above idle. Another port, known as the *off-idle, transitional,* or *transfer* port, is located just above the idle discharge port **[Figure 21-9]**. During idle conditions, this port acts as an air bleed to provide extra air and assist atomization.

As the throttle begins to open vacuum, however, it uncovers the off-idle port and applies vacuum to

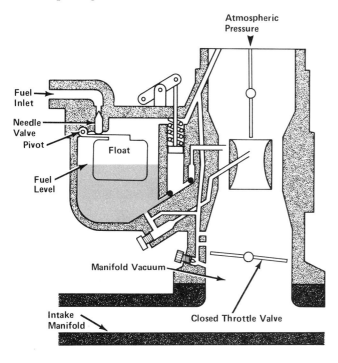

Figure 21-7. A float controls fuel level in a carburetor.

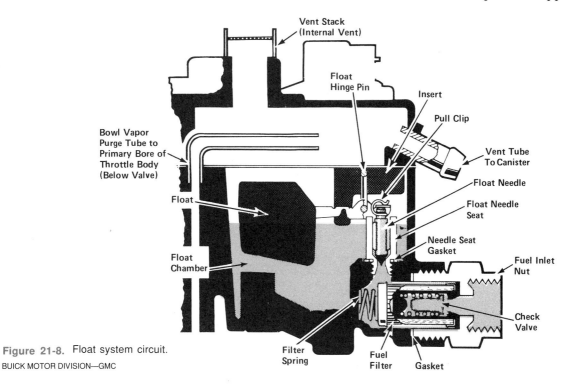

Figure 21-8. Float system circuit.
BUICK MOTOR DIVISION—GMC

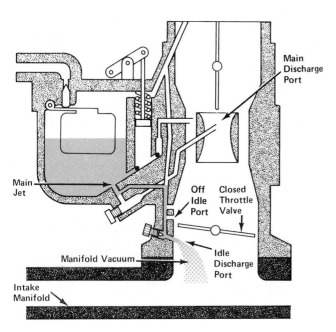

Figure 21-9. An idle discharge port draws fuel from the carburetor to allow the engine to run at idle.

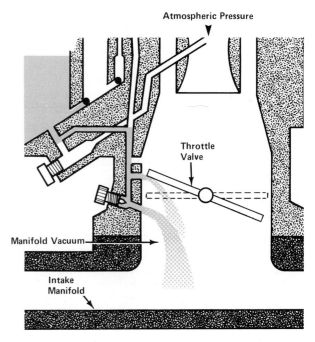

Figure 21-10. The idle discharge port and off-idle port supply fuel to the carburetor at low speeds.

it. At this point, both the idle discharge port and the off-idle port supply fuel to allow smooth engine operation at low speeds [**Figure 21-10**]. The air-fuel ratio at constant low speeds with light engine loads is approximately 14.7:1.

Accelerator Pump Circuit

When the driver presses the accelerator down suddenly to accelerate, the vacuum in the intake manifold drops momentarily. This occurs because the restriction caused by the throttle plate is almost completely removed. Atmospheric pressure forces air into the intake manifold, and the vacuum drops. The discharge of fuel from the idle and off-idle ports slows while a transition, or change, takes place. Fuel had been discharging from the idle circuit. Now, the fuel begins to discharge from the main metering discharge port in the venturi area. This brief period of change causes a lean mixture that can cause the engine to cough or stall.

To richen the mixture quickly, an *accelerator pump,* shown in **Figure 21-11,** is used. An accelerator pump can consist of a synthetic rubber plunger within a well, or cylinder, filled with fuel. Or, it may be a flexible diaphragm next to a small container of fuel. Check balls and springs may be used as inlet and outlet valves for the pump chamber.

The plunger or diaphragm usually is operated by mechanical linkage. When the accelerator is depressed rapidly, the plunger or diaphragm moves, forcing fuel to flow past the check ball and through a discharge port or nozzle into the carburetor bore. This momentary "shot" of fuel enriches the mixture.

This allows the engine to pick up speed smoothly as the air flow speeds up in the venturi. The accelerator pump fuel supply refills with fuel from the float bowl when the accelerator pedal is released. The air-fuel mixture during acceleration ranges from 10 to 12:1.

Main Metering Circuit

At cruising speeds, engine rpm is higher. The air flow through the carburetor bore and venturi creates a low pressure area which draws fuel directly from the float bowl through the *main metering circuit.* The main metering circuit, as discussed in Topic 21.2, operates by drawing fuel through a tube inserted into the float bowl. The *discharge nozzle,* or end of the tube, is placed within the low-pressure area at the venturi for maximum suction [**Figure 21-12**].

The main jet determines the amount of fuel drawn through the tube, as shown in **Figure 21-13.**

At moderate cruising speeds, the air-fuel mixture is approximately 14.7:1 or leaner. Some vehicles operate with fuel as lean as 18:1 during extremely light-load cruising. Heavy engine loads combined with low speeds can produce mixtures as rich as 12 or 13:1.

Enrichment Circuit

For full-power, heavy-throttle operation, the fuel mixture must be made richer. An *enrichment circuit,* shown in **Figure 21-14,** moves a *metering rod,* or tapered needle, within the main jet to allow more fuel to be drawn through the main metering circuit.

The Carburetor

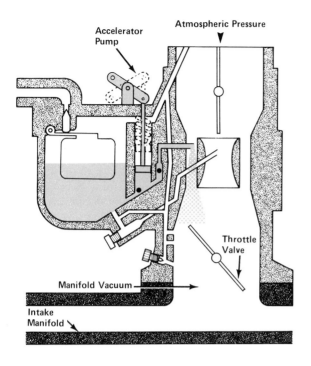

Figure 21-11. Accelerator pump operation.

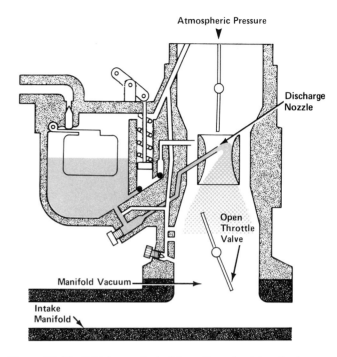

Figure 21-12. At high speeds, fuel flows through a jet and out the discharge nozzle.

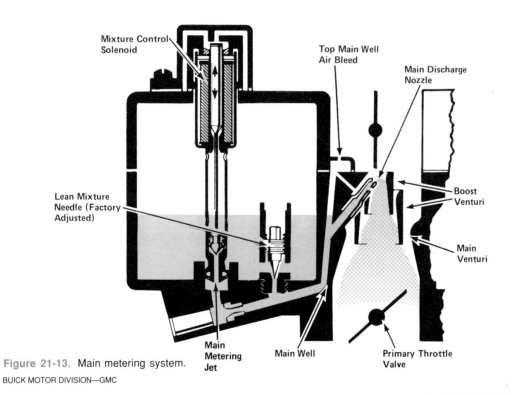

Figure 21-13. Main metering system.
BUICK MOTOR DIVISION—GMC

The needle is tapered or stepped and can be operated by mechanical means, vacuum, or electrical solenoid or stepper motor [Figure 21-15]. This enrichment circuit is sometimes called a power circuit.

During enrichment for full-power operation, the air-fuel mixture is approximately 13:1.

Choke Circuit

When an engine is cold, fuel can collect on the inside of the intake manifold in small droplets. When this condition occurs, fuel is removed from the air-fuel mixture and makes it lean. In addition, cold fuel does not vaporize easily.

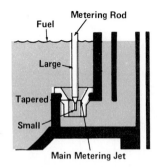

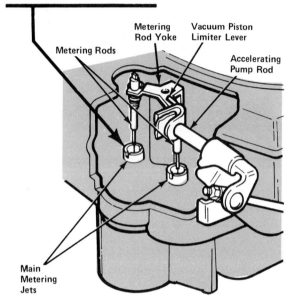

Figure 21-14. An enrichment circuit moves a metering rod to allow more fuel through the main metering circuit.

To supply enough fuel vapor for starting and warmup purposes, another valve is placed at the top of the carburetor bore. Similar to the throttle plate, this valve is called the *choke plate.* On most vehicles, it is operated automatically by a bimetallic (thermostatic) spring, as shown in **Figure 21-16.**

Cold temperatures make the spring wind more tightly, closing the choke plate. When the vehicle is cold, the accelerator pedal must be depressed to the floor to unlock the mechanism.

Heat from the intake manifold, exhaust manifold, electric heater, or circulating coolant is used to operate the bimetallic spring. As the engine warms up, heat causes the spring to unwind and open the choke plate. The choke coil spring can be located on the side of the carburetor or in a well, or depression, in the intake manifold.

When the choke plate is closed, only a small gap exists between the edge of the choke plate and the *air horn.* The air horn is the top of the carburetor bore.

As the starting system begins to turn the crankshaft, the pistons move down, to produce a vacuum in the carburetor bore. With the choke plate closed, this vacuum causes fuel to be drawn from the idle, transfer, and main discharge ports. This action causes a very rich fuel mixture to be drawn into the intake manifold.

As the engine starts, the choke plate must immediately open a slight amount. This allows enough air in to mix with the fuel to prevent flooding. As the engine warms up, the choke coil

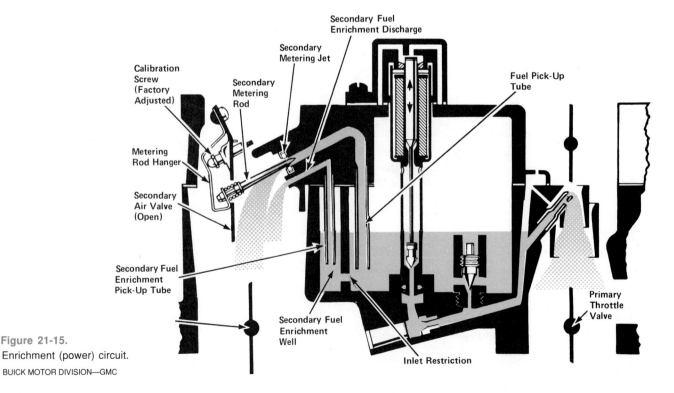

Figure 21-15. Enrichment (power) circuit.
BUICK MOTOR DIVISION—GMC

THE CARBURETOR

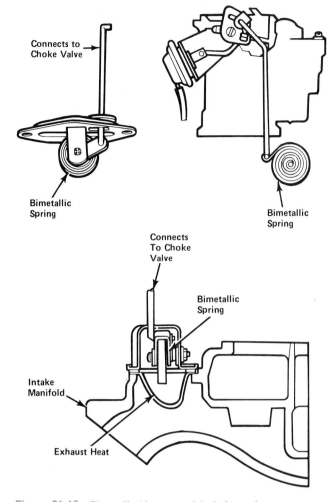

Figure 21-16. Bimetallic (thermostatic) choke spring assemblies. CHRYSLER CORPORATION

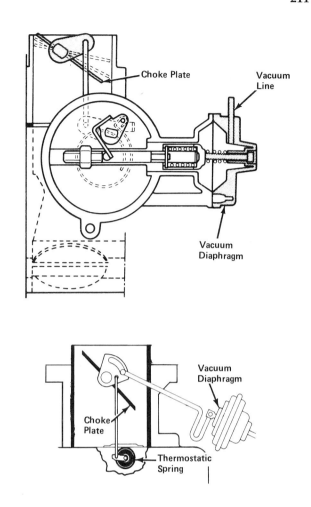

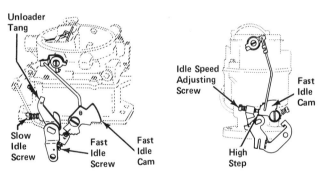

Figure 21-17. Choke pull-off and choke break mechanisms.
CHRYSLER CORPORATION

automatically opens the choke plate to the fully open position. This leans the mixture in response to temperature and engine requirements.

Several methods are used to open the choke plate slightly after starting. The choke plate shaft is offset to one side. The heavier side of the choke plate is positioned so the air flow will help force the choke open slightly.

Choke pull-off or *choke break* mechanisms use engine vacuum or mechanical motion to open the choke slightly after startup [Figure 21-17].

A *fast-idle cam* is connected by small metal rods or levers to the choke plate, as shown in Figure 21-18. This linkage opens the throttle plate to make the engine run faster for quicker and easier warmup.

When the choke is closed, an extremely rich mixture passes through the engine. This rich mixture produces poor fuel economy and large amounts of exhaust emissions.

To help the choke open more quickly, electrical heating elements can be used inside a choke coil housing [Figure 21-19].

When the key switch is on, electrical current can flow to warm the heating element and force the choke plate to open faster.

21.8 MULTIPLE-VENTURI CARBURETORS

The more carburetor bores, the more air-fuel mixture can be drawn into the engine to increase power output. Two- and four-venturi (2V and 4V) carburetors are common. Three-venturi carburetors are produced for some vehicles. The third bore provides a rich mixture for emissions control purposes.

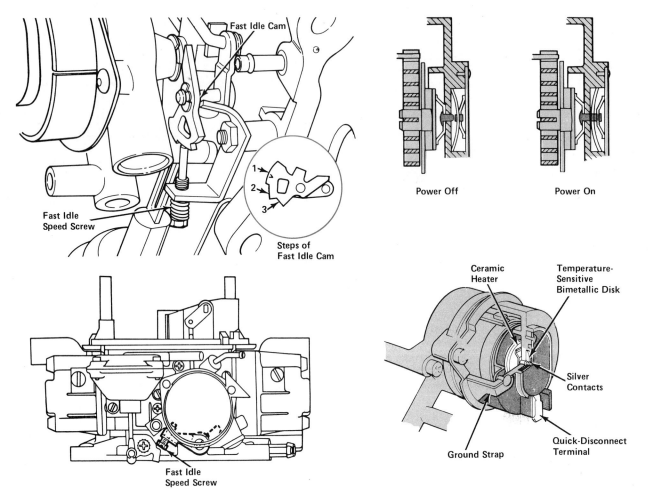

Figure 21-18. Fast-idle cam assembly. FORD MOTOR COMPANY

Figure 21-19. Electrically heated choke assembly.

On 2V carburetors, both bores may operate at all times in the ways discussed above. Alternately, one bore may operate only when extra power is demanded by the driver's pressure on the accelerator. The action and synchronization of the throttle plates can be controlled by vacuum and/or mechanical linkage.

When one bore is used for additional engine power, it may open *progressively* in relation to the *primary,* or first, throttle plate. Such a *secondary progression system* is illustrated in **Figure 21-20**.

Carburetors come in many different sizes and configurations. Carburetors for inline 4-cylinder engines typically include one-, two-, and three-venturi designs. Either two- or four-venturi carburetors are used on V-type engines.

Two-venturi primary/secondary and 4V primary/secondary carburetors were developed to gain a compromise of economy and power. They are "middle-of-the-road" carburetors.

These units have a primary side developed for high economy and a secondary side for additional power. A good example is the General Motors 4V Rochester model known as the Quadrajet, illustrated in **Figure 21-21**.

21.9 ADDITIONAL CARBURETOR EQUIPMENT

To meet complex and stringent fuel economy, driveability, and emission control requirements, modern carburetors have additional equipment, including:

- Temperature-controlled devices
- Fuel-bowl vent
- Altitude-compensation valve
- Throttle-return dashpot
- Vacuum vents
- Throttle positioner solenoid.

Temperature-Controlled Devices

When the engine and carburetor are cold, fuel vaporizes less easily. Fuel is wasted, and high exhaust emissions occur. When the engine and carburetor are hot, fuel may vaporize too easily, possibly causing the engine to stall. To prevent temperature-related problems, devices that respond to temperature are used on current carburetors.

Hot-idle compensator. When the engine begins to overheat, a *hot-idle compensator* opens an air

The Carburetor

213

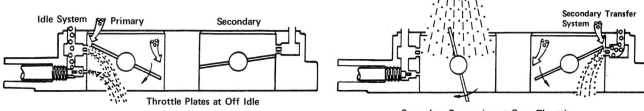

Figure 21-20. Secondary progression system operation. FORD MOTOR COMPANY

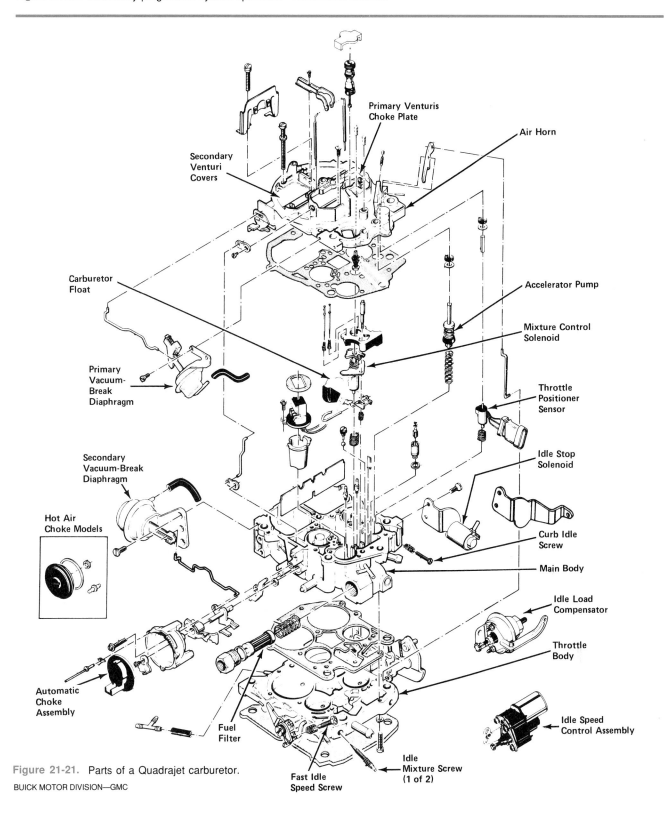

Figure 21-21. Parts of a Quadrajet carburetor.
BUICK MOTOR DIVISION—GMC

passage to lean the mixture slightly. This increases idle speed to help cool the engine, and also helps to prevent too rich a mixture from excess fuel vaporization within the carburetor **[Figure 21-22]**.

Temperature-compensated accelerator pump. A temperature-sensitive device on the accelerator pump can allow more fuel to be pumped when the engine is cold. This helps the engine accelerate well when cold. The device also causes less fuel to be pumped when the engine is hot. An example of such an accelerator pump is shown in **Figure 21-23**.

Other temperature-controlled devices can be used for carburetion problems on specific vehicles.

Fuel-Bowl Vent

Vapors from within the fuel bowl that escape into the air cause pollution. To trap the vapors, a vapor recovery system, discussed in Unit 42, is used. A large, pipe-like vent is attached to the top of the fuel bowl, as shown in **Figure 21-24.** This vent allows

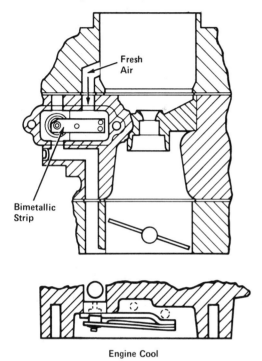
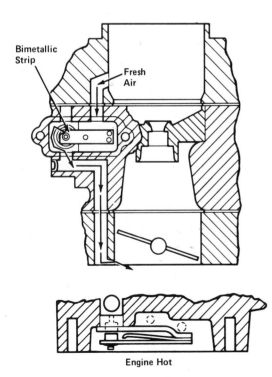

Figure 21-22. Hot-idle compensator operation.

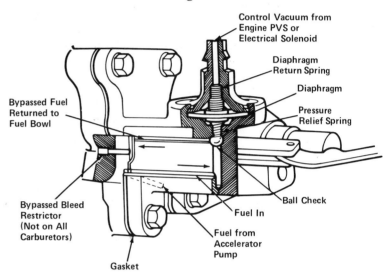

Figure 21-23. Parts of a temperature-compensated accelerator pump. FORD MOTOR COMPANY

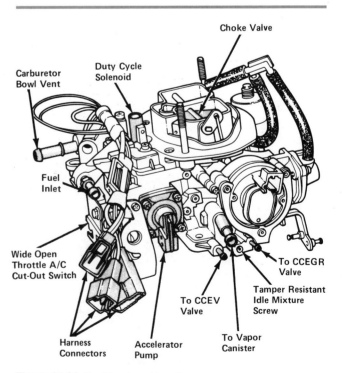

Figure 21-24. Fuel-bowl vent locations. CHRYSLER CORPORATION

THE CARBURETOR

the vapors to be trapped in a vapor-recovery canister and later drawn into the intake manifold.

Modern carburetors usually have an electrically controlled or vacuum-operated vent valve in this location.

Altitude Compensation Valve

At higher altitudes, leaner fuel mixtures are required. An *aneroid bellows* can be used to move a metering needle within a jet to automatically compensate for altitude changes. The bellows responds to atmospheric pressure. An example of such a device is shown in **Figure 21-25.**

Throttle-Return Dashpot

If the throttle is suddenly released, the high vacuum under the throttle plate draws in an excessively rich mixture. This can cause the engine to stall.

To slow the closing of the throttle plate, a *dashpot* can be attached to the throttle linkage. A dashpot is a partially sealed, flexible diaphragm attached to a pushrod. When the accelerator is suddenly released, the throttle linkage hits the pushrod, which slowly moves inward. Some dashpots also use vacuum to control the rate at which the throttle closes **[Figure 21-26].**

Vacuum Vents

Vacuum developed in the venturi and elsewhere in the carburetor bore can be used with vacuum-operated devices to perform work. Ports within the carburetor bore are connected to vacuum motors and other devices by small-diameter pipes and tubes.

Uses of vacuum developed within the carburetor bore can include:

- Making ignition spark occur sooner or later (See Unit 34)
- Operating emission-control mechanisms (See Unit 41).

Ported vacuum. *Ported vacuum* is vacuum from a port placed slightly above the position of the throttle plate at idle. As the throttle plate opens, more of the port is exposed to the flow of air through the bore. This causes a higher vacuum. If the throttle is opened wide suddenly, vacuum drops. Ported vacuum can be used to operate ignition distributor advance and/or retard mechanisms, as discussed in Unit 34. The exhaust gas recirculation system also can be activated by ported vacuum, as discussed in Topic 41.13.

Throttle-Positioner Solenoid

The position of the throttle plate for idle speed operation can be controlled by a simple screw mechanism on the throttle linkage. However, an *electrical solenoid* also can be used to set the correct idle speed when the engine is running. An electrical solenoid is a device that uses electricity to create a magnetic force to push, pull, or hold a mechanical linkage.

Some engines tend to "diesel," or run on, badly when the ignition is shut off. On other engines, turning on accessory units such as an air conditioner will slow engine idle speed. Idle speed may be slowed to a point at which the engine will shake badly or stall. To prevent either condition, idle

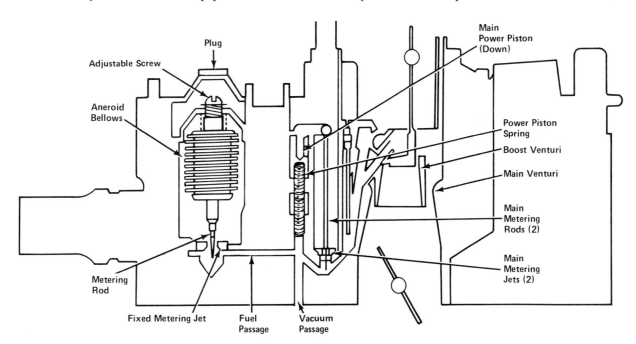

Figure 21-25. Altitude compensation valve assembly. CHEVROLET MOTOR DIVISION—GMC

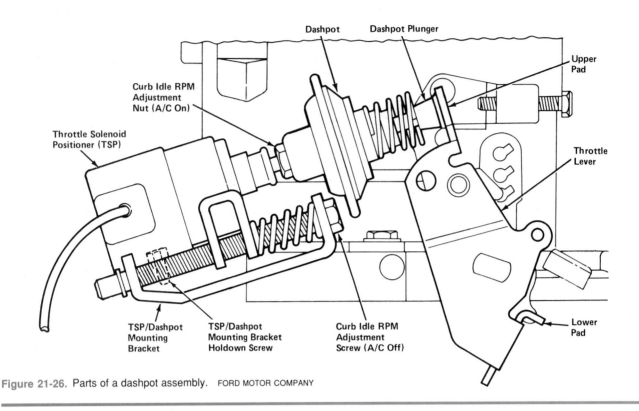

Figure 21-26. Parts of a dashpot assembly. FORD MOTOR COMPANY

speed can be set by use of an *idle-speed solenoid,* or idle positioner solenoid [**refer to Figure 21-26**].

When electricity is applied to the solenoid, its plunger extends. The plunger holds the throttle plate open enough to produce the correct idle speed, often called *curb idle speed.* When electricity is shut off, the plunger retracts, and the throttle plate closes further. This cuts off the fuel supply to prevent run-on or dieseling.

Alternately, turning on the solenoid can increase engine idle speed. This will make up for the slowing effect of parasitic losses from use of engine accessories, such as an air conditioner compressor.

Idle speed control (ISC) motor. Another method of opening the throttle is to use an *idle speed control (ISC) motor.* This device uses electricity to open the throttle plate. The ISC motor also can provide

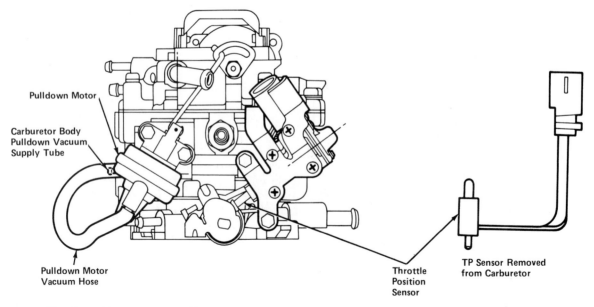

Figure 21-27. Throttle position sensor assembly. FORD MOTOR COMPANY

information to an electronic control device for a feedback carburetor system.

Throttle position sensor. A *throttle position sensor* relays information to an elctronic control system about how far the throttle plate of the carburetor is opened **[Figure 21-27]**. Depending on this information, signals might be sent to other units within the vehicle to modify their operation. The throttle position sensor will also send throttle rate-of-change information to the system controller.

21.10 FEEDBACK CARBURETOR

A *feedback carburetor* is part of a computer-controlled electronic monitoring and operating system (see Unit 38). Electronic control devices can be either input information sensors or output control *actuator* mechanisms. An actuator is a device that responds to a signal by producing an action, such as moving a mechanical device. A feedback carburetor is part of a system that includes an electronic control unit (ECU) or electronic control module (ECM) and an *oxygen sensor* to monitor emissions in the exhaust gases. See Unit 38 for a complete discussion.

Mixture-Control Solenoid

To maintain an optimum air-fuel ratio, a tapered or stepped needle can be positioned in one or more carburetor jets. This needle is similar to that used in a mixture enrichment system. The control unit can send a signal to a solenoid to lower or raise the needle. Lowering the needle restricts or cuts off fuel flow, while raising the needle permits or increases fuel flow. The control unit responds to other signals from the engine about air-fuel ratio, as discussed in Unit 38. A *mixture-control solenoid* unit is shown in **Figure 21-28.**

Feedback carburetors vary mixture ratio in relation to information provided by the oxygen sensor. The raw information is processed by an electronic control unit and "fed back" as control signals to the mixture-control solenoid.

After warmup and during cruising, the needle continuously closes and opens the fuel passage. This provides as close to a stoichiometric (14.7:1) air-fuel ratio as possible for these conditions. This ratio provides optimum engine efficiency during cruising and the lowest exhaust emissions. During warmup and hard acceleration, the electronic control system closes and opens the mixture-control solenoid to provide a richer mixture (see Unit 38).

When electricity flows to the mixture control solenoid, the needle is held in the closed position. The relative amount of time that the mixture-control solenoid closes the fuel passage is called the *duty cycle*. **Figure 21-29** illustrates the results of different duty cycles.

21.11 THE FUTURE OF THE CARBURETOR

To meet increasingly complex and strict emission control, performance, and fuel economy requirements, carburetors have become more complicated and expensive.

Fuel injection systems controlled by electronics can perform air-fuel mixture changes quickly, accurately, and reliably. Thus, vehicle manufacturers are using fuel-injection systems in place of carburetors on many newer vehicles and more expensive models.

However, solenoid-controlled feedback carburetor systems (FBC) continue to be used on inexpensive cars, as well as on trucks and utility vehicles. And carburetor service continues to be a necessary job skill for automotive technicians.

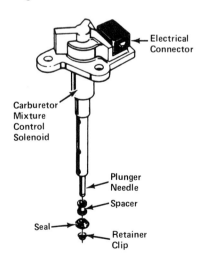

Figure 21-28. Mixture control solenoid assembly.

BUICK MOTOR DIVISION—GMC

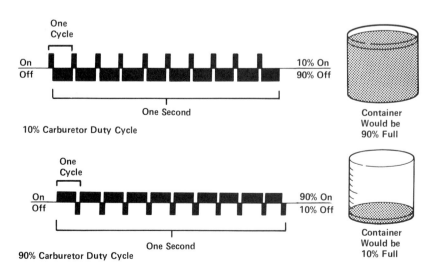

Figure 21-29. Duty cycles.

CHRYSLER CORPORATION

UNIT HIGHLIGHTS

- Carburetion consists of metering, atomization, and vaporization.
- Metering is accomplished by fixed, variable, and air-bleed jets.
- The basic carburetor circuits are the float, idle and low-speed, accelerator pump, main metering, enrichment, and choke circuits.
- Multiple-bore carburetors are used extensively today.
- Additional equipment is added to carburetors to modify their operation for temperature, altitude, emission control requirements, and operation of external devices.
- Feedback carburetors modify air-fuel ratios through information provided by sensors. This information is "fed back" through an electronic control unit to move a mixture-control solenoid.

TERMS

carburetion	metering
jet	needle valve
air-bleed jet	orifice
emulsion tube	atomization
discharge port	circuit
float bowl	float
idle speed	idle discharge port
off-idle port	transitional port
transfer port	accelerator pump
main metering circuit	discharge nozzle
enrichment circuit	metering rod
choke plate	air horn
choke pull-off	choke break
fast-idle cam	hot-idle compensator
aneroid bellows	dashpot
ported vacuum	electrical solenoid
idle-speed solenoid	curb idle speed
actuator	throttle position sensor
feedback carburetor	electronic mixture control
duty cycle	primary throttle plate
oxygen sensor	mixture-control solenoid
idle mixture control screw	secondary progression system
idle speed control (ISC) motor	

REVIEW QUESTIONS

DIRECTIONS: The following questions are similar to those used on automotive technician certification tests. On a separate sheet of paper, write the letter of the correct choice.

1. Two technicians are discussing what happens when a throttle plate is closed suddenly during cruising.
   Technician A says that large amounts of fuel can be drawn into the cylinders and cause increased power output.
   Technician B says that exhaust gases can be drawn into the cylinders during valve overlap and cause increased fuel economy.
   Who is correct?
   A. A only
   B. B only
   C. Both A and B
   D. Neither A nor B

2. Which of the following statements is correct?
   I. Many carburetor circuits are necessary to provide varying air-fuel mixture ratios for different engine temperature, load, and speed conditions.
   II. Many carburetor circuits are necessary to achieve a constant stoichiometric (14.7:1) air-fuel mixture at all times.
   A. I only
   B. II only
   C. Both I and I
   D. Neither I nor II

3. Technician A says ported vacuum can be used to operate ignition distributor mechanisms.
   Technician B says that intake manifold vacuum can be used to draw blowby gases and oil vapors from a PCV system into the cylinders.
   Who is correct?
   A. A only
   B. B only
   C. Both A and B
   D. Neither A nor B

Questions 4 and 5 are not like those above. Each has the word EXCEPT. For each question, look for the choice that does *not* apply. Read each question carefully before you choose your answer.

4. All of the following units are built into a feedback carburetor EXCEPT
   A. throttle position sensor.
   B. tapered or stepped needle in a jet.
   C. mixture-control solenoid.
   D. electronic control unit.

5. All of the following stages are involved in the carburetion process EXCEPT
   A. condensation.
   B. metering.
   C. atomization.
   D. vaporization.

ESSAY QUESTIONS

1. Identify and describe the basic steps involved in carburetion.
2. What are the six basic circuits of a carburetor?
3. What is a feedback carburetor, and how does it operate?

SUPPLEMENTAL ACTIVITIES

1. Examine a carburetor on a vehicle chosen by your instructor. Refer to the vehicle manufacturer's service manual and identify and describe as many parts as possible.
2. Refer to the proper vehicle manufacturer's service manual and locate all adjustments on a shop carburetor. Make a list of the adjustments possible on that particular carburetor.
3. Make a drawing on the classroom blackboard of a needle valve and explain how it could be adjusted to provide more or less fuel flow.
4. Imagine starting a car that has been sitting overnight. Describe the operation of each of the basic carburetor circuits as the vehicle is started, warmed up, and driven from low to highway speeds, including passing other vehicles.
5. Explain the effects of different duty-cycles on a mixture-control solenoid in a feedback carburetor.

22 Carburetor Service

UNIT PREVIEW

Carburetor preventive maintenance consists of replacing filters, checking for leaks, and adjusting automatic choke operation.

Common carburetor problems include lack of fuel at carburetor, stuck or gummy automatic choke, flooding, and driveability problems.

Driveability problems include dieseling, detonation, stalling, rough idle, missing, hesitation, surging, sponginess, poor gas mileage, and cutting out.

On modern carburetors, idle mixture adjustments must be done with the aid of an exhaust gas analyzer. This is done to meet federal and state vehicle exhaust emissions standards.

When attempts to adjust a carburetor indicate that repairs are needed, carburetor overhaul is done.

LEARNING OBJECTIVES

When you have completed your assignments and exercises in this unit, you should be able to:

❏ Perform routine carburetor maintenance.
❏ Adjust an automatic choke.
❏ Adjust idle speed.
❏ Replace carburetor and intake manifold vacuum lines.
❏ Identify and describe possible causes for driveability problems.
❏ Correctly adjust a carburetor float.

SAFETY PRECAUTIONS

Have a fire extinguisher capable of extinguishing class B (flammable liquids) fully charged and readily available when attempting any type of carburetor or fuel system service.

Make sure no sources of possible ignition are present in the area where carburetor or fuel system service is being performed. These sources include lighted cigarettes, electrical switches or tools, and open flames of any kind.

Avoid looking directly down the bore of a carburetor still installed on a vehicle. Wear eye protection at all times in the shop area. If the engine is cranked or started, flames from a backfire can shoot out of the carburetor bore. This can cause serious burns and/or blindness.

Overhauling carburetors involves working with powerful solvents that can cause skin and eye irritation and damage. Wear rubber gloves to avoid any skin contact with carburetor cleaning chemicals.

Wear eye protection when using compressed air to dry carburetor parts.

After carburetor service, crank the engine to allow the fuel pump to deliver fuel to the carburetor for starting. Do not attempt to pour liquid gasoline down the throat of the carburetor. Severe explosions and fires can result.

On vehicles with automatic transmissions, some adjustments are done with the engine running and the gear selector in drive. Set the emergency brake tightly and block the driving wheels before attempting any carburetor adjustment procedures.

Never stand directly in front of a running vehicle while carburetor adjustments are being done. If the throttle is opened suddenly, the vehicle may jump forward over wheel blocks. It can run over or crush anyone standing in front.

Modern carburetors are extremely complicated. They vary greatly among manufacturers and models. Even the same basic carburetor installed on different engines can vary considerably. Do not attempt carburetor maintenance, repairs, or adjustments without the proper manufacturer's service manual.

22.1 CARBURETOR PREVENTIVE MAINTENANCE

Carburetor preventive maintenance consists of the following items:

- Replacing air, PCV, and fuel filters
- Checking for fuel and air leaks
- Checking and/or adjusting automatic choke operation.

Filter Replacement

Refer to Topic 20.5 for fuel, air, and PCV filter replacement. Check the proper shop manual for filter identification.

Fuel and Air Leaks

Refer to Topic 20.4 for procedures to check for leaks. A carburetor is divided into three major sections:

- The *air horn*, or top
- The *main body*, which contains the float bowl and most passages and discharge ports
- The *throttle body*, or base, which contains the throttle plates.

These sections and other attachments are shown in **Figure 22-1**. Leakage can occur between sections or where fittings or external parts are attached.

In some cases, external leakage from parts of the carburetor can be stopped by tightening fittings or attaching fasteners.

Refer to the manufacturer's service manual for correct torque specifications. Carburetors are made of metal that can crack or distort. This can occur if a fitting is overtightened or if covers are tightened in an improper sequence. One example of an air horn tightening sequence is shown in **Figure 22-2**.

Checking an Automatic Choke

In time, fuel varnish and deposits can build on choke plates, cross-shafts and bearing points, and external linkages. Carburetor spray cleaner can be used, as shown in **Figure 22-3**, to remove these deposits. The choke plate can be moved by hand to work the cleaner into tiny crevices and free sticky, gummy linkages.

SAFETY CAUTION

Wear eye protection whenever using spray cans of any kind. Carburetor spray cleaner can cause severe skin and eye irritation, and may lead to blindness.

Carburetor spray cleaner also can be used to clean the external parts of the carburetor.

Internal carburetor problems caused by fuel varnish and deposits require carburetor disassembly and cleaning, as described below.

Adjusting an Automatic Choke

Automatic choke maintenance consists of three basic adjustments:

- Richness
- Choke pull-off
- Fast-idle speed.

Richness adjustment. The amount of fuel delivered during engine cranking depends upon how far the choke plate is closed. An adjustment can be made to make the choke plate close the correct amount. This is done when the engine is cold and the accelerator pedal or linkage has been depressed to unlock the choke. In some cases, a drill bit or other gauge of a specific size is used to measure a specific opening distance. This measurement is made between the top edge of the choke plate and the air horn wall.

One method of adjustment is to physically bend metal parts until they hold the choke plate in the desired position. There should be no pressure on the drill bit or gauge **[Figure 22-4]**.

Modern methods include the use of a choke *protractor*, or angle gauge, that attaches to the choke plate with a magnet **[Figure 22-5]**. Several adjustments can be made accurately with a protractor. These include fast idle cam adjustment, vacuum break adjustments, and unloader adjustments.

Adjustments can be made to automatic choke thermostatic spring mechanisms mounted on the carburetor. Loosen three holding screws and rotate a cover attached to the spring **[Figure 22-6]**.

Choke mechanisms on 1978 and later vehicles may be riveted so that no minor adjustments can be made. The rivets can be drilled out for major carburetor service, as shown in **Figure 22-7**.

CAUTION!!!

Removal of riveted choke covers and adjustment to specifications other than those of the vehicle manufacturer will change exhaust emissions. This is a violation of federal and state air-pollution laws.

Choke pull-off adjustment. Several types of choke pull-off mechanisms can be used. Vacuum-operated pull-off devices may require using a hand vacuum tester. This tool is used to apply vacuum, plug vent holes, and extend plungers and spring mechanisms. These devices are shown in **Figure 22-8**.

Refer to the manufacturer's service manual for correct adjustment procedures. One example of a pull-off adjustment is shown in **Figure 22-9**.

Fast-idle speed adjustment. Adjustments for fast-idle speed are made with the adjusting screw resting on a specified step of the fast-idle cam. To position the fast-idle screw correctly, the throttle must be opened and the choke plate manually closed. This permits moving the fast-idle cam. The correct adjustment is obtained by connecting a tachometer to the engine and adjusting the screw until specified rpm is reached. This adjustment is made while the engine is at normal operating temperature.

22.2 CARBURETOR PROBLEMS

Common carburetor problems include the following:

- Lack of fuel at carburetor
- Stuck or gummy automatic choke
- Flooding.

PART III: AUTOMOTIVE ENGINE SYSTEMS

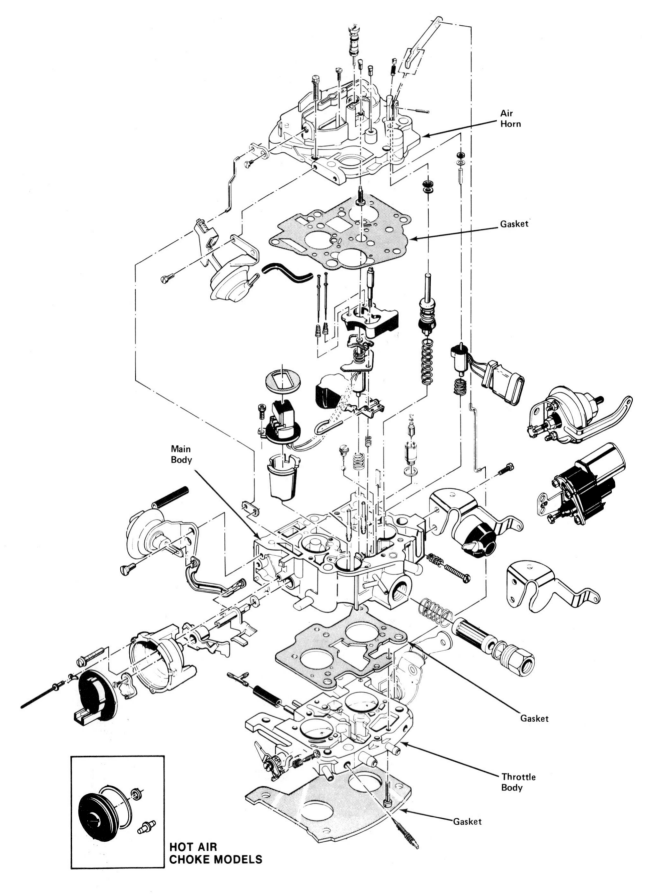

Figure 22-1. Parts of a carburetor. BUICK MOTOR DIVISION—GMC

CARBURETOR SERVICE 223

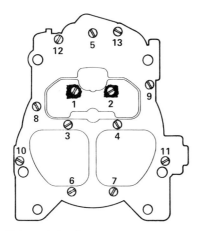

Figure 22-2. Air horn tightening sequence.
BUICK MOTOR DIVISION—GMC

Figure 22-3. Cleaning an automatic choke mechanism.
CHRYSLER CORPORATION

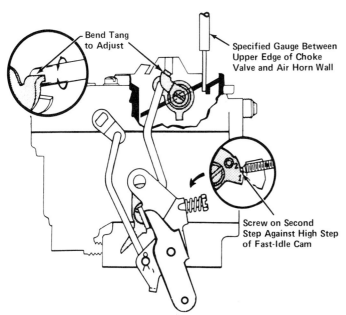

Figure 22-4. Choke rod adjustment. BUICK MOTOR DIVISION—GMC

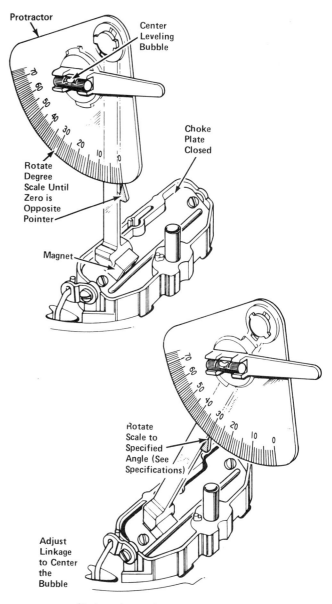

Figure 22-5. Choke valve angle gauge.
BUICK MOTOR DIVISION—GMC

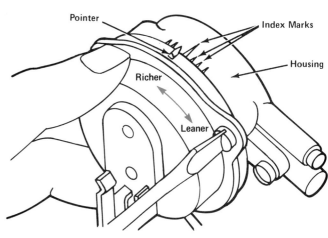

Figure 22-6. Choke spring adjustment.

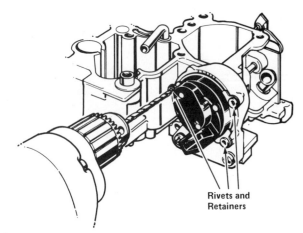

Figure 22-7. Choke cover removal. BUICK MOTOR DIVISION—GMC

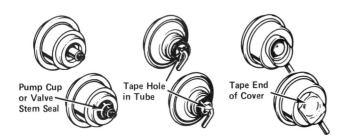

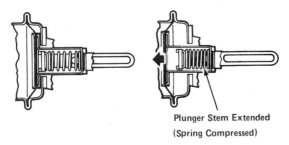

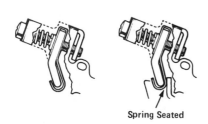

Figure 22-8. Vacuum break adjustment information.
BUICK MOTOR DIVISION—GMC

Lack of Fuel at Carburetor

To determine why there is no fuel in the carburetor, first be sure the vehicle has fuel in its tank. Second, remove the fuel line from the carburetor and place it in a suitable metal container. Crank the engine briefly to see if fuel is flowing from the line. If fuel is not flowing, refer to Unit 20 for diagnosing fuel delivery problems.

If fuel is flowing, the carburetor must be disassembled to check for problems. A new float needle and seat should be installed if necessary **[Figure 22-10]**. Follow these steps to disassemble the carburetor:

1. Remove the air horn.
2. Inspect and clean or replace the needle valve, float parts, and accelerator pump.
3. Properly reinstall the air horn.

If the carburetor must be overhauled, see Topic 22.5.

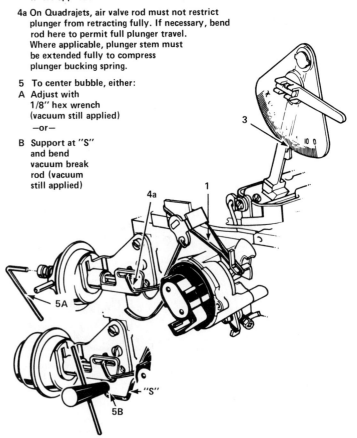

Figure 22-9. Rear vacuum break adjustment.
BUICK MOTOR DIVISION—GMC

CARBURETOR SERVICE

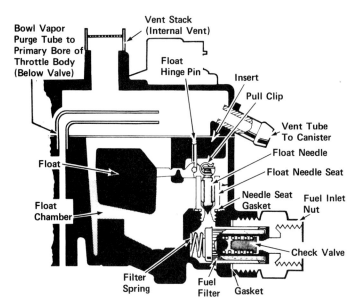

Figure 22-10. Float system assembly. BUICK MOTOR DIVISION—GMC

Stuck Automatic Choke

It is possible for the automatic choke to become stuck fully open, fully closed, or at any point in between. Cleaning a stuck automatic choke is discussed in Topic 22.1.

Flooding

Flooding is excess unvaporized fuel in the intake manifold. Flooding can be caused by heating of fuel in the float bowl after the engine is shut off. (The cooling system is no longer operating.) Flooding occurs most commonly in hot weather or after prolonged stop-and-go driving. Flooding also can occur during unsuccessful attempts at starting in cold weather.

To start a flooded engine, depress the accelerator pedal all the way to the floor and hold it down. This action fully opens the throttle plates and the choke plate for maximum air flow. Under these conditions, only air will enter the engine during cranking. This will vaporize the excess gasoline. Turn the key and crank the engine until it starts. Do not crank longer than 15 seconds. If it won't start, wait five minutes, then try again. This will prevent the starter from overheating.

Percolation. *Percolation* is the name given to the boiling of fuel within the float bowl. Boiling fuel can travel through discharge and vent tubes to the bore of the carburetor and into the intake manifold.

Float and fuel pump problems. Flooding also can be caused by a needle valve that fails to close, or by excess fuel pump pressure. These problems can be detected by observing the fuel bowl vent as the engine is cranked. Fuel will gush out of the vent, as shown in **Figure 22-11**. Flooding also may be caused by a saturated float that has sunk to the bottom of the bowl.

A contaminant particle may lodge between the needle valve and its seat and prevent the valve from closing. Tapping the fuel inlet fitting gently, and briefly may free the valve. If not, the air horn must be removed and the needle valve serviced.

The same problem can be caused by excess fuel pressure. Excess pressure may result from a defective pressure-regulation valve within the fuel pump or a blocked fuel-return line. Checking for correct fuel-pump pressure output is discussed in Unit 20.

22.3 DRIVEABILITY PROBLEMS

Driveability describes the proper operation of a vehicle. The vehicle starts easily, does not stall or die, accelerates well, runs smoothly, and delivers reasonable gas mileage. Driveability problems include:

- Dieseling
- Detonation
- Stalling
- Rough idle
- Missing
- Hesitation
- Surging
- Sluggishness
- Sponginess
- Poor gas mileage
- Cutting out.

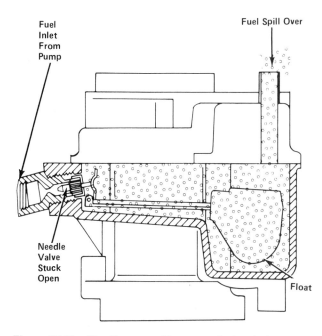

Figure 22-11. Flooding caused by excess fuel or fuel pressure.
CHRYSLER CORPORATION

Dieseling and Detonation

Dieseling and detonation are discussed in Unit 19. Fuel system problems contributing to dieseling and/or preignition are lean fuel mixtures, excessive idle speed, sticking linkages, low octane fuel, and excessive carbon deposits.

An example of a visual flow chart for diagnosis and repair of dieseling and detonation is illustrated in **Figure 22-12**.

Stalling and/or Rough Idle

Stalling, when the engine dies after starting, and/or rough idle can occur when the engine is cold or hot.

Cold stalling or rough idle. Cold stalling can be caused by ignition problems, discussed in Unit 35, or engine mechanical problems, discussed in Unit 43. Fuel system problems that cause too lean a fuel mixture also can produce stalling and rough idle. These problems include:

- Loose, broken, cracked, kinked, or disconnected vacuum hoses
- Leaking carburetor or intake manifold gaskets
- Sticky carburetor throttle or choke linkage
- Stuck needle valve (flooding)
- Defective or misadjusted vacuum pull-off units
- Incorrect choke richness or fast-idle speed
- Defective thermostatic air cleaner (See Unit 42)
- Defective EGR valve (opening too early).

Hot stalling or rough idle. Hot stalling or a rough idle can be caused by ignition problems, discussed in Unit 35, or engine mechanical problems, discussed in Unit 43. Fuel system problems that can produce stalling or a rough idle when the engine is hot include:

- Loose, broken, cracked, kinked, or disconnected vacuum hoses
- Leaking carburetor or intake manifold gaskets
- Worn carburetor throttle shaft and bore
- Incorrect curb-idle speed
- Flooding
- Defective thermostatic air cleaner (See Unit 42)
- Clogged PCV valve (See Unit 42)
- Defective EGR valve (See Unit 42)
- Carburetor idle mixture screws set too rich or too lean.

Missing

Missing is a lack of power from one or more cylinders. Missing can be caused by ignition system and engine mechanical problems, discussed in Unit 43. Fuel system problems that can cause lean misfire include:

- Loose, broken, cracked, kinked, or disconnected vacuum hoses
- Leaking carburetor or intake manifold gaskets
- Worn carburetor throttle shaft and bore.

Hesitation

Hesitation is a momentary lack of response as the accelerator is pressed down, such as when moving away from a stop. Severe hesitation can cause the engine to stall and die. Hesitation can be caused by ignition system problems, discussed in Unit 35. Fuel system-related problems that can cause hesitation include:

- Loose, broken, cracked, kinked, or disconnected vacuum hoses
- Leaking carburetor or intake manifold gaskets
- Sticky carburetor throttle or choke linkage
- Defective accelerator pump linkage or parts
- Incorrect float level
- Defective thermostatic air cleaner system.

Surging

Surging occurs when the engine speeds up and/or slows down with the throttle held steady. Surging can occur at any road speed. Surging can be caused by ignition system problems. In addition, fuel system problems contributing to surging include:

- Loose, broken, cracked, kinked or disconnected vacuum hoses
- Worn carburetor throttle shaft and bore
- Defective thermostatic air cleaner
- Clogged carburetor fuel filter
- Insufficient fuel pump pressure and/or volume output
- Restricted fuel lines
- Low fuel level in the float bowl
- Contaminants or water in fuel tank
- Internal carburetor problems (contaminants, water, defective or damaged parts, and so on).

Sluggishness or Sponginess

Sluggishness occurs when the engine will not deliver sufficient power under load or at high speed. It does not accelerate as quickly as normal, and loses too much speed going up hills.

Sponginess occurs when the engine does not speed up as much as expected when the accelerator is depressed, especially during cruising. If the pedal must be pushed down farther than normal to produce an increase in speed, the engine is said to have a spongy condition.

Carburetor Service

Step/Sequence   Result

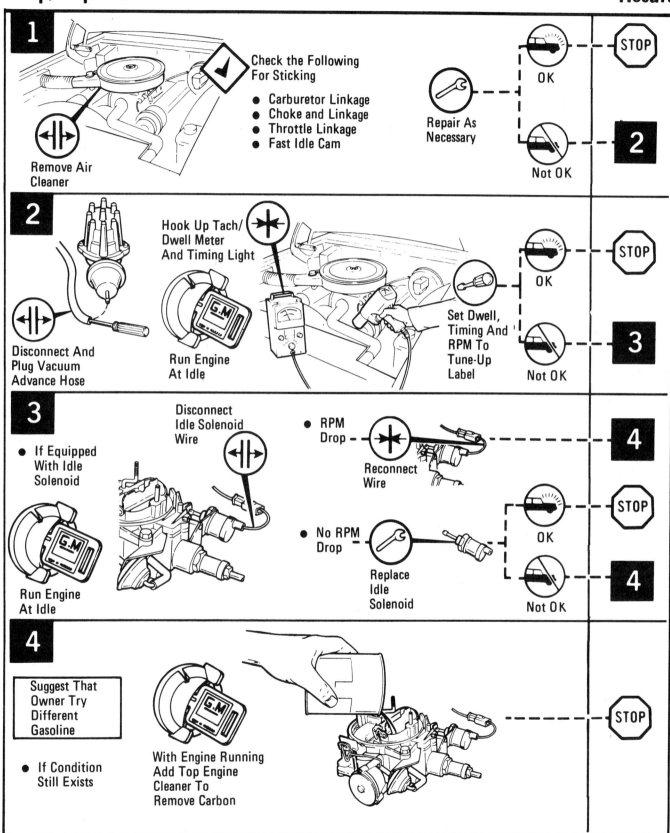

Figure 22-12. Dieseling/detonation diagnosis and repair.   GENERAL MOTORS CORPORATION

Sluggishness or sponginess can be the result of engine problems, as discussed in Unit 43. Incorrect ignition system operation or specifications also can cause these problems, as discussed in Unit 35. Fuel system problems related to sponginess include:

- Dirty air filter
- Throttle and/or carburetor linkage preventing throttle plates from opening fully
- Defective thermostatic air cleaner (See Unit 42)
- Internal carburetor problems (sticky enrichment circuit parts, contaminants in jets, incorrect float adjustment, or bent, damaged, or inoperative metering rods and power valves
- Defective EGR valve (See Unit 42)
- Low fuel pressure.

Poor Gas Mileage

Before assuming that something mechanical is wrong when a vehicle owner complains of poor gas mileage, ask about driving habits. Vehicles driven less than 10 miles before being shut off do not warm up fully. A cold engine will deliver only about half of its gas mileage capability. Best mileage occurs when the engine is fully warmed up and running at constant speeds on the highway. Other factors that lower gas mileage include:

- Extra weight in the vehicle
- Driving on steep mountain grades
- Driving in cold weather
- Underinflated tires
- Hard acceleration (hot rodding)
- Poor wheel alignment
- Driving against the wind.

The EPA (Environmental Protection Agency) new-vehicle car mileage ratings are derived in a laboratory on a chassis dynamometer. The ratings are not achieved by driving vehicles under real-world conditions. Although these ratings are a good indication of vehicle fuel mileage performance, actual driving may yield substantially different numbers. The EPA ratings include a disclaimer, or warning, that driving conditions affect gas mileage.

Mechanical conditions that can affect gas mileage include engine problems, discussed in Unit 43, and ignition system problems, Unit 35. Fuel system problems that can cause lowered gas mileage include:

- Dirty air filter
- Fuel leaks
- Defective oxygen sensor (See Unit 38)
- Loose, broken, cracked, kinked, or disconnected vacuum lines

- Carburetor or intake manifold gasket leaks
- Improper thermostatic air-cleaner operation
- Stuck heat riser valve
- Choke richness and fast-idle settings
- Internal carburetor problems (sticky enrichment circuit parts, contaminants in jets, incorrect float adjustment, or bent, damaged, or inoperative metering rods and power valves).

Cutting Out

Cutting out occurs when the engine stops entirely at irregular intervals and produces no power. The engine usually does not die completely, but the loss of power occurs repeatedly and intermittently. Cutting out is usually worst under heavy acceleration.

Ignition system problems, discussed in Unit 35, are strongly associated with cutting out. Fuel system problems that can cause cutting out include:

- Clogged carburetor fuel filter
- Insufficient fuel pump pressure and/or output volume
- Clogged or leaking fuel lines
- Clogged fuel tank pickup filter or screen
- Contaminants or water in fuel tank.

22.4 CARBURETOR ADJUSTMENTS

Before attempting any carburetor adjustments, correct any fuel leakage problems. Consult the manufacturer's service manual for specific adjustment procedures.

SAFETY CAUTION

Have a fire extinguisher capable of extinguishing class B fires fully charged and readily available if an explosion or fire should occur. Wear eye protection when you work on a running engine.

Start the vehicle and let it warm to normal operating temperature. Check for proper choke operation during warmup.

Adjust curb idle speed by turning idle speed adjustment screws or *solenoid* adjustments [Figure 22-13]. A solenoid uses electricity to provide force to hold the throttle open during normal running.

Carburetors may include both an electrically operated solenoid adjustment and a secondary curb idle speed screw adjustment. A secondary curb idle speed screw allows the throttle plate to close more fully, aiding in shutting off the engine quickly.

Solenoid Adjustment

Refer to the vehicle manufacturer's service manual for specific solenoid adjustment procedures. In many

CARBURETOR SERVICE

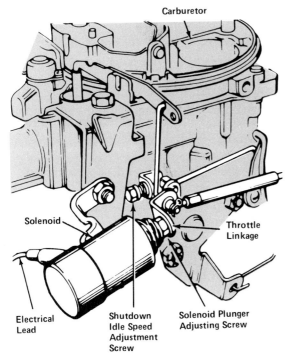

Figure 22-13. Solenoid and secondary curb idle speed screw locations.

cases, other electrical switches—for example, air conditioner controls—must be switched on first.

To set a specified slower engine speed, wiring connections to the solenoid are disconnected. The secondary curb idle speed screw is then turned to position the throttle plate.

The solenoid wire is then reconnected and the throttle is opened to allow the solenoid to push outward and hold the throttle at a higher speed. A threaded shaft positions the solenoid for correct idle speed with the solenoid activated.

Secondary curb idle speed screws are a "backup" system for solenoids. If the solenoid fails, the secondary curb idle speed screw prevents the engine from dying.

Vehicles without solenoid idle speed adjustments have a single curb idle speed screw adjustment. After making any idle speed adjustment, open the throttle momentarily and let it snap back under spring tension. Recheck idle speed again and readjust if necessary.

Idle Mixture Screws

Idle mixture screws control air-fuel mixture at idle. Manufacturers insert steel plugs over idle mixture adjustment screws to prevent tampering. No adjustment is necessary during routine maintenance and service.

However, during carburetor overhaul or to meet emissions standards, these caps are removed so that idle mixture adjustments can be made. After adjustments, the caps are reinserted or the holes plugged again.

To conform to clean-air standards, most idle mixture adjustments require a propane adjustment procedure and an exhaust gas analyzer (refer to Unit 8). Insert the analyzer probe into the tailpipe (see Unit 42). Follow the vehicle manufacturer's instructions to prepare the vehicle properly for testing. This may include ignition system maintenance, explained in Unit 35.

Turn adjustable mixture screws to obtain the proper readings for hydrocarbons (HC) and carbon monoxide (CO).

Float Level Adjustment

An incorrect float level can cause many running problems. If idle speed and mixture adjustments do not cure a problem, refer to the manufacturer's service manual for float adjustment. Remove the air horn and check the float level adjustment [Figure 22-14]. Bend the float tang or make adjustments as indicated in the service manual to obtain the correct float level. After adjustment, install a new air horn gasket and replace air horn correctly.

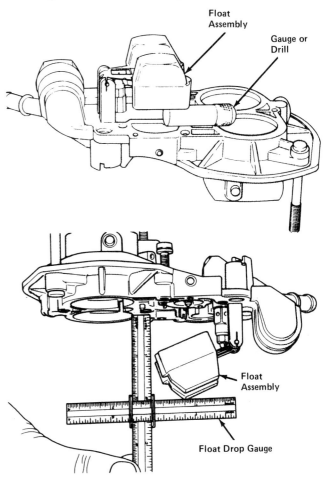

Figure 22-14. Float adjustments. CHRYSLER CORPORATION

22.5 CARBURETOR OVERHAUL

Although vehicle manufacturers are installing an increasing number of fuel injection systems on new vehicles, many cars on the road today have carburetors. Debris and gasoline deposits can partially close or plug tiny jets and passages within the carburetor. Gaskets can crack or leak. Needle valves can become stuck. Floats can become partially or completely soaked with gasoline and sink downward.

When attempts at adjustment indicate that carburetor repairs are necessary, an overhaul is done. The ability to disassemble, overhaul, and reassemble carburetors remains a necessary and valuable job skill for general automotive technicians.

The outline below covers the general steps involved in carburetor overhaul operations. Consult the manufacturer's service manual for detailed information. Carburetor overhaul includes, but is not limited to:

1. Carburetor identification
2. Disconnecting attachments to the carburetor
3. Carburetor removal
4. Carburetor disassembly
5. Removal and testing of electrical sensors, solenoids, and stepper motors
6. Carburetor cleaning and inspection
7. Replacement of parts and gaskets, including necessary new parts
8. Carburetor reassembly
9. Carburetor bench adjustments
10. Carburetor replacement
11. Replacing linkage, hoses and tubes, fuel lines, and electrical connections
12. Carburetor adjustments on a running engine.

Some technicians and shops specialize in carburetor work exclusively. These technicians or shops supply rebuilt carburetors to small garages and service stations that lack the time, tools, or expertise required to rebuild carburetors.

UNIT HIGHLIGHTS

- Many safety precautions must be observed when performing fuel system or carburetor service.
- Carburetor preventive maintenance includes replacing air, PCV, and fuel filters. It also involves checking for fuel and air leaks, and checking and/or adjusting automatic choke operation.
- Automatic choke adjustment concerns richness, fast-idle speed, and choke pull-off operation.
- Common carburetor problems include lack of fuel at carburetor, stuck or gummy automatic choke, flooding, and driveability problems.
- In addition to engine mechanical and ignition system problems, driveability problems can be caused by carburetor and fuel system defects.
- Driveability problems include dieseling, detonation, stalling, rough idle, missing, hesitation, surging, sluggishness, sponginess, poor gas mileage, and cutting out.
- Carburetor overhaul includes several steps. The first steps include carburetor identification, disconnection of attachments, and carburetor removal. The next steps include carburetor disassembly, cleaning and inspection, replacement of parts and gaskets, bench adjustments, and reassembly. The last steps include carburetor replacement, replacement of all linkage and other connections, and carburetor adjustments.
- On modern carburetors, adjustments must be made with the aid of an exhaust gas analyzer to meet clean-air standards.

TERMS

air horn	main body
throttle body	protractor
flooding	percolation
driveability	stalling
sluggishness	sponginess
cutting out	missing
hesitation	surging
solenoid	

REVIEW QUESTIONS

DIRECTIONS: The following questions are similar to those used on automotive technician certification tests. On a separate sheet of paper, write the letter of the correct choice.

1. Technician A says that the choke pull-off closes the choke plate during starting.
   Technician B says that the fast-idle cam opens the choke plate during starting.
   Who is correct?
   A. A only
   B. B only
   C. Both A and B
   D. Neither A nor B

2. A car will not start, and a heavy odor of gasoline is present in the engine compartment. When the air cleaner is removed and the engine is cranked, fuel spurts out of the float bowl vent tube. What is the most likely cause?
   A. Defective fuel pump diaphragm
   B. Needle valve stuck closed
   C. Needle valve stuck open
   D. Automatic choke plate stuck closed

3. Which of the following statements is correct?
   I. Loose, broken, cracked, kinked, or disconnected vacuum lines can cause many driveability problems.
   II. Ignition system defects can cause many driveability problems.
   A. I only
   B. II only
   C. Both I and II
   D. Neither I nor II

Questions 4 and 5 are not like those above. Each has the word EXCEPT. For each question, look for the choice that does *not* apply. Read each question carefully before you choose your answer.

4. Carburetor preventive maintenance includes all of the following EXCEPT
   A. replacing filters.
   B. resetting the float level.
   C. checking for fuel and air leaks.
   D. checking and/or adjusting automatic choke operation.

5. All of the following are part of a carburetor overhaul EXCEPT
   A. identification of the carburetor.
   B. cleaning, inspection, and replacement of parts.
   C. replacement of all main carburetor sections.
   D. bench and running adjustments.

ESSAY QUESTIONS

1. What does carburetor preventive maintenance include?
2. Describe common carburetor problems.
3. What procedures are performed during carburetor overhaul?

SUPPLEMENTAL ACTIVITIES

1. On a vehicle chosen by your instructor, replace air, PCV, and fuel filters, and check for fuel and air leaks.
2. On a vehicle chosen by your instructor, refer to the manufacturer's service manual and adjust choke richness, fast idle, and choke pull-off operation.
3. On a vehicle chosen by your instructor, check and/or replace any defective vacuum lines.
4. Ask at least two relatives or friends about driveability problems with their vehicles. Refer to the text, and make a list of possible causes for each of the driveability problems mentioned.
5. On a disassembled shop carburetor, check and/or make float level adjustments.

23 Fuel Injection

UNIT PREVIEW

An electronic fuel injection system can provide more efficient gasoline engine operation than a carbureted fuel system. An electronically controlled fuel injection system feeds sensor information to a computer. The computer determines how much fuel is to be injected and produces signals that are used to operate the injectors.

Depending on the system, gasoline can be injected twice per crankshaft revolution, once per revolution, or once every two revolutions. Injecting gasoline immediately before the opening of the intake valves is the most advanced form of gasoline fuel injection in production.

In a diesel engine, fuel must be injected near the end of the compression stroke and directly into the cylinder.

LEARNING OBJECTIVES

When you have completed your assignments and exercises in this unit, you should be able to:

❑ Explain the advantages of electronic fuel injection.
❑ Explain how a diesel fuel injection system operates.
❑ Identify and describe the two main types of diesel fuel injection pumps.
❑ Identify and describe the basic parts of a gasoline fuel injection system.
❑ Identify all visible parts of a fuel injection system.
❑ Identify and describe the physical and operating differences possible among fuel injection systems.

23.1 ADVANTAGES OF FUEL INJECTION

A gasoline fuel-injection system atomizes fuel in the amount necessary to provide the correct air-fuel ratio for efficient engine operation. Fuel injection systems have several advantages over carburetors.

A fuel injection system does not need a venturi in the air intake. This means that air can pass through more easily to increase engine volumetric efficiency (its ability to "breathe").

The intake air charge in a fuel injection system does not need to be heated to help vaporize the fuel. When air is heated, it becomes less dense and contains less oxygen. Less oxygen in the air results in less efficient combustion. A cooler, denser intake air charge in a fuel injection system results in better combustion.

Better combustion results in better fuel economy, more power, better acceleration, and less harmful exhaust gases. Thus, a fuel injection system can provide:

- Increased volumetric efficiency
- More efficient combustion
- Better fuel economy
- More power and performance than a carbureted fuel system
- Less harmful exhaust emissions.

The first gasoline fuel injection control mechanisms were extremely complicated and expensive mechanical units. For passenger vehicles, carburetors were less efficient but much cheaper. For many years, the only vehicles that used mechanical fuel injection systems were airplanes, racing and sports cars, and vehicles with diesel engines.

23.2 GASOLINE ELECTRONIC FUEL INJECTION

Electronic controls, like those used in radios and televisions, can also be applied to control mechanical devices. Simple electrical devices, such as solenoids and stepper motors, can be operated by electronic controls.

As early as 1932, crude electronic controls were used for fuel injection on diesel truck engines.

In 1961, the Bendix Corporation was granted patents for all forms of gasoline *electronic fuel injection (EFI)* systems. A gasoline EFI system operates by creating an electromagnetic field to move a solenoid connected to a fuel-injector nozzle. The nozzle is held closed by the force of a spring **[Figure 23-1]**. Pressurized fuel is supplied behind the nozzle from one or more pumps, and fuel is sprayed when the nozzle opens. When electricity to the solenoid is shut off, the spring closes the nozzle, stopping fuel flow.

The first production vehicles with electronic fuel injection were produced by Volkswagen in 1968. This port fuel-injection system, shown in **Figure 23-2,** was known as *D-Jetronic* (D for the German word, *druck*, meaning pressure). Other systems are discussed in 23.3 and 23.4.

Fuel Injection

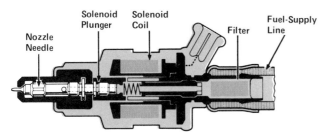

Figure 23-1. Solenoid-operated fuel injector.
ROBERT BOSCH CORPORATION

Microprocessors, or computer chips, have lowered the cost of electronic control systems sharply. Thus, the cost of modern electronic fuel injection systems is just slightly more than that of modern feedback carburetor systems.

Basic Parts of an Electronic Fuel Injection System

A fuel injection system must provide the correct air-fuel ratio for all engine loads, speeds, and temperature conditions. Unlike a carburetor, a fuel injection system uses the same basic single system to provide different air-fuel ratios. This basic system includes:

- Fuel tank and connecting lines
- Fuel pumps
- Fuel pressure regulator
- Fuel filters
- Electronic control unit
- Input information sensors
- Fuel injectors.

Fuel Tank and Connecting Lines

The fuel tank and connecting lines are like those of a carbureted system. One difference is that excess fuel is returned to the tank from a fuel-pressure regulator. This return line is indicated in **Figure 23-3**.

Fuel Pumps

Fuel pressures for EFI systems range from a low of 10 psi (69 kPa) to 79 psi (537 kPa). A transfer, or supply,

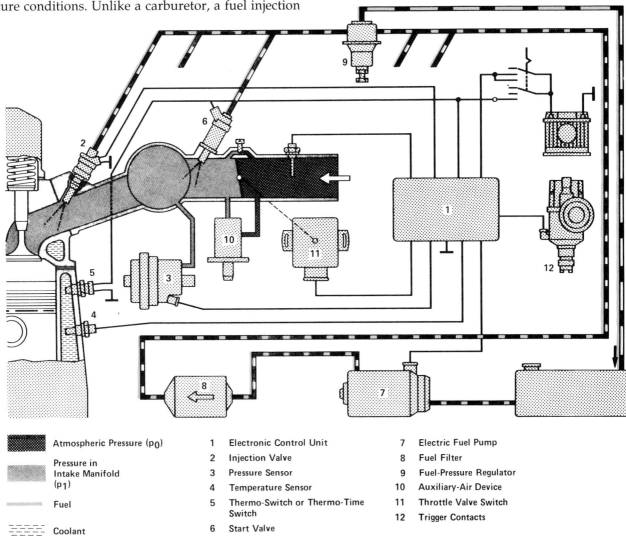

Atmospheric Pressure (p_0)	1 Electronic Control Unit	7 Electric Fuel Pump
Pressure in Intake Manifold (p_1)	2 Injection Valve	8 Fuel Filter
	3 Pressure Sensor	9 Fuel-Pressure Regulator
Fuel	4 Temperature Sensor	10 Auxiliary-Air Device
	5 Thermo-Switch or Thermo-Time Switch	11 Throttle Valve Switch
Coolant		12 Trigger Contacts
	6 Start Valve	

Figure 23-2. Bosch D-Jetronic electronic fuel injection system. ROBERT BOSCH CORPORATION

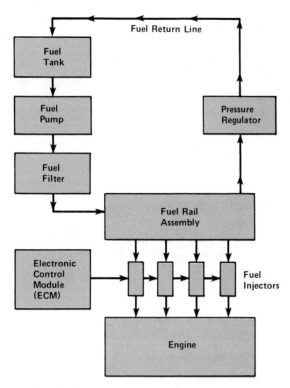

Figure 23-3. Fuel supply system. BUICK MOTOR DIVISION—GMC

pump can be used to help bring fuel from the tank to another high-pressure pump. In other cases, a single pump brings fuel from the tank and supplies the system with sufficient pressure to operate the injectors. Three types of pumps are used for gasoline EFI systems:

- Diaphragm pump
- Rotary roller pump
- Turbine-type pump.

Diaphragm pump. A diaphragm pump is the same type of mechanical pump used on many carbureted fuel systems. Such a pump is used only as a transfer pump to supply fuel to a high-pressure pump.

Rotary roller pump. A *rotary roller pump* consists of rollers in an offset mounted rotor that rotates within a housing **[Figure 23-4]**.

As the rotor turns, the space between the rollers increases in volume because of offset between the rotor and the housing. The increasing volume draws in fuel which is carried around to the pump outlet port. At this point, the volume between the rollers begins to decrease, thus forcing the fuel from this area through the pump discharge port.

Turbine-type pump. A *turbine-type pump* is a centrifugal pump similar to a water pump or turbocharger compressor wheel. Fuel is drawn in at the center of

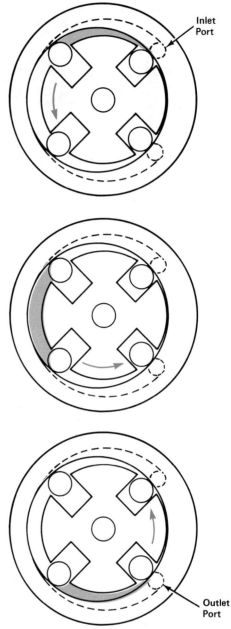

Figure 23-4. Rotary roller pump operation.

the wheel and forced outward through centrifugal force. The centrifugal force creates pressure.

Fuel Pressure Regulator

A fuel pressure regulator is similar to an oil-pressure regulator and is used to maintain a constant pressure for a uniform spray from the injectors.

A diaphragm and spring hold a relief opening closed. When fuel pressure exceeds a preset limit, the diaphragm moves upward, uncovering the relief passage to the fuel tank. Excess fuel is returned to the tank.

In normal operation, the regulator acts as a controlled leak, maintaining a constant fuel pressure.

FUEL INJECTION 235

Up to 85 percent of the fuel may be returned to the tank.

Fuel Filters

Small contaminant particles can lodge in an injection nozzle, blocking it partially open or closed. Water also can corrode the closely machined parts of the injectors. One or more fuel filters are mounted in the system to remove contaminants and small amounts of water. Fuel filters for EFI systems are larger and usually enclosed in a pressed steel shell to withstand the fuel pump pressure.

Electronic Control Unit

The heart of the system is the *electronic control unit (ECU)*, a small computer mounted *on-board*, on the vehicle. The ECU also may be called an *electronic control module (ECM)*, or *combustion control computer (CCC)*. It can be mounted within the passenger compartment, away from the heat and vibration of the engine, or in the engine compartment.

The ECU receives signals from sensors on the engine and determines the amount of fuel to be injected [Figure 23-5]. On current vehicles, this same computer also determines ignition spark timing, as discussed in Unit 35.

The amount of fuel injected is determined by how long the injector nozzle remains open. The nozzle opens and closes in response to signals from the electronic control unit. The time during which the signal is on to open the injector is known as the *pulse width*, measured in milliseconds, or thousandths of a second [Figure 23-6].

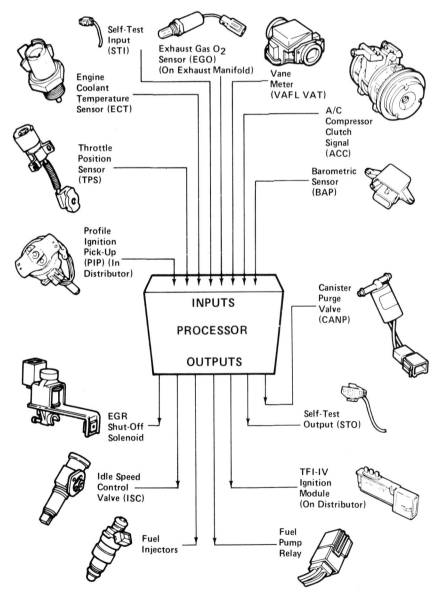

Figure 23-5. On-board computer system. FORD MOTOR COMPANY

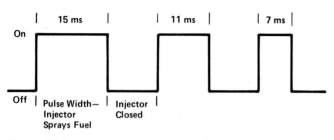

Figure 23-6. Pulse width.

Input Information Sensors

Input information on which the computer determines the amount of fuel to be injected can be based on signals from many sensors. These sensors provide information on engine conditions that can include:

- Engine rpm
- Throttle position
- Coolant temperature
- Crankshaft position
- Camshaft position
- Intake airflow
- Timing of ignition spark
- Air conditioner operation
- Gearshift lever position
- Battery voltage
- Amount of oxygen in exhaust gases
- Emission control device operation
- Power steering operation.

Signals to the computer can be in *analog* or *digital* form. Analog signals vary in strength with changing conditions. For example, a coolant temperature sensor will have more electrical resistance when cold and less when hot.

Digital signals are presented as a series of on-off pulses. The pulses are counted by a computer to determine a condition. For example, a sensor on a crankshaft can produce an electrical signal each time the crankshaft revolves. By counting the number of pulses in a given time, the computer can determine engine rpm.

Some sensor information to and/or control commands from the computer may be in the form of vacuum, rather than electrical, signals. Vacuum hoses, connectors, and other devices are used to transmit, relay, and receive these signals.

Sensors provide information to the computer about engine load, speed, temperature, and other conditions affecting vehicle operation. The computer is *programmed*, or provided with instructions, that produce correct air-fuel mixtures and throttle openings for given conditions. A solenoid or stepper motor is used to open the throttle for fast idle conditions.

For example, starting a cold engine produces signals that provide relatively long pulse widths. This results in a rich air-fuel mixture and a fast idle. Cruising on level roads at moderate speeds would produce shorter pulse widths and a leaner air-fuel mixture.

For reduced exhaust emissions, a stoichiometric (14.7:1) air-fuel mixture is required to allow emission control devices to operate properly. Emission control and other electronic devices are discussed in Units 38 and 39.

Current ECU units include what is known as a "limp-in" or "fail-safe" mode. If the ECU should fail, the vehicle will still run, though poorly. The vehicle will run well enough to be driven to a service garage.

Fuel Injectors

In operation, the opening pulse width may vary from 7 to 15 milliseconds. The pulse width depends on the leanness or richness of the air-fuel mixture needed.

Some confusion may arise regarding the idea of *duty cycle.* A duty cycle is a period of time when power is applied to do work, divided by a period of time when power is off, as shown in **Figure 23-7.**

In a feedback carburetor, power is applied to *close* a fuel metering jet and *shut off* the flow of fuel.

In an EFI system, power is applied to *open* a fuel injector nozzle and *turn on* the flow of fuel.

Therefore, the term duty cycle has opposite meanings as applied to feedback carburetors and EFI systems.

Basic Types of EFI Systems

Fuel can be injected either at a single point, or at each intake port. The following discussions contain examples of both types.

Single-point injection. Also called *throttle-body injection (TBI), single-point injection* can be thought of as a compromise between a carburetor and a complete EFI system. Throttle-body injection provides a more correct air-fuel ratio than a carburetor [**Figure 23-8**]. TBI is simpler and less expensive than multi-point injection, but less efficient. Like a carburetor system, fuel is not distributed equally to all cylinders.

Chrysler system. Chrysler Corporation uses a different type of single-point injection. Fuel is sprayed continuously from *fuel bars,* or hollow tubes above the throttle body assembly [**Figure 23-9**]. Varying the fuel pressure changes the amount of fuel sprayed. The Chrysler system combines the functions of fuel injection and ignition spark timing in a single electronic control unit.

FUEL INJECTION

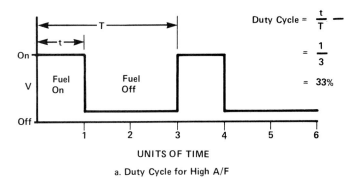

a. Duty Cycle for High A/F

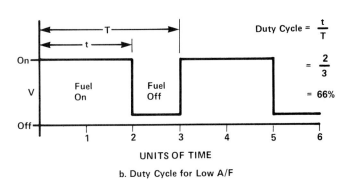

b. Duty Cycle for Low A/F

Figure 23-7. Electronic fuel injection duty cycle.

Multi-point injection. A multi-point, or *port-type injection* system, provides a more accurate and efficient delivery of fuel. However, multi-point systems are more expensive and complicated. Because fuel is injected behind each intake valve, as shown in **Figure 23-10**, *fuel wetting* of the manifold is minimized. Fuel wetting is the accumulation of excess fuel on the walls and floors of the intake manifold and ports. Fuel wetting results in unequal fuel distribution.

Port-type injectors receive fuel from a *fuel rail*, or metal pipe, that connects to all of the injectors.

Injection grouping. On most EFI systems, the injectors spray fuel in groups. Half of the injectors spray once each crankshaft revolution, and the other half spray on the next revolution **[Figure 23-11]**.

Injection timing. Because the 4-stroke cycle requires two complete revolutions to occur, the air-fuel mixture is not drawn into the cylinders immediately. Consequently, some of the mixture "waits" in the manifold until an individual intake valve opens.

Sequential fuel injection (SFI) On these systems, fuel is sprayed just before the opening of each individual intake valve. Thus, each injector sprays once every two crankshaft revolutions **[Figure 23-12]**. The mixture is drawn into the cylinder immediately.

Sequential fuel injection ECUs can be programmed to provide a more precise air-fuel mixture than systems that pulse alternate injector groups.

Sequential fuel injection currently is the most highly developed form of fuel injection. Such a system provides increased performance, better fuel economy, and improved exhaust emissions over a grouped injection system.

Top-feed and bottom-feed injectors. Two types of injectors currently are in use **[Figure 23-13]**. In a *top-feed injector*, fuel is fed in under higher pressure, up to 79 psi (537 kPa) to help prevent vaporization. This higher pressure requires a more expensive fuel pump. In addition, fuel vapors tend to rise into the stream of incoming fuel and may prevent fuel delivery.

Bottom-feed injectors are able to use fuel pressures as low as 10 psi (69 kPa). Fuel vapors flow upward through the injector to a return line to the tank. With bottom-feed injectors, a less expensive pump can be used, and fuel vapors do not tend to block fuel flow.

23.3 COMMON FUEL INJECTION SYSTEMS

There are several common types of fuel injection systems, introduced on European and Japanese vehicles. These include:

- D-Jetronic
- K-Jetronic
- L-Jetronic and its variations
- Motronic.

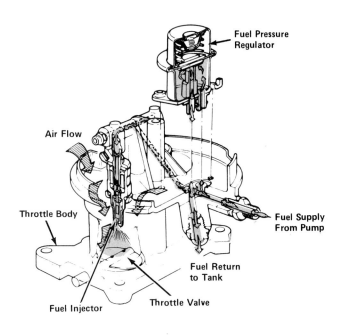

Figure 23-8. Throttle body injection system.
FORD MOTOR COMPANY

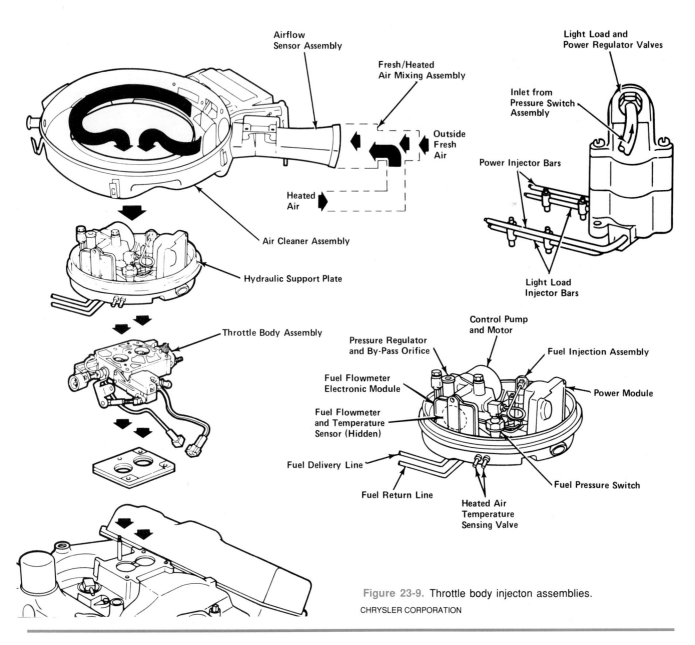

Figure 23-9. Throttle body injecton assemblies.
CHRYSLER CORPORATION

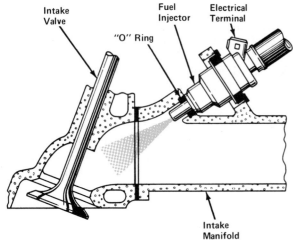

Figure 23-10. Fuel injection behind the intake valve.
BUICK MOTOR DIVISION—GMC

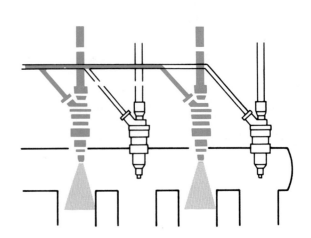

Figure 23-11. Injector grouping.

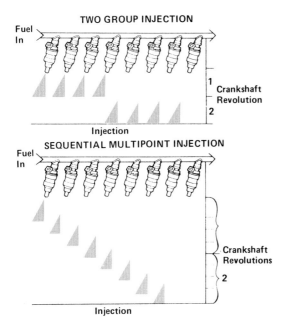

Figure 23-12. Grouped and sequential fuel injection.

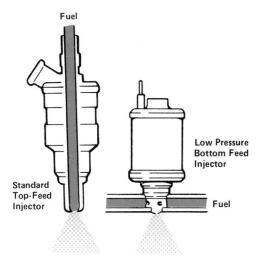

Figure 23-13. Top-feed and bottom-feed injectors.

D-Jetronic

The Robert Bosch D-Jetronic system [refer to Figure 23-2] measures airflow by using a sensor to gauge the pressure of the air in the intake manifold. This measurement is known as *manifold absolute pressure* (*MAP*).

When a higher pressure (less vacuum) exists, a denser air charge will flow relatively slowly into the cylinder. The system responds by injecting more fuel to maintain a correct air-fuel ratio.

However, this system does not allow for the presence of increased amounts of exhaust gases in the cylinder for modern smog controls. The presence of exhaust gases lowers the combustion temperature, and reduces oxides of nitrogen (NO_X), a harmful gas.

K-Jetronic

A different type of fuel injection system allows fuel to be sprayed constantly from all injectors. This type is called **K-Jetronic** (K from the German word **konstanter**, constant). The amount of fuel sprayed is controlled by changing the pressure of the fuel going to the injectors.

The volume of air entering the engine is sensed by the movement of a flat plate within a tapered, venturi-like opening [**Figure 23-14**]. The more air that flows into the engine, the more the plate moves. The movement of an arm connected to the plate operates a fuel-pressure regulator. More fuel is injected to maintain the correct air-fuel ratio.

L-Jetronic

The **L-Jetronic** (L from the German word **luft**, air) system [**Figure 23-15**] measures airflow with a pivoting flap in the airstream. Fuel is injected in proportion to the fresh air entering the intake manifold. All of the injectors squirt half the fuel required *twice* each crankshaft revolution.

Except for these differences, the L-Jetronic system is very similar to the D-Jetronic.

LH-Jetronic. The **LH-Jetronic** EFI system is a variation of the L-Jetronic. LH-Jetronic uses microprocessors (integrated circuit computer chips) instead of transistors to control the injector timing and duration.

Motronic

The Bosch **Motronic** system combines electronic ignition timing control with fuel injection control in a single ECU. Motronic is the most advanced and sophisticated of all Bosch systems. It is particularly efficient in controlling emissions levels.

23.4 EFI AIR MEASUREMENT

Possibly the greatest difference in EFI systems is how the intake air charge is measured. Most current systems measure air volume. The latest designs measure air mass. LN-Jetronic and Buick SFI systems are discussed below. These air-measuring systems include:

- Manifold absolute pressure (MAP) sensor
- Movable plate in a tapered opening
- Movable flap
- Vortex flow
- Heated film
- Heated wire.

The MAP sensor, movable plate, and flap are discussed above.

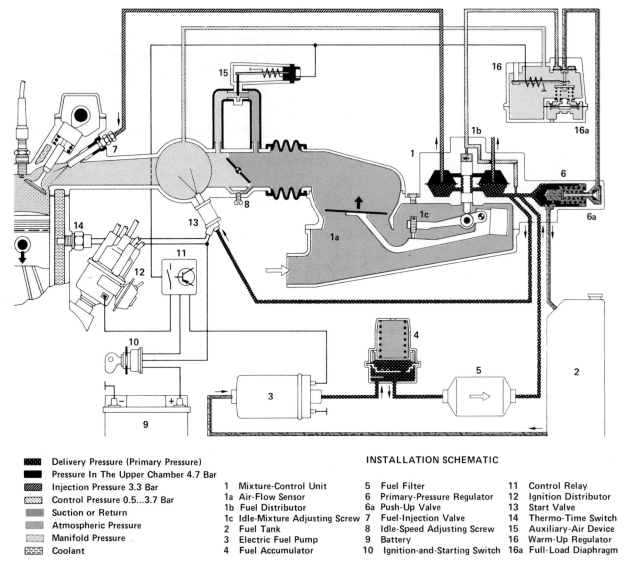

Figure 23-14. Bosch K-Jetronic fuel injection system.   ROBERT BOSCH CORPORATION

Vortex Flow

Chrysler Corporation uses a principle called *vortex flow* to determine the volume of air entering the intake manifold. Vortex flow means a swirling, twisting motion.

Air entering the airflow sensor assembly passes through vanes arranged around the inside of a tube. As the air flows through the vanes, it begins to swirl [Figure 23-16]. The outer part of the swirling air exerts high pressure against the outside of the housing. There is a low-pressure area in the center.

The low-pressure area moves in a circular motion as the air swirls through the intake tube. Two pressure-sensing tubes near the end of the tube sense the low-pressure area as it moves around. An electronic sensor counts how many times the low-pressure area is sensed.

The faster the airflow, the more times the low-pressure area will be sensed. This is translated into a signal that indicates to the combustion control computer how much air is flowing into the intake manifold. The computer then adjusts the pressure of the fuel flowing through the fuel bars to maintain the correct air-fuel ratio.

Mass Air-Flow Sensor

In the mid-1980s, manufacturers introduced a new type of sensor called a *mass air-flow sensor*. This sensor measures the *mass*, not the volume, of the air flowing into the intake manifold. Mass means the quantity, or amount, in absolute terms.

Remember, the ideal 14.7:1 ratio is based on the weight of air and gasoline. Weight, in scientific terms, is the gravitational attraction of the Earth for a given mass.

FUEL INJECTION

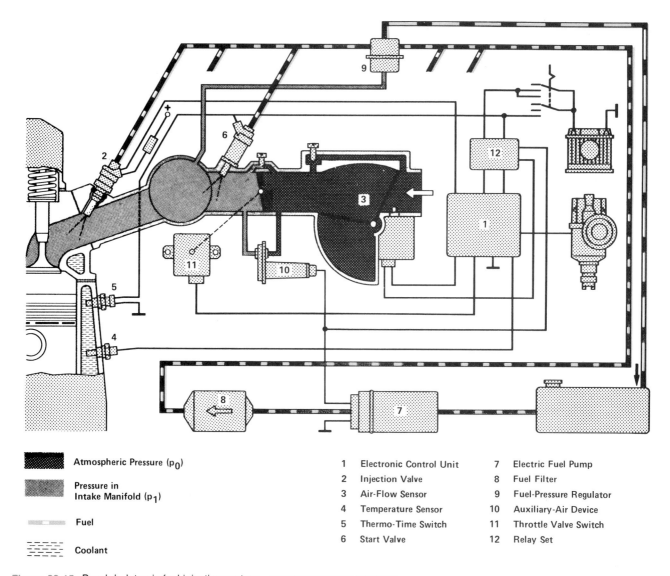

Figure 23-15. Bosch L-Jetronic fuel injection system. ROBERT BOSCH CORPORATION

The volume of air may be misleading, because air can be more or less dense, depending upon temperature and atmospheric pressure. However, being able to measure the mass of air directly means that the stoichiometric air-fuel ratio can be maintained much more accurately.

A mass air-flow sensor [**Figure 23-17**] uses an electrically heated conductive metal foil film placed in the intake airstream. The amount of electricity required to keep the film at a constant 75 degrees C (167 degrees F) is measured. This information indicates how much air mass is flowing over the heated film to cool it.

LN-Jetronic. A variation of the L-Jetronic is the *LN-Jetronic* system that uses low-pressure bottom-feed fuel injectors. A different type of air flow sensor, called a *hot-wire* sensor, is used [**Figure 23-18**]. Electricity is used to heat one wire in the intake airstream. Another wire is unheated. A comparison of the amount of electrical resistance in the heated and unheated wires indicates the mass of air flowing over the heated wire.

23.5 DIESEL MECHANICAL FUEL INJECTION

Diesel fuel systems may be divided into two categories: *direct injection* and *indirect injection*. In direct injection systems, the fuel is injected directly into the combustion chamber, on top of the piston. Indirect injection systems are those in which fuel is injected into a *precombustion chamber*, also called a *prechamber*. The combustion process begins in the prechamber and blows through a port to the top of the piston [**Figure 23-19**].

Ignition in a diesel engine is dependent upon the heat of the compressed air charge. For starting when the cylinder is cold, the cylinder or prechamber contains a *glow plug*, or electrical heating element.

242　　　　　　　　　　　　　　　　　　　　　　　　　　　　　　PART III: AUTOMOTIVE ENGINE SYSTEMS

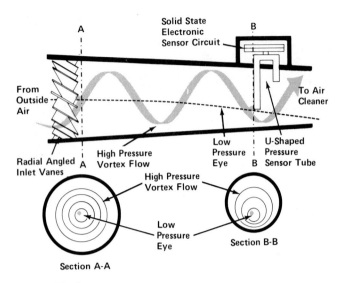

Figure 23-16. Vortex flow.

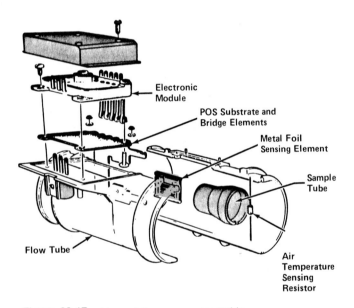

Figure 23-17. Mass airflow sensor assembly.
GENERAL MOTORS CORPORATION

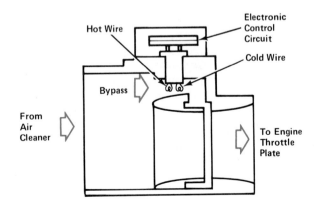

Figure 23-18. Hot-wire sensor operation.

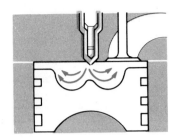

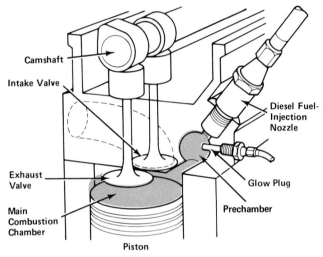

Figure 23-19. Diesel mechanical fuel injection systems.

Before the engine is cranked, the glow plug preheats the air to a temperature sufficient to ignite the fuel. The relative positions of a glow plug and an injection nozzle can be seen in **Figure 23-20**.

Refer to Unit 19 for a diagram of an automotive diesel fuel injection system. Most modern diesel engines have no throttle plates in the air intake. Thus, there is no venturi to restrict air flow. This increases diesel engine volumetric efficiency. Diesel engine power output is determined by the amount of fuel injected. The more fuel injected, the faster the engine runs.

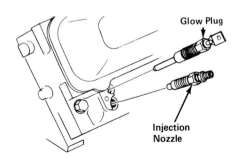

Figure 23-20. Glow plug and injection nozzle locations.
BUICK MOTOR DIVISION—GMC

A supply pump brings fuel under relatively low pressure to the injection pump. The injection pump boosts the pressure to overcome the strong spring pressure holding the diesel fuel injector nozzle closed. Automotive diesel fuel injection pumps produce operating pressures of 800 to 3,000 psi (5,516 to 20,685 kPa). Diesel injection pressures also are measured in *bars.* One bar equals 14.5 psi (99.98 kPa).

Two types of injection pumps are used on automotive diesel engines: *inline pumps* and *rotary pumps.* Inline pumps have a separate pumping plunger for each cylinder. Rotary pumps have a common pumping chamber for all cylinders.

In diesel injection pumps, fuel metering and delivery timing are determined by complicated mechanical and hydraulic devices. These parts must be machined precisely to tolerances sometimes measured in millionths of an inch (hundreds of thousandths of a millimeter). Thus, a mechanical fuel injection pump is complicated and a very expensive part. Pump repairs are best left to trained experts.

UNIT HIGHLIGHTS

- Fuel injection can provide increased volumetric efficiency, more efficient combustion, better fuel economy, and more power than a carbureted fuel system. Fuel injection also can produce less harmful exhaust gases.
- Mechanical fuel injection pumps are complicated and expensive.
- Electronic fuel injection systems are comparable in cost to modern carbureted fuel systems.
- An EFI system uses information from sensors fed through a computer to determine the amount of fuel injected.
- The amount of fuel injected in most EFI systems depends on how long an injector nozzle is held open by a solenoid.
- On the K-Jetronic system, the amount of fuel injected is determined by varying the fuel pressure.
- Depending on the system, EFI can inject fuel twice per crankshaft revolution, once per revolution, or once every two revolutions.
- Sequential fuel injection provides fuel just before individual intake valves open, and is more precise and accurate than other systems.
- Diesel fuel systems may be divided into two categories: direct injection and indirect injection.

TERMS

D-Jetronic	microprocessor
rotary roller pump	turbine-type pump
fuel rail	pulse width
analog	digital
programmed	duty cycle
fuel bars	port-type injection
fuel wetting	top-feed injector
bottom-feed injector	K-Jetronic
L-Jetronic	LH-Jetronic
Motronic	vortex flow
mass air-flow sensor	LN-Jetronic
direct injection	indirect injection
prechamber	glow plug
precombustion chamber	hot wire sensor
bar	single-point injection
electronic fuel injection (ECU)	electronic control unit (EFI)
electronic control module (ECM)	combustion control (CCC)
sequential fuel computer injection (SFI)	manifold absolute pressure (MAP)
throttle body injection (TBI)	inline diesel injection pump
rotary distributor type pump	multi-point injection

REVIEW QUESTIONS

DIRECTIONS: The following questions are similar to those used on automotive technician certification tests. On a separate sheet of paper, write the letter of the correct choice.

1. Which of the following statements is correct?
   I. Fuel injection can provide more power and better fuel economy than a carbureted fuel system.
   II. The cost of EFI makes it suitable only for racing cars and expensive vehicles.
   A. I only   C. Both I and I
   B. II only   D. Neither I nor II

2. The purpose of a glow plug in a diesel engine is to
   A. heat the air in a cold cylinder.
   B. inject fuel directly into the cylinder.
   C. inject fuel into a prechamber.
   D. heat the surrounding metal in a combustion area.

3. Technician A says that gasoline EFI systems can operate with fuel pressures of 10 psi (69 kPa).

   Technician B says that diesel fuel injection systems operate with pressures of up to 3,000 psi (20,685 kPa).

   Who is correct?
   A. A only
   B. B only
   C. Both A and B
   D. Neither A nor B

4. Which of the following supplies input data signals to help the computer determine the air-fuel ratio produced by an EFI system?
   A. Sensors
   B. Mixture control solenoid
   C. Fuel injector
   D. Either B or C

Question 5 is not like those above. It has the word EXCEPT. For question 5, look for the choice that does *not* apply. Read the entire question carefully before you choose your answer.

5. All of the following are advantages of sequential fuel injection EXCEPT
   A. more precise control of air-fuel ratio.
   B. less fuel wetting.
   C. less harmful exhaust emissions.
   D. single fuel injector.

ESSAY QUESTIONS

1. Describe the basic parts of an electronic fuel-injection system.
2. What is the difference between port fuel injection and throttle-body injection?
3. Describe the devices and methods used to measure airflow into a gasoline EFI system.

SUPPLEMENTAL ACTIVITIES

1. Explain the differences between sensing mass airflow and the volume of airflow, and which is preferable for maintaining correct air-fuel ratios.
2. Explain why EFI systems can maintain a more correct air-fuel ratio than a carburetor.
3. Visit the parts department of a new-car dealer who sells vehicles with EFI. Find out the replacement cost of an EFI electronic control unit.

24 Fuel Injection Service

UNIT PREVIEW

Preventive maintenance for fuel injection systems consists of checks and adjustments, where possible. Manufacturer's service manuals must be used to find the correct cause of fuel injection problems.

Fuel injection service includes identification, reference to manufacturer's service manuals, and replacement and/or repair of fuel pumps, sensors, injectors, and ECUs.

LEARNING OBJECTIVES

When you have completed your assignments and exercises in this unit, you should be able to:

- ❏ Perform fuel injection preventive maintenance.
- ❏ Check for fuel delivery on a fuel injection system.
- ❏ Identify and describe sounds of a correctly operating fuel injector.
- ❏ Perform a multi-point fuel injection balance test.
- ❏ Correctly identify a throttle-body injection system.

SAFETY PRECAUTIONS

Have a fire extinguisher capable of extinguishing class B fires (flammable liquids) fully charged and readily available. This type of extinguisher should be nearby whenever you attempt any type of fuel injection or fuel system service.

Make sure no sources of possible ignition are present in the area where fuel injection or fuel service is being performed. These sources include lighted cigarettes, electrical switches and tools, disconnected ignition cables, and open flames of any kind.

Wear eye protection at all times when working on fuel injection or fuel system parts.

Fuel injection systems are extremely complicated and vary greatly between manufacturers, models, and even between similar systems installed on different engines. Do not attempt any fuel injection maintenance, repairs, or adjustments without the proper manufacturer's service manual.

After fuel injection service, crank the engine to allow the system to deliver fuel for starting. Do not attempt to pour liquid gasoline down the throat of the air intake. Severe explosions and fires can result.

In many cases, air must be removed from injection lines after service procedures. "Bleeding," or air removal procedures, will allow fuel to leak onto and around the engine. Wipe up excess fuel with a shop towel and place the towel in a safety container. Allow all fuel to dry before attempting to start the vehicle.

On vehicles with automatic transmissions, some adjustments are done with the engine running and the selector in drive. Set the emergency brake tightly and block the driving wheels before attempting these adjustments. Never stand directly in front of a running vehicle while adjustments are being done. If the throttle is opened suddenly, the vehicle may jump forward over wheel blocks and run over or crush anyone standing in front of the vehicle.

24.1 PRELIMINARY CHECKS

Before assuming that a fuel injection system is the cause of starting and/or running problems, all other systems should be checked. Make a thorough visual inspection for cracked, disconnected, pinched, or kinked vacuum lines (see Unit 42). Check the battery and starting system (see Units 29 and 31). Then, check the charging system (see Unit 33), ignition system and ignition timing (see Unit 35), and engine mechanical operation (see Unit 43).

24.2 FUEL INJECTION PREVENTIVE MAINTENANCE

Preventive maintenance for fuel injection systems includes checking and repairing or replacing the following:

- Fuel and air leaks
- Loose, broken, cracked, kinked, or disconnected vacuum hoses
- Loose, corroded, or grounded electrical connections
- Loose or improperly supported fuel rails, lines, and injectors
- Dirty fuel and air filters
- Incorrect idle speed and/or mixture.

Fuel and Air Leaks

Fuel and air leaks are corrected in basically the same manner as for carbureted systems. However, fuel lines for multi-point fuel injection on many cars are steel or plastic tubing with compression, or flare-type, fittings. A loose connection that leaks fuel can also allow air into the lines, which will cause erratic running problems.

CAUTION!!!

Extremely small amounts of contaminants in a fuel injection system can cause expensive damage. Care must be taken to make sure that all parts to be replaced are extremely clean. Clean fuel lines and other parts thoroughly before removal and again before reassembly. Make sure that tools and hands are clean during reassembly procedures.

Bleeding fuel lines. If lines must be replaced, some fuel injection systems require that air be bled from the fuel lines. A fuel line is loosened and fuel is pumped. At first, both fuel and air bubbles will spurt out. When fuel appears without air bubbles, the line can be tightened. Fittings must be tightened to the manufacturer's specified torque **[Figure 24-1]**.

SAFETY CAUTION

Because liquid fuel and fuel vapors will be present during bleeding, make sure that no possible ignition sources are present. If necessary, push the vehicle away from electrical switches and tools, heaters, and so on. Also, if cold fuel sprays onto a hot drop light bulb, the bulb will crack and the filament will ignite the fuel. Keep droplights as far away as possible during bleeding. Perform bleeding only in a well-ventilated area. Have a fully-charged fire extinguisher capable of extinguishing class B fires available.

Some diesel vehicles include a hand priming pump for bleeding purposes. Electrical connections to the fuel pump on vehicles with gasoline engines can be made to pump fuel through the lines.

Air leaks. Carefully inspect all rubber hoses and connections in the air-induction system. A length of

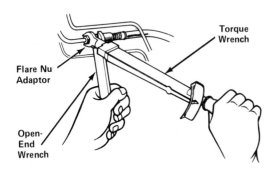

Figure 24-1. Tightening fuel line fittings.

rubber tubing can be used to listen for the hissing of an intake manifold air leak. Even a small air leak will result in a lean air-fuel mixture. Air leaks around metal parts, such as throttle-body shafts, can be detected by listening. A better method is to use propane enrichment (see Topic 42.2).

Fuel injection systems are sensitive to air leaks. Engine air consumption is measured by the mass air flow sensor. Any air getting by the sensor and into the engine (loose dipstick, open PCV valve, or leaking manifold gasket) will upset the computer-controlled balance between air and fuel.

Loose, Broken, Cracked, Kinked, Or Disconnected Vacuum Hoses

Rubber and plastic vacuum hoses are subject to heat and vapors from fuel and oil. These elements tend to dry, deteriorate, and crack the hoses. The ends of the hoses become expanded where they have been pushed onto small-diameter metal or plastic tube connections.

Vacuum hoses can become kinked or pulled loose during other service procedures. Because many signals for some fuel injection systems are in vacuum form, disconnected or switched hoses can cause confusing problems. A typical vacuum hose routing diagram is shown in **Figure 24-2**. Some vehicles have numbered or color-coded hoses to aid in correct replacement. Inspect and/or replace hoses one end at a time to avoid incorrect connections.

Loose, Corroded, or Grounded Electrical Connections

Shake, wiggle, and twist electrical connectors to make sure they are tightly connected **[Figure 24-3]**.

Loose and/or corroded connections. The connections for electrical sensors and fuel injectors can become loose and/or corroded in time. If not properly supported, wires can rub against metal parts and wear through insulation. Any of these conditions can cause a lack of signals or improper signals to be sent to or from the computer control unit.

Connections on sensors and actuators should be removed and checked for corrosion. Clean connectors have shiny metal male and female plug parts.

Some manufacturers use special conductive fluids or greases to help prevent corrosion. Refer to the manufacturer's service manual to determine if the connector pins or tabs should be coated with any special material.

Connectors can be pulled apart to check the condition of metal tabs and pins **[Figure 24-4]**.

CAUTION!!!

Always make sure the ignition switch is in the OFF position before making disconnections, connections, or checks for looseness. In some cases, the battery ground

FUEL INJECTION SERVICE

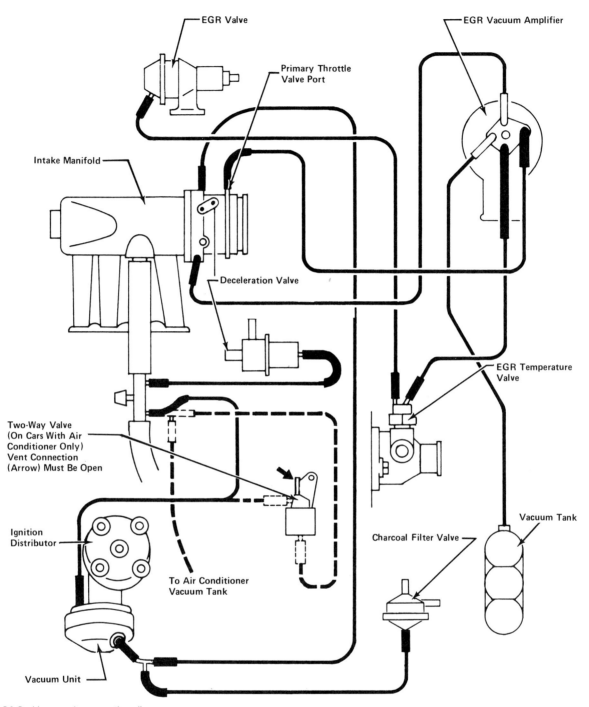

Figure 24-2. Vacuum hose routing diagram. VOLKSWAGON OF AMERICA

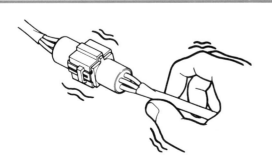

Figure 24-3. Checking electrical connectors for looseness.

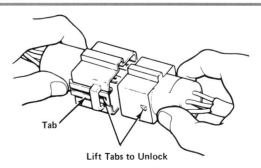

Figure 24-4. Disconnecting electrical connectors.

connection also must be removed. The manufacturer's instructions regarding electrical connector service must be followed exactly. Accidentally touching a screwdriver or other metal tool from one terminal to another can damage or destroy electronic components. Water or other liquids in connectors can cause grounding and short-circuiting. Permanent damage to expensive ECU units can result from improper procedures.

Fuel injector connectors. Some injectors for multi-point fuel injection have metal wire clips to hold the electrical connector securely. Gently pry the wire open while pulling on the connector to remove it **[Figure 24-5]**. Grasp the connector and pull gently. Never pull on a wire; always pull on the connector itself.

Grounded or shorted wiring. Check along the length of wires for rubbing or fraying. If the metal conductor of the wire is visible, wrap it with electrical tape or replace the wire.

A digital ohmmeter with 10 megOhms of impedance also can be used to check for grounded wires. Refer to the manufacturer's instructions for specific directions.

CAUTION!!!

Disconnect the battery ground cable before using an ohmmeter to check electrical wires. If battery power is present in any wire being checked, the ohmmeter will be damaged.

Loose or Improperly Supported Fuel Rails, Lines, and Injectors

Fuel rails and metal or plastic lines should be properly supported to protect them from engine vibration. Vibration can loosen or *fatigue* such parts. Fatigue is cracking or breaking damage that results from repeated flexing and bending. Metal or plastic mounting clips and ties should be in place and securely fastened.

On some vehicles, individual fuel injectors are attached by being pushed into a tight-fitting rubber grommet. In other cases, a bolt or nut holds the injector in a mounting bracket **[Figure 24-6]**.

Check the injector by attempting to move it forward and backward and up and down. Excessive movement can indicate a worn-out rubber grommet or a loose bolt or nut. These problems can result in a lean air-fuel mixture.

Dirty Fuel and Air Filters

Refer to 20.5 for replacement of fuel and air filters. Replacement of PCV filters also is important and is discussed in Topic 20.5.

Incorrect Idle Speed and/or Mixture

Some vehicles equipped with electronic controls also control idle speed and mixture electronically. Periodic adjustment is not required. Adjustment, if possible, is made only during major service, as discussed in Topic 24.5. Other vehicles have easily adjustable idle speed controls. Idle mixture controls, however, are usually hidden and require special tools for adjustment to prevent tampering **[Figure 24-7]**. Provision for fast idle is usually made with an electronically controlled solenoid to open the throttle farther when the engine is cold. Refer to the manufacturer's service manual for correct adjustment procedures.

Correct mixture-control adjustment requires the use of an exhaust gas analyzer. Refer to Unit 22.

CAUTION!!!

Idle speed and mixture adjustments to other than the federal and/or state standards will change the exhaust emissions and may result in violation of federal and/or state laws. In some states, such procedures may result in fines and/or loss of a technician's license or certification.

24.3 TROUBLESHOOTING FUEL INJECTION SYSTEMS

A general procedure for troubleshooting modern vehicles with electronic controls is shown in **Figure 24-8**. *Trouble codes* are numbers that the on-board computer stores to indicate specific problems. The electronic control unit can indicate to the technician what specific problems exist, as discussed in Unit 39.

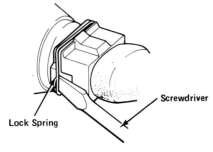

Figure 24-5. Removing the fuel injector connector.

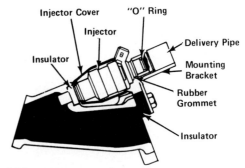

Figure 24-6. Injector mounting.

FUEL INJECTION SERVICE

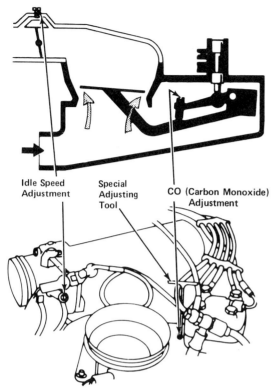

Figure 24-7. Idle speed and CO adjustments.
VOLKSWAGEN OF AMERICA

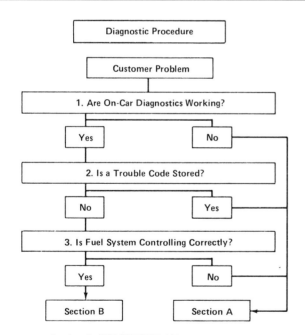

Figure 24-8. Diagnostic procedure for electronically controlled fuel injection. BUICK MOTOR DIVISION—GMC

The technician can activate the electrical control unit to display problem codes by making temporary electrical connections between certain connectors. Refer to the manufacturer's service manual for specific procedures.

If all these systems are operating correctly, refer to the manufacturer's service manual for troubleshooting charts. A troubleshooting chart is similar to a map with different directions at crossroads. Start at the top and follow the directions downward.

After following the beginning instructions, there will be two or more choices. Find the condition that matches your problem. Then follow the chart downward again for further checks, explanations, or instructions for performing repairs [**Figure 24-9**].

Manufacturer's service manuals for modern vehicles contain literally hundreds of such charts. Different charts are needed for specific vehicles, varying engine and transmission combinations, and so on. The service manuals list the meanings for abbreviations such as CKT (circuit) and ECM (electronic control module).

No technician can memorize all of these charts, so it is necessary to find and use the correct chart in the proper service manual. To obtain and hold a job, today's technician must be able to read, understand, and correctly follow these charts.

Driveability Problems

Driveability problems, similar to those of carbureted fuel systems, can occur with fuel injection systems. However, manufacturer's service manuals must be consulted to determine the general or specific causes for particular systems and vehicles. An example of such a chart is shown in **Figure 24-10**.

24.4 CHECKING FOR FUEL DELIVERY

If all other basic systems appear to be working properly, the fuel injection system should be checked for fuel delivery.

Simple Checks

On single-point fuel injection, fuel should be seen spraying down the throat of the throttle-body during cranking.

SAFETY CAUTION

Wear eye protection, and never look directly down the throat of the throttle body. If the engine begins to start, a backfire explosion can cause severe burns or other injury. Have a fire extinguisher capable of extinguishing class B fires fully charged and readily available.

On a multi-point injection system, a technician's stethoscope can be used to listen for the pulsing of the

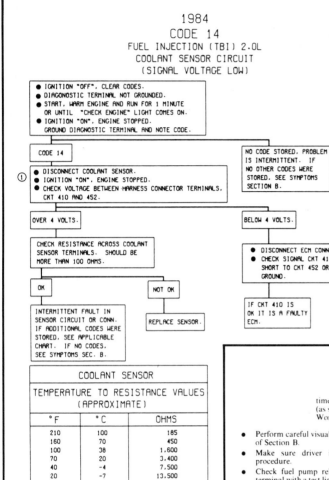

Figure 24-9. Coolant sensor signal troubleshooting chart.
BUICK MOTOR DIVISION—GMC

HARD START

Definition: Engine cranks OK, but does not start for a long time. Does eventually run. If the engine starts but immediately dies (as soon as key is released from "start" position), see "Cranks But Won't Run", **CHART A-3**, page 6E3-39.

- Perform careful visual check as described at start of Section B.
- Make sure driver is using correct starting procedure.
- Check fuel pump relay: Probe fuel pump test terminal with a test light to ground. Turn ignition off for 10 seconds, then turn ignition on. Test light should light for 2 seconds. If not OK, see CHART A-5 TEST POINT 1.
- Check TPS for sticking or binding.
- Check injectors for leaking. See **CHART A-7**.
- Check for high resistance in coolant sensor circuit or sensor itself. See **CODE 15 CHART**, page 6E3-55.

- A faulty in-tank fuel pump check valve will allow the fuel in the lines to drain back to the tank after the engine is stopped. See **CHART A-7**.
- Check ignition system - Section 6D. Check distributor for:
  - Proper Output with ST-125.
  - Worn shaft.
  - Bare and shorted wires.
  - Pickup coil resistance and connections.
  - Loose ignition coil ground.
  - Moisture in distributor cap.
  - Check fuel pressure. See **CHART A-7**.
- Remove spark plugs. Check for wet plugs, cracks, wear, improper gap, burned electrodes, or heavy deposits. Repair or replace as necessary.

STALL AFTER START

Definition: The engine starts OK, but 1) dies after brief idle; 2) dies as soon as any load is placed on engine (such as A/C turned on or transmission engaged); or 3) dies on initial driveaway.

- Perform careful visual check as described at start of Section B.
- Check for proper operation of IAC System. See **CHART C-2C**, page 6E3-187.
- Check PCV valve for proper operation by placing finger over inlet hole in valve end several times. Valve should snap back. If not, replace valve.

- Check for overcharged A/C system.
- Check for high A/C head pressure - could be caused by inoperative engine cooling fan.
- Check for plugged or restricted fuel lines. See **CHART A-5**.
- Check for weak spark from faulty ignition coil (see Section 6D).

HESITATION, SAG, STUMBLE

Definition: Momentary lack of response as the accelerator is pushed down. Can occur at all car speeds. Usually most severe when first trying to make the car move, as from a stop sign. May cause the engine to stall if severe enough.

- Perform careful visual check as described at start of Section B.
- Check fuel pressure. See **CHART A-7**. Also check for water contaminated fuel.
- Check vacuum hose to MAP sensor for leaks or restrictions.
- Check for fouled spark plugs.

- Check for correct PROM number. Also check Service Bulletins for latest PROM.
- Check TPS for binding or sticking.
- Check ignition timing. See Emission Control Information label.
- Check ECM controlled idle speed. See "Fuel Control" system.
- Check generator output voltage. Repair if less than 9 or more than 16 volts.
- Check canister purge system for proper operation.

Figure 24-10. Driveability troubleshooting chart.
BUICK MOTOR DIVISION—GMC

FUEL INJECTION SERVICE

injector solenoid. In some cases, it is possible to feel the pulsing by placing a finger on the injector.

Port-type Injector Problems

Although it is relatively rare, some driveability problems with port-type fuel injectors may occur. A vehicle with clogged port-type fuel injectors does not accelerate or perform well.

The heat generated by the cylinder head may cause the fuel injector nozzle to become clogged with baked-on gasoline deposits. These deposits form after the engine is shut off, during "hot soak" conditions.

Special solutions and tools are available to allow technicians to run a powerful cleaning solvent through the fuel injectors to clean them. Also, gasoline manufacturers have begun adding cleaning chemicals to unleaded regular and premium gasolines.

If driveability problems persist and all other vehicle systems seem to be in order, remove fuel injectors and check for spray patterns. Cleaning or replacement of fuel injectors may be necessary.

After disabling the ignition, multi-point injectors can be removed and the spray observed for proper size and shape [Figure 24-11].

The volume of fuel delivered in a given amount of time also can be checked by spraying the fuel into a graduated container. Refer to the manufacturer's service manual for correct fuel delivery volume.

Instrument Checks

Special tools are needed for accurate diagnosis of fuel injection system problems. Examples of such tools are shown in **Figure 24-12**.

Multi-point injector balance test. A test similar to a cylinder-balance test (see Unit 35) can be performed on some EFI systems [Figure 24-13]. Fuel pressure changes should be almost identical on all injectors. Too low or too high a reading indicates a defective injector.

24.5 FUEL INJECTION SYSTEM REPAIRS

Fuel injection system repairs include the following:

- Identification of system and components
- Reference to manufacturer's troubleshooting charts
- Reference to manufacturer's repair procedures
- Replacement of fuel pumps
- Replacement of sensors
- Replacement of injectors
- Replacement of electronic control units

System Identification

Identification codes for fuel injection system parts may be attached to a tag or plate. In other cases, numbers may be stamped or cast into the parts themselves [Figure 24-14].

Troubleshooting Charts

Reference always should be made to the proper troubleshooting charts [refer to Figures 24-9 and 24-10]. Fuel injection systems are too complicated for "guesswork."

Manufacturer's Service Procedures

Follow the manufacturer's instructions on fuel injection system repairs. Do not attempt to bypass or alter the system in any way. Such repairs may result in increased exhaust emissions and violations of federal and/or state laws.

Fuel Pump Service

Diaphragm-type fuel transfer pumps and electrical pumps for gasoline fuel injection systems are replaced as described in Unit 20.

Repairs to, or replacement of, diesel fuel injection pumps is an automotive specialty. Diesel pump repair requires specialized training, tools, and reference materials.

CAUTION!!!

Diesel fuel injection pumps are highly precise, expensive components. Do not attempt any adjustments or repairs to diesel fuel injection pumps without proper training, tools, reference materials, and calibration equipment. Improper adjustments or repairs may result in failure of the pump.

Sensor Replacement

Sensors are replaced as discussed in Units 16, 17, and 42.

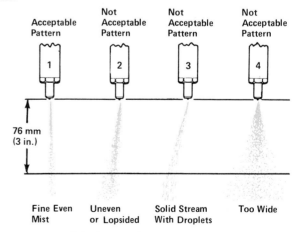

Figure 24-11. Injection nozzle spray pattern.
FORD MOTOR COMPANY

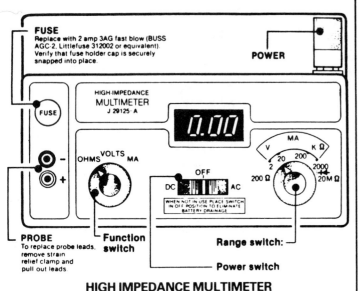

VOLTMETER—Voltage Position Measures amount of voltage. Connected parallel to exiting circuit. A digital high impedance voltmeter is used because some circuits require accurate low voltage readings, and some circuits have a very high resistance in the ECM. This meter also accurately measures extremely low current flow. Refer to meter for more information.

- Both function and range switch must be set properly, and the DC or AC position selected. DC is used for most measurements.

OHMMETER—Resistance Position Measures resistance of circuit directly in ohms. Refer to meter for more information.

- $\boxed{1.}$ display in all ranges indicates open circuit.
- Zero display in all ranges indicates a short circuit.
- Intermittent connection in circuit may be indicated by digital reading that will not stabilize on circuit.
- Range Switch.

HIGH IMPEDANCE MULTIMETER
J29125-A

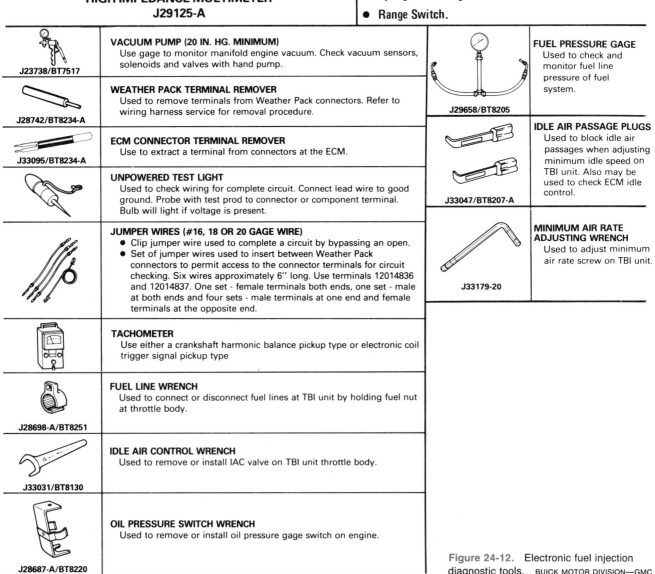

Figure 24-12. Electronic fuel injection diagnostic tools. BUICK MOTOR DIVISION—GMC

Fuel Injection Service

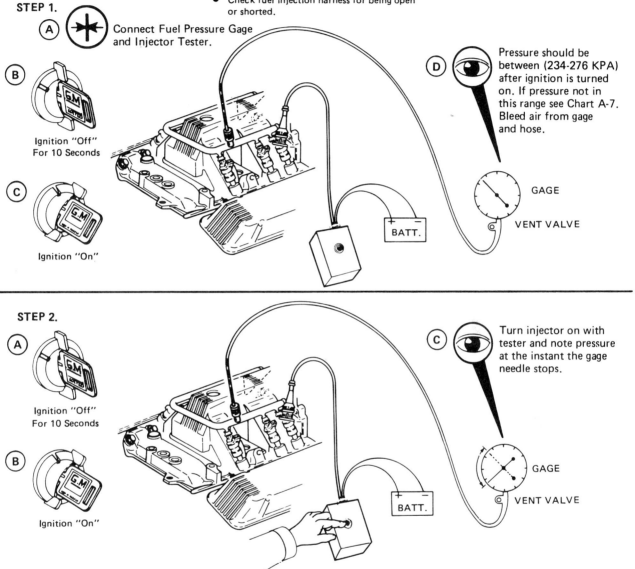

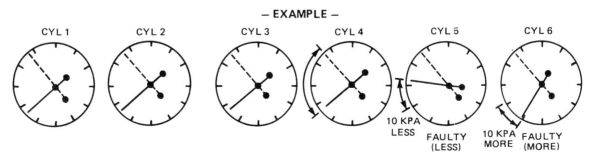

Figure 24-13. Injector balance test. BUICK MOTOR DIVISION—GMC

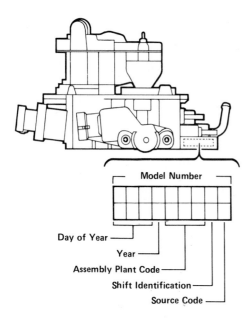

Figure 24-14. Throttle body fuel injection identification.
BUICK MOTOR DIVISION—GMC

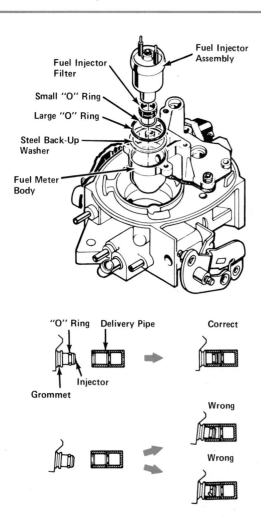

Figure 24-15. Fuel injector grommet and O-ring locations.
BUICK MOTOR DIVISION—GMC

CAUTION!!!

Before attempting to replace any electrical components of a fuel injection system, be sure the ignition switch is in the OFF position. Some vehicles also require that the battery ground cable be disconnected. Simply disconnecting or connecting a wire when the switch is on may damage or destroy expensive electronic equipment.

Injector Replacement

Whenever a fuel injector (single- or multi-point) is replaced, new rubber O-rings and grommets must be used **[Figure 24-15]**.

CAUTION!!!

When refitting O-rings, make sure the injector is aligned correctly, seats squarely, and does not pinch or damage the O-ring.

Electronic Control Unit Replacement

Refer to Unit 39 for a discussion of ECU replacement procedures. Replacement of electronic control units may require making specialized wiring connection repairs, as shown in **Figure 24-16**.

CAUTION!!!

Before attempting to replace any electrical components of a fuel injection system, the ignition switch must be in the OFF position. Some vehicles also require that the battery ground cable be disconnected. Simply disconnecting or connecting a wire when the switch is on may damage or destroy expensive electronic equipment.

Refer to the vehicle or injection system manufacturer's service manual for all major repairs to gasoline fuel injection systems.

24.6 DIESEL FUEL SYSTEM SERVICE

Diesel fuel system service is more specialized than gasoline system service. However, there are several preventive and/or corrective diesel fuel service procedures that can be performed, including:

- Removing water from the system
- Replacing fuel filters
- Testing injection nozzles.

Removing Water From the System

Most diesel-powered vehicles have a provision for dealing with a common diesel problem: water in the fuel. Some vehicles have a special water separator

FUEL INJECTION SERVICE

WEATHER PACK CONNECTORS REPAIR PROCEDURE

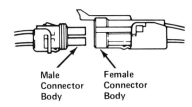

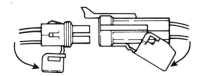

1. Open secondary lock hinge on connector

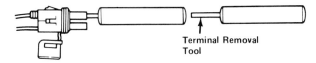

2. Remove terminals using special tool

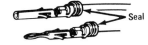

3. Cut wire immediately behind cable seal
   A. Slip new seal onto wire
   B. Strip 5.0 mm (0.2″) on insulation from wire
   C. Crimp terminal over wire and seal

TWISTED/SHIELDED CABLE

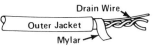

1. Remove outer jacket.
2. Unwrap aluminum/mylar tape. Do not remove mylar.

3. Untwist conductors, strip insulation as necessary.

4. Splice wires using splice clips and rosin core solder. Wrap each splice to insulate.
5. Wrap with mylar and drain (uninsulated) wire.

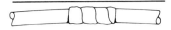

6. Tape over whole bundle to secure as before.

TWISTED LEADS

1. Locate damaged wire.
2. Remove insulation as required.

3. Splice two wires together using splice clips and rosin core solder.

4. Cover splice with tape to insulate from other wires.
5. Retwist as before and tape with electrical tape and hold in place.

Figure 24-16. Electronic control unit replacement and wiring diagram. BUICK MOTOR DIVISION—GMC

in the fuel system (refer to Unit 19). Water separators usually are drained by opening a valve on the bottom of the separator **[Figure 24-17].** Follow manufacturer's service recommendations for these procedures.

Water separators often are installed as aftermarket equipment on vehicles that are not so equipped. Some diesel systems are equipped with a combination fuel pickup/fuel level indicator/water sensor in the fuel tank. This unit will detect the presence of water when it reaches a level of 1 to 1-1/2 gallons in the tank. The unit will light a dashboard light to warn the driver.

The vehicle can be driven some distance under this condition before water is drawn into the fuel system. However, it is advisable to drain water from the tank as soon as possible.

SAFETY CAUTION

Be extremely careful when working with open fuel lines. Hot exhaust system parts, sparks, or open flames can cause fires. Keep a fire extinguisher capable of extinguishing class B fires fully charged and readily available.

Replacing Fuel Filters

All diesel fuel systems are fitted with fuel filters that have very fine filtering elements. These filters can remove particles that measure 10 *microns* in

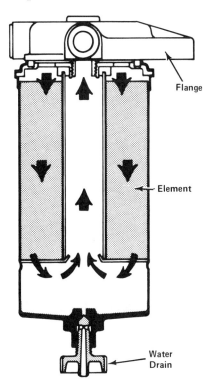

Figure 24-17. Diesel fuel filter/water separator assembly. VOLKSWAGEN OF AMERICA

diameter. A micron is a very small metric measurement, equal to 39 millionths (0.000039) of an inch. These filters must be changed routinely to prevent erratic engine operation or fuel system damage.

Diesel fuel filters may be either self-contained units or separate spin-on units. Always follow the manufacturer's service instructions for correct replacement procedures. These instructions provide detailed information on bleeding the fuel system to eliminate air. In some cases, a hand priming pump is provided for this purpose.

Testing Injection Nozzles

The injection nozzle is an important part of the diesel fuel system. The injection pump develops very high pressures—800 to 3,000 psi (55.2 to 206.9 bars). These pressures are routed through thick steel injection lines to the nozzles [Figure 24-18].

Check the manufacturer's service manual before proceeding with nozzle removal.

CAUTION!!!

Be sure the nozzles and lines are clean before proceeding with nozzle removal. Any contaminants in the fuel system can cause problems.

Once the injection nozzles have been removed, they can be checked for the following:

- Correct spray pattern
- Noise
- Correct opening pressure
- Leakage.

A nozzle or "pop" tester is used to evaluate nozzle condition [Figure 24-19].

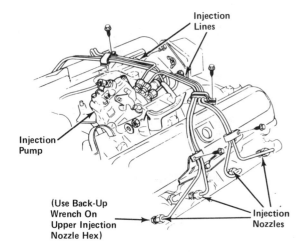

Figure 24-18. Diesel injection fuel lines.
CHEVROLET MOTOR DIVISION—GMC

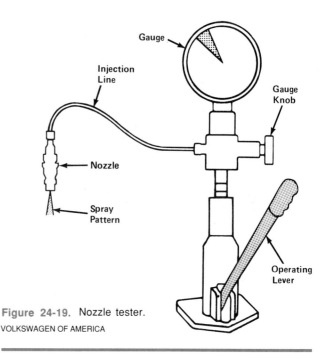

Figure 24-19. Nozzle tester.
VOLKSWAGEN OF AMERICA

SAFETY CAUTION

Always follow the nozzle manufacturer's test procedures and safety cautions. Never pass your hands near the nozzle when testing. The high pressure spray can penetrate the skin and cause blood poisoning.

Carefully follow the equipment manufacturer's test procedures to check for proper spray pattern. **Figure 24-20** shows acceptable and unacceptable patterns for one type of nozzle. Refer to service recommendations for the correct pattern.

While moving the operating handle rapidly, a "creaking" sound should be heard if the nozzle is clean and in good condition. Refer to service recommendations for correct noise.

Open the gauge handle a small amount. Observe the gauge while slowly operating the lever to determine the opening pressure of the nozzle. Refer to service specifications for the correct pressure.

Again, slowly operate the lever, carefully bringing the pressure to 200 psi (13.8 bars) below the opening pressure. Observe the nozzle tip.

SAFETY CAUTION

Do not touch the tip or place your hand near it. The pressure from the spray can cause the fluid to penetrate the skin and cause blood poisoning.

Figure 24-21 shows acceptable and unacceptable amounts of leakage for one type of nozzle. Refer to service recommendations for proper specifications.

FUEL INJECTION SERVICE 257

ACCEPTABLE NOZZLE SPRAY PATTERNS

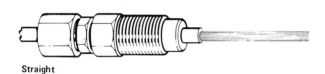
Converging

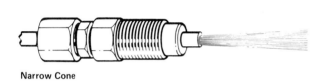
Straight

Narrow Cone

UNACCEPTABLE NOZZLE SPRAY PATTERNS

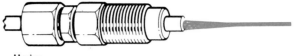
Hosing

Partial Cone

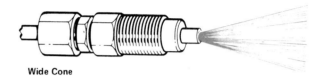
Wide Cone

Figure 24-20. Diesel fuel injector nozzle spray patterns.

1 — No Signs of Any Fuel

2 — No Visible Fuel But Damp

3 — Visible Fuel and Wet

NOT ACCEPTABLE
4 — Drop Forms But Does Not Fall or Run Along the Bottom of the Tip

NOT ACCEPTABLE
5 — Drop Falls or Runs Along Bottom of the Tip

Figure 24-21. Checking poppet nozzle seat tightness.

UNIT HIGHLIGHTS

- Fuel injection preventive maintenance consists of checking for several types of problems. These include leaks, vacuum hose defects, poor electrical connections, dirty filters, incorrect idle specifications, and loose fuel rails, lines, or injectors.
- Troubleshooting fuel injection systems and driveability problems requires reference to the manufacturer's service manuals.
- Simple checks can be made for fuel delivery.
- A multi-point fuel injector balance test can be conducted with a pressure gauge of the proper range.
- Diesel fuel injection pressures are extremely high and can cause physical injury.
- Fuel injection service includes identification, reference to manufacturer's service manuals, and replacement and/or repair of fuel pumps, sensors, injectors, and ECUs.
- Diesel fuel system service includes removing water from an injection system, replacing fuel filters, and testing injection nozzles.

TERMS

fatigue
micron
trouble code

REVIEW QUESTIONS

DIRECTIONS: The following questions are similar to those used on automotive technician certification tests. On a separate sheet of paper, write the letter of the correct choice.

1. Technician A says that fuel injection systems never cause running problems.

   Technician B says that poor engine compression and weak ignition system performance can cause many running problems.

   Who is correct?
   A. A only
   B. B only
   C. Both A and B
   D. Neither A nor B

2. Which of the following statements is correct?
   I. The ignition switch must be in the ON position before any electrical components of a fuel injection system are disconnected or connected.
   II. Electronic parts are not easily damaged by grounding or shorting.

   A. I only
   B. II only
   C. Both I and II
   D. Neither I nor II

3. After removing and replacing a diesel fuel line, which of the following procedures must be done before starting the engine?
   A. Change the fuel filter and engine oil filter.
   B. Drain water from the water separator.
   C. Check injection timing.
   D. Bleed the fuel lines to remove air.

4. Before attempting work on any fuel injection system, what should the technician do?
   A. Talk to the owner about running problems.
   B. Test drive the vehicle.
   C. Check for fuel delivery.
   D. Refer to the manufacturer's service manual for troubleshooting procedures and charts.

Question 5 is not like the first four questions. It has the word EXCEPT. For question 5, look for the choice that does not apply. Read the entire question carefully before you choose your answer.

5. Fuel injection system preventive maintenance includes checking, replacing, and/or repairing all of the following EXCEPT
   A. leaks, vacuum hose defects, and poor electrical connections.
   B. loose or improperly supported fuel rails, lines, and injectors.
   C. air and fuel filters.
   D. diesel fuel injection pump delivery volume and timing.

ESSAY QUESTIONS

1. Describe fuel injection preventive maintenance checks.
2. Why are manufacturer's troubleshooting charts necessary when servicing fuel injection systems?
3. Describe fuel injection system repairs.

SUPPLEMENTAL ACTIVITIES

1. Perform fuel injection preventive maintenance on a vehicle chosen by your instructor.
2. Check for fuel delivery on a fuel injected vehicle chosen by your instructor. Report to your class what method you used.
3. Listen to a fuel injector with a technicians stethoscope on a running engine. Describe the sounds you hear to your class in terms of loudness and frequency. What happens to the sounds if the engine is run at higher or lower rpm?
4. If suitable equipment and a vehicle are available, perform a fuel injector balance test on a multi-point fuel injection system.
5. Refer to the manufacturer's service manual and identify a TBI system. Report to your class where the numbers were found and what they indicate about the system.

25 The Exhaust System

UNIT PREVIEW

The exhaust system directs harmful gases away from the vehicle and reduces noise. Exhaust noise is caused by pressure waves leaving the exhaust port.

The exhaust system also carries away heat from the engine. Almost one-third of the energy in gasoline is converted into unwanted heat, which is routed away safely through the exhaust system.

Finally, pollution control devices associated with the exhaust system help to reduce harmful exhaust emissions.

LEARNING OBJECTIVES

When you have completed your assignments and exercises in this unit, you should be able to:

❏ Identify and describe modern exhaust system parts and their functions.
❏ Explain what devices are used to lessen harmful exhaust gas emissions.
❏ Describe how a muffler cuts down the noise of the exhaust.
❏ Identify and describe the internal parts of a muffler.
❏ Identify and describe the basic types of catalytic converters.

25.1 BASIC EXHAUST SYSTEM

The basic purpose of the exhaust system is to direct harmful exhaust gases and heat away from the engine and underside of the vehicle. Exhaust gases contain *carbon monoxide (CO)*. Carbon monoxide poisoning, caused by accidental venting of exhaust gases into the passenger compartment, results in many deaths each year.

The exhaust system also reduces noise caused by the combustion process within the cylinders.

On current vehicles, the exhaust system also helps to change harmful exhaust gases to relatively harmless compounds through chemical action.

A typical modern exhaust system, shown in **Figure 25-1,** consists of the following parts:

- Exhaust manifold and gasket
- Exhaust pipe and seal
- Catalytic converter
- Intermediate pipes
- Muffler
- Resonator
- Tailpipe
- Heat shields
- Clamps and hangers.

Conventional exhaust systems have been made of mild steel or of steel with an aluminum or zinc coating. Newer vehicles have stainless steel exhaust systems which are much more resistant to rust and corrosion.

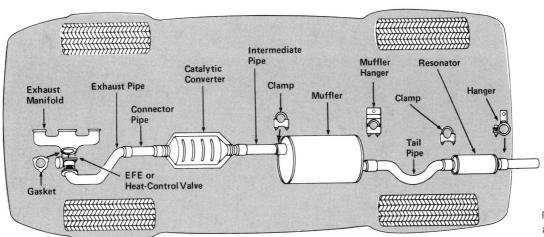

Figure 25-1. Parts of an exhaust system.

Exhaust Manifold and Gasket

Near the end of the power stroke, the exhaust valve opens and exhaust gases begin to leave the cylinder. The piston continues on its exhaust stroke, pushing the burned gases into the exhaust manifold.

Every time the exhaust valve opens, waves of high pressure gases are forced into the exhaust manifold. Sound, or noise, is produced by these pressure waves.

One exhaust manifold design factor is the length of passages within the manifold. If an exhaust manifold is designed correctly, pulses from the cylinders do not interfere with one another. Thus, exhaust gases can flow easily through the manifold.

To assure a good flow of gases with minimum backpressure, the exhaust manifold has large tubular sections. In addition, engineers try to avoid sharp bends that can slow the passage of gases.

However, the exhaust manifold often has to fit into narrow, cramped spaces between the engine and the body of the vehicle. Access to spark plugs and other parts also helps to determine the shape of the exhaust manifold.

Thus, exhaust manifold shapes are often compromises between good gas flow and limited mounting areas.

Exhaust manifolds for higher-performance versions of an engine often are noticeably different in shape for better gas flow **[Figure 25-2]**.

Exhaust manifolds for most passenger vehicles are made of cast iron. An exhaust manifold gasket is sometimes inserted between the manifold and the cylinder head to prevent exhaust leakage **[refer to Figure 25-2]**. Some newer vehicles have stamped, heavy-gauge sheet metal exhaust manifolds.

Air injection. To make exhaust gases less harmful, fresh air can be injected into the exhaust manifold. This helps to oxidize the unburned fuel still remaining in the exhaust gases. Metal connector pipes and fittings to carry and distribute fresh air are screwed into the exhaust manifold **[Figure 25-3]**. Air injection systems are discussed and illustrated in Unit 40.

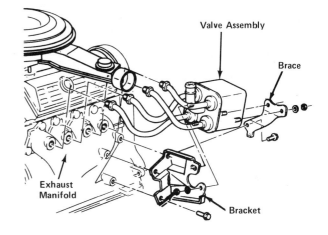

Figure 25-3. Air injection mountings.
CHEVROLET MOTOR DIVISION—GMC

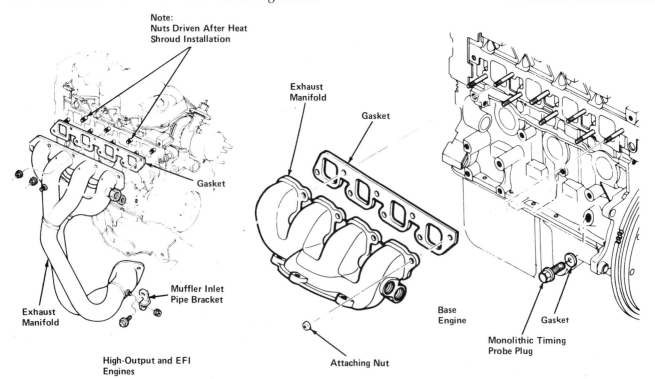

Figure 25-2. High-performance and normal exhaust manifolds. FORD MOTOR COMPANY

EGR. A process called *exhaust gas recirculation (EGR)* is used to lower combustion temperatures for control of NO_x (oxides of nitrogen). The EGR process routes small amounts of exhaust gas into the intake air-fuel charge [**Figure 25-4**]. Exhaust gas recirculation is discussed in Unit 40.

Exhaust Pipe and Seal

The *exhaust pipe* is connected to the exhaust manifold. A beveled seal, made of wire-reinforced asbestos or solid steel, is inserted between the exhaust pipe and manifold. If an EFE system is used, it usually is mounted between the exhaust manifold and exhaust pipe [**Figure 25-5**].

V-type engine exhaust connections. The exhaust gases from most current V-type engines are routed into a single collector pipe. Either a Y-shaped pipe or a *crossover pipe* connects the two sides of the engine [**Figure 25-6**]. A crossover pipe can be routed behind or below the engine.

High-performance vehicles with V-type engines sometimes have two separate exhaust systems, one for each bank of the engine. This is known as "dual exhausts."

Turbocharged engine exhaust connections. Exhaust gases that drive a turbocharger must be routed from the exhaust pipe into and out of the exhaust turbine housing. Examples of exhaust pipe configurations, or shapes, to accomplish this routing are shown in **Figure 25-7**.

A high-performance *header* is a combination of an exhaust manifold and an exhaust pipe, custom-formed from thin steel tubing. This "tuned exhaust" system allows for easier passage of gas from the engine. This results in better volumetric efficiency and greater horsepower.

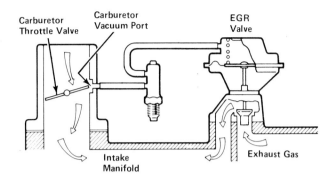

Figure 25-4. Exhaust gas recirculation (EGR) operation.

1 Exhaust Pipe 5 Spring
2 Right Manifold 6 Left Manifold
3 Crossover Pipe 7 EFE Valve
4 22 Ft. Lbs. (30 N·m)

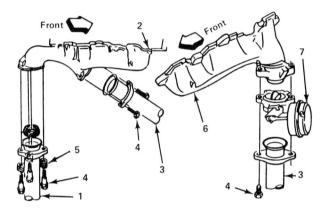

Figure 25-5. Early fuel evaporation (EFE) mounting.
BUICK MOTOR DIVISION—GMC

1 Right Hand Manifold 7 Left Pipe
2 Seal (Engine Code "N") 8 40 Ft. Lbs. (55 N·m)
3 EFE Valve 9 Spring
4 Spring (Engine Code "N") 10 Seal
5 22 Ft. Lbs. (30 N·m) 11 Left Hand Manifold
6 Crossover Pipe 12 Right Pipe

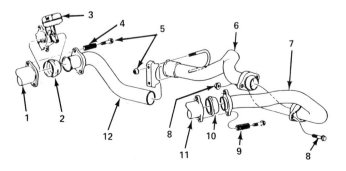

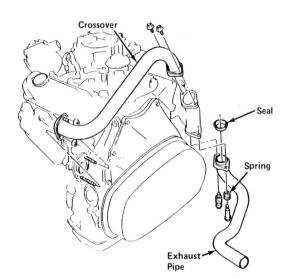

Figure 25-6. Exhaust pipe connections for V-type engines. BUICK MOTOR DIVISION—GMC

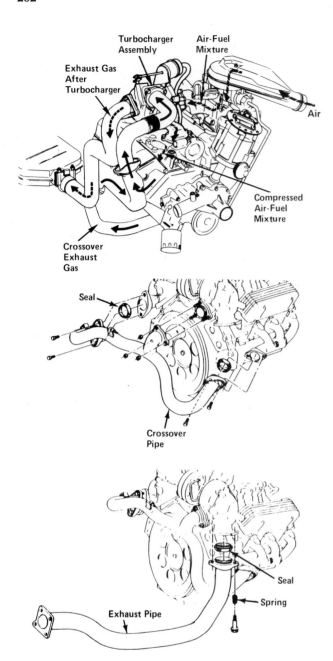

Figure 25-7. Exhaust pipe configurations.
BUICK MOTOR DIVISION—GMC

Catalytic Converter

Catalytic converters have been included in exhaust systems since 1975. A catalytic converter is a part that converts, or changes, harmful exhaust gases into less harmful gases through chemical action. Vehicles with catalytic converters must use unleaded fuel only. The operation of catalytic converters is explained in Unit 40.

One or two catalytic converters may be used in an exhaust system. In a system using two converters, one may be a "light-off" converter that operates during warmup. Such a converter pretreats and heats the exhaust gases for treatment by the main converter, as shown in **Figure 25-8**.

In other systems, one of the converters changes CO (carbon monoxide) and HC (unburned hydrocarbons) into carbon dioxide and water. Technicians refer to this catalytic converter (CO and HC) as a "two-way cat." Another, separate converter changes NO_x (oxides of nitrogen) into nitrogen and oxygen.

Newer vehicles combine the functions of both converters into a single unit, called a *three-way catalytic converter* (CO, HC, and NO_x). This type of catalytic converter also is called a "three-way cat" or a *dual bed converter.* Some three-way catalytic converters have dual beds, or sections, where gases are treated **[Figure 25-9]**.

Other three-way catalytic converters combine all exhaust gas treatment functions into a single unit **[Figure 25-10]**.

The catalytic converter is enclosed within a shell, usually made of stainless steel, to resist corrosion and heat.

Intermediate Pipes

Intermediate, or connecting pipes, can be used at any point past the exhaust pipe to join exhaust system components. These intermediate pipes can be shaped to fit around axles and suspension components, away from shock absorbers, fuel tanks, and so on.

Pipe connections. Gaskets and seals are used to join the exhaust manifold and exhaust pipe to the

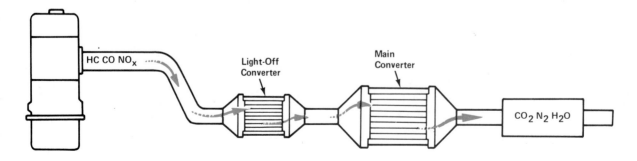

Figure 25-8. Separate catalytic converters.

cylinder head and to each other. In most other parts of the exhaust system, pipes usually are either clamped or welded together **[Figure 25-11]**.

Clamps can be used when the two connecting pipes are formed in such a way that one slips inside the other. This design makes a close fit. The clamp then holds this connection tight.

Welding is a process in which heat is used to melt metal and allow the molten portions to flow together. When cooled, a welded joint is as strong, or sometimes stronger, than a single piece of metal. For mild steel or coated steel, gas welding typically is used. For stainless steel exhaust system parts, special welding techniques such as tungsten inert gas (TIG) and metal inert gas (MIG) must be used. Ask your instructor about welding classes offered at your school or area.

Another type of pipe connection is a *ball joint* **[Figure 25-12]**. A ball joint is a type of connection with a rounded, ball-shaped part that fits into a matching rounded, hollow socket. Ball joints are used where connections cannot be made in a straight line. Nuts and bolts are used to hold the ball joint together. Ball joints are used extensively in front-wheel drive vehicles to join the exhaust pipe to the manifold.

Muffler

The *muffler* muffles, or softens, noise. The exhaust gases, which travel in pulses, would produce a loud roar without a muffler in the system.

A conventional muffler consists of an outer shell and inner resonance chambers. The interior is a series of chambers, holes, and *baffles* through which the exhaust gases must pass **[Figure 25-13]**. A baffle is a wall, plate, or screen to control the flow of fluids, such as liquids or gases.

Passing through the muffler, the gases and pressure waves are slowed down, and much of their energy is absorbed. This process reduces the intensity of exhaust gas pressure waves and resultant noise.

Even a well-designed conventional muffler will produce some backpressure in the system. Backpressure reduces an engine's volumetric efficiency, or ability to "breathe." Excessive backpressure caused by defects in a muffler or other exhaust system part can slow or stop the engine.

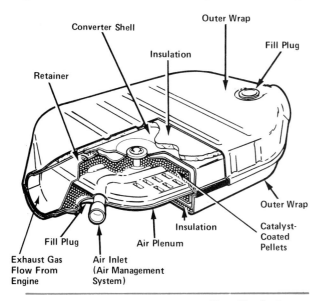
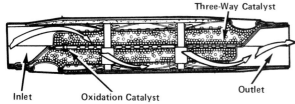

Figure 25-9. Dual-bed catalytic converter.
CHEVROLET MOTOR DIVISION—GMC

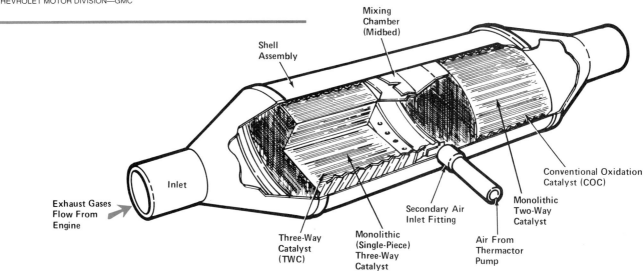

Figure 25-10. Dual-function catalytic converter. FORD MOTOR COMPANY

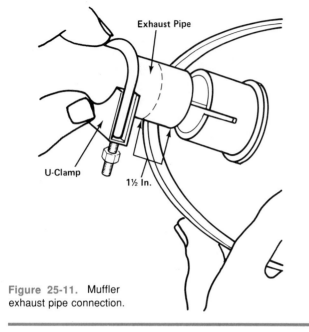

Figure 25-11. Muffler exhaust pipe connection.

A small amount of backpressure can be used intentionally to allow a slower passage of exhaust gases through the catalytic converter. This slower passage results in more complete conversion to less harmful gases, and also may be required for the operation of some types of EGR valves.

Resonator

Another muffler-like device, a *resonator*, is added to some exhaust systems to absorb additional sound frequencies. The resonator helps to cut down objectionable noise **[refer to Figure 25-13]**.

Electronic muffler. Sample models of *electronic mufflers* **[Figure 25-14]** have been produced. In an electronic muffler, there are no baffles. Instead, there is a central, hollow pipe surrounded by noise generators. A noise generator is a device similar to an audio speaker used in radio systems. The noise generators are specially made of materials that withstand exhaust heat and corrosion.

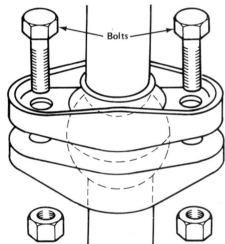

Figure 25-12. Ball joint exhaust pipe manifold connection.

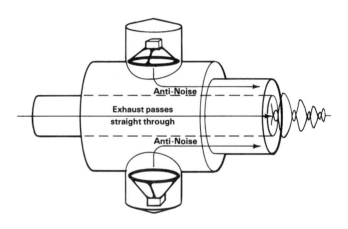

Figure 25-14. Electronic muffler uses a computer to produce anti-noise to cancel out existing noise.

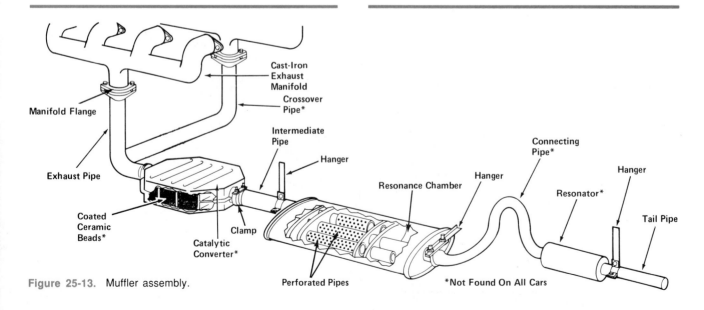

Figure 25-13. Muffler assembly.

The Exhaust System

The principle is simple: noise is created by waves of energy pulsing from a source. If the trough, or low part, of another wave of the same tone combines with a noise wave, it cancels the noise out. In other words, +1 and -1 together make zero. Thus, by creating sound waves of the proper frequency (tone) and amplitude (loudness), exhaust noise can be eliminated. If electronic mufflers are perfected and produced, the major benefit will lie in reduction or total elimination of backpressure. Another benefit is the ability to change the exhaust note to a pleasant tone.

Tailpipe

The *tailpipe* is the endmost part of the exhaust system. Exhaust gases pass out of the tailpipe into the atmosphere.

Heat Shields

Heat shields are made of pressed or perforated sheet metal. They are used to protect vehicle parts from the heat of the exhaust system and catalytic converter **[Figure 25-15]**.

Clamps and Hangers

Clamps and hangers, shown in **Figures 25-11** and **25-13**, serve two purposes. Clamps help to secure exhaust system parts to one another as well as to the bottom of the vehicle. Clamps are bolts formed into a half-circular shape, with a bar that bolts across the open "U."

Hangers help to isolate noise, preventing its transfer through the frame or body to the passenger compartment. Hangers are made from rubber, sometimes reinforced with fabric. One type of hanger resembles a large, thick rubber band stretched between hooks on the underbody and exhaust system **[Figure 25-16]**.

Together, clamps and hangers correctly support and position exhaust system components.

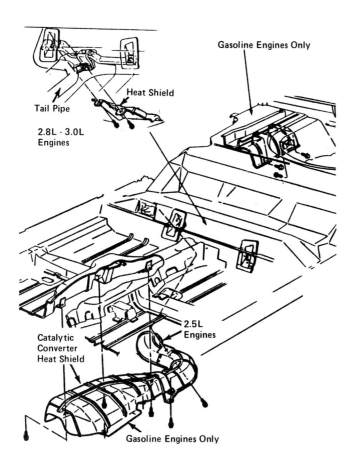

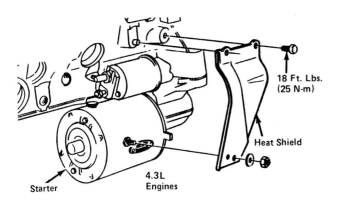

Figure 25-15. Heat shields. BUICK MOTOR DIVISION—GMC

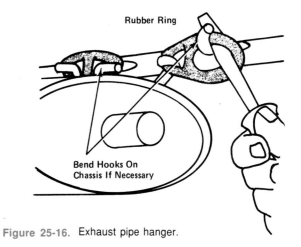

Figure 25-16. Exhaust pipe hanger.

UNIT HIGHLIGHTS

- Exhaust systems on current vehicles direct harmful gases and heat away from the engine and underside of the vehicle. They also reduce noise and harmful exhaust gases.
- Exhaust systems consist of many parts, extending from the engine to the rear of the vehicle.
- Exhaust noise is caused by pressure waves leaving the exhaust port.
- A muffler reduces the intensity and pressure of exhaust gases, thus reducing noise.
- Systems included in the exhaust system to reduce harmful exhaust gases include air injection, EGR, and catalytic converters.

TERMS

exhaust pipe
header
tailpipe
ball joint
baffle
heat shield
electronic muffler
exhaust gas recirculation (EGR)
crossover pipe
catalytic converter
dual bed converter
muffler
resonator
welding
carbon monoxide (CO)
three-way catalytic converter

REVIEW QUESTIONS

DIRECTIONS: The following questions are similar to those used on automotive technician certification tests. Answer each question by circling the letter of the correct choice.

1. Technician A says that exhaust systems direct exhaust gases away from the underside of the vehicle.
   Technician B says that exhaust systems change carbon dioxide and nitrogen into CO, HC, and NO_x.
   Who is correct?
   A. A only
   B. B only
   C. Both A and B
   D. Neither A nor B

2. What causes exhaust noise?
   A. Ignition spark
   B. Pressure waves formed as the exhaust valve opens
   C. Backpressure in the muffler and catalytic converter
   D. HC, CO, and NO_x

3. Which of the following statements is correct?
   I. EGR introduces fresh air into the exhaust manifold.
   II. Air injection introduces exhaust gases into the intake manifold.
   A. I only
   B. II only
   C. Both I and II
   D. Neither I nor II

Questions 4 and 5 are not like those above. Each has the word EXCEPT. For each question, look for the choice that does *not* apply. Read each question carefully before you choose your answer.

4. All of the following are parts of a conventional muffler EXCEPT
   A. perforated pipe.
   B. a resonating chamber.
   C. baffles.
   D. catalyst beds.

5. All of the following are types of catalytic converters EXCEPT
   A. light-off.
   B. dual-bed.
   C. three-way cat.
   D. one-way cat.

ESSAY QUESTIONS

1. Identify and describe the parts and functions of modern exhaust system parts.
2. Describe the parts and functions of a conventional muffler.
3. Identify and describe the basic types of catalytic converters.

SUPPLEMENTAL ACTIVITIES

1. Examine relatives' or friends' vehicles. How many have catalytic converters in the exhaust systems?
2. Examine several V-type engine's exhaust systems. Sketch the location and type of pipes that connect from one side of the engine to the other.
3. Explain why engineers might design an exhaust header or muffler assembly that produces more backpressure than absolutely necessary.

26 Exhaust System Service

UNIT PREVIEW

Exhaust system parts are subject to chemical and physical damage. Leaks from a damaged exhaust system can cause dizziness, nausea, or in extreme cases, death.

Physical damage to exhaust system connector pipes or other components can cause inefficient engine operation and increased air pollution and noise.

Preventive maintenance for the exhaust system consists of inspections and minor repairs. Muffler and connector pipe replacement may require the use of special tools and/or welding equipment.

As with mufflers, catalytic converters can be replaced. Some bead-type converters can be serviced to restore chemical efficiency.

LEARNING OBJECTIVES

When you have completed your assignments and exercises in this unit, you should be able to:

❏ Explain why exhaust leaks are dangerous.
❏ Perform an inspection of an exhaust system from the engine to the tail pipe and identify any defects.
❏ Use a vacuum gauge to check engine operation for evidence of an exhaust restriction.
❏ Replace exhaust manifold gaskets, seals, hanger brackets and/or clamps.
❏ Free a stuck heat riser valve or replace an EFE valve.
❏ Replace a muffler and/or connector pipe.

SAFETY PRECAUTIONS

Exhaust gases are harmful. Work on running vehicles only in well-ventilated areas outdoors, or indoor areas with exhaust hose connections to remove fumes.

Whenever possible, wait until the vehicle has cooled down before attempting exhaust system inspection, preventive maintenance, or repairs. Hot exhaust system parts can cause severe burns.

Working on the exhaust system requires that the entire underside of the vehicle be accessible. Safely raise and support the vehicle before attempting inspection or repair procedures.

Wear eye protection at all times when working under the vehicle. Because they are heated during operation, exhaust system parts rust quickly. Loose rust particles can cause pain and/or permanent blindness if they become lodged in the eyes.

Removal and/or replacement of some exhaust system components may require heating, cutting, or welding with gas, TIG, or MIG equipment. Do not attempt any of these procedures without proper training, safety equipment, and permission from your instructor.

Refer to the manufacturer's service manual for correct replacement procedures, parts, clearances between parts, and recommended sealants.

26.1 EXHAUST SYSTEM PROBLEMS

Exhaust system parts are subject to rust, corrosion, and physical damage. Gas leaks can cause harm to the vehicle's passengers and vehicle parts. Exhaust restrictions can cause the engine to run inefficiently, stall, and/or die.

Gas Leaks

Heat speeds up the chemical processes of rust and corrosion. In addition, acid forms within the pipes. As the exhaust system cools after the engine is off, cool air from the atmosphere is drawn up the pipe. This cool air contains moisture. Moisture combines with sulfur and other minerals deposited in the exhaust system by the flowing exhaust gases, and forms sulfuric and nitric acids. These acids eat away pipes and other components from the inside.

If a vehicle is driven longer distances, the moisture that collects inside the system is turned to vapor and evaporates. This vapor, or "steam," can be seen at the tail pipe when the vehicle is started after being parked overnight.

Vehicles driven only short distances retain large amounts of moisture and produce large amounts of acid. Thus, mild steel or coated steel exhaust system components on such vehicles are rusted and eaten through very quickly.

The underbody of the vehicle is not gas-tight. Exhaust gases from damaged pipes and other components can enter the passenger compartment through openings and seams in the floor pan.

In cold climates, people sometimes become trapped in snow drifts in their vehicles. They close all the windows and leave the engines running to obtain heat from the heater. If the exhaust system has holes in it, exhaust gases can enter the passenger compartment. Every year, many people die from carbon monoxide gas poisoning.

Exhaust Restrictions

A restricted exhaust system will reduce engine efficiency and/or cause the engine to stall or stop.

Exhaust pipes can become bent, flattened, or kinked. In some cases, an inner pipe within an outer shell can collapse. Baffles within a muffler can become loose and partially block the passage of gases. If catalytic converters are damaged or become coated with lead from leaded gasoline, they also can block the flow of exhaust gases.

A restricted exhaust can be detected by using a vacuum gauge connected to a source of engine intake manifold vacuum. Refer to Unit 43 for the complete procedure. Check for a vacuum reading that drops almost to zero as the engine is accelerated, then slowly climbs back to normal. This usually indicates a restricted exhaust system part.

26.2 EXHAUST SYSTEM PREVENTIVE MAINTENANCE

Preventive maintenance for the exhaust system consists of the following items:

- Inspecting the system for leaks, holes, and physical damage
- Inspecting the hangers and brackets for damage
- Replacing hangers and clamps
- Replacing leaking gaskets and seals
- Checking heat riser/EFE valves for sticking
- Servicing stuck heat riser/EFE valves.

Inspecting for Leaks, Holes, and Physical Damage

The vehicle should be safely raised and supported. After the exhaust system has cooled thoroughly, the entire system should be inspected from the tail pipe to the cylinder head. **Figure 26-1** illustrates some of the checks discussed below.

Visual inspection. Mufflers, resonators, and connector pipes should be inspected for evidence of rusted, jagged holes. Rust spots may be only surface rust. Suspected deep rust should be jabbed with a screwdriver. If the screwdriver penetrates the rusty spot, the part is defective and will fail soon.

A wooden block or even your knuckles can be used to knock and pound on pipes and mufflers to detect weak spots. A good part will be solid and firm. A defective part has a tinny, thin sound and feel. Loose baffles within a muffler sometimes can be heard moving around in the muffler.

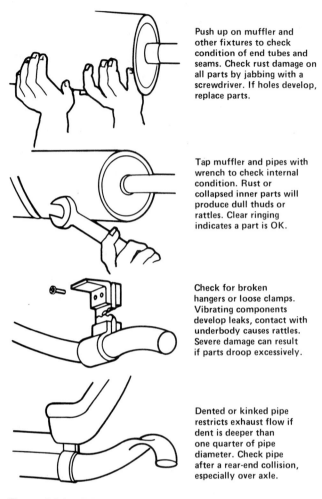

Figure 26-1. Exhaust system inspection points.

The exhaust manifold and exhaust pipe connections should be checked for loose or missing bolts or nuts.

Catalytic converters can overheat. This may be seen in a bluish or brownish discoloration of the outer stainless steel shell. In addition, undercoating or paint above and near the converter may appear blistered or burned. Catalytic converters will usually be surrounded by a heat shield.

Pipes and mufflers should be checked for signs of bending, scraping, or other physical damage. Flattened pipes and mufflers should be replaced.

Leaks at connections sometimes can be spotted by the presence of exhaust deposits (dark or grayish soot). If such a leak is suspected, a check should be made with the engine running.

Noise checks. Exhaust leaks often can be heard as an intermittent hissing or blowing sound when the engine is running. To locate the source, use a length of rubber tubing held near your ear to magnify the sound. Move the rubber tubing near and around the suspected leak. When found, the sound will be loud and clearly defined. In some cases, simply replacing a missing bolt or nut or tightening a loose clamp can stop the leakage.

SAFETY CAUTION

Have a helper in the car to start and stop the engine. Make sure the emergency brake is set tightly and the transmission is in Park (automatic) or neutral (manual). Keep hands, arms, and other body parts away from heated exhaust system and drivetrain parts that could turn during testing.

Inspecting Hangers and Brackets for Damage

Rubber hangers should be inspected for deep cracking or tearing. Small cracks are normal after a few years in service. The tail pipe and connector pipes should be grasped and vigorously wiggled. More than one inch (25.4 mm) of motion usually indicates a loose, broken, or disconnected hanger or support.

Pipes and other components should be inspected for shiny or rubbed spots. Such areas can indicate defective straps or hangers, or improper positioning near moving parts.

Replacing Hangers and Clamps

Hangers and clamps should be replaced if deeply cracked or broken [Figure 26-2]. Clearance around suspension parts and fuel tanks should be maintained as indicated in the manufacturer's service manual. *Penetrating solvent* should be used on bolt and nut connections before removal is attempted. Penetrating solvent is a thin, oily solvent that flows between threads and helps to lubricate parts for removal.

Replacing Leaking Gaskets and Seals

The most likely spots for leaking gaskets and seals are between the exhaust manifold and the cylinder head, and between the exhaust pipe and the exhaust manifold.

Exhaust system nuts and bolts are usually rusty and difficult to unscrew. Penetrating solvent should be used before attempting to unscrew such fasteners. The solvent should be sprayed or squirted on the fasteners. The fasteners then should be tapped sharply with a small hammer on the ends or sides (not on the threads). After a few minutes, the solvent should penetrate and make removal easier.

If penetrating solvent will not free a locked nut or bolt, it will have to be heated with a torch. This will usually free rusted parts, but should only be used as a last resort.

SAFETY CAUTION

Do not attempt to use welding equipment without proper training, safety equipment, and your instructor's permission. The light from gas or electric-arc welding can cause permanent eye damage or blindness. In addition, the heat from welding can ignite fuel vapors or damage parts such as shock absorbers and carpets.

Room must be obtained to move the exhaust manifold away from the engine or the exhaust pipe away from the manifold. This usually makes it necessary to disconnect brackets and hangers. The exhaust system then can be lowered gently and supported during service [Figure 26-3].

Old gaskets should be discarded and replaced with new gaskets during reassembly [Figure 26-4]. In some locations, the manufacturer's service manual may recommend the use of special high-temperature sealants to prevent gas leaks.

Checking for warped manifolds. Exhaust manifolds sometimes warp because of heat or improper torquing of fasteners. Use a straightedge and feeler gauge to check the machined surfaces of the manifold, across the surfaces indicated in Figure 26-5. If the manifold

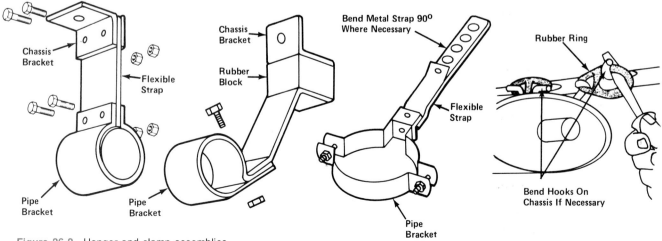

Figure 26-2. Hanger and clamp assemblies.

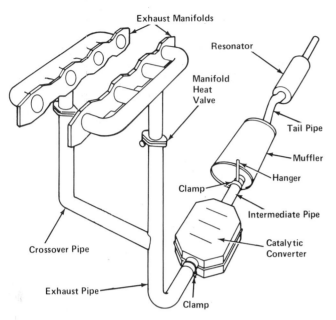

Figure 26-3. Parts of an exhaust system to be removed.

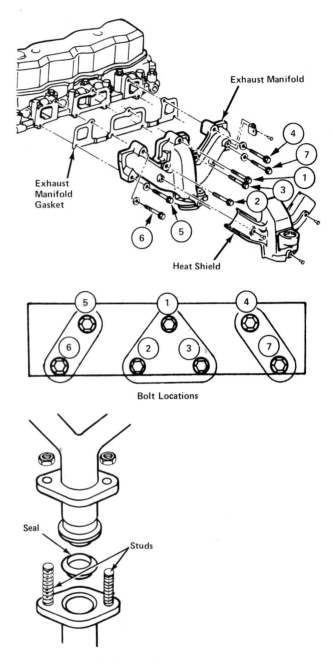

Figure 26-4. Exhaust system gasket locations.

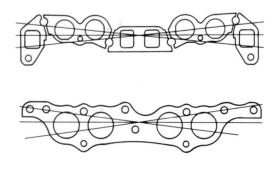

Figure 26-5. Exhaust manifold surfaces are checked for warpage.

is warped beyond the manufacturer's specifications, replace it.

When replacing exhaust manifolds, use a torque wrench to tighten the nuts or bolts to the manufacturer's specifications (as low as 10 to 15 lb.-ft.). Also make sure that you use special heat-resistant nuts and bolts designed for exhaust manifolds. If common fasteners are used, they may work loose or break because of exhaust heat and vibration.

Checking Heat Riser/EFE Valves

Heat riser and EFE valves mounted between the exhaust manifold and exhaust pipe should operate freely **[Figure 26-6]**. Vacuum-type valves are tested by connecting a vacuum pump to the diaphragm connection. As vacuum is applied, the shaft should move. Mechanically operated valves can be checked by moving the balance weight. It should move easily. When rusted, these parts may be frozen, or stuck, in position.

SAFETY CAUTION

Allow the exhaust system to cool before attempting to check or service heat riser or EFE valves.

Servicing Heat Riser/EFE Valves

A stuck vacuum-operated valve must be replaced. A mechanical heat riser valve often can be serviced by spraying it with *graphited oil* and physically moving the balance weight. Graphited oil is a thin lubricant containing graphite particles. After the engine starts, the oil will burn off, but the graphite, an excellent dry lubricant, will remain.

Pliers are used to move the balance weight and valve shaft slowly and carefully. Graphited oil is applied liberally during this process until the shaft moves freely and easily. Another technique is to tap the shaft in and out lightly with a ball peen hammer to loosen the valve.

26.3 MUFFLER AND CONNECTOR PIPE REPLACEMENT

Replacing exhaust system parts requires the use of tools to cut, shape, and expand the steel tubing used **[Figure 26-7]**. To remove a rusted-on muffler, the ends must be slit with a hand or power cutter **[Figure 26-8]**.

Exhaust and intermediate pipes can be cut with a special tool for replacement **[Figure 26-9]**.

An expander tool is used to enlarge the size of pipes made smaller by cutting or by having been clamped **[Figure 26-10]**.

Small sections of damaged pipe can be replaced, or pipes can be joined with couplings **[Figure 26-11]**.

26.4 CATALYTIC CONVERTER PROBLEMS

Catalytic converters can become damaged from being bumped or scraped. In addition, the illegal use of leaded fuel can lead to clogging and overheating. If the engine is operated with one or more of the spark plugs misfiring, it may result in the melting of the catalyst inside the converter.

Physical Damage

Certain types of converters have both an outer and inner shell. External physical damage to the outer shell of the converter may indicate internal damage. The outer shell of the converter is cut apart to inspect the inner converter **[Figure 26-12]**. If damaged, the entire converter must be replaced. If only the outer shell is damaged, it can be replaced separately.

CAUTION!!!

Care must be used when cutting dual-bed converters to avoid damaging or puncturing the air inlet.

If no inner physical damage is evident, a special high-temperature sealant is applied. The outer shell is clamped on with channels and retaining clamps **[Figure 26-13]**.

Bead Replacement

Some catalytic converters are filled with chemically coated beads or pellets. These pellets can be replaced if they are coated with lead or are otherwise not functioning. A plug is removed from such converters. A vibrator is attached to the catalytic converter to vibrate and shake the pellets loose and into a container **[Figure 26-14]**.

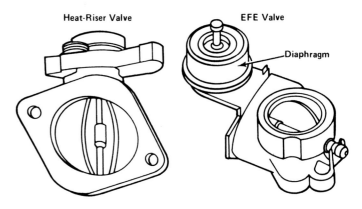

Figure 26-6. Heat riser and EFE valves.

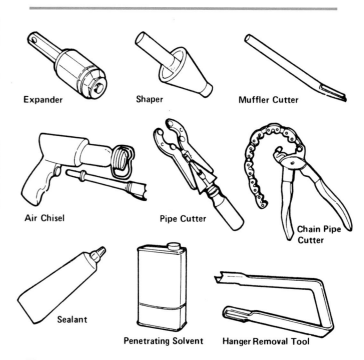

Figure 26-7. Exhaust system tools.

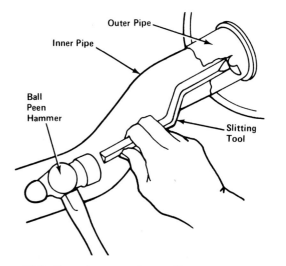

Figure 26-8. Removing a rusted-on muffler.

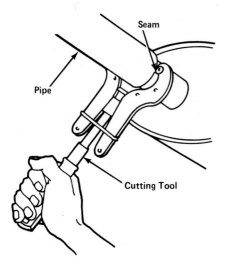

Figure 26-9. Cutting an exhaust pipe.

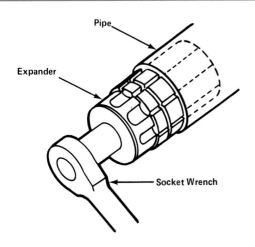

Figure 26-10. Expanding an exhaust pipe.

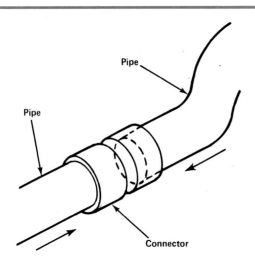

Figure 26-11. Exhaust pipe coupling.

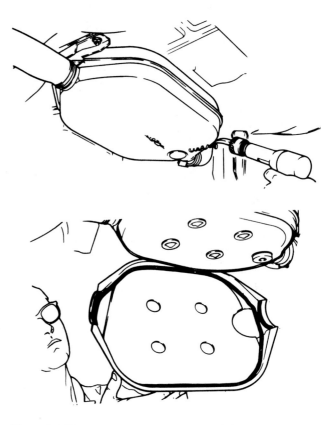

Figure 26-12. Cutting and resealing a catalytic converter.
BUICK MOTOR DIVISION—GMC

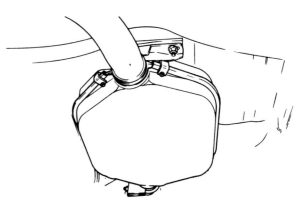

Figure 26-13. Catalytic converter with channel clamps.
BUICK MOTOR DIVISION—GMC

To refill the converter, new beads are placed in the can. The air hose is then attached to the vibrator unit, and a vacuum pump is attached to the tail pipe. New beads will be drawn into the converter. When full, the tools are removed and the converter plug is coated with nickel-based anti-seize compound and replaced.

NOTE: If beads come out through the tail pipe during this procedure, the entire converter must be replaced.

EXHAUST SYSTEM SERVICE

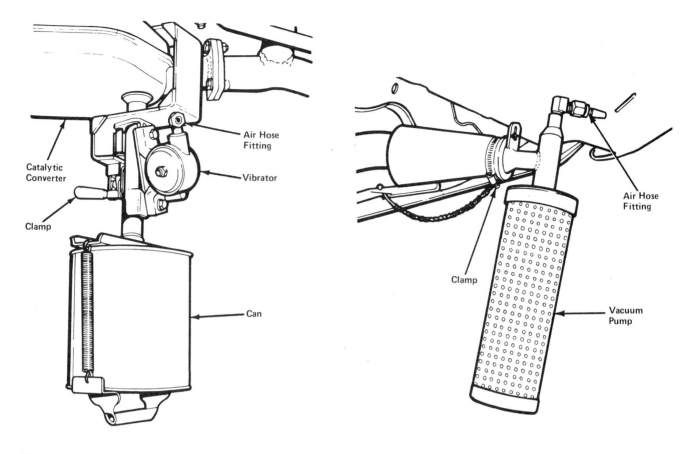

Figure 26-14. Catalytic converter bead replacement. BUICK MOTOR DIVISION—GMC

UNIT HIGHLIGHTS

- Exhaust system parts are subject to rust, corrosion, and physical damage.
- Exhaust gas leaks can be fatal to vehicle passengers.
- Exhaust system restrictions can cause inefficient and erratic vehicle operation.
- Exhaust system preventive maintenance includes inspections for leaks, holes, physical damage, and freeing stuck heat riser/EFE valves.
- Exhaust system part connections may be sealed with gaskets, seals, ball joints, or by welding.
- Muffler and connector pipe replacement may require cutting, expanding pipe sections, or welding.
- Some catalytic converter covers may be removed to inspect the converter. On some vehicles, chemically coated beads can be replaced.

TERMS

penetrating solvent   graphited oil

REVIEW QUESTIONS

DIRECTIONS: The following questions are similar to those used on automotive technician certification tests. On a separate sheet of paper, write the letter of the correct choice.

1. A vehicle used in a humid climate had an entire exhaust system replaced one year ago. Inspection reveals small spots rusted through on the bottom of the muffler and at the bottom of bends in connector pipes. What is the most likely cause?
    A. Wrong muffler installed
    B. Catalytic converter overheating
    C. Vehicle driven only short distances
    D. Improper installation procedures

2. Technician A says that the use of leaded gasoline in a system with a catalytic converter is illegal.
   Technician B says that the use of leaded gasoline in a vehicle with a catalytic converter will lead to converter damage.
   Who is correct?
   A. A only
   B. B only
   C. Both A and B
   D. Neither A nor B

3. Which of the following statements is correct?
   I. An important function of the exhaust system is to vent gases away from the engine.
   II. An important function of the exhaust system is to vent gases away from the passenger compartment.
   A. I only
   B. II only
   C. Both I and II
   D. Neither I nor II

Questions 4 and 5 are not like those above. Each has the word EXCEPT. For each question, look for the choice that does *not* apply. Read each question carefully before you choose your answer.

4. A restricted exhaust system could cause all of the following EXCEPT
   A. loss of power during acceleration.
   B. stalling and dying.
   C. poor gas mileage.
   D. excessive noise.

5. All of the following could occur if a heat riser or EFE valve is stuck in the closed position EXCEPT
   A. slow engine warmup.
   B. exhaust restriction.
   C. engine overheating.
   D. boiling and percolation of fuel in carburetor.

ESSAY QUESTIONS

1. Describe exhaust system preventive maintenance.
2. What types of tools are available for exhaust system service?
3. Describe catalytic converter service and repair.

SUPPLEMENTAL ACTIVITIES

1. Perform a visual inspection of an exhaust system on a vehicle selected by your instructor. Talk to the driver of the vehicle to determine what type of use the vehicle receives. Report to your class on the condition of the exhaust system and possible causes of defects. Which part of the system seemed most solid, toward the engine or toward the tail pipe? Why?
2. On a vehicle selected by your instructor, use a vacuum gauge to check for possible exhaust restrictions.
3. Replace an exhaust manifold gasket and/or exhaust pipe seal on a vehicle selected by your instructor.
4. On a vehicle selected by your instructor, replace hanger brackets and/or clamps.
5. Lubricate a heat riser valve or replace an EFE valve on a vehicle selected by your instructor.
6. Replace a muffler and/or connector pipe on a vehicle selected by your instructor.

PART IV: Automobile Electrical and Electronic Systems

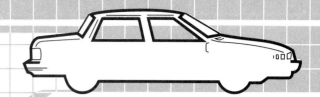

Careers in electrical and electronic servicing can be found at all levels, from specialty shops to dealerships, general repair garages, and fleet service garages. This part of the book begins with an overview of electrical and electronic fundamentals. Following that are knowledge and service units that deal with various electrical systems and electronic devices. This technician is checking electrical wiring on a fuel injected car.

27 Electrical and Electronic Fundamentals

UNIT PREVIEW

All matter is composed of atoms. Electricity is the flow of tiny atomic particles through a conductor. Electricity is a form of energy that can be used to create light, heat, or motion.

Electricity can be used to produce magnetism, and magnetism can produce electricity, depending upon how electrical devices are connected.

Some materials make good electrical conductors, while others act as good insulators. Still others will pass current easily under certain conditions, yet act as insulators, blocking current flow, under other conditions.

When properly connected, a single source can supply electrical energy to many units and systems on an automobile.

As an aid to problem diagnosis, electrical diagrams, or schematics, are used. Schematics help the troubleshooter trace electrical connections between components.

Electronics differs from the study of electricity. Electronics deals with small amounts of current as they operate semiconductor circuits.

LEARNING OBJECTIVES

When you have completed your assignments and exercises in this unit, you should be able to:

❑ Identify and describe the elements of a typical automotive electrical circuit.
❑ Explain what causes a flow of electricity.
❑ Explain how magnetism and electricity can be related.
❑ Identify common electrical conductors, insulators, and semiconductor materials.
❑ Explain the difference between the conventional and electron theories of electrical flow.
❑ Identify and describe series, parallel, and series-parallel circuits.
❑ Explain the difference between direct current and alternating current.
❑ Use basic electrical symbols to draw a simple electrical circuit.
❑ Use basic equations to figure out electrical values.
❑ Conduct basic electrical experiments with an unpowered test light.

27.1 BASIC AUTOMOTIVE ELECTRICAL CIRCUITS

Electricity is a form of energy. Electricity can be used to create light, heat, or force to cause motion. In an automobile, electricity is used for all these purposes. For example, electricity provides the power to operate lights, heating elements, and electrical motors. Electricity also powers other accessories, such as radios, instruments, and computer control units.

Sources, Circuits, and Conductors

A source of electricity, such as a car battery, produces electricity through chemical action. You probably know that batteries have two *terminals*, or connection points, labeled Positive (+) and Negative (-).

Electricity provides energy to do work when it can flow to an electrical *load*, through the load, and back to the source. A load is a device that uses electricity. Electricity flows from one terminal of a source, through conductors and a load, and back to the other terminal and through the source again.

This circular pattern of movement is called an electrical *circuit* **[Figure 27-1]**. Electricity travels through *conductors.* A conductor is a material that allows the passage of electricity. Common conductive materials include most metals (copper, silver, gold, iron, aluminum, and so forth) and some liquids (including water that contains minerals). Copper is the most commonly used metal for wires.

A *switch* can be used to form a complete circuit or to interrupt the flow of electricity, thus shutting off power to the load. The switch can be in an open (off) position or a closed (on) position **[Figure 27-2]**.

Ground Terminal

In a vehicle, one battery terminal, usually the negative, is connected to the metal of the vehicle body. Since metal is a conductor, the entire vehicle can be used as part of an electrical circuit. This common electrical path back to the source of electricity is known as a *ground.*

ELECTRICAL AND ELECTRONIC FUNDAMENTALS 277

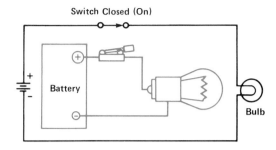

Figure 27-1. Basic electrical circuit: schematic (black) and pictorial (color).

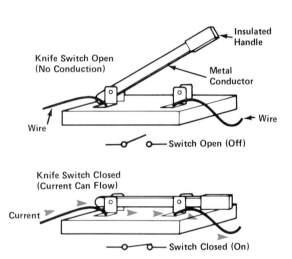

Figure 27-2. Electrical switch positions.

Using a common ground connection eliminates the need for two separate wires to every electrical load. Thus, vehicle electrical systems can be cheaper and simpler to build and service. This way of wiring a vehicle is known as a *single-wire* system.

Before 1956, some vehicles had ground connections made to positive battery terminals, although most were made to negative terminals. In 1956, it was decided to standardize automotive electrical systems by connecting the negative battery terminal to ground. This standard is known as *negative ground*. Some early vehicles (especially British-made) may have *positive ground*. The use of a common ground for an electrical circuit is shown in **Figure 27-3**.

In addition, since 1967, all passenger vehicles have used 12-volt batteries. Before that year, some vehicles had 6-volt batteries.

Schematic Drawings

A special form of drawing, known as an *electrical schematic*, is used to indicate electrical circuits. A schematic is a "map" that shows the routes that electrical current can take. A table of symbols for commonly used automotive components is shown in **Figure 27-4.**

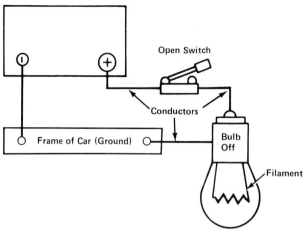

Figure 27-3. Use of ground for circuit.

Insulators

Electricity will not pass through an *insulator*. Common insulating materials include most plastics, enamel, paper, rubber, and most *ceramics*. Ceramics are materials like most household dishes, made from clay baked in special high-temperature ovens.

Electrical conductors in a vehicle, such as copper wires, are covered with an insulator to prevent electricity from flowing to any other conductor **[Figure 27-5]**.

If the wires were not insulated, electricity could flow through any conductor and back to its source. If one conductor touches another conductor or ground, electrical energy can be misused, wasted, or create heat and fire.

27.2 MATTER AND ATOMS

Everything physical is *matter*. Metal, stone, glass, plastic—even air—are all matter. Matter is made of *atoms*. An atom is the smallest portion of a given type of matter that still retains its unique set of characteristics.

Individual atoms are so small that they cannot be seen even with the most powerful microscope. Knowledge about atoms comes from scientific experiments and mathematical calculations.

Matter can be composed entirely of the same type of atoms. Such matter is said to be an *element,* or specific type of material, for example, iron or copper. There are 92 naturally occurring elements. These range from the simplest, hydrogen, to uranium. Other elements have been made in scientific laboratories.

The smallest physical unit of a substance that contains the properties of that substance is called a *molecule.* A molecule may contain a single atom, two or more atoms of the same element, or two or more different atoms. Different elements or molecules can be combined as compounds or mixtures.

ELECTRICAL SYMBOLS			
SYMBOL	REPRESENTS	SYMBOL	REPRESENTS
(Alt)	Alternator	Horn	Horn
(A)	Ammeter		Lamp or Bulb (Preferred)
—⊣⊢—	Battery - One Cell		Lamp or Bulb (Acceptable)
—⊣⎮⊢—	Battery - Multicell	(MOT)	Motor-Electric
12V +⊣⎮⊢−	Where required, battery voltage and/or polarity may be indicated as shown. The long line is always positive polarity.	−	Negative
		+	Positive
Bat	Battery-Voltage Box		Relay
—⎍⎍⎍—	Bi-Metal Strip	—⋀⋀⋀—	Resistor
—♦—	Cable-Connected		Resistor-Variable
—⊥—	Cable-Not Connected	IDLE STOP	Solenoid-Idle Stop
—⟩⊢—	Capacitor		
	Circuit Breaker	SOL / Starting Motor	Starting Motor
—<	Connector-Female Contact		
—>	Connector-Male Contact		
—>>—	Connectors-Separable-Engaged	—o⁄ o—	Switch-Single Throw
—▶⎮—	Diode	—o⁄ o—	Switch-Double Throw
HEI	Distributor	(TACH)	Tachometer
⌒⌒	Fuse	—•—	Termination
(FUEL)	Gauge-Fuel	(V)	Voltmeter
(TEMP)	Gauge-Temperature	⦾⦾⦾ or ⌒⌒⌒	Winding-Inductor
—⏚	Ground-Chassis Frame (Preferred)		
—⎮⎮	Ground-Chassis Frame (Acceptable)		

Figure 27-4. Electrical symbols. GENERAL MOTORS CORPORATION

Atoms

Atoms consist of several parts, as illustrated in **Figure 27-6.** The center, or *nucleus*, of the atom consists of parts called *protons* and *neutrons*. Protons have a positive *charge*, or quality. Neutrons are neutral and have no charge. Spinning around the nucleus in a spherical fashion are layers, or shells, of tiny *electrons*. A single electron weighs only about about 1/1,800 as much as a proton or a neutron. Electrons have a negative charge. All these parts can also exist separately from an atom.

Attraction and Repulsion

In electrical and magnetic applications, electrical charges can attract (pull toward) or repel (push away from) one another. Like charges repel. Unlike charges attract one another.

Two negatively or two positively charged particles will exert a force against each other. Two particles, one negatively charged and the other positively charged, will attract each other.

This can be demonstrated with two bar magnets. The ends of the bar magnets are coded N and S, for

ELECTRICAL AND ELECTRONIC FUNDAMENTALS

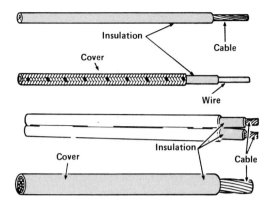

Figure 27-5. Insulated wires.

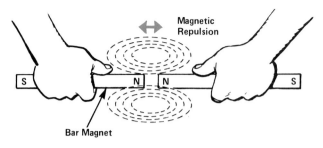

Figure 27-7. Magnetic repulsion.

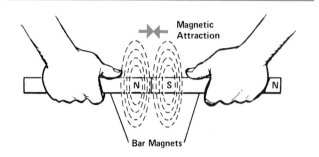

Figure 27-8. Magnetic attraction.

north (positive) and south (negative). Trying to push two north or two south ends together results in a repulsive force you can feel [**Figure 27-7**]. However, opposite charges will attract and pull toward one another [**Figure 27-8**].

Electrons are negatively charged. They are attracted to the positively charged protons in the atom's nucleus. Normally, the positive and negative charges are in balance, creating a *binding force* that literally holds matter together.

Bound and Free Electrons

The element hydrogen, normally a gas, is the simplest atom [**Figure 27-9**]. Only one electron orbits around the nucleus.

An atom of copper is much more complicated [**Figure 27-10**]. Four distinct electron *shells* orbit spherically around the nucleus.

Each shell can contain a certain maximum number of electrons. The closer to the nucleus, the more tightly bound, or held, the electrons are to the atom. The fourth shell of an atom could contain a maximum of 8, 18, or 32 electrons, depending on the element.

Because the fourth, outer shell of copper contains only one electron, the electron is relatively "free," or unbound. Such an electron can be dislodged easily from its orbit. This outer electron can travel to another copper atom and "bump" the electron in its outer shell. This bumping can continue from atom to atom [**Figure 27-11**].

Such a movement of electrons constitutes an *electrical current*, or flow. When an element loses or gains an electron, the number of electrons no longer

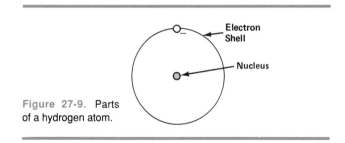

Figure 27-9. Parts of a hydrogen atom.

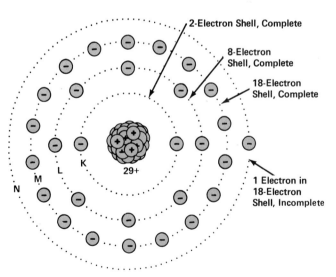

Figure 27-10. Parts of a copper atom.

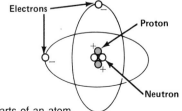

Figure 27-6. Parts of an atom.

balance the number of protons in the nucleus. Such an unbalanced element is known as an *ion*. When an atom or a portion of an atom gains an electron, it is said to be positively *ionized*, or charged. When an electron is lost, the atom is negatively ionized.

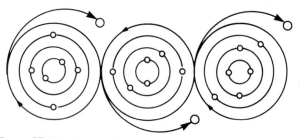

Figure 27-11. Electron flow through atoms.

Electrons that nearly fill the outer shell of a particular element cannot be dislodged easily. These electrons are known as *bound electrons.*

Elements whose outer electron shells are relatively full can be used as insulators or *semiconductors,* discussed in 27.9. Such electron shells are composed of bound electrons that cannot be dislodged easily to create an electrical flow.

27.3 DIRECTION OF ELECTRICAL FLOW

You will encounter two opposite ways of thinking about the direction in which electricity flows. These views on electricity are called:

1. Conventional theory
2. Electron theory.

Conventional Theory

Before the existence and properties of atoms were known, early scientists tried to determine how the world and the universe worked.

Benjamin Franklin, the American statesman/scientist/journalist, conducted an experiment with lightning to determine if it was a form of electricity. On a rainy day, he flew a kite with a metal key attached to the string. Lightning struck the kite and traveled down the string to the key, where it jumped toward the ground. Like other forms of electricity, it flowed through a conductor (the wet string). Had the lightning struck Franklin instead, we might not have a source of confusion over the direction of electrical flow.

Franklin reasoned that electricity was a form of force and, as such, was positive in its nature. He believed that electricity must always travel from a positive source to something that lacked such force. He called the lack of force negative. Thus, Franklin believed that electricity always flowed from positive to negative, or ground.

The *conventional electrical theory* states that electricity flows from positive to negative. Technicians often speak of the positive terminal of a vehicle battery as the source of electricity.

The ground connection, through the vehicle frame, is considered the path for electricity to return to the battery.

Electron Theory

When atomic theory was first introduced, the conventional electrical theory seemed backwards. Electrons have a negative charge, and are attracted to a positive charge.

The *electron electrical theory* states that electricity flows from the negative terminal of a source to a load, and returns to the positive terminal. The chassis ground connection is seen as the source of electrical power, and the conductor to the positive battery post as the return line.

Electrical Flow and Vehicle Service

Either theory, conventional or electron, can be used when working with automotive electrical circuits, as long as it is used consistently. Most technicians like to "think" their way through electrical circuits while troubleshooting or diagnosing problems. In these cases, many find it easier to use the conventional (positive to negative) theory. The main problem you will encounter in working on the electrical system of a vehicle is finding out whether a proper circuit exists.

There may be an *open,* or break in electrical flow at some point in a circuit.

In some cases, a conductor accidentally touches another conductor to produce a *short circuit,* or electricity flows to a different load than the one intended and completes a circuit.

In other cases, a conductor accidentally touches a ground connection, to produce an *accidental ground.* Such an accidental ground can overheat a wire or cause a fuse or circuit breaker to interrupt electrical flow.

Batteries and electrical testers have a positive connection (usually red) and a negative connection (usually black). As long as a circuit exists among an electrical source, conductors, and a load, electricity will flow.

CAUTION!!!

It is important to retain the polarity (direction of electrical flow) of the electrical system intended by the vehicle manufacturer. Switching the battery terminals will result in damage to vehicle components, especially motors and electronic equipment.

Fuses

A *fuse* is a safety device that conducts electricity. A fuse is made of special metal that melts when a certain amount of current flows through it. When the fuse blows, or melts, the circuit is incomplete, or open, and no current can flow. Without protective fuses in an electrical circuit, wires could overheat, burn off insulation, and cause a fire [Figure 27-12]. Typical fuses are shown in Figure 27-13.

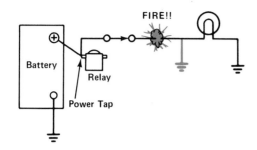

Figure 27-12. Accidentally grounded circuit causing fire.

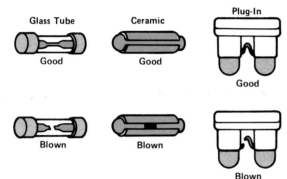

Figure 27-13. Types of fuses.

27.4 DC AND AC

Electrons will move when they are attracted more strongly by an electrical charge than when they are bound to atoms of a conductor. This process occurs in any complete circuit. Electrical current flows from a location with an excess of electrons to a location with a shortage of electrons.

The connection from an electrical source, through conductors and a load, and back to the source, creates such a flow. Thus, electrical current flows when all the elements of a circuit are properly connected:

- Electrical source
- Conductors
- Load
- Switch (optional).

Direct and Alternating Current

Two types of electrical current are used:

1. Direct current (DC)
2. Alternating current (AC).

Direct current. *Direct current (DC)* flows in one direction from one terminal of an electrical source to the other terminal. Batteries produce direct current. The electricity flows in one direction only. The direction of current flow can be reversed by switching conductor connections at a battery.

A simple device, such as a light bulb, will operate with the connections made either way. Motors and other more complicated devices may be damaged or may not operate with *reversed polarity,* or switched connections.

Alternating current. *Alternating current (AC)* is the type of electricity used in homes, shops, and factories. Alternating current changes direction, or switches polarity, many times per second. In some countries, it changes directions 50 or more times per second. In the United States, it changes directions 60 times per second. These direction changes are called *cycles per second (cps).* In the metric system, this is stated as 60 *hertz (Hz),* named after a German scientist.

Some electrical devices will work on either DC or AC. Other units are designed specifically to work on only one or the other. The automobile electrical system is designed to operate on DC electrical flow.

27.5 ELECTRICAL MEASUREMENTS

The electrical qualities of conductors, circuits, insulators, and other equipment can be measured. There are several basic measurements whose meaning you will need to know:

- Amperes (amps)
- Ohms
- Volts
- Watts.

All of these terms come from the names of scientists who conducted early experiments with electricity. A basic analogy, or comparison, can be made between the flow of water through pipes and the flow of electricity through wires. Thus, these terms and their relationships can be understood as relatively simple concepts.

Amperes

The *ampere,* or *amp,* is named after Andre M. Ampere (1775-1836), a French scientist. An ampere is a certain number of electrons flowing each second past a given point. An ampere is similar to the amount of water, measured in drops, flowing each second past a point in a pipe.

Because electrons are extremely small, a tremendous number of electrons must flow to make one ampere. One ampere equals 6,280,000,000,000,000,000 (six quintillion 280 quadrillion) electrons flowing past a given point each second.

Electrical current travels at the speed of light, approximately 186,000 miles in one second. Obviously, no one can count electrons one by one as they pass by. A measuring instrument, called an *ammeter,* is used to measure current flow in amps.

Ohms

Electricity flows easily through materials that are good conductors, such as silver, copper, and aluminum. Electrical flow through other materials is more difficult. Think of water flow, which is more difficult through a small pipe than through a large one. A material through which current flow is difficult is said to have high *resistance.* For example, different sizes of wire have different amounts of resistance. A large, thick wire, like a large pipe, has little resistance. A small, thin wire, like a tiny pipe, has more resistance. Materials with extremely high resistance are used as insulators **[Figure 27-14].**

Resistance is measured in Ohms, symbolized by Ω (the Greek letter *omega*). In equations, resistance is written as R. George S. Ohm (1787-1854) was a German scientist who studied electrical resistance.

If electricity can flow through either a high resistance or a lower resistance, it will flow through the lower resistance. The saying is, "Electricity always follows the path of least resistance."

Volts

Voltage, named after Alessandro Volta (1745-1827), an Italian scientist, is similar to the pressure pushing water through a pipe. The higher the pressure, the greater the flow of water. Thus, more voltage will push greater amounts of current through a circuit.

One volt is defined as the electrical pressure necessary to make one ampere of electricity flow through one ohm of resistance.

Ohm's Law

The symbol for volts is E, short for EMF (electromotive force). The symbol for amps is I, from the French word *intensité* (remember, Mr. Ampere was French!). The symbol for resistance is R. The relationship among amps, ohms, and volts can be summarized in an equation or a drawing **[Figure 27-15].**

Ohm's law allows you to find an unknown value by manipulating known values. For example, in **Figure 27-16** the source voltage in this circuit is given as 12 volts. The total resistance is given as 118 + 2 = 120 (Ohms). We can determine current flow in this circuit using Ohm's Law by solving for I:

$$I = (E \div R) = (12 \div 120)$$
$$= 0.1 \text{ amp}$$

Where:
I = amperes
E = volts
R = resistance

NOTE: Ohm's law, as stated above, applies to DC circuits only. More complicated equations are needed for circuits with AC current. In addition, the resistance of a device, such as a light bulb, changes with temperature. A light bulb might have very little resistance when measured with an ohmmeter. However, when the filament is hot enough to produce light, the resistance may be many times greater.

Watts

Power, as in an engine, is the rate at which work can be done. Forcing a current to flow through a resistance is work. Power is measured in *watts,* named after James Watt (1736-1819), the Scottish scientist who devised the horsepower rating.

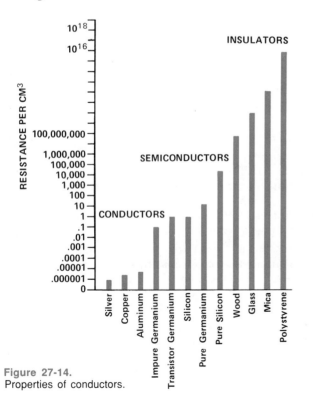

Figure 27-14.
Properties of conductors.

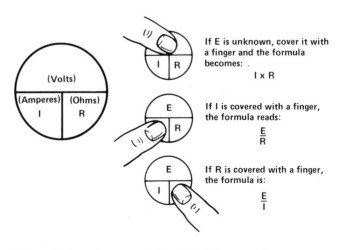

Figure 27-15. Diagrams illustrate Ohm's Law equations.

Electrical power, or the rate of work, is determined by multiplying the amount of voltage by the amount of amperes flowing. The symbol for power (watts) is P. The symbol for voltage is E. The symbol for amperage is I. Thus, the equation to determine electrical power is P = E x I. One horsepower equals 746 watts.

27.6 ELECTRICAL CIRCUITS

Electrical circuits fall into three general categories:

1. Series
2. Parallel
3. Series-parallel.

Series Circuits

In a *series circuit*, electrical current passes through loads one at a time, in order, or in series **[Figure 27-17]**. Electrical current is shared among all loads in a series circuit. Each separate load forms part of the conductive path of the circuit.

To illustrate, strings of inexpensive Christmas tree lights often are wired in series. The current has only one path through which it can flow. If one bulb fails, the entire string goes out. To find the burned-out bulb, each bulb must be checked or replaced one at a time.

Series Circuit Laws

There are three laws that apply to series circuits. These laws are:

1. Current Law: Current in a series circuit is the same value in all parts of the circuit.
2. Resistance Law: Total resistance (ohms) in a series circuit is equal to the sum of all individual resistances.
3. Voltage Law: Source voltage in a series circuit is equal to the sum of the individual voltage drops. This is also known as Kirchhoff's Voltage Law and is expressed as:

$$V_S = V_{D1} + V_{D2} + V_{D3} + ... \text{ (and so on).}$$

Where:
V_S = voltage source
V_D = voltage drop

Voltage drop. Each separate light bulb has resistance. A brighter or more powerful bulb has less resistance than a smaller bulb. The more resistance a load has, the greater the voltage required to push current through it.

Thus, the *voltage drop*, or voltage measurement across a higher-resistance load, is greater **[refer to Figure 27-16]**. The voltage drop can be determined by using Ohm's law if the resistance is known. Alternately, the resistance can be determined by measuring the voltage drop and applying Ohm's law.

Parallel Circuits

In a *parallel circuit*, electricity takes several paths through separate loads to complete a circuit **[Figure 27-18]**.

Consider a string of Christmas tree lights wired in parallel. If one of the bulbs is burned out, the other bulbs will still burn. The voltage drop across each load is the same. However, the total resistance of the circuit is *less* than the resistance of any single load.

Each load in a parallel circuit provides an additional path for electrical current, making it easier for current to complete a circuit. This effect produces less resistance than any individual load.

However, the amount of current flowing does increase according to the number of loads added. Too many loads can create excessive current flow that can cause heat and fire, or blow a fuse.

Parallel Circuit Laws

There are three laws that apply to parallel circuits. These laws are:

1. Current Law: Total current flow in a parallel circuit is equal to the sum of the individual branch current flows. This is also known as Kirchhoff's Current Law and is expressed as:

$$I_T = I_1 + I_2 + I_3 + ... \text{ (and so on)}$$

Where:
I_T = total current flow
I = individual branch current flow

2. Resistance Law: Total resistance in a parallel circuit is less than the lowest individual resistance.
3. Voltage Law: Voltage in a parallel circuit is the same in each individual branch.

Series-Parallel Circuits

It is possible to combine series and parallel circuits into a single circuit **[Figure 27-19]**.

Most circuits in an automobile are *series-parallel circuits*. Ohm's law is used to work out the values for current, voltage drops, and overall resistance values shown, although the process is complicated.

27.7 MAGNETISM

Electricity and magnetism are related. No one really knows what causes certain materials to function as

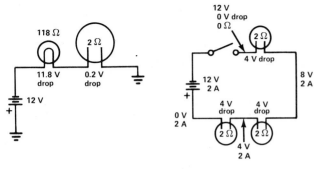

Figure 27-16. Voltage drops.

Figure 27-17. Series circuit.

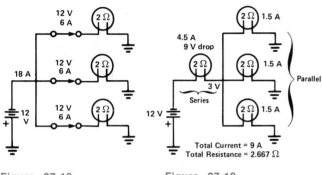

Figure 27-18. Parallel circuit.

Figure 27-19. Series-parallel circuit.

magnets. One theory is that the alignment of molecules allows a natural flow of electrons through and around the magnet [Figure 27-20].

Ferrous, or iron-containing, materials are attracted to this flow. To demonstrate, a sheet of paper can be placed over a magnet. Small iron filings then are sprinkled lightly onto the surface. A pattern of the continuous force fields of magnetism will be formed.

Electromagnetism

The flow of electrons in an electrical circuit can be demonstrated in a similar manner. Iron filings on paper at right angles to a conductor will arrange themselves in a circular pattern around the wire. A compass near the wire will indicate the direction of flow [Figure 27-21].

The direction of current flow can be changed in such a circuit by reversing the wiring connections [Figure 27-22]. This process is also known as reversing the polarity of the circuit. Polarity describes how the ends of a circuit are connected to the electrical source connections. One end is connected to the positive connection, and the other end is connected to the negative connection. The polarity of the circuit determines the direction in which the compass needle will point.

As current flow is shut off, the magnetic field around the conductor collapses, or moves, rapidly inward until it no longer exists.

Coil

To produce a strong magnetic field, conductors can be arranged parallel and next to each other. If the current in the conductors flows in the same direction, the magnetic fields of force will combine and become stronger. If the current in the conductors flows in

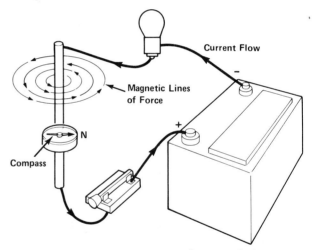

Figure 27-21. Magnetic field surrounding a conductor.

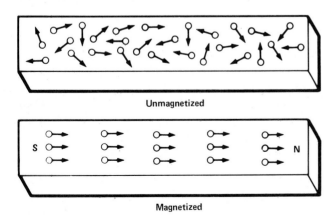

Figure 27-20. Unmagnetized and magnetized steel.

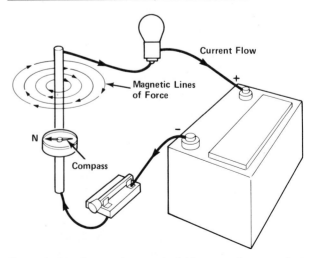

Figure 27-22. Reversed magnetic field surrounding a conductor.

opposite directions, the magnetic fields will repulse each other and tend to cancel out [Figure 27-23].

The shape of the magnetic field surrounding the loops in a coil is shown in **Figure 27-24.**

Electromagnet

A simple way to construct a device that will produce a strong magnetic field is to wrap wire in a coil around a material, such as iron, that conducts magnetic force fields well. An electrical source then can be connected to the ends of the wire. Such a device is known as an *electromagnet* [Figure 27-25].

Solenoid

If an iron bar is placed next to a coil in which current is flowing, the bar will be pulled into the coil. A device that produces such a mechanical effect is known as a *solenoid* [Figure 27-26].

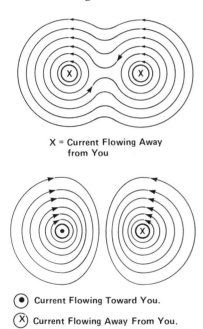

Figure 27-23. Magnetic fields of force.

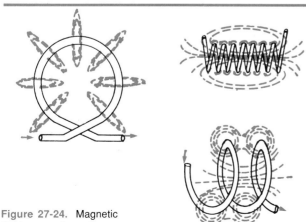

Figure 27-24. Magnetic fields around a coil.

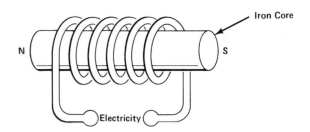

Figure 27-25. Electromagnet.

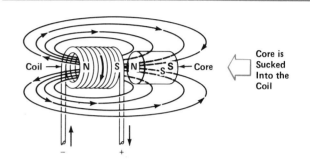

Figure 27-26. Magnetic fields around a solenoid.

Solenoids are used in several applications in modern vehicles, including fuel injectors and starter circuits. Starter circuits are discussed in Unit 30.

27.8 MAGNETIC INDUCTION

A flow of electrons in a conductor can produce magnetism. Magnetism also can produce electrical flow in a conductor.

An ammeter can be connected across the ends of a conductor. Wrapping the conductor in a coil shape will make the process of creating electricity more efficient.

A magnet moved next to the coil of wire will produce a flow of electrons within the conductor. The meter will register how much electrical current is *induced*, or created by magnetic force. The stronger the magnet and the faster it is moved, the more electrical current will be produced [Figure 27-27].

Transformers

The magnetic field produced in one conductor can be used to induce current in a second conductor placed close to, and parallel to, the first. Such an arrangement that uses two parallel coils is known as a *transformer.*

If two coils are close together, and the magnetic field of one is moved, an electrical current will flow through the other.

Shutting off the current in the first coil will cause the magnetic field to collapse and move. This will induce a momentary pulse of current in the second coil [Figure 27-28].

Wrapping both coils around a reluctor, or iron core, will increase the efficiency of this process **[Figure 27-29]**.

The voltage induced in the second coil is proportional to the number of turns of wire in each coil.

If the second coil has more turns of wire, the induced voltage will be greater than that in the first coil, but the current will be proportionally lower. Thus, voltage can be increased through the use of a transformer. This type of transformer is known as a *step-up transformer.*

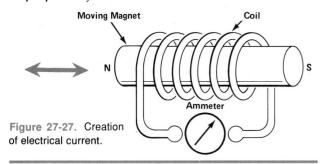

Figure 27-27. Creation of electrical current.

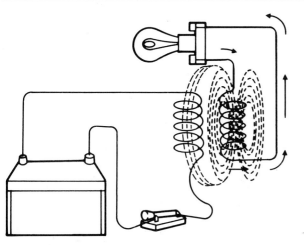

Figure 27-28. Principles of a transformer.

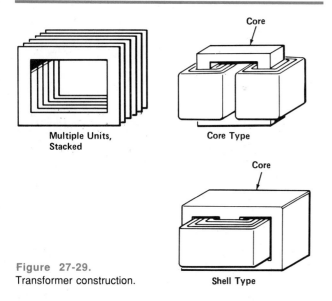

Figure 27-29. Transformer construction.

If the second coil has fewer turns of wire than the first, the voltage induced will be lower, but the current will be higher. Thus, voltage can be decreased through the use of a transformer. This type of transformer is known as a *step-down transformer.*

Except for losses within the transformer, the input power (E x I) is the same as the output power. Basically, what is gained in voltage is lost in current, or vice versa.

27.9 ELECTRONICS AND SEMICONDUCTORS

Electronics is a branch of physics that deals with the behavior and effects of small amounts of electrons in semiconductor circuits. Electrical equipment uses larger amounts of current.

Thus, a television or a computer is considered to be an electronic device, whereas a light bulb or a motor is an electrical device.

Semiconductors

Semiconductors are materials that partially conduct electricity. Semiconductors can act either as conductors or as insulators, depending upon the circumstances.

Silicon, the ingredient in ordinary sand, can be used as a semiconductor when mixed with *trace,* or small, amounts of other minerals.

Doping

Mixing minerals with a semiconductor material is known as *doping.* Minerals used for doping include antimony, arsenic, phosphorus, aluminum, boron, gallium, and indium. Depending upon the type of mineral used, the semiconductor can have a positive (P) or negative (N) characteristic.

Diode

Electrons will flow from a negatively doped semiconductor material (N) toward a positively doped material (P). Joining N and P semiconductor materials will create a junction. Such a junction creates a simple form of electrical device called a *diode* **[Figure 27-30]**. A diode is a device with two (from the Greek *di-*) poles, or a cathode and an anode.

Electricity will flow through a diode in one direction only. Both positive and negative diodes are used in automobiles. Diodes can be used to change alternating current into pulsating DC, as in an automotive alternator.

Transistor

A *transistor* is made of three pieces of semiconductor material in a "sandwich," either NPN or PNP **[Figure 27-31]**. Transistors can be used to control or amplify current flow.

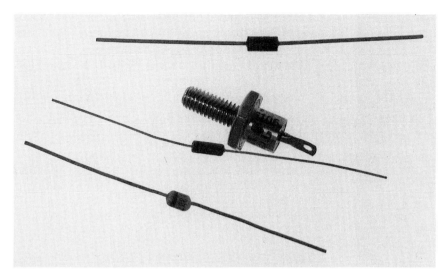
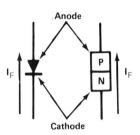

Figure 27-30. Diode (left) and diode junction (right).

Figure 27-31. Transistors.

A transistor is capable of conducting relatively large amounts of current from the *emitter* to the *collector*. The emitter discharges, or emits, electrons. These electrons flow to the collector. However, a transistor will perform in this manner only if a small amount of electricity passes from the emitter to the base. This, in effect, "switches on" the transistor.

A transistor can be used as a switch to turn current flow on or off. A transistor also can act as an amplifier to increase and magnify a given current flow.

Integrated Circuit

Transistors, resistors, and other electrical components can be combined into circuits to perform complicated functions. These components can be greatly miniaturized into what is known as an *integrated circuit*. **Figure 27-32** shows such an integrated circuit, or computer chip, so small that it can pass through the eye of a needle. Such integrated circuits also are known as microchips.

Integrated circuits are made of thin wafers of silicon, on which materials have been deposited. These deposited materials form conductors, resistors, diodes, transistors, and other types of electrical components to form circuits.

Integrated circuits are used as regulators for alternator voltage output, electronic control units (ECUs), radio circuits, and other electronic devices.

Automotive servicing does not require detailed knowledge of electronics. In general, all a technician is required to do is determine whether an electronic device is operating. The technician checks for proper connections, a source of power, and a good ground. If the device is properly connected and not working, it is replaced. Electronic devices and servicing procedures are discussed in Units 38 and 39.

288   PART IV: AUTOMOBILE ELECTRICAL AND ELECTRONIC SYSTEMS

- Electrical current is caused by the movement of electrons from one atom of a conductor to another atom of a conductor.
- Electrical circuits can be hooked up in series, parallel, or series-parallel.
- Magnetism can be used to produce electricity, and electricity can be used to produce magnetism.
- Magnetic induction can produce magnetism or electrical current in a conductor.
- Semiconductors are materials that can be made to conduct electricity. Diodes conduct electricity in one direction only. Transistors can act as switches or as amplifiers.

TERMS

electricity	terminal
load	circuit
conductor	switch
ground	single-wire circuit
negative ground	positive ground
electrical schematic	insulator
ceramic	matter
atom	element
molecule	ionized
bound electron	nucleus
proton	neutron
charge	shell
electron	binding force
electrical current	open
short circuit	accidental ground
fuse	direct current (DC)
reversed polarity	ampere
ammeter	resistance
series circuit	voltage drop
ferrous	hertz (Hz)
electromagnet	solenoid
induce	transformer
step-up transformer	step-down transformer
electronics	ion
semiconductor	trace
doping	diode
transistor	emitter
collector	integrated circuit
alternating current (AC)	cycles per second (cps)
watt	parallel circuit
electron electrical theory	conventional electrical theory
series-parallel circuit	

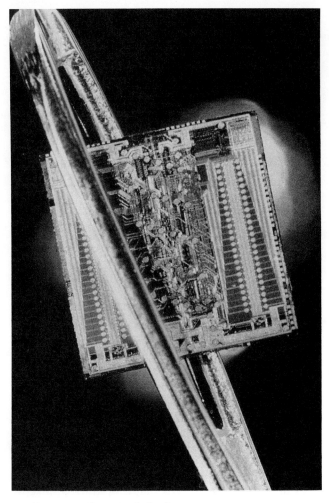

Figure 27-32. Integrated circuit.

UNIT HIGHLIGHTS

- Electricity is a form of energy that can be used to create heat, light, or force to cause motion.
- A typical electrical circuit includes an electrical source, conductors, and a load.
- Modern vehicles all have negative ground.
- Electrical schematics are like maps that use symbols to indicate the position of electrical sources, conductors, loads, and other devices.
- Conductors allow the passage of electricity; insulators prevent the passage of electricity.

REVIEW QUESTIONS

DIRECTIONS: The following questions are similar to those used on automotive technician certification tests. On a separate sheet of paper, write the letter of the correct choice.

1. Technician A says that electrical current consists of a flow of electrons from one atom of a conductor to another.

   Technician B says that electrical current will flow from one pole of an electrical source to the other pole.

   Who is correct?
   - A. A only
   - B. B only
   - C. Both A and B
   - D. Neither A nor B

2. Which of the following statements is correct?
   - I. The conventional theory states that electrical current flows from negative to positive.
   - II. The electron theory states that electrical current flows from positive to negative.
   - A. I only
   - B. II only
   - C. Both I and II
   - D. Neither I nor II

3. Each time a fuse is replaced, it immediately blows. What is the most likely cause?
   - A. Open in the circuit
   - B. Excessive resistance
   - C. Accidental ground
   - D. Reversed polarity

4. In a new car, a current of 10 amperes is measured in a circuit. Using Ohm's law, what is the resistance of the load?
   - A. 0.833 Ω
   - B. 120 Ω
   - C. 1.2 Ω
   - D. Cannot be determined from information given.

Question 5 is not like those above. It has the word EXCEPT. For question 5, look for the choice that does *not* apply. Read the entire question carefully before you choose your answer.

5. A normal electrical circuit contains all of the following EXCEPT
   - A. an electrical source.
   - B. an open.
   - C. a load.
   - D. conductors.

ESSAY QUESTIONS

1. Briefly describe electrical current flow.
2. Describe the three types of circuits found in automobiles.
3. In general, what knowledge of the inner workings of electronic devices does an automotive technician need?

SUPPLEMENTAL ACTIVITIES

1. Trace the battery cable connections on a vehicle. Report to your class where the positive and negative cables are connected.
2. Discussion topic: From what you have learned in this unit, can you figure out how turning an alternator with a drive belt could produce electricity? (Hint: When a vehicle is running, a small amount of electricity from the battery flows through coils in the alternator.)
3. Using the symbols in **Figure 27-4**, draw a schematic of a circuit with two headlight bulbs, four sidemarker lights, and two taillights. Include a battery and a switch. Use the symbol for ground and draw a parallel circuit.
4. Under the supervision of your instructor, connect an unpowered test light to the positive terminal of the battery. Scratch through paint and touch the probe to metal objects in the engine compartment. What does this experiment demonstrate?
5. Under the supervision of your instructor, connect an unpowered test light to a grounded part of a vehicle near the fuse box. Touch the probe to both ends of each fuse and note your results. Explain why the test light might not illuminate on both ends of all fuses.

28 Batteries

UNIT PREVIEW

The battery provides current to start the engine, provides a reserve electrical capacity, and aids in electrical system regulation.

A basic battery consists of two dissimilar metals immersed in an acid. A chemical reaction within the battery produces the potential for electrical current flow. Automotive batteries are classified as lead-acid batteries.

Battery construction may be classified as either conventional or maintenance-free, or wet or dry charged.

Battery ratings indicate the battery's ability to produce electrical flow on demand.

To provide long periods of dependable service, the battery and electrical system must be maintained routinely.

LEARNING OBJECTIVES

When you have completed your assignments and exercises in this unit, you should be able to:

❏ Describe the function, construction, and operation of a vehicle battery.
❏ Identify and describe problems that can shorten battery life.
❏ Explain the meaning of different battery ratings.
❏ Determine the coded group number of a sample battery.
❏ Determine what differences exist between available models of the same group size battery.

28.1 BATTERY FUNCTION

The battery is one of the two electrical sources in a vehicle. The other source is the alternator, discussed in Unit 32. The battery provides current to turn the starter motor and power the ignition system during starting.

After the engine starts, the alternator provides the current needed by the electrical systems. The alternator also replaces electricity used by the starting motor.

Any time the electrical load exceeds alternator output, the battery supplies the rest of the current needed.

The battery also acts as a voltage stabilizer to help even out voltage *spikes*. Spikes are surges of voltage that occur when loads are switched on or off.

28.2 ELECTROCHEMICAL ACTION

A battery does not store electricity. It provides the opportunity, or potential, for electrons to flow. Current flows only when a load is connected across the battery terminals. This process is an example of *electrochemical action.*

A battery uses chemicals to produce electrical potential. A method of constructing a small battery was described in the Supplemental Activities of Unit 17. A penny and a dime were placed in a slice of citrus fruit. A small voltage was created between the two coins.

To create a battery, different metals are placed side by side in an acid solution. A voltmeter (a test instrument for electrical potential) is connected between the metals. Electrical potential can be measured between the metals **[Figure 28-1]**.

Electrolyte

Electrolyte is a material whose atoms become ionized in solution. As in a conductor, electrons can flow easily through electrolyte.

Electrolyte used in automotive batteries is a solution of 36 percent sulfuric acid and 64 percent

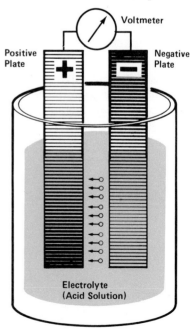

Figure 28-1.
A simple battery.

water, by weight. By volume, sulfuric acid is approximately 25 percent of the electrolyte.

SAFETY CAUTION

Always wear safety glasses when you work around batteries. Battery electrolyte (acid) is extremely corrosive and can cause skin burns, blindness, and damage to auto paint and other parts.

Battery Elements and Cells

A *battery element* consists of two dissimilar metal plate groups separated by porous *separator strips,* or insulators. When the element is immersed in electrolyte, a *battery cell* is created **[Figure 28-2]**.

Lead paste is spread on battery plate grids and baked to form battery plates **[Figure 28-3]**.

Positive Plate

One of the two grids in a battery element is coated with *lead dioxide (PbO_2)*, commonly called "lead peroxide." Lead dioxide is a porous, chocolate-brown, crystalline substance.

Electrons (negatively charged) leave the porous lead dioxide plate and enter the electrolyte. This leaves the lead dioxide with a positive charge, and this grid forms the positive plate of a battery element.

Negative Plate

The other grid in a battery element is coated with porous *sponge lead (Pb).* Electrons travel through the electrolyte and into the sponge lead, causing it to have a negative charge.

When a load is connected across the terminals, electrons flow from the negative terminal, through the load, and back to the positive terminal. Electricity flows to do work, as shown in **Figure 28-4.**

Battery Construction

Each battery cell can produce approximately 2.1 volts. When connected in series, six cells produce approximately 12.6 volts. Construction of an individual cell (called an element before electrolyte is added) is illustrated in **Figure 28-5.**

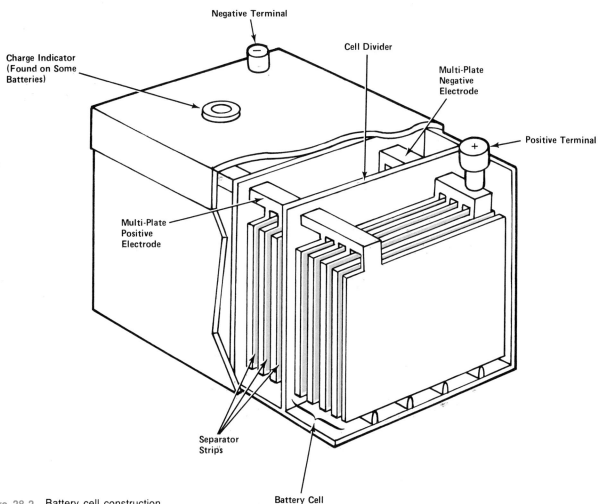

Figure 28-2. Battery cell construction.

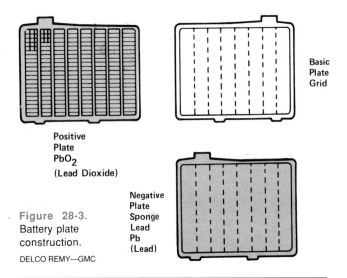

Figure 28-3. Battery plate construction.
DELCO REMY—GMC

Each of the cells is sealed with an individual supply of electrolyte. The cells are enclosed in a polyethylene battery case. A top encloses the cells. Openings to the cells may be permanently sealed or openable. Small spaces, or vents, are designed into the top to allow hydrogen and oxygen gases to be released when the battery charges **[Figure 28-6]**.

If a negative and positive plate touch, an *internal short* is created and the battery is ruined.

In time, material from the plates flakes and falls off. In some batteries, a space at the bottom of the case is provided for this material to collect. This space is called a *sediment chamber*.

Connections to the positive and negative plates are made through plate straps and connectors. These in turn are connected to *side terminals* or internally threaded *battery posts* **[Figures 28-6 and 28-7]**. The battery is sealed around posts or side terminals to prevent electrolyte from leaking. Batteries with top posts have different-sized posts. The positive post is always larger than the negative post **[Figure 28-7]**.

Flanges or *recesses* can be molded into the outer plastic case to aid in holding the battery securely in the vehicle. A flange is a projecting lip or edge. A recess is a shaped hollow space **[Figure 28-8 and refer to Figure 28-6]**.

28.3 BATTERY CHEMISTRY

The supply of electrons in the negative plate of the battery is not unlimited. As electricity discharges, or flows out of the battery, the electrons return to the ions of the positive and negative plates.

Discharging and Sulfation

The positive plate loses oxygen (O_2) to the electrolyte. At the same time, sulfate (SO_4) from the electrolyte combines with the lead of both the positive and negative plates. This chemical process, known as *sulfation*, forms *lead sulfate ($PbSO_4$)*.

Eventually, both the positive and negative plates become coated with lead sulfate, and the percentage of water in the electrolyte increases **[Figure 28-9]**.

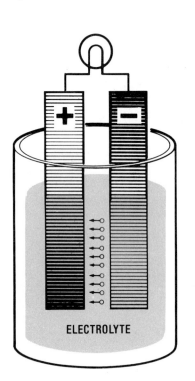

Figure 28-4. Electrical battery current flow.

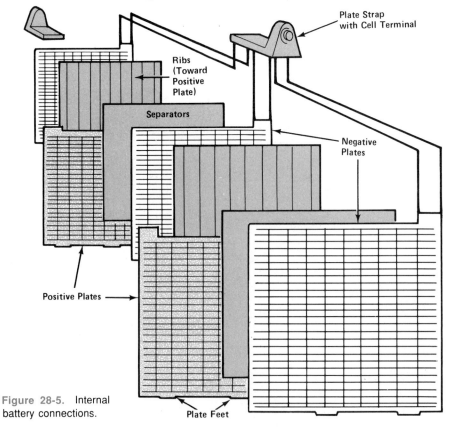

Figure 28-5. Internal battery connections.

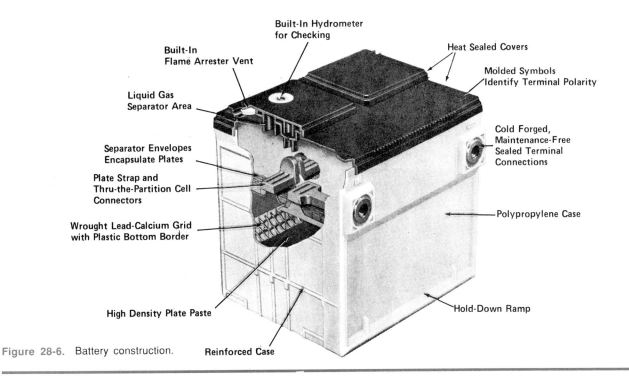

Figure 28-6. Battery construction.

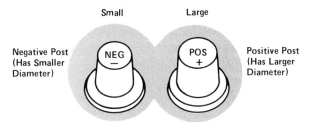

Figure 28-7. Battery posts.

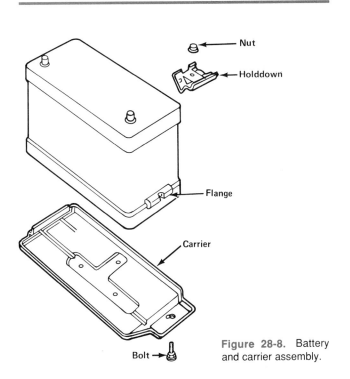

Figure 28-8. Battery and carrier assembly.

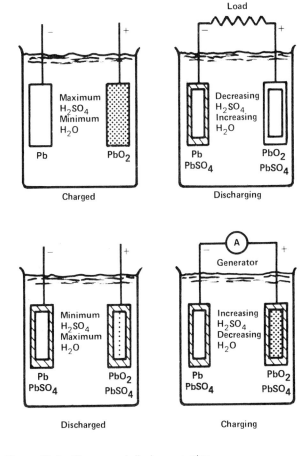

Figure 28-9. Charge and discharge cycles.

Charging

In normal operation, the process of discharging and temporary sulfation is reversed by *charging*. Charging is done by connecting a source of DC (direct current) electricity to the battery terminals.

The charging voltage must be higher than the voltage produced by the battery to reverse the direction of electron flow. In practice, between 13.2 and 15.2 volts are used to charge a 12 volt battery. Charging can be done at a high current input (fast charging) or a low current input (slow charging).

During charging, electrons combine with the ions of the positive plate to create atoms of lead.

Charging causes oxygen and hydrogen gases to be formed from the watery electrolyte. Oxygen from the weak electrolyte combines with the lead on the positive plate to form lead dioxide again.

Charging also causes sulfate to leave the positive and negative plates and return to the liquid. The electrolyte becomes strong sulfuric acid again.

When the battery is fully charged, electrons again are present in the negative plates **[refer to Figure 28-9]**. The battery plates return to being lead dioxide and sponge lead.

SAFETY CAUTION

During charging, an explosive mixture of hydrogen and oxygen gases is released into the atmosphere. If a source of ignition is present, the battery can explode. Battery explosions splatter sulfuric acid and can lead to blindness, skin burns, and damage to parts. Be sure to keep all sources of ignition away from the battery at all times.

Battery Life

With proper maintenance, discussed in Unit 29, batteries should give at least three to four years of good service. No battery lasts forever. In time, hard sulfation, loss of material from plates, and internal shorts can render a battery unserviceable.

Other than problems of age, several factors can shorten battery life:

- Remaining in a discharged state over a long period of time
- Deep discharging and charging cycles
- Dirt and acid buildup
- Inadequate charging
- Overcharging
- Vibration and movement
- Physical damage
- Extremely cold or hot temperatures
- Improper proportions of acid and water in electrolyte
- Loose or corroded terminal connections
- Ignition source for hydrogen gas
- Contaminated electrolyte.

When a battery is left discharged over a long period of time, sulfation can coat the plates permanently. Hard sulfation prevents electrolyte from penetrating the plates properly.

Deep discharging, such as that caused by prolonged cranking when the engine does not start, can shorten battery life.

Dirt and acid buildup on the outer surface of the battery can form a conductive current path between the terminals. Electrical current flow between the terminals can discharge the battery.

Inadequate charging can result in early battery failure. Inadequate charging can be caused by a loose alternator belt, a faulty voltage regulator, limited driving, or starting the vehicle frequently during short-trip driving.

High-current "fast" charging can cause too high an electrolyte temperature and *gassing*. Gassing is the discharge of large amounts of hydrogen gas and loss of water and acid during charging.

If a battery is not held down securely, the internal connections between plates can be broken or separated. In addition, material can be shaken loose from the grids to cause an internal short at the bottom of the plates.

Physical damage, such as a cracked, bulged, or leaking case, will shorten battery life. Such damage can be caused by vibration or loose holding devices, or by overtightening the holding devices.

Chemical reactions occur more readily when the chemicals are heated. Because battery operation depends on chemical reactions, cold temperatures reduce the flow of electricity. In addition, engine lubricants are more viscous, or thick, at cold temperatures. Thus, it is more difficult to turn the engine crankshaft for starting. The combination of these two factors results in difficult starting in cold weather **[Figure 28-10]**.

Severely hot weather conditions or overcharging can result in evaporation and loss of water, resulting in high acid concentrations.

Too much sulfuric acid in the electrolyte can damage the metal plates and grids of the battery and shorten their life.

Corroded battery connections can cause improper charging and hard starting. Electricity cannot pass easily, in or out, through corroded connections.

Loose battery connections can result in sparks that can cause a battery explosion.

Battery life can be shortened by careless servicing. If baking soda enters the cell, it will neutralize the electrolyte. If water containing high amounts of minerals is added to the electrolyte, it will promote internal self-discharging.

28.4 BATTERY RATINGS

Each cell of a battery is designed to produce approximately 2.1 volts. Older vehicles had batteries with three cells, making approximately six volts. All passenger vehicles built since 1967 have 12 volt batteries. Battery size does not change the battery voltage. However, battery plate area does affect the capacity of the battery.

Electrical Capacity

A given battery has a specific *electrical capacity.* More and larger plates increase this capacity. There are several ways to measure and indicate battery electrical capacity:

- Cold cranking performance
- Reserve capacity
- Ampere-hour rating
- Watt-hour rating.

Cold cranking performance is measured with a fully charged 12-volt battery that has been cooled to 0 degrees F (-17 degrees C). The battery is tested for 30 seconds to determine maximum current flow. The *cold cranking amperes (CCA)* rating is the current, measured in amperes, that can be drawn from the battery before the voltage drops below 7.2 volts.

Replacement batteries should have a CCA rating at least as high as the number of cubic inches of engine displacement. For additional starting power, a battery with more cold cranking amps can be used.

Reserve capacity is the length of time a 25-ampere current can be maintained before voltage falls to 10.5 volts. This test is conducted with a fully charged battery at 80 degrees F (27 degrees C).

This test approximates how many minutes a vehicle could continue running satisfactorily if the alternator did not produce electricity. A typical rating for a battery might be 90 minutes of reserve capacity.

Ampere-hour rating refers to a test performed with a fully charged battery at a constant 80 degrees F (27 degrees C). The battery is discharged at a constant rate. At the end of the test, the voltage remaining should be 10.5 volts.

A battery that can supply 4 amperes for 20 hours is rated as an 80 ampere-hour battery. Amperes x hours = ampere-hours. An ampere-hour rating is useful knowledge for some service operations, such as slow charging and battery capacity testing.

Watt-hour is another method of rating a battery. The number of ampere-hours x voltage = watt hours. Thus, a 12-volt battery rated at 80 ampere-hours would be rated at 960 watt-hours. The equation: 80 x 12 = 960.

28.5 MAINTENANCE-FREE BATTERIES

Virtually all new passenger vehicles come equipped with *maintenance-free batteries.* In normal operation, it should not be necessary to add water to the cells of a maintenance-free battery. Older, conventional batteries required periodic addition of water to the cells to keep the plates covered with electrolyte.

Lead-Calcium Grid

The grids on modern long-life batteries are made of a lead calcium alloy rather than the lead antimony alloy used in conventional batteries. Antimony in the negative plates of conventional batteries prevents the flow of electricity and contributes to gassing during charging.

Lead-calcium grids allow lower current flows to be used for charging, which reduces electrolyte temperature. Lower temperatures mean that less water is lost from the electrolyte during charging.

Envelope-Type Separator

The separators used in maintenance-free batteries wrap around the bottom of the plates, preventing loss of plate material. Plate material on older batteries could fall to the bottom of the case and eventually build up to cause an internal short.

Amount of Electrolyte Above Plate

If battery plates become dry because of lack of electrolyte, the battery can be damaged or ruined.

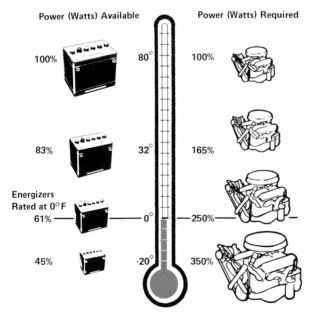

Figure 28-10. Battery power availability and requirements.
DELCO REMY—GMC

Maintenance-free batteries have a larger space above the plates for additional amounts of electrolyte **[refer to Figure 28-6]**. Thus, the plates can remain submerged in the electrolyte. Also, because the plates are enclosed in envelopes, the sediment chamber has been eliminated. This space is used for extra plate area, allowing higher output.

28.6 PHYSICAL SIZE

Batteries are classified as to physical size by *group numbers.* Typical group numbers for top-terminal batteries are 22, 24, 24F, 27, and 27F.

Side-terminal batteries of the same dimensions are 72, 74, 74F, 77, and 77F. The initial "7" indicates side-terminal design **[Figure 28-11]**.

The F designation indicates batteries for Ford vehicles. These batteries have positive and negative terminal posts in reverse position from other batteries.

Grp. Size	Vlt.	Cold cranking power— amps for 30 secs. at 0°F*	No. of mo. warranted	Size of battery container in inches (incl. terminals)		
				Lgth.	Wd.	Ht.
17HF	6	400	24	7¼	6¾	9
21	12	450	60	8	6¾	8½
22F	12	430	60	9	6⅞	8⅛
	12	380	55	9	6⅞	8⅛
	12	330	40	9	6⅞	8⅛
22NF	12	330	24	9½	5½	8⅞
24	12	525	60	10¼	6⅞	8⅝
	12	450	55	10¼	6⅞	8⅝
	12	410	48	10¼	6⅞	8⅝
	12	380	40	10¼	6⅞	8⅝
	12	325	36	10¼	6⅞	8⅝
	12	290	30	10¼	6⅞	8⅝
24F	12	525	60	10¼	6⅞	8⅝
	12	450	55	10¼	6⅞	8⅝
	12	410	48	10¼	6⅞	8⅝
	12	380	40	10¼	6⅞	8⅝
	12	325	36	10¼	6⅞	8⅝
	12	290	30	10¼	6⅞	8⅝
27	12	560	60	12	6⅞	8⅝
27F	12	560	60	12	6⅞	9
41	12	525	60	11 9/16	6 13/16	6 15/16
42	12	450	60	9⅝	6⅞	6¾
	12	340	40	9⅝	6⅞	6¾
45	12	420	60	9½	5½	8⅞
46	12	460	60	10¼	6⅞	8⅝
48	12	440	60	12	6⅞	7½
49	12	600	60	14½	6⅞	7½
56	12	450	60	10	6	8⅜
	12	380	48	10	6	8⅜
58	12	425	60	9¼	7¼	6⅞
71	12	450	60	8	7¼	8½
	12	395	55	8	7¼	8½
	12	330	36	8	7¼	8½
72	12	490	60	9	7¼	8¼
	12	380	48	9	7¼	8¼
74	12	585	60	10¼	7¼	8¾
	12	525	60	10¼	7¼	8¾
	12	505	60	10¼	7¼	8¾
	12	450	55	10¼	7¼	8¾
	12	410	48	10¼	7¼	8¾
	12	380	40	10¼	7¼	8¾
	12	325	36	10¼	7¼	8¾

*Meets or exceeds Battery Council International rating standards.

Figure 28-11. Battery size chart.

UNIT HIGHLIGHTS

- The battery provides current to operate the starter motor and to power the ignition system during starting.
- After starting, the alternator produces current to run the vehicle. The battery supplies current when electrical demand exceeds the alternator's output.
- The battery acts as an electrical stabilizer.
- Electrochemical action within the battery produces the potential for electrical flow. Electrical flow does not occur until a load is connected to a battery.
- A basic battery consists of two different metals immersed in electrolyte, an acid material that becomes ionized in solution.
- The materials in a vehicle battery are lead dioxide (PbO_2), sulfuric acid (H_2SO_4) electrolyte, and sponge lead (Pb).
- In a discharged battery, the plates become lead sulfate ($PbSO_4$) and the electrolyte becomes mostly water.
- In normal operation, charging restores the battery's potential for producing electricity.
- Many problems can shorten battery life.
- The basic ratings for battery capacity are cold cranking amperes, reserve capacity, and watt-hours. One battery capacity rating useful for service procedures is ampere-hours.
- Battery physical size and terminal type is indicated by group numbers.

TERMS

electrochemical action	electrolyte
separator strip	battery element
battery cell	lead dioxide (PbO_2)
sponge lead (Pb)	internal short
battery post	side terminal
flange	recess
lead sulfate ($PbSO_4$)	sulfation
charging	gassing
electrical capacity	spike
sediment chamber	group number
maintenance-free battery	cold cranking amperes (CCA)

REVIEW QUESTIONS

DIRECTIONS: The following questions are similar to those used on automotive technician certification tests. On a separate sheet of paper, write the letter of the correct choice.

1. Technician A says that the battery is a source of electrical power in a vehicle.
   Technician B says that the alternator is a source of electrical power in a vehicle.
   Who is correct?
   A. A only
   B. B only
   C. Both A and B
   D. Neither A nor B

2. Which of the following statements is correct?
   I. A battery stores electricity.
   II. When a load is connected, a battery produces electricity.
   A. I only
   B. II only
   C. Both I and II
   D. Neither I nor II

3. The most maintenance-free batteries are superior to conventional batteries because
   A. a different material is used in the battery plate grids.
   B. a different type of electrolyte is used.
   C. a different type of plastic case is used.
   D. the top of the battery is sealed and cannot be opened.

4. Technician A says that battery electrolyte can explode if ignited.
   Technician B says that the gases produced during charging can explode if ignited.
   Who is correct?
   A. A only
   B. B only
   C. Both A and B
   D. Neither A nor B

Question 5 is not like the previous four questions. It has the word EXCEPT. For question 5, look for the choice that does *not* apply. Read the entire question carefully before you choose your answer.

5. All of the following are part of a battery EXCEPT
   A. positive and negative plates.
   B. electrolyte.
   C. cables.
   D. case and terminals.

ESSAY QUESTIONS

1. Describe the construction of a battery.
2. Describe the process of discharging and charging in a lead-acid battery.
3. What are the important differences between conventional and maintenance-free batteries?

SUPPLEMENTAL ACTIVITIES

1. Inspect the battery on a vehicle owned by a relative or a friend. Note any problems evident in visual inspection and make a report to your class.
2. Measure the physical dimensions of a battery with a plastic ruler and determine its group number. Refer to the chart in **Figure 28-11.**
3. Examine the battery on a vehicle chosen by your instructor. Report to your class its group number and all other information that can be found on the battery.
4. Visit a department store or auto parts store that sells batteries. Ask the prices, warranty periods, cold cranking amps, and reserve capacities of different models of the same group size battery. Make a chart of this information and report to your class what differences are found.

29 Battery Service

UNIT PREVIEW

An important part of battery service is safety. The electrolyte can cause serious burns and blindness. Explosive gases, produced during battery operation, can explode if safety considerations are not followed carefully.

Automotive batteries that are properly sized for the vehicles in which they are installed should provide many years of dependable service. The engine and charging system must be maintained in good operating order to promote long battery life. The battery and its connections also must be serviced routinely to ensure long battery life.

Engine starting problems should be diagnosed carefully before replacing the battery. The problems may be caused by the condition of engine or the charging system. Proper test procedures will help pinpoint the problem.

A discharged battery may be recharged and restored to serviceable condition if it has not been damaged or worn out. There are no magic potions or chemicals that can restore a damaged or worn-out battery back to useful condition. Damaged or worn-out batteries must be replaced.

LEARNING OBJECTIVES

When you have completed your assignments and exercises in this unit, you should be able to:

❏ Describe safety precautions necessary when working around batteries.
❏ Correctly and safely use jumper cables.
❏ Perform battery maintenance.
❏ Correctly and safely use a battery hydrometer.
❏ Correctly and safely perform load tests on batteries.
❏ Correctly and safely charge a battery.

SAFETY PRECAUTIONS

Always wear eye protection when working around batteries. Batteries contain sulfuric acid electrolyte that can cause blindness, skin burns, and metal corrosion. Wear protective rubber or plastic gloves to protect your hands.

Always have a source of fresh, clean water nearby to wash acid off.

Do not tip the battery over or on its side when performing battery service. Electrolyte will flow through the gas vents.

Remove all metal jewelry such as rings, watches, and neck chains when working around batteries. Metal objects conduct electricity and can lead to burns or sparks that cause battery explosions.

The hydrogen and oxygen gases created during battery operation are highly explosive. When a battery explodes, sulfuric acid is splattered over everything nearby.

To prevent battery explosions, shocks, burns, and blown fuses, remove the battery ground terminal before service.

When removing a battery, take off the ground terminal first. When replacing a battery, put the ground terminal on last.

Work in a well-ventilated area when performing battery service and avoid breathing fumes from charging batteries. These fumes contain sulfuric acid that can cause lung and breathing problems. Also be careful not to inhale loose particles of corroded materials when you clean the terminals and tray.

29.1 JUMP-STARTING PROCEDURES

Jump starting is a process of starting a vehicle with a weak or dead battery. *Jumper cables* are connected between the electrical system of the vehicle with a weak battery and a good battery. Jumper cables are heavy electrical cables with color-coded ends.

SAFETY CAUTION

Always wear eye and skin protection when jump starting a vehicle. A battery explosion caused by improper connections can cause blindness or severe skin burns. Keep all jumper cables and clamps away from the fan, belts, and pulleys of both engines.

CAUTION!!!

Consult both vehicle manufacturer's service manuals and service bulletins before attempting jump starting. Some vehicles with on-board computers may not be jump started or used to jump start another vehicle. If

BATTERY SERVICE

such a vehicle's battery is weak, it must be disconnected and recharged or the battery must be replaced before starting is attempted.

The correct hookup sequence for safe jump starting is shown in **Figure 29-1.** Notice especially that the last negative connection (to the vehicle with the weak battery) is *not* made to the battery terminal. Rather, it is made only to the engine or to an engine bracket.

CAUTION!!!

The last connection must only be made on the engine or engine bracket. A connection to a body brace or other non-engine parts, may result in melting the body to battery connections and possibly cause a fire.

When a connection is completed to a good battery, a spark will occur. If the connection is made at the battery, there will be a spark at the battery. This spark could ignite residual gas in the area.

Therefore, the last negative connection always is made away from the battery. A spark there will not cause a battery explosion.

After the connections have been safely made, the vehicle with the good battery is started. The vehicle with the weak battery then is started. Both vehicles are run at a moderately high idle. The cables are removed carefully in the exact reverse order of connection. The last connection, the negative connection on the engine, always is the first to be removed after the engine has started.

SAFETY CAUTION

Roll up long sleeves and keep hands and arms away from moving fans, belts, and pulleys when removing jumper cables.

The charging system of the vehicle with the weak battery then should be checked to find the problem (see Unit 33).

29.2 BATTERY PREVENTIVE MAINTENANCE

Basic battery preventive maintenance consists of the following four areas:

1. Visual inspection
2. Instrument testing
3. Removal and cleaning
4. Replacement, reconnection, and protecting terminals.

29.3 VISUAL INSPECTION

Check the battery for obvious problems. Such problems include:

- Bulging or cracked case or cover
- Physically damaged case or cover

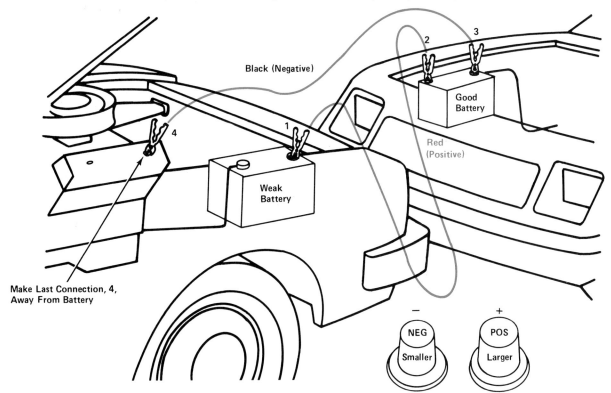

Figure 29-1. Battery jumper cable hookup.

- Leakage
- Corrosion on battery terminals and clamps
- Terminal tightness
- Holding device tightness
- Battery cable insulation condition
- Proper electrolyte level (if openable).

Make a list of all problems found during visual inspection. Parts that are badly damaged require servicing or replacement **[Figure 29-2]**.

29.4 INSTRUMENT TESTING

Instrument testing includes using hydrometers (where possible), test lights, and electrical test instruments.

Hydrometer

A battery hydrometer is illustrated in **Figure 29-3**. The hydrometer measures the relative *specific gravity*, or density, of electrolyte, compared to pure water. A good-quality hydrometer also includes a thermometer to measure electrolyte temperature.

A table of specific gravity readings and necessary temperature compensations is shown in **Figure 29-4**. To test electrolyte, squeeze the hydrometer bulb and carefully insert the tube into the battery. Slowly release the bulb just until it begins to float. Hold the bulb and take a reading. Make any temperature compensations and record the results. Move to the next cell and repeat this test.

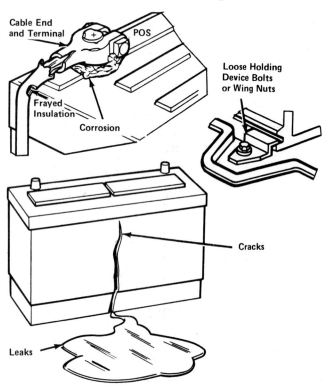

Figure 29-2. Battery inspection. CHRYSLER CORPORATION

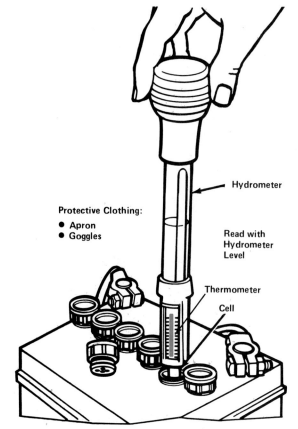

Figure 29-3. Battery testing with a hydrometer.
CHRYSLER CORPORATION

Hydrometer tests must be made of each individual cell. If water recently has been added to the cells, the test is invalid. The battery should be charged before testing (see Topic 29.7).

A charged battery with more than 0.050 specific gravity variance between cells must be replaced.

Maintenance-free batteries have permanently sealed tops. However, a built-in hydrometer allows a visual check of one of the battery's cells **[Figure 29-5]**. When the cell is more than 65 percent charged, a colored ball (usually green or blue) is visible in the plastic window. Some batteries may show orange or red colors when the battery is less than 65 percent charged. This visual check is valid for the cell in which the indicator is located. A battery load test is a more accurate way of determining overall battery condition.

Battery Capacity Testing

If the battery electrolyte specific gravity is above 1.220 or the visual indicator indicates that it is sufficiently charged, electrical testing can be performed. Two types of instruments can be used to test batteries:

1. Voltmeter
2. Battery load tester.

BATTERY SERVICE 301

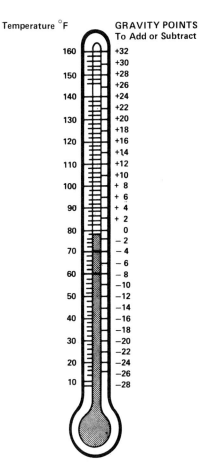

Figure 29-4. Hydrometer readings and temperature corrections.
CHRYSLER CORPORATION

SPECIFIC GRAVITY vs STATE OF CHARGE

% Charged	Dry Charge or Wet Process (1.265 Full Charge)	Spin Processed (1.250 Full Charge)	Tropical (1.225 Full Charge)	Arctic (1.290 Full Charge)
100	1.265	1.250	1.225	1.290
95	1.250	1.235	1.210	1.275
75	1.230	1.220	1.180	1.255
50	1.200	1.190	1.130	1.225
25	1.170	1.160	1.090	1.195
Discharged	1.140	1.130	1.045	1.165

Voltmeter. A voltmeter can be used to test the battery while it is still in the vehicle [Figure 29-6].

The ignition system must be disabled, as discussed in Unit 43. The engine is cranked, without starting, for 15 seconds continuously.

At 70 degrees F (21 degrees C) or more, the voltage should not drop below 9.6 volts during the test. Figure 29-7 shows temperature corrections necessary for colder temperatures.

Battery-starter tester. A *battery-starter tester* can be used for a more accurate check of battery condition

CHART FOR LOAD SETTING

Battery Group Size	Test Load (Amps)
22	200
72	215
24	250
74	250
27	280
77	280

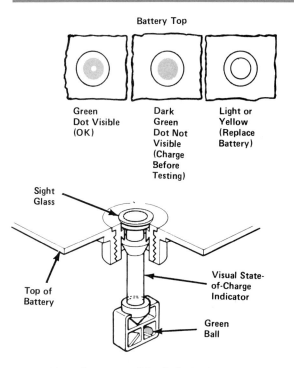

Figure 29-5. Battery condition indicator.
CHRYSLER CORPORATION

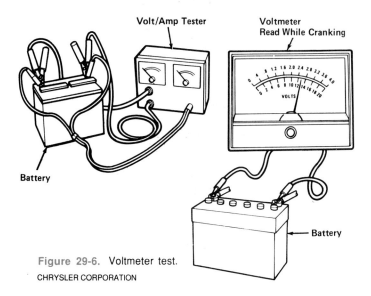

Figure 29-6. Voltmeter test.
CHRYSLER CORPORATION

Temperature (F°)	Minimum Voltage Acceptable	
	12-Volt	6-Volt
70 (or more)	9.6	4.8
60	9.5	4.75
50	9.4	4.7
40	9.3	4.65
30	9.1	4.55
20	8.9	4.45
10	8.7	4.35
0	8.5	4.25

Figure 29-7. Temperature correction table. FORD MOTOR COMPANY

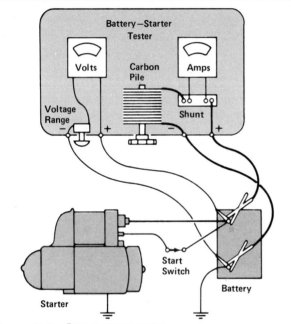

Figure 29-8. Battery-starter tester.

[Figure 29-8]. A battery-starter tester has a voltmeter, an ammeter, and a variable electrical resistance load.

To perform this test accurately, the ampere-hour rating of the battery must be known. For a 12-volt battery, if the watt-hour rating is known, the ampere-hour rating can be determined by the following equation:

$$ahr = whr \div 12$$

Where:
ahr = ampere-hour rating
whr = watt-hour rating

In the event that neither the amp-hour nor the watt-hour ratings can be found, divide the cold cranking amperes (CCA) rating by two. This number is used as the test load described below.

The positive (red) and negative (black) leads are connected to the battery. The load is adjusted to three times the ampere-hour rating and held for 15 seconds continuously. During the test, the voltage should not drop below 9.6 volts. Apply the temperature corrections shown in **Figure 29-7**.

29.5 BATTERY REMOVAL AND CLEANING

The battery should be removed and cleaned once per year, or whenever corrosion and acid buildup are noticed.

Removing the Battery

In removing the battery, first disconnect the grounded terminal [**Figure 29-9**]. Position the grounded cable well away from the battery.

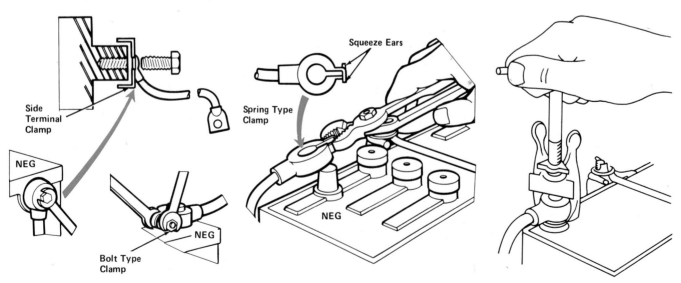

Figure 29-9. Battery cable removal. CHRYSLER CORPORATION

SAFETY CAUTION

Always disconnect the grounded terminal from the vehicle first. Otherwise, a wrench or other tool can connect positive to the negative circuit causing a direct short and sparks. Sparks can produce battery explosions.

Disconnect the battery positive terminal last. Then remove the holding devices (clamps, bars, or frames). Use a battery carrier to lift and remove the battery from the vehicle **[Figure 29-10]**.

Cleaning the Battery

Before cleaning, the battery caps (where used) must be tightened securely. A mild cleaning solution of 1 tablespoon baking soda mixed with one quart of water is used to clean and neutralize acid. Brush the solution over the battery and flush with water **[Figure 29-11]**. Dry the battery with paper towels and throw the towels away.

Cleaning the Battery Tray and Holding Devices

Corrosion and rust on the battery tray, holding clamps, and nearby metal parts must be removed by scraping and brushing.

Use scrapers, putty knives, and wire brushes to clean metal parts down to bare metal.

After physically removing all corrosion, use a mild solution of baking soda and water to neutralize acid **[Figure 29-12]**. Flush with water and dry with paper towels. Throw the towels away.

After the parts have air-dried thoroughly, paint them with rust-resistant paint and allow it to dry. When it is completely dry, coat the parts with a special grease- or silicone-base spray to protect them from battery acid.

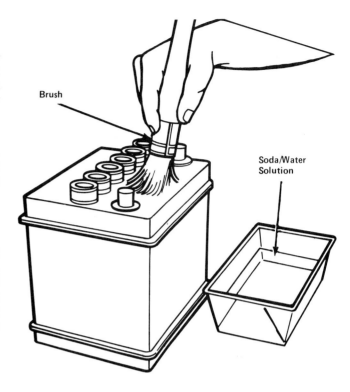

Figure 29-11. Cleaning the battery.
CHRYSLER CORPORATION

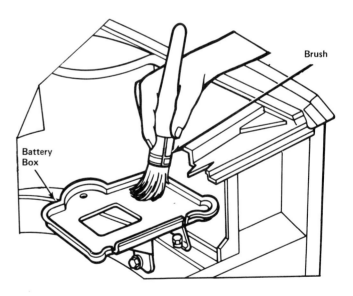

Figure 29-12. Servicing the battery tray.
CHRYSLER CORPORATION

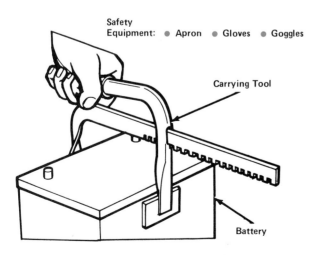

Figure 29-10. Battery carrier is used to lift and remove the battery. CHRYSLER CORPORATION

Cleaning Clamps and Cable Ends

Clean battery clamps, terminal ends, and battery connections with a wire brush **[Figure 29-13]**. After proper cleaning, the metal should be shiny.

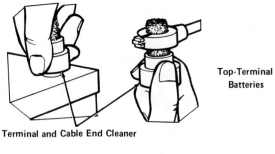

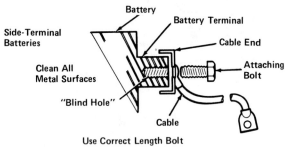

Figure 29-13. Cleaning battery posts and cables. CHRYSLER CORPORATION

If a cable end is badly corroded or physically damaged, cable ends or entire cables should be replaced.

29.6 BATTERY REPLACEMENT

After all cleaning operations are complete, the battery is replaced in its tray. Holding devices are tightened.

CAUTION!!!

Tighten battery holding devices securely, but do not overtighten. Overtightening can lead to distortion and bulging of the battery case, and battery case leaks.

Replacing Cables

To replace top-type battery clamps, loosen the holding bolt and spread the clamp with special post pliers. Always replace the positive cable first. Replace the grounded (negative) cable last.

SAFETY CAUTION

The grounded (negative) cable must be replaced last to prevent shocks, sparks, and possible battery explosions.

Protective Sprays and Liquids

After replacing the clamps, special battery protective spray or liquids may be used on the outer parts of the connections.

CAUTION!!!

Avoid getting battery protective materials between the cable and the battery terminal. Such materials are insulators and prevent the flow of electricity.

29.7 BATTERY CHARGERS

A vehicle's own charging system should keep the battery properly charged. However, if the vehicle is started frequently and driven only short distances, the battery can become discharged. A faulty charging system also can cause problems (see Unit 32).

Batteries can be charged with a *battery charger.* A battery charger is a device that changes line current (110V AC) to low-voltage DC. Charging can be done with the battery in or out of the vehicle. However, if the battery is left in the vehicle, remove the battery cables before charging.

SAFETY CAUTION

Under no circumstances should a frozen battery be charged. Be sure the battery is thawed out before charging.

Fast Charging

Fast charging (at high current) can be done to "boost" a weak battery. This procedure may be necessary to get a vehicle started **[Figure 29-14]**. If the vehicle's charging system is operating properly, normal driving should bring the battery up to full charge.

CAUTION!!!

Battery electrolyte temperature must not exceed 125 degrees F (52 degrees C), and excessive gassing must not occur during fast charging. If possible, distilled water must be added if the electrolyte level drops below the proper level. Be sure not to overfill when adding water [Figure 29-15].

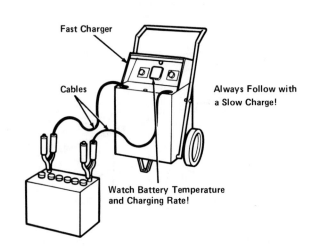

Figure 29-14. Fast charging. CHRYSLER CORPORATION

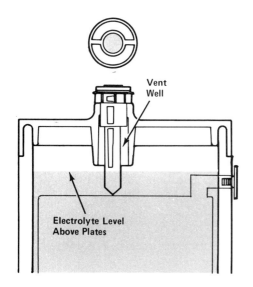

Figure 29-15. Electrolyte fill level.

Refer to the manufacturer's service manual for recommended fast charging rates and times. In general, 12-volt batteries should not be charged at a rate over 35 amperes. One manufacturer's recommendations are shown in **Figure 29-16**.

Multiple batteries can be connected in parallel for charging **[Figure 29-17]**.

Slow Charging

Slow charging, or charging at rates less than 3 amperes, can be used to bring a battery to full charge. One definition of "fully charged" is proper specific gravity reading after 24 hours of slow charging.

Periodic slow charging also can be used to keep stored batteries charged before sale or installation.

Specific Gravity Reading	Charge Rate Amperes	Battery Capacity—Amp Hours			
		45	55	70	80
		High Rate Charging Time			
1.125* to 1.150	35	65 min.	80 min.	100 min.	115 min.
1.150 to 1.175	35	50 min.	65 min.	80 min.	95 min.
1.175 to 1.200	35	40 min.	50 min.	60 min.	70 min.
1.200 to 1.225	35	30 min.	35 min.	45 min.	50 min.
Above 1.225	5	Note: Charge at Low Rate Only Until Specific Gravity Reaches 1.250 at 80°F.			

*If Specific Gravity is Below 1.125, Use Indicated High Rate, Then Follow with Low Rate of Charge (5 Amperes) Until Specific Gravity Reaches 1.250 at 80°F.

Figure 29-16. Fast charging guidelines. FORD MOTOR COMPANY

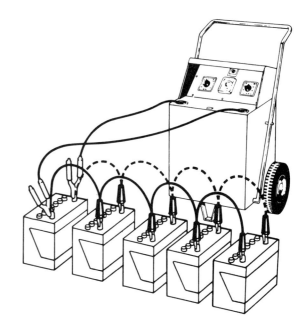

Figure 29-17. Multiple-battery charging hookup.
FORD MOTOR COMPANY

CAUTION!!!

Continuous overcharging, even with a slow charger, can cause positive plate damage.

29.8 "MAGIC" BATTERY CHEMICALS

Chemical products that are claimed to bring old batteries back to life "magically" are available on the market. However, these chemicals do nothing but increase battery electrolyte temperature temporarily. This causes a temporary increase in electrical output. The battery will appear to be "rejuvenated" for a short time but then will fail again.

SAFETY CAUTION

Sulfuric acid added to weak electrolyte can cause an explosion. Weak electrolyte is caused by internal battery chemistry that cannot be corrected by addition of sulfuric acid.

No permanent, magic cure exists for a badly sulfated, internally shorted, or physically damaged old battery. Batteries that fail the 0.050 variance hydrometer test and load test must be replaced.

UNIT HIGHLIGHTS

- Jumper cable starting hookups must be made in a specific order to prevent battery explosions. The last connection is to an engine ground.
- Basic battery preventive maintenance consists of visual inspection, instrument testing, removal, cleaning, replacement, and terminal protection.
- Hydrometer readings must be corrected for electrolyte temperature.
- Load testing can be done by cranking the engine, or by using a battery-starter tester.
- Battery electrolyte level and temperature must be checked during fast charging. Fast charging must be followed by slow charging to bring the battery to a fully charged state.
- "Magic" battery chemicals are a waste of money.

TERMS

jump starting
specific gravity
battery charger
fast charging
jumper cables
battery-starter tester
slow charging

REVIEW QUESTIONS

DIRECTIONS: The following questions are similar to those used on automotive technician certification tests. On a separate sheet of paper, write the letter of the correct choice.

1. Technician A says that jumper cables are connected battery-to-battery only.
   Technician B says that the last connection is made to a ground away from the weak battery.
   Who is correct?
   A. A only
   B. B only
   C. Both A and B
   D. Neither A nor B

2. How much variance is acceptable in battery cell readings with a hydrometer?
   A. 0.050
   B. 0.500
   C. 5.00
   D. 0.005

3. Which of the following statements is correct?
   I. The maximum test load on a 12V, 720 watt-hour battery should be 240 amps.
   II. The maximum test load on a 12V, 600 watt-hour battery should be 150 amps.
   A. I only
   B. II only
   C. Both I and II
   D. Neither I nor II

4. After fast charging a 12-volt battery at 75 amps, the battery will not hold a charge. Which of the following is the most likely cause?
   A. Battery has died due to old age
   B. Electrolyte contains too much water
   C. Battery damaged during charging
   D. Excess sulfation prevented full charging

Question 5 is not like those above. It has the word EXCEPT. For question 5, look for the choice that does *not* apply. Read the entire question carefully before you choose your answer.

5. Basic battery preventive maintenance includes all of the following EXCEPT
   A. visual inspection and instrument testing.
   B. removal and cleaning.
   C. fast charging.
   D. replacement, reconnection, and protecting terminals.

ESSAY QUESTIONS

1. Describe jump-starting procedures and safety procedures to be followed.
2. What methods can be used to check battery charge and condition?
3. Describe the safety procedures to be followed when using a battery charger.

SUPPLEMENTAL ACTIVITIES

1. Correctly and safely connect and disconnect jumper cables on two vehicles for jump starting.
2. Use a hydrometer and perform a check of battery electrolyte. Compensate properly for temperature. Make notes and report to your class on the readings noted.
3. Correctly and safely perform a load test using the starter of the vehicle as a load. Compensate properly for temperature as shown in **Figure 29-7**. What was the final voltage reading, and what does it indicate?
4. Correctly and safely perform a load test on a battery using a battery-starter tester. Compensate properly for temperature. What was the final voltage reading, and what does it indicate?
5. Charge a discharged battery by either fast or slow charging. Perform a load test and/or a hydrometer test on the charged battery. What readings are obtained, and what do they indicate?

30 The Starting System

UNIT PREVIEW

The battery supplies electrical power to a motor that cranks the engine for starting. The starter motor, like other electrical motors, uses electro-magnetism to produce motion.

A basic starting system consists of several elements. These parts are used to control and protect the starting motor, engine, and vehicle.

Different types of starter and control mechanisms can be used on passenger vehicles.

LEARNING OBJECTIVES

When you have completed your assignments and exercises in this unit, you should be able to:

❑ Explain how an electric motor operates.
❑ Identify and describe the basic parts of a starter motor.
❑ Identify and describe the parts of a basic starting system.
❑ Explain how magnetic switches operate.
❑ Describe devices used to prevent vehicles from starting in gear.
❑ Locate visible parts of a vehicle's starting system.

30.1 STARTING AN ENGINE

To start an engine, the crankshaft is turned. As it turns, the pistons and valves begin to move and the fuel system supplies fuel. The cylinders draw in an air-fuel mixture. The ignition system (see Unit 34) ignites the mixture, and the engine begins to run on its own. On early vehicles, the crankshaft was turned by hand, using a metal crank that attached to the front crankshaft pulley.

In 1912, an engineer for Cadillac named Charles F. Kettering invented the self-starter. The self-starter uses a battery, a switch, and a powerful electric motor to turn the crankshaft. The flywheel, attached to the end of the engine crankshaft, has gear teeth. The gear teeth may be cut into the flywheel, or may be a separate, attached *ring gear.* The starter motor has a small drive gear called a *pinion gear.* The pinion gear meshes with the flywheel gear teeth to turn the crankshaft [Figure 30-1]. The gear teeth mesh only during starting, then are disengaged.

30.2 ELECTRIC MOTOR OPERATION

Electromagnetism supplies the turning power in an electric motor. Magnetic fields oppose each other to produce a turning force. This turning force, or torque, is used to turn a starter pinion gear.

Electromagnetic Fields

Figure 30-2 shows a conductor within a horseshoe, or U-shaped, magnet. The continuous lines of magnetic force from the magnet move from north to south poles.

When electrical current flows through the conductor, concentric circles of magnetic force are produced around the conductor. However, the flow of magnetic force on one side of the conductor is opposite to that of the magnet. The opposing magnetic forces cause the conductor to be unbalanced, magnetically, in relation to the magnetic field of the magnet. This unbalanced condition causes the conductor to move **[Figure 30-3].**

Simple Electric Motor

The conductor is shaped like a loop, or coil, and placed within a magnet. When electricity flows through the loop, one side will be pushed upward, and the other side downward. This action causes rotational, or circular, movement **[Figure 30-4].**

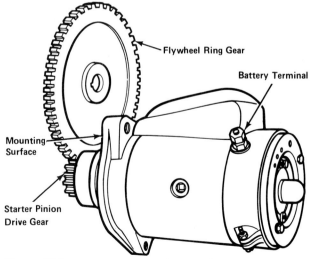

Figure 30-1. Starter and flywheel connection.
CHRYSLER CORPORATION

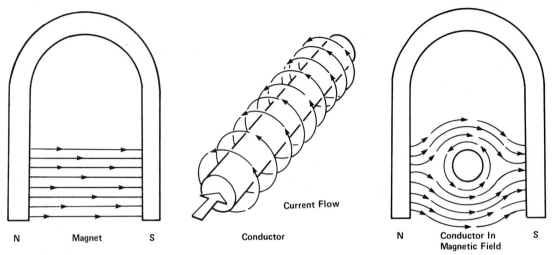

Figure 30-2. Electromagnetic starter fields.

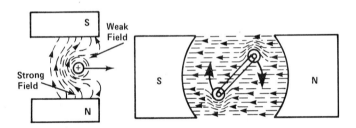

Figure 30-3. Unbalanced electromagnetic field.

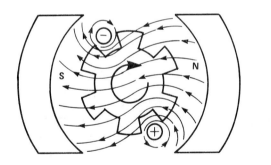

Figure 30-4. Simple electric motor principle.

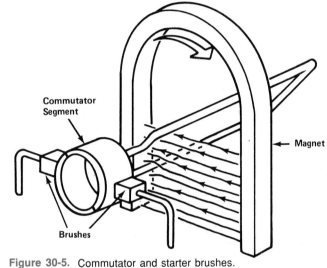

Figure 30-5. Commutator and starter brushes.

Armature

The ends of the loop can be connected to semicircular *segments,* or partial rings, that rotate. Together, the segments form a *commutator.* *Brushes* that ride against the segments can conduct electricity to the segments as they rotate [Figure 30-5]. Brushes can be made of carbon or soft copper. Starter motor brushes are usually copper impregnated with graphite. The graphite serves as a lubricant.

When the loops are parallel to the magnetic field, the magnetic effect is strongest. As the loops move farther away from the magnet ends, the magnetic effect is weaker.

30.3 STARTER MOTOR

The simple motor described above would not run very well. The loop would stop turning at the center position.

To maintain a constant turning force, many *windings,* or coils of wire, and segments are used [Figure 30-6]. To increase magnetic force, the windings are looped around a laminated, or layered, core made of iron discs. Such an arrangement is known as an *armature.*

A real motor often has hundreds of loops of wire to create a constant turning force. A starter motor is a specialized form of electrical motor that produces high torque to turn the flywheel ring gear.

Field Coils

Magnets are not used in most starter motors. Instead, coils of wire, called *field coils,* create magnetic

The Starting System

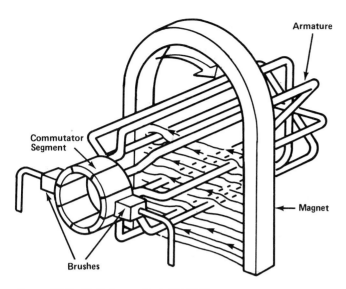

Figure 30-6. Multiple armature windings.

fields for the armature when electricity flows through them. The field coils are wrapped around metal cores called *pole pieces* or *pole shoes* **[Figure 30-7]**.

Frame Assembly

The field coils are mounted within a heavy metal frame. **Figure 30-8** shows a frame and field coil assembly.

30.4 STARTING SYSTEM

A basic starting system consists of the following elements **[Figure 30-9]**:

- Battery
- Ignition switch
- Battery cables
- Magnetic switch
- Starter drive:
  - Pinion gear
  - Overrunning clutch
  - Mesh spring drive flange.

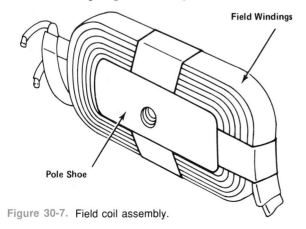

Figure 30-7. Field coil assembly.

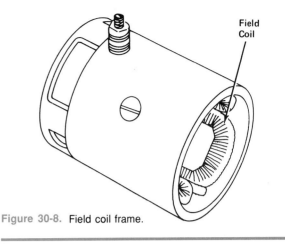

Figure 30-8. Field coil frame.

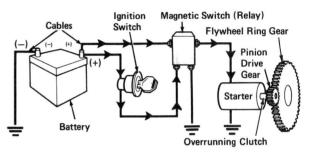

Figure 30-9. Basic starting system.
CHRYSLER CORPORATION

Battery

The battery supplies current for the starter motor. In cold weather, required starting current for a large engine can approach 300 amperes or more.

Ignition Switch

The ignition switch is a light-duty switch that operates the heavy-duty magnetic switch when the ignition key is turned to the START position **[Figure 30-10]**. Generally, the switch/key can be turned clockwise to a spring loaded "crank" position. This position energizes the starting motor and continues to do so until the key is released. At that point, the spring loaded switch automatically returns to the "run" position and the starting motor is de-energized.

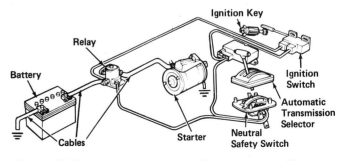

Figure 30-10. Ignition switch and starting system assemblies.
CHRYSLER CORPORATION

Battery Cables

Large, heavy-duty copper cables are used to conduct the high amounts of electrical current needed for cranking the engine from the battery to the starting motor [refer to Figure 30-10].

Magnetic Switch

Because turning the engine crankshaft is hard work, the starter uses more current than any other component. To connect the battery to the starter, a heavy-duty magnetic switch, called a *relay*, is used.

Relay. A small amount of electricity passes through the ignition switch to a coil within the relay. The coil pulls the plunger and holds it in place. Connected to the plunger is a thick copper *contact disk*, or washer [Figure 30-11].

When pulled in and held, the contact disk connects two terminals. One of the terminals is attached to the battery positive cable. The other terminal is attached to the starter motor positive terminal. The starter is grounded through its connection to the engine block. Electricity is supplied to the field coils and the armature, forcing the armature to turn. When the engine has started, the key is released and the magnetic relay is de-energized. A spring inside the relay pushes the contact disc to the "off" position.

Some vehicles use starter relays, and others use a combination of a relay and a *solenoid*. Still others use only a solenoid [Figure 30-12].

Solenoid. Instead of a relay mounted close to the battery, a solenoid can be mounted on the starter motor. A solenoid is similar to a starter relay. However, the motion of the plunger also moves a *shift lever* which causes the starter pinion gear to mesh with the flywheel ring gear [Figure 30-13].

A diagram of a starting system using a solenoid is shown in **Figure 30-14**. Most vehicles use such a starting system.

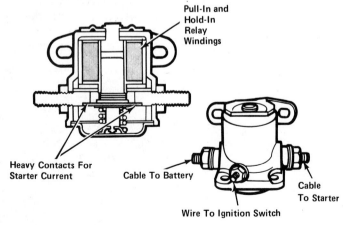

Figure 30-11. Starter relay assembly.
CHRYSLER CORPORATION

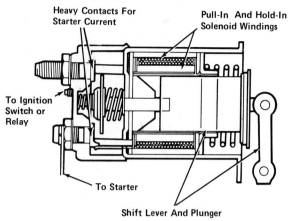

Figure 30-12. Starter solenoid assembly.
CHRYSLER CORPORATION/FORD MOTOR COMPANY

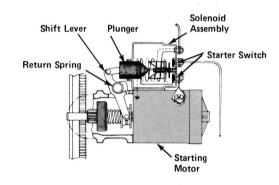

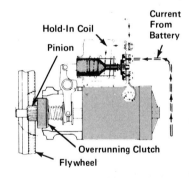

Figure 30-13. Solenoid and overrunning clutch operation.
DELCO REMY—GMC

Starter Drive

The *starter drive* is an assembly consisting of the following main elements:

- Pinion drive gear
- Overrunning clutch
- Mesh spring and drive flange.

The pinion gear is connected to the starter motor armature. When the armature turns, the pinion gear

THE STARTING SYSTEM

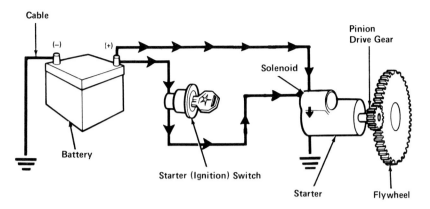

Figure 30-14. Solenoid starting system. CHRYSLER CORPORATION

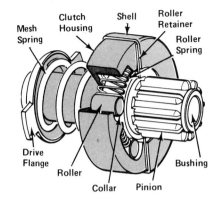

Figure 30-15. Overrunning clutch assembly.

turns. The connection between the armature shaft and the pinion gear is called the *overrunning clutch* [Figure 30-15].

When the starting motor is turning the engine ring gear, the overrunning clutch is locked up. However, when the engine starts, the ring gear begins to drive the pinion gear. When this happens, the overrunning clutch allows the pinion gear to disengage, or overrun, the armature shaft until the key is released and the pinion gear is unmeshed from the ring gear. This prevents starter motor damage.

The gear ratio between the pinion gear and the ring gear is around 20:1. Thus, with an engine speed of up to 2,000 rpm at start-up, for example, the starter armature would be driven at 40,000 rpm. The speed would destroy the starter if the pinion gear didn't overrun the armature shaft.

A cutaway view of a complete starter and solenoid is shown in **Figure 30-16**.

30.5 TYPICAL STARTER MOTORS

Figure 30-17 shows three types of starters that are in general use:

1. Solenoid direct drive
2. Movable pole
3. Solenoid gear reduction.

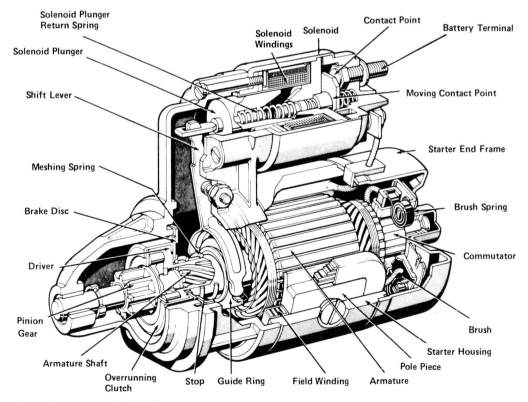

Figure 30-16. Parts of a starter and solenoid.
ROBERT BOSCH CORPORATION

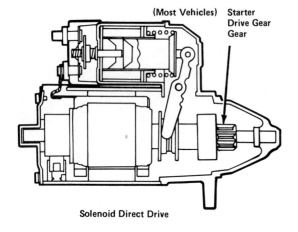

Solenoid Direct Drive

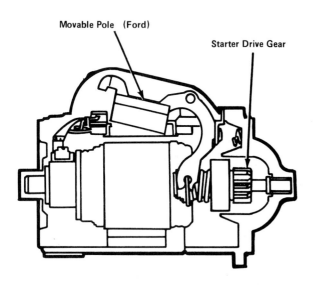

Movable Pole (Ford)

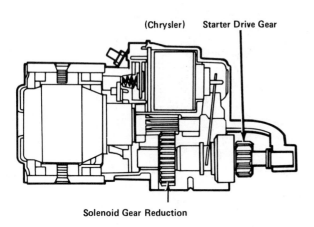

Solenoid Gear Reduction

Figure 30-17. Typical starter motors. CHRYSLER CORPORATION

Solenoid Direct Drive

The most common type of starter is the solenoid direct drive. Such starters are used on most vehicles.

Movable Pole

Some Ford vehicles use a movable pole piece to engage the starter drive. A shift lever attached to the pole moves the starter gear into mesh with the flywheel gear.

Solenoid Gear Reduction

Some vehicles use a system of gears to increase starter torque. See Unit 49 for a discussion of torque increase. The additional set of gears produces a distinctive, high-pitched sound during starting.

30.6 STARTER MOTOR WIRING

Different combinations of series, parallel, and series-parallel circuits can be used within a starter. The circuit design depends on many factors. These factors include engine cranking speed and torque requirements, battery cable size, battery capacity, and motor brush and switch capacity **[Figure 30-18]**.

Series and Shunt Field Coils

Two types of field coils can be used: series and *shunt*. A shunt is an electrical bypass **[Figure 30-18]**. Series-wound field coils are wound with heavy, flat copper ribbon. Shunt coils contain more turns of smaller wire.

Current that flows through series-wound field coils also flows through armature windings. These types of motors tend to have a rather high free speed and are thus a bit noisy. Also, when the motor turns freely, without load, it can spin too rapidly. Too great a rotational speed can cause the armature windings to be thrown outward from their slots.

Shunt-wound field coils are used to prevent the starter motor from turning too rapidly. Current through a shunt coil bypasses the armature and

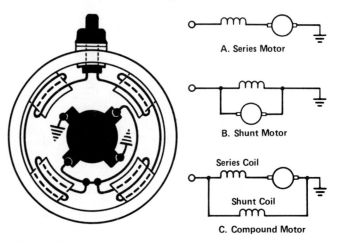

Figure 30-18. Starter circuits.

30.7 STARTER MOTOR CIRCUITS

To prevent vehicles from jumping forward in gear when the starter operates, a *neutral safety switch* is used. The switch is part of the starter control circuit [Figure 30-19]. The neutral safety switch will complete the circuit only if an automatic transmission is in park or neutral. For vehicles with automatic transmissions, the switch can be located on the transmission itself or on the linkage to the gearshift lever.

If the vehicle is equipped with a manual transmission, a safety switch is operated by the clutch linkage, and is known as a "clutch switch." The switch allows the starter to operate only if the clutch pedal is depressed.

Many newer vehicles with automatic transmissions use an *ignition interlock* to prevent the key from being removed except in Park [Figure 30-20]. An ignition interlock is a theft-prevention device. It also locks the steering wheel if the key is not inserted and turned.

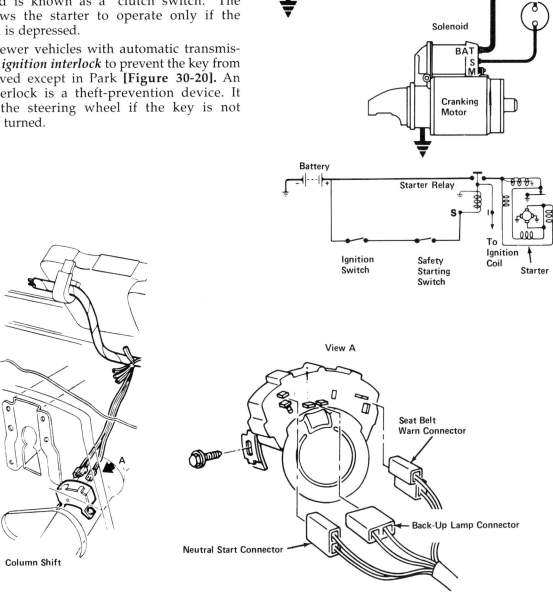

Figure 30-19. Neutral safety switch assemblies.
GENERAL MOTORS CORPORATION

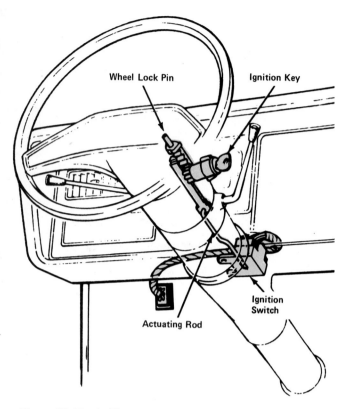

Figure 30-20. Ignition switch interlock assembly.

UNIT HIGHLIGHTS

- A battery supplies electrical power to a motor to crank the engine for starting.
- A motor uses electromagnetism to produce opposing lines of magnetic force, and thus produce motion.
- In a motor, an armature turns in the magnetic lines of force produced by field coils.
- A basic starting system consists of several elements. These include the battery, ignition switch, magnetic switch, starter motor, and overrunning clutch.
- A magnetic switch can be a relay or a solenoid.
- An overrunning clutch prevents the starter from being turned too rapidly by the flywheel when the engine starts.
- Three types of starter are commonly used: direct drive solenoid, gear drive solenoid, and movable pole.
- Shunt-wound field coils prevent the starter from turning too rapidly when disengaged.
- A neutral safety switch, clutch switch, or an ignition interlock prevents a vehicle from starting in gear.

TERMS

segments	commutator
brush	winding
armature	field coil
pole piece	pole shoe
relay	contact disk
starter drive	solenoid
shift lever	overrunning clutch
ignition interlock	shunt
pinion gear	neutral safety switch
ring gear	

REVIEW QUESTIONS

DIRECTIONS: The following questions are similar to those used on automotive technician certification tests. On a separate sheet of paper, write the letter of the correct choice.

1. Electrical motor operation is being discussed.
   Technician A says that the magnetic lines of force oppose each other to produce motion.
   Technician B says that separate armature windings are used to increase magnetic force.
   Who is correct?
   A. A only   C. Both A and B
   B. B only   D. Neither A nor B

2. Which of the following statements about magnetic switches is correct?
   I. A coil produces electromagnetic force to pull a plunger.
   II. A coil produces electromagnetic force to hold a plunger.
   A. I only   C. Both I and II
   B. II only   D. Neither I nor II

3. A neutral safety switch is
   I. in series with the ignition switch.
   II. closed when the transmission is in Drive.
   A. I only   C. Both I and II
   B. II only   D. Neither I nor II.

Questions 4 and 5 are not like those above. Each has the word EXCEPT. For each question, look for the choice that does *not* apply. Read each question carefully before you choose your answer.

4. All of the following are part of a starting system EXCEPT
    A. ignition interlock.
    B. solenoid.
    C. overrunning clutch.
    D. alternator armature.

5. All of the following are types of starters EXCEPT
    A. solenoid and movable pole.
    B. solenoid gear reduction.
    C. movable pole.
    D. solenoid direct drive.

ESSAY QUESTIONS

1. Describe the parts and functions of a simple electric motor.
2. What are the functions of the parts of a basic starting system?
3. Describe the parts and operation of the three types of starter motors.

SUPPLEMENTAL ACTIVITIES

1. Trace battery positive cables on several different vehicles. To what are they attached?
2. Examine a vehicle's starting system. Note the make and year of the vehicle. Report to your class what type, or combination, of magnetic switch or switches is used.
3. Explain how a starter relay works, and how a solenoid differs from a starter relay.
4. Examine a vehicle's starter. Report what difficulties would be encountered in removing the starter from the vehicle.
5. Locate the neutral safety switch on a vehicle equipped with an automatic transmission. Is the switch adjustable? If so, how?

31 Starting System Service

UNIT PREVIEW

Starter preventive maintenance consists of checking and cleaning connections and terminals.

On-car and off-car tests may be necessary to diagnose starting problems. These tests include electrical tests, visual inspection, and simple physical manipulation.

Faulty solenoids or starter relays are replaced. Faulty starters can be repaired, or an exchange starter can be installed.

LEARNING OBJECTIVES

When you have completed your assignments and exercises in this unit, you should be able to:

❑ Safely remove, clean, and replace battery and starter cables and clean connections.
❑ Perform on-car electrical tests.
❑ Check the pull-in and hold-in windings of a solenoid.
❑ Remove, disassemble, test, and replace a starter.

SAFETY PRECAUTIONS

Remove all rings, watches, neck chains, and any other metal objects when servicing starting system parts. Be sure that no metal tools connect from grounded metal to parts connected to the battery positive terminal.

The battery must be fully charged and in good condition before starter tests can be made. Refer to Unit 29.

Before attempting to remove starter motors or connecting cables and wires, disconnect the grounded battery terminal. Secure the cable end away from the battery. Injury can result from sparks and red-hot metal caused by grounded positive leads if the negative terminal is not removed.

Starter motors are heavy. Plan how to get the starter around obstructions before starting removal procedures. During removal, support the starter as the last mounting bolt is taken out. If the starter falls, it can crush fingers, hands, or other parts of your body. If the starter falls to the floor, it can be damaged.

31.1 STARTER PREVENTIVE MAINTENANCE

Starter motors are reasonably durable. Preventive maintenance consists of the following items:

• Checking and cleaning connections and wiring
• Replacing damaged or frayed cables or terminal ends.

Checking and Cleaning Connections and Wiring

Loose, corroded, or damaged positive connections at the battery, solenoid, or starter motor may cause starting problems.

Battery terminals. Check the battery connections for corrosion and, if necessary, clean as discussed in Unit 29.

SAFETY CAUTION

Remove the grounded battery terminal first. Remove the positive terminal last. Replace the positive terminal first. As the last step, replace the grounded battery terminal. Battery explosions, sparks, and burns may result if the terminals are not removed and replaced in this order.

Also check negative cable connections and ground straps for tight and clean connections. If cleaning or tightening is needed, remove the battery negative terminal clamp. Then remove fasteners, scrape, and wire brush connections until they are clean and shiny. Replace fasteners and tighten to the manufacturer's specified torque.

Cables. Check cables for proper size, broken wires, loose or frayed insulation, and corrosion. Proper cable sizes for 12-volt batteries are shown in **Figure 31-1**. An undersize cable can overheat, resulting in insufficient current flow. Cables or wires that are routed near an exhaust manifold or pipe may appear good at both ends but be corroded internally. The outside insulator may appear to be serviceable, but the inner wires can be corroded and conduct poorly. The voltage drop test, described in 31.3, will locate defective cables.

Cable ends. Battery cable connections at the battery, relay or solenoid, starter, and engine or frame ground

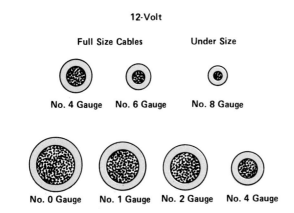

Figure 31-1. Battery cable sizes.

must be clean and tight. Check cables both visually and manually for looseness and corrosion. Attempt to rotate and wiggle the connections. Tighten loose cables.

SAFETY CAUTION

Remove the grounded battery terminal before attempting to tighten the positive battery cable at either end. A metal tool, ring, or metal watch band will conduct electricity from ground to any metal part of a positive battery terminal. Battery explosions, sparks, and physical injury can result.

Corrosion is removed by scraping and using a wire brush on the connections until they are clean and shiny.

Starter mounting bolts. Attempt to wiggle and move the starter motor itself. If the mounting bolts are loose, a poor or intermittent ground will cause starter problems.

Replacing Damaged or Frayed Cables

Broken individual wires in a cable will effectively reduce cable size and increase resistance. Cables with broken wires should be replaced.

SAFETY CAUTION

Remove the grounded battery terminal before attempting to replace the positive battery cable. Metal tools can ground the positive cable at any point. Such grounding can result in sparks, battery explosions, burns, or damage to parts.

Loose or frayed insulation on the positive cable can cause grounding. If grounding occurs near the battery, a battery explosion can result. Damage to parts can result from grounding the positive cable at other points.

Frayed or missing insulation on either the positive or grounded cable can lead to corrosion under the insulation. Corrosion causes resistance and hampers electrical current flow. Replace cables that have visible, thick corrosion between the wires of the conductor.

If cable wires are clean and unbroken, electrical tape can be used to insulate cable wires. Wrap several layers around and past the damaged portion. *Heat-shrink tubing* can be used in place of tape to make a neater and more permanent repair. The tubing can be slipped over the end of the cable and heated to shrink tightly against the conductor.

31.2 STARTING SYSTEM TROUBLESHOOTING

A failure to crank may originate at any point in the starting system circuit. **Figure 31-2** shows potential trouble spots. Cranking problems generally fall into four distinct categories:

1. No cranking
2. Engine cranks slowly, does not start
3. Starter spins, does not crank engine
4. Excessive noise during cranking.

NOTE: It is a good idea to make sure the engine is in good mechanical shape before you begin starter tests. If there is mechanical damage inside the engine, the crankshaft may be unable to turn for starting. For example, it can become jammed by a frozen bearing, a broken valve, or a broken connecting rod. With the engine OFF, test for mechanical damage by turning the crankshaft pulley nut clockwise at least two full turns with a large socket wrench. If the crankshaft rotates, then proceed with starter tests.

No cranking and slow cranking can be diagnosed with the starter installed on the vehicle. An operating starter that fails to engage the flywheel gear teeth may require testing on and off the vehicle. Excessive noise during cranking can be caused by gear misalignment, incorrect gear tooth clearance, a cracked engine flex plate or flywheel, or starter internal problems. Gear tooth clearance can be checked with a wire gauge between starter pinion and flywheel gear teeth [**Figure 31-3**].

Internal starter problems include a defective starter drive, worn bearings, a bent armature shaft, dragging pole shoes, worn brushes and bushings, and electrical problems.

Troubleshooting procedures for basic cranking problems on Ford Motor Company vehicles are summarized in **Figure 31-4**. Refer to the manufacturer's service manual for troubleshooting procedures for specific vehicles.

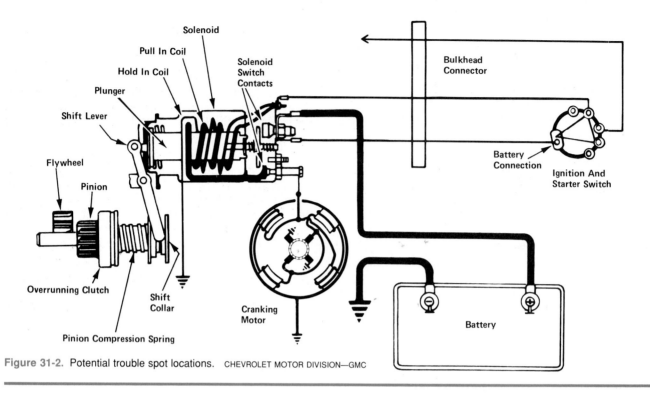

Figure 31-2. Potential trouble spot locations. CHEVROLET MOTOR DIVISION—GMC

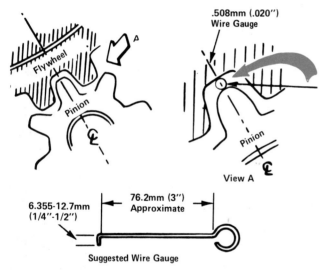

Figure 31-3. Measuring starter and flywheel teeth clearance. CHEVROLET MOTOR DIVISION—GMC

31.3 ON-CAR SERVICE

Several tests can be made with the starter installed in the vehicle. Other problems, discussed below, require starter removal and repair.

No Cranking

Starter system troubleshooting generally proceeds from the most accessible parts to the least accessible. The items to be checked, shown in **Figure 31-5**, are:

1. Ignition/starter switch
2. Neutral safety switch or clutch start switch
3. Starter relay or solenoid
4. Starter motor.

Troubleshooting a no-cranking situation is illustrated in **Figure 31-6**.

Slow Cranking, Failure to Start

In Unit 27, it was explained that a voltage drop can be measured across a given load, or resistance. Excessive resistance in battery cables can be determined with a *voltage drop test*.

A voltage drop test measures the amount of voltage drop, or loss, across a conductor while current is flowing. The higher the voltage reading, the greater the resistance.

Battery cables in good condition should have very low resistance. The voltage drop across cables during cranking should be less than 0.2 volt.

Two things always must be done to perform a voltage drop test. First, the voltmeter must be connected on each side of the circuit, or component, to be measured. Second, electrical current must be flowing through the circuit or component being measured. Although you may get a voltage reading as soon as the connections are made, that by itself does not mean current is flowing. Current flows through the battery cable to the starter motor *only* during cranking.

To perform a voltage drop test, a voltmeter is connected to the ends of a cable. The ignition system is disabled so that the vehicle cannot start. The

STARTING SYSTEM SERVICE

STARTING SYSTEM TROUBLESHOOTING CHART

Complaint Condition	Perform Test*	Results Indicate	Corrective Action
A. Starter spins, but does not crank engine.	Starter drive test.	Slipping starter drive.	Replace starter drive.
	Inspect starter drive components.	Worn or broken starter drive components.	Replace starter drive.
	Inspect flywheel ring gear.	Worn or broken ring gear teeth.	Replace ring gear.
B. Engine cranks slowly.	Starter cranking circuit test.	Excessive resistance in cranking circuit.	Clean and tighten connections; repair or replace components as necessary.
	Starting motor load (current draw) test.	Out-of-spec readings indicate internal starter motor problems.	Make bench tests. Repair or replace starter motor.
C. Engine won't crank.	Starting control circuit test.	Excessive resistance in starter control circuit.	Clean and tighten connections; repair, replace or adjust wires or switches as necessary.
	Relay bypass test. (Positive engagement only)	Inoperative starter relay.	Replace starter relay.
	Starter solenoid test. (Solenoid starter only)	If no pull-in or over 10 volts required, solenoid is damaged, or not operating properly.	Replace solenoid.
	Starter no-load tests.	Internal starter problems.	Disassemble and perform further bench tests.
	Armature open circuit test Armature and field grounded test.	Specific starter problems.	Repair or replace starter motor.
D. Noisy starter cranking.	Inspect starter mounting and drive components.	Improper mounting or misalignment condition.	Replace worn components; correct alignment problems. If problem persists, repair or replace starter motor.

*See manual for description of test.

Figure 31-4. Starting system troubleshooting chart. FORD MOTOR COMPANY

starter is operated, and the voltage drop across the cable is measured while the engine is cranking **[Figure 31-7]**.

CAUTION!!!

Catalytic converters may receive an overrich air-fuel mixture during prolonged cranking. Electronic ignition systems also may be damaged by incorrect starting system test procedures. Refer to the manufacturer's service manual for specific starting system troubleshooting procedures.

Starter Spins, Does not Crank Engine

This condition can be caused by a faulty starter drive mechanism or a gear mesh problem. Some manufacturers' service manuals recommend an on-car check for testing starter drive mechanisms. Other manufacturers require that the starter be removed from the vehicle for testing. Refer to specific service manuals for proper testing procedures.

However, if the starter makes an electric motor sound, rather than a gear-grinding sound, the starter drive probably is at fault.

31.4 STARTER SOLENOID AND RELAY TESTING

A typical solenoid wiring diagram and connections are shown in **Figure 31-8**. Most starter solenoids and relays have two heavy duty terminals for current in and current out connections. They also have at least one smaller terminal. This is the "S" terminal and is used to control or turn on the solenoid or relay.

320 PART IV: AUTOMOBILE ELECTRICAL AND ELECTRONIC SYSTEMS

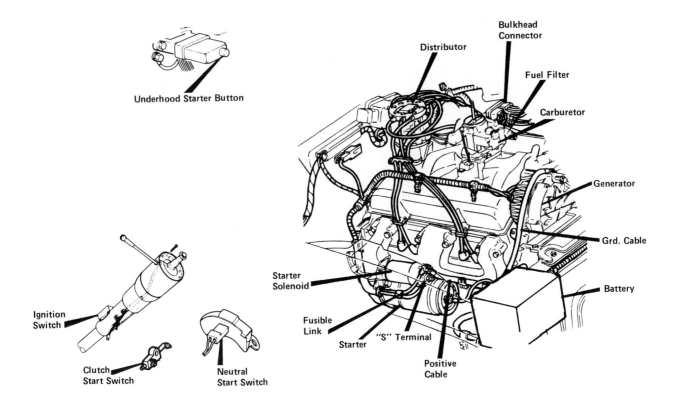

Figure 31-5. A typical locator chart that indicates a no-cranking situation. GENERAL MOTORS CORPORATION

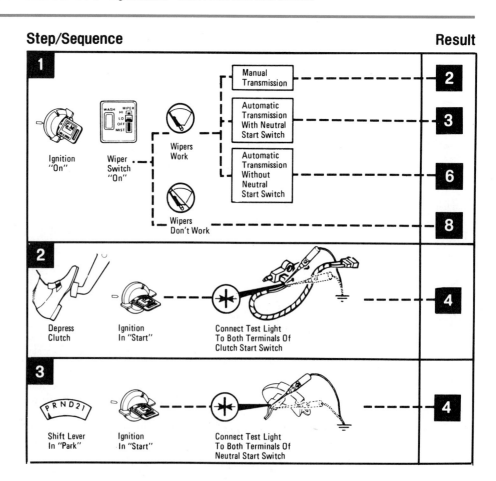

Figure 31-6. Troubleshooting a no-cranking situation.
GENERAL MOTORS CORPORATION

STARTING SYSTEM SERVICE 321

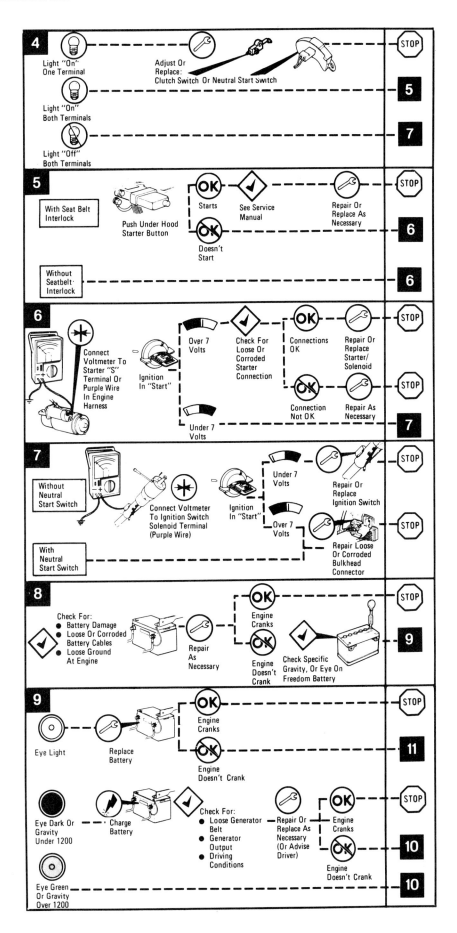

Figure 31-6. Continued.

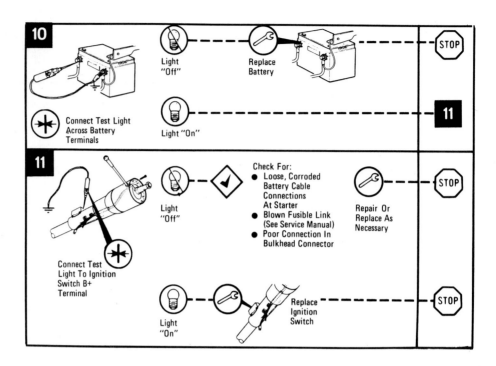

Figure 31-6. Concluded.

Tests

1. Positive lead to battery "+" terminal;
   Negative lead to solenoid "Starter" terminal.
   Reading should be no more than 0.5 volts.

2. Positive lead to solenoid "Bat" terminal;
   Negative lead to solenoid "Starter" terminal.
   Reading should be no more than 0.3 volts.

3. Positive lead to battery "+" terminal;
   Negative lead to solenoid "Bat" terminal.
   Reading should be no more than 0.2 volts.

4. Negative lead to battery "−" terminal.
   Positive lead to engine ground.
   Reading should be no more than 0.1 volt.

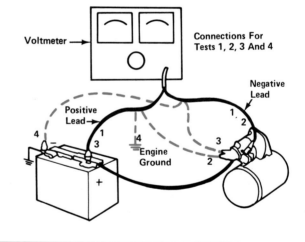

Figure 31-7. Cranking circuit test.

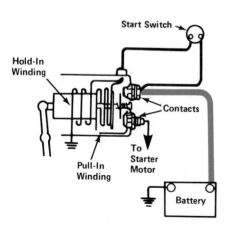
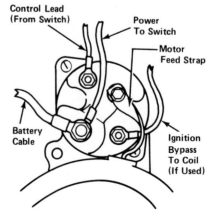

Figure 31-8. Basic solenoid circuit and connections.

Starting System Service

The majority of solenoid/relay problems have to do with the electrical conductivity of the internal copper disc and associated contacts. In most problem cases, the driver will hear a clicking sound but the engine will not crank when the driver tries to start the automobile. The clicking sound indicates the solenoid/relay is operating. If it is not clicking, connect a jumper wire from the battery positive terminal to the "S" terminal on the relay. If the unit still does not click it must be replaced. If it clicks but does not crank the engine, or cranks it very slowly, it must be given a voltage drop test.

To conduct this test, connect a voltmeter across the two large terminals and attempt to crank the engine. When the voltmeter is first connected, it will read battery voltage. The voltage drop reading will be displayed when the engine is cranking. A reading higher than 0.3 volts indicates excessive internal resistance and the unit must be replaced.

Refer to the manufacturer's service manual for specific solenoid testing procedures and test result interpretation.

31.5 STARTER REMOVAL

The greatest problem in removing starters is gaining access to the starter and mounting fasteners, and disconnecting attached wires and cables. On some vehicles, the starter can be removed only from the bottom of the engine compartment [Figure 31-9]. In some cases, suspension components, exhaust components, and other parts must be removed first.

SAFETY CAUTION

Remove the grounded battery terminal before any starter or cable removal procedure. Place the cable in a position where it cannot accidentally touch the negative battery terminal.

SAFETY CAUTION

Make sure the exhaust system is cool before you attempt to remove the starter. Hot exhaust system parts can cause severe burns.

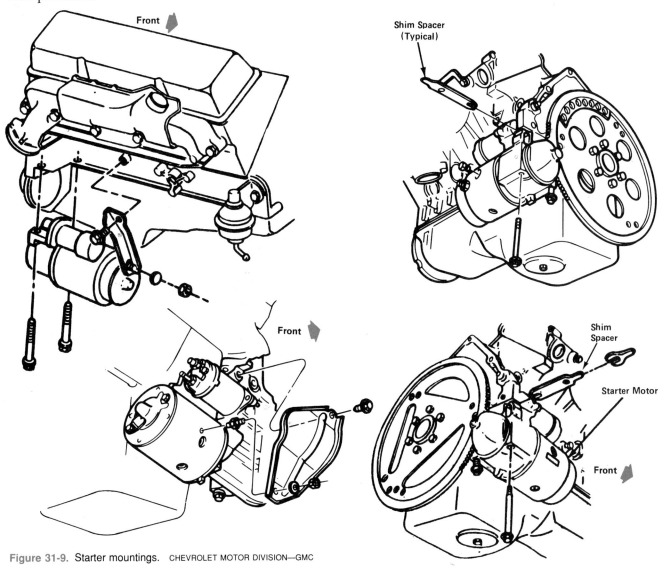

Figure 31-9. Starter mountings. CHEVROLET MOTOR DIVISION—GMC

31.6 STARTER INSPECTION

If a starter engagement problem is noted, inspect the starter pinion gear and flywheel ring gear teeth **[Figure 31-10]**.

A damaged starter pinion gear can be replaced. Major damage to flywheel gear teeth requires that the flywheel ring gear, or flexplate, be replaced.

Simple Checks

The starter pinion gear should turn freely in one direction only. The armature should turn freely when the pinion gear is pryed with a screwdriver in the opposite direction. **NOTE:** You may not be able to do this on gear reduction starters. Tight bearings, a bent shaft, or a loose pole shoe will bind the armature. If the armature does turn freely, a no-load rpm test should be performed.

No-Load RPM Test

A *no-load rpm test* checks the free rotational speed of the armature with a special tachometer **[Figure 31-11]**.

The variable resistance shown in the figure adjusts battery voltage to the same value as during actual cranking.

Many problems can be diagnosed using a no-load rpm test. Such problems include:

- Excess friction from worn bushings
- Shorted armature
- Grounded armature
- Direct ground in terminal or field coils
- Open field circuit
- Open armature coils

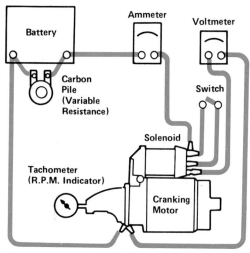

Figure 31-11. No-load rpm test hookup. DELCO REMY—GMC

- Poor connections between the brushes and commutator
- High internal resistance
- Shorted field coils.

Refer to the manufacturer's service manual for specific tests and interpretation of results.

31.7 STARTER REPAIR

Some service facilities simply exchange a faulty starter for a new or rebuilt unit. Other shops rebuild starters. The major steps in rebuilding a starter include:

1. Disassembly
2. Replacing starter drive
3. Replacing bushings or bearings
4. Testing armature and field coils
5. Resurfacing commutator
6. Checking brush spring tension
7. Replacing brushes
8. Checking and/or replacing solenoid
9. Reassembly
10. Testing
11. Replacement.

Before a starter is disassembled, punch marks to aid in reassembly must be made. Punch marks are made at each end of the starter end housings and frame **[Figure 31-12]**. These marks identify the proper locations and orientation of each end housing in relation to the frame. Starter end housings and frames must be reassembled in the same positions as found during disassembly.

Some special tools are required for complete starter testing. For example, an electrical tester

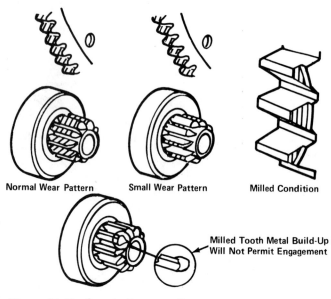

Figure 31-10. Gear tooth wear patterns.

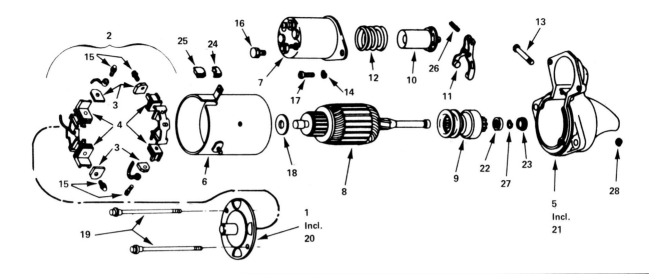

1. Frame—Commutator End
2. Brush And Holder Pkg.
3. Brush
4. Brush Holder
5. Housing—Drive End
6. Frame And Field Asm.
7. Solenoid Switch
8. Armature
9. Drive Asm.
10. Plunger
11. Shift Lever
12. Plunger Return Springer
13. Shift Lever Shaft
14. Lock Washer
15. Screw—Brush Attaching
16. Screw—Field Lead To Switch
17. Screw—Switch Attaching
18. Leather Washer—Brake
19. Thru Bolt
20. Bushing—Commutator End
21. Bushing—Drive End
22. Pinion Stop Collar
23. Thrust Collar
24. Grommet
25. Grommet
26. Plunger Pin
27. Pinion Stop Retainer Ring
28. Lever Shaft Retaining Ring

Figure 31-12. Parts of a starting motor. CHEVROLET MOTOR DIVISION—GMC

called a *growler* is used to detect shorts in armature windings. A growler produces strong magnetic lines of force. This strong magnetic field can induce current flow and magnetism in a conductor. The growler "hums" with 60-cycle AC current and produces a "growling" noise.

A hacksaw blade is placed parallel to the armature, and the armature is rotated. If the armature is shorted, the hacksaw blade will vibrate, indicating the location of the short [Figure 31-13].

Most growlers include a powered 110-volt AC continuity test light. When connected to a conductor, the light will shine. An armature can be checked for grounds by placing the leads on a commutator segment and the laminated core segments [Figure 31-14].

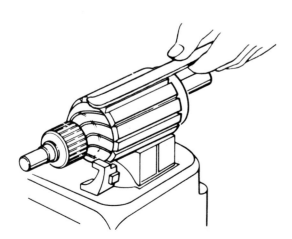

Figure 31-13. Using a growler to check an armature for short circuits. DELCO REMY—GMC

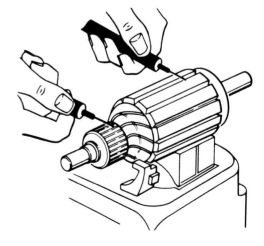

Figure 31-14. Testing an armature for grounds. DELCO REMY—GMC

SAFETY CAUTION

The powered 110-volt continuity tester leads will produce a dangerous shock if touched. Make sure your hands are dry and clean, and hold the test probes only by the insulated handles during continuity checking.

Commutators can become out-of-round and worn as the brushes ride against them. The commutator can be cut in a lathe, as illustrated in **Figure 31-15**, to restore its concentricity.

Brush wires may be soldered to coil leads or screwed to ground or positive terminals **[Figure 31-16]**. Ford starters also may include a set of *contact points*, also shown in **Figure 31-16**. Refer to the manufacturer's service manual for special testing procedures for such starters.

To replace soldered brushes, the old leads must be cut. New brush leads must be soldered to the coil leads and insulated with heat-shrink tubing or *spaghetti tubing*. Spaghetti tubing is a hollow, straw-shaped flexible insulating material that is slipped over wires before solder connections are made. After soldering, the spaghetti tubing is positioned to cover the soldered joint.

Reassembly of the end housings and frame must be performed according to previously made punch marks.

The starter must be tested, as discussed above, before installation. When tests indicate that the starter is working properly, it is reinstalled in the vehicle. After installation, the last steps are to reconnect the battery ground terminal and check starter operation in the vehicle.

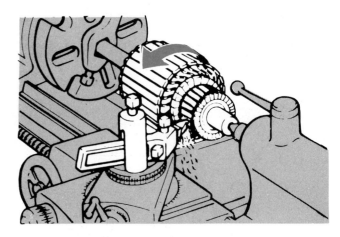

Figure 31-15. Turning an armature on a lathe.

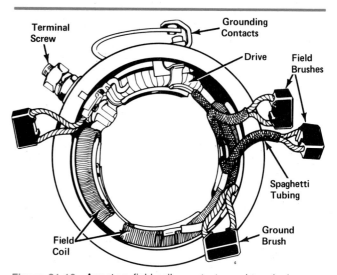

Figure 31-16. Armature field coils, contacts, and terminal assembly.

STARTING SYSTEM SERVICE

UNIT HIGHLIGHTS

- The battery must be fully charged and in good condition before on-car starter tests are made.
- Starter preventive maintenance consists of checking and cleaning connections and terminals.
- Cranking problems include no cranking, slow cranking without starting, starter spinning without cranking, and excessively noisy cranking.
- On-car and off-car tests may be necessary to diagnose some starting problems.
- Faulty solenoids or starter relays are replaced.
- Faulty starters can be repaired, or an exchange starter can be installed.

TERMS

voltage drop test
growler
contact point
no-load rpm test
spaghetti tubing
heat-shrink tubing

REVIEW QUESTIONS

DIRECTIONS: The following questions are similar to those used on automotive technician certification tests. On a separate sheet of paper, write the letter of the correct choice.

1. Which of the following statements is correct?
   I. The battery must be fully charged and in good condition before on-car starter tests are made.
   II. The grounded battery terminal must be disconnected before attempting starter removal.
   A. I only
   B. II only
   C. Both I and II
   D. Neither I nor II

2. After replacing faulty battery cables with new No. 8 gauge cables, a vehicle fails to start in cold weather. What is the most likely cause?
   A. Insufficiently charged or low-capacity battery
   B. Battery damaged or shorted during cable replacement
   C. Defective starter solenoid, relay, or starter
   D. Wrong size cables installed

3. Technician A says that a rotating starter motor that does not crank an engine can have a defective drive mechanism.
   Technician B says that the problem may be in the starter pinion and flywheel gear mesh.
   Who is correct?
   A. A only
   B. B only
   C. Both A and B
   D. Neither A nor B

4. A voltage drop test checks for
   A. insufficient resistance in a conductor.
   B. excessive resistance in a conductor.
   C. correct battery voltage.
   D. correct cranking voltage.

Question 5 is not like those above. It has the word EXCEPT. For question 5, look for the choice that does *not* apply. Read the entire question carefully before you choose your answer.

5. Common starter problems can include all of the following EXCEPT
   A. no cranking.
   B. engine cranks slowly, does not start.
   C. engine cranks quickly, does not start.
   D. excessive noise during cranking.

ESSAY QUESTIONS

1. Describe starter preventive maintenance.
2. What are the four categories and symptoms of cranking problems?
3. Describe the steps in off-the-car starter repair.

SUPPLEMENTAL ACTIVITIES

1. Safely clean and replace battery and starter cables and connections.
2. Demonstrate to your class how to conduct a voltage drop test on both battery cables.
3. Demonstrate to your class how to conduct a voltage drop test across the solenoid or starter relay.
4. Remove a starter from a vehicle. Refer to the manufacturer's service manual and perform a no-load rpm test on the starter. Report test findings to your class.
5. Refer to the manufacturer's service manual and disassemble the starter removed in Supplemental Activity 4, above. Perform inspection and tests on the armature and field coils. Report test results to your class. Reassemble and reinstall the starter.

32 The Charging System

UNIT PREVIEW

The charging system provides for the electrical needs of the vehicle when the engine is running. The battery provides additional current as needed.

A charging system includes several parts to produce, control, and monitor electrical current.

Alternating electrical current can be produced by magnetic induction and is changed to DC through the use of diodes.

The output of the alternator is controlled by a switching device, which can be controlled either electromechanically or electronically.

Warning lights, ammeters, or voltmeters can be used to monitor charging system operation.

LEARNING OBJECTIVES

When you have completed your assignments and exercises in this unit, you should be able to:

❏ Identify and describe the parts of a charging system.
❏ Explain how AC current is produced and changed to DC current in an alternator.
❏ Identify and describe the parts of an alternator.
❏ Describe the basic operation of a voltage regulator.
❏ Identify and describe the types, functions, and basic circuits used for charging indicators.

32.1 CHARGING SYSTEM FUNCTION

The charging system supplies the vehicle's electrical needs when the engine is operating. In addition, the alternator supplies electrical current to recharge the battery. The battery supplies electrical current beyond the capacity of the charging system.

In addition to the battery, a typical charging system, illustrated in **Figure 32-1**, includes the following components:

- Alternator
- Voltage regulator
- Charging indicator
- Wiring connections.

32.2 ALTERNATOR

An *alternator* produces alternating electrical current (AC) when driven by a belt turned by engine power. The AC is *rectified*, or changed, to direct current (DC) for use in vehicle systems and for recharging the battery. Aluminum is used for the alternator case because it is nonmagnetic, light in weight, and dissipates heat well **[Figure 32-2]**. Most alternators use a fan to pull cooling air through the alternator.

Generating Current

A simplified alternator consists of a rotating magnet within a coil **[Figure 32-3]**. As the magnet is rotated, its lines of magnetic force pass through the coil, inducing electrical current flow. The rotating magnet is called a *rotor*. The stationary coil in which current is produced is called a *stator*.

When the rotor is midway through its revolution (perpendicular to the coil), the induced current drops to zero. However, as the magnet continues to rotate, the opposite pole approaches the coil. This produces electrical flow in the opposite direction. A graph, or picture, of such a current flow shows what is called a *sine wave* **[Figure 32-4]**.

In an alternator, the rotor consists of a *field coil* with many turns of wire. Magnetism is produced by connecting the ends of the coil to battery voltage. Smooth rings, called *slip rings*, and brushes are used to form the moving connection.

Coils of wire with many turns also form the stator, shown in **Figure 32-5**.

Electrical Output

The electrical output of the alternator must be great enough even at low speeds to satisfy the electrical requirements of the vehicle. Also, the electrical output must be smooth, not pulsing.

Too low an output would result in an undercharged battery and poor starting and electrical component performance. Irregular output could damage sensitive electronic components of the vehicle.

Increasing electrical output. An alternator can produce more electrical output if the strength of the rotor's magnetic field is increased. Magnetic field strength is increased by raising the current flow passing through the rotor field coil.

The Charging System

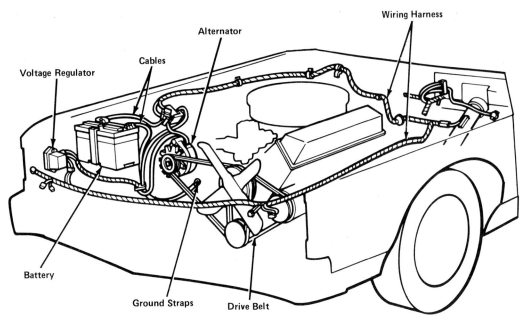

Figure 32-1. Charging system and wiring harness. CHRYSLER CORPORATION

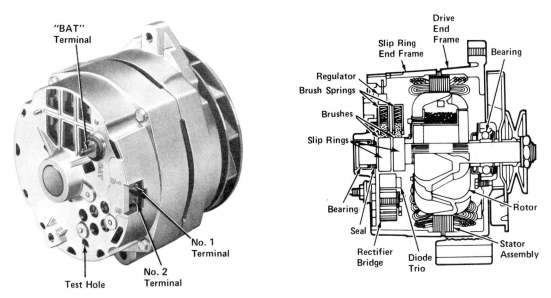

Figure 32-2. Alternator assembly. CHEVROLET MOTOR DIVISION—GMC

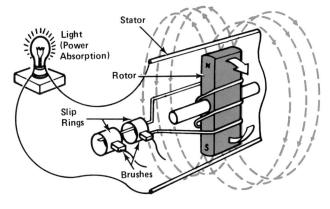

Figure 32-3. Simplified alternator.

Increasing alternator efficiency. To increase alternator efficiency, the rotor has several "fingers," or *poles*, that form alternating north and south magnetic poles. As the rotor turns, these separate poles form separate continuous lines of magnetic force. The more poles, the more even is the voltage output.

Three-Phase Current

To further increase alternator output, the stator consists of more than one coil. Thus, each time the rotor's magnetic force lines pass through a separate coil, electricity is produced. An automotive stator

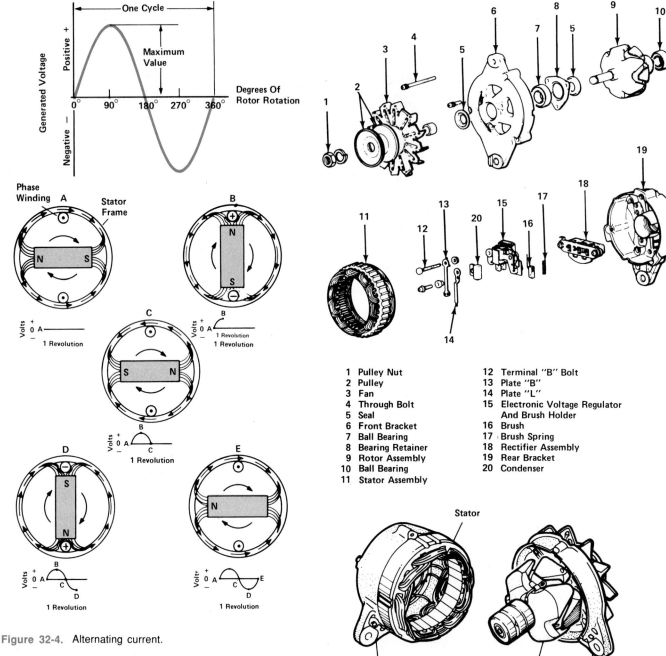

Figure 32-4. Alternating current.

has three separate windings, each consisting of many coils in series. The three windings can be connected in a "Y" or a "Delta" configuration **[Figure 32-6]**.

The three windings overlap within the laminated core sections of the stator **[Figure 32-7]**.

Voltage is induced in each of three overlapping coils at slightly different times by the rotor poles **[Figure 32-8]**.

A graph of the overlapping sine waves produced by this arrangement is shown in **Figure 32-9**. Each sine wave is at a slightly different point, or *phase*, of its cycle at a given time. Thus, such alternating current produced by the three coils is known as *three-phase AC current*.

1 Pulley Nut
2 Pulley
3 Fan
4 Through Bolt
5 Seal
6 Front Bracket
7 Ball Bearing
8 Bearing Retainer
9 Rotor Assembly
10 Ball Bearing
11 Stator Assembly
12 Terminal "B" Bolt
13 Plate "B"
14 Plate "L"
15 Electronic Voltage Regulator And Brush Holder
16 Brush
17 Brush Spring
18 Rectifier Assembly
19 Rear Bracket
20 Condenser

Figure 32-5. Alternator assembly. CHRYSLER CORPORATION

Rectifying AC Current

AC must be rectified to DC to charge the battery and to operate vehicular electrical systems. The shape of AC and DC current graphs indicate that DC flows only in one direction **[Figure 32-10]**.

Diodes are used to change alternator output from AC current to DC current. A *diode,* as discussed in

The Charging System

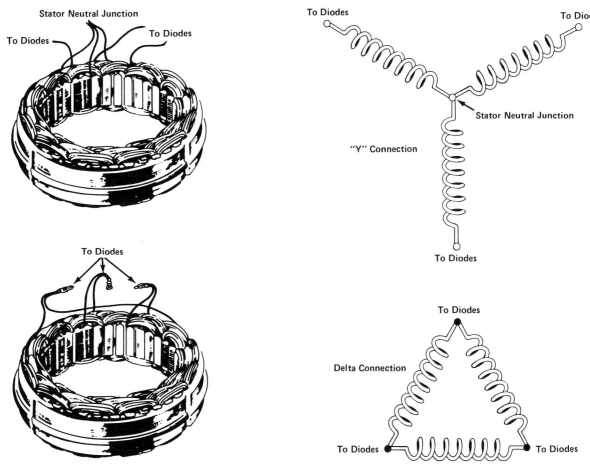

Figure 32-6. Stator coil connections.

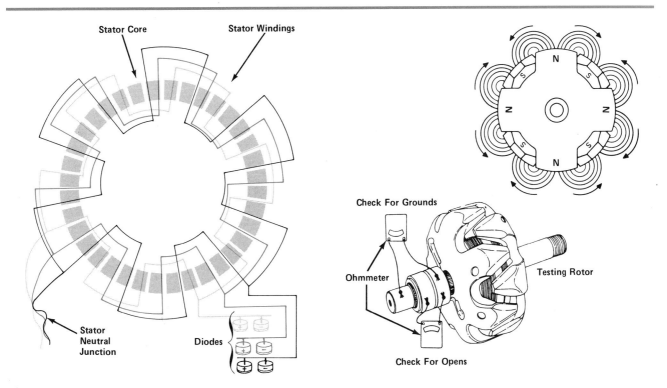

Figure 32-7. Stator windings and diodes.

Figure 32-8. Rotor poles and coils. FORD MOTOR COMPANY

332　PART IV: AUTOMOBILE ELECTRICAL AND ELECTRONIC SYSTEMS

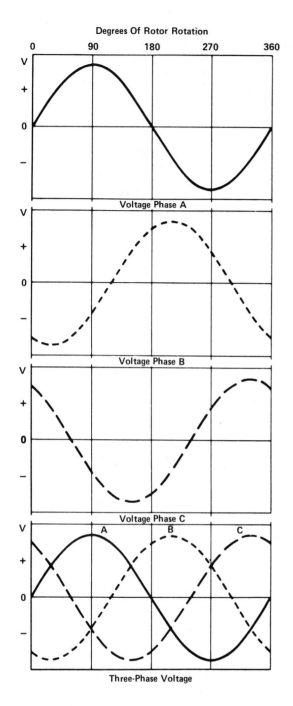

Figure 32-9. Electrical current sine waves.

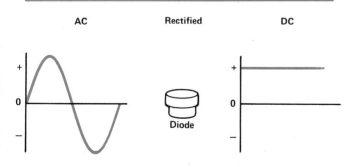

Figure 32-10. AC and DC waveforms. CHRYSLER CORPORATION

Unit 27, allows current to pass in one direction only [Figure 32-11].

This principle can be used to prevent current from flowing in the opposite direction. In a multi-phase alternator, a *bridge rectifier circuit* is used to provide reasonably constant DC voltage [Figure 32-12].

An alternator bridge rectifier circuit has two diodes for each phase winding. Thus, six diodes are used to allow current flow in only one direction.

As can be seen from the illustrations, although rectified alternator output is DC, it does have an

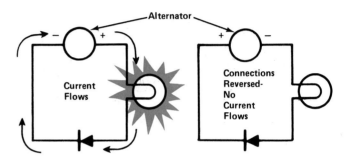

Figure 32-11. Diode current flow.

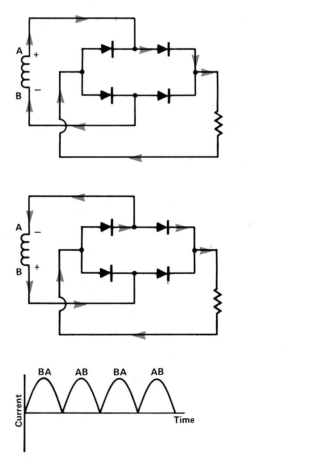

Figure 32-12. Bridge rectifier circuit.

insignificant, slight pulsation. This "ripple" can be further "smoothed" by using a *capacitor*. A capacitor is an electronic device that can hold an electrical charge, then release it. A capacitor can thus act as a "surge tank" to smooth out electrical pulses. In practice, the slight pulsations that remain after current passes through the capacitor do not interfere with battery charging.

During rectification of AC to DC, the diodes produce heat. Thus, rectifier bridges are mounted on finned aluminum *heat sinks* to dissipate heat **[Figure 32-13]**.

32.3 ALTERNATOR REGULATION

If the voltage applied to the rotor coil was unregulated, the alternator output would be uncontrolled. This could result in excessive battery voltage, battery failure, and other electrical system problems. To regulate alternator output, a ***voltage regulator*** is used. A voltage regulator is a device that monitors electrical system voltage levels. It switches current on and off to the rotor field to regulate current output.

Voltage regulators and alternators include devices to prevent battery voltage from draining back to ground and discharging the battery when the system is not operating. In most modern systems, diodes provide for this anti-drain-back feature.

Alternator regulators also include temperature-sensitive devices. These devices provide higher charging rates during cold weather and lower charging rates during hot weather.

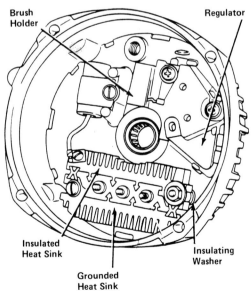

Figure 32-13. Alternator (rectifier) heat sink.

CHEVROLET MOTOR DIVISION—GMC

Figure 32-14 shows the basic types of regulators in use on modern cars. These regulators can be divided into three basic categories:

- Separate electromechanical
- Separate electronic
- Integral electronic.

Separate Electromechanical Regulator

Electromechanical regulators use *relays* **[Figure 32-15]**. A relay is a specialized form of switch. These relays use electromagnetic force created by a small current flow. This closes contacts through which a larger current flows.

A relay consists of a coil, a flat spring, a movable plate, and a set of electrical contacts. Spring tension

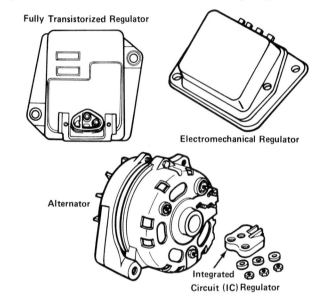

Figure 32-14. Types of voltage regulators.

CHRYSLER CORPORATION

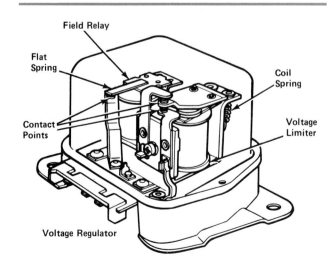

Figure 32-15. Electromechanical voltage regulator.

CHRYSLER CORPORATION

can hold the electrical contacts open or closed. When sufficient current is applied to the coil, a magnetic field moves the plate against the spring. This opens or closes electrical contacts, depending on the design of the relay.

Voltage limiter relay. A small amount of current flows through a set of contacts to provide an initial magnetic field when the key is switched to the ON position. This action provides a small magnetic field to "prime" the rotor coil for current generation. Voltage also flows through an indicator lamp on the dash to warn the driver that the charging system is not operating. When the vehicle starts, the indicator lamp should go out.

The voltage limiter relay is connected to the field circuit of the charging system. During operation, when system voltage decreases below acceptable levels, the voltage limiter acts to supply additional current to the rotor field, boosting alternator output.

As system voltage rises, the voltage limiter, sensing this rise, acts to reduce the amount of current flowing to the rotor field. The cycle of connecting and disconnecting battery current to the rotor field may occur as often as 100 times per second.

Field relay. When the engine is not running, a spring holds the rotor field relay contacts open. As the engine begins to start, voltage is applied to the field relay. The contacts close, allowing voltage to travel from the voltage relay to the rotor field coil. When the engine stops, a spring pulls the field relay contacts open and prevents battery discharge through the rotor coil.

Separate Electronic Regulator

As discussed in Unit 27, transistors can be used to control current flows. Small amounts of current across the emitter-base connections will cause the transistor to conduct electricity. Larger amounts of current can be conducted across the emitter-collector connections **[Figure 32-16]**.

The small amount of current used to make the transistor conduct is known as *bias current.* Transistors can thus be *forward-biased* (conducting across emitter and collector), when bias current is in the correct direction.

Transistors can also be *reverse-biased* (nonconducting), when bias current is in the opposite direction.

Thus, transistors can be used to perform the functions of both the field and voltage regulator relays without any moving parts. However, transistors operate much more rapidly than relays. They are not subject to mechanical problems and the burning, pitting, and sticking common with contact points in electromechanical relays.

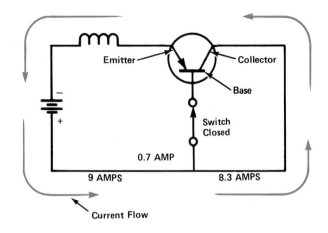

Figure 32-16. Basic transistor circuit.

By using transistors, the rotor field coil can be energized, on and off, up to several hundred times a second. An electronic voltage regulator can be mounted separately from the alternator.

Some older voltage regulators combine an electromechanical field relay with a transistorized voltage regulator. However, most voltage regulators in use today are fully transistorized **[Figure 32-17]**.

Integral Electronic Regulator

An *integrated circuit (IC)* can contain hundreds or thousands of transistors, resistors, capacitors, and other circuit parts. IC "chips" can be extremely small **[Figure 32-18]**. All the functions of the regulator can be performed by an extremely small integrated circuit.

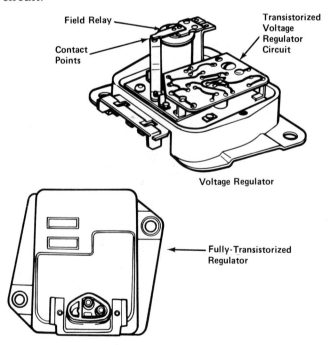

Figure 32-17. Transistorized voltage regulator.
CHRYSLER CORPORATION

The Charging System

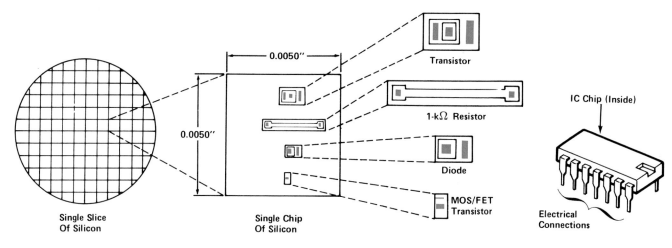

Figure 32-18. Integrated circuit.

In most vehicles built within the last few years, an IC regulator is built into the alternator housing **[Figure 32-19]**.

Some alternators use a ***diode trio.*** A diode trio consists of three diodes connected in parallel. Diode trios supply the rotor field coil with rectified DC current from the stator. In addition, diode trios are used to apply bias current to regulator transistors.

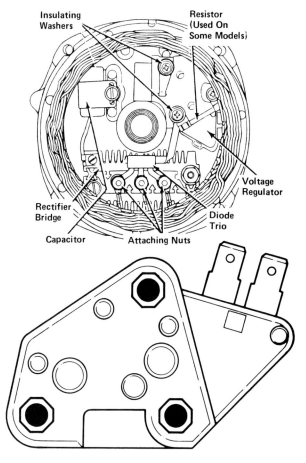

Figure 32-19. Integrated circuit regulator. DELCO REMY—GMC

32.4 CHARGING INDICATOR

Three types of charging indicators currently are used:

1. Indicator light
2. Ammeter
3. Voltmeter.

Indicator Light

An indicator light for a charging system is shown in **Figure 32-20**. When the charging system is operating properly, battery and alternator output voltage should be equal. Electrical flow occurs only from a source with a surplus of electrons to a source with a lack of electrons. Thus, when equal electrical potential is present at both sides of the bulb, current does not flow.

When alternator output falls below battery voltage, current flows from the battery rather than toward it. The indicator light bulb lights, warning the driver that the charging system is not operating properly.

Ammeter

An ammeter can be connected in series in the alternator output circuit. The ammeter usually is calibrated to 60 amperes, plus and minus. The ammeter circuit is designed so that the current used

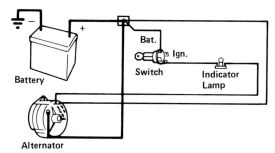

Figure 32-20. Indicator lamp circuit.

by the starter does not pass through the ammeter. Otherwise, starter current (75 to 300 amps) would damage the ammeter.

The ammeter indicates the current flowing into (+) or out of (-) the battery. Thus, after starting, the ammeter should indicate a relatively high charging rate to replace the energy used by the starter.

As the battery becomes recharged, the charging rate should taper off. If current is flowing out of the battery, the ammeter will show a discharge. A discharge is normal if, for example, the headlights are turned on while the engine is not running.

Normal alternator charging system operation is indicated by a slight charging rate when the engine is running above specified idle speed.

Voltmeter

Many current vehicles use a voltmeter to measure charging system operation. After starting and while running, battery voltage should remain between 13.2 and 15.2 volts. If it drops below 13.2 volts at normal road speeds, the battery can become discharged. If the voltage climbs above 15.2 volts, damage to the battery and electrical and electronic parts can result.

A voltmeter is connected between the battery positive and negative terminals **[Figure 32-21]**.

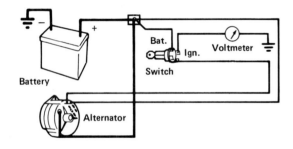

Figure 32-21. Voltmeter circuit.

UNIT HIGHLIGHTS

- The charging system provides for the electrical needs of the vehicle when the engine is running. The battery provides additional current as needed.
- A charging system includes a battery, alternator, voltage regulator, wiring connections, and charging indicator.
- DC passing through the rotor creates a moving magnetic field. AC is produced in the stator coils by this magnetic field.
- AC produced in the stator coils is rectified by a rectifier bridge that consists of six diodes.
- A voltage regulator turns rotor field current on and off to control alternator output.
- Mechanical relays or transistors can be used to switch current on and off in a voltage regulator.
- Lights, ammeters, or voltmeters can be used to monitor charging system operation.

TERMS

alternator	rotor
stator	sine wave
field coil	slip ring
phase	three-phase AC current
diode	rectify
bridge rectifier circuit	capacitor
heat sink	voltage regulator
relay	bias current
forward-biased	reverse-biased
integrated circuit (IC)	diode trio
pole	phase

REVIEW QUESTIONS

DIRECTIONS: The following questions are similar to those used on automotive technician certification tests. On a separate sheet of paper, write the letter of the correct choice.

1. Technician A says an alternator produces AC current in the rotor coil.

   Technician B says DC current is induced in the stator coils.

   Who is correct?

   A. A only
   B. B only
   C. Both A and B
   D. Neither A nor B

2. Which of the following controls alternator output?

   A. Rectifier bridge
   B. Diode trio
   C. Voltage regulator
   D. Brushes

THE CHARGING SYSTEM

3. Which of the following statements is correct?
   I. Most current vehicles use electromagnetic voltage regulators.
   II. Most current vehicles use transistorized IC voltage regulators.
   A. I only
   B. II only
   C. Both I and II
   D. Neither I nor II

Questions 4 and 5 are not like those above. Each has the word EXCEPT. For each question, look for the choice that does *not* apply. Read each question carefully before you choose your answer.

4. All of the following statements about charging systems are correct EXCEPT
   A. Batteries can be charged only by direct current (DC).
   B. Alternators produce pulsating DC.
   C. Rectifier bridges change AC into pulsating DC.
   D. Voltage regulators regulate alternator output by periodically energizing the stator field coil.

5. All of the following can be used to monitor charging system operation EXCEPT
   A. a voltmeter.
   B. an indicator light.
   C. an ohmmeter.
   D. an ammeter.

ESSAY QUESTIONS

1. Describe the parts of the charging system and their functions.
2. Describe how an alternator produces current to charge the battery.
3. Describe the functions of a voltage regulator relay and a field relay.

SUPPLEMENTAL ACTIVITIES

1. Examine a vehicle's charging system and identify all visible parts. Does the vehicle have a separate or integrated voltage regulator? Does the alternator have a fan? What type of circuit indicator does the system have? Report to your class the location of charging system parts.
2. Describe how the rotor of an alternator is similar to the field coils of an electrical motor, and how they are different.
3. Identify and describe the basic internal parts of an alternator.
4. Explain the purpose of a rectifier bridge.
5. Describe how a field relay and a voltage regulator relay operate to control and protect the charging system.
6. Draw basic charging system indicator circuits, and explain how such indicators operate.

33 Charging System Service

UNIT PREVIEW

Most charging systems are designed to give years of trouble-free operation. Preventive maintenance for the charging system is quite minimal.

However, because it is more electrical than mechanical, many technicians are reluctant to service charging systems. There are no mysterious components or magical procedures associated with charging system service.

Undercharging or overcharging may be caused by something as simple as a loose drive belt or a defective regulator. On the other hand, a problem may involve extensive bench testing and repair. However, simple and accurate step-by-step troubleshooting and diagnosis, followed by careful servicing, will result in quality repair work.

LEARNING OBJECTIVES

When you have completed your assignments and exercises in this unit, you should be able to:

- ❏ Identify and describe causes of undercharging and overcharging.
- ❏ Perform charging system preventive maintenance.
- ❏ Perform a charging voltage/alternator output test.
- ❏ Perform a circuit-resistance test on a charging system.
- ❏ Safely and correctly remove, disassemble, reassemble, and replace an alternator.

SAFETY PRECAUTIONS

When you work on a running engine, roll up long sleeves and remove watches, rings, neck chains, and all other jewelry. Wrap or tie long hair in a ponytail and pin it to the top of your head under a cap. Be especially careful to keep fingers, hands, and arms away from moving parts.

Before attempting to remove an alternator or voltage regulator, disconnect the grounded battery terminal. Place the cable in a position where it cannot accidentally contact the battery terminal.

CAUTION!!!

Do *not* disconnect the battery terminals from a vehicle with an alternator when the engine is running. Without the battery to act as a load, the alternator can produce up to 100 volts. This excessive voltage will damage all electrical parts and destroy all electronic devices on the vehicle.

After alternator or voltage regulator service or replacement procedures, double-check all wiring connections. Reconnect the grounded battery cable only after you are sure all wiring connections are correct.

33.1 CHARGING SYSTEM PREVENTIVE MAINTENANCE

Routine charging system preventive maintenance includes the following procedures:

- Checking charging system indicators and battery charge
- Checking, adjusting, and/or replacing alternator drive belts
- Checking, inspecting, and/or cleaning wiring connections and battery cables
- Visually inspecting the alternator and voltage regulator
- Checking for alternator bearing noise.

Undercharging and Overcharging

Warning lights, ammeters, or voltmeters can indicate charging system problems. Charging or a normal voltage reading (approximately 13-15 volts) should be indicated when the engine is running. However, problems in indicator circuits can be misleading. A burnt-out warning light bulb or non-functioning meter will not indicate a discharge condition. Bulbs and/or meters are replaced if defective (see Unit 37). Indicator circuit wiring connections should be checked as described below.

If indicators are functioning properly and indicating discharge or insufficient charging with the engine idling, engine idle speed should be checked. Engine idle speed below the manufacturer's specifications can cause an indicator to show undercharging.

Before charging system components are suspected of failure, engine idle speed should be checked and/or adjusted.

If idle speed is correct, battery checks are made (see Unit 29). Overcharging is the result of a faulty voltage regulator. Undercharging can be caused by many things, including:

- Frequent starting and short-trip driving
- Faulty starters
- Loose alternator drive belts
- Loose or corroded connections
- Alternator problems
- Voltage regulator problems.

Frequent starting and short-trip driving may not allow the charging system to operate sufficiently to fully charge a battery, especially if the climate is cold.

Faulty starters may draw excessive amounts of current. This may show up as a discharged battery, especially during frequent short-trip driving. In such a case, a starter overhaul may solve the undercharged/discharged battery problem.

Loose alternator drive belts also can cause undercharging. A loose belt slips and does not transmit turning power properly. Belts should be inspected and replaced if worn or damaged **[Figure 33-1]**. However, properly tightening a belt that is in good condition may correct an undercharging condition.

Loose or corroded wiring connections at the alternator and/or regulator can cause charging problems. A loose alternator or regulator also can have the same effect. Some alternators which have special rubber mounts must have a grounding strap or cable properly connected.

CAUTION!!!

Do not overtighten a drive belt. Check for proper tension after tightening (see Unit 18). Overtightened belts will cause bearing failure in driven units.

Alternator problems include worn or dirty brushes; shorted, open, or grounded rotor or stator windings; shorted or open diodes; and bearing problems. These also can cause undercharging. Alternator service is discussed in 33.4.

Voltage regulator problems can cause undercharging. After long service, electromechanical regulator contact points can become burned, pitted, or "welded" together. In addition, spring tension can decrease

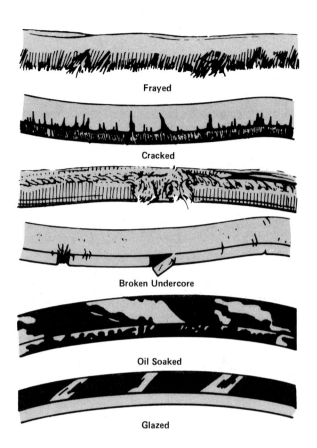

Figure 33-1. Drive belt damage.
FORD MOTOR COMPANY

with heat and time. Any of these problems can cause improper alternator output.

Some electromechanical regulators are adjustable to increase or decrease alternator output to specifications. Partially transistorized voltage regulators also may be adjustable. However, such service requires specialized tools and training.

In general, fully transistorized IC regulators are not adjustable. If faulty, the entire unit is replaced. Alternators with integral IC regulators may require disassembly for such replacement.

A voltage regulator also can cause overcharging. Overcharging symptoms include rapid loss of electrolyte, excessive corrosion near the battery, and frequent burned-out bulbs from excessive voltage. A faulty field relay can cause battery discharge when the engine is not running.

Alternator and Regulator Wiring Connection Checks

Alternator and regulator connections are checked in a manner similar to that for starter connections. However, caution must be taken not to damage small wires or plastic connectors. Connectors are gently shaken, twisted, and pulled. Corroded connections can be cleaned and coated with a light lubricant spray. Loose screws, nuts, or bolts can be tightened.

SAFETY CAUTION

Remove the battery ground cable before using metal tools on alternator or regulator connection fasteners.

Slightly cracked plastic connectors or frayed wiring can be wrapped with electrical tape. If badly damaged, plastic connectors and/or wiring must be replaced. The electrical integrity of the charging circuit may be checked by performing a circuit resistance voltage drop test. This test is discussed in Topic 33.2.

Visual Checks of Alternator and Voltage Regulator

Alternators and voltage regulators should be inspected for physical damage. Burn marks or traces of carbon can indicate shorting or grounding. Cracked housings can indicate physical abuse that can cause internal problems.

Wiring and/or insulating parts can be wrapped with electrical tape or replaced. However, damage that appears to come from inside the units requires unit disassembly or replacement.

Checking for Alternator Bearing Noise

A short length of hose or a technician's stethoscope can be used to check for bearing noise.

SAFETY CAUTION

Noise checks are made with the engine running. Use caution around the fan, belts, and other moving parts of the alternator.

Carefully place the end of the hose near the alternator pulley. The probe of a technician's stethoscope is placed on the alternator housing near the bearing. Excessive bearing noise requires that the alternator be disassembled and the bearing replaced. Refer to Topic 33.4.

Charging System Troubleshooting

Charging system problems may include a single test or a combination of tests. The alternator, regulator, battery, indicator, and wiring all may need to be checked to determine the cause of a problem. Troubleshooting charts can be helpful in tracking down a charging system problem. **Figure 33-2** shows a troubleshooting chart for Ford Motor Company vehicles.

33.2 CHARGING SYSTEM ELECTRICAL TESTS

Charging system electrical tests can include the following:

- Charging voltage/alternator output

BASIC TYPES OF TROUBLE	PROBABLE CAUSES	SERVICE PROCEDURES
Alternator noise Note: Water pump noise is sometimes confused with alternator noise. A sound detecting device, such as a stethoscope, will eliminate indecision in this respect.	Alternator drive belt (Squealing noise).	Adjust or replace belt, as required. (An application of belt dressing may eliminate noise caused by minor surface irregularities.)
	Alternator bearing (Squealing noise).	Replace bearing if found to be out-of-round, worn, or causing shaft scoring.
	Alternator diode (Whining noise).	Test alternator output. (A shorted diode causes a magnetic whine and a reduction in output.) Test diodes and replace, as required. Replacing rectifier assembly may be most feasible fix.
Indicator gauge fluctuates —or— Indicator light flickers.	Charging system wiring.	Tighten loose connections. Repair or replace wiring, as required.
	Regulator contacts	Oxidized or dirty regulator contacts. Replace regulator, if necessary.
	Brushes	Check for tightness and wear. Replace, if necessary.

Figure 33-2. Troubleshooting chart. FORD MOTOR COMPANY

CHARGING SYSTEM SERVICE

BASIC TYPES OF TROUBLE	PROBABLE CAUSES	SERVICE PROCEDURES
Indicator light stays on.	Broken, loose, or slipping drive belt.	Adjust or replace, as required.
	Battery cables, charging system wiring.	Clean battery cables and terminals. Tighten loose connections. Repair or replace as required.
Ammeter registers constant discharge. Battery will not hold charge.	Battery specific gravity.	If unsatisfactory, replace battery.
	Alternator output low, or no alternator output.	Perform the voltage output tests to determine if trouble is in the regulator, wiring harness or the alternator itself.
	Alternator drive belt.	Adjust or replace belt, as required.
	Battery cables, charging system wiring	Clean battery cables and terminals. Tighten loose connection. Repair or replace, as required.
Battery low in charge. Headlights dim at idle. Note: A history of recurring discharge of the battery, which cannot be explained, suggests the need for checking and testing the complete charging system.	Electrolyte (specific gravity)	Test each cell and evaluate condition: All readings even at 1.225 or above — Battery OK. All readings even, but less than 1.225 — Recharge and retest. High-low variation between cells less than 50 gravity points — Recharge and retest. High-low variation between cells exceeds 50 gravity points — Replace battery.
	Battery (capacity)	Test capacity and evaluate condition: Minimum voltage 9.6 for 12 volt battery or 4.8 for 6 volt battery. (Both values under specified test load conditions.) If capacity is under minimum specifications, perform 3 minute charge test. If below maximum (15.5 volts for 12 volt battery or 7.75 volts for 6 volt battery at 40 and 75 amps, charge rate, respectively) battery is OK — recharge. If above maximum, battery is sulfated. Slow charge at 1 amp./positive plate. Replace battery if it doesn't respond to slow charge.
Lights and fuses fail prematurely. Short battery life. Battery uses excessive water. Burning of distributor points. Burning of resistor wire. Coil damage. High charging rate.	Charging system wiring, including regulator ground wire.	Tighten loose connections. Repair or replace wiring, as required.
	Voltage limiter setting.	Perform the voltage output tests to verify the condition of the regulator.

Figure 33-2. Concluded.

- Oscilloscope trace
- Circuit resistance (voltage drop).

Before charging system electrical tests are made, obvious problems are corrected. Loose or damaged belts are tightened or replaced. Corroded battery connections are cleaned. Engine idle speed is adjusted. In addition, the battery is tested and recharged (see Unit 29).

Charging Voltage and Alternator Output Tests

To isolate the cause of undercharging, two tests are performed: charging voltage and alternator output.

Charging voltage test. A charging voltage test is used to indicate whether the charging system will boost battery voltage. With the engine off, connect a voltmeter across the battery terminals. If the voltage is above 11 volts, turn on the headlights until battery voltage drops to approximately 11 volts, then shut the headlights off. Start the engine and run it at a fast idle (1,500 to 2,000 rpm) for three minutes. The voltage should increase by approximately two volts over the original reading [Figure 33-3]. If the voltage increase is not approximately two volts, an alternator output test is made. To check the charging system under load, turn the air conditioning or heater blower on high and the high beam headlights. Again, operate the engine at fast idle (1,500 to 2,000 rpm) and note the battery voltage reading. This reading should be at least 0.5 volts above the original reading. If not, an alternator output test is made.

Alternator output test. Alternator output is the ability of an unregulated alternator to produce electricity. Different methods, depending on the charging system, are used to *full field* the alternator. Full-fielding means fully, or constantly, energizing the alternator field windings with full battery current to produce maximum output. Alternator current output then is checked. Be sure to check the manufacturer's instructions before proceeding with this test.

An *inductive ammeter* that clamps around a battery cable should be used. An inductive ammeter does not require disconnection of wires that can cause *voltage spikes*, or pulses. Voltage spikes can damage or destroy sensitive electronic equipment on the vehicle.

Two types of circuits are used: A and B circuits [Figure 33-4].

On Delco alternators, a small screwdriver is used to ground a test tab inside the rear of the alternator [Figure 33-5]. This grounding bypasses the internal regulator. At the same time, the voltmeter and ammeter readings are noted.

CAUTION!!!

Do not force the screwdriver farther than 1 in. (25.4 mm) into the test hole. Damage to alternator internal components, fuel injectors, computers, meters, and other display units can result. The tab is kept grounded less than 10 seconds, just long enough to complete the test.

If the battery voltage increases to between 14 and 16 volts, the problem is in the regulator. If

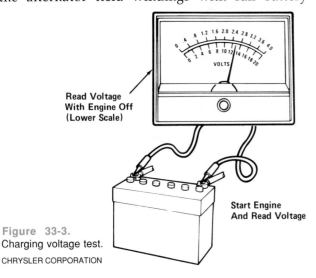

Figure 33-3. Charging voltage test.
CHRYSLER CORPORATION

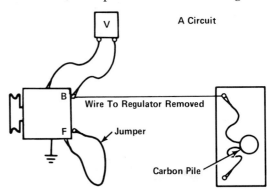

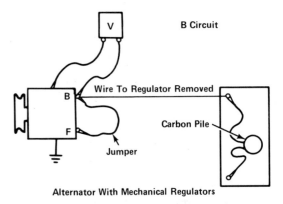

Figure 33-4. Connections for testing internal field ground B-circuit and external ground A-circuit alternators.

Figure 33-5. Generator test hole. BUICK MOTOR DIVISION–GMC

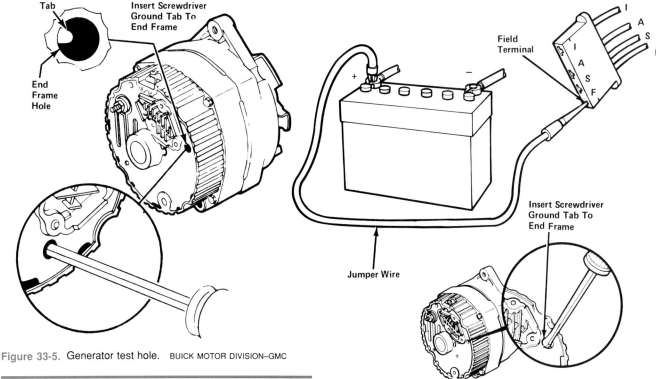

Figure 33-6. Alternator output test procedures. CHRYSLER CORPORATION

battery voltage does not increase to this level, the problem is in the alternator. Alternator current output should come to within five amps of the manufacturer's rating. Typical full alternator output varies according to alternator design. On most vehicles, it ranges from 55 to 80 amps. Recent vehicles may have 90, 100, or 110-amp alternators.

Full-fielding some charging systems requires momentarily connecting a jumper wire to a disconnected regulator plug. The other end of the jumper is connected to ground or to the battery positive terminal [**Figure 33-6**].

CAUTION!!!

Refer to the manufacturer's service manual to determine the correct test locations and connections. Improper connections can instantly damage or ruin internal alternator components.

A complete test sequence for General Motors Corporation vehicles is shown in **Figure 33-7**.

Oscilloscope Trace Test

An *oscilloscope* is a device that resembles a television set. Different electrical signals will produce characteristic *traces,* or patterns, on a screen.

Oscilloscope test leads usually are connected to ignition system primary terminals (see Unit 34). When the engine is running, the characteristic pattern of slightly pulsing DC current should be seen [**Figure 33-8**].

Any other pattern indicates problems that require the alternator be removed from the vehicle and repaired. Oscilloscope traces are the quickest way to determine alternator problems, although simpler instruments can be used.

Circuit Resistance (Voltage Drop) Test

This test is similar to that performed on the battery cables during starter problem diagnosis. A voltmeter is connected across the insulated charging system, one connection on the battery positive terminal and the other on the alternator battery terminal. A carbon pile is used to load the alternator to create a current output of 20 amps.

The insulated charging circuit should have less than 0.7 volt (ammeter indicator) or 0.3 volt drop (warning light indicator). **Figure 33-9** shows the correct connections for a voltage drop test.

33.3 FUSIBLE LINK SERVICE

A *fusible link* is a wire about four gauge sizes smaller in diameter than the wire used in the circuit it is to protect. Like a fuse, excessive current will cause a fusible link to burn out and open the circuit. Fusible links can be used at several locations on a vehicle to protect circuits, including the charging system circuit [**Figure 33-10**].

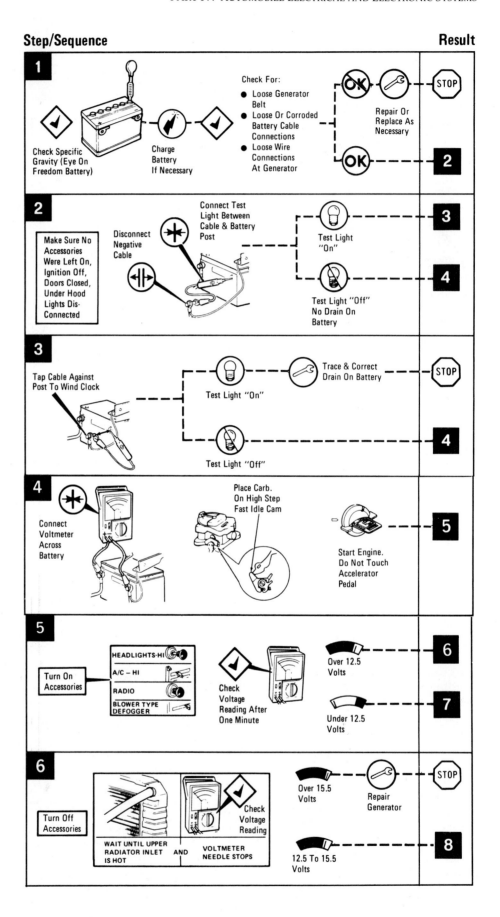

Figure 33-7. Battery charging problem diagnosis.

GENERAL MOTORS CORPORATION

CHARGING SYSTEM SERVICE 345

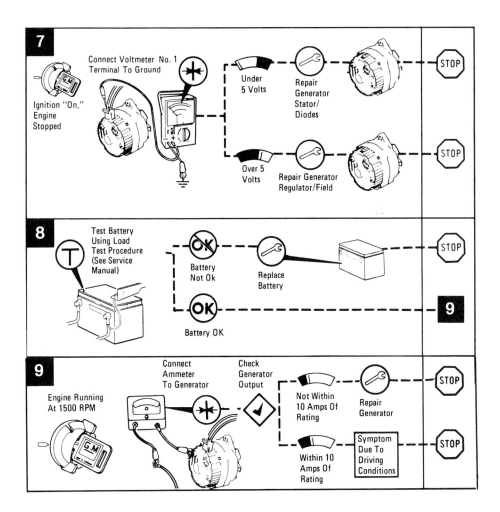

Figure 33-7. Concluded.

Figure 33-8. Normal alternator oscilloscope trace pattern.

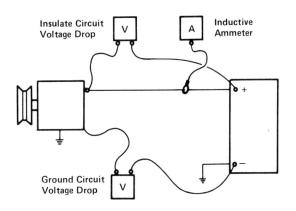

Figure 33-9. Charging system voltage drop test.

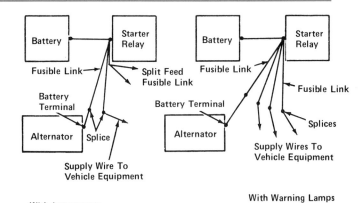

Figure 33-10. Fusible link locations. FORD MOTOR COMPANY

Fusible links usually are identified by color and/or by a molded tag in the insulation. A burned-out fusible link may have bare wire ends protruding from the insulation, or expanded or bubbled insulation.

If excessive current flows, such as during incorrect booster cable hookup, the fusible link will burn out to protect the alternator or wiring. Fusible link replacement is illustrated in **Figure 33-11.**

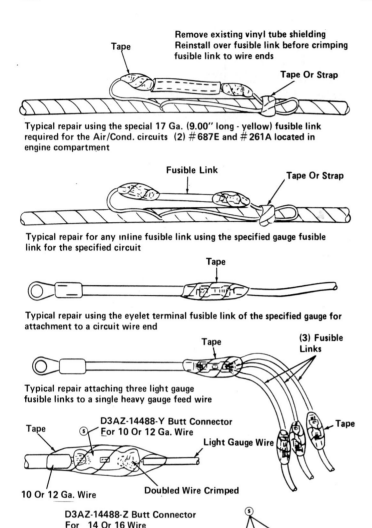

Figure 33-11. Fusible link service procedure.
FORD MOTOR COMPANY

33.4 ALTERNATOR OFF-CAR SERVICE

If tests indicate alternator and/or internal regulator problems, the alternator must be removed from the vehicle for service.

SAFETY CAUTION

Remove the grounded battery cable from the battery before you begin alternator removal procedures.

The alternator is mounted with brackets to the engine, as shown in **Figure 33-12**. Wiring connections and the alternator belt are removed. Then the pivot and adjusting bolts are removed. Other fasteners may need to be loosened or removed if the alternator cannot be lifted from the vehicle.

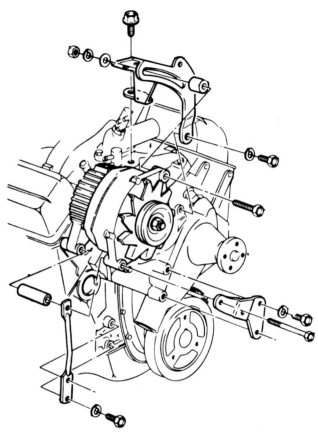

Figure 33-12. Alternator mounting.
CHEVROLET MOTOR DIVISION—GMC

Alternator bench tests can be performed with appropriate test equipment. An electric motor is used to drive the alternator during testing.

Basic repair steps for General Motors Corporation alternators are illustrated in **Figure 33-13**. Refer to the vehicle manufacturer's service manual for specific procedures. Be sure to mark the end frames for correct reassembly. After service procedures are performed, the alternator should be bench tested before reinstallation.

UNIT HIGHLIGHTS

- Routine charging system preventive maintenance includes checking charging indicators, alternator drive belts and wiring connections. In addition, visual checks for physical damage are made.

- Undercharging can be caused by low engine idle rpm, loose belts, or alternator or regulator problems.

CHARGING SYSTEM SERVICE

Thru-Bolt Location

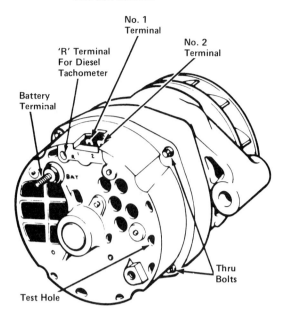

End Frame View

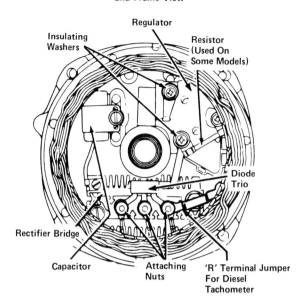

1. Make scribe marks on end frames to facilitate reassembly.
2. Remove four thru-bolts and separate drive end frame assembly from rectifier end frame assembly.

3. Remove three attaching nuts and three regulator attaching screws.
4. Separate stator, diode trio and regulator from end frame. The regulator cannot be tested on the work bench except with a regulator tester.

Testing Stator

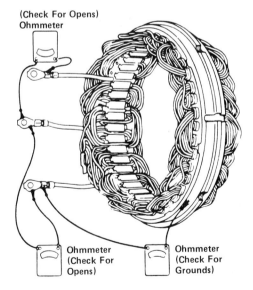

Testing Rotor

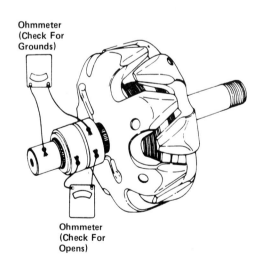

5. Check stator for opens with ohmmeter (two checks). If either reading is high (infinite), replace stator.
6. On all series, check stator for grounds. If reading is low, replace stator.

7. Check rotor for grounds with ohmmeter. Reading should be very high (infinite). If not, replace rotor.
8. Check rotor for opens. Should read 2.4 – 3.5 ohms. If not, replace rotor.

Figure 33-13. Alternator disassembly and repair procedures.
BUICK MOTOR DIVISION—GMC

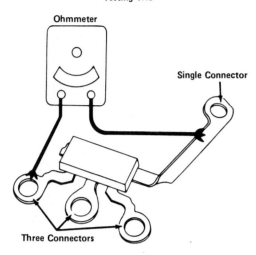

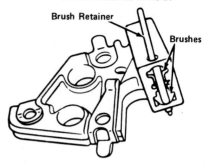

14. Clean brushes with soft, dry cloth.
15. Put brushes in holder and hold with brush retainer wire.

9. To check diode trio, connect ohmmeter as shown, then reverse lead connections. Should read high and low. If not, replace diode trio.
10. Repeat same test between single connector and each of other connectors.

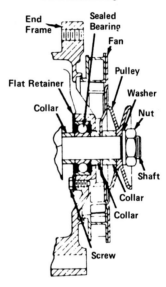

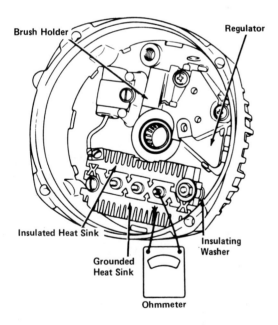

16. Observe stack-up of parts in both illustrations. To remove rotor and drive end bearing, remove shaft nut, washer and pulley, fan and collar. Push rotor from housing.
17. Remove retainer plate inside drive end frame and push bearing out. Clean all parts with soft cloth.
18. Press against outer race to push bearing in. The 12SI uses a sealed bearing — no lubricant is required. Assemble retainer plate.
19. Press rotor into end frame. Assemble collar, fan pulley, washer and nut. Torque shaft nut to 54 - 82 N-m (40 - 60 ft. lbs.).

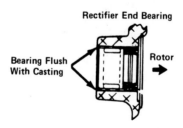

11. Check rectifier bridge with ohmmeter connected from grounded heat sink to flat metal on terminal, not the stud. Reverse leads. If both readings are the same, replace rectifier bridge.
12. Repeat test between grounded heat sink and other two flat metal clips.
13. Repeat test between insulated heat sink and three flat metal clips, not the stud. To replace bridge, remove attaching screws.

20. Push slip ring end bearing out from outside toward inside of end frame.
21. Place flat plate over new bearing, press from outside toward inside until bearing is flush with end frame.
22. Assemble brush holder, regulator, resistor, diode trio, rectifier bridge and stator to slip ring end frame.
23. Assemble end frames together with thru-bolts. Remove brush retainer wire.

Figure 33-13. Concluded.

- Charging system troubleshooting includes checking the battery, alternator, regulator, and wiring connections.
- Charging system electrical tests include charging voltage/alternator output tests, oscilloscope trace tests, and circuit resistance tests.
- Fusible links may be used to protect a charging system alternator and/or wiring.
- Alternator off-car service includes removal, disassembly, checking and/or replacing parts, replacement, and final testing before reinstallation.

TERMS

full field
voltage spike
trace
inductive ammeter
oscilloscope
fusible link

REVIEW QUESTIONS

DIRECTIONS: The following questions are similar to those used on automotive technician certification tests. On a separate sheet of paper, write the letter of the correct choice.

1. Which of the following statements is correct?
   I. Fully transistorized IC regulators usually can be adjusted.
   II. Electromechanical regulators usually cannot be adjusted.
   A. I only
   B. II only
   C. Both I and II
   D. Neither I nor II

2. Technician A says that an oscilloscope trace pattern can be used to check for normal alternator operation.
   Technician B says that a charging voltage/alternator output test can determine whether the alternator or regulator causes undercharging.
   Who is correct?
   A. A only
   B. B only
   C. Both A and B
   D. Neither A nor B

3. A charging voltage test indicates less than 1.2 volts rise above battery voltage. An alternator output test indicates rated alternator current output.
   What is the most likely cause of undercharging?
   A. Shorted alternator field windings
   B. Shorted diode in rectifier bridge
   C. Defective regulator
   D. Defective diode trio

4. Technician A says that fusible links are located only in a vehicle's charging system.
   Technician B says that the conductor in a fusible link is smaller in diameter than the circuit wire it is to protect.
   Who is correct?
   A. A only
   B. B only
   C. Both A and B
   D. Neither A nor B

Question 5 is not like those above. It has the word EXCEPT. For question 5, look for the choice that does *not* apply. Read the entire question carefully before you choose your answer.

5. Undercharging can be caused by all of the following EXCEPT
   A. excessively tight alternator belt or excessively high engine rpm.
   B. frequent starting or defective starter.
   C. low idle speed or loose alternator belt.
   D. alternator or regulator problems.

ESSAY QUESTIONS

1. What are common causes of undercharging?
2. What problems can be found during a charging system visual inspection?
3. Describe charging voltage and alternator output tests.

SUPPLEMENTAL ACTIVITIES

1. Perform charging system preventive maintenance on a vehicle. Report to your class what problems, if any, were found.
2. Perform a charging voltage test on a vehicle. Report to your class what readings are obtained.
3. Perform an alternator output test on the same vehicle used in Supplemental Activity 2, above. Report to your class on the readings obtained.
4. Perform a circuit-resistance (voltage-drop) test on a vehicle. Report the readings obtained to your class.
5. Demonstrate to your class how to correctly connect and perform an oscilloscope trace test, if available, on a charging system.
6. Referring to the manufacturer's service manual, correctly remove and disassemble an alternator. Perform tests and replace parts as necessary. Correctly reassemble, install, and replace the alternator in the vehicle.

34 The Ignition System

UNIT PREVIEW

The purpose of the ignition system is to ignite the air-fuel mixture in the combustion chambers of the engine.

A gasoline engine ignition system increases the relatively low voltage of the battery to as much as 50,000 volts. This high voltage is timed and directed to individual cylinders at the proper time to begin combustion.

Proper engine operation depends upon accurate ignition system adjustment and service. Ignition timing affects both engine performance and emissions.

LEARNING OBJECTIVES

When you have completed your assignments and exercises in this unit, you should be able to:

❏ Explain how an ignition system operates and identify what factors influence correct ignition timing.
❏ Identify and describe ignition system parts on a vehicle.
❏ Determine what type of switching device is used on a vehicle.
❏ Describe a direct, or distributorless, ignition system (DIS).

34.1 PURPOSE OF THE IGNITION SYSTEM

The ignition system produces a spark in a cylinder at the proper point in a gasoline engine 4-stroke cycle. The spark must be sufficiently strong to jump the spark plug air gap and ignite the air-fuel mixture, which begins combustion.

Lean mixtures and/or high compression pressures require higher voltage, or stronger, sparks. Richer mixtures and/or lower compression pressures can be ignited with lower-voltage sparks.

34.2 IGNITION SYSTEM PARTS

An ignition system includes a low-voltage *primary circuit* and a high-voltage *secondary circuit*. Mechanical and/or vacuum controls can be used to make the spark occur sooner or later than an initial setting. In addition, most recent passenger vehicles use sensors and computers to control the ignition system (see Unit 38).

However, the basic operation of all spark-ignition systems is the same. In 1908, Charles F. Kettering, who also invented the electric starter, designed a basic spark ignition system. A basic system, shown in **Figure 34-1,** consists of a battery, ignition coil, switch, primary and secondary wiring connections, and a spark plug.

The following discussion of modern automotive ignition systems includes several key components:

• Ignition coil
• Distributor
• Primary circuit
• Secondary circuit.

34.3 IGNITION COIL

An ignition coil is a *pulse transformer.* A pulse transformer produces periodic surges of high voltage. A transformer, as discussed in Unit 27, consists of two

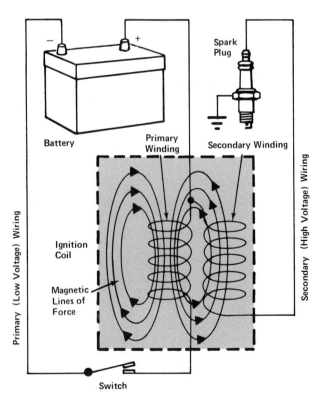

Figure 34-1. Basic ignition system.

The Ignition System

coils wrapped around an iron core. Magnetic lines of force formed by current flow in one coil will induce current flow in the second coil. The voltage output of the second coil is proportional to the number of turns in the first and second coils.

Both coils in an ignition coil are wrapped around a laminated iron core. The first, or primary, outer coil consists of from 100 to 150 turns of 20-gauge wire. This wire is approximately 0.05 in. (1.27 mm) thick. The inner, or secondary, coil is made of 15,000-20,000 turns of fine copper wire thinner than human hair. Windings are separated by paper or plastic insulation and can be enclosed in an oil-filled canister or exposed **[Figure 34-2]**.

When a switch is turned on, battery current energizes the primary coil to form magnetic lines of force. When the switch is turned off, the magnetic field collapses, or moves inward toward the laminated iron core. As it collapses, the magnetic lines of force move across the secondary coil. This action induces electric current in the secondary coil and produces a current flow. In modern ignition systems, secondary voltage can reach 40,000 volts or more. High voltage is required to cause a spark to jump across the *air gap*, or gap between electrodes, at the end of a spark plug.

34.4 DISTRIBUTOR

An ignition system must produce a spark for each cylinder every time a piston completes a compression stroke. Typical passenger vehicle ignition systems must be capable of producing and distributing up to 350 sparks per second. To accomplish this task, a *distributor* can be used. A distributor, shown in **Figure 34-3**, controls primary current and distributes secondary current. A distributor includes the following parts:

- Housing
- Shaft and drive mechanism
- Switching or sensing device
- Rotor
- Cap.

The outer housing of a distributor is made of aluminum. The distributor shaft is either turned directly by the camshaft or by a drive mechanism attached to the camshaft. Thus, the distributor shaft turns with the camshaft at one-half crankshaft speed.

The lower part of the distributor contains either a switching device and wiring connections for the primary circuit, or a sensing device. This is used to control current flow to the primary winding in the ignition coil.

The upper part of the distributor contains a *rotor* and a cap with metal terminals for the secondary circuit. A rotor conducts secondary current from a central, carbon coil terminal to metal terminals

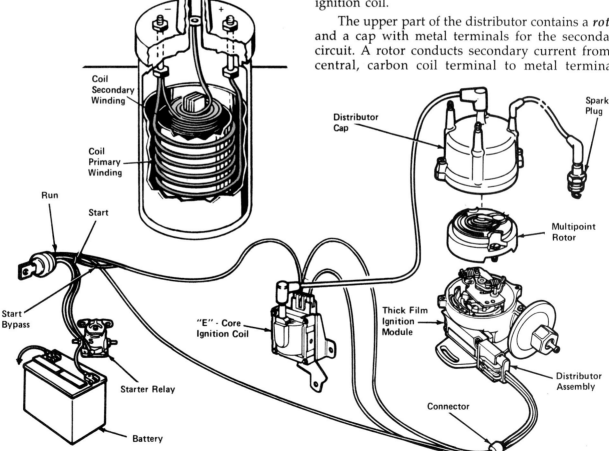

Figure 34-2. Ignition coil systems. FORD MOTOR COMPANY

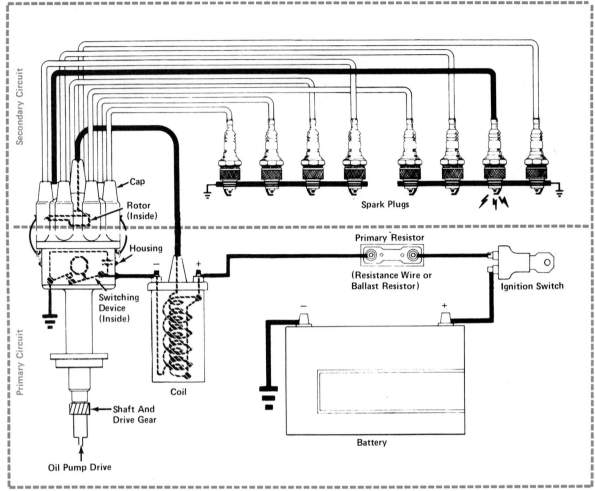

Figure 34-3. Distributor system. PRESTOLITE

outside the *distributor cap.* The rotor is attached to the distributor shaft, which is turned by the crankshaft. A *coil cable* connects the ignition coil to the carbon terminal in the center of the distributor. Spark plug cables attach from the metal cap terminals to spark plugs screwed into the cylinder head **[refer to Figure 34-3]**.

The action of the rotor is coordinated with that of the switching device in the distributor. Each time the rotor nears a metal terminal, the primary circuit is opened and secondary voltage is produced **[Figure 34-4]**.

The high voltage travels from the coil to the central terminal of the distributor cap, then to the rotor. A spark jumps an air gap between the tip of the rotor and the metal terminals inside the cap. The high voltage continues through the spark plug cables to the center electrode of the spark plug. The high voltage again produces a spark at the spark plug's air gap to ignite an air-fuel mixture.

The primary circuit is then closed, or completed, in preparation for production of another spark. This action is repeated as many as 350 times per second in passenger vehicles.

Each cylinder must receive ignition spark at the proper time for the engine to run efficiently.

34.5 PRIMARY CIRCUIT

A basic primary circuit consists of a current source, a primary ignition coil, and a switching device that controls current flow. Two types of switching devices are common:

- Breaker points
- Electronic sensor/switching circuits.

Mechanical Switching Circuit

A set of *breaker points,* or ignition points, is a mechanically operated switch. Two electrical contacts are held together by spring tension. One of the contacts can move against the spring. The other contact is stationary, or fixed. A *rubbing block* is attached to the movable contact point. A rubbing block rubs, or is held, against a *distributor cam* which is attached to the rotating distributor shaft. A distributor cam is a multi-lobed shape that opens the breaker points.

The Ignition System

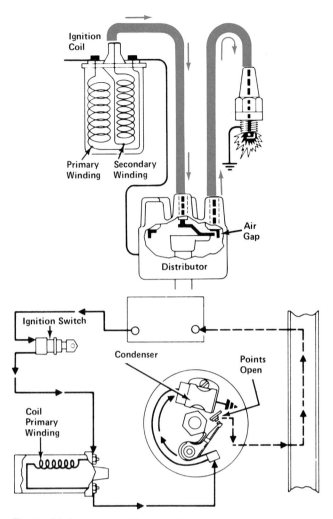

Figure 34-4. Spark ignition circuit.

When closed, current can flow from one contact point to another. The movable contact is connected to one of the primary coil terminals. Thus, when the points are closed, the primary coil is energized **[Figure 34-5]**.

The lobes, or high points, of the distributor cam push the breaker points open momentarily. A 4-cylinder engine's distributor would have four lobes, a 6-cylinder would have six, and so on.

Each time the points are opened, a high-voltage pulse is produced in the secondary coil. High voltage from the secondary coil flows, in order, to:

1. Coil cable
2. Distributor cap center terminal
3. Rotor
4. Distributor cap side metal terminals
5. Spark plug cable
6. Spark plug
7. Ground.

A flat spring in the breaker point assembly closes the points as the distributor cam rotates. A *condenser*, or

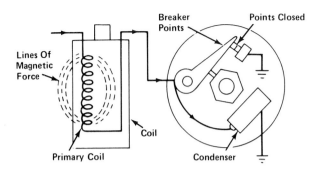

Figure 34-5. Breaker points.

capacitor, prevents excessive arcing across the breaker points as they open.

The points must remain closed long enough to *saturate*, or fully develop, the primary coil's magnetic field. The amount of distributor shaft rotation through which the distributor points remain *closed* determines magnetic saturation. This rotation is known as the *dwell angle* **[Figure 34-6]**.

Breaker points eventually wear and may burn. This deterioration changes the dwell angle, reduces primary coil current, and changes the point at which ignition spark occurs. In addition, incorrect spring tension and/or high distributor rpm can cause other problems.

Current available to energize the primary coil is limited by the contact points. Thus, secondary voltage in a conventional breaker-point ignition is limited to approximately 20,000 volts.

A *ballast resistor* can be used to protect the primary circuit from excessive voltage. A ballast resistor changes resistance in response to dwell angle and engine rpm. At relatively slow engine speeds, less current is needed to ignite the mixture. At high rpm, more current is used to prevent the spark from being "blown out" by high combustion pressures.

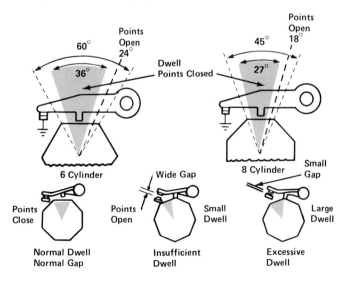

Figure 34-6. Dwell angle.

During starting, a ballast resistor is bypassed to provide maximum current flow to the primary coil **[Figure 34-7]**. A ballast resistor may be a separate unit or a specially made wire.

Since 1973, mechanical breaker point systems have been phased out by automobile manufacturers in search of improved fuel economy and reduced emissions. However, many vehicles with breaker-point ignitions are still being driven and require periodic servicing.

Electronic Sensor/Switching Circuit

To eliminate breaker-point problems, non-wearing electronic sensors and transistorized *electronic ignition control units* can be used. An electronic ignition control unit switches current to the ignition primary coil on and off with transistorized circuits. Three types of electronic sensors are used on recent vehicles:

- Magnetic position sensor
- Hall-effect sensor
- Optical LED sensor.

Magnetic position sensors consist of a central *armature* and a *pickup coil assembly*. An armature is a device that conducts lines of magnetic force. A pickup coil assembly consists of a permanent magnet and a small coil **[Figure 34-8]**.

The armature is attached to the rotating distributor shaft with an air gap between the teeth and the pickup coil. The number of "teeth," or pole pieces, of the armature correspond to the number of cylinders. When a tooth aligns with the core of the pickup coil, magnetic lines of force flow through the core and pickup coil. This magnetic field induces a signal current in the pickup coil. The ends of the pickup coil are connected to a transistorized IC control unit.

The control unit shuts off battery current to the ignition coil which creates a high-voltage pulse.

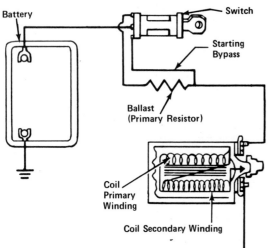

Figure 34-7. Ballast resistor circuits.

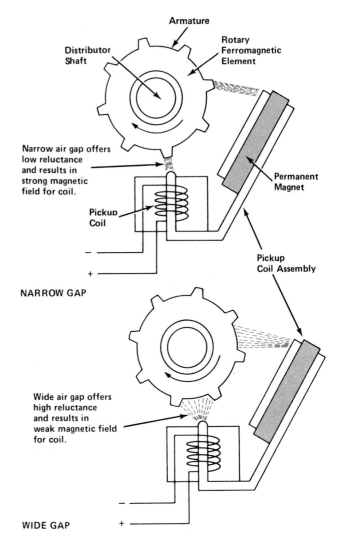

Figure 34-8. Magnetic position sensor.

General Motors products use parts similar in function to the above description. However, General Motors' High-Energy Ignition (HEI) parts are slightly different in appearance **[Figure 34-9]**.

Magnetic position sensors and electronic ignition control units eliminate the physical wear and deterioration problems of breaker points. However, magnetic position sensors can cause ignition timing to vary slightly at high rpm.

Hall-effect sensors use sophisticated electronic principles to create a signal voltage for a transistorized control unit.

A source of constant voltage is applied in one direction across a flat piece of semiconductor material. A magnet brought close to the semiconductor will cause current to flow at right angles to the applied voltage **[Figure 34-10]**.

In a distributor, a U-shaped magnet faces a similar steel or iron core. A Hall-effect semiconductor element is placed between one end of the

The Ignition System

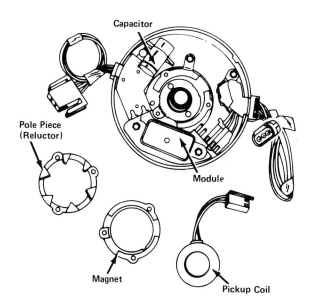

Figure 34-9. General Motors High Energy Ignition (HEI) parts.
CHEVROLET MOTOR DIVISION—GMC

facing "U"s. A steel disc passes between the lower ends of the "U"s, without touching them. As in the magnetic position sensor, magnetic lines of force flow in a circular path. The magnetic lines of force cause voltage to be created across the Hall element **[Figure 34-11]**.

This voltage is used as a signal by a transistorized control unit, which shuts off primary current to produce high voltage.

Hall-effect sensors, unlike magnetic position sensors, are not affected adversely by rotational speed. Thus, ignition timing can remain more accurate as engine speed increases.

Optical sensors use *light-emitting diodes (LEDs)* to produce a signal light. Light-emitting diodes are special-purpose diodes that produce light when current flows through them.

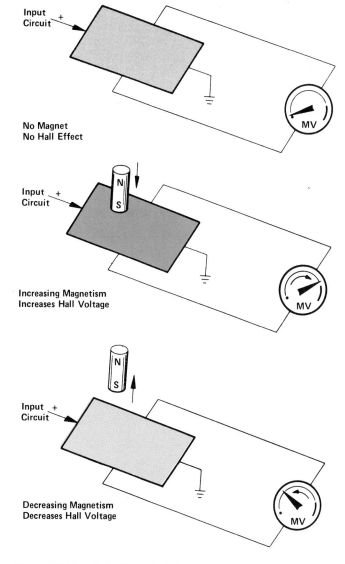

Figure 34-10. Hall-effect principles.

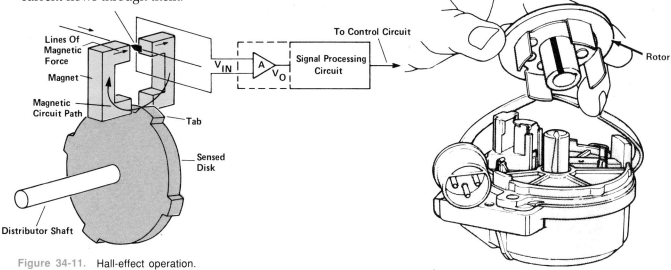

Figure 34-11. Hall-effect operation.

A rotating, slotted plate is placed between the LEDs and optical pickup sensors **[Figure 34-12]**. When the slot rotates, light from the LEDs passes through to photo diode sensors. The sensors produce a signal for a control module. The control module switches off the current to the ignition coil to produce a high-voltage pulse. The pulse travels from the coil through the distributor to a spark plug.

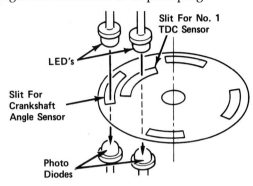

Figure 34-12. Optical sensor with LEDs.
MITSUBISHI MOTOR SALES OF AMERICA, INC.

Control modules, or control units, may be enclosed in housings separate from the distributor. In other cases, control modules may be located within, or attached to, a distributor **[Figure 34-13]**.

Control modules contain transistors and other semiconductors to open and close the primary circuit. A simplified diagram of such a circuit is shown in **Figure 34-14**. The use of transistors allows higher current flows to be used to produce greater magnetic saturation. Secondary voltages of approximately 50,000 volts or more can be produced by some late-model electronic ignition systems and ignition coils.

Dwell time is controlled by additional circuits within the control unit and is not totally dependent on distributor shaft rotation. In addition, ballast resistor functions are taken over by current-limiting circuits within the control unit.

34.6 SECONDARY CIRCUIT

The secondary circuit consists of the secondary coil winding, coil cable, distributor rotor, spark plug cables, and spark plugs.

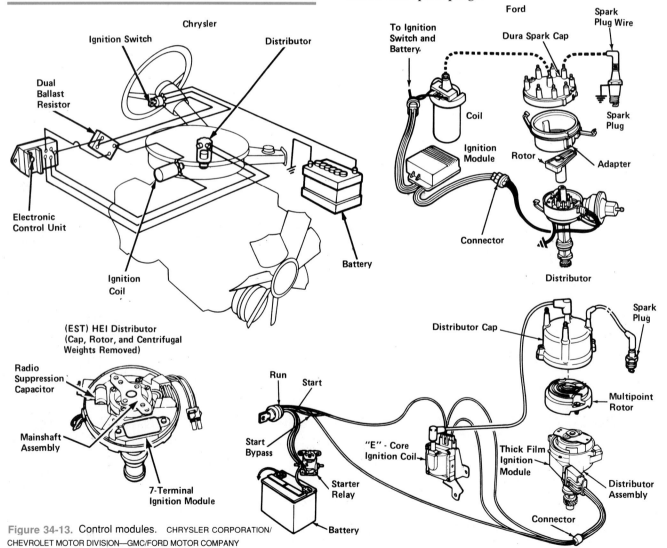

Figure 34-13. Control modules. CHRYSLER CORPORATION/ CHEVROLET MOTOR DIVISION—GMC/FORD MOTOR COMPANY

THE IGNITION SYSTEM

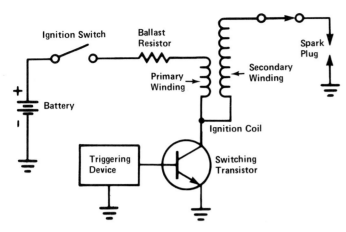

Figure 34-14. Simplified control module.

Distributor Rotor and Cap

Many different sizes and shapes of distributor caps and rotors are common [**Figure 34-15**]. Plastic used for rotors and caps must be able to insulate against high voltage and prevent it from reaching ground.

Spark Plug Cables and Coil Cable

Spark plug cables and the coil cable are made of strands of carbon-impregnated fiberglass. The strands are surrounded by layers of insulation and an outer jacket [**Figure 34-16**].

Electronic ignition coil cable and spark plug cable insulation must be able to insulate against higher voltages than a breaker-point system. Wires for electronic ignition systems are 8 mm in diameter. Cables for breaker-point ignitions are slightly smaller, 7 mm in diameter.

General Motors was the first to incorporate the ignition coil within the plastic distributor cap [**Figure 34-17**]. The ignition secondary contact is directly below the coil and eliminates the need for a separate coil wire.

Spark Plugs

Spark plugs are made of several separate parts [**Figure 34-18**]. High-voltage current from the secondary coil completes a circuit across the air gap between the center and side electrodes.

To fit properly, spark plugs must be of the proper size and *reach.* Reach refers to the length of the threaded part of the spark plug [**Figure 34-19**].

The center electrode tip of the spark plug must reach a temperature hot enough to burn off deposits. Yet the firing tip must remain cool enough to prevent rapid oxidation and electrode wear. The optimum firing tip temperature is between 650 and 1,500 degrees F (344 and 816 degrees C).

Spark plugs are available in different *heat ranges.* A heat range indicates how well a spark plug can conduct heat away from the tip. A "colder" plug

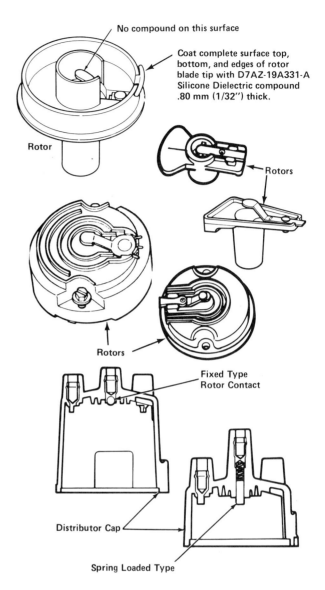

Figure 34-15. Rotors and caps. CHRYSLER CORPORATION/FORD MOTOR COMPANY/GENERAL MOTORS CORPORATION

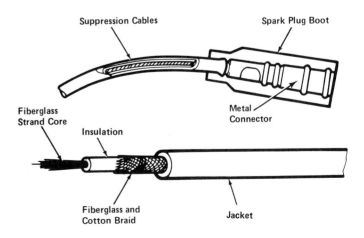

Figure 34-16. Spark plug cables. CHRYSLER CORPORATION

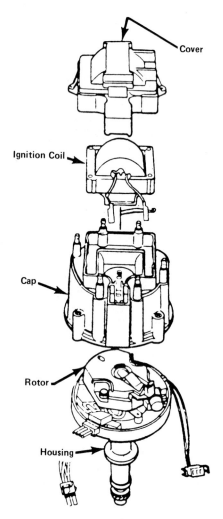

Figure 34-17. General Motors HEI distributor.
BUICK MOTOR DIVISION—GMC

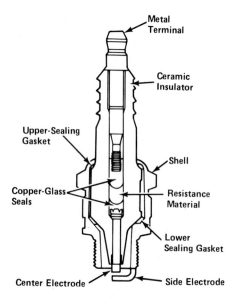

Figure 34-18. Parts of a spark plug.
AC SPARK PLUG DIVISION—GMC

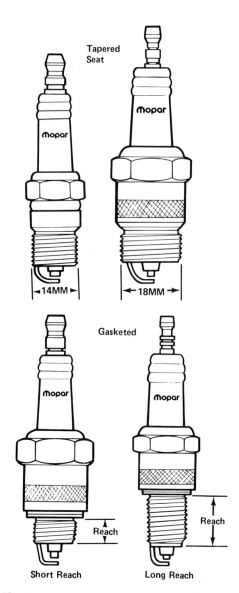

Figure 34-19. Types of spark plugs. CHRYSLER CORPORATION

will transfer heat away from the tip rapidly, resulting in lower tip temperatures. A "hotter" plug will transfer heat away slowly, resulting in higher tip temperatures [Figure 34-20]. The shape of the porcelain insulator and its contact with the outer metal shell determines spark plug heat range. The heat range is indicated by a code imprinted on the side of the plug, usually on the porcelain insulator.

A "normal" heat range plug is specified for a mixture of stop-and-go and highway driving. Hotter or colder plugs are available in steps for specialized driving conditions.

34.7 COMBUSTION AND ENGINE POWER OUTPUT

Combustion of the air-fuel mixture requires a short period of time, usually measured in thousandths of a second.

THE IGNITION SYSTEM

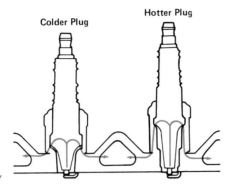

Figure 34-20. Heat transfer through a spark plug. CHAMPION SPARK PLUG COMPANY

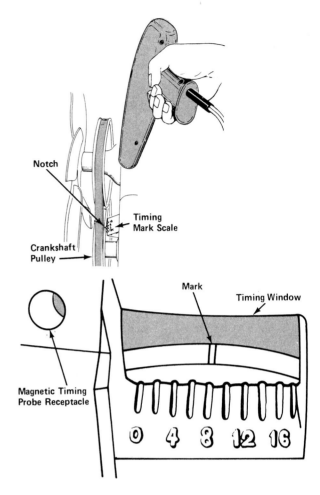

Figure 34-21. Timing mark.

Higher compression pressures tend to speed up combustion. Higher-octane gasolines and lower-cetane diesel fuels ignite less easily and require increased burning times. Increased vaporization and turbulence tend to decrease combustion times. Other factors, including intake air temperature, and humidity also affect combustion.

Just as combustion is completed, maximum pressure is exerted against the piston top. For most efficient engine operation, maximum pressure should occur about 10 degrees ATDC on the power stroke.

Ignition Timing

Ignition timing refers to the precise time when spark occurs. Ignition timing is specified by referring to a piston position in relation to crankshaft rotation from top dead center. External *ignition timing marks* can be located on a pulley or the flywheel to indicate piston position **[Figure 34-21]**. Vehicle manufacturers specify initial, or basic, ignition timing.

When the marks are aligned at top dead center, or 0, the piston in the No. 1 cylinder is at TDC (compression stroke). Additional numbers on a scale indicate a number of degrees of crankshaft rotation *before TDC (BTDC)* or *after TDC (ATDC)*. In a majority of engines, initial timing is specified at a point between TDC and 20 degrees BTDC. A few manufacturers specified initial timing of from 1 to 5 degrees ATDC for vehicles built during the 1970s.

Two major engine factors influence operational ignition timing:

- Speed (rpm)
- Load.

Engine RPM and Ignition Timing

At higher rpm, the crankshaft turns through more degrees in a given period of time. If the combustion is to be completed by 10 degrees ATDC (power stroke), ignition timing must vary according to rpm. Higher rpm requires that ignition spark must occur sooner, or be *advanced.*

However, air-fuel mixture turbulence (swirling) increases with rpm. Increased turbulence causes the mixture to burn faster. Increased turbulence in some engines requires less ignition advance than other engines. Thus, increasing engine speed requires ignition advance that varies with engine speed. A mechanical device can be used to move the rotor tip and switching mechanism ahead as rpm increases. This *centrifugal advance mechanism* causes ignition spark to occur sooner **[Figure 34-22]**.

Engine Load and Ignition Timing

The load on an engine is the work it must do. Driving up hills or pulling extra weight increases vehicle engine load. Under load, the engine operates less efficiently. A good indication of engine load is the amount of intake manifold vacuum.

Under light loads, intake manifold vacuum is relatively high. But the amount of air-fuel mixture drawn into the manifold and cylinders is small because the throttle is opened only part way. On compression, this "thin" mixture will produce less combustion pressure and will allow the ignition timing to be advanced.

Under heavy loads, however, intake manifold vacuum is low. But because throttle opening is

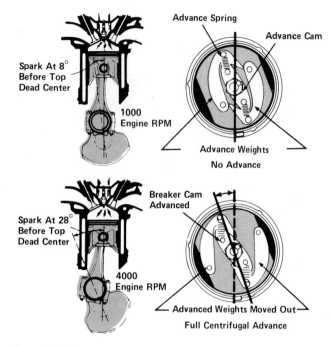

Figure 34-22. Centrifugal advance. DELCO REMY—GMC

increased, a larger mass of air-fuel mixture can be drawn in. High combustion pressure and rapid burning will result. In fact, as the flame front travels across the combustion chamber, the unburned "end gases" are additionally heated and compressed. If these end gas conditions are severe, they will self ignite or detonate and cause knocking or pinging noise. In such cases, the ignition timing must be *retarded* to avoid these conditions and promote normal combustion.

A *vacuum advance mechanism* can advance and retard ignition timing in response to engine load and throttle position. A vacuum advance unit consists of a flexible diaphragm within a metal housing. The diaphragm is connected to the metal plate of the distributor on which the switching device is located [**Figure 34-23**]. A spring pushes the diaphragm toward the switching device.

The other side of the diaphragm is connected to a source of *ported vacuum.* Ported vacuum is a vacuum source on the carburetor slightly above the throttle plate.

When the throttle plate is closed during idle or deceleration, no vacuum can reach the vacuum diaphragm. No ignition advance due to vacuum occurs.

When the throttle is partially open, manifold vacuum acts on the diaphragm. The switching device is pulled in a direction opposite to distributor shaft rotation. This action causes ignition timing to advance.

When the throttle is fully open, low manifold vacuum allows the spring to push the diaphragm back. The effect of this action is to retard total ignition advance.

Vacuum advance operates together with centrifugal advance. The portion of total ignition advance due to vacuum is shown in **Figure 34-24**.

A *vacuum retard mechanism* applies vacuum to the opposite side of a sealed diaphragm. Such devices are used to retard ignition timing for emission control requirements (see Unit 41).

Electronic Controls and Spark Advance

Engineers have found that the optimum spark advance for a given engine is not a simple curve. A more correct way of picturing optimum spark advance for a particular engine is to use a *map* [**Figure 34-25**]. A map is a three-dimensional graph that resembles a mountain range.

For any given combination of engine vacuum and speed, a point can be found on the map that corresponds to best ignition advance. A computerized control unit can vary spark advance according to the coordinates of a map stored in the computer. Sensors are used to monitor engine vacuum, speed, and other factors to determine the ignition advance necessary [**Figure 34-26**]. Engines equipped with electronic spark advance may not have centrifugal or vacuum advance units.

34.8 CYLINDER NUMBERING AND FIRING ORDER

Cylinders receive ignition spark in firing order, not cylinder number order. Examples of firing orders are shown in **Figure 34-27**.

34.9 DETONATION CONTROL

Abnormal combustion, or detonation, can damage pistons and other engine parts. Extreme detonation, or "knocking" can be heard at low speeds. However, inaudible, or high-speed, detonation usually is not heard because of engine and road noise. High-speed detonation can cause great engine damage.

Knocking can be reduced by retarding ignition timing slightly. However, retarded ignition timing reduces engine efficiency. For most efficient operation, an engine should operate at maximum possible advance before detonation occurs.

To prevent detonation, a *detonation sensor* can be part of an electronic ignition control circuit. A detonation, or knock, sensor acts like a phonograph cartridge. It picks up high-intensity sound wave vibrations in an engine and converts them to an electrical signal. This signal is sent to a control unit that retards ignition timing just enough to eliminate detonation [**Figure 34-28**].

Because of higher pressures developed in the cylinders, turbocharged and supercharged engines are more likely to experience detonation. However, naturally-aspirated engines also produce better performance when they run at the maximum possible

The Ignition System

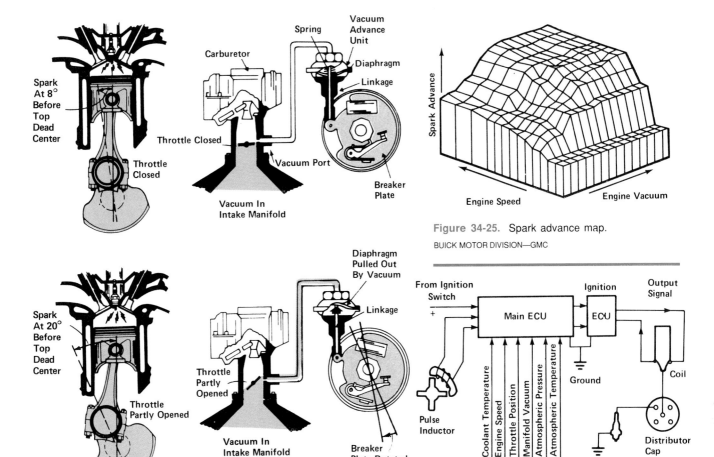

Figure 34-23. Vacuum advance. DELCO REMY—GMC

Figure 34-25. Spark advance map.
BUICK MOTOR DIVISION—GMC

Figure 34-26. Computer-controlled ignition advance system.

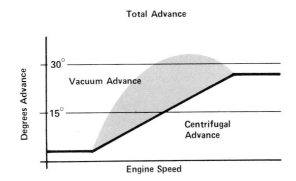

Figure 34-24. Vacuum and centrifugal advance curves.
DELCO REMY—GMC

spark advance. Thus, recent electronic ignition systems can include detonation sensors for turbocharged, supercharged, and naturally-aspirated engines.

Electronic controls also can be used to monitor and adjust air-fuel ratio. Thus, very precise control of many engine functions can be included in an electronic control device. Electronic controls are discussed in Units 38 and 39.

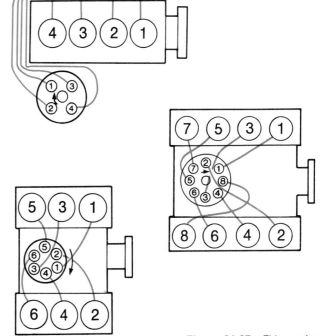

Figure 34-27. Firing orders.

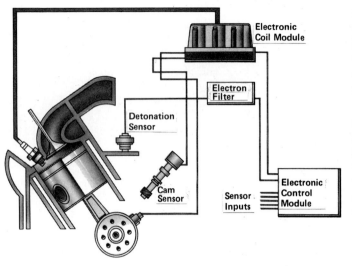

Figure 34-28. Distributorless ignition detonation sensor.
FORD MOTOR COMPANY/BUICK MOTOR DIVISION–GMC

34.10 DISTRIBUTORLESS IGNITION

The plastic materials used for distributor caps and rotors can break down under the high voltage produced by electronic ignitions. In severe cases, no ignition spark reaches the spark plugs, and the engine cannot run.

Several manufacturers, including Ford and General Motors, have produced larger distributor caps to increase separation between terminals. However, the distributor cap and rotor still remain the weak spots of electronic ignition.

In 1985, Buick was the first automobile manufacturer to introduce a production version of an automotive ignition system without a distributor. *Distributorless, or direct, ignition systems (DIS)* use special double-ended ignition coil units. Each end of a secondary coil is attached directly to a spark plug cable **[Figure 34-29]**. Each double-ended coil produces sparks for *two* spark plugs. For a four-cylinder engine, two coil units are used; for a 6-cylinder engine, three coil units are used, and so on.

With a direct ignition system, dwell time, the time that the primary coil is energized, can be increased for a higher-voltage, more intense spark. This produces better engine performance, especially at high engine speeds.

When a sensor signals the ECU, the ECU switches off current to the primary coil. The magnetic field collapses and high voltage travels to both spark plugs.

Different types of sensing, or triggering, mechanisms are used with direct ignition. Some use devices driven off the camshaft **[refer to Figure 34-28]**. Others pick up pulses with a magnetic position sensor similar to that used inside a distributor. But the reluctor consists of teeth cut into a crankshaft pulley or the teeth of the flywheel ring gear. An electrical pulse is produced each time a tooth passes the sensor. The computer counts the pulses and determines the position of pistons in the cylinders and where they are in the four-stroke cycle.

One of the plugs will fire to begin the power stroke in its cylinder. The other plug is in an alternate-firing cylinder on its exhaust stroke. The next time the coil fires, the alternate-firing cylinder will begin its power stroke, and the first cylinder will be on its exhaust stroke. The spark that occurs during the exhaust stroke, known as "wasted spark," does not interfere with normal engine operation.

SAFETY CAUTION

Direct ignition systems produce much higher secondary voltage and current output than conventional systems. A shock from a direct ignition system may cause great physical injury or loss of consciousness. Use safe working procedures to avoid shocks from any type of ignition system.

34.11 DIESEL IGNITION TIMING

Diesel ignition timing is called *injection timing* and is determined by the moment at which injection occurs. It is specified by reference to piston position and degrees of crankshaft rotation. Diesel injection advance may be provided either hydraulically or electronically in the injection pump.

UNIT HIGHLIGHTS

- A basic ignition system consists of a battery, an ignition coil, a switch, a distributor, cables, and spark plugs.
- A distributor controls primary current and distributes secondary current to spark plugs at the proper time for efficient engine operation.
- Ignition coil primary current can be switched on and off mechanically or through electronic sensors and control units.
- Initial ignition timing is specified by the vehicle manufacturer. Correct engine advance is most accurately illustrated by a vacuum advance map.
- Mechanical and/or electronic devices may be used to advance and/or retard ignition timing for different speed and load conditions.
- Detonation sensors can be used with electronic control units to operate engines at maximum advance before detonation.
- Distributorless ignition systems can be used to prevent arcing and insulation problems in distributors.

The Ignition System

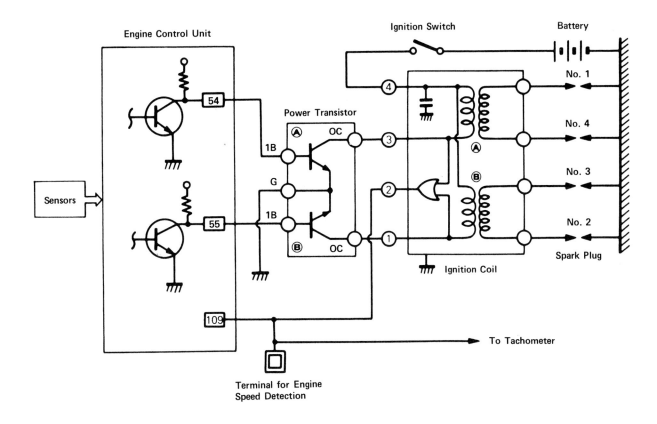

Figure 34-29. Four-cylinder direct ignition circuit. MITSUBISHI MOTOR SALES OF AMERICA, INC.

TERMS

primary circuit
pulse transformer
distributor
coil wire
rubbing block
condenser
dwell angle
ported vacuum
armature
reach
ignition timing
before TDC (BTDC)
advance
detonation sensor
distributor cap
electronic ignition control unit
vacuum advance mechanism
light-emitting diode (LED)

secondary circuit
air gap
rotor
breaker points
distributor cam
saturate
ballast resistor
reluctor
pickup coil assembly
heat range
ignition timing marks
after TDC (ATDC)
retard
map
injection timing
centrifugal advance mechanism
vacuum retard mechanism
direct ignition system (DIS)

REVIEW QUESTIONS

DIRECTIONS: The following questions are similar to those used on automotive technician certification tests. On a separate sheet of paper, write the letter of the correct choice.

1. Technician A says that an ignition spark is produced when the switching device turns on current to the primary coil.
   Technician B says that an ignition spark is produced when the switching device turns off current to the primary coil.
   Who is correct?
   A. A only
   B. B only
   C. Both A and B
   D. Neither A nor B

2. A vehicle will not start. The coil wire is removed from the distributor cap and held 0.25 in. (6.3 mm) from the engine block. When the engine is cranked, ignition spark will jump from the coil cable to ground. The coil wire is replaced and spark plug cables checked in the same manner. No sparks will jump from any spark plug cable to ground.

Which of the following defective components is the most likely cause?
A. Coil
B. Switching mechanism
C. Rotor
D. Spark plug cables

3. What does an ignition advance map indicate?
A. Ignition advance for maximum performance
B. Ignition advance necessary at different rpm
C. Ignition advance necessary at different vacuum (load) conditions
D. All of the above

4. Which of the following will help to prevent detonation?
A. Advancing ignition timing
B. Retarding ignition timing
C. Detonation sensor
D. Distributorless (direct) ignition

Question 5 is not like those above. It has the word EXCEPT. For question 5, look for the choice that does *not* apply. Read the entire question carefully before you choose your answer.

5. Any of the following can be used in a distributor as part of a switching device EXCEPT
A. breaker points.
B. detonation sensor.
C. magnetic position sensor.
D. Hall-effect sensor.

ESSAY QUESTIONS

1. Describe the parts and functions of both distributor-type and direct ignition systems.
2. Describe the methods and parts used to switch primary current on and off.
3. What factors affect ignition timing, and how is ignition timing varied?

SUPPLEMENTAL ACTIVITIES

1. Remove the distributor cap from a vehicle and determine what type of switching device is present. If it is an electronic ignition, determine which type of switching device is used. Refer to the manufacturer's service manual and locate the ignition control module.
2. Discussion topic: what effect does changing the dwell time (the time the coil is energized) have on the ignition timing?

35 Ignition System Service

UNIT PREVIEW

The ignition system has a great influence on engine performance, economy, emissions, and driveability. The ignition system must deliver high-voltage electrical current to the correct cylinder at the proper instant.

Such a system requires careful maintenance and service. A tune-up includes checks and maintenance of the charging, fuel, ignition, and emission control systems.

Ignition system preventive maintenance includes troubleshooting, replacement of defective parts, and adjustments on ignition timing. Basic checks to determine why an engine won't start include checks for fuel and for ignition spark. Commonly used ignition system diagnostic instruments include engine analyzers, timing lights, hand-operated vacuum pumps, and oscilloscopes.

Setting ignition timing properly may involve following carefully detailed procedures as listed on emission control specification stickers.

LEARNING OBJECTIVES

When you have completed your assignments and exercises in this unit, you should be able to:

- ❏ Check for the presence of spark.
- ❏ Perform a visual inspection of ignition system parts and identify problems found.
- ❏ Correctly use a remote starter, engine analyzer, timing light, hand vacuum pump, and oscilloscope to diagnose ignition system problems.
- ❏ Remove, color read, and replace spark plugs and set ignition timing properly.
- ❏ Remove a distributor and dead time an engine.

SAFETY PRECAUTIONS

Keep hands, arms, and other body parts away from moving fans, belts, and pulleys when the engine is cranking or running. Roll up long sleeves and remove all jewelry from hands, arms, and around necks. Tie long hair in a ponytail and place inside your shirt.

Ignition timing specifications are checked and adjusted with the engine running. Keeps hands, arms, and other body parts away from hot engine exhaust manifolds, radiator hoses, and other parts.

Ignition systems can produce up to 50,000 volts. This high voltage can produce painful electrical shocks. When testing for spark, hold spark plug and coil cables with insulated pliers.

Wear eye protection when using compressed air to clean debris from spark plug wells.

When replacing breaker points, make sure the ignition switch is in the off position.

35.1 CHECKING FOR SPARK

Basic tests to determine why a vehicle won't start include checks for fuel (see Unit 20) and for ignition spark. Refer to the vehicle manufacturer's service manual for correct and safe methods to check for ignition spark.

SAFETY CAUTION

Direct ignition systems produce much higher secondary voltage and current output than conventional systems. A shock from a direct ignition system may cause great physical injury or loss of consciousness. Use safe working procedures to avoid shocks from any type of ignition system.

The following is a general description of the procedure used to check for ignition spark. Always refer to the manufacturer's service manual for specific procedures.

To check for ignition spark, with the ignition off grasp any spark plug cable by the boot then twist and pull to remove the cable.

After removal, a metal object, such as an insulated screwdriver, is inserted into the spark plug boot. The metal shank is placed approximately 1/4 inch (6.3 mm) from the engine block or cylinder head. When the engine is cranked, a spark should jump the gap from the screwdriver shaft to the engine block or cylinder head [Figure 35-1].

CAUTION!!!

Some manufacturers require the use of a special spark tester. The spark plug cable is connected to the tester.

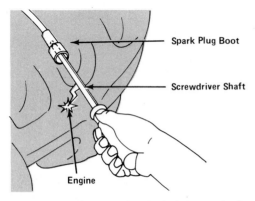

Figure 35-1. Checking for spark at the spark plug cables.

A grounding wire on the tester is connected to a good ground in the engine compartment. When the engine is cranked, a spark should be observed between the electrodes of the spark tester. Refer to the manufacturer's service manual for details and test procedures.

The spark should continue to jump the gap as the cable is moved up to 1/2 inch (12.2 mm) away. If not, inspect the cables, distributor cap, and rotor as described below.

If no spark is observed, replace the spark plug cable and try other cables. Ignition spark should be present at all cables.

If no spark is observed from spark plug cables, remove the coil cable (if present) from the distributor cap. Check for spark from the coil cable when the engine is cranked **[Figure 35-2]**.

NOTE: General Motors HEI (High Energy Ignition) units and those of some other manufacturers do not have coil cables. Refer to the manufacturer's service manual for specific ignition coil test procedures.

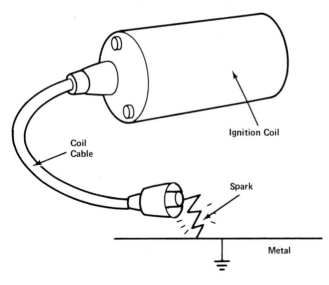

Figure 35-2. Checking for spark at the coil.

Spark output from the coil but not from the spark plug cables indicates a distributor rotor or cap problem. No spark output from the coil cable indicates a primary system problem, bad coil cable, or bad coil. Too-slow engine cranking that speeds up when the coil cable is removed indicates over-advanced initial ignition timing.

Breaker point primary ignition systems are subject to wear and burning. In many cases, primary system problems are due to worn rubbing blocks and/or burned contact points. Breaker point replacement is covered in 35.7. For primary electronic ignition circuits, consult the manufacturer's service manual for correct diagnostic and service procedures.

35.2 IGNITION SYSTEM SERVICE

Ignition system service usually is performed as part of a tune-up which includes checks and maintenance of:

- Battery and charging systems
- Fuel system
- Ignition system
- Automotive emissions-related equipment.

Maintenance of the first three items in this list have always formed the basis for a tune-up. Ignition system service aims at providing most efficient ignition system operation. In previous years, maximum engine power and/or economy was the goal of a tune-up.

However, the need to control automotive emissions has changed the emphasis of tune-up procedures. Federal and state laws require that harmful automotive exhaust emissions be reduced (see Unit 41). Proper ignition system operation is necessary for both efficient engine operation and minimal emissions. The following items form the basis for ignition system service:

- Diagnostic procedures to locate defective parts
- Removal and replacement of defective parts
- Adjustments to ignition timing.

Troubleshooting procedures include inspections, basic tests for ignition spark, and diagnostic procedures done during tune-ups. Advanced diagnostic techniques include the use of electronic instruments.

35.3 VISUAL INSPECTION

If an engine runs poorly, the ignition system is functioning well enough to provide high-voltage sparks. Before making basic or advanced tests, the ignition system is checked for obvious problems. Such problems can include:

- Disconnected, loose, or damaged secondary cables
- Loose, or corroded primary wiring

IGNITION SYSTEM SERVICE

- Loose distributor cap
- Damaged distributor cap and/or rotor
- Worn and/or damaged primary system switching mechanism
- Improperly mounted electronic control unit.

Secondary and Primary Wiring Connections

Spark plug and coil cables should be clean and pushed tightly into the distributor cap and coil, and onto spark plugs. Loose or corroded primary coil connections should be cleaned and tightened. Frayed or cracked primary wires or secondary cables are replaced. Replacement of distributor spark plug cables on Chrysler products may require the use of longnose pliers **[Figure 35-3]**.

Secondary cables must be connected in correct *firing order*. Firing order is the sequence in which ignition occurs in the numbered cylinders of an engine. Many firing orders are common **[Figure 35-4]**. Refer to

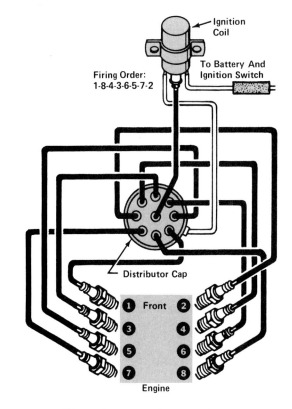

Figure 35-4. Firing order and spark plug routing.

the manufacturer's service manual to determine correct firing order.

White or grayish powdery deposits on secondary cables where they cross or near metal parts indicate faulty insulation. High-voltage electricity has "burned" collected dust. In the dark, such faulty insulation may produce a spark that sometimes can be heard and seen. An occasional glow around the spark plug cables, known as a *corona effect*, is not harmful. To test for faulty insulation, a grounded test lead is moved along a spark plug cable. A spark jumping to the test lead indicates faulty cable insulation.

Spark plug cables from consecutively firing cylinders should cross rather than run parallel to one another **[Figure 35-5]**. Parallel secondary cables can induce firing voltage in one another and cause spark plugs to fire at the wrong time.

Primary ignition system wiring should be checked for tight connections, especially on vehicles with electronic ignitions. Electronic circuits operate on very low voltage. Resistance caused by corrosion or dirt can cause running problems. Separate connectors and check them for dirt and corrosion. Clean the connectors according to the manufacturer's service manual recommendations. Some late-model vehicles require the use of a special silicone dielectric grease on connector terminals. This is done to seal the connections from outside elements.

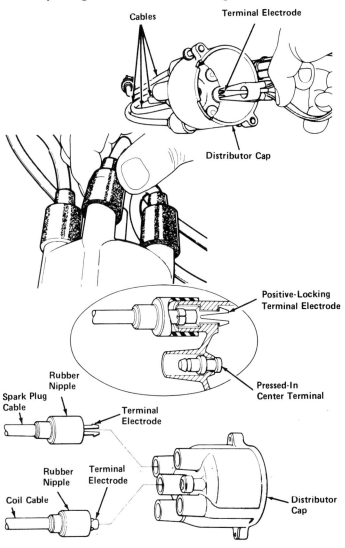

Figure 35-3. Removing and installing distributor cap wires.
CHRYSLER CORPORATION

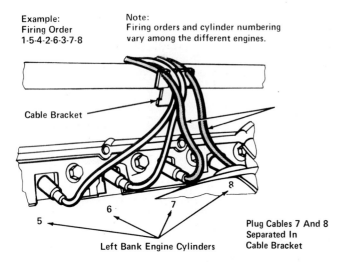

Figure 35-5. Routing of spark plug cables.
CHRYSLER CORPORATION

Distributor Cap

The distributor cap should be properly indexed and firmly seated on its base. All clips or screws should be tightened securely [**Figure 35-6**]. Also check for corrosion inside the cap terminals and carefully scrape them clean if necessary.

The distributor cap and rotor also should be removed for visual inspection. Physical damage is easily recognizable, electrical damage sometimes is harder to locate.

Electrical damage from high voltage can include corroded or burned metal terminals and *carbon tracking* inside distributor caps. Carbon tracking is formation of a line of carbonized dust between distributor cap terminals, or between a terminal and the distributor housing. Carbon tracking indicates that high-voltage electricity has found an improper, low-resistance conductive path over or through the plastic. The result is a cylinder that fires at the wrong time, or a misfire. Damaged and/or carbon-tracked distributor caps and/or rotors are replaced [**Figure 35-7**]. Carbon tracking frequently takes on the appearance of a crack.

Check the outer cap towers and metal terminals for defects. Cracked plastic requires replacement of the unit. Lightly corroded outer metal contacts sometimes can be cleaned with a special wire brush [**Figure 35-8**].

The rotor plastic should be inspected carefully for discoloration, especially on vehicles with electronic ignition. Inspect the top and bottom of the rotor carefully for grayish, whitish, or rainbow-hued discolored spots. Such discoloration indicates that the rotor plastic has lost its insulating qualities. High-voltage electricity is being conducted to ground through the plastic [**Figure 35-9**].

Check separate ignition coils for damage around the center tower. Cracks or burned spots require coil replacement [**Figure 35-10**].

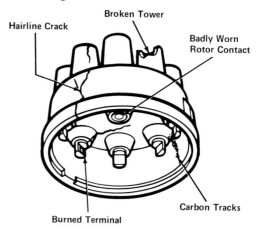

Figure 35-7. Distributor cap and rotor damage.
CHRYSLER CORPORATION

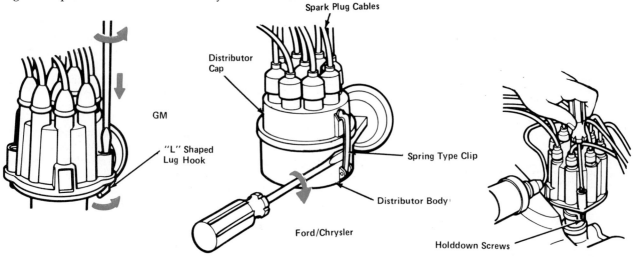

Figure 35-6. Distributor cap holding devices.
CHRYSLER CORPORATION

IGNITION SYSTEM SERVICE

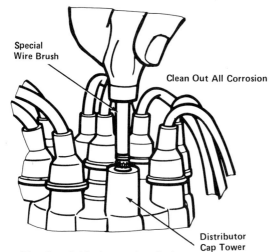

Figure 35-8. Cleaning distributor cap terminals.
CHRYSLER CORPORATION

Primary System Switching Mechanisms

With the distributor cap and rotor off, inspect the switching mechanism inside the distributor base. Primary system problems are most common in vehicles equipped with breaker point ignitions. The points and condenser or electronic switch components are checked for correct routing and secure fastening. Loose components can cause erratic ignition system operation.

Vacuum advance mechanism linkage should be attached securely. Disconnected linkage can cause stalling and dying.

Breaker points should be pushed open with a screwdriver to check the condition of the contact surfaces **[Figure 35-11]**. Badly oxidized (blackened) or pitted points should be replaced. Badly pitted surfaces indicate a mismatched condenser. If the points are pitted, both the points and the condenser must be replaced.

Magnetic pulse generators are relatively trouble-free. The reluctor or pole piece is replaced only if broken or cracked.

Pickup coil wire leads can become grounded if insulation rubs off when vacuum advance and/or retard mechanisms operate. Inspect these leads carefully **[Figure 35-12]**. If worn insulation is found, a temporary repair can be made by wrapping the wire with electrical tape. The wires also should be placed so that they do not rub the switch plate as it moves. Sometimes the pickup coil leads harden from the constant action of the vacuum advance unit and break. The leads can be tested with an ohmmeter. If they are open, the pickup coil must be replaced. See Topic 35.4, Hand Vacuum Pump.

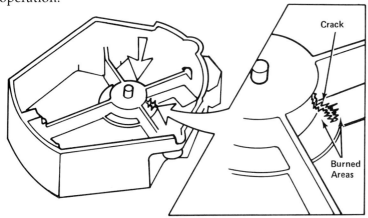

Figure 35-9. Distributor/rotor inspection.

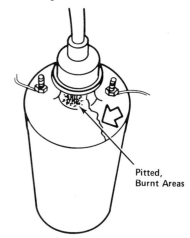

Figure 35-10. Cracked coil tower.

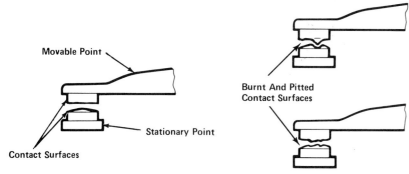

Figure 35-11. Breaker point inspection.

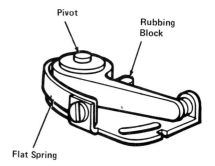

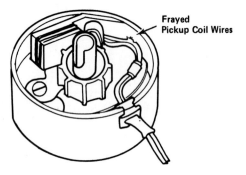

Figure 35-12. Coil wire inspection.

Hall-effect switch problems are similar to those of magnetic pulse generators. Connecting wires that can move and rub as the vacuum advance/retard mechanisms move the switch plate should be inspected **[Figure 35-13]**.

Electronically controlled advance mechanisms eliminate centrifugal and vacuum advance/retard mechanisms within the distributor. However, all wiring connections and parts should be thoroughly inspected for damage.

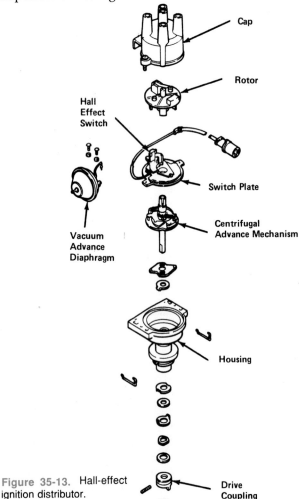

Figure 35-13. Hall-effect ignition distributor.
CHRYSLER CORPORATION

Distributorless ignitions are controlled by an electronic circuit within the coil/control unit. Refer to the manufacturer's service manual for detailed diagnosis and service procedures.

Centrifugal advance mechanisms can be checked for free motion. The rotor on the distributor shaft is rotated clockwise and counterclockwise. The rotor should rotate in one direction approximately 1/4 inch (6.3 mm), then spring back. If not, the centrifugal advance mechanism may be binding or rusty. It may be necessary to lubricate advance mechanism pivots and rubbing surfaces.

Electronic Control Units

Heat can damage or destroy sensitive electronic control units. Control units must be tightly mounted to clean surfaces. Loose mounting can cause a heat build-up that can destroy transistors and other electronic components. Some manufacturers require the use of a special heat-conductive silicone grease between the control unit and its mounting **[Figure 35-14]**.

35.4 TROUBLESHOOTING WITH INSTRUMENTS

Instruments can be used to diagnose running problems and the need for tune-up procedures. The instruments most commonly used include:

- Engine analyzer
- Timing light
- Hand vacuum pump
- Oscilloscope.

Specialized instruments, including ignition coil testers, condenser testers, and others, can be used for specific tests. However, the instruments discussed here can be used to diagnose almost all ignition system problems.

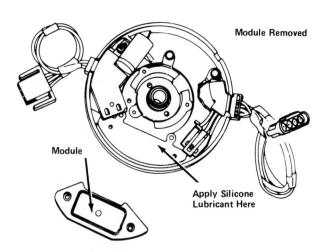

Figure 35-14. Module replacement.
CHEVROLET MOTOR DIVISION—GMC

Ignition System Service

Engine Analyzer

An engine analyzer [**Figure 35-15**] includes several test functions. These functions can include testing for:

- Low and high rpm
- Dwell
- Voltage
- Resistance
- Amperage
- Pickup coil output
- Alternator ripple current.

Instruments may require batteries or may have cables to use battery voltage to power certain test functions. Ohmmeter functions require a separate battery.

The instrument shown in **Figure 35-15** has an *inductive pickup* for rpm functions. An inductive pickup is a device that contains a pickup coil which clamps around a wire or cable. A current passing through the conductor induces a signal voltage. The signal voltage can be modified and interpreted by the instrument to produce a specified reading, such as rpm. An inductive pickup for rpm clamps around any spark plug cable.

Other instruments require connections to an ignition coil primary terminal and ground to pick up coil pulses. Such instruments require setting a dial for the number of cylinders or reading a separate scale for correct rpm figures.

Diagnostic meters can have an *analog* or *digital* readout. A meter is a form of analog readout. In response to higher rpm, for example, the needle moves up the scale farther. A digital readout would be in the form of numbers. Instead of a scale, the meter simply displays a number, such as 650. As rpm changes, the number changes.

Engine analyzers can have many test leads and cords for various functions. Refer to the instrument manufacturer's instructions for proper test hookups.

Many ignition system problems can be found by using an ohmmeter [**Figure 35-16**]. High resistance will lower current flow. Low resistance will increase current flow. Manufacturers' service manuals list proper resistance specifications for given components. Resistance in some units, such as electronic ignition pickup coils, can change greatly with temperature. In such cases, temperature compensations must be made when taking resistance readings.

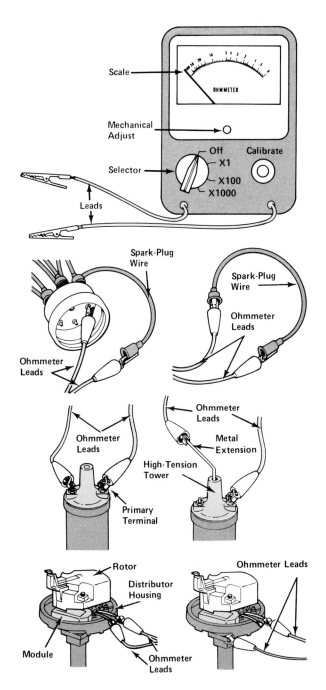

Figure 35-16. Ohmmeter checks.

Figure 35-15. Engine analyzer.

CAUTION!!!

Ohmmeters can be ruined instantly by attaching test leads to powered circuits. Disconnect the grounded battery terminal to prevent damage to ohmmeters during resistance tests.

Secondary cables are constructed to have electrical resistance primarily to decrease radio interference. This resistance can increase over time and thus decrease firing current. A general recommendation is that no ignition secondary cable, regardless of length, should have more than 50,000 Ω (ohms) resistance.

Timing Light

A *timing light* is an instrument that produces a brief (1/1000 second), repeated flash of bright light that appears to "freeze" a moving pulley or flywheel. The flash is triggered by current flowing to the No. 1 spark plug. Thus, timing marks on the engine can be seen for checking and/or adjustment.

An *inductive timing light* has an inductive pickup that is clamped around the No. 1 spark plug cable. Two other connections to the battery terminals power the flash and sensing circuitry inside the timing light **[Figure 35-17]**.

SAFETY CAUTION

On most vehicles, timing marks are located on and near the power pulley, near the fan and moving belts. The fan becomes almost invisible as it spins. Be extremely careful to avoid the spinning fan, belts, and pulleys as the timing light is aimed at timing marks. Personal injury or damage to the timing light can result if it is caught by moving parts.

Vacuum and centrifugal advance mechanisms can be checked with an *advance timing light* **[Figure 35-18]**. A calibrated dial on the instrument can be set to, in effect, "move" the timing marks. This function allows a technician to determine ignition advance by reading a number scale on the timing light.

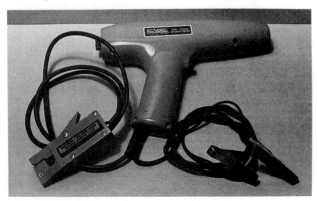

Figure 35-17. Timing light.

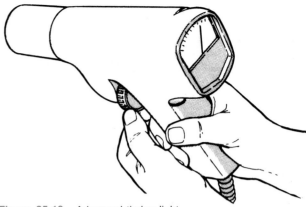

Figure 35-18. Advanced timing light.

Manufacturers' service manuals specify vacuum and centrifugal advance at different engine rpm. By removing and plugging the vacuum advance and/or retard hoses, centrifugal advance alone can be determined. Subtracting the centrifugal advance from the total advance with hoses connected will indicate vacuum advance alone.

Magnetic probe timing adapters are meant for use with special instruments. A probe inserted into the hole can detect the presence of a notch on a crankshaft balancer pulley. Some vehicles have provision for using either a timing light or a magnetic probe timing instrument **[Figure 35-19]**.

Hand Vacuum Pump

A *hand vacuum pump* is useful for checking vacuum-diaphragm operated devices for proper operation and/or leakage **[Figure 35-20]**. After securely attaching a rubber tube, the hand vacuum pump is squeezed until the gauge registers vacuum. At least 15 inches (381 mm) of vacuum tubing should be used for testing. At the same time, the mechanical linkage is checked for movement.

If the linkage moves, the vacuum is held for at least 30 seconds. A loss of vacuum indicates a leaking diaphragm assembly that must be replaced. A leaking or broken vacuum advance diaphragm can cause hesitation, stalling, and dying on acceleration.

To check for internal breakage in primary switching mechanism wires, such a test can be performed with the engine running. If the engine dies or stalls, the primary switching mechanism and leads should be doublechecked with an ohmmeter while repeating the vacuum test.

Oscilloscope

An *oscilloscope* is a sophisticated electronic testing instrument. An oscilloscope displays changing levels of voltage versus time in an electrical system in relation to time. Characteristic patterns, or traces,

IGNITION SYSTEM SERVICE

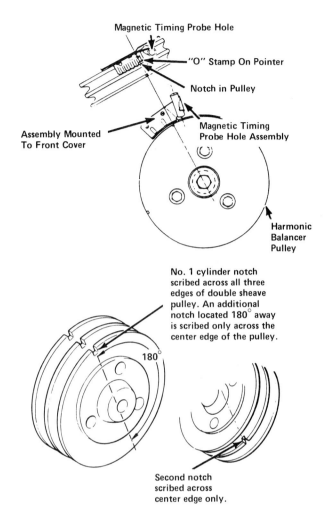

Figure 35-19. Magnetic probe timing adapter.
BUICK MOTOR DIVISION—GMC

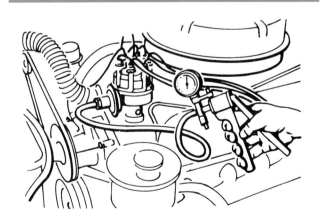

Figure 35-20. Using a hand vacuum pump.

are recognizable on the face of a screen similar to that of a television set [**Figure 35-21**]. Patterns for both the ignition primary and secondary systems can be displayed. A typical firing trace for a single spark plug in a breaker point ignition system is

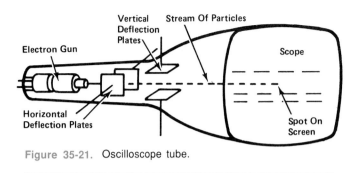

Figure 35-21. Oscilloscope tube.

shown in **Figure 35-22**. Deviations from the typical pattern shown can pinpoint ignition system problems.

Spark plug firings can be displayed in several ways [**Figure 35-23**]. For example, all firing patterns can be displayed one on top of the other, or *superimposed*. A *raster* pattern displays all firing order traces separately, in a stack. A normal *display* shows the firings one after the other, in firing order.

Patterns for electronic ignition systems differ. Typical patterns for General Motors, Chrysler, and Ford systems are shown in **Figure 35-24**.

Problems can be spotted on a display [**Figure 35-25**]. Single cylinder firings then can be displayed for more detailed analysis.

Properly used, an oscilloscope can be an important tool in determining the cause of engine operation problems. Experience in recognizing ignition system problems from oscilloscope patterns is a valuable job skill for a technician.

35.5 TESTING ELECTRONIC IGNITION CONTROL UNITS

Test procedures for electronic ignition control units may involve multiple steps and measurements with voltmeters and ohmmeters. Manufacturers also may specify the use of special testers for electronic ignition modules. *Testing by substitution* may be the recommended procedure. Testing by substitution means substituting a part that is known to be good for a suspected bad part. If the good part functions properly, the problem is in the old part.

Refer to the vehicle manufacturer's service manual for correct servicing procedures. Improper servicing procedures can result in damage to sensitive electronic components.

35.6 SPARK PLUG REPLACEMENT

On older models, spark plugs were replaced at intervals of 10,000-12,000 miles (16,000-19,200 km), or approximately once a year. However, high-voltage electronic ignition systems, the use of unleaded gasoline, and platinum-tipped spark plugs have increased this interval. On current vehicles, spark plug replacement intervals of 15,000-45,000 miles (24,000-72,000 km) are recommended.

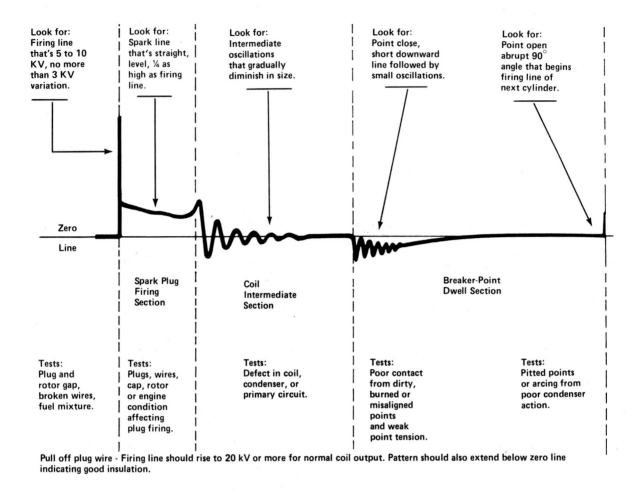

Figure 35-22. Ignition system firing pattern. SUN ELECTRIC CORPORATION

Removing Spark Plugs

Before attempting spark plug removal, mark the spark plug wires for correct replacement. Compressed air can be used to blow out dirt and debris from around spark plugs before removal [Figure 35-26]. Such materials could fall into the cylinder through the spark plug hole and cause internal damage. Small chips of rock also can lodge between the valve face and seat, preventing valve closure.

SAFETY CAUTION

Wear eye protection whenever working with compressed air. Pain or blindness can result from debris or liquids blown into the eyes.

Obstructions often cause problems in spark plug replacement. Different combinations of extensions and universal joints may be necessary to reach some spark plugs. On some vehicles, spark plugs can be removed from below the engine through wheel wells when the vehicle is raised.

Use a ratchet or breaker bar to loosen the spark plug, then turn it out by hand. Keep the spark plugs in order as they are removed. Inspect the spark plug for damage, wear, and indications of engine problems [Figure 35-27].

Spark Plug Color Reading

An engine that is operating properly with the specified spark plugs will produce characteristic colored deposits on the spark plugs. Color reading is done at the firing tip and insulator nose only. Correct colors are medium grays, browns, and tans.

Extremely light deposits (white, yellow, pink) indicate combustion temperatures that are too high. Possible causes include vacuum leaks, lean mixtures, cooling system problems, and spark plugs of incorrect (too hot) heat range.

Extremely dark deposits (dark brown or sooty black) indicate combustion temperatures that are too low. Possible causes include rich air-fuel mixtures, short trip/cold weather operation, cooling system problems, and spark plugs of incorrect (too cold) heat range. Spark plug color reading can be done accurately only on plugs from an engine that has been brought to normal operating temperature.

IGNITION SYSTEM SERVICE

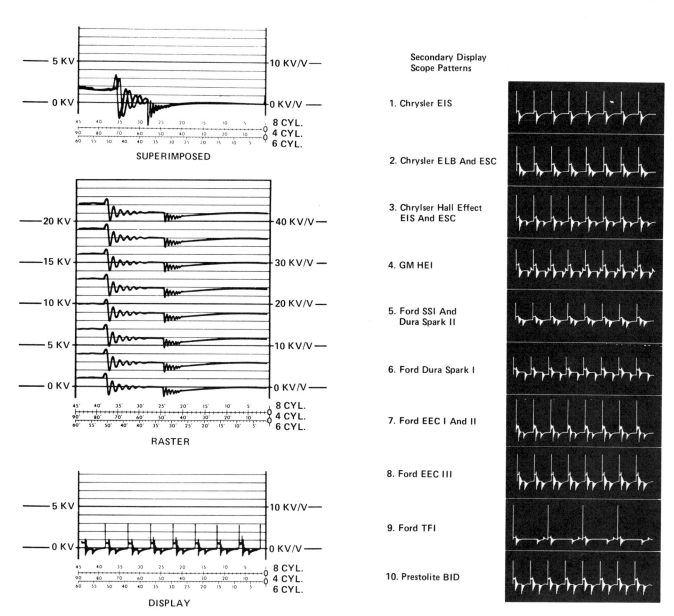

Figure 35-23. Superimposed, raster, and display patterns.
SUN ELECTRIC CORPORATION

Figure 35-24. Secondary display patterns.
SUN ELECTRIC CORPORATION

Replacing Spark Plugs

The plugs found installed in an engine may not have the correct heat range or reach. Always refer to the manufacturer's service manual for correct replacement spark plug numbers.

New plugs always must be checked for a correct air gap. If incorrect, the plug must be regapped. To install, start the plug by hand, then tighten to the manufacturer's torque specification [Figure 35-28]. As a final step, the correct spark plug cable is pushed firmly onto the spark plug.

35.7 REPLACING BREAKER POINTS

Breaker points require regular replacement at intervals of 10,000-12,000 miles (16,000-19,200 km).

If no pitting of the contact point surfaces is evident, the condenser need not be replaced.

Leads from the coil primary terminal and condenser may be screwed on or held behind the point spring. Screws holding the point set must be loosened and/or removed [Figure 35-29].

To adjust a replacement set of points, the point rubbing block must rest on a corner of the distributor cam. To position the distributor cam, the engine must be rotated. If the crankshaft pulley nut is accessible, the engine can be turned by hand with the proper-sized wrench. Another method of positioning the crankshaft is to connect a *remote starter* to the starter relay or solenoid [Figure 35-30].

A remote starter is an electrical switch with leads. Either lead is connected to battery power. The

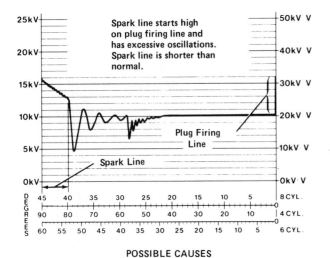
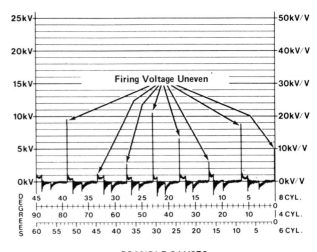

POSSIBLE CAUSES (left)

A. Higher than normal.
1. Worn spark plug.
2. Open plug wire.
3. Distributor cap.
4. Unbalanced (lean) fuel mixture.

B. Lower than normal.
1. Closely-gapped or fouled spark plugs.
2. Shorted or broken spark plug.
3. Cylinder compression low.

POSSIBLE CAUSES (right)

A. Present on all cylinders.
1. Coil tower.
2. Coil tower wire.
3. Corroded coil terminals.
4. Distributor cap.
5. Rotor.

B. Present on one or more cylinders, but not all.
1. Open plug wire.
2. Distributor cap.
3. All plug wire connections corroded.

Figure 35-25. Firing problems. HEATH COMPANY

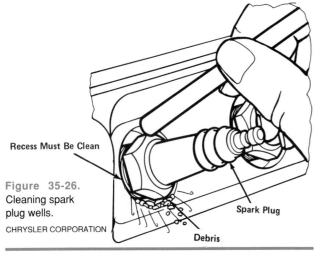

Figure 35-26. Cleaning spark plug wells. CHRYSLER CORPORATION

other end is connected to the key switch terminal. When the button is pushed, the starter cranks the engine. A remote starter switch also can be used to position the engine for other service operations.

If the vehicle is equipped with a manual transmission, put the transmission in high gear and gently push or pull the vehicle to position the ignition rubbing block on the tip of a distributor cam lobe.

When the points are fully opened, the breaker point air gap must be set correctly [Figure 35-31]. Point and condenser holding screws are tightened and the point gap checked again. Tightening point holding screws may change the adjustment. Connecting wires from the coil and condenser are tightened securely.

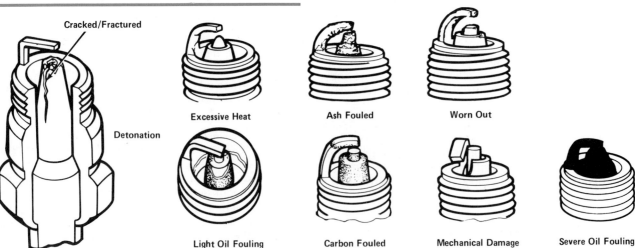

Figure 35-27. Inspecting spark plugs. CHRYSLER CORPORATION

IGNITION SYSTEM SERVICE

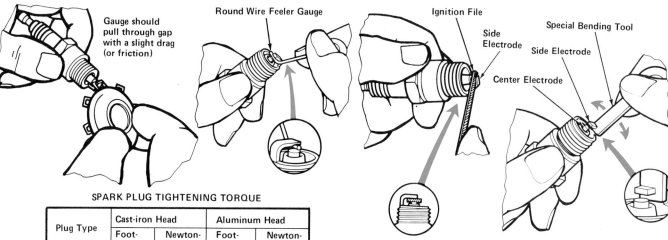

SPARK PLUG TIGHTENING TORQUE

Plug Type	Cast-iron Head		Aluminum Head	
	Foot-Pounds	Newton-Meters	Foot-Pounds	Newton-Meters
14-mm Gasketed	25-30	34-40	15-22	20-30
14-mm Tapered Seat	7-15	9-20	7-15	9-20
18-mm Tapered Seat	15-20	20-27	15-20	20-27

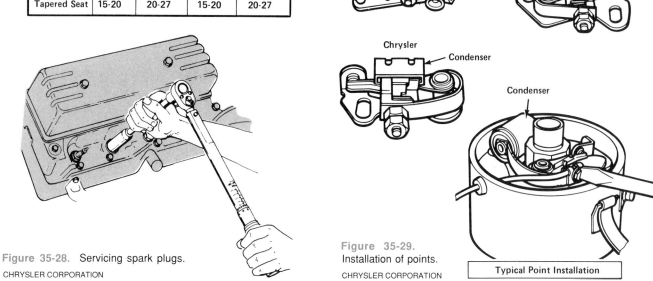

Figure 35-28. Servicing spark plugs.
CHRYSLER CORPORATION

Figure 35-29. Installation of points.
CHRYSLER CORPORATION

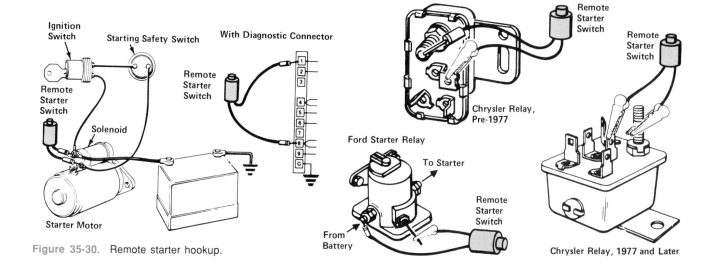

Figure 35-30. Remote starter hookup.

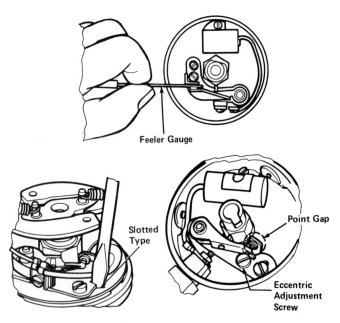

Figure 35-31. Adjustment of points. CHRYSLER CORPORATION

Most vehicles built since 1969 include emissions control instructions on a sticker or decal in the engine compartment. Step-by-step instructions must be followed to tune an engine correctly. It may be necessary to disconnect and plug several vacuum hoses before ignition timing can be set properly. Emission control systems are discussed in Unit 41.

Refer to the underhood sticker or manufacturer's service manual for proper ignition timing procedures.

To check ignition timing, timing light leads are attached to the battery and No. 1 cylinder spark plug cable. All instrument leads are positioned away from moving parts at the front of the engine. Vacuum advance hose is disconnected and plugged. Clean and mark the timing marks with white chalk before checking timing.

The engine is started and the timing checked. Be sure the idle speed is correct. If it is necessary to correct timing, the hold-down clamp beneath the distributor is loosened. The distributor is rotated slightly until the marks are aligned correctly [Figure 35-32]. Rotating the distributor changes the relationship of the rotor and switching mechanism to the

CAUTION!!!

Air gaps on electronic ignition systems normally do not need adjustment or checking. Electronic ignition systems, which can be adjusted, require the use of nonmagnetic brass feeler gauges. Use of steel gauges can cause a pickup coil or reluctor to become magnetized and no longer function properly.

After setting points, a dwell reading is taken with an engine analyzer. If the dwell angle is incorrect, the point gap must be readjusted. Too wide a gap will produce a low dwell angle. Too narrow a gap will produce a high dwell angle.

35.8 SETTING IGNITION TIMING

Ignition timing is dependent upon many factors. These factors include:

- Spark plug air gap
- Dwell
- Engine rpm
- Emission control devices.

Both spark plug gap and dwell must be correct before ignition timing is set. Also, the engine must be at full operating temperature. On electronic ignitions, dwell is determined electronically and is not adjustable, but must be within the manufacturer's specifications.

In some cases, engine rpm must be adjusted slower or faster than normal for ignition timing purposes. Engine rpm is changed by turning fuel-system idle speed adjustments. Be sure to follow manufacturer's instructions.

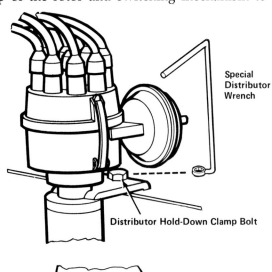

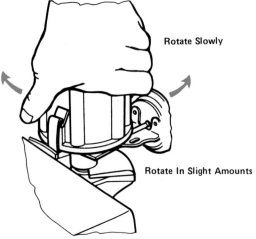

Figure 35-32. Turning distributor to set timing.
CHRYSLER CORPORATION

distributor shaft. After the hold-down clamp bolt is tightened, timing is rechecked. Tightening the clamp may cause the timing to shift slightly. All disconnected vacuum hoses are replaced.

The timing mark on some late-model vehicles appears not as a single notch, but as a wide pattern [Figure 35-33]. Refer to the manufacturer's service manual for correct ignition timing procedures.

35.9 DISTRIBUTOR REMOVAL AND REPLACEMENT

For some service operations, the distributor must be removed. Most distributors are driven by a helical gear that meshes with another gear on the camshaft. If the distributor is removed, the relationship between ignition timing and valve timing is changed.

The distributor rotor must be properly positioned in relationship to crankshaft and camshaft position when the distributor is replaced. Misalignment by only one gear tooth will change ignition timing enough to keep the engine from starting or running.

Before removing the distributor, reference marks are made when the rotor is pointing toward the engine [Figure 35-34]. Because the gear is curved, the rotor will rotate as the distributor is removed and replaced. When the distributor is reinstalled, the reference marks must be realigned by correctly positioning the gear. The engine must not be cranked when the distributor is removed. After installation, always recheck ignition timing.

Dead Timing

Dead timing can be done if the engine is accidentally cranked, or when reinstalling a distributor on an overhauled engine. Dead timing is a process of replacing a distributor correctly in an engine for starting purposes. After starting, ignition timing can be reset properly. Dead timing consists of two steps:

1. Position the engine crankshaft at the No. 1 cylinder's firing position. This can be determined from ignition timing marks and pressure from the No. 1 cylinder's spark plug hole. As the piston moves on its compression stroke, pressure is created. Placing a thumb *over* the spark plug hole will help you to feel compression. Then align the timing marks to initial timing.
2. Place the distributor back in its mounting hole. The rotor must be aligned with the distributor cap terminal leading to the No. 1 spark plug wire.

After the engine has been started, use a timing light to set ignition timing.

Refer to the manufacturer's service manual for firing order and approximate distributor positioning.

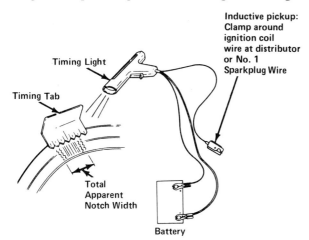

Figure 35-33. Timing notch width. BUICK MOTOR DIVISION—GMC

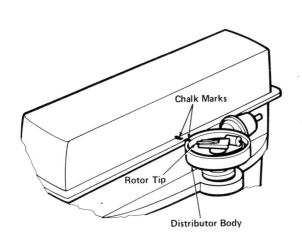

Figure 35-34. Removing and replacing distributor.
CHRYSLER CORPORATION

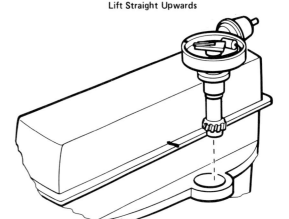

Do Not Crank Engine When Distributor Is Removed

UNIT HIGHLIGHTS

- Basic checks to determine why an engine won't start include checks for the presence of fuel and ignition spark.
- A tune-up includes checks and maintenance of the charging, fuel, ignition, and emission control systems.
- Visual inspection can locate the causes of many ignition system faults. Diagnostic instruments are used to help locate other ignition system problems.
- Ignition system service includes diagnostic procedures, removal and replacement of defective parts, and adjustments to ignition timing.
- Commonly used ignition system diagnostic instruments include the engine analyzer, timing light, hand vacuum pump, and oscilloscope.
- Experience in recognizing proper and improper oscilloscope patterns is a valuable job skill for technicians.
- Ignition timing is set at a specified rpm. The engine must be at full operating temperature, and spark plug gap and dwell must be correct. Emission control vacuum lines may need to be disconnected and plugged before timing is set.

TERMS

firing order
carbon tracking
analog
timing light
advance timing light
oscilloscope
raster
testing by substitution
dead timing

corona effect
inductive pickup
digital
inductive timing light
hand vacuum pump
superimposed
display
remote starter

REVIEW QUESTIONS

DIRECTIONS: The following questions are similar to those used on automotive technician certification tests. On a separate sheet of paper, write the letter of the correct choice.

1. With the coil cable connected, an engine cranks very slowly and will not start. With the coil cable disconnected, an engine cranks quickly and easily. The most likely cause is:
   A. a weak battery.
   B. a defective rotor or cap.
   C. the ignition timing is advanced too far.
   D. the ignition timing is retarded too far.

2. Technician A says that ignition system problems can be found during visual inspections.
   Technician B says that diagnostic instruments can be used to find ignition system problems.
   Who is correct?
   A. A only   C. Both A and B
   B. B only   D. Neither A nor B

3. Which of the following statements is correct?
   I. Ignition timing procedures may require changing idle speed.
   II. Ignition timing procedures may require disconnecting and plugging vacuum hoses.
   A. I only   C. Both I and II
   B. II only   D. Neither I nor II

4. When is dead timing necessary?
   A. During every tune-up
   B. When ignition timing is adjusted
   C. When the distributor is removed and replaced
   D. When the crankshaft has been rotated while the distributor is out

Question 5 is not like those above. It has the word EXCEPT. For question 5, look for the choice that does *not* apply. Read the entire question carefully before you choose your answer.

5. All of the following are checked and/or adjusted during a tune-up EXCEPT
   A. belts, hoses, and tire pressures.
   B. battery and charging system.
   C. fuel and ignition systems.
   D. emission control system.

ESSAY QUESTIONS

1. What ignition system problems can be found during a visual inspection?
2. Identify and describe instruments used to troubleshoot ignition system problems.
3. What problems can be found by "reading" spark plugs?

SUPPLEMENTAL ACTIVITIES

1. Check a vehicle for spark at both the spark plug and coil cables. Demonstrate to your class proper operation and/or any problems found.
2. Perform a visual inspection of all ignition system parts. Make notes and report any problems found to your class.
3. Form small groups. Choose one diagnostic instrument and read the instrument manufacturer's instructions for use. Demonstrate to the rest of the class, one group at a time, how to connect and use diagnostic instruments. Choose either the remote starter, engine analyzer, timing light, hand vacuum pump, or oscilloscope.
4. Remove, color read, and replace spark plugs and set ignition timing properly on a vehicle. Report to your class on problems found and conditions necessary for setting ignition timing.
5. Remove a distributor and crank the engine of a vehicle. Perform dead timing, start the engine, and set ignition timing properly.

36 Electrical Devices

UNIT PREVIEW

Wires for vehicle circuits are grouped together in harnesses, or bundles. Individual wires are color-coded to aid in identification. Several wires may be connected to a common source of power or ground.

Circuit protection devices include fuses, fusible links, and circuit breakers. Relays are used in place of heavy-duty switches to conduct large current flows.

A wiring diagram is like a road map to indicate the path of current flow for individual circuits.

Warning devices and instruments are used to alert the driver to problem conditions. Sending units are used with either lights or gauges. Some indicating devices that operate mechanically, rather than electrically.

Electrical devices include lighting, warning or indicating devices, and comfort accessories.

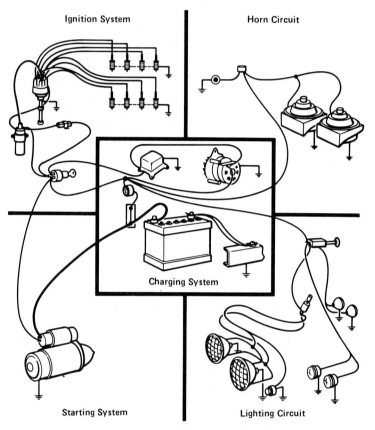

Figure 36-1. Simplified electrical system.

LEARNING OBJECTIVES

When you have completed your assignments and exercises in this unit, you should be able to:

❑ Describe how wiring connections are made in a vehicle.
❑ Identify and describe parts on vehicle wiring diagrams.
❑ Trace wires and locate connectors on a vehicle.
❑ Locate fuse blocks and turn-signal flashers on a vehicle.
❑ Locate warning light or gauge sending units on a vehicle.
❑ Check and replace burned-out bulbs on a vehicle.

36.1 WIRING CIRCUITS

As discussed in Unit 27, electricity flows from a source, through conductors to an electrical load, and back to the source. The roughly circular path through which electricity flows is called a circuit. In addition to charging and starting systems, electrical system circuits include circuits for lights, indicator lights, and accessories. A simplified diagram of an automotive electrical system is shown in **Figure 36-1**.

Each electrical unit on a vehicle is part of a specific circuit. Lights, radios, windshield wiper motors, blower motors, and other units each may have a separate circuit. Even though physically far apart, series, parallel, and series-parallel circuits may be electrically interconnected **[Figure 36-2]**.

Common Connections

A shared source of power or a common ground can be used to decrease separate connections **[Figure 36-3]**. Wherever several units are physically close, such as instruments and lights in an instrument panel, common connections can be used.

Wiring Harness

Wires of different sizes run throughout a vehicle **[Figure 36-4]**. Wires are grouped together in *harnesses*. A harness is a bundled group of wires. At specific points, wires lead from a harness to electrical units.

382

ELECTRICAL DEVICES

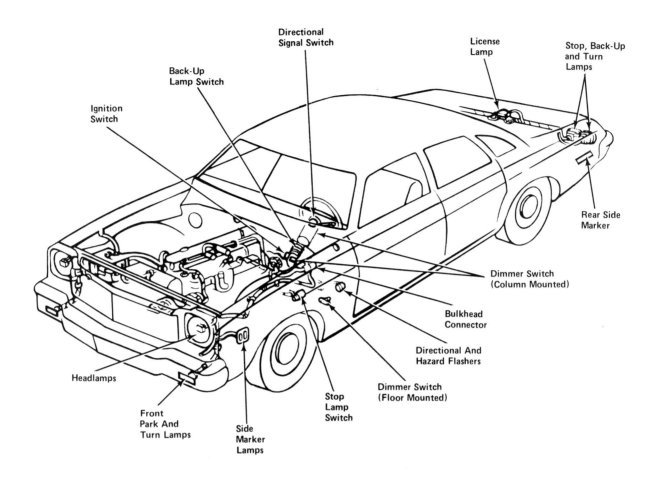

Figure 36-2. Typical locator.
GENERAL MOTORS CORPORATION

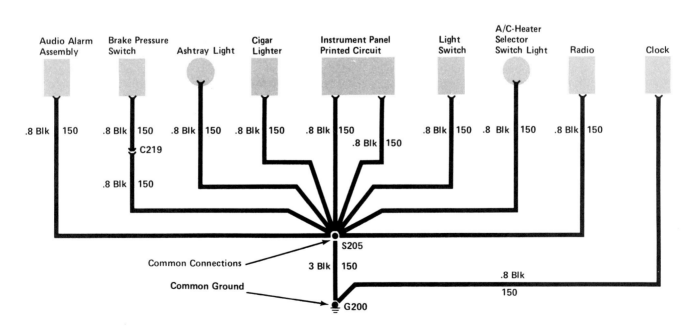

Figure 36-3. Instrument panel ground distribution.
BUICK MOTOR DIVISION—GMC

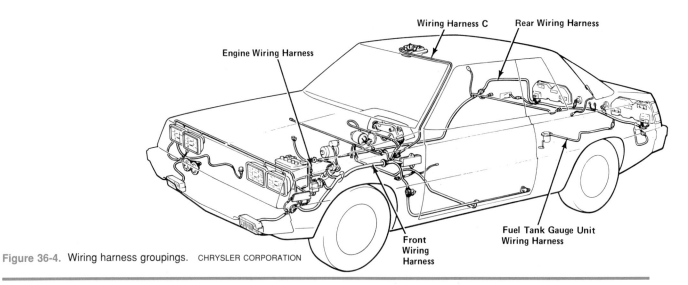

Figure 36-4. Wiring harness groupings. CHRYSLER CORPORATION

Wiring harnesses can contain many wires, especially in the area of the dash and firewall **[Figure 36-5]**.

Color Coding of Wires

Colored plastic insulation on wires helps a technician to trace wires through connections. The plastic insulation may be a solid color or may have a small *tracer*, or stripe, of another color. Wires often run to connectors that can be disassembled for checking and/or replacement of electrical units **[Figure 36-6]**. Finding the same color wire on both sides of a connector helps to trace a wire if a problem occurs.

Printed Circuits

A *printed circuit* is a thin sheet of nonconductive plastic material on which conductive metal has been deposited. Parts of the metal are etched, or eaten away, by acid. The remaining metal lines form conductors for separate circuits **[Figure 36-7]**.

A wiring connector can be plugged into a common point **[see numbers 1 and 2 in Figure 36-7]**. Power or ground circuits are completed through connected wires.

Wiring Diagram

A *wiring diagram* is a drawing similar to a road map. Wiring diagrams show how wires are connected in circuits **[Figure 36-8]**. Insulation colors may be reproduced on a diagram. In other cases, black lines are used to indicate all wires. In such cases, insulation colors may be written next to lines indicating wires, for example: "pink with black stripe." Insulation colors may also be abbreviated (pnk/blk) or coded with letters (p/b). When coded with letters, a guide is included with a wiring diagram (p = pink, b = black).

Wiring diagrams may include all or only a portion of a vehicle's electrical system. Because of the complexity of vehicle electrical systems, partial wiring diagrams are easier to follow and understand.

36.2 CIRCUIT PROTECTION

The vehicle must be protected from fire hazards that occur if a powered circuit is accidentally shorted or grounded. Fuses, fusible links, and *circuit breakers* can be used to protect circuits. A circuit breaker interrupts an excessive flow of current but does not require replacement.

Fuses and Fusible Links

The functioning of fuses and fusible links is discussed in Unit 27. Fuses and fusible links are rated by current-carrying capacity. For example, a fuse for a given circuit might be rated at 5, 10, 15, or 20 or more amperes. When more than the rated current attempts to flow, the fuse or fusible link conductor melts to open the circuit. Burned-out fuses and fusible links must be replaced. Fusible links are usually located in the engine compartment near the battery, alternator, or separate voltage regulator.

Although separate fuses also are used, main circuit fuses are usually grouped together in *fuse blocks*. Fuse blocks contain fuses, circuit breakers, *turn signal flashers*, and relays **[Figure 36-9]**. A turn signal flasher contains a bimetallic spring that is heated by current flowing through it. As it heats, the spring flexes to open the circuit. When the spring has cooled, it flexes back to make contact and conduct current to the flasher bulbs again.

Fuse blocks often are located under the dash near the headlight switch. However, fuse blocks also can be located elsewhere under the dash, in the glove compartment, or in the engine compartment.

Circuit Breaker

A circuit breaker also is a device that interrupts, or opens, a circuit when too much current flows.

ELECTRICAL DEVICES

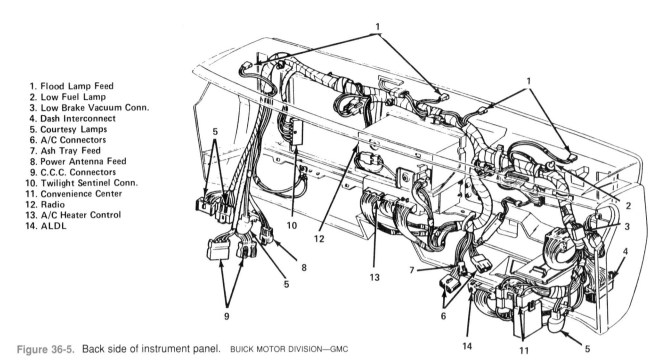

1. Flood Lamp Feed
2. Low Fuel Lamp
3. Low Brake Vacuum Conn.
4. Dash Interconnect
5. Courtesy Lamps
6. A/C Connectors
7. Ash Tray Feed
8. Power Antenna Feed
9. C.C.C. Connectors
10. Twilight Sentinel Conn.
11. Convenience Center
12. Radio
13. A/C Heater Control
14. ALDL

Figure 36-5. Back side of instrument panel. BUICK MOTOR DIVISION—GMC

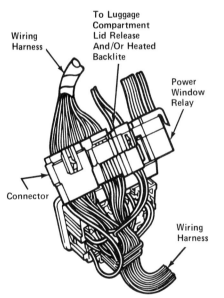

Figure 36-6. Connections from instrument panel to fuse panel. FORD MOTOR COMPANY

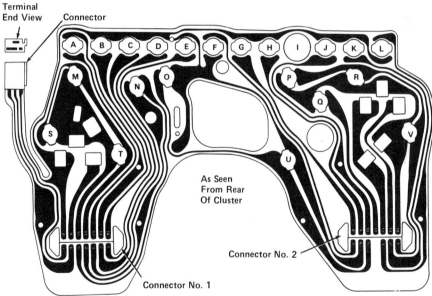

Figure 36-7. Printed circuit. BUICK MOTOR DIVISION—GMC

However, a circuit breaker does not need replacement like a fuse or fusible link.

A simple form of circuit breaker contains only a bimetallic spring and a set of contact points [**Figure 36-10**]. The bimetallic spring heats as current passes through it. When excessive current flows, the spring bends upward to interrupt a circuit. While the circuit is interrupted, the spring cools. When the spring has cooled sufficiently, it bends downward to complete a circuit again. The turn signal flasher is a small form of a circuit breaker.

Some circuit breakers must be reset by hand. A locking device must be released to close contact points and complete a circuit again.

36.3 RELAYS

A large current flow can be switched by a heavy-duty switch. For example, older vehicles used a foot-operated headlight dimmer switch to switch from low to high beams and back.

386 PART IV: AUTOMOBILE ELECTRICAL AND ELECTRONIC SYSTEMS

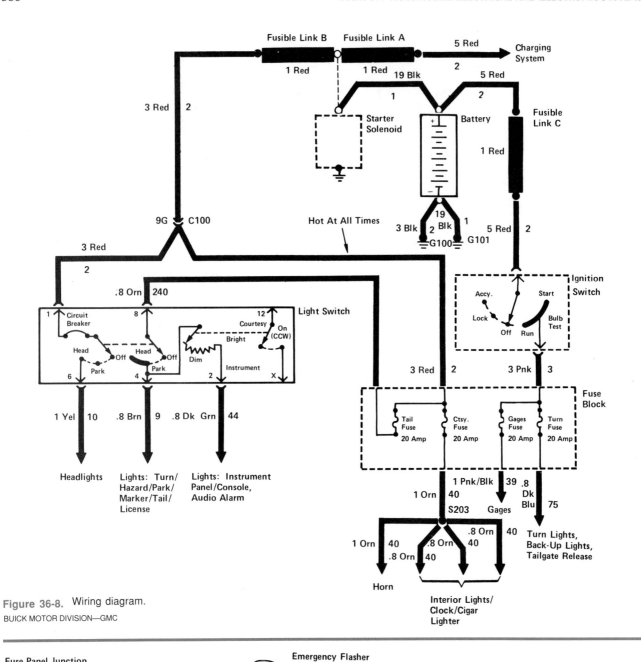

Figure 36-8. Wiring diagram.
BUICK MOTOR DIVISION—GMC

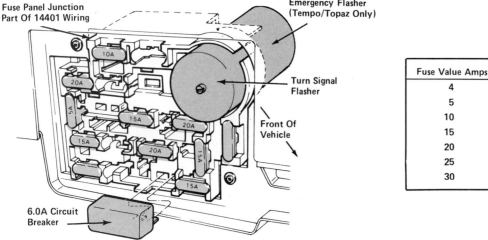

Figure 36-9. Fuse block and fuse locations.

Fuse Value Amps	Color Code
4	Pink
5	Tan
10	Red
15	Light Blue
20	Yellow
25	Natural
30	Light Green

ELECTRICAL DEVICES

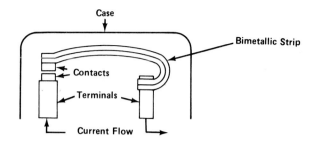

Figure 36-10. Simple circuit breaker.

Newer vehicles may use a small switch mounted on a *stalk*, or lever, next to the steering wheel. The small switch is connected to a relay that operates like a starter relay or solenoid **[Figure 36-11]**. When the small switch is operated, relay contacts close to direct a large current flow to the headlights.

Relays are used wherever high current flows are necessary. Other examples of such circuits include the starter motor, horns, and electrically operated window defroster units.

36.4 WARNING DEVICES

To alert a driver quickly to problem conditions, warning lights or sound warning devices can be used. These devices operate only when a problem condition occurs.

Warning Lights

Warning lights are located on the dash to indicate problem conditions to a driver. Standard warning lights include:

- Overheating
- Battery discharging
- Low oil pressure
- Brake system failure/handbrake applied
- Turn signal operation
- High-beam headlight operation.

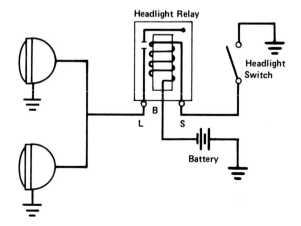

Figure 36-11. Headlight relay system.

When the engine is running and all monitored systems are operating normally, lights remain off. Circuits that include warning light bulbs and/or sensors are incomplete until a problem condition occurs. When a monitored system is malfunctioning, a power or ground connection is made to complete a circuit **[Figure 36-12]**.

Lights also may be used to warn of other problem conditions, such as burned-out bulbs, low fluid levels, and water in the fuel **[Figure 36-13]**.

Sound Warning Devices

Buzzers or chimes are used to alert a driver to certain conditions. These can include doors that are ajar, keys left in an ignition switch, and lights left on. **Figure 36-14** illustrates a tone generator system schematic and a typical location of a warning device.

Graphic Displays

Graphic displays are translucent drawings or pictures of a vehicle. Lights operate to indicate problem locations. **Figure 36-15** illustrates examples of graphic warning displays.

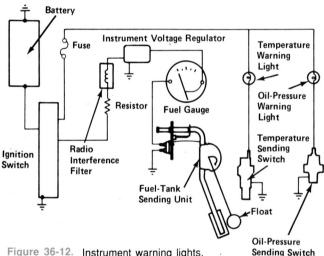

Figure 36-12. Instrument warning lights.

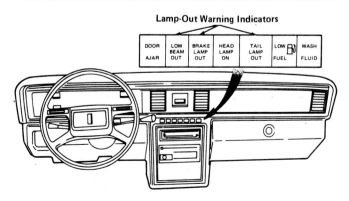

Figure 36-13. Lamp-out warning indicators.
FORD MOTOR COMPANY

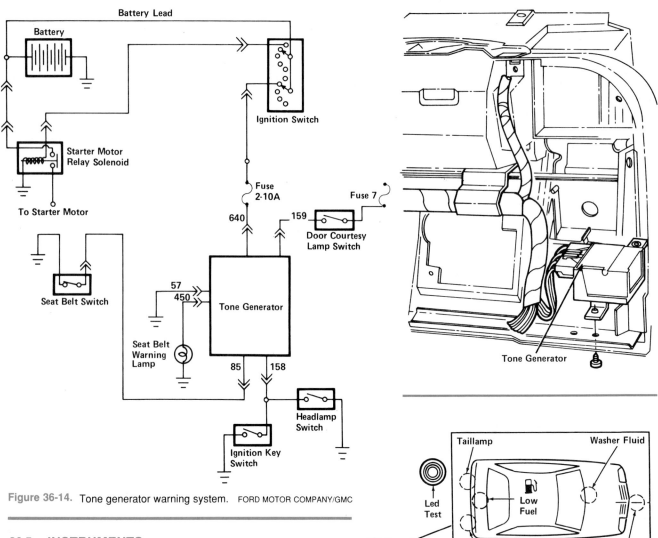

Figure 36-14. Tone generator warning system. FORD MOTOR COMPANY/GMC

36.5 INSTRUMENTS

Instruments, or gauges, are used to monitor vehicle systems. Instruments can indicate a relative position within a scale of values or an exact number. Those that indicate a relative scale of values are known as analog instruments. Those that indicate an exact number for a measured quantity are known as digital instruments.

Gauges

Common electrical analog instruments include gauges for:

- Fuel level
- Engine oil pressure
- Coolant temperature
- Charging system operation
- Tachometer.

Instrument Voltage Regulator

The fuel, oil pressure, and coolant temperature gauges require a stable voltage source for proper

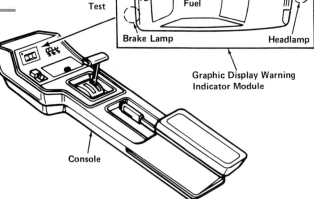

Figure 36-15. Graphic display warning system.
FORD MOTOR COMPANY

operation. An *instrument voltage regulator* is used to stabilize and limit voltage for accurate instrument operation [Figure 36-16].

Sending Units

Sending units are devices with variable electrical resistance based on physical movement. Movement

ELECTRICAL DEVICES

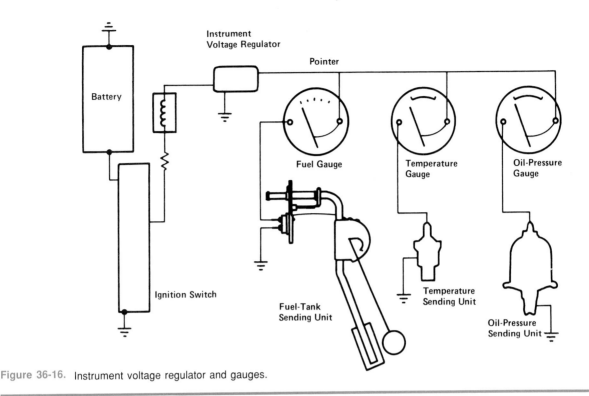

Figure 36-16. Instrument voltage regulator and gauges.

may be caused by pressure against a diaphragm, heat, or motion of a float as liquid fills a fuel tank. Two types of electrical analog gauges commonly are used with sending units:

- Magnetic
- Thermal.

Magnetic Gauges

The simplest form of a *magnetic gauge* is a simple ammeter. A permanent magnet attracts a ferrous indicator needle connected to a pivot point and holds it centered. An armature, or coil of wire, is wrapped around the base of the needle, near the pivot point. Current can flow through a conductor beneath the armature [Figure 36-17].

When current flows, a magnetic field around the conductor induces magnetism in the armature. This magnetism opposes that of the permanent magnet. Attractive or repulsive magnetic forces cause the needle to swing left or right. The direction the needle swings depends on the direction of current flow in the conductor.

Balancing Coil Gauge

A *balancing coil gauge* also operates on principles of magnetic attraction and repulsion. However, no permanent magnet is used. The base of an indicating arm is pivoted and includes an armature. Two coils are used to create magnetic fields [Figure 36-18].

A sending unit has a variable resistance. Electricity always follows the path of least

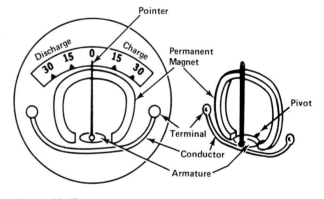

Figure 36-17. Simple ammeter.

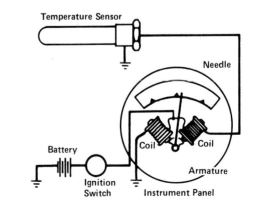

Figure 36-18. Balancing coil gauge.

resistance. The two coils are connected so that electricity can flow through either one. When the resistance of a sending unit is low, the right-hand coil receives more current than the left-hand coil. More magnetism is created in the right-hand coil, attracting the armature. Thus, a gauge needle moves to the right.

When resistance of a sending unit is high, the left-hand coil receives more current. More magnetic force is created in the left-hand coil, and a gauge needle swings to the left.

Thermal Gauge

A *thermal gauge* operates through heat created by an electrical flow [Figure 36-19]. A variable-resistance sending unit causes different amounts of current to flow through a heating coil within a gauge. The heat acts on a bimetallic spring attached to a gauge needle. When more heat is created, the needle swings farther up the scale. When less heat is created, the needle moves down the scale.

Nonelectrical Gauges

Not all instruments on a vehicle are electrical. Some gauges operate mechanically, without electrical power. Examples include:

- Speedometers
- Odometers
- Vacuum/pressure gauges.

Speedometer

On most vehicles, speedometers are mechanical, not electrical, instruments. A drive cable attached to a gear in the transmission turns a magnet inside a cup-shaped metal piece [Figure 36-20]. The cup is attached to a speedometer needle and held at zero by a *hairspring*, a fine wire spring. As the cable rotates faster with increasing speed, magnetic forces act on the cup and force it to rotate. The speedometer needle, attached to the cup, moves up the speed scale.

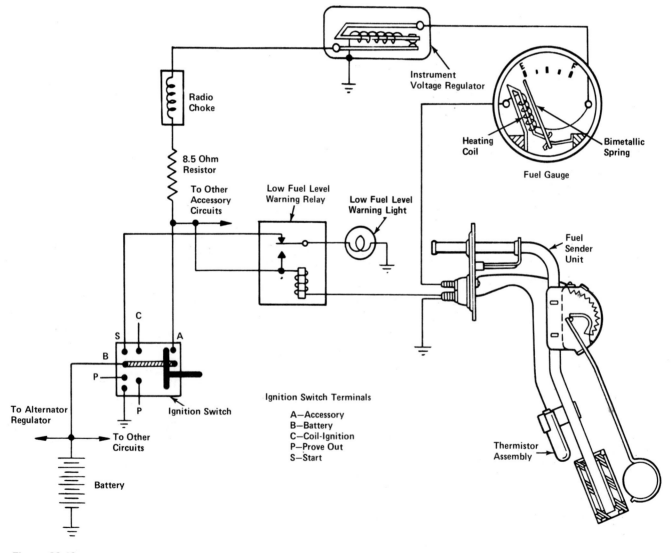

Figure 36-19. Thermal gauge.

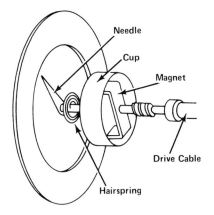

Figure 36-20. Speedometer.

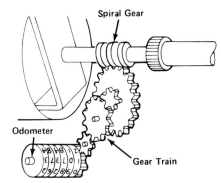

Figure 36-21. Odometer.

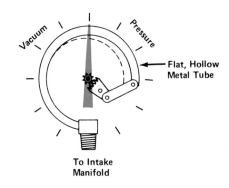

Figure 36-22. Bourdon tube gauge.

Odometer

An *odometer* is a digital instrument that records how far a vehicle has traveled. An odometer displays numbers that indicate distance in miles or kilometers. Odometer number drums are turned by a gear arrangement driven off a speedometer cable **[Figure 36-21]**.

Vacuum/Pressure Gauge

A vacuum/pressure gauge can be used to indicate how much manifold vacuum is present as an engine runs. Such gauges often are used as a rough indicator of fuel mileage because engine load is reflected in manifold vacuum readings.

For turbocharged or supercharged engines, *boost pressure gauges* indicate when manifold pressure increases or decreases.

A vacuum/pressure gauge is operated by a **Bourdon tube**. A Bourdon tube is a flat, hollow, coiled tube. A gauge needle is attached to the tube. A source of vacuum or pressure is connected to one end of the tube. The other end is sealed. A Bourdon tube coils or uncoils in response to vacuum or pressure, thus moving a gauge needle **[Figure 36-22]**. Some oil pressure gauges are Bourdon tube units.

36.6 LIGHTS

Lighting systems to improve visibility at night and to make a vehicle visible to other drivers include:

- Headlights, parking lights, and taillights
- Turn signal, cornering, and side marker lights
- Stop lights
- License plate lights
- Courtesy and convenience lights.

Headlights

Headlights are usually large, *sealed-beam* bulbs with single or double *filaments*. A sealed beam bulb is a large glass lens and reflector assembly that contains a filament. A filament is a metal element within a lightbulb. This unit is filled with an inert gas to prevent filament burnout. When electricity passes through a filament, the filament becomes white-hot and thus produces light. Headlights may have one or two filaments, depending upon their use. A headlight with two filaments is used for both bright and dim operation. A single-filament lamp is used only for one mode or the other.

Halogen lights with replaceable small bulbs also may be used as headlights. A halogen light contains a small quartz-glass bulb, inside of which is a filament surrounded by halogen gas. The small bulb fits within a larger reflector and lens element. Halogen lights produce a whiter, brighter light than conventional sealed-beam headlights **[Figure 36-23]**.

When burned out, covers and *bezels* must be removed to replace headlight bulbs **[Figure 36-24]**. A bezel is a retainer around a light or instrument. Bezels may be decorative or functional. Screws or springs are used to fasten a metal retainer ring that holds a bulb.

Headlight adjusting screws are located near lights. One screw tilts the bulb upward and downward. The other tilts the bulb left and right. When bulbs are removed, care must be taken not to

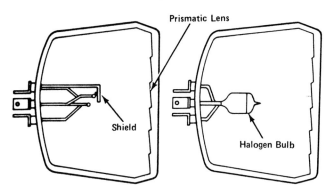

Figure 36-23. Headlights.

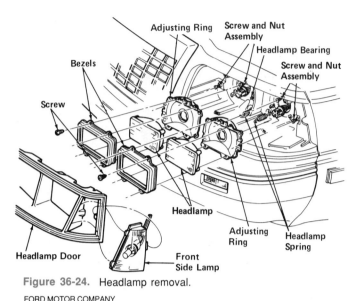

Figure 36-24. Headlamp removal. FORD MOTOR COMPANY

mistake headlight aiming screws for retaining screws. Misadjusted headlights can create poor visibility at night and/or create glare that temporarily "blinds" drivers of oncoming vehicles.

Taillights, Turn Signal, Side Marker, and Cornering Lights

Taillights burn when the headlights are on. Taillights have replaceable bulbs that usually can be reached from inside the trunk or by removing a plastic lens [Figure 36-25].

Turn signals and *side marker lights* may be separate units or combined in wrap-around plastic lenses. Side marker lights are located on the sides of a vehicle so that it can be seen more easily at night. Taillights, parking lights, side marker lights, and license plate lights operate when headlights are turned on. *Cornering lights* burn steadily when turn signals operate to light up an area in the direction of a turn.

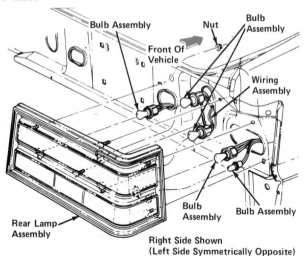

Figure 36-25. Taillights. FORD MOTOR COMPANY

A typical parking light, side marker light, and cornering light circuit is shown in Figure 36-26.

Stop Lights

Stop lights operate when a brake pedal is pushed. Most late-model vehicles have motion-operated switches attached to a brake pedal [Figure 36-27]. Older vehicles may have pressure-operated switches operated by brake master cylinder pressure. The switch illustrated in Figure 36-27 is constructed so that the contacts move with the pedal and the actuating pin is connected to the pushrod.

Courtesy and Convenience Lights

Courtesy and convenience lights include dome and courtesy lights that light when a door is opened. A switch is held in the open position by a closed door. When a door is opened, a spring pushes the switch closed to complete a circuit to dome or courtesy lights.

Underhood and trunk lights may be operated by a *mercury switch*. A mercury switch consists of two separated contacts within a glass tube that also

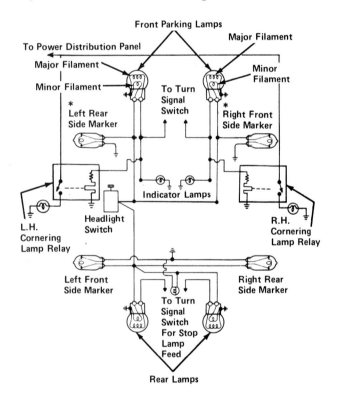

* Add cornering lamps into circuit to function with park and turn lamps when turn signal is activated.

Note: If parking lamp or taillamp has a three-wire socket, the black wire is the ground. If parking lamp or taillamp has a two-wire socket, the bulb is grounded through the lamp assembly.

Figure 36-26. Parking, side marker, and cornering light circuit. FORD MOTOR COMPANY

ELECTRICAL DEVICES

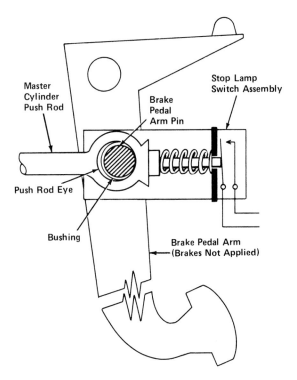
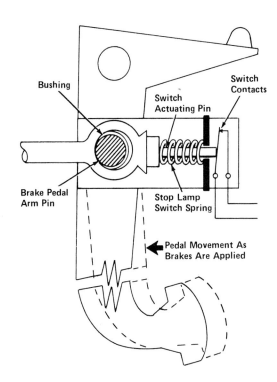

Figure 36-27. Brake pedal switch. FORD MOTOR COMPANY

contains liquid metal mercury. When the glass tube is tilted, mercury flows to contacts and completes a circuit. When the tube is tilted in the opposite direction, mercury flows away from contacts and opens a circuit.

Bulbs

Bulbs fit into sockets and are held by spring tension or mechanical force [Figure 36-28]. Bulbs are coded with numbers for replacement purposes. Bulbs with different code numbers may appear physically similar but have different wattage ratings. Smaller bulbs may simply push into a socket. Most bulbs must be pushed inward and turned counterclockwise to be removed. Replacement is accomplished by pushing in and turning clockwise.

36.7 WINDSHIELD WIPER AND WASHER

Windshield wipers are powered by a small single- or multi-speed electric motor. A switch on a steering

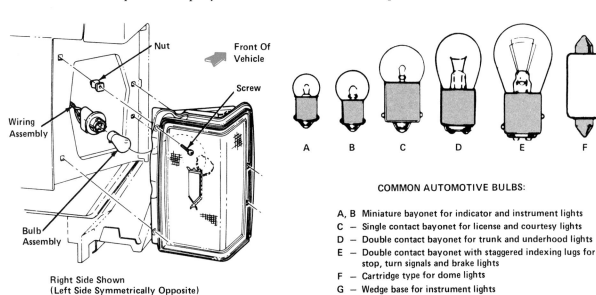

Figure 36-28. Types of bulbs and installations.

column lever or on the dash activates the motor. A separate plastic reservoir, motor, and pump force liquid through tubing to adjustable nozzles **[Figure 36-29]**. The nozzles spray liquid on the windshield to wash it. The wash and wiper functions often are combined, so that both occur when the washer is activated.

36.8 HORN

Horn switches are mounted in the steering wheels of most vehicles. Some vehicles have a horn switch mounted on a steering column lever **[Figure 36-30]**. A horn relay is part of the circuit, usually mounted in a small metal box in the engine compartment.

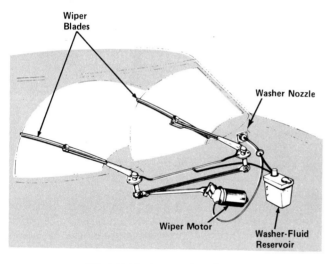

Figure 36-29. Windshield wipers and washers.
TRICO PRODUCTS CORPORATION

36.9 COMFORT ACCESSORIES

Comfort accessories are provided for the comfort and convenience of vehicle occupants. Such accessories can include:

- Power seats
- Power windows
- Electrically positioned outside mirrors
- Automatic leveling devices
- Cruise control
- Automatic headlight dimmers.

Power seats are operated by small electric motors **[Figure 36-31]**. Power window motor switches are located at each window, with a master control near the driver.

Some devices, such as cruise control, automatic leveling devices, and so on may be electronically controlled by computer units on a vehicle. Unit 38 discusses functions of electronic devices.

Vehicle manufacturers offer many comfort accessory options. Refer to the manufacturer's service manual for information on specific vehicle systems.

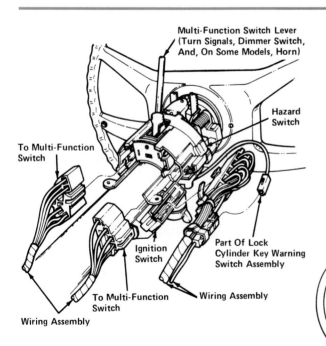

Figure 36-30. Horn assemblies. FORD MOTOR COMPANY

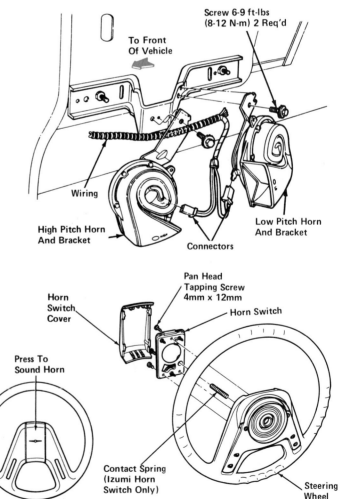

ELECTRICAL DEVICES

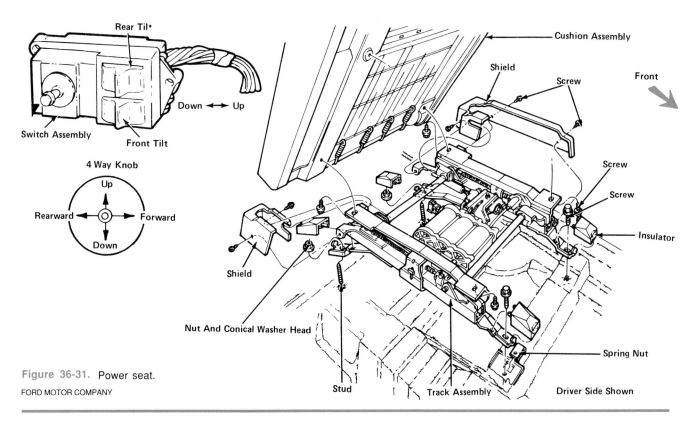

Figure 36-31. Power seat. FORD MOTOR COMPANY

36.10 AUDIO SYSTEMS

Audio systems include radios, cassette tape players, compact disc players, graphic equalizers, amplifiers, and speaker systems. Radios are installed in the dash, with wiring harnesses to door and rear shelf speakers [Figure 36-32].

36.11 ANTI-THEFT SYSTEM

Each year, thieves steal or break into tens of thousands of automobiles. In response to this problem, car manufacturers have offered optional or standard-equipment anti-theft systems.

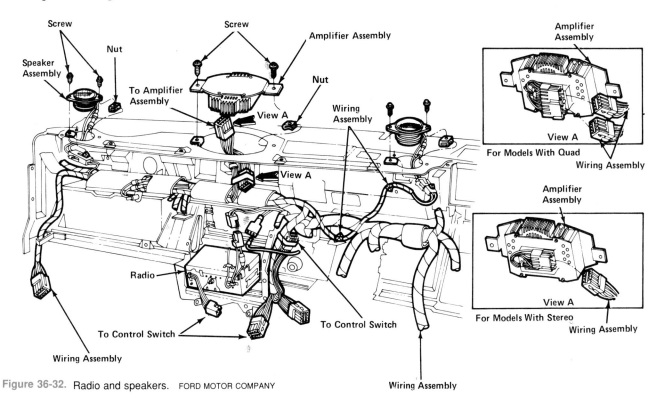

Figure 36-32. Radio and speakers. FORD MOTOR COMPANY

Several components make up an anti-theft system [Figure 36-33]. These parts include:

- Electronic control unit
- Front and rear door switches
- Trunk key cylinder switch
- Hood switch
- Starter inhibitor relay
- Horn relay
- Theft-alarm horn.

Anti-theft systems disrupt the ignition and/or starter systems, sound the horn intermittently, and flash headlights. These actions alert the car owner and attract attention to the thief.

To arm the system, lock the driver's door. A security light on the dash lights for approximately 30 seconds to signal that the system will function. When the light goes out, the alarm and other functions are armed.

If a door, hood, or trunk is opened before the key is used to unlock the driver's door, the alarm sounds. The system will continue to sound and flash a warning for two to three minutes. Then, if the door, hood, or trunk has been closed, the system will shut itself off and reset the alarm automatically.

The trunk can be opened with the key to load packages or cargo. The alarm will not be triggered and will reset automatically, 30 seconds after the trunk lid is closed.

To disarm the system, open the driver's door with the key.

UNIT HIGHLIGHTS

- Wiring harnesses are necessary to complete circuits in all parts of the vehicle. Wires are color-coded for identification purposes. Several units may share a common connection for power or ground.
- Circuits can be protected with fuses, fusible links, or circuit breakers. Relays are used in place of heavy-duty switches to conduct large current flows.
- A wiring diagram indicates color coding of wires and circuit connections.
- Warning devices include lights, buzzers, chimes, and graphic displays.
- Most common electrical gauges are analog instruments. Sending units vary resistance to change the amount of current in a gauge circuit. Gauges can be permanent-magnet, balancing-coil, or thermal-type.
- Speedometers, odometers, and vacuum/pressure gauges are usually operated mechanically rather than electrically.
- Lights, horns, windshield wipers and washers, and many comfort and convenience accessories operate electrically.
- Anti-theft systems disrupt the ignition and/or starting systems, sound the horn, and flash headlights to discourage thieves.

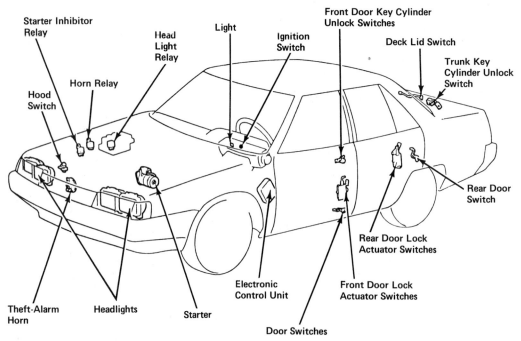

Figure 36-33. Major parts of an anti-theft system.
MITSUBISHI MOTOR SALES OF AMERICA, INC.

TERMS

harness	tracer
printed circuit	wiring diagram
circuit breaker	fuse block
stalk	graphic displays
magnetic gauge	sending unit
balancing coil gauge	thermal gauge
hairspring	odometer
boost pressure gauge	Bourdon tube
sealed-beam	filament
turn signal flasher	halogen light
bezel	side marker light
cornering light	mercury switch
instrument voltage regulator	

REVIEW QUESTIONS

DIRECTIONS: The following questions are similar to those used on automotive technician certification tests. On a separate sheet of paper, write the letter of the correct choice.

1. Technician A says colored stripes or tracers on wire insulation indicate the size of wires.
   Technician B says colored stripes or tracers on wire insulation indicate a specific conductor.
   Who is correct?
   A. A only
   B. B only
   C. Both A and B
   D. Neither A nor B

2. Which of the following statements is correct?
   I. Headlight adjusting screws are located near headlight retaining screws.
   II. Every time a headlight is replaced, it must be readjusted for correct road illumination.
   A. I only
   B. II only
   C. Both I and II
   D. Neither I nor II

3. After replacing a turn-signal flasher bulb, the turn signals blink faster on one side of the vehicle than the other.
   What is the most likely cause?
   A. Short circuit
   B. Grounded circuit
   C. Wrong number bulb installed
   D. Bad fuse

Questions 4 and 5 are not like the first three questions. Each has the word EXCEPT. For each question, look for the choice that does *not* apply. Read each question carefully before you choose your answer.

4. All of the following are used in vehicle electrical circuits to protect the vehicle and electrical components EXCEPT
   A. flashers.
   B. circuit breakers.
   C. fuses.
   D. fusible links.

5. All of the following are types of electrical analog instruments EXCEPT
   A. permanent-magnet ammeter.
   B. warning light.
   C. balancing-coil meter.
   D. thermal gauge.

ESSAY QUESTIONS

1. Describe how wiring connections are made in a vehicle.
2. What are some possible uses of an automobile electrical schematic diagram?
3. Describe the purpose and operation of circuit protection devices.

SUPPLEMENTAL ACTIVITIES

1. Examine and follow wiring harnesses in a vehicle. Make a simple diagram of where connectors are located. Check for color coding of wires on both sides of connectors.
2. Locate fuse blocks and turn-signal flashers on a vehicle.
3. Refer to a manufacturer's service manual for wiring diagrams of a vehicle. Trace wiring connections for headlights, taillights, radio, turn signals, horn, cigarette lighter, and windshield wipers. Make a list of the type and size of protective devices used for each device.
4. Locate sending units on a vehicle for warning devices or gauges for coolant temperature, oil pressure, and fuel. Make a simple diagram of their location and color coding of connecting wires.
5. Check all exterior and interior lights, horn, and windshield wipers and washers for proper operation. Remove any burned-out bulbs and make a list of the code numbers on the bulbs.

37 Electrical Service

UNIT PREVIEW

Regular checks are performed to make sure that all electrical equipment is working properly. Basic checks include visual and physical inspections of the battery, electrical units, circuit protection devices, and wiring connections.

Electrical system service includes both troubleshooting and repair procedures. Troubleshooting requires a thorough understanding of basic principles, logic, and common sense. Troubleshooting procedures can include visual inspections, use of wiring diagrams and test instruments, and testing by substitution.

The most difficult part of electrical troubleshooting or service can be gaining access to wiring and to electrical units. Wires may be run under carpeting, seats, and through body panels.

Breaks or shorts in wires can be found with special test equipment, and wires can be bypassed to make repairs.

In general, defective electrical units are replaced, not repaired.

LEARNING OBJECTIVES

When you have completed your assignments and exercises in this unit, you should be able to:

❏ Explain which electrical checks are done on a regular basis.
❏ Identify and describe techniques used in electrical troubleshooting and service.
❏ Safely remove, clean, and replace exterior light bulbs and clean bulb sockets.
❏ Safely make and solder electrical wire joints.
❏ Safely and correctly use solderless connectors and quick-splice connectors to terminate and join wires.
❏ Remove and replace dash warning light bulbs and/or a radio unit.

SAFETY PRECAUTIONS

Accidental shorting or grounding of powered circuits can cause sparks, fires, and/or personal injury. In addition, expensive electrical or electronic components can be damaged or destroyed. Remove and replace electrical units when no power is connected to the circuit. Make sure all switches are off before replacing electrical units.

For unfused circuits, disconnect the grounded battery terminal from the battery. After repair or replacement procedures, doublecheck all wiring connections before replacing the grounded terminal on the battery.

Testing procedures may require working with "live," or powered, electrical circuits. Use extreme care when working with jumper leads or test wires connected to sources of electrical power and ground. Make sure such leads do not touch each other and do not cause an accidental short or ground.

Use extreme care when working in cramped, crowded areas with many connections. Conductive metal tools can cause an accidental electrical short or ground when working near fuse blocks and connectors. Be sure to remove all jewelry before doing any electrical work.

TROUBLESHOOTING AND MAINTENANCE

No specific preventive maintenance is done to electrical wiring and circuit protective devices. However, the battery and charging system affect the operation of all other parts of the electrical system. Battery and charging system preventive maintenance are discussed in Units 29 and 33.

Regular checks are performed to make sure that safety, lighting, and other electrical equipment is functioning properly. Common electrical units include such items as:

- Light bulbs
- Turn signal flasher units
- Electrical motors
- Horns
- Radios and audio systems
- Electrical heating elements.

Electrical units and circuit protective devices are replaced when defective.

ELECTRICAL SERVICE

37.1 BASIC ELECTRICAL CHECKS

To find defects that cause electrical system malfunctions, basic checks are done. These checks include inspections of:

- Battery connections and electrical system voltage
- Electrical bulbs and units
- Circuit protection devices
- Wiring connections.

Battery Connections and Electrical System Voltage

If no electrical units operate, turn on headlights to check if they operate. If the headlights do not operate, attempt to wiggle and move the battery cable connections. On top terminal batteries, a thin screwdriver can be inserted between the battery terminal and the clamp. If lights and other electrical units work when these checks are done, the terminals must be removed and cleaned (refer to Unit 29). Battery connections must be clean and fit securely for maximum current to flow in or out of the battery. If the vehicle can be started, connect a voltmeter across the battery terminals. After the vehicle has been running at a high idle for a few minutes, the system voltage should read 13.2 volts or more. If not, refer to Topic 33.2 and perform charging system checks and service as necessary.

Electrical Bulbs and Units

Electrical bulbs, motors, and other units can become defective. For example, a single nonworking bulb in a circuit may need replacing. An example of a single circuit with multiple bulbs is the taillight, parking light, and side marker light circuit.

If only one bulb is not burning, remove the bulb and inspect it. Check for broken filaments. Check the brass base and bottom contacts for corrosion. Gently attempt to turn the glass element in the brass base. Replace bulbs with broken filaments or loose glass elements. Clean corrosion from the base and/or contacts, or obtain a new bulb.

With all switches OFF, also check and/or clean the socket by scraping or using a small wire brush **[Figure 37-1]**. Apply a light film of grease to the connector to help reduce corrosion. If none of the bulbs in a circuit, such as the one mentioned above, operate, then the switch or circuit protection devices must be checked.

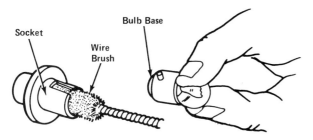

Figure 37-1. Cleaning a light socket.

CAUTION!!!

Make sure all light switches are off before attempting to clean a socket. Cleaning a socket with power present can ground the circuit and blow a fuse.

Circuit Protection Devices

Depending on the vehicle, several devices may be protected by one fuse, fusible link, or circuit breaker. If no devices operate in a common circuit, the most likely cause is a blown fuse or fusible link, or an open circuit breaker. Special tools are available for removing and replacing fuses **[Figure 37-2]**.

If fuses or fusible links burn out again when they are replaced or a switch is closed, the circuit is grounded or shorted. A grounded circuit will cause a circuit breaker to click rapidly and constantly. A grounded conductor must be repaired or bypassed, as discussed in Topic 37.4.

Wiring Connections

Inspect accessible wiring for bare spots or frayed insulation. Electrical tape can be wrapped around wires with worn insulation to insulate them. Gently push and pull on terminals and connections to see if wires are loose or disconnected.

37.2 ELECTRICAL SYSTEM SERVICE

If basic checks and repairs do not solve a problem, electrical system service is necessary. Electrical system service consists of:

- Troubleshooting
- Service and/or repair procedures.

37.3 ELECTRICAL TROUBLESHOOTING

Electrical problems can occur at any point in an electrical circuit. For example, consider the oil pressure warning light circuit shown in **Figure 37-3**. Suppose that the light fails to operate when the key

Figure 37-2. Fuse puller and replacer.

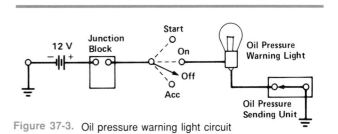

Figure 37-3. Oil pressure warning light circuit

switch is in the "on" position, engine off. The problem could be at any of the following points in the circuit:

- Battery
- Battery positive connection or cable
- Wire from battery positive to junction block
- Junction block
- Wire from junction block to ignition switch
- Ignition switch
- Wire from ignition switch to indicator lamp
- Indicator lamp socket
- Indicator bulb
- Wire to oil pressure switch
- Connector terminal on wire
- Oil pressure switch
- Engine ground
- Battery ground connection or cable.

Troubleshooting involves thinking and checking until a trouble source is found. The more knowledge and experience a technician has, the more rapidly and easily a problem can be found and fixed. Electrical system troubleshooting can involve:

- Thorough basic knowledge of electricity
- Visual inspection
- Use of wiring diagrams
- Logical approach
- Use of test instruments
- Testing by substitution.

Basic Knowledge of Electricity

Electrical and electronic fundamentals are discussed in Unit 27. Knowledge of how series, parallel, and series-parallel circuits operate is necessary to troubleshoot interconnected circuits.

Basic causes of electrical problems include opens, shorts, and grounds. Opens can be caused by burned-out fuses or fusible links or tripped circuit breakers. Burned-out electrical units, or disconnected wiring also create opens. Shorts and grounds can be caused by physical damage to wires or incorrect connections.

Inspection

Before proceeding to more complicated troubleshooting, make a thorough inspection for obvious problems. Burned-out bulbs and disconnected wires are obvious when found.

It is estimated that more than 90 percent of all "electrical problems" are not electrical, but mechanical. That is, most of these problems involve loose or dirty connections, not defects in electrical parts.

In some cases, pulling a connector apart and reconnecting it may restore an electrical circuit. Light scraping and/or polishing of electrical contacts may be necessary to restore a corroded connection **[Figure 37-4]**. Use your eyes and hands to find simple problems. If wires are bundled into harnesses, it may be necessary to trace them with the help of an electrical diagram.

Use of Wiring Diagrams

When attempting troubleshooting, wiring diagrams can be a valuable help in tracing circuit conductors and interconnections. Small diagrams may require the use of a magnifying glass and a ruler to trace specific circuits. In some cases, a wire of one color may be connected to another color of wire at a *junction block* or plug-in connector.

A junction block is an area where common wire connections are made. A junction block usually consists of a plastic base with metal terminals and connectors. Wires run from remote locations to common junction block connections, then continue. A fuse block is a junction block that includes circuit protection devices in addition to common connections.

Common connections can cause confusion when attempting to find defects. Tracing a circuit with a wiring diagram may be necessary to find the ultimate source of a problem.

Logical Approach

In order to find the cause of a problem, it is necessary to approach the problem logically. As mentioned above, if only one bulb in a common circuit is out, the cause is most likely in the bulb itself. If all bulbs are out, the problem is most likely in a fuse, fusible link, or circuit breaker.

However, it is necessary to separate problems and the reasons for those problems. In the following example, the problem is that the vehicle's parking lights are not working. The chart is an example of cause-and-effect diagnosis in a step-by-step format:

Problem (effect)	Reason (cause)
No parking lights	Burned-out fuse
Burned-out fuse	Grounded wire
Grounded wire	Physical damage to wire in trunk
Physical damage to wire in trunk	Loose jack in trunk

Cause-and-effect diagnosis thus identified an unlikely cause to the parking light failure: A jack had not been secured after a tire was changed. To fix electrical or other problems, it often is necessary to find the underlying cause.

ELECTRICAL SERVICE

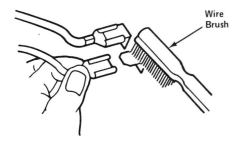

Figure 37-4. Cleaning a terminal and connection.

Consider the oil pressure circuit in **Figure 37-3**. A logical approach to determining why the light doesn't work would be to eliminate the least likely possibilities.

If other electrical parts of the vehicle function properly, the battery and its connections can be eliminated. The next check would be to find if other circuits with a common power connection operate. That is, are other circuits that draw power from the same terminal of the ignition switch operating? If such circuits work properly, all connections up to and including the switch terminal can be eliminated.

Another logical step is to consider the probability of failure for given parts. For example, oil-pressure sending units fail more often than light bulbs used for warning systems. Thus, it would be more logical to look for a problem at the sending unit than at the dash light.

The next step would be to check the most accessible parts of the system. In the warning-light system above, these parts would be the oil-pressure sender and connecting wire. Parts most difficult to access are checked last and/or repaired last. Removal of dash and/or instrument panel parts for dash light access would be done last. Thus, logic and common sense are used to find and fix the problem in the quickest, easiest way.

Use of Test Instruments

Most electrical problems can be found with an *unpowered test light* and *jumper wires* **[Figure 37-5]**. An unpowered test light contains a small bulb that will light when connected to power and ground. Jumper wires are wires used to "jump over" existing wiring and make connections.

Other electrical instruments such as voltmeters and ohmmeters can be used to measure voltage drops and resistance.

Unpowered test lights can be used to check for power or ground. One contact of the test light bulb is connected to a wire and a clip. The other contact of the bulb is connected to a sharp, ice pick-like probe. The probe can be used to scratch through corrosion or paint to make a good connection.

If the clip is attached to ground and the probe touched to a source of power, the bulb will light. If the connections are reversed, the bulb will light when ground is touched. Thus, the test light can be used to find or confirm a source of power or a good ground **[Figure 37-6]**.

If an electrical unit has a source of adequate power and ground, it should operate. If not, the unit itself may be defective.

Jumper wires can be used to connect a unit to a known source of power and/or ground. This is done to determine whether the problem lies in the unit itself **[Figure 37-7]**. In some cases, wiring or connectors may be faulty. If a unit operates when connected with jumper wires, the problem is in a connection, not in the unit.

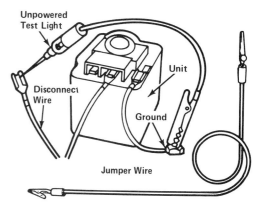

Figure 37-5. Unpowered test light and jumper wires.

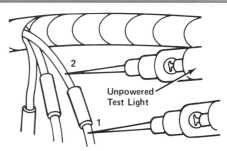

Figure 37-6. Using a test light.

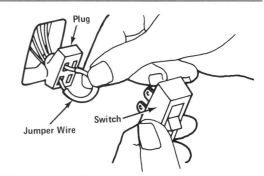

Figure 37-7. Using a small jumper.

Voltmeters and ohmmeters can be used to determine resistance in conductors and/or electrical units. Excessive voltage drops indicate excessive resistance. Excessive resistance can be checked directly with an ohmmeter.

CAUTION!!!

Using an ohmmeter on a powered circuit will destroy the ohmmeter. Disconnect the grounded battery terminal to prevent damage when performing resistance checks with an ohmmeter.

Testing by Substitution

After simple checks have been made, testing by substitution can be used to determine if an electrical unit is defective. A unit that works can be substituted for one that does not. If the substitute unit functions properly, the problem is in the old unit. This approach often works when other checks indicate connections are all correct but a unit still does not function.

Testing by substitution can have drawbacks. In most cases, electrical parts are not returnable. If a substitute unit does not function, money is wasted. In addition, an improper connection that damaged an existing unit can also damage a substitute unit.

However, if other methods fail, testing by substitution may be the simplest and quickest way to determine if a unit is defective.

37.4 ELECTRICAL SERVICE

The main difficulty in performing electrical service is gaining access to wiring and electrical units. During assembly, wiring harnesses and many electrical units are installed before the dash, seats, carpeting, headliner, and upholstery are fitted. Thus, wiring harnesses, connectors, and electrical units may be in areas that are accessible only if many parts are removed. Such removal can greatly increase the cost of a simple repair. For wiring under carpets and panels, other service techniques are available.

Gaining Access to Wiring And Electrical Units

It may be necessary to remove air conditioning units, radios, dash pads, and many other parts to locate problems. Such work is often more difficult than the actual troubleshooting or electrical repair necessary.

Wiring harnesses in the area of the fuse block contain many tightly bundled wires. On older vehicles, wiring harnesses were formed by wrapping wires with plastic ribbon. Newer vehicles also may use removable plastic sleeves to contain, protect, and bundle wires **[Figure 37-8]**.

Replacement Wiring

In many cases, it is not practical to inspect nonconductive wiring that runs beneath carpeting and inside body panels. Special test instruments, called

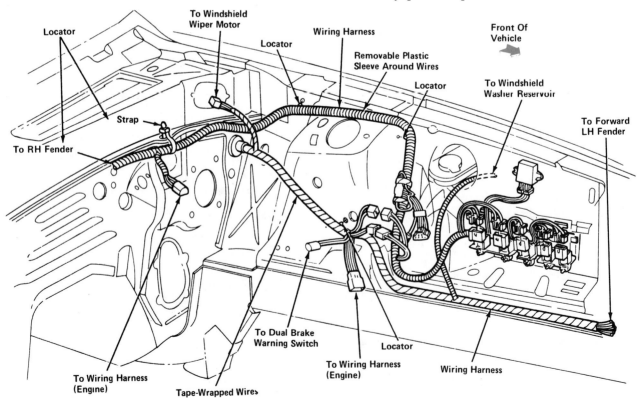

Figure 37-8. Engine compartment wiring. FORD MOTOR COMPANY

short finders, are available to pinpoint the location of shorted or grounded wires in such locations.

Short finders induce current in a wire through induction. A powered coil within the short finder can induce current in a conductor. An inductive ammeter detects this current. At a point where a wire is broken or shorted, no electrical energy is present. The short finder indicates this point with a meter and/or buzzer **[Figure 37-9]**.

Splicing and Terminating Wires

The most permanent way of joining electrical wires is by soldering. A heated soldering iron is used to apply heat to wires. The heated wires melt solder, which flows to form an electrically conductive and physically secure joint **[Figure 37-11]**. After soldering, joints are insulated by wrapping them securely with electrical tape or by using heat-shrink tubing.

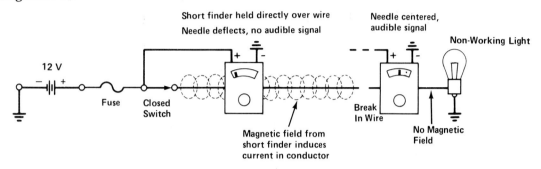

Figure 37-9. Short finder operation.

The quickest and simplest way to replace a wire in an inaccessible location is to physically bypass the wire. *Bypass wiring*, or jumper wiring, is used to replace physically inaccessible defective wires.

A defective wire is cut at both ends. A jumper, or replacement, wire of the correct size is *spliced* on the cut ends. Splicing means joining. There are three common types of joints: pigtail, T-type, and Western Union **[Figure 37-10]**.

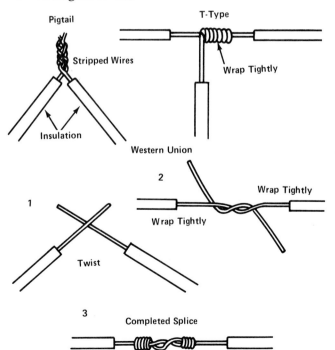

Figure 37-10. Types of wire joints.

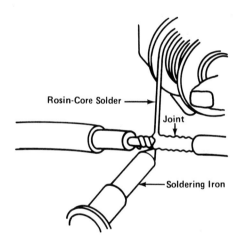

Figure 37-11. Soldering procedure.

SAFETY CAUTION

Soldering irons in use and molten solder are extremely hot and can cause serious burns. Use only rosin-core solder in electrical circuits. Wear eye and face protection when soldering.

Solderless connectors are hollow, tubular metal parts covered with insulating plastic. The terminals may be at the end of a wire, or in a splice. *Stripped*, or bare, wire conductors are inserted into a connector. The connector is then *crimped*, or deformed, to hold the wire securely **[Figure 37-12]**. *Quick-splice connectors* also can be used to quickly splice or tap into wires. A quick-splice connector is a device specifically intended for making quick, easy splices.

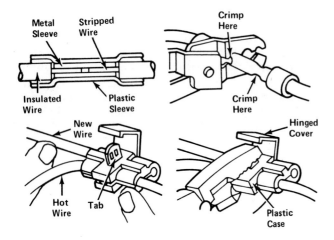

Figure 37-12. Crimped connectors.

Electrical current bypasses the nonconductive wire and flows through the new wire **[Figure 37-13]**. Such replacement may involve running a jumper wire under the floor of the vehicle. Replacement wiring must be of sufficient size to carry current expected in a circuit, as illustrated in **Figure 37-14**.

Removing and Replacing Defective Units

Although some electrical units can be repaired, in most cases replacement is recommended. For service procedures, any electrical device can be considered as a replaceable unit, regardless of function.

Electrical power to the unit is shut off. If battery power is not switched, a fuse or circuit breaker can often be removed. If a circuit is not fused, as some safety systems are not, the battery must be disconnected.

Unit replacement of a radio is illustrated in **Figure 37-15**.

SAFETY CAUTION

Disconnect the grounded battery cable to avoid sparks, shocks, and damage to electrical units. Reconnect the cable after all wiring connections have been made and doublechecked.

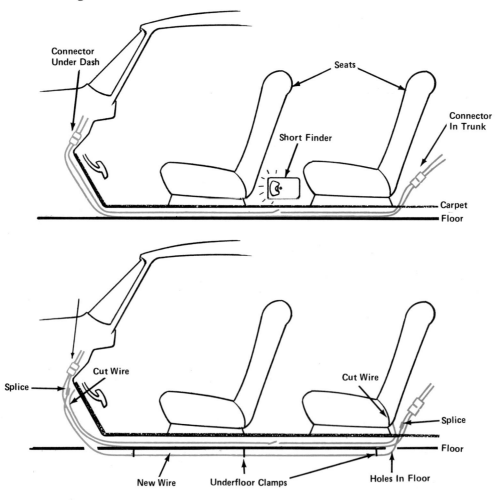

Figure 37-13. Replacement wiring.

ELECTRICAL SERVICE

Suggested AWG Cable Size 12 V				METRIC SIZE	AWG SIZE
Ammeter to Fuse Block	12	Heater to Fuse Block	16	.22	24
Ammeter to Switch	14	Horn Button Circuit	16	.35	22
Back Up Lights	18	Horn to Relay and Battery	12	.5	20
Battery to Ammeter	12	Ignition Switch Feed	12	.8	18
Battery to Fender Ground	14	Instrument Lamps/Sending		1.0	16
Cigar Lighter	14	Units	18	2.0	14
Coil Wire (Low Tension)	16	Interior Lights	18	3.0	12
Dome Light	16	Light Switch to Relay		5.0	10
Directional Signals	16	or Fuse	18	8.0	8
Fuel Gauge	18	Marker (Fender) Lights	18	13.0	6
Fuse Lights	See Note	Parking Lights	18	19.0	4
Generator/Alternator		Radio to Ammeter or Fuse	18	32.0	2
to Regulator	16	Taillights	18		
Generator/Alternator		Stop Lights	18		
to Starter Relay	12	Windshield Wiper and Washer	16		
Headlight Circuit	16	High Tension Spark Plug			
Headlight Relay	12	and Coil Wire	7 mm Suppression Cable		

Note: For fuse links
Where circuit is 10 gauge, use 14 gauge for fuse link.
Where circuit is 14 gauge, use 18 gauge for fuse link.
Fuse links are usually 4 gauges smaller than circuit in which they are found.

Figure 37-14. Replacement wiring sizes. BUICK MOTOR DIVISION—GMC

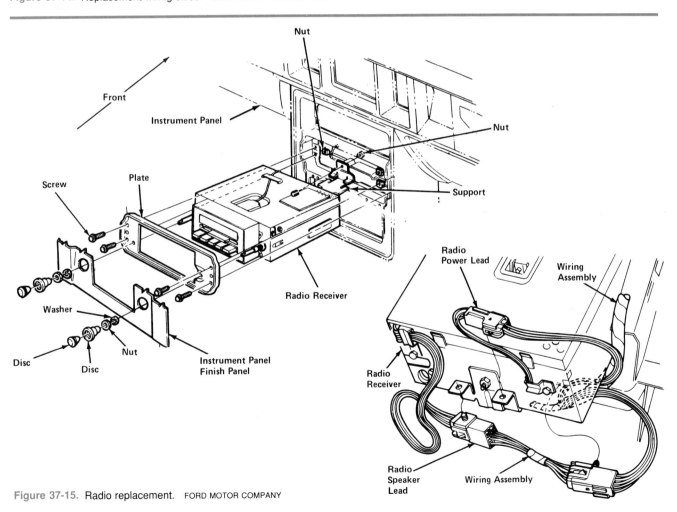

Figure 37-15. Radio replacement. FORD MOTOR COMPANY

UNIT HIGHLIGHTS

- Regular checks are done to make sure that lighting, safety, and other electrical equipment is working properly.
- Basic checks include inspection of the battery, electrical units, circuit protection devices and wiring connections.
- Electrical system service includes troubleshooting and repair procedures. Troubleshooting requires a thorough understanding of basic principles, logic, and common sense.
- Electrical system troubleshooting procedures can include visual inspections, use of wiring diagrams and test instruments, and testing by substitution.
- Gaining access to wiring and to electrical units can be the most difficult part of electrical system repair procedures. Breaks or shorts in wires can be found with special test equipment, and wires can be bypassed.
- In most cases, electrical units are replaced, not repaired.

TERMS

junction block
jumper wire
bypass wiring
solderless connector
crimp
unpowered test light
short finder
splice
strip
quick-splice connector

REVIEW QUESTIONS

DIRECTIONS: The following questions are similar to those used on automotive technician certification tests. On a separate sheet of paper, write the letter of the correct choice.

1. Two taillight bulbs do not burn in a common circuit with a total of eight bulbs. What should be checked first?
   A. Fuse
   B. Bulbs
   C. Sockets
   D. Wiring

2. The wire in a fuse looks okay. A test light indicates there is power at one end of the fuse but not at the other end.
   What is the most likely cause?
   A. Test light defective
   B. Nonconductive fuse
   C. Accidental ground in wiring from fuse to circuit
   D. Excessive resistance in the electrical unit

3. Which of the following should be done last?
   A. Check battery
   B. Replace bulb
   C. Replace fuse
   D. Replace wiring

4. Which of the following statements is correct?
   I. Electrical troubleshooting requires knowledge, logic, and common sense.
   II. Electrical troubleshooting is done when simple checks do not locate the cause of a problem.
   A. I only
   B. II only
   C. Both I and II
   D. Neither I nor II

Question 5 is not like those above. It has the word EXCEPT. For question 5, look for the choice that does *not* apply. Read the entire question carefully before you choose your answer.

5. Basic electrical checks include all of the following EXCEPT
   A. voltage drop across an electrical unit.
   B. battery and electrical system voltage.
   C. wiring connections.
   D. fuses, fusible links, and circuit breakers.

ESSAY QUESTIONS

1. What are some basic electrical checks?
2. Describe electrical troubleshooting.
3. What procedures may be necessary for electrical service?

SUPPLEMENTAL ACTIVITIES

1. Remove all exterior light bulbs from a vehicle, one at a time. Inspect and clean bulb contacts and safely clean all socket connections.
2. Locate the fuse block of a vehicle. Use a test light to check both ends of all fuses for power. Replace any defective fuses found.
3. Strip lengths of wire and make neat examples of T-type, pigtail, and Western Union joints. Safely and correctly solder the joints.
4. Strip lengths of wire and use solderless connectors and quick-splice connectors to terminate and splice wires.
5. Refer to the manufacturer's service manual for procedures to remove and replace dash light bulbs and a radio unit. Safely and properly perform the procedures on a vehicle.

38 Electronic Devices

UNIT PREVIEW

Electronic devices can monitor and/or control virtually all of an automobile's major parts and systems. Display devices alert the driver to problems. Electronic computers can produce signals to operate mechanisms and devices that control vehicle functions. Because they operate so quickly and precisely, electronic devices are the best and least expensive way to control complicated vehicle functions.

LEARNING OBJECTIVES

When you have completed your assignments and exercises in this unit, you should be able to:

❏ Describe how electronic displays can be used in warning systems.
❏ Identify parts of a speed-control system.
❏ Explain how a basic feedback-control system operates.
❏ Describe the difference between closed-loop and open-loop computer modes.
❏ Identify vehicle systems that use electronic controls.

38.1 AUTOMOTIVE ELECTRONICS

Many electronic devices are used on modern automobiles. They can be broken down into two general groups: those that present information and those that control vehicle functions.

Warning and Indication Systems

Electronic circuits and *display devices* can be used in place of conventional gauges to present information [**Figure 38-1**]. When information is presented as digits, or numbers, it is known as *digital* information.

A conventional fuel gauge needle moves up or down a scale, from E to F. A sending unit, or *sensor,* in the fuel tank sends an electrical signal to the gauge based upon the amount of fuel in the tank. Such a conventional type of instrument is an *analog* display device.

More complicated information can be presented in other forms. For example, a warning device that

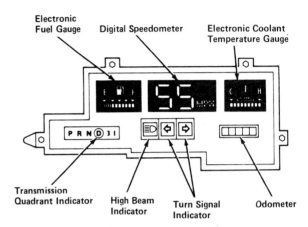

Figure 38-1. Digital dashboard display. FORD MOTOR COMPANY

plays or creates a message in human language can be used. Such a *voice warning device* is shown at the bottom left of **Figure 38-2.** When a door is ajar or the ignition key is left in the lock, a message is played to alert the driver. More sophisticated units have *voice simulators* that produce language sounds to form sentences and warnings.

Parts for display and voice warning devices are not available separately. If defective, the device must be replaced as a unit.

Computers

Facts, or *data,* can be gathered by sensors. For example, a coolant temperature sensor can produce a signal (data) that indicates the temperature of an engine. Data also can be added to other data, used in mathematical calculations, or used as is. A *computer* is a device that *processes,* or changes, raw data into useful information. The processing is directed by a set of instructions in the computer known as a *program.* Computers receive and process data, and deliver information in the form of small electrical currents.

The human brain can be thought of as an extremely complicated and sophisticated form of computer. It processes many types of data rapidly.

When you are driving down the street and see a red light at an intersection, you step on the brakes and bring the vehicle to a safe stop. Data (red light ahead) is gathered by the eyes (sensor input signals) and sent to the brain (computer). The brain then processes this data, based on your memory (your programming). The result is a signal (output signals)

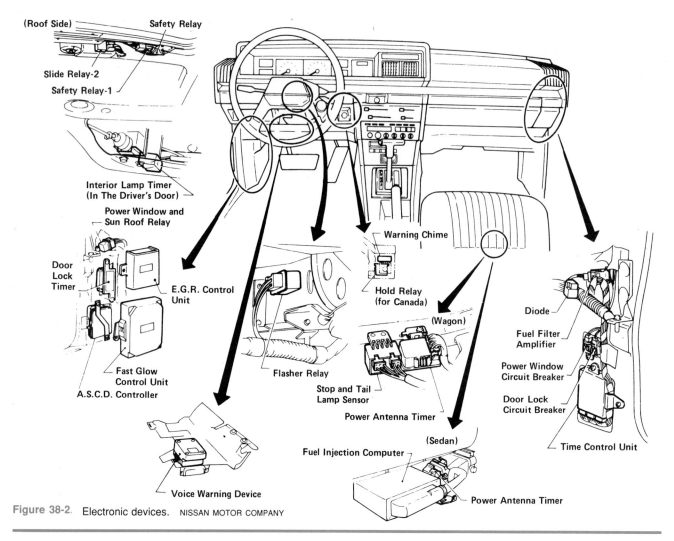

Figure 38-2. Electronic devices. NISSAN MOTOR COMPANY

to the muscles in the legs and feet to push on the brake pedal.

An automotive example of this is an electronic engine control system that uses a microprocessor or computer. As the automobile is driven at a steady speed, the ignition timing also is fairly steady. As the accelerator is depressed, the manifold pressure sensor detects a drop in vacuum. The sensor sends an electrical signal to the computer, telling the computer that the vacuum is dropping. The computer is programmed to respond to this input signal by producing an output signal to retard ignition timing enough to prevent detonation.

An automotive trip computer can receive data about fuel level, fuel usage, vehicle speed, and distance to a destination. The trip computer processes these *inputs,* or data units. The results, or *outputs,* of the processing are displayed for driver use. For example, a trip computer can compute and display a fuel consumption figure in miles per gallon, or an estimated time of arrival. Using such information, the driver can determine how to drive for best fuel economy or an early arrival time. **Figure 38-3** shows a trip computer and related parts.

38.2 CONTROL SYSTEMS

Computers can also be used to control vehicle functions. The data from various sensors is processed to produce an output signal. The output signal is used to operate switches, motors, solenoids or other *output devices* to control vehicle functions. The main parts of a basic computer control system are:

* Input sensors
* Computer processing unit
* Output devices
* Wiring connections.

The computer receives inputs from sensors located at various parts of the vehicle. One type of sensor is shown in **Figure 38-4**. The computer output is used to operate relays and other devices to control various functions. An output device, or actuator, for a cruise control system is shown in **Figure 38-5**.

An automotive computer can be thought of as a *"black box,"* as shown in **Figure 38-6**. If a given input signal is presented to the computer, it should produce a certain output. The black box simply processes the

ELECTRONIC DEVICES

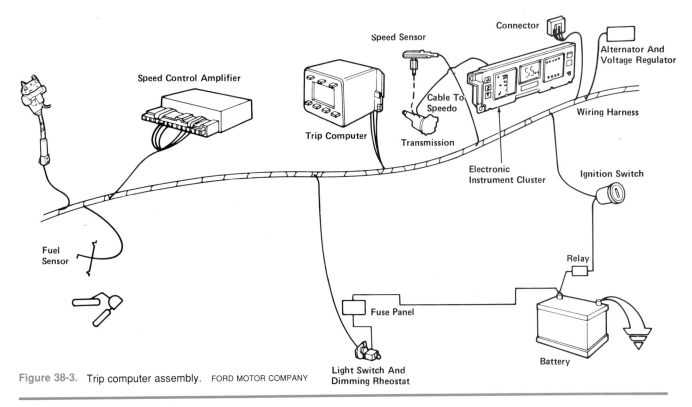

Figure 38-3. Trip computer assembly. FORD MOTOR COMPANY

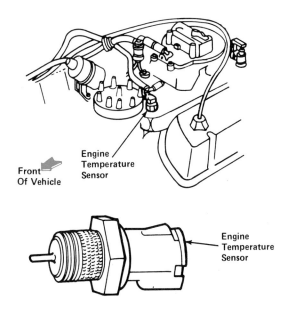

Figure 38-4. Engine temperature sensor. FORD MOTOR COMPANY

input and produces an output. It is not necessary for a technician to fully understand the inner parts and functions of the computer. The technician needs only to be able to determine if the unit is working properly.

Test procedures for particular computer processing units, or *modules,* are given in manufacturers' service manuals. With a given input signal, the computer should produce a certain output signal. If not, the unit is defective.

In most cases, computer modules are not serviceable. If a computer module is defective, it must be replaced with a new unit. A typical computer module is shown in **Figure 38-7.**

Feedback Control

Computers, sensors, and output devices can be connected to monitor and adjust vehicle functions continuously. For example, a cruise control unit can maintain a set speed. When vehicle speed falls below the set speed, the computer output signal operates a *servomechanism,* or output device, to move the throttle linkage. The servo opens the throttle to increase vehicle speed. When the set speed is reached, the servo eases off the throttle. The speed control unit continuously monitors and adjusts vehicle speed.

A block diagram of such an arrangement, known as *feedback control,* is shown in **Figure 38-8.** Feedback control means that data concerning the effects of the computer's output are fed back into the computer as input signals.

These input data are compared to a set value, such as the speed selected on a cruise control. If the input data do not match the selected speed, the computer changes the output signal. The servo moves the throttle linkage. The sensor reports the new speed, and the input data are compared again to the set speed. The output signal is changed until vehicle speed matches the selected speed.

The computer receives a constant flow of data about the process being controlled. The computer

410   PART IV: Automobile Electrical and Electronic Systems

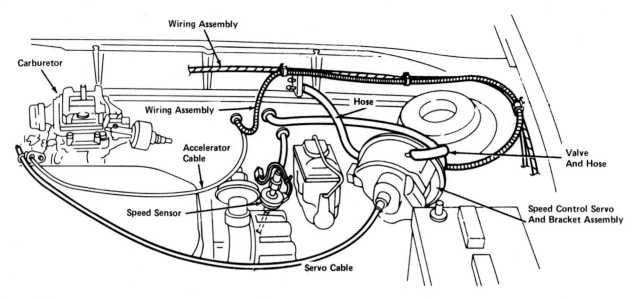

Figure 38-5. Cruise control assembly. FORD MOTOR COMPANY

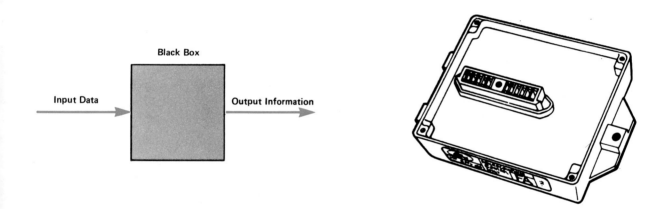

Figure 38-6. A black box, or computer, receives inputs, processes them, and produces usable outputs.

Figure 38-7. Computer processing control unit. FORD MOTOR COMPANY

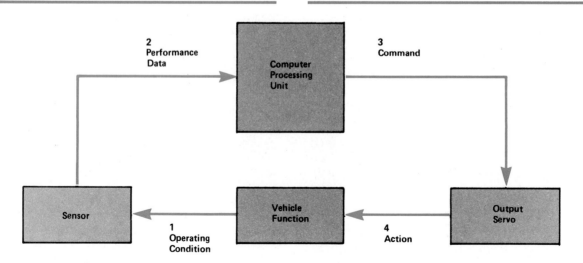

Figure 38-8. Closed-loop feedback control.

compares the input data to a set figure in its memory. The computer then changes its output. Whenever the process being controlled changes, the computer changes the output to bring everything back into line. This type of feedback control is also known as a *closed-loop* mode of operation. In **Figure 38-8,** the line starting at 1 and returning at 4 forms a continuous, or closed, loop.

Engine Controls

A more important and necessary use of electronic controls is for engine operation. Automobiles must offer better fuel economy and improved emissions controls in each new model year. Electronic controls react quicker and more precisely than the mechanical controls used in the past. Another significant advantage of electronic controls is that they do not wear out.

Since electricity travels at the speed of light (approximately 186,000 miles (300,000 km) per second), electronic controls operate rapidly. In addition, the mass production of *integrated circuits* has brought the cost of electronic controls to a reasonable level. An integrated circuit, or IC, contains many thousands of miniaturized electronic components, such as resistors, capacitors, and transistors. Using sensors, computer modules, and output devices in feedback-controlled circuits, it is possible to accurately control an engine's operation.

Closed-loop systems. The two main areas of engine control at present are ignition timing and air-fuel ratio. **Figure 38-9** shows the parts of a closed-loop system for adjusting carburetor air-fuel ratio. An electronic ignition control system is shown in **Figure 38-10.**

By adjusting ignition timing and air-fuel ratio precisely for running conditions, the vehicle can be made to run very efficiently. The vehicle will start easily, run well, have good power and acceleration, and deliver good fuel economy. In addition, the combustion process can be adjusted so that the vehicle's exhaust gases are less harmful.

Open-loop systems. For starting, warmup, and hard acceleration, the engine needs a steady air-fuel ratio richer than 14.7:1. The closed-loop system that continually adjusts air-fuel mixture is not needed. In *open-loop* operation, the computer does not continually monitor and adjust the air-fuel ratio. It simply provides a constant, rich mixture for smooth engine operation during warmup and hard acceleration. After the engine warms up or the driver stops pushing hard on the accelerator pedal, the system returns to closed-loop operation. An open-loop mode for a fuel supply system is shown in **Figure 38-11.** The line connecting the oxygen sensor, control module, and feedback carburetor does not form a closed loop. The loop is open.

Emission Controls

The ideal, or *stoichiometric*, air-fuel ratio of 14.7 to 1 is also ideal for reducing harmful exhaust gas emissions. By using electronic controls, the exhaust emissions can be reduced significantly.

To monitor the air after the engine is warmed up, an exhaust gas *oxygen sensor* is used. An oxygen sensor is a device that senses the amount of oxygen in the exhaust gases. The oxygen sensor, shown in **Figure 38-12,** is screwed into the exhaust manifold or header pipe.

Carbon monoxide results from incomplete burning of fuel. Exhaust gases containing high amounts of carbon monoxide also contain much unburned fuel. The unburned fuel is a major cause of air pollution. The oxygen sensor sends an input signal to the computer module based upon how much oxygen is present.

The computer output signal is sent to the fuel injection system or a special feedback carburetor. After the engine warms up, the exhaust gases are sampled continuously, and the air-fuel mixture is adjusted continuously. The air-fuel mixture stays at, or very near, 14.7:1.

Several emission control devices directly affect the amount of other harmful gases in vehicle exhausts. Typical locations for these devices are shown in **Figure 38-13.** Some of these devices are used to provide input data or affect the combustion process. Electronic controls are used to monitor or control these emission control devices (see Unit 42).

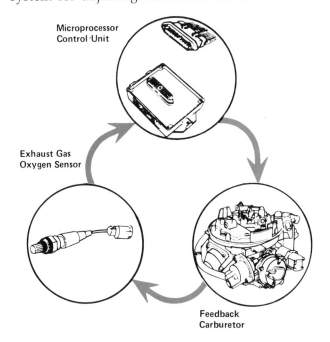

Figure 38-9. Closed-loop fuel control. FORD MOTOR COMPANY

PART IV: AUTOMOBILE ELECTRICAL AND ELECTRONIC SYSTEMS

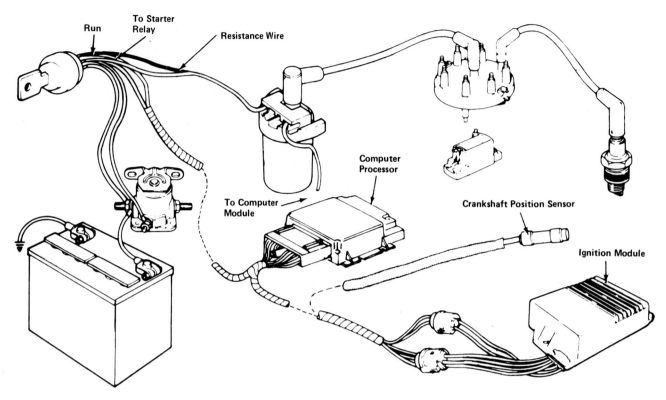

Figure 38-10. Electronic ignition system. FORD MOTOR COMPANY

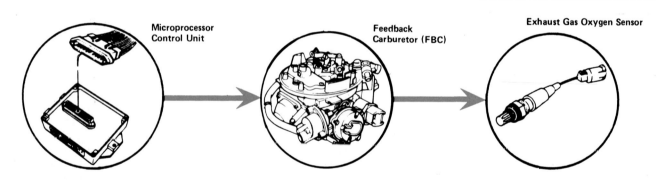

Figure 38-11. Open-loop fuel control. FORD MOTOR COMPANY

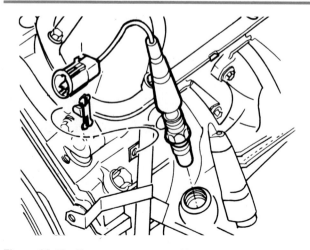

Figure 38-12. Oxygen sensor assembly. FORD MOTOR COMPANY

38.3 SELF DIAGNOSTICS

Computers located in, or "on board," the vehicle also can be used to store data or information about vehicle operation. For example, the computer can store information codes about vehicle malfunctions. When proper electrical test connections are made, the computer will flash the dashboard "CHECK ENGINE" light in a coded sequence **[Figure 38-14]**. Flashes separated by pauses indicate numbers. One flash, followed by a pause, and then two more rapid flashes, indicate a code 12 malfunction. Manufacturer's service manuals list the problems indicated by the code numbers. Vehicles with trip computers often use the trip computer display to show trouble code numbers during test procedures.

Electronic Devices

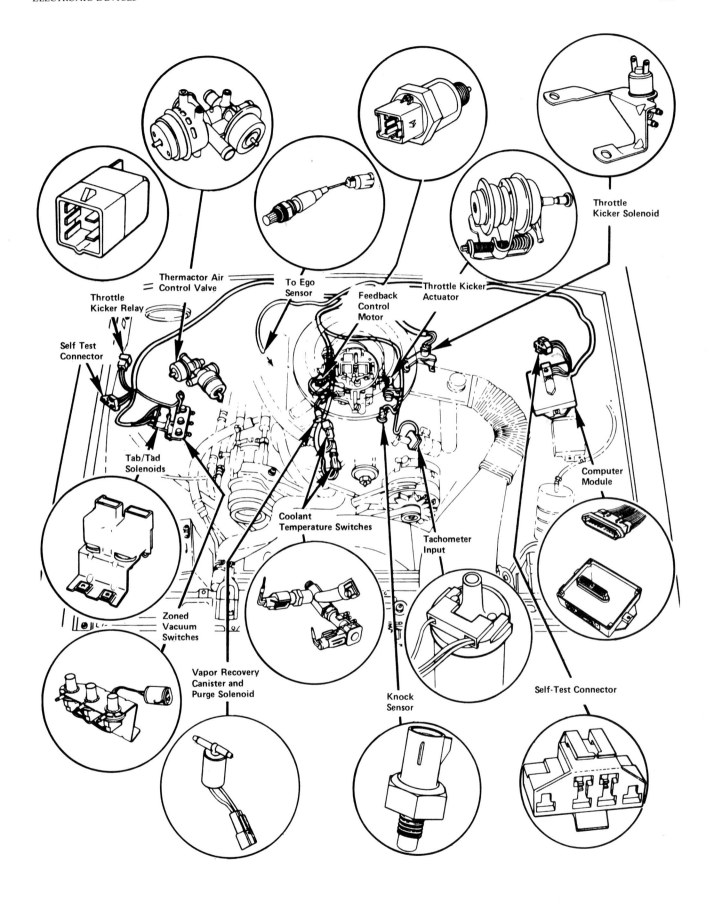

Figure 38-13. Emission control device locations. FORD MOTOR COMPANY

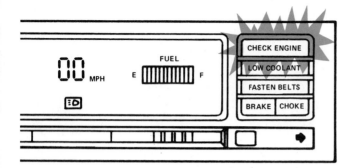

Figure 38-14. Dashboard computer code readout.

38.4 OTHER USES OF ELECTRONIC CONTROLS

Electronic control systems can be used to monitor and adjust virtually every area of vehicle operation. For example, vehicles currently being manufactured use computers and electronics to control the following systems and functions:

- Ignition (timing advance and retard)
- Fuel (amount and delivery timing)
- Automatic transmission (shifting)
- Power steering (variable assist)
- Power brakes (anti-skid system)
- Suspension (height, damping, and motion control)
- Heating and air conditioning (automatic temperature control)
- Comfort and convenience (lighting, locking, and anti-theft).

The future uses of electronic controls are unlimited. For example, Porsche of Germany has a prototype "drive-by-wire" system that uses electronically controlled accelerator linkage. No mechanical connection exists between the accelerator pedal and the throttle plate. Instead, an electrical stepper motor positions the throttle plate according to input from a sensor on the accelerator pedal linkage. Since 1982, Rolls-Royce of England has used a similar system to control the brake system.

UNIT HIGHLIGHTS

- Automotive electronics can be used to display data or information or control vehicle functions.
- A computer is a device that processes data, or facts, into usable information.
- A computer can be thought of as a "black box" that produces a certain output from a given input.
- A basic computer control system consists of input sensors, a computer processing unit, and output devices.
- Computers can store data and/or information about vehicle malfunctions.
- Electronic controls can be used to monitor or control virtually all vehicle functions.

TERMS

display device	digital
analog	sensor
voice warning device	voice simulator
data	computer
process	program
input	output
output device	black box
module	servomechanism
feedback control	closed-loop
integrated circuit	open-loop
oxygen sensor	stoichiometric

REVIEW QUESTIONS

DIRECTIONS: The following questions are similar to those used on automotive technician certification tests. On a separate sheet of paper, write the letter of the correct choice.

1. Which of the following statements is correct?
   I. Automotive electronics can be used to operate gauges or warning devices.
   II. Automotive electronics can be used to control vehicle operation.
   A. I only C. Both I and II
   B. II only D. Neither I nor II

2. Which of the following statements is correct?
   I. For starting and warmup, the computer that operates ignition timing and the fuel delivery system operates in closed-loop mode.
   II. After warmup, the computer that operates ignition timing and the fuel delivery system operates in open-loop mode.
   A. I only C. Both I and II
   B. II only D. Neither I nor II

3. Technician A says that a technician needs to know how a computer operates internally to correct automotive electronics problems.

   Technician B says that a technician needs to know how to determine if a computer is working.

   Who is correct?

   A. A only
   B. B only
   C. Both A and B
   D. Neither A nor B

Questions 4 and 5 are not like those above. Each has the word EXCEPT. For each question, look for the choice that does *not* apply. Read each question carefully before you choose your answer.

4. A computer is capable of doing all of the following EXCEPT

   A. receiving input data.
   B. processing input data according to a program, and producing output information.
   C. moving linkage to affect vehicle operation.
   D. storing data and/or information.

5. All of the the following statements about feedback control are correct EXCEPT

   A. sensor input data are sent to the computer.
   B. computer output is used to operate servos.
   C. the computer changes output in response to input data.
   D. computer output signals are fed back into the computer as input data.

ESSAY QUESTIONS

1. What are the parts and functions of a computer control system?
2. Describe feedback control, using the air-fuel ratio as an example.
3. What are self-diagnostics?

SUPPLEMENTAL ACTIVITIES

1. If available at your library, research computer controls in magazines such as MOTOR and the SAE Journal. Report on service and developments in computer controls.
2. Read the proper section in a manufacturer's service manual for an instructor-chosen vehicle and check if indicating devices are operating correctly.
3. Locate and identify sensors and mechanical output devices on an instructor-chosen vehicle.

39 Electronic Device Service

UNIT PREVIEW

Typical tasks for electronic device service include checking for proper wiring connections and operation, replacing defective units, and making adjustments on mechanical output devices. Self-diagnostic features aid the technician in finding problems. Test procedures and equipment vary among manufacturers, and from model to model. Technicians must refer to the vehicle manufacturer's service manual instructions for specific troubleshooting.

LEARNING OBJECTIVES

When you have completed your assignments and exercises in this unit, you should be able to:

❏ Explain the precautions necessary when checking and replacing electronic systems components or modules.

❏ Inspect and clean electrical connector plugs.

❏ Check electronic indicating devices for proper operation before attempting service procedures.

❏ Activate self-diagnostic functions on computer-controlled systems.

❏ Determine whether a vehicle is operating in the open-loop or closed-loop mode.

❏ Replace defective sensors or computer modules, and adjust mechanical output devices.

SAFETY PRECAUTIONS

Personal injury and unnecessary damage to electronic components can be avoided by observing the following precautions.

Do not attempt to check or replace electronic parts without the manufacturer's specific test or replacement procedures. These procedures change from year to year and from model to model. Make sure you have the correct service manual for the vehicle being checked.

Expensive damage to electronic systems can result from common checking procedures that are used for checking simpler systems. For example, to check a simple warning light system, a technician might normally connect battery voltage to the bulb. On some vehicles with a low-fuel warning light for an electronic fuel gauge, this procedure will destroy the gauge.

Use the manufacturer's special test equipment when necessary. Use of other than the correct test equipment may result in false readings.

If necessary, remove the battery negative terminal to disconnect electrical power from the vehicle. Remove rings, watches, and any other metal jewelry items before working on electrical or electronic components.

When procedures call for connecting test leads, or wires, to electrical connections, use extreme care and follow the manufacturer's instructions. Identify the correct test terminals before attempting to connect test leads.

Accidentally touching a metal clip lead or test probe between metal terminals can cause a short circuit. Expensive computer modules or sensors can be destroyed instantly, without warning, by incorrect test wiring hookups.

Many test procedures require that electrical power or ground be supplied to the circuit being tested. Avoid grounding powered circuits with metal tools. Avoid touching live electrical leads to grounded metal parts of the automobile. Personal injury from sparks may result, or the unit being tested may be damaged or destroyed.

CAUTION!!!

Computer "chips" can be destroyed by static electricity generated by your clothing, hair, or shoes. To discharge static electricity from your body, simply touch a clean, unpainted ground connection anywhere on the vehicle before you touch or unplug computers or replaceable chips.

TROUBLESHOOTING AND MAINTENANCE

In most circumstances, no preventive maintenance is required for electronic sensors or modules. However, in areas of extreme humidity or dust, electrical connections can become degraded by oxidation or contaminated by dust. These poor connections can cause erratic operation and/or damage to sensitive electronic components. Preventive maintenance in such areas might consist of periodically cleaning

ELECTRONIC DEVICE SERVICE

electrical terminals. Mechanical output units operated by electronic devices may require periodic lubrication, adjustment, and/or other maintenance. The following general procedures are provided as a general guide to the maintenance of electronic system components.

Cleaning Electrical Connector Plugs

It has been estimated that more than 90 percent of the problems with electrical or electronic units are caused by poor electrical connections. In most cases, electrical connections between vehicle electronic units are made with the use of *electrical connector plugs.* Figures 39-1 and 39-2 show several types of connectors.

Disconnect the grounded battery terminal before disconnecting electrical leads or connectors. Plastic locking tabs on some connector bodies must be released before the connector can be pulled apart. Some tabs are released by pressing inward. Other types of tabs must be lifted outward to release. Release the tabs and pull the connector apart, as shown in **Figure 39-2.**

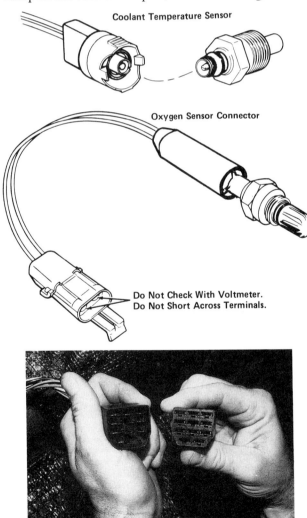

Figure 39-1. Electrical wire connectors.

Simply pulling the connector apart and reconnecting it may restore an electrical connection. However, the terminals may need to be cleaned or replaced.

Clean the terminals as recommended by the manufacturer. Never use metal tools, such as screwdrivers, to scrape the terminals. Metal tools can cause a short circuit between the terminals.

After cleaning, protect the terminals as the manufacturer recommends to prevent further corrosion or oxidation. Some manufacturers recommend using a *dielectric compound,* such as silicone grease. Line up any alignment tabs, and push the connector plug back together firmly. Seat the locking tabs firmly into position so that the connector cannot pull apart from vibration.

39.1 INDICATING DEVICES

If a problem is suspected with an electronic indicating device, preliminary checks can be made to isolate the source or cause. Refer to the manufacturer's service manual for the correct checking procedures.

Figure 39-3 shows what is known as a check-out sequence, or *prove-out sequence,* for an electronic speedometer. A prove-out sequence is a display of information indicating that the basic operation of an indicating device is correct. When the key switch of this vehicle is turned on, all *segments,* or lighted areas, of the display come on. The display reads "188 km/h MPH." This proves that none of the segments are burned out. Next, all segments go off, proving that none of them remain lit incorrectly. Finally, the speedometer displays "0 MPH," and the display remains on. This proves that the speedometer reads accurately when the vehicle is stopped. Each manufacturer may have a different check-out or prove-out sequence for each instrument. If an electronic indicating device is defective, the unit must be removed and replaced with a new unit. No field repairs should be attempted on electronic indicating devices.

39.2 SELF DIAGNOSTICS

Most recent vehicles with electronic controls include features to help technicians locate malfunctions. The control modules have *memory* features that can store, or save, information about problems. To activate these *self-diagnostic* features, jumper wires are connected to special connector plugs located under the dash or hood. Connecting the jumper wires correctly will cause the control module to display a *trouble code.*

Activating Self-Diagnostic Functions

Figure 39-4 shows how to check a self-diagnostic connector with jumper wires and a voltmeter probe. This illustration applies to some Ford MCU systems.

PART IV: AUTOMOBILE ELECTRICAL AND ELECTRONIC SYSTEMS

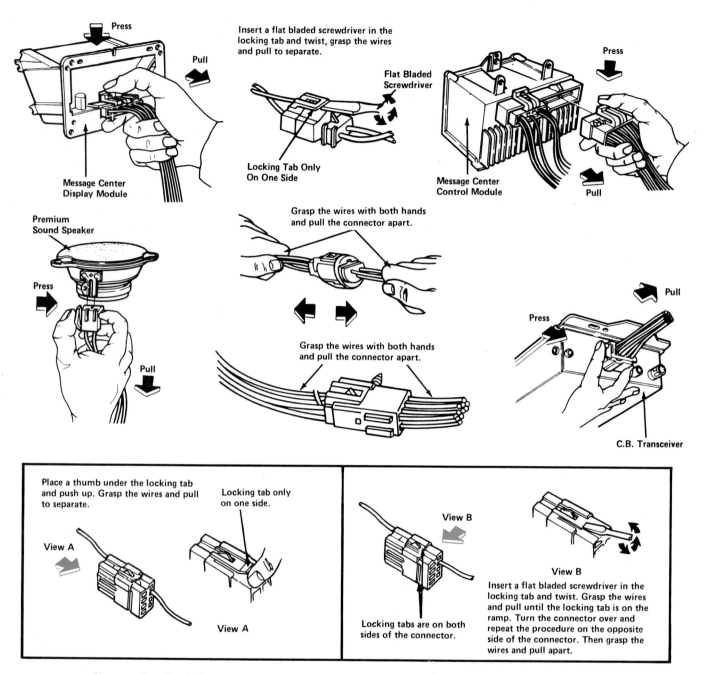

Figure 39-2. Disconnecting electrical connectors. FORD MOTOR COMPANY

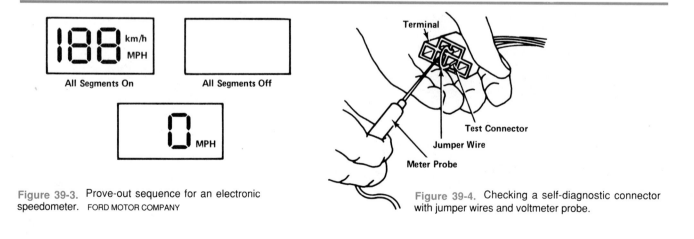

Figure 39-3. Prove-out sequence for an electronic speedometer. FORD MOTOR COMPANY

Figure 39-4. Checking a self-diagnostic connector with jumper wires and voltmeter probe.

The simplest display is a dash light that blinks on and off in coded sequence. Some vehicle manufacturers require that a voltmeter be connected to certain terminals. Others require that a vacuum/pressure gauge be inserted in certain hoses. The movement of the meter needle or gauge then indicates code numbers. **Figure 39-5** shows how to read a typical light code, used on General Motors vehicles.

Trouble Codes

Figure 39-6 shows a table of codes used for problems with General Motors vehicles. A code 12 would indicate that no rpm signal is being received from the distributor when the engine is off. This checking code is used to confirm that the diagnostic system is operating properly before other tests are performed. Each vehicle manufacturer uses a different set of codes.

Hand-held Diagnostic Testers

Vehicle manufacturers may specify the use of special hand-held diagnostic testers that connect to electrical sockets on the vehicle **[Figure 39-7]**. These testers communicate with vehicle computers to identify system malfunctions. These testers can provide a display of trouble codes, read input sensor information, and operate actuators. In addition, the testers can perform troubleshooting functions and provide a readout of problem components.

Transmitting Troubleshooting Data

Some domestic vehicle manufacturers provide sockets on the vehicle that a technician can connect to a phone *modem*, or transmitting device. Information from the vehicle's computer is transmitted over phone lines to a central computer. The computer, in the manufacturer's service head-quarters, logs the input data, then compares it to similar problems in its memory banks. A computer screen displays diagnosis information to the technician at his workplace. Refer to manufacturers' service manuals for specific testers, connections, and instructions on their use.

39.3 OPEN-LOOP AND CLOSED-LOOP OPERATION

As explained in Unit 38, engine control systems can operate in either a closed-loop or an open-loop mode. For General Motors vehicles, a dwell meter can be used to determine if a feedback carburetor is operating in the proper mode.

Determining Open-Loop and Closed-Loop Operation

Connect the dwell meter leads to the mixture control solenoid on the feedback carburetor, as shown in **Figure 39-8**.

NOTE: This procedure requires that the dwell meter be set to the six-cylinder scale, even if the engine being checked is a 4- or 8-cylinder engine. Some dwell meters may not work properly for these procedures. Do not use a dwell meter that causes engine speed to change when connected.

In the open-loop mode, an on-off signal of fixed length is sent to the carburetor mixture control solenoid. This action provides a steady, rich mixture to

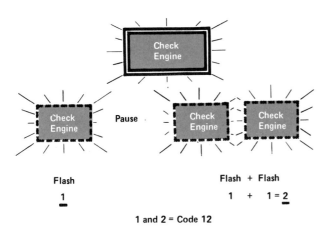

Figure 39-5. Diagnostic code display.
GENERAL MOTORS CORPORATION

GENERAL MOTORS TROUBLE CODES	
Code	Problem
12	Computer not receiving rpm signal
13	Defect in oxygen sensor's wiring
14	Short circuit in coolant-temperature sensor's wiring
15	Open circuit to coolant temperature sensor
21	Defect in throttle-position sensor
23	Defect in carburetor solenoid wiring
24	Defect in vehicle-speed sensor
32	Defect in barometric-pressure sensor
35	Short circuit in idle-speed control switch
42	Ground in ignition timing bypass circuit
44	Defect in oxygen sensor
51	Defects in computer module
54	Defect in carburetor mixture-control solenoid and/or computer module
55	Defect in throttle-position sensor, oxygen sensor, or computer module (depending on engine model)

Figure 39-6. Trouble code identification.
GENERAL MOTORS CORPORATION

The "CHECK ENGINE" light will only be "ON" if the malfunction exists under the conditions listed below. It takes up to five seconds minimum for the light to come on when a problem occurs. If the malfunction clears, the light will go out and a trouble code will be set in the ECM. Code 12 does not store in memory. If the light comes "ON" intermittently, but no code is stored, go to the "Driver Comments section. Any codes stored will be erased if no problem reoccurs within 50 engine starts.

THE TROUBLE CODES INDICATE PROBLEMS AS FOLLOWS:

Code	Problem	Code	Problem
12	No distributor reference pulses to the ECM. This code is not stored in memory and will only flash while the fault is present.	32	Barometric pressure sensor (BARO) circuit low, or altitude compensator low on J-car.
13	Oxygen sensor circuit — The engine must run up to five minutes at part throttle, under road load, before this code will set.	35	Idle speed control (ISC) switch circuit shorted. (Over 50% throttle for over 2 sec.)
		42	Electronic spark timing (EST) bypass circuit or EST circuit grounded or open.
14	Shorted coolant sensor circuit — The engine must run up to five minutes before this code will set.	44	Lean exhaust indication — The engine must run up to five minutes, in closed loop, at part throttle and road load before this code will set.
15	Open coolant sensor circuit — The engine must run up to five minutes before this code will set.	44 & 55	(At same time) — Faulty oxygen sensor circuit.
21	Throttle position sensor circuit — The engine must run up to 25 seconds, at specified curb idle speed, before this code will set.	51	Faulty calibration unit (PROM) or installation. It takes up to 30 seconds before this code will set.
23	Open or grounded M/C solenoid circuit.	54	Shorted M/C solenoid circuit and/or faulty ECM.
24	Vehicle speed sensor (VSS) circuit — The car must operate up to five minutes at road speed before this code will set.	55	Grounded V ref. (terminal "21", faulty oxygen sensor or ECM.

Figure 39-6. Concluded.

the engine. In the open-loop mode, the dwell meter will indicate a fixed reading of between 21 and 35 degrees.

At idle, after warmup, the engine control system should be in the closed-loop mode. The control module constantly modifies the signal to the carburetor mixture control solenoid in response to exhaust emissions. In the closed-loop mode, the dwell meter indicates a dwell reading that varies constantly between 5 and 55 degrees. **Figure 39-9** shows the range of dwell change during closed-loop operation.

To double check that the system is in closed-loop operation, momentarily "choke" the carburetor air horn with your hand. The dwell meter should

ELECTRONIC DEVICE SERVICE

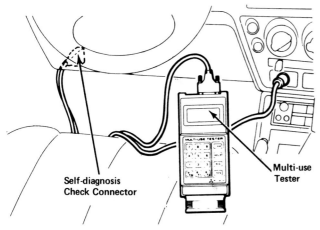

Figure 39-7. Hand-held diagnostic computer.
MITSUBISHI MOTOR SALES OF AMERICA, INC.

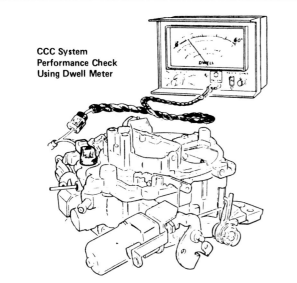

Figure 39-8. Dwell meter connection.
GENERAL MOTORS CORPORATION

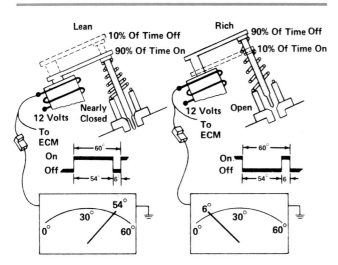

Figure 39-9. Dwell range during closed-loop operation.
GENERAL MOTORS CORPORATION

indicate a slightly higher reading if the system is in closed-loop operation. Create a vacuum leak by disconnecting a vacuum hose to the intake manifold. The dwell meter should indicate a slightly lower reading if the system is in closed-loop mode.

39.4 SENSORS

Many types of sensors are used with electronic engine controls. Examples of sensors can include:

- Engine coolant temperature
- Exhaust gas oxygen sensor
- Throttle position sensor (TPS)
- Barometric pressure (BARO)
- Manifold absolute pressure (MAP)
- Vacuum
- Intake air mass (MAF)
- Intake air temperature
- Transmission gear
- Ambient temperature.

These sensors provide information to the control module about engine and vehicle functions and atmospheric conditions. Based on this information, the control module sends output signals. These signals control air-fuel mixture, spark advance, and some emissions control devices. Signals from the system also can be used to control many other functions (refer to Unit 38).

Each manufacturer provides specific test and replacement information for each type of sensor used on its vehicles. The following information is provided as a general guide to the replacement of the exhaust gas oxygen sensor used on General Motors vehicles.

Oxygen Sensor Replacement

An oxygen sensor is shown in **Figure 39-10**.

SAFETY CAUTION

The exhaust gas oxygen sensor is located in the exhaust manifold. Make sure the exhaust system is cool before attempting to check or replace the sensor. Severe burns can result from touching a hot exhaust manifold.

CAUTION!!!

If an oxygen sensor is handled roughly or dropped, it will no longer provide accurate input information to the control module. Handle the oxygen sensor carefully. Do not allow grease, lubricants, or cleaning solvents of any kind to touch the sensor end or the electrical connector plug. Apply the manufacturer's special anti-seize compound to the threads before installing the sensor.

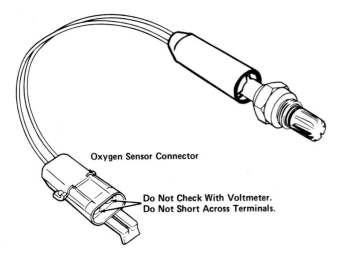

Figure 39-10. Oxygen sensor. GENERAL MOTORS CORPORATION

Disconnect the electrical connector plug. Unscrew the sensor with the manufacturer's special wrench or a flare-nut wrench. Coat the threads of the new sensor with the manufacturer's special anti-seize compound. Thread it in by hand. Tighten the sensor to the manufacturer's specified torque. Make sure that the silicone rubber boot on the connector wire does not touch the exhaust manifold after tightening.

39.5 CONTROL MODULES

Some manufacturers provide special diagnostic test equipment for their own electronic engine control systems. In other cases, normally used shop testers are used to check the operation of electronic control modules. **Figure 39-11** shows the types of tools needed to check a General Motors system.

Vehicles can be equipped with a variety of engines, transmissions, differential ratios, and wheel and tire sizes. Because of these variations, one control module will not work for all vehicles. Some manufacturers provide separate modules with programs for each individual vehicle. Others provide a separate, smaller module that plugs into the main control module. This smaller module is known as an *engine calibration unit* or *programmable read-only memory (PROM)* [Figure 39-12].

Replacing a Control Module

To replace the main control module, lift the locking tabs and gently rock the connectors to unplug them. Remove the main control module. Remove the access cover over the PROM.

CAUTION!!!

Computer "chips" can be destroyed by static electricity generated by your clothing, hair, or shoes. To discharge static electricity from your body, simply touch a clean, unpainted ground connection anywhere on the vehicle before you touch or unplug computers or replaceable chips.

Unplug the connector terminals and gently rock the PROM to remove it. Note the alignment tabs and depressions. Check the stock numbers on the new main control module to make sure they match the numbers on the old module. Carefully replace the PROM and seat it in the new module. Replace the access cover and main control module. Plug the connectors into the module and make sure that the locking tabs seat properly.

39.6 MECHANICAL OUTPUT DEVICES

Mechanical output devices are operated by engine vacuum, electrical solenoids, or electrical motors. A typical flowchart showing the diagnosis procedure for a cruise-control unit is shown in **Figure 39-13**.

Adjusting a Mechanical Output Device

Some mechanical output devices require adjustment. **Figure 39-14** shows the adjustment procedure for a cruise-control servo that operates throttle linkage.

UNIT HIGHLIGHTS

- There are many differences in servicing procedures for electronic devices. Therefore, the technician must always refer to the proper manufacturer's service manual for correct checking and/or replacement procedures.
- Many electronic system problems are caused by poor connections.
- Self-diagnostic functions can help the technician to find problems.
- A dwell meter can be used to determine if an engine control system is operating in open-loop or closed-loop mode.
- Service procedures for electronic devices consist of cleaning connections, replacing modules, sensors, and indicating devices, and maintaining mechanical output devices.

TERMS

electrical connector plug	prove-out sequence
segment	memory
self-diagnostic	trouble code
engine calibration unit	dielectric compound
programmable read-only memory (PROM)	modem

ELECTRONIC DEVICE SERVICE

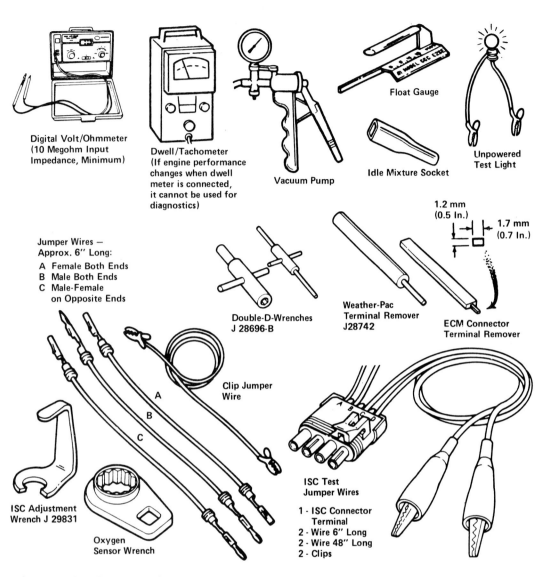

Figure 39-11. Control module diagnostic tools. GENERAL MOTORS CORPORATION

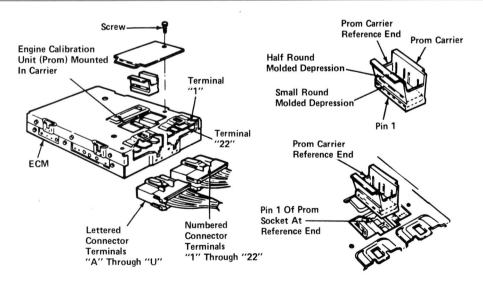

Figure 39-12. Electronic control module.
GENERAL MOTORS CORPORATION

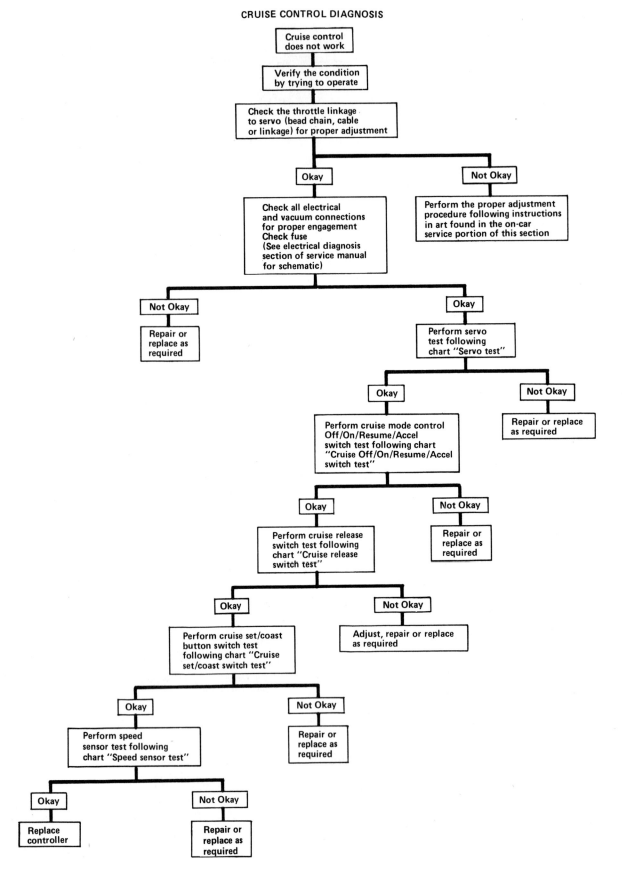

Figure 39-13. Cruise control diagnostic chart. PONTIAC MOTOR DIVISION—GMC

ELECTRONIC DEVICE SERVICE

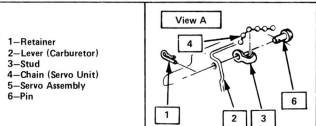

Figure 39-14. Throttle linkage adjustment procedure. PONTIAC MOTOR DIVISION—GMC

REVIEW QUESTIONS

DIRECTIONS: The following questions are similar to those used on automotive technician certification tests. On a separate sheet of paper, write the letter of the correct choice.

1. Technician A says that the procedures for repairing electronic devices are the same as for electrical devices.

   Technician B says that special precautions and specific instructions from the manufacturer's service manual are necessary for servicing electronic devices.

   Who is correct?
   A. A only
   B. B only
   C. Both A and B
   D. Neither A nor B

2. Which of the following statements is correct?
   I. A prove-out sequence shows how accurate an indicating device is throughout its range of measurement.
   II. A prove-out sequence checks the basic operation of an indicating device.
   A. I only
   B. II only
   C. Both I and II
   D. Neither I nor II

3. During self-diagnostic functions of an electronic control module, two flashes or pulses followed by one flash or pulse indicate
   A. no rpm signal being received (engine off).
   B. exhaust gas oxygen sensor faulty.
   C. a code 21.
   D. a code 15.

4. When checking for open- or closed-loop operation, a dwell reading that constantly *varies* between 5 and 55 degrees indicates
   A. open-loop operation.
   B. closed-loop operation.
   C. dwell meter not set to 6-cylinder scale.
   D. dwell meter is changing engine speed when connected.

Question 5 is not like those above. It has the word EXCEPT. For question 5, look for the choice that does *not* apply. Read the entire question carefully before you choose your answer.

5. When replacing an exhaust gas oxygen sensor, all of the following must be done EXCEPT
   A. making sure the exhaust system is cool.
   B. checking the old sensor by connecting a voltmeter across its terminals.
   C. coating the threads of the replacement sensor with the manufacturer's special antiseize compound.
   D. making sure the silicone rubber boot does not touch the exhaust manifold.

ESSAY QUESTIONS

1. What procedures and devices may be involved in activating self-diagnostics and reading trouble codes?
2. Describe how to replace an oxygen sensor.
3. Describe how to replace a PROM.

SUPPLEMENTAL ACTIVITIES

1. Locate, disconnect, clean, and reconnect connector plugs on a shop vehicle.
2. Use a dwell meter to determine whether a shop vehicle's engine control system is operating in open-loop or closed-loop mode.
3. Read the proper section in a manufacturer's service manual for an instructor-chosen vehicle and activate the computer's self-diagnostic functions.
4. Read trouble codes on the vehicle used for Supplemental Activity 3 and diagnose a "bug," or problem, created by your instructor.

PART V: Emissions Systems

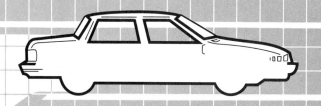

To troubleshoot emissions control systems, a technician must have a thorough background in all automotive systems. Many states require special testing and certification for emissions-control technicians. The first unit provides an introduction to the sources and causes of natural and man-made pollution. The next knowledge unit details the construction and operation of emissions-control systems. This technician is working on an exhaust system.

40 Emission Control Fundamentals

UNIT PREVIEW

Unhealthful air can result from natural or man-made causes. Incomplete burning of fuels to produce heat and energy can produce chemicals that pollute the air. Air pollution can cause irritation or diseases in living things and damage to objects.

Atmospheric conditions can trap and concentrate polluted air. Smog, or polluted air, may be visible or invisible. Invisible pollution can be more harmful than visible pollution.

Major sources of air contaminants from automobiles include the exhaust, fuel system, and crankcase. U.S. government regulations limit the amount of certain pollutants produced by new vehicles.

The combination of clean-air and fuel-economy regulations has stimulated vehicle manufacturers to use advanced technology. This technology includes feedback carburetors, fuel injection, electronic ignition, catalytic converters, and computer controls.

The technology of emission controls is not simple. However, technicians who have a good attitude and are willing to learn new knowledge and skills always are in demand. This is true especially in the area of emission control service.

LEARNING OBJECTIVES

When you have completed your assignments and exercises in this unit, you should be able to:

❑ Identify and describe sources and effects of air pollution.
❑ Identify and describe the products of complete and incomplete combustion of a hydrocarbon fuel.
❑ Locate emission control information labels on late-model vehicles and describe specific tune-up procedures required for limiting emissions.
❑ Identify and describe trends in engine size, horsepower output, and ignition timing of vehicles in recent years.
❑ Describe the type of training necessary for technicians who wish to perform emission control service.

40.1 AIR POLLUTION

Pollution means a process of contamination. *Air pollution* is the contamination of air by harmful substances. Air pollution may occur through natural processes or through the contamination of air by man-made substances and gases. Air pollution is a problem because it damages and harms plants, animals, people, and objects.

Air is made up of nitrogen (N_2), oxygen (O_2), and a very small amount of other natural gases **[Figure 40-1]**. Air pollution occurs when harmful gases and particles are added to air.

Natural Pollution

Natural pollution comes from many sources in nature. Natural pollutants include:

* Smoke from naturally caused forest and grass fires
* Gases given off by growing and decomposing vegetation
* Dust and dirt blown by the wind
* Smoke, ashes, dust, and gases from volcanic eruptions.

Natural pollution was present in the atmosphere before humans appeared on Earth. However, natural forces such as wind and rain tended to dilute and spread pollutants. The "balance of nature" maintains an atmosphere that allows plants and animals to grow and live in a healthful manner.

Man-Made Air Pollution

Man-made air pollution is caused mainly by the burning of fuel to produce heat and power. Refineries and other industrial operations also produce chemicals and gases that pollute the air. Under certain conditions, the concentration of automobiles and

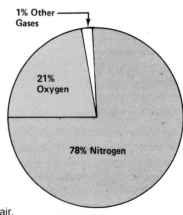

Figure 40-1. Composition of air.

factories in cities can produce air pollution that causes serious health problems.

Air pollution can irritate and damage breathing passages in the nose, throat, and lungs. Acids are formed from a combination of water vapor and sulfur, nitrogen, and other chemical pollutants.

Taking a deep breath can be painful because of the acids' effect on the lungs. Coughing is another symptom noticed during periods of high pollution. Eyes become watery from irritating particles and acids. Air pollution has been linked to many health problems, including:

- Asthma
- Cancer of the lungs and stomach
- Emphysema
- Heart and circulatory system problems
- Skin problems.

Smog

Smog is a combination of the words "smoke" and "fog." However, air pollution is not always caused by a combination of *particulate* matter and fog. Particulates are extremely small particles of solid material, like bits of burned material in smoke. However, the slang term "smog" is widely used to mean "air pollution" of all sorts.

Hundreds of years ago, the burning of soft coal in European and British cities caused foul-smelling, dirty air. As recently as the 1950s, many people in London died because of a "killer fog." This deadly fog was caused by a combination of stagnant air, carbon monoxide gas, coal dust, and sulfuric acid from the smoke.

Smog became a health and visibility problem in Los Angeles as early as the late 1930s. Hundreds of years before, native Indians in Southern California called the area "valley of smokes." They had noticed that smoke from campfires would rise to a certain level, then spread out in a flat layer.

This atmospheric phenomenon is known as an *inversion layer,* a condition in which normally cooler upper air becomes warmer. This warmer air acts as a "lid" to trap and concentrate air pollution [Figure 40-2]. When geographic conditions combine with a lack of wind, stagnant air can intensify air pollution.

Photochemical Smog

The action of sunlight can change contaminants into more harmful pollution. *Photochemical smog* is air pollution produced through the action of sunlight on chemicals and moisture in air.

Incomplete Combustion and Air Pollution

If combustion of a hydrocarbon fuel is 100 percent efficient, the products are water vapor (H_2O) and

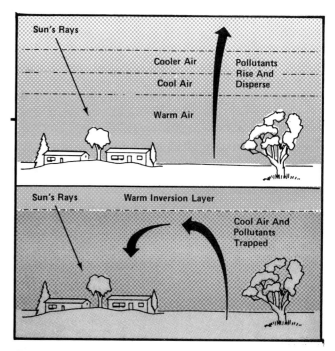

Figure 40-2. Inversion layer.

carbon dioxide (CO_2). Both of these substances are part of the natural environment. Carbon dioxide is one of the colorless, odorless gases exhaled when you breathe.

As discussed in Unit 10, 100 percent combustion is never achieved in an internal combustion engine. Quench areas within the combustion chamber put out the flame, and some fuel remains unburned. Thus, even in a new, perfectly tuned vehicle, internal combustion engine exhaust gases always contain some unburned *hydrocarbon (HC)* fuel. In addition, incomplete combustion produces *carbon monoxide (CO),* a toxic gas. Both of these pollutants usually are colorless.

Bluish-white smoke from a tailpipe indicates that engine oil, a hydrocarbon, is being burned. Thus, unburned petroleum hydrocarbons are being released into the air. Black smoke indicates an air-fuel mixture too rich to burn completely. Large amounts of hydrocarbons are released into the air. However, even vehicles that produce no smoke at all are releasing pollutants into the air.

Invisible Pollution

Small amounts of hydrocarbon pollutants and carbon monoxide are not visible. *Invisible pollution* takes the form of invisible gases and vapors that cause health problems.

For example, carbon monoxide (CO) gas from auto exhausts can displace oxygen in the bloodstream, resulting in *carbon monoxide poisoning.*

The concentration of a gas or particulate in air can be measured in *parts per million (ppm).* That is, in one million parts of air, a certain number of the parts are composed of a certain gas or particulate.

Below 50 ppm of CO in air, no symptoms are noticeable. Mild CO poisoning is noticeable as a headachy, drowsy feeling. Increasingly severe stages of CO poisoning include throbbing headaches, vomiting and collapse, and coma. At a concentration of 600 ppm, death occurs **[Figure 40-3]**.

Thus, even though it cannot be seen, invisible air pollution can cause serious health problems. Technicians must be especially aware of the dangers of carbon monoxide poisoning when working on vehicles in an enclosed garage **[Figure 40-4]**.

- Fuel system emissions
- Crankcase emissions.

Exhaust Emissions

Emissions, or pollutants, in the exhausts of internal-combustion engines account for a large percentage of pollutants caused by man **[Figure 40-6]**. Automotive exhaust emissions include:

- Hydrocarbons (HC)
- Carbon monoxide (CO)
- Oxides of nitrogen (NO_x)

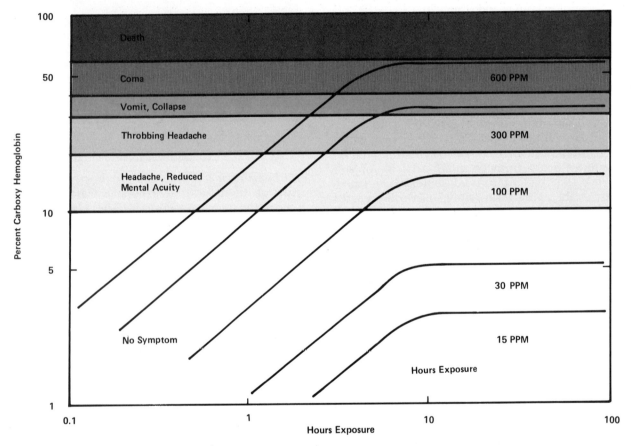

Figure 40-3. Carbon monoxide exposure graph.

SAFETY CAUTION

Always have an adequate source of ventilation when operating an engine indoors. Open the garage doors or turn on ventillating fans whenever engines are running inside an enclosed area. Carbon monoxide poisoning can cause serious health problems.

40.2 AUTOMOTIVE EMISSIONS

Sources of air pollution from vehicles are illustrated in **Figure 40-5.** These sources include:

- Exhaust emissions

Figure 40-4. Carbon monoxide warning sign.

EMISSION CONTROL FUNDAMENTALS

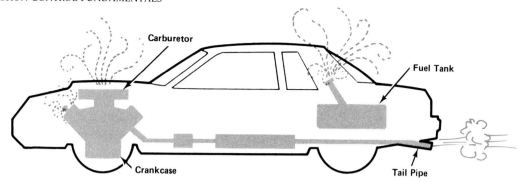

Figure 40-5. Sources of automotive pollution.

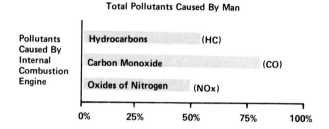

Figure 40-6. Internal combustion pollutants.

- Oxides of sulfur (SO_x)
- Ozone (O_3)
- Particulate matter.

Hydrocarbons are molecules that contain hydrogen and carbon atoms, such as petroleum fuels and lubricating oils. Unburned hydrocarbons are the major source of automotive air pollution.

Carbon monoxide is a colorless, odorless, toxic gas produced by incomplete combustion. As discussed in Topic 40.1, carbon monoxide can be deadly.

Oxides of nitrogen are produced when combustion temperatures and pressures are raised. To reduce HC and CO emissions, engines can be designed to run on leaner mixtures and at higher combustion temperatures. Higher combustion temperatures, however, result in the combination of nitrogen and oxygen to form NO_x. NO_x contributes to the formation of nitric acid, a powerfully corrosive substance.

Oxides of sulfur are produced by the chemical reaction in a catalytic converter. (Refer to Unit 25). Sulfur combines with water vapor to form sulfuric acid.

Ozone can be produced by sunlight acting on oxygen in the air, or by an electrical spark. Thus, the secondary ignition system is a source of ozone. Ozone has a sharp, pungent smell. It is harmful to the lungs and can cause rubber to deteriorate.

Particulates in exhaust gases can include carbon soot and lead particles (in older vehicles using leaded fuel). Diesel engines produce large amounts of soot, which is a *carcinogen*. A carcinogen is a substance that causes cancer.

Fuel System Emissions

Petroleum-based motor fuels are pure hydrocarbons. When fuel is allowed to evaporate, massive amounts of HC are released into the air. When temperatures rise, fuel in vehicle tanks expands and vaporizes. In addition, fuel in the carburetor float bowl also can evaporate.

On earlier vehicles, neither of these sources of pollution was controlled. It was known that fuel that became heated in a fuel tank produced vapors and pressure. Vented gas caps allowed pressure buildup and vapors to be vented to the atmosphere to prevent rupturing the fuel tank.

Before pollution controls, fuel within the carburetor float bowl was allowed to evaporate into the air to produce HC vapors [Figure 40-7].

Crankcase Emissions

Crankcase emissions include *blowby gases* and fumes from heated oil. Blowby gases are gases that leak past piston rings during combustion. Blowby gases contain water vapor (H_2O), HC, CO, and sulfur (S). These pollutants combine to form acids and sludge in lubricating oil. To prevent pressure buildup from blowing out oil pan and valve cover gaskets, blowby gases must be controlled.

Before emission controls, blowby gases were vented directly to the atmosphere through a *road draft tube*. A road draft tube is an open tube leading from the crankcase. Blowby gases and crankcase vapors leave the crankcase area through this tube and are vented to the atmosphere.

Positive crankcase ventilation (PCV) systems were tried as early as the 1920s to prevent contamination of lubricating oil. A PCV system uses intake

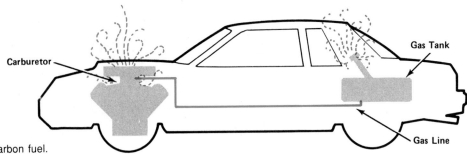

Figure 40-7. Evaporation of hydrocarbon fuel.

manifold vacuum to draw the blowby gases from the crankcase and blend them with the intake charge. Crankcase emissions are thus "recycled" for further combustion **[Figure 40-8]**.

40.3 GOVERNMENT REGULATION AND AIR POLLUTION

In 1959, California became the first state to enact automobile emission regulations. A state Air Resources Board (ARB) was formed to set standards for automotive emissions as well as testing procedures.

In 1968, based on the work of California's ARB, the U.S. Congress passed the Clean Air Act, mandating emission controls for all vehicles. As a result, allowable levels of the three pollutants considered most harmful were to be controlled. Those three pollutants are:

- Hydrocarbons (HC)
- Carbon monoxide (CO)
- Oxides of nitrogen (NO_x).

Federal standards for the reduction of these pollutants, measured in *grams per mile (gpm)* were established **[Figure 40-9]**. A gram is a metric measurement of mass, equal to 0.035 ounce. Grams per mile is a way of measuring a mass of pollutant emitted over a distance traveled.

Other states, primarily high-altitude western states, passed their own laws and applied for

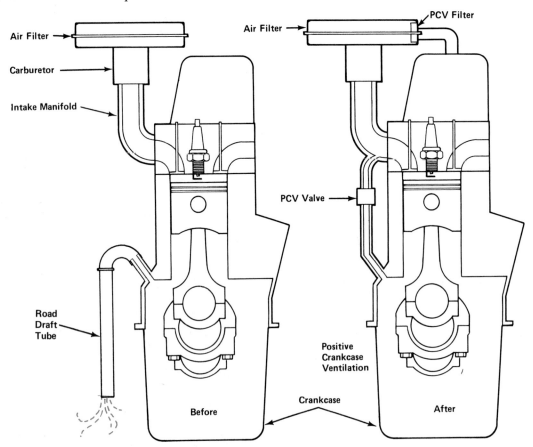

Figure 40-8. Road draft tube and basic PCV system.

Federal Standards—New Vehicle Summary
Passenger Cars[1]

The following standards, up to 1975, apply only to gasoline-fueled light-duty vehicles. Standards for 1975 and later apply to both gasoline-fueled and diesel light–duty vehicles.

Year	Test Procedure[2]	Hydrocarbons	Carbon Monoxide	Oxides Of Nitrogen	Diesel Particulates	Evaporative Hydrocarbons[3]
Prior	7-mode	850 ppm	3.4%	1000 ppm	–	–
to	7-mode	11 gpm	80 gpm	4 gpm	–	–
controls	CVS-75	8.8 gpm	87.0 gpm	3.6 gpm	–	–
1968-69	7-mode					
	50-100 CID	410 ppm	2.3%	–	–	–
	101-140	350 ppm	2.0%	–	–	–
	over 140	275 ppm	1.5%	–	–	–
1970	7-mode	2.2 gpm	23 gpm	–	–	–
1971	7-mode	2.2 gpm	23 gpm	–	–	6 g/test[4]
1972	CVS-72	3.4 gpm	39 gpm	–	–	2 g/test
1973-74	CVS-72	3.4 gpm	39 gpm	3.0 gpm	–	2 g/test
1975-76	CVS-75	1.5 gpm	15 gpm[5]	3.1 gpm	–	2 g/test
1977[6]	CVS-75	1.5 gpm	15 gpm	2.0 gpm	–	2.0 g/test
1978-79	CVS-75	1.5 gpm	15 gpm	2.0 gpm	–	6.0 g/test
1980	CVS-75	0.41 gpm	7.0 gpm	2.0 gpm	–	6.0 g/test
1981	CVS-75	0.41 gpm	3.4 gpm[7]	1.0 gpm[8,9]	–	2.0 g/test
1982[10, 11]	CVS-75	0.41 gpm (0.57)	3.4 gpm[7] (7.8)	1.0 gpm[8,9] (1.0)[8]	0.60 gpm –	2.0 g/test (2.6)
1983[10,11]	CVS-75	0.41 gpm (0.57)	3.4 gpm (7.8)	1.0 gpm[8] (1.0)[8]	0.60 gpm –	2.0 g/test (2.6)
1984[11,12]	CVS-75	0.41 gpm (0.57)	3.4 gpm (7.8)	1.0 gpm[8] (1.0)[8]	0.60 gpm –	2.0 g/test (2.6)
1985-86[13]	CVS-75	0.41 gpm	3.4 gpm	1.0 gpm	0.60 gpm	2.0 g/test
1987 & later[13]	CVS-75	0.41 gpm	3.4 gpm	1.0 gpm	0.20 gpm[14]	2.0 g/test

Light Duty Vehicles:

gpm - grams per mile / ppm - parts per million / CVS - Constant Volume Sampler

1. Standards do not apply to vehicles with engines less than 50 CID from 1968 through 1974. Diesel particulate standards apply only to diesels.
2. Different test procedures have been used since the early years of emission control which vary in stringency. The appearance that the standards were relaxed from 1971 to 1972 is incorrect. The 1972 standards are actually more stringent because of the greater stringency of the 1972 test procedure.
3. Evaporative emissions determined by carbon trap method through 1977, SHED procedure beginning in 1978. Applies only to gasoline-fueled vehicles.
4. Evaporative hydrocarbon standard does not apply to off-road utility vehicles for 1971.
5. Carbon monoxide standard for vehicles sold in the State of California is 9.0 gpm.
6. Cars sold in specified high altitude counties required to meet standards at high altitude.
7. Carbon monoxide standard can be waived to 7.0 gpm for 1981-82 by the EPA Administrator (Appendix 2).
8. Oxides of nitrogen standard can be waived to 1.5 gpm for innovative technology or diesel (Appendix 2).
9. Oxides of nitrogen standard can be waived to 2.0 gpm for American Motors Corporation.
10. Standards in parentheses apply to vehicles sold in specified high altitude counties. Vehicles eligible for a carbon monoxide waiver for 7.0 gpm at low altitude are eligible for a waiver to 11 gpm at high altitude.
11. Exemptions from the high-altitude standards are provided for qualifying low-performance vehicles.
12. These same numerical standards apply to vehicles sold in high-altitude areas. Standards in parentheses apply to heavy passenger cars sold in high-altitude areas for the 1984 model year only.
13. These same numerical standards apply to vehicles sold in high-altitude areas. Exemptions from the high-altitude standards are provided for qualifying low-performance vehicles.
14. Emissions averaging may be used to meet this standard.

Additional Requirements

No crankcase emissions are permitted (applies to gasoline-fueled only).

Emission control component maintenance is restricted to **specified** intervals (See Appendix 1).

Figure 40-9. Federal standards for new automobiles.

waivers of the federal standards. As a result, since the 1970s gasoline vehicles have been built to three different emission standards:

- California vehicles
- Federal (49-state) vehicles
- High-altitude vehicles.

California's standards allowed less pollution than federal standards. Federal controls were less stringent. Because there is less oxygen at high altitudes to promote combustion, high-altitude standards were almost as strict as California's. To identify the vehicle and to aid technicians, emission control information labels were put in the underhood area [Figure 40-10].

Diesel engines must meet additional regulations. Diesel soot is a carcinogen. Also, because of high combustion pressures and temperatures, diesel engines produce large amounts of NO_x. In recent years, a partial or total ban on the sale of passenger-car diesels has been considered by California's ARB. Appropriate technology for trapping particulates without restricting exhaust flow has been difficult to achieve.

In 1977, the federal Clean Air Act was amended to require vehicle emission control inspection and maintenance (I/M) programs. Dec. 31, 1987, was the date chosen by which all states would be required to meet national air quality standards.

The Environmental Protection Agency (EPA) was given broad powers to enforce the Clean Air Act. The EPA can limit federal grants or cancel federal projects, such as highways, until states comply with clean air standards.

The EPA also has established a pollutant standards warning system [Figure 40-11]. Health warnings are broadcast on television and radio station weather reports.

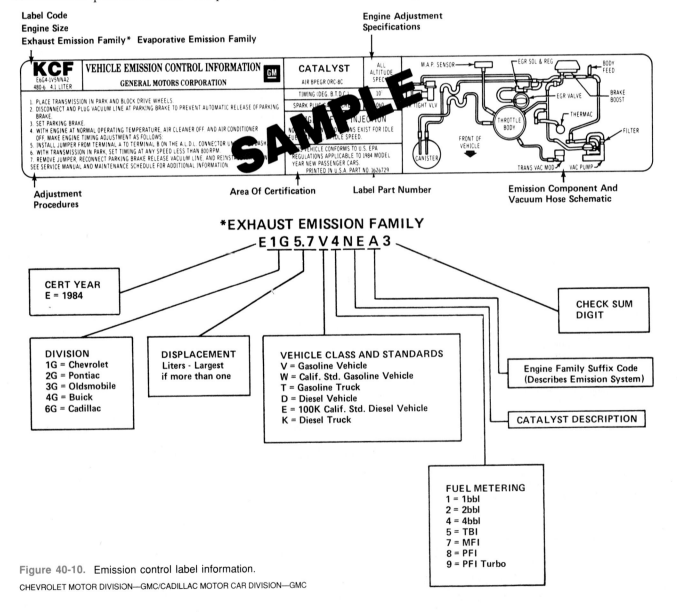

Figure 40-10. Emission control label information.

CHEVROLET MOTOR DIVISION—GMC/CADILLAC MOTOR CAR DIVISION—GMC

EMISSION CONTROL FUNDAMENTALS

Stages		Health Effects	Health Warnings
0-50	Good	None	None
50-100	Moderate	None	None
100-200	Air Quality Goal Unhealthful	Mild aggravation of symptoms in susceptible persons, with irritation symptoms in the healthy population.	Persons with existing heart or respiratory ailments should reduce physical exertion and outdoor activity.
200-300	1st Stage Alert Very Unhealthful	Significant aggravation of symptoms and decreased exercise tolerance in persons with heart or lung disease, with widespread symptoms in the healthy population.	Elderly and persons with existing heart or lung disease should stay indoors and reduce physical activity.
300-400	2nd Stage Alert Hazardous	Premature onset of certain diseases in addition to significant aggravation of symptoms and decreased exercise tolerance in healthy persons.	Elderly and persons with existing diseases should stay indoors and avoid physical exertion. General population should avoid outdoor activity.
400-500	3rd Stage Alert Hazardous	Premature death of ill and elderly. Healthy people will experience adverse symptoms that affect their normal activity.	All persons should remain indoors, keeping windows and doors closed. All persons should minimize physical exertion and avoid traffic.

Figure 40-11. Pollutant standards index.

The technology for achieving a constant reduction in levels of pollutants was not specified by the Clean Air Act. Manufacturers are free to choose methods and devices for achieving these reductions. As a result, general similarities exist among different manufacturers' methods, but there are many variations in hardware and devices.

To correctly diagnose and find problems on literally hundreds of different pollution control systems is a problem for technicians. As a result, technicians must refer to specific manufacturer's service manuals for correct inspection and test procedures of pollution controls.

40.4 GOVERNMENT REGULATION AND FUEL MILEAGE

During late 1973 and early 1974, the oil-producing countries of the Middle East refused to ship petroleum products. Motorists waited in gasoline lines for hours or days, and the price of gasoline and diesel fuel doubled. In mid-1973, premium (leaded) gasoline cost approximately 30 to 36 cents per gallon. Diesel fuel cost approximately 24 cents per gallon. Since that time, fuel prices have risen and fallen dramatically to present levels.

To encourage a reduction in fuel usage, Congress enacted the Energy Policy and Conservation Act. This act set standards for *corporate average fuel economy (CAFE)*. Overall fuel mileage requirements for an average of all vehicles produced by a manufacturer were part of this legislation **[Figure 40-12]**.

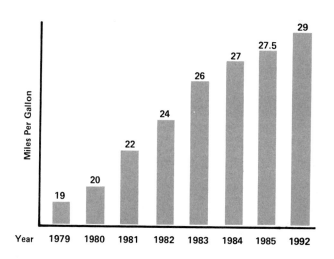

Figure 40-12. CAFE standards.

Technology, Air Pollution, and Performance

Prior to the CAFE regulations, emissions control technology tended to lessen fuel mileage. To meet clean air and CAFE regulations, vehicle manufacturers applied advanced technology to vehicle controls. Sensors, computer modules, and actuator devices became standard equipment on vehicles.

Several late-model vehicles produce a combination of good fuel mileage, driveability, low exhaust emissions, and brisk acceleration. These improvements have been brought about by several relatively recent applications of high technology to automobiles:

- Feedback carburetors and electronic fuel injection
- Electronic ignition
- Catalytic converters
- Computer controls
- Lockup torque converters for automatic transmissions (See Unit 53)
- Turbocharging and supercharging (See Unit 19).

Computer-controlled fuel injection, ignition, and automatic transmissions increase performance while lowering automotive exhaust pollutants. Fuel system controls, including catalytic converters, allow engines to be tuned for performance without sacrificing air quality. Turbocharging and supercharging permit a smaller engine to have as much, or more, horsepower than a normally aspirated larger engine.

In addition to these devices, specific pollution control equipment has been developed. Emission control systems are discussed in Unit 41.

All these approaches and devices have benefitted each of us, in terms of improved air quality and automotive performance. However, for the technician they have meant the necessity of learning new technology and work skills.

In the future, automotive technology will continue to change and improve. As an automotive technician, you must be ready and able to learn new skills and approaches. In any field of work, as in life, a good attitude and a willingness to learn are important. With such an attitude and outlook, you can satisfy yourself, your employer, and the customer whose car you service.

- Major sources of automotive pollutants include the exhaust, fuel system, and crankcase.
- Exhaust emissions include HC, CO, NO_x, SO_x, O_3, and particulates. U.S. government regulations govern new-car emissions of HC, CO, and NO_x.
- There are different regulations for California and federal (49-state) vehicles.
- The combination of CAFE regulations and Clean Air Act requirements stimulated vehicle manufacturers to use advanced technology. This technology includes feedback carburetors, fuel injection, electronic ignition, catalytic converters, and computer controls.
- Technicians who have a good attitude and are willing to learn new knowledge and skills will always be in demand.

TERMS

pollution
smog
inversion layer
carbon dioxide (CO_2)
carbon monoxide (CO)
emissions
blowby gases
grams per mile (gpm)
carbon monoxide poisoning
corporate average fuel economy (CAFE)

air pollution
particulate
photochemical smog
hydrocarbon (HC)
invisible pollution
parts per million (ppm)
carcinogen
road draft tube
positive crankcase ventilation (PCV)

UNIT HIGHLIGHTS

- Air pollution can be caused by natural or man-made pollutants. Man-made pollutants result from the burning of fuels to produce heat and energy. Air pollution causes health problems for living things and damage to objects. Atmospheric conditions can trap and concentrate air pollution in geographical areas.
- Complete combustion of a hydrocarbon fuel results in the formation of water vapor (H_2O) and carbon dioxide (CO_2). Incomplete combustion releases unburned hydrocarbons (HC) and carbon monoxide (CO).
- Air pollution can be visible as "smog" or invisible. Invisible pollution can be more harmful to health than visible pollution.

REVIEW QUESTIONS

DIRECTIONS: The following questions are similar to those used on automotive technician certification tests. On a separate sheet of paper, write the letter of the correct choice.

1. Which of the following statements is correct?
   I. Complete combustion of a hydrocarbon fuel produces HC and CO.
   II. Incomplete combustion of a hydrocarbon fuel produces H_2O and CO_2.
   A. I only
   B. II only
   C. Both I and II
   D. Neither I nor II.

2. Which of the following mandates fuel-economy standards?
   A. Clean Air Act
   B. CAFE requirements
   C. California Air Resources Board
   D. Environmental Protection Agency

3. Technician A says that ozone, sulfur, and particulate emissions levels are controlled by federal and state laws.
   Technician B says that HC, CO, and NO_x emissions levels are controlled by federal and state laws.
   Who is correct?
   A. A only
   B. B only
   C. Both A and B
   D. Neither A nor B

Questions 4 and 5 are not like those above. Each has the word EXCEPT. For each question, look for the choice that does *not* apply. Read each question carefully before you choose your answer.

4. Major sources of automotive pollutants include all of the following EXCEPT
   A. crankcase.
   B. fuel system.
   C. exhaust.
   D. transmission.

5. Different U.S. emission control requirements and equipment are applied to all of the following EXCEPT
   A. vehicles sold in Alaska and Hawaii.
   B. high-altitude vehicles.
   C. 49-state vehicles.
   D. California vehicles.

ESSAY QUESTIONS

1. What are the sources and types of automotive pollution?
2. Describe regulations that have led to current automotive emission control systems.
3. What technical advances are used on current automobiles to reduce harmful emissions?

SUPPLEMENTAL ACTIVITIES

1. Inspect several vehicles and locate their emission control information labels. Make up a table that indicates the locations of different manufacturers' labels.
2. Copy all information from a vehicle's emission control information label. Report to your class what adjustments and conditions must be met to correctly tune-up the vehicle.
3. Refer to general repair manuals for tune-up specifications for a given make and model of vehicle. Note engine size, horsepower output, and ignition timing specifications from 1967 to the present. Report to your class what general trend can be noticed.
4. Visit a local auto dealership and talk to the service manager. Ask him or her what training is provided for technicians in emission control servicing. Report to your class on what training is done, and how often technicians must attend training sessions.
5. Visit an independent garage and talk to a technician or shop owner. Ask about the ways that independent technicians keep informed about the latest pollution control system technology and service.

41 Emission Control Systems

UNIT PREVIEW

Virtually every part or system that changes vehicle performance and driveability affects vehicle exhaust emissions. Emission control measures are not add-on devices but integral parts of total vehicle systems.

Vacuum and/or electricity are used to control many engine and emissions control parts and systems. Vacuum-operated devices can be used to control other mechanical or electrical devices.

Electronic control is more precise than vacuum control. Electronically controlled functions operate in either an open-loop or closed-loop mode.

Specific emission control equipment includes devices to control fuel evaporation, crankcase emissions, and exhaust emissions. These devices include vapor recovery, positive crankcase ventilation, air injection, exhaust gas recirculation, and catalytic converters.

Catalytic converters permit engines to have both relatively clean exhaust emissions and improved performance and driveability.

LEARNING OBJECTIVES

When you have completed your assignments and exercises in this unit, you should be able to:

- Identify and describe parts and systems that affect exhaust emissions.
- Explain how vacuum controls operate and what sources of vacuum may be used.
- Identify what specific systems are used for emission-control purposes.
- Locate and identify vacuum-operated control mechanisms on a vehicle.
- Locate and identify PCV, air injection, EGR, and catalytic converter system parts.

41.1 FACTORS THAT AFFECT EMISSIONS

An automobile is a complex group of interrelated systems and parts, like a human body. A single change can affect many other parts and systems.

For example, a change in idle speed can affect idle air-fuel mixture and ignition timing, and thus affect engine performance. Decreased engine performance can change automatic transmission shift points, thus decreasing acceleration, overall performance, and economy. These factors, in turn, can change the amount of pollutants emitted by a vehicle.

Vehicle Parts and Systems

Every vehicle part and system that affects performance and driveability also can influence automotive exhaust emissions. Thus, emission-control measures are not merely add-on parts or systems, but are inseparable from total vehicle operation. These parts and systems include:

- Internal engine parts (crankshaft, pistons, valves)
- Engine systems (cooling, lubrication, fuel, ignition, exhaust, electrical, and electronic control)
- Combustion chamber shape and design
- Camshaft profile
- Emissions control equipment
- Transmission gear ratios and shift points
- Final drive gear ratios
- Tire and wheel size.

Vehicle Operating Mode

Automotive emissions also vary according to vehicle *operating mode.* An operating mode is a specific way in which a vehicle might operate. These include:

- Warmup
- Idling
- Acceleration
- Wide-open throttle (WOT)
- Cruising
- Deceleration.

41.2 ENGINE CONTROLS

Controlling an engine during different operating modes requires both a source of force and a method of control. The most commonly used sources of force for emissions control purposes are:

- Vacuum
- Electricity.

Vacuum

Vacuum can be used as a signal or as an actuating force. That is, vacuum can be used to signal another device to operate, or to operate a mechanical device directly. Valves that block or open vacuum passages can be operated mechanically and/or electrically. For example, a temperature-operated valve can direct or block vacuum to a given device.

Electricity

Electricity can be used to operate solenoids and stepper motors to push, pull, rotate, or hold mechanical devices. For example, a solenoid valve can open or block a vacuum passage for different engine operating modes. Electricity may be controlled by mechanical switches or by electronic control units.

Figure 41-1 shows the interrelationship among vacuum, electricity, and electronic controls for one type of emission control.

41.3 VACUUM-OPERATED CONTROLS

Vacuum-operated controls use one of three sources of vacuum [**Figure 41-2**]:

- Manifold vacuum
- Ported vacuum
- Venturi vacuum.

Manifold Vacuum

Manifold vacuum, as described in Unit 10, is created in an intake manifold when an engine cranks or runs. Manifold vacuum is high when a throttle plate is closed or partially closed during idling and deceleration. Manifold vacuum drops when a throttle plate is opened wide during acceleration and high-speed operation.

Thus, manifold vacuum is related to throttle position, engine load, and engine volumetric efficiency. Manifold vacuum will vary from 15 to 22 inches of mercury (38.1 to 55.88 cm) Hg at idle.

When a throttle plate is momentarily opened to *wide-open throttle (WOT)*, manifold vacuum will drop to zero.

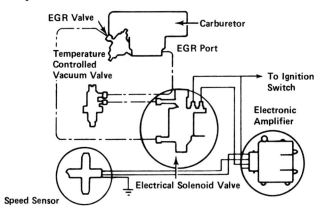

Figure 41-1. Emission control system using temperature, electronics, and manual actuation.

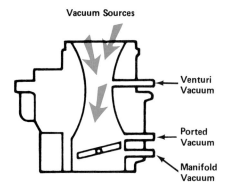
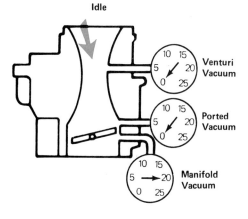
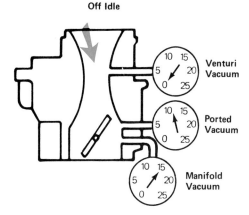
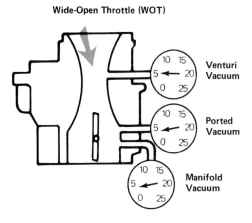

Figure 41-2. Sources of vacuum.

Ported Vacuum

Ported vacuum, as discussed in Unit 21, is created in a throttle body bore. When a throttle plate is moved to increase engine speed above idle, the vacuum port is exposed to manifold vacuum. Ported vacuum does not exist at idle, but is created and increases as the throttle is opened. Ported vacuum may vary between zero and 22 in. (55.88 cm) Hg. Ported vacuum, as explained in Topic 34.7, can be used to operate devices such as distributor vacuum advance mechanisms. Another example of a ported vacuum device is shown in **Figure 41-3**.

Venturi Vacuum

Venturi vacuum is created at a venturi, or restriction, in a carburetor bore. At idle and when a throttle plate is only partially open, air velocity is too low to develop vacuum. However, as a throttle plate is opened and air flow increases, venturi vacuum is developed. Venturi vacuum varies from zero to 4 in. (10.16 cm) Hg.

Mechanically Operated Valves

Mechanically operated controls can take many forms. Distributor centrifugal advance mechanisms are mechanical controls. A heat-operated valve also can direct or block vacuum to vacuum-operated controls.

Electrically Operated Controls

Electrically operated controls, such as solenoid valves, can be used to direct or block vacuum to vacuum-operated controls. The unit shown in **Figure 41-4** uses mechanical actuation and/or a solenoid to direct vacuum to a distributor vacuum advance unit.

41.4 VACUUM-OPERATED CONTROL SYSTEM

The use of vacuum-operated engine and emissions control equipment may require the use of multiple vacuum hoses. See **Figure 41-23**. If a single vacuum

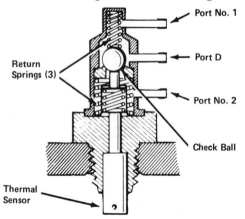

Figure 41-3. Typical ported vacuum switch.
AMERICAN HONDA MOTOR CO., INC.

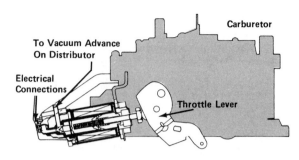

Figure 41-4. Vacuum-electric solenoid assembly.

hose becomes loose, disconnected, or broken, elements of a total emissions control system will not function properly. Manufacturer's service manuals contain vacuum hose diagrams for specific models, years, engines, and optional equipment.

Vacuum-operated systems operate on basic mechanical principles and thus are relatively durable. However, vacuum-operated controls do not act immediately. In addition, leaks and accidental disconnections can occur.

The need for controls that respond rapidly and reliably has led vehicle manufacturers to develop electronically operated systems.

41.5 ELECTRONICALLY OPERATED CONTROL SYSTEMS

Sensors, electronic control units, and actuator units can be used to adjust air-fuel ratios, exhaust gas recirculation, and ignition timing. This is discussed in Unit 38. As necessary, other actuator-operated devices can be combined with basic systems to form overall electronic control systems. A diagram of inputs and outputs that include emissions control devices is shown in **Figure 41-5**.

41.6 EMISSION CONTROL EQUIPMENT

Specific emission control equipment **[Figure 41-6]** is included as part of an overall group of vehicle systems. Emission control subsystems include:

- Evaporative emission controls
- Positive crankcase ventilation (PCV) system
- Fuel system modifications and controls
- Combustion control
- Engine and intake manifold modifications
- Ignition timing modifications
- Exhaust gas recirculation (EGR)
- Air injection
- Catalytic converters.

41.7 EVAPORATIVE EMISSION CONTROLS

Evaporative emission controls are parts and systems designed to prevent fuel vapors from escaping directly into the air. ***Vapor recovery systems*** trap and

Emission Control Systems

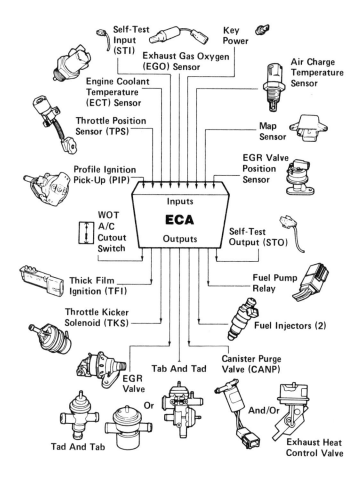

Figure 41-5. Emissions system parts. FORD MOTOR COMPANY

hold fuel vapors until they can be drawn into the engine with an intake charge. A vapor recovery system [Figure 41-7] can include:

- Nonvented fuel tank filler cap
- Domed fuel tank
- Vapor/fuel separator
- Connecting hoses and tubes
- Vapor recovery canister
- Check valves
- Source of manifold vacuum.

Nonvented Fuel Tank Filler Cap

Filler caps on late-model vehicles are sealed to prevent escape of fuel vapors [Figure 41-8]. These caps include vacuum and pressure relief valves to prevent damage to a fuel tank from excessive pressure or vacuum.

As fuel cools, it contracts. A partial vacuum is formed in the tank. Atmospheric pressure acting on the walls of a tank could collapse it. A *vacuum relief valve*, however, opens to allow outside air into the tank. This action balances pressure against inside and outside surfaces of a tank.

When a fuel tank is heated by outside temperatures, fuel and vapors expand, producing pressure that could rupture fuel tank seams. Normally, this pressure buildup is slight, and pressure is dissipated throughout a vapor recovery system. However, if hoses or tubes become clogged or accidentally pinched, pressure could damage a tank. A *pressure relief*

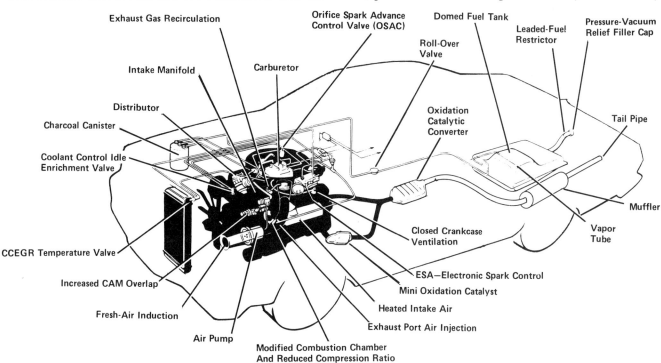

Figure 41-6. Emissions system parts locations. CHRYSLER CORPORATION

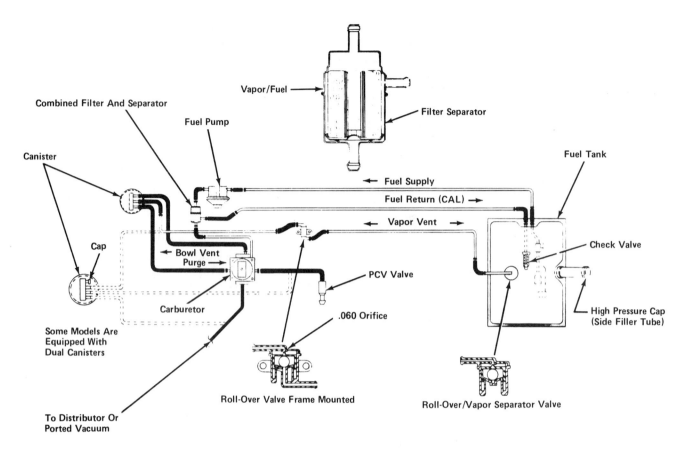

Figure 41-7. Vapor recovery system. CHRYSLER CORPORATION

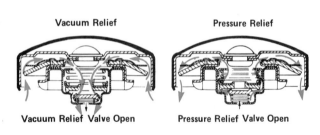

Figure 41-8. Sealed fuel tank cap.

valve opens to allow excessive pressure to escape. Normally, up to 1 psi (6.895 kPa) of pressure can exist in a tank and vapor recovery system. Pressure over that limit will open the pressure relief valve.

Domed Fuel Tank

Fuel vapors rise to the upper portion of a tank. An *expansion dome* collects fuel vapors from the tank [**Figure 41-9**]. An expansion dome is a raised portion of an upper fuel tank wall.

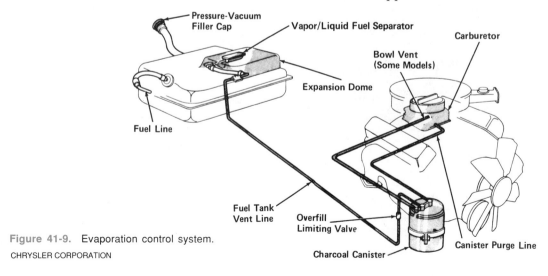

Figure 41-9. Evaporation control system.
CHRYSLER CORPORATION

Vapor/Fuel Separator

A *vapor/fuel separator* allows fuel vapors to pass into connecting hoses. The vapor/fuel separator collects droplets of liquid fuel and directs them back into the tank [Figure 41-9].

Connecting Hoses and Tubes

Hoses and tubes connect parts of a vapor-recovery system. Special fuel/vapor rubber tubing must be used. Ordinary rubber hoses would rot and crack quickly from effects of fuel. Metal tubing is used under a vehicle.

Vapor Recovery Canister

A *vapor recovery canister* is a container that stores fuel vapors. The vapor recovery canister is filled with *activated charcoal.* Activated charcoal is a highly porous type of charcoal that can store large amounts of vapors. During engine operation, fresh air is drawn in through the charcoal by means of intake manifold vacuum. This air flow purges, or strips, the charcoal of the gasoline vapors. These vapors then are mixed with the incoming air-fuel mixture.

Check Valves

Check, or one-way, valves keep vapors confined. When an engine runs, vacuum and/or electrically operated valves open to allow both fresh air and vapors to be drawn into the intake manifold.

Source of Manifold Vacuum

Ported vacuum often is used to control the canister purge circuit. This is done to prevent purge operation during idle. If vapors were drawn in during idle, a rough idle could result along with excessive exhaust emissions, especially during hot operation.

41.8 POSITIVE CRANKCASE VENTILATION SYSTEM

A *positive crankcase ventilation (PCV)* system removes blowby gases, oil, and fuel vapors from crankcase and air passages within an engine. A typical closed PCV system is shown in Figure 41-10.

A PCV, or check, valve meters flow of blowby gases and fresh air through a PCV system. The PCV valve also prevents backfire explosions in an intake manifold from travelling back to internal engine spaces. Operation of a PCV valve during different engine modes is shown in Figure 41-11.

41.9 FUEL SYSTEM MODIFICATIONS

The first approach to reducing HC and CO exhaust emissions was to create a leaner than stoichiometric mixture. A lean mixture increases combustion temperatures. In this way, HC and CO emissions can be

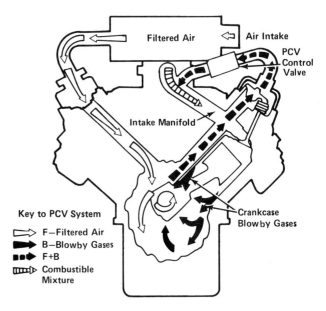

Figure 41-10. PCV system.

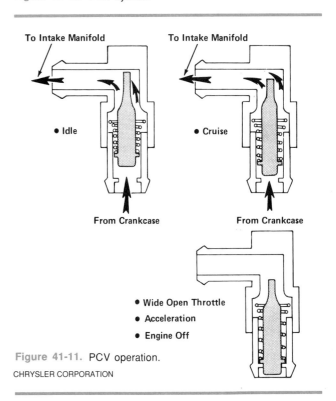

Figure 41-11. PCV operation.
CHRYSLER CORPORATION

reduced. However, as can be seen from the graph in Figure 41-12, NO_x emissions climb sharply as the air-fuel mixture becomes leaner.

Stoichiometric Mixtures

A stoichiometric mixture is best for both engine efficiency and smog control. A catalytic converter operates most efficiently with a stoichiometric (14.7:1) air-fuel mixture [Figure 41-13].

Computer-operated feedback carburetors can vary jet openings to provide a nearly stoichiometric

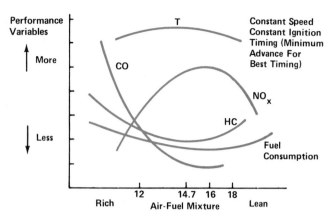

Figure 41-12. Air-fuel ratio and emissions variations.

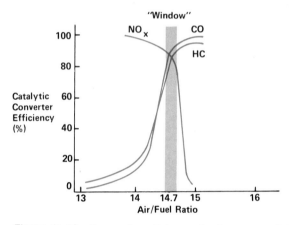

Figure 41-13. Conversion efficiency of a three-way catalyst.

mixture. Electronic fuel-injection systems operate in a similar way. Thus, computer controls can help reduce exhaust emissions when the system operates in closed-loop mode.

Tamper-Proof Adjustments

To prevent tampering, idle mixture screws may be factory-preset and plugged. On late-model vehicles, technicians can make only minor idle speed adjustments.

Thermostatically Controlled Air Cleaner

A *thermostatically controlled air cleaner* can be used to promote intake fuel charge vaporization. Depending on underhood air temperature, a thermostatically controlled air cleaner can provide hot, warm, or *ambient* air. Ambient air is air at the surrounding, or ambient, temperature. Heated air aids in fuel vaporization. The source of heated air is a *heat stove*. A heat stove is a sheet metal enclosure around an exhaust manifold, through which air can flow and be heated.

The mixing of heated and cold air is controlled by a temperature sensor inside an air cleaner. Heated air maximizes vaporization and minimizes unburned HC exhaust emissions [Figure 41-14].

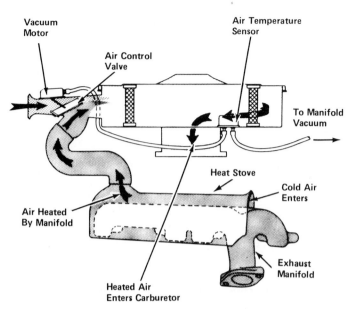

Figure 41-14. Thermostatic air cleaner. CHRYSLER CORPORATION

A temperature sensor with a bimetallic spring opens and closes a vacuum valve. The valve bleeds off or passes manifold vacuum to a *vacuum motor* on an air cleaner intake. A vacuum motor consists of a vacuum diaphragm attached to a pull rod. The pull rod is attached to an air control valve, or damper door, in an air intake. A spring pulls the door closed. When vacuum is applied to the vacuum motor, intake manifold vacuum pulls the damper open [Figure 41-15].

When intake temperature is low, heated air is directed into an air cleaner to promote fuel vaporization. As intake temperature rises, part heated and part cool air are mixed for proper vaporization. After warmup the sensor and vacuum motor act to provide correct inlet air temperature. During hard acceleration, the damper closes to provide cool, dense air [Figure 41-16].

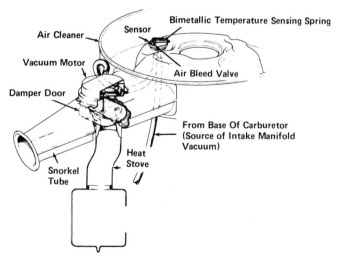

Figure 41-15. Parts of a thermostatically controlled air cleaner. CHEVROLET MOTOR DIVISION—GMC

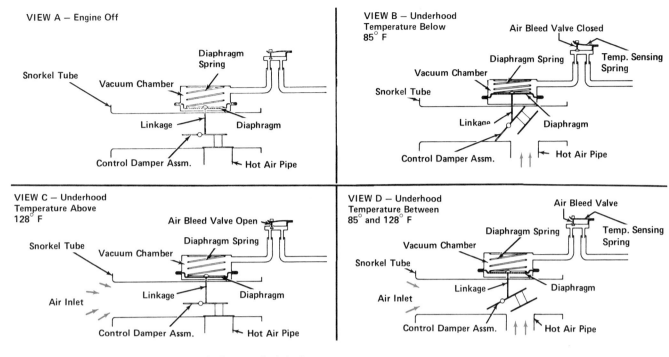

Figure 41-16. Operation of a thermostatically controlled air cleaner. CHEVROLET MOTOR DIVISION—GMC

Throttle Positioners

When a throttle plate is closed suddenly during deceleration, a high vacuum is created in the intake manifold. This high vacuum draws in a rich mixture from idle discharge ports. It also will cause any liquid fuel in the manifold to flash off into a vapor. A sudden "spike," or high reading, of high HC and CO emissions is created.

To prevent this problem, some carburetors use *deceleration valves* and *throttle positioners*. A deceleration valve is a valve that operates for a specific length of time on deceleration. A throttle positioner holds a throttle open. Throttle positioners use a small vacuum diaphragm and an air bleed to gradually reduce throttle position to idle. These devices are often called dashpots. These actions eliminate temporary high readings of HC and CO.

41.10 COMBUSTION CONTROL

The quality of combustion within an engine directly affects engine performance and exhaust emissions. If combustion is efficient, maximum power is produced from minimum fuel. In addition, efficient combustion means that less unburned HC and CO emissions are produced. Controlling the temperature of the combustion process can help reduce NO_x emissions. Methods of controlling the combustion process include:

- Engine and intake manifold modifications
- Fuel system modifications
- Ignition timing modifications
- Exhaust gas recirculation

41.11 ENGINE AND INTAKE MANIFOLD MODIFICATIONS

Engine and intake manifold modifications are made to improve fuel vaporization, intake gas flow and turbulence, and exhaust gas flow. In addition, exhaust gases can be mixed with an intake charge to decrease burning temperatures.

Combustion Chamber Shape

Combustion chamber shape can be changed to decrease quench areas. Spark plug holes can be placed so that a combustion flame front travels rapidly and evenly through the combustion chamber. Modifications to combustion chamber shape can help increase gas flow and promote vaporization, as discussed in 14.1. Increasing intake valve size and repositioning intake and exhaust valves can result in reduced exhaust emissions. Intake and exhaust valves also can be made to operate at steeper angles to each other, which increases turbulence.

Late-model vehicles may use a *pentroof* combustion chamber, as shown in **Figure 41-17** and **41-18**. A pentroof combustion chamber is shaped like the roof of a house. It is an angular version of a hemispherical combustion chamber.

Additional Intake and/or Exhaust Valves

Two intake and/or two exhaust valves can be used in each cylinder to improve both turbulence and volumetric efficiency **[Figure 41-18]**. Two smaller valves are better than one large valve because gas flow rate must increase through smaller openings. Higher flow

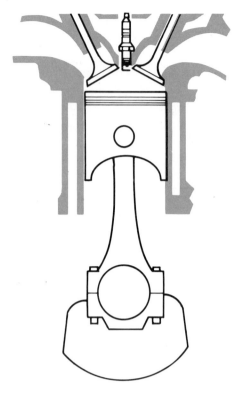

Figure 41-17. Pentroof combustion chamber.

Figure 41-18. Four-valves-per-cylinder cylinder head.

rates increase turbulence and promote smoother operation. Thus, engine efficiency increases, providing better power output and lower exhaust emissions.

Stratified Charge

Stratified charge engines, such as those used in some Honda-built automobiles, have an additional special, smaller intake valve. A rich mixture from a third carburetor bore is drawn in directly below a spark plug. Below this, a lean mixture is drawn in through a normal, larger intake valve [**Figure 41-19**]. Thus, the charge is layered, or stratified, with a rich mixture on top and a leaner mixture on the bottom. When the spark plug fires, the rich upper mixture ignites easily and begins the burning of the hard-to-ignite lean mixture beneath.

Cylinder Head and Intake Manifold Design

Redesigned cylinder head intake ports and intake manifolds can increase turbulence. Some late-model vehicles have a variable induction system that opens additional intake ports at high rpm [**Figure 41-20**]. This allows high rates of gas flow at both low and high rpm, for increased volumetric efficiency, torque, and turbulence.

Increased Valve Overlap

Valve overlap, as discussed in Topic 14.10, occurs when both intake and exhaust valves are open during exhaust and intake strokes. Camshafts are

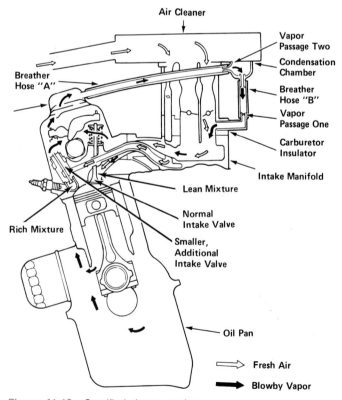

Figure 41-19. Stratified charge engine.
AMERICAN HONDA MOTOR CO., INC.

EMISSION CONTROL SYSTEMS

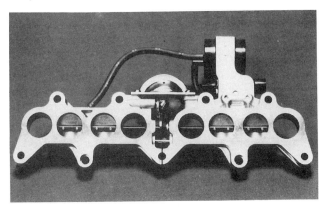

Figure 41-20. Variable induction system.

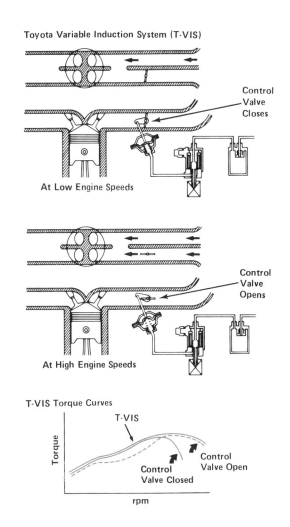

designed differently for emission control purposes. Intake and exhaust camshaft lobes are repositioned relative to each other and reshaped to provide increased valve overlap [Figure 41-21]. This allows more exhaust gases to mix with an intake charge.

Because exhaust gases contain little oxygen, the mixed intake charge and exhaust gases do not burn rapidly. A slower, lower-temperature burning takes place, thus reducing NO_X exhaust emissions. Many engines may use an EGR valve to accomplish this. See Topic 41.13.

41.12 IGNITION TIMING MODIFICATIONS

Ignition timing directly affects engine efficiency and emissions. However, the dual goals of efficient engine operation and low HC and CO emissions are not always compatible.

Timing for Engine Power and Emissions

For maximum performance, ignition timing should occur so that maximum pressure is produced at a point approximately 10 degrees ATDC. This normally requires that initial timing be set some degrees BTDC. Retarding spark so that ignition occurs ATDC will lower pressure, engine torque, and power. However, as pressure decreases, a more thorough, slow burning occurs.

More fuel is burned when the burning process is slow and thorough. Retarded timing can help to burn a rich air-fuel mixture, such as that which occurs during deceleration, more completely.

The graph in Figure 41-22 illustrates a basic problem with ignition timing for good fuel economy and minimum exhaust emissions. Timing that produces lowest fuel consumption (and maximum engine torque) increases HC and NO_X emissions. If ignition timing remains constant, an engine will produce either low emissions or good fuel economy/high torque, but not both. Thus, ignition timing must advance or retard both for fuel economy and emissions considerations.

Vacuum-Operated Ignition Advance/Retard Controls

Mechanical devices can be used to control vacuum advance/retard diaphragm operation [Figure 41-23].

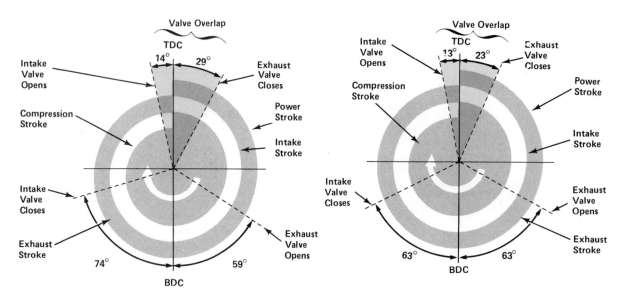

Figure 41-21. Valve overlap changes.

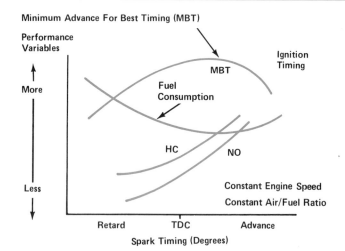

Figure 41-22. Variation of performance with spark timing.

Such devices can include:

- Thermo-vacuum valves
- Vacuum delay valves
- Vacuum check valves.

Thermo-vacuum valves can be used to block or direct ported or manifold vacuum. In this way, vacuum can control distributor advance and/or retard mechanisms for maximum power or more complete combustion. These valves also are temperature sensitive.

Vacuum delay valves are small plastic units in a vacuum line that contain a calibrated restriction. Vacuum delay valves delay vacuum buildup that operates a device. A vacuum delay valve can be used to delay spark advance until an engine reaches higher rpm.

Vacuum check valves allow vacuum to travel in only one direction through a tube or hose. Vacuum check valves can be used to isolate a vacuum-operated unit from specific sources of vacuum.

Computer-Controlled Ignition

Computer-controlled ignition timing, as discussed in Unit 34, can control ignition timing according to a "spark map." A computer program thus can select the best ignition timing compromise for good power and low exhaust emissions.

41.13 EXHAUST GAS RECIRCULATION

As discussed in 41.11, dilution of an intake charge with exhaust gases can lower NO_x emissions. *Exhaust gas recirculation (EGR)* is a method of reusing exhaust gases in an intake charge to reduce NO_x.

An *EGR valve* is a vacuum-operated valve that can open a passage between exhaust and intake manifold areas. Many types and combinations of EGR control systems are possible [Figure 41-24].

Temperature-Controlled EGR

Exhaust gas recirculation reduces power output and may cause a vehicle to stall during warmup. Temperature-sensitive vacuum control valves can be used to prevent EGR valve operation until an engine warms up [Figure 41-24].

Backpressure Transducer

In previous years, EGR operated only in an on or off mode. Late-model vehicles' EGR systems are *modulated,* or controlled. Thus, an EGR valve can be closed, partially opened, or fully opened.

EMISSION CONTROL SYSTEMS

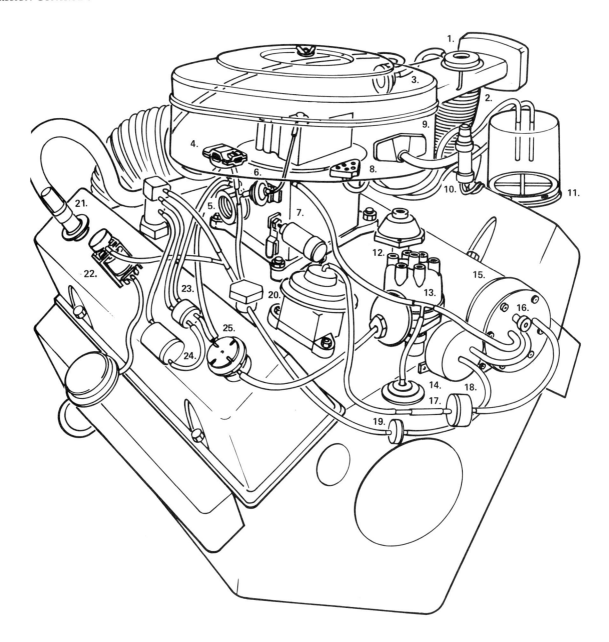

1	Vacuum Motor	9	Breather Filter (Inside Housing)	17	Wide-Open Throttle (WOT) Valve
2	Thermostatically-Controlled Air Cleaner Manifold Stove Heater Duct	10	Carburetor Float Bowl Solenoid Valve	18	Thermal Control Valve
		11	Vapor Recovery Canister Replacement Filter	19	Distributor Check Valve
3	Temperature Control Valve	12	Deceleration Valve	20	EGR Valve
4	Temperature Sensor	13	Vacuum Hose Tee Connector	21	PCV Valve
5	Automatic Choke Thermostat	14	Vacuum Reservoir	22	Spark Delay Valve
6	Choke Pull-Off	15	Vacuum Amplifier	23	Fuel/Vapor Separator
7	Idle Speed Solenoid	16	EGR Air Bleed Filter	24	Vacuum Regulator
8	Choke Air			25	Three-Way Spark Control

Figure 41-23. Vacuum control system.

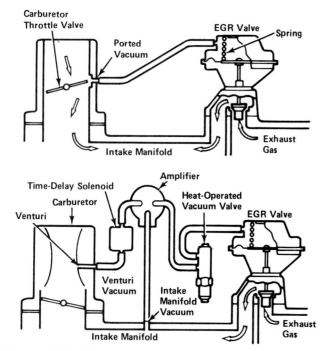
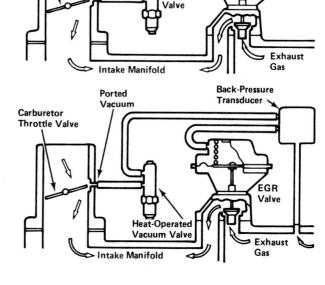

Figure 41-24. EGR variations.

A *backpressure transducer* can be used to modulate, or change, the amount an EGR valve opens [Figure 41-24]. A backpressure transducer is a device that senses exhaust gas pressure and regulates a vacuum-control valve. Exhaust gas pressure is dependent on engine rpm. The EGR valve may be closed or partially opened at different engine speeds.

Ported Vacuum Control

In many cases, the vacuum control line leading to the EGR valve vacuum diaphragm will have ported vacuum. If EGR would operate at idle, it would cause rough idle or stalling. For this reason, ported vacuum is used to control EGR to ensure EGR only above engine idle speed.

Electronically Controlled Solenoid Valve

A computer-controlled solenoid can open or close an EGR valve in relation to any sensor-provided information [Figure 41-25]. Thus, many engine operating conditions can influence EGR.

41.14 AIR INJECTION

Air injection is an introduction of fresh air into an exhaust manifold. In previous years, air injection was done to promote additional burning, or oxidation, of HC within the exhaust manifold. For a burning process to occur, there must be fuel, heat, and oxygen. Small amounts of fuel (HC) are present in hot exhaust gases. Introducing air into an exhaust manifold provides oxygen, and additional burning occurs.

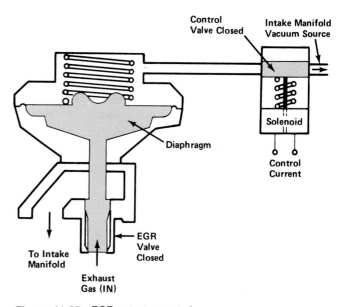

Figure 41-25. EGR actuator control.

Late-model vehicles use air injection to provide additional air for an oxidizing catalytic converter. An oxidizing catalytic converter, described in 41.15, promotes the combination of CO and HC with oxygen. Relatively less-harmful carbon dioxide (CO_2), water (H_2O), and oxygen (O_2) are formed. Two types of air injection are common:

- Engine-driven pump system
- Exhaust gas pulse system

Engine-Driven Pump

An engine-driven pump air injection system is illustrated in **Figure 41-26**. This type of system can include the following parts:

- Air pump
- Diverter valve
- Air-switching valve
- Check valves
- Distribution manifold
- Temperature-sensitive vacuum valves
- Vacuum source
- Connecting tubes and hoses.

An air pump draws air in and forces pressurized air out. A diverter valve diverts, or detours, air during deceleration. Excess air in exhaust that is rich with fuel can produce a backfire or explosion in a muffler during deceleration. A vacuum signal operates the diverter valve during deceleration. Compressed air is diverted to the atmosphere.

Late-model vehicles include diverter/bypass valves that can direct air to an exhaust manifold and/or a catalytic converter **[Figure 41-27]**.

Catalytic converter air can be injected into the exhaust pipe before the catalytic converter, or between elements of the converter.

An air-switching valve switches the location of injected air to lessen emissions of NO_X after warmup. Check valves prevent exhaust gases from reaching a diverter valve or air pump.

Exhaust Gas Pulse System

Exhaust gases leave an engine in pulses, as explained in Unit 25. These pulses produce noise, which is lessened by a muffler. However, these pulses also can be used to provide force to draw fresh air into an exhaust manifold.

A *pulse air valve* directs air into distribution manifold pipes **[Figure 41-28]**. A pulse air valve is a container with *reed valves*. A reed valve is a device with a flexible diaphragm covering an opening. After an exhaust pulse has passed the valve, a low pressure is developed, which pulls a reed valve open. Air is drawn in through the reed valve to an exhaust manifold port.

41.15 CATALYTIC CONVERTER

Catalytic converters have been part of new vehicles' exhaust systems since 1975, as discussed in Unit 25. Vehicles that are designed to use catalytic converters must use only unleaded fuel. A catalytic converter contains a chemical *catalyst*. A catalyst is a chemical that promotes a chemical reaction between other substances, without being consumed itself.

The complete combustion of HC fuel in air produces water and carbon dioxide. However, as discussed in Unit 40, internal combustion engines do not completely burn all the fuel in an intake charge. The result of incomplete combustion is HC and CO pollutants. In addition, high combustion temperatures produce NO_X.

Early attempts to reduce hydrocarbons involved raising combustion temperatures, using lean mixtures, and retarding ignition timing. Such practices produced vehicles that stalled and died repeatedly when cold, ran poorly, and gave poor gas mileage. In addition, high amounts of NO_X were produced.

The use of catalytic converters allows richer mixtures and advanced ignition timing for improved driveability, performance, and fuel mileage. Thus, catalytic converters promote both cleaner air and reasonable performance.

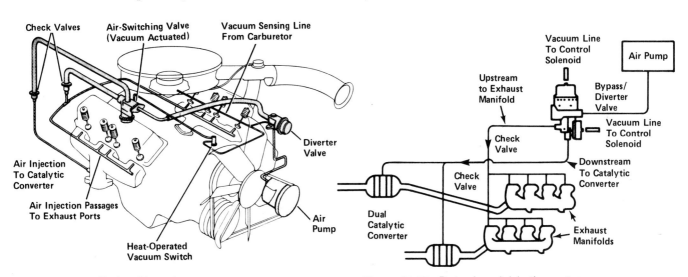

Figure 41-26. Engine-driven air pump.
CHRYSLER CORPORATION

Figure 41-27. Parts of an air injection system.
FORD MOTOR COMPANY

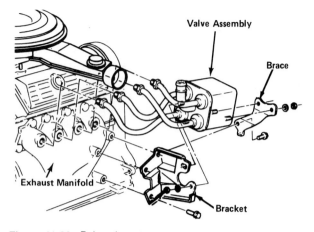

Figure 41-28. Pulse air system. CHEVROLET MOTOR DIVISION—GMC

Catalytic converters change harmful pollutants into relatively harmless products through two processes:

- Oxidation
- Reduction.

Oxidation

If HC and CO could be combined with extra oxygen, the result would be H_2O and CO_2. The catalyst in a catalytic converter promotes a combining action of oxygen with both HC and CO. This action is known as chemical *oxidation,* or combination with oxygen.

Such a *two-way catalyst* was used in catalytic converters in 1975 and subsequent years. The chemicals used for two-way catalysts are *platinum* and *palladium.* Platinum and palladium are expensive, rare metals. Ceramic beads or a ceramic *substrate* can be coated with a thin layer of catalyst. A substrate is a honeycomb grid structure [Figure 41-29].

Exhaust gases pass through beads or a substrate, and oxidation occurs. Thus, HC and CO are combined with oxygen from excess air provided by an air-injection system. The resulting compounds are H_2O and CO_2, both part of the natural environment [Figure 41-30].

Reduction

Another chemical reaction that can be promoted by a catalyst is *reduction.* Reduction is the removal of oxygen from a compound. If oxygen can be removed from NO_x, nitrogen (N_2) and oxygen (O_2), both part of air, will be formed.

Chemical reduction of NO_x is accomplished with catalysts of platinum and *rhodium.* Rhodium is another rare metal, often used for plating jewelry. Platinum and rhodium also are effective for oxidizing HC and CO. A *three-way catalyst* thus changes HC, CO, and NO_x into H_2O, CO_2, N_2 and O_2 [Figure 41-31].

Catalytic converters may contain *dual beds,* or two areas where catalytic reactions occur. Dual beds may be used in two-way or three-way catalytic converters. A three-way (HC, CO, NO_x) catalytic converter is shown in Figure 41-32.

SO_x Compounds

During oxidation, sulfur (S) present in fuel and lubricating oil combines with oxygen (O_2). The resulting compounds are known as oxides of sulfur (SO_x).

In addition, sulfur dioxide gas (SO_2) is formed. Catalytic converters in some vehicles produce an unpleasant, "rotten egg" smell. SO_2 is the same gas formed in rotting eggs. Although unpleasant, SO_2 is not considered a major pollutant at present.

Oxides of sulfur also combine with water vapor (H_2O) to produce small amounts of sulfuric acid (H_2SO_4). The long-term effects of adding small amounts of sulfuric acid to air are thought to contribute to *acid rain.*

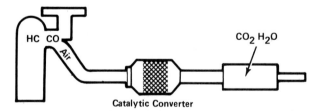

Figure 41-30. Exhaust conversion from a two-way catalytic converter.

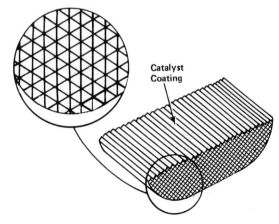

Figure 41-29. Substrate structures.

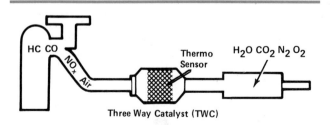

Figure 41-31. Exhaust conversion from a three-way catalytic converter.

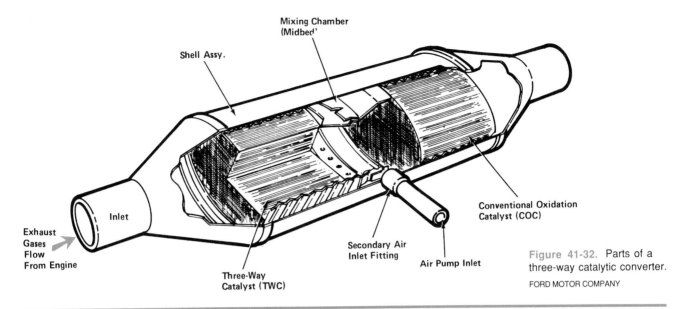

Figure 41-32. Parts of a three-way catalytic converter.
FORD MOTOR COMPANY

However, acid rain and *acid fog* are caused by large amounts of sulfur added to air through industrial processes. Acid rain and acid fog contain large amounts of sulfuric acid, which pollutes rivers and streams, kills fish, and harms vegetation.

Misfueling

Misfueling is adding leaded gasoline to vehicles with catalytic converters. Exhaust fumes containing lead will coat catalyst beads or substrate, "poisoning" a converter. Exhaust emissions rise dramatically, and if the practice continues, will probably plug the converter.

More than one tankful of leaded gasoline will poison a converter and make converter replacement necessary.

UNIT HIGHLIGHTS

- Virtually every part and system that affects vehicle performance and driveability also influences vehicle exhaust emissions. Thus, emission control measures are not merely add-on devices but integral parts of total vehicle systems.
- Vehicle operating modes also influence exhaust emissions.
- Vacuum and/or electricity are used to control many engine and emission control functions. Three types of vacuum are used: manifold, ported, and venturi. Vacuum controls are relatively durable but do not respond quickly.
- Electronically controlled functions operate in either an open-loop or closed-loop mode. Electronic control is more precise than vacuum control.
- Specific emission control equipment includes devices to control fuel evaporation, crankcase emissions, and exhaust emissions. These devices include vapor recovery, PCV, air injection, EGR, and catalytic converters.
- Catalytic converters permit engines to use richer fuel mixtures and advanced ignition timing for improved performance.

TERMS

operating mode	vapor recovery system
vacuum relief valve	pressure relief valve
expansion dome	vapor/fuel separator
vapor recovery canister	activated charcoal
heat stove	vacuum motor
deceleration valve	throttle positioner
pentroof	stratified charge
EGR valve	ambient
modulate	backpressure transducer
air injection	pulse air valve
reed valve	catalyst
oxidation	reduction
two-way catalyst	platinum
palladium	substrate
rhodium	three-way catalyst
dual bed	misfueling
acid rain	acid fog
evaporative emission controls	exhaust gas recirculation (EGR)
positive crankcase ventilation (PCV)	wide-open throttle (WOT)
thermostatically controlled air cleaner	

REVIEW QUESTIONS

DIRECTIONS: The following questions are similar to those used on automotive technician certification tests. On a separate sheet of paper, write the letter of the correct choice.

1. Which of the following statements is correct?
   I. Specific emission control devices influence vehicle emissions.
   II. Virtually any part or system that affects vehicle operation also affects vehicle emissions.
   A. I only
   B. II only
   C. Both I and II
   D. Neither I nor II

2. Technician A says that vacuum can be used to control engine parts that affect vehicle operation and emissions.
   Technician B says that electricity can be used to control vehicle parts that affect vehicle operation and emissions.
   Who is correct?
   A. A only
   B. B only
   C. Both A and B
   D. Neither A nor B

3. EGR stands for
   A. exhaust gas recombination.
   B. exhaust gas recirculation.
   C. exhaust gas reduction.
   D. exhaust gas restriction.

4. A three-way catalyst
   A. treats HC, CO, and SO_x emissions.
   B. has two areas for oxidation reactions.
   C. treats HC, CO, and NO_x emissions.
   D. has three areas for reduction reactions.

Question 5 is not like those above. It has the word EXCEPT. For question 5, look for the choice that does *not* apply. Read the entire question carefully before you choose your answer.

5. All of the following statements about air injection systems are true EXCEPT
   A. they can use an air pump.
   B. they can use a pulse-air valve.
   C. they can prevent additional burning of HC in the exhaust manifold.
   D. they can provide oxygen for an oxidation catalyst.

ESSAY QUESTIONS

1. Describe parts and operating modes that affect emissions.
2. What types of control systems can be used to reduce emissions?
3. Describe specific emission control equipment used on current automobiles.

SUPPLEMENTAL ACTIVITIES

1. Refer to the manufacturer's service manual. Locate and identify vacuum-operated control mechanisms and individual parts on a vehicle.
2. Locate and identify electrically operated solenoid valves on a vehicle's vacuum control system.
3. Locate vacuum taps on a carburetor and/or intake manifold. Mark vacuum tubing for reconnection, and test each tap with a vacuum gauge. Test each tap at idle, during acceleration, WOT-operation, and deceleration. Make notes of the readings for each tap during tests.
4. Locate and trace vacuum and vapor lines and components for PCV, air injection, EGR, and catalytic converter systems. Demonstrate to your class where components are located on a vehicle.
5. Refer to manufacturers' service manuals and determine where exhaust gases are to be sampled during air-fuel mixture adjustment. Report to your class on what differences are found.

42 Emission Control Systems Service

UNIT PREVIEW

Late-model vehicles are designed to deliver good economy and reasonable performance while greatly reducing harmful exhaust emissions. They do this best when all systems, including emission control devices, are tuned and adjusted to manufacturers' specifications.

Recommended service for emission control systems includes careful inspection, testing, and diagnostic procedures as outlined by the manufacturer. Refer to the appropriate manufacturer's service manual for correct emission control system service.

Intentionally disconnecting, bypassing, or disabling emission control devices will result in lost performance, economy, and customer dissatisfaction. These practices also are prohibited by law and can result in fines and loss of licenses or certifications.

LEARNING OBJECTIVES

When you have completed your assignments and exercises in this unit, you should be able to:

- Identify the problems caused by tampering with emission control devices.
- Describe the systems that must be maintained to reduce exhaust emissions to legal levels.
- Inspect and replace defective vacuum control lines.
- Perform checks on heat and solenoid-operated vacuum valves and vacuum delay valves.
- Check PCV valves, thermostatically operated air cleaners, air pumps, diverter valves, and EGR valves for proper operation.
- Correctly connect and use an exhaust gas analyzer to determine exhaust emission levels and diagnose the causes of emission problems.

SAFETY PRECAUTIONS

Many ignition, fuel system, and emission checks and adjustments are done with the engine running. Clothing, jewelry, or hair can be caught in moving fans, belts, and pulleys and cause serious personal injury.

Remove all watches, rings, neck chains, and other jewelry. Roll up long sleeves or wear a short-sleeved shirt or blouse. Tie long hair in a ponytail, pin it to the top of your head, and wear a cap. Keep hands, arms, and face away from moving parts.

Hot exhaust manifolds and other parts can cause serious burns. Avoid touching exhaust manifolds, radiator surfaces, and other heated metal parts. Allow the vehicle to cool fully before attempting service procedures on exhaust or engine system parts.

Exhaust gas analysis is performed with the engine running. Make sure there is proper ventilation before working on a running engine in an enclosed area. Carbon monoxide poisoning can cause serious health problems or death.

42.1 EMISSION CONTROL DEVICES AND TAMPERING

Federal and/or state laws prohibit intentional *tampering* with equipment that can change vehicle emissions levels. Tampering means removing, modifying, disconnecting, or disabling such devices.

Legal Sanctions

Licensed or certified automotive technicians can lose licenses or certification and/or be fined for tampering with emission control devices. Additional penalties may be imposed by specific states.

In many states, vehicles that cannot pass a strictly monitored emissions equipment inspection and test cannot be registered.

In California, tests and inspections are performed on a regular basis to detect tampering and/or high emissions levels. Emission control equipment that has been tampered with must be returned to original condition, no matter what the replacement cost. Vehicles must be equipped with all original, manufacturer-supplied emission control equipment.

Emission Control and Performance

Late-model engines and vehicles are built for proper operation with all emission control equipment in place and functioning. Tampering often can cause serious and expensive damage.

For example, disconnecting an EGR valve increases engine power output and thus increases manifold vacuum. However, such tampering can cause detonation and piston damage. In addition, vacuum controls on automatic transmissions respond to increased vacuum by reducing hydraulic pressures. Lowered hydraulic pressures can cause component failures within automatic transmissions.

Health and Environmental Pollution

Even if no legal, performance, or engine durability problems were created by tampering, health and environmental problems still exist. Everyone, including automotive technicians, breathes the same air and eventually can suffer health problems from air pollution.

As a technician, you may gain the knowledge and skill necessary to tamper with emission controls. However, you know that such tampering is wrong and contributes to health and environmental problems.

42.2 EMISSION CONTROLS PREVENTIVE MAINTENANCE

As discussed in Unit 35, the main purpose of a modern tune-up is to reduce exhaust emissions and maximize fuel economy. An emissions-legal tune-up includes checks and maintenance of the following systems:

- Fuel
- Charging
- Ignition
- Emission control.

Refer to the manufacturer's emission control information label in the engine compartment for proper tune-up procedures. Refer to the manufacturer's service manual for location of emissions-related parts and systems [Figure 42-1].

Initially, federal laws required that vehicle manufacturers warranty emissions-related equipment for the first 50,000 miles of vehicle use. Proposals to increase this to the first 100,000 miles of use have been made. Electronic ignition, fuel injection, and electronic controls all contribute to reliable and durable emission control equipment.

Refer to Units 20, 22, 24, 29, 33, and 35 for preventive maintenance for the fuel, charging, and ignition systems. Refer to Unit 39 for electronic device service.

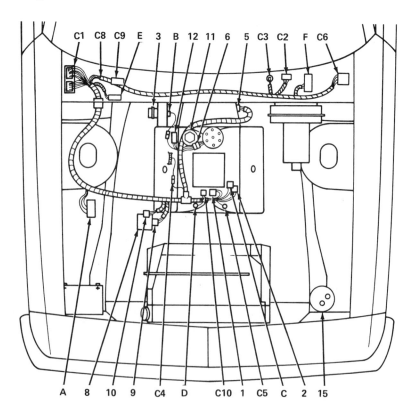

Figure 42-1. Locations of emissions-related parts. CHEVROLET MOTOR DIVISION—GMC

Checking Emission Control Devices

Checks for proper emission system operation can be performed with simple test equipment for some systems, such as vapor-recovery and PCV systems. However, for some devices, such as a catalytic converter, an *exhaust gas analyzer* may be necessary to determine proper operation. An exhaust gas analyzer is an electronic instrument that samples exhaust gases for the presence of HC, CO, and CO_2. Specific emission control devices that can be checked include:

- Vacuum, vapor, and fuel lines
- Vacuum control systems
- Thermostatically controlled air cleaner
- Vapor recovery system
- PCV system
- Air injection system
- EGR system
- Catalytic converter.

Integrated Control Systems

Electronic controls on late-model vehicles may operate several emission control system parts. In addition, information about emission control device operation can be "fed back" into a computer control module. Thus, *integrated control systems* can monitor and adjust almost every aspect of vehicle performance and emissions. An integrated control system combines many functions into a single electronic control unit.

Vacuum Leaks

Vacuum leaks create an excessively lean mixture in the intake manifold and cylinders. In addition, a feedback carburetor or fuel-injection system will react to a "lean" oxygen sensor output signal. Additional fuel will be delivered, which can cause an increase in HC and CO emissions. Thus, a lean mixture can create driveability problems. *Propane enrichment* is a useful and efficient technique to locate difficult-to-find vacuum leaks.

Propane is a hydrocarbon fuel. When additional fuel is added to a lean intake charge, engine speed will increase. In addition, oxygen sensor output voltage and exhaust emissions will change. These changes will occur if propane gas enters the intake charge through a vacuum leak.

A standard propane enrichment tester for checking and adjusting air-fuel mixtures is modified to check for vacuum leaks **[Figure 42-2]**. A long metal tube is added to the end of the rubber hose. This allows the technician to direct a small amount of propane gas toward parts and locations where vacuum leaks may occur. A tachometer or other test instrument is used to check whether the running engine reacts by changing speed and so on.

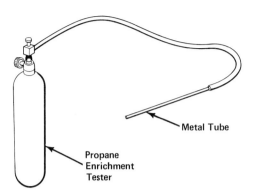

Figure 42-2. Propane enrichment tester.

Attach a one-foot long, 1/8-inch diameter copper or aluminum tube to the end of the tester's rubber hose.

SAFETY CAUTION

Propane is highly flammable. Fire or explosions can occur if the testing equipment is not used properly. Avoid directing propane gas around the distributor cap or near spark plug cables. Have a fully charged class B and C fire extinguisher ready for use.

Before you start the test, disable the air injection system. Disconnect the output hose from a pump or block the air inlet tube to a pulse air unit.

Start the engine and let it warm up to normal operating temperature and slow to normal curb idle speed. Open the valve of the propane enrichment tester and move the end of the metal tube toward suspected vacuum leaks. Watch your test instrument for indications that the air-fuel mixture is being enriched. Direct propane gas toward each of the following areas on all engines:

- Carburetor/throttle body base (around throttle shafts)
- Intake manifold fittings and gaskets
- All vacuum hoses and vacuum-operated devices connected to the intake manifold (especially in areas where they are affected by heat)
- Vacuum actuated power brake booster system
- Threaded base of air temperature sensor
- Air injection system plumbing.

For fuel injection systems, additional tests are necessary. Also direct propane gas toward the following areas:

- Ducting and hoses between the air metering box and the intake manifold
- Around the bases of each fuel injector
- Around the base of a separate cold enrichment injector

- Around the joints of vacuum-operated fuel pressure regulators.

Use one of the following methods to check your results. For RPM readings, use a low-RPM tachometer. For oxygen sensor output voltage, use only a high-impedance (10 megOhm or higher) digital voltmeter connected to the oxygen sensor output wire. Or, for CO and HC readings, use an exhaust gas analyzer. Check for the following indications:

- An increase in engine RPM
- An increase and fluctuation in oxygen sensor output voltage
- An increase in exhaust gas CO and HC outputs (allow 20 seconds for the exhaust gas analyzer to react).

Increases in idle speed, oxygen sensor output voltage, or CO and/or HC outputs indicate the source of a leak. Replace tubing or vacuum-operated units as necessary to correct vacuum leaks. After you have made repairs, make sure you have found and corrected all vacuum leaks. Reconnect the air-injection system output hose or unblock the pulse-air inlet tube when you finish.

42.3 VACUUM, VAPOR, AND FUEL LINES

Carefully examine all vacuum and vapor hoses and fuel lines for hardening, cracks, breaks, kinks, and loose or disconnected connections. Rubber hoses should be flexible, not stiff and hard. Typical vacuum line defects are shown in **Figure 42-3**. Replace defective lines one at a time to avoid misrouting. Refer to the manufacturer's service manual for correct vacuum hose routing if more than one hose is disconnected **[Figure 42-4]**.

Replace all vapor and fuel lines with special fuel/vapor hose. Use of heater hose or vacuum tubing will result in rapid cracking, leaks, and breakage.

42.4 VACUUM CONTROL SYSTEMS

To check vacuum controls, refer to the manufacturer's service manual for correct location and identification of components. As discussed in Unit 41, several types of vacuum control valves may be used.

Troubleshooting Vacuum Controls

Vacuum control troubleshooting is similar to all other troubleshooting. Troubleshooting procedures can include:

- Visual inspection
- Thorough basic knowledge of operating principles
- Use of vacuum hose diagrams
- Logical approach
- Use of test instruments.

Faulty operation can be caused by any part of the system or connecting lines. Check all parts thoroughly.

Checking Vacuum Valves

When a vacuum valve is closed, vacuum will be blocked from passing through the valve. When a valve opens, vacuum will pass through. Refer to the manufacturer's service manual for correct valve operating conditions and test procedures.

A hand-held vacuum pump and vacuum gauge can be used to check for proper opening and closing. Use at least 15 in. (38.1 cm) Hg of vacuum for checking purposes.

Replace vacuum control valves that do not operate properly.

Vacuum Delay Valve

To check a vacuum delay valve, connect a vacuum gauge on the output side of the delay valve. Vacuum

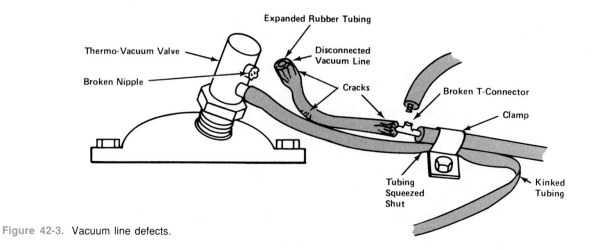

Figure 42-3. Vacuum line defects.

Emission Control Systems Service

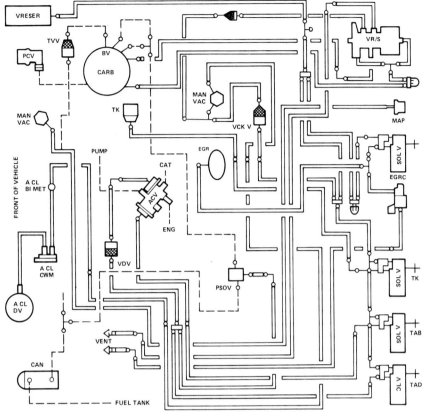

Figure 42-4. Vacuum hose routing diagram. FORD MOTOR COMPANY

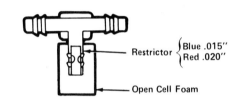

Figure 42-5. Checking the vacuum delay valve.

Figure 42-6. Vacuum vent air bleed with filter assembly. FORD MOTOR COMPANY

is applied on the vacuum supply side of the delay valve. Use a watch to determine how many seconds are required for full vacuum to appear **[Figure 42-5]**. Refer to the manufacturer's service manual for correct delay times.

On-Car and Off-Car Checks

Checks can be done on the vehicle or off. Checking vacuum control devices on the vehicle may require watching for movement of an operated device with the engine off. In other cases, movement or a change in an instrument reading may be checked when the engine is running.

To check vacuum control valves removed from a vehicle, use a vacuum gauge and a hand vacuum pump **[Figure 42-5]**.

Air Bleeds

Some vacuum-operated devices include a calibrated *air bleed* to limit vacuum actuation **[Figure 42-6]**. An air bleed is a calibrated *orifice*, or opening, through which air is drawn into a vacuum line. An air bleed limits the amount of vacuum applied to a device.

When testing vacuum-operated units, check for air bleeds in lines. Air bleeds will cause vacuum readings and/or actuation to decrease as air enters through the orifice.

Heat-Operated Vacuum Valves

Heat-operated controls include heat-operated vacuum valves and bimetallic temperature sensors. A heat-operated vacuum valve screws into a water jacket passage in the cylinder head. Checking a removed valve is illustrated in **Figure 42-7**.

Solenoid Vacuum Valves

Check electrically operated solenoid vacuum valve connections with a test light or voltmeter for power. Use jumper wires to provide electrical power to actuate a solenoid. Use hand vacuum pumps and vacuum gauges to verify proper valve operation.

42.5 THERMOSTATICALLY CONTROLLED AIR CLEANER

Check thermostatically controlled air cleaners for proper hose and vacuum line connections. Replace missing or torn hot-air and/or cool-air ducts.

Checking Bimetallic Temperature Sensors

Check bimetallic temperature sensors in air cleaner housings by applying heat or cold. Such sensors should allow vacuum to pass through when cold, and bleed vacuum off when hot. A hair dryer or heat gun can be used to heat the sensor.

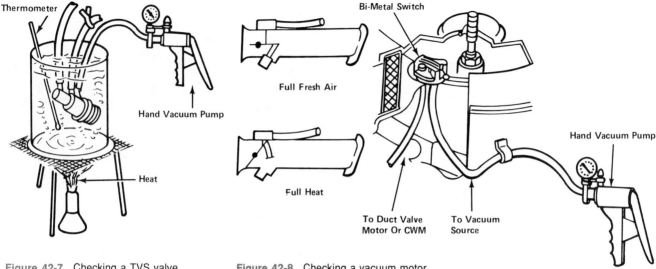

Figure 42-7. Checking a TVS valve.

Figure 42-8. Checking a vacuum motor.

An ice cube wrapped in a shop rag can be used to cool the sensor. Connect a hand vacuum pump to the vacuum source tube and check the operation of the damper door [Figure 42-8]. If the damper door does not operate, check the vacuum motor.

Checking Vacuum Motor Operation

Apply vacuum directly to the vacuum motor and watch the damper door to check its operation. If the motor is defective, drill out spot welds and replace the vacuum motor [Figure 42-9]. Expanding rivets are used to hold the new vacuum motor in place.

If the vacuum motor only operates properly when checked directly, the bimetallic temperature sensor is at fault and must be replaced.

42.6 VAPOR RECOVERY SYSTEM

Check vapor recovery systems for proper hose connections and hose conditions. Replaceable fiberglass air filters may be incorporated in vapor recovery canisters [Figure 42-10].

Filter Replacement

To replace vapor recovery canister filters, pull out the old filter and carefully insert the new filter under the bar. Replacement filters should be installed when the fuel system air filter is replaced.

Late-model canisters may not have replaceable filters. The entire canister must be replaced if defective. Canisters are held in place by sheet metal straps and screws. Loosen the screws, mark vapor and vacuum lines for replacement, and remove the canister. Install a replacement canister and reconnect vapor and vacuum lines correctly.

Vapor Recovery System Problems

Plugged or gasoline-soaked vapor recovery canisters, faulty purge valves, and/or clogged and pinched

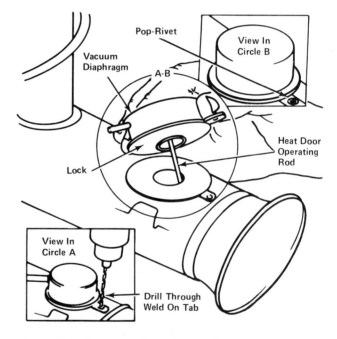

Figure 42-9. Vacuum motor replacement.

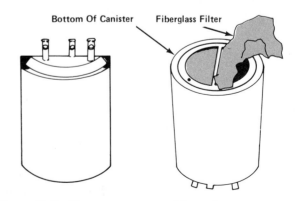

Figure 42-10. Vacuum recovery canister replacement.

lines can produce excessive pressure in fuel tanks. Sometimes when a defective filler cap is opened, a strong hiss can be heard as vapors are released. Gasoline blends with a high vapor pressure also may contribute to hissing and excessive pressure in the fuel tank.

Refer to the manufacturer's service manual for proper checking procedures for vapor recovery canisters and purge valves.

Lines to the fuel tank can be checked by applying low-pressure (3 psi or less) compressed air at the vapor recovery canister end [Figure 42-11]. A strong hissing should be heard through the fuel tank filler opening with the filler cap removed.

breather, filter, if used, is fitted in the air cleaner housing [Figure 42-12].

PCV System Problems

PCV filters should be changed when the air filter is changed. Oil in the air cleaner housing and an oily PCV filter can indicate problems. Even a light film of oil on the paper air cleaner element will decrease air flow and result in a richer mixture. A richer

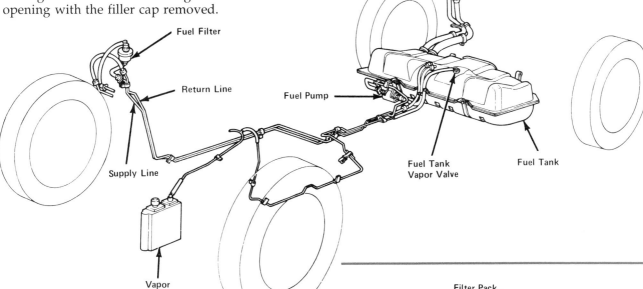

Figure 42-11. Fuel pump and fuel lines. FORD MOTOR COMPANY

SAFETY CAUTION

Always wear eye protection when working with compressed air. Eye injury or permanent blindness can result from liquid or solid contaminants blown into the eyes. Excessive air pressure can rupture the fuel tank, split hoses, or damage other parts. Use care when applying air pressure.

Fuel odors in the passenger compartment can be caused by broken, leaking, or disconnected vapor and/or fuel lines. Gasoline vapors noticeable during hard cornering may indicate faulty operation of a vapor/fuel separator in the fuel tank.

42.7 PCV SYSTEM

Check PCV systems for proper hose connections, clogged filters, and proper PCV valve operation. When an engine is running, vacuum should be present at oil filler openings on the valve cover. A PCV, or

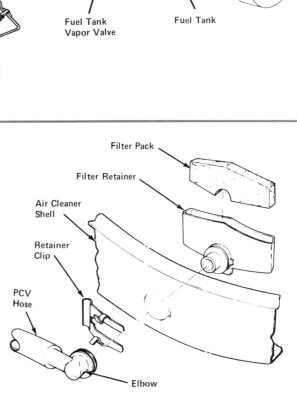

Figure 42-12. PCV filter installation. FORD MOTOR COMPANY

mixture will waste fuel and cause an increase in exhaust HC and CO emissions. Excessive HC exhaust emissions also may lead to overheated and damaged catalytic converters.

Oil from a PCV system can be caused by two problems. First, the PCV valve or hose may be clogged, allowing oil fumes to back up into the air filter housing. Second, excessive blowby from worn piston rings and cylinders can overload the PCV system and force oily vapors through the PCV breather filter tube (see Unit 43).

Checking PCV Valves

To check a PCV valve, remove it from its rubber grommet and shake it. A valve that is not clogged will rattle, indicating that the check valve is free. To check for proper operation, start the engine. Note the tachometer reading, then block the engine end of the PCV valve with your thumb. A strong suction should be felt, and engine speed should drop at least 50 rpm **[Figure 42-13]**. If not, the valve or line is clogged.

A defective PCV valve that remains open can cause rough idling, stalling, and the introduction of oil to the air cleaner. These problems also can affect on-board computer controlled fuel and ignition systems and catalytic converters. PCV valves should be replaced approximately every two years or 24,000 miles, or as recommended by the manufacturer.

42.8 AIR INJECTION SYSTEM

Air injection system checks include vacuum hose, air hose, and distribution manifold conditions, drive belt operation, and diverter valve operation. Replace damaged or defective parts. Tighten loose drive belts.

NOTE: Air pumps on some late-model Chrysler products are driven off the camshaft. Refer to specific Chrysler Corporation service manuals for proper servicing procedures.

Exhaust gas pulse air injection systems are relatively trouble-free. Check for evidence of scorched paint and/or burned-through metal pipes and valve units. Replace exhaust pulse valves and manifolds if defects are found.

Checking Air Pump Output

Air pump output can be checked by removing output hoses and holding your hand over the ends. A strong, continuous gush of air should be felt from either or both outlets when the engine is accelerated. If not, the pump may need replacement. Air pumps are *vane-type pumps*, as illustrated in **Figure 42-14**. A vane-type pump uses vanes, or blades, to compress air. A fanlike assembly behind the drive pulley uses centrifugal force to remove contaminant particles from intake air **[Figure 42-15]**.

Most shops install a new or rebuilt air pump rather than servicing defective pumps. If an air pump is defective, loosen and remove the drive belt. Mark hoses for proper reassembly. Loosen and remove pivot and adjusting bolts, braces, and hoses. Remove and replace the air pump **[Figure 42-16]**.

Checking Diverter Valve Operation

Diverter valve operation can be checked by starting the engine and holding the throttle at a high idle (1,500-2,000 rpm). When the throttle is released suddenly, air should be forced out of diverter/bypass valve outlets **[Figure 42-17]**. This rush of air can be heard as well as felt with the fingers.

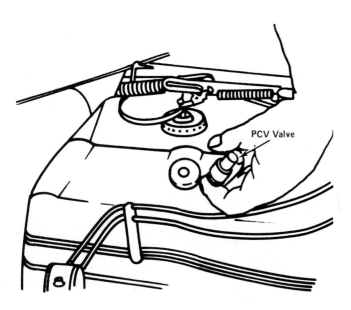

Figure 42-13. Checking a PCV valve.

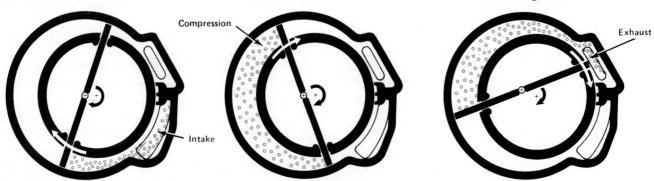

Figure 42-14. Vane-type air pump operation.   CHEVROLET MOTOR DIVISION—GMC

EMISSION CONTROL SYSTEMS SERVICE

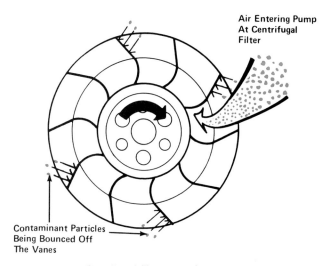

Figure 42-15. Centrifugal filter operation.

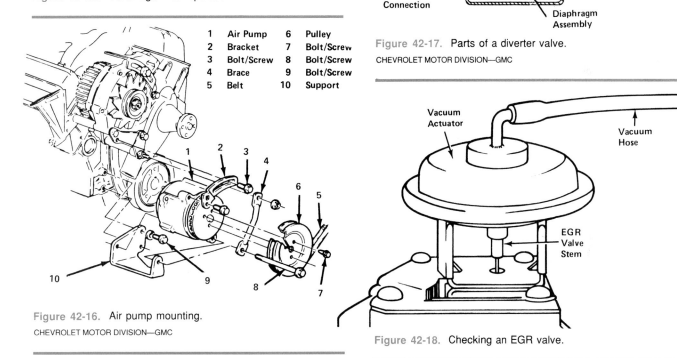

Figure 42-17. Parts of a diverter valve.
CHEVROLET MOTOR DIVISION—GMC

1	Air Pump	6	Pulley
2	Bracket	7	Bolt/Screw
3	Bolt/Screw	8	Bolt/Screw
4	Brace	9	Bolt/Screw
5	Belt	10	Support

Figure 42-16. Air pump mounting.
CHEVROLET MOTOR DIVISION—GMC

Figure 42-18. Checking an EGR valve.

The diverter valve also can be checked with a hand vacuum pump. Apply vacuum to the control diaphragm at the signal line connection and check for diverted air from the bypass holes. If a diverter valve fails these tests, replace it.

42.9 EGR SYSTEM

Check EGR systems for vacuum hose condition and valve operation. EGR valves may be checked visually and/or with a hand vacuum pump. Usually, a first indication of EGR malfunction is stalling at idle or audible spark knock on acceleration.

EGR Valve Checks

Start the engine and let it warm up to normal operating temperature. Increase engine speed to about 2,500 rpm and watch the EGR valve stem [Figure 42-18]. It should move upward.

Allow the engine to return to idle. The stem should move downward. Apply vacuum to the diaphragm. The stem should move upward, and the engine should die. If the EGR valve fails these tests, remove the valve [Figure 42-19]. Clean passages in the intake/exhaust manifolds as recommended by the manufacturer.

Replacing and/or Cleaning EGR Valves

Some vehicle manufacturers recommend a cleaning procedure if exhaust deposits have jammed an EGR valve [Figure 42-20]. Other manufacturers recommend replacement with a new valve if there are any functional problems.

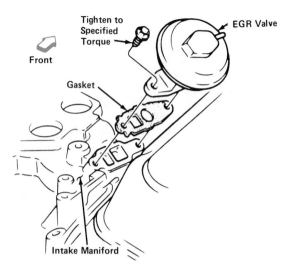

Figure 42-19. EGR valve mounting.
CHEVROLET MOTOR DIVISION—GMC

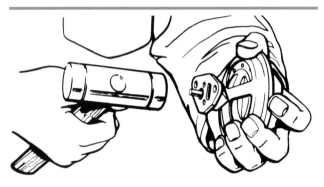

Figure 42-20. Cleaning an EGR valve.
CHEVROLET MOTOR DIVISION—GMC

CAUTION!!!

Do not wash EGR valve assemblies in solvents or degreaser. Permanent damage to the diaphragm will result. Also, sandblasting may cause defective valve operation. Refer to the manufacturer's service manual for proper cleaning procedures.

42.10 CATALYTIC CONVERTER

Catalytic converters do not require maintenance. However, if an engine is poorly tuned or has fuel or ignition system problems, a catalytic converter can be damaged or destroyed. Excess unburned fuel from an incorrectly adjusted carburetor or misfiring spark plug, can cause heat buildup within the converter. The heat generated can melt substrate or pellet bed retainers and destroy the catalyst.

"Rotten egg smell" from the exhaust is caused by the creation of sulfur dioxide gas (SO_2) as a byproduct of converter operation. The smell is not harmful in terms of vehicle function or emissions. A thorough tune-up will restore the efficiency of the engine and may reduce or eliminate this smell.

Heat shields around the catalytic converter should be tightly secured in their proper location.

SAFETY CAUTION

Allow all exhaust system components to cool completely before attempting work on catalytic converters. Heat in excess of 1,800 degrees F (968.8 degrees C) is generated by the catalytic converter. Severe burns can result from touching or brushing against a hot catalytic converter.

Check heat shield and catalytic converter fasteners. Tighten or replace fasteners and/or shields as required.

Exhaust Gas Analyzer

After all other tune-up and emissions control service has been performed, exhaust emissions can be checked with an exhaust gas analyzer. A *sampling probe* is inserted into the tailpipe of the vehicle. A sampling probe is a tube device that allows small amounts of exhaust gas to enter an exhaust gas analyzer. A sampling probe also can be used to find fuel vapor and exhaust gas leaks by moving the probe in the area of a suspected leak.

Refer to the vehicle manufacturer's service manual for correct placement of the sampling probe. Probes may be inserted directly into a tailpipe. In other cases, a plug may need to be removed from the exhaust system to insert the probe. Sampling locations may be located in the exhaust pipe leading to a catalytic converter, or in the body shell of the converter itself.

With the sampling probe inserted in the proper location, readings are taken at idle and under cruise (2,500 rpm) conditions for HC, CO, and CO_2. A table of analyzer readings and possible causes is shown in **Figure 42-21**. Standards for NO_x emissions are also planned.

Late-model computerized exhaust gas analyzers have memories that store data on allowable levels of emissions. These emissions levels can be compared with sampled gases for different vehicle makes, models, and years **[Figure 42-22]**.

The technician enters the year, make, and model of vehicle on a keyboard. The entered information is displayed on a screen. After sampling, results appear on the screen and on a slip of paper printed by the analyzer **[Figure 42-23]**.

If a vehicle's exhaust has higher levels of pollutants than allowed by law, some analyzers will indicate probable causes and recommended servicing.

Catalytic converters that have failed due to heat buildup must be serviced or replaced. Bead-type converters may be refillable. Substrate-type converters must be replaced as a unit. Catalytic converter service and bead replacement is covered in Unit 26.

EMISSION CONTROL SYSTEMS SERVICE

DIAGNOSIS OF EXHAUST GAS ANALYZER READINGS AT IDLE

HC	CO	SYMPTOMS	POSSIBLE CAUSES

High	Normal	Rough Idle	1. Faulty ignition:
			   a. Incorrect timing
			   b. Defective condenser or points
			   c. Fouled, shorted or improperly gapped plugs
			   d. Open or crossed ignition wires
			   e. Cracked distributor cap
			2. Leaking exhaust valves
			3. Leaking cylinder

High	Low	Rough Idle
		
		Fluctuating HC reading
			1. Vacuum leak:
			   a. Vacuum hose
			   b. Intake manifold
			   c. Head gasket
			2. Lean mixture

High	High	Rough Idle
		
		Black smoke fron exhaust
			1. Restricted air filter
			2. Plugged PCV Valve
			3. Faulty carburetion:
			   a. Idle mixture too rich
			   b. Faulty choke action
			   c. Incorrect float setting
			   d. Leaking needles or seats
			   e. Leaking power valve

Figure 42-21. Diagnosis of exhaust gas analyzer readings.

Figure 42-22. Exhaust gas analysis on a computer screen.

Figure 42-23. Technician removes analysis printout from computerized engine analyzer.

UNIT HIGHLIGHTS

- Intentional tampering with parts and/or systems that affect emissions is prohibited by federal and state laws. Automotive technicians can lose licenses and/or certifications and be fined for tampering.
- Late-model vehicles are designed to operate efficiently with all manufacturer-supplied emission control equipment in place and functioning. Tampering with emission control devices can cause expensive damage.
- An emissions-legal tune-up includes checks and maintenance of fuel, charging, ignition, and emission control devices.
- Vacuum, vapor, and fuel lines are checked for proper connections and hose conditions. Vapor canisters, PCV valves, air pumps and diverter valves, and EGR valves can be replaced if defective.
- A hand vacuum pump is a useful tool to check vacuum controls and operate many emissions-related parts. Many checks consist of simple manipulations and visual observation. Other checks may require the use of instruments.
- After all other tune-up and emission control checks have been performed, exhaust emissions are checked. Late-model computerized exhaust gas analyzers can produce visual and printed records of test results and recommended maintenance.

TERMS

tampering
orifice
propane enrichment
sampling probe
exhaust gas analyzer
air bleed
vane-type pump
integrated control system

REVIEW QUESTIONS

DIRECTIONS: The following questions are similar to those used on automotive technician certification tests. On a separate sheet of paper, write the letter of the correct choice.

1. Which of the following statements is correct?
   I. It is both illegal and counterproductive to tamper with emission control devices.
   II. Emission control devices on late-model vehicles contribute to cleaner, healthier air.
   A. I only
   B. II only
   C. Both I and II
   D. Neither I nor II

2. Technician A says plugging an EGR vacuum control line can increase engine power.
   Technician B says plugging an EGR vacuum control line can result in engine and automatic transmission damage.
   Who is correct?
   A. A only
   B. B only
   C. Both A and B
   D. Neither A nor B

3. Technician A says a diverter valve can be checked by watching for stem movement as the engine is speeded up.
   Technician B says a diverter valve can be checked by listening or feeling for air output as the throttle is released to slow down the engine.
   Who is correct?
   A. A only
   B. B only
   C. Both A and B
   D. Neither A nor B

Questions 4 and 5 are not like those above. Each has the word EXCEPT. For each question, look for the choice that does *not* apply. Read the questions carefully before you choose your answer.

4. An exhaust gas analyzer determines the levels of all the following EXCEPT
   A. HC (hydrocarbons).
   B. CO (carbon monoxide).
   C. SO_2 (sulfur dioxide).
   D. CO_2 (carbon dioxide).

5. Defects in vacuum controls for emissions-related devices can include all of the following EXCEPT
   A. excessive manifold vacuum.
   B. vacuum lines connected to incorrect locations.
   C. cracked, loose, leaking, disconnected, or pinched vacuum lines.
   D. defective heat or solenoid-operated vacuum valves.

ESSAY QUESTIONS

1. Describe the problems that can be caused by tampering with emissions control systems.
2. What devices can be checked during emission control system preventive maintenance?
3. Describe the checks that are made to check thermostatically controlled air cleaners, vapor recovery, PCV, air injection, and EGR systems for proper operation.

SUPPLEMENTAL ACTIVITIES

1. Inspect a vehicle's vacuum lines for defects. Replace lines, one at a time, if defects are found.
2. Check a vehicle's PCV valve for proper operation both by shaking it and by checking rpm drop.
3. Refer to the manufacturer's service manual and perform vacuum checks on heat and solenoid-operated vacuum valves and vacuum delay valves.
4. Check an air pump for proper output.
5. Use a hand-operated vacuum pump and check a thermostatically controlled air cleaner, diverter valve, and EGR valve for proper operation.
6. Refer to the instrument manufacturer's instructions and set up an exhaust gas analyzer correctly for sampling. Refer to the emission control information label and check for proper emission levels. Make notes of high emission levels, and write your recommendations for service of any problems found.

PART VI: Automotive Engine Service

Career opportunities for automotive engine technicians are found in dealerships, fleet, garages, and independent repair shops. The first unit covers preliminary tests to determine the need for engine rebuilding. The following service units detail the tasks necessary to rebuild 4-stroke piston engines, such as measurement inspections, and replacement of worn parts. In this photo, a technician is using a torque wrench to assemble a cylinder head.

43 Engine Problem Diagnosis

UNIT PREVIEW

This chapter discusses gasoline engine problem diagnosis. Before work is done to replace or repair mechanical engine parts, tests are performed to determine specific engine problems. Checks can be made for intake vacuum, cylinder compression pressure and leakage, engine lubricating oil pressure, and engine noise sources.

Some tests are more specific than others. The combination of results from several tests can confirm suspected engine problems. Once the problem has been diagnosed carefully, repair procedures can be performed.

Engine repair and service procedures are discussed in Units 44 through 46.

LEARNING OBJECTIVES

When you have completed your assignments and exercises in this unit, you should be able to:

- ❑ Explain how engine tests can be used to locate specific engine problems.
- ❑ Identify and describe the most common types of tests used to diagnose engine mechanical problems.
- ❑ Perform engine tests and make a diagnosis of engine problems.

SAFETY PRECAUTIONS

Almost all engine diagnostic procedures involve working on engines that are at normal operating temperatures. Hot exhaust and cooling system parts can cause severe burns. Avoid touching such areas.

Some testing procedures involve cranking, or turning over, the engine using the starter motor. Keep hands, arms, and loose clothing away from moving belts, pulleys, and fan blades. Tie up long hair and wear a cap.

Remove all rings, watches, neck chains, and other jewelry.

Wear short sleeved clothing or roll up long sleeves when working near moving belts and pulleys.

Wear safety goggles or safety glasses at all times in the shop area. Avoid pointing compressed air nozzles and blowguns at the face. Serious injury from metal or dirt particles or sprayed liquids can result.

Do not attempt to use any power machinery until you have received proper safety instructions and operating directions from your instructor. Obtain your instructor's permission before operating any power tools or equipment.

43.1 DIAGNOSTIC PROCEDURES

To diagnose engine mechanical problems, several tests can be made. Some tests are better than others for locating specific problems. In addition, two or more tests may be used to confirm a problem diagnosis. Engine tests generally include:

- Vacuum
- Compression
- Cylinder leakage
- Engine oil pressure.

43.2 VACUUM TESTS

When an engine in good condition is running, the low pressure created by the pistons on the intake stroke produces a steady vacuum, or suction, in the intake manifold. During cranking, less vacuum is produced. Vacuum readings during cranking and at idle can be compared to determine problems. A vacuum reading taken while the engine is running can locate many engine problems including leaking piston rings, leaking valves, worn valve guides, weak valve springs, and engine tuning problems.

Cranking Vacuum Test

To perform a cranking vacuum test, first run the engine until it has reached normal operating temperature. Connect the vacuum gauge hose securely to a source of engine *manifold vacuum*, as shown in **Figure 43-1**.

Disable the ignition system so that the engine will not start. To do this on vehicles with a separate ignition coil, remove the coil wire from the distributor cap. Connect an electrical jumper lead from the coil wire to a good ground connection in the engine compartment.

To disable electronic ignition systems without a separate ignition coil, disconnect the electrical connector that supplies battery voltage to the system. **Figure 43-2** shows the connector for the General Motors High Energy Ignition (HEI) disconnected.

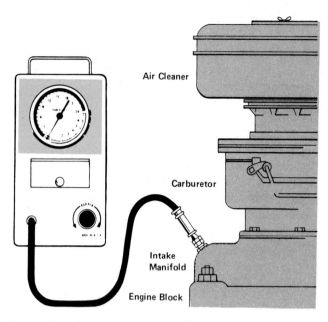

Figure 43-1. Vacuum gauge connection.

SAFETY CAUTION

Make sure the vacuum gauge hose is well away from the belts, pulley, and fan before cranking or running the engine.

Using the key switch or a remote starter switch, crank the engine and watch the vacuum gauge. (Do not depress the accelerator pedal.) Make a note of the vacuum reading produced and the steadiness of the reading. In general, engines in good condition should produce cranking vacuum readings of 5 in. (126 mm) of vacuum or more.

Less than 5 inches of cranking vacuum indicates leakage. Leakage could be external, such as from an intake manifold or carburetor gasket, or a broken or disconnected vacuum line.

Inspect external engine components for signs of leakage or disconnection. If everything appears correct, the leakage may be from internal engine components. To locate internal engine vacuum problems, perform a vacuum test with the engine running.

Running Vacuum Test

Reconnect the ignition system so that the engine can be started.

SAFETY CAUTION

Hot exhaust system, cooling system, and other engine parts can cause severe burns. Avoid touching hot surfaces during test procedures. Also be careful to avoid moving engine parts.

Figure 43-2. Disconnecting High Energy Ignition (HEI) system.

Start the engine and let it warm up. After the engine has reached normal operating temperature, note the vacuum reading produced while the engine is idling. Observe the steadiness of the reading. **Figure 43-3** illustrates different vacuum readings and what they indicate.

NOTE: Late model engines with high-lift camshafts and increased valve overlap may have lower and more uneven readings than illustrated in Figure 43-3. Emission control systems may also lower intake manifold vacuum readings.

43.3 COMPRESSION TESTS

A *compression test* measures compression pressure directly and can be used to determine piston ring or valve problems. A compression tester is shown in **Figure 43-4.**

To conduct a compression test, run the engine until it is at normal operating temperature.

SAFETY CAUTION

Hot exhaust system, cooling system, and other engine parts can cause severe burns. Avoid touching hot surfaces during test procedures. Also be careful to avoid moving engine parts.

Engine Problem Diagnosis

Steady reading between 15 and 22 in.-Hg with the engine warmed up and idling is normal (left, top). Snap the throttle plate suddenly. The needle should drop to 5 in.-Hg or lower before stabilizing at the normal reading between 15 and 22 in.-Hg (left, bottom).

Fluctuating reading that periodically drops 2 to 6 in.-Hg below normal indicates worn points or low compression.

Needle floats over a range of about 14 to 16 in.-Hg. Suggests that spark plugs may be gapped too close after incorrect servicing. Adjust gap.

Regular fluctuation between a low reading of about 5 in.-Hg and a slightly lower-than-normal reading means the head gasket is leaking.

Needle that swings erratically between about 10 and 20 in.-Hg when the engine is accelerated smoothly may indicate weak valve springs.

Reading drifts back and forth over a range of 4 to 5 in.-Hg within the normal range to indicate incorrect carburetor adjustment.

Steady high reading that holds above 21 in.-Hg indicates restricted air intake. Check for clogged air filter or a stuck choke.

Low reading that holds steady between 8 and 14 in.-Hg suggests that ignition timing is off or that piston rings are leaking. Check timing and compression.

Rapid needle vibration between 14 and 19 in.-Hg indicates that worn valve guides are letting intake valves chatter as they seat.

Needle drops to near zero when the engine is accelerated, then climbs back almost to normal level. Exhaust system may be blocked or kinked.

Figure 43-3. Analyzing vacuum readings.

Make sure the battery is fully charged so that the engine will crank quickly and easily.

Remove the spark plug cables and identify them for replacement in correct order. Next, loosen the spark plugs several turns before blowing any loose dirt from around the base of each plug. This will prevent dirt from falling into the cylinders when the plugs are removed.

Remove spark plugs. Disable the ignition system, as discussed in Topic 43.2, to prevent high-voltage sparks at the spark plug wires. Block the accelerator pedal or throttle linkage fully open to allow air to be drawn easily into the engine.

Screw the compression tester into cylinder No. 1. Crank the engine through the same number of revolutions for each cylinder, at least four full revolutions for each. Note how the pressure reading rises in steps, and note the final pressure reading obtained. Repeat the test for each cylinder.

Normal Compression Reading

Pressure readings for a cylinder in good condition will be built up in large jumps at first, and then smaller jumps. Refer to the manufacturer's service manual for correct compression pressure specifications. In general, all cylinder readings should be between 100 and 180 psi (690 and 1240 kPa). The lowest reading cylinder should be within 40 psi (27.6 kPa) of the highest reading. Excessive pressure readings can indicate carbon buildup in the cylinders.

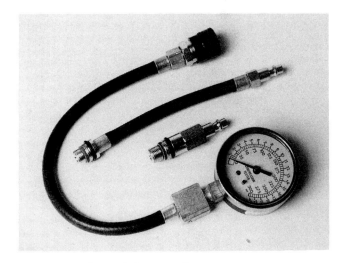

Figure 43-4. Compression gauge/tester.

Piston Ring Problems

Pressure readings for a cylinder with leaking piston rings will build up in smaller jumps and not reach the manufacturer's specified reading. To check for piston ring problems, perform a *wet compression test.* Remove the compression tester and squirt one tablespoon of engine oil through the spark plug hole. Replace the tester and repeat the test. If the readings for this test rise to nearly normal, the piston rings are defective. The added engine oil temporarily forms a seal around the piston.

NOTE: One broken compression ring on a piston will result in a low or zero compression reading for that cylinder. In this case, a wet compression test will not greatly improve the compression reading.

Valve Problems

Compression readings for an engine with leaking valves will build up in nearly equal, small jumps. The final reading will be below the manufacturer's specification. A wet compression test is not meaningful in checking for defective valves. The compression tester will indicate only that valve problems exist. A compression test will not specify whether the problem is due to an intake or an exhaust valve. This can be determined with a cylinder leakage test, discussed in Topic 43.4.

Head Gasket Problems

The small areas of a head gasket between cylinders can rupture, or be blown, by combustion pressure. A blown head gasket between cylinders will result in those two cylinders having very low, but equal final readings. As one piston moves up on the compression stroke, pressure leaks into the adjacent cylinder. On that cylinder's compression stroke, pressure leaks into the first cylinder. Both cylinders produce similar low readings.

A *blown head gasket* or a crack in a combustion chamber also can be detected by using an exhaust gas analyzer. Remove the radiator cap and place the analyzer probe near the radiator neck. If the analyzer records the presence of exhaust gases, a blown head gasket or combustion chamber leak or crack is indicated.

43.4 CYLINDER LEAKAGE TEST

To perform a cylinder leakage test, a special instrument called a *leak-down tester* is used with compressed air. A leak-down tester is shown in **Figure 43-5**. A leakdown tester can determine piston ring, intake or exhaust valve, and head gasket problems quickly and specifically.

To perform a cylinder leakdown test, first remove the radiator pressure cap. Then, run the engine until it has reached normal operating temperature.

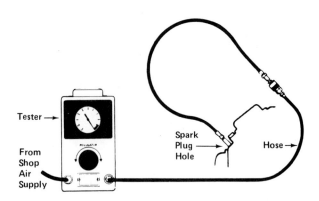

Figure 43-5. Leak-down tester. SUN ELECTRIC CORPORATION

SAFETY CAUTION

Hot exhaust system, cooling system, and other engine parts can cause severe burns. Avoid touching hot surfaces during test procedures. Also be careful to avoid moving engine parts.

Remove the spark plug from cylinder No. 1. Disable the ignition system. Crank the engine until the piston in cylinder number one is at TDC on its compression stroke. The timing marks should be on zero. Remove the air cleaner cover and oil filler cover.

Connect air pressure to the leakdown tester. Connect a blowgun to the output hose of the tester. Zero the tester dial. Operate the blowgun then release the trigger. Rezero the tester as necessary. Screw the *air-hold adapter* into the spark plug hole and connect the hose from the tester to the adapter. An air hold adapter is a hollow metal fitting of the same size as a spark plug [Figure 43-6].

Note the reading on the leakdown tester. The tester reads the amount of air leaking from the cylinder, expressed as a percentage. No leakage at all would result in a reading of zero. A cylinder in normal condition will have up to 10 percent leakage. More than 10 percent leakage indicates problems.

To determine the source of the problem, use a short length of rubber tubing held to your ear as a primitive "stethoscope."

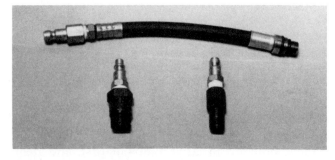

Figure 43-6. Air-hold adapters.

Hissing heard through the carburetor or throttle body indicates a leaking intake valve. Hissing from the oil filler opening indicates leaking piston rings. Bubbling heard or seen in the radiator indicates a blown head gasket, cracked cylinder head, or cracked block. Hissing from the tail pipe indicates a leaking exhaust valve.

Repeat the leakage test for all cylinders. Make a note of each cylinder's leakage and where you hear the greatest leakage.

NOTE: You can move the crankshaft to TDC for each cylinder with a remote starter. Connect the remote starter. Watch the tester dial and "bump" the engine by tapping the starter switch until the leakage reading is at its minimum value.

43.5 ENGINE OIL PRESSURE TEST

The lubricating system of an engine in good condition supplies pressurized oil to the critical moving parts and bearing surfaces within the engine. Oil is supplied to the main and connecting rod bearings on the crankshaft, the valve train, and the lower cylinder walls. The faster the engine turns, the more pressure is produced. At a preset pressure, a relief valve in the lubrication system opens to release excess pressure.

Worn engine main and connecting rod bearings, as explained in Unit 10, will cause a loss of oil pressure. With good maintenance and frequent oil changes, engine bearing wear should not be excessive until the vehicle has traveled at least 100,000 miles (160,900 km).

If high-mileage wear problems are suspected, a check of engine oil pressure at idle and at higher rpm can help determine lubrication system and bearing wear problems.

To conduct an engine oil pressure test, first check that the engine oil level is correct. Add oil if necessary. Then locate the oil pressure sensor, shown in **Figure 43-7**, on the engine block.

NOTE: Gasoline from a leaking fuel pump diaphragm can contaminate engine oil and make it thinner. Thin oil will cause false oil pressure readings. Check for gasoline contamination by smelling the oil on the dipstick. You also can attempt to light the oil on the dipstick with a match. If it lights easily, it is contaminated with gasoline.

Remove the wiring connection. Using a special socket, remove the sensor.

Thread the end of an oil pressure gauge hose into the engine block **[Figure 43-8]**. Start the engine and let it run at idle until normal operating temperature is reached.

Figure 43-7. Oil pressure sensor.

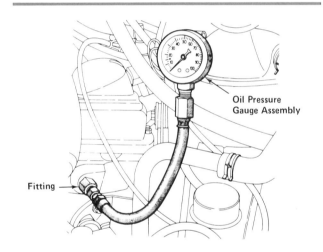

Figure 43-8. Oil pressure gauge. CHRYSLER CORPORATION

SAFETY CAUTION

Hot exhaust system, cooling system, and other engine parts can cause severe burns. Avoid touching hot surfaces during test procedures. Also be careful to avoid moving engine parts.

In general, idle oil pressure should be at least 20 psi (138 kPa). Slowly depress the throttle linkage until the engine is running at approximately 2,000 rpm. At 2,000 rpm, engine oil pressure should be approximately 45 to 60 psi (310 to 415 kPa). Refer to the manufacturer's service manual for correct oil pressure readings.

Oil pressures below specifications can indicate any of several problems, including excessive clearance between bearings or a worn oil pump. Other possible problems include leaking seals, a clogged oil pump screen, or a defective oil pump pressure relief valve. To determine the exact cause, it is necessary to remove the oil pan. Then, inspect the main and connecting rod bearings, rear main bearing seal, and oil pump.

Quick Check of Main Bearing Clearance

While the oil pressure gauge is connected and the engine is running, a quick check of main bearing clearance may be performed. While watching the pressure gauge, quickly open the throttle fully and then let it snap shut. The high vacuum during idle tends to "pull" or "hold" the crank up against the main bearing oil discharge holes. This tendency keeps oil pressure artificially high. The quick throttle opening will drop intake manifold vacuum. This will allow the crankshaft to drop down, if bearing clearance is excessive, and produce a momentary drop in oil pressure.

43.6 LOCATING ENGINE NOISES

A length of rubber tubing or a *technician's stethoscope*, can be used to find engine noise sources. Hold one end of the rubber hose to your ear and move the other end of the hose close to suspected noise sources. Touch the metal probe of the stethoscope to stationary metal parts.

SAFETY CAUTION

Be extremely careful when listening for noises around moving belts and pulleys at the front of the engine. Keep the end of the hose or stethoscope probe away from moving parts. Physical injury can result if the hose or stethoscope is pulled inward or flung outward by moving parts.

Start the engine and begin checking for noise sources. Some noises are more pronounced on a cold engine because clearances are greater when parts are not expanded by heat.

Open the oil filler cover and listen through the opening. A tapping noise indicates either a clogged hydraulic lifter or excess valve clearance. Tapping that stops after the engine warms up indicates a partially closed oil passage on a hydraulic lifter.

Touch the stethoscope to each intake and exhaust manifold *runner*, or hollow tube, near the cylinder head. The tapping noise will be loud and distinct at the affected valve.

Engine noises that seem to occur at a rate slower than crankshaft rotation are related to camshaft rotation. Distributors, fuel pumps, and oil pumps are accessories usually driven by the camshaft.

Engine accessories such as alternators, water pumps, power-steering pumps, air-conditioning compressors, and air pumps all can cause noise. Noises from the extreme ends of these units are usually caused by bad bearings or bushings.

Listen at several locations on the block for noises that seem to occur at the same speed as crankshaft revolution. Noises heard from the block can be caused by piston slap, loose piston pins, or a *rod bearing knock.*

Listen at the oil pan near the crankshaft front pulley and at the extreme rear of the oil pan. Noises heard from these locations are usually caused by main bearing problems.

Remember that aluminum and iron expand at different rates as temperatures rise. For example, a cold knock that disappears as the engine warms up probably is piston slap or knock. An aluminum piston expands more than the iron block, allowing the piston to fit more closely as engine temperature rises. A knock heard only at operating temperature may be caused by loose piston pins.

UNIT HIGHLIGHTS

- More than one engine test can be performed to confirm suspected engine problems.
- Vacuum tests can indicate many engine problems, but are not as specific as compression and cylinder leakage tests.
- A compression test can determine leaking or broken piston rings, leaking valves, or a blown head gasket.
- A cylinder leakage test determines problems more specifically than a compression test.
- Engine oil pressure tests can indicate engine bearing wear problems, oil pump and lubrication system problems, and leaking seals.
- Simple procedures can be used to locate the source of engine noises.

TERMS

manifold vacuum	runner
compression test	wet compression test
leak-down tester	air-hold adapter
technician's stethoscope	rod bearing knock
blown head gasket	

REVIEW QUESTIONS

DIRECTIONS: The following questions are similar to those used on automotive technician certification tests. On a separate sheet of paper, write the letter of the correct choice.

1. Technician A says that a wet compression test can determine whether piston rings or valves are leaking.
   Technician B says that a wet compression test can determine whether the intake or the exhaust valve is leaking.
   Who is correct?
   A. A only
   B. B only
   C. Both A and B
   D. Neither A nor B

2. Which of the following statements is correct?
   I. A cylinder leakage test can locate specific engine problems more quickly than a compression test.
   II. A cylinder leakage test can provide more information than a compression test.
   A. I only
   B. II only
   C. Both I and II
   D. Neither I nor II

3. Which of the following can be diagnosed with a technician's stethoscope or length of rubber tubing while the engine is running?
   A. Worn piston rings
   B. Plugged hydraulic lifter
   C. Leaking intake valve
   D. Weak valve springs

Questions 4 and 5 are not like those above. Each has the word EXCEPT. For each question, look for the choice that does *not* apply. Read each question carefully before you choose your answer.

4. An oil pressure test can indicate problems with all of the following items EXCEPT
   A. main and connecting rod bearings.
   B. oil pump.
   C. oil intake screen.
   D. piston rings.

5. Vacuum tests can determine all of the following EXCEPT
   A. blown head gaskets.
   B. worn valve guides.
   C. restricted exhaust system.
   D. worn main bearings.

ESSAY QUESTIONS

1. What can cranking and running vacuum tests indicate?
2. Describe gasoline engine compression test procedures and how to interpret them.
3. Describe cylinder leakage test procedures and how to interpret them.

SUPPLEMENTAL ACTIVITIES

1. Perform vacuum tests on a vehicle selected by your instructor. Report your findings to the class.
2. Perform a compression test on the same vehicle tested in Activity 1. Report your findings to the class.
3. Perform a cylinder leakage test on the same vehicle tested in Activity 1. Report your findings to the class.
4. Perform an engine oil pressure test on the same vehicle tested in Activity 1. Report your findings to the class.
5. Locate engine noises on the same vehicle tested in Activity 1. Report your findings to the class.
6. Based on your own tests and on the reports of your classmates, list problems found in the vehicle tested.

44 Cylinder Head Service

UNIT PREVIEW

Some service procedures for the engine cylinder head do not require removing the head from the engine. These service procedures include valve clearance adjustment and replacement of valve springs, valve stem seals, and retaining parts. Measurements of critical valve train operational dimensions also can be done with the cylinder head in place.

Most overhead camshaft engines require removal of the camshaft for cylinder head repair procedures. Cylinder head repair procedures for most pushrod engines can be performed without major engine disassembly.

After the head has been removed, major resurfacing and machining operations are done by an automotive machinist.

LEARNING OBJECTIVES

When you have completed your assignments and exercises in this unit, you should be able to:

❏ Safely and correctly perform cylinder head service procedures.
❏ Identify and describe the steps that must be followed to perform valve clearance checks and adjustments.
❏ Perform a valve adjustment.
❏ Remove and replace valve retainers, springs, and valve stem seals.
❏ Check for proper valve guide clearance and camshaft lobe lift.
❏ Identify the problems involved in resurfacing and machining operations on the cylinder head.

SAFETY PRECAUTIONS

Always refer to the manufacturer's service publications for correct service procedures. Incorrect procedures may result in damage to expensive parts.

Some procedures require working on engines at normal operating temperatures. Avoid touching hot exhaust and cooling system parts.

Sharp metal edges on parts can inflict serious cuts. When inspecting parts, move your hands and fingers slowly around the edges to prevent cuts. Use a shop rag when lifting or moving parts.

Remove all rings, watches, neck chains, and other jewelry before beginning work. Tie long hair in a ponytail and pin to the top of your head or stuff it down the back of your shirt or blouse.

Wear short-sleeved clothing or roll up long sleeves when working near a running engine.

Wear safety goggles or safety glasses at all times in the shop area.

Use caution when working with compressed air. Do not point blowguns or other air tools at other persons.

Do not attempt to use any power machinery until you have received proper instruction and training from your instructor. Obtain your instructor's permission before operating any power tools or equipment.

44.1 ON-CAR CYLINDER HEAD SERVICE

Service procedures that can be performed with the cylinder head still attached to the cylinder block include:

- Inspection for worn valve train parts
- Measurement of valve spring installed height
- Valve adjustment
- Checking for valve guide wear
- Checking camshaft lobe lift
- Checking for valve spring damage
- Replacing valve springs, retainers, and valve stem seals.

Before on-car service procedures can be performed, the engine and cylinder head must be cleaned. Dirt and grease can fall through oil drain holes in the cylinder head and contaminate the engine's lubrication system.

Cleaning the Engine

Scrape heavy grease and dirt deposits from the engine with a dull scraper. Use solvent and a parts brush to dissolve grease in hard-to-reach areas. Use a steam-cleaning machine or high-pressure spray to wash the engine.

CYLINDER HEAD SERVICE

SAFETY CAUTION

Hot steam from the steam-cleaning machine can cause severe burns. Wear protective clothing, gloves, and safety goggles during steam cleaning.

Thoroughly and carefully clean the areas around covers, such as valve covers, timing gear cover, and oil pan. Rinse the engine with cool water and dry off with compressed air. Dry the spark plug wires and ignition distributor thoroughly.

SAFETY CAUTION

Wear eye protection whenever compressed air is used. Severe eye injury or permanent blindness can result from particles or liquids blown into the eyes.

44.2 INSPECTING THE CYLINDER HEAD

To aid in reassembly, make drawings of parts that are removed to allow access to the valve cover or covers. Remove parts to allow access to the cylinder head cover. Tag and number disconnected wires and hoses with masking tape. Identify and save attaching fasteners and brackets.

Remove the valve cover bolts or nuts in a crisscross pattern, from the ends toward the center [Figure 44-1]. Identify and save the attaching bolts or nuts.

Inspect the cylinder head for cracks and leaks. (Final inspection for cracks and leaks should be performed after the head has been removed.) Cracked heads require removal from the engine for service or replacement.

Inspect *soft plugs*, also called "core plugs," for leaking. Soft plugs are cup-shaped, stamped metal plugs inserted into cast metal parts to fill holes. As shown in **Figure 44-2**, a soft plug can be either a *cup plug* or an *expansion plug*.

If accessible, replacement soft plugs can be installed with the cylinder head attached to the block.

To replace soft plugs, drive a center punch or small chisel through the plug. Pull the old plug out with pliers. Coat the new soft plug with sealant and drive it in with a special tool [Figure 44-2].

44.3 INSPECTING THE VALVE TRAIN

Visually inspect all visible parts of the valve train for looseness, wear, or damage. Check for the following problems:

- Worn or loose timing chain or sprockets
- Worn or loose timing belt
- Loose mounting bolts, studs, and nuts
- Worn or pitted OHC lobes [Figure 44-3]

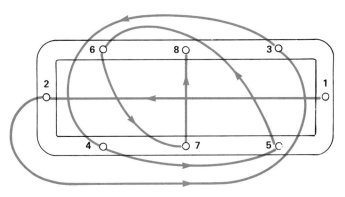

Figure 44-1. Valve cover bolt removal pattern.

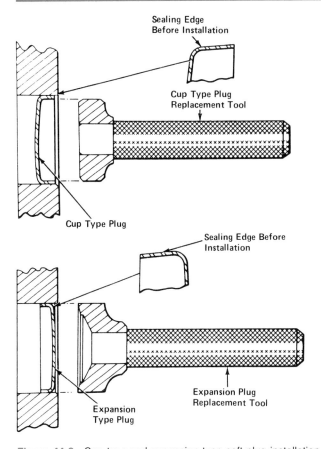

Figure 44-2. Cup-type and expansion-type soft plug installation.
FORD MOTOR COMPANY

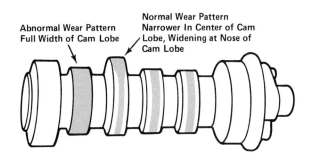

Figure 44-3. Even and uneven wear on camshaft lobes.

- Damaged OHC valve train parts
- Worn rocker arms or rocker followers
- Broken valve springs
- Bent pushrods
- Improperly seated valve locks and retainers
- Missing or damaged valve stem seals.

Tighten loose mounting bolts, studs, and nuts. Refer to the manufacturer's service manual for OHC removal procedures. Camshafts can be replaced on OHC engines without removing the cylinder head. On pushrod engines and some OHC engines, a number of parts can be replaced without removing the cylinder head. These parts include damaged rocker arms or followers, valve springs, retainers, valve stem seals, and pushrods. Check for bent pushrods by rolling each pushrod across a flat surface.

Check for plugged oil drain holes in the cylinder head by inserting a wire or thin metal probe gently into the holes.

After the cylinder head has been removed from the engine, clogged oil drain holes can be unplugged by cleaning with screwdrivers, metal rods, and wire brushes. Be sure that gasket sealing surfaces are not damaged or deeply scratched. Minor damage to gasket sealing surfaces can be carefully dressed with mill files or fine emery cloth. Make sure that abrasive particles do not get into the engine.

44.4 VALVE CLEARANCE ADJUSTMENT

Valve clearance adjustments are required at specified intervals on engines with solid or mechanical lifters. Engines with hydraulic valve lifters or hydraulic lash adjusters, normally require adjustment (if possible) only when an engine is being reassembled. Hydraulic lifters also can be collapsed, or compressed, to check for proper positioning and functioning.

Valve clearance for mechanical lifters must be checked as specified by the vehicle manufacturer, with the engine hot or cold. Refer to the manufacturer's service manual for correct adjustment procedures. The following is intended as a general guide to performing valve clearance adjustment.

Positioning the Crankshaft

Valve clearance on a cylinder is checked when both valves are in their fully closed positions. To achieve this condition, position the crankshaft as close as possible to TDC on the compression stroke.

The piston in a given cylinder can be positioned by cranking the engine with the starter motor. Another method is to turn the crankshaft pulley nut manually, using a breaker bar and socket. To avoid damage to the valve train, never attempt to turn the engine by rotating the camshaft.

The ignition distributor rotor can be used to indicate which cylinder is near TDC on its compression stroke. When the distributor rotor is aligned under one of the distributor cap terminals, the piston in that cylinder is very near its *firing position.* The firing position is the point at which the spark ignites the air-fuel mixture. This point is near TDC on the compression stroke.

Crank or position the engine until the pulley timing marks indicate TDC. Remove the distributor cap. Find the terminal on the cap with which the distributor rotor tip is aligned. Follow that spark plug wire to its cylinder. Valve clearance can be checked and/or adjusted at that cylinder **[Figure 44-4]**. To adjust any other cylinders' valves, reposition the engine to align the distributor rotor with that cylinder's spark plug wire terminal.

Checking and Adjusting Valve Clearance

Valve clearance usually is different for intake and exhaust valves. Valves aligned with intake manifold runners are intake valves; valves aligned with exhaust manifold runners are exhaust valves, as shown in **Figure 44-5**.

Rocker arm. Valve clearance on an engine with rocker arms is checked between the valve stem tip and the end of the rocker. Slide a feeler gauge of the proper size between the valve tip and the rocker. A light drag should be felt if the clearance is correct. If not, an adjustment is necessary. Loosen the locknut with a wrench, if so equipped, and turn the adjusting screw to adjust the clearance **[Figure 44-6]**. When the clearance is correct, tighten the locknut. Tightening the locknut can rotate the adjusting screw, so recheck the clearance and readjust if necessary.

Cam follower. Valve clearance on overhead camshaft engines with cam followers is checked between the camshaft lobe base circle and the adjustment shim, as shown in **Figure 44-7**. If adjustment is necessary, the shim must be removed with special tools, as shown in **Figure 44-8**. Replacement shims of different thicknesses are available to reduce or increase the valve clearance to specifications **[Figure 44-9]**.

Hydraulic lifter. To check hydraulic valve lifter assembly clearance, collapse the lifter according to instructions in the manufacturer's service manual. Measure the clearance between the valve stem and the tip of the rocker arm, as shown in **Figure 44-10**.

If the clearance is incorrect, adjust it by turning the pivot point adjusting nut **[Figure 44-11]**.

44.5 REMOVING VALVE SPRINGS

On most engines, damaged valve springs, retainers, and valve stem seals can be replaced without

Cylinder Head Service

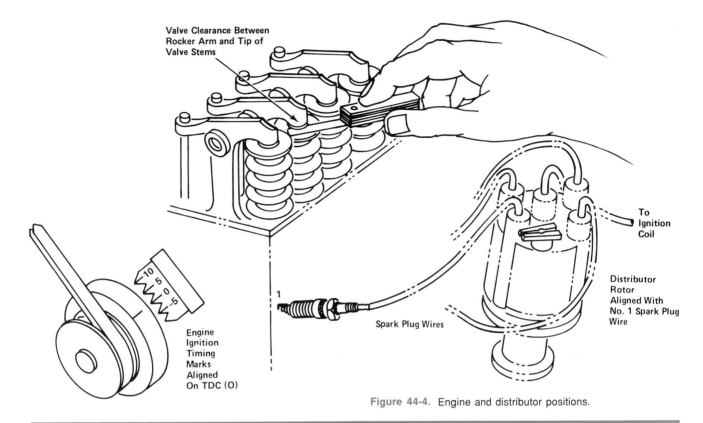

Figure 44-4. Engine and distributor positions.

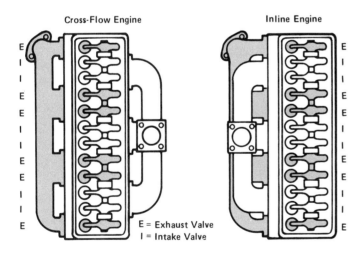

Figure 44-5. Valve positions.

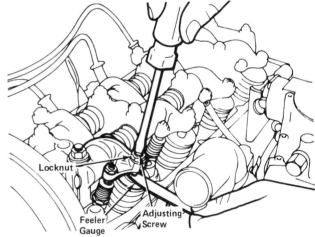

Figure 44-6. Valve clearance adjustment.
CHRYSLER CORPORATION

removing the cylinder head from the engine. Overhead camshaft engines require that the camshaft and timing belt or chain be removed. Refer to the manufacturer's service manual for correct service procedures.

Engines with rocker arms require that the rockers be removed. The rocker shaft and rockers or individual rocker pivots must be unbolted and removed.

To remove valve springs and related parts, position the piston in its cylinder at BDC. Remove the spark plug and screw in an air-hold adapter. Connect an air hose to supply air pressure of at least

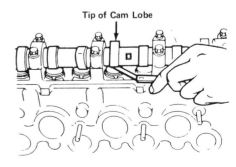

Figure 44-7. Checking valve clearance. FORD MOTOR COMPANY

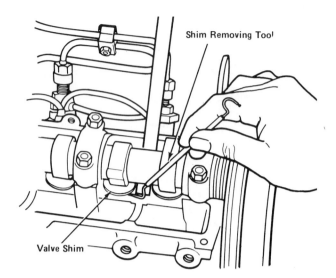

Figure 44-8. Shim removal. FORD MOTOR COMPANY

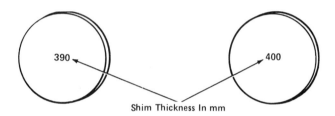

Figure 44-9. Valve shim sizes. FORD MOTOR COMPANY

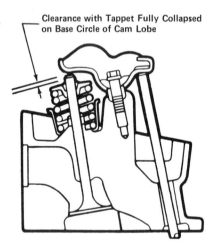

Figure 44-10. Checking collapsed valve lifter clearance.
CHEVROLET MOTOR DIVISION—GMC

Figure 44-11. Turning pivot point adjusting nut.
CHEVROLET MOTOR DIVISION—GMC

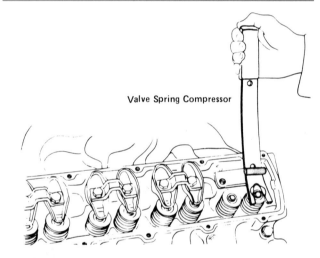

Figure 44-12. Removing valve springs.
CHEVROLET MOTOR DIVISION—GMC

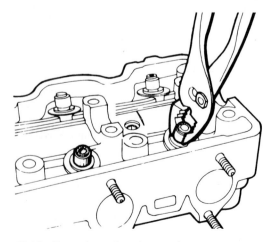

Figure 44-13. Removing valve stem seals.
CHRYSLER CORPORATION

75 psi (425 kPa). The air pressure will keep the valves in their closed positions so that they cannot drop down into the cylinder.

Use a special tool, such as shown in **Figure 44-12**, to compress the valve springs. Have a helper or magnet ready to help remove the valve locks, retainer, and spring. Remove the valve stem seals with a pair of pliers as shown in **Figure 44-13**.

CYLINDER HEAD SERVICE

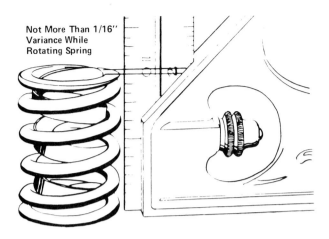

Figure 44-14. Checking a valve spring.
CHEVROLET MOTOR DIVISION—GMC

44.6 CHECKING FOR VALVE SPRING DAMAGE

Bent valve springs can cause uneven valve seating. To check a spring, stand it on a *surface plate* or a flat surface. A surface plate is a steel plate that has been accurately machined until it is very flat. Rotate the spring on the plate, next to a metal square. The clearance between the end of the spring and the square should be 1/16 inch (1.6 mm) or less as the valve is turned, as shown in **Figure 44-14.**

Valve springs can weaken and lose spring tension, and cause the engine to run poorly at higher rpm or become damaged. Valve springs can be checked for proper tension with a special fixture and a torque wrench [**Figure 44-15**]. Refer to the manufacturer's service manual for proper valve spring checking procedures.

44.7 MEASURING VALVE STEM CLEARANCE

Before valve stem seals are replaced, valve stem clearance should be checked with a dial indicator, as shown in **Figure 44-16.**

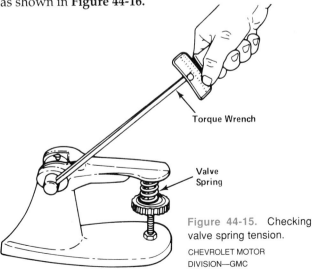

Figure 44-15. Checking valve spring tension.
CHEVROLET MOTOR DIVISION—GMC

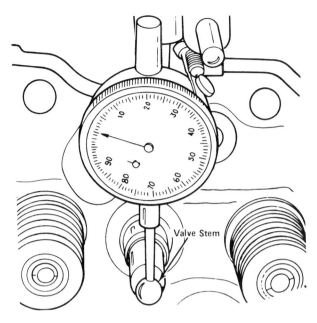

Figure 44-16. Measuring valve stem clearance with a dial indicator. FORD MOTOR COMPANY

Wrap adhesive tape around the mid-part of the valve stems so that the valves cannot drop into the cylinder. Release the air pressure so that the valve is free to move. Position the dial indicator so that the plunger is resting against the side of the valve tip. Move the valve down about 1/16 in. (1.6 mm) off its seat. Move the valve stem back and forth in line with the dial indicator plunger and note the readings. Subtract the lower reading from the higher. The difference indicates a rough approximation of valve guide clearance. If clearance is greater than manufacturer specifications, the cylinder head and valves must be removed from the engine for servicing.

44.8 MEASURING CAMSHAFT LOBE LIFT

Pushrod engine camshaft lobes can be checked for wear and correct lift by using a dial indicator mounted on the cylinder head [**Figure 44-17**].

Position the dial indicator plunger against the pushrod tip. Hold the pushrod in position. Zero the dial indicator. Using a breaker bar and socket on the crankshaft pulley nut, slowly rotate the engine. Note the dial indicator reading, this reading indicates camshaft lift. If camshaft lobe lift is not within manufacturer specifications, the camshaft must be replaced.

Overhead camshaft lobe lift can be determined by measuring the lobes directly with a micrometer, as shown in **Figure 44-18.** In many cases, camshaft lobes also can be measured while the camshaft is still installed. Camshaft service includes:

- Lobe inspection and measurement
- Journal inspection and measurement

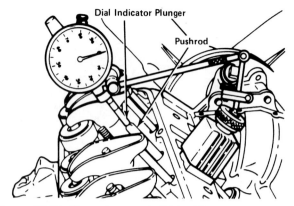

Figure 44-17. Measuring camshaft lobe lift.
CHEVROLET MOTOR DIVISION—GMC

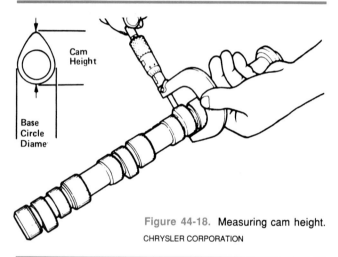

Figure 44-18. Measuring cam height.
CHRYSLER CORPORATION

Camshaft and lifter condition are directly linked to each other. Defective lifters may be replaced individually if the camshaft is in good condition.

The camshaft lobe surface should be straight or parallel with the shaft. The surface of the lobe should be free of metal pitting or flaking. If the lobe tip is worn convex or the metal surface in this area is not smooth, the camshaft must be replaced.

If the camshaft is replaced, the lifters must also be replaced. Failure to do this will result in the old lifters causing early failure of the new camshaft.

The camshaft must be removed, in most cases, to check the journal surfaces. They should not be discolored, scored, or pitted. A micrometer is used to determine if the journal diameter has been worn past specifications. The amount of lobe wear is determined by measuring across the lobe tip to determine the major diameter. When the smaller base circle measurement is subtracted from the major diameter, you have determined lobe lift. If this lift measurement does not meet specifications, the camshaft and lifters must be replaced.

If no other problems require removal of the cylinder head, reapply air to the air-hold adapter. Lift the valves to the closed positions and apply air pressure. Then, install new valve stem seals and/or other needed parts. Compress the springs and replace the retainers and valve locks. Remove the air-hold adapter. Replace the rockers and/or camshaft. Adjust valve clearance.

44.9 OFF-CAR CYLINDER HEAD SERVICE

Off-car cylinder head service includes:

- Removal of the cylinder head
- Cleaning and visual inspection
- Checking for physical damage
- Measuring valve and head parts for wear
- Machining and repair operations.

Before attempting to remove a cylinder head from an engine, drain the cooling system and engine oil. Disconnect the grounded battery terminal to prevent electric shock. Disconnect electrical wires to sensors, spark plugs, or other devices attached to the head.

All parts that would interfere with head removal must be moved or disconnected. Make simple drawings of how parts are to be replaced. Tag and identify all removed parts and fasteners.

44.10 CYLINDER HEAD REMOVAL

Before removing the cylinder head(s), follow the procedures described in 44.8 to determine if the camshaft meets manufacturer's specifications. If it does not, the camshaft and lifters (followers) must be replaced.

Some inline-engine cylinder heads can be removed with intake and/or exhaust manifolds attached. Unbolt and removed connections to the manifolds, then remove the cylinder head.

On V-type or flat engines and some inline engines, it is necessary to remove manifolds before the head is removed. Unbolt and remove the exhaust manifold, shown in **Figure 44-19**, or the exhaust pipe from the manifold.

Disconnect the fuel lines and throttle linkage and remove the carburetor or fuel injection throttle body. Unbolt and remove the intake manifold, shown in **Figure 44-20**. Unbolt and remove the intake manifold with the carburetor attached, if possible.

A V-type engine's intake manifold must be unbolted from both heads before the cylinder heads can be removed **[Figure 44-21]**.

On overhead camshaft engines, align the timing marks, loosen the belt or chain tensioner, and remove the timing belt. Timing marks for one type of belt-driven overhead camshaft are shown in **Figure 44-22**.

Remove the rocker shaft or individual rockers. Remove pushrods, in order, from their openings. Tag and identify the pushrods so that they can be reinstalled in their original locations.

Cylinder Head Service

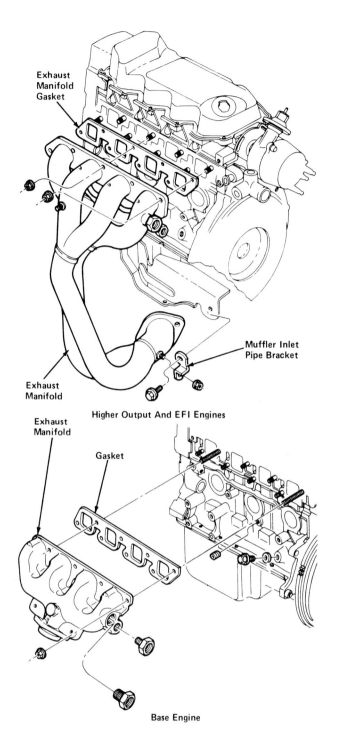

Figure 44-19. Exhaust manifold removal. FORD MOTOR COMPANY

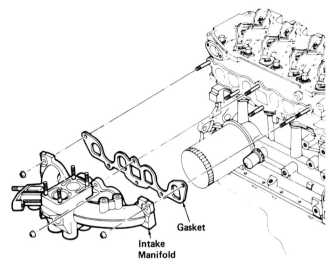

Figure 44-20. Intake manifold removal. FORD MOTOR COMPANY

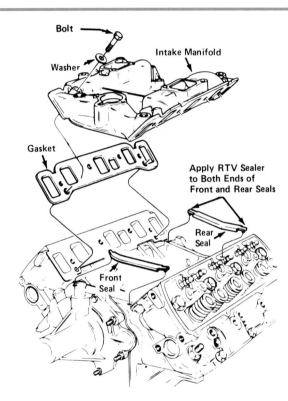

Figure 44-21. Removing a V-type intake manifold and gasket. CHEVROLET MOTOR DIVISION—GMC

Remove the spark plugs from the cylinder head. Unbolt the cylinder head attaching bolts in the pattern recommended by the manufacturer. Be sure that all head bolts have been removed. On some engines, head bolts are located under plugs in the intake manifold.

If the cylinder head is stuck, pry gently against its edges until it becomes loose. Remove the cylinder head.

CAUTION!!!

Do not insert tools into the intake or exhaust ports or other head openings. Prying against the ports can crack the cylinder head.

Use a special tool to remove the valve lifters [Figure 44-23]. Place the lifters in separate clean plastic bags, and identify them so that they can be reinstalled in their original locations.

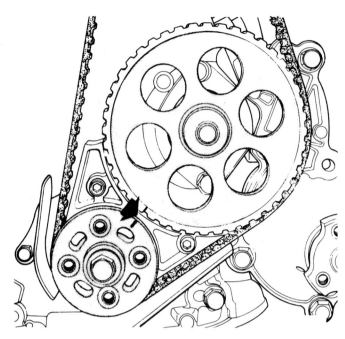

Figure 44-22. Timing marks. CHRYSLER CORPORATION

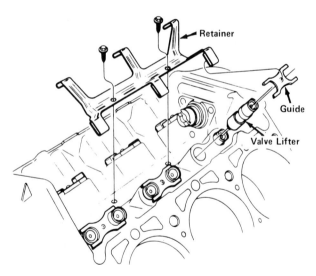

Figure 44-23. Removing valve lifters, guides, and retainers.
CHEVROLET MOTOR DIVISION—GMC

Figure 44-24. Inspecting for corrosion and wear.
BLACK & DECKER

44.11 CLEANING AND VISUAL INSPECTION

After the head has been removed, look for evidence of leakage or damage around the block sealing surface, intake and exhaust ports, and valve seats, as shown in **Figure 44-24**.

Scrape the sealing surfaces of the head with a flat gasket scraper or putty knife. Clean the combustion chamber walls with a scraper and a wire brush. Use a wire brush and drill motor to remove heavy deposits.

CAUTION!!!

Wear eye protection and use only a brass bristle brush on aluminum cylinder heads. Use care during this cleaning process so that no gasket sealing surface is deformed.

Remove the valves, springs, and retainers. Check for heavy carbon deposits and pitting on the valve head and fillet area. Remove any threaded plugs or other covers. Remove all soft plugs.

Place the head in a *hot tank* or a *jet spray booth* to remove rust, corrosion, and deposits. A hot tank is a metal tank filled with corrosive solvent, used to clean iron or steel parts. A jet spray booth is a closed container in which large parts are cleaned with multiple high-pressure spray jets **[Figure 44-25]**.

CAUTION!!!

Do not place aluminum heads, or iron heads with aluminum attachments, into a hot tank. The caustic cleaning solution will damage aluminum parts.

Steam clean the head after removal from the hot tank. Rinse in cool water and dry with compressed air.

Use a drill motor, or manual valve guide cleaner to remove varnish deposits from the valve guides, as shown in **Figure 44-26**.

Check the cylinder head carefully for cracks, especially around the valve seats and spark plug holes. Small cracks may be repairable by welding.

44.12 CHECKING FOR PHYSICAL DAMAGE

Cracks that may not be visible can be detected by spraying special liquids onto the metal surface of the block. This may involve using more than one liquid, one after another. Chemical reactions between the substances cause the cracks to stand out in color. Alternately, special chemicals that fluoresce, or glow, under ultraviolet (black) light can be used to detect hairline cracks. Special attention must be paid to areas around valve seats, spark plug holes, and ports.

CYLINDER HEAD SERVICE

Figure 44-25. High-pressure spray cleaner booth.
STORM VULCAN

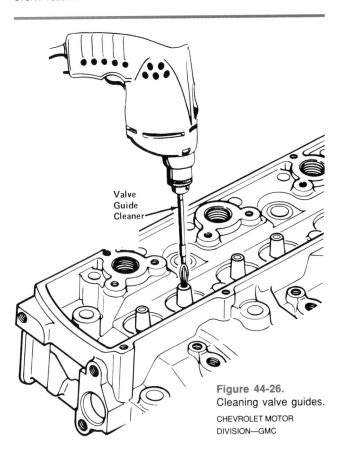

Figure 44-26.
Cleaning valve guides.
CHEVROLET MOTOR
DIVISION—GMC

Warpage, or bending, of the cylinder block sealing surface can be detected by using a *straightedge* and feeler gauges. A straightedge is a heavy metal bar that has been machined so that its edges are straight and parallel. The head sealing surface must be absolutely clean and free of dirt and old gasket material before this measurement can be made. Lay the straightedge across the head in seven places, and slide increasingly larger feeler gauges under the bar [**Figure 44-27**]. The largest feeler gauge that will slide under the bar with a light drag indicates the amount of warpage. In general, no more than 0.004 in. (0.1 mm) of warpage is allowable.

If the cylinder head is warped more than 0.004 in. (0.1 mm), it must be machined flat or replaced.

44.13 MEASURING VALVES AND VALVE GUIDES

Valves are measured with a micrometer and precision rulers at the points indicated in **Figure 44-28.**

If a valve stem is not badly worn or damaged, the valve stem tip and valve face can be ground smooth, as discussed in Topic 44.14.

Valve guides should be checked for wear with a small hole gauge and micrometer, as shown in **Figure 44-29.** Subtract the outside diameter of the valve stem from the inside diameter of the valve guide to determine valve guide clearance accurately.

Worn valve guide inserts can be pressed or driven out and new guides inserted with a hydraulic press. Worn integral valve guides can be reamed to a larger size, and valves with larger stems can be installed [**Figure 44-30**]. Damaged guides can be completely drilled or reamed out and new guides pressed in, if necessary. This operation is usually completed by an automotive machine shop.

44.14 MACHINING AND REPAIR OPERATIONS

Machining and repair operations are usually done by an *automotive machinist.* An automotive machinist specializes in precision machining of automotive

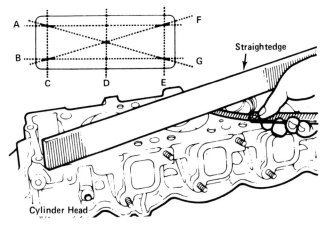

Figure 44-27. Distortion checkpoints. CHRYSLER CORPORATION

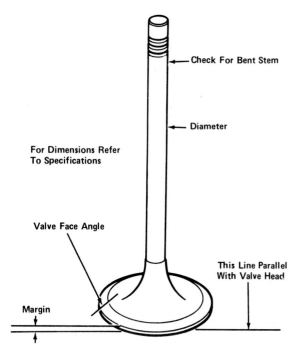

Figure 44-28. Valve measurement points.
FORD MOTOR COMPANY

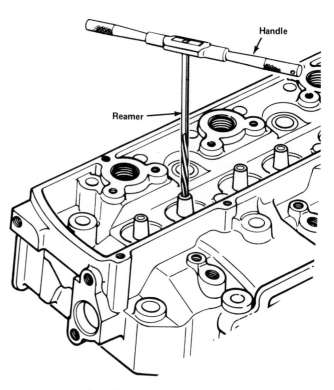

Figure 44-30. Reaming out valve guides.
CHEVROLET MOTOR DIVISION—GMC

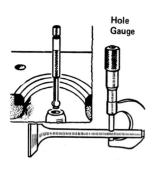

Figure 44-29. Checking valve guide wear.
BLACK&DECKER

parts, and must have a good background in auto mechanics, machine shop practices, mathematics, and geometry.

When material is machined or ground away, other parts may not fit properly. An automotive machinist must know how to machine other affected parts so that everything fits together properly again. For example, when one head from a V-type engine is refinished, the other cylinder head also must be machined the same amount. Otherwise, the intake manifold will not line up properly with the intake ports and bolt holes. The machinist must be able to figure out where to remove material from the intake manifold and how much to remove to restore proper alignment.

Slightly worn valve seats can be machined smooth with an electric seat grinder [Figure 44-31]. Badly worn valve seat inserts must be replaced. If an integral valve seat is badly damaged, the seat may be machined to accept a valve seat insert. In some cases, the entire head must be replaced.

Figure 44-31. Machining valve seats smooth.
SIOUX TOOLS INCORPORATED

After the valve seats have been machined, the seat width must be checked [Figure 44-32].

If the resurfaced seat is too wide, it must be narrowed by grinding, as shown in Figure 44-33.

After the valve seats have been ground, the valve will protrude farther through the head than previously. The valve spring installed height must be checked and corrected [Figure 44-34]. Shims are added under the valve springs to keep them under proper tension and correct installed height.

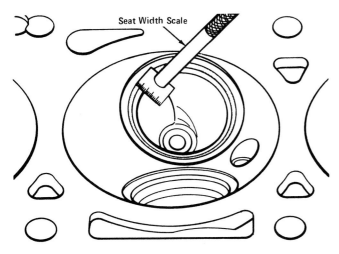

Figure 44-32. Checking valve seat width. FORD MOTOR COMPANY

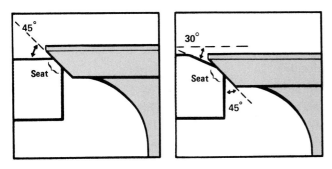

Figure 44-33. Valve seat machining. SIOUX TOOLS INCORPORATED

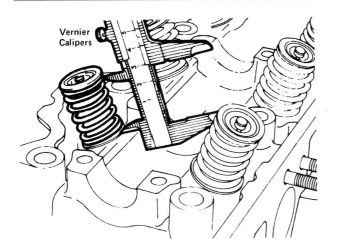

Figure 44-34. Checking the length of the valve in the seat. CHRYSLER CORPORATION

If a cylinder head is not badly cracked, it usually can be repaired by drilling, threading, and screwing in metal rods. The rods are then ground off flush and welded to the head. This process is repeated several times to repair larger cracks.

After all machining operations are completed, the cylinder head is scrubbed with hot water and detergent. Bottle brushes are used to clean ports and head passages. All traces of metal chips, abrasive dust, or other contaminants must be removed before reassembly. Rinse the head with clean water and dry with compressed air.

After cleaning, the machined surfaces are swabbed with engine oil applied to a lint-free cloth. Valves, springs, seals, and retainers are installed. Rocker shafts or pivots, and lifters or hydraulic lash adjusters are replaced. New soft plugs are coated with the manufacturer's recommended sealant and installed. The valve lifters are checked for excessive wear. If a lifter is noticeably *concave* (dished or curved inward) on the bottom (cam) side, both the lifters and camshaft must be replaced. If new lifters are required, a new camshaft also should be installed.

Use a tap of the correct size to clean block attaching bolt holes.

44.15 CYLINDER HEAD REASSEMBLY AND INSTALLATION

When the head is ready for reinstallation, install valve lifters in lifter bores. Install reusable pushrods in their original locations. Install new parts as necessary. Reassemble all upper valve train parts as specified by the manufacturer.

Dowels or *pilots* are used to align the head to the block. A pilot is a shaft or other part that guides two attaching parts into correct alignment.

Place a new head gasket between the head and the block. Do not apply gasket sealers to the head gasket unless recommended by the gasket manufacturer. Tighten the head bolts, in the manufacturer's specified order, to the correct torque.

CAUTION!!!

Head bolts must be oiled before installation, and partially tightened by stages, in the correct sequence, to the proper torque. Many current engines have head bolts that are tightened to a specific torque, then rotated an additional number of degrees. These "torque-to-yield" fastener bolts cannot be reused after they have been tightened. Always use new torque-to-yield fasteners when you reassemble an engine.

A typical cylinder head bolt tightening sequence is shown in **Figure 44-35**. After the head is in place, adjust the valve clearance and replace the rocker cover. Align crankshaft and overhead camshaft timing marks and install the timing chain or belt. Tighten the chain or belt tensioner as recommended by the manufacturer.

Replace the exhaust and intake manifolds, the carburetor or fuel injection throttle body, all hoses, brackets, and all other removed parts.

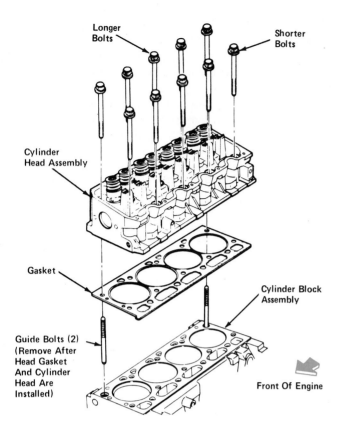

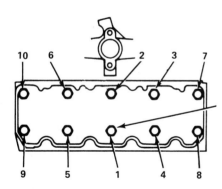

Figure 44-35. Cylinder head bolt tightening sequence.
FORD MOTOR COMPANY

Refill the cooling system. Change the oil filter and add the recommended amount of oil to the engine.

Reconnect the battery and start the engine. Let it run until it has reached normal operating temperature. On diesels and on some imported gasoline engines, the following procedures are necessary. Shut the engine off and allow it to cool. Follow the procedure discussed in Topic 44.2 to remove the valve cover. Tighten the cylinder head bolts, in sequence, to the proper torque again. Readjust the valves. Reinstall the valve cover and connect all parts that were removed.

UNIT HIGHLIGHTS

- Specific procedures in the vehicle manufacturers' service manuals must be followed to correctly perform cylinder head service.
- All of the following can be done without removing the cylinder head from the engine: valve clearance adjustment; replacement of valve springs, retainers, and valve stem seals; a rough check of valve guide clearance; and checking camshaft lobe lift.
- Overhead-camshaft engines may require that the camshaft be correctly removed and replaced for service procedures.
- An automotive machinist specializes in precision machining operations, such as valve seat and cylinder head resurfacing.

TERMS

soft plug
expansion plug
surface plate
jet spray booth
warpage
straightedge

cup plug
firing position
hot tank
pilot
concave
automotive machinist

REVIEW QUESTIONS

DIRECTIONS: The following questions are similar to those used on automotive technician certification tests. On a separate sheet of paper, write the letter of the correct choice.

1. What must be done before proceeding with any cylinder head service?
   A. Observe all safety precautions and refer to the manufacturer's service manual for correct procedures.
   B. Inspect for cracks.
   C. Measure valve spring installed height.
   D. Adjust valve clearance.

2. Technician A says that valve clearance can be adjusted when the piston in a cylinder is exactly at TDC of its compression stroke.
   Technician B says that valve clearance can be adjusted when the piston in a cylinder is at its firing position.
   Who is correct?
   A. A only
   B. B only
   C. Both A and B
   D. Neither A nor B

3. Which of the following statements is correct?
   I. The cylinder head must be removed to check approximate valve guide clearance.
   II. The cylinder head must be removed to check camshaft lobe lift.
   A. I only
   B. II only
   C. Both I and II
   D. Neither I nor II

4. Which of the following is usually done by an automotive technician?
   A. Machining cylinder head sealing surfaces
   B. Adjusting valve clearance
   C. Machining intake manifolds
   D. Grinding valve stems and adjusting valve spring installed height

Question 5 is not like those above. It has the word EXCEPT. For question 5, look for the choice that does *not* apply. Read the entire question carefully before you choose your answer.

5. All of the following require cylinder head removal for repair EXCEPT
   A. visible cracks in the head.
   B. excessive valve guide clearance.
   C. broken, damaged, or missing valve stem seals.
   D. damaged valve stems.

ESSAY QUESTIONS

1. Describe on-car cylinder head service.
2. What procedures are necessary to make valve adjustments on an engine with mechanical (solid) lifters?
3. Describe off-car cylinder head service.

SUPPLEMENTAL ACTIVITIES

1. Perform a valve adjustment on a vehicle chosen by your instructor.
2. Remove valve retainers, springs, and valve stem seals from a vehicle chosen by your instructor.
3. Measure approximate valve guide clearance and camshaft lobe lift on a vehicle chosen by your instructor.
4. Explain what other machining must be done if one cylinder head on a V-type engine must be resurfaced.
5. Make a report to the class on the shop equipment available at your school for major cylinder head machining operations.

45 Piston, Ring, And Rod Service

UNIT PREVIEW

Before pistons, rings, and rods can be serviced, the engine should be removed from the vehicle. The oil pan and cylinder head must be removed to allow access to cylinders and connecting rod bolts.

Ridges in the upper cylinder area, formed through wear, must be removed before the pistons can be pushed out the top. Pistons, connecting rods, and rod bearing caps must be marked for proper re-installation.

Pistons are removed from connecting rods by pressing out the piston pins with a hydraulic press.

Piston service includes cleaning, inspection, measurement, and machining operations. Connecting rod service includes cleaning, inspection, measurement, and surface finishing. Connecting rod bearings are inspected for problems and wear.

Before reinstallation, cylinder block and lower engine services are performed.

LEARNING OBJECTIVES

When you have completed your assignments and exercises in this unit, you should be able to:

- Safely and properly remove ridges from engine cylinders.
- Remove and replace piston rings.
- Measure pistons for wear.
- Safely and properly clean pistons.
- Safely and properly remove pistons from connecting rods.

SAFETY PRECAUTIONS

Always refer to the manufacturers' service manuals for correct service procedures. Incorrect procedures may result in damage to expensive parts.

Engine removal procedures require that heavy assemblies and parts be securely fastened before removal. Before transmission and engine fasteners are loosened, the assemblies must be safely and properly supported.

Sharp metal edges on piston rings can inflict serious cuts. When inspecting rings and other parts, use care to avoid cuts.

Wear safety goggles or safety glasses at all times in the shop area.

Do not attempt to use any power machinery until you have received proper instruction and training from your instructor. Obtain your instructor's permission before operating any power tools or equipment.

45.1 ENGINE REMOVAL

Before attempting to service pistons, rings, and connecting rods, the engine should be removed from the vehicle. Engine removal can include the following procedures:

- Disconnecting the battery
- Steam cleaning or power spraying the exterior of the engine
- Draining the cooling and lubrication systems
- Removing the fan and radiator
- Removing all external engine system accessories
- Disconnecting exhaust pipes
- Disconnecting hoses and other plumbing connections
- Disconnecting and/or removing linkage
- Disconnecting and/or removing instrument wiring and cables
- Disconnecting the exhaust system
- Supporting the transmission and removing transmission mounting bolts
- Disconnecting the driveline
- Supporting the engine and loosening motor mounts
- Lifting the engine from the vehicle with an engine hoist.

Refer to the manufacturer's service manual for specific procedures.

SAFETY CAUTION

Do not attempt engine removal procedures without safety instruction, specific training, and permission from your instructor.

45.2 PREPARING FOR PISTON REMOVAL

Before pistons and connecting rods can be removed, the cylinder head must be removed from the engine, as discussed in Unit 44. Other procedures necessary before piston removal include:

- Removing the oil pan
- Marking connecting rods and bearing caps
- Removing connecting rod bearing caps
- Measuring connecting rod bearing clearance
- Removing cylinder ridges.

Marking Connecting Rods and Bearing Caps

Remove the oil pan and oil pump from the cylinder block. The connecting rod and rod bearing cap must be replaced in their original locations. If the rods and bearing caps are not marked by the manufacturer for reinstallation, the technician must make numbering and alignment marks. Use a centerpunch, small chisel, or numbering die to number the rods and bearing caps. Mark the rods and bearing caps as shown in **Figure 45-1** to indicate which part should face the front of the engine. Always mark before disassembly has started. Failure to do so may cause metal distortion resulting in improper reassembly fit. Remove the connecting rod nuts. Tap the bearing caps with a soft hammer to loosen them. Remove the connecting rod bearing caps and inspect the insert bearings for defects **[Figure 45-2]**.

Measuring Connecting Rod Bearing Clearance

To measure connecting rod bearing clearance, lay a strip of *Plastigage* across the bearing surface, parallel to the crankshaft as shown in **Figure 45-3**. Perfect Circle Plastigage is a thin strip of wax-like material that flattens when it is compressed. The amount of flattening indicates the clearance between two parts.

Replace the bearing cap and lower insert bearing. Tighten the connecting rod nuts to the manufacturer's specified torque. Loosen and remove the nuts and bearing cap. Use the paper gauge supplied with the Plastigage to find the bearing clearance, as shown in **Figure 45-4**.

Figure 45-2. Bearing defects. FORD MOTOR COMPANY

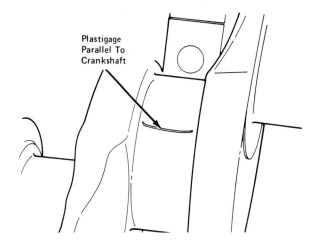

Figure 45-3. Using Plastigage on a journal.
CHEVROLET MOTOR DIVISION—GMC

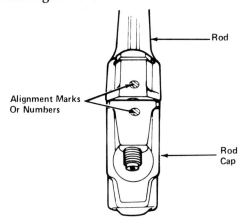

Figure 45-1. Connecting rod alignment marks.
FORD MOTOR COMPANY

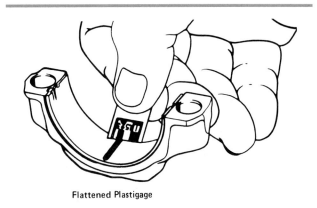

Figure 45-4. Measuring bearing clearance with Plastigage.
CHEVROLET MOTOR DIVISION—GMC

Refer to the manufacturer's service manual for acceptable bearing clearance. Different readings at the ends of the Plastigage indicate one of three conditions. The first is a tapered journal, larger at one end than the other. Another is a spool-shaped crankshaft journal (smaller in the middle than at the ends). The last is a barrel-shaped journal (larger in the middle than at the ends).

Removing Cylinder Ridges

After many thousands of miles of operation, the area of the cylinders where the rings travel become worn and enlarged. But because the rings do not travel all the way to the top of the cylinder bore, *cylinder ridges* are formed at the top of the cylinder [Figure 45-5].

Cylinder ridges must be removed before the piston can be pushed out the top of the cylinder. A *cylinder ridge remover*, or ridge reamer, is used to remove the ridge [Figure 45-6]. Turn the crankshaft until the piston in the cylinder is at BDC. Insert a clean shop towel to catch metal shavings.

CAUTION!!!

Refer to the cylinder ridge remover manufacturer's service manual to correctly adjust and use the tool. Improper use can damage the cylinder beyond repair.

Cover the other cylinders to protect them from shavings. Insert the ridge remover and adjust the cutting bit. Rotate the tool by using hand tools. After the ridge has been cut, remove the tool and clean the shavings from the cylinder. Repeat this procedure for all cylinders.

45.3 REMOVING PISTONS AND CONNECTING RODS

Rotate the crankshaft so that the piston/rod to be removed is at the bottom of its stroke. Unbolt and remove the rod bearing caps. Place short lengths of

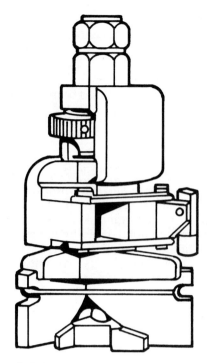

Figure 45-6. Cylinder ridge remover.

rubber hose over the connecting rod bolts to protect the crankshaft journals and cylinder walls [Figure 45-7].

If the piston is to be removed from the connecting rod, mark the piston top for cylinder number and proper mounting on the connecting rod so that correct piston pin offset is maintained. Pistons are often notched or otherwise marked to indicate which part is to face toward the front of the engine, as shown in **Figure 45-8.**

Place a length of wooden dowel or broomstick against the bottom of the piston. Carefully drive out the piston with a mallet.

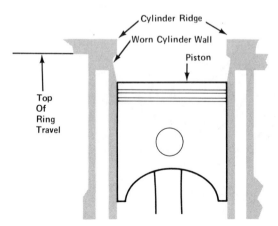

Figure 45-5. Cylinder ridges are caused by piston ring wear.

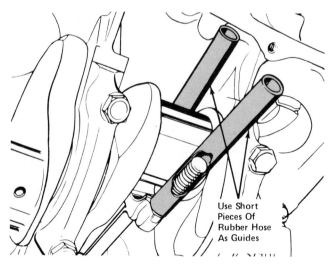

Figure 45-7. Connecting rod-bolt guide.
CHEVROLET MOTOR DIVISION—GMC

Piston, Ring, and Rod Service

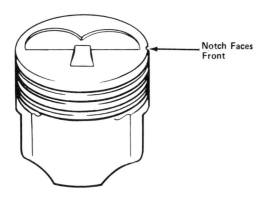

Figure 45-8. Piston identification.
CHEVROLET MOTOR DIVISION—GMC

Removing Piston Rings

A *piston ring expander* [Figure 45-9], is used to remove cast rings, such as compression rings and some oil-control rings, from the piston without breakage.

Oil control rings made of expanders and side rails can be removed carefully by hand, as shown in Figure 45-10.

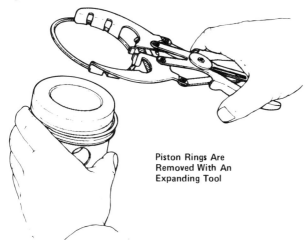

Figure 45-9. Piston ring expander.
CHRYSLER CORPORATION

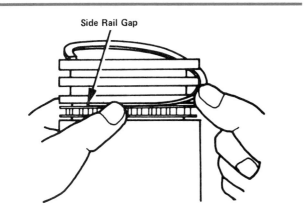

Figure 45-10. Removing oil-control ring.
CHRYSLER CORPORATION

45.4 REMOVING PISTONS FROM CONNECTING RODS

To remove piston pins that are press-fitted to connecting rods or pistons, a *hydraulic press* is used. A hydraulic press is an assembly that uses the pressure of a hydraulic jacking unit or hydraulic cylinder to produce pushing force.

SAFETY CAUTION

Do not attempt to use the hydraulic press without safety instruction, specific training, and permission from your instructor. The force generated by the press can cause metal to shatter suddenly and throw sharp metal parts outward.

Remove pin retaining clips, if used, before attempting to press out the piston pin. During the pressing operation, the piston pin boss area of the piston must be supported. A press setup for removing piston pins is shown in Figure 45-11. Do not attempt to remove a press fit piston pin unless proper piston support tooling is available. Failure to have the proper support may result in piston breakage.

45.5 PISTON SERVICE

If not badly damaged or worn, pistons can be reused. Piston service can include:

- Cleaning
- Inspection
- Measurement
- Machining operations.

Cleaning Pistons

Carbon, varnish, and deposits are removed from pistons by soaking the pistons in carburetor cleaner or

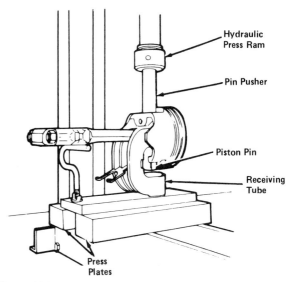

Figure 45-11. Pressing out piston pin. FORD MOTOR COMPANY

by *bead blasting.* Bead blasting is a process that uses compressed air to blow small glass beads at metal parts. The glass beads gently abrade, or wear off, deposits.

SAFETY CAUTION

Do not attempt bead blasting procedures without safety instruction, specific training, and permission from your instructor. Bead blasting must be done in a properly enclosed area. Safety hoods, goggles, gloves, and protective clothing must be worn.

After cleaning, the piston ring grooves are scraped with a groove cleaner **[Figure 45-12]**.

Oil return holes in the oil ring groove can be cleaned by hand with a small drill. During cleaning operations, no metal is to be removed from the piston. Deposits and varnish must be cleaned off without damage to the piston.

Inspecting and Measuring Pistons

Measure the piston as shown in **Figure 45-13**. Compare this reading with the bore size of the cylinder block. Refer to the manufacturer's service manual for recommended piston-to-bore clearance.

Piston taper. Pistons can become *tapered,* or narrower at one end. When measuring taper, the largest reading will be at the bottom of the skirt. Refer to the manufacturer's service manual for allowable piston taper. Pistons can be expanded, or made slightly larger, by machining operations described below.

CAUTION!!!

Handle pistons with care. Do not attempt to force pistons into or through cylinders until the cylinders

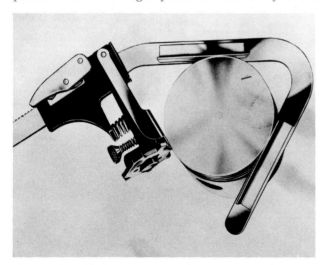

Figure 45-12. Cleaning piston ring groove. AMMCO TOOLS, INC.

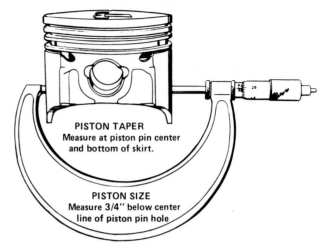

Figure 45-13. Measuring a piston.
CHEVROLET MOTOR DIVISION—GMC

have been machined to the correct size. Pistons can be damaged through careless handling.

Ring groove wear. The ring grooves can become enlarged through wear. To check ring groove clearance, use feeler gauges inserted between a new ring and the piston groove **[Figure 45-14]**.

Refer to the manufacturer's service manual for allowable clearance measurements. If clearance on the top compression ring groove is excessive, the groove can be machined approximately 0.025 in. (0.64 mm) larger. A flat steel groove spacer then can be inserted above the ring to reduce the clearance to specifications **[Figure 45-15]**.

Piston pin bore wear. If the piston pin hole, or bore, is worn, the pin can move excessively within the piston. Such wear generally requires that the pin be replaced. The piston and connecting rod small end are rebored and a new, larger pin installed.

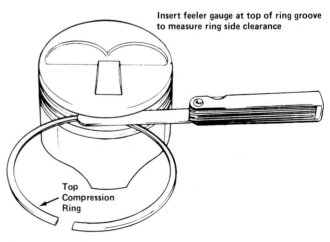

Figure 45-14. Piston ring side clearance.
CHEVROLET MOTOR DIVISION—GMC

Piston, Ring, and Rod Service

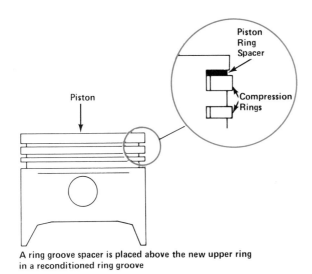

Figure 45-15. Ring groove spacer.

Machining Pistons

A piston skirt may be slightly expanded by *knurling*, as shown in **Figure 45-16**. Knurling is a machining process. The piston is mounted in a lathe. A knurling bit is slowly forced into the spinning skirt. The bit produces a raised pattern, or knurl. This raised pattern enlarges the diameter of the skirt.

45.6 CONNECTING ROD BEARING AND JOURNAL SERVICE

Connecting rods that are not badly twisted or worn can be reconditioned and reused. Connecting rod service can include:

- Cleaning
- Inspection
- Measurement.

Cleaning

Connecting rods can be cleaned after removal from the piston and pin by soaking them in a hot tank of cleaning solution. Blow compressed air through the oil squirt hole to make sure it is open.

SAFETY CAUTION

Wear eye protection whenever you use compressed air. Liquids or particles can be blown into the eyes with great force and cause injury or blindness.

Inspection

Connecting rods can become bent or twisted because of the forces exerted against them by the piston pin and the crankshaft journal **[Figure 45-17]**.

Special measurement tools and fixtures are available to check the rod for bending and twisting. If only slightly twisted, connecting rods can be straightened by inserting a bar and twisting. The limits for correctable connecting rod twist are shown in **Figure 45-18**.

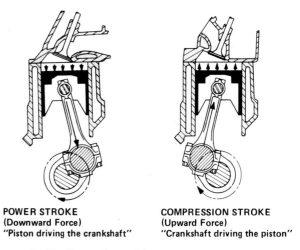

Figure 45-17. Connecting rod forces.

Figure 45-16. Knurling a piston. PERFECT CIRCLE

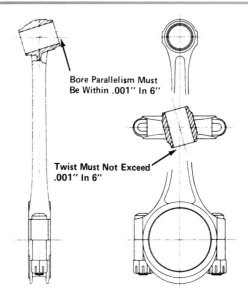

Figure 45-18. Connecting rod twist. FEDERAL-MOGUL

The connecting rod big end bore can become enlarged through stretching **[Figure 45-19]**. Occasionally, the bearing itself will spin or rotate inside the rod, effectively increasing the diameter. Connecting rods with enlarged bores should be remachined by an automotive machine shop or be replaced.

Inspecting Connecting Rod Bearings And Crankshaft Journals

Inspect the crankshaft rod journals for matching defects and *scoring*. Scoring consists of grooves worn around the inside or outside circumference of a part. A rough check for unacceptable scoring is to run the tip of a clean fingernail across the journal. If your fingernail drops into any grooves, they are too deep.

45.7 CONNECTING ROD BEARING INSPECTION

Connecting rod bearings are subject to forces from the power stroke, spinning, or centrifugal force, and forces due to the weight being moved **[refer to Figure 45-17]**.

Bearing damage from these forces can take many forms **[refer to Figure 45-2]**. The most frequent cause of damaged bearings is contaminated engine oil. Small particles of dirt or sand, metals, sludge, or other contaminants left in the engine can destroy new bearings quickly.

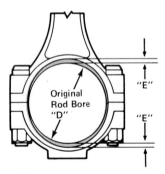

Figure 45-19. Stretched connecting rod bore.

There are four major factors that contribute to rapid bearing failures in newly rebuilt engines. These factors are: improper fit, excess or insufficient crush, lack of lubrication, and contamination. Be sure to have clean work habits—especially clean hands. Always use good quality engine oils with the highest API service ratings.

45.8 INSTALLING PISTONS ON CONNECTING RODS

After piston and connecting rod service, piston pins are pressed into the connecting rods. The press setup for this operation is shown in **Figure 45-20**.

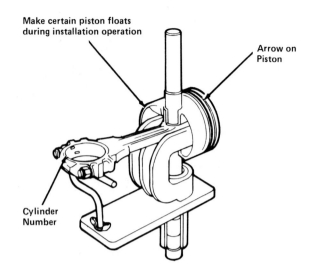

Figure 45-20. Piston pin installation. FORD MOTOR COMPANY

45.9 INSTALLING PISTON RINGS

Refer to the manufacturer's specifications for proper types of compression and oil control rings to be fitted. However, before rings are installed on the piston, the rings must be checked in the cylinder for proper fit. Other procedures, such as refinishing crankshaft journals, are necessary before installing the piston and connecting rod assembly in its cylinder. These procedures are discussed in Unit 46.

UNIT HIGHLIGHTS

- The engine should be removed from the vehicle before pistons, rings, and connecting rods are serviced.
- Cylinder ridges must be removed from the cylinders before the pistons can be pushed out.
- Connecting rods, bearing caps, and pistons must be marked for proper reinstallation.
- Pistons are removed by pushing them out and through the top of the cylinders.
- Piston rings are removed before piston pins are pressed out.
- Piston machining operations include enlarging the top compression ring groove for spacer installation and knurling the piston skirt.
- Small amounts of twist in connecting rods can be corrected.
- Before new piston rings are fitted and the piston and connecting rod assembly reinstalled, other machining and fitting operations are necessary.

TERMS

scoring
cylinder ridge
piston ring expander
bead blasting
knurl

Plastigage
cylinder ridge remover
hydraulic press
taper

REVIEW QUESTIONS

DIRECTIONS: The following questions are similar to those used on automotive technician certification tests. On a separate sheet of paper, write the letter of the correct choice.

1. Technician A says that the connecting rods must be marked to indicate in which cylinder they are to be reinstalled.
   Technician B says that the connecting rods must be marked to indicate which way they must face in their cylinders.
   Who is correct?
   A. A only
   B. B only
   C. Both A and B
   D. Neither A nor B

2. Which of the following must be done before pistons can be removed?
   A. Check connecting rod twist
   B. Remove cylinder ridges
   C. Measure cylinder taper
   D. Remove the main bearing caps

3. Which of the following statements is correct?
   I. Knurling produces a better gripping surface for the piston against the cylinder wall.
   II. Knurling expands the piston skirt diameter.
   A. I only
   B. II only
   C. Both I and II
   D. Neither I nor II

Questions 4 and 5 are not like those above. Each has the word EXCEPT. For each question, look for the choice that does *not* apply. Read each question carefully before you choose your answer.

4. All of the following must be done to remove or replace a piston on a connecting rod EXCEPT
   A. receive safety instruction and specific training from your instructor.
   B. wear eye protection.
   C. support the piston pin boss.
   D. use a larger-diameter piston pin to drive out the old piston pin.

5. All of the following contribute to early bearing failures EXCEPT
   A. improper fit.
   B. excessive or insufficient crush.
   C. lack of lubrication.
   D. cleanliness during reassembly.

ESSAY QUESTIONS

1. What steps are necessary to prepare for piston removal?
2. Describe piston service.
3. Describe connecting rod service.

SUPPLEMENTAL ACTIVITIES

1. Read the instruction manual for your shop's cylinder ridge remover. Ask your instructor about safety precautions and correct use of the tool. Demonstrate to your class how to avoid cutting too much material from the cylinder wall.
2. Ask your instructor about safety precautions and correct use of your shop's bead blaster or other compressed air cleaning device. Demonstrate to your class proper piston cleaning procedures.
3. Inspect and measure a shop piston for taper with an outside micrometer. Refer to the manufacturer's service manual and report to your class why the piston can or cannot be reused.
4. Demonstrate to the class the correct removal and replacement procedures for cast compression and oil-control rings.
5. Ask your instructor about the safe and proper use of your shop's hydraulic press. Demonstrate for your class proper safety precautions and use of the press to remove and replace piston pins.

46 Cylinder Block And Crankshaft Service

UNIT PREVIEW

Before block and crankshaft service can be completed, the components must be cleaned. After a thorough cleaning, measurements are taken. Then, a careful inspection of all parts and components must be performed.

Following the cleaning, measuring, and inspection, it will be possible to make accurate service recommendations. These may include replacement, remachining, and/or reconditioning of components.

After engine reassembly, special procedures are followed for initial startup, break-in, and the first 100 miles of driving.

LEARNING OBJECTIVES

When you have completed your assignments and exercises in this unit, you should be able to:

❏ Measure main bearing clearance and crankshaft end play.
❏ Determine if a cylinder is misshapen or worn.
❏ Determine the condition of crankshaft and camshaft journals and lobes.
❏ Fit piston rings correctly to a piston.
❏ Explain the importance of correctly tightening cylinder heads and other parts in sequence to the manufacturer's specified torque.

SAFETY PRECAUTIONS

Always refer to the manufacturer's service manual for correct service procedures. Incorrect procedures may result in damage to expensive parts.

Wear safety goggles or safety glasses at all times in the shop area.

Parts to be reused must be identified as to location and position. Handle parts, especially precision bearing inserts, carefully. Store such parts wrapped in soft cloths.

Do not attempt to use any power machinery until you have received proper safety training and operating instructions from your instructor. Obtain your instructor's permission before operating any power tools or equipment.

New gaskets and the manufacturer's recommended sealants must be used when parts are reinstalled.

The following are general procedures for cylinder block and crankshaft service. Refer to the manufacturer's service manual for specific procedures.

46.1 PRE-DISASSEMBLY MEASUREMENTS

Before proceeding with cylinder block and crankshaft service, check crankshaft end play and main bearing clearances.

Measuring Crankshaft End Play

End play is the movement of a shaft in a back-and-forth direction parallel to its length. *Crankshaft end play* is the movement of the crankshaft forward or rearward in the block. Excessive end play usually indicates worn flanges on the crankshaft thrust bearing.

To measure crankshaft end play, use a screwdriver to pry carefully between the crankcase and a crankshaft counterweight. Pry the crankshaft toward the front of the engine. Measure the clearance between the crankshaft flange surface and the rear main bearing with a feeler gauge [Figure 46-1].

If the clearance is not to the manufacturer's specifications, remove and examine the thrust bearing. All main bearings should be replaced during an engine overhaul. Reassembling the engine is discussed in Topic 46.8.

NOTE: In most engines, the thrust bearing and main bearing are parts of a set. The thrust bearing cannot be replaced separately.

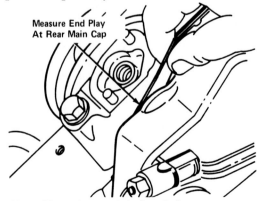

Figure 46-1. Measuring crankshaft end play.
CHEVROLET MOTOR DIVISION—GMC

Measuring Main Bearing Clearance

To check main bearing clearance, mark the main bearing caps for location and alignment with punchmarks and/or numbering dies. Unbolt the main bearing caps. Position Plastigage on the lower main bearing insert, parallel to the crankshaft.

CAUTION!!!

The crankshaft must not rotate during the gauging operation. If the crankshaft rotates, the plastic material will be damaged and will not indicate correct clearance measurements.

Replace the main bearing caps and bearing inserts. Tighten the cap nuts or bolts to the manufacturer's specified torque. Loosen the bolts, carefully remove the main bearing caps and bearing inserts. Use the width gauge supplied with the Plastigage to judge main bearing clearance, as shown in **Figure 46-2**. The plastic measuring material may be stuck to the journal or to the bearing insert.

Different readings at the ends and/or middle of a journal can indicate defects. The journal may be tapered, spool-shaped, or barrel-shaped. These defects can be corrected by regrinding the crankshaft journals to a smaller size. Thicker main bearings then are used to return the clearance to specifications.

Inspect the crankshaft main journals for signs of damage or scoring. In most cases, the crankshaft can be reused if the following conditions apply:

- The main journals are not severely damaged.
- The crankshaft is not cracked
- The crankshaft clearance is within specifications.
- The engine was removed for repairs other than main bearings.

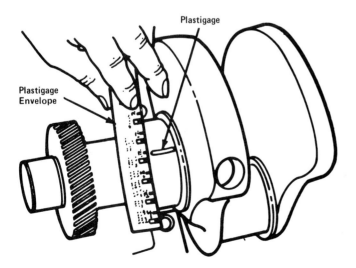

Figure 46-2. Checking main bearing clearance.
FORD MOTOR COMPANY

46.2 CYLINDER BLOCK SERVICE

Engine overhaul procedures generally include refinishing cylinders, and cleaning the cylinder head sealing surfaces and bearing surfaces. Cylinder block service and repair procedures can include:

- Cleaning and inspection
- Measuring cylinders and bearing surfaces
- Reboring and honing cylinders
- Replacing cylinder sleeves
- Reboring and refinishing camshaft and main bearing bores
- Replacing soft plugs
- Repairing cracks
- Repairing damaged threads.

46.3 CYLINDER BLOCK CLEANING AND INSPECTION

All major attached and inserted parts must be removed from the cylinder block before the block can be thoroughly cleaned.

SAFETY CAUTION

The crankshaft is heavy. Have a helper ready to assist you in lifting the crankshaft from the crankcase area. If the crankshaft drops, it will be damaged. Physical injury also can result.

Parts Removal

Unbolt and remove the water pump. Unbolt and remove the engine flywheel and any cover plates, as shown in **Figure 46-3**.

The crankshaft pulley or damper is removed with a special puller, as shown in **Figure 46-4**. On pushrod-type engines, the timing cover **[Figure 46-5]** is removed to expose the timing chain and/or gears. Check for looseness in a timing chain.

The timing chain and/or gears are disconnected and removed **[Figure 46-6]**. Pushrod-type valve lifters are removed as discussed in Unit 44.

Mark the crankshaft and flywheel or flex plate for reassembly in the original location. The crankshaft must be removed from the cylinder block and inspected for damage and cracks prior to cleaning operations. If main bearing caps are not marked, use a punch and/or numbering dies to make identification and alignment marks before removing them. Remove the main bearing caps and lower bearing inserts **[Figure 46-7]**.

Remove the crankshaft and inspect it for damage or cracks. To check for cracks quickly, carefully hang the crankshaft from a chain hoist with rags around an end counterweight. Strike any other counterweight with a hammer. A cracked crankshaft will produce a dull thunking sound. A ringing sound indicates a crankshaft without cracks.

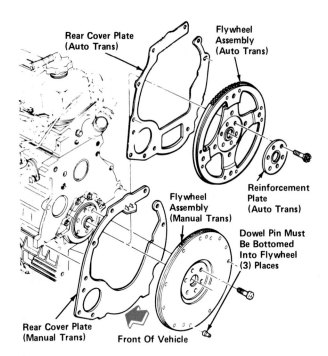

Figure 46-3. Flywheel removal. FORD MOTOR COMPANY

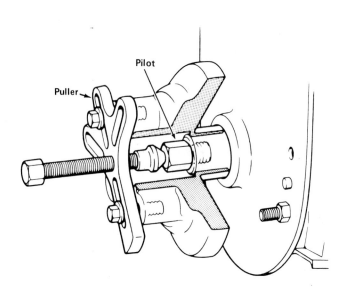

Figure 46-4. Damper removal. CHEVROLET MOTOR DIVISION—GMC

The camshaft then is removed carefully from the block, as shown in **Figure 46-8.** Camshaft removal on some engines may require the use of special puller tools. Special camshaft retainers or thrust plates must be removed on some engines.

Camshaft bearings on pushrod engines are removed with a special puller, shown in **Figure 46-9.** As the nut is tightened, the camshaft bearing is pulled from its bore. If the camshaft bearings are to be reused, number them and make alignment marks for reinstallation.

The thermostat housing and all other covers and attachments then are removed from the cylinder block. All soft plugs are removed so that the interior passages of the block can be cleaned.

Cleaning

After all attached parts have been removed, the block is cleaned. A hot tank is used to remove sludge, rust, and other deposits from cast iron blocks. A high-pressure or jet spray booth is used to clean aluminum blocks and also may be used to clean a cast iron block. Then the block is steam cleaned. Circular wire brushes are used to clean oil and coolant passages **[Figure 46-10].**

Taps are used to clean the threads of the cylinder block attaching bolt holes. Dirt in the holes can cause false torque readings when head bolts are tightened during cylinder head replacement.

Inspection

After cleaning, the block is thoroughly inspected for evidence of cracks and other damage. The cylinder head sealing surfaces are checked for warpage. This check is performed by using a straightedge and feeler gauges across the surfaces **[Figure 46-11].**

Refer to Topic 44.12 for a guide to checking for cracks in metal surfaces. Large cracks in sealing surfaces or cylinder walls usually require that the block be discarded. Small cracks in cylinder walls may be repaired by *sleeving,* or machining the cylinder and fitting a metal sleeve (see Topic 46.6).

Porosity, or tiny holes caused by air bubbles during casting, can be patched with *epoxy cement.* Epoxy cement is a plastic resin that hardens to form a patch. The porous areas are ground down and filled with epoxy **[Figure 46-12].** Heat is applied to dry the epoxy. After repair, the epoxy may be sanded and painted.

46.4 CYLINDER MEASUREMENT

Use a cylinder bore gauge to measure the cylinders at the top, middle, and bottom of the ring travel area. Measure the cylinder at points parallel and perpendicular (at 90 degrees) to the crankshaft as shown in **Figure 46-13.**

Different measurements between points A and B indicate an out-of-round, or oval, cylinder. Increasingly smaller measurements at the top, middle, and bottom of the ring travel indicate *cylinder taper.* Refer to the manufacturer's service manual for allowable cylinder dimensions.

46.5 BEARING SURFACE MEASUREMENT

To measure the main bearing bore, the main bearing caps are replaced in the correct order and position.

CYLINDER BLOCK AND CRANKSHAFT SERVICE 501

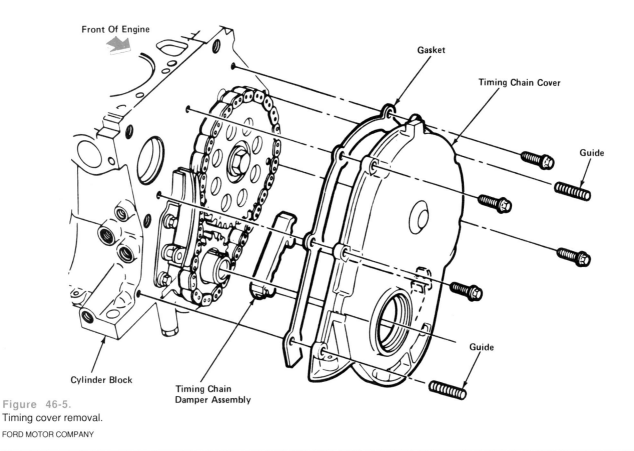

Figure 46-5.
Timing cover removal.
FORD MOTOR COMPANY

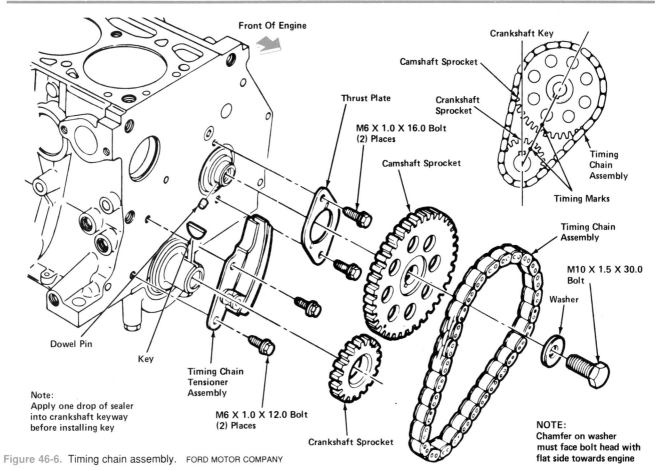

Figure 46-6. Timing chain assembly. FORD MOTOR COMPANY

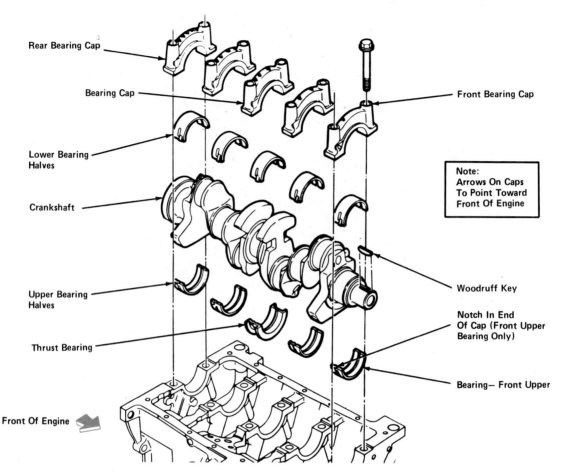

Figure 46-7. Removing main bearing caps. FORD MOTOR COMPANY

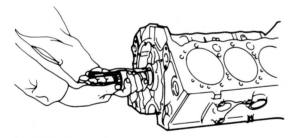

Figure 46-8. Camshaft removal. CHEVROLET MOTOR DIVISION—GMC

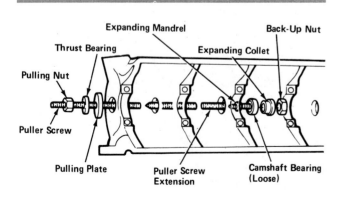

Figure 46-9. Camshaft bearing removal. FORD MOTOR COMPANY

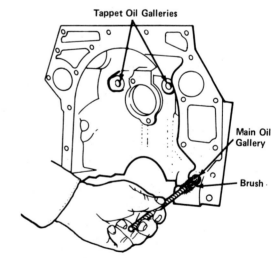

Figure 46-10. Cleaning oil galleries. FORD MOTOR COMPANY

The caps then are tightened to the manufacturer's torque specification. A telescoping gauge and outside micrometer are used to measure the bore.

The bores can become misaligned, as shown in **Figure 46-14,** due to crankcase warpage or a bent crankshaft.

Cylinder Block and Crankshaft Service

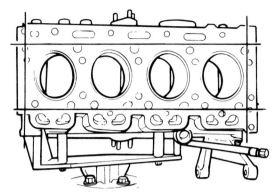

Figure 46-11. Checking cylinder block sealing surface distortion. FORD MOTOR COMPANY

Note — Portions of the front surface that are not machined or not part of the water jacket are repairable. Shaded areas may be repaired with epoxy.

Front And Left Side

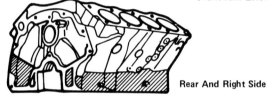

Rear And Right Side

Shaded areas may be repaired with epoxy.

Figure 46-12. Repairable block areas. FORD MOTOR COMPANY

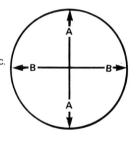

A — At right angle to center line of engine.
B — Parallel to center line of engine.

Top Measurement:	Make 12.70mm (½") below top of block deck.
Bottom Measurement:	Make within 12.70mm (½") above top of piston—where piston is at BDC.
Bore Service Limit:	Equals the average of "A" and "B" when measured at the center of the piston travel.
Taper:	Equals difference between "A" top and "A" bottom.
Out-of-Round:	Equals difference between "A" and "B" when measured at the center of piston travel.

Refer to manufacturer's specification tables.

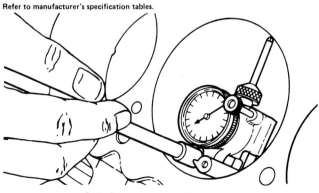

Figure 46-13. Cylinder measurement. BUICK MOTOR DIVISION—GMC/FORD MOTOR COMPANY

To check for main bearing bore misalignment, a specially ground *arbor*, or bar, is placed in the crankcase bore [Figure 46-15]. Feeler gauges are used to check for clearance under the bar at each bearing housing. Clearance indicates misaligned bearing housings.

Such misalignment can be corrected by *line boring* the bore. The ends of the main bearing caps are ground down to create a smaller than normal diameter hole when they are installed and torqued to specifications. A special machine is used to cut new, larger bores in the block. Oversized bearings are fitted to the new bore. Line boring is done by an automotive machinist.

Camshaft bearing bores are measured with a telescoping gauge and outside micrometer. Out-of-round or oversize conditions can be corrected by machining operations.

46.6 CYLINDER BORING AND REFINISHING

Cylinders can be rebored *oversize*, usually in steps of 0.010 inch (0.25 mm) up to 0.060 inch (1.52 mm). Larger pistons and piston rings, available in matching oversize steps, are installed. Cylinder reboring usually is done by an automotive machinist.

After boring, the cylinders must be honed, or finished, to the proper size and smoothness to a "crosshatch" pattern with a honing tool, shown in **Figure 46-16.**

Cylinder Sleeves

In some cases, it is possible to repair cylinder wall cracks or excessive wear by boring the cylinder to accept a dry liner. The *overbore*, or larger bore diameter, is usually 0.125 inch (3.18 mm) or more. The dry liner is then pressed or driven into the block, bored to the correct size, and honed.

Wet cylinder sleeves that are in contact with engine coolant are removed from blocks and replaced when worn past the manufacturer's limits. A wet liner is one which has coolant in direct contact with it. They are usually sealed to the block with "O" rings at the top and bottom.

46.7 CRANKSHAFT AND CAMSHAFT JOURNAL MEASUREMENT

If crankshaft journals have not been badly scored or damaged, the crankshaft can be reused. Use an outside micrometer to measure the crankshaft journals as indicated in **Figure 46-17**.

Inspect and measure camshaft lobes as discussed in Topic 44.8. Measure camshaft bearing journals and compare the measurements to the manufacturer's specifications.

Crankshaft and camshaft journals can be *reground*, or refinished, to a smaller size. Camshaft lobes, if worn but not damaged, also can be reground.

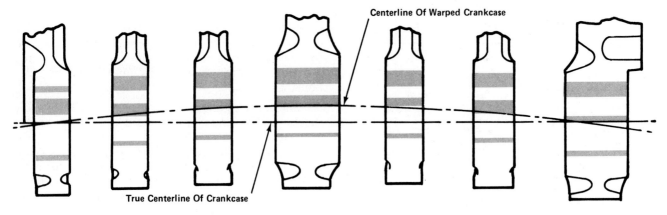

Figure 46-14. Bearing surface measurements.

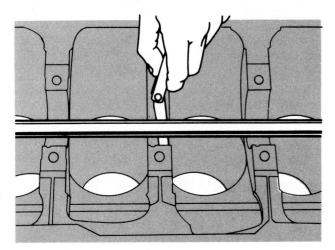

Figure 46-15. Checking crankcase alignment.

Figure 46-16. Cylinder honing tool.

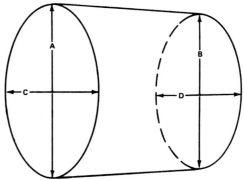

Figure 46-17. Crankshaft journal measurement.
FORD MOTOR COMPANY

Note that the difference in insert thickness is made up in the flexible steel backing itself, while the thin alloy lining layer remains constant.

Figure 46-18. Standard (left) and oversize (right) insert bearings.

Thicker bearings, known as "undersized bearings," are used on undersize crankshaft and camshaft journals **[Figure 46-18]**.

46.8 CYLINDER BLOCK REASSEMBLY

After all machining procedures on the cylinder block have been performed correctly, the engine is ready for reassembly. Cylinder block reassembly can include the following procedures:

- Preparing the cylinder block for reassembly
- Replacing camshaft bearings and camshaft
- Replacing soft plugs
- Replacing crankshaft seals
- Replacing main bearings
- Checking crankshaft end play
- Fitting piston rings
- Replacing piston/connecting rod assemblies
- Replacing connecting rod bearings
- Replacing the oil pump and oil pan
- Replacing the flywheel
- Checking connecting rod side clearance

Cylinder Block and Crankshaft Service

- Replacing the cylinder head
- Replacing timing gears and/or chains on pushrod engines
- Replacing parts and covers.

Preparing the Cylinder Block for Reassembly

The cylinder block is cleaned by scrubbing it with hot water and detergent. A stiff bristle brush and bottle brushes are used to thoroughly scrub the cylinders, crankcase, camshaft bore, and all other openings.

CAUTION!!!

The cylinder block must be thoroughly cleaned. Any traces of abrasives or metal chips will quickly wear cylinder bores, crankshaft and camshaft journals, and bearings. Do not use gasoline, kerosene, or solvent to clean the cylinder block. These materials will not remove traces of dirt and abrasives from the "pores" of the metal.

After washing, swab the bores repeatedly with clean engine oil on a lint-free cloth. Do this until no trace of iron or abrasive particles are left.

Replacing Camshaft Bearings and Camshaft

Camshaft bearings for pushrod engines are available for standard and undersize journals. Undersize journals are the result of machining operations to remove scoring and other problems from the camshaft.

In some cases, the removal tool also can be used to pull or press the camshaft bearings into the block. In other cases, special tools are used to push the camshaft bearings into place.

The camshaft bearing oil holes are aligned with the cylinder block oil galleries before insertion. After bearing installation, check for oil hole alignment [Figure 46-19].

NOTE: Apply a special break-in lubricant to camshaft lobes and bearing journals before installation.

The camshaft timing gear is attached to the camshaft, and the shaft is inserted through the bore. Retaining devices are connected and tightened to the manufacturer's specified torque.

Replacing Soft Plugs

Cylinder block soft plugs are replaced in the same manner as cylinder head soft plugs. Refer to Topic 44.2 for replacement procedures.

Replacing Crankshaft Seals

Several types of rear main bearing oil seals are used in modern automobiles. A *rope-type seal* is used on many engines. After being coated with oil, this type of seal is pressed or rolled into a groove behind the rear main bearing [Figure 46-20]. The corresponding half of this seal is rolled into the rear main bearing cap. A razor blade cutting tool is used to cut off excess seal material flush with the housing.

Replacing Main Bearings

Replacement precision bearing inserts are available for standard or undersize crankshaft main journals. Before installation, the bearing inserts and crankshaft journals are coated with the recommended oil for lubrication when the engine is first started. The tangs and oil holes are aligned correctly, and the upper bearing insert is snapped into place. The lower main bearing inserts also are installed in the main bearing caps. The crankshaft then is lowered carefully into place, taking care to lower it in evenly. Finally, the main bearing caps are installed in their proper location and alignment.

The main bearing cap bolts or nuts are installed finger tight. A block of wood is used to align the thrust bearing flanges, as shown in **Figure 46-21**. One by one from the center outward, each bearing cap is tightened to the manufacturer's specified torque. Rotate the crankshaft to check for proper fit. It should turn smoothly. If not, remove the bearing caps for inspection. Turning the shaft as each cap is removed will help isolate the problem.

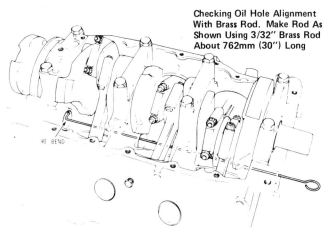

Figure 46-19. Checking oil hole alignment.

CHEVROLET MOTOR DIVISION—GMC

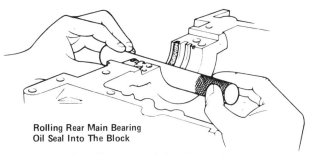

Figure 46-20. Installing rear main bearing seal.

CHEVROLET MOTOR DIVISION—GMC

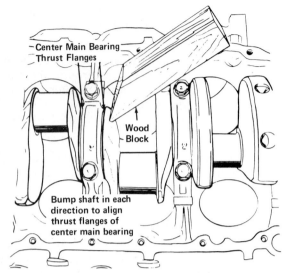

Figure 46-21. Aligning the thrust bearing flanges.
CHEVROLET MOTOR DIVISION—GMC

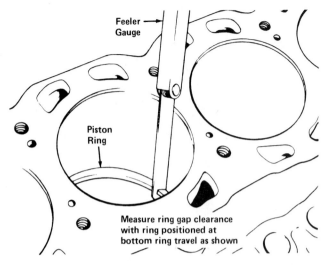

Figure 46-23. Measuring piston ring gap.
CHEVROLET MOTOR DIVISION—GMC

Checking Crankshaft End Play

After main bearing installation, check for proper crankshaft end play, as explained in Topic 46.1. End play also can be checked by using a dial indicator [**Figure 46-22**].

Fitting Piston Rings

Before installation on the pistons, an important check is performed on the piston compression rings and oil-control ring rails. Insert them squarely in the cylinder at the bottom of the ring travel area. Check the end gap of the rings for the proper dimension [**Figure 46-23**].

Also check the piston rings in the piston ring grooves for proper ring side clearance [**Figure 46-24**].

Replacing Piston/Connecting Rod Assemblies

Position the piston ring end gaps so that they are staggered to lessen blowby [**Figure 46-25**]. Refer to the manufacturer's service manual for the correct positioning of ring gaps.

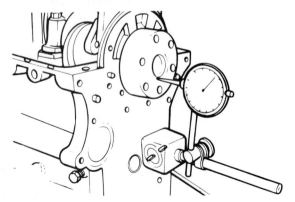

Figure 46-22. Checking crankshaft end play with a dial indicator.
FORD MOTOR COMPANY

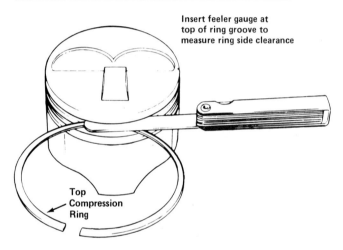

Figure 46-24. Checking ring side clearance.
CHEVROLET MOTOR DIVISION—GMC

Coat the pistons and rings with engine oil. Insert lengths of rubber hose over the connecting rod bolts to protect the cylinder walls and crankshaft journals. Use a *piston ring compressor,* shown in **Figure 46-26,** to squeeze the rings tightly for insertion into the cylinder. Place the upper connecting rod bearing insert in the connecting rod and coat it with lubricant, then install the bearing insert in the rod cap. Align the numbered piston/connecting rod assembly properly over the crankshaft throw and insert into the cylinder. Position offset connecting rods properly. Use the wooden handle of a hammer or mallet to tap the top of the piston until it is inserted in the cylinder.

Replacing Connecting Rod Bearings

Properly align and position the numbered bearing cap. Screw the rod nuts on by hand. Rod nuts must be

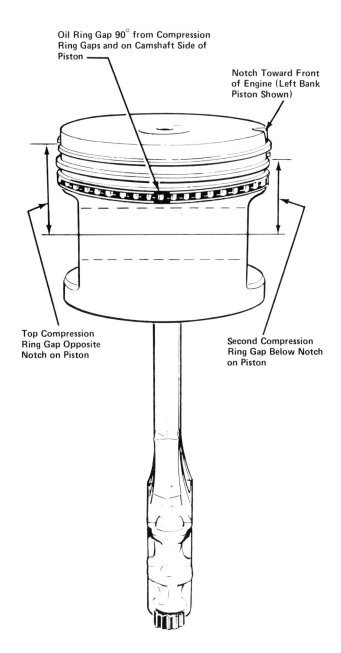

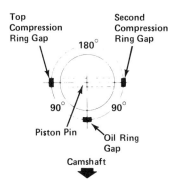

Figure 46-25. Piston ring gap positioning. CHEVROLET MOTOR DIVISION—GMC

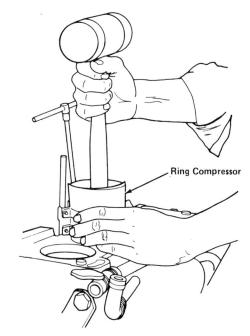

Figure 46-26. Piston ring compressor. FORD MOTOR COMPANY

torqued to the manufacturer's specifications. At this point, the rod should be free to slide back and forth on the journal with finger pressure. If no movement, no matter how slight, is visible, this indicates the bearing is too tight or has foreign material in it and must be disassembled for inspection.

Checking Connecting Rod Side Clearance

When two connecting rod assemblies are mounted on a single crankpin, the side clearance between the rods must be checked, as shown in **Figure 46-27**. Improper clearance can indicate incorrectly installed offset connecting rods.

Replacing the Flywheel

Align the marks so that the flywheel and crankshaft are in their original positions as shown in **Figure 46-28**. Coat threads with thread sealant as recommended by the vehicle manufacturer. Torque the attaching bolts as specified in the service manual.

Replacing the Oil Pump and Oil Pan

Install a new oil pump, oil pump drive shaft if used, and gasket. Install a new oil pan gasket and/or sealant as recommended by the manufacturer. Then bolt the oil pan into place. Torque the oil pan bolts in the proper sequence to the manufacturer's specification.

Replacing the Cylinder Head

Replace the cylinder head and upper valve train parts. Tighten the attaching bolts or nuts in the

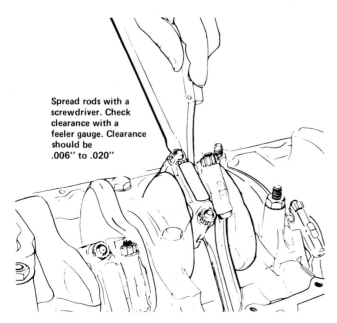

Figure 46-27. Checking connecting rod side clearance.
CHEVROLET MOTOR DIVISION—GMC

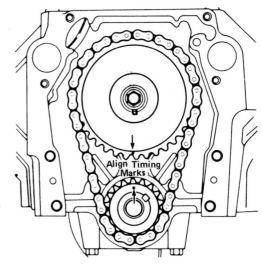

Figure 46-29. Aligning timing gear marks.
CHEVROLET MOTOR DIVISION—GMC

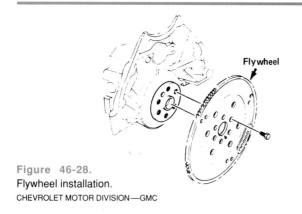

Figure 46-28. Flywheel installation.
CHEVROLET MOTOR DIVISION—GMC

specified order to the manufacturer's specified torque. Perform a valve adjustment when the engine is cold. Refer to Unit 44 for correct procedures.

Replacing Timing Gears and/or Chain

Attach the crankshaft timing gear to the crankshaft. Rotate the crankshaft and camshaft until the timing gear marks are aligned, as shown in **Figure 46-29**. Then install the timing chain.

Replacing Parts and Covers

Use new gaskets and the manufacturer's recommended sealants when removed parts are replaced. Install all parts that do not interfere with engine reinstallation, such as water and fuel pumps. Install the intake manifold and seals **[Figure 46-30]**.

Torque the intake manifold bolts in the proper sequence **[Figure 46-31]**. Install the exhaust manifold in a similar manner.

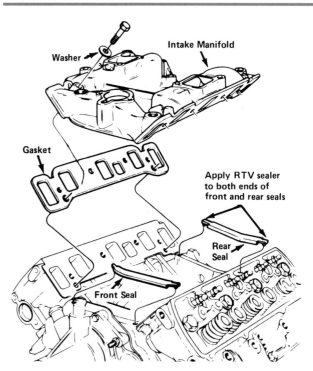

Figure 46-30. Intake manifold installation.
CHEVROLET MOTOR DIVISION—GMC

Apply the manufacturer's recommended sealants to seals and threaded plugs. Tighten covers in the proper sequence to the manufacturer's recommended torque specifications.

46.9 ENGINE REINSTALLATION

Engine reinstallation is essentially the reverse of removal. Lower the engine carefully into place. Connect the motor mounts and tighten them to the manufacturer's recommended torque specification.

CYLINDER BLOCK AND CRANKSHAFT SERVICE

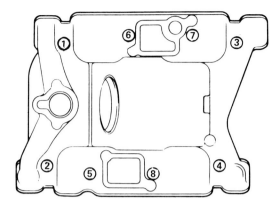

Figure 46-31. Intake manifold torque sequence.
CHEVROLET MOTOR DIVISION—GMC

SAFETY CAUTION

Do not attempt engine reinstallation procedures without proper safety instruction and training from your instructor. The engine is heavy enough to injure you or another person fatally if it falls. Refer to the manufacturer's service manual for correct installation procedures.

Install all remaining engine accessories, such as alternators, power steering pumps, and air conditioning compressors. Reconnect hoses and other plumbing parts. Reinstall electrical wiring and instrument connections.

Fill the engine with the correct amount of oil, and install a new oil filter. Refill the cooling system to the proper level. Install belts on engine accessories and tighten them to the proper tension.

Remove the oil pressure sensor and turn the oil pump drive through the distributor opening with a drill motor until oil appears at the sensor bore. Replace the oil pressure sensor and distributor. Reconnect the ignition system.

Install new spark plugs and reconnect the spark plug wires. Install the distributor and set the ignition timing (dead timing) as accurately as possible, as explained in Topic 35.10.

After a final check to make sure that everything has been reconnected and installed properly, reconnect the battery ground terminal.

Starting Procedure

Start the engine and allow it to run at fast idle until it reaches normal operating temperature. Shut off the engine. Tighten the cylinder head attaching bolts or nuts again to the correct torque specification as necessary. Check the manufacturer's service manual for information.

Check and readjust valve clearance if necessary. Start the engine again, and make final ignition timing and carburetor or fuel injection adjustments.

46.10 ENGINE BREAK-IN PROCEDURE

To properly *seat* piston rings against the cylinder walls, a break-in procedure must be followed. Seating is a process in which parts that *mate*, or fit together, rub against each other until their mating surfaces match almost perfectly. Until the piston rings seat against the cylinder walls, there will be more blowby and higher oil consumption.

To correctly break in the engine, drive the vehicle and accelerate gently to 30 mph (48 km/hr). Then accelerate rapidly until a speed of approximately 50 mph (80 km/hr) is reached. Allow the vehicle to coast down to 30 mph (48 km/hr) again. The acceleration and deceleration procedure is repeated at least 10 times.

This procedure applies a load rapidly to the engine for short periods after the engine has reached operating temperature. The procedure pushes the piston rings against the cylinder walls and speeds the break-in process. In addition, the deceleration creates a vacuum above the piston that draws oil up to the rings.

After the break-in procedure, the vehicle can be driven at normal speeds. However, sustained high speeds must be avoided for the first 100 miles (160 km). After the first 100 miles, the vehicle can be driven normally. The oil and oil filter should be changed after about 500 miles.

UNIT HIGHLIGHTS

- Before the cylinder block is disassembled, main bearing clearance and crankshaft end play are checked.
- Before cleaning, all attached and inserted parts are removed from the cylinder block.
- In general, cylinders in cast-iron cylinder blocks can be overbored in steps of 0.010 inch (0.25 mm), up to a maximum of 0.060 inch (1.52 mm).
- Crankshafts and camshafts may be reground to undersize measurements. Thicker bearings are used to make up for removed material.
- The manufacturer's service manual procedures must be followed to correctly reinstall engines.
- After overhaul, specific procedures must be followed during initial startup, after warmup, and during the first 100 miles of driving.

TERMS

end play	crankshaft end play
sleeve	porosity
epoxy cement	cylinder taper
arbor	line boring

oversize
regrind
rope-type seal
seat

overbore
piston ring compressor
mate

REVIEW QUESTIONS

DIRECTIONS: The following questions are similar to those used on automotive technician certification tests. On a separate sheet of paper, write the letter of the correct choice.

1. Technician A says that main bearing clearances that are smaller in the middle than at either end indicate tapered crankshaft journals.
   Technician B says that main bearing clearances that are smaller at one end than another indicate barrel-shaped or spool-shaped journals.
   Who is correct?
   A. A only
   B. B only
   C. Both A and B
   D. Neither A nor B

2. Which of the following statements is correct?
   I. Different cylinder bore measurements at the top and bottom of the ring travel indicate cylinder taper.
   II. Different cylinder bore measurements at 90 degrees to each other at the top of the ring travel indicate an out-of-round condition.
   A. I only
   B. II only
   C. Both I and II
   D. Neither I nor II

3. How is residue from honing and cylinder machining operations removed from the cylinder block?
   A. By washing with gasoline.
   B. By washing with kerosene.
   C. By washing with solvent.
   D. By washing with water and detergent, then swabbing with clean cloths and motor oil.

4. Technician A says that the valve adjustment must be corrected after a rebuilt engine is allowed to warm up.
   Technician B says that the main and connecting rod bearings must be retorqued after a rebuilt engine is allowed to warm up.
   Who is correct?
   A. A only
   B. B only
   C. Both A and B
   D. Neither A nor B

Question 5 is not like the previous questions. It has the word EXCEPT. For question 5, look for the choice that does *not* apply. Read the entire question carefully before you choose your answer.

5. All of the following must be done before the cylinder block is cleaned EXCEPT
   A. check main bearing clearances.
   B. check crankshaft end play.
   C. measure cylinder block warpage.
   D. remove all attached and inserted parts from the cylinder block.

ESSAY QUESTIONS

1. What steps are involved in cylinder block service?
2. Describe how to measure cylinder bore and what problems can be indicated.
3. Identify and describe the steps that are part of cylinder block reassembly.

SUPPLEMENTAL ACTIVITIES

1. Measure main bearing clearances on a shop engine and inspect the bearings and crankshaft journals. Compare your readings with the manufacturer's specifications. What, if anything, must be done to the crankshaft and/or bearings?
2. Measure end play on a shop engine by using feeler gauges and a dial indicator. Make a report on what differences, if any, you find between the two methods.
3. Measure cylinder dimensions on a disassembled shop engine. Check for out-of-round and cylinder taper conditions. Compare your readings to the manufacturer's service manual specifications. Make a report to your class about the condition of the cylinders and what needs to be done.
4. Measure a shop engine's crankshaft journals. Make a report to your class on the condition of the journals and what needs to be done to the crankshaft.
5. Insert a telescoping gauge into a disassembled shop engine's cylinder, parallel to the crankshaft axis. Lock the gauge. Using both hands, press hard on the sides of the block parallel to the cylinder. What happens, and why? What importance does this experiment have in regard to engine reassembly?
6. Measure the clearance between rings and the ring grooves on a shop piston. Determine what must be done for proper fit.

PART VII: Automotive Drivetrains

Career opportunities for technicians skilled in manual and automatic transmission service and four-wheel drive drivetrains are found in dealerships, fleet service garages, and independent repair shops. Topics covered in this part include drivelines, manual transmissions and transaxles, automatic transmissions and transaxles, differentials, driving axles, and four-wheel drive transfer cases. Here, the technician is reassembling a transmission.

47 Drivelines

UNIT PREVIEW

The torque generated by the engine must be transmitted to the driving wheels to move the vehicle. This can be done in any of several ways, depending upon the design of the vehicle. Engine torque can be used to drive, or rotate, the front wheels, rear wheels, or all four wheels.

Rear-wheel drive (RWD) vehicles transmit torque from the transmission to a differential and driving axles at the rear of the vehicle through a driveline. Truck and utility-type four-wheel drive (4WD) vehicles also transmit torque from a transfer case attached to the transmission through a driveline to a front differential and driving axles.

Most RWD vehicle drivelines use a hollow metal tube and flexible joints to transfer the torque from the transmission to the differential and driving axles. Another RWD driveline design uses a solid metal shaft inside a hollow tube to transfer torque.

Different types of flexible joints can be used to align and connect driveline parts, and transmit torque.

LEARNING OBJECTIVES

When you have completed your assignments and exercises in this unit, you should be able to:

❑ Describe how the driveline operates.
❑ Identify and describe the parts of the driveline.
❑ Describe the operation of a conventional universal joint.
❑ Explain how a constant-velocity universal joint operates.
❑ Describe the operation and purpose of a slip joint.

47.1 DRIVELINE DESIGN

The parts of a rear-wheel *drivetrain* are the clutch or torque converter, manual or automatic transmission, *driveline*, differential, and driving axles. A basic rear-wheel drivetrain is shown in **Figure 47-1**.

The purpose of the driveline is to transfer torque, or turning force, from the transmission to the differential. Most drivelines consist of a hollow tube

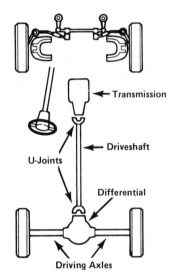

Figure 47-1. Conventional rear-wheel drivetrain.
CHEVROLET MOTOR DIVISION—GMC

or solid metal shaft with one or more flexible joints, as shown in **Figure 47-2**.

Driveshaft

The *driveshaft*, sometimes called the *propeller shaft,* usually is a hollow metal tube. Vehicles with a torque-tube drive system, discussed in Topic 47.3, use a solid metal shaft. Solid metal shafts also are used on vehicles with a front-engine, rear-transaxle design such as the German Porsche 944 and 928 models. (Pontiac used a similar system for the Tempest in the early 1960s.)

As the vehicle moves down the road, the driveline must be able to transfer torque smoothly and continuously. However, the wheels and suspension of the vehicle must move up and down with the contours of the road surface. As the rear suspension moves, the differential also moves up and down. The differential allows the inner and outer driving wheels to rotate at different speeds as the

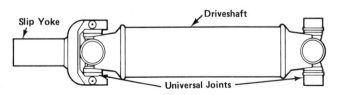

Figure 47-2. Driveline. OLDSMOBILE DIVISION—GMC

vehicle turns corners. Differentials are discussed in Unit 55.

The upward and downward movement of the differential with the suspension causes both the distance and the angle between the transmission and the differential to change. To transfer torque smoothly and continuously, the driveline must perform these three functions:

- Rotate
- Move up and down
- Compensate for varying distances between the transmission and differential.

Universal Joint

The driveline must be able to perform all of these tasks at the same time. However, the driveshaft tube itself can only transmit torque in a straight line. To transfer torque as the angle between the transmission and the differential changes, *universal joints,* or U-joints, are used. A simplified universal joint is shown in **Figure 47-3.**

A U-joint can transfer torque at an angle while rotating. The front U-joint accepts torque from the transmission and transmits the torque to the driveshaft. The rear U-joint accepts torque from the driveshaft and transmits the torque to the differential. Thus, the driveline can move as required by road and driving conditions without interrupting torque transfer.

Slip Joint

The distance between the transmission and the differential changes as the angle between them varies. However, the driveshaft and universal joints cannot stretch or shrink to make up for this changing distance. To allow the driveline to maintain a connection as the distance changes, a *slip joint* is used. A slip joint is formed of two parts. One part is the *splined* end of the transmission output shaft. Splines are machined ridges around or inside a part. The other part of a slip joint is called the *slip yoke.* The slip yoke consists of a hollow, splined metal tube at one end. The other end is the yoke, which connects to the front U-joint. The splined end of the transmission output shaft fits inside the hollow end of the slip yoke **[Figure 47-4].**

Splines act much like gears. When the transmission output shaft rotates, the two sets of splines mesh, or connect. This causes the slip yoke to rotate. At the same time, the slip yoke can slide, or slip, back and forth on the transmission shaft. This sliding action maintains the connection as the distance between the transmission and the differential changes. Slip joints also may be found at connecting points in split driveshafts.

47.2 THE EFFECTS OF TORQUE

As torque is applied by the driveline to the differential, the entire differential and driving axle housing rotates. This effect is known as *rear-end torque.* This torque presents certain problems in the operation of the driveline, differential, and driving axles.

Rear-end torque always moves in a direction that is opposite to rear wheel rotation. The axle housing also tries to turn in a direction opposite to pinion gear movement.

As the vehicle begins to move forward, the front of the differential moves upward. This upward motion could cause the U-joints to bind, or stick, and be damaged **[Figure 47-5].**

To prevent excessive movement of the differential and axle housing, some form of bracing must be used. Common methods of bracing used on modern vehicles are:

- Torque-tube drive
- Hotchkiss drive.

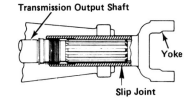

Figure 47-4. Slip joint.

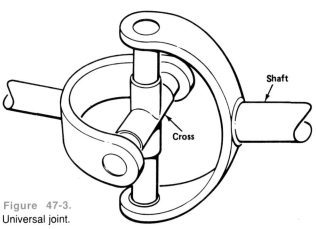

Figure 47-3. Universal joint.

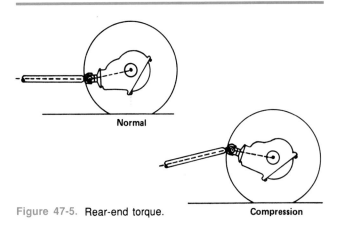

Figure 47-5. Rear-end torque.

47.3 TORQUE-TUBE DRIVE

The idea of the *torque tube* is to keep the driveshaft in perfect alignment with the differential at all times. To do this, a driveshaft housing, or torque tube, is bolted between the differential housing and the transmission or vehicle frame. This direct, rigid connection helps to control rear-end torque. A typical torque-tube system is shown in **Figure 47-6.**

A solid steel driveshaft runs inside the torque tube. This driveshaft attaches to the differential drive pinion by way of a flexible drive coupling, as shown in **Figure 47-7.**

Only one U-joint is used with a torque-tube drive. This U-joint is mounted at the front of the driveshaft, with a slip joint that connects it to the rear of the transmission. In addition, a support bearing is often mounted near the front of the torque tube. The support bearing helps to control whip, or uneven rotation of the driveshaft. Additional braces called *truss rods,* are sometimes used to further locate and hold the torque tube [refer to Figure 47-6].

A torque-tube driveline is extremely heavy. In addition, the entire torque tube, differential, and rear axle assembly must be removed to service the transmission.

Because of its disadvantages, torque-tube drive is used infrequently on modern vehicles. Recent examples of torque-tube drive systems include the Chevrolet Chevette and some Japanese-made RWD passenger vehicles.

47.4 HOTCHKISS DRIVE

The *Hotchkiss drive* uses an open driveshaft. It eliminates the disadvantages of torque-tube drive and uses the vehicle's rear suspension to control rear-end torque. In a Hotchkiss drive system, the differential and driving axle housing can twist slightly because of rear-end torque. However, this motion is kept within safe limits by the rear suspension of the vehicle.

Driveshafts used with Hotchkiss type drives are usually hollow tubes. To help reduce vibration and noise, the hollow driveshaft may have cardboard or rubber elements inside. The universal joints allow torque to be transferred smoothly.

A Hotchkiss drive is more adaptable than torque-tube drive. A Hotchkiss drive system can use longer driveshafts. In addition, Hotchkiss-drive driveshafts can be made in separate sections, connected by additional universal and slip joints.

A Hotchkiss drive can be used either with leaf springs, shown in **Figure 47-8,** or with coil springs. When Hotchkiss drive is used with coil springs, additional braces, called *control arms* must be used [Figure 47-9].

Leaf springs are attached to the frame or subframe by brackets located ahead of and behind the rear axle housings. The leaf springs also attach to the rear axle housings with U-shaped bolts.

When rear-end torque begins, the axle housing rotates up at the front. This action flexes, or bends,

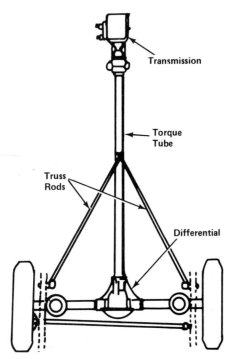

Figure 47-6. Torque-tube drive.

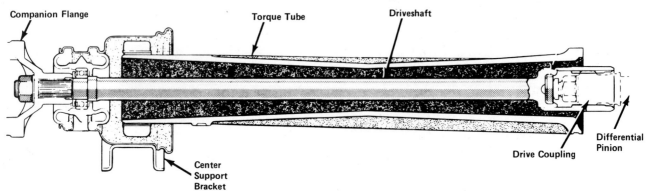

Figure 47-7. Torque-tube assembly. PONTIAC MOTOR DIVISION—GMC

DRIVELINES 515

the springs. When the springs are flexed to their limits, rear axle movement stops, and rear-end torque is controlled. Force is transferred from the axle housings to the vehicle's frame through leaf spring mounts or control arms. This force moves the vehicle forward.

47.5 DRIVESHAFT

The lengths and diameters of driveshafts vary, simply because all vehicles are not alike. Many factors determine what type of driveshaft system is used. The physical size and *wheelbase* of the vehicle are prime considerations. Wheelbase is the distance from the axle center of the front wheel to the axle center of the rear wheel.

While most vehicles have one-piece driveshafts, some vehicles have *split driveshafts* that consist of two or more connected shorter shafts. A split driveshaft can be angled upward or downward from the transmission to reduce the size of the hump that runs down the middle of the vehicle floor. Larger vehicles, such as trucks, use split driveshafts because of greater distances between their transmissions and differentials [Figure 47-10].

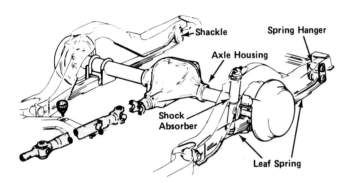

Figure 47-8. Hotchkiss drive with leaf spring.
CHEVROLET MOTOR DIVISION—GMC

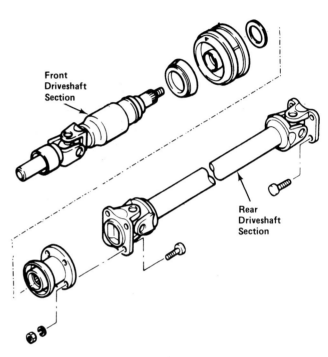

Figure 47-10. Split driveshaft. NISSAN MOTOR CORPORATION

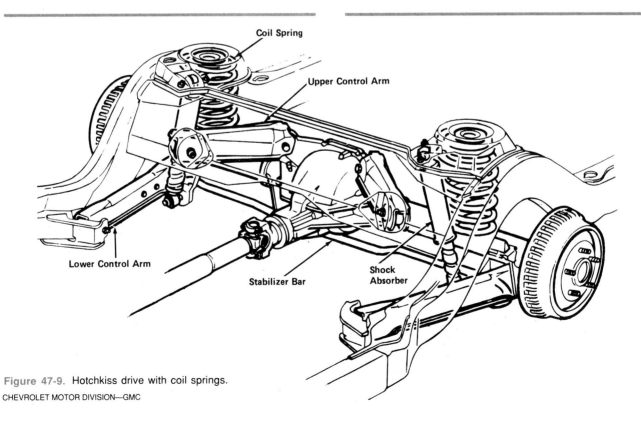

Figure 47-9. Hotchkiss drive with coil springs.
CHEVROLET MOTOR DIVISION—GMC

A driveshaft must be strong and well balanced. The driveshaft receives and transfers great twisting forces from the transmission to the differential. Because the driveshaft spins rapidly, it must be in balance.

Small pieces of metal are welded to the driveshaft to correct imbalances. An unbalanced driveshaft will vibrate, cause noise, and wear out other drivetrain parts.

47.6 CENTER SUPPORT BEARING

Vehicles with split driveshafts must use a center support bearing to support and align the sections of the driveshaft.

A center support bearing also may be used with a one-piece driveshaft. The bearing minimizes whip and vibration and allows the driveshaft to rotate smoothly. Center support bearings are used on a variety of vehicles, from expensive luxury sedans to imported economy cars. A typical center support bearing is shown in **Figure 47-11**.

47.7 UNIVERSAL JOINT

The conventional universal joint, or *Cardan joint*, is a double-hinged joint. This means that the U-joint can swivel in two different directions. A U-joint consists of two metal *yokes* and a cross, or spider. A yoke is a U-shaped piece used to attach one part to another **[Figure 47-12]**.

Two *trunnions*, or arms, of the cross-shaped spider connect to the driving yoke. The other two trunnions connect to the driven yoke. The driving yoke applies torque. The driven yoke accepts torque.

Small, lubricated roller bearings, called needle bearings, allow the joint to swivel and transfer power smoothly. A seal between the bearings and the spider helps to keep contaminants out and lubricant in **[Figure 47-13]**.

Conventional Universal Joint

An undesirable characteristic of a conventional universal joint is that as it changes angle during rotation, the torque applied to the driven yoke changes. The driving yoke rotates at a constant speed. However, the driven yoke speeds up and slows down twice during each rotation.

This change in speed could cause pulsing and vibration in the driveline. As the angle of the yokes increases, the changes in speed occur more frequently. However, the second U-joint at the differential end of the driveshaft cancels out the uneven rotation. The rear U-joint produces a reaction that is opposite to the reaction produced at the front U-joint. That is, the rear U-joint slows down as the speed increases in the front U-joint. Also, the rear U-joint will speed up when the front U-joint slows down. The final result is a fairly constant rate of driveshaft rotation.

Double Cardan (Constant Velocity) Universal Joint

As U-joint angles increase, driveline speed variations and vibration increase. In addition, conventional U-joints are limited as to the angle at which they can operate.

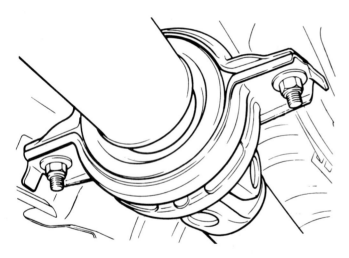

Figure 47-11. Center support bearing. CHRYSLER CORPORATION

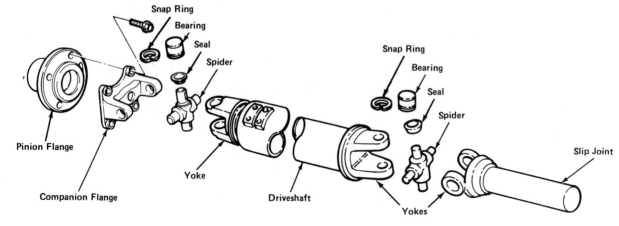

Figure 47-12. Universal joint assembly. CHRYSLER CORPORATION

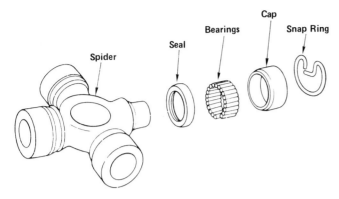

Figure 47-13. Parts of a universal joint.
CHEVROLET MOTOR DIVISION—GMC

When driveshaft angles become too great for conventional U-joints, a *constant-velocity U-joint* is used. A constant-velocity joint maintains a relatively constant speed during rotation, unlike a conventional U-joint. One or more constant-velocity U-joints may be used anywhere along the driveshaft.

The type of constant-velocity U-joint used on most rear-drive vehicles is called a *double Cardan joint*. The double Cardan joint is actually two conventional U-joints that are closely connected. **Figure 47-14** shows a double Cardan joint. In a double Cardan joint, each U-joint flexes only half as much as a conventional U-joint to transmit torque through a given angle.

Since the constant-velocity U-joint has a driving yoke and a driven yoke, it still changes speed as it rotates. However, because it contains two conventional U-joints, this speed change cancels itself out.

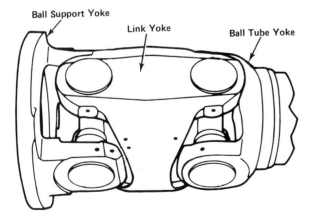

Figure 47-14. Double Cardan universal joint assembly.
CADILLAC MOTOR CAR DIVISION—GMC

UNIT HIGHLIGHTS

- Engine torque can be used to drive, or rotate, the front, rear, or all four wheels, depending on the design of the vehicle.
- A RWD drivetrain accepts and controls engine torque, and transfers it through a driveline to the driving wheels. A 4WD drivetrain also transmits torque from a transfer case attached to the transmission through a driveline to two additional driving wheels.
- A RWD driveline consists of the driveshaft and universal joints, or a torque tube assembly.
- A universal joint transfers torque at an angle while rotating.
- Some drivelines use a center support bearing for support, alignment, and reduction of vibration.
- A slip joint allows two rotating parts to remain connected as the distance between them changes.
- A double Cardan joint is used when the angle between the differential and transmission is too great for conventional universal joints.
- Some vehicles use split drivelines to transmit torque through greater distances or at a sharper angle.

TERMS

drivetrain
driveshaft
slip joint
torque-tube drive
wheelbase
yoke
trunnion
spline
control arm
constant velocity U-joint

driveline
universal joint
rear-end torque
Hotchkiss drive
split driveshaft
slip yoke
double Cardan U-joint
truss rod
Cardan joint

REVIEW QUESTIONS

DIRECTIONS: The following questions are similar to those used on automotive technician certification tests. On a separate sheet of paper, write the letter of the correct choice.

1. Hotchkiss drive
    A. uses the vehicle's rear suspension to control rear-end torque.
    B. uses a driveshaft inside a rigid tube.
    C. never uses more than two U-joints.
    D. is never used with long or split driveshafts.

2. Two conventional U-joints are used on most drivelines because
    A. all drivelines have two U-joints.
    B. two U-joints reduce uneven rotation.
    C. no driveline has more than two U-joints.
    D. two U-joints have to bend only half as much as a single U-joint.

Questions 3, 4, and 5 are not like those above. Each has the word EXCEPT. For each question, look for the choice that does *not* apply. Read each question carefully before you choose your answer.

3. All of the following statements are correct EXCEPT
    A. the function of the driveline is to transfer torque.
    B. drivelines are used on RWD and 4WD vehicles.
    C. the driveshaft is either a solid steel shaft or a hollow tube.
    D. all drivelines have two U-joints.

4. All of the following statements about torque-tube drive are correct EXCEPT
    A. the torque tube is bolted directly to the differential.
    B. a torque-tube drive system is extremely heavy.
    C. a hollow steel driveshaft runs inside the torque tube.
    D. only one U-joint is used with a torque-tube drive.

5. All of the following statements are correct EXCEPT
    A. the driveshaft rotates.
    B. the driveshaft moves up and down.
    C. the driveshaft remains connected as the distance between the transmission and the differential changes.
    D. the driveshaft transfers power at an angle.

ESSAY QUESTIONS

1. Identify and describe the parts of a driveline.
2. Describe the methods used to control and apply torque to move a rear-wheel drive vehicle.
3. What are the differences between a conventional universal joint and a double Cardan joint?

SUPPLEMENTAL ACTIVITIES

1. Identify the parts of the drivetrain on a vehicle in the shop.
2. Explain why two universal joints are used on Hotchkiss drive systems.
3. Examine a driveline on a vehicle in the shop. Identify the parts of the driveline, including any balancing weights.
4. Name the parts of a universal joint.
5. Explain why a double Cardan universal joint is used.
6. Explain why split drivelines are used.

48 Driveline Service

UNIT PREVIEW

Preventive inspections and service are important to the safe and reliable operation of the drivetrain. The driveline is a relatively simple and rugged part of the drivetrain. However, its parts are subject to wear and damage.

The driveline must be inspected at regular intervals to spot defects and prevent potential problems. Driveline preventive maintenance includes cleaning, inspection, and lubrication. Driveline service includes universal joint replacement and rebalancing.

LEARNING OBJECTIVES

When you have completed your assignments and exercises in this unit, you should be able to:

- ❏ Describe the precautions necessary when servicing the driveline.
- ❏ Clean, inspect, and lubricate a driveline.
- ❏ Check driveshaft inclination and runout.
- ❏ Remove and replace a driveline.
- ❏ Balance a driveshaft.

SAFETY PRECAUTIONS

Personal injury and unnecessary damage to parts can be avoided by following the precautionary steps discussed here.

Driveshaft preventive maintenance and service require that the vehicle be raised and supported safely. *Never* work under a vehicle unless you are absolutely sure that it cannot move or fall. No type of jack is meant to support the weight of a vehicle safely. Safety stands, or jack stands, must be used if the vehicle is not on a hoist.

The underside of the vehicle is dirty. Dirt, grime, or foreign objects can fall into the eyes during service procedures. Safety glasses or goggles must be worn to protect the eyes from injury.

Hot exhaust system parts on vehicles can cause severe burns. Be careful when working near exhaust system parts, and let the vehicle cool off, whenever possible, before servicing.

The driveline is heavy. If it drops, it can cause severe personal injury or damage to the parts. Have a helper ready to assist in removing or replacing the driveline.

Some driveline service procedures involve working under a running vehicle, near the spinning driveline. Loose clothing or long hair can be caught by the moving driveline, causing serious injury. Use extreme caution near the spinning driveline and/or wheels.

 ## TROUBLESHOOTING AND MAINTENANCE

Driveline preventive maintenance includes:

- Cleaning
- Inspection
- Lubrication.

SAFETY CAUTION

When working under a vehicle, the vehicle must be supported properly by jackstands or on a hoist. *Never* work under a vehicle supported only by a jack.

Cleaning the Driveline

Shut off the engine. Safely raise and support the vehicle. Shift the transmission of the car into neutral, and release the emergency brake so that the driveline can be rotated by hand during inspection and service.

SAFETY CAUTION

Be sure the vehicle to be inspected or serviced is cooled down. Hot exhaust system parts can cause severe burns. Driveline parts can have sharp edges that will cut. Move your hands slowly and carefully over the driveline during inspection to prevent cuts and scrapes.

Remove any dirt, mud, undercoating, or other foreign material from the driveline with a putty knife or scraper and a wire brush. These accumulations can cause the driveline to become unbalanced. Turn the shaft by hand to clean all areas of the driveline.

Driveline Inspection

Use a droplight to inspect the driveline. Grasp the driveshaft near the front universal joint, or U-joint. Try to move the U-joint and slip joint up and down. Excess movement indicates a bad transmission bushing or U-joint **[Figure 48-1]**.

Shake and twist the driveshaft at both ends. Check for erratic, binding rotation and/or clicking as you try to move the U-joint.

Look for evidence of powdery red rust around the U-joint trunnion grease seals. Rust indicates that the seals have failed and that the bearing surfaces have become rusted. A rusted U-joint must be replaced, as discussed in 48.2.

Examine the seal in the transmission around the slip joint. Check for leaks, cracks, or other signs of seal damage.

Check that the driveshaft balancing weight is not loose or missing **[Figure 48-2]**. The balancing weight usually is located at the rear of the driveshaft.

Check visually and manually for dents, wrinkles, or cracks in the driveshaft tube. Very shallow dents are not necessarily serious. However, deep dents, wrinkles, or cracks in the driveshaft are signs of serious damage. Finally, check the center and rear universal joints for looseness, cracks, or damage.

Lubricating Universal Joints

Factory-installed U-joints on passenger vehicles usually are sealed, and cannot be lubricated with a grease gun. Grease fittings are included on most replacement U-joints **[Figure 48-3]**.

Look for grease fittings on the front U-joint, behind the transmission, and on the rear U-joint, ahead of the differential. Some vehicles may have a third U-joint and grease fittings in the middle of the driveshaft **[Figure 48-4]**.

Inspect each universal joint for grease fittings. Refer to a lubrication chart or the proper service manual for specific lubrication procedures. Some U-joints may have plugs that can be removed to insert a nozzle for greasing. Sealed U-joints are lubricated before installation and cannot be serviced. Lack of lubrication in a sealed U-joint requires that the U-joint be replaced.

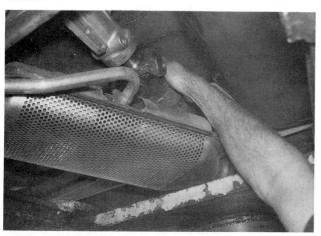

Figure 48-1. Twisting and shaking the front universal joint to check it for looseness.

Figure 48-3. A universal joint with grease fittings

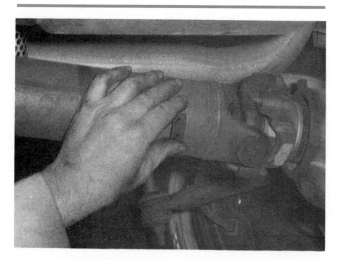

Figure 48-2. Checking to make sure driveshaft balancing weight is still attached.

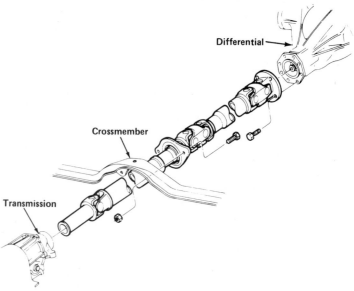

Figure 48-4. Split driveshaft with three universal joints.
CADILLAC MOTOR CAR DIVISION—GMC

Use a grease gun to pump grease slowly into grease fittings on the universal joints. When grease begins to appear at all four bearing cups, stop pumping. Too much pressure can damage the U-joint seal. No grease at one or more bearing cups may indicate clogged grease passages within the trunnions. (Clogged passages cannot be cleared without removing the U-joint. Removal and replacement of U-joints is discussed in 48.2.) Wipe off any excess grease with a rag.

Grease all of the fittings in the driveline in the same manner. If grease cannot be pumped through a fitting, the fitting may be clogged with dirt or other foreign material. Replace clogged fittings and attempt to grease the unit again. If no grease can be pumped through a unit, the unit must be replaced.

Make a thorough inspection for grease fittings. Lack of lubrication can cause U-joint and/or driveline failure.

48.1 DIAGNOSING DRIVELINE PROBLEMS

The technician is responsible for properly diagnosing problems before expensive work is begun. Driveshaft mechanical problems include:

- Incorrect U-joint angles
- Excessive driveshaft runout.

U-Joint Angle

U-joints transmit power through a limited range of angles. Manufacturers specify the correct driveshaft angle, or inclination, for each vehicle. To check front or rear U-joint angles, a special tool, called an *inclinometer* is used [Figure 48-5].

Clean all surfaces where the inclinometer will be placed. Hold the inclinometer on the driveshaft near the U-joint to be checked and read the angle indicated.

If the angle is incorrect, shims, or thin spacers, can be added to or removed from transmission and rear axle housing mounts. See the vehicle manufacturer's service manual for specific procedures.

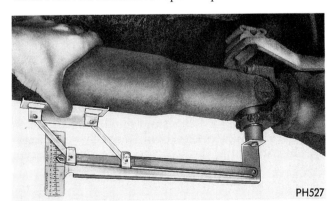

Figure 48-5. Driveshaft angle (inclination) is checked with an inclinometer. CHRYSLER CORPORATION

CAUTION!!!

Do *not* use force when positioning the inclinometer. Hold the instrument firmly but gently to obtain correct readings.

Driveshaft Runout

A bent or damaged driveshaft can cause vibration and noise. A driveshaft in good condition will have very little *runout* during rotation. Runout is the distance that an edge or surface of a rotating part moves in or out during rotation. Excessive runout can cause vibration and noise.

Figure 48-6 shows a setup for measuring driveshaft runout. Runout readings are taken at the center, and near both ends, of the driveshaft with a dial indicator.

Before attempting to measure driveshaft runout, clean all dirt and/or foreign material from the areas where the plunger of the dial indicator will ride. Mount the dial indicator base on the frame or underbody of the vehicle, according to the vehicle manufacturer's specifications. Position the dial indicator so that the plunger contacts a smooth and clean area of the driveshaft. Rotate the driveshaft slowly by hand.

The plunger of the dial indicator will move in or out as the driveshaft moves toward, or away from, the indicator. As the plunger is moved, the dial indicator will record how much runout is present. An acceptably true, or straight, driveshaft will move a few thousandths of an inch (hundredths of a millimeter) toward or away from the dial indicator.

Refer to the vehicle manufacturer's service manual for specific runout limits. If the readings are greater than the manufacturer's allowed runout specifications, the driveshaft must be replaced.

Figure 48-6. Driveshaft runout is checked with a dial indicator.
NISSAN MOTOR CORPORATION

48.2 DRIVELINE SERVICE

The most frequent driveline service is replacement of universal joints, which requires that the driveline be removed from the vehicle.

SAFETY CAUTION

Driveline assemblies are heavy and awkward to carry. Use proper care and have a helper to assist when you remove and replace the driveline. Place an oil catch pan at the rear of the transmission to prevent lubricant from dripping on the shop floor. Install a special transmission rear plug or spare slip joint to keep the lubricant from draining while you service the driveline.

Raise and support the vehicle safely on a hoist. Mark the parts for reassembly. If the driveline parts are not reassembled in the same positions relative to each other and to the transmission and differential, vibration can occur. Use paint or scratchmarks to mark the front and rear parts of the driveline for reassembly **[Figure 48-7]**.

Remove Center Bearing Support Bolts

To remove the driveshaft of a vehicle equipped with a center bearing, the bearing support assembly must be removed **[Figure 48-8]**.

NOTE: Make matchmarks on the separate parts of a split driveline for reassembly.

Support the driveline as the center support bolts are removed. Use a jackstand to support the driveline, or have a helper hold it.

SAFETY CAUTION

Support the front of a split driveline to prevent it from separating and falling. If the driveline falls, it can be damaged or cause injury.

Remove Driveline

The rear universal joint is held to the differential companion flange by U-bolts or straps. Loosen the nuts or bolts and slide the driveline forward toward the transmission. Then lower the driveline and pull it off the transmission output shaft.

Place a plug on the transmission output shaft to prevent the lubricant from leaking **[Figure 48-9]**.

Remove and Replace Center Bearing/Support Assembly

Some center support bearings are made in one piece with the rubber support assembly, and must be replaced as a unit. Other center support bearings are separate from the support assembly, and can be replaced. A hydraulic press is used to push out the old bearing and push in the new one **[Figure 48-10]**.

Remove and Replace Universal Joint

Scratch or paint matchmarks on the U-joint bearing caps, trunnions, and the driveshaft yoke for reassembly. Remove all snap rings from the yoke openings.

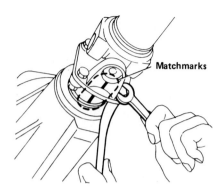

Figure 48-7. Matchmarks on the driveshaft assembly indicate how it is to be reassembled. NISSAN MOTOR CORPORATION

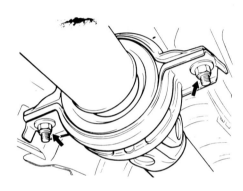

Figure 48-8. Remove the center bearing by unfastening the nuts and bracket. CHRYSLER CORPORATION

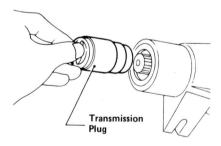

Figure 48-9. Installing a plug at the rear of the transmission extension housing to prevent fluid from leaking out. NISSAN MOTOR CORPORATION

Press out the U-joints from the driveshaft yokes, using a vise and sockets. Using a small socket, push the bearing cups through the yoke, into a larger socket, as the vise is tightened **[Figure 48-11]**.

The vise and sockets also are used to push the new U-joint into the yoke. After reassembly, check to see that each U-joint swivels easily in all directions. Then install new snap rings in the yoke openings.

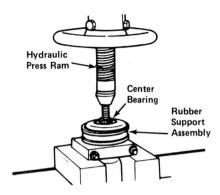

Figure 48-10. A support fixture holds the center bearing assembly while a press is used to remove and install the bearing. NISSAN MOTOR CORPORATION

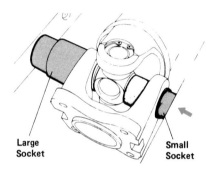

Figure 48-11. A vise and different size sockets are used to press universal joints out of the yoke. CHRYSLER CORPORATION

A driveline with a double Cardan joint requires special tools and procedures for disassembly and reassembly. In addition, whenever a double Cardan joint is disassembled, the U-joints must be replaced. Consult the vehicle manufacturer's service manual for required tools and specific procedures.

Replace Driveline

After both U-joints have been installed, the driveline can be replaced in the vehicle. Align matchmarks on split driveshafts and slide the sections together.

Support split driveshafts to prevent them from bending during installation. Slide the slip joint yoke into the transmission end. Align the matchmarks on the differential companion flange and the rear driveshaft yoke. Thread the fasteners a few turns to hold the parts loosely.

Support the driveline, and thread the center bearing support bolts a few turns to hold the support loosely. Tighten the differential companion flange bolts to the vehicle manufacturer's torque specification. Finally, tighten the center support bearing bolts to the vehicle manufacturer's torque specification.

48.3 DRIVELINE BALANCING

During driveline manufacture, small variations in thickness of parts can occur. These variations can cause one side of the driveline to be heavier than the other side. During rotation, centrifugal force will cause the heavy side to be pulled outward and create vibration.

The driveshaft portion of the driveline is balanced during manufacture by welding small balancing weights to the lighter side of the driveshaft. However, balancing weights can come off, causing the driveshaft to become unbalanced.

An unbalanced driveshaft will cause vibration and noise in the driveline. If a technician notices that a balancing weight is missing, the driveshaft must be rebalanced. Mark driveline parts for reassembly, then remove the driveline from the vehicle and remove the U-joints.

Driveshaft balancing is done by an automotive machinist in a driveshaft specialty shop. The machinist welds steel weights to the shaft to bring it back into balance. After balancing, serviceable or new U-joints are installed, and the driveline is replaced in the vehicle.

UNIT HIGHLIGHTS

- Visual inspection can reveal driveline problems, such as worn or damaged universal joints and driveshafts, and missing balance weights.
- Regular lubrication can prevent universal joint failure.
- The angle of the front and rear U-joints is important for reliability and smooth operation of the driveline.
- Driveline vibration can be caused by damage, imbalance, or excessive runout.
- Driveshafts should be rebalanced by an automotive machinist.

TERMS

inclinometer   runout

REVIEW QUESTIONS

DIRECTIONS: The following questions are similar to those used on automotive technician certification tests. On a separate sheet of paper, write the letter of the correct choice.

1. Which of the following statements is correct?
   I. Damage to the driveshaft can result in vibration.
   II. A lost balancing weight can result in vibration.
   A. I only
   B. II only
   C. Both I and II
   D. Neither I nor II

2. Technician A says that some center bearings are replaced as a unit with their rubber support assembly.
   Technician B says that some center bearings can be pressed out from rubber support assemblies and replaced separately.
   Who is correct?
   A. A only
   B. B only
   C. Both A and B
   D. Neither A nor B

3. Which of the following is done FIRST?
   A. Remove U-joints
   B. Mark driveline parts for reassembly
   C. Balance driveline
   D. Remove driveline

Questions 4 and 5 are not like those above. Each has the word EXCEPT. For each question, look for the choice that does *not* apply. Read each question carefully before you choose your answer.

4. To find a driveline problem, all of the following might have to be done EXCEPT
   A. lubrication.
   B. visual inspection.
   C. checking U-joint angles.
   D. measuring driveshaft runout.

5. Driveline preventive maintenance includes all of the following EXCEPT
   A. cleaning.
   B. inspection.
   C. U-joint replacement.
   D. lubrication.

ESSAY QUESTIONS

1. What steps are involved in driveline preventive maintenance?
2. Describe measurement tests that can indicate driveline problems.
3. What steps are necessary to replace universal joints?

SUPPLEMENTAL ACTIVITIES

1. Perform a visual inspection and lubrication of a vehicle's driveline.
2. Measure front and rear U-joint angles on a vehicle.
3. Measure driveshaft runout on a vehicle.
4. Under the supervision of your instructor, mark parts for reassembly, then remove a driveline.
5. Under the supervision of your instructor, remove and replace universal joints and reinstall a driveline.

49 Manual Transmissions And Transaxles

UNIT PREVIEW

A transmission is used in a rear-wheel drive (RWD) vehicle. In a manual transmission, the driver controls the meshing of different sets of gears by hand. The gears increase or decrease engine torque for different road speeds and load conditions.

Front-engine, front-wheel drive (FWD) automobiles use manual transaxles in place of transmissions. A transaxle combines transmission and differential assemblies in a single unit (see Unit 55).

LEARNING OBJECTIVES

When you have completed your reading assignments and exercises in this unit, you should be able to:

❏ Explain how gears are used to increase or decrease torque and speed.
❏ Identify and describe the main parts and operation of a manual transmission and transaxle.
❏ Follow the flow of power through transmission gears.
❏ Explain how torquing-over affects powertrain operation.
❏ Describe the effects of torque steer and how it can be corrected.
❏ Identify and describe different types of constant-velocity joints.

49.1 TRANSMISSION DESIGN

An automobile must be able to perform well under many different conditions. Automobiles must be able to move away from a resting position smoothly and easily and accelerate well. They also must be able to carry extra weight or passengers when necessary. Automobiles must be able to go up or down steep grades and to travel in reverse.

To allow the vehicle to do all these jobs, varying amounts of engine torque must be applied to the drivetrain. A transmission allows engine torque to be increased or decreased before it is transmitted to the rest of the drivetrain.

A manual transmission is an arrangement of shafts and gears inside a case, which is made of aluminum for passenger vehicles, and may be made of cast iron for trucks and utility vehicles. A cutaway view of a transmission is shown in **Figure 49-1**.

Torque

To move a vehicle, sufficient engine torque, or twisting force, must be transmitted through the drivetrain to the driving wheels. However, the torque produced by an engine at low rpm is not enough to move the vehicle easily.

If the engine were coupled directly to the driveline, it would be difficult to move the vehicle from rest. As power was engaged to move the vehicle, the engine would tend to stall and die. High crankshaft rpm would be necessary to provide enough torque to move the vehicle away from rest easily.

Acceleration would be jerky and poor, and the vehicle would perform well at only one speed. The vehicle would not go up steep hills or grades. In addition, there would be no way to make the vehicle move in reverse.

To make the vehicle perform well under many speed and load conditions, a transmission is used. A transmission allows the driver to increase or decrease engine torque, obtain different road speeds, and move a vehicle in reverse.

Gear Ratios

A *gear* is a toothed wheel that fits into another toothed wheel **[Figure 49-2]**. Suppose two gears with the same number of teeth are meshed, or connected. If one gear (the driving gear) is turned, the other gear (the driven gear) will rotate at the same speed. For each revolution of the driving gear, the driven gear will rotate one revolution **[Figure 49-3]**.

The comparison between the relative number of turns and turning speed of two gears is known as the *gear ratio.* For two gears that have equal numbers of teeth, the gear ratio is 1:1 (read: one to one). For each turn of one gear, the other gear will make one turn. In addition, both gears will turn at the same speed.

However, suppose two gears with different numbers of teeth are meshed. These gears do *not* rotate at the same speed, or through an equal number of rotations. In **Figure 49-4,** a smaller driving gear of 10 teeth is meshed with a larger driven gear of 20 teeth. The smaller driving gear must make two revolutions to turn the larger gear through a single revolution. The smaller gear must turn twice as fast as the larger gear. The gear ratio of such a set of gears is 2:1.

To determine the gear ratio of two meshed gears, first count the number of teeth on each gear. Then

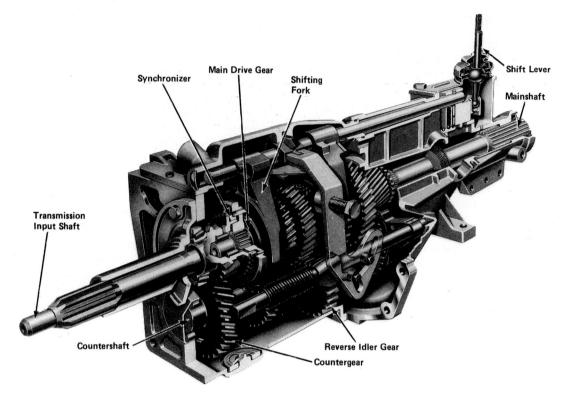

Figure 49-1. Parts of a manual transmission.

divide the number of teeth on the driving gear into the number of teeth on the driven gear. The answer tells you the gear ratio. For example, say the driving gear of a meshed set has 10 teeth and the driven gear has 25 teeth. Divide 10 into 25, which gives a quotient of 2.5 (25 ÷ 10 = 2.5). Thus, the gear ratio is 2.5:1. The smaller (driving) gear will rotate 2.5 times to turn the larger (driven) gear around once.

Notice that the value of the driving gear in a gear ratio is expressed as the first number. A ratio simply is another form for expressing a fraction. The ratio 2.5:1 has the same value as the fraction 2.5/1, or 25/10.

Torque on Gears

Torque on a gear is measured in a straight line from the center of a tooth to the gear *axis*. The axis is the center of the gear, around which the gear rotates. Say that 10 pounds of force is exerted on a gear tooth one foot from the axis of the gear. In this case, 10 pounds-feet (lb.-ft.) of torque is exerted at the axis of the gear. More precisely, 10 lb.-ft. of torque is exerted at the axis of the shaft holding the gear [Figure 49-5].

Figure 49-2. A spur gear.

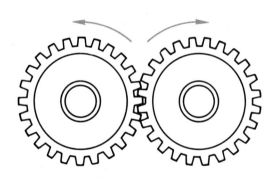

Figure 49-3. Meshing gears of the same size.

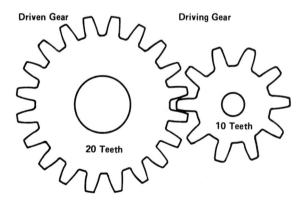

Figure 49-4. Meshing gears of different sizes.

There is an easy method for finding the torque on a shaft in customary units. Simply multiply the pounds of force on a gear tooth by the distance from the tooth to the shaft axis. If the distance is measured in feet, the torque is expressed in pounds-feet (lb.-ft.). If the distance is measured in inches, the torque is expressed in pounds-inches (lb.-in.).

Similarly, force can be applied to the teeth of a gear mounted on a shaft by turning the shaft. Say that 10 lb.-ft. of torque are applied to such a shaft. Also say that the distance from the axis of the shaft to the gear teeth is one foot. In this case, 10 pounds of force will be exerted on each gear tooth **[Figure 49-6]**.

Meshing another gear with this driving gear results in 10 pounds of force being exerted on each tooth of the second gear. If both gears are of the same size, the torque on each gear shaft is the same.

However, if a driving gear is meshed with a larger driven gear, torque on the driven gear shaft is greater. Suppose that the distance from the teeth to the axis of the larger driven gear is 2 feet. The resulting torque on the driven gear shaft is 20 lb.-ft. To determine torque exerted on a shaft, multiply the force on the gear teeth by the distance to the axis **[Figure 49-7]**:

10 lb-ft **x** 2 = 20 lb.-ft. of torque.

In this example, the torque on the larger gear's shaft is twice the torque on the smaller gear's shaft. Using larger driven gears in this way, it is possible to greatly multiply, or increase, an original amount of torque.

In this same example, the larger gear and its shaft turn at one-half the speed of the smaller gear. However, the shaft of the larger gear is turned with twice the torque of the smaller gear. A reduction in speed from a smaller driving gear to a larger driven gear increases torque.

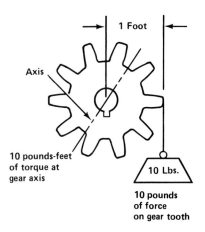

Figure 49-5. Torque on gears.

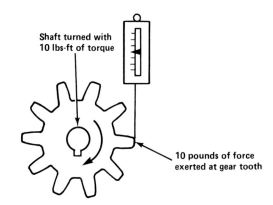

Figure 49-6. Torque applied on a single gear tooth.

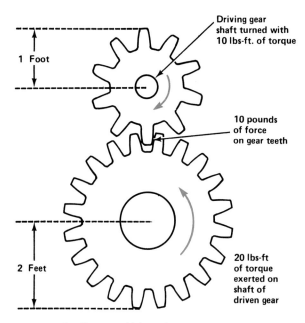

Figure 49-7. Smaller gear driving a larger gear.

Overdrive Ratios

It is also possible to make the larger of two gears the driving gear instead of the smaller. For example, in **Figure 49-8** the larger driving gear has 20 teeth, and the smaller driven gear has 10 teeth. Dividing 20 into 10 gives the gear ratio: 0.5:1. This is known as an *overdrive* gear ratio. Overdrive means that the speed of the driven gear is greater than, or over, the speed of the driving gear.

In an overdrive gearset, the torque on the driven gear is reduced rather than increased. The torque on the driven gear shaft in the example above is only half the torque on the driving gear.

However, the speed of the driven gear is twice that of the driving gear. Overdrive gearing is used to increase the speed of the driven gear and shaft.

An increase in speed from a larger driving gear to a smaller driven gear reduces torque.

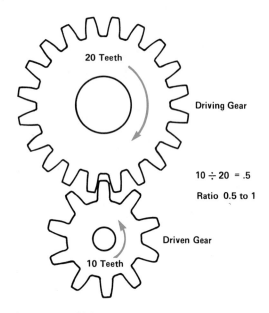

Figure 49-8. Larger gear driving a smaller gear.

49.2 MANUAL TRANSMISSION CONSTRUCTION

A manual transmission consists of parallel sets of metal shafts on which meshing gearsets of different ratios are mounted. By moving the shift lever, gear ratios can be selected to produce different amounts of torque multiplication. The shafts and gears are mounted inside a metal case made of aluminum or cast iron [Figure 49-9]. The case contains lubricant to reduce friction between the meshing gears and between shaft bearing surfaces.

Vehicle manufacturers specify different lubricants for different vehicles. Current lubricants for a manual transmission include multi-weight gear oil, automatic transmission fluid, and engine oils. Always refer to the manufacturer's service manual for correct lubricant recommendations.

Engine torque is connected to, or disconnected from, the transmission by engaging or disengaging the clutch. The construction and operation of the clutch are discussed in Unit 51. When the clutch is engaged, engine torque turns the *transmission input shaft* [Figure 49-10].

A *main drive gear* is attached to the rear of the input shaft. The main drive gear meshes with and drives a larger *countergear*. A countergear is one of four gears on the *countershaft* of the simplified sliding-gear transmission shown in Figure 49-11. Three shafts can be seen: the input shaft, the countershaft (the lower shaft), and the *mainshaft* (the upper shaft). The mainshaft, sometimes called the output shaft, is connected at the rear to the driveline.

When the input shaft and main drive gear turn, the counter shaft drive gear rotates. From the front of the transmission to the rear, the gears on the countershaft become smaller. From front to rear, these gears are: countergear, intermediate (second) gear, low (first) gear, and reverse gear.

Constant-Mesh Gears

Modern manual transmissions are of the *constant-mesh* type. Constant mesh means that all countershaft gears and mainshaft gears are constantly in

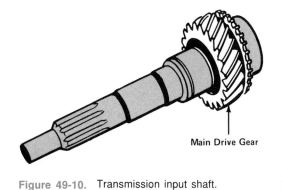

Figure 49-10. Transmission input shaft.

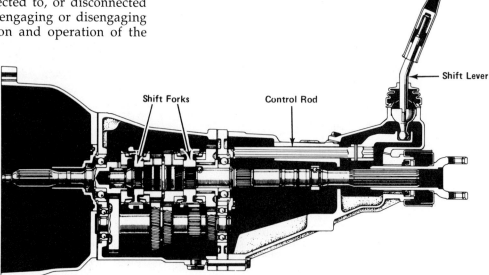

Figure 49-9. Transmission assembly.

CHRYSLER CORPORATION

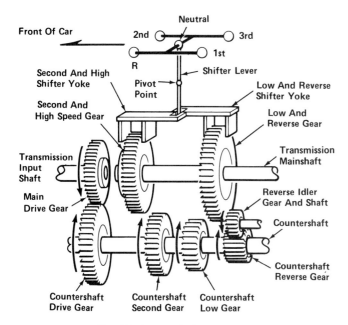

Figure 49-11. Simplified transmission.

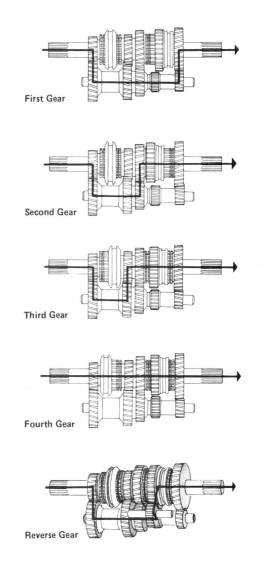

Figure 49-12. Transmission power flow diagram.
VOLVO CAR CORPORATION

mesh with each other. During operation, however, only one of the gears on the mainshaft is engaged, or locked, to the mainshaft. The other gears turn freely on the mainshaft and are not engaged to it.

Study the power flow diagram in **Figure 49-12**. All the gears on the mainshaft and countershaft are meshed. Because each gearset has a different gear ratio, only one gearset can be engaged at a time to turn the mainshaft.

The gears in a modern transmission or transaxle are helical cut gears, which transfer power more quietly and smoothly than spur gears **[Figure 49-13]**.

Approximate gear ratios in a typical four-speed manual transmission are: first, 2.5:1; second, 1.9:1; third, 1.5:1; and high, 1:1. The numerically higher the gear ratio, the greater the torque multiplication.

In high (fourth) gear, which has a 1:1 gear ratio, there is no torque multiplication. That is why fourth gear is not used to get the vehicle rolling. Fourth gear is used for high speeds and light loads.

Shift Rails

When the driver moves the shift lever, linkage moves *shift forks* inside the transmission. The shift forks are mounted on metal shafts called *shift rails* **[Figure 49-14]**. The shift forks determine which gearset drives the mainshaft.

Synchronizers

The action of the shifting forks moves the *synchronizers.* A synchronizer is a metal assembly that can lock a selected gear to the mainshaft. The action of the synchronizers prevents gear clashing when the

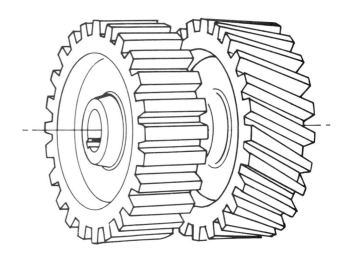

Figure 49-13. Comparing a helical gear (right) with a spur gear.

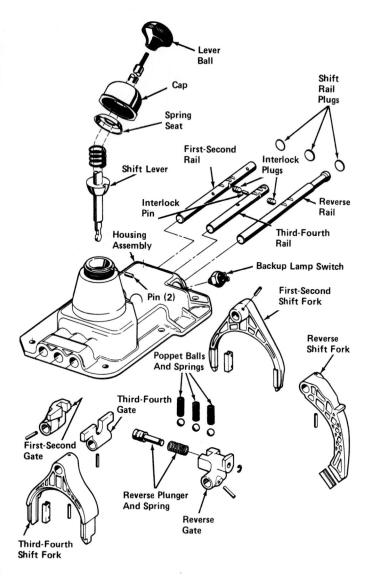

Figure 49-14. Shifting forks and shafts. CHRYSLER CORPORATION

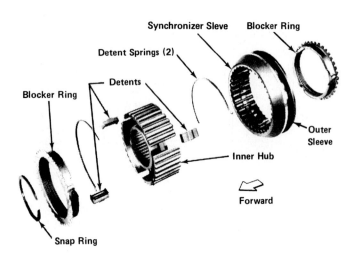

Figure 49-15. Parts of a synchronizer. CHRYSLER CORPORATION

driver shifts gears. Two synchronizers, or synchros, are used in a four-speed transmission.

When the driver selects a gear, linkage moves the shift forks inside the transmission case. The shift forks move parts of the synchronizers back and forth.

When a synchronizer moves toward a gear, it transfers rotating motion to the gear to prevent gear clashing. The gear locks, or engages, and turns the mainshaft. The synchronizer can move forward or backward to engage either of two gears.

Between these two positions, the synchronizer does not engage either gear in its neutral position. When both synchronizers are in their neutral positions, the countershaft and mainshaft gears mesh and turn. However, no mainshaft gear is engaged to turn the mainshaft. **Figure 49-15** shows the parts of a synchronizer assembly, which include:

- Inner hub
- Detent springs
- Detents, or inserts
- Outer sleeve
- Clutch gear sleeve
- Blocker rings
- Snap ring.

Refer to **Figures 49-11, 49-12, and 49-15** while reading the following discussion of synchronizer operation.

The synchronizer hub has splines inside that mate with splines on the mainshaft to make both parts rotate together. The outside of the synchronizer hub also is splined to the synchronizer sleeve **[refer to Figure 49-15]**.

As a transmission operates in a forward gear, first for example, torque from the engine turns the input shaft and main drive gear. The main drive gear turns the countershaft drive gear and the entire countershaft. All the gears on the countershaft turn together **[refer to Figure 49-11]**. First gear on the mainshaft is engaged, or locked, to the mainshaft. Thus, torque flows from the countershaft first gear through the mainshaft first gear to turn the mainshaft. The mainshaft turns the rest of the drivetrain to move the vehicle forward in first.

All the gears on the countershaft rotate at the same speed. The mainshaft gears are meshed with the countershaft gears and are turning at different speeds, depending on the gear ratios. In first gear, only the mainshaft first gear is engaged to the mainshaft and turns it.

When the driver shifts up to second gear, the clutch pedal is depressed to interrupt the flow of torque through the transmission. The shift lever is moved from first to second gear position **[refer to Figure 49-12]**.

As the shift lever is moved, the first-second shift fork moves the synchronizer to disengage the mainshaft first gear so that it no longer turns the mainshaft. Then the synchronizer moves toward the mainshaft second gear. As it moves toward second gear, the synchronizer rotates the gear at the same speed as the mainshaft first gear. When the clutch pedal is released, engine torque flows through the input shaft/main drive gear, countershaft, and mainshaft second gear to turn the mainshaft.

To engage a different gear ratio smoothly, the synchronizer assemblies match, or synchronize, the rotating speeds of the gears as one gear is disengaged and another is engaged to turn the mainshaft. Thus, the mainshaft continues to be turned at the same speed. As the clutch pedal is released, the flow of torque is applied smoothly from the main drive gear through the countergear to the mainshaft.

If the synchronizer did not provide this action, excessive force would be exerted against the transmission driving gears and shafts as the clutch was released. This force would create noise and could damage the gear teeth and mainshaft. In this way, the synchronizer prevents damage to the transmission gears and shafts.

Synchronizers also work in a similar way to slow down the selected gear as a downshift is made from a higher to a lower gear. The synchronizer assemblies thus allow upshifts or downshifts to be made without clashing, grinding, noise, or damage to transmission parts.

Synchronizing begins when the shift fork moves the synchronizer sleeve toward the selected gear. The detents force the blocker ring toward the gear. A cone-shaped clutch surface on each gear creates friction against the cone-shaped clutch part of the blocker ring **[refer to Figure 49-15]**. The friction slows or speeds up the main drive gear/input shaft and clutch disc.

When the speeds are matched, the sleeve slips over the small teeth on both the blocker ring and the gear. Until the hub and gear rotate at the same speed, the blocker ring acts as its name implies. It blocks, or prevents, the small mainshaft gear teeth from fully engaging the inner teeth of the sleeve. When hub and gear rotation speeds are the same, the gear is forced to turn with the synchronizer and mainshaft.

When the driver selects reverse, both forward gear synchronizer sleeves are moved to their neutral positions. On most transmissions, reverse gear does not mesh directly with a gear on the countershaft. The reverse-gear shift fork moves as the shift lever is moved toward reverse. The reverse-gear shift fork slides reverse gear toward the reverse *idler gear.*

The idler gear is an intermediate gear positioned between the countershaft and mainshaft reverse gears on a separate shaft **[refer to Figure 49-11]**. The reverse idler gear reverses the direction of rotation of the mainshaft reverse gear. Thus, the mainshaft turns in the opposite direction and the vehicle moves in reverse.

On some recent manual transaxles used with foreign front-wheel drive vehicles, a synchronizer has been added to reverse gear. This reverse synchronizer helps to prevent gear clash and noise as the reverse idler gear begins to turn reverse gear.

Gearshift Linkage

Shift linkage on modern transmissions is connected to the bottom of the shift lever. A control rod mounted inside the transmission housing connects to shift forks **[refer to Figure 49-9]**.

49.3 FOUR-SPEED TRANSMISSION POWER FLOW

The movement of the synchronizers, discussed above, controls the flow of power through the transmission gears. Refer to **Figures 49-12** and **49-16** to follow the power flow through a four-speed transmission as you read the following discussion.

When the driver selects first gear, the shifting fork moves the first/second synchronizer sleeve to engage first gear.

When the driver selects second gear, the fork moves the first/second synchronizer sleeve away from first and toward second. When the synchronizer sleeve engages second gear, second gear rotates the mainshaft.

When the driver selects third gear, two movements occur inside the transmission. As the shift lever begins to move, the first/second synchronizer sleeve is moved away from second gear. The sleeve is moved into the neutral position between first and second gears. As the driver continues to move the lever, the second movement occurs in the transmission. The other shifting fork moves the third/fourth synchronizer sleeve toward third gear until it is engaged with the mainshaft.

When the driver selects fourth (high) gear, the forward shifting fork moves the forward synchronizer sleeve away from third gear. The synchronizer sleeve moves toward the clutch gear on the input shaft. When the clutch gear is engaged, the input shaft and the mainshaft are locked to rotate at the same speed. The countergear continues to turn but no longer transmits any torque. In fourth gear, there is no speed change or torque multiplication. The transmission is in direct drive.

When the driver selects reverse gear, both synchronizer sleeves are moved to their neutral positions. The reverse-gear linkage rod from the shifter moves the reverse-gear shifting arm. The shifting arm slides the reverse gear forward until it engages the reverse idler gear. Between the rearmost countershaft gear and reverse gear is an idler gear **[Figure 49-17]**. The idler gear rides on a separate, short

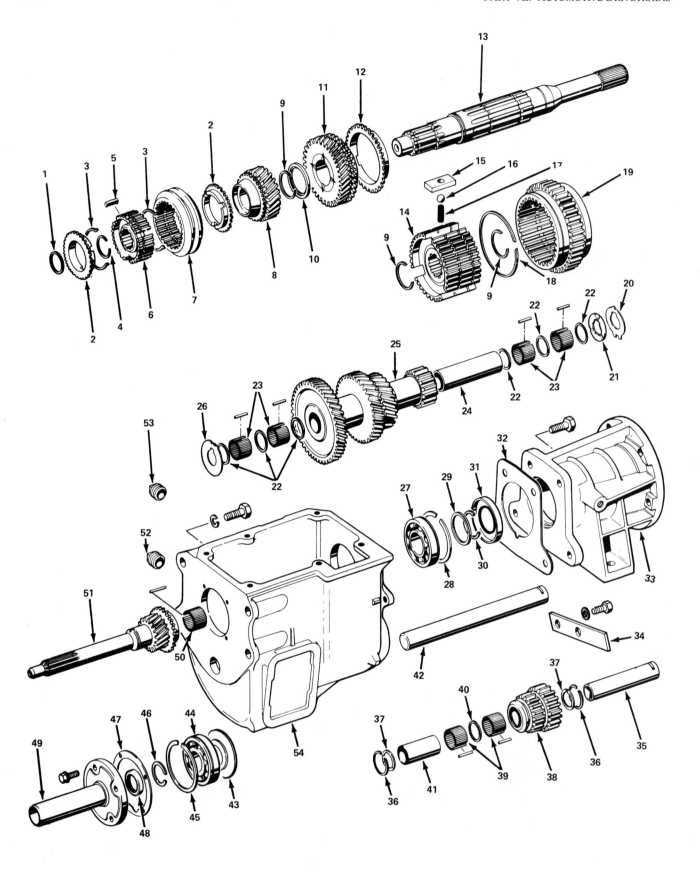

Figure 49-16. Parts of a four-speed transmission. CHRYSLER CORPORATION

1	Mainshaft Pilot Bearing Roller Spacer	
2	Third-Fourth Blocking Ring	
3	Third-Fourth Retaining Ring	
4	Third-Fourth Synchronizer Snap Ring	
5	Third-Fourth Shifting Plate (3)	
6	Third-Fourth Clutch Hub	
7	Third-Fourth Clutch Sleeve	
8	Third Gear	
9	Mainshaft Snap Ring	
10	Second Gear Thrust Washer	
11	Second Gear	
12	Second Gear Blocking Ring	
13	Mainshaft	
14	First-Second Clutch Hub	
15	First—Second Shifting Plate (3)	
16	Poppet Ball	
17	Poppet Spring	
18	First-Second Insert Ring	
19	First-Second Clutch Sleeve	
20	Countershaft Gear Thrust Washer (Steel) (Rear)	
21	Countershaft Gear Thrust Washer (Steel Backed Bronze) (Rear)	
22	Countershaft Gear Bearing Washer	
23	Countershaft Gear Bearing Rollers (88)	
24	Countershaft Gear Bearing Spacer	
25	Countershaft Gear	
26	Countershaft Gear Thrust Washer (Front)	
27	Rear Bearing	
28	Rear Bearing Locating Snap Ring	
29	Rear Bearing Spacer Ring	
30	Rear Bearing Snap Ring	
31	Adapter Plate Seal	
32	Adapter Plate to Transmission Gasket	
33	Adapter to Transmission	
34	Countershaft - Reverse Idler Shaft Lockplate	
35	Reverse Idler Gear Shaft	
36	Reverse Idler Gear Snap Ring	
37	Reverse Idler Gear Thrust Washer	
38	Reverse Idler Gear	
39	Reverse Idler Gear Bearing Rollers (74)	
40	Reverse Idler Gear Bearing Washer	
41	Reverse Idler Shaft Sleeve	
42	Countershaft	
43	Front Bearing Retainer Washer	
44	Front Bearing	
45	Front Bearing Locating Snap Ring	
46	Front Bearing Lock Ring	
47	Front Bearing Cap Gasket	
48	Front Bearing Cap Seal	
49	Front Bearing Cap	
50	Mainshaft Pilot Bearing Rollers (22)	
51	Clutch Shaft	
52	Drain Plug	
53	Filler Plug	
54	Transmission Case	

Figure 49-16. Concluded.

shaft. The idler gear makes the reverse gear rotate in the opposite direction. When reverse is engaged and the clutch pedal is released, the mainshaft rotates backwards and the vehicle moves in reverse.

Many modern transmissions use *all-indirect gearing.* In such a transmission, power always flows through the countergear and countershaft gears to the mainshaft gears. A 1:1 gear ratio is obtained by selecting countershaft and mainshaft gears with the same number of teeth.

Some four-speed transmissions have an overdrive ratio fourth gear. This overdrive ratio allows the mainshaft of the transmission to turn faster than engine speed for highway driving. The engine turns slower for better gas mileage.

The gear to be used depends upon load and speed requirements. Lower gears are used when the load is heavy or when speed is low. The more gears a transmission has, the more easily it can meet a vehicle's load and speed requirements. Current automobiles with a manual transmission usually have four or five forward speeds. The 1989 Chevrolet Corvette was the first production passenger vehicle with a six-speed transmission. The number of speeds refers to forward speeds only. All automotive transmissions have a reverse gear.

In a normal start from a standstill, shifting progresses from first gear through second and third and on into fourth. This process is called *upshifting.*

If the driver encounters an increased load, such as a hill, or slow traffic, a lower gear must be used. When climbing a hill, the driver may shift from fourth down to third, second, or even first gear. This process is called *downshifting.*

Overall Transmission Gear Ratio (OTR)

Transmissions contain many gears and several shift positions. The *overall transmission gear ratio (OTR)* in each gear position can be determined by using the following formula:

$$OTR = (FDNG \div FDG) \times (RDNG \div RDG)$$

Where:

OTR = overall transmission gear ratio

FDNG = front driven gear

FDG = front drive gear

RDNG = rear driven gear

RDG = rear drive gear

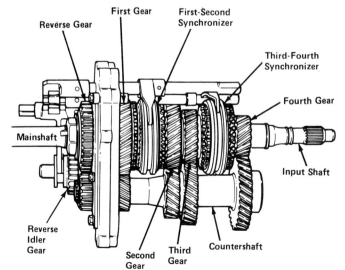

Figure 49-17. Gear locations. FORD MOTOR COMPANY

Sometimes this is also called a *transmission torque multiplication factor (TMF)*. As an example, suppose a transmission has 16 teeth on the front drive gear, 22 teeth on the front driven gear, 24 teeth on the rear driven gear and 11 teeth on the rear drive gear. The OTR for this transmission in first gear is:

$$\begin{aligned} OTR &= (22 \div 16) \times (24 \div 11) \\ &= 1.375 \times 2.18 \\ &= 2.99 : 1 \text{ or } 3 : 1 \end{aligned}$$

Drive Axle Ratio (DAR)

Differentials have a driving pinion gear and a driven ring gear. Thus, a *drive axle ratio (DAR)* also can be calculated. Assume a differential has 12 teeth on the driving pinion gear and 34 teeth on the driven ring gear. The gear ratio for the differential then could be calculated as:

$$DAR = (34 \div 12) = 2.8 : 1$$

An axle with this ratio might be called an economy axle, because it allows the engine to turn relatively slowly at highway speeds. This contributes to increased fuel mileage.

For better acceleration but limited top speed, a differential with a numerically higher gear ratio, such as 4.11:1, can be used. This also causes the engine to turn more revolutions for a given road speed, so fuel mileage decreases.

Overall Drivetrain Gear Ratio (ODR)

You can calculate the *overall drivetrain gear ratio (ODR)* by taking into account both the transmission OTR and differential DAR. The ODR is calculated by using this formula:

$$ODR = OTR \times DAR$$

Where:
ODR = overall drivetrain gear ratio
OTR = overall transmission gear ratio
DAR = drive axle ratio

If we used the OTR example in section 49.1 of 3:1 and the above DAR of 2.8:1, you would calculate the ODR to be:

$$\begin{aligned} ODR &= 3 \times 2.8 \\ &= 8.4 : 1 \end{aligned}$$

This would be the torque multiplication factor for the first gear.

For passenger vehicles, engineers calculate these factors for each gear and different drive axles. They select individual transmission gear ratios and drive axle ratios to give a good compromise of both acceleration and fuel economy.

49.4 FIVE-SPEED TRANSMISSION

Five-speed transmissions are used in many modern vehicles. The extra gearset makes smaller engines more flexible in handling different loads and road conditions. Some five-speed transmissions have overdrive gearing for both fourth and fifth gears to improve fuel economy.

Some five-speeds have cases that are constructed in two halves. These cases are split down the middle **[Figure 49-18]**.

49.5 OTHER TRANSMISSION PARTS

Other parts and devices found in manual transmissions include:

- Ball and needle roller bearings (with separate or caged balls or rollers)
- A gear arrangement to drive the speedometer cable **[Figure 49-19]**
- A switch to turn on backup lights when the vehicle is shifted into reverse
- Sensors that indicate to computer control mechanisms which gear is engaged (refer to Unit 38).

49.6 TRANSAXLE DESIGN

A manual transaxle is a combination of two drivetrain units: a transmission and a differential in one housing. (See Unit 55 for a discussion of differential functions.) Because only one lightweight aluminum housing is used, the weight of a transaxle is less than the combined weight of a separate transmission and differential. The engine and transaxle can be mounted together at the front or rear of the vehicle **[Figure 49-20]**.

Most transaxle automobiles are of front-wheel drive (FWD) design. In a FWD automobile, the engine and drivetrain are combined at the front to drive the front wheels. The powertrain usually is mounted transversely, or sideways **[Figure 49-21]**. In longitudinal mounting, used with transmissions and a few FWD transaxles, the engine is mounted in a line from front to back of the automobile.

Passenger room and trunk space are increased when the engine and drivetrain are mounted in the front. The large transmission hump and driveline tunnel in the floorpan are smaller, and may be used to route exhaust system parts. Front-wheel drive automobiles have smaller humps and tunnels, which are designed to stiffen the floorpan. A stiffened floorpan strengthens the body.

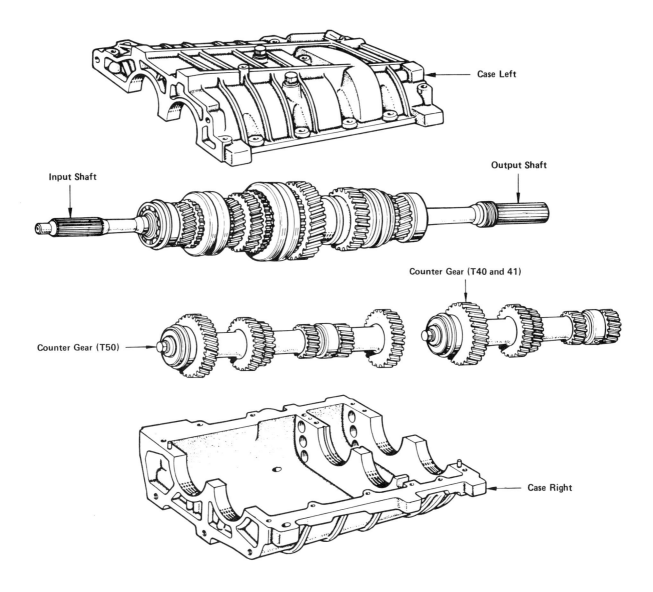

Figure 49-18. Split-case, five-speed transmission assembly. NISSAN MOTOR CORPORATION

Figure 49-19. Removing the speedometer drive gear.

49.7 TRANSAXLE DRIVETRAIN

No driveline or driveshaft is used with front-wheel drive (FWD) drivetrains. Manual transaxle drivetrain parts include:

- Clutch
- Transaxle
- Driving axles.

Clutch

The same type of clutch is used with manual transaxles as with manual transmissions [Figure 49-22]. The clutch is discussed in Unit 51.

Transaxle

The entire powertrain of a FWD vehicle is located under the hood of the automobile. Parts of a typical

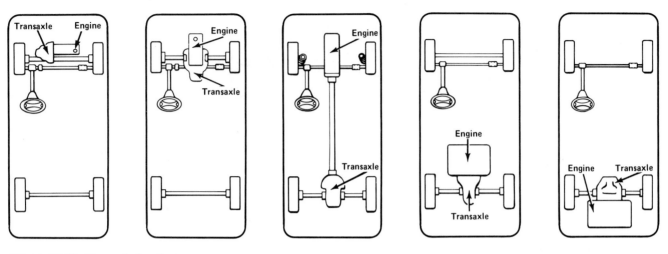

Figure 49-20. Transaxle locations.

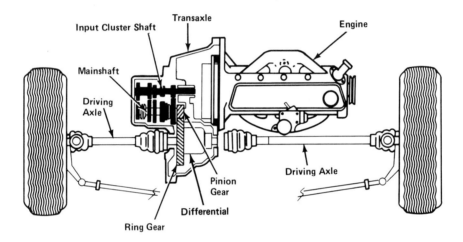

Figure 49-21. Transverse-mounted transaxle.

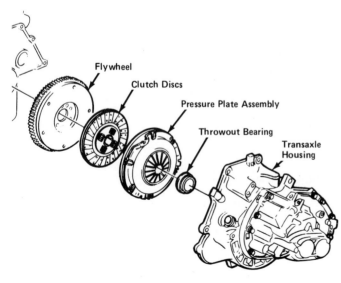

Figure 49-22. Clutch assembly.
PONTIAC MOTOR DIVISION—GMC

FWD transaxle are shown in **Figure 49-23.** The differential ring gear is driven by the output shaft of the transmission, instead of by a pinion gear, as discussed in Unit 55.

The engine, transaxle, and driving axles are in a relatively small space, which can make servicing more difficult. In addition, during acceleration and deceleration, the engine twists slightly in the engine mounts. This twisting is called *torquing-over.*

Torquing-over can twist shift linkage, which could damage the linkage or cause the transaxle to shift into another gear. To prevent these problems, braces or struts often are connected between the body and the powertrain, parallel to the shift linkage [**Figure 49-24**].

Power flow and all-indirect gearing. In a rear-wheel drive automobile, power flows to the transmission through the input shaft. Power then flows from the transmission through the output shaft. Power flows in at one end and out the opposite end.

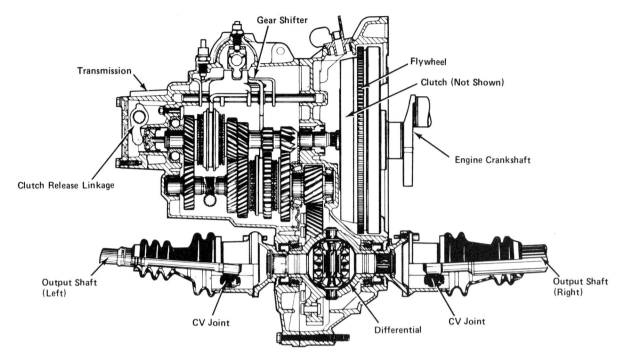

Figure 49-23. Transaxle assembly. CHRYSLER CORPORATION

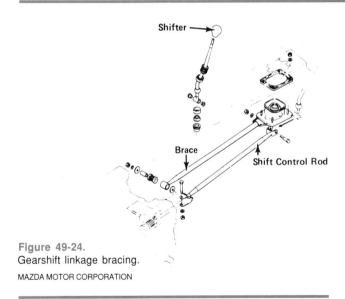

Figure 49-24. Gearshift linkage bracing. MAZDA MOTOR CORPORATION

In many FWD automobiles with transverse-mounted powertrains, torque flows from the end of the engine into the transmission portion of the transaxle. The differential ring gear is a helical gear that is driven by the output shaft of the transmission [refer to Figure 49-23]. The driving axles are splined into the differential side gears and are located below and parallel to the engine crankshaft. See Unit 55 for a discussion of differential operation.

As in modern transmissions, modern transaxles use all-indirect gearing. Power flow through the input shaft always passes through the countershaft gear cluster. From the countershaft gear cluster, power is coupled to the main, or output, shaft. Power flow through a transaxle never directly couples the input shaft and the mainshaft.

Gear ratios for economy. As in modern transmissions, modern transaxle final transmission gear ratios are overdrive ratios (refer to 49.1). A typical fourth-gear ratio might be 0.87:1, while a fifth-gear ratio might be 0.78:1 or less.

Shift and clutch linkages. As explained above, in transversely mounted powertrains, the transaxle is *not* parallel to the front-to-back line of the automobile. It is perpendicular, or at a 90-degree angle, to the front-to-back line. This means that the shift linkage must change the normal front-to-back motions of the shift pattern into side-to-side motions. One solution is to use flexible cables [Figure 49-25]. Rod-and-lever linkage systems, with braces, also are used. Cable and hydraulic clutch linkages are used with transaxles (see Unit 52).

Driving Axles

Driving axles on a FWD automobile are not enclosed in a housing, as they would be in a conventional differential (see Unit 55). The inner parts of the driving axles are connected to the differential. The outer parts of the axles are connected to a wheel hub.

When the powertrain is mounted transversely, the transaxle is offset to one side of the engine compartment. Because of this offset, one of the driving axles must be longer than the other. This can cause problems.

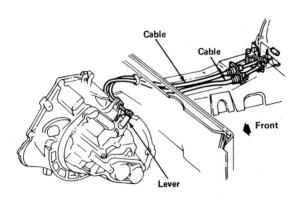

Figure 49-25. Cable gearshift linkage.
PONTIAC MOTOR DIVISION—GMC

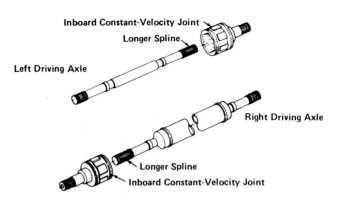

Figure 49-26. Solid and hollow driving axle assemblies.
FORD MOTOR COMPANY

For example, when torque from the transaxle begins to turn the driving axles, the axles begin to twist before they turn. The longer axle shaft will twist more than the shorter shaft before it begins turning. This added twisting, or torquing, of the longer axle makes it begin to turn slightly later than the shorter axle.

Thus, the wheel connected to the longer axle begins to turn slightly later than the opposite driving wheel. When this happens, the vehicle will pull toward the side with the longer driving axle. This is known as *torque steer*. During deceleration, torque steer also can cause the vehicle to pull toward the side with the shorter driving axle. **Figure 49-26** shows a driving axle system with solid and hollow axle shafts.

The most common way used to reduce torque steer on recent FWD vehicles is to make both outer drive axles the same length. A third, or connecting, shaft is used between the outer driving axle on the side farthest from the wheel **[Figure 49-27]**.

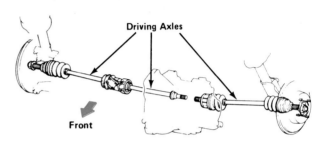

Figure 49-27. Driving axles with three shafts.
MAZDA MOTOR CORPORATION

Constant-velocity joints. Driving axles used with front-drive automobiles must steer from side to side as well as move up and down. Universal joints (discussed in Unit 47) are not flexible enough to do the job. Power must be transmitted smoothly through the sharp angles necessary when steering the front wheels. To do this, a *constant-velocity joint* is used.

The most common constant-velocity (CV) joint is the *Rzeppa joint*. A Rzeppa joint has large-diameter ball bearings that run in grooves in the inner and outer parts of the joint **[Figure 49-28]**. This type of joint can be used at the inner and outer ends of the drive axles.

Another type of joint may be used on the inner, or transaxle, ends of the axles. This is called a *tripod*, or *tripot*, joint. A tripod joint has a triangular-shaped center spider and needle bearings. It also has circular races that run in grooves machined into the outer portion of the joint **[Figure 49-29]**.

UNIT HIGHLIGHTS

- Gearsets can be used to increase or decrease torque or speed.
- A transmission uses gearsets to control and multiply engine torque.
- The three shafts of a manual transmission are the input shaft, countershaft, and mainshaft.
- Although all the gearsets on the mainshaft and countershaft are meshed at the same time, only one gearset drives the mainshaft at any one time.
- Synchronizers engage the gears to the mainshaft to select and engage a chosen gear.
- Automobile transmissions commonly have four or five forward speeds, plus reverse.
- A manual transaxle combines a transmission and a differential in one unit.
- Most front-drive powertrains are mounted transversely.
- A manual transaxle drivetrain consists of a clutch, transaxle, and driving axles.
- Torquing-over is engine movement that is in a direction opposite to crankshaft rotation.

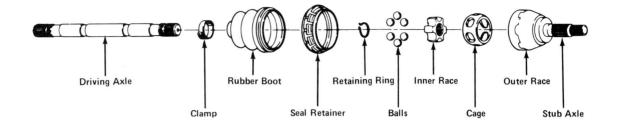

Figure 49-28. Rzeppa joint. PONTIAC MOTOR DIVISION—GMC

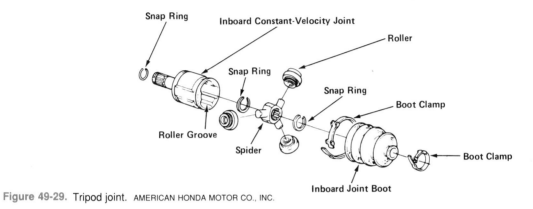

Figure 49-29. Tripod joint. AMERICAN HONDA MOTOR CO., INC.

- Shift linkage may be operated by cables or by rod-and-lever mechanisms in transverse-mounted powertrains.
- Clutch linkages for FWD transaxles are hydraulic or cable-operated.
- Driving axles in transaxle powertrains usually are of different lengths.
- A constant-velocity (CV) joint transmits power smoothly through sharp angles.
- A tripod joint has a triangular-shaped center spider and needle bearings. It is used on the inner ends of axles.

TERMS

gear	gear ratio
overdrive	transmission input shaft
synchronizer	main drive gear
countergear	countershaft
mainshaft	constant-mesh gear
shift forks	shift rails
idler gear	axis
upshifting	downshifting
all-direct gearing	tripod joint
constant-velocity joint	torquing-over

drive axle ratio (DAR)
overall transmission ratio (OTR)
Rzeppa joint
torque steer
overall drivetrain gear ratio (ODR)
transmission torque multiplication factor (TMF)

REVIEW QUESTIONS

DIRECTIONS: The following questions are similar to those used on automotive technician certification tests. On a separate sheet of paper, write the letter of the correct choice.

1. Twenty-five lb.-ft. of torque is applied to a driving gear with a radius of 1 foot. That gear is meshed with a gear with a radius of 3 feet. What is the torque on the shaft of the larger gear?
   A. 75 lb.-ft.   C. 37.5 lb.-ft.
   B. 450 lb.-ft.   D. 25 lb.-ft.

2. A driving gear with nine teeth is meshed with a driven gear of 37 teeth. Which of the following numbers represents the gear ratio?
   A. 4.11:1   C. 333:1
   B. 0.24:1   D. 49:1

3. Technician A says that the driving gear of an overdrive gearset is larger than the driven gear.

   Technician B says that the driving gear of any gearset is always the smaller gear.

   Who is correct?
   A. A only
   B. B only
   C. Both A and B
   D. Neither A nor B

Questions 4 and 5 are not like those above. Each has the word EXCEPT. For each question, look for the choice that does *not* apply. Read each question carefully before you choose your answer.

4. All of the following statements are true EXCEPT
   A. a manual transaxle combines a transmission and a differential.
   B. in an FWD automobile, the engine and drivetrain are combined at the front to drive the front wheels.
   C. a transverse-mounted powertrain usually is mounted in a front-to-back direction.
   D. front-wheel drive automobiles have small humps and tunnels in the floorpan.

5. All of the following statements about constant-velocity joints are true EXCEPT
   A. torque is transmitted at different rates to eliminate torque steer.
   B. a Rzeppa joint is the most common CV joint in use.
   C. a tripot joint uses a triangular spider.
   D. a tripod joint is used on the inner ends of driving axles.

ESSAY QUESTIONS

1. Describe how gears can be used to increase torque or rotational speed.
2. Describe the parts of a manual transmission.
3. What are the differences between a manual transaxle and a manual transmission?

SUPPLEMENTAL ACTIVITIES

1. List the main parts of a manual transmission.
2. Describe how gearing affects rotational speed and torque in a gearset.
3. Describe how you might be able to figure out the gear ratios on a manual transmission removed from a vehicle without disassembling it.
4. Explain how a synchronizer matches the speed of a mainshaft gear to the speed of the mainshaft during upshifting.
5. Discussion topic: Synchronizers are not used on the transmissions of large tractor-trailer rigs. How do the drivers manage to shift gears without clashing?
6. Describe the problems that can be caused in a FWD vehicle if the engine torques over, and identify methods for controlling engine movement.
7. Describe the transaxle shift and clutch linkage systems on a vehicle selected by your instructor.
8. Explain what is meant by torque steer and describe how driving axles may be designed to reduce torque steer.
9. Describe the flow of power from the engine crankshaft to the tires on a transaxle-equipped automobile.
10. Explain the difference between a 1:1 ratio in a direct-drive transmission and in a transmission with all-indirect drive.
11. Determine the overall transmission gear ratio (OTR) for a transmission.
12. Determine the drive axle ratio (DAR) for a differential.
13. Determine the overall drivetrain gear ratio (ODR) for each gear on a vehicle selected by your instructor.

50 Manual Transmission And Transaxle Service

UNIT PREVIEW

Inspection of transmission problems usually requires removing and reinstalling the transmission or transaxle. This unit discusses preventive maintenance, diagnosis, and servicing of manual transmissions and transaxles.

Limited space is available for servicing front-wheel drive manual transaxles. The technician may need to remove several parts of the suspension, brake system, and steering to service a transaxle.

Service procedures for manual transaxles are almost identical to service procedures for manual transmissions. Service for transaxle differentials are similar to RWD differential service, except that the differential is driven by spur gears instead of a hypoid gearset (see Unit 56).

LEARNING OBJECTIVES

When you have completed your assignments and exercises in this unit, you should be able to:

❏ Describe how preventive maintenance is performed on a manual transmission.
❏ Identify problems often found in manual transmissions and transaxles and the causes of those problems.
❏ Identify the parts that must be removed before a transmission or transaxle can be removed from a vehicle.
❏ Describe how manual transmissions and transaxles are removed, disassembled, cleaned, inspected, and reassembled.
❏ Remove and replace driving axles on a FWD vehicle.
❏ Disassemble, reassemble, and inspect constant-velocity joints.
❏ Safely remove and replace a transaxle in a FWD vehicle.

SAFETY PRECAUTIONS

Wear eye protection whenever you work underneath a vehicle. Transmissions and transaxles are extremely heavy. Engine support fixtures and a transmission jack are required when removing a transmission or transaxle.

A helper is required to assist in removal and replacement of transmissions, transaxles, drivelines, and driving axles.

Hot transmission or transaxle lubricant, transaxle cases, exhaust systems, engines, and brakes can cause severe burns. When possible, allow the automobile to cool before you begin work.

Disconnect the grounded battery terminal before attempting to service a transaxle. Physical injury and/or electrical system damage may result if electrical connectors are pinched or squeezed by tools or heavy transaxle and drivetrain parts. Also, electric cooling fans controlled by thermostats may run and cause injury even when the ignition switch is off.

Clutch and brake assemblies contain asbestos dust, which causes lung cancer. Always use an approved special vacuum cleaner or liquid removal system to remove asbestos dust. Wear a breathing mask designed specifically for use with asbestos fibers.

TROUBLESHOOTING AND MAINTENANCE

Preventive maintenance includes checking and draining manual transmission and transaxle gear lubricant. External gear linkage on manual transaxles is lubricated. A visual inspection is made to locate possible problem areas.

Check Transmission or Transaxle Lubricant Level

Transmission lubricant level should be checked at regular intervals, usually during chassis lubrication. Raise the automobile on a hoist to check the transmission lubricant level. The vehicle must be level and the engine must be OFF.

To check lubricant level, locate the *fill plug* on the side of a transmission [Figure 50-1]. Clean the area around the plug with a shop towel. Loosen and remove the plug. Insert a finger or a bent rod into the fill hole. Lubricant should be level with, or not more than one-half inch below, the bottom of the fill hole.

If lubricant level is low, add the proper lubricant. Add lubricant with a filler pump [Figure 50-2]. If you overfill the transmission accidentally, let the lubricant drain until it is flush with the

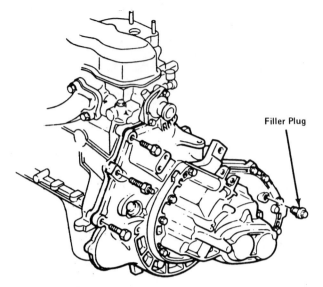

Figure 50-1. Filler plug location. BUICK MOTOR DIVISION—GMC

Figure 50-2. Refilling transmission with fluid.

Checking and Adjusting Shift and Clutch Linkage

Transaxle shift and clutch linkages should be lubricated at the pivot points whenever a chassis lubrication is performed. Refer to the manufacturer's service manual and bulletins for details.

Replace Transmission Lubricant

Manual transmission/transaxle lubricant should be drained and refilled periodically according to the manufacturer's recommendations. Change the lubricant at least every two years or 24,000 miles, whichever comes first. If a vehicle is used for trailer towing, off-road operation, or continuous stop-and-go driving, the lubricant should be changed more frequently.

To drain and refill lubricant, drive the vehicle to warm up the lubricant so that it will drain quickly. Then shut off the engine and safely raise the vehicle on a hoist.

Clean the areas near the fill plug and the drain plug or bolt. Place a catch pan under the transmission or transaxle [Figure 50-3]. Remove the fill plug to allow air to enter the transmission and help the draining process. Remove the drain plug or bolt and its washer (if used), and drain the lubricant into the catch pan.

Inspect the draining lubricant for a large amount of metal particles. Small metal particles often appear as a shiny, "metallic" color in the old lubricant. Large amounts of metal particles in the lubricant indicate severe bearing, synchronizer, gear, or housing wear.

When the lubricant has drained, replace the washer or apply a recommended sealant to the threads on the drain plug or bolt. Tighten the drain plug or bolt to the manufacturer's specified torque.

Fill the transmission or transaxle with the proper lubricant. Fill only until the lubricant is even

Figure 50-3. Draining transmission fluid.

bottom of the check hole. Then replace and tighten the fill plug.

NOTE: Some foreign vehicles' transaxles include a dipstick and/or filler tube accessible from under the hood. Refer to the manufacturer's service manual for specific information.

CAUTION!!!

Manual transmission/transaxle lubricants in use today include single-and multi-viscosity gear oils, engine oils, and automatic transmission fluid. Refer to the vehicle manufacturer's service manual to determine the correct lubricant for the transmission.

with the lower edge of the hole. Do not overfill. Replace and tighten the fill plug to the manufacturer's specified torque.

Visual Inspection

A visual inspection of transmission and transaxle drivetrains should be made at regular intervals. Check for damage, looseness, and leaks.

SAFETY CAUTION

Allow the vehicle to cool before inspection.

Inspecting case and mounts. Check all parts for looseness and leaks. Also check the transmission or transaxle case for any casting porosity that shows up as leaks or seepage of lubricant.

Push up and pull down on the transmission or transaxle case. Watch the mounts to see if the rubber separates from the metal plates [**Figure 50-4**]. Also watch to see if the transaxle case moves up but not down. If the housing does not move down, the mount must be replaced.

Inspecting the linkages. Move the clutch and transmission shift linkages around to check for looseness. Look for excessive or jerky movement, and damaged or missing parts. Cable linkage should have no kinks or sharp bends. Check the fluid level in the clutch master cylinder if a hydraulic clutch system is used.

Inspecting driving axles and CV joints. Driving axles should be checked for cracks, deformation, or damage.

The constant-velocity (CV) joints are enclosed in rubber boots to keep contaminants and moisture out and lubricant in [**Figure 50-5**]. Inspect the boots for

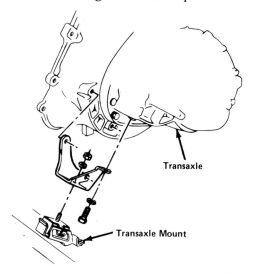

Figure 50-4. Transaxle mounting. PONTIAC MOTOR DIVISION—GMC

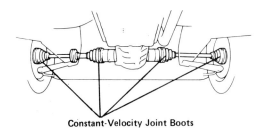

Figure 50-5. Constant-velocity joint boot locations.

cracks, tears, or splits. Look for indentations in the boots. Check for leakage between the CV joint boot ridges.

To check a CV joint, grasp the axle and try to move it up and down. If it moves, the CV joint should be serviced.

Split or cracked CV joint boots must be replaced. Otherwise, dirt, salt, or other debris will enter the CV joint and cause it to fail. In most cases, the drive axle must be removed to replace a bad boot. Some aftermarket manufacturers supply split boots that do not require axle removal for replacement. An indentation in a boot usually can be removed by grasping the dimpled portion on both sides. Pull on the boot in opposite directions until the indentation pops out.

50.1 DIAGNOSING TRANSMISSION AND TRANSAXLE PROBLEMS

Diagnosing a transmission problem involves troubleshooting, or locating the source of a problem. A reported transmission problem actually might be a problem in the clutch, driveline, or differential. The following is a discussion of the most common transmission/transaxle problems.

Noises

Most manual transmission/transaxle complaints involve noises. However, noises can reverberate through the driveline parts. The first step is to attempt to isolate the source of the noise.

Make a visual inspection of the transmission or transaxle housing. Check for loose mounting bolts and crossmembers. Check the level and condition of the transmission lubricant. If the level is low, check for leaks. Large amounts of metal in drained lubricant may indicate worn or broken parts. Some small metal particles usually will be present due to normal wear.

Noises also may indicate worn or damaged bearings, gear teeth, or synchronizers.

To determine the location of the noise, you can listen with a technician's stethoscope or a length of rubber tubing. (On transaxle-equipped cars, determining the source of a noise can be difficult because the transmission and differential share the same housing.) Listen under the car with a stethoscope or length of rubber tubing held to your ear. Have a helper operate the transmission in different gears. A noise that changes or disappears in different gears can point to specific problem areas in a transmission.

SAFETY CAUTION

When the transmission or transaxle is in gear and the engine is running, the driving wheels and related parts will turn. Avoid touching moving parts. Severe physical injury can result from contact with spinning drive wheels, drivelines, and so on.

The type of noise also is important. For example, a loud clicking or knocking sound indicates the possibility of a missing gear tooth. A humming or grinding noise indicates a bad bearing.

Gear Clash

Gear clash is a grinding noise that occurs during shifting. The noise occurs when one gearset remains partly engaged while another gearset attempts to turn the mainshaft. Gear clash can be caused by incorrect clutch adjustment or binding of clutch and/or gearshift linkage. Internal transmission or transaxle problems that can cause gear clash include damaged, worn, or defective synchronizer blocker rings. Use of an improper gear lubricant also can contribute to gear clash.

Hard Shifting

If the shift lever is hard to move from one gear to another, check clutch linkage adjustment. Clutch service is discussed in Unit 52. Hard shifting also may be caused by damage inside the transmission or transaxle, or by an improper lubricant that is too thick. Common hard-shifting problems include badly worn bearings and damaged clutch gears, control rods, shift rails, shift forks, and synchronizers.

Jumping Out of Gear

If the car jumps out of gear into neutral, especially when decelerating or going down hills, first check the shift lever and internal gearshift linkage. Excessive clearance between gears and the input shaft can cause jumping out of gear. Still another cause is badly worn bearings. Other internal transmission/transaxle parts to inspect are the clutch pilot bearing, gear teeth, shift forks, shift rails, and springs or detents.

Locked in Gear

If a transmission or transaxle is locked in one gear and cannot be shifted, first check the gearshift lever linkage for misadjustment or damage. Low lubricant level can cause needle bearings, gears, and synchronizers to seize and lock up the transmission.

If the problem is not found, the transmission or transaxle must be removed from the vehicle and disassembled. After disassembly, inspect the internal countershaft gear, clutch shaft, reverse idler, shift rails, shift forks, and springs or detents for damage. Also check for worn support bearings.

50.2 DRIVELINE AND DRIVING AXLE REMOVAL

Before a transmission can be removed, the driveline first must be removed. Refer to Unit 48 for driveline removal and service procedures.

Remove Transaxle Driving Axles

Driving axle removal is necessary when the axles are serviced or when the transaxle is removed. Use the following brief description as a general guide to transaxle removal.

Place the transaxle shift lever in first or reverse. Set the emergency brake securely. Instruct a helper to depress the brake pedal firmly to prevent the wheel from turning. Loosen the *hub nut,* as shown in **Figure 50-6,** with a breaker bar and socket. Raise the vehicle on a hoist.

Remove the wheel and hub nut. Remove brake, steering, and suspension parts as the service manual indicates. Avoid pulling or bending the driving axle, which can separate and damage the CV joints.

Use a special tool to remove the driving axle from the transaxle **[Figure 50-7]**. Support the outer end of the driving axle so that the axle and CV joints will not be damaged.

Servicing Driving Axles

Replacing constant-velocity joints and boots are the most common services performed on driving axles. Always inspect the shafts and splines for damage and wear when the driving axles have been removed. **Figure 50-8** shows a driving axle assembly.

Outer constant-velocity joint service is shown in **Figures 50-9** and **50-10.** Inner constant-velocity joint service is shown in **Figures 50-11** and **50-12.**

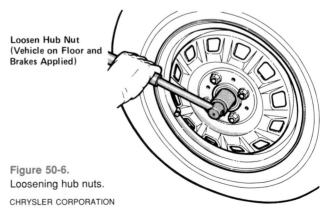

Figure 50-6.
Loosening hub nuts.

CHRYSLER CORPORATION

Manual Transmission and Transaxle Service

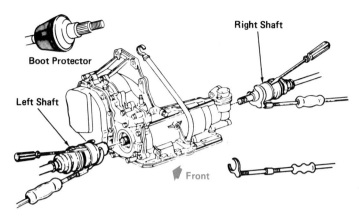

Figure 50-7. Removing driving axle shafts.
CHEVROLET MOTOR DIVISION—GMC

50.3 REMOVE TRANSMISSION PARTS AND LINKAGE

To remove the transmission or transaxle, any connecting or obstructing parts must be removed. These include, but are not limited to:

- Clutch linkage
- Speedometer cable
- Backup light switch
- Shift linkage
- Brake system
- Suspension system
- Exhaust system
- Transmission crossmember.

Refer to the manufacturer's service manual for correct removal procedures for each of these parts. Use the following as a general guide for removal of these parts.

Remove clutch linkage. Refer to the manufacturer's service manual and remove rod-and-lever, cable, or hydraulic clutch linkage (see Unit 52).

Remove speedometer cable. The speedometer is operated by a *speedometer cable* that is connected to the transmission or transaxle. The speedometer cable housing may be screwed or bolted to the transmission housing [**Figure 50-13**]. Disconnect the speedometer cable from the transmission or transaxle and wire it out of the way.

Remove backup light switch connector. A backup light switch is located near, and activated by, the shift lever [**Figure 50-14**]. Disconnect the switch by pulling the connector apart. Secure the wire out of the way.

Remove shift linkage. The shift-lever assembly and the shift linkage may need to be removed [**Figure 50-15**]. This means working inside and under the vehicle.

Remove emergency brake linkage. Metal cables operate the rear brakes to act as an emergency and parking brake. In many cases, these must be removed to allow access for transmission removal. Refer to the manufacturer's service manual and disconnect any

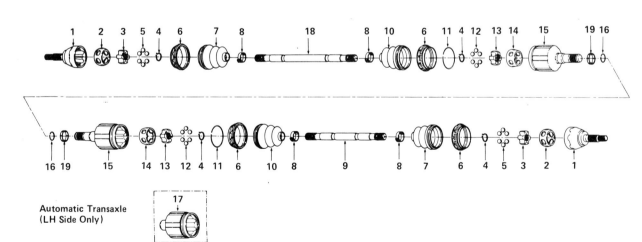

1. Race, CV Joint Outer
2. Cage, CV Joint
3. Race, CV Joint Inner
4. Ring, Race Retaining
5. Ball (6)
6. Retainer, Seal
7. Seal, CV Joint
8. Clamp, Seal Retaining
9. Shaft, Axle (LH)
10. Seal, D/O Joint
11. Ring, Ball Retaining
12. Ball (6)
13. Race, D/O Joint Inner
14. Cage, D/O Joint
15. Race, D/O Joint Outer
16. Ring, Joint Retaining
17. Race, D/O Joint Outer
18. Shaft, Axle (RH)
19. Slinger

Figure 50-8. Driving axle assembly. CHEVROLET MOTOR DIVISION—GMC

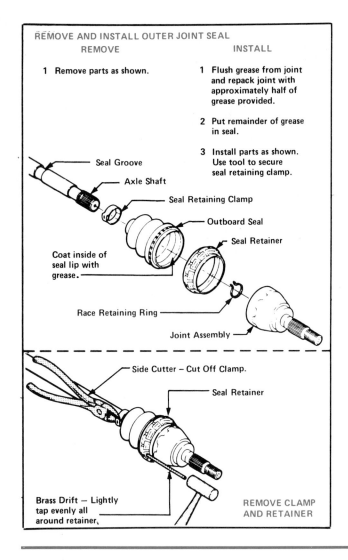
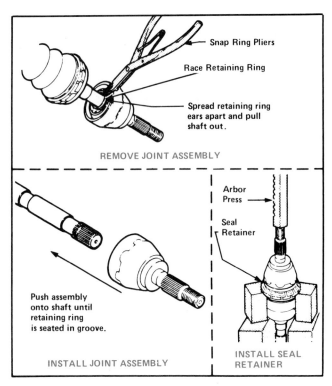

Figure 50-9. Removing and installing outer joints.
CHEVROLET MOTOR DIVISION—GMC

emergency brake linkage that would interfere with transmission or driveline removal (see Unit 63).

Remove brake system parts. For front-wheel drive transaxle removal, brake system parts may need to be removed. Refer to the manufacturer's service manual and to Unit 64.

Remove suspension system parts. FWD transaxle removal may require disconnecting and/or removing some front suspension parts. Refer to the manufacturer's service manual and to Unit 58 for procedures.

Remove exhaust system. If the exhaust system pipes will interfere with transmission, transaxle, or driveline removal, they must be removed. Disconnect the header pipe from the exhaust manifold, and remove hangers and straps as necessary (refer to Unit 26).

Remove transmission crossmember. The transmission crossmember is a steel frame that supports the transmission and the engine. Before unbolting the crossmember from the chassis, the engine must be safely and securely supported with a jack or jack stand [Figure 50-16].

SAFETY CAUTION

Failure to support the engine before transmission removal can result in serious injury and vehicle damage.

50.4 TRANSMISSION AND TRANSAXLE REMOVAL

If a visual inspection does not identify a problem, the transmission or transaxle must be removed from the vehicle. Use care to avoid injuries. When the engine is safely supported, secure the transmission or transaxle to a transmission jack with safety chains [Figure 50-17]. It may be necessary to lift the engine slightly with a jack to aid in transmission or transaxle removal. Special fixtures are used to support the engine and/or transaxle for removal and service [Figure 50-18].

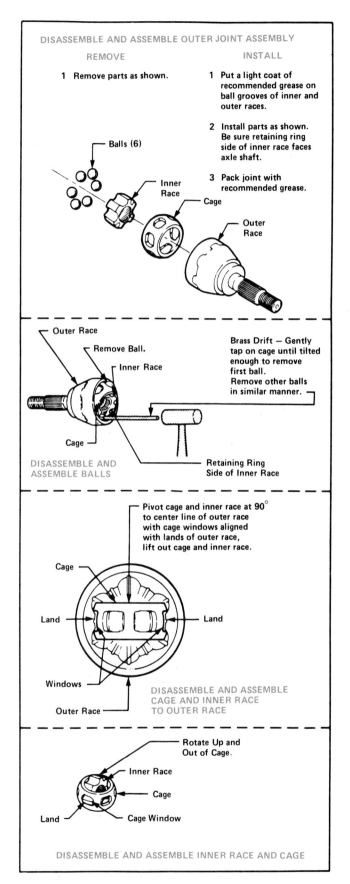

Figure 50-10. Disassembling and assembling outer joints.
CHEVROLET MOTOR DIVISION—GMC

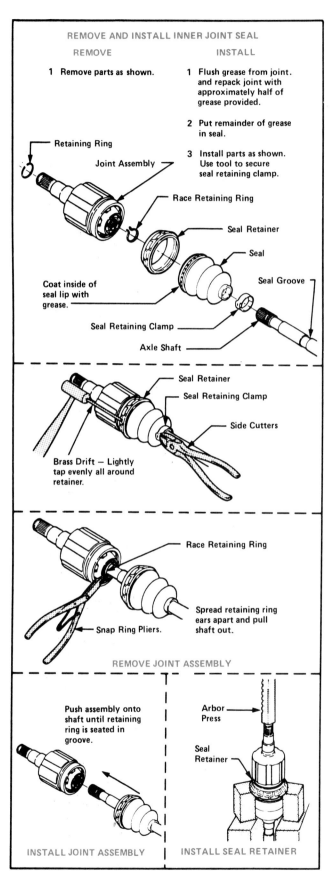

Figure 50-11. Removing and installing inner joints.
CHEVROLET MOTOR DIVISION—GMC

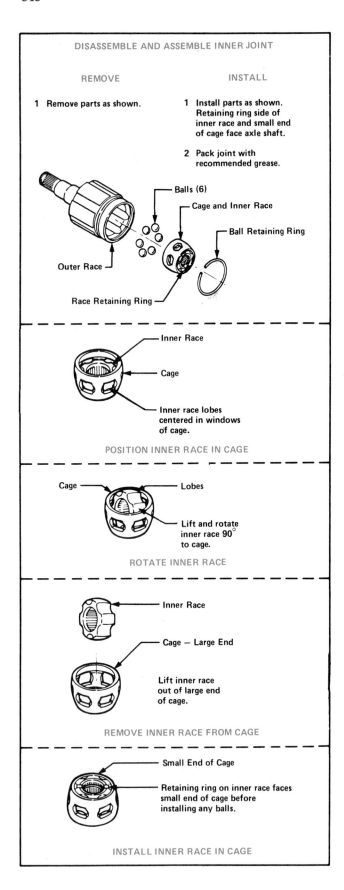

Figure 50-12. Disassembling and assembling inner joints.
CHEVROLET MOTOR DIVISION—GMC

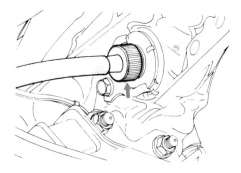

Figure 50-13. Speedometer cables screwed (top), bolted (bottom) to transmission case. CHRYSLER CORPORATION/FORD MOTOR COMPANY

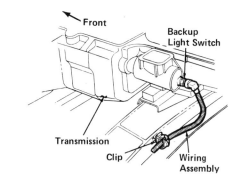

Figure 50-14. Backup light switch. FORD MOTOR COMPANY

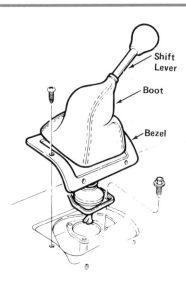

Figure 50-15. Floor shifter assembly.
AMERICAN MOTORS CORPORATION

MANUAL TRANSMISSION AND TRANSAXLE SERVICE

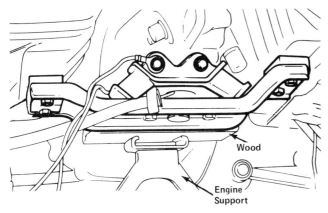

Figure 50-16. Supporting the engine. CHRYSLER CORPORATION

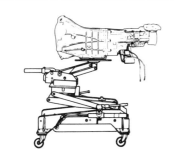

Figure 50-17. Transmission jack. CHRYSLER CORPORATION

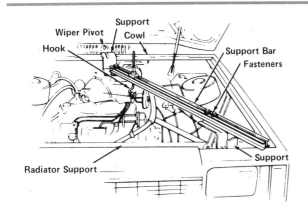

Figure 50-18. Engine support fixture.
CHEVROLET MOTOR DIVISION—GMC

For specific removal procedures, refer to the manufacturer's service manual. If a manual transmission can be separated from the bellhousing, it is not necessary to remove the bellhousing. Then the transmission or transaxle, supported by the transmission jack, is pulled away from the engine.

SAFETY CAUTION

Severe physical injuries can result if a transmission falls. Raise and support the vehicle safely on a hoist. Secure the transmission to a transmission jack with safety chains before attempting removal or replacement procedures. Use caution and work with a helper to remove and replace the transmission slowly and carefully.

Before you attempt to remove a transmission or transaxle, place an oil catch pan under the transmission and drain the lubricant. Alignment dowels, or threaded metal shafts, often are installed in place of fastening bolts to help guide the transmission off of and back onto the engine. Refer to the manufacturer's service manual for specific procedures.

Disassemble Transmission or Transaxle

Manual transmission disassembly procedures vary considerably among manufacturers and models. Refer to the manufacturer's service manual for specific, step-by-step procedures. The following is a generalized list of parts that are removed for transmission inspection:

- Covers and gaskets
- Shifting forks
- Front bearing retainer
- Input shaft
- Extension housing (if used)
- Reverse idler gear
- Speedometer drive gear
- Mainshaft
- Needle bearings
- Countergear assembly.

A hydraulic press is used to remove gears and synchronizers from the mainshaft. After disassembly, all parts are cleaned and inspected for defects. **Figure 50-19** shows parts of a transaxle that are visible after the clutch cover is removed. Refer to Unit 56 for transaxle differential service procedures.

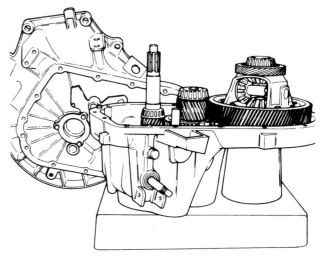

Figure 50-19. Transaxle with clutch cover removed.
CHEVROLET MOTOR DIVISION—GMC

Clean and Inspect Parts

Clean all parts thoroughly in clean solvent. Remove all traces of old gaskets. Wash roller bearings in solvent and wipe them dry with a lint-free cloth. Do *not* spin-dry bearings with compressed air.

Rotate the outer *bearing race* against the inner race. A race is part of a bearing, the machined circular metal surface against which roller or ball bearings ride. Listen carefully for noise, and feel for tightness or roughness. Visually check the rollers and races for wear. Lubricate bearings with the manufacturer's recommended lubricant before reassembly. Check all needle bearings and needle-bearing bushings for signs of wear or roughness. Replace any defective bearings.

Synchronizer blocker rings should remain with the gears they mate with. Do *not* mix the blocker rings between gears. Check blocker rings for fit and wear according to the manufacturer's service manual.

Check all parts of the transmission, including brass bushings (for the output shaft, countergear, etc.), for wear or damage. Replace parts as necessary.

Lubricate each transmission part after inspection with the manufacturer's recommended lubricant. Install new oil seals and gaskets and lubricate the seal lips. Install new snap (retaining) rings.

Figure 50-20 shows internal parts of a transaxle in an exploded view.

Reassemble Transmission or Transaxle

Reassembly procedures usually are in reverse order from disassembly. New parts are used as necessary. All new gaskets and seals always are used.

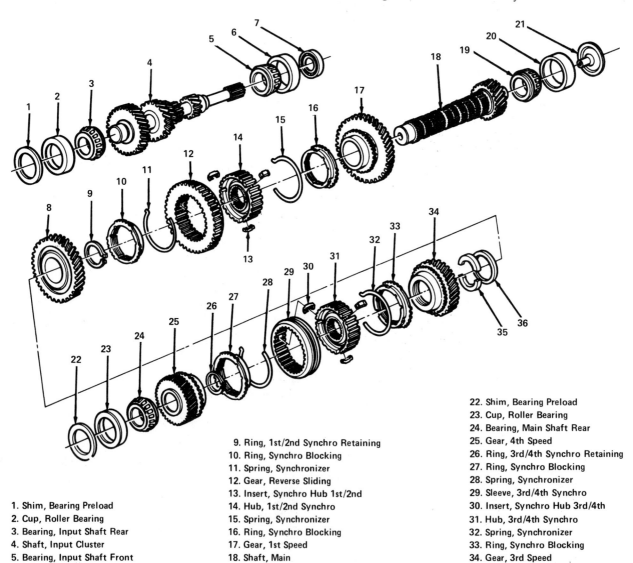

1. Shim, Bearing Preload
2. Cup, Roller Bearing
3. Bearing, Input Shaft Rear
4. Shaft, Input Cluster
5. Bearing, Input Shaft Front
6. Cup, Roller Bearing
7. Seal Assembly, Input Shaft
8. Gear, 2nd Speed
9. Ring, 1st/2nd Synchro Retaining
10. Ring, Synchro Blocking
11. Spring, Synchronizer
12. Gear, Reverse Sliding
13. Insert, Synchro Hub 1st/2nd
14. Hub, 1st/2nd Synchro
15. Spring, Synchronizer
16. Ring, Synchro Blocking
17. Gear, 1st Speed
18. Shaft, Main
19. Bearing, Main Shaft Front
20. Cup, Roller Bearing
21. Funnel, Mainshaft
22. Shim, Bearing Preload
23. Cup, Roller Bearing
24. Bearing, Main Shaft Rear
25. Gear, 4th Speed
26. Ring, 3rd/4th Synchro Retaining
27. Ring, Synchro Blocking
28. Spring, Synchronizer
29. Sleeve, 3rd/4th Synchro
30. Insert, Synchro Hub 3rd/4th
31. Hub, 3rd/4th Synchro
32. Spring, Synchronizer
33. Ring, Synchro Blocking
34. Gear, 3rd Speed
35. Washer, 2nd/3rd Gear Thrust
36. Ring, 2nd/3rd Thrust Washer Retaining

Figure 50-20. Parts of a transaxle assembly. FORD MOTOR COMPANY (Continued on next page.)

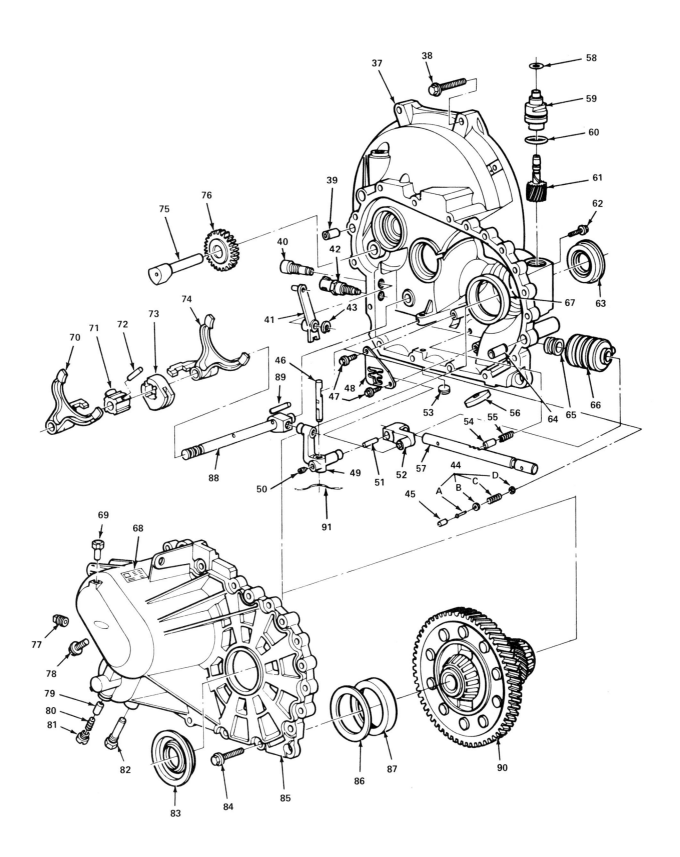

Figure 50-20. Continued.

37. Case, Clutch Housing
38. Bolt, Transaxle to Engine Attaching
39. Dowel, Trans Case to Clutch Housing
40. Pin, Reverse Relay Lever Pivot
41. Lever, Reverse Relay
42. Switch Assembly, Back Up Lamps
43. Ring, Retaining External
44. Spring and Retaining Assembly, Reverse Inhibitor
    A. Pin
    B. Washer
    C. Spring
    D. Ring
45. Plunger, Reverse Inhibitor
46. Shaft, Shift Lever
47. Bolt, Selector Plate Attaching
48. Plate, Selector
49. Lever, Shift
50. Screw, Shift Lever Shaft Set
51. Pin, Spring
52. Arm, Input Shift Shaft Selector Plate
53. Plug, Expansion
54. Plunger, Input Shift Shaft Detent
55. Spring, Input Shift Shaft Detent
56. Magnet, Case Ceramic
57. Shaft, Input Shift
58. Seal, 5.16mm x 1.6 O-Ring
59. Retainer, Speedo Driven Gear
60. Seal, Speedo Retainer to Case
61. Gear, Speedo Driven
62. Screw, Speedo Retaining
63. Seal Assembly (RH), Differential
64. Dowel, Trans Case to Clutch Housing
65. Seal Assembly, Shift Shaft Oil
66. Boot, Input Shift Shaft
67. Cup, Differential Bearing
68. Tag, Transaxle Identification
69. Vent, Case
70. Fork, 3rd/4th
71. Arm, Fork Selector
72. Pin, Spring
73. Sleeve, Fork Interlock
74. Fork, 1st/2nd
75. Shaft, Reverse Idler
76. Gear, Reverse Idler
77. Plug, Fill
78. Bolt, Reverse Shaft Retaining
79. Plunger, Main Shift Shaft Detent
80. Spring, Main Shift Shaft Detent
81. Screw, Detent Plunger Retaining
82. Pin, Fork Interlock Sleeve Retaining
83. Seal Assembly (LH), Differential
84. Bolt, Trans Case
85. Case, Trans
86. Shim, Differential Bearing Preload
87. Cup, Differential Bearing
88. Shaft, Main Shift
89. Pin, Reverse Relay Actuating Lever
90. Differential and Final Drive Ring Gear
91. Shift Bias Spring, 3rd/4th

Figure 50-20. Concluded.

Serviceable gears are pressed onto the mainshaft. Separate needle bearings are held in place with heavy grease so that shafts can be inserted into place. During reassembly, shaft end play is measured and adjusted with shims, spacers, or snap rings of different thicknesses. A torque wrench is used to tighten all fasteners to the manufacturer's specifications.

The counter gear is then put into place. Soft-faced mallets are used to tap shafts and other parts into place. The reverse idler gear is installed. Then the mainshaft and geartrain are installed.

The input shaft, front bearing retainer, and extension housings are installed. All used gaskets are discarded and new gaskets are installed.

After reassembly, the transmission is secured to a transmission jack with safety chains and raised into place. Before the transmission is installed, the clutch is inspected and serviced as necessary (see Unit 52).

Replace Transmission or Transaxle

Transmission or transaxle replacement is essentially the reverse of removal. Raise the transmission on the transmission jack slowly and carefully. If alignment dowels are used, raise the unit and guide the case holes carefully onto the dowels.

Rock the transmission and push it forward until the input shaft splines engage with the clutch hub. Insert and tighten the lower transmission or transaxle bolts, remove the dowels, and insert the upper bolts. Install any support crossmembers and tighten all fasteners with a torque wrench to the manufacturer's specifications.

Install all parts you removed for transmission or transaxle removal. Connect the brake, suspension, exhaust system parts correctly and according to the directions in the manufacturer's service manual. Install the transmission driveline (refer to Unit 48) or driving axles, as discussed below. Doublecheck your work and tighten all fasteners with a torque wrench to the manufacturer's specifications.

After the transmission or transaxle has been reinstalled and all other parts and systems are correctly replaced, make a road test. The road test will determine if the problem has been corrected.

SAFETY CAUTION

Before beginning a road test, buckle your safety belt and test the brakes to be sure they are functioning properly. Drive carefully and only in a manner required to determine if the problem has been corrected. Obey all traffic laws.

UNIT HIGHLIGHTS

- Transmission and transaxle preventive maintenance and visual inspections are conducted at regular intervals.
- Most transaxle and clutch problems can be identified by noises and shifting difficulties.
- Diagnosis is done to determine whether a problem is being caused by the transmission or transaxle or by another drivetrain assembly.
- Driving axles are removed whenever the axles and transaxle are serviced.

MANUAL TRANSMISSION AND TRANSAXLE SERVICE 553

- Several parts usually must be removed for access to the transmission.
- Transmission or transaxle removal requires special jacks and supporting fixtures.
- A manual transmission or transaxle must be disassembled in a specific order, as specified by the manufacturer.
- Cleaning and inspection of transmission parts is a key to proper diagnosis.

TERMS

fill plug
speedometer cable
bearing race
gear clash
hub nut

REVIEW QUESTIONS

DIRECTIONS: The following questions are similar to those used on automotive technician certification tests. On a separate sheet of paper, write the letter of the correct choice.

1. Preventive maintenance for manual transmissions and transaxles include
   A. checking and replacing lubricant.
   B. replacing seals and gaskets.
   C. inspecting transmission gear teeth for defects.
   D. using a hydraulic press to remove and replace gear assemblies.

2. Which of the following statements is correct?
   I. Diagnosing transmission or transaxle problems always requires disassembly.
   II. Diagnosing transmission or transaxle problems never requires making visual inspections, listening to noises, or making a test drive.
   A. I only
   B. II only
   C. Both I and II
   D. Neither I nor II

3. Technician A says that after cleaning and inspection, all old manual transmission or transaxle parts except gaskets and seals are reused.
   Technician B says that after cleaning and inspection, defective parts, gaskets, and seals are replaced.
   Who is correct?
   A. A only
   B. B only
   C. Both A and B
   D. Neither A nor B

Questions 4 and 5 are not like the previous questions. Each has the word EXCEPT. For each question, look for the choice that does *not* apply. Read each question carefully before you choose your answer.

4. All of the following statements are correct EXCEPT
   A. most transaxle and clutch problems can be identified by shifting difficulties.
   B. driving axle removal is necessary when removing a transaxle.
   C. a special tool may be needed to remove a CV joint from a transaxle.
   D. the outer CV joint must be removed before loosening the hub nut.

5. To remove a transaxle, all of the following parts may need to be removed EXCEPT
   A. brake parts.
   B. shift linkage.
   C. U-joints and slip yokes.
   D. suspension parts.

ESSAY QUESTIONS

1. What procedures are part of manual transmission and transaxle preventive maintenance?
2. Describe procedures for diagnosing manual transmission or transaxle problems.
3. Explain the general steps involved in removing and overhauling a manual transmission or transaxle.

SUPPLEMENTAL ACTIVITIES

1. Check the lubricant level of a manual transmission or transaxle.
2. Inspect transaxle mounting and CV joint boots on a vehicle selected by your instructor.
3. Drain and fill the transmission on a vehicle selected by your instructor.
4. Identify the type of transmission or transaxle noise and possible cause of that noise on a vehicle selected by your instructor.
5. Remove a transmission or transaxle from a shop vehicle.
6. Disassemble a manual transmission or transaxle.
7. Clean and inspect the parts of a manual transmission or transaxle.
8. Reassemble a transmission or transaxle and reinstall it in an automobile.
9. Remove, disassemble, inspect, reassemble, and replace a CV joint on a vehicle selected by your instructor.

51 The Clutch

UNIT PREVIEW

A clutch provides a mechanical coupling between an engine flywheel and a manual transmission input shaft. The clutch is operated by linkage that extends from the passenger compartment to the bellhousing between the engine and transmission. This unit discusses how the clutch operates.

LEARNING OBJECTIVES

When you have completed your assignments and exercises in this unit, you should be able to:

- Describe how engine torque is transferred by the clutch disc to the transmission.
- Describe the construction of a clutch plate.
- Identify and describe the parts of a pressure plate assembly.
- Describe coil-spring and diaphragm-spring pressure plate assemblies.
- Describe the operation of the throwout bearing and clutch fork.
- Identify and describe three different types of clutch linkages.

51.1 CLUTCH DESIGN

The clutch is located between the transmission and the engine. The clutch allows the driver to connect or disconnect the flow of torque from the engine to the transmission. This action allows easy shifting between gears. The functional parts of clutch assembly include:

- Flywheel
- Clutch disc
- Pressure plate assembly
- Throwout bearing
- Clutch fork.

A clutch assembly is shown in **Figure 51-1.** All parts of the clutch, except linkage and pedal, are enclosed in a *bellhousing.* A bellhousing is a protective metal case shaped like a bell. The large end of the bellhousing is connected to the engine block. The small end is connected to the transmission housing.

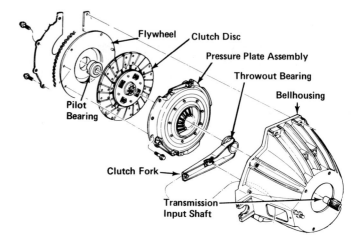

Figure 51-1. Clutch assembly. FORD MOTOR COMPANY

Flywheel

The *flywheel* is an important part of the engine (see Units 10 and 12). The flywheel also is the main driving member of the clutch. The rear surface of the flywheel provides a machined frictional surface for proper clutch operation. Holes are drilled and tapped in the flywheel surface for pressure plate attachment.

Clutch Disc

A *clutch disc,* also called a *clutch plate,* **[Figure 51-2]** is a driven member that transfers engine torque from the flywheel to the transmission input shaft. The clutch disc is made from a thin steel disc that is covered with a friction material, also called a "friction facing." The friction material is gripped by

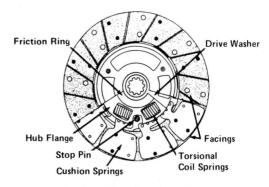

Figure 51-2. Parts of a clutch disc assembly.
CHEVROLET MOTOR DIVISION—GMC

The Clutch

the flywheel and pressure plate when the clutch is coupled, or engaged. The flywheel and clutch disc separate when the clutch is uncoupled, or disengaged.

The friction facing on the clutch disc is riveted or *bonded* to the metal clutch disc. Bonded means held in place with a strong epoxy adhesive. Separate metal parts under the friction material, called *cushioning springs,* absorb shock when the clutch is engaged. *Torsional coil springs* around the hub absorb and cushion torque forces and help the clutch disc to rotate smoothly. Stop pins limit this torsional movement.

The splined input shaft from the transmission passes through the central, splined *hub* of the clutch disc. The forward end of the input shaft fits into a pilot bearing or bushing in the flywheel end of the crankshaft.

Pressure Plate Assembly

The purpose of the *pressure plate assembly* is to press the clutch disc tightly against the flywheel. The pressure plate assembly also must be able to release, or disengage, the clutch disc so that it can stop rotating, even though the flywheel and pressure plate continue to rotate [Figure 51-3]. The pressure plate assembly includes the following parts:

- Cover
- Pressure plate
- Release levers
- Spring assembly.

The steel cover bolts to the flywheel and acts as a housing to hold the parts together.

The *pressure plate* is a heavy, flat ring made of cast iron. The machined surface of the pressure plate contacts the clutch disc to provide a good gripping surface.

Release levers are positioned around the cover. Three release levers usually are used. Release levers "release" the holding force of the springs.

Two types of springs are used in a pressure plate assembly: coil springs and diaphragm springs.

Coil spring pressure plate assembly. Helical springs, or *coil springs,* are made of wire wound into rings or spirals. Coil springs in a pressure plate assembly are spaced evenly around the inside of the cover [Figure 51-4]. As the release levers pivot, the pressure plate moves in and out against the spring tension. A coil spring pressure plate also is known as a "long style" pressure plate.

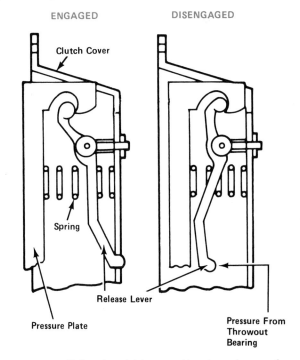

Figure 51-4. Coil-spring clutch assembly release lever action.

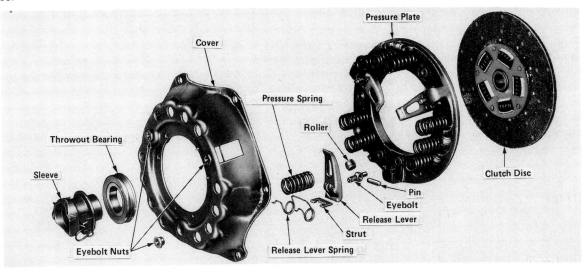

Figure 51-3. Pressure plate assembly. CHRYSLER CORPORATION

The springs exert pressure to hold the pressure plate tightly against the clutch disc. This forces the clutch disc against the flywheel. To disengage the clutch, the release levers move to compress the springs. As it disengages, the pressure plate releases pressure and no longer holds the clutch plate against the flywheel. When the clutch is disengaged, torque no longer is transmitted through the clutch disc to the transmission input shaft.

Diaphragm spring pressure plate assembly. A *diaphragm spring* is a single, thin sheet of metal that works in the same manner as the bottom of an oil can. When pressure is applied to the bottom of the oil can, the metal yields to the pressure. When pressure is released, the bottom of the oil can springs back to its original position.

On a pressure plate assembly, the diaphragm spring is placed between the cover and the pressure plate. The center portion of the diaphragm is slit into numerous fingers that act as release levers [**Figure 51-5**].

When the clutch is engaged, the diaphragm spring is almost flat at the center. At the outer rim, the spring is moved outward and the pressure plate is forced against the clutch disc.

To disengage this style of clutch, the fingers move against the diaphragm spring and force the outer rim inward. This action pulls the pressure plate away from the clutch disc and releases holding pressure.

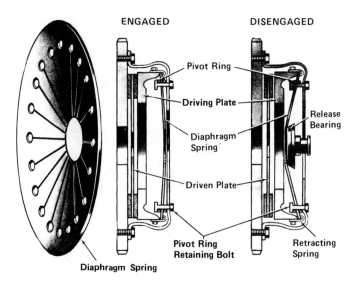

Figure 51-5. Diaphragm-spring clutch assembly.
CHEVROLET MOTOR DIVISION—GMC

Throwout Bearing Assembly

The release bearing, or *throwout bearing*, is a sealed, prelubricated ball bearing. The throwout bearing is pressed onto a sleeve, or collar, that fits into the yoke of the clutch fork [**Figure 51-6.**]. Throwout bearing movement is controlled by the clutch fork, which is operated by the clutch pedal and linkage.

When the clutch pedal is depressed to disengage the clutch, the throwout bearing presses against the release levers to operate the pressure plate springs. As the release levers rotate with the pressure plate, the throwout bearing rotates.

When the clutch is disengaged, the throwout bearing is pushed against the release levers. When the clutch is engaged, the throwout bearing does not touch the release levers.

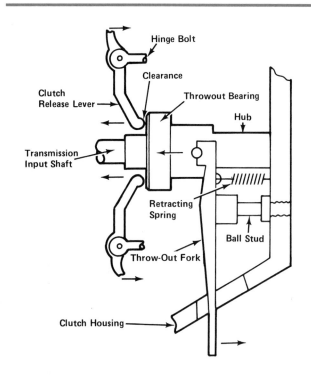

Figure 51-6. Throwout bearing assembly action. Arrows indicate clutch assembly movement.

Clutch Fork

The *clutch fork* is a forked lever that pivots in an opening in the bellhousing, as shown in **Figure 51-7**. The small end of the clutch fork protrudes from the bellhousing and is connected to the clutch linkage and clutch pedal.

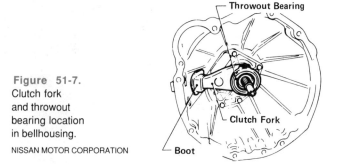

Figure 51-7. Clutch fork and throwout bearing location in bellhousing.
NISSAN MOTOR CORPORATION

THE CLUTCH

51.2 CLUTCH VARIATIONS

Two other types of clutches are found in some vehicles. These variations are the semi-centrifugal clutch and the double-disc clutch. Double-disc clutches, used primarily in heavy-duty trucks, are not discussed here.

Semi-Centrifugal Clutch

A *semi-centrifugal clutch* assembly uses *centrifugal force* to increase the force exerted by the pressure plate against the clutch plate. Centrifugal force is an outward pull from the center of a rotating axis.

The release levers on a semi-centrifugal clutch pressure plate include a weighted end **[Figure 51-8]**. The weight acts to increase centrifugal force as the rotational speed of the pressure plate increases. The centrifugal force adds to the spring pressure to produce greater holding force against the clutch plate. This allows the pressure plate springs themselves to be made less stiff, which results in easier clutch pedal operation.

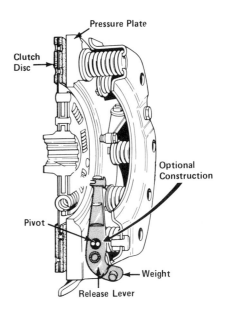

Figure 51-8. Semi-centrifugal clutch pressure plate assembly. FORD MOTOR COMPANY

51.3 CLUTCH LINKAGE

The *clutch linkage* is a series of parts that connects the clutch pedal to the clutch fork. The linkage transfers the motion of the pedal to the clutch fork and provides increased leverage to help compress the pressure plate springs.

Clutch linkage may be mechanical or hydraulic. The basic types of clutch linkage are:

- Rod and lever
- Cable
- Hydraulic.

Rod and Lever Linkage

Rod and lever linkage includes parts to transfer the pedal motion and apply force to move the clutch fork **[Figure 51-9]**.

A pedal rod connects the clutch pedal to a lever on the *lever and shaft assembly,* or "bellcrank." When the upper lever in **Figure 51-9** is moved by the clutch pedal, the shaft rotates and moves the lower lever.

The lower lever is connected to a pushrod that is attached to the clutch fork. The lever and shaft assembly is located between the chassis and bellhousing, near the lower rear part of the engine block.

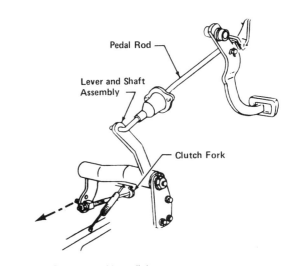

Figure 51-9. Rod and lever linkage. BUICK MOTOR DIVISION—GMC

Cable Linkage

Cable linkage includes a flexible, protective housing that covers a wire cable. At one end, the wire cable is connected to the clutch pedal. At the other end, the cable is connected to the clutch fork **[Figure 51-10]**.

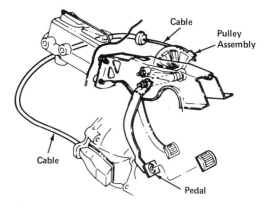

Figure 51-10. Cable linkage. FORD MOTOR COMPANY

Clutch pedal linkage transfers motion from the clutch pedal to the clutch fork, which in turn operates the pressure plate. The pressure plate springs and springs on the clutch pedal lever provide force to push the cable back when the clutch pedal is released.

Hydraulic Linkage

In a *hydraulic linkage,* hydraulic (liquid) pressure transmits motion from one sealed cylinder to another through a connecting tube, or hydraulic line **[Figure 51-11]**.

When the clutch pedal is depressed, the piston inside the clutch "master" cylinder is moved forward by the pushrod. Hydraulic fluid under pressure is pushed from the master cylinder through the hydraulic line and into the "slave" cylinder. When fluid enters the slave cylinder, it applies force against the slave cylinder piston. The piston moves against a pushrod connected to the clutch fork.

When the clutch pedal is released, the pressure plate springs and clutch pedal lever springs push the linkage backwards. Hydraulic fluid is forced back into the master cylinder reservoir.

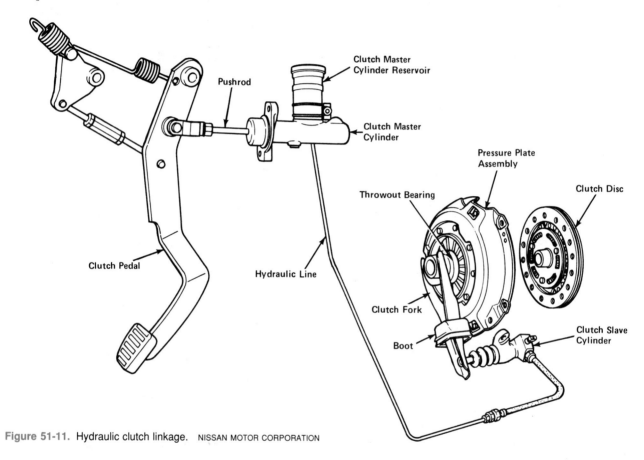

Figure 51-11. Hydraulic clutch linkage. NISSAN MOTOR CORPORATION

UNIT HIGHLIGHTS

- A clutch allows the driver to connect and disconnect engine torque from a manual transmission.
- The main parts of the clutch are contained in a bellhousing.
- A clutch disc transfers torque from the flywheel to the transmission input shaft.
- A pressure plate assembly presses the clutch disc against the flywheel.
- Coil springs and diaphragm springs are used for a pressure plate assembly.
- The clutch fork moves a throwout bearing against pressure plate release levers.
- Clutch linkage connects the clutch pedal to the clutch fork.
- Rod and lever, cable, and hydraulic clutch linkages are used.

TERMS

bellhousing	clutch disc
clutch plate	bonded

cushioning spring
pressure plate assembly
release lever
diaphragm spring
clutch fork
centrifugal force
rod and lever linkage
cable linkage
hub

torsional coil spring
pressure plate
coil spring
throwout bearing
semi-centrifugal clutch
clutch linkage
lever and shaft assembly
hydraulic linkage
flywheel

REVIEW QUESTIONS

DIRECTIONS: The following questions are similar to those used on automotive technician certification tests. On a separate sheet of paper, write the letter of the correct choice.

1. The pressure plate contacts the
   A. transmission mainshaft.
   B. throwout bearing.
   C. clutch plate.
   D. flywheel.

2. Technician A says that a diaphragm-spring pressure plate has no release levers.
   Technician B says that slits in a diaphragm spring create fingers that act as release levers.
   Who is correct?
   A. A only
   B. B only
   C. Both A and B
   D. Neither A nor B

3. Which of the following statements is correct?
   I. A semi-centrifugal clutch has weighted release levers.
   II. Double-disc clutches are used primarily in large trucks.
   A. I only
   B. II only
   C. Both I and II
   D. Neither I nor II

Questions 4 and 5 are not like those above. Each has the word EXCEPT. For each question, look for the choice that does *not* apply. Read each question carefully before you choose your answer.

4. A clutch disc includes all of the following parts EXCEPT
   A. a splined hub.
   B. a throwout bearing.
   C. a torsional coil spring.
   D. a facing.

5. All of the following statements about clutch linkages are true EXCEPT
   A. electrical solenoids and switches are used in some clutch linkages.
   B. hydraulic cylinders and lines are used in some clutch linkages.
   C. metal cables are used in some clutch linkages.
   D. rod-and-lever mechanisms are used in some clutch linkages.

ESSAY QUESTIONS

1. Describe the parts of a clutch assembly.
2. What differences are found in a semi-centrifugal clutch assembly?
3. What types of clutch linkage are used?

SUPPLEMENTAL ACTIVITIES

1. Identify the separate parts of a clutch disc.
2. Describe how coil-spring and diaphragm-spring pressure plates operate.
3. Describe how a throwout bearing operates when the clutch is engaged and disengaged.
4. Discussion topic: What damage can result if the clutch plate is not held securely against the flywheel during driving?
5. Discussion topic: What damage to the clutch can result from downshifting to slow a vehicle instead of using the brakes?

52 Clutch Service

UNIT PREVIEW

Preventive maintenance for the clutch consists of periodic clutch linkage checks, lubrication, and adjustment.

Clutch service requires removing the transmission and other drivetrain parts. This unit discusses clutch preventive maintenance and clutch repairs.

LEARNING OBJECTIVES

When you have completed your assignments and exercises in this unit, you should be able to:

- Measure and adjust clutch pedal free travel.
- Diagnose clutch problems including clutch pedal pulsation, slipping, drag, and chatter.
- Disassemble and reassemble a clutch assembly.
- Inspect clutch parts for defects.

SAFETY PRECAUTIONS

Always wear eye protection when you work underneath a vehicle. Clutch service involves two major hazards: asbestos fibers and the removal of several very heavy assemblies. The dust inside the bellhousing and on the clutch assembly contain asbestos fibers. Asbestos dust causes lung cancer.

Remove asbestos dust only with a special, approved vacuum collection system or an approved liquid cleaning system. Never use compressed air or a brush to clean off asbestos dust. Dispose of collected asbestos dust or liquid containing asbestos dust in accordance with all federal, state, and local laws.

Clutch preventive maintenance and service require that the vehicle be raised and supported safely. *Never* work under a vehicle unless you are absolutely sure that it cannot move or fall. No type of jack is meant to support the weight of a vehicle safely. Safety stands, or jackstands, must be used if the vehicle is not raised on a hoist.

Jackstands and special jacks are used to support the engine and transmission. The transmission is very heavy, so it is necessary to have a helper assist in its removal. Use care during removal to avoid personal injuries and damage to parts.

Many clutch troubleshooting procedures require that the automobile be operated in the shop area. Always place *wheel chocks* against the wheels that remain on the ground. Be sure the work area is properly ventilated, or attach a ventilating hose to the vehicle's exhaust system whenever you run an engine indoors. Do *not* allow anyone to stand in front of or behind the automobile while the engine is running.

Hot engine and exhaust system parts can cause severe burns. Avoid touching hot surfaces and whenever possible let the vehicle cool down before beginning work.

TROUBLESHOOTING AND MAINTENANCE

For vehicles with external clutch linkage, periodic lubrication is required. All clutches require checking and/or adjustment of linkage at regular intervals.

Lubricating External Clutch Linkage

External clutch linkage should be lubricated on a regular basis, usually during a chassis lubrication. The same chassis grease used for suspension parts and U-joints is used on many clutch linkages. Refer to the vehicle manufacturer's service manual to determine the proper lubricant.

Lubricate clutch linkage at all sliding surfaces and pivot points in the linkage **[Figure 52-1]**. After lubrication, the linkage should move freely.

On automobiles with hydraulic clutch linkage, check the clutch master cylinder reservoir fluid level. It should be approximately 1/4 in. (6.35 mm) from the top of the reservoir. Use approved brake fluid to refill if necessary. The clutch master cylinder does not "consume" fluid. If the fluid is low, check for leaks in the master cylinder, connecting flexible line, and slave cylinder.

Clutch Linkage Adjustment

When the clutch is disengaged (pedal up), the throwout bearing must not touch the pressure plate release levers. Clearance between these parts prevents premature clutch plate, pressure plate, and throwout bearing wear. As the clutch disc becomes thinner with wear, this clearance will change.

To make sure there is clearance, the clutch linkage is adjusted so that the pedal has a specified

CLUTCH SERVICE

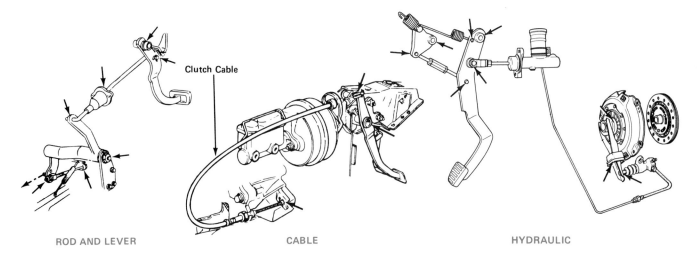

Figure 52-1. Clutch linkage lubrication points. BUICK MOTOR DIVISION—GMC/FORD MOTOR COMPANY/NISSAN MOTOR CORPORATION

amount of free play, or *free travel.* Free travel is slack, or the distance a clutch pedal moves when depressed, before the clutch fork begins to move the throwout bearing.

Some vehicles have automatic clutch adjusting mechanisms. For correct free travel measurements and adjustments, refer to the manufacturer's service manual. To measure free travel, use a tape measure or ruler. Place one end of the tape measure or ruler against the floor of the automobile [Figure 52-2]. Note the reading. This measurement is the height of the fully extended clutch pedal.

Hold the tape measure or ruler and push down with one finger until resistance is felt. Note the reading on the tape measure or ruler. The difference between the two measurements is the amount of free travel.

Clutch pedal free travel should be checked on a regular basis. Adjustment should be performed when free travel is not correct, or when the clutch does not engage or disengage properly.

To adjust clutch free travel, refer to the vehicle manufacturer's service manual to locate adjustment points. Usually, a threaded fastener can be turned to increase or decrease free travel. The threaded fastener may be located under the dash at the clutch pedal, or where the linkage attaches to the clutch fork [refer to Figure 52-1].

Clutch adjustments are made from under the dash, hood, or from under the vehicle. Follow all shop safety rules to safely raise and support the vehicle.

Clean the linkage with a shop towel and solvent, if necessary. Check linkage and replace any damaged or missing parts or cables. On automobiles with hydraulic linkage systems, check for leaks at the clutch master cylinder, hydraulic hose, and slave cylinder. Adjust linkage to provide the manufacturer's specified clutch pedal free travel.

Some newer automobiles use a constant duty throwout bearing. This bearing normally operates while against the clutch release levers. Clutches using this type of release mechanism will have no clutch pedal free travel and therefore will not need any adjustments.

52.1 DIAGNOSING CLUTCH PROBLEMS

Before attempting to diagnose clutch problems, check and attempt to adjust clutch pedal free travel. A clutch that is worn out (friction lining too thin) cannot be adjusted successfully. The most common clutch problems are the following:

* Slippage
* Drag and binding
* Chatter
* Pedal pulsation
* Vibration
* Noises.

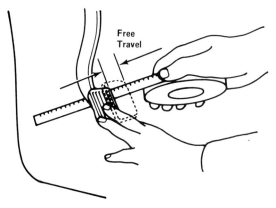

Figure 52-2. Measuring free travel.

Slippage

Clutch slippage is occurring when the engine races but the vehicle does not keep up. Slippage will be worst during acceleration, upshifts, and/or downshifts. It is usually most noticeable in higher gears. Check for slippage by driving the vehicle.

SAFETY CAUTION

Perform a test drive only with the permission and supervision of your instructor. Before beginning a test drive, buckle the safety belt and check brake operation. Drive in a manner that will allow you to correctly diagnose the problem. Drive safely and obey all traffic laws.

Normal acceleration from a stop and several gear changes will indicate whether the clutch is slipping. Slippage also can be checked in the shop.

SAFETY CAUTION

Perform this test only with the permission and supervision of your instructor. Block the front and rear wheels with wheel chocks. Set the parking brake securely and place the gearshift lever in neutral. Be sure that no one is standing in front of the automobile during this test because the vehicle could jump forward.

Refer to the manufacturer's service manual for correct procedures. The following is a general procedure for checking clutch slippage.

Depress the clutch pedal, shift the transmission into high gear, and increase engine speed to about 2,000 rpm. Release the clutch pedal slowly until the clutch engages. The engine should stall immediately.

If the engine does not stall within a few seconds, depress the clutch pedal to end the test quickly. Then raise the vehicle safely on a hoist and check the clutch linkage. If no linkage problems are found, the transmission must be removed and the clutch disassembled and inspected to locate the cause of the slippage.

Clutch slippage is caused by an oil-soaked or worn disc facing, warped pressure plate, weak diaphragm spring, or the throwout bearing contacting and applying pressure to the release levers. Both the clutch disc and pressure plate should be replaced. Always replace the throwout bearing whenever you replace major internal clutch components.

NOTE: Severe or prolonged clutch slippage also causes grooving and/or heat damage to the pressure plate.

Drag and Binding

Clutch drag occurs when the clutch disc is not fully released when the clutch pedal is depressed fully. Clutch drag will cause gear clash, especially when shifting into reverse. Clutch drag also can cause hard starting because the engine attempts to turn the transmission input shaft.

SAFETY CAUTION

Perform this test only with the permission and supervision of your instructor. Before checking clutch drag in the shop, block all wheels with wheel chocks. Set the parking brake securely and place the gearshift lever in neutral. Be sure that no one is standing in front or in back of the automobile during this test because the vehicle could move.

The clutch disc, input shaft, and transmission gears should require no more than five seconds to come to a complete stop after disengagement. This is known as clutch *spindown time*. It is normal and should not be mistaken for clutch drag.

Start the engine. Depress the clutch pedal completely. Shift the transmission into first gear. Do *not* release the clutch. Now, shift the transmission into neutral and wait 5-10 seconds. Then attempt to shift the transmission smoothly into reverse.

If the shift into reverse causes gear clash, raise the vehicle on a hoist and check clutch linkage. If no defects are found, the clutch must be disassembled for inspection.

Clutch drag can be caused by a warped disc or pressure plate, a loose disc facing, or a defective release lever. Incorrect clutch pedal adjustment that results in excessive free travel also can cause clutch drag.

Binding can occur because the splines in the clutch disc hub and/or on the transmission input shaft are damaged. Problems with the release levers also can cause binding.

Chatter

Clutch chatter is a shaking or shuddering that is felt in the vehicle as the clutch is engaged. Chatter usually occurs when the pressure plate first makes contact with the clutch disc, and will stop when the clutch is fully engaged.

SAFETY CAUTION

Perform this test only with the permission and supervision of your instructor. If the clutch is being tested in the shop area, be sure no one stands in front of or behind the automobile. The automobile could move. Place chocks in front of and in back of all wheels. Set the emergency brake securely and place the gearshift lever in neutral.

Start the engine and depress the clutch completely. Shift the transmission into first gear. Increase engine speed to about 1,500 rpm. Slowly release the clutch pedal and check for chattering as it begins to engage. Depress the clutch pedal immediately and reduce engine speed to prevent damage to the clutch parts.

SAFETY CAUTION

Perform this test only with the permission and supervision of your instructor. Do not release the clutch pedal completely, or the automobile might jump and cause serious injury. As soon as the clutch is partially engaged, depress the clutch pedal immediately.

Clutch chatter usually is caused by a glazed clutch facing and/or a mirror-like shine on the pressure plate. Liquid leaks into the clutch can lead to clutch glazing and pressure plate damage. Sources of oil and/or clutch hydraulic fluid leaks include the engine rear main bearing seal, transmission input shaft seal, and the clutch slave cylinder and hydraulic line. Also look for broken engine mounts, loose bellhousing bolts, and damaged clutch linkage.

During disassembly, check for a burned or glazed disc facing, warped pressure plate or flywheel, and worn input shaft splines. If the chattering is caused by an oil-soaked clutch disc and no other parts are damaged, only the disc need be replaced. The source of the oil leak must also be found and repaired.

Pedal Pulsation

A rapid up-and-down pumping movement of the clutch pedal as the clutch disengages or engages is known as *clutch pedal pulsation*. Pedal movement usually is slight but can be felt through the clutch pedal. Pulsation occurs when the throwout bearing first touches, or is in contact with, the release levers.

SAFETY CAUTION

Perform this test only with the permission and supervision of your instructor. Before checking for clutch pedal pulsation in the shop, block the wheels with wheel chocks. Set the parking brake securely and place the gearshift lever in neutral. Make sure that no one stands in front of or behind the automobile because the vehicle could move.

Start the engine. Depress the clutch pedal slowly until the clutch just begins to disengage. Stop briefly. Continue depressing the clutch pedal slowly, and check for pulsation as the pedal is depressed to a full stop.

Minor pulsation is considered normal on many automobiles. If excessive pulsation is felt, the clutch must be removed and disassembled for inspection.

Pedal pulsation is caused by misalignment of parts. Broken, bent, or warped release levers create misalignment. Check for a misaligned bellhousing, bent flywheel, and warped clutch disc, pressure plate, or clutch cover.

Vibration

Clutch vibration can occur at any clutch pedal position. Usually, vibration occurs at normal engine operating speeds—over 1,500 rpm. Clutch-related vibrations are different from pedal pulsations. Clutch-pedal vibrations can be felt throughout the automobile.

To check for clutch vibration, raise the automobile on a hoist. Check the engine mounts and look for any indication that engine parts are rubbing against the body or frame. Look for a damaged crankshaft damper pulley. Lower the automobile.

Remove the drive belts one at a time. Place the transmission in neutral, set the emergency brake securely, and follow all safety precautions. Start the engine after each belt is removed to see if accessories may be the cause. Do *not* run the engine for more than one minute when checking for vibrations with the belts removed.

The transmission will have to be removed before other clutch parts can be examined. After removal, check for loose flywheel bolts, excessive flywheel runout, and pressure plate cover balance problems.

Noises

Many clutch noises come from bearings and bushings. Throwout bearing noise is a whirring, grating, or grinding sound. It happens when the clutch pedal is depressed. Throwout bearing noise stops when the pedal is fully released.

The noises are most noticeable in neutral with the clutch engaged. However, these noises also may be heard when the clutch is engaged and the transmission is in gear.

Pilot bushing noises are squealing, howling, or trumpeting sounds. The noises are most noticeable in cold weather. They usually occur when the pedal is being depressed and the transmission is in neutral.

52.2 CLUTCH SERVICE

Removing and replacing the clutch in an automobile also requires removal of the driveline or driveshafts and transmission or transaxle.

SAFETY CAUTION

Always wear safety goggles or a face mask when working under an automobile. Extreme caution must be used when working with any heavy parts. Special jacks are used to support the engine and transmission. These parts are very heavy, so you must have a helper to assist in their removal. Exhaust systems become very hot. Let the vehicle cool down before starting work.

Removing the Clutch

Raise the vehicle on a hoist. Clean excessive dirt, grease, or debris from around the clutch and transmission. Disconnect and remove the clutch linkage.

On rear-wheel drive automobiles, remove the driveline (see Topic 48.3) and remove the transmission (see Topic 50.3). Depending on the automobile, the bellhousing is removed either with the transmission or after the transmission is removed.

On front-wheel drive vehicles with transaxles, drive axles, parts of the engine, brake and/or suspension system, or body parts may need to be removed first. These parts may interfere with transaxle removal. Refer to the vehicle manufacturer's service manual for specific procedures.

SAFETY CAUTION

Remove asbestos dust only with a special, approved vacuum collection system or an approved liquid cleaning system. Never use compressed air or a brush to clean off asbestos dust. Dispose of collected asbestos dust or liquid containing asbestos dust in accordance with all federal, state, and local laws.

After the bellhousing has been removed, the clutch assembly will be accessible. Remove asbestos dust and dirt from the clutch assembly with a special vacuum cleaner or liquid cleaning system. Mark the flywheel and clutch plate for proper reassembly [Figure 52-3]. Unbolt and remove the clutch assembly by turning each bolt approximately 1/4 turn at a time to prevent warping and damage.

Inspection

Look for evidence of oil leaks on the clutch assembly, flywheel, and bellhousing.

After disassembly, examine the flywheel surface closely for discoloration, a mirror-like shine, scoring, or uneven wear. Any of these defects indicate that the flywheel must be resurfaced or replaced. A cracked pressure plate must be replaced with a new unit.

The pilot bearing or bushing in the flywheel end of the crankshaft should be replaced when the clutch has been disassembled. Pilot bearings or bushings should be smooth, with no signs of wear.

Avoid touching the friction surface of the clutch disc. Grease from hands can contaminate the clutch lining. The friction facings must be uniform over the entire contact area. Check the splined hub for wear. Check for loose or broken hub springs, rivets, and cushioning plates.

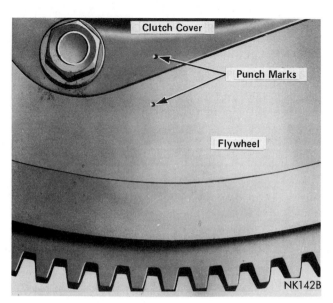

Figure 52-3. Clutch cover and flywheel matchmarks.
CHRYSLER CORPORATION

Measure the thickness of the lining above the rivet heads for wear [Figure 52-4]. Bonded facing is measured for total thickness. Check the proper service manual for acceptable lining thickness. Always replace discs that are in questionable condition.

If either the clutch disc or pressure plate are defective, replace both with new units. If one is damaged, both should be replaced.

Inspect the pressure plate surface for grooving, discoloration, mirror-like shine, cracks, and warping. To check a pressure plate for warping, set a straightedge across the surface [Figure 52-5]. Use a feeler gauge to determine the largest gap, and compare this gap to the manufacturer's specifications.

Release levers should be worn evenly where they contact the throwout bearing. The throwout bearing should be replaced whenever the clutch is disassembled. To check a throwout bearing, turn the bearing with your fingers. Check for smooth operation and damage.

Inspect the clutch fork for wear on the throwout-bearing mount and on the linkage pivot. Also look for distortion or bending. Replace damaged forks and linkage parts as necessary.

Assembling and Replacing the Clutch

Grease the pilot bearing or bushing and tap it into the center of the flywheel with the special tool recommended in the manufacturer's service manual. Place the clutch disc against the flywheel. Be sure the correct side of the clutch disc is against the

CLUTCH SERVICE

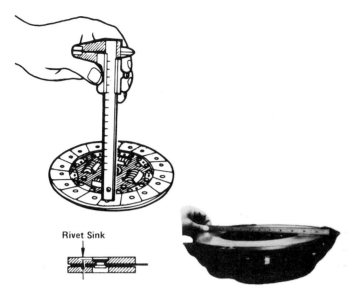

Figure 52-4. Measuring clutch lining. CHRYSLER CORPORATION

Figure 52-5. Checking for pressure plate warpage. FORD MOTOR COMPANY

flywheel. Insert an alignment tool or old transmission input shaft to support the disc [**Figure 52-6**].

Line up the marks on the clutch cover and flywheel. Insert the bolts and tighten to the manufacturer's specified torque. Remove the alignment tool or input shaft.

Throwout bearings usually are mounted in a sleeve, either with spring clips or by being press-fitted to the sleeve. For press-fit bearings, support the sleeve properly and use a hydraulic press to press the bearing in [**Figure 52-7**]. Place small amounts of lubricant inside the hub and on the surface that contacts the release levers or fingers.

Put a small amount of lubricant on the clutch fork pivot points. Install the clutch fork.

Install the bellhousing if it has been removed. Install the transmission or transaxle (see Unit 50), driveline (see Unit 48), and linkage systems. Install and torque all parts that were removed to remove the clutch and transmission. Check and adjust clutch pedal freeplay. Road test the vehicle to check for proper clutch operation.

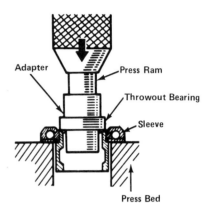

Figure 52-7. Installing a throwout bearing.

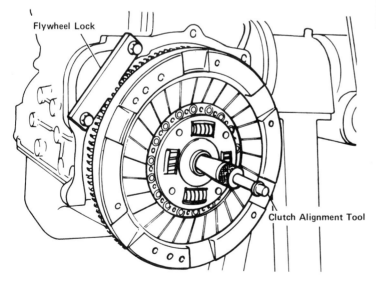

Figure 52-6. Clutch alignment tool. MAZDA MOTOR CORPORATION

UNIT HIGHLIGHTS

- Clutch preventive maintenance includes lubricating external clutch linkage.
- Clutch pedal free travel is measured to determine whether clutch linkage adjustment is needed.
- Correct clutch linkage adjustment prevents excessive clutch wear and makes shifting gears easier.
- A road test can help the technician to determine clutch problems.
- Testing for clutch problems includes blocking wheels and observing safety precautions, then starting the engine and operating the clutch.
- To remove a clutch assembly, the transmission or transaxle and other parts must be removed.
- Inspection of a disassembled clutch assembly includes visual and measurement checks.

TERMS

wheel chocks
clutch slippage
clutch drag
clutch pedal pulsation
free travel
clutch chatter
spindown time
clutch vibration

REVIEW QUESTIONS

DIRECTIONS: The following questions are similar to those used on automotive technician certification tests. On a separate sheet of paper, write the letter of the correct choice.

1. When making a clutch adjustment, it is necessary to
   A. measure clutch pedal free travel.
   B. lubricate the clutch linkage.
   C. check hydraulic fluid level.
   D. place the transmission in reverse.

2. Which of the following statements is true?
   I. Clutch slippage is most noticeable during acceleration and gear shifts.
   II. Clutch drag makes a transmission easy to shift into reverse.
   A. I only   C. Both I and II
   B. II only   D. Neither I nor II

3. Technician A says that the clutch disc can be replaced without removing the transmission.
   Technician B says that the clutch linkage can be adjusted without removing the transmission.
   Who is correct?
   A. A only   C. Both A and B
   B. B only   D. Neither A nor B

Questions 4 and 5 are not like those above. Each has the word EXCEPT. For each question, look for the choice that does *not* apply. Read each question carefully before you choose your answer.

4. During reassembly and clutch replacement, all of the following must be done EXCEPT
   A. lubricate the release bearing hub and release lever contact surfaces.
   B. align marks on the clutch cover and clutch disc.
   C. support the clutch disc with one of the mounting bolts.
   D. inspect and measure the clutch disc and pressure plate.

5. Clutch inspection requires the following procedures EXCEPT
   A. inspecting clutch disc springs.
   B. measuring the depth of friction material.
   C. looking for a mirror-like shine on the flywheel and pressure plate.
   D. placing a straightedge across the clutch disc.

ESSAY QUESTIONS

1. Why is clutch adjustment important?
2. Describe common clutch problems and how to diagnose them.
3. Describe the steps in clutch service.

SUPPLEMENTAL ACTIVITIES

1. Demonstrate how to remove asbestos fibers safely from a clutch assembly.
2. Measure and adjust clutch pedal free travel on a vehicle selected by your instructor.
3. Diagnose and describe possible causes of a clutch noise.
4. Remove and replace a clutch from a vehicle selected by your instructor.
5. Describe and list defects that were found in a clutch assembly during disassembly and inspection.

53 Automatic Transmissions And Transaxles

UNIT PREVIEW

A conventional automatic transmission is used in a rear-wheel drive drivetrain. The automatic transmission transfers torque from the engine to the rest of the drivetrain by using a fluid coupling instead of a clutch. An automatic transmission upshifts and downshifts through forward gears automatically.

Automatic transmissions are designed to function mechanically and hydraulically. In addition, computer controls can be used to shift the transmission for better acceleration or greater fuel economy.

Conventional automatic transaxles used in front-wheel drive vehicles are similar to automatic transmissions. As with manual transaxles, automatic transaxles include a differential unit.

Another design of automatic transaxle used in front-wheel drive transaxles for smaller vehicles is known as a continuously variable transmission (CVT). The CVT has no torque converter and greatly simplifies the job of creating different gear ratios.

LEARNING OBJECTIVES

When you have completed your assignments and exercises in this unit, you should be able to:

❑ Describe how a fluid coupling and a torque converter work.
❑ Describe how a planetary gearset operates.
❑ Describe the function of apply devices and holding and driving members.
❑ Describe the functions of a valve body.
❑ Identify and describe the four types of pressures used to control automatic transmission gear shifting.
❑ Describe the operation of an automatic transmission cooling system.
❑ Describe how computer controls are used to shift an automatic transmission or transaxle.
❑ Describe the differences between an automatic transaxle and an automatic transmission.
❑ Describe the parts and operation of a continuously variable transmission in a transaxle.

53.1 CONVENTIONAL AUTOMATIC TRANSMISSION DESIGN

A conventional automatic transmission includes a fluid coupling instead of a mechanically operated clutch. Different types of gears are used to control torque and speed output. Based on driver demands, engine load, and road speed, automatic transmissions upshift and downshift through forward gears automatically. The five main assemblies of a conventional automatic transmission are:

1. Torque converter (fluid coupling)
2. Planetary gearset
3. Apply devices
4. Hydraulic controls
5. Computer controls.

The parts of an automatic transmission perform the same functions as a clutch and manual transmission. Torque from the engine is applied to gearsets for different torque multiplication factors and vehicle road speeds. However, in most automatic transmissions, the mechanisms used to do these tasks are very different from those in a manual transmission.

53.2 FLUID COUPLING

An automatic transmission accepts power from the engine through a fluid connection, or coupling. A simple *fluid coupling* consists of an *impeller* and a *turbine* [Figure 53-1]. The impeller is connected to the engine flywheel through the fluid coupling metal housing. The turbine is placed closely behind the impeller and separated by a small space.

The impeller and turbine resemble two fans with blades, or vanes, that face each other. When one fan

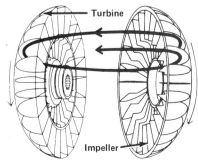

Figure 53-1. Fluid coupling.

(impeller) is turned, it pushes fluid through the vanes of the second fan (turbine). The fluid passing through the vanes of the second fan force it to turn.

The engine turns the impeller to create a flow of fluid. The fluid pressure generated by the revolving impeller turns the turbine.

All of the torque of the turning impeller is not transferred to the turbine. Some *slippage,* or *slip,* occurs because a fluid coupling is not as efficient as a mechanical clutch coupling. To reduce slippage, a simple fluid coupling is modified to create a torque converter, as used in modern automatic transmissions.

53.3 TORQUE CONVERTER

A *torque converter* is a series of parts that reduce slip and multiply torque. As shown in **Figure 53-2,** a torque converter may have six separate parts:

1. Torque converter housing
2. Impeller
3. Turbine
4. Stator
5. Overrunning clutch
6. Lockup system.

Torque Converter Housing

All parts of a torque converter are enclosed and sealed in a *torque converter housing.* The torque converter housing bolts to a *flexplate,* which bolts to the crankshaft. A flexplate/torque converter combination serves as the flywheel on cars with automatic transmissions **[Figure 53-3]**.

Impeller

In a torque converter, the impeller is part of the *rear* half of the torque converter housing. The torque converter housing is bolted to the flexplate, which is attached to the crankshaft. Thus, the impeller turns at engine speed.

The impeller is often called "the pump" by technicians because it pushes, or pumps, fluid inside the torque converter housing. The fluid is directed forward and toward the turbine. The impeller, however, is *not* the automatic transmission oil pump. The oil pump is a separate part that builds up hydraulic pressure in an automatic transmission.

Turbine

The turbine spins inside the *front* half of the torque converter housing. Fluid from the impeller is pumped into the curved vanes on the outside of the turbine. The moving fluid strikes the turbine, causing the turbine to rotate **[Figure 53-4]**. The input shaft also rotates because it is splined to the turbine hub. The input shaft is used to transmit torque from the torque converter to the gear assemblies.

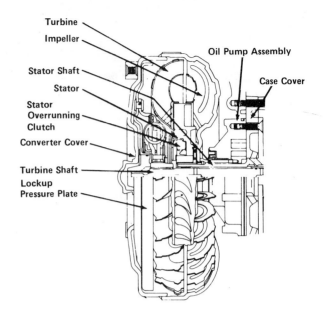

Figure 53-2. Parts of a torque converter assembly.
OLDSMOBILE DIVISION—GMC

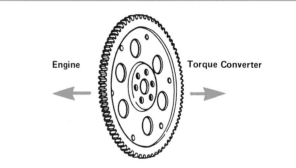

Figure 53-3. Flexplate.

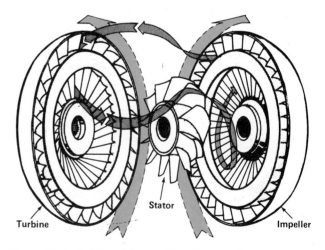

Figure 53-4. Turbine, stator, and impeller operation.
AMERICAN MOTORS CORPORATION

As the turbine rotates, the fluid that caused it to rotate falls to the center of the turbine. At the center of the turbine, the fluid passes through the stator.

Stator

When fluid is directed from the impeller to the turbine, the fluid flow is reversed. To correct the flow, a *stator*, **[Figure 53-5]**, is placed between the impeller and turbine. The vanes on the stator redirect fluid from the turbine to the impeller's direction of rotation. This change in fluid flow direction to match that of the impeller increases force and creates torque multiplication.

Overrunning Clutch

An *overrunning clutch*, or one-way clutch, allows the stator to rotate in only one direction. This assures that the stator will redirect fluid flow during low-speed operation. During higher-speed operation, the returning fluid from the turbine strikes the back side of the stator vanes. This causes the stator to rotate in a clockwise direction. The overrunning clutch is built into the hub of the stator.

Lockup System

Even during high-speed operation, the torque converter has a certain amount of slippage. To eliminate slip and improve efficiency, a *lockup system* may be used. A lockup system locks the turbine and the impeller together to form a mechanical coupling like a clutch.

The parts of a lockup system include clutch friction material, a sliding clutch piston, and torsion springs **[Figure 53-6]**.

During low-speed operation, automatic transmission fluid flows through the turbine shaft and forces the piston away from the lockup position. To engage lockup, the converter ATF flow is reversed, which forces the piston against the clutch surface. Some systems use a special silicone fluid that thickens when hot to create lockup, as in a fan clutch mechanism.

53.4 PLANETARY GEARSET

An automatic transmission is equipped with one or more *planetary gearsets* to create different forward

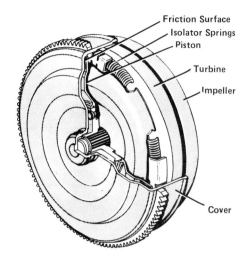

Figure 53-6. Lockup torque converter. CHRYSLER CORPORATION

gear ratios and reverse operation. A simple planetary gearset **[Figure 53-7]** includes:

- Sun gear
- Ring gear with internal teeth
- Planet gears
- Planet carrier.

Planet gears orbit, or circle, the sun gear. Planet gears are held in place by the planet carrier and the ring gear. The ring gear surrounds, and meshes with, the planet gears.

In a planetary gearset, the teeth of each gear are meshed at all times with the teeth of another gear. This is known as a *constant-mesh gearset*. Whenever one gear, or part, is turned, all other gears and parts in the system are affected.

To transmit power through a planetary gearset, one gear, or part, turns as the *drive member*. Another part, called the *reaction member*, is held and prevented from moving. A third part becomes the *driven member*, or output member. The driven member transmits torque to the driveline.

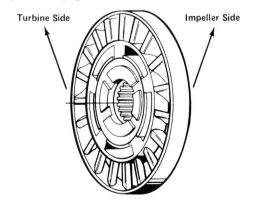

Figure 53-5. Stator. HYDRA-MATIC DIVISION—GMC

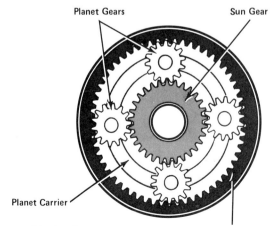

Figure 53-7. Simple planetary gearset.

All modern automatic transmissions have either two or three planetary gearsets linked together. These gearsets are called *compound planetaries*. A three-speed automatic has two planetary gearsets. A four-speed overdrive automatic has three planetary gearsets. Each gearset is controlled by a common or separate sun gear. **Figure 53-8** shows a sun gear that actually is a long tube splined to accept planetaries at both ends.

Planetary Gearset Power Flow

Planetary gears are used to change torque, change speed, and reverse direction of rotation. Planetary gears also provide neutral and act as a coupling for direct drive, a 1:1 ratio between the transmission input and output shafts.

Multiplying torque generally is known as reduction, because there is always a decrease in the speed of the output member. With a constant input speed, the output torque increases as the output speed decreases.

The following discussions describe the common forms of planetary gearset power flow.

Simple reduction. Simple reduction occurs when the sun gear is held and torque is applied to the ring gear in a clockwise direction **[Figure 53-9]**. In this case, the planetary pinions rotate in a clockwise direction and "walk" around the stationary sun gear. This causes the carrier assembly to rotate clockwise, in reduction.

Direct drive. Direct drive occurs when any two members of the planetary gearset rotate in the same direction at the same speed. This forces the third member to turn at the same speed **[Figure 53-10]**. In this condition, the pinions do not rotate on their axles. The pinions lock the entire unit together to form one rotating part.

Neutral. If none of the members are held, torque cannot flow through the gearset. It is in its neutral position.

Overdrive. When the sun or ring gear is held and the carrier is driven, the remaining member turns more

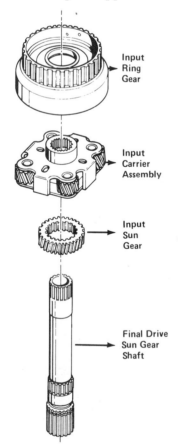

Figure 53-8. Compound planetary sun gear.

HYDRA-MATIC DIVISION—GMC

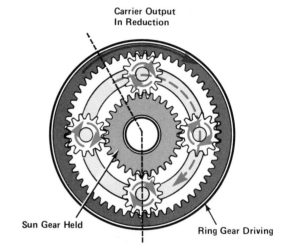

Figure 53-9. Simple reduction.

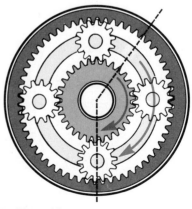

Figure 53-10. Direct drive.

rapidly than the driving member with an overdrive (less than 1:1) gear ratio.

Reverse. Reverse occurs whenever the carrier is held and power is applied to either the sun gear or the ring gear. This causes the planet pinions to drive the output member in the opposite direction, as shown in **Figure 53-11**.

53.5 APPLY DEVICES

A planetary gearset is controlled by *apply devices*. An apply device applies force that actuates, holds, or drives other parts. Depending upon which apply device is activated, one member of a planetary gearset is held and another is driven. Apply devices can be separated into two categories:

1. Hydraulic actuators
2. Holding and driving devices.

Hydraulic actuators, such as servos and piston clutches, apply hydraulic pressure to activate holding and driving devices. The holding and driving devices, such as multiple-disc clutches, transmission bands, and overrunning clutches, apply mechanical force to gearsets.

Multiple-Disc Clutch

A *multiple-disc clutch* consists of a series of friction discs, or plates, sandwiched between steel discs **[Figure 53-12]**. Friction discs have composition surfaces. Steel discs have smooth surfaces.

To lock a multiple-disc clutch, fluid pressure pushes a piston to compress a return spring and force the clutch plates against the pressure plate. When friction and steel discs are compressed against the pressure plate, the discs, input shaft, and drum rotate as a unit. The clutch pack is released when fluid pressure is released.

Piston Clutch

A *piston clutch* is an actuating device. It has a large cylinder fitted with a large flat piston. When pressure is applied to the piston, force is applied directly to the pressure plate **[Figure 53-13]**.

Transmission Band

A *transmission band,* also called a brake band, is a holding member. A transmission band is a flexible piece of metal wrapped around a clutch drum **[Figure 53-14]**. The inside of a transmission band has a composition friction surface that grips the clutch drum. The band is tightened to hold a clutch drum and is released to allow the drum to rotate freely.

Servo

A *servo* is an actuating device that consists of a piston and rod in a hydraulic cylinder **[Figure 53-15]**. Hydraulic pressures operate a servo to clamp or release a transmission band around a clutch drum.

Accumulator

An *accumulator* acts as a hydraulic shock absorber when a clutch, piston, or band is applied. When pressure enters a servo, the pressure also enters the accumulator, which cushions the application of clutches and bands **[Figure 53-16]**.

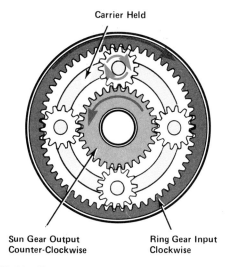

Figure 53-11. Reverse gear.

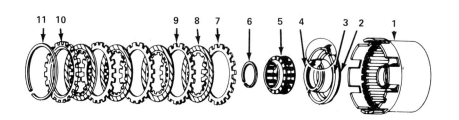

1. Reverse Input Housing
2. Reverse Input Outer Seal
3. Reverse Input Piston
4. Reverse Input Inner Seal
5. Reverse Input Spring Assembly
6. Reverse Input Spring Snap Ring
7. Reverse Input Plate (Waved)
8. Reverse Input Plate (Composition)
9. Reverse Input Plate (Steel)
10. Reverse Input Backing Plate
11. Reverse Input Snap Ring

Figure 53-12. Multiple-disc clutch assembly. CHEVROLET MOTOR DIVISION—GMC

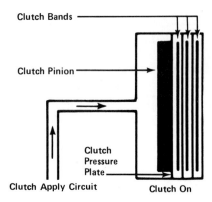

Figure 53-13. A piston clutch.

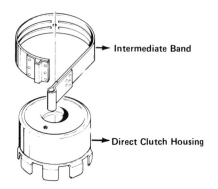

Figure 53-14. Transmission band assembly. HYDRA-MATIC DIVISION—GMC

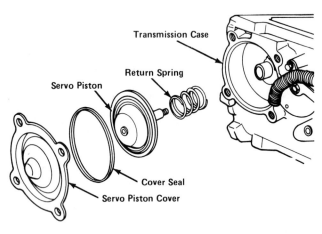

Figure 53-15. Servo assembly. FORD MOTOR COMPANY

Overrunning Clutches

An overrunning clutch is used with a planetary gearset system. It holds or applies planetary gearset members as the vehicle starts to move. During coasting, it allows the transmission shaft to turn faster than the engine during coasting. An overrunning clutch has small, spring-loaded rollers, or *spags*, s-shaped metal parts between an inner drum and an outer hub. When the clutch is applied, these parts are wedged between the drum and the hub to prevent counterclockwise rotation [Figure 53-17].

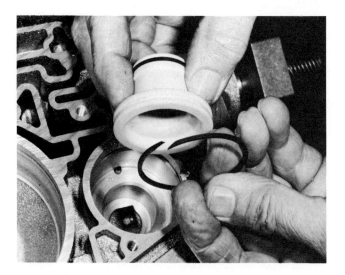

Figure 53-16. Accumulator assembly.

53.6 HYDRAULIC PRINCIPLES

Liquids cannot be compressed, or squeezed into a smaller volume. For example, place a plunger into a tightly sealed cylinder filled with water [Figure 53-18]. When force is applied to the plunger, the plunger cannot squeeze the liquid into a smaller volume. The plunger pushes against the liquid, causing *pressure* to build up. Pressure is the amount of force pushing on each unit of the liquid's surface area. Pressure is applied to the bottom surface of the plunger and to the inner surfaces of the container.

Pressure is measured by dividing the force pushing on a liquid by the amount of surface area to which the force is applied. If the force on the

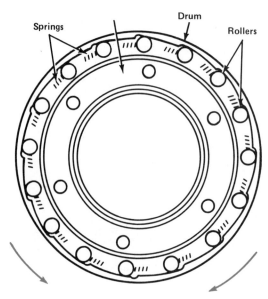

Figure 53-17. A roller-type overrunning clutch.

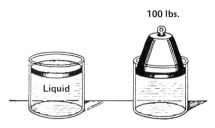

Figure 53-18. Liquid cannot be compressed even when weight is added. PONTIAC MOTOR DIVISION—GMC

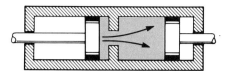

Figure 53-19. Liquid under pressure can be pushed from one cylinder to another by the use of pistons.

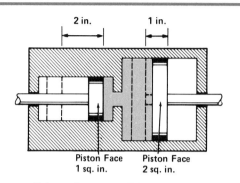

Figure 53-20. If the surface area of the two pistons is different, the larger piston will move a shorter distance when the smaller piston is moved.

plunger is 100 pounds, and the surface area of the plunger is 10 square inches, the pressure on the liquid is 100 pounds divided by 10 square inches, or 10 *pounds per square inch (psi).* To find pressure, divide the force by the area of the plunger, or piston, face.

Pressure is applied equally to every surface touched by the liquid. This relationship is known as Pascal's Law. This law states: Pressure applied on a confined liquid is transmitted equally in all directions and acts with equal force on equal areas.

Motion Transfer

Liquid under pressure can be made to do work. Connecting two piston and cylinder units together will allow liquid pressure to push against the second piston, causing it to move [Figure 53-19]. This is called *motion transfer.*

Moving one piston will push liquid into the second cylinder. If both pistons and cylinders are the same size, moving one piston will cause the other piston to move the same distance.

However, if one piston has half the area of a connected piston, the motion of the second piston will be less. Moving the smaller piston two inches will cause the larger piston to move only one inch [Figure 53-20].

Pushing on a larger piston would transfer more liquid to the smaller cylinder. If the larger piston is pushed 2 inches, the smaller piston will move twice as far, or 4 inches.

The distance between cylinders has no effect. If the connecting tube and both cylinders are completely filled with fluid, both fluid and motion will be transferred.

Pressure Transfer

Pressure transfer occurs in a fluid-filled system. The pressure in a closed, sealed system is the same everywhere. If pressure gauges are attached at various points, they will all register the same pressure [Figure 53-21].

Force Transfer

In a hydraulic system, pressure in connected cylinders is the same. However, if the cylinders are of

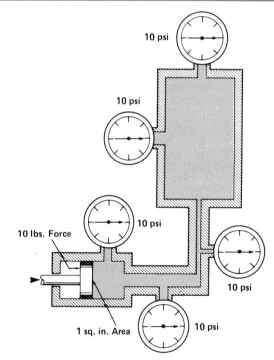

Figure 53-21. Pressure is the same throughout a sealed system.

different sizes, different amounts of force will be exerted on the pistons. This is called *force transfer.*

Remember, pressure is measured in force per unit area, commonly in pounds per square inch (psi) or, in metric terms, kilopascals (kPa). Pounds per square inch are converted to kilopascals by multiplying psi by 6.895. Thus, a pressure of 10 psi is equal to 68.95 kPa in metric terms.

Suppose a 1-square-inch piston is connected by a passage or a tube to a 2-square-inch piston. The smaller piston has a 10-pound force acting on it. The pressure acting against the larger piston is 10 pounds per square inch. However, because the larger piston has 2 square inches of area, the pushing force is doubled. The total force pushing the larger piston is 10 pounds x 2, or 20 pounds **[Figure 53-22]**.

Pressure and Flow

Pressure can be generated only if there is some resistance to the flow of liquid. In a closed container, pressure is generated by the resistance of the walls of the container to the movement of the fluid. If a leak develops, pressure within the container drops.

In a system with separate cylinders connected by tubes or passages, the resistance of the cylinder pistons to movement and leakage causes pressure to build up.

53.7 HYDRAULIC CONTROLS

A hydraulic control system is used to control and distribute force. The following discussions describe the various parts, components, and functions of the hydraulic control system.

Gearshift Controls

The gearshift lever for an automatic transmission may be located on the steering column or on the floor. An indicator near the gearshift lever always points to one of the shift positions, which may include:

- Park (P)
- Reverse (R)
- Neutral (N)
- Overdrive (O)
- Drive (D)
- Second gear (2)
- First gear (marked either 1 or L).

This series of letters, P-R-N-O-D-2-L, is called the *shift quadrant*. This is one type of quadrant for an automobile equipped with a four-speed overdrive transmission **[Figure 53-23]**. Another transmissions quadrant marking that indicate overdrive is "D4."

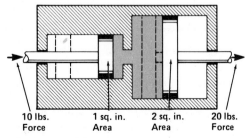

Figure 53-22. Pressure acting against a larger piston will increase the amount of force transferred.

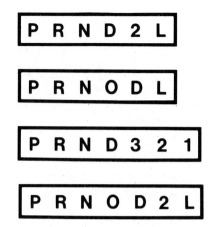

Figure 53-23. Shift quadrant variations.

Gearshift linkage for an automatic transmission is connected from the gearshift lever to a lever arm on the transmission. Either rod and lever or cable linkage is used with an automatic transmission **[Figure 53-24]**.

A separate linkage system, called *kickdown linkage*, is used for downshifting the transmission without moving the gearshift lever. Kickdown linkage, or *throttle valve cable*, is connected to the throttle, or throttle pedal, and the transmission. Kickdown linkage can be mechanical **[Figure 53-25]**, cable **[Figure 53-26]**, or electronic **[Figure 53-27]**. Kickdown occurs when the throttle pedal is depressed to the floor, such as in highway passing situations.

Automatic Transmission Fluids

The liquid used to operate hydraulic controls should have certain qualities. It must be able to transmit pressure, lubricate, cool, and have the proper *viscosity* to flow through the narrow orifices. Viscosity is the thickness or thinness of a fluid, and its ability to resist flowing.

Automatic transmission fluid (ATF) is a thin lubricant oil, similar to engine oil, and may be made from either petroleum or from synthetic lubricants. Special additives are mixed with the fluid to maintain viscosity and to help resist foaming, sludge and varnish buildup, and corrosion. Three common types of petroleum-based automatic transmission fluid are:

- DEXRON II
- Ford Mercon
- Ford Type F
- Ford Type H.

Refer to the vehicle manufacturer's service manual and service bulletins for the correct fluid to be used in a specific transmission. Do not mix different types of automatic transmission fluid.

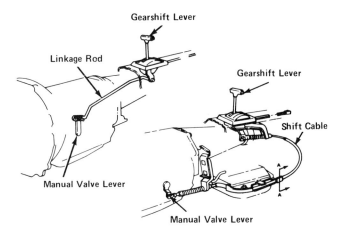

Figure 53-24. Shift linkage systems. FORD MOTOR COMPANY

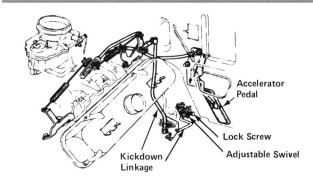

Figure 53-25. Mechanical kickdown linkage.
CHRYSLER CORPORATION

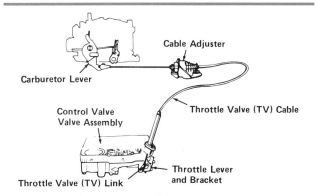

Figure 53-26. Cable kickdown linkage.
CHEVROLET MOTOR DIVISION—GMC

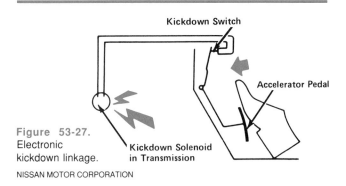

Figure 53-27. Electronic kickdown linkage.
NISSAN MOTOR CORPORATION

Oil Pump

The transmission *oil pump* creates pressure to operate apply devices and to circulate fluid throughout the system.

On modern conventional automatic transmissions, one oil pump is used. It is located just behind and is driven by the torque converter. The pump operates only when the engine runs.

An oil pump is made up of a drive member, a driven member, and two pump body halves. The pump body halves contain passages that direct fluid to various parts of the transmission. Some oil pumps have other controls built into them.

All transmission oil pumps draw fluid from a reservoir and distribute the fluid throughout the hydraulic system. The three common types of oil pumps are:

1. Gear and crescent [Figure 53-28]
2. Rotor [Figure 53-29]
3. Vane type [Figure 53-30].

Control Valves

A *valve body* is a housing that contains a maze of passages with valves to control and direct hydraulic pressures. These pressures are used to control the actions of apply devices and shift the planetary gearsets. A valve body is located inside the transmission oil pan [Figure 53-31]. A *separator plate*, or transfer plate, is placed between the two parts of the valve body. The separator plate helps to direct and connect fluid pressure [Figure 53-32].

Control valving consists of pressure-regulating, balancing, and shifting valves located inside machined bores in the valve body. Hydraulic pressure is regulated for use in a particular hydraulic circuit. Four main types of fluid pressure are used in automatic transmissions:

1. Mainline pressure
2. Throttle pressure
3. Vacuum modulator pressure
4. Governor pressure.

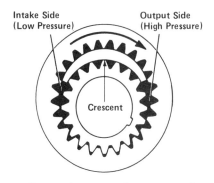

Figure 53-28. Gear and crescent oil pump operation.
CHEVROLET MOTOR DIVISION—GMC

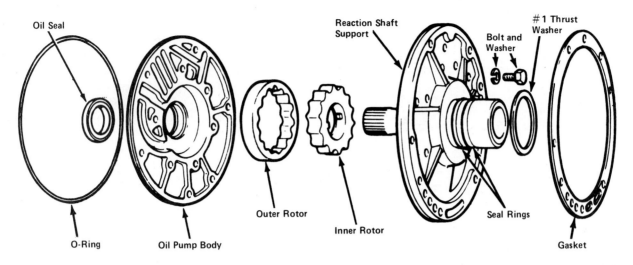

Figure 53-29. Rotor oil pump. AMERICAN MOTORS CORPORATION

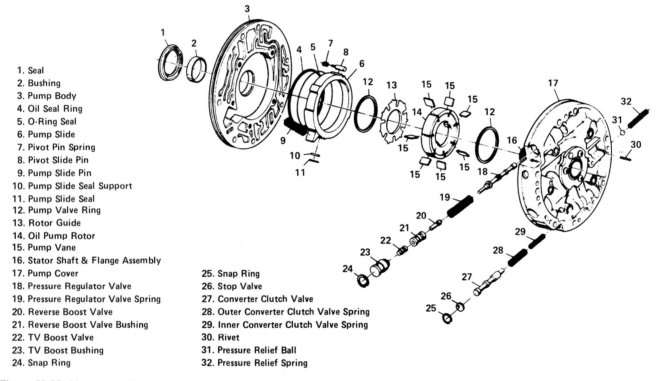

1. Seal
2. Bushing
3. Pump Body
4. Oil Seal Ring
5. O-Ring Seal
6. Pump Slide
7. Pivot Pin Spring
8. Pivot Slide Pin
9. Pump Slide Pin
10. Pump Slide Seal Support
11. Pump Slide Seal
12. Pump Valve Ring
13. Rotor Guide
14. Oil Pump Rotor
15. Pump Vane
16. Stator Shaft & Flange Assembly
17. Pump Cover
18. Pressure Regulator Valve
19. Pressure Regulator Valve Spring
20. Reverse Boost Valve
21. Reverse Boost Valve Bushing
22. TV Boost Valve
23. TV Boost Bushing
24. Snap Ring
25. Snap Ring
26. Stop Valve
27. Converter Clutch Valve
28. Outer Converter Clutch Valve Spring
29. Inner Converter Clutch Valve Spring
30. Rivet
31. Pressure Relief Ball
32. Pressure Relief Spring

Figure 53-30. Vane-type oil pump. CHEVROLET MOTOR DIVISION—GMC

Mainline pressure. Hydraulic pressure from the oil pump is regulated to prevent damage. This regulated pressure is called *mainline pressure.*

A *pressure regulator valve* controls pressure from the oil pump. The valve balances between a force exerted by a spring and a hydraulic force against the piston [Figure 53-33].

When the gearshift lever is moved, it moves a *manual control valve* [Figure 53-34]. It directs fluid to other shifting and pressure-regulating valves and to apply devices.

An extra load on the engine requires higher hydraulic pressure for clutches and bands to ensure a stronger grip. A *booster valve* is used to raise mainline pressure during these conditions. A booster valve is a balancing valve with a pressure-regulating spring.

Shifting valves route oil pressure to apply devices to engage the correct gear ratio for driving conditions. Shifting valves are operated by a combination of throttle pressure and governor pressure.

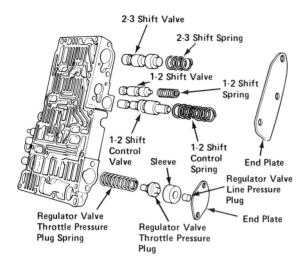

Figure 53-31. Valve body. AMERICAN MOTORS CORPORATION

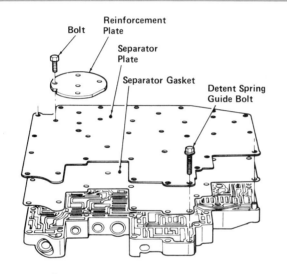

Figure 53-32. Separator plate. FORD MOTOR COMPANY

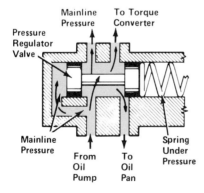

Figure 53-33. Pressure regulator valve assembly.

Throttle pressure. *Throttle pressure* is used to increase mainline pressure and to help control shifting. Throttle pressure increases and decreases with throttle movement.

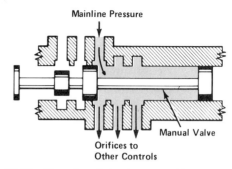

Figure 53-34. Manual control valve assembly.

Throttle pressure is regulated by a *throttle valve* [Figure 53-35]. A throttle valve is a balancing valve with a lever-regulated spring. A throttle valve can be either connected to the throttle linkage or operated by engine vacuum.

Vacuum modulator pressure. Engine vacuum also is used to exert pressure on the throttle valve spring. A vacuum-controlled diaphragm, or *vacuum modulator*, is controlled by vacuum from the engine intake manifold [Figure 53-36]. As the throttle is opened and closed, the vacuum modulator modulates, or changes, throttle pressure in response to engine load. High vacuum results in lower modulator pressure. Low vacuum results in higher modulator pressure.

Other valves also are used to control pressure. These valves may be called downshift, detent, or kickdown valves.

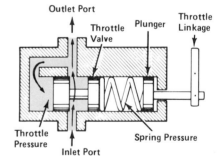

Figure 53-35. Throttle valve assembly.

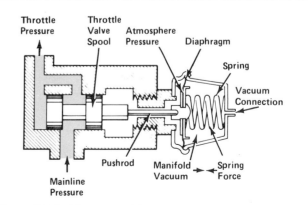

Figure 53-36. Vacuum modulator.

Governor pressure. *Governor pressure* is used to help upshift the transmission into the proper gear. Governor pressure and throttle or modulator pressure work against each other to produce upshifts or downshifts at the correct time.

A governor reduces mainline pressure in response to output shaft speed **[Figure 53-37]**. Higher output shaft speed results in higher governor pressure.

In recent automatic transmissions, some or all of the functions listed above are replaced by electrically controlled solenoid valves.

53.8 COMPUTER CONTROLLED AUTOMATIC TRANSMISSIONS

On recent vehicles, computers are used to control automatic transmission shifting functions. Input signals for engine and road speed, manifold vacuum, engine operating temperature, gear selection, throttle position, and other factors are provided to the computer. The computer produces output signals that operate relays, which activate *electrical solenoid valves.* Refer to Unit 38 for a discussion of how computers operate.

An electrical solenoid valve moves a spool valve inside a valve body. Instead of hydraulic pressures and springs, the motion of the solenoid controls the valve position. When activated, the solenoid moves the valve to control fluid pressures in a valve body. The operation of a solenoid for a starter system is discussed in Unit 30.

Computer controls can "hold" the transmission in each gear longer for better acceleration, when a driver selects a "Power" range (computer program) for automatic shifting. Alternately, the computer creates upshifts sooner for improved gas mileage in a "Normal" or "Economy" mode. Computer controls also are applied to solenoid valves to create torque converter lockup for better fuel economy (refer to 53.3).

53.9 TRANSMISSION OIL COOLING

The actions of the torque converter, shafts, and gears heat the automatic transmission fluid. Excessive heat and friction will quickly damage an automatic transmission. An internal or external cooler, or combination of both, is used to dissipate this heat.

Internal Transmission Cooler

The engine radiator outlet tank contains a small, sealed heat exchanger that operates as an *internal transmission cooler.* Heated transmission fluid is cooled and returned to the transmission, as shown in **Figure 53-38**.

External Transmission Cooler

A separate, small radiator may be used for an *external transmission cooler* **[Figure 53-39]**. This cooling system is common on trucks and other vehicles that are subjected to heavy loads, such as towing. The transmission radiator is mounted outside the engine radiator, frequently just in front of it. Heated fluid usually runs through the internal cooler first and then through the external cooler before returning to the transmission.

53.10 CONVENTIONAL AUTOMATIC TRANSAXLE DESIGN

An automatic transaxle housing encloses an automatic transmission and a differential. A manual

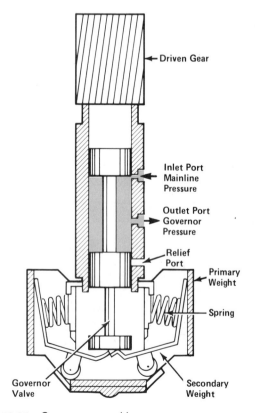

Figure 53-37. Governor assembly.

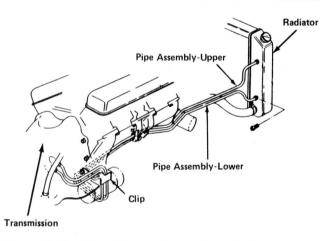

Figure 53-38. Internal transmission cooler routing lines.
BUICK MOTOR DIVISION—GMC

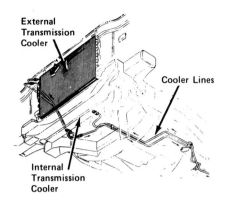

Figure 53-39. External transmission cooler assembly.
CHRYSLER CORPORATION

transaxle drivetrain is similar in design and mounting to a manual transaxle [Figure 53-40]. Both use the same type of driving axles and constant-velocity (CV) joints (refer to Unit 50).

53.11 CONVENTIONAL AUTOMATIC TRANSAXLE OPERATION

The transmission portion of an automatic transaxle uses planetary gears, just as in a conventional automatic transmission. The differential of an automatic transaxle operates in the same manner as its manual transaxle counterpart (see Unit 55).

Engine torque is routed through two 90-degree turns in a transaxle. Power is transferred from the output end of the turbine shaft to the differential input shaft through a chain, shaft, or gears. When a chain is used, it is connected to sprockets at each shaft [Figure 53-41]. Note the location of the valve body in the transaxle shown.

Torque Converter and Planetary Gearsets

A torque converter is mounted inside the transaxle housing and behind the engine flywheel. Transaxle planetary gear assemblies are mounted inside the same transaxle body, but either behind or to one side of the torque converter. Figure 53-42 shows a transaxle gear assembly mounted to one side of the torque converter. In a rear-drive automatic transmission, the transmission is mounted directly behind a torque converter.

Hydraulic Controls

Once the gearshift lever is set in the Drive position, automatic shifting is controlled by engine speed, load, and throttle position. When the torque converter is operating, the oil pump creates hydraulic pressure. The valve body and other valve systems control hydraulic pressure and regulate shifting.

Computer-controlled Transaxles

Computer controls are used extensively on recent transaxles (refer to 53.8, above, and to Unit 38). Input signals for engine and road speed, manifold vacuum, engine operating temperature, gear selection, throttle position, and other functions are used. The computer produces output signals that activate electrical relays. The relays supply current to operate solenoid valves.

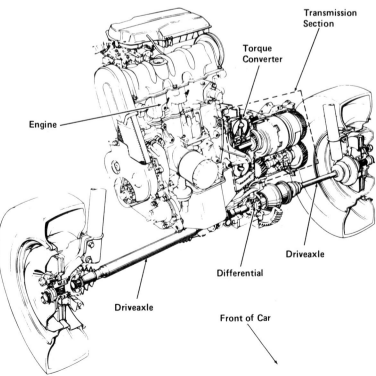

Figure 53-40. Automatic transaxle assembly.
CHRYSLER CORPORATION

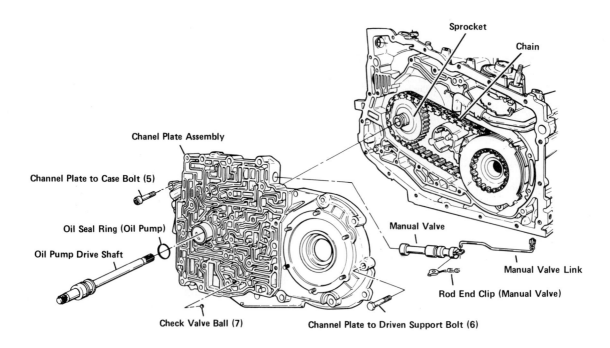

Figure 53-41. Sprocket and chain drive assembly. CADILLAC MOTOR CAR DIVISION—GMC

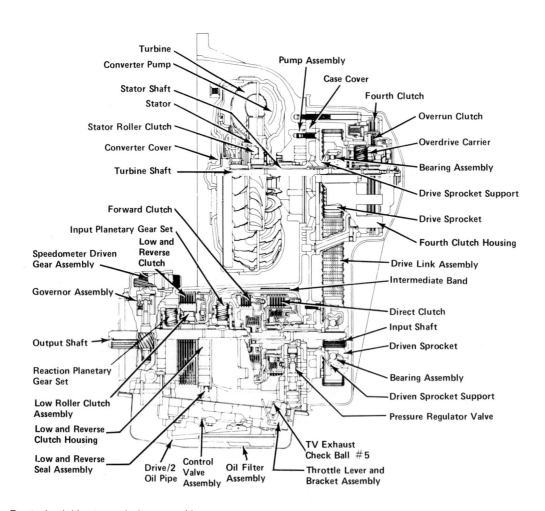

Figure 53-42. Front-wheel drive transmission assembly. BUICK MOTOR DIVISION—GMC

ATF Reservoir

Automatic transmission fluid (ATF) in some automatic transaxles may be stored in more than one oil pan, or reservoir. In some cases, an oil pan is mounted on the side of the transaxle. A thermostatic valve is used to regulate fluid flow between the two reservoirs, depending on transaxle temperature.

Automatic Transaxle Design Differences

Space and mounting requirements dictate transaxle design. Examples of important differences between automatic transmissions and transaxles include:

- Valve body locations may be on the top, side, or bottom of a transaxle.
- A valve body may be integral with the transaxle housing.
- A drive pinion gear may be integral with a transfer shaft.

As discussed above, recent transaxle designs include computer-controlled solenoid valves in place of hydraulically controlled units. Refer to Unit 38 for information on computer control systems.

53.12 CONTINUOUSLY VARIABLE TRANSMISSION (CVT)

In 1989, Subaru introduced a new type of automatic transmission for front-wheel drive transaxles. This transmission is known as a *continuously variable transmission (CVT)* **[Figure 53-43]**. A continuously variable transmission provides a stepless transition from the lowest to the highest forward gear ratios. The CVT includes a computer-controlled hydraulic system like that in current transaxles. A magnetic clutch assembly couples the CVT to the engine.

Electromagnetic Powder Clutch

In place of a torque converter, the Subaru CVT uses an *electromagnetic powder clutch* **(refer to Figure 53-43).** This clutch mechanism uses magnetic force to hold metal powder rigidly in place between a driving and a driven member to lock them together mechanically.

The clutch housing between the driving and driven members is filled with powdered stainless steel. A switch on the accelerator linkage signals the computer when the driver begins to press the accelerator. The computer operates a relay that sends a

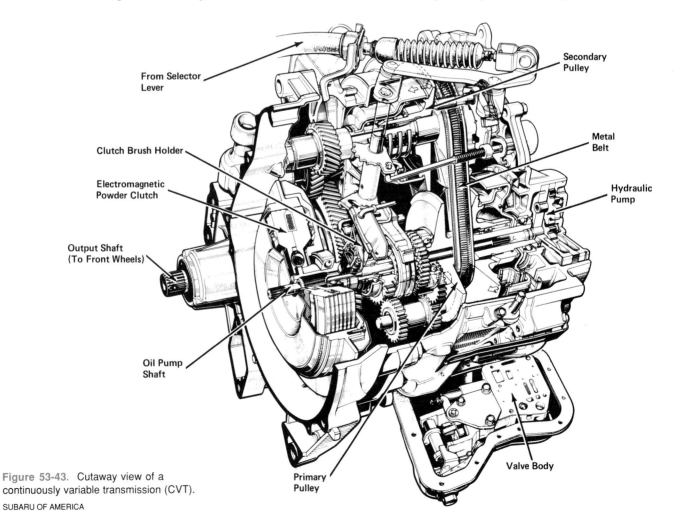

Figure 53-43. Cutaway view of a continuously variable transmission (CVT).
SUBARU OF AMERICA

small amount of current to the coils in the electromagnetic clutch. This creates enough magnetism to hold some of the powder rigidly between the driving and driven members. This mode allows a slight slippage in the transmission to move the vehicle forward gradually.

As the accelerator is depressed further, the computer responds by energizing a relay to deliver a large amount of current to the coils. The powder then locks the engine directly to the transmission input shaft.

When the vehicle comes to a full stop, a reverse DC current is sent to the coils to de-energize and uncouple the electromagnetic clutch completely.

Variable-Width Pulley System

In place of a complex planetary gear system, two *variable-width pulleys* are used to create a range of gear ratios. Hydraulic pressure is used to move one side of either pulley inward or outward. This action make the pulleys narrower or wider. A special, non-stretchable metal belt composed of multi-layered flexible bands and individual steel links, is used between the two pulleys **[Figure 53-44]**.

Either pulley can be made wider or narrower **[Figure 53-45]**. The belt remains taut at all times between the pulleys. When a pulley is widened, the belt will ride down toward the bottom of the groove. When the pulley is narrowed, the belt will ride up toward the top of the groove. This creates different size pulleys, which, like gears of different sizes, creates different gear ratios (refer to Unit 49).

When the primary, or drive pulley is widest, and the secondary, or driven pulley is narrowest, a "low" gear ratio (approximately 2.5:1) is created. When the pulleys are the same size, a 1:1 direct drive gear ratio is created.

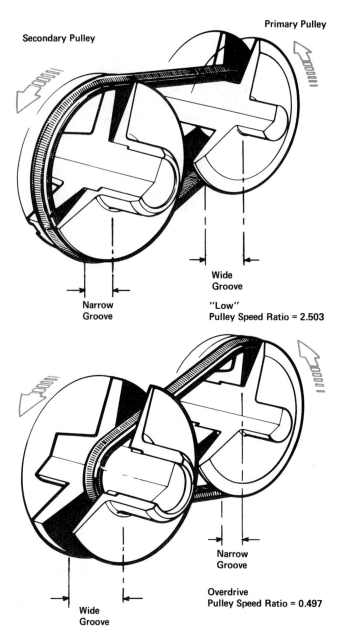

Figure 53-45. CVT variable pulleys. SUBARU OF AMERICA

When the drive pulley is narrowest and the driven pulley is widest, an overdrive gear ratio (approximately 0.5:1) is created. Between these extremes, the CVT provides a stepless range of gear ratios. This allows the engine to run in its most efficient speed range for good fuel mileage and maximum torque output.

A separate idler gear setup is used to create reverse rotation, as in a manual transmission or transaxle.

CVT Control System

The Subaru CVT includes a computerized-hydraulic control system **[Figure 53-46]**. Spool valves and

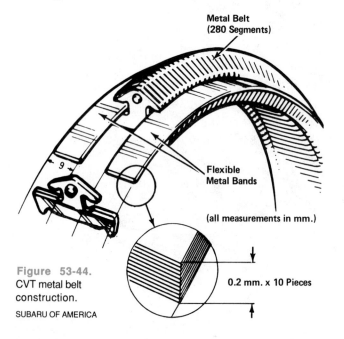

Figure 53-44. CVT metal belt construction. SUBARU OF AMERICA

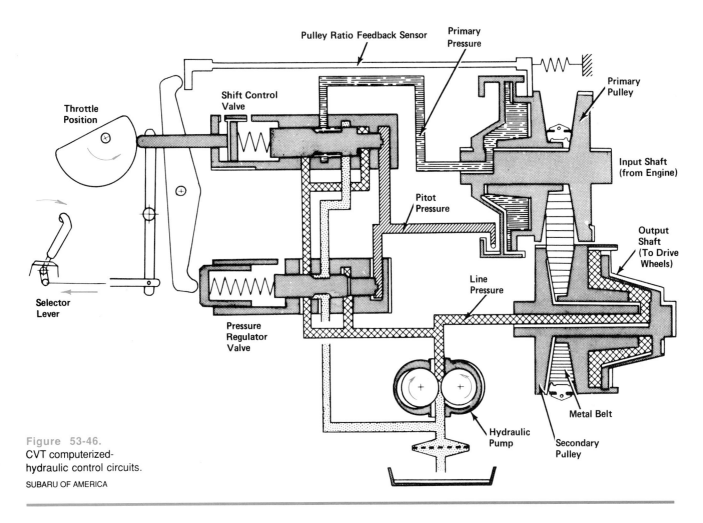

Figure 53-46. CVT computerized-hydraulic control circuits. SUBARU OF AMERICA

solenoid valves are used in the valve body. Dexron II is the recommended replacement transmission fluid.

A metal cable from the throttle linkage moves a throttle valve, just as in some conventional automatic transmissions. A road speed signal is created by hydraulic pressure in the primary pulley hydraulic circuit.

The CVT computer receives input signals for engine speed, coolant temperature, vehicle speed, barometric/manifold pressure, and fuel system operation. Based on a program, or set of instructions, the computer creates output signals to energize relays that provide current to the electromagnetic clutch and pulley solenoid valve control system.

UNIT HIGHLIGHTS

- A fluid coupling transfers engine power to a conventional automatic transmission or transaxle.
- A torque converter reduces slip and increases performance in an automatic transmission.
- A planetary gearset can change gear ratios and shift while it is meshed.
- Apply devices are holding and driving assemblies.
- Different types of hydraulic valves supply pressure to shift the planetary gearset.
- A valve body is a housing that contains a maze of passages and valves to control and direct hydraulic pressure.
- Four main types of fluid pressures are used in automatic transmissions.
- A transmission is cooled by circulating its fluid through a heat exchanger in the radiator.
- The transmission part of a conventional automatic transaxle operates in much the same way as an automatic transmission.
- A torque converter is mounted to the side of the transmission assembly in a conventional automatic transaxle.
- A continuously variable transmission (CVT) is used in transaxles on small FWD vehicles.
- A CVT has an electromagnetic clutch in place of a torque converter and uses variable-width pulleys and a steel belt.
- A CVT can create a range of gear ratios from low through overdrive.

TERMS

fluid coupling	impeller
turbine	slippage
torque converter	torque converter housing
flexplate	stator
overrunning clutch	lockup system
planetary gearset	drive member
reaction member	driven member
apply device	multiple-disc clutch
constant mesh gearset	compound planetaries
transmission band	servo
accumulator	pressure
motion transfer	pressure transfer
force transfer	shift quadrant
kickdown linkage	throttle valve cable
viscosity	piston clutch
oil pump	valve body
mainline pressure	pressure regulator valve
manual control valve	booster valve
shifting valve	throttle pressure
throttle valve	vacuum modulator
governor pressure	sprags
electrical solenoid valve	separator plate
electromagnetic powder clutch	variable width pulley
pounds per square inch (psi)	continuously variable transmission (CVT)
internal transmission cooler	external transmission cooler
automatic transmission fluid (ATF)	

REVIEW QUESTIONS

DIRECTIONS: The following questions are similar to those used on automotive technician certification tests. On a separate sheet of paper, write the letter of the correct choice.

1. A planetary gearset
   A. is not a constant-mesh gearset.
   B. requires a reverse idler gear to provide reverse rotation.
   C. includes a ring gear, a carrier and planetary gears, and a sun gear.
   D. turns the held member during reverse rotation.

2. Which of the following statements is correct?
   I. Apply devices include multiple disc clutches and bands.
   II. A servo acts as a shock absorber.
   A. I only   C. Both I and II
   B. II only   D. Neither I nor II

3. Technician A says that pressure varies in different parts of a sealed system according to the surface area of the fluid passages.
   Technician B says that pressure can be generated only if there is a resistance to the flow of a liquid.
   Who is correct?
   A. A only   C. Both A and B
   B. B only   D. Neither A nor B

Questions 4 and 5 are not like those above. Each has the word EXCEPT. For each question, look for the choice that does *not* apply. Read each question carefully before you choose your answer.

4. Conventional automatic transaxles and transmissions share all the following features EXCEPT
   A. a torque converter.   C. a valve body.
   B. a planetary gearset.   D. a differential.

5. All of the following statements about a continuously variable transmission (CVT) used in a FWD automatic transmission are true EXCEPT
   A. a CVT uses a variable-width pulley system to create different gear ratios.
   B. a CVT uses a torque converter to couple the engine to the input shaft.
   C. a CVT includes a hydraulic control system.
   D. a CVT includes computer-controlled solenoid valves in the valve body.

ESSAY QUESTIONS

1. What are the functions of the parts of a conventional automatic transmission?
2. What four types of control pressures are used in an automatic transmission valve body?
3. Describe how a CVT changes forward gear ratios and creates reverse rotation.

SUPPLEMENTAL ACTIVITIES

1. Identify the parts of a torque converter.
2. Describe how a lockup system operates.
3. Identify and describe the parts of a planetary gearset.
4. Describe how a transmission band and servo work together.
5. Describe how force, motion, and pressure is transferred in a hydraulic system.
6. Locate the oil reservoir(s) on a conventional automatic transmission.
7. If available, safely test drive a vehicle equipped with a continuously variable transmission. Report to your class the differences you noticed in acceleration, engine speed, and general performance, compared to both a conventional automatic transaxle and a manual transaxle.

54 Automatic Transmission And Transaxle Service

UNIT PREVIEW

Regular checks of automatic transmission and transaxle fluid level and condition should be made. Preventive maintenance for these units includes lubricating shift linkage and draining and replacing automatic transmission fluid. As part of the fluid change, a disposable filter is replaced or a reusable screen is cleaned. Where applicable, adjustments to transmission bands are made.

Specialized tests can be performed to diagnose automatic transmission or transaxle problems. These tests include measurements of hydraulic operating pressures and air pressure checks of the valve body. Transmission linkage is checked and adjusted, if necessary. In addition, road tests usually are part of problem diagnosis.

Service procedures for transaxle differentials are essentially the same as those for rear-wheel drive (RWD) vehicle differentials.

Most preventive maintenance procedures for continuously variable transmissions (CVTs) in transaxles are similar to conventional automatic transaxles. CVT service also includes specific mechanical services related to the pulleys and metal link belts.

LEARNING OBJECTIVES

When you have completed your assignments and exercises in this unit, you should be able to:

- Check for proper automatic transmission or transaxle fluid level.
- Diagnose the condition of transmission fluid.
- Change automatic transmission or transaxle fluid and replace a filter or clean a screen.
- Perform a pressure test and explain how it helps in diagnosing transmission or transaxle problems.
- Change a vacuum modulator.
- Describe adjustments that can be made to linkage systems.
- Adjust transmission bands.
- Describe common service procedures for continuously variable transaxles.

SAFETY PRECAUTIONS

Refer to the manufacturer's service manual and literature for specific work procedures and safety cautions.

Always wear eye protection in the shop area. *Never* work under a vehicle supported only by a jack. Proper jacks and supporting tools, as well as a helper, are essential when servicing transmissions.

Hot automatic transmission fluid, transmission and transaxles housings, exhaust systems, engines, and brakes can cause severe burns. When possible, allow the automobile to cool before you begin work.

If you perform a road test to diagnose transmission problems, fasten safety belts and drive in a safe manner.

Many automatic transmission and transaxle diagnostic tests require that the engine be running. Be sure that proper ventilation equipment is used to remove dangerous exhaust fumes from an enclosed shop area. Block the wheels. Set the parking brake securely and have a helper apply the service brakes. Do not allow anyone to stand in front of, or behind, the automobile during these tests.

If a procedure requires that the transmission be running and in gear with the wheels off the ground, use extreme care. Contact with a spinning driveline or wheel can cause serious injury.

Always disconnect the grounded battery terminal before attempting to service a transaxle to prevent personal injury and damage to parts.

Do not wipe internal transmission or transaxle parts with a cloth or paper towel. Lint can cause internal valves to stick and result in major transaxle damage. Clean parts with solvent, then dry with compressed air. However, do not use compressed air to spin-dry bearings.

Transaxle removal on some vehicles requires removing brake assemblies. Brake systems contain asbestos dust. Use an approved special vacuum cleaner or liquid cleaning system to remove asbestos before performing any service work on brake assemblies.

If suspension parts must be disconnected before servicing a transaxle, always refer to the manufacturer's service manual for correct procedures and wheel alignment specifications.

AUTOMATIC TRANSMISSION AND TRANSAXLE SERVICE

TROUBLESHOOTING AND MAINTENANCE

Automatic transmission and transaxle preventive maintenance includes lubricating the gearshift linkage, checking automatic transmission fluid level and condition, changing fluid, and replacing the filter and screen. A visual inspection also is made.

Lubricating and Checking Gearshift Linkage

Gearshift linkage usually is lubricated during chassis lubrication service. Clean the linkage with a shop towel. Check for missing or damaged parts. The linkage is lubricated where parts pivot or slide.

Checking Transmission Fluid Level

Automatic transmission fluid (ATF) level should be checked at regular mileage and time intervals. The dipstick is located on the transmission housing, at the end of the engine opposite the belts and pulleys.

Make sure the vehicle is level. On most automobiles, the ATF level can be checked accurately only when the transmission is at operating temperature. For most vehicles, the engine must be running and the shift lever placed in either Park or Neutral, *as specified by the vehicle manufacturer*, with the parking brake applied. Markings on a dipstick indicate full, safe, or low levels **[Figure 54-1]**.

Some dipsticks have readings on both sides **[Figure 54-2]**. Others have readings on only one side **[Figure 54-3]**. On some vehicles with automatic transaxles, the cold fluid level may be higher than the hot fluid level **[Figure 54-4]**. Refer to the manufacturer's service manual and service bulletins for specific information.

To check fluid level, start the engine and bring it to operating temperature. Remove the dipstick and wipe it clean with a lint-free cloth or paper towel. Reinsert the dipstick fully. Remove it again and note the reading.

Refer to the manufacturer's service manual for the recommended ATF. A special funnel with a long, narrow neck is inserted into the transmission dipstick hole to add fluid **[Figure 54-5]**. Avoid overfilling the transmission. Remove excess fluid with a suction gun.

Diagnosing Transmission Fluid

Uncontaminated automatic transmission fluid is pinkish or red in color. A dark brownish or blackish color and/or a burnt odor indicate overheating. If the fluid has overheated, the ATF and the filter must be changed and the transmission inspected. A milky color can indicate that engine coolant is leaking into the transmission cooler in the radiator outlet tank.

Bubbles on the dipstick indicate the presence of air. The bubbles usually are caused by a high-pressure leak.

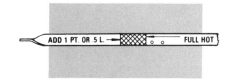

Figure 54-1. Typical automatic transmission dipstick.
PONTIAC MOTOR DIVISION—GMC

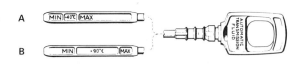

Figure 54-2. Dipstick with readings on both sides.
VOLVO OF AMERICA CORPORATION

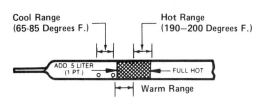

Figure 54-3. Hot and cold dipstick readings on one side.
CHEVROLET MOTOR DIVISION—GMC

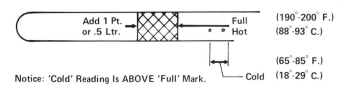

Figure 54-4. Transaxle dipstick with hot reading above cold reading. CHEVROLET MOTOR DIVISION—GMC

Figure 54-5. Adding automatic transmission fluid.

Wipe the dipstick on absorbent white paper. Look at the fluid stain. Dark particles in the fluid indicate band or clutch material. Silvery metal particles indicate excessive wear on metal parts or housings. Varnish or gum deposits on a dipstick indicate the need to change the ATF and the transmission filter.

Changing Transmission/Transaxle Fluid

Change the fluid when the engine and transmission or transaxle are at normal operating temperatures. Contaminants flow out more completely when the fluid is warm. Raise and support the automobile properly on a hoist. Place a catch pan under the transmission.

On most units, it is necessary to remove the oil pan to drain the fluid. Some transmission pans on recent vehicles include a drain plug **[Figure 54-6]**. After draining, the pan must be removed for inspection and to replace the filter.

To remove an oil pan without a drain plug, start by removing bolts near the lowest edge of the pan. Position the catch tank under the edge. Then start to loosen, but do not remove, the bolts along the sides that are nearest to that edge. Separate the pan gently from the case by tapping it with a rubber mallet or by prying gently with a putty knife. Take care not to damage the sealing surface of the transmission case. As the pan separates from the transmission, the ATF will drain through the gap. Then remove all of the pan bolts **[Figure 54-7]**.

Remove the oil pan and check the bottom for deposits and metal particles. A small amount of blackish deposits from transmission clutches and bands, and a few small metal particles are normal. Large amounts of deposits or metal particles indicate major damage to friction materials or metal parts in the transmission. Clean the inside of the pan with solvent and blow it dry with compressed air.

SAFETY CAUTION

Wear eye protection whenever you work with compressed air. Do not use any sort of rag or paper towel to wipe internal transmission parts. Lint from the rag can clog small passages and valves and cause major transmission problems.

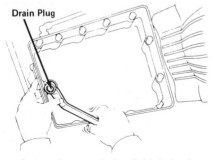

Figure 54-6. Automatic transmission fluid drain plug.

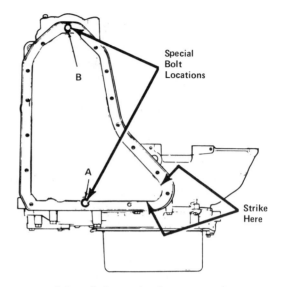

Figure 54-7. Automatic transaxle oil pan removal.
BUICK MOTOR DIVISION—GMC

Draining the torque converter. Some fluid remains in the torque converter during draining. This remaining fluid must be removed if the ATF is badly contaminated.

Most recent automobile torque converters are not manufactured with a torque converter drain plug. However, aftermarket kits are available to install drain plugs in converters that do not have them. Special care must be taken when drilling and tapping the converter housing to install a drain plug. No metal particles must enter the converter housing. Refer to the manufacturer's service manual and bulletins for recommended procedures to drain the torque converter.

Place a catch pan under the torque converter. If installed, the drain plug is located behind an access-hole plate at the bottom of the transmission housing **[Figure 54-8]**. Remove the plug to drain the fluid.

Replacing the filter and screen. A filter or screen usually is attached to the bottom of the valve body **[Figure 54-9]**. A disposable filter or reusable screen on the oil pump suction side is used to remove particles from the ATF. Replace a filter or clean a screen according to manufacturer's recommendations.

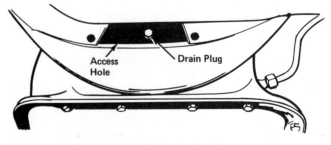

Figure 54-8. Access cover drain plug location.
CHRYSLER CORPORATION

AUTOMATIC TRANSMISSION AND TRANSAXLE SERVICE

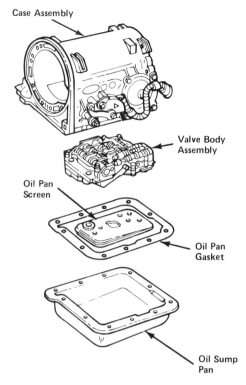

Figure 54-9. Oil screen location. FORD MOTOR COMPANY

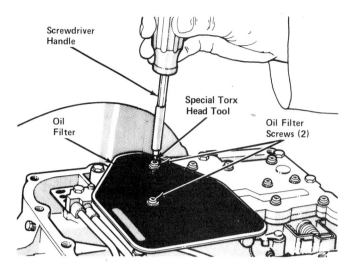

Figure 54-11. Removing a transaxle filter with a Torx driver.
CHRYSLER CORPORATION

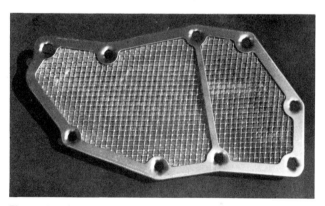

Figure 54-12. Transmission fluid screen.

Filters are made of paper or fabric **[Figure 54-10]**. A filter may be held in place by screws, clips, or bolts. Remove the fasteners and the filter **[Figure 54-11]**. A filter kit contains a new filter and a transmission pan gasket. Install the filter.

Screens are removed in the same way as filters **[Figure 54-12]**. Some transmissions have valves and springs between the valve body and screen. Do *not* loosen or remove valves or springs. Clean the screen with fresh solvent and a stiff brush. Replace the screen.

Remove any traces of the old gasket on the case housing and oil pan by scraping with a gasket scraper or putty knife. Do not damage the sealing surface of the housing or oil pan. If the pan is dimpled, straighten it as described in Unit 16, under Replacing Valve Cover Gaskets. Install the oil pan and new gasket.

Lower the vehicle and refill with the recommended type and amount of new ATF. Be careful not to overfill. When ATF has been added to the manufacturer's specifications, move the shift lever slowly through all the shift positions. Allow the transmission to shift into forward, reverse, and Park positions. Then place the shift lever in Park or Neutral, as recommended in the manufacturer's service manual, and recheck the ATF level.

54.1 AUTOMATIC TRANSMISSION AND TRANSAXLE DIAGNOSIS

Automatic transmission or transaxle identification numbers are stamped into the case **[Figure 54-13]** or on an attached metal identification tag. To identify and diagnose specific transmission or transaxle problems correctly, refer to the manufacturer's service manual and troubleshooting charts.

NOTE: Transmission or transaxle problems can be caused by engine malfunctions. A poorly tuned or worn-out engine that does not produce enough vacuum

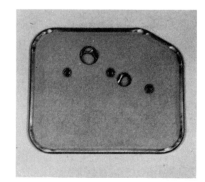

Figure 54-10.
Transmission fluid filter.

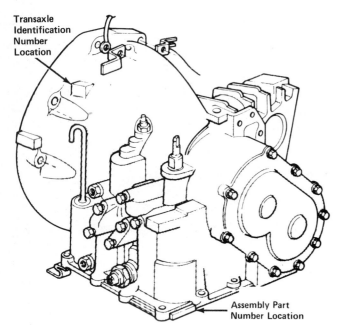

Figure 54-13. Transaxle identification number location. CHRYSLER CORPORATION

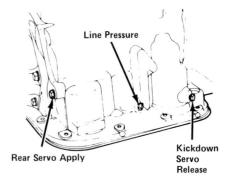

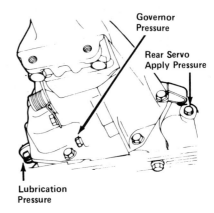

Figure 54-14. Pressure test locations. CHRYSLER CORPORATION

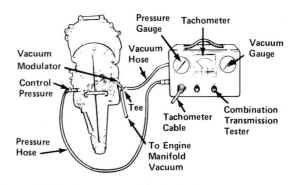

Figure 54-15. Pressure combination tester.
FORD MOTOR COMPANY

can cause shifting problems. A clogged engine cooling system can cause the transmission fluid to overheat. Worn-out engine bearings may exert force against the input shaft, damaging or destroying internal transmission parts. Any problem that affects engine operation can affect transmission or transaxle operation.

Make a visual inspection of the housing and cooler lines for damage. Check fluid level and condition. Look for leaks, broken lines, and misadjusted linkage. Check electrical connections and look for damaged wiring. Be sure all vacuum lines are connected.

A properly conducted road test can indicate how the apply devices are operating. The technician also can check for slippage, harsh or delayed shifting shifts at incorrect road speeds.

54.2 TESTING

Specific tests are made to diagnose automatic transmission or transaxle problems. Hydraulic fluid *pressure tests* and *air pressure tests* on valve bodies are helpful aids in diagnosing problems.

Hydraulic Pressure Tests

Operating pressures inside an operating transmission or transaxle can be checked through pressure test openings, or ports, on the transmission case [**Figure 54-14**].

Fittings and test instruments with gauges [**Figure 54-15**] are attached to the ports, and the transmission or transaxle is shifted into a specific gear with the engine running. Pressure readings are recorded during the test procedures. Some testers include long hoses that allow the tester to be used inside the car during a road test.

Procedures for hydraulic pressure tests vary considerably. Refer to the manufacturer's service manual for specific procedures.

Air Pressure Test

After the valve body has been removed, air pressure can be directed through specified case holes and passages to test the operation of apply devices. A blowgun with a conical rubber nozzle is used to apply

Automatic Transmission and Transaxle Service

low air pressure. Refer to the manufacturer's service manual for correct test procedures and test hole locations.

SAFETY CAUTION

Keep hands and fingers away from valves and servos when you apply air pressure. Moving parts can cause painful injuries. Wear eye protection at all times when you use compressed air.

When the air pressure is applied, the operation of servos and clutches can be seen, heard, or felt as vibrations through the case.

Some manufacturers recommend the use of special metal plates that are bolted to the transmission to seal the apply passages **[Figure 54-16]**.

54.3 ADJUSTMENT AND LIGHT SERVICE

Servicing an automatic transmission or transaxle may involve only on-car service, such as an adjustment. Off-car service, such as an overhaul, is defined as a major job. This section covers on-car adjustments and light service.

Linkage Adjustment

There are three types of linkage adjustments: quadrant, gearshift, and kickdown. Quadrant adjustment is made to center the pointer **[Figure 54-17]**.

Gearshift linkage adjustment varies between manufacturers. Usually, the manual valve lever on the side of the transmission or transaxle is positioned in a specific location. Adjustments are made to linkage that is connected to this lever **[Figure 54-18]**. Always refer to the manufacturer's service manual for correct procedures.

Several types of linkage systems can be controlled by the accelerator pedal. These include: accelerator, kickdown rod, and throttle valve. Many automobiles use more than one of these linkage systems. **Figure 54-19** shows a typical kickdown linkage system on an automatic transmission.

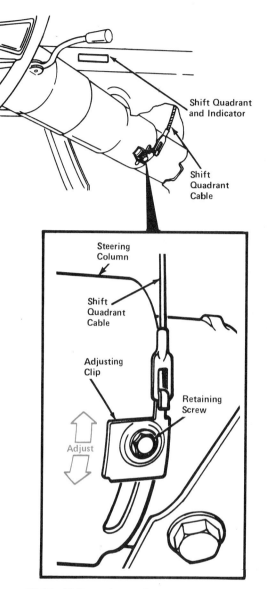

Figure 54-17. Shift quadrant adjustment.
AMERICAN MOTORS CORPORATION

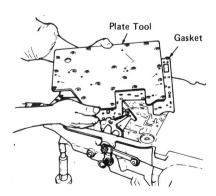

Figure 54-16. Air pressure plate. FORD MOTOR COMPANY

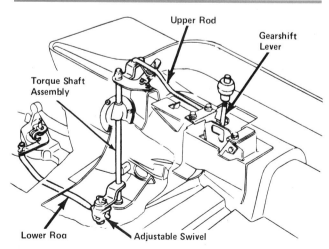

Figure 54-18. Gearshift linkage adjustment locations.
CHRYSLER CORPORATION

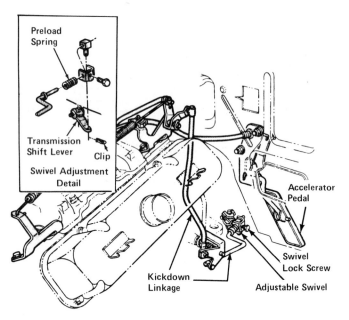

Figure 54-19. Kickdown linkage system. CHRYSLER CORPORATION

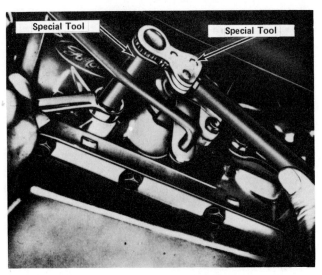

Figure 54-20. Band adjustment. FORD MOTOR COMPANY

Band Adjustment

Adjusting transmission bands may or may not be part of a scheduled maintenance program. Some transmissions have no provision for band adjustments. In other cases, band adjustments may require draining the ATF and removing the oil pan.

Figure 54-20 shows an external intermediate band adjustment on a Ford transmission. To adjust the band, hold the adjuster screw from turning and loosen the locknut one or two turns. Tighten the adjusting nut to the manufacturer's specifications with the special torque wrench shown in the illustration. Then carefully back off the adjuster the number of turns as specified. Then hold the adjuster from turning and retighten the locknut.

Vacuum Modulator Service

The vacuum modulator usually is screwed or clamped into place at the side or rear of the transmission **[Figure 54-21]**.

Improper shifting may be caused by a defective vacuum modulator or by defective or missing modulator vacuum connectors **[Figure 54-22]**. Engine vacuum leaks can produce the same symptoms. Follow the diagnostic procedure outlined in the manufacturer's service manual to determine the exact cause.

A repeated low fluid level in the transmission with no visible external leaks and the sudden appearance of excessive blue-white smoke from the tailpipe usually indicates a ruptured diaphragm in the modulator. To check for this condition, pull the vacuum hose off the modulator connection. If ATF is present in the hose, the vacuum modulator must be replaced.

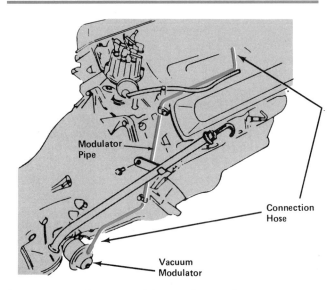

Figure 54-21. Vacuum modulator and connections.

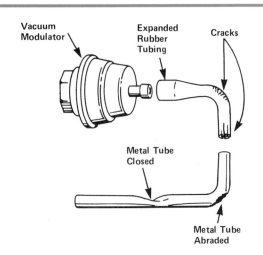

Figure 54-22. A damaged vacuum line can cause improper shifting.

Clean the area around the modulator with solvent. Then disconnect the vacuum hose. The modulator itself is either unscrewed or unclamped from the transmission housing and removed carefully. Special tools may be necessary. Take care that the modulator valve does not fall out during this removal. Then install the new replacement unit. Refer to the manufacturer's service manual for more complete replacement procedures.

Governor Service

Improper shift points may be caused by a malfunction in the governor or governor drive gear system **[Figure 54-23]**. Governor operation is discussed in Unit 53.

Although all governors are mounted internally, some require removal of the extension housing or oil pan for access.

Another type of governor is serviced by removing an external retaining clamp and then removing the unit **[Figure 54-24]**. Follow the manufacturer's recommended diagnostic and service procedures.

Oil Leaks

There are many possible sources for ATF leakage. One of these, the pump seal, can be replaced only if the transmission is removed from the vehicle.

Leaks caused by worn or defective gaskets or seals are repaired by replacing the defective part. The transmission need not be removed for this service. Follow the replacement recommendations of the manufacturer. **Figure 54-25** shows possible sources of leaks on an automatic transmission.

Case porosity, tiny holes caused by trapped air bubbles during the casting process, may occur. Leakage of ATF through the cap can be seen and felt. Case porosity may be repaired using an epoxy-type sealer. All traces of oil must be removed before attempting this repair, which may require transmission removal.

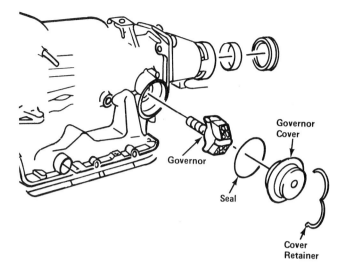

Figure 54-24. Governor held in place by retaining clamp.

Computer and Solenoid Valve Tests

Refer to the manufacturer's service manual procedures for specific transmission computer tests. Also refer to Unit 39 for general procedures for computer testing. Be aware that improper test procedures can damage or destroy computer units immediately.

Specific tests for input and output signals are made. Special hand-held computer testers may be required. If input signals are correct and output signals are incorrect, the computer unit must be replaced.

Solenoid valves are tested by removing the electrical connector. Then the resistance of the solenoid coil can be checked between the connector and ground. Another way of testing a solenoid is to apply current with a small jumper wire to the disconnected solenoid valve. Correct operation can be heard or felt when the solenoid moves sharply. Refer to the manufacturer's service manuals and service bulletins for specific tests.

54.4 AUTOMATIC TRANSMISSION AND TRANSAXLE OVERHAUL

Overhauling an automatic transmission or transaxle requires removing the unit from the vehicle. Refer to Unit 50 for general procedures to remove transmissions and transaxles.

The unit then is disassembled, cleaned, and inspected. Worn or damaged parts are replaced, and internal adjustments are made. Always refer to the manufacturer's service manual for specific automatic transmission or transaxle overhaul procedures. An exploded view of internal automatic transaxle parts is shown in **Figure 54-26**. As with manual transaxles, automatic transaxle service includes differential checks and repairs. See Unit 56 for differential service procedures.

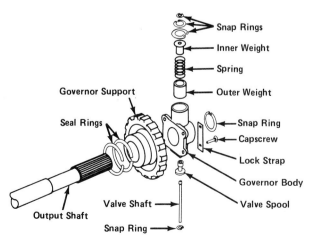

Figure 54-23. Governor valve assembly.

AMERICAN MOTORS CORPORATION

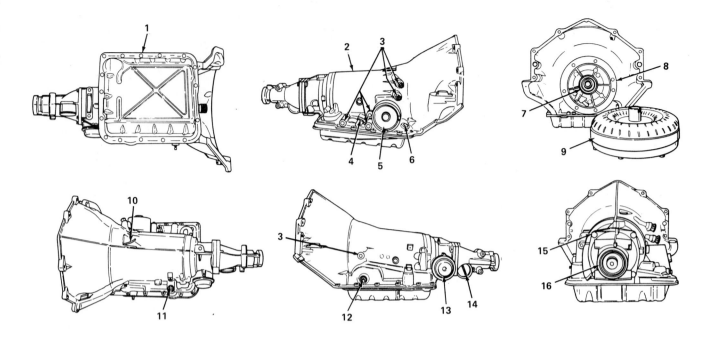

1	Oil Pan	7	Oil Pump Seal Assembly	12	Manual Shaft Seal
2	Case	8	Oil Pump to Case Seal	13	Governor Cover
3	Cooler Connectors and Plugs	9	Converter	14	Speedo Seal
4	T.V. Cable Seal	10	Vent	15	Extension to Case Seal
5	Servo Cover	11	Electrical Connector Seal	16	Extension Oil Seal Assembly
6	Oil Fill Tube Seal				

Figure 54-25. Possible sources of transmission fluid leaks. PONTIAC MOTOR DIVISION—GMC

54.5 CONTINUOUSLY VARIABLE TRANSMISSION DIAGNOSIS AND SERVICE

The hydraulic controls on a CVT are similar to those on automatic transmissions, and can have the same problems. Specific internal problems with CVTs include:

- Broken or damaged drive belt
- Pulley damage
- Belt slippage due to low pulley control oil pressure
- Leakage in seals that control pulley width
- Improperly adjusted throttle cable.

To replace the belt, spread the secondary (driven) pulley with a special tool. Remove the belt and install a new belt. Replace damaged parts as necessary. Always consult the vehicle manufacturer's service manual and service bulletins for CVT preventive maintenance, diagnostic service, and specific repair procedures.

UNIT HIGHLIGHTS

- Transmission fluid level is checked with a dipstick when the fluid is at operating temperature.
- Most vehicles require that the engine be running and the transmission be in Park or Neutral to check ATF fluid level correctly.
- ATF fluid color and condition can help the technician to diagnose internal transmission problems.
- Transmission fluid usually is drained by removing the oil pan bolts and pan.
- An automatic transmission/transaxle filter or screen is located on the valve body.
- Hydraulic and air pressure tests help to pinpoint transmission and transaxle problems.
- Continuously variable transmissions used in transaxles can have familiar hydraulic control problems and specific mechanical problems.

Automatic Transmission and Transaxle Service

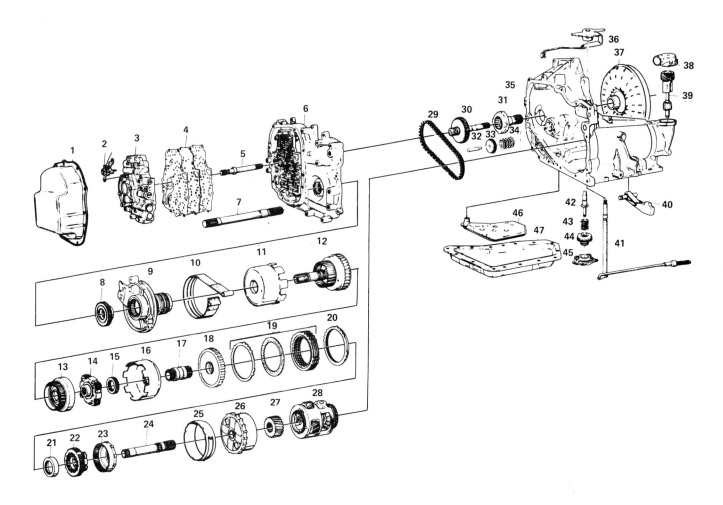

1. Valve Body Cover
2. Throttle Lever Bracket & Link
3. Control Valve & Oil Gasket
4. Spacer Plate & Gaskets
5. Oil Pump Shaft
6. Case Cover
7. Output Shaft
8. Driven Sprocket
9. Driven Sprocket Support
10. Intermediate Band
11. Direct Clutch Housing
12. Forward Clutch Housing
13. Input Internal Gear
14. Input Carrier
15. Input Sun Gear
16. Input Drum
17. Reaction Sun Gear
18. Low/Reverse Housing
19. Low/Reverse Clutch Plates
20. Low/Reverse Clutch Backing Plate
21. Low Overrunning Clutch Race
22. Reaction Carrier
23. Reaction Internal Gear
24. Final Drive Sun-Gear Shaft
25. Final Drive Internal-Gear Spacer
26. Final Drive Internal Gear
27. Final Drive Sun Gear
28. Differential Carrier
29. Drive Link
30. Drive Sprocket & Turbine Shaft
31. Drive Sprocket Support
32. 1-2 Accumulator Pin
33. 1-2 Accumulator Piston
34. 1-2 Accumulator Spring
35. Case
36. Manual Detent Lever & Rod
37. Converter
38. Governor Cover
39. Governor
40. Parking Pawl
41. Actuator Rod & Manual Shaft
42. Intermediate Band Apply Pin
43. Intermediate Servo Cushion Spring
44. Intermediate Servo Piston
45. Intermediate Servo Cover
46. Oil Filter
47. Oil Pan

Figure 54-26. Parts of an automatic transaxle. CHEVROLET MOTOR DIVISION—GMC

TERMS

pressure test
air pressure test

REVIEW QUESTIONS

DIRECTIONS: The following questions are similar to those used on automotive technician certification tests. On a separate sheet of paper, write the letter of the correct choice.

1. Technician A says fluid levels are marked on all automatic transmission and transaxle dipsticks.

   Technician B says ATF color and condition can indicate internal problems.

   Who is correct?
   - A. A only
   - B. B only
   - C. Both A and B
   - D. Neither A nor B

2. Which of the following statements is correct?
   - I. Hydraulic fluid pressure checks help diagnose transmission and transaxle problems.
   - II. Air pressure checks are part of normal preventive maintenance.
   - A. I only
   - B. II only
   - C. Both I and II
   - D. Neither I nor II

3. Which of the following statements about automatic transmission and transaxles is correct?
   - A. A linkage adjustment positions the shift valves in the valve body.
   - B. Linkage adjustments include quadrant, gearshift, and kickdown adjustments.
   - C. All transmissions require band adjustments when fluid is changed.
   - D. The pump seal can be replaced with the transmission installed in the car.

Questions 4 and 5 are not like those above. Each has the word EXCEPT. For each question, look for the choice that does *not* apply. Read each question carefully before you choose your answer.

4. All of the following precautions should be followed when servicing automatic transaxles EXCEPT
   - A. use special support fixtures to hold the engine and transaxle safely.
   - B. disconnect the battery before you service the unit.
   - C. block the wheels and set the emergency brake before servicing.
   - D. blow asbestos dust from brake units with an air hose before servicing.

5. All of the following may be required to diagnose a conventional automatic transaxle EXCEPT
   - A. making hydraulic pressure tests.
   - B. checking the pulleys for damage.
   - C. checking differential bearings and gears.
   - D. making an air-pressure test of the valve body and apply devices.

ESSAY QUESTIONS

1. Describe a conventional automatic transaxle overhaul.
2. Describe service procedures for CVTs (continuously variable transmissions) used in transaxle.
3. What differences in servicing procedures are there between conventional automatic transmissions and automatic transaxles?

SUPPLEMENTAL ACTIVITIES

1. Check the automatic transmission fluid level on a vehicle and describe the condition of the fluid.
2. Make a visual inspection of a transmission installed in a vehicle and describe any defects that are found.
3. Perform a fluid and filter change on an automatic transmission or transaxle.
4. Demonstrate how to make a hydraulic or air pressure test on a transmission.
5. Adjust the quadrant adjustment linkage on a vehicle.
6. Refer to a service manual and describe diagnostic procedures for an automatic transmission or transaxle chosen by your instructor.
7. Remove, disassemble, inspect, reassemble, and reinstall an automatic transmission or transaxle in a vehicle chosen by your instructor.
8. If available, disassemble and inspect a CVT. Report on any damage or defects found.

EMISSION CONTROLS

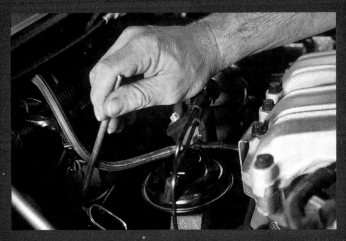

(1) Use a mirror to check if the EGR valve is working properly.

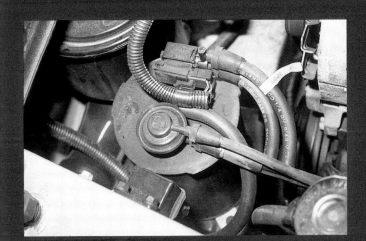

(2) The carbon canister is the heart of the evaporative emission system.

(3) Checking the fuel delivery system with a scan tool.

(4) Computer-controlled throttle body-injection system.

ENGINE TESTING TOOLS

(1) Checking the manifold pressure with a vacuum gauge.

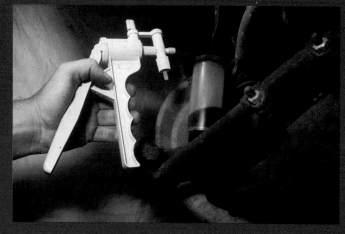
(2) Using a hand vacuum pump to check a brake system.

(3) Vacuum gauge with control ranges printed on its face.

(4) Vacuum pump with built-in gauge.

(5) Checking exhaust with an infrared analyzer.

(6) Checking emissions with computerized analyzer.

(7) Checking air/fuel mixer with a scan tool.

(8) Various methods—from simple strobe light to highly accurate magnetic probe devices—are available for checking and adjusting initial timing.

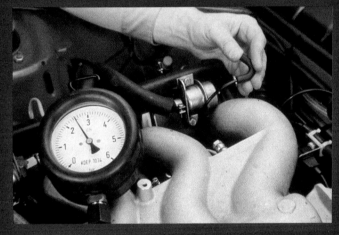
(9) Metric vacuum gauge marked in bars, each of which is equal to 14.5 psi.

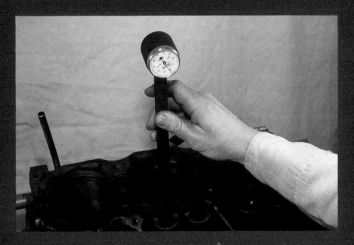
(10) Engine bore indicator in use.

CHECKING FOR IGNITION AND INJECTOR PROBLEMS

(1) Checking for ignition vacuum leak problems with an electric device that responds to the sound frequency they produce.

(2) Typical ignition system in Japanese vehicles.

(3) A cause for ignition problems is a defective carbon button found in the distributor cap.

(4) To maximize available voltage, use a spark tester. You can create your own spark tester by using a modified gap spark plug and grounding wire.

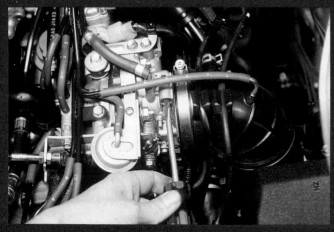

(5) On some fuel-injected vehicles, idle speed adjustment can be accomplished by a twist of the throttle stop screw.

(6) Injector power balance procedure requires disconnecting the injectors one by one.

COMPUTERIZED ENGINE CONTROLS AND CHECKING DEVICES

(1) Most multiport fuel injectors are actually easier to service than most carbureted fuel systems.

(2) The use of portable recorders, such as this one, captures and stores the on-board data.

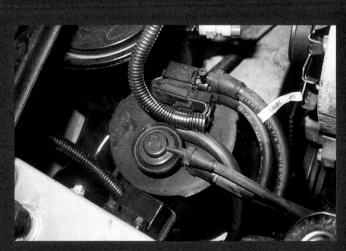

(3) The engine control computer is able to indirectly control the vapor canister's purge solenoid via an electrically operated solenoid located in the vacuum line.

(4) To save time and prevent lost or overlooked data, use this scan tool, which displays all information with a built-in thermal printer.

(5) Diagnostic readout tester that is very helpful in troubleshooting.

(6) Typical engine computer power module.

(7) Connecting an engine computer diagnostic driveability tester.

(8) Typical driver computer testers.

ENGINE COMPONENTS

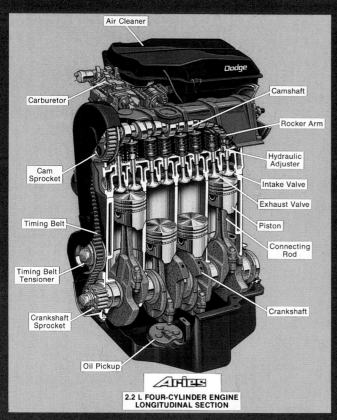

(1) Components of a four-cylinder engine.

(2) Turbocharged engine in place under a hood.

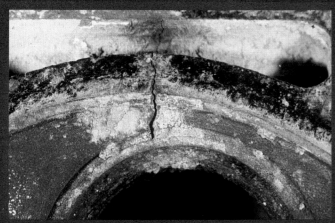

(4) Typical crack found in a cylinder head.

(3) Welding a cracked cylinder head.

(5) Use one of the new replacement rings to check if the grooves are clean.

(6) Typical aluminum head.

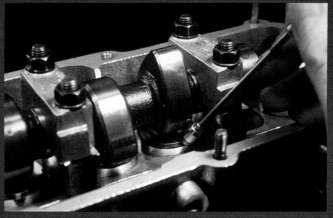

(7) Typical crankcase.

SPARK PLUG ANALYSIS

(1) Normal, good spark plug.

(2) Lead-fouled spark plug.

(3) Broken insulator.

(4) Oil-fouled plug.

(5) Carbon-fouled plug.

(6) Proignition.

(7) Overheating.

55 Differentials, Driving Axles, and 4WD

UNIT PREVIEW

When a vehicle turns, the inner and outer driving wheels must travel different distances. The driving wheels must turn at different speeds to reduce tire wear on turns.

A conventional differential is a gear mechanism that can transfer different amounts of torque to rotate the driving axles at different speeds during turning. Driving axles transfer torque from the differential to the driving wheels.

Differentials in front-wheel drive vehicles are usually mounted in the same case as the transmission components, as discussed in Units 57 and 59. This unit discusses separately mounted differentials, as used in rear-wheel drive and four-wheel drive (4WD) vehicles.

Transfer cases are used in 4WD vehicles to supply torque to an additional pair of driving wheels. Four-wheel drive vehicles may be based on rear-wheel drive or front-wheel drive vehicle designs.

LEARNING OBJECTIVES

When you have completed your assignments and exercises in this unit, you should be able to:

❏ Describe how a conventional differential operates.
❏ Describe the operation of a limited-slip differential.
❏ Describe the three common types of driving axles.
❏ Identify the rear axle parts involved in power flow.
❏ Describe the functions of transfer cases and interaxle differentials for four-wheel drive (4WD) vehicles.

55.1 DIFFERENTIAL DESIGN

The differential is a geared mechanism located between two driving axles. It rotates the driving axles at different speeds when the vehicle is turning, and at the same speed when the vehicle is traveling in a straight line.

The rear drive axle assembly provides for several important functions. First, it directs driveline torque to the driving wheels. The rear axle assembly redirects this driveline torque power flow so it is perpendicular to the driveline.

Second, the gear ratio between the drive pinion and ring gear is used to increase engine torque and drivability. Third, the gears within the differential unit serve to establish a state of balance of forces/torques between the drive wheels. The driving torque to each wheel is no greater than the torque required by the wheel with the least traction.

Fourth, the differential allows the drive wheels to turn at different speeds as the vehicle turns a corner.

Numerically high differential gear ratios (for example, 4.11:1) will give high torque multiplication and generally faster acceleration rates. Lower ratios (2.75:1) will provide less torque multiplication, but reduce fuel consumption by allowing the engine to turn more slowly for given road speeds.

Differential Function

When the vehicle is moving in a straight line, the differential delivers equal torque to the two driving axles. When the driving wheels go around a turn, the differential provides unequal torque to the driving axles.

Torque enters the differential through the *drive pinion gear* [Figure 55-1]. A pinion gear is the smaller of two meshing gears. The drive pinion gear is connected to the driveline at the rear U-joint yoke.

The drive pinion gear meshes with a larger *ring gear* inside the differential housing. The ring gear transfers torque, through a series of parts, to the driving wheels.

When driveline torque reaches the differential, it must be redirected at an angle of 90 degrees to turn the driving wheels.

The drive pinion gear and ring gear are *hypoid gears* that resemble beveled gears. A hypoid gear contacts more than one tooth at a time and makes contact with a sliding motion. Also, the centerlines of the ring and pinion gears do not match. The drive pinion meshes with the ring gear at a point below its centerline [Figure 55-2].

To allow a different rotation for each driving wheel, the differential contains four assemblies:

1. Differential case or carrier
2. Pinion shaft

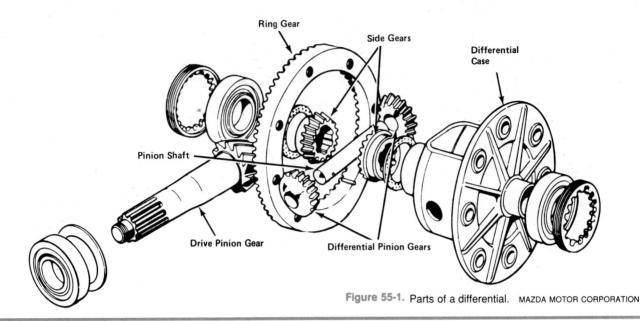

Figure 55-1. Parts of a differential. MAZDA MOTOR CORPORATION

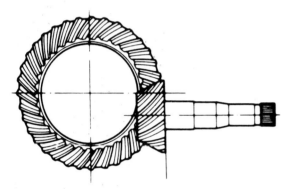

Figure 55-2. Hypoid gearset. NISSAN MOTOR CORPORATION

3. Differential pinion gears
4. Differential side gears.

The ring gear is bolted to the *differential case.* As the ring gear rotates, the carrier rotates. The remaining differential parts are located inside the carrier.

The *pinion shaft* is mounted in holes in the case. *Differential pinion gears* are mounted on the pinion shaft to mesh with the *side gears.* This relationship is illustrated in **Figure 55-3.**

The pinion shaft is mounted rigidly to the differential carrier. The carrier rotates, carrying the pinion shaft around with it.

When driving in a straight line, the differential pinion gears do not rotate on the pinion shaft, but remain stationary. As the pinion shaft moves with the case, the gears are meshed with, and exert pressure on, the side gears. This pinion shaft movement is a tumbling motion. The side gears turn at the same speed as the ring gear to drive both wheels at the same speed.

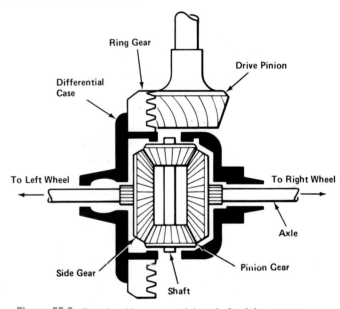

Figure 55-3. Relationship among pinion shaft, pinion gears, and side gears in a typical differential.

The partial circular paths that the wheels must travel around a curve are of different lengths. The outer wheel must travel a greater distance. Thus, when the automobile turns a corner, or rounds a curve, the outer driving wheel must turn faster than the inner driving wheel.

As the vehicle begins to turn, the inner wheel requires greater torque to turn. As a result, the side gear connected to the inner wheel slows down.

The pinion gears then begin to turn, pivoting on, or "walking around," the inner wheel side gear teeth. The rotating action of the pinion gears applies more torque to the opposite side gear and driving axle, connected to the outside wheel. The outside wheel turns faster than the inside wheel.

Differentials, Driving Axles, and 4WD

Calculating Differential Gear Ratios

To determine a gear ratio, as discussed in Unit 49, divide the number of teeth on the driving gear into the number of teeth on the driven gear. In a differential, the ring gear is the driven gear. The pinion gear is the driving gear. So, *drive axle ratio (DAR)* is calculated by:

$$DAR = NRG \div NPG$$

Where:
DAR = drive axle ratio
NRG = number of ring gear teeth
NPG = number of pinion gear teeth

If the drive pinion has 12 teeth and the ring gear has 34 teeth, the DAR is:

$$34 \div 12 = 2.8$$

If the drive pinion has 9 teeth and the ring gear has 37 teeth, the DAR is:

$$37 \div 9 = 4.11$$
$$= 4.11 : 1$$

In Unit 49, the *overall transmission gear ratio (OTR)* for a particular transmission in 1st gear was determined to be approximately 3:1. (Refer to Unit 49 for the calculation of OTR.)

To find the *overall drivetrain gear ratio (ODR)*, multiply the overall transmission ratio (OTR) by the drive axle ratio (DAR). The ODR indicates the overall torque multiplication of the combination of the transmission and differential together. For a vehicle with an OTR of 3:1 and a DAR of 2.8:1, the ODR is calculated by:

$$ODR = OTR \times DAR$$
$$= 3.0 \times 2.8 = 8.4$$
$$= 8.4 : 1$$

Where:
ODR = overall drivetrain gear ratio
OTR = overall transmission gear ratio
DAR = drive axle ratio

Thus, the engine torque transferred through the transmission 1st gear and differential to the driving axles is increased 8.4 times.

Suppose you have an engine that can produce a maximum of 240 lb.-ft. of torque. In a vehicle with an ODR of 8.4:1, the maximum torque that can be delivered to the drive axles in 1st gear is calculated by:

$$TDA = mt \times ODR$$
$$= 240 \text{ lb.-ft.} \times 8.4$$
$$= 2,016 \text{ lb.-ft.}$$

Where:
TDA = torque to driving axles
mt = maximum torque
ODR = overall drivetrain gear ratio

If you substitute a differential with a numerically high 4.11:1 gear ratio, the torque applied to the drive axles in 1st gear becomes:

$$ODR = OTR \times DAR$$
$$= 3.0 \times 4.11 = 12.3 : 1$$

$$TDA = mt \times ODR$$
$$= 240 \text{ lb.-ft.} \times 12.3$$
$$= 2,959 \text{ lb.-ft.}$$

Calculating maximum top speed. If you assume the OTR in high gear is 1:1, and you know the DAR and other information, you can calculate the theoretical top speed of a vehicle. Use this formula:

$$mph = (Max\ RPM \times r) \div (168 \times DAR)$$

Where:
mph = miles per hour
Max RPM = maximum rpm (engine speed)
r = rolling radius of tires
DAR = drive axle ratio

(In the formula, 168 is a mathematical constant that is needed to make the equation produce accurate results.)

You can use this formula to figure out the top speed of a drag racing car in high gear (1:1 OTR). Suppose your car has the following specifications:

maximum engine rpm = 5,400
r (rolling radius of tires) = 12"
DAR (drive axle ratio) = 4.11:1

Substituting these values into the equation, you can calculate the following:

$$mph = (5,400 \times 12) \div (168 \times 4.11)$$
$$= 64,000 \div 690.48$$
$$= 93.8 \text{ mph}$$

Thus, at the end of the 1/4-mile strip, if your engine was running at 5,400 rpm in top gear (1:1), your car theoretically would be going 93.8 mph.

Calculating maximum engine rpm. In drag racing, top end speed also is important. Suppose you know that the winners in your racing class have top end speeds of approximately 100 mph at the end of the quarter mile. To win your class, suppose that you want your car to be running at a speed of 102 mph in top gear (1:1) at the end of the drag strip. Using algebra, the formula above can be rearranged to determine what engine rpm is necessary for a given speed:

$$rpm = (168 \times DAR \times mph) \div r$$

Where:
rpm = revolutions per minute
DAR = drive axle ratio
mph = miles per hour
r = rolling radius of tire

Assume that you have a car with the following specifications:

DAR = 3.56:1
r = 12.5"

To find out how fast your engine must be running at the end of the drag strip to produce a top end speed of 102 mph in high gear (1:1), substitute these factors into the formula:

$$rpm = (168 \times 3.56 \times 102) \div 12.5$$
$$= 4,880$$

Calculating drive axle ratio (DAR). You also can rearrange the formula to find out what top end speed a given drive axle ratio (DAR) would produce at the finish line. The formula becomes:

$$DAR = (rpm \times r) \div (168 \times mph)$$

If you knew the engine would develop peak power at 6,000 rpm in top gear (1:1) and you wanted to aim for 102 mph at the end of the drag strip, you would need to use the correct drive axle ratio (DAR). Calculate as follows:

$$DAR = (6,000 \times 12.5) \div (168 \times 102)$$
$$= 4.37$$
$$= 4.37 : 1$$

Calculating rolling radius of tire (r). A differential with a driving axle ratio of 4.37:1 would be difficult to find, or would have to be customed machined. However, a differential with a 4.11:1 ratio is relatively common and easy to find.

Because the 4.11:1 ratio will change the results of the formula, you could substitute different size tires to keep your engine at its 6,000 rpm power peak.

Using algebra, you can rearrange the formula to solve for r, the rolling radius of the new tires you would need with a 4.11:1 DAR:

$$r = (168 \times DAR \times mph) \div rpm$$

By substituting the 4.11 DAR and using our desired top end speed of 102 mph and engine speed of 6,000 rpm, you would calculate:

$$r = (168 \times 4.11 \times 102) \div 6,000$$
$$= 11.74"$$

Thus, you would try to find a tire/wheel combination which would give a rolling, or loaded, radius of about 11.74, or approximately 11 3/4 inches.

55.2 DIFFERENTIAL CONSTRUCTION

The differential can be constructed in either of two ways: as an integral-carrier differential and as a removable-carrier differential.

Integral-Carrier Differential

An *integral-carrier differential* assembly is part of the rear axle housing **[Figure 55-4]**. The differential must be serviced through the rear of the housing. The differential bearings must be disassembled and the axles must be removed before it can be removed from the automobile.

Removable-Carrier Differential

A *removable-carrier differential* can be unbolted and removed, intact, from the rear axle housing after the axles have been removed **[Figure 55-5]**. The entire differential can be set on a workbench and serviced as a unit.

55.3 LIMITED-SLIP DIFFERENTIAL

With a conventional differential, described above, wheel alignment is directed to the wheel that spins most easily, after traction has been broken, even if that wheel will spin without traction. A *limited-slip differential* eliminates this problem. If one wheel begins to turn faster than the other, the limited slip differential locks both driving axles together to deliver torque even if one wheel begins to slip. When the vehicle is moving straight ahead, a limited-slip differential operates in the same manner as a standard differential.

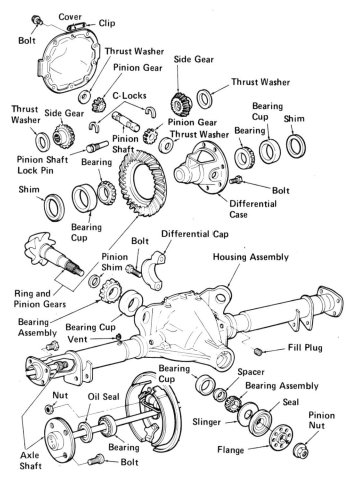

Figure 55-4. Parts of an integral-carrier differential assembly.
FORD MOTOR COMPANY

In addition to the gears found in a conventional differential, a limited-slip differential includes an additional clutch assembly. The clutch assembly locks both driving wheels together even when one wheel may not have good traction. Basically, the clutch "locks" the axles to the differential case.

Two types of limited-slip differentials are common: the clutch plate type and the cone type.

Clutch Plate Limited-Slip

The *clutch plate limited-slip* differential consists of a series of clutch plates, called a *clutch pack.* Two clutch packs may be located inside the differential case—one at each side gear. A clutch pack has steel plates connected to the case and friction plates connected to the side gear that are stacked alternately **[Figure 55-6]**. Most limited-slip differentials include preload springs to apply beginning pressure on the clutch plates. Increasing drive torque applies increased pressure on the clutch.

During normal operation, there is enough slippage in the clutch pack to prevent lockup. When one wheel begins to spin, the torque is transmitted through the clutch pack to the wheel with traction. Driving torque causes side thrust of the side gears. The side gear, in turn, applies an added force against the clutch packs. This locks the axles to the differential case with added force to resist single wheel spinning.

Cone Limited-Slip

Many automobiles use *cone limited-slip* assemblies at each side gear **[Figure 55-7]**. The cones have

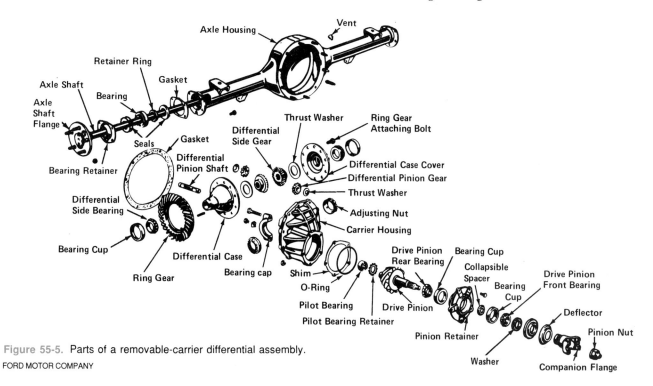

Figure 55-5. Parts of a removable-carrier differential assembly.
FORD MOTOR COMPANY

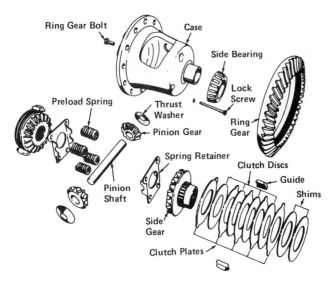

Figure 55-6. Limited-slip clutch plate assembly.
CHEVROLET MOTOR DIVISION—GMC

used for each driving wheel. For vehicles with a separately mounted differential, the axle shafts are enclosed in a housing that extends from the differential assembly toward the driving wheels [Figure 55-8].

The inner ends of the driving axle are splined into and are driven by the differential side gears.

The driving wheels are turned by the axle shafts. For the most common form of rear-wheel drive vehicles, a flange is formed on the end of the axle. Threaded studs are pressed into the flange, and the wheel lugs tighten on the studs to hold the wheel tightly against the axle flange [Figure 55-9]. Three types of axles are commonly used on vehicles with drivelines:

1. Semi-floating axles
2. Three-quarter floating axles
3. Full-floating axles.

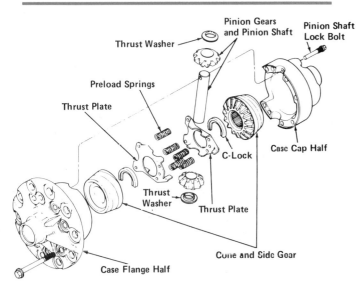

Figure 55-7. Limited-slip cone assembly. FORD MOTOR COMPANY

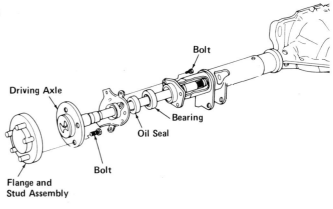

Figure 55-8. Differential housing and drive axle assembly.
FORD MOTOR COMPANY

friction surfaces that apply force against the differential case when limited-slip operation is necessary. The cones are splined to the side gear hubs.

The cone-type system operates in essentially the same way as the clutch plate design. The cones are forced against the case by preload springs and side gear thrust. At this point, the cone rotates with the case and locks up both axles.

Special lubricants must be used in limited-slip differentials. Refer to the manufacturer's service manual for correct lubricant recommendations.

55.4 DRIVING AXLES

Driving axles transfer torque from the differential to the driving wheels. A separate steel axle shaft is

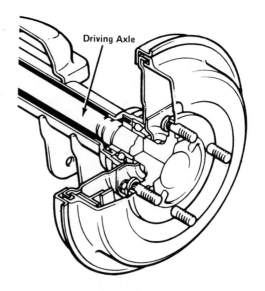

Figure 55-9. Driving wheel mounting. CHRYSLER CORPORATION

Semi-Floating Axle

A *semi-floating axle* [Figure 55-10] has three functions. First, it transfers torque to drive the wheel. Second, the axle flange helps to attach the wheel. Third, it helps to support the weight of the vehicle. The rear weight of the vehicle is supported through the outer wheel bearing and axle shaft. Semi-floating axles are most commonly found on rear-wheel drive passenger vehicles.

Three-Quarter Floating Axle

A *three-quarter floating axle* has only two functions: it turns and helps to attach the wheel [Figure 55-10]. However, a three-quarter floating axle does not help support the vehicle. The weight of the vehicle is supported through a hub, outer bearing, and the axle housing. Three-quarter floating axles are used on older passenger vehicles and on some trucks.

Full-Floating Axle

A *full-floating axle* [Figure 55-10] has only one function: to turn the wheel. Bearings in the housing and hub support the weight of the vehicle. Fasteners on the hub assembly attach the wheel. Full-floating axles are used mainly on large, heavy-duty trucks. If a full-floating axle breaks, the wheel remains attached to the hub assembly.

Independently Suspended Axles

This type of rear driving axle system is found mostly on rear-wheel drive luxury cars and sports cars. The driving axles are open instead of being enclosed in an axle housing.

Two types of independently suspended rear driving axles are common: the De Dion system and the swing axle.

The *De Dion axle* driving axle resembles a driveshaft, with U-joints at each end of the axle [Figure 55-11]. A slip joint is attached to the inboard, or innermost, U-joint. The outboard, or outermost, U-joint is connected to the driving wheel. This allows the driving axle to move up and down as it rotates.

On automobiles with *swing axles*, the driving axles may be open or enclosed. An axle fits into the differential by way of a ball-and-socket system. The ball-and-socket system allows the axle to pivot up and down. As the axle pivots, the driving wheel "swings" up and down [Figure 55-12].

55.5 FOUR-WHEEL DRIVE (4WD)

To apply torque at more locations for improved traction on wet, uneven, or slippery surfaces, all four wheels can be driven. This arrangement is known as *four-wheel drive (4WD)*. Four-wheel drive vehicles require additional drivetrain parts to transmit torque to additional driving wheels.

Part-Time Four-Wheel Drive

Most 4WD trucks, utility vehicles, and automobiles operate in two-wheel drive (2WD) for highway driving. Then, as required, the driver shifts the vehicle's transfer case into a 4WD position. This arrangement is known as *part-time 4WD*.

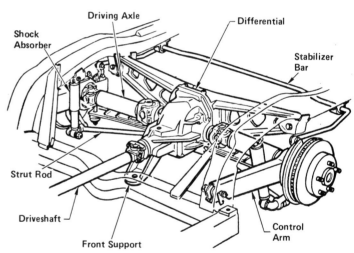

Figure 55-11. De Dion axle assembly.
CHEVROLET MOTOR DIVISION—GMC

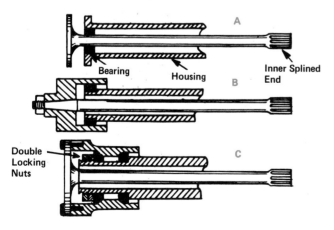

Figure 55-10. Semi-floating (A), three-quarter floating (B), and full-floating (C) axle assemblies.

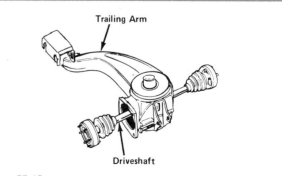

Figure 55-12. Swing axle assembly. VOLKSWAGEN OF AMERICA

4WD Design Variations

Most 4WD pickup trucks and utility vehicles are designed as rear-wheel drive vehicles. To create a 4WD vehicle, a *transfer case,* front drive-shaft, and front differential are added to transmit torque to the front wheels [Figure 55-13]. A transfer case is a housing that contains a gear and/or chain mechanism. The transfer case mechanism transmits engine torque to an additional driveline, differential, and set of driving wheels.

However, many smaller 4WD automobiles are designed as front-wheel drive vehicles. A transfer case or coupling mechanism, rear driveline, and rear differential are added to transmit torque to the rear wheels [Figure 55-14]. Thus, 4WD vehicles have two differentials and two sets of driving axles, at both front and rear.

Transfer Case

To switch to four-wheel drive, the driver engages the transfer case, attached to the vehicle's transmission [Figure 55-15].

Gearsets and/or a chain drive transfers torque from the transmission to an additional driveline and differential [Figure 55-16].

When shifted into a 4WD position, the transfer case sends torque from the transmission or transaxle output shaft to two additional wheels. All four wheels turn when driving straight ahead to provide four-wheel drive.

Most transfer cases include a 2WD "High" (normal speed and torque) gear position, Neutral, a 4WD "High" gear position, and a 4WD "Low" (lower speed, increased torque) gear position. The internal parts of some transfer cases include a single planetary gearset, similar in function to that used in an automatic transmission (refer to Unit 53). The planetary gearset provides both gear reduction ("Low") and direct drive ("High").

In many applications, the vehicle must be brought to a complete stop before the driver can shift the transfer case lever. In addition, the transmission must be in Neutral or Park so that the transmission output shaft is not turning.

In other cases, to shift into 4WD it may be necessary to get the vehicle moving slowly, then quickly shift into Neutral and move the transfer case lever.

On some 4WD vehicles, the driver can shift the transfer case into a 4WD position while the vehicle is being driven. This design is called "shift on the fly" capability.

Hub Locking System

Most 4WD truck and utility vehicles operate as rear-wheel drive vehicles during highway driving. With these vehicles, the front hubs must be locked manually to the front driving axles while the vehicle is stopped. To lock the hub mechanism, a rotating selector at the center of each front wheel hub is turned to a "lock" or "4WD" position.

More recent designs include "auto locking" hub mechanisms that engage automatically when the 4WD differential begins to turn the drive axles. On some of these systems, unlocking the hubs requires that the driver stop, shift the transfer case back to 2WD, then drive a short distance in reverse.

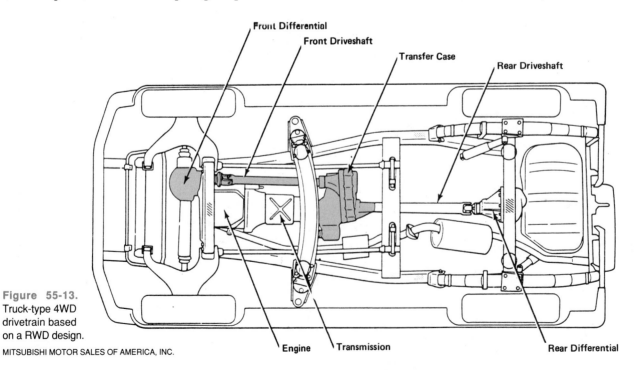

Figure 55-13. Truck-type 4WD drivetrain based on a RWD design.

MITSUBISHI MOTOR SALES OF AMERICA, INC.

DIFFERENTIALS, DRIVING AXLES, AND 4WD

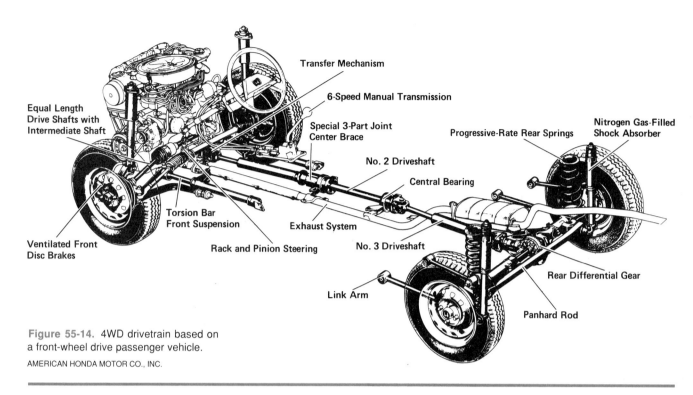

Figure 55-14. 4WD drivetrain based on a front-wheel drive passenger vehicle.
AMERICAN HONDA MOTOR CO., INC.

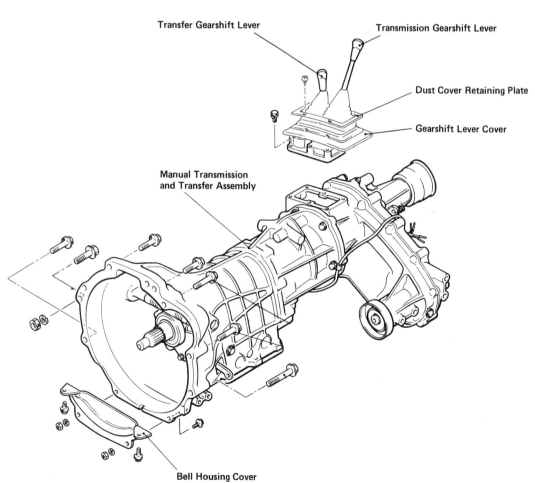

Figure 55-15. Transmission and transfer case for 4WD.
MITSUBISHI MOTOR SALES OF AMERICA, INC.

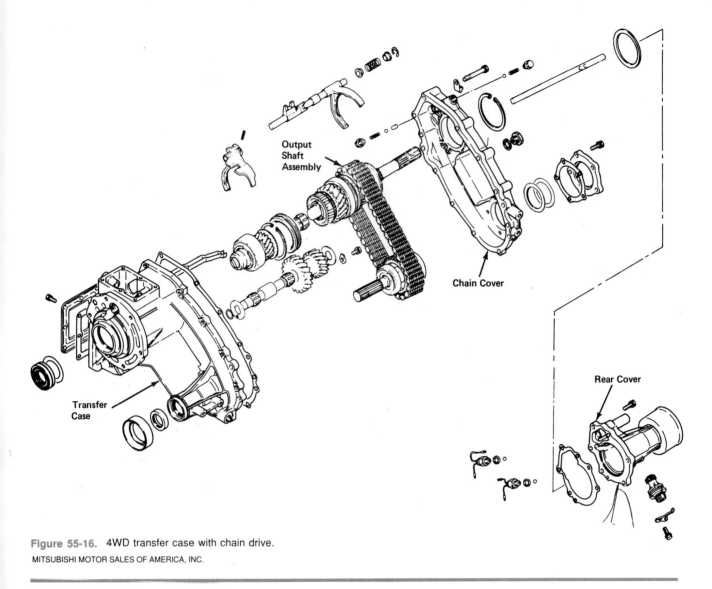

Figure 55-16. 4WD transfer case with chain drive.
MITSUBISHI MOTOR SALES OF AMERICA, INC.

Full-Time Four-Wheel Drive

Some vehicles provide *full-time 4WD*. Instead of a shiftable transfer case, full-time 4WD vehicles include a third differential in the drivetrain. For highway driving, this *interaxle differential*, or center differential, compensates for different distances travelled by the front and rear driving wheels on turns.

For maximum traction during low-speed driving, the driver locks the interaxle differential by shifting into the "Low" range or a "Lock" position. When the interaxle differential is locked, equal torque is transferred to both the front and rear differentials, as if there were no interaxle differential.

Some current high-performance sports sedans are equipped with full-time 4WD. These vehicles use full-time 4WD for added traction during driving at highway speeds in rain, snow, or on icy roads. Driver- or computer-controlled interaxle differentials provide varying amounts of torque to the front and rear driving wheels.

Viscous Coupling

Recent developments in 4WD drivetrains for passenger vehicles include special *viscous couplings* that eliminate the need for the driver to shift into 4WD. The viscous coupling is similar in function to that used on a cooling system fan clutch (refer to Topic 17.6). When the non-driving wheels turn at the same speed as the driving wheels, the viscous coupling does not transfer torque and the vehicle remains in two-wheel drive.

However, when the driving wheels begin to spin on slick surfaces, the viscous coupling fluid becomes heated, locks up, and transfers torque to the remaining two wheels. Only enough torque is transmitted to keep the normal driving wheels from slipping. When both sets of wheels turn at the same speed, the viscous coupling stops transferring torque. Thus, the system automatically shifts the vehicle into and out of 4WD as necessary.

Differentials, Driving Axles, and 4WD

UNIT HIGHLIGHTS

- Hypoid gears mesh with a sweeping action and make contact below the ring gear centerline.
- A conventional differential mechanism transfers more torque and turning speed to the wheel that is easiest to turn.
- The differential can have an integral or a removable carrier.
- A limited-slip differential locks both axles together through the differential case for better traction.
- Driving axles transfer power from the differential to the driving wheels.
- A part-time 4WD system typically includes a transfer case, driveline, and front or rear differential.
- A full-time 4WD system requires a third, interaxle differential mounted in the transfer case.
- Viscous couplings can be used to shift a vehicle into or out of 4WD automatically.

TERMS

drive pinion gear
hypoid gear
pinion shaft
side gear
clutch plate limited-slip
semi-floating axle
swing axle
clutch pack
viscous coupling
full-time 4WD
integral-carrier differential
three-quarter floating axle
interaxle differential
ring gear
differential case
differential pinion gear
limited-slip differential
cone limited-slip
De Dion axle
transfer case
full-floating axle
four-wheel drive (4WD)
part-time 4WD
removable-carrier differential
drive axle ratio (DAR)

REVIEW QUESTIONS

DIRECTIONS: The following questions are similar to those used on automotive technician certification tests. On a separate sheet of paper, write the letter of the correct choice.

1. Which of the following statements is correct?
   I. When turning, a conventional differential transfers the most torque to the wheel that is more difficult to turn.
   II. When traveling in a straight line, the differential transfers unequal amounts of torque to each driving axle.
   A. I only
   B. II only
   C. Both I and II
   D. Neither I nor II

2. Technician A says that an integral-carrier differential can be serviced while installed in the vehicle.
   Technician B says that a removable-carrier differential is part of the rear axle housing.
   Who is correct?
   A. A only
   B. B only
   C. Both A and B
   D. Neither A nor B

3. Technician A says that full-time 4WD systems have an interaxle (center) differential.
   Technician B says that part-time 4WD vehicles have an interaxle differential.
   Who is correct?
   A. A only
   B. B only
   C. Both A and B
   D. Neither A nor B

Questions 4 and 5 are not like those above. Each has the word EXCEPT. For each question, look for the choice that does *not* apply. Read each question carefully before you choose your answer.

4. All of the following statements about a limited-slip differential are correct EXCEPT
   A. the mechanism locks the driving axles together when one wheel spins.
   B. a clutch pack is made up of several steel discs next to each other.
   C. side thrust applies force that causes the clutch mechanism to lock.
   D. cones can be used in place of clutch packs to create lockup.

5. All of the following are types of driving axles EXCEPT
   A. one-quarter floating axles.
   B. semi-floating axles.
   C. three-quarter floating axles.
   D. full-floating axles.

ESSAY QUESTIONS

1. Describe how the parts of a conventional differential operate.
2. Describe how different types of limited-slip differentials operate.
3. What are the functions of additional parts in a 4WD drivetrain?

SUPPLEMENTAL ACTIVITIES

1. Describe how hypoid gears mesh and transfer force.
2. Identify the type of differential on a vehicle selected by your instructor.
3. Calculate the drive axle ratio (DAR) on a vehicle selected by your instructor.
4. Refer to a service manual to find maximum engine torque output. Calculate the overall transmission ratio (OTR) on the same vehicle used in 3, above. Calculate the torque applied to the driving axles.
5. Measure the loaded radius of the tires on the same vehicle used in 3, above. Calculate its theoretical top speed.
6. Identify different types of driving axles.
7. Locate and identify all the parts of a part-time or full-time 4WD system.
8. Demonstrate to your class how to engage and disengage the hubs and transfer case on a part-time 4WD vehicle.

56 Differential, Driving Axle, and 4WD Service

UNIT PREVIEW

Preventive maintenance for separately mounted differentials on rear-wheel drive and four-wheel drive vehicles includes checking for leaks, adding lubricant, and changing lubricant.

Differential problems often cause specific types of noises that can be recognized by the technician. Differential service procedures include listening for noises, making gear tooth contact pattern tests, axle removal, disassembly, cleaning and inspection of parts, and reassembly and adjustment.

Transfer case service includes removal, disassembly, cleaning and inspection, replacement of defective parts, reassembly, and replacement of the unit in the vehicle.

LEARNING OBJECTIVES

When you have completed your assignments and exercises in this unit, you should be able to:

❏ Remove driving axles from C-lock and retainer-type axles.
❏ Disassemble integral-carrier and removable-carrier differentials.
❏ Make a gear tooth contact pattern test.
❏ Remove a differential drive pinion assembly.
❏ Make differential bearing preload and backlash adjustments.

SAFETY PRECAUTIONS

Differential and driving axle preventive maintenance and service require that the automobile be raised and supported safely. *Never* work under an automobile unless you are sure that it cannot move or fall. Safety stands, or jackstands, must be used if the vehicle is not on a hoist.

Always wear eye protection when you work underneath a vehicle. Dirt, grime, or foreign objects can fall into the eyes during service procedures.

Hot exhaust system parts on vehicles can cause severe burns. Whenever possible, allow the vehicle to cool off before servicing.

The differential is heavy. If it falls, it can cause personal injury and damage to parts. Have a helper ready to assist in removing and replacing the differential.

Diagnosing differential and driving axle problems may require a test drive. Drive safely and only in a manner that allows you to diagnose the problem.

To remove asbestos from the brake assemblies, use an approved vacuum cleaner or liquid collection system. Do *not* use compressed air to blow out a brake assembly.

 ## TROUBLESHOOTING AND MAINTENANCE

A visual inspection of the differential and driving axles should be made whenever the automobile is on a hoist. The differential oil level must be checked and refilled, if necessary, during chassis lubrication.

Checking for Leaks

Inspect the differential housing for leaks, cracks, or damage. Check for leaks at the filler plug, differential housing, and drive pinion seal.

Differential housing leaks caused by casting porosity, tiny holes formed during the manufacturing process, can be sealed with special epoxy compounds.

Look inside each rear wheel for evidence of leaking brake fluid or differential lubricant. Brake fluid can leak from defective rear-wheel brake cylinders or calipers. Differential lubricant can leak from defective rear-axle bearing seals.

Checking and Adding Correct Lubricant

The differential has a fill plug similar to that on a manual transmission. The fill plug may be located at either the back or the front side of the differential **[Figure 56-1]**.

To check differential lubricant level, wipe off the fill plug. Use the proper wrench to remove the plug. Place a finger or a bent rod into the hole.

SAFETY CAUTION

Do not turn the tires or the drive shaft while your finger is inside the check hole. Rotating differential gears can pinch or cut your finger severely.

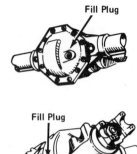

Figure 56-1. Fill plug locations.

Lubricant level should be even with or no more than one-half inch below the fill hole. Do not overfill. To add lubricant, use a pressure gun or suction gun. After lubricant has been added, replace and tighten the fill plug and wipe off any spilled lubricant.

CAUTION!!!

Refer to the vehicle manufacturer's service manual for the correct lubricant. Special lubricants must be used in limited-slip differentials.

Draining and Refilling

To drain differential lubricant, remove a drain plug (if present), rear cover and gasket, or use a suction gun [Figure 56-2]. If the lubricant is in good condition, replace the drain plug or rear cover and new gasket, and refill with the recommended lubricant. If the drained lubricant has a shiny, "metallic" appearance, it indicates severe damage to bearings and/or gears. Inspect and service the differential as described in 56.2

56.1 DIAGNOSING DIFFERENTIAL PROBLEMS

Noises from the differential during a test drive can indicate specific problems. However, sounds tend to reverberate through the entire drivetrain and may not originate in the differential. For specific information, refer to the troubleshooting charts in the manufacturer's service manual.

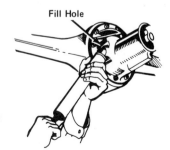

Figure 56-2. Removing differential lubricant with a suction gun.

Gear Noises

Two types of gear problems cause unusual noises: damaged gears, and gears that do not mesh properly.

Noises from a damaged gear are heard over the entire speed range. Noises caused by gears that don't mesh properly usually are heard at specific speeds. The noise may be different during acceleration and deceleration. Worn or damaged side gears and differential pinion gears rarely make noise during normal driving. These noises will be more noticeable when the vehicle is turning.

Bearing Noises

Differential bearings usually make a growling or scraping noise. The noise remains constant but will vary in intensity with vehicle speed. Bearing noise also may be louder when the vehicle is turning.

Limited-Slip Diagnosis

Improper limited-slip operation can cause a whirring, snapping sound, usually heard while turning a corner. Chattering caused by intermittent lockup also can be felt as the vehicle turns. Swerving during acceleration is another problem indication.

To make sure the limited-slip is operating properly, place the transmission in neutral. Block the front wheels so that the vehicle won't move. Raise one rear wheel off the ground. Place a jackstand under the vehicle. Block the front of the opposite wheel. Remove two lugnuts from the raised wheel.

Place a special tool over the wheel lugs [Figure 56-3]. Replace and tighten the lug nuts. Attach a torque wrench to the special tool. Rotate the torque wrench slowly until the wheel begins to turn. Take a torque reading just as the wheel begins to rotate. This is called the *breakaway torque*. Continue rotating the torque wrench slowly. The axle shaft should turn with even pressure without binding.

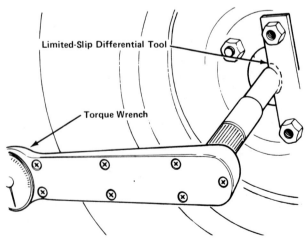

Figure 56-3. Checking limited-slip operation.
FORD MOTOR COMPANY

If breakaway torque is less than specifications, or the axle shaft does not rotate with even pressure, check the limited-slip differential.

SAFETY CAUTION

Never attempt to start or run the engine when making this check. A limited-slip differential will direct power to the wheel on the ground, causing it to rotate. The car can run off the jackstand, possibly causing serious injury.

56.2 DIFFERENTIAL SERVICE

Before servicing, the size, type, and rear-axle ratio of the differential must be identified. To locate this information, look for an identification code tag **[Figure 56-4]**.

An identification code may be stamped on the front side of the rear-axle housing. Sometimes the code is stamped on a metal tag that is attached to a differential cover retaining bolt. On some automobiles, the code is included in the vehicle identification number or on the driver's side door jamb.

Ring and Pinion Contact

This gearset consists of the ring gear and drive pinion gear. Teeth on the drive pinion can mesh the ring gear at either the same or different places after several revolutions. Three terms describe gearset contact:

1. Hunting gearset
2. Nonhunting gearset
3. Partial nonhunting gearset.

The number of teeth on the ring gear and drive pinion gear determine the type of gearset. A nonhunting or partial nonhunting gearset will be marked with special timing marks to ensure correct assembly.

Hunting gearset. When one drive pinion gear tooth contacts every ring gear tooth after several revolutions, it is a *hunting gearset.*

Nonhunting gearset. When one drive pinion gear tooth contacts only certain ring gear teeth, it is a *nonhunting gearset.*

Partial nonhunting gearset. On a *partial nonhunting gearset,* one pinion gear tooth contacts twice as many teeth as a nonhunting gearset. On the first revolution, it may contact three ring gear teeth. On the second ring gear revolution, it will contact three different ring gear teeth.

Gear Tooth Contact Pattern Test

A *gear tooth contact pattern test* indicates how ring and pinion gears mesh. Proper mesh will prevent gear damage and gear mesh noises.

Gear tooth contact pattern tests determine gear mesh problems and corrective actions that may be necessary. The test can be made before disassembling the differential and again after reassembly. Drive pinion depth and backlash are sometimes adjusted based on the results of a contact pattern test.

Before this test is made, clean each tooth of the ring and pinion gearset. Use a stiff brush to apply a gear marking compound, as shown in **Figure 56-5.**

SAFETY CAUTION

Use a gear marking compound that does not contain lead. Lead is poisonous and can lead to many physical ailments.

After the gear teeth have been coated with the compound, rotate the drive pinion with a torque wrench and note the torque necessary to turn the pinion. First rotate the drive pinion in the normal direction of rotation to produce a drive pattern. Then rotate the drive pinion backwards to produce a coasting pattern.

In some cases, wedging a block of wood lightly between the differential housing and the back surface of the ring gear will help. The slight force on the gear teeth produces a clearer, more distinct test pattern.

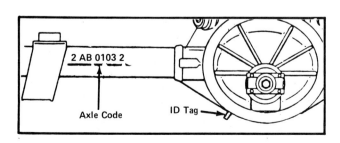

Figure 56-4. Differential identification code.
OLDSMOBILE DIVISION—GMC

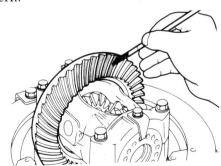

Figure 56-5. Applying a coat of gear marking compound.
FORD MOTOR COMPANY

Analyze the shape and position of the contact pattern. A correct pattern is centered on the gear tooth surface. Incorrect patterns are not centered **[Figure 56-6]**.

If the ring and pinion gear are to be reinstalled and the pattern is incorrect, make notes for adjustments to be made during reassembly. Refer to the manufacturer's service manual for correct adjustment procedures.

Removing Axles

The driving axles must be removed before the differential can be taken out. Two different driving-axle assemblies are found in modern automobiles: the C-lock type and the retainer-plate type.

The *C-lock axle assembly* is held in place by a C-lock that is located inside the differential housing. The axle bearing is a press fit in the axle housing. Whenever the axle is removed, a new axle seal must be used.

On a *retainer-plate axle assembly*, the axle shaft is held in place by a retaining plate. The retaining plate is bolted to the end of the axle, and the bearing is a press-fit on the axle shaft.

Whenever the axle assembly is removed, new gaskets and oil seals must be used.

C-lock axle removal. Raise the automobile on a hoist to a comfortable working height. Remove the rear wheels. Release the parking, or emergency, brake and remove the brake drums (see Unit 64).

Use a wire brush and a rag dipped in solvent to clean all dirt from the differential rear cover seal area. Place an oil catch pan under the differential and remove the drain plug or the rear cover to drain the differential.

Rotate the differential carrier by hand to expose the differential pinion lockscrew or lock bolt **[Figure 56-7]**. Remove the lockscrew or bolt and pull out the pinion shaft. If the bolt is broken, use screwdrivers, ice picks, and other tools to unscrew and remove it.

CAUTION!!!

Do not rotate the axle shafts while the pinion shaft is out. The differential pinion gears may fall out.

Push each axle shaft inward. Remove the C-locks from the grooves in the ends of the axle shaft **[Figure 56-8]**.

Remove the axle shaft by pulling on the flanged end of the shaft **[Figure 56-9]**. Pull slowly, being careful not to damage the bearing, which will remain in the housing. Remove the oil seal and axle bearing with a slide hammer **[Figure 56-10]**.

Retainer-plate type axle removal. Raise the automobile on a hoist to a comfortable working height. Remove the rear wheels. Release the parking, or emergency, brake and remove the brake drums (see Unit 64). A retainer-plate type axle assembly is shown in **Figure 56-11**.

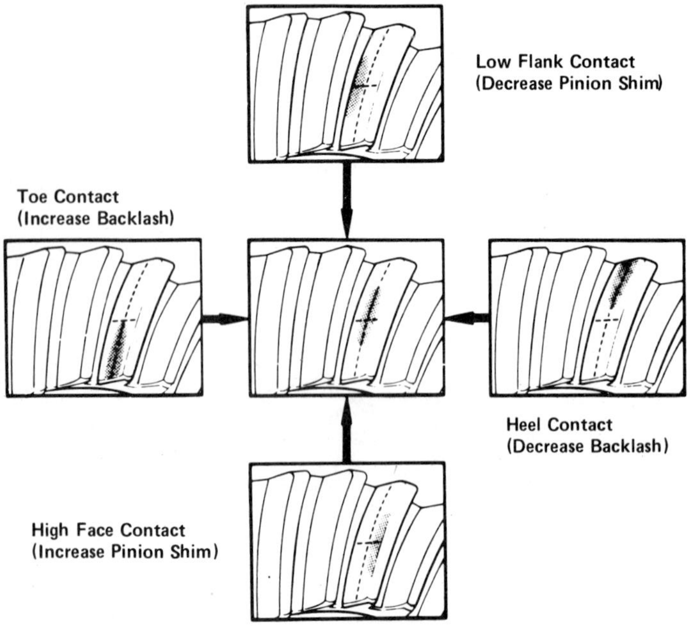

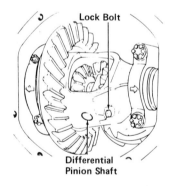

Figure 56-7. Drive pinion lockscrew locations.
FORD MOTOR COMPANY

Figure 56-6. Gear tooth contact patterns.
CHEVROLET MOTOR DIVISION—GMC

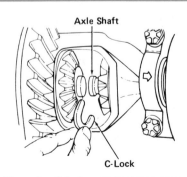

Figure 56-8. Removing C-locks. FORD MOTOR COMPANY

DIFFERENTIAL, DRIVING AXLE, AND 4WD SERVICE

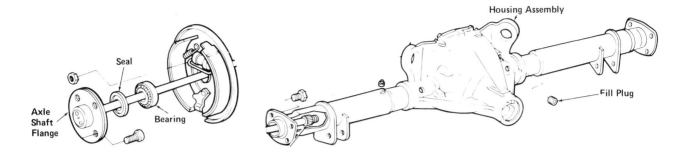

Figure 56-9. Parts of a C-lock axle assembly. FORD MOTOR COMPANY

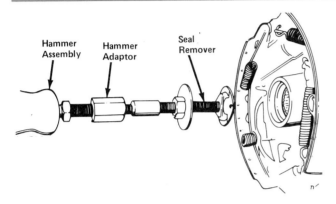

Figure 56-10. Removing an oil seal with a slide hammer.
BUICK MOTOR DIVISION—GMC

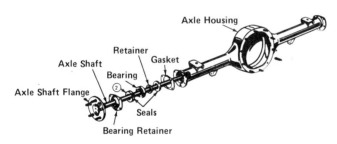

Figure 56-11. Parts of a retainer-plate type axle assembly.
FORD MOTOR COMPANY

The flange on the axle shaft has one or more access holes. Rotate the flange by hand to align the access holes with the nuts on the retainer plate. Place the appropriate socket wrench and extension through the access hole [Figure 56-12]. Remove the retainer plate nuts. If no access holes are provided, the nuts are removed from the back side of the retainer plate.

In some cases, the axle shaft can be pulled out carefully by hand. More often, a special axle puller must be used [Figure 56-13]. A hydraulic press must be used to remove the bearing if it is to be replaced.

Use a piece of wire to attach the brake backing plate to the frame of the vehicle and prevent damage to hydraulic brake lines.

Removing the Differential

Two types of differential assemblies are used: integral carrier and removable carrier.

Integral carrier. The *integral-carrier differential* is removed, part by part, from the rear of the axle housing. An integral-carrier differential is shown in **Figure 56-14**.

First, remove the driveline (see Unit 48), wheels, brake drums, pinion shaft, and axle shafts as described above. Reinstall the pinion shaft through the differential pinion gears and thrust washers.

Check the bearing caps to make sure they are marked "R" and "L" for right and left. If not, mark them. Loosen the bearing cap bolts until only a few

Figure 56-12. Removing a retainer plate assembly.
CHRYSLER CORPORATION

Figure 56-13. Removing an axle shaft with an axle puller.
FORD MOTOR COMPANY

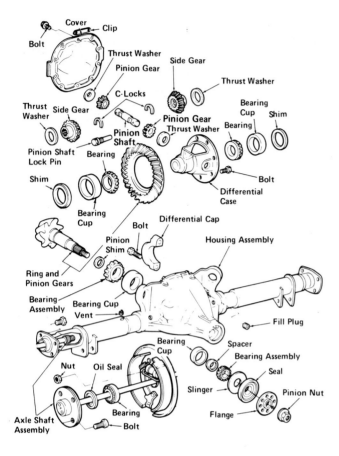

Figure 56-14. Parts of an integral-carrier differential assembly.
FORD MOTOR COMPANY

threads hold them in place. Place a pry bar behind the differential case [**Figure 56-15**]. A special spreader tool may be necessary to spread the differential housing enough to remove the carrier. Pry out the case until it falls free against the bearing caps.

Hold the differential assembly with one hand. Remove the bolts and other differential parts. Make a note of where shims are located and their thicknesses. Remove the case and place it on a workbench. Be sure to keep the bearing races in their original locations.

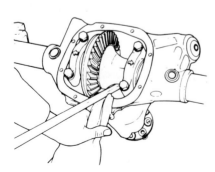

Figure 56-15. Prying out a differential case.
FORD MOTOR COMPANY

Drive pinion removal. Measure pinion bearing preload, if specified by the manufacturer, before removing the drive pinion assembly. Place a pound-inch (Nm) torque wrench on the drive pinion nut [**Figure 56-16**]. Rotate the wrench and record the torque that is required to maintain rotation.

Mark the positions of the pinion nut, companion flange, and drive pinion shaft for reassembly [**Figure 56-17**].

Place a special holding tool on the companion flange, as shown in **Figure 56-18**. Hold the special tool and, with a socket wrench, remove the drive pinion nut and companion flange.

Remove the oil seal with an oil seal remover. Hold the drive pinion with one hand and carefully tap out the drive pinion with a soft-faced mallet [**Figure 56-19**]. Remove the rear cover, drive pinion assembly, and all differential parts.

Removing the removable carrier. A *removable-carrier differential* is taken off the axle housing as an assembly and disassembled on a workbench. Parts of a removable-carrier differential are shown in **Figure 56-20**.

Clean the differential thoroughly, using a wire brush. Drain the lubricant. Remove the rear wheels, brake drums, axle shafts, and driveline (see Unit 48).

Support the removable-carrier housing on a floorjack or transmission jack. Have a helper hold the carrier housing while the nuts or bolts and

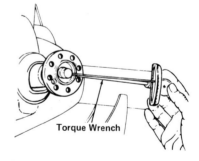

Figure 56-16. Measuring pinion shaft preload.
FORD MOTOR COMPANY

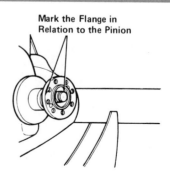

Figure 56-17. Marking the pinion flange for reassembly.
FORD MOTOR COMPANY

Differential, Driving Axle, and 4WD Service

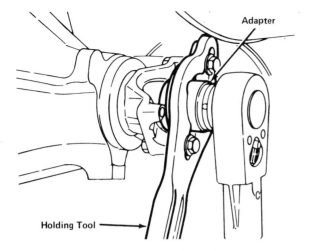

Figure 56-18. Companion flange holding tool.
AMERICAN MOTORS CORPORATION

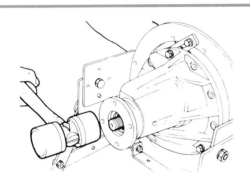

Figure 56-19. Driving out drive pinion with a soft-faced mallet.
CHRYSLER CORPORATION

washers are removed. Remove the carrier housing from the axle housing. Place the carrier housing in a holding fixture on a workbench **[Figure 56-21]**.

SAFETY CAUTION

The carrier housing is extremely heavy. Use care and have a helper to assist in moving and positioning the carrier housing.

Removable carrier disassembly. Disassembling the removable carrier is similar to disassembling the integral carrier.

Figure 56-21. Removable-carrier holding fixture.
FORD MOTOR COMPANY

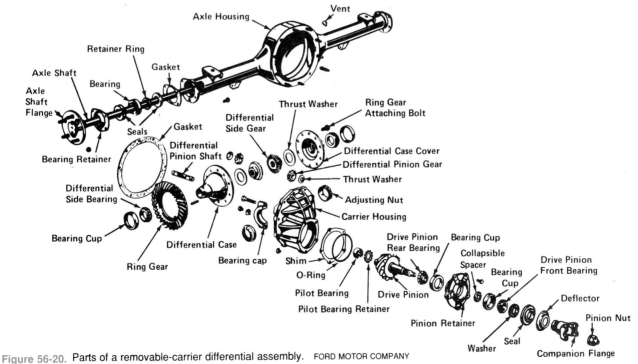

Figure 56-20. Parts of a removable-carrier differential assembly. FORD MOTOR COMPANY

Mark each bearing cap. Remove the bearing adjuster lock bolts and brackets, caps, and bearing cups. Lift the differential assembly out of the carrier housing. Place the differential assembly on a clean bench.

Make an alignment mark on the differential case and ring gear. Remove the bolts that hold the ring gear and case together. Hold the ring gear to prevent it from falling and becoming damaged.

Drive out the differential lockpin with a brass drift. Separate the halves of the differential case. Drive out the pinion shaft with the drift. Remove the differential pinion gears. Remove the differential side gears.

Two different drive pinion assemblies are used with removable carriers. These are the *retainer-type drive pinion* **[Figure 56-22]** and the *nonretainer-type drive pinion* **[Figure 56-23]**. Drive pinions are disassembled in much the same way as integral-carrier assemblies. The inner drive pinion bearing is a press-fit on the pinion gear and must be removed with an arbor press or hydraulic press.

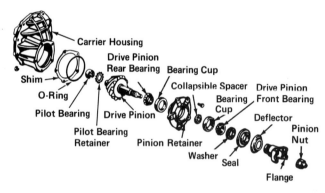

Figure 56-22. Parts of a retainer-type drive pinion assembly.
FORD MOTOR COMPANY

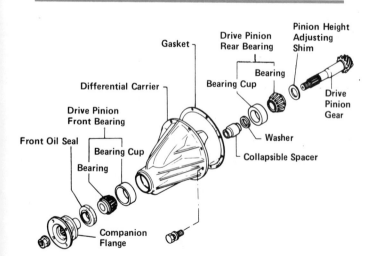

Figure 56-23. Parts of a nonretainer-type drive pinion assembly.
FORD MOTOR COMPANY

Retainer drive pinion disassembly. Remove the bolts that hold the pinion retainer and remove the drive pinion assembly from the carrier housing. Take care not to lose the retainer shims that are between the retainer and the housing. Remove the companion flange using a holding tool and socket wrench. Remove the drive pinion gear and oil seal with a slide hammer **[refer to Figure 56-10]**. Remove the front and rear pinion bearings.

Nonretainer drive pinion disassembly. A nonretainer drive pinion is disassembled in two steps. First, the companion flange, front oil seal, and front bearing assembly are removed. The second step is performed after the differential has been disassembled: The drive pinion and rear bearing assembly are removed from the rear of the differential housing.

Limited-slip removal and disassembly. Special precautions must be observed when working on clutch-type limited-slip differentials. For example, to avoid clutch disc misalignment, neither axle should be rotated unless both axles are in position. To remove clutches and cones from limited-slip differentials, refer to the manufacturer's service manual for specific precautions and instructions.

Cleaning and Inspection

Improper cleaning of differential parts can lead to early failure. Use the following guidelines.

Bearings can be cleaned in a suitable solvent. Lubricate the bearings immediately after cleaning. *Never* spin a bearing with compressed air. Use compressed air to blow it dry from the unsealed side.

The remaining parts, including the bores, can be cleaned with solvent. These parts can be blown dry with compressed air.

If the ring and pinion gears are scored with grooves or are chipped, they must be replaced and the axle housing and axle tubes must be thoroughly cleaned.

Limited-slip clutch packs are best cleaned with fast-evaporating mineral spirits or a dry-cleaning solvent. Clutch packs can be blown dry with compressed air.

SAFETY CAUTION

Safety glasses or goggles should always be worn when working with compressed air. Compressed air will blow metal chips, dirt, and liquids in all directions. Serious injury can result.

Inspect all parts. Look for signs of damage or wear. Check the case and housing for cracks, nicks, burrs, or scoring. Scoring may occur where the bearing is pressed onto the case. Scoring may be caused by the bearing race spinning on the case.

Differential, Driving Axle, and 4WD Service

Replace any parts that are worn or damaged. The ring and pinion gearset must be replaced as a matched set, even though only one may be damaged. Oil seals and drive pinion nuts are always replaced. Check the proper service manual for parts that must be replaced after disassembly.

Reassembly and Adjustment

Many checks and adjustments are necessary when reassembling any differential. The drive pinion and ring gear are always replaced as a set. If any of these checks or adjustments are not made, differential noise or early failure can occur. Reassembling the differential requires the following procedures:

- Drive pinion depth setting
- Drive pinion reassembly
- Differential reassembly.

Drive pinion depth setting. A dial indicator and special tools or fixtures may be required to check and adjust *pinion depth setting* [Figures 56-24 and 25]. Measurements are made to determine the thickness of the shim to be used, and to recheck that the pinion depth is correct [Figure 56-26]. Refer to the manufacturer's service manual for specific instructions and tools to be used.

A special installation tool is used to replace any bearings [Figure 56-27].

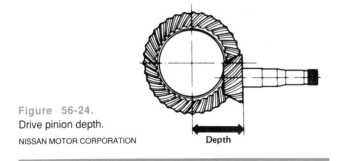

Figure 56-24. Drive pinion depth.
NISSAN MOTOR CORPORATION

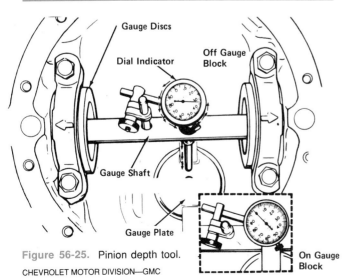

Figure 56-25. Pinion depth tool.
CHEVROLET MOTOR DIVISION—GMC

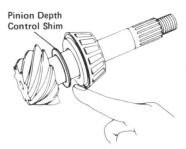

Figure 56-26. Depth-control shim location.
NISSAN MOTOR CORPORATION

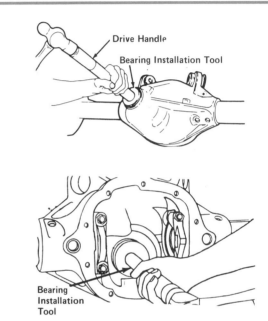

Figure 56-27. Pinion bearing installations.
OLDSMOBILE DIVISION—GMC/CHEVROLET MOTOR DIVISION—GMC

Integral-carrier drive pinion reassembly. Use all new shims, spacers, washers, gaskets, and oil seals when reassembling the differential. Refer to the manufacturer's service manual. Be sure all parts are lubricated properly.

Reassembly begins with the drive pinion assembly. Use an arbor press and a pinion bearing installation tool to press on the correct shim and bearing [Figure 56-28]. Place the front bearing cone in the housing. Install the pinion oil seal with a seal installation tool. Insert the drive pinion, and install the companion flange, using the same special tool that was used to remove it. Tighten the pinion nut according to the manufacturer's recommendations. Measure pinion bearing preload with a torque wrench.

Removable-carrier drive pinion reassembly. The nonretainer drive pinion assembly is reassembled and adjusted in the same manner as the integral-carrier drive pinion assembly [Figure 56-29].

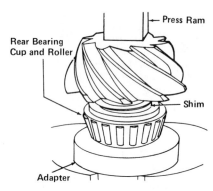

Figure 56-28. Press on new pinion bearing.
FORD MOTOR COMPANY

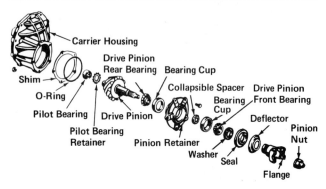

Figure 56-29. Parts of a retainer-type drive pinion assembly.
FORD MOTOR COMPANY

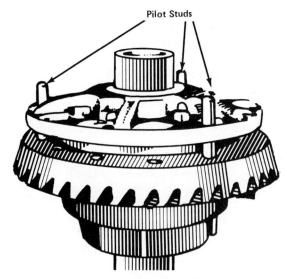

Figure 56-30. Pilot bolts are used to position the ring gear correctly on the differential case.

On the retainer drive pinion assembly, pinion bearing preload adjustment is made before pinion depth setting.

Press the rear and front bearing cones and rollers into place. Be sure the new collapsible spacer is against the rear bearing. Place the drive pinion assembly into the carrier. Install the oil deflector and oil seal. Install the companion flange and washer. Torque the pinion nut according to manufacturer's specifications, checking pinion bearing preload several times as you tighten the drive pinion nut. Avoid tightening the nut so far that the bearing preload is excessive. If this happens, obtain another new collapsible spacer and pinion nut and repeat the procedure.

Integral-carrier differential reassembly. Install thrust washers and side gears. Install the differential pinion gears between the side gears. Install the pinion shaft and the lockscrew or bolt.

Make pilot studs by cutting off the heads of bolts with the proper size thread, then cut a screwdriver slot in the ends with a hacksaw. Screw in the pilot studs, then position the ring gear over them on the case [Figure 56-30]. Remove the pilot studs and thread new bolts into the ring gear. Torque the bolts to specifications. Press side-bearing cones in place.

Removable-carrier differential reassembly. Procedures for reassembling the removable-carrier differential are similar to reassembling an integral-carrier differential. The main difference is that removable-carrier reassembly is done on a workbench. Reassembly of an integral-carrier differential is done on the vehicle. Refer to the manufacturer's service manual for instructions on measuring and adjusting pinion bearing preload with a torque wrench.

Limited-slip differential reassembly. The only difference in reassembling a limited-slip differential is when the clutch packs or cones are installed. Refer to the manufacturer's service manual for these procedures. If the differential contains preload springs, be careful and wear safety glasses during installation.

Final Checks and Adjustments

After reassembly, two checks are made to determine whether the differential is operating within specifications:

1. Differential bearing preload and backlash
2. Gear tooth contact pattern test.

Differential bearing preload and backlash. Adjustments for *differential bearing preload* and *backlash* usually are done at the same time. Differential bearing preload refers to the load or pressure the bearings are under when installed. Backlash is the proper meshing and free play of the ring and pinion gear teeth.

Differential bearing preload adjustment varies considerably from automobile to automobile. Refer to the manufacturer's service manual for correct procedures. Adjusting bearing preload often involves using shims or adjusting nuts at each side of the carrier [Figure 56-31].

Backlash on the ring gear is checked with a dial indicator [Figure 56-32]. Backlash is corrected by changing the shims or adjusting nuts on each side of the carrier [Figure 56-33].

Gear tooth contact pattern test. After you reassemble the differential, make a gear tooth contact pattern test (refer to Topic 56.2). Make drive pinion depth and backlash adjustments to produce a centered gear mesh contact pattern. Repeat the test until a correct, centered pattern is produced. When gear mesh is correct, thoroughly clean all marking compound from the differential parts.

Differential Installation

An integral-carrier differential is installed during reassembly. The driving axles (see Topic 56.3) and

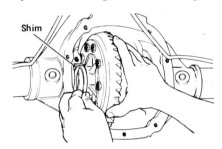

Figure 56-31. Using shims to adjust bearing preload.
FORD MOTOR COMPANY

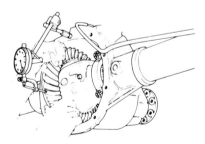

Figure 56-32. Checking backlash with a dial indicator.
FORD MOTOR COMPANY

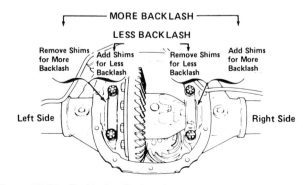

Figure 56-33. Backlash adjustments. FORD MOTOR COMPANY

rear cover are installed. Replace the rear cover using a new gasket or sealant. Refill the differential with lubricant according to manufacturer's specifications.

Install the removable-carrier assembly using a gasket and floorjack to support the carrier. The carrier is heavy, so a helper will be needed. Tighten the carrier bolts to specifications. Install the driving axles (see Topic 56.3), brake assemblies (see Unit 64), and wheels (see Unit 62).

56.3 DRIVING AXLE SERVICE

Premature failure of an oil seal or axle bearing frequently is caused by poor inspection or reassembly. Inspect the inside of the axle housing, where the seal is mounted, for rough spots, burrs, rust, and corrosion. Axle shaft splines should be free of burrs, wear, and damage. Check the retainer and studs for damage. New oil seals must be used.

Unsealed bearings can be cleaned in solvent. Look for signs of pitting, wear, and other damage [Figure 56-34]. Also check the axle shaft bearing surface. Sealed bearings can be checked by rotating them slowly by hand. Listen for noise and feel for roughness. Replace defective bearings.

Driving Axle Runout

Improper runout at any part of the driving axle can cause vibration. Runout is checked with a dial indicator at three different areas of the driving axle assembly:

1. Pilot
2. Flange face
3. Axle shaft.

The *pilot* is a machined area at the outside end of the axle shaft on C-lock axle assemblies. Place a dial indicator on the pilot, and rotate the shaft [Figure 56-35].

On retainer axle assemblies, check the flange and wheel lugs for runout [Figure 56-36].

C-Lock Bearing and Seal Installation

Lubricate the new bearing. Install the bearing squarely and correctly and align it in the housing bore. Use a special bearing installation tool to seat the bearing [Figure 56-37].

CAUTION!!!

Never hit the bearing with a hammer or apply heat to install it. Do not try to use the oil seal to straighten the bearing in the bore.

Use an oil seal drive, or the recommended special tool, to install the oil seal [Figure 56-38]. Spread grease on the seal after installation.

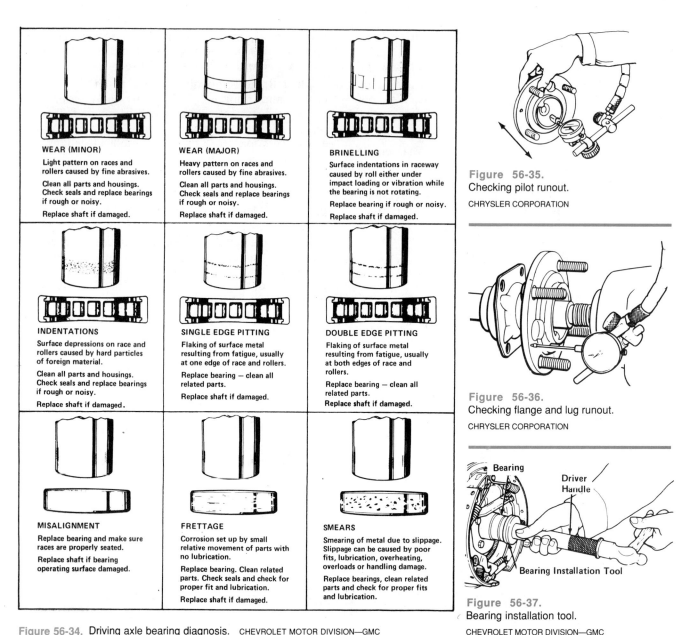

Figure 56-34. Driving axle bearing diagnosis. CHEVROLET MOTOR DIVISION—GMC

Figure 56-35. Checking pilot runout. CHRYSLER CORPORATION

Figure 56-36. Checking flange and lug runout. CHRYSLER CORPORATION

Figure 56-37. Bearing installation tool. CHEVROLET MOTOR DIVISION—GMC

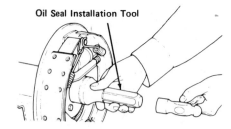

Figure 56-38. Oil seal installation tool. CHEVROLET MOTOR DIVISION—GMC

Retainer-Type Bearing and Seal Installation

There are many variations for installing a bearing and seal on different automobiles. Always check the proper service manual for the correct procedure.

Lubricate the bearing and seal. Install the bearing and retaining collar on an arbor press or with special tools. Press the new bearing onto the axle shaft, as shown in **Figure 56-39**. Press the new seal in place with a seal installation tool.

C-Lock Axle Reassembly

Remove the lockscrew and pinion shaft. Slide the axle shaft into the housing. Be careful not to damage the oil seal and axle bearing. Start the splines into the side gears. Push the axle shaft in until the button end of the shaft can be seen in the differential case. Install the C-locks and pull out on the axle. Replace the pinion shaft and lockscrew.

C-lock end-play adjustment. End play is controlled by the C-locks and is checked at the outer axle end.

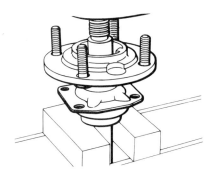

Figure 56-39. Pressing on the axle bearing.
CHRYSLER CORPORATION

End play is the in-and-out movement of the axle shaft inside the axle housing.

Strike the ends of both axle shafts with a soft-faced mallet. Install an end-play tool and dial indicator **[Figure 56-40]**. Push and pull on the axle shaft and read the movement on the dial indicator. Shims are used at the C-locks to correct end play.

Retainer-Type Axle Reassembly

Position the gaskets, brake backing plate, and brake assembly over the axle housing. Lubricate the axle splines. Slide the axle shaft into the housing. Push the axle splines into the side gears. Install retainer plate nuts and torque to specifications.

Retainer-type end-play adjustment. End play adjustment is done in much the same way as C-lock adjustment. On retainer-type assemblies, end play usually is adjusted at the left axle shaft only. Adjustments are made to shims located behind the retainer plate.

Complete reassembly by installing the brake drums (see Unit 64), retainer clips, and wheel assembly.

56.4 TRANSFER CASE SERVICE

Transfer case mechanisms are similar to the internal parts of manual and automatic transmissions. Thus, transfer case service is similar to transmission service (refer to Units 50 and 54).

Preventive Maintenance

Correct lubrication is important for long transfer case life. Preventive maintenance jobs include:

- Making visual inspections for leaks and damage to the transfer case housing.
- Checking and adding lubricant as necessary.

CAUTION!!!

Always refer to the manufacturer's service manual for recommended transfer case lubricants. Many transfer cases require EP (extreme pressure) lubricants as used in differentials and in some manual transmissions.

Transfer Case Overhaul

The following is a brief overview of transfer case services. Refer to the manufacturer's service manual for specific repair and overhaul procedures.

Transfer Case Removal

To remove the transfer case, first disconnect and remove driveline assemblies. Always mark the relative positions of the yokes before disassembly so that driveline balance can be maintained. Disconnect linkage to the transfer case shift lever. Also disconnect wires to switches for 4WD dash indicator lights.

Support the transfer case with a transmission jack and safety stands for removal. Remove fasteners holding the transfer case, then move the case away from the transmission or transaxle and lower the unit.

Transfer Case Disassembly

Disassembly includes removing covers and electrical switches for 4WD dash indicator lights. Loosen and carefully remove plugs that secure springs and detent balls. Drive out pins that secure shift forks.

Some transfer cases use spur or helical cut gearsets to transfer torque from the transmission to the output shaft. Other gearsets use chain drives **[refer to Figure 55-15]**. Planetary gearsets provide gear reduction functions in some transfer cases.

Remove front and rear output shafts and gears or transfer chains and planetary gearsets from the case **[Figure 56-41]**.

Remove, clean, and inspect all parts. Inspect chains for wear and stretching. Carefully examine sprockets for damage and broken teeth. Replace any defective parts. If you find defects in any part of a planetary gearset, replace the entire unit.

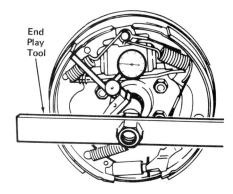

Figure 56-40. End play tool. AMERICAN MOTORS CORPORATION

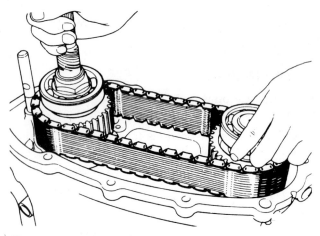

Figure 56-41. Transfer case chain drive and sprocket wheels.
MITSUBISHI MOTOR SALES OF AMERICA, INC.

Transfer Case Reassembly

Measurements for shaft assembly end play may be necessary, and selective snap rings or shims may be used to adjust end play of shaft assemblies. Use new gaskets between covers when you reassemble the unit. Replacement of the transfer case is essentially the reverse of removal. Refer to the vehicle manufacturer's service manual for specific reassembly and replacement procedures.

UNIT HIGHLIGHTS

- Preventive maintenance includes a visual inspection and checking differential lubricant level.
- Specific noises indicate differential problems.
- The ring and pinion gears can make contact in three different ways.
- A gear tooth contact pattern test indicates how the ring and pinion gears mesh.
- Driving axles are held in place by C-locks or retainer plates.
- An integral carrier is serviced from the rear of the differential and is removed part by part.
- A removable carrier is bolted to the front of the differential and is removed as a unit before disassembly.
- Drive pinion depth setting is the in-and-out movement of the drive pinion gear.
- Differential bearing preload is the load or pressure on bearings after installation.
- Runout is checked at three places on a driving axle.
- Axle bearings and seals are installed with special tools.
- End play is the in-and-out movement of the axle shaft in the axle housing.
- Preventive maintenance and service for transfer cases is similar to service for manual transmissions.
- Transfer case mechanisms include gears, drive chains, sprocket wheels, and planetary gearsets.

TERMS

pilot	breakaway torque
nonhunting gearset	hunting gearset
C-lock axle assembly	backlash
pinion depth setting	end play
removable-carrier differential	differential bearing preload
gear tooth contact pattern test	nonretainer-type drive pinion
partial nonhunting gearset	retainer-plate axle assembly
retainer-type drive pinion	integral-carrier differential

REVIEW QUESTIONS

DIRECTIONS: The following questions are similar to those used on automotive technician certification tests. On a separate sheet of paper, write the letter of the correct choice.

1. Which of the following statements is correct?
   I. An integral carrier case is bolted to the rear of the axle housing and is removed in one piece.
   II. Pinion bearing preload can be measured before removing the drive pinion gear.
   A. I only
   B. II only
   C. Both I and II
   D. Neither I nor II

2. Technician A says that drive pinion depth setting adjusts the in-and-out movement of the drive pinion gear.

   Technician B says that differential bearing preload is the in-and-out movement of the axle shaft.

   Who is correct?
   - A. A only
   - B. B only
   - C. Both A and B
   - D. Neither A nor B

3. Which of the following driving axle service procedures is correct?
   - A. Bearings are driven in with hammers and heat from a welding torch.
   - B. Old seals are reused if they look good.
   - C. End play is controlled by shims.
   - D. Oil seals are installed with a hammer and a drift punch.

Questions 4 and 5 are not like those above. Each has the word EXCEPT. For each question, look for the choice that does *not* apply. Read each question carefully before you choose your answer.

4. When diagnosing a differential problem, all of the following are correct EXCEPT
   - A. consult the troubleshooting chart in the manufacturer's service manual.
   - B. worn or damaged bearings rarely make noise.
   - C. noises from a damaged gear are heard over the entire speed range.
   - D. swerving during acceleration may indicate a limited-slip differential problem.

5. When servicing transfer cases, all of the following are correct EXCEPT
   - A. visual inspections for leaks and damage are necessary.
   - B. follow the manufacturer's recommendations for lubricants.
   - C. all transfer cases include both a chain drive and a planetary gearset.
   - D. overhauling transfer cases is similar to overhauling transmissions.

ESSAY QUESTIONS

1. What preventive maintenance procedures are needed for differentials, driving axles, and 4WD transfer cases?
2. Describe how to diagnose differential problems.
3. Describe service procedures for transfer cases.

SUPPLEMENTAL ACTIVITIES

1. Drain and refill a differential.
2. Remove, disassemble, clean, and inspect differential parts for defects.
3. Remove and replace C-lock and retainer-type driving axles.
4. Refer to a service manual and adjust gear tooth contact on a differential.
5. Refer to a service manual to remove, disassemble, clean and inspect a transfer case. Report what defects are found. Reassemble the transfer case and replace it in the vehicle.
6. Perform a gear tooth contact pattern test.
7. Measure and adjust drive pinion depth and differential bearing preload and backlash.

PART VIII: The Automotive Chassis

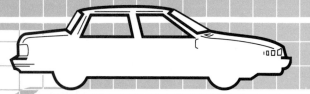

Independent shops that specialize in tires, brakes, and front-end alignment provide excellent career opportunities for skilled automotive technicians. The suspension, steering, and brake systems directly affect the stability, control, and overall safety of a vehicle. The knowledge and service units cover suspension, steering and wheel alignment, and tires and wheels. This technician is beginning to perform a computerized, front-end wheel alignment.

57 The Suspension System

UNIT PREVIEW

A suspension system supports an automobile and keeps its wheels in contact with uneven road surfaces. Modern suspensions produce a smooth, stable ride as well as predictable cornering and stopping characteristics. The suspension system determines how an automobile "handles" and how much weight can be carried safely.

LEARNING OBJECTIVES

When you have completed your assignments and exercises in this unit, you should be able to:

❑ Describe suspension functions.
❑ Describe the different types of springs used in suspensions.
❑ Explain the function and operation of a shock absorber.
❑ Describe the operation of an independent suspension system.
❑ Identify and describe the differences between passive, semi-active, and active suspension systems.

57.1 SUSPENSION FUNCTIONS

The suspension system supports the weight of the vehicle's chassis and provides a comfortable and safe ride over many types of road surfaces. In addition, the suspension minimizes irregular and excessive tire wear.

A vehicle is most stable and controllable when its body is level and the forces acting on the tire *contact patches* are evenly distributed. Tire contact patches are the areas of the tires on which the vehicle rests on the road.

Forces created by acceleration, braking, and cornering cause a vehicle's body to tilt front-to-back and lean from side-to-side. As the body tilts or leans, weight shifts from one end or side of the vehicle to another. As a result, the traction of the tires becomes uneven and the vehicle is less controllable. **Figure 57-1** illustrates the range of conditions that occur during driving.

57.2 SPRINGS

The main parts of a suspension system are the *springs*. Springs are used to support the automobile's frame, body, engine, and drivetrain above and between the wheels. Springs also permit the wheels to move upward and downward. As the wheels go over bumps and uneven road surfaces, the springs compress and expand. The basic types of springs used on modern automobiles are:

- Coil springs
- Leaf springs
- Torsion bars
- Air suspensions.

All four corners of an automobile are supported by spring assemblies. However, many automobiles have different spring assembly designs at the front and rear. Most of the weight of an automobile is supported by the springs. *Sprung weight* refers to parts of the automobile suspended by the springs, such as the body, engine, and transmission. *Unsprung weight* identifies those parts that are *not* suspended by the springs, such as wheels, brakes, and axles.

Spring Function

Springs are designed to compress a certain amount when they support the sprung weight of an automobile. Enough additional movement is built into the springs to absorb road bumps when the automobile is driven. When traveling over uneven surfaces, wheels and other unsprung parts act to compress and expand springs, and cushion the ride.

Coil Springs

A *coil spring* is a thick coil of special steel alloy. Weight and/or force compresses a coil spring. This action stores energy. When the weight and/or force is removed, the spring expands and energy is released. Coil springs are used in both front and rear suspension assemblies. *Control arms* are connected from the frame to the bottom of a coil spring and to the wheel. Control arms allow a coil spring to expand and compress as road conditions affect the wheels. When wheel action stops, the coil spring returns to its original position. **Figure 57-2** shows a coil spring suspension.

625

Visual Reference	Suspension Terminology	Definition
	Vehicle Attitude	The relative levelness and/or tilt of the vehicle body, both side-to-side and front-back
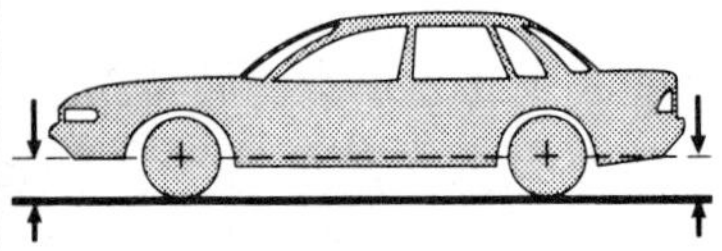	Vehicle Height	The height of the vehicle body in relation to the centers of the front and rear wheels
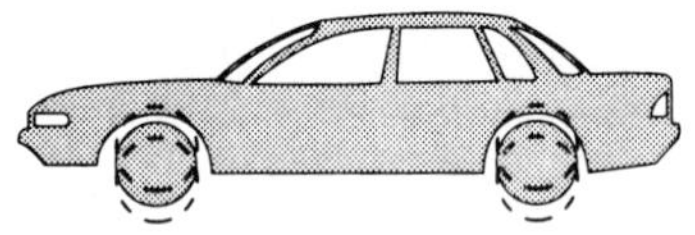	Damping	The resistance created as a shock absorber plunger moves through hydraulic fluid during compression or expansion
	Dive	The downward movement of the front, and upward movement of the rear of a vehicle when the brakes are applied*
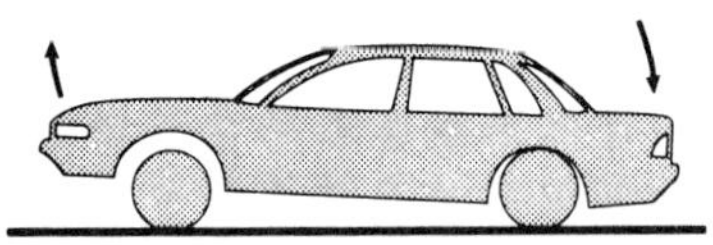	Squat	The upward movement of the front, and downward movement of the rear of a vehicle during acceleration*
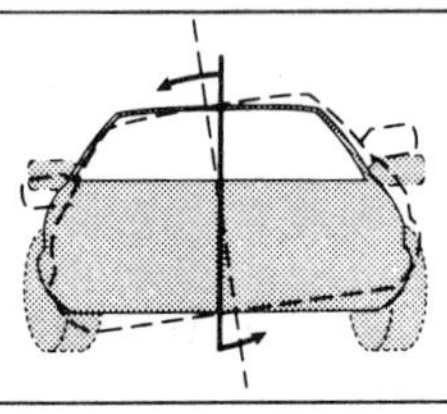	Roll	The leaning movement of the vehicle body when turning
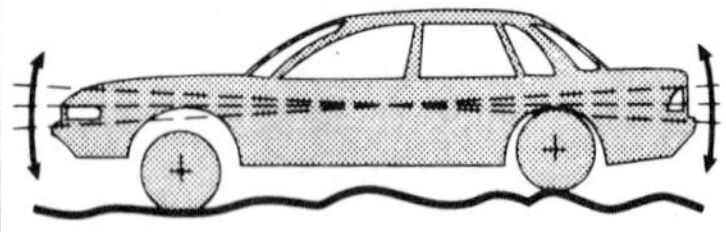	Pitching and Bouncing	The alternate rising and dipping of the front and rear of the vehicle body over uneven road surfaces

* When traveling in Reverse, opposite ends of the vehicle are affected.

Figure 57-1. Suspension conditions and terminology.
MITSUBISHI MOTOR SALES OF AMERICA, INC.

Leaf Springs

A *leaf spring* is made of one or more long strips of metal. The metal strips, called *leaves,* are laid one atop another to form a curved assembly. Leaves bend and move against another when the spring is compressed or expanded.

Leaf springs can be mounted longitudinally or transversely at the front or rear of an automobile.

The Suspension System

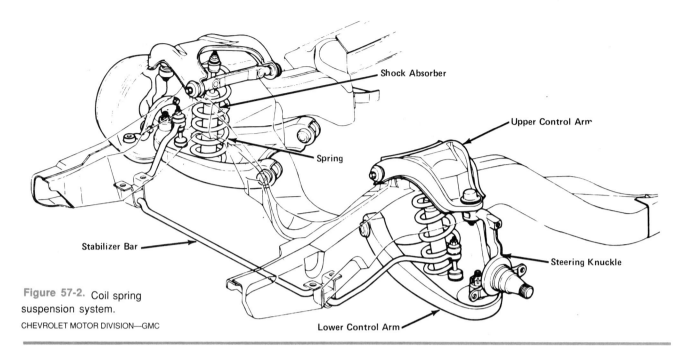

Figure 57-2. Coil spring suspension system.
CHEVROLET MOTOR DIVISION—GMC

Leaf springs usually are held in place at the center and at the eyes, or loops, at each end. Shackles attach the spring loops to the frame [Figure 57-3].

Torsion Bars

A *torsion bar* is a specially made steel bar that is capable of twisting to provide spring action. A torsion bar is attached to a control arm and to a frame crossmember [Figure 57-4]. Torsion bars are used as front suspension systems on light trucks and some passenger vehicles. As wheel movement causes the control arm to move up and down, the torsion bar is twisted. The rear of the torsion bar is held firmly in the crossmember. The torsion bar resists the twisting force, eventually untwisting to return the control arm to its original position. Torsion bars can be mounted longitudinally or transversely in an automobile.

Air Suspensions

An *air suspension* system consists of air spring assemblies which can replace or supplement conventional springs [Figure 57-5]. An air spring consists of a flexible rubber chamber that is filled with compressed air. The suspension may be controlled by a computer that automatically increases or decreases air pressure in the air spring chambers to raise, lower, or level the vehicle as needed.

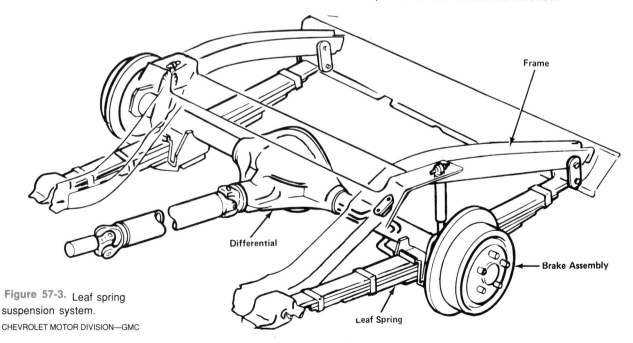

Figure 57-3. Leaf spring suspension system.
CHEVROLET MOTOR DIVISION—GMC

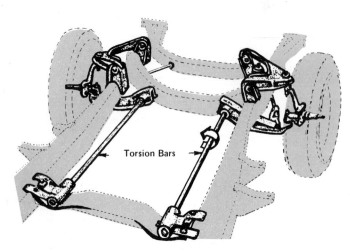

Figure 57-4. Torsion bar suspension system.
CHRYSLER CORPORATION

57.3 SHOCK ABSORBERS

A *shock absorber* is a device that works with the springs to control movements of the body, wheel, and axle. A spring that is not controlled by a shock absorber will continue to expand and compress, or oscillate, several times. A shock absorber dampens, or limits, the number and extent of spring oscillations.

Four shock absorbers usually are used on an automobile, one at each wheel. Shock absorbers are said to be *double-acting.* Double-acting means that the shock absorber controls both upward and downward suspension movements.

When a wheel travels over a bump, the spring is compressed, or shortened, and the wheel moves upward. This *compression,* or *jounce,* builds up energy in the spring. As the energy is released, the spring expands beyond its normal length. This expansion is called *rebound.* A spring also expands when a wheel travels over a low spot in a road surface or falls into a hole. This expansion is followed by compression and extension.

A tubular shock absorber lengthens and shortens, like a telescope, as it operates. A piston inside the shock absorber moves up and down in a fluid-filled chamber [Figure 57-6]. The piston has small orifices, or holes, through which fluid flows. This slows the movement of the piston and the telescoping action of the shock absorber. This action controls and dampens spring movement.

Bubbles can be produced on extremely bumpy roads as the plunger moves up and down rapidly through the fluid. These bubbles cause jerky, erratic motions and reduce the damping effect of the shock absorber. *Gas pressure shock absorbers* use a high-pressure gas (usually nitrogen) above the fluid to prevent fluid bubbling and foaming. The gas pressure maintains shock absorber damping characteristics even on the roughest roads.

Automatic Level Control

Some automobiles use an *automatic level control,* or *load leveling,* system. An automatic level control keeps the rear of the automobile at a normal level when a heavy load is added. An automatic level control is activated by a *height sensor* at the rear of the automobile. The height sensor signals an electronic control module when the rear suspension is compressed by a heavy load. Normal up-and-down movement from driving does not affect an automatic level control. A delay switch in the electronic control module prevents premature activation of the system. The signal must be on for several seconds before the load leveling system begins to operate.

In operation, the electronic control module activates an electrically operated compressor under

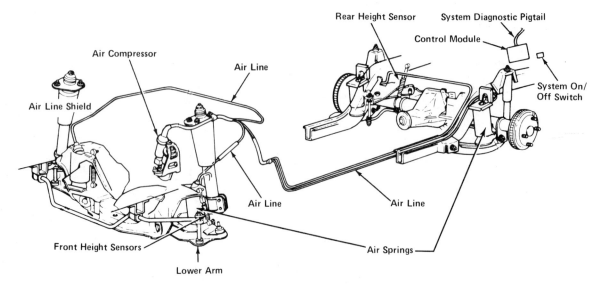

Figure 57-5. Air suspension system. FORD MOTOR COMPANY

THE SUSPENSION SYSTEM   629

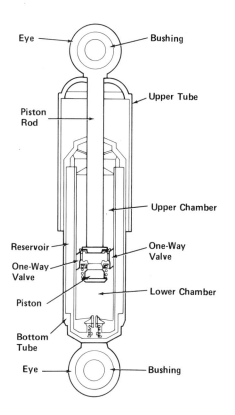

Figure 57-6. Parts of a shock absorber.

the hood. The compressor pumps air into lines that lead to air chambers in the rear hydraulic shock absorbers. The air chambers fill with compressed air, raising the rear of the automobile to a predetermined height. **Figure 57-7** shows an automatic load leveling system.

57.4 FRONT SUSPENSION

Modern automobiles have *independent suspension* systems at the front wheels. An independent suspension system is designed to allow each wheel to move up and down separately, or independently. Each wheel assembly has its own suspension system. This type of suspension usually has coil springs or torsion bars **[Figure 57-8]**. Other parts of an independent front suspension include:

- Control arms
- Steering knuckle and spindle assembly
- Ball joint
- Stabilizer bar.

Control Arms

Two control arms at each wheel direct the actions of the suspension system. One end of a control arm is connected to a frame member. The other end is connected to a steering knuckle **[Figure 57-9]**. As the

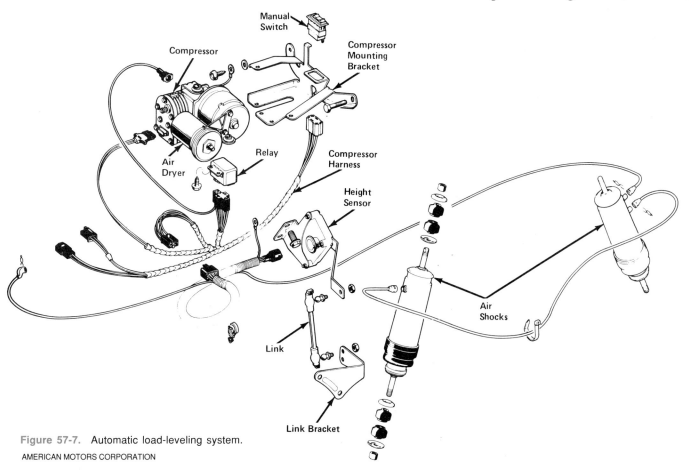

Figure 57-7. Automatic load-leveling system.
AMERICAN MOTORS CORPORATION

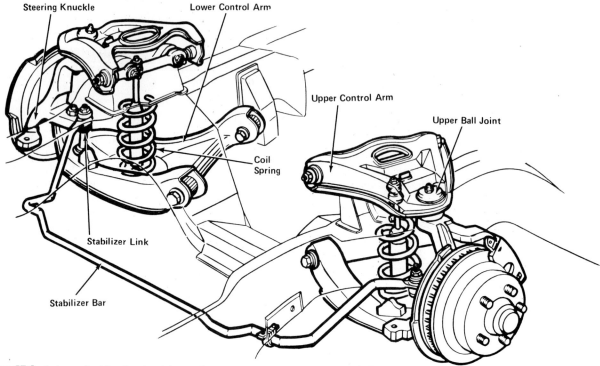

Figure 57-8. Independent front suspension system.   CHEVROLET MOTOR DIVISION—GMC

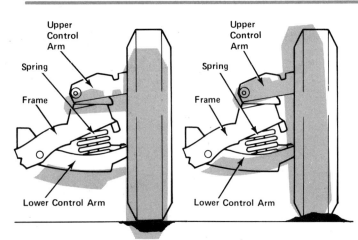

Figure 57-9. Front suspension movement.

wheel moves up and down, it moves the control arms and the spring.

A control arm often is shaped like the letter A. Because of this, a control arm may be called an "A-arm."

Upper and lower front control arms are made in different lengths. Short and long control arms, called "SLA type," are used at each wheel. The upper control arm is shorter. As the wheels move up and down, this arrangement allows stable tire contact with the ground.

Figure 57-9 shows a suspension system with a spring located between the frame and lower control arm. It also shows how control arms operate when the wheel moves.

Steering Knuckle and Spindle Assembly

A *steering knuckle and spindle assembly* consists of a *wheel spindle* and a *steering knuckle* [**Figure 57-10**]. A spindle is connected to a wheel through wheel bearings. The spindle is the point at which the wheel hub and wheel bearings are connected. A steering knuckle is connected to control arms. In most cases, a steering knuckle and wheel spindle are forged in a single piece.

Ball Joint

A steering knuckle and control arms are connected by a *ball joint* [**Figure 57-11**]. A ball joint has a ball stud that is positioned in a seal retainer, as shown in **Figure 57-12**. The ball stud pivots in a socket as the control arms are moved up and down. Ball joints may be riveted or bolted in place.

Ball joints help to support the weight of an automobile. Ball joints also allow rotary (side-to-side steering) movement and vertical (up and down) movement.

Stabilizer Bar

A *stabilizer bar*, or sway bar, is a long steel rod. The ends of the rod are connected to the lower control arms. The center section of the rod is connected to the frame or subframe [**Figure 57-13**].

A stabilizer bar controls *body roll*. Body roll occurs when the automobile is turning and the body leans, or sways, toward the outer suspension. Body roll shifts weight to the outside and moves the lower

The Suspension System

Figure 57-10. Steering spindle assembly.
CHEVROLET MOTOR DIVISION—GMC

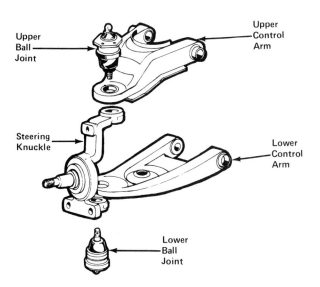

Figure 57-11. Ball joint locations. MOOG AUTOMOTIVE

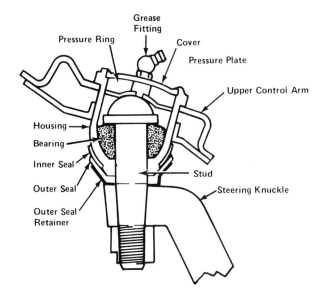

Figure 57-12. Ball joint assembly.

control arm upward to compress the spring. When the outer lower control arm moves upward, it twists the stabilizer bar. The stabilizer bar resists the twisting and limits movement of the control arm. This action reduces body roll and improves driver control.

Types of Independent Front Suspensions

There are many variations of independent front suspensions. The most common systems are:

- Double A-arm
- Straight control arm and strut rod
- MacPherson strut
- Modified strut
- Longitudinal torsion bar
- Transverse torsion bar.

Double A-arm. The *double A-arm* suspension is one of the oldest and best designs still in use **[Figure 57-14]**. This design is used mostly on larger automobiles and light trucks. Parts of this design were highlighted earlier in this section.

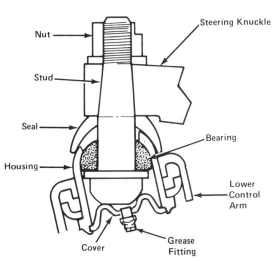

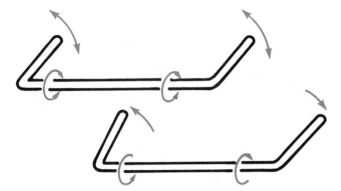

Figure 57-13. Stabilizer bar operation. Stabilizer bar does not offer torsional resistance when both parts of the frame move simultaneously (top). When frame is tipped on one side (bottom), it is pressed down on one end and is lifted up on the other end. This creates torsional resistance.

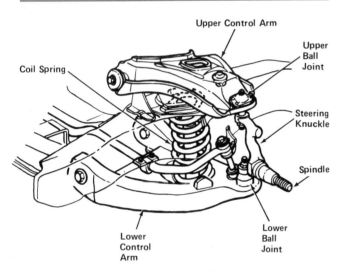

Figure 57-14. Double A-arm suspension.
CHEVROLET MOTOR DIVISION—GMC

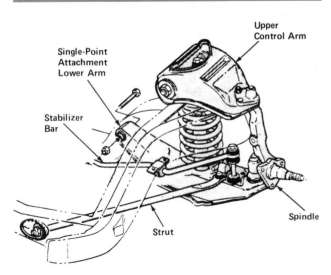

Figure 57-15. Straight control arm and strut-rod suspension.
FORD MOTOR COMPANY

Straight control arm and strut rod. The *straight control arm and strut rod* design is a variation of the double A-arm system [Figure 57-15]. The upper control arm has an A-shape. The lower control arm is straight. The lower control arm is attached to the frame at a single point. It is stabilized by a strut rod. The strut rod is connected to the outer end of the control arm and to a front frame member.

MacPherson strut. The *MacPherson strut* suspension has only a lower control arm [Figure 57-16]. The remainder of the suspension parts act as an upper control arm and a shock absorber. The coil spring is mounted at the top of the strut and must be compressed for strut removal.

The bottom of the strut is connected to the steering knuckle/wheel spindle assembly. At the top, the coil spring is connected to a tower in the body sheet metal. The tower is part of a reinforced inner fender. The top of the strut assembly rotates in the tower as the wheels pivot during turning. Most recent automobiles are designed with MacPherson strut front suspensions or modifications of this design.

Modified strut. The *modified strut* has a separately mounted coil spring [Figure 57-17]. In this type of suspension, only the shock absorber is combined with a strut. This design makes servicing easier.

Longitudinal torsion bar. A *longitudinal torsion bar* suspension system is similar to the straight control arm and strut system. However, the coil spring is replaced by a torsion bar [Figure 57-18]. The torsion bar is connected to the lower control arm at its pivot point on the frame. This type of suspension is used extensively on small pickup trucks.

Transverse torsion bars. *Transverse torsion bars* operate on the same principle as longitudinal torsion bars. However, the torsion bars are positioned across the front of the automobile [Figure 57-19].

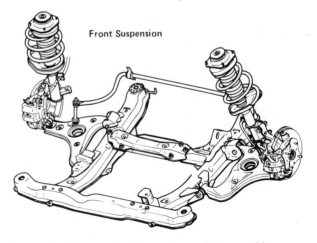

Figure 57-16. MacPherson strut suspension assembly.
MAZDA MOTOR CORPORATION

THE SUSPENSION SYSTEM

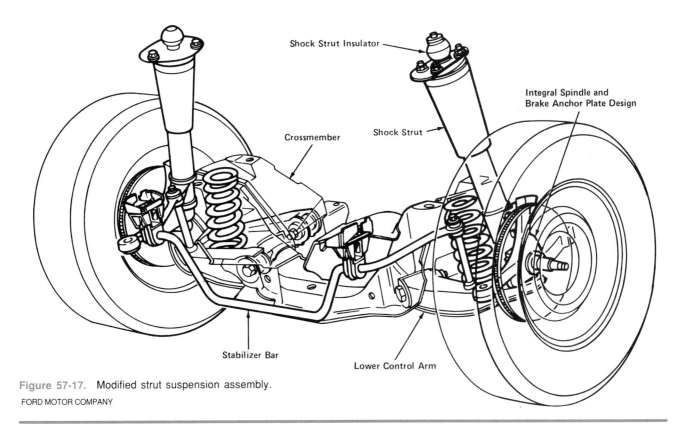

Figure 57-17. Modified strut suspension assembly.
FORD MOTOR COMPANY

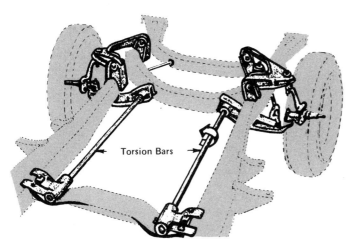

Figure 57-18. Longitudinal torsion bar system.
CHRYSLER CORPORATION

Fiberglass Springs

Chevrolet uses a transverse fiberglass leaf spring design on the independent front suspension of recent Corvettes. The ends of the leaf spring are connected to each lower control arm, or A-arm **[Figure 57-20]**.

57.5 REAR SUSPENSION

Designs similar to front suspension systems are used at the rear suspension. However, because the rear wheels on most vehicles are not steered, a rear suspension system usually does not include ball joints.

Coil Springs

Coil springs are located between brackets on the rear axle housing and *spring seats* in the frame or unit body. Spring seats are saucer-like brackets that position the springs. Springs are held in the spring seats by the weight of the automobile and by the shock absorbers **[Figure 57-21]**.

Coil springs can be flexed or moved in all directions. Control arms are used to control this random movement. Rear suspension control arms usually are made of channeled steel and mounted with rubber bushings to prevent damage from shock. Upper and lower control arms usually are used.

Upper control arms control *rear-end torque* and sideways movement of the axle housing assembly. Rear-end torque occurs when the driveshaft causes the differential and driving axles to twist (refer to Unit 47). Upper control arms are connected to the frame or unit body and to the differential housing.

Lower control arms are mounted between the axle assembly and the frame. Lower control arms maintain the fore-and-aft relationship of the axle housing to the chassis. The rigid axle housing holds the rear wheels in proper alignment.

The Volvo coil-spring rear suspension uses only lower control arms. Upper control arms are replaced with a wishbone-shaped subframe. Rear-end torque is transferred to the subframe by two torque arms.

A single, long torque arm also is used on some General Motors automobiles **[Figure 57-22]**. The front of the torque arm is connected to the transmission.

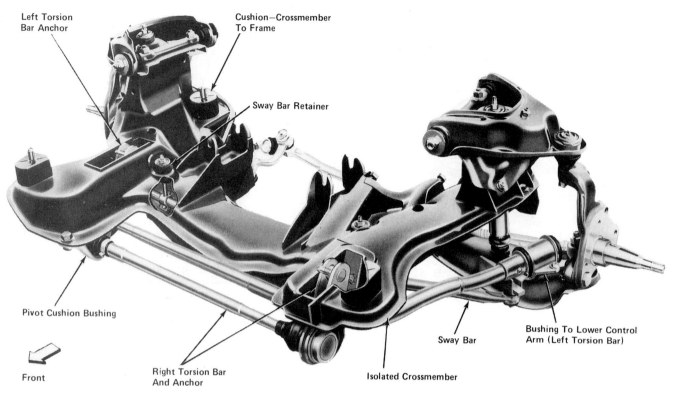

Figure 57-19. Transverse torsion bar system. CHRYSLER CORPORATION

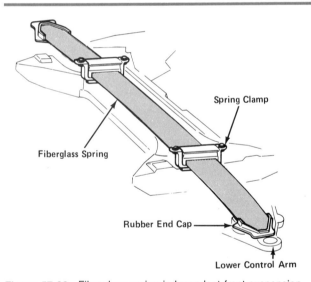

Figure 57-20. Fiberglass spring independent front suspension.
CHEVROLET MOTOR DIVISION—GMC

The rear of the torque arm is connected to the differential housing.

Some recent rear-wheel drive and full-time four-wheel drive automobiles have open axle shafts at the rear. The coil spring suspension used on these automobiles has only lower control arms. A suspension assembly crossmember supports the control arms [**Figure 57-23**]. The tops of the shock absorbers are mounted to the unit body. The springs are positioned on seats at the top and bottom.

MacPherson Strut

Many automobiles use MacPherson strut assemblies at the rear, as shown in **Figure 57-24.** Rear suspension MacPherson strut assemblies, also called "Chapman struts," operate in the same way as front suspension strut assemblies. A *radius rod* may be connected to the unit body and to the wheel spindle assembly. The radius rod keeps the lower control arm properly aligned.

MacPherson struts also are used at the rear in some mid-engine and rear-engine automobiles.

Another type of strut assembly also is used. On this type of suspension, shown in **Figure 57-25,** only the shock absorber is part of the strut assembly. The coil spring is located elsewhere on the lower control arm. A stabilizer bar also may be used with this suspension.

Semi-Independent Rear Suspension

Many automobiles with front transaxles have a *semi-independent suspension* system at the rear. A semi-independent system is one in which the two rear wheels are connected by a crossmember. In addition, each rear wheel is independently suspended by a spring.

One semi-independent rear suspension system uses a *trailing arm* design. A trailing arm extends rearward from the actual suspension mounting points on the body. This allows more interior room for

The Suspension System

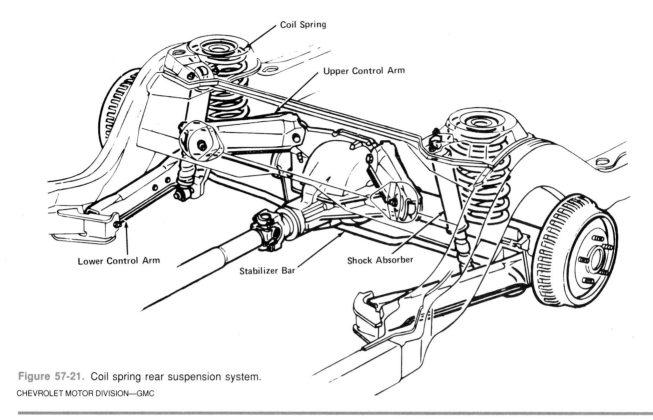

Figure 57-21. Coil spring rear suspension system.
CHEVROLET MOTOR DIVISION—GMC

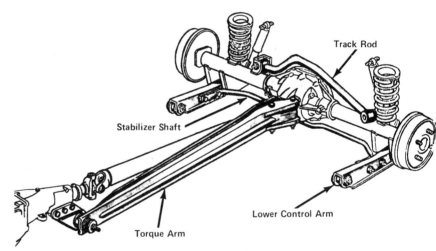

Figure 57-22. Rear suspension torque arm assembly.
OLDSMOBILE DIVISION—GMC

occupants. A wheel spindle is attached to the trailing arm. The crossmember between the trailing arms may twist and act as a stabilizer bar. **Figures 57-26 and 57-27** show trailing arm suspension systems.

A torsion bar also is used on some trailing arm suspensions. For example, Renault has used a transverse torsion bar built into the crossmember. When a torsion bar is used, only a shock absorber and wheel spindle are connected to the trailing arm.

Leaf Springs

Leaf spring rear suspensions are used on some larger rear-wheel drive vehicles and trucks. This type of rear suspension usually has longitudinally mounted springs. Springs and shock absorbers are positioned below the rear-axle housing and are connected to the frame or unit body [refer to Figure 57-3].

The front eye of a leaf spring is attached to the frame. The rear eye is attached to a *spring shackle* [Figure 57-28]. A spring shackle allows the leaf spring to change length as the leaves bend. The center of each leaf spring is connected to the rear-axle housing with U-bolts. Rubber bumpers are located between the rear-axle housing and frame or unit body to dampen severe shocks.

Transverse-mounted leaf springs sometimes are designed to act as independent suspensions. On these

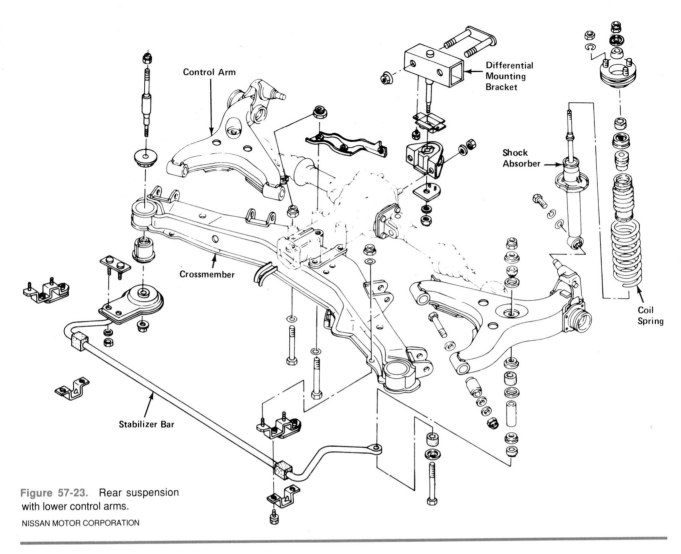

Figure 57-23. Rear suspension with lower control arms.
NISSAN MOTOR CORPORATION

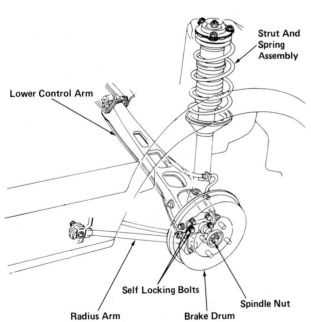

Figure 57-24. Rear suspension with MacPherson struts.
AMERICAN HONDA MOTOR CO., INC.

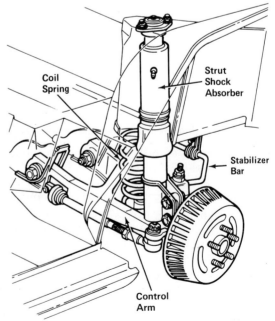

Figure 57-25. Non-MacPherson strut assembly.
CADILLAC MOTOR CAR DIVISION—GMC

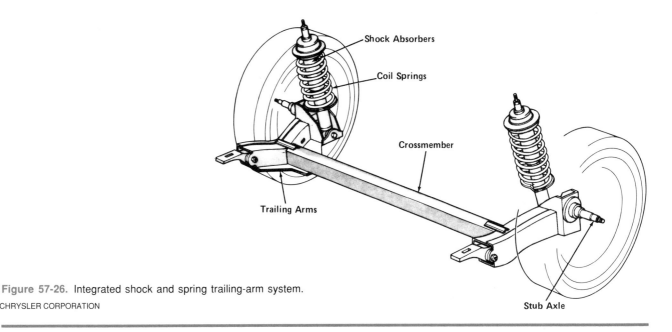

Figure 57-26. Integrated shock and spring trailing-arm system.
CHRYSLER CORPORATION

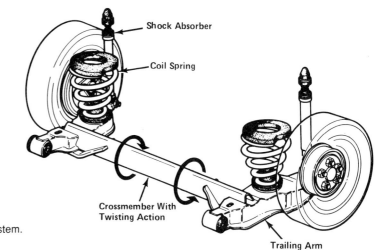

Figure 57-27. Separate shock and spring trailing-arm system.
PONTIAC MOTOR DIVISION—GMC

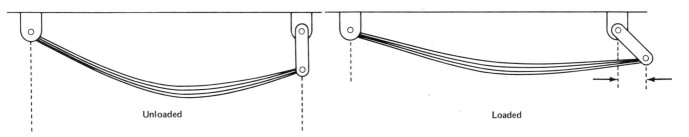

Figure 57-28. Rear leaf-spring shackle in operation.

designs, a multi-leaf [**Figure 57-29**] or a single-leaf [**Figure 57-30**] spring system may be used. The transverse leaf spring is mounted to the differential housing. The spring eyes are connected to wheel spindle assemblies.

On rear-drive automobiles, extra bracing is required for the suspension because of rear-end torque [**Figure 57-31**]. For leaf spring suspensions, the common methods of bracing are Hotchkiss drive and control rods.

Hotchkiss drive. Automobiles with rear leaf spring suspensions have *Hotchkiss drive*. Hotchkiss drive is a design that allows spring action to control rear-end torque. When rear-end torque begins, the axle housing twists upward at the front. This action lifts

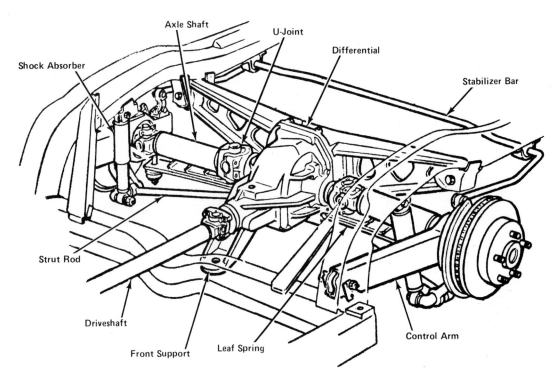

Figure 57-29. Transversely mounted multi-leaf rear springs. CHEVROLET MOTOR DIVISION—GMC

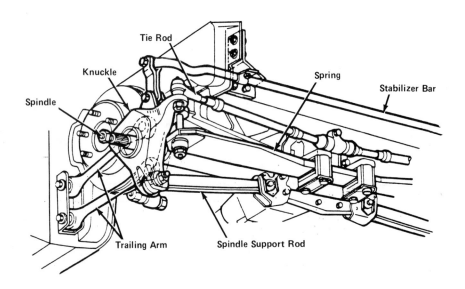

Figure 57-30. Transversely mounted single-leaf rear spring. CHEVROLET MOTOR DIVISION—GMC

the front of the leaf springs and lowers the rear of the leaf springs. Hotchkiss drive limits this movement to control rear-end torque.

Control rods. The *control rods* are used to maintain the proper front-to-back relationship of the rear axle in relation to the body. In some cases, a *torque tube* is used to control rear-end torque. This is a hollow tube containing the rear portion of the driveline [Figure 57-32]. This tube is fastened to the rear axle housing and a sub-frame member under the car. Any rear-end torque is controlled by this torque tube arrangement.

57.6 PASSIVE, SEMI-ACTIVE, AND ACTIVE SUSPENSION

To deal with changes in vehicle attitude that occur during driving, three types of suspension systems are used: *passive, semi-active,* and *active* suspension.

THE SUSPENSION SYSTEM

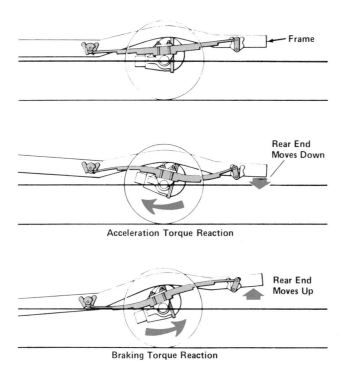

Figure 57-31. Rear-end torque reaction. FORD MOTOR COMPANY

Passive Suspension

Passive suspension systems are found on most vehicles. Vehicle height and damping depend on fixed, nonadjustable springs and shock absorbers or struts. When weight is added, the vehicle lowers as the springs are compressed. Vehicle body motion and tire traction varies with the road surface and driving conditions.

The characteristics of the springs and shock absorbers or struts used determine how stiffly or softly the vehicle rides. The amount of traction that keeps the vehicle under control changes during cornering and braking.

For example, the suspension design and parts of some luxury cars produces a high, "soft" ride on smooth roads for maximum passenger comfort. Soft springs and soft shock absorber damping allow the wheels to move upward over small bumps without transferring this motion to the body.

However, this choice of parts allows the body and wheels to move excessively during sudden maneuvering or hard braking. The vehicle may experience excessive body roll under these conditions.

At the other extreme, sports cars are built low to the ground and have stiff springs and shock absorbers. This design and choice of parts produces a "hard," stiff ride on most surfaces. Stiff springs and shock absorbers control body lean and wheel motion rapidly and efficiently. Thus, sports cars corner well at high speeds and are more controllable during hard braking and quick emergency maneuvers. However, sports cars are not well known for soft, comfortable rides.

Semi-Active Suspension

To provide some of the benefits of both "soft" and "hard" suspensions, *semi-active* suspensions were first introduced during the mid-1980s. These systems use computer-controlled adjustable springing devices and adjustable shock absorbers.

Computer Control. Like other computer-controlled systems, semi-active suspensions receive input data in the form of electrical signals from sensors (refer to Unit 38). Sensors provide input data for front and rear suspension height, *G-forces* (acceleration), steering wheel movement, vehicle speed, and other conditions. The computer unit processes the data and controls output units—air suspension units and adjustable shock absorbers.

Semi-active suspensions control some vehicle body motions. The system controls overall vehicle

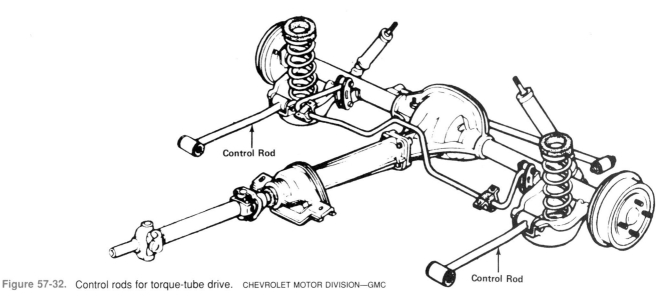

Figure 57-32. Control rods for torque-tube drive. CHEVROLET MOTOR DIVISION—GMC

height, shock absorber or strut stiffness or softness, and spring stiffness.

Springing Control. Spring stiffness is the resistance of a springing device to compression or extension. To adjust spring stiffness rapidly in a semi-active suspension system, flexible air chambers can be used.

As shown in **Figure 57-33,** the top of a coil spring rests against flexible air chambers. In effect, the flexible air chambers form an adjustable "cushion" between the spring and the vehicle body. The computer unit controls an air valve system to increase or decrease air pressure in the chambers.

When air pressure is decreased, the flexible chambers become softer and more flexible, just as removing air from a tire makes it softer. When the spring is compressed as the wheels move upward over a bump, it presses against the air chambers. The soft, flexible chamber expands as it absorbs some of the force of spring compression. This action provides a smooth ride.

Pumping air into the flexible chambers makes them stiffer. Then, when the spring compresses, the force is transmitted more directly to the body and the springing becomes "stiffer" for greater vehicle control.

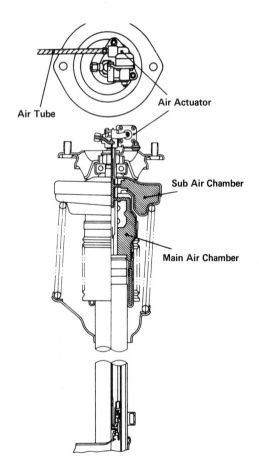

Figure 57-33. Flexible air chambers for semi-active suspension.
MITSUBISHI MOTOR SALES OF AMERICA, INC.

Height Control. The flexible air chambers also can raise or lower the vehicle's height. The chambers expand as air is pumped into them. Because the body is resting on top of the chambers, it rises as air pressure increases. To lower the vehicle, air pressure is reduced in the chambers.

For example, if vehicle load is increased by adding passengers and cargo, semi-active suspension reacts to height sensor information. The system increases air pressure in the chambers to raise the vehicle and compensate for the increased load.

Another use of height control is to improve aerodynamics at highway speeds. As vehicle speed increases, the vehicle lowers, with the front end angled downward. This action reduces wind resistance for better gas mileage and greater stability. When the driver switches the headlights ON, the system lowers the rear of the vehicle to level the vehicle and maintain correct headlight aiming.

As the vehicle slows, the semi-active suspension raises the body to normal height and level conditions.

Damping Control. *Damping* refers to the actions of shock absorbers or struts in controlling the bouncing motions, or oscillations, of springing devices. Special adjustable shock absorber or strut units control spring damping in semi-active suspensions.

Shock absorbers dampen spring oscillations by providing a resistance to the bouncing movement of the springs. Within the shock absorber, a piston with orifices, or holes, is forced to move through hydraulic fluid when the shock absorber is compressed or expanded. This restricted movement through fluid provides damping resistance.

For semi-active suspension systems, a computer controls the motion of an actuator (an air motor or an electrical stepper motor). The actuator turns a control rod that extends through the shock absorber.

On the control rod is a *rotary valve* through which the shock absorber fluid flows during compression or expansion. A rotary valve includes orifices that can be aligned with holes in a surrounding hollow enclosure **[Figure 57-34]**.

When the holes are fully aligned, fluid can flow through the passages easily. If the fluid passages are not fully aligned, fluid flow is partially blocked. The rotary valve varies the size of the shock absorber orifices through which hydraulic fluid flows. The larger the passageway, the more easily the fluid flows and the less damping resistance the shock absorber provides.

The rotary valve controls the amount of damping by providing a large passageway (soft damping) or a small passageway (hard damping). Intermediate positions produce medium damping.

THE SUSPENSION SYSTEM

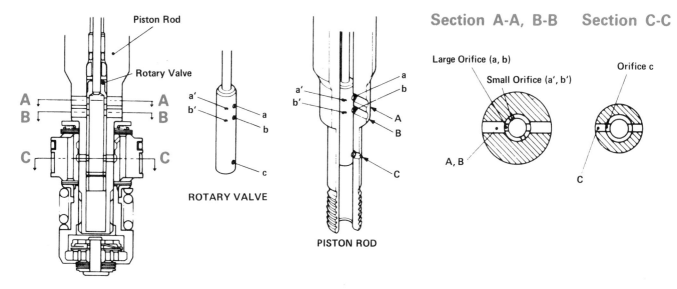

Figure 57-34. Rotary valve for adjustable damping control. MITSUBISHI MOTOR SALES OF AMERICA, INC.

Active Suspension

In addition to all of the functions provided by semi-active suspension, fully *active suspension* reacts to many other types of body motion. Fully active suspension also controls body tilt from front-to-back and body lean from side-to-side during cornering, braking, and acceleration. Additional solenoid air valves are used to transfer air pressure and control body motion.

Major parts [Figure 57-35] of one type of fully active suspension include:

- Adjustable strut/shock absorber units with air chambers
- Solenoid air valves
- Air compressor
- Front and rear height sensors
- Vehicle speed sensor
- G-sensor for side-to-side acceleration
- Throttle-position sensor
- Steering wheel angular velocity (turning speed) sensor
- Information/control panel
- Electronic control unit.

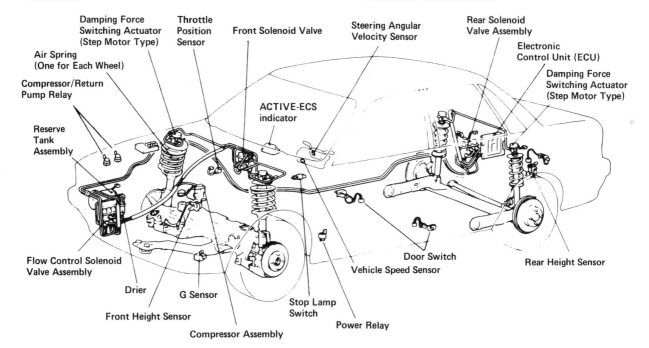

Figure 57-35. Major parts of a fully active suspension system. MITSUBISHI MOTOR SALES OF AMERICA, INC.

In this fully active suspension design, an air control system similar to that of semi-active suspension systems is used. However, additional air control valves are added so that air pressure is transferred from side-to-side and from front-to-back. This transfer of air pressure changes suspension height and counteracts body motions caused by turning, braking, and acceleration.

Roll Control. During cornering, the side of the vehicle nearest the inside of a turn tends to lift upward. As the vehicle leans toward the outside of the turn. This motion is known as *roll*. Roll reduces traction on the tires at the inside of a turn.

To counteract these forces, the active suspension system increases air pressure in the air chambers at the outside wheels. At the same time, valves open to reduce air pressure in the inside wheel air chambers. These actions counteract the motions caused by turning. The vehicle body remains relatively level for better traction and driver control **[Figure 57-36]**.

After the turning motion is completed, valves are opened to equalize the pressures on both sides of the vehicle and return it to a level position.

Dive Control. During hard braking, weight transfer tends to push the front of the vehicle downward and lift the rear upward. This motion is known as *dive*. Dive reduces traction on the rear end of the vehicle and may cause it to slide or spin during emergency braking.

During hard braking, the active suspension increases air pressure in the front air chambers and reduces air pressure in the rear chambers. These actions minimize dive to keep the vehicle level and make it easier for the driver to control. After braking, valves operate to equalize air pressures in front and rear air chambers and level the vehicle again.

Squat Control. When the driver depresses the accelerator quickly during hard acceleration, the front end of the vehicle tends to lift up. The rear end lowers. This motion is known as *squat*.

The active suspension system controls squat by operating valves that increase air pressure in rear wheel air chambers and reduce air pressure in front wheel air chambers. When the vehicle is no longer accelerating quickly, the control system operates valves to equalize air pressures and level the vehicle.

Thus, an active suspension system changes the height of the front, rear, or either side of the vehicle to counteract tilting and leaning. These active attitude control functions improve vehicle stability and increase tire traction and driver control.

UNIT HIGHLIGHTS

- The suspension system helps to keep the vehicle body level for good traction and control.
- Coil springs, leaf springs, torsion bars, and air springs are used in current suspension systems.
- A shock absorber dampens and controls spring action.
- An independent suspension allows each wheel to move independently of the others.
- Control arms position the wheels and pivot to follow the motion of the wheels over bumps and into depressions.
- A wheel spindle assembly connects the wheel to the suspension.
- Ball joints provide pivot points at the steering knuckle.
- A MacPherson strut combines a spring and a shock absorber into one assembly.
- A stabilizer bar controls body roll.
- Rear suspension systems must control rear-end torque.
- Passive suspensions use nonadjustable springs and shock absorbers.
- Semi-active suspensions use adjustable air chambers and shock absorbers to control vehicle height and damping.
- An automatic level control raises and lowers the rear suspension automatically.
- Fully active suspensions control vehicle height, shock absorber damping, roll, dive, and squat for better traction and vehicle control.

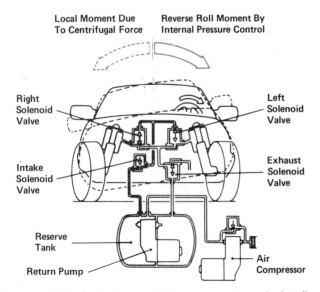

Figure 57-36. Active suspension systems counteract body roll.
MITSUBISHI MOTOR SALES OF AMERICA, INC.

TERMS

springs	sprung weight
unsprung weight	coil spring
control arm	leaf spring
torsion bar	air suspension
shock absorber	double-acting
compression	jounce
rebound	automatic level control
load leveling	height sensor
independent suspension	wheel spindle assembly
wheel spindle	steering knuckle
ball joint	stabilizer bar
body roll	MacPherson strut
rear-end torque	spring seat
trailing arm	spring shackle
Hotchkiss drive	control rods
double A-arm	contact patch
passive suspension	semi-active suspension
G-force	leaves
rotary valve	active suspension
roll	dive
squat	modified strut
transverse torsion bar	longitudinal torsion bar
semi-independent suspension	straight control arm and strut rod
damping	radius rod
torque tube	Gas pressure shock absorber

REVIEW QUESTIONS

DIRECTIONS: The following questions are similar to those used on automotive technician certification tests. On a separate sheet of paper, write the letter of the correct choice.

1. Which of the following statements is true?
   I. A shock absorber dampens spring action.
   II. A shock absorber absorbs shock.
   A. I only
   B. II only
   C. Both I and II
   D. Neither I nor II

2. Technician A says that a MacPherson strut assembly can be used as a front or rear suspension assembly.
   Technician B says that a MacPherson strut assembly uses only an upper control arm.
   Who is correct?
   A. A only
   B. B only
   C. Both A and B
   D. Neither A nor B

Questions 3, 4, and 5 are not like the previous questions. Each has the word EXCEPT. For each question, look for the choice that does *not* apply. Read each question carefully before you choose your answer.

3. All of the following are used as a spring in a suspension EXCEPT
   A. control arm.
   B. leaf.
   C. torsion bar.
   D. air spring.

4. All of the following statements are correct EXCEPT
   A. a non-independent suspension uses a rigid axle.
   B. an independent suspension allows each wheel to move independently.
   C. upper control arms are longer than lower control arms.
   D. ball joints allow vertical and rotary motion.

5. A fully active suspension system controls all of the following EXCEPT
   A. roll during cornering.
   B. vehicle weight during driving.
   C. dive during braking.
   D. squat during acceleration.

ESSAY QUESTIONS

1. What are the functions of a suspension system, and what forces can cause uneven traction and a reduction in driver control?
2. Identify and describe common front and rear suspension designs.
3. Describe passive, semi-active, and active suspension systems.

SUPPLEMENTAL ACTIVITIES

1. Identify and describe leaf springs, coil springs, torsion bars, and air suspension bags.
2. Describe how a shock absorber dampens spring motions.
3. Describe how a leaf-spring suspension controls rear-end torque.
4. Describe how rear-end torque is controlled on vehicles with coil-spring rear suspensions.

58 Suspension System Service

UNIT PREVIEW

Suspension system inspection and lubrication usually is performed at the same time as an oil change. Suspension and driveline grease fittings must be serviced regularly. Non-driving wheel bearings also should be disassembled, inspected, repacked with grease, and properly torqued. This service usually is performed at the same time as brake service, discussed in Unit 64.

Suspension system service and repair procedures include diagnosing suspension system problems, replacing shock absorbers, stabilizer bar bushings, ball joints, springs, control arm bushings, and MacPherson strut assemblies.

LEARNING OBJECTIVES

When you have completed your assignments and exercises in this unit, you should be able to:

❏ Lubricate all grease fittings on a vehicle.
❏ Repack wheel bearings.
❏ Inspect the parts of a suspension system for wear and damage.
❏ Replace a shock absorber.
❏ Replace a ball joint.
❏ Replace stabilizer bar and control arm bushings.
❏ Replace a MacPherson strut assembly.

SAFETY PRECAUTIONS

Always refer to the manufacturer's service manual and bulletins for correct suspension service procedures. An improperly repaired suspension system can lead to an accident and serious injury.

Use extreme caution when making road tests with an automobile that may have suspension problems. Before starting, buckle up the seat belt and check the brakes. Drive carefully and only in a manner that will help to diagnose any problems.

Raise and support vehicles safely. If the vehicle is not being serviced on a hoist, safely position and support the vehicle with jack stands.

Suspension springs exert great pressure. Follow all safety recommendations in the manufacturer's service manual when you use a spring compressor. Improper use can lead to severe personal injury or death.

Suspension parts can move while being serviced. Keep your fingers and hands clear of parts that may move to prevent serious injury.

A fastener must be replaced with one of the same part number or hardness grade. Use only the fasteners recommended in the vehicle manufacturer's service manual.

Always torque fasteners to the proper specifications. Improperly torqued fasteners can loosen, fall off, or fail.

Always replace worn or damaged suspension parts with new parts. Never heat or attempt to straighten any suspension part. These procedures weaken parts and make them subject to cracking and breaking.

TROUBLESHOOTING AND MAINTENANCE

Refer to the manufacturer's service manual for preventive maintenance schedules. Preventive maintenance for passive suspension systems includes periodic lubrication of the ball joints, tie rod ends, and other types of joints in the front suspension. This service usually is performed at the same time that steering parts are checked and lubricated.

Ball Joint Seal Inspection and Lubrication

To inspect ball joint seals, first clean any dirt or grease from the outside surface. If a ball joint seal is damaged, the ball joint may fail due to contaminants and moisture [Figure 58-1]. Seals are not available separately. New ball joint assemblies include new seals.

Ball joints may have a lubrication plug or grease fitting at the top or bottom of the ball joint housing [Figure 58-1]. Clean the plugs and fittings to remove dirt from the grease inlet. Remove the lubrication plugs, if necessary, and install grease fittings. Use a low-pressure lubrication gun and the proper lubricant. Too much lubricant will damage the seals. Stop filling when either of the following occurs:

• The seal begins to fill with grease and expand

SUSPENSION SYSTEM SERVICE

- Grease begins to flow from the bleed areas at the base of the seal.

Tie rod ends, other steering joints, and driveline fittings are lubricated in the same manner. Refer to Unit 48 and see Unit 60.

Repacking Non-Driving Wheel Bearings

Wheel bearings on non-driving wheels should be inspected and repacked with grease approximately every 30,000 miles, or whenever the brake system is inspected or serviced.

Wheel bearing removal. Raise the automobile and safely support it so that the wheels to be packed are off the ground. Remove the wheel cover and wheel. If the wheel has a disc brake assembly, remove the brake caliper. Hang the caliper out of the way with wire. Make sure that the brake hose is not stretched or damaged by the weight of the caliper. Remove the bearing dust cover and the cotter pin. Remove the locking device and the adjusting nut and washer **[Figures 58-2, 58-3, and 58-4]**.

Remove the thrust washer and outer bearing cone. Pull the drum or disc assembly off the spindle. Remove the inner grease seal and inner bearing according to the manufacturer's service manual instructions. Discard the old grease seal. A disc brake wheel bearing assembly is shown in **Figure 58-5**.

Cleaning and inspection. Clean the hub and drum assembly. Use kerosene, mineral spirits, or a similar low-flammability solvent.

CAUTION!!!

Dry the bearing by using compressed air from the side or ends of the rollers. Do not spin the wheel bearings dry with compressed air. This will damage the bearings.

Examine the bearing cups for pitting, scoring, or other damage. If the cups are damaged, remove them from the hub with a soft steel drift punch. The bearing cup areas in the hub should be smooth, with no scored or raised metal. Bearing cones and rollers should be smooth and free of pits, chipping, or other damage. Any evident damage on any part of the wheel bearing means that the entire bearing assembly must be replaced. Refer to Unit 56 for illustrations of typical bearing defects.

Each non-driving wheel has two wheel bearing assemblies. There is an inner, large bearing that carries the weight of the vehicle, and an outer, smaller bearing that helps to align the assembly **[refer to Figure 58-2]**.

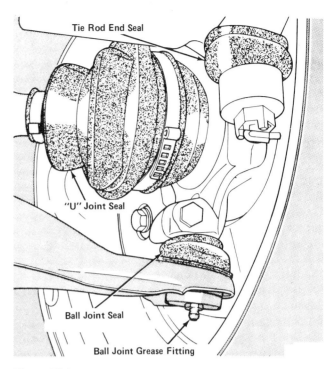

Figure 58-1. Ball joint grease fitting. CHRYSLER CORPORATION

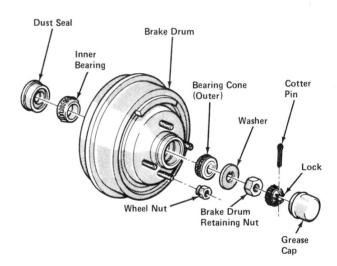

Figure 58-2. Parts of a drum brake wheel bearing assembly.
CHRYSLER CORPORATION

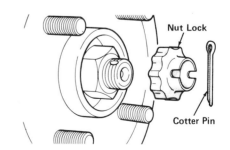

Figure 58-3. Removing cotter pin and nut lock.
CHRYSLER CORPORATION

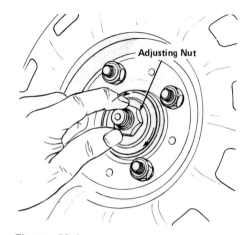

Figure 58-4. Removing adjusting nut and washer. CHRYSLER CORPORATION

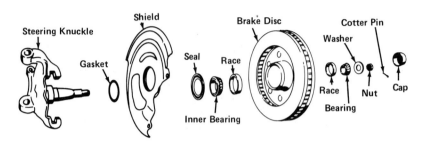

Figure 58-5. Parts of a disc brake wheel bearing assembly. BUICK MOTOR DIVISION—GMC

Each bearing assembly includes a tapered roller bearing and a race, or "cup," as shown in **Figure 58-6**. These are mated, or matched, and must not be interchanged from one side of the vehicle to the other.

When a wheel bearing is replaced, *both* the tapered roller bearing and its matching race must be replaced. Races are removed and driven into brake discs or drums with a drift punch and a ball peen hammer.

SAFETY CAUTION

Always wear eye protection when you use a hammer and punch.

If both wheel bearings are serviceable, they must be repacked with wheel bearing grease and reinstalled. It is a good practice to use high-temperature disc brake wheel bearing grease for both disc and drum brakes.

Wheel bearing installation. Force the recommended lubricant between the cleaned bearing cone rollers by hand, or use a pressure bearing packer. Coat the outer rollers of both bearings, the inner and outer races, and the spindle with a liberal layer of wheel bearing grease.

Install the inner bearing, with the small end facing into the recess. Drive the seal in squarely with a seal driver or suitable tool.

Place the drum or disc on the spindle. Install the smaller outer wheel bearing with the small end facing into the recess. Install the thrust washer and nut. The next step is to adjust the wheel bearings to the correct torque.

Wheel bearing adjustment. Tighten the wheel bearing nut tightly as you rotate the drum or disc.

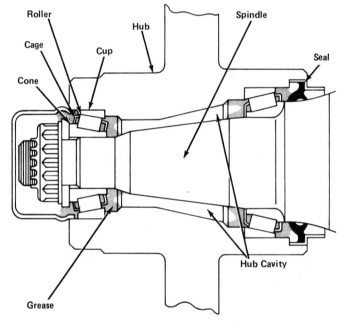

Figure 58-6. Assembled wheel bearing. CHRYSLER CORPORATION

Then back the nut off until it is loose. Use a lb.-in. torque wrench to tighten the adjusting nut to specifications **[Figure 58-7]**. Rotate the adjusting nut with a wrench until the cotter pin can be inserted. Install the nut lock and a new cotter pin **[Figure 58-8]**.

Clean and install the grease cap. On disc brake systems, install the caliper. Install the tire and wheel assembly.

58.1 DIAGNOSING SUSPENSION SYSTEM PROBLEMS

Refer to the manufacturer's service manual and bulletins for correct service procedures and safety precautions. Tire wear tells you when suspension and alignment service are needed (see Unit 62). The

SUSPENSION SYSTEM SERVICE

Figure 58-7. Non-driving wheel bearing adjustment.
SAVERIO BONO

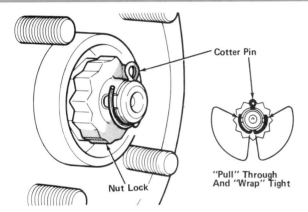

Figure 58-8. Nut lock and cotter pin installation.
CHRYSLER CORPORATION

driver also may complain of noises, bouncing, and/or erratic movements by the car while driving or cornering.

Mechanical Checks

A suspension inspection should be made when tire wear and/or driver complaints indicate a problem. It is a good practice to inspect suspension system parts whenever the vehicle is raised on a lift for any type of service.

Safely raise and support the vehicle on a hoist or jackstands. Make a visual inspection to locate broken, bent, or worn suspension parts.

Check tire wear patterns for specific problem indications (see Unit 62). Unit 59 discusses steering and wheel alignment service.

Wheel bearings. Rotate each tire by hand. The wheel should rotate quietly. If not, the wheel bearing should be checked for adjustment or removed and inspected for damage. However a slight drag from brake components is considered normal.

Grasp each wheel at the top and bottom. Attempt to move it in at the bottom and out at the top. Reverse this motion. Looseness may indicate a bad wheel bearing or improper bearing adjustment. Adjust the wheel bearing as described above and recheck for looseness. If there is still movement, check for worn ball joints.

Ball joints. Before checking ball joints for excessive movement, the wheel bearings must be in proper adjustment.

Refer to the manufacturer's service manual to determine where to support the suspension to check ball joint wear. Place safety stands at the indicated locations **[Figure 58-9]**.

SAFETY CAUTION

Be sure the vehicle is stable and does not rock on the safety stands. An improperly secured vehicle could fall and cause serious injury.

Grasp the tire and attempt to move it in and out. Then attempt to raise it. Watch at the ball joint for movement.

Some ball joints have visible plastic *wear indicators* that move inward or outward if the ball joint is worn excessively **[Figure 58-10]**. Refer to the manufacturer's service manual for checking procedures.

To measure ball joint movement accurately, use a dial indicator. Position the plunger at the position

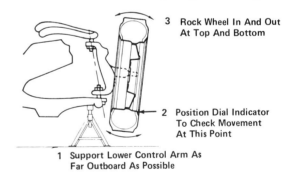

Figure 58-9. Checking upper ball joint.
BUICK MOTOR DIVISION—GMC

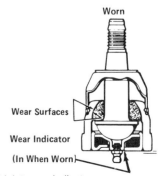

Figure 58-10. Ball joint wear indicator.
BUICK MOTOR DIVISION—GMC

shown in **Figure 58-9**. Grasp the wheel. Push in on the bottom of the tire and pull out on the top. Read the dial indicator. Reverse the motion and read the dial indicator again. The difference should not exceed the manufacturer's specifications for acceptable movement.

Also check the ball stud portion of the ball joint. If it has become disconnected from the steering knuckle, or if any looseness is detected, replace the ball joint as described in 58.2.

Shock absorbers. Check each shock absorber for loose or broken mounts or mounting brackets. Look for leaks. A black, gritty, grease-like deposit of dirt and leaking shock absorber fluid on the outside of the shock absorber indicates a defective unit. Make sure that the fluid leak is not grease from a differential, wheel bearing, driveline, or CV joint. Also check that it is not brake fluid or engine oil.

If no fluid is leaking from the shock absorber, another check can be made. Lower the vehicle so that it rests on its wheels. Push down on the corner of the vehicle nearest the shock absorber to be checked. Attempt to get the car bouncing up and down as much as possible. At the bottom of a downward stroke, step away and watch the bouncing motion. The vehicle should rebound upward, then downward, then come to a stop. If not, the shock absorber may be defective. Disconnect or remove the shock absorber to make a further check (see Topic 58.2).

Disconnect the lower end of the shock absorber. Grasp the bottom end and fully extend and compress the shock absorber several times. If no resistance is felt, a worn shock absorber is indicated. If only light resistance is felt, compare it with the action of a new shock absorber. Shock absorbers must always be replaced as pairs, never individually. Either both fronts, both rears, or all four shock absorbers must be replaced.

Control arms, strut rods, and bushings. Rubber and/or metal bushings are used at control arm pivot points and around strut rods [Figure 58-11]. Excessive wear on one side of a tire (camber wear), cupping, irregular wear, and squeaking noises can be caused by defective bushings.

Upper and lower control arms, strut rods, and their bushings can be inspected when the wheel is removed. Check rubber bushings for cracking, breaking, and separation. Check metal bushings for excessive wear and clearance around shafts. Check all metal parts for damage or wear. Special attention must be given to the rubber isolator bushings. If they are showing signs of deterioration or movement, they must be replaced. Also, check the strut rods for straightness.

If a control arm or strut rod moves excessively within the bushing or is noisy, replace the bushing.

SAFETY CAUTION

Removal of control arms and bushings requires compressing the spring, which exerts great pressure. Never attempt to compress a spring or remove control arms without proper training and supervision from your instructor.

Stabilizer bar. Defective bushings, broken or disconnected stabilizer bars or mounting hardware can cause excessive swaying and loss of control on turns. Inspect the stabilizer bar for damage. Check the rubber bushings and mounting hardware for proper connections and damage or wear [refer to Figure 58-11]. Replace any defective bushings or mounting hardware.

Springs. Sagging vehicle height and suspension bottoming over bumps are the most obvious signs that new springs are needed. However, weak springs can be detected in other ways. Rapid tire wear, early failure of suspension parts, and excessive bouncing and swaying are reasons to suspect weak springs. It is recommended that springs always be replaced as pairs on an axle, not separately.

To detect sagging springs, measure the *trim height* and inspect all suspension parts for damage [Figure 58-12]. Trim height is a manufacturer's recommended height at different parts of the chassis. Place the automobile on a level surface. Make sure the fuel tank is full, the trunk is empty, and the tires are properly inflated.

Figure 58-12 shows typical trim height measurement locations. The numbers in the following list correspond to the numbers in **Figure 58-12**. Types of information that can be obtained from trim height measurements include:

- The distance (1) between the lower control arm and the frame. The difference between control arm measurements should be no more than 1/4 inch (6.35 mm).
- The differences between distance 2 on both sides and between distance 3 on both sides should not exceed 3/4 inch (19.05 mm).
- The distances (4) between the ends of the bumper and the ground should not exceed 3/8 inch (9.525 mm). Before this measurement is made, be sure the bumper and its brackets and chassis mounting points are straight.

Other measurements are taken from the road surface to various body points, front and rear. Refer to the vehicle manufacturer's service manual when measuring trim height for specific vehicles.

Inspect the leaf and coil spring assemblies for any problems. Check leaf spring systems for wear or damage around the hangers, shackles, bolts, and bushings. Also check for broken leaves.

SUSPENSION SYSTEM SERVICE

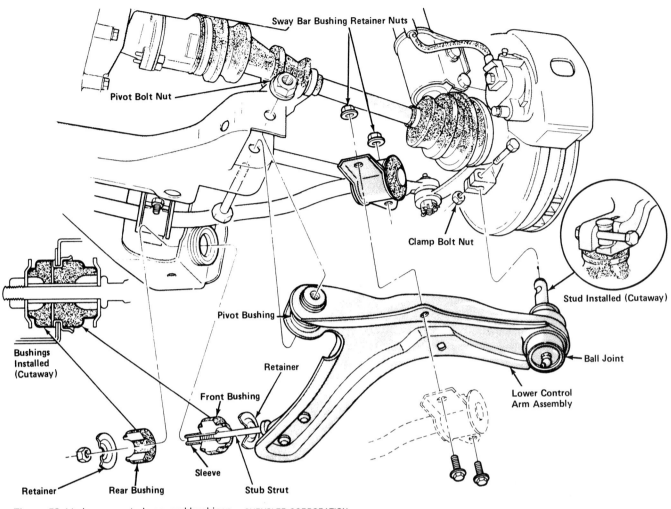

Figure 58-11. Lower control arm and bushings. CHRYSLER CORPORATION

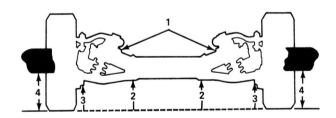

Figure 58-12. Measuring trim height.

Check for worn spring shackle bushings, which can cause a car to wander left or right. Insert a pry bar between the spring eye and shackle. Pull down on the pry bar. If movement occurs between the spring eye and hanger bolts, the bushings should be replaced. Little or no movement means the bushing is satisfactory.

SAFETY CAUTION

A special compressor tool or spring jack is needed to compress coil or leaf springs safely for spring and bushing replacement. Technicians have been severely injured and killed by expanding springs. Do not attempt these procedures without proper training and supervision from your instructor.

Check coil spring assemblies for wear, damage, or missing parts. If the vehicle's trim height is low and other parts are not defective or modified, both front, rear, or all four springs must be replaced to restore proper height.

MacPherson struts. Diagnosing MacPherson strut suspension problems is similar to checking coil spring suspensions. In some cases, the shock absorber portion of the strut and the springs can be replaced separately. In other cases, the entire strut may have to be replaced as a unit. As with shock absorbers and springs, always replace MacPherson strut parts or units as pairs on an axle.

58.2 SUSPENSION SYSTEM SERVICE

Coil spring and MacPherson strut suspensions are found most often on modern automobiles. This section discusses common service procedures for these types of suspension systems.

Coil Spring Suspension

Figure 58-13 shows a typical General Motors front coil spring suspension. Service procedures for this suspension system include replacing:

- Shock absorbers
- Stabilizer bars, bushings, and mounting hardware
- Ball joints
- Coil springs
- Control arms, strut rods, and associated bushings

Shock Absorber Replacement

Always follow the instructions enclosed with replacement shock absorbers because service procedures can vary. The following is a general procedure for replacing shock absorbers.

Locate the upper control arm. If the threads are rusted, apply penetrating oil to them and let it soak in for a few minutes. Then place a wrench or special socket on the upper stem of the shock absorber to keep it from turning **[Figure 58-14]**. Use an open-end wrench to remove the upper "pal nut," or sheet metal lock nut, metal retainer, and the retaining nut. Remove the retainer, and rubber shock grommets. One type of mounting hardware and shock absorber is shown in **Figure 58-15**. When you remove the shock absorber, be careful not to damage mounting bolts or studs that must be reused.

Raise the automobile on a hoist. Remove the two bolts holding the bottom of the shock absorber to the lower control arm. Pull out the shock absorber assembly from the bottom.

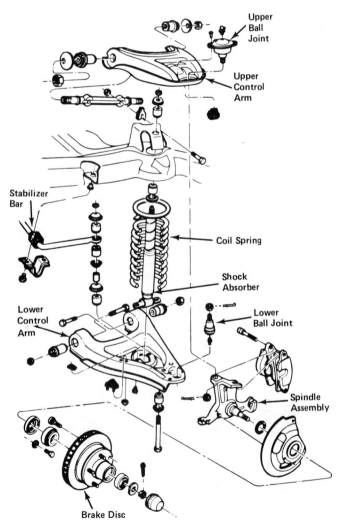

Figure 58-13. Front coil spring suspension.
OLDSMOBILE DIVISION—GMC

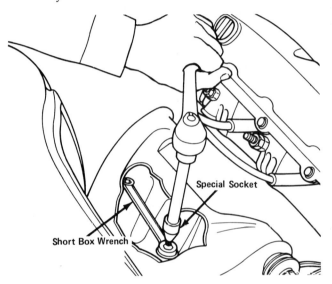

Figure 58-14. Removing nut at the top of the shock absorber.
CADILLAC MOTOR CAR DIVISION—GMC

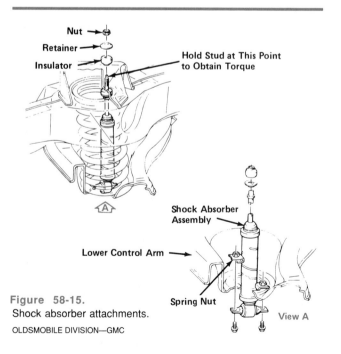

Figure 58-15. Shock absorber attachments.
OLDSMOBILE DIVISION—GMC

SUSPENSION SYSTEM SERVICE

Before installing the new shock absorber, carefully read all information on the instruction sheet. Hold the shock absorber upright and collapse and expand it through its full stroke several times to bleed out air bubbles.

To install, place the retainer and grommet over the upper stem of the new shock absorber. Install the shock absorber, fully extended, up through the lower control arm and spring. Continue to insert the upper stem until it passes through the mounting hole in the upper control arm frame bracket. Install the upper rubber grommet, retainer, and nut. Hold the upper stem from turning with an open-end wrench. Tighten the nut to specifications. In general, the nut must be tightened until the rubber grommet expands to the same diameter as the retainer. Install the pal nut, if supplied. Install the bolts attaching the shock absorber to the lower control arm and torque them to specifications.

Stabilizer Bar Replacement

Stabilizer bar service usually requires replacing the bushings and/or mounting hardware. Use **Figure 58-16** as a guide during replacement. Raise the automobile. If using a jack, position safety stands in the proper spots under the automobile before disassembly.

Disconnect the stabilizer linkage at each end of the bar. Remove the nuts, retainers, and grommets. Pull out the bolts and remove additional retainers, grommets, and spacers.

Remove the bolts that hold the mounting brackets to the frame. Remove the brackets and bushings. Remove the stabilizer bar.

To replace the stabilizer bar, position the bar under the front frame. Slide the rubber bushings into position on the stabilizer bar with the rounded part matching the mounting bracket shape. Install the mounting brackets over the bushings and torque the bolts to specifications.

Install the grommets, retainers, spacers, and bolts on the ends of the stabilizer bar. Be sure all parts are arranged correctly. Install the upper grommets, retainers, and nuts. Tighten the nuts to the manufacturer's torque specifications.

Ball Joint Replacement

Refer to the manufacturer's service manual for correct ball joint service procedures and safety precautions. It is not necessary to replace lower and upper ball joints at the same time. However, it is recommended that either both upper, both lower, or all four ball joints be replaced. The following general procedures are for lower and upper ball joint replacement on a double A-arm front suspension with coil springs.

SAFETY CAUTION

A floor jack must remain under the lower control-arm spring seat during removal and installation. The jack holds the spring and control arm. The spring is under great tension between the two control arms. In most cases, a special spring compressor tool is used to hold the spring. If the spring should come loose, it can cause serious injury. Do not attempt these procedures without proper training and supervision from your instructor.

The following is a generalized list of procedures to replace a lower ball joint. To replace the joint:

1. Raise and support the vehicle safely and securely.
2. Remove the wheel and tire.
3. If necessary for access, remove the brake caliper from the steering knuckle and hang it with a wire to prevent damage to the brake hose.
4. Remove the shock absorber.
5. Remove the grease fitting from the old ball joint.
6. Safely and carefully use a spring compressor to compress the spring **[Figure 58-17]**. Use only hand tools to operate a spring compressor.
7. Position a jackstand squarely under the lower control-arm spring seat.
8. Remove the cotter pin and loosen the lower ball joint nut a few turns.
9. Use a special ball joint tool to break the ball joint loose from the knuckle **[Figure 58-18]**.
10. Remove the stud nut completely.
11. Drill out rivets **[Figure 58-19]** or remove bolts or use a special tool to press out the old ball joint from the control arm **[Figure 58-20]**.
12. Install the ball joint to the control arm **[Figure 58-21]**. Torque attaching bolts and nuts to the manufacturer's specified torque.
13. Install the new ball joint stud into the steering knuckle and install the retaining nut. Torque the nut to the manufacturer's specifications and install a new cotter pin.

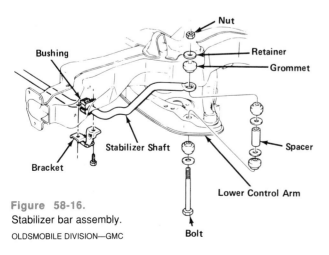

Figure 58-16.
Stabilizer bar assembly.
OLDSMOBILE DIVISION—GMC

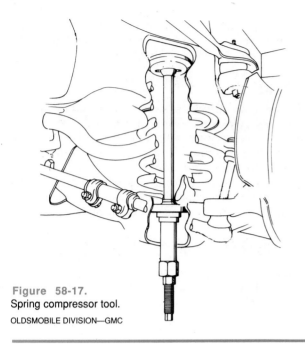

Figure 58-17.
Spring compressor tool.
OLDSMOBILE DIVISION—GMC

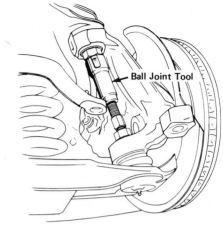

Figure 58-18.
Breaking the ball joint loose.
OLDSMOBILE DIVISION—GMC

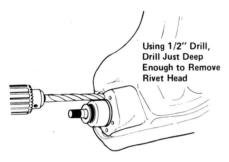

Figure 58-19.
Drilling out rivet heads.
OLDSMOBILE DIVISION—GMC

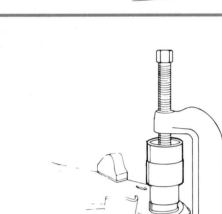

Figure 58-20.
Removing a pressed-in lower ball joint.
OLDSMOBILE DIVISION—GMC

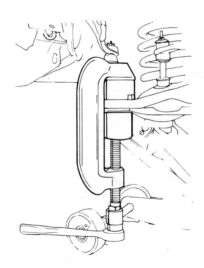

Figure 58-21.
Installing a pressed-in lower ball joint.
OLDSMOBILE DIVISION—GMC

14. Install the grease fitting and lubricate the ball joint.
15. Slowly and carefully release the spring compressor and guide the spring into its seat.
16. Install the tire and wheel assembly.
17. Check/adjust wheel alignment (see Units 59 and 60).

Spring removal. To remove a spring, remove the two lower control arm pivot bolts. Disengage the lower control arm from the frame. Rotate the lower control arm rearward, with the spring. Remove the spring from the arm [**Figure 58-22**]. Then complete spring removal by following these steps:

- Slowly and carefully release tension on the old spring by turning the compressor screw with a hand tool.
- Insert the compressor into a new spring and compress it.

SUSPENSION SYSTEM SERVICE

- Install the compressed new spring in place and lift the lower control arm up into position.
- Install the ball joint stud into the steering knuckle and install the retaining nut. Torque the nut to the manufacturer's specifications and install a new cotter pin.
- Slowly and carefully release the spring compressor and guide the spring into its seat.
- Install the tire and wheel assembly.
- Check/adjust wheel alignment (see Units 59 and 60).

Control Arm/Bushing Replacement

Refer to the manufacturer's service manual for specific safety precautions and service information. Bushing replacement is the most common service procedure performed on a control arm. Steps for replacing upper and lower control arms are similar. The upper or lower ball joint is disconnected to replace the appropriate control arm. Alignment shims are used to position the upper control arm only [**Figure 58-23**]. An upper control arm and related parts are shown in **Figure 58-24**. The following steps are a general procedure for replacing control arms and bushings:

- Remove the upper control arm attaching bolts and the control arm assembly.
- Press out old bushings. Inspect the control arm bushing holes for wear and damage. Replace the control arm if worn or damaged.
- Press in new bushings.
- Position the upper control arm bolts loosely in the frame.
- Install the pivot shaft and alignment shims for an upper control arm. Install the shaft or bolt for a lower arm. Torque the nuts to specifications.

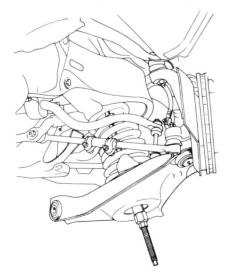

Figure 58-22. Rotate arm, with spring rearward, and remove spring from arm. OLDSMOBILE DIVISION—GMC

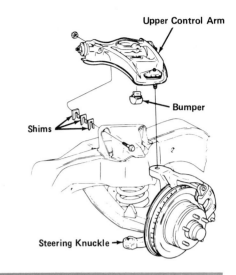

Figure 58-23. Upper control arm assembly. OLDSMOBILE DIVISION—GMC

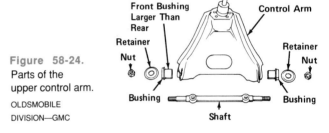

Figure 58-24. Parts of the upper control arm. OLDSMOBILE DIVISION—GMC

- Connect the ball joint to the steering knuckle and torque the nut to the manufacturer's specifications. Install a new cotter pin.
- Install the tire and wheel assembly.
- Check/adjust wheel alignment (see Units 59 and 60).

MacPherson Strut Replacement

The only service possible for many MacPherson strut suspensions is replacement of the shock absorber cartridge or spring. Always refer to the manufacturer's service manual when servicing any MacPherson strut assembly.

The following general list of procedures applies to suspensions similar to that shown in **Figure 58-25**:

1. Raise and support the vehicle safely and securely.
2. Remove the wheel and tire.
3. Mark the adjustment cams for replacement [**Figure 58-26**], then remove the two bolts that attach the strut to the steering knuckle.
4. Remove the brake line from the strut.
5. Remove the nuts holding the mounting assembly at the top of the strut.
6. Remove the strut/spring assembly.

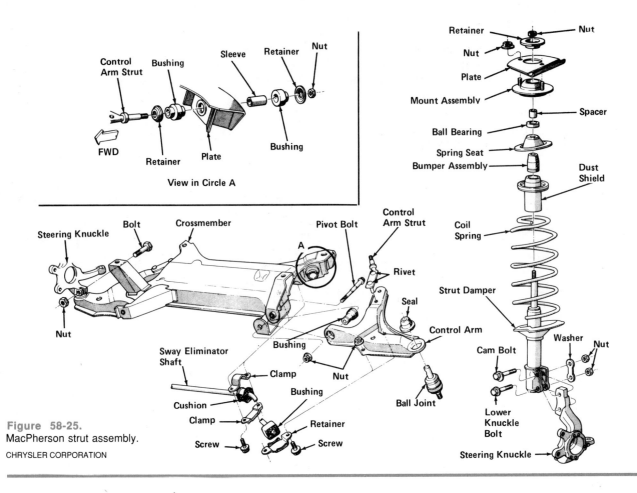

Figure 58-25. MacPherson strut assembly. CHRYSLER CORPORATION

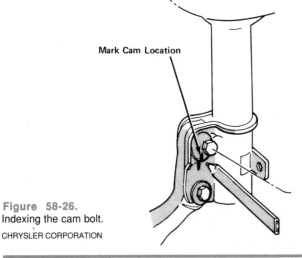

Figure 58-26. Indexing the cam bolt. CHRYSLER CORPORATION

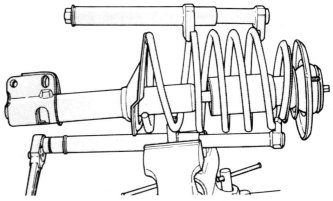

Figure 58-27. Spring compressor tool. CHRYSLER CORPORATION

Strut disassembly. Strut disassembly is completed on the bench. Place the strut in a vise and slowly and carefully compress the spring, using hand tools only **[Figure 58-27]**. Mark the coil spring for reassembly in its original location.

SAFETY CAUTION

Follow the manufacturer's instructions when you use a spring compressor. A compressed spring stores great amounts of energy and can cause serious injury if the spring is released suddenly.

Use the following list of procedures to disassemble the strut:

- Remove the strut rod nut while holding the strut rod **[Figure 58-28]**.
- Remove the strut assembly.
- Slowly and carefully release the spring compressor tool.

SUSPENSION SYSTEM SERVICE

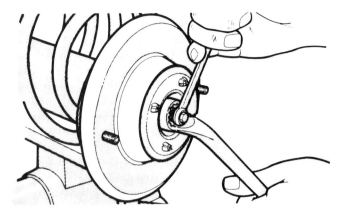

Figure 58-28. Removing strut-rod nut. CHRYSLER CORPORATION

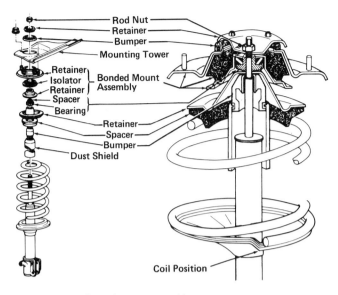

Figure 58-29. Strut damper assembly. CHRYSLER CORPORATION

- Inspect the mount assembly **[Figure 58-29]**. Check for cracks, distortions, and other wear or damage. Replace any defective parts of the strut and mounting assemblies.

Strut reassembly. Refer to the vehicle manufacturer's service literature for specific procedures and safety precautions. The following list is a general list of reassembly procedures:

- Using hand tools, slowly and carefully compress the spring with the compressor.
- Place the compressed spring over the new strut and onto the lower spring seat.
- Assemble the bumper, dust shield, upper spring retainer, bearing, and spacer on the strut rod.
- Mount the assembly over the strut, along with the rebound bumper, retainer, and rod nut.
- Align the spring, as shown in **Figure 58-30.**
- Use a special tool **[Figure 58-31]** to tighten the rod nut to specifications.
- Using hand tools, slowly and carefully release the spring compressor tool.

Strut installation. Always refer to the manufacturer's service manual for correct procedures. The following list is a general procedure for MacPherson strut installation:

- Place the top of the strut assembly into the fender reinforcement. Install and tighten the mounting nuts to specifications.
- Place the lower end of the strut into the steering knuckle. Install the lower bolt that holds the parts together, but do not tighten.
- Attach the brake hose retainer to the strut and tighten to specifications.
- At the steering knuckle, rotate the top cam bolt to the index mark you made during removal.

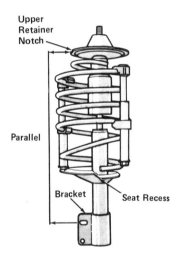

Figure 58-30. Align the spring. CHRYSLER CORPORATION

Figure 58-31. Tightening strut-rod nut. CHRYSLER CORPORATION

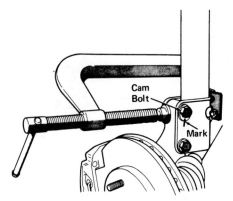

Figure 58-32. Attach C-clamp and connect strut.
CHRYSLER CORPORATION

- Place a C-clamp on the strut and knuckle **[Figure 58-32]**. Tighten the clamp just enough to eliminate any looseness between the knuckle and the strut.
- Realign the cam bolt and index marks, if necessary. Torque the bolts to specifications.
- Remove the C-clamp.
- Install the tire and wheel assembly.
- Check/adjust wheel alignment (see Units 59 and 60).

Semi-active and Active Suspension Service

Semi-active and active suspension systems use air suspension units and adjustable shock absorbers. Service procedures include checking electrical, mechanical, and air pressure system problems.

Electrical malfunctions. Semi-active and active suspension systems include many electrical parts to control system functions. Problems include malfunctions in:

- Sensors
- Relays
- Switches
- Solenoid air valves
- Strut shock absorber stepper motors
- Electrical air compressor
- Electronic control unit (computer).

Units 37 and 39 of this book provide a general guide to electrical and electronic component service. *Self-diagnosis* codes produced by a system computer help to pinpoint the cause of problems. Unit 39 provides a general guide to electronic device service and reading self-diagnosis codes.

Some vehicle manufacturers specify the use of special hand-held diagnostic computers that connect to electrical sockets on the vehicle. These testers communicate with the suspension system computer to identify system malfunctions. Refer to the manufacturer's service manual for specific testers and instructions on their use.

Mechanical malfunctions. Just as on passive suspension systems, mechanical malfunctions may occur on semi-active and active suspension systems. These malfunctions include problems with:

- Front and rear struts or shock absorbers
- Controls arms
- Ball joints
- Body and chassis mountings.

Refer to the material in Topic 58.2 for general information on mechanical suspension system service.

Air pressure system malfunctions. Air pressure operates air suspension units in both semi-active and active suspension systems to raise, lower, and level the vehicle. Air system problems include malfunctions and defects in:

- Joints and seals
- Air tubing and lines
- Solenoid air valves
- Air compressor pumps.

Refer to the manufacturer's service manual for specific information on troubleshooting and repairing semi-active and active suspension systems.

UNIT HIGHLIGHTS

- Ball joints should be inspected for wear and damage regularly and lubricated at least once a year.
- Wheel bearings should be repacked whenever the brakes are serviced, or at least every 30,000 miles or three years.
- Check wheel bearings by shaking a wheel for looseness.
- A dial indicator is used to measure ball joint movement.
- Some lower ball joints have a wear indicator.
- Broken mounts and leaks are common shock absorber problems.
- Weak springs can be detected by measuring trim height.
- Follow the manufacturer's instructions when installing new shock absorbers.

- Ball joints must be broken free from the steering knuckle and control arm before they can be removed.
- When a coil spring is replaced, the opposite spring also must be replaced.
- Old control arm bushings are pressed out and new bushings are pressed in.
- MacPherson strut service usually involves replacing the shock absorber strut assembly.
- Service for semi-active and active suspensions includes electrical, mechanical, and air pressure system diagnosis and repair.

TERMS

wear indicator
self-diagnosis
trim height

REVIEW QUESTIONS

DIRECTIONS: The following questions are similar to those used on automotive technician certification tests. On a separate sheet of paper, write the letter of the correct choice.

1. Which of the following statements is correct?
   I. Wheel bearings on non-driving wheels are permanently lubricated.
   II. Wheel bearings on driving wheels must be repacked with grease.
   A. I only
   B. II only
   C. Both I and II
   D. Neither I nor II

2. Technician A says that, when replacing shock absorbers, original mounting bolts and other parts are never reused.
   Technician B says that shock absorber bolts should be torqued to specifications.
   Who is correct?
   A. A only
   B. B only
   C. Both A and B
   D. Neither A nor B

3. Which of the following statements is correct?
   I. Semi-active and active suspension problems include electrical, mechanical, and air pressure system malfunctions.
   II. Self-diagnosis codes help to diagnose electrical problems on semi-active and active suspensions.
   A. I only
   B. II only
   C. Both I and II
   D. Neither I nor II

Questions 4 and 5 are not like the previous questions. Each has the word EXCEPT. For each question, look for the choice that does *not* apply. Read each question carefully before you choose your answer.

4. All of the following procedures for replacing a ball joint are correct EXCEPT
   A. supporting the vehicle with safety stands.
   B. disconnecting the ball joint from the steering knuckle.
   C. suspension springs are not strong enough to cause injury.
   D. torquing the bolts on a replacement upper ball joint to the manufacturer's specifications.

5. All of the following statements are correct EXCEPT
   A. ball joints with grease fitting or plugs should be lubricated.
   B. when a ball joint rubber seal begins to expand, stop adding lubricant.
   C. some ball joints include wear indicators.
   D. ball joints must be replaced every 30,000 miles.

ESSAY QUESTIONS

1. What checks should be made to diagnose suspension system problems?
2. What safety precautions are necessary when using a spring compressor?
3. Describe common sources of problems in electronically controlled suspension systems.

SUPPLEMENTAL ACTIVITIES

1. Lubricate ball joints.
2. Clean and repack non-driving wheel bearings with grease.
3. Make a visual inspection of the suspension system and record any problems that are found.
4. Make a trim height inspection.
5. Replace shock absorbers.
6. Replace the bushings on a stabilizer bar.
7. Remove and replace a ball joint on a vehicle selected by your instructor.
8. Replace upper control-arm bushings on a double A-arm coil spring suspension.
9. Replace a MacPherson strut assembly.
10. Refer to the manufacturer's service manual and call up trouble codes for a semi-active or active suspension system.

59 Steering and Wheel Alignment

UNIT PREVIEW

A steering system provides directional control of a vehicle. Gears and a linkage system transfer force from the steering wheel to road wheels which pivot around a vertical axis.

Steering systems can be either manually operated or power assisted. Wheel alignment must be to the manufacturer's specifications to provide safe steering and handling characteristics and to reduce tire wear.

Traditionally, only the front two wheels of a vehicle were steered. However, recent vehicles include four-wheel steering (4WS) systems that steer all four wheels.

LEARNING OBJECTIVES

When you have completed your assignments and exercises in this unit, you should be able to:

- ❑ Locate and identify the parts of a steering linkage.
- ❑ Describe the operation of a recirculating ball steering gearbox.
- ❑ Describe the operation of a rack and pinion steering gear.
- ❑ Describe how an integral power steering system works.
- ❑ Describe the functions and operation of a four-wheel steering (4WS) system.
- ❑ Identify and describe steering column designs.
- ❑ Describe wheel alignment service.

59.1 STEERING SYSTEM OPERATION

The steering wheel is connected to a *steering shaft* that rotates with the steering wheel inside a stationary *steering column.*

The steering shaft extends from the steering wheel into the engine compartment to a *steering gearbox.* The steering gearbox gears transfer force to *steering linkage* rods that move from side to side, perpendicular to the length of the vehicle.

Steering Linkage

The steering linkage is a system of metal rods and flexible joints. As the steering wheel is turned, the linkage exerts a pulling and pushing motion to pivot both front wheels for steering. Two types of steering systems are commonly used, *parallelogram steering linkage* and *rack and pinion* steering linkage.

Parallelogram steering linkage is used on larger, heavier vehicles to reduce steering effort. The major parts of a parallelogram steering system as shown in **Figure 59-1** include:

- Pitman arm
- Relay rod
- Idler arm
- Tie rod
- Steering arm.

Refer to **Figures 59-1** and **59-2** as you read the following descriptions of steering linkage parts.

Pitman arm. A *Pitman arm* extends from the steering gearbox and transmits gear movement to the relay rod.

Relay rod. A *relay rod* transmits, or relays, steering movement toward both front wheels. A relay rod also is called a *center link.*

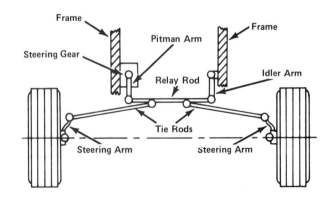

Figure 59-1. Parallelogram steering linkage.
AMERICAN MOTORS CORPORATION

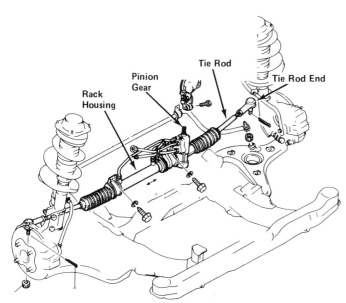

Figure 59-2. Rack and pinion steering assembly.
MAZDA MOTOR CORPORATION

Idler arm. An *idler arm* is connected to the relay rod and to the frame of the automobile. An idler arm supports the end of the relay rod that is opposite the Pitman arm. The length of the idler arm is the same as the Pitman arm. It forms the right side of the parallelogram.

Tie rod. A *tie rod* is connected between the relay rod and the steering arm. There are two tie rods in a steering linkage system, one on each side. Either the outer end, or both ends, of a tie rod have a ball-and-socket joint. Tie rods usually are connected with *adjusting sleeves.* These sleeves are adjusted during wheel alignment (see Topic 59.5).

Steering arm. Tie rods connect to the steering knuckle through a *steering arm*, usually formed as a part of the steering knuckle (see Unit 57). A steering arm transmits force to pivot the steering knuckle and turn a front wheel.

Notice that, in a parallelogram steering linkage system, the Pitman and idler arms are always parallel. These two arms also keep the relay rod in a position parallel to an imaginary line drawn through the pivot points of the idler arm and the Pitman arm. Thus, these parts make up three sides of a parallelogram.

In a rack and pinion steering system, tie rods and steering arms are the only steering linkage parts needed to connect the end of the rack to the steering knuckle arms.

Rack and Pinion Steering

Rack and pinion steering is used on smaller, lighter vehicles and sports cars to provide quick steering responsiveness. The major parts of a rack and pinion system [Figure 59-2] include:

- Pinion gear
- Rack (internal)
- Tie rod
- Steering arm.

Pinion gear. The pinion gear is connected to the steering shaft/wheel. Whenever the steering wheel is turned, the pinion gear also turns.

Rack. The pinion gear is meshed with the *rack*. The rack is a steel bar mounted within the rack housing. The rack has teeth machined into it which mesh with the pinion gear. Thus, the rotary motion of the steering wheel is changed to side-to-side motion for steering control.

Tie rod. Tie rods are connected at each end of the rack. These rods transfer the rack movement to the steering arms. The rods include a feature for length adjustment during wheel alignment.

Steering arms. The steering arms in a rack and pinion system, serve the same function as in a parallelogram steering system.

Steering Ratio

Moving the steering wheel from extreme left to extreme right, is called stop-to-stop, or *lock-to-lock*, movement. During this steering wheel movement, the front wheels usually turn a total of about 60 degrees, from all the way left to all the way to the right.

If the *steering ratio* were 1:1, this slight movement would steer the automobile all the way left or right. A steering ratio expresses the amount of angular turning of the steering wheel to the angular turning of the front wheels as they steer.

For example, think of 12 o'clock as the position in which the automobile's front wheels are pointing straight ahead. With a 1:1 steering ratio, moving the steering wheel from 11 o'clock (30 degrees left) to 1 o'clock (30 degrees right) would steer the wheels all the way from the left to the right.

A 1:1 steering ratio is too extreme. During driving, the slightest movement of the steering wheel would cause the automobile to swerve. For more accurate steering control, a slower gear ratio is needed.

A steering ratio of 15:1 is acceptable. This means the driver must turn the steering wheel through 15 degrees of angular motion for both front wheels to turn 1 degree.

With a 15:1 steering ratio, 2-1/2 turns (lock-to-lock) of the steering wheel are required to turn the front wheels from all the way left to all the way right (60 degrees). To work out the ratio:

$$2\text{-}1/2t \times 360 \text{ degrees} = 900 \text{ degrees swm}$$
$$900 \text{ degrees swm} \div 60 \text{ degrees wm} = 15$$

Where:
t = turns
swm = steering wheel movement
wm = wheel motion

A movement of 2-1/2 to 3-1/2 turns of the steering wheel to steer the front wheels all the way from lock to lock is considered normal. Thus, steering ratios in the range of about 15:1 to 21:1 usually are used.

59.2 MANUAL STEERING

An automobile may have either a manual or a power steering system. A *manual steering* system requires the driver to provide all of the force needed to move the steering linkage and pivot the front wheels for steering. A *power steering* system adds force to move the steering linkage that is created by a hydraulic pump driven by the engine.

Both manual and power steering systems are named for the types of steering gears that are used. Most steering gears have recirculating ball or rack and pinion systems.

Recirculating Ball

A *recirculating ball* gearbox has two gears: a driving gear and a driven gear. The driving gear, called a *worm gear*, is connected to the steering shaft. A worm gear has spiral threads.

The driven gear is a *sector gear* that is connected to the Pitman arm. A sector gear has teeth in a semicircle, or on a section of a circle. The name "recirculating ball" comes from the system of moving ball bearings that reduce friction inside the gearbox **[Figure 59-3]**.

The gearbox is attached to the frame and is filled with lubricant, either single-weight gear oil or grease. Inside the gearbox, the worm gear is threaded into a *ball nut.* A ball nut is a nut with an internal spiral groove. Ball bearings fit into the internal thread, or groove, between the worm gear and the ball nut. When the driver turns the steering wheel, the steering shaft rotates. As it turns, ball bearings move through tubes, or *ball return guides,* that are connected to each end of the groove. These tubes provide a continuous loop that collects and recycles the ball bearings.

When the steering wheel is turned, the steering shaft rotates the worm gear. The ball bearings transmit turning force from the worm gear to the ball nut and help to reduce friction between these parts. Teeth on the edge of the ball nut mesh with and move the sector gear and shaft, which in turn swivel the Pitman arm. Parallelogram linkage is used with recirculating ball steering gears.

Rack and Pinion

A rack and pinion steering system is much simpler in design and operation **[Figure 59-4]**. A rack housing, usually made of aluminum, encloses a steering rack and a pinion gear.

The steering shaft from the steering wheel is connected to the pinion gear. The pinion gear meshes with teeth cut into a metal bar, or rack. As the pinion gear turns, it forces the rack to move to the left or right, depending upon the rotation of the pinion gear. Each end of the rack is connected to a tie rod assembly. The ends of the tie rods are connected to steering knuckle arms, which transmit force to pivot the front wheels.

59.3 POWER STEERING

Power steering systems are classified as hydromechanical servo systems. Hydraulic pressure is used to operate a power steering system, either Pitman or rack-and-pinion **[Figure 59-5]**. Parts necessary for power steering, or "power assisted steering," include:

- Pump
- Reservoir
- Hydraulic lines
- A special gearbox or assist assembly on the linkage.

Two types of power steering systems are used: *integral power steering* and *linkage power steering.* Current passenger vehicles use integral power steering systems.

An integral power steering system applies hydraulic pressure to a piston within the steering gearbox. The differences in linkage power steering are discussed at the end of this topic.

Power Steering Pump

Hydraulic fluid flow for power steering is provided by a *power steering pump,* driven off the engine crankshaft pulley by a belt **[Figure 59-6]**. Hydraulic fluid for the power steering pump is stored in a reservoir. The reservoir is either part of the pump body itself or is a separate tank, connected by hoses to the pump. Fluid pressure to operate the steering system is routed to and from the pump by hoses and lines. Excess pressure is controlled by a relief valve.

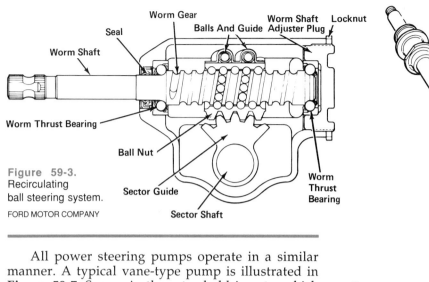

Figure 59-3. Recirculating ball steering system. FORD MOTOR COMPANY

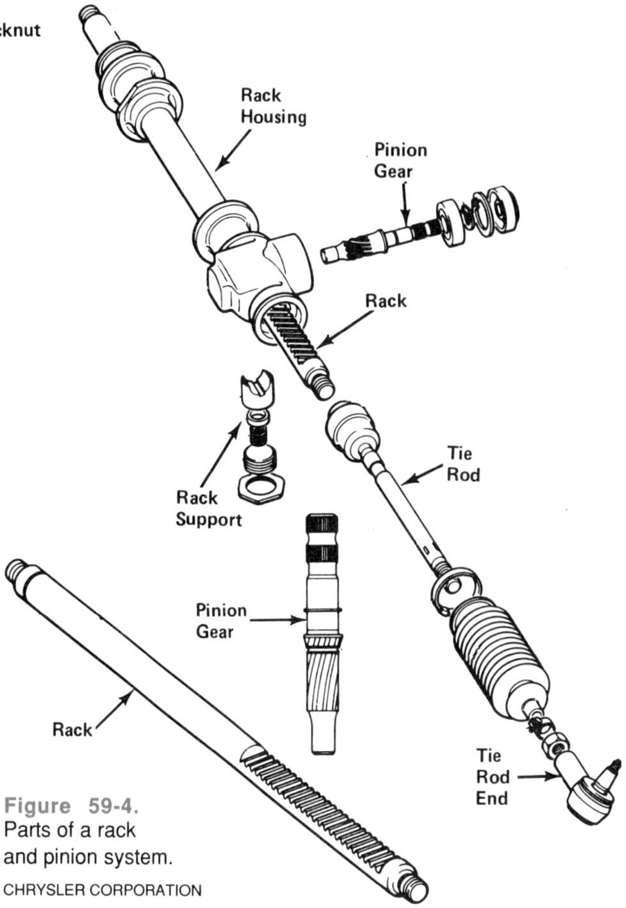

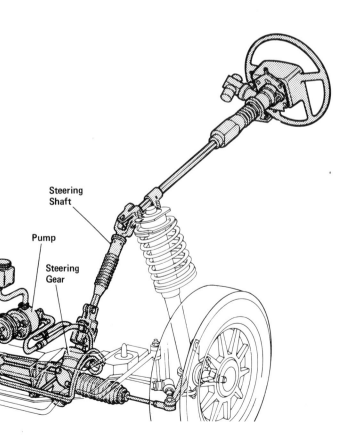

Figure 59-4. Parts of a rack and pinion system. CHRYSLER CORPORATION

All power steering pumps operate in a similar manner. A typical vane-type pump is illustrated in **Figure 59-7**. Spaces in the rotor hold inserts, which may be vanes, slippers, or rollers **[Figure 59-8]**. As the rotor turns, fluid is drawn between the inserts and forced out through an outlet. Centrifugal force holds the inserts tightly against the inner surface as the pump turns.

Power Steering Gearbox

A power steering gearbox is similar to a manual steering gearbox. Hydraulic pressure applies force to the internal gear mechanism and lessens the effort required to turn the steering wheel.

A recirculating ball power steering gearbox is illustrated in **Figure 59-9**. The ball nut separates the gearbox into two chambers. Ring-type seals on the special ball nut allow it to act as a piston. The control valve directs hydraulic fluid to push the ball nut piston. Fluid is applied to one chamber or another, depending on which way the steering wheel is turned. Fluid returns to the power steering pump reservoir through a low-pressure hose from the other chamber.

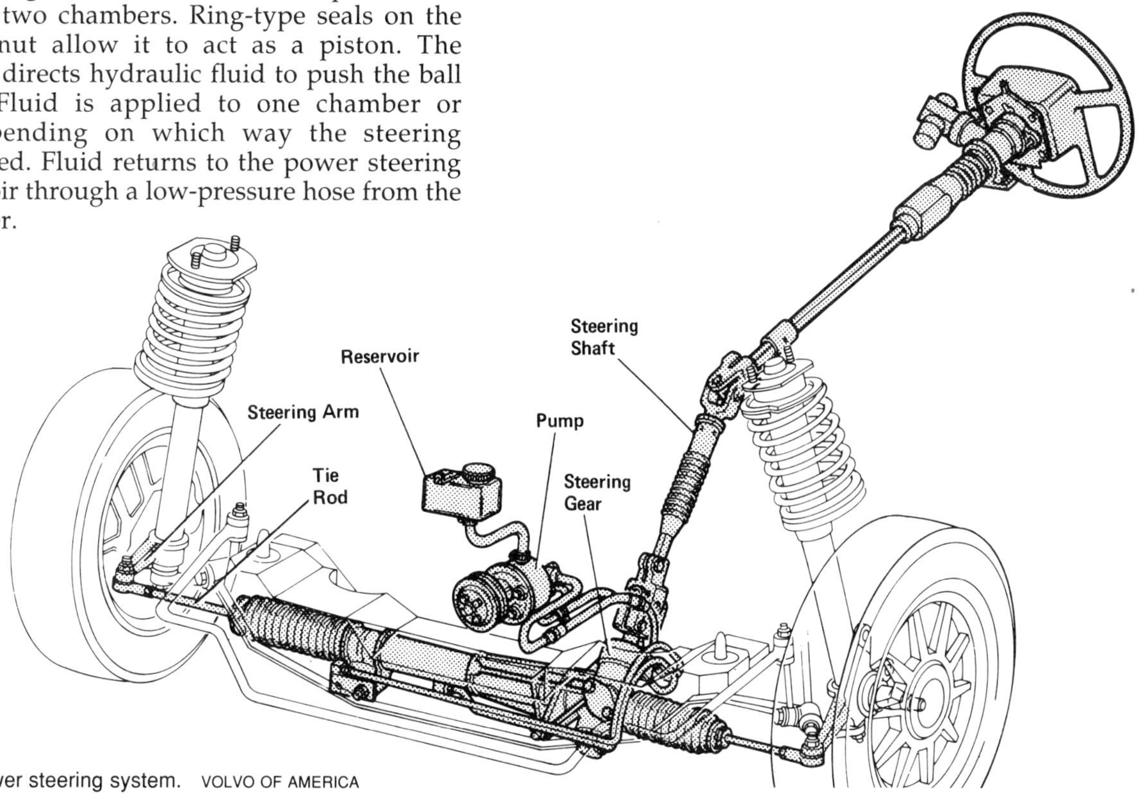

Figure 59-5. Power steering system. VOLVO OF AMERICA

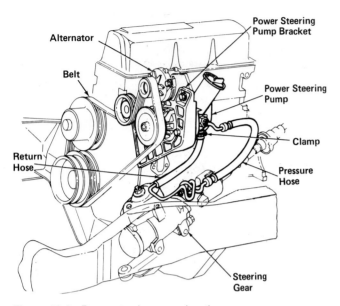

Figure 59-6. Power steering pump location.
FORD MOTOR COMPANY

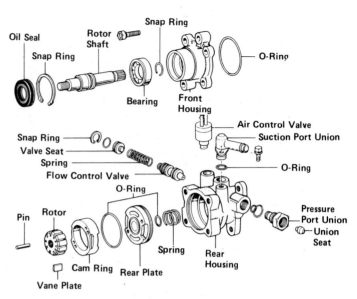

Figure 59-7. Parts of a vane-type pump.

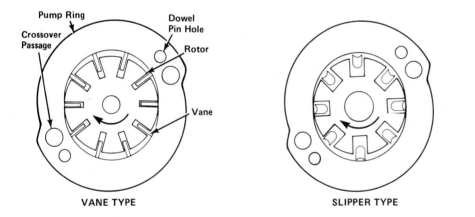

Figure 59-8. Steering pump rotors.

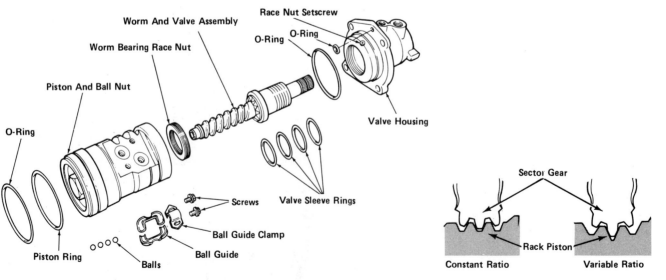

Figure 59-9. Recirculating ball power steering gearbox assembly.
FORD MOTOR COMPANY

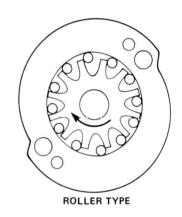

Figure 59-10. Variable-ratio steering.

The control valve directs hydraulic fluid to push the ball nut piston one way or the other. One chamber receives fluid for left turns. The other chamber receives fluid for right turns. Fluid returns to the power steering pump through a low-pressure hose from the other chamber.

The control valve is activated when the steering shaft is turned. The valve opens and closes passages to direct fluid into the proper chamber. The valve also directs return fluid back to the reservoir.

Variable ratio steering. A *variable ratio steering* system allows the steering ratio to be changed as steering wheel rotation is increased. The steering ratio may change from 15:1 to 10:1 to allow faster steering during parking. Many power steering systems are designed with variable ratio steering.

Variable ratio steering is accomplished by redesigning the sector gear in a recirculating ball gearbox. A sector gear with a long middle tooth and two shorter teeth on each side is a variable ratio gear **[Figure 59-10]**. This feature provides for slower steering over center and quicker steering at the ends.

Power Rack and Pinion Steering

A typical rack and pinion power steering assembly is shown in **Figure 59-11.** The rack housing forms the power cylinder inside which hydraulic fluid creates pressure. Seals at each end control leaks. A flange on the rack functions as a piston. Lines connect the rack housing pressure inlets with a rotating control valve, located on the end of the steering shaft. Hydraulic pressure is directed to one side of the flange or the other for right or left turns, as shown in **Figure 59-12.**

Linkage Power Steering

This system uses a separate hydraulic power cylinder mounted on the steering linkage **[Figure 59-13]**. On this type of system, the control valve is mounted between the Pitman arm and the relay rod.

When the Pitman arm is moved by turning linkage, the control valve senses this movement and directs fluid to the proper end of the power steering cylinder. This fluid then provides steering assistance. When steering wheel movement stops, the control valve stops fluid flow to the cylinder.

Electronic Power Steering (EPS)

To provide better steering control and improved steering "feel," computer controls can be applied to power steering mechanisms. *Electronic power steering (EPS)* systems increase the force applied to steering linkage and also change the effort required to turn the steering wheel.

For example, at slow speeds when parking and maneuvering, an EPS system applies greater force to move the steering linkage and reduce steering effort

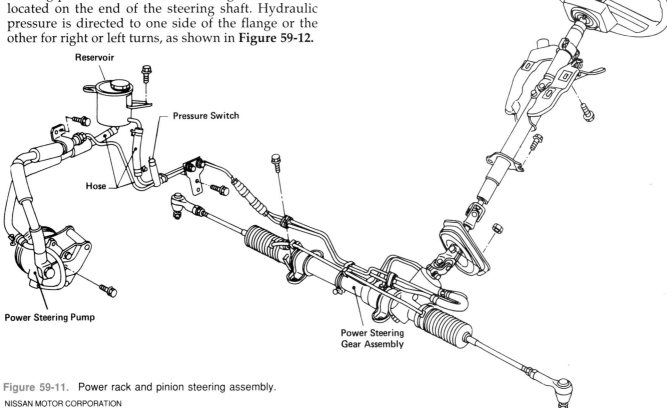

Figure 59-11. Power rack and pinion steering assembly.

NISSAN MOTOR CORPORATION

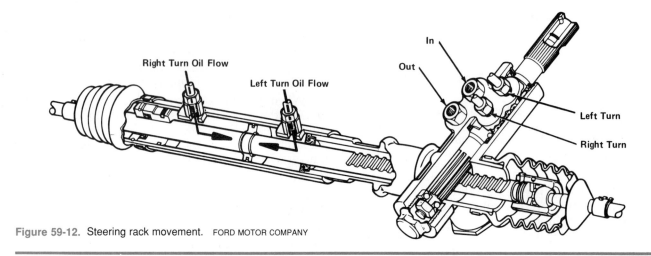

Figure 59-12. Steering rack movement. FORD MOTOR COMPANY

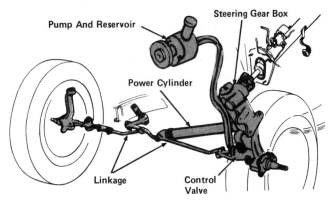

Figure 59-13. Linkage power steering system. FORD MOTOR COMPANY

at the same time. At high road speeds, an EPS system decreases the force applied to steering linkage and increases steering effort to provide more stable steering.

To change the force applied to the steering linkage that moves the front wheels, fluid pressure to the power cylinder changes. In addition, different pressures act on the internal steering gearbox mechanism to provide increased or decreased steering wheel turning effort.

Electronic power steering systems are controlled by computer units. The system shown in **Figure 59-14** includes:

- Vehicle speed sensor
- Normal/Sport steering effort selector switch
- Electronic control unit (computer)
- Pressure control valve
- Solenoid valve
- Reaction force plungers
- Rotary valve.

Fluid flows from the power steering oil pump to a pressure control valve within the steering gearbox. A vehicle speed sensor and manually controlled switch unit provide data to a computer. Based on a program, or set of instructions, the computer controls the operation of a solenoid actuator. The actuator positions a rotary pressure control valve within the rack-and-pinion steering gearbox.

The rotary valve can be positioned to provide higher or lower pressure to the power cylinder that moves the rack. In addition, higher or lower pressures are applied to plungers that restrict the motion of the steering input shaft. These *reaction force plungers* create or relieve resistance to steering wheel motion. The plungers control the effort required to turn the steering wheel.

For low effort steering, greater fluid pressure flows to the power cylinder and less fluid pressure is applied to the reaction force plungers. This action reduces the effort required to turn the front wheels.

To increase steering effort for steering stability, less fluid pressure is applied to the power cylinder and greater pressure is applied to the reaction force plungers.

Steering effort. To provide a good "road feel," EPS changes the effort required to turn the steering wheel. The system increases the effort required to turn the steering wheel with the angle that the wheel is turned. This function helps the driver to judge how sharply and how quickly the vehicle is turning.

Steering response. To provide a crisp, "sporty" steering feel, pressures applied to the power cylinder are increased just as the steering wheel begins to move from the center position. The vehicle responds more quickly to initial steering input and begins to turn rapidly.

Decreased steering kickback. Another function of the EPS system shown is to decrease *steering kickback,* or steering harshness. Steering kickback is force and

STEERING AND WHEEL ALIGNMENT

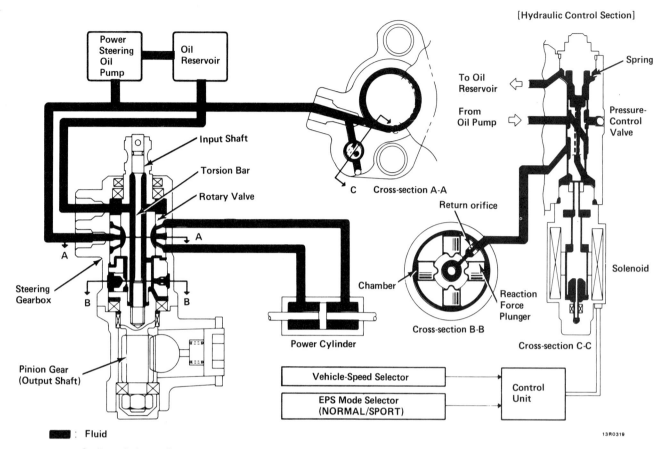

Figure 59-14. Outline of electronic power steering components.
MITSUBISHI MOTOR SALES OF AMERICA, INC.

motion transmitted from the front wheels backwards through steering linkage to the steering wheel. At low speeds on rough roads, the system decreases fluid pressure to the power cylinder and increases fluid pressure to the reaction force plungers. This action reduces steering kickback.

Selectable steering effort. Electronic power steering also allows the driver to select higher or lower steering effort manually. When the driver presses a selector switch, the system responds by routing higher or lower pressures to the power cylinder. At the same time, correspondingly lower or higher pressures act on the reaction force plungers to control steering effort.

Four-Wheel Steering (4WS)

To improve steering responsiveness, all four wheels can be steered. **Four-wheel steering** systems turn the rear wheels slightly to make the vehicle respond more quickly to steering wheel motion. Four-wheel steering systems are operated mechanically or through electrical and/or hydraulic systems.

The system shown in **Figure 59-15** is an all-mechanical system. A gear driven by the rack turns a

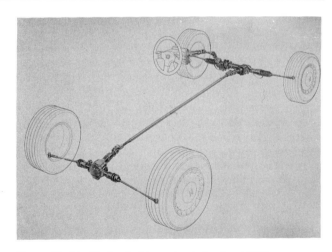

Figure 59-15. Mechanical four-wheel steering (4WS) system.
AMERICAN HONDA MOTOR CO., INC.

long shaft connected to the rear wheel steering gearbox and tie rods.

Within the rear wheel steering gearbox, the input shaft is connected to an eccentric that turns a planetary gear **[Figure 59-16]**. When the amount of rotation transmitted is small, the planetary gear

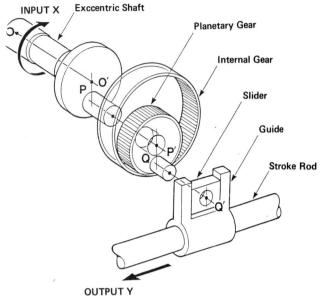

Figure 59-16. Simplified 4WS rear steering gearbox.
AMERICAN HONDA MOTOR CO., INC.

or collapsing. **Figure 59-17** shows how three different types of steering columns are designed to collapse to prevent injury to the driver.

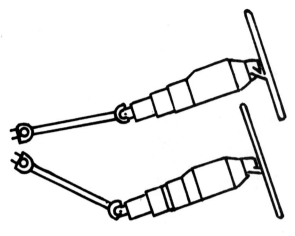

U-Joints Pivot To Absorb Crash Impact

turns the slider and stroke rod to move the rear wheels in the same direction as the front wheels. This action produces a small change in rear wheel angle for sharper turning, as in parallel parking.

When the driver turns the steering wheel sharply, the eccentric rotates and drives the planetary gear to transfer motion in the *opposite* direction. The slider and stroke rod then turn the rear wheels in a direction *opposite* to front wheel movement. The rear wheels trace an arc that is similar to the arc of the front wheels for more responsive turning, as in sharp left or right turns.

Other 4WS systems use a power-assisted mechanism that is similar to part of an electronic power steering system. Power steering fluid from a pump is routed through flexible and rigid lines to a rear-mounted steering rack connected through tie rods to the rear wheels. A computer controls solenoid valves to apply hydraulic force to the rack piston and turn the wheels. Some of these systems also can move the rear wheels in a direction opposite to front wheel motion for sharper turning.

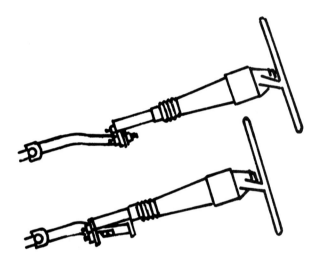

Plate Separates

59.4 STEERING COLUMN

A steering column can house several driver-operated controls, such as a shift lever, turn signal indicator, and hazard warning switch. Additional equipment inside the steering column housing includes windshield wiper/washer, high beam, and horn switches.

Collapsible Steering Column

A *collapsible steering column* is a safety feature, designed to absorb energy from an impact by bending

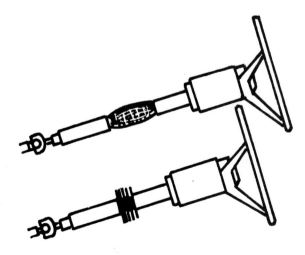

Mesh Compresses

Figure 59-17. Collapsible steering columns.

STEERING AND WHEEL ALIGNMENT 667

Tilt Steering Wheels and Adjustable Columns

A *tilt steering wheel* allows the driver to adjust the angle of the wheel to a more comfortable driving position [Figure 59-18]. The lock/release mechanism for a tilt steering column usually is mounted on a lever below or behind the turn signal lever.

Some automobiles have steering columns that can be adjusted for slight upward or downward adjustment of the entire column.

A steering column that can telescope toward or away from the driver is illustrated in Figure 59-19.

Locking Steering Column

A *locking steering column* is an anti-theft device. The locking mechanism is activated by the ignition switch. When the ignition switch is turned off, the steering wheel and shift lever are locked and cannot be moved.

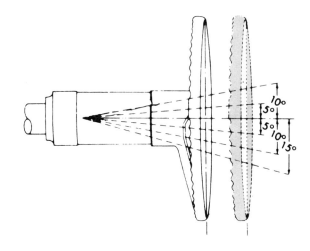

Figure 59-19. Telescoping steering operation.
CADILLAC MOTOR CAR DIVISION—GMC

59.5 WHEEL ALIGNMENT

Wheel alignment refers to the positioning of parts that affect steering action and tire contact with the road. Proper wheel alignment gives the driver good control under many different circumstances and prevents rapid tire wear. Improper wheel alignment causes steering problems, rapid tire wear, and may contribute to damage of suspension and steering components.

Front-End Geometery

Steering geometry, or *front-end geometry*, is a term used to describe angles in the front suspension and steering of an automobile. During wheel alignment, some of these angles can be adjusted. The five angles of front-end geometry are:

- Camber
- Caster
- Steering axis inclination
- Toe
- Toe-out on turns.

Camber. A *camber* angle refers to the inward or outward tilt of a wheel, viewed from the front of an automobile. This tilt is measured in degrees as the angle between the tire centerline and vertical [Figure 59-20]. If the wheel tilts outward at the top, the camber is positive. If the wheel tilts inward at the top, the camber is negative. Incorrect camber will cause one side of the tire to wear faster than the other. Camber angle is not adjustable on some MacPherson strut suspensions.

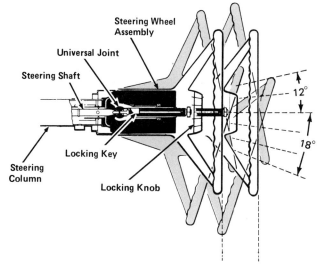

Figure 59-18. Tilt steering operation. CHRYSLER CORPORATION

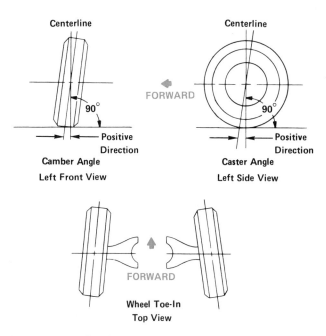

Figure 59-20. Steering geometry.
CHEVROLET MOTOR DIVISION—GMC

Caster. A *caster* angle is a forward or rearward tilt of the steering knuckle when viewed from the side [Figures 59-20 and 59-21]. If the steering knuckle is tilted forward at the top, the caster is negative. If the steering knuckle is tilted rearward at the top, the caster is positive.

Caster does not have a great effect on tire wear but can affect steering effort and return, and other vehicle handling characteristics. The caster angle is adjustable on double A-arm and related suspensions. Caster is not adjustable on most Macpherson strut suspensions.

Steering axis inclination. *Steering axis inclination* refers to an inward slant from vertical of the top of the steering knuckle [Figure 59-21]. The steering axis inclination is designed into the steering knuckle. No adjustments are possible to adjust steering axis inclination on any type of suspension. However, if chassis, steering, or suspension parts become bent or damaged, the steering axis angle may be changed and will cause steering and tire wear problems.

Toe. *Toe-in* is illustrated in Figure 59-20. When viewed from above, the centerlines of the tires are angled inward toward the front. Toe-out means that the wheels are angled outward toward the front. Some toe-in is designed into rear-drive automobiles to assist in handling and to promote long tire life. Incorrect toe settings will cause rapid tire wear. The toe measurement can be made in either fractions of an inch (mm) or degrees and is adjustable on all types of suspensions.

Toe-out on turns. *Toe-out on turns* refers to the difference in angles between the front wheels during turns [Figure 59-22]. Toe-out on turns is designed into the angles of the steering knuckle arms and helps turn the inner wheel at a sharper angle. Toe-out on turns cannot be adjusted. However, if the steering knuckles are bent or damaged, this angle will be incorrect and cause rapid tire wear.

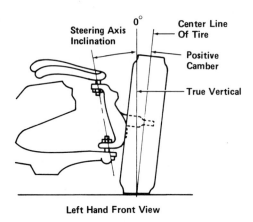

Figure 59-21. Steering axis inclination.
CHEVROLET MOTOR DIVISION—GMC

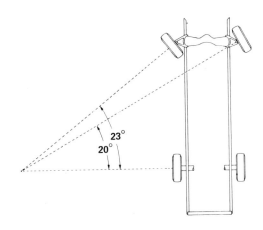

Figure 59-22. Toe-in and toe-out on turns.
CHEVROLET MOTOR DIVISION—GMC

UNIT HIGHLIGHTS

- The steering linkage transfers force to pivot wheels for steering.
- The steering ratio is a comparison of how many degrees the steering wheel must be turned to turn the front wheels one degree.
- A recirculating gearbox contains a worm gear, a sector gear, a ball nut, and ball bearings.
- A rack and pinion steering system includes a rack housing, a pinion gear, and a sliding rack.
- A power steering system provides hydraulic pressure and force to move parts in the steering gearbox or linkage to decrease steering effort.
- A variable ratio steering system has a specially designed sector gear that causes steering ratios to change as the steering wheel is turned.
- Electronic power steering (EPS) systems use vehicle speed sensors and manual switches to control steering force and steering effort.
- Four-wheel steering (4WS) systems are operated mechanically or through power steering systems. These systems can steer the rear wheels in the same or opposite direction as the front wheels for ease in parking or sharp turning.
- Steering columns have features for safety, driver comfort, and anti-theft protection.
- Wheel alignment is the process of adjusting front end geometry.
- Front-end geometry describes angles that affect steering control and tire contact.

TERMS

steering shaft
steering gearbox
rack
relay rod
reaction force plunger
tie rod
steering arm
steering ratio
power steering
worm gear
ball nut
rack and pinion
linkage power steering
variable ratio steering
locking steering column
steering column
steering linkage
Pitman arm
center link
idler arm
adjusting sleeve
lock-to-lock
manual steering
recirculating ball
sector gear
ball return guide
integral power steering
power steering pump
tilt steering wheel
wheel alignment
front-end geometry
caster
toe-in
toe-out on turns
four-wheel steering (4WS)
collapsible steering column
camber
steering axis inclination
toe-out
steering kickback
parallelogram steering linkage
electronic power steering (EPS)

REVIEW QUESTIONS

DIRECTIONS: The following questions are similar to those used on automotive technician certification tests. On a separate sheet of paper, write the letter of the correct choice.

1. The relay rod is connected to the
   A. drag link.
   B. steering arm.
   C. adjusting sleeve.
   D. tie rods.

2. Which of the following statements is correct?
   I. Lock-to-lock is one complete rotation of the steering wheel.
   II. A 20:1 steering ratio requires 3-1/3 rotations of the steering wheel, lock-to lock.
   A. I only
   B. II only
   C. Both I and II
   D. Neither I nor II

3. Technician A says that electronic power steering (EPS) varies the force applied to the steering linkage that turns the wheels.
   Technician B says that EPS varies the force required to turn the steering wheel.
   Who is correct?
   A. A only
   B. B only
   C. Both A and B
   D. Neither A nor B

Questions 4 and 5 are not like those above. Each has the word EXCEPT. For each question, look for the choice that does *not* apply. Read each question carefully before you choose your answer.

4. All of the following angles are adjustable during wheel alignment EXCEPT
   A. camber.
   B. steering axis inclination.
   C. toe-in.
   D. toe-out.

5. All of the following statements about recirculating ball steering gearboxes are correct EXCEPT
   A. a worm gear has rounded threads.
   B. a ball nut shaft is connected to the Pitman arm.
   C. ball bearings transmit force from the worm gear to the ball nut.
   D. ball bearings reduce friction.

ESSAY QUESTIONS

1. Describe parallelogram and rack-and-pinion steering systems.
2. What types of power steering systems are available?
3. What wheel alignment factors influence steering, handling, and tire wear?

SUPPLEMENTAL ACTIVITIES

1. Determine the steering ratio on a vehicle, assuming a total of 60 degrees of front wheel movement.
2. Describe how a recirculating ball steering gearbox operates.
3. Describe how a rack and pinion steering system works.
4. Describe how hydraulic pressure operates a power steering gearbox.
5. Describe the angles in front-end geometry.
6. Describe how an electronic power steering (EPS) system works.
7. Describe how a mechanical four-wheel steering (4WS) system works.

60 Steering and Wheel Alignment Service

UNIT PREVIEW

Steering systems and wheel alignment angles require periodic inspection and adjustments. Damage to steering and suspension parts can be caused by normal wear or by driving over potholes, hitting curbs during parking, or during a collision. Whenever steering or suspension parts are replaced, wheel alignment checks or adjustments are necessary. Steering and wheel alignment problems can cause safety hazards, create excessive tire wear, and reduce fuel economy.

LEARNING OBJECTIVES

When you have completed your assignments and exercises in this unit, you should be able to:

- ❑ Check steering lubricant level.
- ❑ Diagnose steering problems.
- ❑ Adjust a steering gearbox.
- ❑ Replace a power steering pump.
- ❑ Diagnose electronic power steering (EPS) problems.
- ❑ Perform two- and four-wheel alignments.

SAFETY PRECAUTIONS

Always refer to the manufacturer's service manual for correct steering and wheel alignment service procedures and safety precautions.

Raise and safely support the vehicle on a hoist or with jack stands.

Allow the vehicle to cool, whenever possible, before you begin work. Hot parts and fluid around steering system parts can cause burns.

Do not use heat or attempt to straighten any suspension or steering part. Install new parts to replace defective units.

Shut the engine OFF before you attempt to work on a power steering pump. On vehicles with electrically driven cooling fans, disconnect the negative battery terminal to prevent the fan from operating unexpectedly.

Use caution during road tests. A faulty steering system can be dangerous to you and other drivers. The automobile could swerve or make unusual movements. Drive safely and only enough to diagnose a steering or wheel alignment problem.

When using wheel alignment equipment, set the parking brake, place the gearshift lever in Park (automatic transmissions) or in neutral (manual transmissions), and block the wheels. Follow all safety precautions and service procedures recommended by the wheel alignment equipment and vehicle manufacturer.

Performing a wheel alignment requires the technician to work in tight areas around the wheel. Be alert at all times. A hand can become caught in parts, such as a coil spring. Keep your hands and other body parts away from springs that may become compressed. Take care to avoid bumping your head and upper body against chassis, suspension, and other parts under the vehicle.

All wheel alignment fasteners are important attaching parts. Always use the manufacturer's recommended replacement parts and mounting hardware and tighten fasteners to the manufacturer's torque specifications.

TROUBLESHOOTING AND MAINTENANCE

Always refer to the manufacturer's service manual for steering and wheel alignment preventive maintenance recommendations. Preventive maintenance for steering systems usually involves fluid level checks, lubrication, and visual inspection. Some parts of a steering system may be permanently lubricated and do not require maintenance.

Manual Steering Gearbox Lubrication

Both Pitman and rack and pinion steering gearboxes on newer vehicles are filled with grease and permanently sealed. Lubricant can be added or changed only when the gearbox is disassembled for service.

On older vehicles, check and add Pitman steering gearbox lubricant (usually 90 weight EP gear oil) at the fill plug [Figure 60-1].

Refer to the manufacturer's service manual or visually check for a flexible metal universal joint between steering shaft and the steering gearbox. If recommended by the manufacturer, lubricate the joint with light oil.

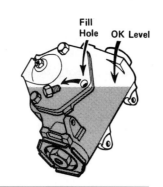

Figure 60-1. Checking manual gearbox lubricant. AMERICAN MOTORS CORPORATION

Power Steering Lubrication

On some automobiles, a power steering fluid dipstick indicates both hot and cold fluid levels. On others, the lubricant must be checked at driving temperature. To warm up power steering fluid, start the engine, then turn the steering wheel from lock-to-lock several times. This expels air from the system and warms the fluid.

Check the dipstick or fluid level and add power steering fluid to the reservoir on the pump, or in a separate tank [Figure 60-2]. Older vehicles use automatic transmission fluid (ATF) in the power steering pump. Newer vehicles may require a special fluid. If low, add the lubricant recommended in the manufacturer's service manual and replace the cap.

Steering Linkage Lubrication

Parts of the steering linkage system that require periodic lubrication are shown in Figure 60-3. Whenever possible, remove plugs from the ball joints, tie rod ends, and other parts and install grease fittings. Search carefully for all grease fittings. Use a grease gun to lubricate all fittings properly (refer to Unit 58).

Visual Inspection

During lubrication, check steering linkage for damage, excessive wear, and looseness by attempting to push and shake connected parts. Also, check for lubricant leaks. Check for missing grease fittings.

Check around suspension and steering mountings, bushings, ball joints, and at gearboxes. Look for leaks at gearboxes, power steering connections, and driving axle boots. Check the power steering belt for cracks, fraying, wear, and belt tension. Install new parts to replace any parts that are worn, damaged, or missing.

60.1 DIAGNOSING STEERING PROBLEMS

Tire wear is the best indication of problems in the steering system and/or with wheel alignment (see Unit 62). Also, the driver may complain of pulling or swerving, hard steering, poor steering wheel return after turns, and other related problems.

Steering problems may involve suspension, wheels, and/or tires. All must be considered when diagnosing a steering condition. Safely conduct a road test to check for problems. Before a road test, make the following preliminary checks:

1. Check wear patterns and tire condition, and properly inflate all tires to the vehicle manufacturer's recommended pressures.
2. Check for free play, or looseness, in the steering linkage. Unlock and move the steering wheel back and forth and watch the motion of the front tire. More than two inches of play on Pitman steering systems indicates a problem. There should be little or no play in a rack and pinion steering system. Move the steering wheel back and forth sharply and listen for clunking, knocking, or banging sounds from loose steering parts. Repeat the test with the engine running for power steering systems. Then shut off the engine.
3. Check couplings and universal joints from the steering shaft to the steering gearbox for loose connections or wear. Replace worn parts.

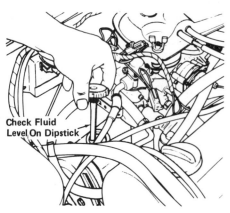

Figure 60-2. Checking power steering lubricant. CHRYSLER CORPORATION

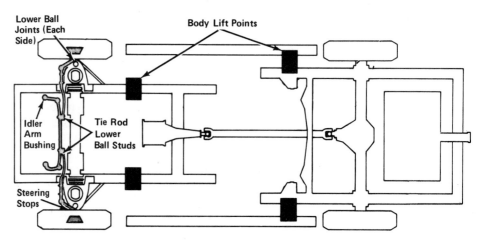

Figure 60-3. Steering linkage lubrication points. AMERICAN MOTORS CORPORATION

4. Safely raise and support the vehicle on a hoist or with jackstands. Check the front and rear suspension and steering linkage for loose or damaged parts. Have a helper move the steering wheel back and forth to help locate looseness and wear in steering parts.
5. While the car is raised, rotate the front and rear wheels by hand. Check for out-of-round tires, out-of-balance wheel assemblies, bent rims, loose and/or rough wheel bearings and loose ball joints.
6. Check for leaks on the power steering system. Check power steering fluid level. Check tension on the drive belt.
7. Check the shock absorbers for proper damping and leakage.

If any defects are found, correct them. Then road test the automobile to determine if the problem has been corrected.

Vibration, steering response, and noise are indications of a possible steering problem. Refer to the vehicle manufacturer's troubleshooting charts in the service manual for proper diagnosis.

The following descriptions are intended as a general guide to troubleshooting steering problems:

Shimmy, shake, or vibration. These are usually caused by worn steering linkage pivots, worn ball joints, improper wheel balance, and/or bad tires.

Hard steering (manual). This can be caused by insufficient lubrication, a bind in steering linkage, front wheel misalignment, improper rack and pinion adjustment, and underinflated or worn out tires.

Hard steering (power). This can be caused by the hydraulic system, improper gearbox or rack and pinion adjustment, a loose belt, or low power steering fluid level. It also can be caused by binding in the steering linkage or in a rack and pinion assembly, or underinflated tires.

Excessive play. Play is the free movement of the steering wheel before the front wheels begin to turn. Excessive play is caused by loose or worn steering and suspension parts. These include worn ball joints, steering gearboxes, steering shaft couplings, universal joints, relay rods, tie-rod ends, and the idler arms.

Pitman steering gearboxes can be adjusted for looseness without disassembly [**Figure 60-4**]. Rack and pinion gearboxes do not have external adjustments for looseness or wear. Rack and pinion bearing preload can be adjusted during overhaul procedures. These adjustments are described below.

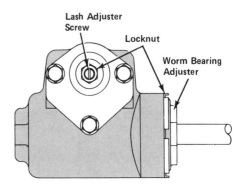

Figure 60-4. Steering gearbox adjustment points.
CHEVROLET MOTOR DIVISION—GMC

Poor returnability (manual). *Returnability* is the tendency of the steering wheel to return to a straight-ahead position at the end of a turning maneuver. Poor returnability may be caused by lack of lubrication at Pitman steering linkage or within a rack and pinion steering gearbox. Other causes include binding in the steering column, incorrect caster, incorrect rack and pinion gearbox adjustments, or damaged steering components.

Poor returnability (power). This can be caused by lack of steering linkage lubrication, improper caster, or a sticking hydraulic valve. Binds can occur in the steering shaft, lower coupling flange, or anywhere in the steering linkage. Adjustments may be required in a steering gearbox or in a rack and pinion unit.

Wander. Wander shows up as poor steering stability. The driver must make constant steering corrections to keep the automobile moving straight ahead. Wander can be caused by a lack of lubrication at the steering linkage or a steering gearbox adjustment that is too tight. It also can indicate front or rear wheel alignment problems, or loose, worn, or damaged parts in the steering and suspension systems.

Pull during braking. This is caused by incorrect or uneven caster, uneven tire inflation, underinflated tires, or loose and/or worn steering and suspension system parts. Brake defects also can cause pulling.

Kickback. *Steering wheel kickback* is a sharp, rapid movement of the steering wheel when the automobile strikes a bump or other obstruction. Excessive kickback can be caused by lack of steering linkage lubrication, loose tie rod ends, or air in a power steering hydraulic system. It also can be caused by loose gearbox or rack and pinion attachments or adjustments. A worn or loose steering shaft coupling or universal joints also could be the problem.

Steering wheel surge. *Steering wheel surge* describes the condition in which the steering wheel moves erratically in the driver's hands. Surge is caused by a power steering hydraulic system problem, sluggish control valve, a loose belt, or low fluid level.

Abnormal power steering pump noise. A groan or whine is caused by low fluid level, air in the fluid, a loose pump mounting, or mechanical pump problems. Most pumps make a low groaning noise when the steering is turned as far as the wheels will turn.

A rattle or chuckle as the steering wheel is turned may indicate loose internal gearbox parts or loose steering linkage. A squeal or chirp usually indicates a loose belt on the power steering pump. Rattles also can be caused by any loose parts that can hit each other during driving or when the steering wheel is turned.

Abnormal front-end noise. Possible causes include lack of lubrication at the steering linkage or worn control arm or strut bushings. Also, check for worn linkage pivots or tie-rod ends, or steering gearbox adjustments. A dry or defective upper strut bearing in the shock/strut tower will also cause abnormal noises, especially during turns over bumps.

60.2 STEERING SYSTEM SERVICE

Steering system service involves a steering gear adjustment and replacing parts. Procedures vary considerably among automobiles. Always consult the manufacturer's service manual.

Recirculating Ball Steering Gear Adjustment

Adjustments are provided for worm bearing preload and lash, or gear mesh, between the ball nut and sector gear. In most cases, preload adjustment must be checked and adjusted first. Always refer to the manufacturer's service manual for steering gear adjustment procedures. Improper adjustment procedures can result in a damaged steering gear or poor steering response.

Most adjustments require that the steering linkage be disconnected and/or the Pitman arm removed. Many power steering gearboxes must be removed from the automobile before adjustments can be made.

On most rack and pinion gearboxes, the pinion bearing preload is checked and adjusted to manufacturer's specifications during overhaul procedures. The gearbox must be removed from the vehicle and disassembled to make adjustments.

Steering Linkage Replacement

The automobile should be raised on a hoist to replace steering linkage. **Figure 60-5** shows parts of a Pitman steering linkage system. Special tools and pullers are used to separate tie rod end and other studs and remove Pitman arms. Use care to avoid damaging seals on tie rod ends that will be reused.

CAUTION!!!

Do not turn the steering wheel hard against the stop when the linkage is disconnected. This could damage the ball return guide in the gearbox.

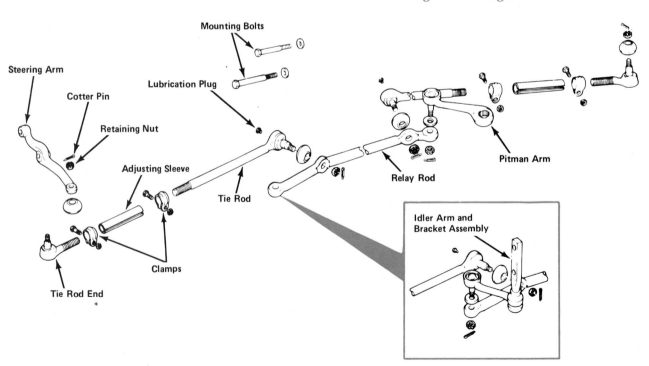

Figure 60-5. Parts of a steering linkage system. AMERICAN MOTORS CORPORATION

Whenever linkage parts are replaced, a wheel alignment must be performed (see Topic 60.3).

Power Steering Pump Replacement

A typical power steering pump assembly is shown in **Figure 60-6**. Parts of the engine fan assembly and air conditioning system may have to be removed for access to the pump.

When the pump is installed, be sure all fasteners are torqued properly, including the hose fittings. Fill the reservoir, tighten the belt, and bleed the system by starting the engine and turning the steering wheel lock-to-lock several times.

Steering Gear Replacement

Always refer to the manufacturer's service manual for specific repair procedures. First disconnect and remove the steering shaft, coupler, and/or universal joint.

Set the front wheels in a straight-ahead position. Remove any power steering hoses. Mark the relationship of the Pitman arm to the Pitman shaft if necessary. Make alignment marks for reassembly and remove the Pitman arm with a special puller **[Figure 60-7]**. Disconnect the steering shaft and remove the bolts holding the gearbox in position and remove the gearbox **[Figure 60-8]**.

When you replace the gearbox, bolt it in place and install the Pitman arm according to the alignment marks you made during removal. Attach the steering shaft coupler or joint. Torque all fasteners to the manufacturer's recommended specifications.

Check and/or adjust wheel alignment to specifications.

Rack and Pinion Replacement

Before removing a rack and pinion assembly, raise the automobile on a hoist and remove the front wheels. Rack and pinion attachment points are shown in **Figure 60-9**. Disconnect the tie rod ends with a special puller and disconnect the steering shaft. If necessary, support the crossmember with a hydraulic jack. Remove all attachments according to the manufacturer's service manual, including power steering parts. Remove the bolts holding the steering assembly, then remove the assembly.

A rack and pinion assembly is installed in the reverse order of removal. Wheel alignment must be checked and adjusted if necessary after the unit is replaced.

Electronic Power Steering (EPS) Service

In addition to mechanical and hydraulic system problems, *electronic power steering (EPS)* systems can have malfunctions in electrical and electronic subsystems. These malfunctions generally fall into two categories:

- Electrical problems with solenoid control valves
- Electronic problems with connections, vehicle speed sensors, or computer control units.

Refer to the manufacturer's service manual for specific information on resistance and current checks for solenoid valves **[Figure 60-10]**. Also refer to the manufacturer's service manual for information on checking EPS wiring connections, vehicle speed sensor inputs, and computer control units.

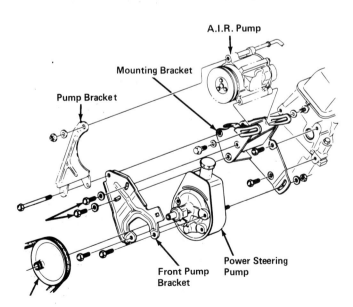

Figure 60-6. Power-steering pump assembly.
BUICK MOTOR DIVISION—GMC

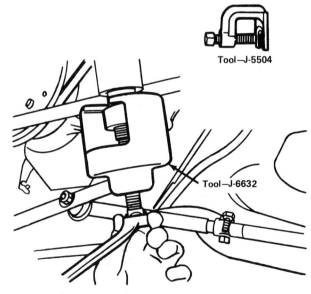

Figure 60-7. Removing Pitman arm.
CHEVROLET MOTOR DIVISION—GMC

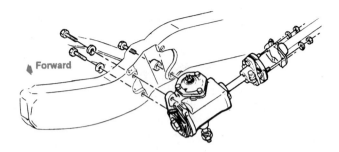

Figure 60-8. Removing bolts that hold gearbox. CHEVROLET MOTOR DIVISION—GMC

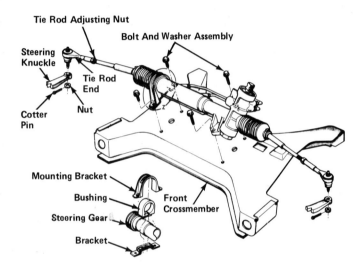

Figure 60-9. Rack and pinion assembly. CHRYSLER CORPORATION

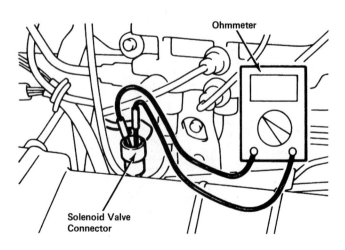

Figure 60-10. Checking EPS solenoid valve coil. MITSUBISHI MOTOR SALES OF AMERICA, INC.

60.3 WHEEL ALIGNMENT

Always refer to the manufacturer's service manual when making a wheel alignment. Wheel alignment must be checked and/or adjusted whenever any steering or suspension part is replaced or when uneven tire wear is noticed. As a preventive measure, wheel alignment frequently is checked and adjusted when new tires are installed.

Front-end alignments are done on all vehicles. Some vehicles also have adjustments for rear-wheel alignment.

On some suspensions, angles such as caster and camber may not be adjustable. If non-adjustable angles are incorrect and caused by damaged suspension parts, those parts must be replaced to restore correct wheel alignment.

Prealignment Checks

Before any checks are made, the automobile must be on a level surface. The fuel tank must be full. All doors must be closed, with no occupants or excess weight inside the automobile.

On most vehicles, all tires should be of the same size. If tires are worn unevenly, rotate the best tires to the wheels to be aligned. If tires are badly worn, install new tires before wheel alignment checks and adjustments.

A thorough inspection must be made of all steering and suspension parts, and any fault corrected before wheel alignment can be done. Check:

1. Tire inflation pressures, condition, size, and even tread wear
2. Front wheel bearings for proper adjustment
3. All suspension and steering parts for wear or damage
4. Shock absorber appearance and condition
5. Wheel and tire runout
6. Vehicle trim height
7. Steering gearbox to chassis attachment.

Wheel Alignment Measurement

Many different types of wheel alignment equipment are used. The equipment can be portable or permanently mounted. **Figure 60-11** shows an example of wheel alignment equipment. Newer units are computerized, like diagnostic engine testers. They provide quick and accurate information about front-end geometry angles and corrections needed.

The method of checking alignment will vary depending on the type of equipment used. Instructions furnished by the equipment manufacturer always should be followed. Check camber, caster, and toe.

Steering and Wheel Alignment Service

Figure 60-11. Wheel alignment equipment.

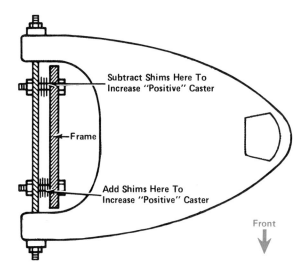

Figure 60-12. Shim caster adjustment.
CHEVROLET MOTOR DIVISION—GMC

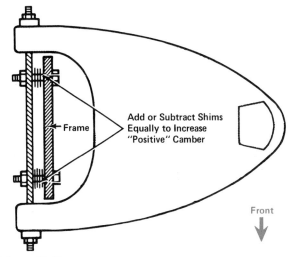

Figure 60-13. Camber adjustment with shims.
CHEVROLET MOTOR DIVISION—GMC

Front-End Alignment

Always refer to the manufacturer's service manual for correct alignment procedures and safety precautions. If both camber and caster must be adjusted, and they will be adjusted separately, always do caster first, then camber. If camber is done first, it may have to be redone after caster has been adjusted.

Caster. Before adjusting caster and camber, bounce the car several times and allow the vehicle to settle to its normal height.

When making a shim adjustment for caster, a different number of shims will be needed at the front and rear bolts that hold the upper control arm pivot shaft. To correct caster, the shims are transferred from front to rear or from rear to front [Figure 60-12].

For vehicles with cam-type caster adjustments, a caster change is made by turning each cam bolt to obtain one-half of the needed correction. Check the caster setting to make sure it is correct. If necessary, repeat the adjustments [see Figure 60-14].

On some vehicles, caster also may be adjusted by turning a nut to reposition a strut rod attached to a lower control arm.

Camber. Camber adjustments are usually made with shims or by turning cam bolts. To adjust camber on a suspension with shims, shims may be added or subtracted to correct the angle [Figure 60-13]. After caster has been adjusted, an equal number of shims must be changed at both the front and rear of the control arm shaft to change camber.

Loosen the frame nuts at the upper control arm. Add or subtract shims according to the proper alignment correction charts. Retorque the nuts.

To make adjustments with cam bolts, the cam bolt is loosened and turned. Use Figure 60-14 as a guide in making the following cam bolt adjustments.

Say that the camber on a wheel is -0.6 degrees. To obtain 0.0 degrees, camber must be corrected by +0.6 degrees. First, hold one of the cam bolts and loosen the nut. Turn the cam bolt to obtain a change that is equal to half of the needed correction, or +0.3 degrees. Hold the cam bolt in this position to maintain the setting, and tighten the nut to specifications. Repeat the same procedure for the other cam bolt to get another correction of +0.3 degrees. The two corrections will total a change of +0.6 degrees.

Toe-in/toe-out. The last wheel alignment adjustment should be toe. Caster and camber angle corrections may affect the toe setting. Toe usually is increased or

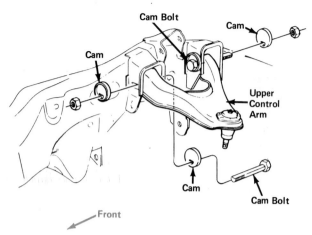

Figure 60-14. Cam bolt adjustments.
CHEVROLET MOTOR DIVISION—GMC

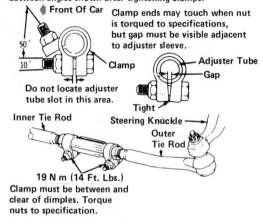

Figure 60-15. Tie rod adjustment. BUICK MOTOR DIVISION—GMC

decreased by turning both tie-rod sleeves to lengthen or shorten the tie rods. The steering wheel is held centered with a special tool. Then both tie rods are adjusted to correct toe. One side is lengthened, and the other side is shortened. If only one tie rod is adjusted, the steering wheel will not be properly centered during driving.

CAUTION!!!

Never bend a tie rod as a method of adjusting toe-in. Bent tie rods will cause steering and safety problems and must be replaced.

If a tie rod adjuster is heavily rusted, discard the nuts and bolts. Apply penetrating oil between the clamp and the tube. Rotate the clamps until they move freely. Install new nuts and bolts of the proper size and strength.

Use **Figure 60-15** as a guide to adjusting toe-in. Loosen the clamp bolts at each end of the adjusting sleeves. With the steering wheel set in a straight-ahead position, turn the adjusting sleeves to obtain the proper toe-in adjustment.

When adjustment is completed, check that the tie-rod end housings are at proper angles to the steering arms. Place tie rod clamps and sleeves in their proper positions.

Before locking the clamp bolts, be sure that both tie-rod ends are in alignment with each other. To check alignment, rotate the tie-rod as far as it will go. The stud to rod angle should be the same at each end. Tighten the clamps to specifications. Make sure the adjuster tubes and clamps are positioned properly. Torque the nuts to specifications.

MacPherson Strut Front-End Alignment

On many MacPherson strut front suspensions, only toe-in adjustments are made. In most cases, there is no provision for caster adjustments on MacPherson strut front or rear suspensions. Camber can be adjusted on some vehicles' MacPherson strut front suspensions.

Camber. Use **Figure 60-16** as a guide for camber adjustment. Loosen the bolts that connect the strut to the steering knuckle. The bolts should be just loose enough to allow movement between the strut and the knuckle.

Grasp the top of the tire firmly and move the tire inward or outward until the correct camber adjustment is obtained. Attempt to tighten one or both bolts firmly to hold the adjustment in place. If the wheel and tire interfere with full tightening, tighten the bolts as much as possible. Then remove the wheel and torque the bolts to the manufacturer's specifications. Finally, reinstall the wheel and tire.

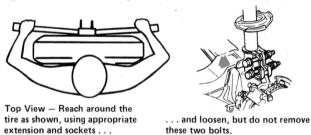

Figure 60-16. MacPherson strut camber adjustment.
BUICK MOTOR DIVISION—GMC

Toe-in. When adjusting toe-in on a rack and pinion steering system, the tie rod boots will be moved. Be careful that the boots are not damaged or twisted. Replace any boots that are worn or damaged.

Use **Figures 60-17** and **60-18** as guides when adjusting toe-in. Loosen the jam nuts or clamp bolts on the tie rod. Rotate the tie rods to adjust the toe to specifications. Torque the jam nuts or clamp bolts to specifications. Reposition the boot correctly.

Rear-Wheel Alignment

In addition to front-wheel alignment, some vehicles with independent rear suspensions require rear-wheel alignment. Rear-wheel alignment procedures are similar to those described above in Front-End Alignment. In most cases, only camber and toe can be adjusted on rear suspensions. Refer to the manufacturer's service manual for rear-wheel alignment specifications and procedures.

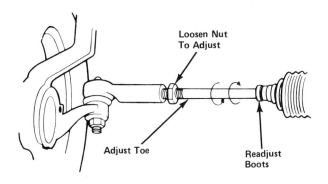

Figure 60-17. MacPherson strut toe adjustment.
BUICK MOTOR DIVISION—GMC

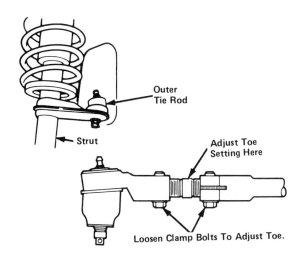

Figure 60-18. MacPherson strut toe adjustment.
BUICK MOTOR DIVISION—GMC

UNIT HIGHLIGHTS

- Preventive maintenance for steering systems can include steering gearbox and steering linkage lubrication and visual inspection.
- Unusual tire wear is the best indication of steering and alignment problems.
- Inspection and repairs may be necessary before a vehicle can safely be taken for a test ride.
- In addition to mechanical and hydraulic checks, service for electronic power steering (EPS) includes checking electrical and electronic parts and systems.
- A wheel alignment is necessary whenever any steering or suspension system part is replaced or adjusted.
- Camber and caster adjustments are made at control arms or at MacPherson strut mountings.
- Toe-in adjustments are made by turning both tie rod adjusting sleeves.
- MacPherson strut suspensions usually cannot be adjusted for caster or camber.
- Rear-wheel alignment usually includes camber and toe adjustments.

TERMS

returnability
steering wheel kickback
steering wheel surge
electronic power steering (EPS)

REVIEW QUESTIONS

DIRECTIONS: The following questions are similar to those used on automotive technician certification tests. On a separate sheet of paper, write the letter of the correct choice.

1. Which of the following statements is correct?
   I. All steering gearboxes are checked for lubricant level.
   II. Power steering lubricant level is checked at the reservoir.
   A. I only   C. Both I and II
   B. II only   D. Neither I nor II

2. Technician A says that a recirculating ball steering gearbox has external adjustments.
   Technician B says that rack and pinion steering gearboxes have external adjustments.
   Who is correct?
   A. A only
   B. B only
   C. Both A and B
   D. Neither A nor B

3. After caster has been adjusted, camber is adjusted by
   A. turning tie-rod sleeves.
   B. installing new tie-rod ends.
   C. adding or subtracting the same number of shims at front and rear.
   D. turning a cam bolt to obtain twice the necessary correction.

4. Which of the following statements is correct?
   I. Toe-in is adjusted by turning both adjusting sleeves.
   II. Toe-in on a MacPherson strut suspension cannot be adjusted.
   A. I only
   B. II only
   C. Both I and II
   D. Neither I nor II

Question 5 is not like those above. It has the word EXCEPT. For question 5, look for the choice that does *not* apply. Read the entire question carefully before you choose your answer.

5. All of the following statements are correct EXCEPT
   A. steering play can be checked by moving the steering wheel and watching the motion of the front tire.
   B. returnability is the ability of a steering wheel to return to a straight-ahead position after a turn.
   C. kickback is the bending of a steering shaft during impact.
   D. steering wheel surge causes the steering wheel to move erratically in the driver's hands.

ESSAY QUESTIONS

1. Describe checks that should be done before you attempt to diagnose steering problems.
2. What should you check before you do a wheel alignment?
3. Describe front-wheel alignment on vehicles with adjustable double A-arm and MacPherson strut suspensions.

SUPPLEMENTAL ACTIVITIES

1. Perform steering system preventive maintenance.
2. Diagnose a steering problem presented by your instructor.
3. Make a steering gearbox or rack and pinion adjustment.
4. Replace a power steering pump.
5. Remove, inspect, and replace steering system linkage and a steering gearbox.
6. Adjust camber, caster, and toe-in on a vehicle selected by your instructor.

61 Tires and Wheels

UNIT PREVIEW

Tires and wheels are responsible for the final actions of the powertrain, suspension, steering, and brakes. Tire construction and inflation determines how much traction, or frictional contact, exists to keep a vehicle safely in contact with the road surface. Information molded into the tire sidewall includes size, maximum pressure and load information, construction details, and safety warnings.

LEARNING OBJECTIVES

When you have completed your assignments and exercises in this unit, you should be able to:

❏ Identify the parts of a tire.
❏ Identify and describe the three different types of tires.
❏ Describe how to determine tire size.

61.1 TIRE FUNCTION

Tires are constructed of rubber, fabric, and reinforcing materials and are filled with air to hold them tightly against a metal wheel rim. Tire construction and inflation pressure determine how much *traction* it will provide. Traction refers to a tire's ability to "grip" the road surface when the vehicle is moving.

A tire functions best when it is inflated to the vehicle manufacturer's recommended pressures. The listing molded into the sidewall of a tire lists the *maximum* pressure and load. Always refer to the manufacturer's service manual or sticker on the vehicle for the correct recommended tire pressures. Recommended pressures for front and rear tires may differ.

A tire's *contact patch*, or "footprint" is the area that touches the road surface. While driving, each tire's contact patch is about the size of your outstretched hand and palm. The traction created by the tire against the road determines how well a vehicle handles, corners, and stops.

61.2 TIRE CONSTRUCTION

Air pressure can be held inside a tire in two ways. On most passenger vehicles, a *tubeless* tire is constructed to form an air-tight seal against the metal wheel without a tube [Figure 61-1]. Another way is to use a separate *inner tube*, or sealed, hollow rubber enclosure, between a tire and the metal wheel rim, as on a bicycle.

A tire is designed to hold a specified maximum amount of air pressure. The air in a tire is measured in pounds per square inch (psi) or in kilopascals (kPa). A typical passenger car tire may have a maximum capacity of 35 psi (approximately 240 kPa). Heavy-duty tires, such as those used on trucks, may be designed to accept a maximum air pressure of between 60 and 100 psi (approximately 414-690 kPa).

Air is forced into a tire through a *tire valve* [Figure 61-1]. A small spring and air pressure inside the tire keep the tire valve closed. When compressed air is used to inflate a tire, it acts against air pressure inside the tire and against the spring to

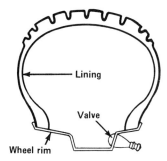

TUBELESS TIRE has a soft inner lining to keep air from leaking between the tire and rim. This inner lining often forms a seal around a nail or other object that punctures the tread. A self-sealing tire holds in air after a nail is removed.

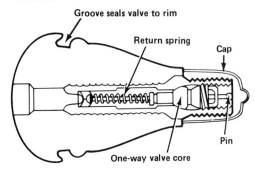

TIRE VALVE has a central core that is spring-loaded to allow air to pass only inward, unless the pin is depressed. If the core becomes defective, it can be unscrewed and replaced. The airtight cap on the end of the valve provides extra precaution against valve leakage.

Figure 61-1. Tire valves.

open the valve. A tire valve can be molded as part of an inner tube, or can be a separate, replaceable rubber plug that fits into a wheel rim. A removable cap is used to protect the valve from dirt and damage.

Parts of a Tire

Tires are constructed to combine the properties of ride, safety, and durability. Parts of a tire include [Figure 61-2]:

- Tread
- Sidewalls
- Carcass
- Plies
- Belts
- Beads
- Inner liner.

Tread. *Tread* is formed by molding rubber into a thick pattern of raised areas around the outer circumference of the tire. The spaces between raised tread areas allow water to escape for better traction on wet surfaces. Most tread patterns are designed to work well on both wet and dry road surfaces. Tread patterns also can be designed specifically for use on snow and ice.

Sidewalls. *Sidewalls* are the sides of the tire. Sidewalls must flex to absorb road shocks and loads. Size and identification information, maximum pressure and load rating, and decorative patterns are molded into the sidewall.

Carcass. The *carcass* is the strong, inner "body" of the tire. The carcass is constructed of *plies*, or layers, of bonded rubber and fabric that are arranged in specific patterns. When filled with air, the carcass supports the weight of the vehicle.

Plies. Layers of rubber and fabric are *vulcanized*, or bonded through heat and pressure, to form the carcass plies. Typically, automobile tires have 2, 4, or 6 plies for sufficient strength. Heavy-duty truck or heavy equipment tires may have 10 or more plies.

Belts. *Belts* are layers of woven material placed between the carcass and the tread. Belts reinforce the tread to keep it flat on the ground, reducing wear and improving traction.

Beads. *Beads* are the thick edges of the tire that are fitted onto a wheel rim and form a tight seal. A bead is formed at the edge of a sidewall. Metal wires are molded into the beads to increase strength.

Inner liner. The inside of the tire has a soft rubber *inner liner* to ensure against air leaks. Some tires have an additional soft, sticky material to self-seal tread punctures.

Tire Materials

The fabric material used with rubber to form the plies is known as the tire *cord.* Today's tires use body plies made from rayon, polyester, nylon, or other synthetic fibers.

Rayon and polyester are the least expensive cord materials and help to provide a comfortable ride, but are affected by heat. Nylon is more heat-resistant and has greater impact resistance, but produces a slightly firmer ride.

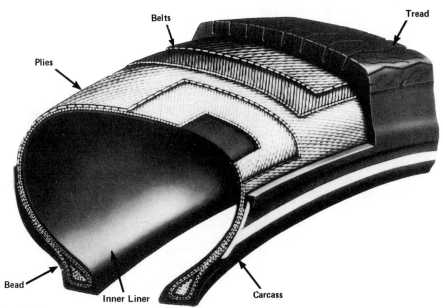

Figure 61-2. Parts of a tire.

Under the tread area, reinforcing belts woven of stronger materials can be used. Belts can be made of woven threads of steel, fiberglass, or synthetic materials such as Aramid® or Kevlar®. Kevlar® also is used to make bulletproof vests. These reinforcing belts also help to provide protection against cuts and punctures.

Types of Tires

Tires are classified by the ways in which the cords are arranged. Refer to **Figure 61-3** as you read the following discussion. Types of tires in common use are:

- Radial ply
- Bias belted
- Bias ply.

Radial. The plies on this type of tire extend around the tire from bead to bead. This *radial*, or circular, arrangement creates a supple sidewall that can flex to keep the tread in contact with the road surface. To reinforce and strengthen the tire, additional belts are placed directly under the tread area. The belts encircle the entire tire under the tread, just as a belt goes around your waist. Tread reinforcing belts are woven from steel strands, fiberglass, Kevlar®, or Aramid®.

Radial tires are standard equipment on all new passenger cars. The "footprint," or contact patch, of a radial tire during cornering typically is greater than a comparable bias ply tire **[Figure 61-4]**.

Radial tires provide superior traction in cornering and braking, as well as longer tire life. Radials also offer improvements in fuel mileage for most vehicles, due to decreased rolling resistance.

Bias-belted tire. In a *bias-belted* tire, the plies crisscross at an angle, or on a *bias*, from bead to bead around the tire carcass. Relatively few bias plies are used to create a flexible sidewall. This results in a soft tread area. Belts are used to stiffen the tread, as in a radial tire. This type of construction is considered a compromise between a true radial-ply tire and the oldest design of tire, the *bias-ply*.

Bias ply. The plies of material crisscross from bead to bead on this tire. To create a relatively stiff tread area, the sidewall also must be stiff. During hard cornering, the stiff sidewall tends to tilt with the rest of the car. This tilting shifts the vehicle's weight from the full tread area to the outside edge of the tire, which results in decreased traction.

SAFETY CAUTION

For best performance and driver control, do *not* intermix radials with bias-belted or bias-ply tires. Ideally, all of the tires on the car should be of the same type. However, if there are only two radial tires out of the four total, the radials must be mounted at the *rear* of the vehicle. Radial tire traction is greater than non-radial tires, and a rear-wheel skid is a dangerous and difficult-to-control situation. Thus, the radial tires—those with the greatest traction—must be mounted at the rear.

Tread Design

Treads can be designed for use on dry pavement, rain and snow conditions, or a mixture of both. **Figure 61-5** shows the three most common types of current tread designs.

Tire Information

Coding marks embossed on the tire's sidewall describe each tire and its performance ratings. Tire speed rating information also may be given on the sidewall. It is added as a letter, usually H, S, or V following the aspect ratio number. The speed rating indicates the highest sustained speed at which the tire may be operated without danger of a speed related failure. For example, 215/65HR15 indicates

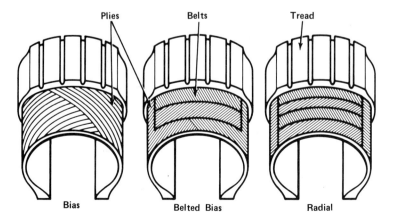

Figure 61-3. Types of tires.

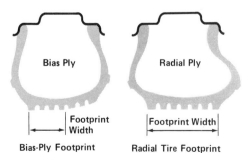

Figure 61-4. Tire footprints.

Figure 61-5. Tread designs.

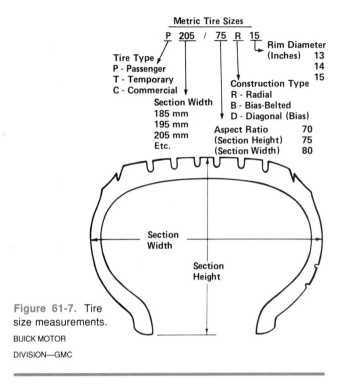

Figure 61-7. Tire size measurements.
BUICK MOTOR DIVISION—GMC

a tire has a speed rating of "H," or 130 mph (210 km). An "S" rating indicates a top speed of 112 mph (180 km), and "V," indicates a top tire speed of over 130 mph (210 + km).

Current domestic tire sizes. Tire size coding, serial numbers, maximum load, and maximum pressure recommendations and other information is molded into the sidewall of a tire [**Figure 61-6**].

The numbers on a typical new tire meant for a domestic American car might be: P205/75R-15. The P stands for passenger car use. If the tire were intended for use as a temporary spare only, the first letter would be a T. The number 205 represents the tire width, from sidewall to sidewall, in millimeters. Thus, this tire has a *section width* of 205 millimeters (approximately 8 inches) [**Figure 61-7**].

The number after the slash or dash, in this case 75, indicates the *aspect ratio* or "profile" [**Figure 61-8**]. The aspect ratio indicates the height of a tire as a percentage of its width. In this case, the tire's height is 75 percent of its width. Higher numbers indicate taller, narrower tires; lower numbers indicate wider, "fatter," tires.

The next letter, R, stands for a radial-ply tire. (B would indicate bias-belted and D would mean bias-ply.)

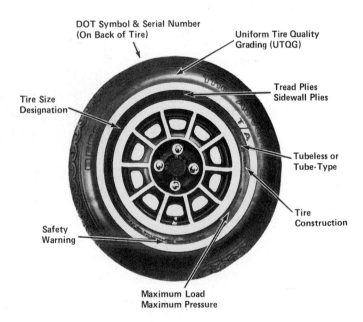

Figure 61-6. Sidewall markings.

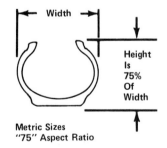

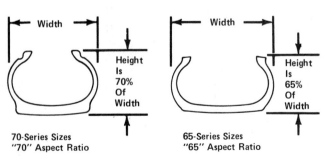

Figure 61-8. Tire size and aspect ratio.
AMERICAN MOTORS CORPORATION

The last number, 15, indicates the diameter, in inches, of the wheel rim on which the tire will be mounted.

Domestic tire sizes. Some tires might have a designation such as JR78-15. The letter J indicates a tire's weight-carrying capacity. The further into the alphabet, the higher the tire's load-carrying capacity. The R indicates radial-ply design. The 78 indicates an aspect ratio. The number after the dash indicates that the tire fits a 15-inch metal rim.

Imported car tire sizes. Another system of tire sizing is used in Europe and Asia. These numbers typically might read: 175SR-14. In this sizing system, the first number, 175, indicates the height of the tire from tread to bead, in millimeters. Thus, this tire is 175 millimeters (approximately 6.9") in section height [refer to Figure 61-7]. The "S" is a speed rating, as described earlier.

The R stands for radial-ply tire construction. The last number, 14, indicates the rim diameter, in inches, on which the tire will be mounted.

Passenger car tires must conform to federal safety requirements. Tires are graded by the Department of Transportation (DOT) for traction and for temperature qualities. These ratings are called the *Uniform Tire Quality Grading (UTQG)* system. They are based on tests conducted by DOT. The grades are indicated on a tire by encircled letters or numbers.

Truck tire sizes. For truck and utility vehicle tires, another tire-sizing system is used. One example of a truck tire size is 30X950R-16.5. The first number, 30, indicates the total height of the wheel-and-tire combination in inches. The X indicates "by." The second number, 950, indicates that the tire is 9.5 inches wide, sidewall-to-sidewall. The R means that this is a radial tire. The last number, 16.5, indicates that the wheel is to be mounted on a 16.5-inch diameter rim.

61.3 TEMPORARY USE TIRES

A compact spare tire and wheel assembly is used on many automobiles. This tire is smaller and narrower than the automobile's normal driving tires [Figure 61-9]. It is mounted on a special, smaller wheel, and is to be used only for emergency situations.

Compact spares typically must be inflated to 60 psi to support the weight of the vehicle safely. Some newer compact spares are to be inflated to the same pressure, usually 35 psi, as the normal tires. The correct inflation pressure is molded in inch-high large block letters on the side of the temporary spare.

Another type of temporary tire is the collapsible spare. This tire is deflated, or collapsed, to save space. It is inflated from a supplied canister of compressed air before use.

Figure 61-9. Compact spare tire.

SAFETY CAUTION

Temporary spare tires must *never* be used as permanent substitutes for the normal tires. They can be driven only up to a maximum speed of 50 mph, and for a maximum of only one hour. After one hour of driving, they should be allowed to cool before further driving.

61.4 RETREADED TIRES

Retreaded tires are made from bald, used tire carcasses on which a new tread has been installed. Also called a *recapped tire*, a retreaded tire is less expensive than a new tire, but usually is not as durable. Owners of semi-trucks/trailer combinations often use recapped tires as an economy measure. The large sections of rubber abandoned by the side of most interstate highways are the remains of treads from recapped tires.

Passenger-car recapping should be limited to tire carcasses that are in good condition. The vulcanizing (bonding through heat) process used for recaps is the same as that used on new tires. The difference can occur in the quality of the application. Many unsatisfactory or unsafe recaps originate in poor recapping shops. On the other hand, a good-quality, factory-guaranteed recap from a major company may deliver good service.

Although the original tire casing may still bear a UTQG rating, recaps are not subject to government quality-control testing.

61.5 WHEELS

A wheel must be strong enough to move and support the weight of the car. It must be able to withstand forces caused by hard driving, braking, and engine torque. For tubeless tires, it must be air-tight.

Most passenger car wheels are made of pressed steel. Steel wheels usually consist of a stamped center section welded to an outer section, called the rim [Figure 61-10].

Figure 61-10. Parts of a wheel.

To fit properly and safely with a tire, the inside of a wheel rim has a *drop center* and a *safety rim*. A drop center is a well inside the wheel rim. A safety rim has raised sections on both sides of the drop center [Figure 61-11]. When the tire is inflated, the bead is forced over the raised sections of the safety rim. The raised sections help to hold the tire on the wheel if a loss of pressure occurs.

The wheel fits over studs, or *wheel lugs*, that are pressed into a *wheel hub*. *Lug nuts* thread onto the studs, or lugs, to hold the wheel to the wheel hub. Lugs are tapered on one side to properly index the wheel to the axle. The wheel hub is mounted to the wheel spindle assembly, or axle.

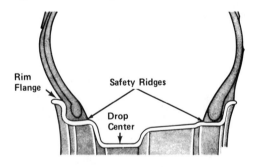

Figure 61-11. Safety rim. CHRYSLER CORPORATION

Some automobiles have custom wheels. Custom wheels usually are aluminum alloy wheels that have been added to improve the appearance of an automobile and to decrease unsprung weight (refer to Unit 57). Stud holes in aluminum wheels are damaged easily, so special nuts and washers often are used to mount these wheels. In many cases, spacers are used to provide a correct fit or proper positioning of the wheel.

UNIT HIGHLIGHTS

- Tire construction and inflation determines how much traction, or frictional contact, exists to keep a vehicle safely in contact with the road surface. A tire cushions the ride of an automobile and provides traction.
- Most current passenger vehicle tires are tubeless.
- Three basic types of tires are used on automobiles—bias-ply, bias-belted, and radial.
- Tire sizing information, UTQG gradings, maximum pressure and load ratings, and DOT serial numbers are molded into the tire sidewall.
- Compact, temporary-use tires and wheels are used as spares on most modern cars.
- A safety rim helps hold a tire on the wheel rim.

TERMS

traction	inner tube
tubeless tire	tire valve
tread	sidewall
carcass	ply
belt	bead
cord	bias ply
radial ply	bias-belted
section width	aspect ratio
retreaded tire	contact patch
recapped tire	drop center
safety rim	wheel lugs
lug nuts	wheel hub
vulcanized	inner liner
Uniform Tire Quality Grading (UTQG)	

REVIEW QUESTIONS

DIRECTIONS: The following questions are similar to those used on automotive technician certification tests. On a separate sheet of paper, write the letter of the correct choice.

1. Technician A says that tires always should be inflated to the maximum pressure molded into the tire sidewall.

   Technician B says that tires should be inflated to the vehicle manufacturer's recommended pressures.

   Who is correct?

   A. A only   C. Both A and B
   B. B only   D. Neither A nor B

2. Which of the following statements is correct?
   I. Plies on a bias-ply or bias-belted tire criss-cross at an angle.
   II. Plies on a radial tire run straight across from bead to bead.
   A. I only
   B. II only
   C. Both I and II
   D. Neither I nor II

3. Which of the following statements about sidewall markings is correct?
   A. The UTQG ratings indicate maximum safe inflation pressure and load.
   B. Section width is the distance between the tread and the bead.
   C. Aspect ratio, or profile, indicates a grading for tread life.
   D. Important information and safety warnings are indicated.

4. Technician A says that you can drive on a temporary spare tire indefinitely, as long as you are careful.
   Technician B says that there are strict limitations on how fast and how long you may drive on a temporary spare tire.
   Who is correct?
   A. A only
   B. B only
   C. Both A and B
   D. Neither A nor B

Question 5 is not like those above. It has the word EXCEPT. For question 5, look for the choice that does *not* apply. Read the entire question carefully before you choose your answer.

5. All of the following statements are correct EXCEPT
   A. most tread patterns are designed to work well on both wet and dry roads.
   B. the carcass of a tire is made up of plies, layers of cloth and rubber.
   C. belts are reinforcing materials that encircle the tire under the tread.
   D. the bead is the decorative pattern at the outer edge of the tread.

ESSAY QUESTIONS

1. Describe tire construction and types of tires.
2. What do UTQG ratings indicate?
3. What does the tire size designation P175/80R13 mean?

SUPPLEMENTAL ACTIVITIES

1. Explain why tire inflation pressures are important.
2. Identify and describe the parts of a tire.
3. Identify and describe the types of materials used for tire body cords and reinforcing belts.
4. Check friends' or relatives' spare tires for correct pressure and type. Report to your class on what you find.

62 Tire and Wheel Service

UNIT PREVIEW

Tire wear is the best indication of suspension, steering, tire inflation pressure, and tire balance problems. Tire and wheel preventive maintenance includes checking and correcting tire inflation pressures, visual inspections, tire rotation, and wheel balancing. Technicians also need to be able to identify suspension, steering, and tire pressure problems by examining treadwear. Service procedures include checks for tire and wheel runout, using a tire machine to remove and mount tires, and repairing tire punctures.

LEARNING OBJECTIVES

When you have completed your assignments and exercises in this unit, you should be able to:

- ❏ Check and adjust tire pressures to the vehicle manufacturer's recommendations.
- ❏ Rotate tires.
- ❏ Diagnose possible steering, suspension, or tire problems by identifying wear patterns.
- ❏ Balance tires.
- ❏ Check tire and wheel runout.
- ❏ Repair tire punctures.

SAFETY PRECAUTIONS

Always wear eye protection when you work on a vehicle. Compressed air, used for air tools and for inflating tires, can cause serious injury if dust, dirt, or contaminants are blown into your eyes.

Raise and safely support vehicles on a hoist or use jack stands.

Never sit on, stand on, or straddle a tire that is being inflated. Use a tire cage and other safety equipment as recommended by your instructor. Serious injuries can result if a tire and/or wheel combination explodes.

Follow the equipment manufacturer's directions and safety precautions when using balancers, tire machines, and all other types of power equipment.

Perform road tests safely and only in a manner to diagnose problems.

TROUBLESHOOTING AND MAINTENANCE

The following preventive maintenance procedures are required to keep tires and wheels in good operating condition. These steps are:

- Checking and correcting tire inflation pressures
- Visual inspection of tread and sidewalls
- Tread wear
- Tire rotation
- Wheel balancing.

Proper Inflation

Maintaining recommended tire pressures is the most important procedure for ensuring long tire life. Because the materials of the tire carcass and liner are porous, tires tend to lose air over time. Each tire may lose air at a different rate. In general, losses of 1/2 to 3 psi per month are considered normal. Low and incorrect tire pressures cause:

- Shortened tire life
- Lower gas mileage
- Increased heat buildup that can lead to blowouts
- Longer stopping distances
- Less traction on turns.

Uneven tire pressures, from side-to-side, front-to-rear, or tire-to-tire can cause a car to swerve or spin during emergency braking. Inflation pressures directly affect the tire's contact with the road, indicated by the letter "C" in **Figure 62-1.**

Recommended tire pressures. Consult the owner's manual, service manual, or stickers on the car for recommended tire pressures. The information on the

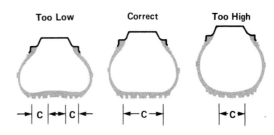

Figure 62-1. Inflation and footprint patterns.

tire sidewall indicates the *maximum* inflation pressure, not the *recommended* pressure.

Typically, recommended tire pressures are slightly less than the maximum pressure. Also be aware that recommended pressures for front and rear tires may be different to improve braking, cornering, and weight-carrying performance.

Too little tire pressure, or **underinflation**, will cause the outer edges of the tread to wear quickly. Underinflated tires can cause hard steering, pulling to one side, tire squeal, front-end shimmy, and other steering and braking problems. Underinflation also can damage sidewalls, plies, and, in extreme cases, wheel rims that come in contact with the road surface.

Excessive tire pressure, or **overinflation**, will cause the center of the tread to wear quickly. Excessive pressure causes a harsh ride and loss of traction and makes tires susceptible to cuts and punctures. Overinflation also can damage the tire carcass and cord. Overinflation also reduces a tire's "footprint," or contact patch, the amount of tire tread that makes contact with the road. This reduces traction.

Uneven pressures among the tires on an automobile can cause the vehicle to wander while driven or to pull or swerve when the brakes are applied.

Checking tire pressure. Check tire pressures when the tires are cold. Remove the cap from the tire valve. Use an accurate tire gauge to check the pressure. All tire pressures should be to the vehicle manufacturer's specifications. Front and rear tire pressures may not be the same due to automotive design or load requirements. Be sure to check the spare tire also. Inflate tires or release pressure as necessary.

CAUTION!!!

Always check inflation pressures when the tires are cold. Tire pressures can increase considerably as heat builds up during driving. Never lower the pressure of a hot tire, because pressure will drop when it cools. If the automobile has been driven a long distance, wait about three hours before checking inflation.

If air is needed, use a compressed-air hose with the proper adapter. If air is to be removed, press in on the valve stem until the desired pressure is reached. Replace the valve cap.

Visual Inspection

Tires must be checked for unusual wear and damage regularly. Unusual wear can indicate specific problems with inflation pressures, wheel alignment and balance, and worn suspension and steering parts. Ideally, all tires should wear in a flat and even manner across and around the circumference of the tread.

Visually check for badly worn or bald tires, damage or cuts in the sidewall and tread, and nails or other items stuck in the tire. Check the tire valve for cracks and damage.

Inspect wheels for damage and cracks. Make sure that wheel lugs are tight and that none are missing. Also look for clean spots on the metal wheel that can indicate lost balance weights.

Tread Wear

Automobile problems, road conditions, and poor driving habits can show up on tire treads. Tread wear can serve as a diagnostic "chart" for a technician [**Figure 62-2**]. Close your eyes and run your fingertips in and out across the tread. Then move

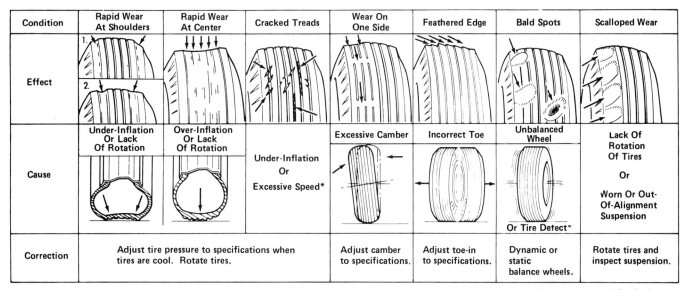

Figure 62-2. Tire tread wear chart. CHRYSLER CORPORATION

*Have tire inspected for further use

your fingertips around the outer circumference of the tread. Feel carefully for uneven wear that is not yet visible.

New tires have a *tread wear indicator* molded into the tire tread. A tread wear indicator, shown in **Figure 62-3,** alerts the owner that only 1/16 in. of tread remains. In most cases, this is the legal limit of tread wear. The tire should be replaced when the indicator bands become visible.

Tire Rotation

Tires should be rotated approximately every 5,000 to 6,000 miles (approximately 9,600 km) or whenever uneven tire wear is noticed. Greater tire wear may occur at the *traction wheels,* or driving wheels. On front-wheel drive vehicles, the front tires, which both drive and steer, wear most quickly.

Tire rotation means moving tire-and-wheel assemblies to different locations on an automobile to even out wear and increase tire life. **Figure 62-4** shows typical tire rotation patterns. This rotation sequence allows the tires to rotate in the same direction at all times. Some manufacturers may recommend other tire rotation patterns. Only normal sized spare tires can be used in rotation. Compact and collapsible spare tires cannot be used for tire rotation.

See Topic 62.2 for removing and replacing tire-and-wheel assemblies. Always check and correct tire pressures after rotation.

CAUTION!!!

Wheel lug nuts or bolts should be tightened to the manufacturer's recommended torque specifications. Lugs that are too loose or too tight can cause damage to metal wheels, brake parts, tires, and suspension parts, and also may result in unsafe vehicle conditions. Use a torque wrench and tighten lugs in a crisscross pattern, not sequentially in a circular pattern.

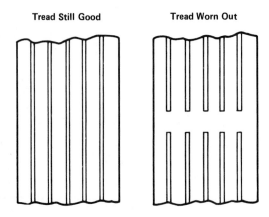

Figure 62-3. Tread wear indicator. AMERICAN MOTORS CORPORATION

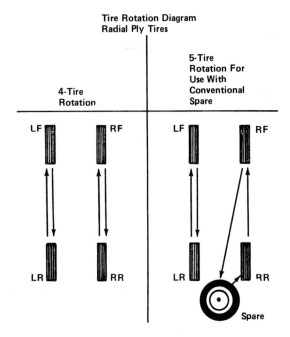

Figure 62-4. Tire rotation patterns. FORD MOTOR COMPANY

Wheel Balancing

When tires are manufactured, the carcass is not perfectly even in all areas. Heavier spots exist in the tire. Centrifugal force will lift the heavy spot upward as the tire rotates. This will cause the tire to bounce and vibrate, which can result in abnormal tire wear, shock absorber damage, and suspension and steering damage.

To balance a tire, an equally heavy weight is placed opposite the heavy spot of the tire, on the metal wheel rim. Centrifugal force then acts on both the heavy spot and the balance weight to balance and eliminate bouncing.

Typically, tire balancing is done when new tires are mounted, when punctures are repaired, or whenever a balancing weight has been lost or removed.

There are two types of tire and wheel balance: static and dynamic. *Static balance* is the equal distribution of weight around the circumference of a wheel **[Figure 62-5].** A statically unbalanced tire and wheel will hop up and down, or *tramp,* as it rotates.

Dynamic balance is the equal distribution of weight on each side of the vertical centerline of a wheel **[Figure 62-6].** A dynamically unbalanced tire and wheel will *shimmy* from side to side at medium and high driving speeds.

The most common form of dynamic and static wheel balancing is done with the wheel and tire are removed from the car and mounted on an electronic balancer. The balancer spins the assembly and identifies the heavy spots and the amount of weight needed to correct the imbalance. The technician clips a lead weight to the rim at a location opposite that of the heavy spot **[Figure 62-7]** or uses adhesive

TIRE AND WHEEL SERVICE

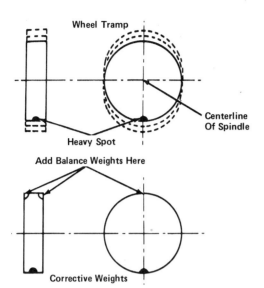

Figure 62-5. Static balance. FORD MOTOR COMPANY

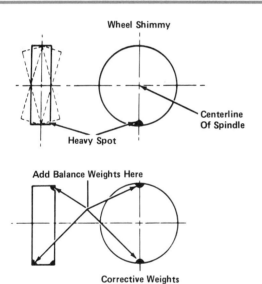

Figure 62-6. Dynamic balance. FORD MOTOR COMPANY

Figure 62-7. Adding lead weights for balance.

weights on custom wheels. Follow wheel balancer manufacturers' instructions for best results.

62.1 DIAGNOSING TIRE PROBLEMS

As discussed above, tread wear is the best indication of steering and suspension problems. Check inflation pressures and correct them if necessary. Raise the car safely and rotate tires by hand to visually inspect for treads that are not centered on the tire carcass, bulges or bubbles in the tire, and objects stuck in the tread or sidewall.

Some noises result from tire design. Tires with wide spaces between treads, such as all-weather tires or snow tires, can be noisy. The road surface also can cause unusual noises.

Driving Problems

Information from the driver and making a test drive also can help to troubleshoot a problem. Ask the driver about specific driving conditions such as, when the noise or problem occurred, how the vehicle was loaded, and so on. To troubleshoot the problem, you must try to get specific information about what the vehicle did and under what conditions.

Most vibrations at highway speeds are caused by tire and wheel problems. An unbalanced wheel assembly, or improper tire or wheel runout, will cause vibration.

Tire and Wheel Runout

Excessive radial and lateral runout of a tire or wheel can cause excessive noise, vibration, and wear. *Radial runout* is a measurement that determines how much the wheel assembly is out of round. *Lateral runout* is a measurement of the in-and-out movement, or wobble, of the assembly.

Make runout measurements with a dial indicator while the wheel assembly is on the automobile. Inflate the tires properly and torque the wheel bearings to the manufacturer's specifications.

The automobile must be driven approximately seven miles (11.3 km). This will eliminate false readings caused by temporary flat spots in the tires.

Tire runout is measured at the points indicated in **Figure 62-8**. Use chalk to mark the runout high points on the tire. If runout exceeds specifications, check wheel runout to determine whether the tire or the wheel is the problem.

Wheel runout is measured at the points indicated in **Figure 62-9**. Mark the runout high points. If wheel runout is not within specifications, install a new wheel. If tire runout exceeds specifications and wheel runout is satisfactory, relocate the tire on the rim. Deflate the tire and turn it on the wheel. Position the high-point mark of the tire 180 degrees from its original location on the rim. If tire runout cannot be reduced to specifications, install a new tire.

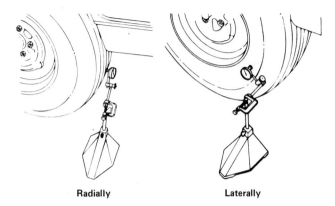

Figure 62-8. Tire runout measurement locations.
AMERICAN MOTORS CORPORATION

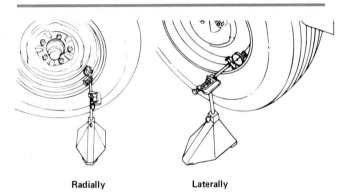

Figure 62-9. Wheel runout measurement locations.
AMERICAN MOTORS CORPORATION

62.2 TIRE AND WHEEL SERVICE

Tire and wheel service includes removing wheel/tire assemblies, removing tires from rims, checking tires for problems, repairing punctures, and checking wheels.

Removing Wheels

Remove wheel covers before the automobile is lifted on a hoist or jack. Support the automobile securely. Place a wheel cover or hubcap near each wheel. Loosen the lug nuts with a lug wrench or an impact wrench.

NOTE: Some 1970s and earlier automobiles, especially Chrysler products, have *left* and *right* hand threads. A left hand thread tightens to the *left*, or counterclockwise. A right hand thread tightens to the right, or clockwise. On these vehicles, the ends of the studs will be marked "L" and "R."

Remove the lug nuts and place them inside the wheel cover or hubcap near each wheel to prevent loss. Remove the wheels.

Removing Tire From the Rim

If the tire is to be remounted after it is serviced, make chalk or crayon marks across the sidewall and the rim. The tire should be replaced as close as possible to its original position to make wheel rebalancing easier.

Unscrew and remove the valve core to release air pressure. Do not try to break the bead loose from the wheel rim until the tire is deflated completely. Always inspect the tire valve stem for cracks or damage.

Using a Tire Machine

Follow the equipment and vehicle manufacturer's directions for using a tire machine. Mount the tire on the center shaft with conical adapters. Use a soap-and-water or special lubricating solution to coat the tire next to the wheel rim. The solution reduces friction so that the tools can be forced under the bead more easily.

CAUTION!!!

Special adapters may be required to mount cast aluminum, "mag," and chromed wheels. The machine will bend, break, or scratch these wheels if you do not use the correct adapters.

An air-powered clamp on the machine breaks the bead loose. Special tire irons are mounted over the shaft in the middle, which turns when the technician steps on a foot switch. The special tire iron is forced around and under the bead so that the sidewall can be lifted over the tire rim, as shown in **Figure 62-10**.

Remove the tire from the wheel. Repair or replace the tire as described below.

Before installing the tire, remove all rust scale from the wheel rim. Repaint the wheel, if necessary. Install a new valve stem, if necessary, and valve core. Coat the wheel rim and tire beads with the lubricant for easier mounting. If necessary, align the tire and wheel chalk or crayon marks that were made before removal.

Install the tire, using the special tire iron. Clip the air hose onto the valve stem. Push and pull upwards and downwards on the tire until the air begins to fill and expand the tire on the rim. When the beads "pop" into the safety rims, inflate the tire to approximately 40 psi to make sure it is seated. After it seats, reduce tire pressure to the vehicle manufacturer's recommended pressure.

CAUTION!!!

Do not stand on or over a tire during inflation. Excessive pressure can cause a tire to explode and

TIRE AND WHEEL SERVICE

Figure 62-10. Tire-changing machine.

cause serious injury. Use a tire cage and safety equipment as directed by your instructor.

Balance the tire and wheel assembly. Place the wheel on the hub, and tighten the lugs to the vehicle manufacturer's recommended torque. Replace the wheel cover.

Repairing Punctures

Most punctured tires should be removed from the wheel. Repairs usually are recommended from the inside of a tire, using a combination repair plug and patch.

Only punctures in the tread area are repairable. Never try to repair punctures in the shoulders or sidewalls. Never try to repair a tire with the following types of damage:

- Bulges or blisters
- Ply separation
- Broken or cracked beads
- Fabric cracks or cuts
- Tires worn to the fabric
- Visible wear indicators
- Punctures larger than 1/4 inch (6.35 mm).

Many tire manufacturers do not recommend using externally applied repair plugs, blowout patches, and aerosol sealants except for emergency repair.

CAUTION!!!

If a tire is to be replaced, manufacturers usually recommend replacement with the same type of tire. The replacement tire should be of the same size, load range, and type of construction as the original tire. Never mix radial and bias-ply or bias-belted tires on the same axle. Different types or sizes of tires on opposing wheels can result in damage to the powertrain, suspension, and body. If there are only two radial tires, always mount them on the rear wheels.

Plug repair. A permanent *tire plug* has a flat plug head at one end [Figure 62-11]. The narrow end of the plug is inserted into the puncture. The head adheres to the inside of the tire. Other types of plugs are recommended only for emergency use.

To install a tire plug, remove the object that caused the puncture. Clean the puncture. Use a buffing tool to roughen up the tire's inner surface where the plug head will make contact. The roughened surface gives the adhesive a better gripping surface. Apply the recommended sealant to the plug and to the plug head. Follow the instructions given by the tire plug manufacturer for exact installation procedures.

Cold patch repair. A tire repair patch is placed over a puncture on the inside of a tire. The area around the puncture should be cleaned and buffed. The proper adhesive should be spread around the puncture, and the patch should be placed over the puncture. A stitching tool that makes small indentations is run over the patch to help bind, or stitch, the patch to the tire.

Only special patches approved for radial tires can be used with radials. When using these patches, be sure the arrow on the patch is parallel to the radial plies.

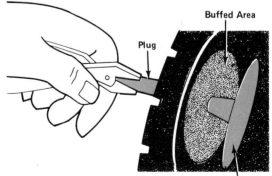

Figure 62-11. Plug for a radial tire.

Hot patch repair. A tire hot patch is applied much as a cold patch. After cleaning and buffing, clamp the hot patch over the puncture. Light the flammable material on the back of the patch. Allow it to burn completely, then let it cool thoroughly. Remove the clamp.

Tube repair. Cold or hot patches are used to repair punctures in tubes. Since the tube holds the air, it is not necessary to seal a small puncture in the tire.

Puncture-sealing tires. Some tires are made with a sealing substance on the inside of the tread area. If a nail punctures the tread, the nail can be removed without loss of air. The sealing compound will plug the puncture from the inside **[Figure 62-12]**.

Checking Wheels

Inspect the wheels for bending, dents, elongated bolt holes, and heavy rusting. Check for excessive runout and for air leaking through the welds. If any of these conditions is present, install a new wheel. Manufacturers do not recommend repairing wheels.

Replacement wheels should be equivalent to the original wheels. Check wheel load capacity, diameter, rim width, lug pattern, and *offset.* Offset is the measurement between the mounting point of the center section and the center of the rim. Offset is especially critical on front-drive cars. **Lug pattern**, or **bolt pattern**, refers to the spacing and position of holes in the wheel that fit over the wheel studs.

CAUTION!!!

Wheels and tires that are not of the same size or design as original equipment units can cause problems. They may decrease wheel bearing life, brake cooling, change speedometer readings, and affect ground clearance and the clearance between the tire and chassis or body.

Servicing Compact Spare Tires

A compact spare tire can be mounted and removed from its special wheel rim with the procedures discussed above. Although most compact spare tires are meant to be inflated to 60 psi, the beads will seat at 40 psi. This seating pressure is comparable to regular-size tires. Compact spare tires can be repaired with the same materials used on normal tires.

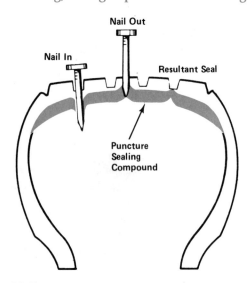

Figure 62-12. Puncture-sealing tire. BUICK MOTOR DIVISION—GMC

UNIT HIGHLIGHTS

- Inflation affects tire wear and traction qualities.
- Visual inspection can be used to diagnose suspension, steering, and tire defects.
- Tire rotation is performed regularly to increase tire life.
- Wheel balancing corrects an imbalance caused by a heavy spot.
- Tire problems often show up as vibrations.
- A tread wear indicator becomes visible when tires are worn excessively.
- Tire and wheel runout are checked with a dial indicator.
- Lug nuts on wheels may have right- or left-handed threads.
- Tires are mounted and replaced on a tire machine.
- Small damaged areas in the tread of the tire can be repaired.
- Tire repairs are made with plugs and patches.
- A new wheel and/or tire may be required to correct a defect.

TERMS

underinflation	shimmy
overinflation	tire rotation
traction wheels	tramp
static balance	tread wear indicator
dynamic balance	lateral runout
radial runout	offset
tire plug	bolt pattern
lug pattern	

TIRE AND WHEEL SERVICE

REVIEW QUESTIONS

DIRECTIONS: The following questions are similar to those used on automotive technician certification tests. On a separate sheet of paper, write the letter of the correct choice.

1. Technician A says that tire rotation is done to even out wear on the front and rear tires.
   Technician B says that tire wear is always greatest on the rear wheels.
   Who is correct?
   A. A only
   B. B only
   C. Both A and B
   D. Neither A nor B

2. Which of the following statements about tires is correct?
   A. Underinflated tires wear more quickly at both outside edges.
   B. Recommended tire pressures are molded into the sidewall.
   C. Excessive camber causes a feathered edge on the tread.
   D. Scalloped wear is caused by incorrect toe settings.

3. Which of the following statements is correct?
   I. Lateral runout on a tire causes an in-and-out wobbling movement.
   II. Radial runout on a tire causes a side-to-side squirming movement.
   A. I only
   B. II only
   C. Both I and II
   D. Neither I nor II

Questions 4 and 5 are not like those above. Each has the word EXCEPT. For each question, look for the choice that does *not* apply. Read each question carefully before you choose your answer.

4. All of the following statements are correct EXCEPT
   A. recommended tire pressures are often lower than maximum pressures.
   B. recommended tire pressures for front and rear tires may be different.
   C. recommended tire pressures are molded into the sidewall of the tire.
   D. tire inflation pressure directly affects traction.

5. All of the following statements are true EXCEPT
   A. wheel balancing is required only when new tires are installed.
   B. a tread puncture that is less than 1/4 inch in diameter is repairable.
   C. for best results, punctures should be repaired from inside the tire.
   D. a cold patch should be stitched to the inner surface of the carcass.

ESSAY QUESTIONS

1. What procedures are part of preventive maintenance for wheels and tires?
2. Describe how to diagnose tire problems.
3. What procedures are part of tire and wheel service?

SUPPLEMENTAL ACTIVITIES

1. Check and correct all tire pressures, including the spare, on a vehicle selected.
2. Perform a tire rotation.
3. Balance a wheel.
4. Make runout readings on a tire and wheel assembly furnished by your instructor.
5. Remove a tire from a wheel.
6. Repair a tire, replace it on the wheel, and inflate it to the proper pressure.

63 The Brake System

UNIT PREVIEW

A brake system slows and stops an automobile. Brakes are operated by hydraulic pressure and act at each wheel. Most passenger vehicles have disc brakes on the front wheels and drum brakes on the rear wheels. All vehicles with disc brakes have a power assist device to reduce brake pedal effort. The parking, or emergency, brake is operated by mechanical linkage.

Computerized anti-lock brake systems (ABS) include an additional hydraulic control unit to modify brake system fluid pressures. ABS systems prevent wheel lockup and increase driver control.

LEARNING OBJECTIVES

When you have completed your assignments and exercises in this unit, you should be able to:

❑ Describe how friction is used to slow and stop a vehicle.
❑ Identify and describe the parts of a basic brake system.
❑ Describe how disc and drum brakes operate.
❑ Locate, identify, and describe the functions of metering and proportioning valves in a combination disc and drum brake system.
❑ Describe how a power brake system works.
❑ Describe the function, parts, and operation of an anti-skid brake system (ABS).
❑ Describe the parts and function of the parking, or emergency, brake.

63.1 BRAKE SYSTEM DESIGN

A brake is a device that uses *friction* to slow or stop a mechanism. Friction is the resistance to motion that occurs when two parts rub against each other. An automotive brake system has brakes at all four wheels. The brakes apply friction to parts that rotate with the wheels.

Friction

Three factors determine the amount of friction produced by a brake unit: *pressure, friction material,* and *surface area.* To see how this works, grasp a fairly heavy object, such as a large wrench, just in your fingertips. Held too loosely, the wrench will begin to slip from your grasp. If you do any one or all of the following, you can stop the wrench from slipping:

• Increase the pressure of your grip
• Hold the wrench through rough leather gloves
• Press the palms of both hands against the wrench.

Friction causes heat. An automobile's braking system must be able to withstand extremely high temperatures and still provide braking action.

Automotive Braking Friction

The braking friction of an automobile is created when force from the brake pedal is multiplied and transmitted to the wheels as hydraulic pressure. Pressure moves non-rotating friction materials into contact with a rotating surface connected to the wheel. When these two surfaces make contact, friction is created. This friction slows and eventually stops the rotation of the wheel. The kinetic energy of the moving automobile is changed to heat within the brakes.

Hydraulic Principles

Pressure is transmitted and multiplied through hydraulic principles, like those used in an automatic transmission. To simplify, liquids cannot be compressed. Therefore, if force can be applied to fluid inside a closed system, the fluid will transfer motion, pressure, and force. This will work if there are no gases or bubbles in the liquid. Gases, unlike liquids, can be compressed, and force will not be transmitted.

When the driver steps on the brake pedal, force from the brake pedal operates a piston inside a fluid-filled cylinder **[Figure 63-1]**. This unit is called a *master cylinder,* and acts as a plunger pump. The movement of the piston creates hydraulic pressure and forces brake fluid to flow into fluid-filled hoses and steel tubing, called **brake lines.**

At the ends of the brake lines are two types of cylinders with pistons inside them, called disc brake *wheel calipers* and drum brake *wheel cylinders.*

Hydraulic pressure creates force against the pistons. The pistons move and force brake linings into contact with rotating metal brake discs and drums.

The Brake System

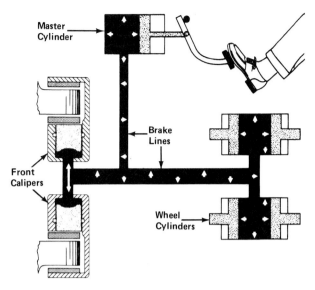

Figure 63-1. Automotive hydraulic system.
CHEVROLET MOTOR DIVISION—GMC

Friction slows the rotating motion of the discs, drums, and the wheels to which they are connected.

63.2 BRAKE SYSTEM CONSTRUCTION

A basic disc/drum combination brake system is shown in **Figure 63-2**. The parts include:

- Brake pedal, linkage, and brake light switch
- Brake fluid
- Power brake booster
- Master cylinder and fluid reservoir
- Brake lines
- Disc brake calipers
- Drum brake wheel cylinders
- Friction materials
- Disc brake rotors and brake drums

Brake Fluid

Friction between the brake linings and metal rotors or drums heats the wheel brake units and the *brake fluid*. If the heat becomes great enough, the brake fluid will boil. Gas bubbles in boiling brake fluid will reduce or eliminate the transfer of pressure, motion, and force, as explained above. Thus, if the brake fluid boils, the brake system will not work.

Brake fluid is glycol, a liquid related to alcohol and ether, and is *not* an oil. The piston seals used inside the master cylinder, disc brake calipers, and wheel cylinders are destroyed by petroleum oil. If the seals fail, the brake system will not work.

The brake fluid used in most vehicles is known as Department of Transportation class 3 (DOT 3) brake fluid. Uncontaminated DOT brake fluid will not boil until it reaches 384 degrees F. However, brake fluid absorbs water vapor from the air. Water boils at only 212 degrees F. Brake fluid contaminated by moisture will boil at a lower temperature. In addition, the moisture will cause rust and corrosion in the brake hydraulic system parts and lines.

Master Cylinder

The master cylinder creates hydraulic fluid pressure. Above the master cylinder is a reservoir of brake fluid [**Figure 63-3**].

For safety purposes, the master cylinder has two pistons so that two separate fluid chambers and plunger pumps are created.

One chamber provides hydraulic fluid and pressure either to the front brakes alone or to a combination of one front and one rear brake. The other chamber provides hydraulic fluid and pressure either to the rear brakes alone or another front-rear brake combination. If the brake lines or cylinders from one chamber fail, the automobile still can be stopped.

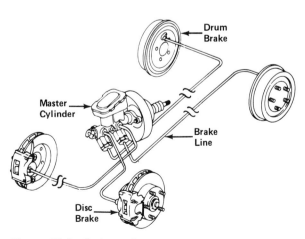

Figure 63-2. Brake system.
DELCO MORAINE DIVISION—GMC

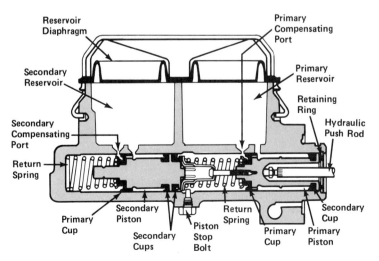

Figure 63-3. Parts of a master cylinder. BENDIX CORPORATION

The force applied by the driver to the brake pedal is multiplied in three ways:

1. By leverage in the brake pedal linkage
2. By hydraulic principles (see Unit 53)
3. By a power assist unit.

The force is applied through a pushrod to the rearmost, or primary, piston of the master cylinder. Each of the two pistons has two sealing devices, called "cups." The front seal is called the primary cup, and helps the piston to develop hydraulic pressure in the cylinder. The back seal is the secondary cup, which prevents fluid leaks **[Figure 63-3]**. In the illustration, the brake pedal is to the right and applies force to the hydraulic pushrod.

The *primary piston* is moved by the hydraulic push rod. As it moves forward, its primary cup covers the primary compensating port from the reservoir above. This traps fluid between the primary cup and the rear of the secondary piston (at left in the illustration). The fluid creates pressure which forces the secondary piston to move forward.

As the *secondary piston* primary cup covers the secondary compensating port, fluid also is trapped in front of the secondary piston. Further pedal movement builds pressure in both primary and secondary pressure chambers and causes brake fluid to flow out into the brake lines. This flow results in brake application, as outlined above.

When the brake pedal is released, expanded springs and other parts on the wheel brake units contract. Fluid flows back up the lines to the master cylinder. Return springs inside the master cylinder move the pistons rearward (to the right in **Figure 63-3**) to a "home" position. This action draws additional fluid into the areas in front of the primary cups so that the brakes can be applied immediately again if needed.

As explained previously, the brake fluid temperature rises as the wheel brake units create friction and heat. The heat causes the brake fluid to expand. When the pedal is released and the pistons move back, the primary cups uncover the compensating ports, or holes into the reservoir above. This allows the expanded volume of fluid to flow upward into the reservoirs.

Brake Lines

The brake lines include both steel tubes and reinforced flexible lines. Steel tubing carries brake fluid from the master cylinder along the chassis. Flexible, reinforced hoses connect the steel tubing to the front wheel brake units to allow movement. One or more flexible lines also are used at the rear to allow for movement of the rear wheels.

Brake fluid flows under pressure from the master cylinder through the brake lines to wheel brake units when the driver depresses the brake pedal. When the driver releases the brake pedal, fluid flows back up into the master cylinder.

Dual system brakes. Since 1967, all passenger vehicles sold in the United States have been equipped with *dual system brakes.* If one half of the system fails due to a loss of pressure, the other half continues to operate. When pressure drops in either part of the system, a dashboard warning light operates.

Separate pistons within the master cylinder provide pressure to operate each half of the brake system. Two patterns of brake line routing are used to divide the brake system in half **[Figure 63-4]**:

- Front and rear system
- Dual-diagonal system.

Front and rear system. In this system, one piston in the master cylinder supplies pressure to both front wheel brakes. The other master cylinder piston supplies pressure to both rear wheel brakes.

Dual-diagonal system. In a dual-diagonal system, one master cylinder piston operates one front wheel brake and the diagonally opposite rear wheel brake. The other piston operates the other front wheel brake and diagonally opposite rear wheel brake.

Wheel Cylinder and Disc Brake Caliper Cylinder

Hydraulic cylinders at each wheel receive fluid and pressure from the master cylinder. A hydraulic cylinder is located at each drum brake assembly **[Figure 63-5]**. A similar hydraulic cylinder is cast and machined into a *disc brake caliper*. Rubber cups in drum brake wheel cylinders and piston seals in disc brake calipers seal the pistons inside the cylinder bores.

Friction Materials

Conventional *organic* brake linings are made of compressed asbestos, soft metal chips, friction modifiers, and a binder material. *Brake linings* are formed to fit specific disc pad backing plates and curved drum brake "shoes."

SAFETY CAUTION

Asbestos dust can cause lung cancer. As asbestos brake linings rub against brake discs or drums, a fine dust is created. If inhaled, severe lung damage can occur. Ordinary breathing masks will not filter out asbestos particles. Wear a breathing mask specifically designed for use with asbestos fibers. Always use an approved vacuum cleaner or liquid collection system to remove asbestos fibers before you work on brake

The Brake System 699

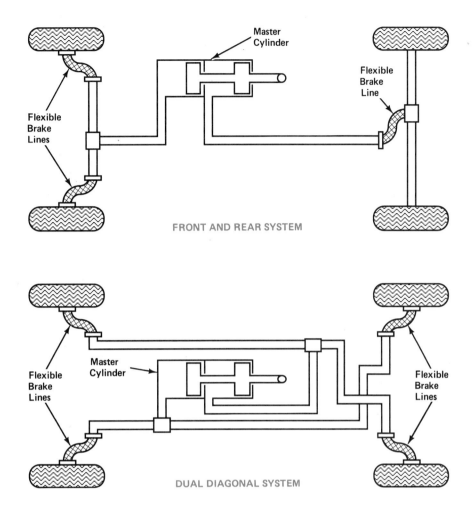

Figure 63-4. Dual system brake line routing.

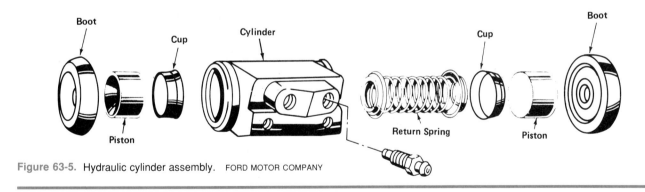

Figure 63-5. Hydraulic cylinder assembly. FORD MOTOR COMPANY

units. Do *not* use compressed air to blow dust off of brake assemblies.

To reduce the unacceptable health hazards posed by asbestos dust, vehicle manufacturers equip virtually all new passenger vehicles with *semi-metallic* linings. Semi-metallic brake linings are made of steel fibers, coarse sponge iron, graphite, friction modifiers, and resin binders. These linings have excellent wear characteristics and also help to dissipate heat, through conduction, away from the brake disc rotors.

Also, replacement linings made of fiberglass and of synthetic materials such as Aramid® and Kevlar® are now used. These brake lining materials are made for both original equipment and replacement use.

63.3 DISC BRAKES

Disc brakes work in the same manner as hand brakes on a bicycle. Pads of friction material are forced against a rotating surface by a caliper. Disc brakes produce good stopping force, dissipate heat well, and are relatively unaffected by water. For these reasons, disc brakes are used on the front of vehicles.

A disc brake system has a metal disc, called a *rotor*, that is mounted on the axle. The rotor rotates with the wheel. A U-shaped gripping mechanism, or *caliper*, is mounted over the disc. The caliper is mounted on the steering knuckle. Brake *pads*, or shoes, are mounted inside a caliper to grip the rotor. Some pads have "wear indicators" which scrape against the edge of the rotor and create noise when the pad lining is worn excessively. A disc brake assembly is shown in **Figure 63-6**.

A ventilated rotor has holes, or vents, between the outer frictional surfaces. The vents help to dissipate heat more quickly. In some cases, the vents are designed like the blades of a fan to force air through and cool the discs. Thus, some rotors must be mounted on either the right side or the left side of the vehicle. Solid discs, without vents, also are used on smaller vehicles.

As outlined above, the caliper includes a hydraulic cylinder. Hydraulic pressure forces pistons to move in the caliper bore. The pistons act to push against the disc brake pads to force them against the rotor to create friction.

During brake application, a square-cut caliper piston seal is stretched and distorted. When the brakes are released, the seal returns to its original position and retracts the brake shoes slightly. In addition, lateral runout (side-to-side motion) of the brake disc also helps to push the pads away from the rotor. Three common designs of disc brake systems are used:

- Fixed caliper
- Floating caliper
- Sliding caliper.

A *fixed caliper* is a brake design in which the caliper is mounted solidly to the steering knuckle. Two or four pistons—one or two on each side of the caliper—operate against the disc brake pads, or friction material assemblies **[Figure 63-7]**.

The two most widely used designs are the sliding and floating caliper. In these designs, there is only one piston. Sliding and floating calipers work in a manner similar to the action of a C-clamp.

The piston moves inward and pushes one pad against the disc. This action forces the other side to move and push the other pad against the disc.

A *floating caliper* assembly slides on bushings and pins **[Figure 63-8]**. The housing of a *sliding caliper* slides in and out **[Figure 63-9]**.

As the pads wear against the disc, the friction materials become thinner. Fluid pressure behind the piston repositions it closer to the disc. Fluid flows from the master cylinder downward to fill the additional space behind the piston. Thus, disc brakes require no manual adjustment to keep linings close to the disc rotor. But remember that a low fluid level in the disc brake fluid reservoir can indicate excessively worn disc brake pads.

63.4 DRUM BRAKES

Drum brakes are used on the rear of most vehicles. A *brake drum* is mounted on an axle between the end of the axle and the wheel. When the brakes operate, two semi-circular pieces of metal covered with friction material, push against the hollow inner surface of the drum. These curved metal pieces are called *brake shoes*. The brake shoes are pressed against the inside of the drum by a wheel cylinder. Drum brake parts are shown in **Figure 63-10**.

A *backing plate* is mounted to the steering knuckle or axle housing and prevents the brake shoes from rotating with the drum during braking. Two brake shoes are used: a *primary brake shoe* and a *secondary brake shoe*. The primary shoe is positioned toward the front of an automobile. The secondary shoe is positioned toward the rear.

The brake friction linings are riveted or bonded (glued) to the brake shoes. A wheel cylinder is mounted at the top of the backing plate, between the two brake shoes. Heavy return springs pull the brake shoes inward toward the wheel cylinder.

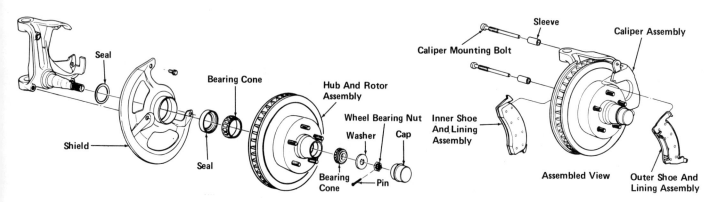

Figure 63-6. Disc brake assembly. CHEVROLET MOTOR DIVISION—GMC

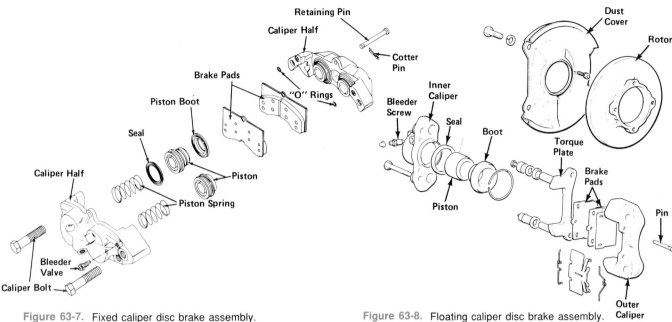

Figure 63-7. Fixed caliper disc brake assembly.
CHEVROLET MOTOR DIVISION—GMC

Figure 63-8. Floating caliper disc brake assembly.
CHRYSLER CORPORATION

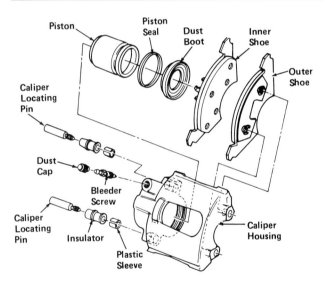

Figure 63-9. Sliding caliper disc brake assembly.
FORD MOTOR COMPANY

A spring and adjuster screw holds shoes together at the bottom. Hold-down springs secure the shoes against the backing plate.

As the brake pedal is depressed, pistons in the wheel cylinder force the shoes to move outward and contact the drum. When the brake pedal is released, the brake shoe return springs pull the shoes back to their original positions. This return movement also compresses the wheel cylinder pistons to push brake fluid back to the master cylinder.

Self-Adjusting Drum Brakes

Self-adjusters are used on many modern drum brake systems to maintain the proper clearance between the brake shoes and drum automatically.

Two types of self-adjusting mechanisms are commonly used. One type is operated by the motion of the brake shoes when the car is driven in reverse and stopped. Another type is operated when the parking brake lever is pulled up. Both types use levers to turn an adjuster screw between the bottoms of the brake shoes to spread them farther apart. This action reduces the clearance between the brake drum and the surface of the shoe linings.

Self-Energizing Drum Brakes

Some drum brakes are designed to be *self-energizing*. Self-energizing means that additional mechanical force is created by the braking action. This also means that less foot pressure is required to operate the brake pedal. Most rear-wheel drive automobiles have self-energizing drum brakes at the rear [**Figure 63-11**].

As the primary shoe contacts the rotating drum, it pivots downward and rearward. The primary shoe acts as a pivot point, and has a shorter lining than the secondary shoe. The force from the pivoting motion of the primary shoe is transferred to the secondary shoe through the adjusting screw mechanism.

At the same time, the secondary shoe is forced upward and forward, and stopped by an *anchor pin* at the top of the backing plate. This forces the secondary shoe outward and toward the rotating drum with great force.

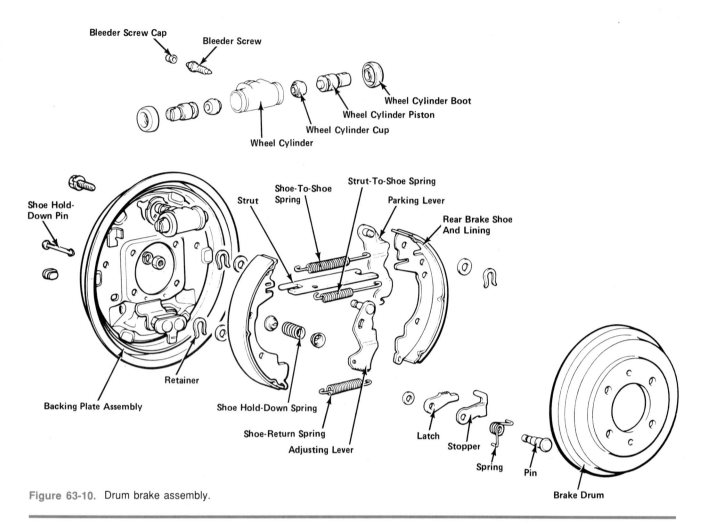

Figure 63-10. Drum brake assembly.

As the brake shoes rub against the drum, they wear down. As the distance between the friction material and the drum increases, the wheel cylinder must move out farther to operate the drum brake. The result is that the driver has to push the brake pedal farther to make the brakes work. Some drum brakes do not have self-adjusting mechanisms. A technician must adjust these brakes by turning the adjuster screw by hand to spread the brake shoes farther apart and closer to the inner surface of the brake drum.

63.5 DISC AND DRUM BRAKE COMBINATIONS

As described previously, most automobiles use a combination of disc brakes at the front and drum brakes at the rear. Because the front brakes must do 75 to 80 percent of the work of stopping the vehicle, more efficient disc brakes are used on the front wheels.

A disc brake dissipates heat better than a drum brake. Only 20 to 25 percent of the braking force is applied at the rear wheels as the rear end lifts under heavy braking. For this reason, and also because a parking, or emergency, brake can be made to operate the rear drum brakes more easily, drum brakes are used on the rear wheels. Because the two brake systems operate differently, control valves are used for proper braking action.

Proportioning Valve

Disc and drum brakes require different operating pressures. Disc brakes are operated at higher pressures. To furnish proper hydraulic pressures to each system, a *proportioning valve* is used. A proportioning valve is placed between the master cylinder and rear brakes **[Figure 63-12]**. The proportioning valve limits maximum pressure to the rear drum brakes during hard braking, effectively preventing rear-wheel lockup.

Metering Valve

A *metering valve* delays hydraulic pressure to the disc brakes. Disc brakes react quicker to braking action. The delay allows both disc and drum systems to operate simultaneously **[Figure 63-13]**.

Combination Valve

A *combination valve* combines a proportioning valve, metering valve, and *failure warning switch* into one

THE BRAKE SYSTEM

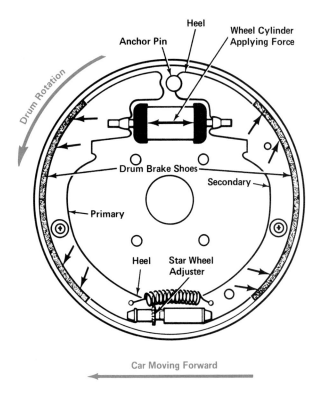

Figure 63-11. Self-energizing drum brake assembly.

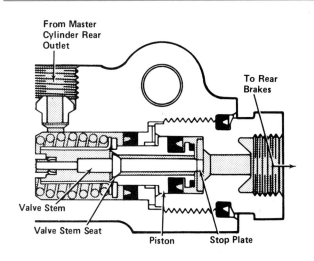

Figure 63-12. Proportioning valve.
CHEVROLET MOTOR DIVISION—GMC

unit [Figure 63-14]. A failure warning switch activates a warning light on the dashboard when a loss of front or rear brake pressure occurs. In most cases, this is the same warning light used to alert the driver that the emergency brake is applied.

Residual Check Valve

Drum brakes require a small amount of residual, or remaining, pressure to minimize leaking at the wheel cylinders. This is accomplished through a *residual check valve*. A residual check valve usually is located in the master cylinder outlet port.

63.6 POWER BRAKES

A power assist is required to provide the operating pressures needed for disc brake operation. A *power brake booster unit* is used between the brake pedal and master cylinder. The unit supplements and increases the driver's foot pressure to reduce braking effort. A power brake system is operated by either vacuum or hydraulic pressure.

Vacuum Power Brakes

Most power brakes use engine vacuum and atmospheric pressure to create force to move the master cylinder plunger. A brake booster unit contains a large diaphragm that separates the housing into two chambers [Figure 63-15].

Two types of units are used. In one type, vacuum is applied to both sides of the diaphragm (vacuum suspended). When the brake pedal is depressed, atmospheric pressure enters the rear chamber and forces the diaphragm and master cylinder pushrod forward.

In the other type, both sides of the diaphragm are open to atmospheric pressure (air suspended). When the brake pedal is depressed, vacuum is applied to the front chamber. As in the first type, atmospheric pressure forces the pushrod forward.

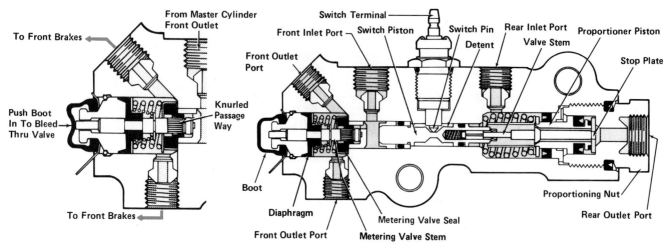

Figure 63-13. Metering valve. CHEVROLET MOTOR DIVISION—GMC

Figure 63-14. Combination valve. CHEVROLET MOTOR DIVISION—GMC

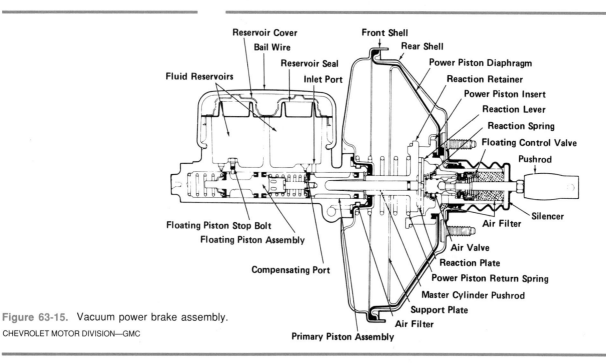

Figure 63-15. Vacuum power brake assembly.
CHEVROLET MOTOR DIVISION—GMC

Hydraulic Pressure Power Brakes

Power steering fluid pressure is used to assist braking on some General Motors and foreign vehicles. Fluid under pressure from the power steering pump flows to a brake booster. The booster is mounted between the brake pedal and master cylinder, on the firewall [Figure 63-16]. Fluid enters the booster and is controlled by a *spool valve*. The spool valve directs fluid to help force the pushrod forward.

63.7 ANTI-LOCK BRAKE SYSTEM (ABS)

The function of an *anti-lock*, or *anti-skid*, *braking system (ABS)* is to prevent the wheels from locking under hard braking. The tires produce maximum braking just before the wheels lock and skid. Also, if the wheels lock and skid, steering control is lost.

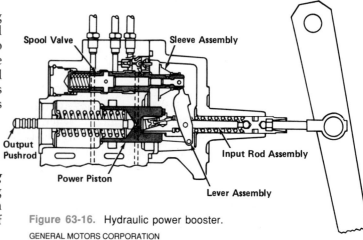

Figure 63-16. Hydraulic power booster.
GENERAL MOTORS CORPORATION

When the wheels begin to skid on dry pavement, the rubber of the tires heats, burns, and liquefies. The liquefied rubber acts as a lubricant between the tire and the road surface. Thus, the vehicle does not slow down as fast as it would before the wheels lock.

Anti-lock braking systems help to prevent wheel lockup. They also apply and release the brakes rapidly, as an experienced driver would, for maximum braking force. ABS systems **[Figure 63-17]** include:

- Master cylinder (power assisted)
- Speed sensors at each wheel
- A computer-controlled hydraulic unit
- An electronic control unit (computer module)
- Relays.

Brake fluid under pressure passes through the hydraulic control unit. When the brakes are applied, the computer determines whether the hydraulic unit should be activated to modify fluid pressures.

The hydraulic unit contains fluid passageways and control mechanisms similar to those found in an automatic transmission valve body **[Figure 63-18]**. Solenoid valves, spool valves, and accumulators control brake fluid pressure and flow.

The solenoid valves open or close to modify the fluid pressure flowing from the master cylinder to the wheel calipers. The hydraulic unit decreases, maintains, or increases hydraulic pressure to either of the front two brakes independently, or to both rear brakes together.

Wheel speed sensors provide information to the computer on the rotational speed of each wheel. The sensor assembly includes a toothed metal rotor that

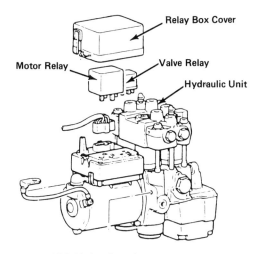

Figure 63-18. ABS Hydraulic unit.
MITSUBISHI MOTOR SALES OF AMERICA, INC.

rotates with the wheel and a stationary magnetic pickup unit **[Figure 63-19]**.

A wheel speed sensor uses the same magnetic operating principles as a magnetic position sensor in an ignition distributor (refer to Unit 34). When the wheel turns, an AC signal is generated. The frequency of the current indicates the wheel's rotational speed.

Anti-Lock Brake System Operation

The computer compares and averages input signals from all four wheels to determine the vehicle's road speed. If any wheel begins to turn slower than the average speed, the computer determines that the wheel is beginning to skid. The computer operates the hydraulic unit to decrease brake fluid pressure slightly to allow the wheel to unlock.

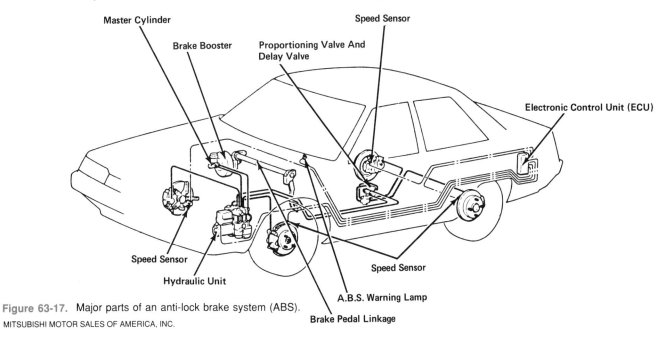

Figure 63-17. Major parts of an anti-lock brake system (ABS).
MITSUBISHI MOTOR SALES OF AMERICA, INC.

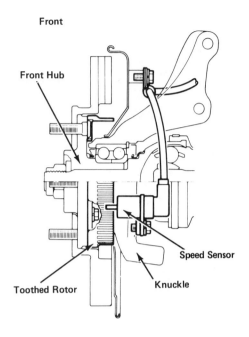

Figure 63-19. Anti-lock braking system wheel speed sensor.
MITSUBISHI MOTOR SALES OF AMERICA, INC.

If any wheel turns faster than the others during braking, the computer reacts. Output signals from the computer cause the hydraulic unit to increase brake fluid pressure to that wheel brake. The increased pressure slows the wheel to the speed of the other wheels. When the computer determines that the wheel is rotating at the same speed as the rest of the wheels, hydraulic pressure is maintained at a constant level.

In operation, ABS "pumps" the brakes rapidly to maintain maximum braking force without wheel lockup. This action is similar to the way that an experienced driver would operate the brakes on an icy road to prevent wheel locking.

The entire course of events described above occurs in about one second. The result is that the wheels do not lock or skid. Braking distances are reduced and directional control is increased, even on an icy curve.

If any part of the system should fail to work, the system goes into a "fail-safe" mode. The hydraulic unit does not modify master cylinder output pressures. The master cylinder controls brake application, just as on a vehicle that is not equipped with ABS.

63.8 PARKING BRAKE OPERATION

On most vehicles, the *parking brake* is a mechanically operated backup system that operates only the rear brake assemblies. The parking brake locks the rear brake assemblies so that the vehicle cannot move while it is parked. On a few vehicles, such as Subarus and Saabs, the parking brake operates the front brakes only.

The parking brake also functions as a backup "emergency brake" if the entire hydraulic brake system should fail. A metal cable and/or rods transfer force and motion when the driver operates the parking brake [Figure 63-20]. This action operates the rear brake units to force brake shoes against drums, or brake pads against disc rotors.

The leverage required to operate a rear disc parking brake is difficult to achieve. For this reason, some luxury and sports vehicles with 4-wheel disc brakes include small drum brake units within the rear brake assemblies to function as a parking brake.

UNIT HIGHLIGHTS

- A brake uses friction to slow and stop an automobile.
- When the brake pedal is depressed, hydraulic pressure is applied to force friction materials into contact with brake discs or drums.
- A master cylinder operates like a simple plunger pump.
- Steel and flexible brake lines carry hydraulic pressure from the master cylinder to disc brake calipers and drum brake wheel cylinders.
- Hydraulic cylinders in disc brake calipers and drum brake wheel cylinders, move brake pads or brake shoes against rotating parts to create friction.
- A disc brake includes a rotor, caliper, and pads.
- A drum brake includes a drum, wheel cylinder, and brake shoes.
- Drum brakes can be designed to be self-energizing.
- Valves are used to control and coordinate the actions of disc and drum brakes.
- Power brakes can be operated by vacuum and atmospheric pressure or by hydraulic pressure from a power steering pump.
- Anti-lock brake systems (ABS) use computer-controlled hydraulic units to modify brake pressures.
- Anti-lock brake systems decrease, increase, or maintain brake system pressures to prevent skidding and loss of steering control.
- A mechanically operated parking brake locks the rear brakes on most vehicles.

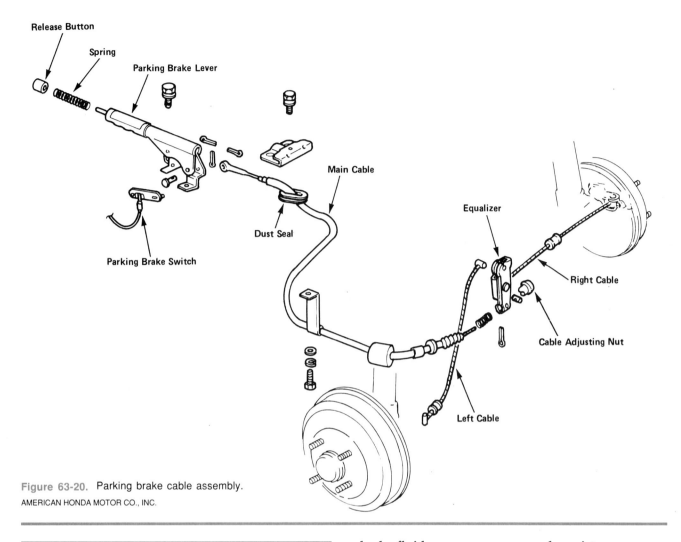

Figure 63-20. Parking brake cable assembly.
AMERICAN HONDA MOTOR CO., INC.

TERMS

friction
brake lines
rotor
drum brakes
wheel caliper
brake pads
floating caliper
primary brake shoe
brake linings
pressure
self-energizing
frictional material
metering valve
failure warning switch
power brake booster unit
parking brake
organic brake linings
brake drum
master cylinder
wheel cylinder
disc brakes
brake shoes
fixed caliper
sliding caliper
backing plate
secondary brake shoe
anchor pin
caliper
self-adjuster
proportioning valve
combination valve
residual check valve
surface area
dual system brakes
primary piston
spool valve
brake fluid
anti-lock braking system (ABS)
disc brake caliper
secondary piston
semi-metallic brake linings

REVIEW QUESTIONS

DIRECTIONS: The following questions are similar to those used on automotive technician certification tests. On a separate sheet of paper, write the letter of the correct choice.

1. Which of the following statements is correct?
   I. Modern master cylinders have two chambers to create pressure for each half of a brake system.
   II. Disc brakes are most often found on the rear of vehicles.
   A. I only
   B. II only
   C. Both I and II
   D. Neither I nor II

2. Which of the following statements is correct?
   A. A disc brake is self-energizing.
   B. A disc brake caliper applies hydraulic pressure to a wheel cylinder.
   C. A fixed caliper is a rebuilt caliper.
   D. A floating caliper pivots on flexible bushings to clamp the pads against the rotor.

3. Technician A says that an anti-lock brake system (ABS) includes sensors, a computer, and a hydraulic unit to modify brake pressures.

   Technician B says that ABS can increase, decrease, or maintain brake system pressures to prevent tire skidding and loss of steering control.

   Who is correct?
   A. A only   C. Both A and B
   B. B only   D. Neither A nor B

Questions 4 and 5 are not like those above. Each has the word EXCEPT. For each question, look for the choice that does *not* apply. Read each question carefully before you choose your answer.

4. All of the following statements are true EXCEPT
   A. friction is created between stationary brake linings and rotating discs or drums.
   B. braking friction is created between the brake pedal and the master cylinder.
   C. a hydraulic system is activated when a driver depresses the brake pedal.
   D. the brake lines transfer fluid and pressure to a wheel cylinder.

5. All of the following statements are correct EXCEPT
   A. a power brake system applies extra force to the master cylinder pushrod.
   B. brake fluid fills the engine side of a vacuum brake booster.
   C. the diaphragm in a power brake booster is moved by atmospheric pressure.
   D. a hydraulically operated brake booster uses pressure from a power steering pump to apply force to the master cylinder pushrod.

ESSAY QUESTIONS

1. Describe the operation of a basic brake system.
2. What are the functions of a combination valve?
3. Describe the parts, function, and operation of an anti-lock brake system.

SUPPLEMENTAL ACTIVITIES

1. Identify and describe three different types of disc brakes.
2. Describe how self-energizing drum brakes operate.
3. Locate and identify the valves in a disc and drum brake combination system.
4. Describe what occurs in different types of power brake assist units when the brakes are applied.
5. Describe how the parking, or emergency, brake operates.

64 Brake System Service

UNIT PREVIEW

Brake service includes making regular brake inspections for leaks, lining wear, damage to brake lines, and disc and drum brake components. After inspections or service, a road test is performed to check that the brake system operates properly.

A complete brake job includes rebuilding or replacing the master cylinder, disc brake calipers, and drum brake wheel cylinders. All brake linings are replaced. Disc rotors and drums are resurfaced or replaced. New parts are installed to replace any defective mechanical parts.

Anti-lock brake system (ABS) service includes checking electrical and/or electronic sensors and parts. Special tools and testers may be required.

LEARNING OBJECTIVES

When you have completed your assignments and exercises in this unit, you should be able to:

❏ Inspect the brake system and identify defects.
❏ Safely conduct a road test to check for brake operation.
❏ Bleed air from a brake system.
❏ Service disc and drum brake systems.
❏ Adjust drum brakes and emergency brake mechanical linkage.
❏ Replace vacuum power brake boosters.
❏ Diagnose anti-lock brake system (ABS) problems.

SAFETY PRECAUTIONS

Refer to the vehicle manufacturer's service manual and bulletins for specific directions and safety precautions to be followed during brake service.

Safely raise and support the vehicle on a hoist or with jack stands. Always wear eye protection when working underneath a vehicle or with compressed air.

Many brake linings contain asbestos. Inhaling asbestos dust can lead to lung cancer. Wear a special breathing mask approved for use with asbestos fibers. Always use either a special, approved vacuum collection system or liquid cleaning system to remove asbestos dust from brake units. Never use compressed air or a brush to remove asbestos dust. Dispose of collected asbestos dust or liquid waste in accordance with all federal, state, and local laws.

Never use any petroleum product to clean brake system parts. Even the slightest amount of oil, gasoline, kerosene, or any related mineral oil product will damage rubber seals in the brake system and cause complete brake system failure.

Grease and other foreign material must be kept off brake parts, especially disc and drum friction surfaces and brake lining materials.

 TROUBLESHOOTING AND MAINTENANCE

Regular preventive maintenance includes checking and adding brake fluid. To check wheel brake assemblies, remove wheels and brake parts as described below.

Brake Fluid Inspection

A master cylinder brake fluid reservoir is located under the hood and near the firewall, usually on the driver's side. Wipe the outside of the cover and reservoir to remove dirt and grime. Check for visible leaks. Older master cylinder covers may be held on with a bail-type wire retainer **[Figure 64-1]** or an attaching bolt. Other covers pull off or unscrew. Remove the cover and inspect the gasket and cover for damage or plugged vent holes. Replace the cover and/or gasket or clean vent holes as necessary.

The reservoir should be up to the MAX (maximum) mark on translucent plastic containers. On containers without indicator lines, the level should be filled to within 1/4 in. (6.35 mm) of the top. If fluid must be added, either the brake linings are worn or there is a leak.

To check for contaminated fluid, place a small amount of brake fluid in a clear glass or plastic jar. New DOT 3 brake fluid is a clear, honey-colored liquid. If the fluid is dirty or separates into layers, it is contaminated. Contaminated fluid must be flushed from the entire brake system, including the brake lines, new brake fluid added, and the entire system bled of air bubbles.

PART VIII: THE AUTOMOTIVE CHASSIS

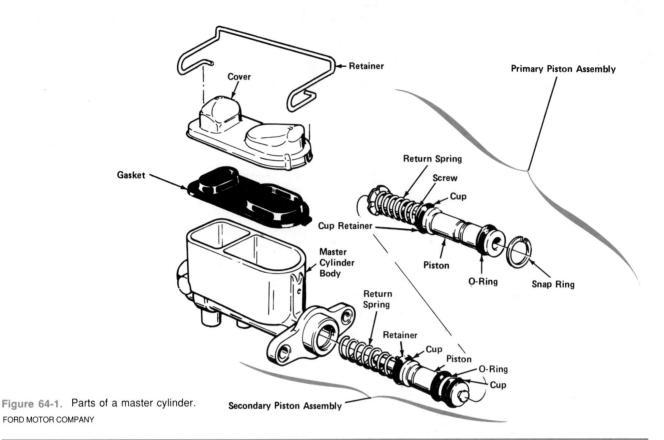

Figure 64-1. Parts of a master cylinder.
FORD MOTOR COMPANY

CAUTION!!!

Brake fluid can damage paint. Avoid spilling it on the vehicle. Always use fresh DOT 3 or higher grades of brake fluid as recommended in the vehicle manufacturer's service manual.

Check all around the master cylinder for dampness and leaks, especially at the brake line fittings and at the rear. To check the rear of the brake booster, unbolt the master cylinder and move it forward gently just enough to see the rear of the unit. Leaks at the rear indicate a defective rear piston secondary seal. The master cylinder must be rebuilt or replaced.

If the outside of the master cylinder is dry, but fluid must be added to the master cylinder more often than once a month, there is an external leak in the brake system. The leak must be located and repaired.

Brake Line Inspection

Check all tubing, hoses, and connections from under the hood to the wheels for leaks and damage. Flex flexible brake hoses to check for flexibility, bulges, and cracks, especially near the metal end fittings. General procedures for flexible brake line replacement are described below. Steel brake lines must be double lap flared. Wheels and tires also must be inspected for signs of brake fluid that has leaked.

Brake Pedal Inspection

Depress and release the brake pedal several times. Have a helper check for proper stoplight operation. Check for friction and noise. Pedal movement should be smooth, with no squeaks from the pedal, linkage, or power brake unit. The pedal should rise quickly when it is released.

Test for brake system pressure loss. On a mechanically operated brake system, apply heavy foot pressure to the brake pedal. Check for a *spongy pedal* and *pedal reserve*. Pedal action should feel firm. A spongy pedal feels soft yet springy. Pedal reserve is the distance between the brake pedal and the floor after the pedal has been depressed fully. The pedal should not go lower than 1 inch (25.4 mm) above the floor.

With the engine off, apply and hold heavy foot pressure on the pedal for 30 seconds. The pedal should remain firm to indicate that the system is holding pressure. Downward pedal movement indicates loss of pressure. Repeat the procedure with the engine running for power brakes.

SAFETY CAUTION

When operating the engine, be sure the transmission lever is in Park (automatic) or in neutral (manual). Be sure the area is properly ventilated for the exhaust to escape.

If the pedal slowly sinks downward under foot pressure and the fluid level is correct, the master cylinder has internal leakage. It must be rebuilt or replaced.

NOTE: A pedal that will "rise" with repeated pumping may indicate either a defective master cylinder, or drum brakes with excessive clearance between the linings and the drum.

Power brake booster check. Three checks can be made to test the operation of a vacuum power brake booster. These are:

1. Start the engine and run it for two minutes to build vacuum in the booster, then switch the engine OFF. Depress the brake pedal 10 times with normal foot pressure. If the pedal gradually rises and becomes harder to push, the booster is operating properly; if the pedal height does not rise, the booster is defective.
2. With the engine OFF, depress the brake pedal and start the engine. If the pedal drops slightly under your foot as the engine starts, the booster is operating properly; if not, the booster is defective.
3. Start the engine, depress the brake pedal and hold it down with normal foot pressure, then switch OFF the engine. Hold the pedal depressed for 30 seconds. If the pedal height does not change, the booster is operating properly; if the pedal height rises, the booster is defective.

To check the operation of a hydraulically operated brake booster, refer to the manufacturer's service manual for specific instructions. Check around a hydraulic booster unit for leaks or damage. If there are leaks, the booster is defective.

64.1 BRAKE SYSTEM PROBLEM DIAGNOSIS

A road test is necessary to check brakes for safe, quiet operation. Road tests should be conducted on a level road that is dry, clean, and smooth. Use the proper service manual to identify problems and causes during the road test. The following tests will help to evaluate brake performance.

First, check that the stoplights and warning lights are operating properly. Be sure the brake fluid is at the proper level. Check tire pressures and wheel bearing adjustments. A road test should be made only when the operator is sure that the brakes will stop the automobile.

SAFETY CAUTION

Before driving, depress the brake pedal to make sure it will not bottom. Make a few slow-speed stops to determine if the brakes are safe for driving. Do not apply the brakes continually, because they might overheat and *fade,* or lose their holding power.

Pulls, Noises, and Braking Effectiveness

Make light and moderate but complete stops from about 15 mph (24 km/h). Notice whether the foot pressure required to stop the vehicle is excessively heavy or excessively light. Check for pulling. Front brakes pull in the direction of the brake doing most of the work. Rear brake pull may not be noticeable during low-speed stops. Check for noise. Open the windows and turn off all accessories. Listen to determine the type of noise and the wheel from which it is coming.

Make hard stops from 55 mph (88 km/h) and check in the same manner as for the low-speed test. A hard stop is one in which the automobile is stopped quickly, just short of skidding. Do not repeat a hard stop until the automobile has traveled about 2 miles (3.2 km) so that the brakes will cool.

SAFETY CAUTION

Use extreme caution when making high-speed tests. There should be no traffic where these tests are conducted. The area around the road test area should be clear of obstructions. Stay away from residential areas. If a brake pulls or locks up, or the automobile skids, ease up on the brakes and bring the car safely to a stop.

Roughness or Pulsation

Make light stops from 55 mph (88 km/h). Check for pedal vibration felt through your foot as a rough or pulsating movement.

Brake Fade

Brake fade is a temporary loss, or reduction, of brake effectiveness. Fade is caused by heat that causes the lining material to create a gas. The gas builds up between the lining material and the metal surface of a drum or disc, and prevents friction contact. Make three hard stops from 55 mph (88 km/h) at 1/2-mile (0.8 km) intervals. Check for pulling. Notice how much pedal effort is required. Check pedal reserve after each stop.

Delayed Brake Fade

After performing the brake fade test, allow the automobile to stand for about 10 minutes. Then to test for delayed brake fade, accelerate to 55 mph (88 km/h) and make one hard stop. Check for pulling, pedal effort, and pedal reserve.

The results of brake tests, along with a service or diagnosis manual, will assist in pinpointing any problems.

Disc Brake Inspection

Raise the automobile on a hoist or support it with safety stands. Remove all four wheels. (If drum brakes are used at the rear wheels, refer to the discussion of Drum Brake Inspection below.) Whenever wear symptoms appear, or at least every six months or 6,000 miles, *whichever comes first*, inspect the brake linings for wear.

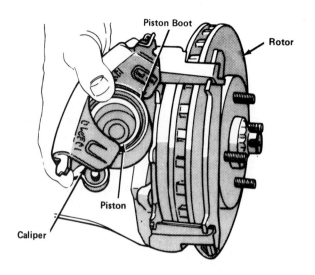

Figure 64-2. Caliper removal. CHRYSLER CORPORATION

SAFETY CAUTION

Use extreme care when working around the brake linings. Organic brake linings contain asbestos. Asbestos can cause lung cancer. Wear a special breathing mask approved for use with asbestos fibers. Use a special, approved vacuum cleaner or liquid cleaning system to remove asbestos dust from the brakes before disassembly. Never use compressed air to blow off brake dust.

Lining wear. Lining should be thicker than the thickness of a coin. It usually is not necessary to remove disc brake calipers to check lining thickness. First, check ends of the linings visible from the outside of the caliper. One or both linings may be visible through an inspection hole in the caliper. If any lining is the approximate thickness of a coin, or less, the linings are worn and must be replaced.

Disc brake pad replacement. In most cases, the caliper must be removed to replace pads. On some sliding-caliper designs, one slide bolt can be removed, the other slide bolt loosened, and the caliper swiveled up to replace pads.

Removal of one type of caliper is shown in **Figure 64-2**. Fastening bolts or wedges are removed to free the caliper housing. Slide the caliper slowly out and away from the rotor.

CAUTION!!!

Support the caliper assembly on an upper control arm with a wire hanger. Do not kink or stretch the brake lines.

Remove the outer brake pad by sliding it out from the rotor **[Figure 64-3]**. Remove the inner brake pad from the caliper housing **[Figure 64-4]**. Notice how the pads are located and positioned within the caliper. If linings are oil-soaked, imbedded with foreign material, too thin, or have loose rivets, new pads must be installed.

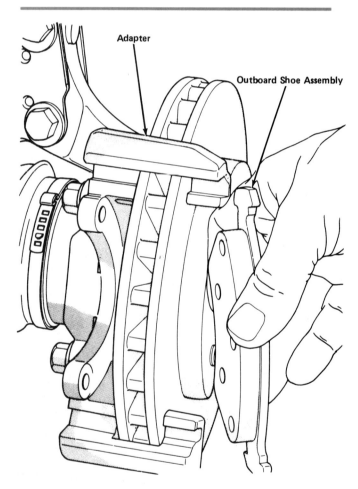

Figure 64-3. Outboard brake pad removal.
CHRYSLER CORPORATION

Lining thickness measurement. Check the brake pads for uneven or excessive wear. If one pad is noticeably thicker than the other in a caliper, the caliper is sticking or binding during operation. Both calipers

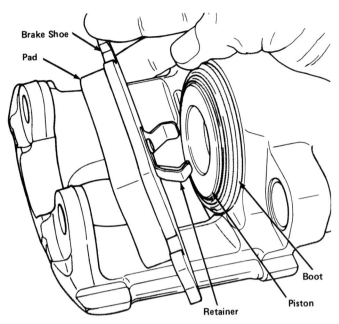

Figure 64-4. Inboard brake pad removal.
CHRYSLER CORPORATION

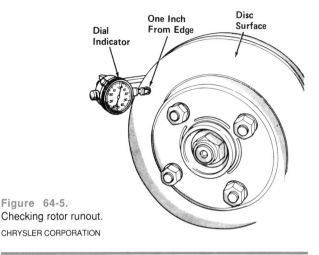

Figure 64-5. Checking rotor runout.
CHRYSLER CORPORATION

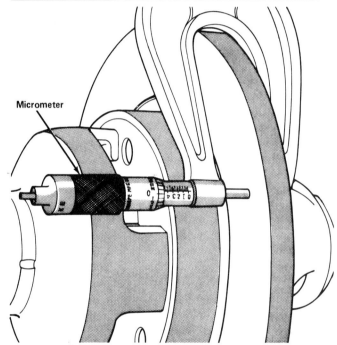

Figure 64-6. Checking rotor parallelism. CHRYSLER CORPORATION

for both wheels on that axle should be rebuilt, as described under Topic 64.4.

Measure the lining thickness. Lining thickness should be no less than the thickness of a coin at any point. If the inner lining is noticeably thinner than the outer lining, it indicates the caliper is not sliding/floating during operation. All machined surfaces must be cleaned and lubricated.

Rotor measurement. A rotor should be checked for wear, runout, parallelism, and flatness. Visual defects in the rotor friction surface include cracks, scoring, rust, imbedded lining material, deep scratches and grooves, or a shiny, mirror-bright surface. Rust causes noise and chatter. Wear and scoring can create an improper contact area with the lining. A shiny, mirror-like surface causes squealing and loss of braking effectiveness. Some discoloration and wear is normal.

Excessive runout, or wobble, can increase pedal travel or cause pedal pulsation, and damage a caliper. Check runout with a dial indicator [Figure 64-5]. Slowly rotate the rotor while watching the dial indicator. A runout of 0.002 to 0.004 in. (0.05 to 0.10 mm) usually indicates a defect. Refer to the manufacturer's service manual for allowable runout specifications.

Parallelism is the thickness variation of a rotor. Parallelism can cause brake pedal pulsation, chatter, and surge. Refer to the manufacturer's service manual for specifications. Check at four places 90 degrees apart with a micrometer [Figure 64-6]. Generally, a difference in the measurement of over 0.005 in. (0.13 mm) indicates the rotor must be resurfaced or replaced.

Dishing and distortion are problems that affect the flatness of a rotor. A straightedge can be used to check flatness.

Reassembly inspection. Before reassembling a disc brake assembly, inspect all parts. Check the linings for looseness, cracks, unusual wear, or imbedded foreign material. Check the pads for distortion. Inspect the caliper and hydraulic hose for leaks or damage, and the mounting bolts and other attachment parts for corrosion and damage. Correct any faults.

Use a C-clamp or special tool to force the caliper piston back into its bore to create clearance for the thickness of new pads. Reassemble the disc brake and install the wheel assembly. Pump the brake pedal with short, gentle strokes several times to force the disc pads into contact with the rotor before you attempt to drive the vehicle.

Drum Brake Inspection

Whenever wear symptoms appear, or at least every six months or 6,000 miles, *whichever comes first*, inspect drum brakes. The drums must be removed from the vehicle.

SAFETY CAUTION

Use extreme care when working around drum brakes. Organic brake linings contain asbestos. Asbestos can cause lung cancer. Wear a special breathing mask approved for use with asbestos fibers. Use a special, approved vacuum cleaner or liquid cleaning system to remove asbestos dust from the brakes before disassembly. Never use compressed air to blow off brake dust.

Drum removal. Different procedures may be required when removing rear drum and front drum assemblies. Always refer to the vehicle manufacturer's service manual. Rear drum removal and inspection are discussed here.

Remove the wheel assembly. Safely block the front wheels or support the vehicle on a hoist. The brake adjuster may have to be backed off through the adjusting hole if drum brakes are adjusted too tightly to remove the drum. A special drum puller tool may be necessary.

For non-driving wheels, remove the grease cap and wheel bearing assembly **[refer to Figure 58-2]**. Carefully remove the drum, shown in **Figure 64-7**.

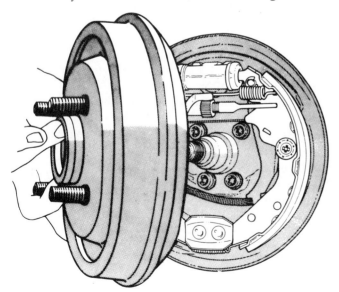

Figure 64-7. Removing the brake drum. CHRYSLER CORPORATION

Clean the brake dust immediately with a special, approved vacuum cleaner or liquid cleaning system.

Brake inspection. Check the linings for looseness, cracks, unusual wear, and imbedded foreign material. No oil, grease, or brake fluid should be on the linings.

Check the shoes for distortion or broken parts. Check the springs, fasteners, and backing plate for damage or excessive wear.

Remove all debris from the brake assembly. Brake parts must be clean to operate properly.

Lining thickness measurement. Check the linings for uneven or excessive wear. Measure the lining thickness. Lining thickness should be no less than the thickness of a coin at any point. If the inner lining is noticeably thinner than the outer lining, this is an indication that the caliper is "frozen." The caliper must be removed, cleaned, lubricated, and reinstalled.

Drum inspection. The drum surface should be clean, relatively smooth, free of damage, and should not have a mirror-like shine. Check for hard spots, scoring, and imbedded foreign material.

Use a brake drum micrometer to check drum inside diameter, and replace any drums that are not within the manufacturer's specifications. Use a *clearance gauge* to measure and adjust drum and brake shoe clearances **[Figure 64-8]**.

Before reassembling brake and non-driving wheel bearing assemblies, replace any defective brake lines. Check the backing plate shoe pads for deep grooves. If they are deeply grooved, replace the backing plates.

Lubricate the brake shoe metal contact pads and anchor pin lightly with special high-temperature lubricant. Repack non-driving wheel bearings and install new grease seals. Replace the drum. Tighten wheel bearings to torque specifications as described above.

64.2 BLEEDING BRAKES

As explained above, there must be no air in brake lines or parts. Brake *bleeding* is a process of forcing brake fluid out through the brake lines, calipers, and wheel cylinders. The fluid carries with it and eliminates any air bubbles that may be in the system.

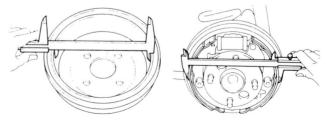

Figure 64-8. Checking brake drum clearances.
BUICK MOTOR DIVISION—GMC

Always follow the procedures described in the vehicle manufacturer's service manual for bleeding brakes.

Bleeder screws or *bleeder valves* thread into the wheel cylinders or calipers. First clean the bleeder screw, then connect a drain hose from the bleeder screw to a transparent plastic jar **[Figure 64-9]**.

Two types of brake bleeding procedures are used: manual bleeding and pressure bleeding. The sequence in which bleeding is performed can be critical. When bleeding a power brake system, remove vacuum from the power unit and plug it. To remove vacuum, the engine must be off. Pump the brake pedal several times.

CAUTION!!!

Always use fresh brake fluid when bleeding the system. Do not use fluid that has been drained. Drained fluid is contaminated and can damage the system.

Manual Bleeding

A *manual bleeding* procedure requires two people. One person operates the bleeder. The other person operates the brake pedal. Bleed only one wheel at a time. Brakes can be bled starting with the wheel closest to, or farthest away from, the master cylinder. Follow the vehicle manufacturer's recommendations in the service manual.

CAUTION!!!

The master cylinder should be full before bleeding is begun. Be sure the bleeder hose is below the surface of the liquid in the drain container at all times. Do not allow the master cylinder to run out of fluid at any time. If these precautions are not followed, air can enter the system, and it must be bled again. The master cylinder cover must be kept in place to prevent splashing during bleeding.

Place the bleeder hose and jar in position. Have a helper pump the brake pedal several times and then hold it down with moderate pressure. Slowly open the bleeder valve by turning it counterclockwise with a wrench. After fluid/air has stopped flowing, close the bleeder valve. Have your helper slowly release the pedal.

Repeat this procedure until fluid that flows from the bleeder is clear and free of bubbles. Repeat this procedure for all four wheels. Discard all used brake fluid. Fill the master cylinder reservoirs. Check the brakes for proper operation.

Pressure Bleeding

A *pressure bleeding* procedure can be done by one person. Pressure bleeding equipment uses air pressure to force brake fluid through a special adapter into the master cylinder and out through the brake lines **[Figure 64-10]**.

The use of pressure bleeding equipment varies with different automobiles and different equipment makers. Always follow the automobile manufacturer's recommendations when using pressure bleeding equipment.

On automobiles with metering valves, the valve must be held open during pressure bleeding. **Figure 64-11** shows one type of special tool used to hold open the metering section of a combination valve.

Open the bleeder valves one at a time until clear, air-free fluid is flowing. Bleed all four wheels in the order recommended in the vehicle manufacturer's service manual.

Do not exceed recommended pressure while bleeding the brakes. Always release air pressure

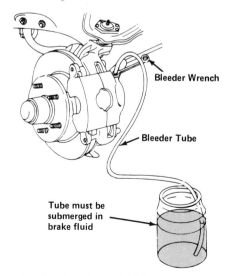

Figure 64-9. Brake bleeding attachments.
BUICK MOTOR DIVISION—GMC

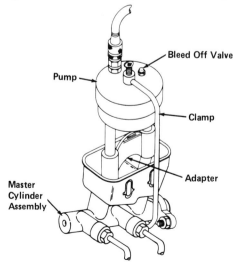

Figure 64-10. Pressure bleeder adapter.
BUICK MOTOR DIVISION—GMC

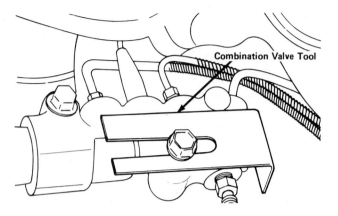

Figure 64-11. Combination valve tool. BUICK MOTOR DIVISION—GMC

before removing the master cylinder adapter. Clean and fill the master cylinder after pressure bleeding. Remove the special tool used to hold the metering valve. Check the brakes for proper operation.

64.3 MASTER CYLINDER REBUILD

A master cylinder is rebuilt to replace leaking seals or gaskets. If a more serious problem exists, the master cylinder should be replaced. Use the following procedures as a general guide to master cylinder service. Always refer to the manufacturer's service manual for correct procedures.

To remove a master cylinder, disconnect the brake lines at the master cylinder. Install plugs in the brake lines and master cylinder to prevent dirt from entering. Remove the nuts that attach the master cylinder to the firewall or power brake unit, and remove the cylinder.

Remove the cover and seal. Drain the master cylinder and carefully mount it in a vise. Remove the piston assembly and seals according to the manufacturer's instructions. **Figure 64-12** shows a master cylinder assembly. New seals always are included in rebuilding kits. Other parts may be included.

CAUTION!!!

Clean master cylinder parts only with brake fluid, brake cleaning solvent, or alcohol. Do not use a solvent containing mineral (petroleum) oil, such as gasoline. Mineral oil is harmful to rubber seals.

Hone and inspect the master cylinder. Damage, cracks, porous leaks, and worn piston bores mean the master cylinder must be replaced. Check for pitting or roughness in the bore after honing and cleaning. If there is any deep pitting, a new cylinder must be installed. After rebuilding, install and bleed the master cylinder according to the manufacturer's directions.

64.4 DISC BRAKE SERVICE

Always refer to the manufacturer's service manual for specific instructions on rebuilding disc brake calipers. Use the following as a general guide to disc brake service.

Disassembly

Raise the automobile on a hoist or safety stands at the proper locations. Remove the wheel and caliper

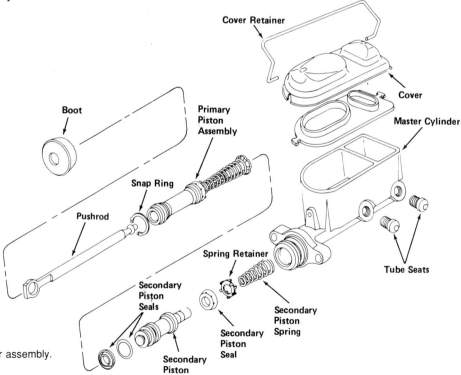

Figure 64-12. Parts of a master cylinder assembly. AMERICAN MOTORS CORPORATION

assemblies. Disassemble the caliper according to the manufacturer's recommendations. A disc brake assembly is shown in **Figure 64-13.**

Cleaning and Inspection

All brake dust should be removed from the disc brake assembly with a special, approved vacuum or liquid brake cleaning system.

Check the linings and rotor for damage and wear. Check the rotor for runout, parallelism, and flatness, as discussed in Topic 64.1

Check for leaks around the piston seal, boot area, and on the brake linings. If you find any leaks, remove the flexible brake line at the chassis end. Then remove the line from the caliper and disassemble the disc brake caliper for inspection and rebuilding.

Rebuilding Disc Brake Calipers

Always refer to the vehicle manufacturer's service manual for specific brake caliper service procedures. Use the following procedures as a general guide to rebuilding disc brake calipers.

Some calipers consist of two halves that must be disassembled for rebuilding operations. Also, rear disc brake calipers may include special ratchet locking mechanisms that position the rear caliper pistons and pads close to the disc. These mechanisms increase the parking brake's clamping efficiency.

The lock mechanisms must be disabled to move the caliper piston back into the bore.

To rebuild a disc brake caliper, first remove the flexible brake line from the caliper. Discard the used copper washer that fits between the end of the brake line and the caliper body.

Place a shop towel or block of wood between the caliper piston and the opposite side of the caliper. Apply low-pressure air (10 psi or less) through the brake fitting hole in the caliper to push out the piston and dust boot **[Figure 64-14].**

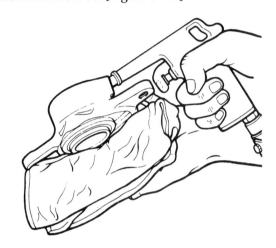

Figure 64-14. Use low-pressure air to remove a caliper piston.
MITSUBISHI MOTOR SALES OF AMERICA, INC.

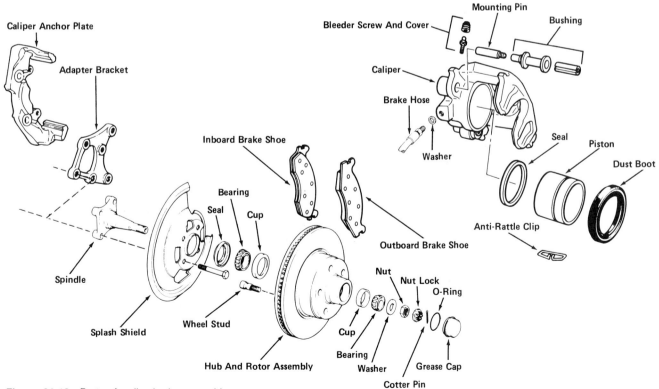

Figure 64-13. Parts of a disc brake assembly.
AMERICAN MOTORS CORPORATION

SAFETY CAUTION

When air pressure is applied to the caliper piston, it will move outward with great force. Keep your hands and fingers away from the caliper piston to prevent injuries during removal.

Remove and discard the old piston seal inside the caliper piston bore **[Figure 64-15]**. Clean the outside of the caliper piston and the inside of the caliper bore with clean brake fluid or alcohol.

Inspect the outside of the caliper piston and the inside of the bore for pitting, corrosion, and rough spots. Some manufacturers recommend honing the bore of cast-iron calipers to remove roughness and light imperfections. Use a disc brake cylinder hone with a drill motor to clean up the inside of the bore. If deep pits remain after 30 seconds of honing, the caliper and piston should be discarded.

Use crocus cloth to lightly clean the outer surface of the piston. If heavy corrosion or deep pits remain, discard the piston and the caliper.

If the caliper piston and bore are serviceable, clean them thoroughly. Install a new piston seal inside the caliper bore **[Figure 64-16]**. Lubricate the bore and the seal with brake fluid.

Place a new dust seal on the caliper piston. Lubricate the metal body of the piston and carefully slide it into the caliper bore. Make sure the piston slides inside the bore seal without distorting it **[Figure 64-17]**. Sometimes you can use thin feeler gauges to help fit the piston inside the seal. Some manufacturers recommend using a special tool to seat the dust seal. Refer to the vehicle manufacturer's service manual for specific procedures.

Grease caliper pins with the manufacturer's recommended lubricant. Fit a new copper washer between the flexible brake line and the caliper body. Tighten the brake line fitting to the manufacturer's recommended torque specification. Notice the lines and markings in the flexible brake line. Reinstall

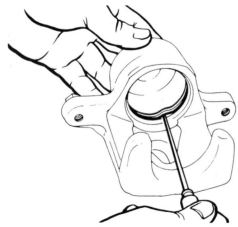

Figure 64-15. Remove and discard the used piston seal.
MITSUBISHI MOTOR SALES OF AMERICA, INC.

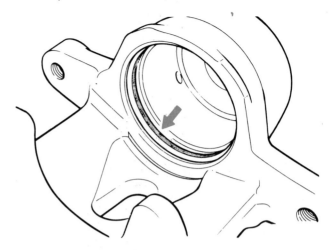

Figure 64-16. Always install a new caliper bore seal.
MITSUBISHI MOTOR SALES OF AMERICA, INC.

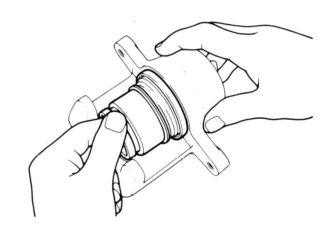

Figure 64-17. Reassemble the brake caliper piston into the bore.
MITSUBISHI MOTOR SALES OF AMERICA, INC.

the flexible brake line and torque the fitting on the caliper to the manufacturer's specifications. Make sure there are no twists in the line when you reattach the flexible line at the chassis end.

Rotor Removal and Replacement

Once the caliper assembly has been removed, the rotor can be pulled off the hub or spindle **[Figure 64-18]**. On some vehicles, the rotor and hub are removed as a unit, then disassembled on a workbench. Refer to Unit 58 for general information on removing non-driving wheel bearings. Refer to the manufacturer's service manual for specific procedures on removing front-wheel drive front wheel disc rotors.

If a rotor is not deeply grooved or thinner than allowed by law, it can be turned, or resurfaced. A brake lathe is used to resurface both friction surfaces of the rotor, remove light scratches, and make the friction surfaces parallel.

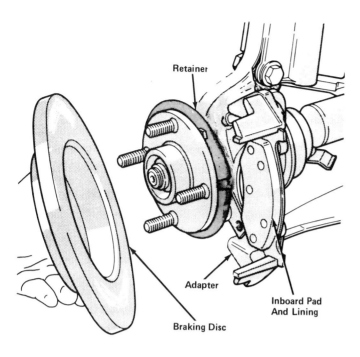

Figure 64-18. Removing the rotor. CHRYSLER CORPORATION

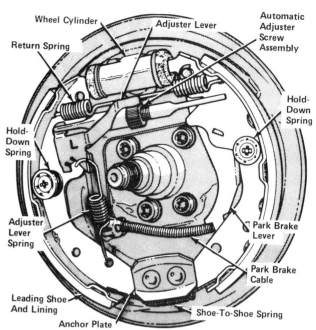

Figure 64-19. Parts of a drum brake assembly. CHRYSLER CORPORATION

SAFETY CAUTION

Always wear safety glasses when resurfacing or refacing a rotor. When using rotor-turning equipment, read the manufacturer's instructions carefully. The rotor must be cleaned of chips or other contamination before it is turned.

To replace the rotor, pack and install the inner wheel bearings. On some front-drive automobiles, a hub and sealed wheel bearings may be one assembly. Install the hub, if it was removed, and the rotor. Be sure that the rotor is installed on the correct side. Some rotors are made for right or left side installation only. Install the outer wheel bearing, thrust washer, and spindle nut. Adjust the wheel bearing and install a nut lock, a new cotter pin, and the grease cap. Install the caliper and wheel assembly.

64.5 DRUM BRAKE SERVICE

Drum brake service includes disassembly, cleaning and inspection, resurfacing the drums, rebuilding wheel cylinders, and replacing linings. Refer to Topic 64.1 for procedures to remove the wheel assembly and drum, and to inspect drum brake parts. Service procedures can vary considerably. Always refer to the manufacturer's service manual. Use the following as a general guide to drum brake service.

Disassembly

To disassemble a rear drum brake assembly [Figure 64-19], first pull up on the self-adjuster lever spring and unhook it. Remove the hold-down springs. Then grasp the brake shoes with both hands and pivot them outward toward you. The shoes and return springs will come off together. Remove the self-adjuster strut and disconnect the parking brake cable from the lever.

SAFETY CAUTION

Use extreme care when disassembling drum brakes. Organic brake linings contain asbestos, which is a health hazard when inhaled. Wear a special breathing mask for use with asbestos fibers. Use a special, approved vacuum cleaner or liquid cleaning system to remove asbestos-containing brake dust from the brakes before disassembly. Never use compressed air to blow off brake dust.

Cleaning and Inspection

Clean all parts, except lining and drums, with brake cleaning solvent. Do not attempt to clean linings contaminated with brake fluid or axle lubricant. Contaminated linings must be replaced with new units. Remove brake fluid contamination on other parts with denatured alcohol. Clean the brake drums only with a soap and water solution.

Pull back the wheel cylinder boots and check for leakage. If leakage is found, the wheel cylinders must be rebuilt, if possible. Some aluminum wheel cylinders cannot be rebuilt. Inspect for brake lining wear. If wear is uneven, the drums and shoes must be checked for distortion and damage. Inspect parts for unusual wear and damage. Replace all defective parts.

Resurfacing Drums

The drums should be inspected for damage and excessive wear. Measure drum runout and diameter with a brake drum micrometer. If only lightly scored, the drum can be resurfaced on a brake drum lathe.

A brake drum cannot be resurfaced if the drum diameter already exceeds, or would exceed, the maximum allowable diameter after resurfacing. The diameter usually is marked on the outside of the drum. Always follow the manufacturer's measurements for resurfacing.

Rebuilding Wheel Cylinders

If the boot on a wheel cylinder has cuts, tears, or heat cracks, or the cylinder is leaking, the wheel cylinder should be removed. Wheel cylinders should be cleaned, honed, and inspected.

To remove a wheel cylinder, first disconnect the brake tube from the rear of the wheel cylinder [Figure 64-20]. Remove the bolts that hold the wheel cylinder in place. Remove the wheel cylinder [Figure 64-21]. Measure the diameter of the wheel cylinder.

Carefully attempt to loosen the bleeder valve, using penetrating oil if necessary. After soaking, try to open the valve. It should open. If the valve breaks off, a new cylinder is needed.

Disassemble the wheel cylinder by prying off the boots. Slide out the piston assembly by pushing in on one end. Remove all wheel cylinder parts [Figure 64-22].

Clean the metal parts thoroughly in brake fluid or alcohol. Do not use a rag when cleaning the bore surfaces. Lint will adhere to the surfaces and contaminate the brake fluid.

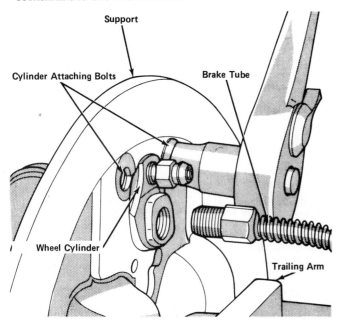

Figure 64-20. Removing brake lines. CHRYSLER CORPORATION

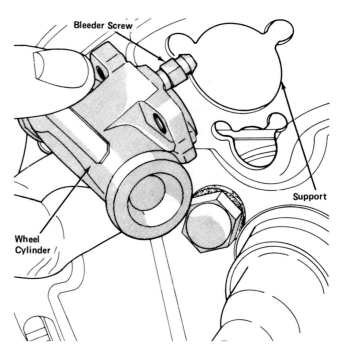

Figure 64-21. Removing a wheel cylinder.
CHRYSLER CORPORATION

Hone and inspect the cylinder bore and piston for damage or excessive wear. Pitting or roughness are reasons to discard the cylinder and install a new one.

A wheel cylinder kit contains replacement cups and other parts necessary for reassembly. Dip the pistons and new cups in clean brake fluid. Coat the cylinder bore with clean brake fluid. Assemble all the parts according to the manufacturer's recommendations.

Install and tighten the cylinder assembly and brake line. Torque the fasteners to specifications.

Brake Springs

Check all brake springs for distortion and overheating. The spring coils should be tight against one another and should have the original paint. An absence of paint may indicate overheating. Replace any springs that are questionable.

Backing Plate

Inspect the backing plate for any looseness or wear. Pay special attention to the pads or bosses. They sometimes have deep grooves worn in them by the brake shoes. Replace deeply grooved backing plates. Lubricate the pads with a thin coat of special high-temperature brake lubricant.

Reassembly

Replace all damaged and excessively worn drums with new ones. Reassemble the brake assembly according to the manufacturer's recommendations.

BRAKE SYSTEM SERVICE

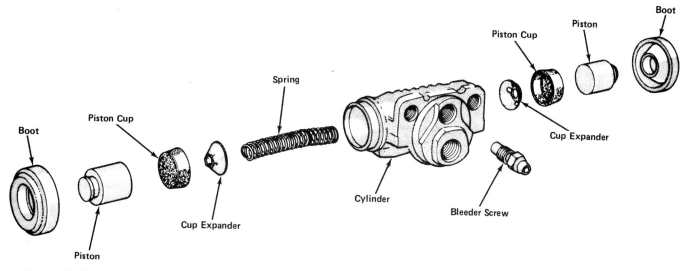

Figure 64-22. Parts of a wheel cylinder assembly. CHRYSLER CORPORATION

Install the drum and wheel assembly as discussed in Topic 64.1.

64.6 POWER BRAKE SERVICE

If diagnosis indicates an internal problem with the power brake unit, replace the entire power unit. Most manufacturers do not recommend disassembly, repair, or adjustment of power brake booster units.

To remove a vacuum power brake unit, first disconnect the vacuum hose. Remove the nuts and lockwashers that hold the power unit to the master cylinder and to the firewall. Pull the master cylinder forward gently until it clears the power booster unit. Remove the power unit. It usually is not necessary to disconnect the brake lines. A typical power unit installation is shown in **Figure 64-23.**

Install a new power unit in the reverse order of disassembly. A lubricant usually is used on the end of the pushrod. Refer to the manufacturer's service manual for hydraulically operated power brake unit service procedures.

64.7 ANTI-LOCK BRAKE SYSTEM (ABS) SERVICE

Anti-lock brake system (ABS) service includes all repairs required for ordinary brake systems. In addition, check the electrical and electronic parts of ABS systems for proper operation. Problems in anti-lock brake systems generally fall into two categories:

- Electrical and electronic components
- Hydraulic system components.

CAUTION!!!

Do not attempt ABS service without the vehicle manufacturer's service manual and special tools

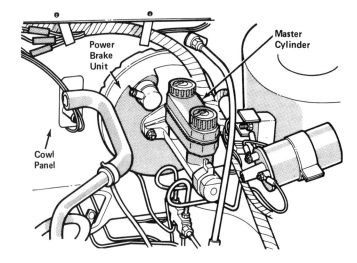

Figure 64-23. Power brake mounting. CHRYSLER CORPORATION

required. Use special tools, electrical harness adapters, and testers as recommended. Improper service procedures can result in failure of the ABS system.

Electrical and Electronic Troubleshooting

The most common problem with ABS brakes is that dirt, rocks, and other debris get between the wheel speed sensor pickups and the toothed hubs and cause damage. Examine the sensor pickup units and the toothed hubs carefully. Clean all dirt and contamination from between sensor teeth. Replace damaged sensors and/or toothed hubs. Also check all wiring and connectors leading to the sensor pickups.

For other types of failures, some ABS systems may be equipped with a self-diagnosis function that operates a dash warning light. If the ABS fails and the brake system is operating in the fail-safe mode, the light will illuminate. Other electrical and electronic troubleshooting includes:

- Checking electrical connections
- Testing switches and relays
- Inspecting and adjusting speed sensor units **[Figure 64-24]**.

Recent ABS systems may include self-diagnosis functions to determine electrical and/or hydraulic system problems. In addition, special electronic testers may be required. Refer to the vehicle manufacturer's service manual for correct test instruments and testing procedures.

Hydraulic System Repairs

In addition to the master cylinder, brake lines, and wheel brake units, the ABS system includes a hydraulic unit. Internal parts of the unit include spool valves, solenoid valves, and accumulators similar to those used in modern automatic transmissions. The hydraulic unit also includes an electric fluid pump.

Some vehicle manufacturers do not recommend servicing the hydraulic unit. If troubleshooting indicates internal problems, the entire hydraulic unit is replaced. Refer to the vehicle manufacturer's service manual for specific troubleshooting and repair procedures.

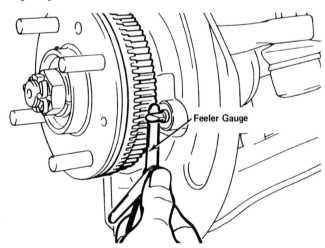

Figure 64-24. Checking ABS wheel speed sensor clearance.
MITSUBISHI MOTOR SALES OF AMERICA, INC.

UNIT HIGHLIGHTS

- Brake fluid is contaminated if it separates into layers in a glass jar.
- The way a brake pedal feels when it is being depressed can be an indication that a brake problem exists.
- Brake pad wear can be checked by looking around and/or through the caliper.
- To change disc brake pads, the caliper must be removed.
- A rotor is measured to determine runout, flatness, and parallelism.
- The drum must be removed to inspect drum brake linings.
- Drum and lining measurements are made with a clearance gauge.
- The brakes should be checked thoroughly before conducting a road test.
- Air is removed from a hydraulic system by a procedure called bleeding.
- If not badly pitted, master cylinders, disc brake calipers, and wheel cylinders can be rebuilt to correct a leakage problem.
- If not deeply grooved or beyond legal sizes, brake rotors and drums can be resurfaced on a brake lathe.
- Power brake units must be replaced when they fail to operate properly.
- Anti-lock brake system (ABS) service may require special tools, testers, and information from the vehicle manufacturer's service manual.
- In addition to typical brake system service, ABS systems require troubleshooting electrical and/or electronic parts.

TERMS

spongy pedal
parallelism
bleeding
bleeder screw
pressure bleeding

pedal reserve
clearance gauge
bleeder valve
manual bleeding
brake fade

REVIEW QUESTIONS

DIRECTIONS: The following questions are similar to those used on automotive technician certification tests. On a separate sheet of paper, write the letter of the correct choice.

1. Technician A says that calipers usually do not have to be removed to check brake pad thicknesses.
   Technician B says that brake drums usually do not have to be removed to check shoe lining thicknesses.
   Who is correct?
   A. A only
   B. B only
   C. Both A and B
   D. Neither A nor B

2. Which of the following statements are correct?
   I. Wheel bearing adjustment must be correct before conducting a road test.
   II. Brake fade is a temporary loss of pedal pressure.
   A. I only
   B. II only
   C. Both I and II
   D. Neither I nor II

3. When manually bleeding a brake system, a technician should
   A. remove a brake line.
   B. have a helper to open and close the bleeder valves.
   C. reuse the brake fluid.
   D. bleed the master cylinder.

4. Technician A says that rotors and drums must be within legal specifications after resurfacing.
   Technician B says that all rotors and drums can be resurfaced and reused.
   Who is correct?
   A. A only
   B. B only
   C. Both A and B
   D. Neither A nor B

Question 5 is not like those above. It has the word EXCEPT. For question 5, look for the choice that does *not* apply. Read the entire question carefully before you choose your answer.

5. All of the following statements are *false* EXCEPT
   A. drum brakes are used on the front wheels of most cars.
   B. deeply pitted wheel cylinders can be rebuilt.
   C. asbestos dust is a severe health hazard.
   D. self-adjusters are used on modern disc brakes.

ESSAY QUESTIONS

1. Describe preventive maintenance procedures for a brake system.
2. Describe brake system problem diagnosis.
3. What steps are involved in rebuilding a disc brake caliper?

SUPPLEMENTAL ACTIVITIES

1. Perform a complete brake inspection on a vehicle selected by your instructor. Note any problems found.
2. Demonstrate special breathing masks and cleaning systems to protect against asbestos dust hazards.
3. Remove and replace disc brake pads on a vehicle.
4. Measure lining thickness on a brake system selected by your instructor.
5. Explain the procedures that are used to perform a road test to check the braking system.
6. Use micrometers to measure disc rotor parallelism, thickness, and drum diameter. Determine whether the units can be resurfaced or must be discarded.
7. Bleed a brake system.
8. Rebuild a master cylinder correctly, bench bleed it, and install it. Check for proper operation.
9. Disassemble, inspect, and reassemble a brake assembly on an automobile selected by your instructor.
10. Disassemble, hone, inspect, and rebuild, if possible, a disc brake caliper and wheel cylinder.
11. Remove and install a power brake unit.
12. Refer to the service manual, inspect and diagnose ABS system faults.

PART IX: Auxiliary Systems

This final part covers heating, air conditioning, and safety systems. Many career opportunities for skilled and certified technicians in these areas are increasing. The first units cover heating and air conditioning systems. The final units provide details on the operation, construction, and service of specific safety systems. The technician in this photo is adding antifreeze to a cooling system.

65 Heating and Air Conditioning Systems

UNIT PREVIEW

Heating and air conditioning systems provide warmed or cooled air to the passenger compartment. Both systems use heat exchangers, or cores, similar to a cooling system radiator. Coolant from the engine is circulated through a heater core. Fresh air is directed over the core, becomes warm, and flows into the interior.

An air conditioning system evaporator core is cooled by the evaporation of a special gas. The gas is circulated through the system by a compressor pump. Fresh air is directed over the evaporator, becomes cool, and flows into the passenger compartment. Blower motors help to force air through the heater core and evaporator core and through ducting to the interior.

LEARNING OBJECTIVES

When you have completed your assignments and exercises in this unit, you should be able to:

❏ Describe the function and operation of heating system parts.
❏ Locate and identify heating system parts on a vehicle.
❏ Describe the function and operation of air conditioning system parts.
❏ Locate and identify air conditioning system parts on a vehicle.
❏ Describe special handling procedures for automotive refrigerants.
❏ Describe how air is directed and forced through distribution plenums.

65.1 HEATING SYSTEM

An automotive heating system uses engine heat to provide occupant comfort. As discussed in Unit 17, heat from the engine coolant warms a heater core, or heat exchanger, that is similar to a small radiator. Air that flows through the core becomes heated and circulates through air ducts and doors into the passenger compartment. The basic assemblies in a heating system, shown in **Figure 65-1,** are:

- Air inlet
- Heater
- Heater controls.

Air Inlet Assembly

An *air inlet assembly* draws in outside air and forces it into the passenger compartment. Outside air usually is drawn into a heating system through a grille assembly in an opening below the windshield.

A duct directs the incoming air to a **blower.** A blower is a fan that pushes, or blows, the incoming air through, or out of, a heater assembly. An electric motor turns the blower fan.

Heater Assembly

The *heater assembly* includes a heater core, distribution plenums and ducts, and control doors. Heated coolant from the cylinder head flows through an inlet heater hose into the heater core. Pressure from the coolant pump moves the liquid through the inlet hose and into the core.

In most heater assemblies, a valve in the inlet hose controls the amount of heated coolant that flows through the core. In other systems, there is no valve in the inlet hose **[Figure 65-2].** Coolant flows through the heater core whenever the coolant pump operates. These systems control the amount of heat that enters the passenger compartment by controlling airflow through the heater core.

The heat from the coolant is transferred into the metal of the heater core. When the heater operates, fresh outside air, or recirculated air from inside the passenger compartment, is directed through vents to flow through the core. As the air passes through the heater core, it is warmed. An electric blower motor forces air through the heater core and provides rapid heating **[Figure 65-3].**

The warmed air flows through control doors into *distribution plenums,* or chambers that accept circulating air and distribute it. From the plenums, the warmed air can be directed toward floor and/or dash vents. Plenum chambers that direct warmed air directly onto the inner surface of the windshield are called *defrosters.*

After passing through the heater core, the coolant has lost some of its heat. This cooler liquid flows through the outlet heater hose back to the suction side of the coolant pump. The cooled liquid is

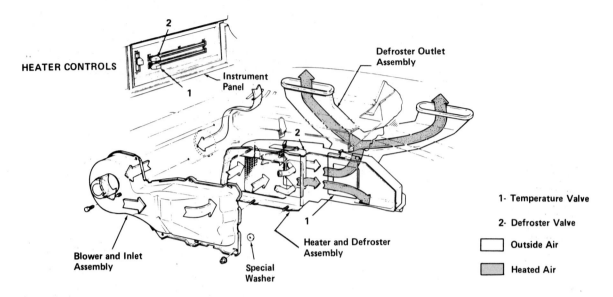

Figure 65-1. Heating system assembly. CHEVROLET MOTOR DIVISION—GMC

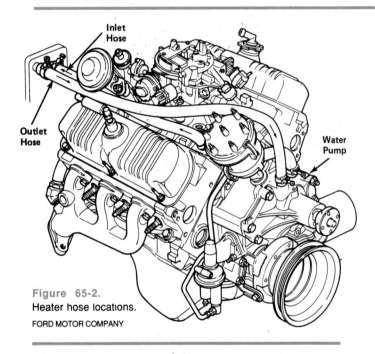

Figure 65-2. Heater hose locations. FORD MOTOR COMPANY

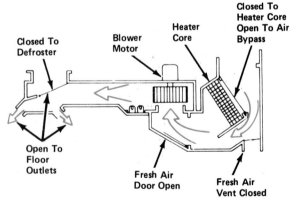

Figure 65-3. Airflow through a heater assembly. AMERICAN MOTORS CORPORATION

drawn into the engine, recirculated, and reheated as it passes through the cylinder block and head.

Many heating systems also are used as *ventilation systems.* A ventilation system can provide unheated outside air or heated air for the comfort of passengers. Ventilation systems include a bypass to divert outside air from flowing through the heater core.

Heater Controls Assembly

A *heater controls assembly* regulates a heating system. Manual cables, vacuum lines and motors, and/or electric motors can be used to open or close a heater valve and move control doors. Parts of a heater control panel are shown in **Figure 65-4.**

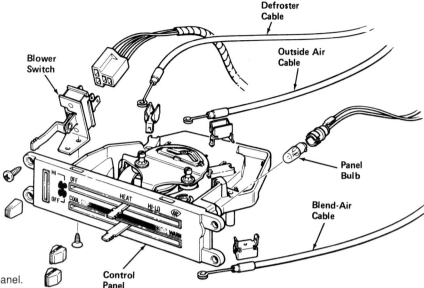

Figure 65-4. Parts of a control panel.
AMERICAN MOTORS CORPORATION

65.2 AIR CONDITIONING SYSTEM

An air conditioning system provides cooled air to the vehicle passenger compartment. Fresh outside air or recirculated air from inside the compartment is cooled, dehumidified, and circulated. Air conditioning systems operate on the principle of *evaporation.*

Evaporation

You can experience cooling produced through evaporation by applying alcohol to the back of your hand. The cooling sensation you feel is caused by the alcohol absorbing heat from your skin as it evaporates.

When a liquid is converted to a gas (as when the alcohol evaporates), it absorbs heat. If this gas can be compressed and liquified again, the heat will be released.

Air conditioning systems operate by evaporating a gas within a closed system. If there are no leaks, the gas remains within the system and is compressed and liquified so that the process can occur repeatedly. The air conditioning system absorbs heat from the air in the passenger compartment and transfers it to the outside air.

Refrigerants

A liquid that is used to cool by way of evaporation is called a *refrigerant.* A refrigerant is a liquid that will begin to boil and evaporate at a very low temperature. These temperatures are well below the freezing point of water, 32 degrees F (0 degrees C).

A good refrigerant also is nonpoisonous, nonexplosive, noncorrosive, and can be mixed with oil. Automotive air conditioning systems commonly use a refrigerant called R-12.

CAUTION!!!

R-12 contains chlorofluorocarbons which damage the ozone layer in the upper atmosphere. Federal, state, or local regulations may apply to the use of refrigerant. Use only approved techniques when charging or discharging air conditioning systems.

Humidity Control

An air conditioning system not only cools the air but also dehumidifies it. *Humidity* is the amount of water vapor that air can hold. A humidity of 100 percent means the air is saturated with water vapor. It can hold no more. A humidity rating of 50 percent means that the air is holding only half the amount of water vapor it is capable of holding.

Cold air holds less moisture than warm air. Consequently, as warm, humid air is cooled by an air conditioning system, it becomes super-saturated with water vapor. This means that the air has more water vapor than it can hold. The excess moisture begins to separate from the air and collect on the cooler air conditioning parts. This process is called *dehumidification.* Dryer, less humid air is more comfortable than humid air.

Air Conditioner Operation

Automotive air conditioning systems, called *air conditioners,* use the principle discussed earlier to cool the passenger compartment. For the refrigerant to vaporize and condense properly, it is routed through many parts in the air conditioner. Refrigerant flow through a basic air conditioning system is shown in **Figure 65-5.** Parts of an air conditioner include:

- Compressor

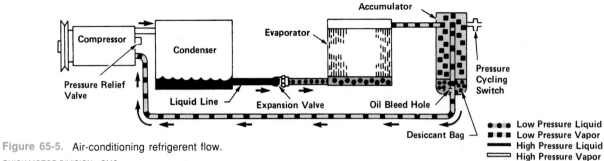

Figure 65-5. Air-conditioning refrigerent flow.
BUICK MOTOR DIVISION—GMC

- Condenser
- Receiver/dryer
- Refrigerant control
- Evaporator
- Controls.

Approximate locations of air conditioner parts are shown in **Figure 65-6.**

Compressor. A *compressor* is a pump that compresses and forces heated refrigerant vapor to flow through the system. From the compressor, the hot vapor flows into the condenser.

Condenser. A *condenser* changes the hot vapor into a liquid, or condenses it. A condenser looks and functions like a radiator. It usually is located just in front or along side of the engine cooling radiator. As hot refrigerant vapor passes through the condenser, cooler air flows through the condenser core. The refrigerant cools and is condensed from a vapor to a liquid. This liquid refrigerant then flows to the receiver/dryer.

Receiver/dryer. A *receiver/dryer*, sometimes called a receiver/dehydrator stores liquid refrigerant and removes moisture through the use of a **desiccant**. A desiccant is a special substance that absorbs moisture. Moisture in the refrigerant will combine with refrigerant to produce hydrochloric acid which will damage or destroy parts of the system. A receiver/dryer is used only in systems that use expansion valves.

Refrigerant control. To control the process as the high-pressure refrigerant evaporates into low-pressure gas inside the evaporator core, different devices are used. If the evaporator core becomes too cold, moisture on it will freeze and block airflow. The air conditioning system will not operate properly.

In older systems, some compressors ran continuously whenever the engine was operating. Cooling was controlled with expansion valves and evaporator pressure regulators.

In more modern systems, the compressor cycles off and on during engine operation to maintain proper cooling. In this case, the refrigerant control is simply a metered hole, or orifice, similar to a carburetor jet.

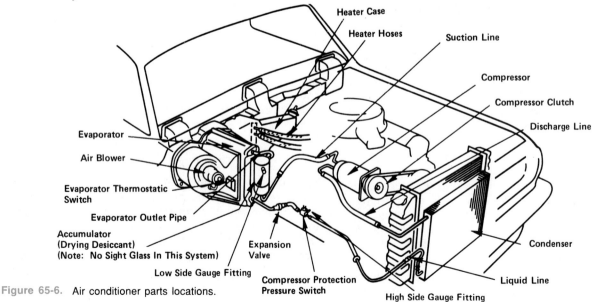

Figure 65-6. Air conditioner parts locations.
CHRYSLER CORPORATION

The hole controls how much refrigerant passes into the evaporator core as the compressor operates.

Evaporator. Outside air, or recirculated air from the passenger compartment, passes over the *evaporator tubes.* An evaporator is a heat exchanger, similar in appearance to a heater core. However, the internal evaporation of refrigerant causes the evaporator core to become cool rather than hot. Thus, air that is forced through the evaporator core gives up its heat and humidity and becomes cooler and drier.

Liquid refrigerant passes through a metered hole into the low-pressure area within the evaporator tubes. This action releases pressure on the liquid and causes it to evaporate rapidly and absorb heat through the metal of the evaporator core. Moisture from the air collects on the evaporator and drains through an opening onto the ground. The air that passes over the evaporator tubes becomes cool and dehumidified.

In some systems, the evaporated refrigerant passes into an *accumulator,* or holding tank. An accumulator holds liquid refrigerant, allowing it to vaporize before allowing it to go back to the compressor. A desiccant in the accumulator helps remove moisture from the refrigerant. The refrigerant continues on to the compressor, where the refrigeration process begins again.

NOTE: **Because the air becomes dehumidified, operating the air conditioner at the same time as the heating system defroster will clear mist from all windows quickly. Some automatic temperature control systems switch both of these units on automatically when the driver selects defrosting.**

Air Conditioner Controls

Control systems for air conditioners usually are connected with heater controls. Most heater and air conditioner systems use the same plenum chamber for air distribution. Two types of air conditioner controls are used: manual and automatic.

Manual controls. Air conditioner manual controls operate in a manner similar to heater controls. Depending on the control setting, doors are opened and closed to direct airflow **[Figure 65-7]**. The amount of cooling is controlled manually through the use of control settings and blower speed.

Automatic temperature control. An *automatic temperature control system* maintains a specific temperature automatically inside the passenger compartment. To maintain a selected temperature, heat sensors send signals to a computer unit that controls compressor, heater valve, blower, and

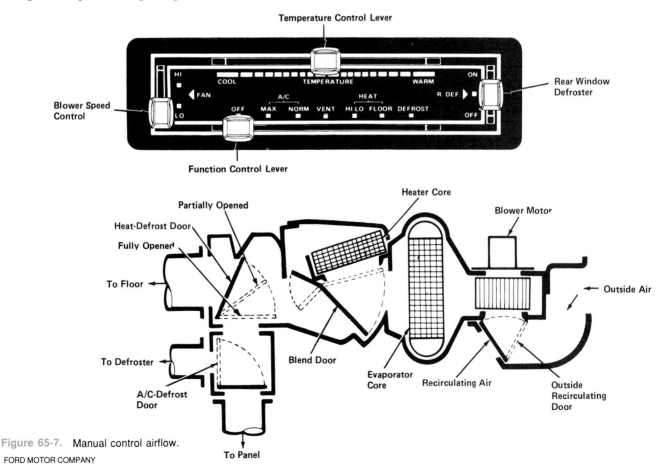

Figure 65-7. Manual control airflow.
FORD MOTOR COMPANY

plenum door operation. Warm and cool air usually are mixed to maintain the proper temperature. Airflow in an automatic temperature control system is shown in **Figure 65-8.**

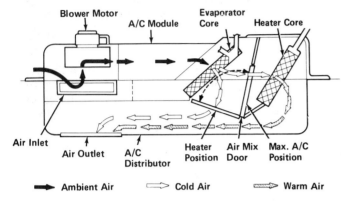

Figure 65-8. Automatic temperature airflow.
CADILLAC MOTOR CAR DIVISION—GMC

UNIT HIGHLIGHTS

- A heater assembly transfers engine coolant heat into air that circulates into the passenger compartment.
- A refrigerant evaporates at very low temperatures.
- An air conditioning system transfers heat from the passenger compartment into the outside air.
- Air conditioning systems use the principle of evaporation to cool and dehumidify air.
- Ventilation, heating, and air conditioning systems are often combined into one unit.
- An automatic temperature control system operates to maintain a set temperature by regulating heater, air conditioner, blower, and plenum door operations.

TERMS

air inlet assembly	blower
heater assembly	distribution plenum
defroster	ventilation system
heater controls system	evaporation
refrigerant	humidity
dehumidification	air conditioner
compressor	condenser
receiver/dryer	desiccant
evaporator tubes	accumulator
automatic temperature control system	

REVIEW QUESTIONS

DIRECTIONS: The following questions are similar to those used on automotive technician certification tests. On a separate sheet of paper, write the letter of the correct choice.

1. Technician A says that used refrigerant may contain hydrochloric acid.
   Technician B says that used refrigerant should be released into the air.
   Who is correct?
   A. A only
   B. B only
   C. Both A and B
   D. Neither A nor B

2. Which of the following is cooled by expanding refrigerant?
   A. Compressor
   B. Receiver/dryer
   C. Evaporator
   D. Condenser

3. Which of the following statements is correct?
   I. The vaporization of refrigerant produces a cooling effect.
   II. Vaporized refrigerant is liquified in the condenser.
   A. I only
   B. II only
   C. Both I and II
   D. Neither I nor II

Questions 4 and 5 are not like those above. Each has the word EXCEPT. For each question, look for the choice that does *not* apply. Read each question carefully before you choose your answer.

4. A heating system includes all of the following parts EXCEPT
   A. a distribution plenum.
   B. a receiver/dryer.
   C. a heater core.
   D. a ventilation system.

5. All of the following statements are *true* EXCEPT
   A. air conditioner/heater systems can be computer-controlled.
   B. heaters and air conditioners use the same plenum chambers.
   C. the evaporator changes heated refrigerant to a liquid.
   D. automatic temperature control operates to maintain a set temperature.

ESSAY QUESTIONS

1. Describe a typical heating system.
2. What are the functions of the parts of an air-conditioning system?
3. How does an air conditioning system produce cool and dehumidified air?

SUPPLEMENTAL ACTIVITIES

1. Locate and identify the functions of all heater system parts on a vehicle selected by your instructor.
2. Describe how doors to plenum chambers distribute heat and air conditioning to different parts of an automobile.
3. Locate and identify the parts of an air conditioning system on an automobile selected by your instructor.
4. Describe how an automatic temperature control system operates to maintain a selected temperature when the outside temperature changes.

66 Heating and Air Conditioning Service

UNIT PREVIEW

Heating and air conditioning systems share distribution plenums and ventilation systems and may share control systems. Heating system service includes engine cooling system checks and repairs, flushing, and repairs to hoses. Air conditioning system service involves checks for damage and proper operation as well as special safety procedures when working with refrigerant.

LEARNING OBJECTIVES

When you have completed your assignments and exercises in this unit, you should be able to:

❑ Perform heating system checks and service.
❑ Remove and replace a heater valve and heater core.
❑ Check for refrigerant level and locate leaks in an air conditioning system.
❑ Use approved methods and equipment to discharge, evacuate, and recharge an air conditioning system.

SAFETY PRECAUTIONS

Wear eye protection at all times when you work. Use extreme care to avoid serious burns from heated engine components and other vehicle parts.

The engine will be running during some tests. Use care to avoid having clothing, hair, tools, or any body part injured by moving fans and belts.

When servicing electrical parts, always disconnect the battery to avoid shocks, burns, and damage to electrical or electronic components.

The R-12 refrigerant used in air conditioning systems can cause serious injuries. Any skin or eye contact will cause instant frostbite and can lead to serious burns or blindness.

Always wear safety glasses or goggles when servicing a refrigeration system. Keep sterile mineral oil and a weak boric acid solution handy when working with refrigerant. Should any liquid refrigerant get into the eyes, use a few drops of sterile mineral oil to wash them out. Then, wash your eyes with the boric acid solution. See a doctor immediately, even though you may not feel irritation after this first-aid treatment.

If R-12 comes into contact with an open flame, it produces poisonous phosgene gas. Avoid using older leak-detector tools that operate with an open flame. Newer, safer electronic leak detectors and dye-chemical leak detectors are recommended.

Air conditioning refrigerant can cause serious damage to the environment. Refrigerant must be collected in accordance with all federal, state, and local regulations. Use only approved equipment and methods when charging, evacuating, and recharging an air conditioning system.

Although ozone at the surface level is considered a pollutant, the upper ozone layer in the Earth's atmosphere protects against harmful ultraviolet radiation from the sun that can cause skin cancer. It is expected that the refrigerant R-12 will be phased out and replaced with a different refrigerant in the future. This may require redesign of air conditioning system parts to use less harmful refrigerants.

Do not heat a refrigerant container or store it in a hot portion of the building. Temperatures above a recommended safety level can increase refrigerant pressure and cause containers to explode. Do not weld or steam clean on or near system parts or refrigerant lines.

 ## TROUBLESHOOTING AND MAINTENANCE

A visual inspection of heating system parts should be made when other underhood services are done. The most common problems are leaks, damaged, loose, or missing drive belts, defective/inoperative compressor clutches, and low coolant or refrigerant levels.

Check the heater hoses. There should be no leaks, frayed ends, cracks, or hardness when a heater hose is squeezed. Hose clamps should be tight and in good condition. If any of these problems are found, replace the hose or clamp. Air conditioning hose inspection is discussed under Topic 66.4.

Check the air inlet and heater assemblies for debris and damage. Remove debris and repair or replace any damaged parts.

Control cables, wires, and vacuum hoses for heating system components are routed through the firewall. Check these cables, wires, and hoses for wear, damage, tightness, or missing parts. Replace or repair them as necessary.

HEATING AND AIR CONDITIONING SERVICE

66.1 DIAGNOSING HEATER PROBLEMS

To check heater output, place a thermometer at an outlet duct in the automobile and set the heater controls according to the manufacturer's recommendations. Some manufacturers' service manuals will list minimum temperature requirements for a heater that is operating properly **[Figure 66-1]**. If heating problems exist, first check the heating system for leaks and damage.

Insufficient Heat

If the heater does not produce sufficient heat, make the following checks:

- Coolant level in the vehicle's radiator
- Proper cooling system thermostat operation
- Flow of coolant through heater hoses/core
- Heater control setting and operation
- Cable control attachment and operation
- Vacuum hose conditions
- Electrically operated connections, switches, and fuses
- Distribution system openings and door operation
- Sensor operation on automatic temperature control systems.

Draftiness and cold spots can be caused by missing or damaged weatherstripping, insulation, seals, or other material. Check around the doors, windows, firewall, and floor areas.

Too Much Heat

When a heater puts out too much heat, check the cooling system thermostat for proper operation. Check heater valve controls and make sure the valve opens and closes properly. Also check thermostatic switches and resistors, when used.

Blower Problems

If a blower is not operating properly, check the following items:

- Fuse circuit and other electrical connections
- Relay switches and all thermostatically controlled switches.

Check for electrical current at the blower motor when the switch is on. Be sure the motor has a proper ground circuit. If current is present and the blower does not operate, replace the motor. If there is no current, check the switch, fuse, wiring, or thermostatic controls. Refer to the manufacturer's service manual for specific troubleshooting procedures.

66.2 HEATER SERVICE

Always refer to the manufacturer's service manual for correct heater servicing procedures. In many cases, wire, cable, and vacuum hose replacement can be completed without removing the heater assembly. The cooling system must be drained before you attempt to remove a heater core. Heater hose, heater valve, and cooling system service is discussed in Unit 18.

Heater core leaks usually produce a wet carpet on the passenger side. To gain access to the heater core, you usually must remove parts of the distribution plenums, dash, and other units **[Figures 66-2 and 66-3]**.

A cable-type heater control assembly usually is removed from the dashboard to check connections and operation **[Figure 66-4]**. If they are accessible, doors, mechanical cables, vacuum hoses, and electrical switches can be checked at the heater housing or from under the dashboard.

66.3 DIAGNOSING AIR CONDITIONER PROBLEMS

Air conditioner problems usually are noticed as lack of cooling, noises, and/or odors. Always refer to the vehicle manufacturer's troubleshooting charts and service procedures for air conditioning system diagnosis.

SAFETY CAUTION

When you work around air conditioning systems, always wear eye protection. Do not disconnect air conditioning lines without proper instruction or without your instructor's permission.

Insufficient/No Cooling

If the air conditioner does not produce enough cold air, check for the following:

- Condenser fin cleanliness
- Compressor belt/operation
- Compressor clutch/electrical connections/dash switch

Temperature Reference Chart			
Ambient Temperature		Minimum Heater Outlet Duct Temperature	
Celsius	Fahrenheit	Celsius	Fahrenheit
15.5°	60°	62.2°	144°
21.1°	70°	63.8°	147°
26.6°	80°	65.5°	150°
32.2°	90°	67.2°	153°

Figure 66-1. Temperature reference chart.
CHRYSLER CORPORATION

734　　　　　　　　　　　　　　　　　　　　　　　　　　　　　　　PART IX: AUXILIARY SYSTEMS

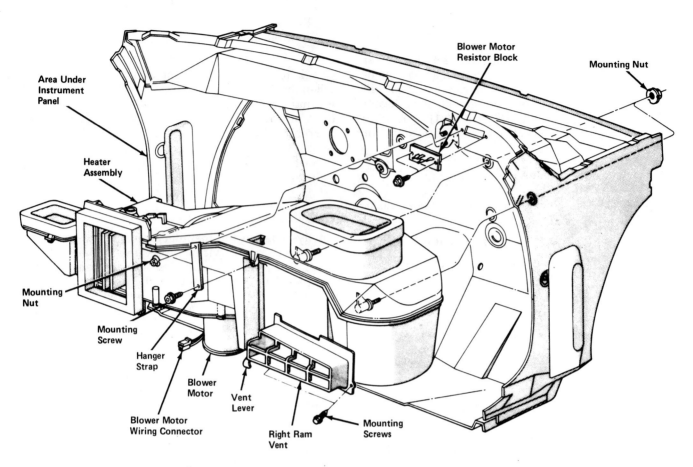

Figure 66-2. Heater assembly removal. CHRYSLER CORPORATION

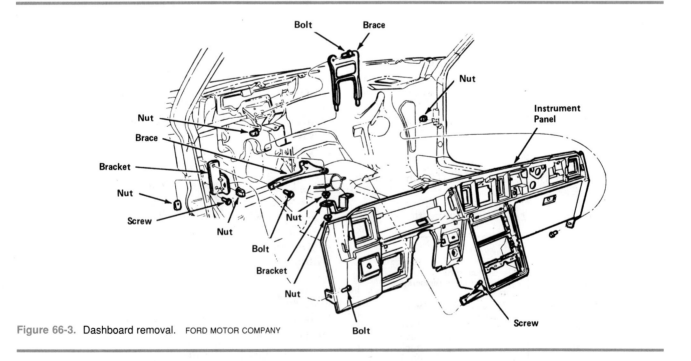

Figure 66-3. Dashboard removal. FORD MOTOR COMPANY

- Electrical connections to compressor clutch
- Refrigerant level/condition
- Vent and duct openings
- Refrigerant level/hoses/leaks (see below)
- Distribution plenum door operation
- Evaporator core fins cleanliness

HEATING AND AIR CONDITIONING SERVICE

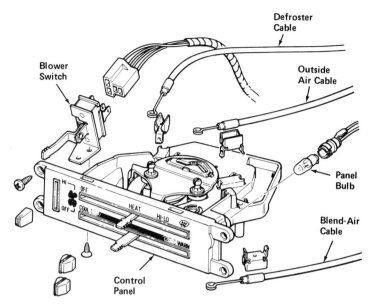

Figure 66-4. Parts of the control panel.
AMERICAN MOTORS CORPORATION

- Evaporator core icing (refrigerant control system problems)
- Receiver/dryer operation.

No/Insufficient Volume of Air

If cold air is felt at the air conditioner outlets, but no air is blowing, check the blower motor circuit. Check the on/off switch, resistor, and wires leading to the blower. No airflow also can mean that the plenum doors are not being opened by control cables, vacuum motors, or electrical motors.

Noises

Most air conditioner noises come from a drive belt or from the compressor itself. A squealing sound usually indicates a loose or badly worn drive belt. A rumbling sound indicates an internal compressor problem. Grinding or clashing noises indicate that compressor shaft or pulley bearings may be defective. The compressor contains oil for lubrication of internal parts. First check the drive belt with the engine OFF. Move back, then have a helper start the engine. Listen and watch to see if the noise only is heard when the air conditioning is switched ON and the compressor clutch engages. This indicates a defective compressor.

Odors

Mold and bacteria can grow in the bottom of the evaporator housing where dirt and debris have collected. First check if the evaporator drain tube is clogged. Use a wooden pencil or dowel to gently probe the drain tube and unclog it so that it can drain.

If odors persist, it may be necessary to disassemble the evaporator housing and clean the evaporator core. Use a mixture of half rubbing alcohol and half household laundry bleach to kill mold and bacteria that cause odors. Clean the core and housing thoroughly, allow to dry, and apply the mixture liberally with a paintbrush or spray bottle.

66.4 AIR CONDITIONER MAINTENANCE

Periodic maintenance usually involves inspection of some of the air conditioning components. This should be done at least once a year to check for proper operation.

Condenser Cleaning

Check the outside of the radiator and condenser cores for blockage caused by dirt, leaves, insects, or other foreign material. To clean out debris, use a hose and water pressure, directing the water from the inside toward the outside. Be sure the core is cleaned thoroughly. Use a paintbrush, if necessary, but be careful not to damage the delicate tubes and fins on the condenser and radiator. Never install a bug screen in front of the condenser, it will restrict air flow to the point of causing possible air conditioning and engine overheating problems.

Drive Belt Inspection

All belts, including the air conditioning compressor belt, should be checked whenever you make other underhood checks. Check the belt for wear, tension, cracks, glazing, and other defects. Tighten a serviceable belt or install a new belt as necessary.

Tubing and Hose Inspection

Check all hose connections and tubing for kinks, cracks, frayed ends, and loose or damaged connectors. Oil deposits on the hoses or tubing indicate a refrigerant leak.

SAFETY CAUTION

Leaking refrigerant can cause painful frostbite and serious injury. To remove or replace a defective air conditioner hose, the system must be discharged completely, as discussed in Topic 66.5.

Loose fittings may be carefully tightened to the manufacturer's torque specifications to stop minor leaks.

Magnetic Clutch Inspection

A *magnetic clutch* rotates with the drive pulley whenever the engine is running [Figure 66-5]. The clutch does not engage to turn the compressor until the air conditioner control switch is turned on.

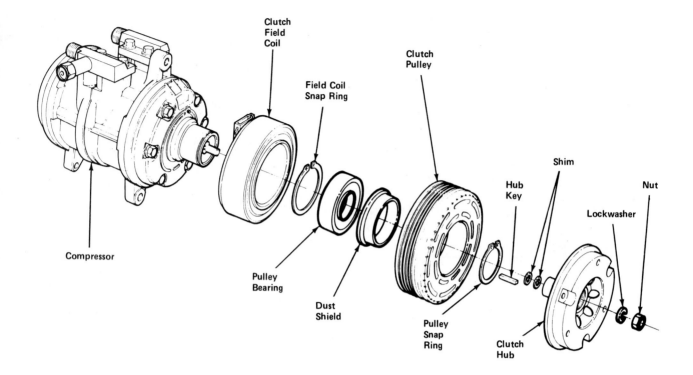

Figure 66-5. Parts of the clutch assembly. FORD MOTOR COMPANY

When a selector switch is set to Air Conditioning and the blower is switched ON, electrical current is sent to the field coil inside the clutch. The coil creates a strong magnetic force that locks the driving pulley and the compressor driveshaft together.

To check the operation of a magnetic clutch, raise the hood and have a helper start the engine and turn the air conditioner on and off. Watch the motion of the clutch hub on the compressor and listen for a noticeable clicking noise when it engages and disengages.

Refrigerant Level Inspection

Manufacturers often provide a *sight glass* that allows the technician to see the flow of refrigerant in the lines **[Figure 66-6]**. It may be located either on the receiver/dryer or between the receiver/dryer and the expansion valve.

To check refrigerant, open the windows and doors and set the controls for maximum cooling, and set the blower on its highest speed. Let the system run for about five minutes. Be sure the automobile is in a well-ventilated area, or connect an exhaust gas ventilation system.

Use care to check the sight glass while the engine is running. If you see oil-streaking, the system is empty. Bubbles, or foam, indicate the refrigerant is low. A sufficient level of refrigerant is indicated by what looks like a flow of clear water, with no bubbles.

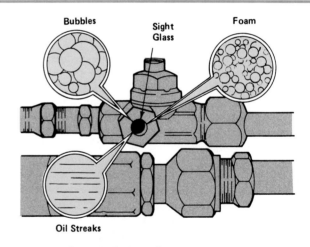

Figure 66-6. Sight glass diagnosis. FORD MOTOR COMPANY

SAFETY CAUTION

You will be working near the engine fan and rotating belts. Keep hands, clothing, and tools away from those moving parts or serious injury can result. Hot engine parts can cause serious burns.

If an air conditioning system has no sight glass, the system charge can be checked with pressure gauges. To connect the gauges, follow the manufacturer's directions carefully. A gauge assembly that is connected to check operating pressures is shown in **Figure 66-7**. Gauge assemblies are discussed later.

Heating and Air Conditioning Service

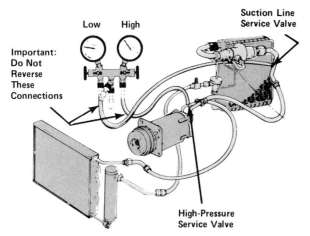

Figure 66-7. Checking refrigerant pressure.
FORD MOTOR COMPANY

66.5 AIR CONDITIONER SERVICE

Always service air conditioning systems according to the manufacturer's service recommendations. Servicing involves working with refrigerant under pressure. Refrigerant leaks can cause severe and painful injuries. Always wear eye protection and follow all safety precautions recommended for air conditioning service. Collect refrigerant only with approved equipment and methods, and in accordance with all federal, state, and local safety regulations.

Checking for Leaks

A system that is low on refrigerant usually has a leak. Most leaks are located at seal connections [**Figure 66-8**]. A common source of leakage is the compressor shaft seal. Recommended types of refrigerant *leak detectors* include:

- Electronic sensor
- Dye charge.

Electronic leak detectors have a sensing tube, or *sniffer*. To locate a leak, move the sensing tube near air conditioning lines. The unit emits a high-pitched sound when refrigerant is detected.

Dye charge is added to the system when recharging. The dye will flow out through leaks. Colored dyes or fluorescent materials are used. To see fluorescent materials, a special ultraviolet, or "black" light is directed toward the refrigerant lines. The fluorescent dye will glow under the black light.

Even a small leak reduces system performance after a period of time. Whenever the system is recharged or serviced, inspect for leaks.

Pressure Testing

A pressure test is conducted with a ***manifold gauge assembly***. A manifold gauge assembly determines

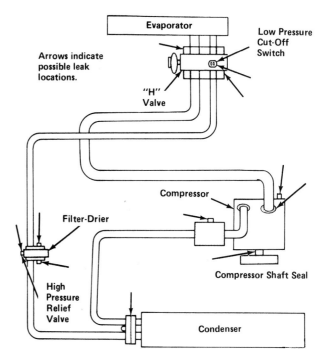

Figure 66-8. Possible refrigerant leak locations.
CHRYSLER CORPORATION

high and low pressures and correct refrigerant charge. It also helps in the diagnosis of system problems. High and low pressures are compared to determine proper system operation. A manifold gauge assembly is shown in **Figure 66-9.**

When connecting a manifold gauge assembly, follow the manufacturer's recommendations. Pressure readings on the gauge assembly are compared with specifications in the manufacturer's service manual.

Refrigerant Recharging

To replace parts or clean an air conditioning system, the system must be discharged, evacuated, and charged. The refrigerant is under pressure, so it must be handled safely and carefully. Always follow the manufacturer's recommendations.

A precisely measured amount of refrigerant is added to an air-conditioning system with a manifold gauge set or a ***charging station.*** A charging station includes a bulk 30-lb cylinder of refrigerant, pressure gauges, and a collection container for used refrigerant. Computerized charging stations that automatically discharge, evacuate, and recharge the system according to the technician's settings are used in many modern shops.

SAFETY CAUTION

Always wear safety glasses. Never open the high-pressure valve on the gauge set when the engine is running. High pressure from the compressor can cause the refrigerant container to explode.

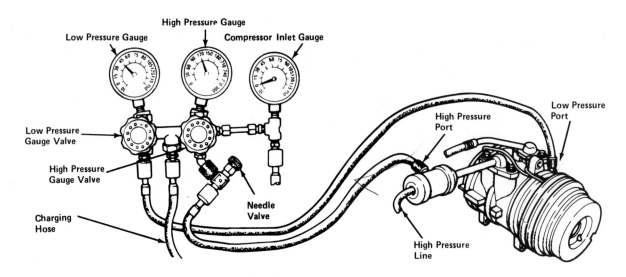

Figure 66-9. Manifold gauge assembly. CHRYSLER CORPORATION

Discharging

Refrigerant chemicals (chlorine) combine with and destroy the Earth's protective ozone layer in the upper atmosphere. Follow all federal, state, and local regulations for the safe handling of refrigerant. Use only approved equipment and methods to discharge, collect, and recycle used refrigerant.

Follow the vehicle or collection station manufacturer's instructions. Make sure that the gauge set valves are closed before attaching hoses to the refrigerant system [Figure 66-10].

Connect the discharge hose securely to the collector container. Open the discharge and suction valves to release the refrigerant into the collection

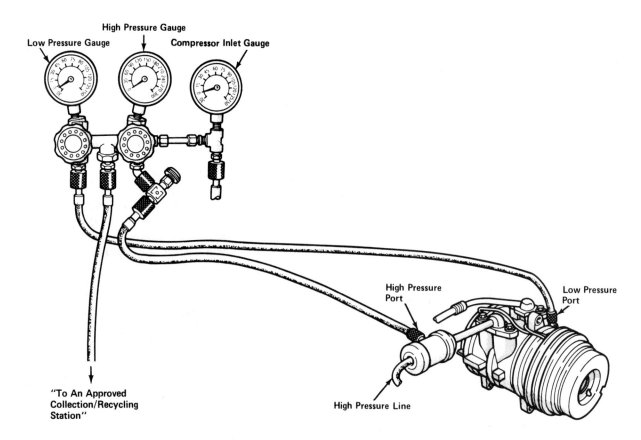

Figure 66-10. Manifold gauge set for discharging refrigerant. CHRYSLER CORPORATION

HEATING AND AIR CONDITIONING SERVICE

container. Follow the manufacturer's recommendations about measuring and adding compressor lubrication oil.

If any air conditioner parts are removed, cap or plug fittings to keep out moisture and dirt. Replace or repair any parts as necessary and add refrigerant and compressor lubricant oil as recommended.

Evacuating

Whenever an air conditioning system has been opened, it must be *evacuated.* Air and moisture in an air conditioning system will cause problems. Air will reduce system performance, and moisture will combine with the R-12 to form hydrochloric acid, which will damage system components. A vacuum pump is used to evacuate, or apply a vacuum to, the system. The vacuum causes refrigerant, oil, moisture, hydrochloric acid, and other contaminants to evaporate rapidly and be drawn from the system.

To evacuate a system, install a gauge set according to the manufacturer's directions [Figure 66-11]. Be sure the gauge set valves are closed before attaching hoses to the refrigerant system. If the gauge set indicates pressure, discharge the system.

Open the gauge set valves and operate the vacuum pump until the suction gauge registers the recommended reading. Continue operating the vacuum pump for at least a half an hour after the pressures on the gauges start dropping, or as recommended by the manufacturer. Close the manifold valves. Turn off the vacuum pump. The vacuum level should remain constant for at least five minutes. If not, the system has a leak. Locate and repair all leaks.

Charging

The system must be evacuated before charging. Use only the recommended refrigerant. Before handling R-12, review the safety precautions discussed at the beginning of this unit. Manufacturers specify the weight of the refrigerant that should be used in the system in ounces (oz.) or grams (gm). There are 16 ounces in a pound, and 1,000 grams in a kilogram. Refrigerant is available in 30-pound cylinders.

Avoid overcharging a system with more than the recommended weight of refrigerant. Overcharging can lead to air conditioning system parts damage as well as poor cooling performance.

Keep the refrigerant supply container upright. If a refrigerant container is on its side or upside down, liquid refrigerant will enter the system and can damage the compressor. A charging hookup for a manifold gauge set is shown in **Figure 66-12.**

SAFETY CAUTION

Carefully follow the manufacturer's recommendations on the refrigerant tank. Improperly handled, refrigerant tanks can explode and cause severe physical injury.

Close the manifold valves. Purge air from the charging line by opening the valve into the

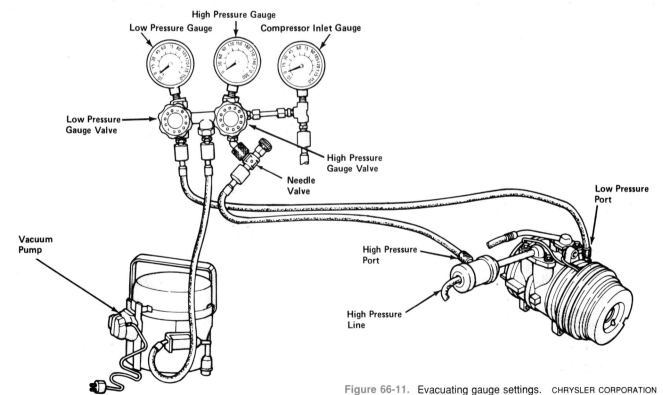

Figure 66-11. Evacuating gauge settings.   CHRYSLER CORPORATION

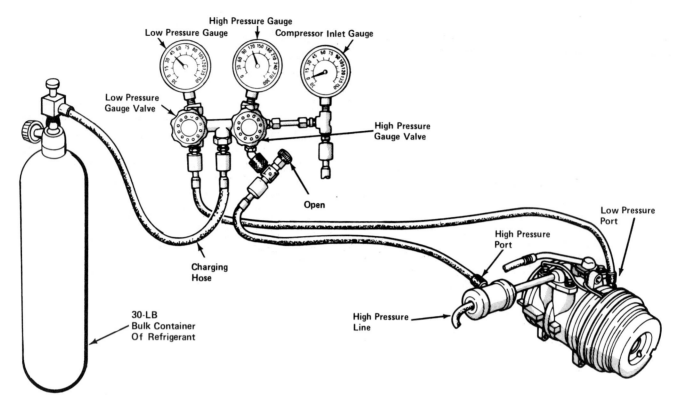

Figure 66-12. Charging hookup for manifold gauge set.
CHRYSLER CORPORATION

collection station as recommended by the manufacturer. Release the recommended amount of refrigerant into the system, according to the manufacturer's instructions.

When refrigerant no longer flows into the system, start the engine. Turn on the air conditioner and blower switches. Continue charging according to instructions.

Use care when removing the charging assembly from the air conditioner. Any refrigerant on the fittings can evaporate and freeze. Place a shop towel on fittings when removing them to protect your hands and eyes.

Manufacturers furnish directions on correcting low refrigerant level. Consult the vehicle manufacturer's service manual for correct amounts of refrigerant to be used in specific vehicles.

Replacing Air Conditioner Parts

If tests indicate only a small amount of refrigerant is needed, first check for leaks. If no leaks are found, small amounts of refrigerant may be added without evacuating the system. However, whenever the air conditioning system lines are disconnected to replace worn or damaged parts, the system must be evacuated and recharged. Check the manufacturer's service manual to determine if additional refrigeration oil must be added with the new components.

UNIT HIGHLIGHTS

- Refrigerant must be handled carefully with approved equipment and in accordance with all federal, state, and local regulations.
- Preventive maintenance for heater systems includes visual inspection of hoses, housings and ducts, and control systems.
- The dashboard and ducting assemblies often must be removed when servicing heater cores and air conditioners.
- Preventive maintenance for air conditioning systems includes visual inspection of hoses and cleaning of condenser fins.
- Refrigerant level sometimes can be checked through a sight glass.
- Electronic and dye check leak detectors are recommended to find refrigerant gas leaks.
- Pressure testing determines air conditioner pressures and the level of charge.
- Discharging a system means removing refrigerant and pressure from the air conditioner.
- Evacuating removes air, oil, moisture, acid, and contaminants from a system.
- Adding refrigerant to an air conditioner is called charging the system.

TERMS

magnetic clutch
leak detector
manifold gauge assembly
charging station
sight glass
sniffer
evacuate

REVIEW QUESTIONS

DIRECTIONS: The following questions are similar to those used on automotive technician certification tests. On a separate sheet of paper, write the letter of the correct choice.

1. Which of the following statements is correct?
   I. Heater operation can be checked with a thermometer.
   II. Too much heat can be caused by a blocked heater hose.
   A. I only
   B. II only
   C. Both I and II
   D. Neither I nor II

2. Technician A says that the compressor operates at all times when the engine is running.
   Technician B says that the compressor clutch uses magnetic force to couple a driving belt and pulley to the compressor driveshaft.
   Who is correct?
   A. A only
   B. B only
   C. Both A and B
   D. Neither A nor B

3. A sight glass is used to check
   A. whether the weight of the refrigerant in the system is correct.
   B. whether the refrigerant level is low.
   C. the temperature of the condenser core.
   D. the temperature of the evaporator core.

4. Which of the following statements is correct?
   I. Evacuating an air conditioning system removes moisture that can cause hydrochloric acid damage.
   II. It is not necessary to evacuate a system to add small amounts of refrigerant.
   A. I only
   B. II only
   C. Both I and II
   D. Neither I nor II

Question 5 is not like the first four questions. It has the word EXCEPT. For question 5, look for the choice that does *not* apply. Read the entire question carefully before you choose your answer.

5. All of the following statements are *true* EXCEPT
   A. to service a heater assembly, the dashboard may need to be removed.
   B. the battery should be disconnected before a blower is serviced.
   C. engine coolant should be drained before a heater assembly is disconnected.
   D. automatic temperature control systems use a combination heating/evaporator core.

ESSAY QUESTIONS

1. What checks can be made if a heater does not produce enough heat?
2. Describe common air conditioning problems and possible causes.
3. What does air conditioner maintenance and service include?

SUPPLEMENTAL ACTIVITIES

1. Describe safety precautions that must be followed when working with refrigerant.
2. Make a visual inspection of a heating system and diagnose a heater problem on a vehicle selected by your instructor.
3. Remove and replace a heater core.
4. Clean condenser fins, check refrigerant level, and check for leaks in an air conditioning system.
5. Safely and correctly discharge, evacuate, collect refrigerant, and recharge an air conditioning system.
6. Refer to the manufacturer's service manual and diagnose an air conditioner problem on a vehicle selected by your instructor.

67 Safety Systems

UNIT PREVIEW

Automobiles include devices and features that help to prevent damage and injury. Safety devices include brakes, wheels, and tires. Lights and indicators alert drivers to potential problems. Electrical circuit protection devices prevent fires and damage to the vehicle. Safety switches prevent vehicle operation under unsafe conditions. Windshield wipers and defrosters help to provide clear vision. Anti-theft devices and systems protect vehicles. Emission control devices help to promote and protect health. Occupant protection systems include both exterior and interior design features. This unit provides a brief overview of the functions and operation of safety systems.

LEARNING OBJECTIVES

When you have completed your assignments and exercises in this unit, you should be able to:

- Identify and describe devices that contribute to automotive safety.
- Describe how automobiles are designed to help prevent injury to occupants.
- Explain how vehicle design features contribute to safety.
- Identify and describe active and passive restraint systems.
- Describe the function and operation of air bags.

67.1 SAFETY DEVICES

Safety features are included in many major systems of the automobile. Examples of devices that contribute to the safety of vehicle occupants and others include:

- Brakes
- Tires and wheels
- Lights and indicators
- Electrical circuit protection
- Safety switches
- Windshield wipers
- Defrosters
- Anti-theft devices
- Emission controls.

Brakes

The brake system is the most important safety system of an automobile. A dual braking system provides for braking action even if one half of the system fails completely.

A brake system warning light on the dashboard alerts the driver if the pressure in either or both halves drops. The switch for this light is located in the combination metering/proportioning valve.

The light will remain on until the problem has been corrected and, in some cases, until the switch has been reset by a technician. The same light also may be used to warn the driver that the parking brake is applied or that the fluid level in the master cylinder reservoir is low.

The combination metering/proportioning valve controls brake fluid flow and application. Some automobiles have anti-lock brake systems to prevent wheel lockup, stop the automobile quicker and straighter, and maintain directional control. Refer to Unit 63 for a discussion of the brake system.

Tires and Wheels

If a tire loses air pressure suddenly, it may lead to an accident. Special sealant materials are used on the interior surfaces of some tires to help prevent flats. Some special tires are designed with special sidewalls that support the rim even if the tire loses pressure.

To prevent separation of the tire from the wheel, wheels are built with a safety rim. Wheels and tires are discussed in Unit 61.

Lights and Indicators

Current vehicles use many electrical lights to increase visibility, both to see and to be seen. Electrical indicators, such as turn signals, can be used to alert other drivers. Others act as indications of problem conditions or of required servicing **[Figure 67-1]**. Refer to Unit 36.

Headlights, stop lights, turn signals. Headlights allow a driver to see the road ahead. A dimmer switch allows a driver to adjust the beams according to traffic conditions. Correctly aimed headlights are angled away from oncoming traffic to allow other drivers to see without glare. Some automobiles have delay switches that turn off the headlights automatically when the engine is not running to prevent

SAFETY SYSTEMS

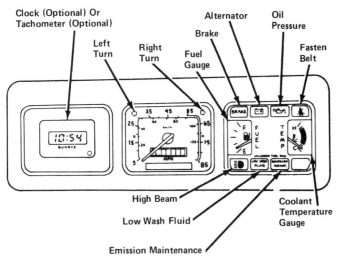

Figure 67-1. Safety indicators and gauges on dashboard. AMERICAN MOTORS CORPORATION

accidental battery drain. A headlight assembly is shown in **Figure 67-2.**

A headlight switch also operates taillights and side marker lights so that the automobile can be seen more clearly at night. Indicators inside the automobile can signal whether an external bulb is burned out.

Stop lights and a third, safety warning light in the rear operate when the brake pedal is depressed. Turn signals alert other drivers and are mounted so that they can be seen from the front, rear, and side of an automobile.

Emergency flashers. An emergency flasher system warns other drivers of dangerous conditions. The emergency flasher operates both front and rear turn signals at the same time.

Indicators. An indicator may signal that fuel is low or that a door is ajar. Indicators also may warn of brake problems, low oil pressure, or an overheated engine or catalytic converter. Indicators also are used to alert the driver to the need for service. Indicators include self-diagnostic codes produced by computers that can help a technician to troubleshoot a problem.

Audible warning devices. To get a driver's attention buzzers, chimes, and voice-warning devices warn of safety problems. If a key is left in the ignition switch and the driver's door is opened, an audible warning device will sound. Other uses for these devices include warnings of unbuckled seat belts. Some vehicles are equipped with voice warning devices (refer to Unit 38).

Other indicators. Not all indicators are operated electrically. Many parts have wear indicators that indicate the part should be replaced. These wear indicators are early warning signs that a damaging failure may result. They may be visual indicators (ball joints, Unit 58) or noise indicators (disc brake pads, Unit 64).

Electrical Circuit Protection

If a short circuit or accidental ground occurs, an excessive flow of electrical current can heat wires

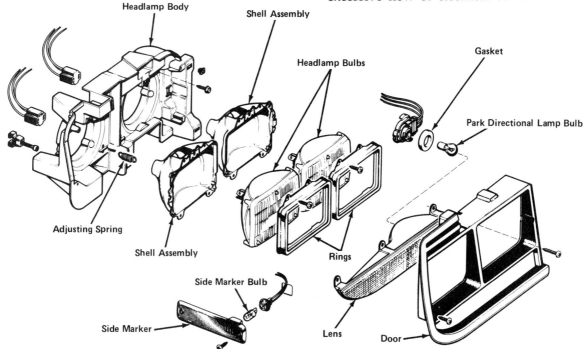

Figure 67-2. Headlight assembly. AMERICAN MOTORS CORPORATION

and lead to a fire. Fuses, fusible links, and circuit breakers are placed in most electrical circuits to protect the circuit components, the vehicle, and its occupants. If excessive current flows, these devices stop the flow of current (refer to Units 27 and 36).

Safety Switches

A neutral-start switch or clutch switch prevents an automobile from being started in gear. A rollover switch shuts off an electric fuel pump if the vehicle turns over in a collision to help prevent fires. An oil-pressure safety switch also may shut off the engine when oil pressure becomes too low.

Windshield Wipers

A windshield wiper and washer unit removes outside dirt and moisture from a windshield **[Figure 67-3]**. The washer sprays a cleaning solvent on the windshield, or rear window, to help remove dirt.

Defrosters

A fogged or steamed window on the inside of an automobile can be cleared by using the defroster and air conditioner at the same time. Refer to Unit 65. Defroster ducts are directed toward a windshield, and sometimes toward rear and side windows.

Most new automobiles use an electric defogger at the rear window. An electric defogger heats conductive paint on the surface of the glass to remove mist buildup from the interior window surface. Some defogger/deicers have electrical wires embedded in the glass to heat the surface and melt snow on the outside of the window, as well as defog the window.

Anti-theft Devices

A locking steering wheel is a simple anti-theft device (refer to Unit 59). It prevents the steering wheel from being turned fully when the ignition key is removed.

More complex anti-theft systems include computers and sensors. These units flash lights and activate the horn if a door, hood, or trunk is opened after the ignition key is removed. In addition, the ignition system is disabled (refer to Unit 36).

Emission Controls

Automobiles produce air pollution from fuel system evaporation, crankcase emissions, and exhaust emissions. Evaporative emission controls reduce the amount of unburned fuel vapors that escape into the atmosphere. Crankcase emissions are controlled by a positive crankcase ventilation system. Exhaust emissions are affected by engine design, fuel and ignition system controls, and exhaust system parts.

Emissions control devices are integral parts of an automobile and contribute to overall engine and vehicle durability and reliability. Refer to Unit 41.

67.2 OCCUPANT PROTECTION

Automotive designers engineer vehicles to protect occupants during a collision. Exterior and interior

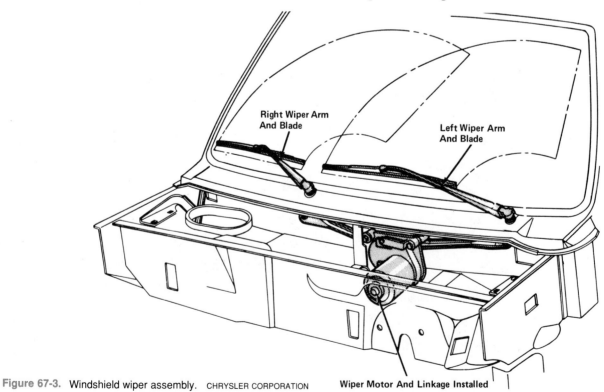

Figure 67-3. Windshield wiper assembly. CHRYSLER CORPORATION

design features work together to decrease the likelihood of injury.

Exterior Design

The front and rear sections of some passenger vehicles are designed to collapse on impact to absorb the force of a collision. The passenger compartment is designed to resist collapse and form a more rigid structure to protect the occupants.

Strong metal beams are built into automobile doors and side panels to protect occupants in side collisions.

Energy-absorbing bumpers, which function in much the same way as gas-filled shock absorbers, are used on many vehicles. These bumpers cushion low-speed impacts (from three to five miles per hour) to prevent damage to passenger cars and injuries from minor impacts **[Figure 67-4]**.

Interior Design

A dashboard is designed to "give" and is padded to cushion an occupant who might strike it. Safety knobs and latches on the doors prevent accidental opening and are shaped to lessen impact injury.

Strong anchors hold the seats firmly in place during an impact. Additional strengthened areas are used for mounting seat belts.

Safety glass is used in all windows. Safety glass crumbles into small pieces when broken. No large pieces of glass protrude to seriously cut an occupant. Some windshields also include a clear plastic inner liner. The liner helps to prevent an occupant's head from going through the windshield during a collision.

Collapsible steering column. To prevent driver injury, steering columns on passenger vehicles are made to collapse or bend on impact. Special connections and joints are used to keep the steering wheel from striking and seriously injuring the driver. Collapsible steering columns are discussed in Unit 59.

Active restraints. An *active restraint* is one that a vehicle occupant must make an effort to use. For example, in most vehicles the passengers must fasten seat belts for crash protection.

Seat and shoulder belts are required for front seat occupants. Shoulder belts for rear seat passengers are available on some automobiles. The *restraint anchor,* or attachment point, is connected securely to a strong part of the body or frame. A belt must be anchored properly so it will not break loose in an accident **[Figure 67-5]**.

A restraint system belt is made from webbed, synthetic material. On many current vehicles, a one-piece belt serves as both lap and shoulder restraint. A sliding "tongue" fixture is connected to a stationary buckle to fasten the harness. The buckle may be attached to the end of a short length of webbed material or may be part of a swiveling anchoring assembly. A release button allows the buckle to be removed.

The restraint belt on recent vehicles is passed through a *retractor*. A retractor is a reel that rotates and locks in place when a belt is tightened **[Figure 67-6]**. A retractor will allow the belt to move out slowly, permitting normal upper body movement by occupants. However, when a sudden movement occurs, as in an impact, a ratchet or pendulum mechanism

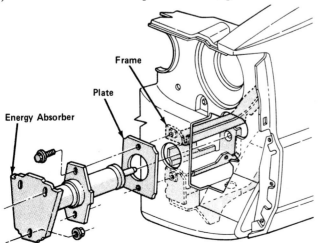

Figure 67-4. Energy-absorbing bumper. CHRYSLER CORPORATION

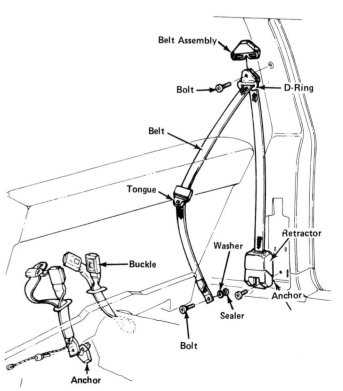

Figure 67-5. Safety belt assembly. FORD MOTOR COMPANY

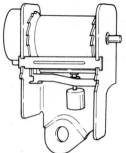

Figure 67-6. Safety belt retractor.
FORD MOTOR COMPANY

locks the belt tightly. This action helps to hold the passengers securely and prevent injuries.

An electrical connection is attached to the buckle end of a belt assembly. The circuit operates a warning light and an audible warning device. These devices operate for several seconds after the ignition is turned on. The audible warning will shut off if the belts are fastened or if the transmission is put into a gear position.

Passive restraints. A *passive restraint* is one that operates automatically. No action is required of the occupant to make it functional. Two types of passive restraints are automatic seat belts and air bags.

An automatic seat and shoulder belt system can be connected to an automatic retractor and to a door [Figure 67-7]. When a door is opened, the belt stretches out with the door. An occupant enters and sits in the seat. When the door is closed, the belt wraps around the occupant. Electric motors are used in some automatic belt systems to position shoulder restraint belts, but the occupant must fasten the lap belt by hand. Both mechanical and electrical passive restraint systems work to position front belts only.

Air bags. The formal name for an air bag is an *air cushion restraint system.* In a frontal impact collision, a sensor activates chemicals that produce a high-pressure gas rapidly. The gas, sodium azide, inflates the air bag in a few thousandths of a second to provide a cushion for the occupant [Figure 67-8]. The bag then deflates in a few seconds.

Most air bag systems are designed for front seat driver/passenger protection only. A driver's side air bag system is mounted in the steering wheel. When an air bag device is included for the front passenger side, it is mounted in the dashboard or glove compartment cover.

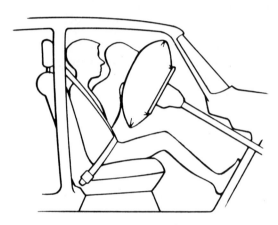

Figure 67-8. Air bag system.

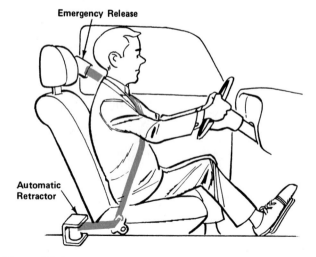

Figure 67-7. Passive restraint system.

UNIT HIGHLIGHTS

- A safety device helps to prevent accidents and to avoid or minimize damage and injury.
- Occupant protection features include both exterior and interior design features.
- An active restraint must be operated by a driver or passenger.
- A passive restraint operates automatically, either by mechanical or electrical means.

TERMS

active restraint
passive restraint
safety glass
restraint anchor
air cushion restraint system
retractor

REVIEW QUESTIONS

DIRECTIONS: The following questions are similar to those used on automotive technician certification tests. On a separate sheet of paper, write the letter of the correct choice.

1. Which of the following statements is correct?
   I. A rollover switch operates when an automobile is upside-down or in a collision.
   II. A rollover switch shuts off fuel flow to help prevent fires.
   A. I only
   B. II only
   C. Both I and II
   D. Neither I nor II

2. A passive restraint system
   A. requires that passengers operate the system.
   B. operates automatically.
   C. consists of an automatic shoulder belt and a manual seat belt.
   D. consists of manual seat belts and an air-bag system.

3. Technician A says that air bags are inflated by high-pressure gas stored in large steel cylinders behind the steering wheel or dash.
   Technician B says that air bags are inflated when chemicals combine to produce gas rapidly.
   Who is correct?
   A. A only
   B. B only
   C. Both A and B
   D. Neither A nor B

Questions 4 and 5 are not like those above. Each has the word EXCEPT. For each question, look for the choice that does *not* apply. Read each question carefully before you choose your answer.

4. All of the following statements are correct EXCEPT
   A. safety devices are designed to help avoid or prevent accidents and injuries.
   B. dual-system brake lines allow half of the system to operate if the other half fails.
   C. safety devices are not part of the design of a vehicle.
   D. indicators and warning devices alert the driver to problems.

5. All of the following are safety features EXCEPT
   A. brakes.
   B. exterior trim.
   C. safety glass.
   D. energy-absorbing bumpers.

ESSAY QUESTIONS

1. What systems contribute to driver and occupant safety?
2. Describe active restraints.
3. What are passive restraints?

SUPPLEMENTAL ACTIVITIES

1. Identify and describe visual and audible safety warning devices in a brake system.
2. Locate fuses, fuse links, and circuit breakers in an automobile selected by your instructor.
3. Locate and identify a rollover switch and brake pressure warning switch.
4. Locate and identify a collapsible steering column on an automobile selected by your instructor.
5. Identify interior design features that contribute to safety on a vehicle selected by your instructor.
6. Locate and identify parts of a passive restraint system on a vehicle selected by your instructor.
7. Locate and identify energy-absorbing bumper cartridges on a vehicle selected by your instructor.

68 Safety Systems Service

UNIT PREVIEW

Many safety device preventive maintenance procedures and repairs are included in service procedures for other systems. These systems include brakes, wheels and tires, electrical systems, ignition, fuel, and emission control systems.

Refer to the manufacturer's service manual and to specific units in this book for safety precautions and procedures. This unit provides an overview of servicing procedures for seat belts, air bags, and energy-absorbing bumper cartridges.

LEARNING OBJECTIVES

When you have completed your assignments and exercises in this unit, you should be able to:

❑ Inspect and replace defective seat belts.
❑ Inspect safety belt mounting areas for damage and distortion.
❑ Safely disarm and inspect an air bag assembly.
❑ Replace an energy-absorbing bumper cartridge.

SAFETY PRECAUTIONS

Always wear eye protection when you work on air bags or energy-absorbing bumper cartridges.

Refer to the vehicle manufacturer's service manual for specific air bag inspection and service procedures. Air bags are inflated by a gas produced in a chemical reaction. The air bag sensor system must be disarmed before you attempt service or inspection procedures to prevent injuries caused by accidental inflation. Some manufacturers may recommend that the battery and/or electrical system connections be disconnected while working on air bag systems.

Cartridges for energy-absorbing bumpers operate under extreme pressures. Do not drill into, strike, or heat a cartridge. If the cartridge explodes or pressurized contents escape, severe injury can result.

TROUBLESHOOTING AND MAINTENANCE

A visual inspection of all safety systems should be made routinely whenever an automobile is being serviced. Check tires, wheels, windshield wipers, and lights. Look for fluid leaks and damaged or worn parts. All safety systems should be in good condition so they will operate properly if needed.

68.1 DIAGNOSING OCCUPANT PROTECTION SYSTEM PROBLEMS

Seat and shoulder belt assemblies should be checked for wear and damage periodically. Replace any belt that shows fraying and/or wear.

Also note the date of manufacture as indicated by labels on the belt *webbing.* As belt material ages, it becomes weaker and may need to be replaced. Refer to the manufacturer's service manual for belt life and replacement guidelines.

Replace any safety belts that have been bleached or dyed. Bleaching and dying will weaken and ruin safety belt webbing material, even though it may appear good. Also replace any seat belt that was worn during a collision. The force of a collision stretches and weakens the webbing material.

Air bag systems should be checked according to the manufacturer's recommendations. Parts of an air bag system that should be inspected are shown in **Figure 68-1.**

Make sure that all warning indicators are operative. Check all anchors, retractors, and connections. Replace any parts that are damaged or that fail to function.

68.2 OCCUPANT PROTECTION SYSTEMS SERVICE

Seat and/or shoulder belts should be replaced after they have been subjected to stretching forces during a collision. Refer to the vehicle manufacturer's service manual for belt replacement guidelines.

Belt Assembly Replacement

Before installing a new belt assembly, the attaching areas should be inspected for damage and distortion. Sheet metal around the attaching points may have to be reworked to its original shape.

To remove belt assemblies, refer to **Figures 68-2** and **68-3.** Disconnect the anchor assembly near the floor and the D-ring that is attached near the ceiling. At the rear seat, remove the seat cushion, then disconnect all assemblies. Replace all worn or damaged parts. Install the assemblies in the reverse order of removal. Torque all fasteners to the manufacturer's specifications.

SAFETY SYSTEMS SERVICE

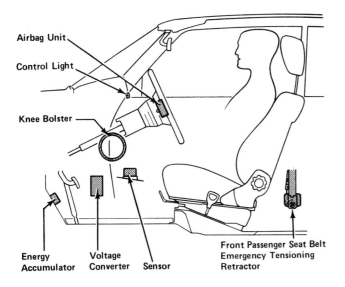

Figure 68-1. Parts of an air bag system.
MERCEDES-BENZ OF NORTH AMERICA

NOTE: Sometimes a new retractor will not operate properly after it has been bolted into a damaged or distorted mounting area. The new retractor could be warped and fail to function. Reshape the sheet metal at the mounting area and install another complete new belt assembly. Be careful when rustproofing an automobile not to spray body sealant onto the belt retracting mechanism. Pull belts out all the way before application to prevent soiling the belts.

Air Bag Replacement

Air bags are not reusable. If they are activated, the air bag and the activating gas-producing cartridge must be replaced.

SAFETY CAUTION

Air bag systems can be activated accidentally if sensors and/or electrical connections are activated. Always disconnect and tape the negative connector before servicing. Wait at least two minutes before proceeding. Refer to the manufacturer's service manual for disarming procedures and other servicing information.

Energy-Absorbing Bumper Cartridge Replacement

A bumper usually must be removed to replace an energy-absorbing cartridge. On some automobiles, front bodywork also must be removed. Always refer to the manufacturer's service manual before replacing an energy-absorbing cartridge.

SAFETY CAUTION

Never drill into an energy-absorbing bumper cartridge to remove pressure. Never strike or heat the cartridge. Cartridge pressures can be 10,000 psi (68,950 kPa). If the cartridge explodes or leaks, severe injuries can result.

To replace a cartridge, remove the push-on fasteners that hold a boot over the unit [Figure 68-4]. Support the bumper properly. Remove the nuts that

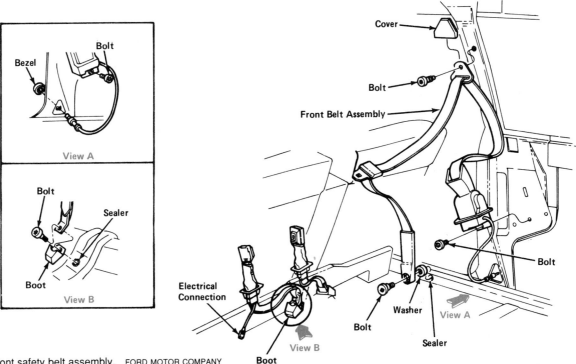

Figure 68-2. Front safety belt assembly. FORD MOTOR COMPANY

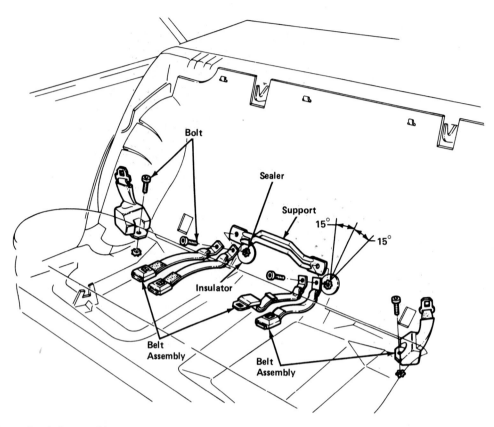

Figure 68-3. Rear safety belt assembly. FORD MOTOR COMPANY

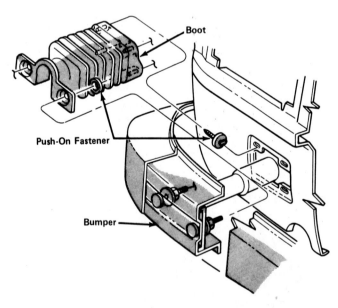

Figure 68-4. Bumper boot removal. CHRYSLER CORPORATION

hold the bumper to the energy absorber. **Figure 68-5** shows parts of the bumper assembly. Lower the bumper to the floor. Replace the cartridge. Parts of a cartridge assembly are shown in **Figure 68-6**.

UNIT HIGHLIGHTS

- Labels on safety belts indicate the date of manufacture; older belts must be replaced at intervals recommended by the manufacturer.
- Safety belts become stretched and weakened during a collision and must be replaced.
- Belt anchor and attaching points may require straightening after a collision.
- Air bag systems must be disarmed before servicing.
- Bodywork and bumpers may require removal and/or straightening to replace an energy-absorbing cartridge.

TERMS

webbing

SAFETY SYSTEMS SERVICE 751

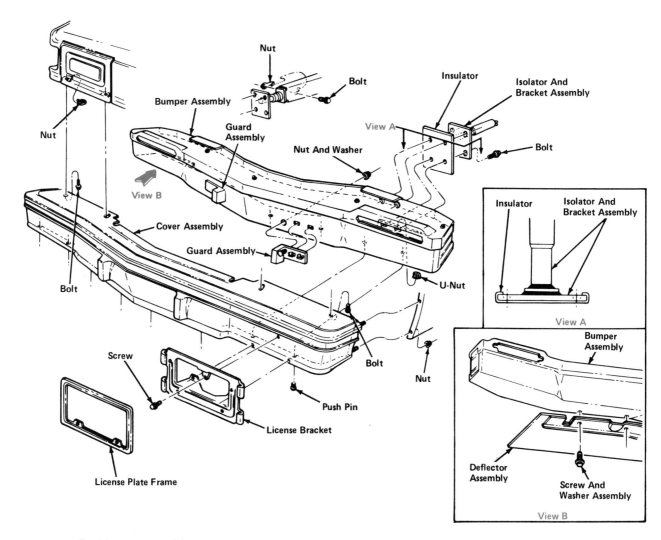

Figure 68-5. Front bumper assembly. FORD MOTOR COMPANY

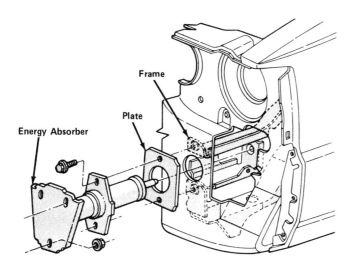

Figure 68-6. Bumper cartridge assembly.
CHRYSLER CORPORATION

REVIEW QUESTIONS

DIRECTIONS: The following questions are similar to those used on automotive technician certification tests. On a separate sheet of paper, write the letter of the correct choice.

1. Technician A says that labels on a belt assembly indicate when the belt was manufactured.
   Technician B says that belts will last the life of the automobile.
   Who is correct?
   A. A only
   B. B only
   C. Both A and B
   D. Neither A nor B

2. Which of the following statements is correct?
   I. An air bag assembly cannot be serviced.
   II. An air bag assembly creates high-pressure gas when activated.
   A. I only
   B. II only
   C. Both I and II
   D. Neither I nor II

3. To remove an energy-absorbing bumper cartridge, you must
   A. hammer on the cartridge.
   B. remove the bumper.
   C. remove pressure from inside the cartridge.
   D. heat the cartridge with a torch.

4. Technician A says that an energy-absorbing bumper cartridge contains high pressure.
   Technician B says that the cartridge is not pressurized.
   Who is correct?
   A. A only
   B. B only
   C. Both A and B
   D. Neither A nor B

Question 5 is not like the first four questions. It has the word EXCEPT. For question 5, look for the choice that does *not* apply. Read the entire question carefully before you choose your answer.

5. All of the following statements are *true* EXCEPT
   A. a belt assembly that was used in a collision should be replaced.
   B. seat belt anchor mounting points often must be straightened after a collision.
   C. a new belt retractor that does not operate properly must be discarded.
   D. anchor fasteners do not have to be torqued to specifications.

ESSAY QUESTIONS

1. What problems require that seat belts be replaced?
2. What safety precautions should be followed when an air bag is replaced?
3. What safety precautions should be followed when an energy-absorbing bumper cartridge is replaced?

SUPPLEMENTAL ACTIVITIES

1. Inspect seat/shoulder belt webbing for damage and wear and replace any defective belts.
2. Inspect seat belt anchor points for damage or distortion on a vehicle selected by your instructor. Make sheet-metal repairs as necessary.
3. Refer to the manufacturer's service manual to safely disarm and inspect an air-bag system on a vehicle selected by your instructor.
4. Refer to the manufacturer's service manual and replace an energy-absorbing bumper cartridge on a vehicle selected by your instructor.

Glossary

A

accelerator pump A small pump in the carburetor that supplies extra fuel when the throttle is depressed quickly for acceleration. It prevents a hesitation or flat spot when the fuel discharge changes from the idle circuit to the main metering circuit.

accumulator A shock absorber for apply devices in an automatic transmission; also, in an air conditioning system, a tank that holds and prevents liquid refrigerant from returning to the compressor. In some cases, the accumulator contains a desiccant, which removes moisture from the refrigerant.

acid fog Fog that contains sulfuric, nitric, and other acids that are harmful to living things.

acid rain Rain that contains sulfuric, nitric, and other acids that are harmful to living things.

activated charcoal A highly porous substance that can store large amounts of vapors.

active restraint A manual seat and shoulder belt assembly that must be activated by an occupant.

active suspension system A suspension system that reacts to, or controls, additional body movements, such as: body tilt (front to back) and body lean (side to side) during cornering, braking, and acceleration.

actuator A device that responds to a signal by producing an action, such as moving a mechanical device.

adhesive A glue used to hold objects together.

adjustable open-end wrench An open-end wrench that has a moveable jaw that can be opened or closed to fit many different diameters.

adjustable steering column A steering column that contains a mechanism to allow the steering wheel to be tilted or telescoped.

adjusting sleeve A part of a tie rod that is used during wheel alignment for an adjustment of tie rod length (toe).

advance To make ignition spark occur sooner.

advance timing light A timing light that allows a technician to determine ignition advance by reading a number scale.

aeration The presence or addition of air bubbles to a liquid.

after TDC (ATDC) A piston position after top dead center, specified in degrees of crankshaft rotation. *See also* top dead center (TDC).

air bleed An orifice that controls the flow of air mixing with fuel.

air-bleed jet A jet through which air can be added progressively to a fuel flow to prevent overenrichment.

air chisel An air-powered hammering tool used to remove bodywork and exhaust system parts.

air conditioner A device used to control the temperature of an enclosed environment, such as the passenger compartment of a car.

air cooled An engine cooling system that operates with air.

air cushion restraint system An air bag.

air drill A spark-resistant drill that operates on compressed air.

air filter A device to trap and hold airborne contaminants in an intake air charge.

air-hold adapter A hollow metal fitting of the same size as a spark plug; used to introduce compressed air into a cylinder.

air horn The top section of the carburetor, through which air enters the barrel(s). Provides a base for the air cleaner housing.

air impact wrench A socket wrench powered by compressed air.

air injection The introduction of fresh air into an exhaust manifold for additional burning; also, to provide oxygen to a catalytic converter.

air inlet assembly The portion of a heating system that draws in outside air and forces it through the heating system with a blower.

air nozzle A device that controls and directs compressed air coming out of a line.

air pressure test Applying air pressure to an automatic transmission case to check operation of parts.

air ratchet An air-powered wrench similar to an air impact wrench but used for lighter parts.

air resistance The resistance to the motion of a body through the air.

air suspension A suspension that uses four air bags instead of conventional springs.

align boring The process of boring crankshaft bearing housings in a straight line.

Allen head screw A screw that has a hollow hexagonal depression in its head.

alloy A mixture of metallic and/or nonmetallic elements. Usually produced by melting the elements and allowing them to fuse together.

alternator An electrical generator that produces alternating electrical current (AC) when turned by engine power. The AC is changed to direct current (DC) for vehicle systems and the battery.

ammeter An electrical measuring device used to measure current flow.

ampere A measurement of electrical current, equivalent to 6,280,000,000,000,000,000,000 electrons passing a point in one second.

analog A non-digital measuring method that uses a needle to indicate readings. A typical dashboard gauge with a moving needle is an analog instrument.

anchor pin A pin that prevents brake shoe rotation during braking.

aneroid bellows A device that responds to atmospheric pressure and can be used to move a metering needle within a jet to automatically compensate for altitude changes.

anti-friction bearing A ball or roller bearing.

anti-lock braking system (ABS) A series of sensing devices at each wheel that control braking action to prevent wheel lockup.

anti-seize compound A type of lubrication that prevents fasteners from seizing on one another.

anti-wear additive An oil additive that coats bearings and other moving parts to help prevent wear.

anvil The stationary surface of a micrometer.

apply device A part that holds or drives the planetary gear system in an automatic transmission.

arbor A shaft to which other parts may be attached.

arbor press A device that applies strong force to objects through a rack and pinion gear arrangement.

arc-joint pliers Slip-joint adjustable pliers that can be adjusted to many different positions.

armature The inner, rotating part of an electrical motor or generator.

aspect ratio The height of a tire, from bead to tread, expressed as a percentage of the tire's section width.

atom The smallest particle of matter that can exist alone and retain its unique characteristics, or in combination with other elements.

atomization The process of reducing fuel to tiny droplets during carburetion.

automatic level control An electronically controlled system that keeps the rear suspension at a proper level when a heavy load is added.

automatic temperature control A ventilation, heating, and air conditioning control system that uses a microprocessor to maintain temperature automatically.

automatic transaxle A transaxle that shifts automatically once placed in the proper drive range.

automatic transmission A transmission that shifts automatically.

automatic transmission fluid (ATF) An oil used in automatic transmissions.

automotive machinist A machinist who specializes in precision machining of automotive parts.

automotive specialty technician A technician who specializes in servicing one part of an automobile.

axle bearing A roller bearing, or ball bearing, that supports the weight of the vehicle through the axle housing and axle shafts.

B

backfire An explosion of air-fuel mixture in the intake manifold.

backing plate A stationary mount that prevents brake shoes from turning when they contact the drum.

backlash The amount of free play between the ring gear teeth and the drive pinion gear teeth.

backpressure Pressure created by restriction in an exhaust system.

backpressure transducer A device that senses exhaust gas pressure and operates a vacuum control valve to regulate the pressure.

balance To adjust the weight of a rotating part so that it rotates without vibration or whip.

balance pad A thick portion of a piston or connecting rod that can be ground down to lighten and balance the part.

balancing coil gauge A gauge that uses two coils of wire to create magnetic fields to move an indicating needle.

ballast resistor A primary ignition circuit electrical resistor that changes resistance in response to dwell angle and engine rpm.

ball joint A connection formed by a rounded, ball-shaped part that fits into a matching hollow socket; connects the control arms to the steering knuckle.

ball nut A nut with gear teeth on one side that fits over a worm gear and is rotated by ball bearings.

ball return guide Part of a recirculating ball gearbox that provides a continuous loop that collects and recycles ball bearings.

bank An area of a V-type or flat engine cylinder block that contains the cylinders. This term applies to engines with more than one bank.

bar A unit of pressure measurement. One bar is equal to 14.5 psi (99.98 kPa).

barrel *See* throttle bore.

battery A chemical device that produces electrical energy to start an engine.

battery cell A group of battery elements connected in series, positive to positive and negative to negative. A single automotive battery cell produces approximately 2.1 volts.

battery charger A machine that changes line current (110V AC) to low-voltage DC current to charge vehicle batteries.

battery element Two battery plates (one positive and one negative) immersed in electrolyte and kept apart by a separator strip.

battery hydrometer A measuring tool used to check the density of electrolyte.

battery load tester An instrument used to measure a battery's electrical capacity.

battery post A round projection to which all negative or positive battery plates are connected.

battery starter tester An electrical testing device that includes a voltmeter, an ammeter, and a variable electrical resistance load.

battery voltage tester A voltmeter.

bead The edge of a tire's sidewall that helps hold the tire to the wheel.

bead blasting A process that uses compressed air to blow small glass beads and chips at metal parts. The glass beads gently abrade, or wear off, deposits.

bearing A component used to hold a rotating part and allow it to turn smoothly.

bearing cap A heavy, machined metal clamp used to attach a connecting rod bearing or main bearing to the crankshaft. Also, the sheet metal protective cap on a wheel bearing assembly, sometimes called a dust cover.

bearing crush The process of compressing a bearing into place as the bearing cap is tightened.

bearing spread The distance a larger bearing insert must be compressed to fit into its smaller machined opening.

before TDC (BTDC) A piston position before top dead center, specified in degrees of crankshaft rotation. *See also* top dead center (TDC).

bellhousing A protective bell-shaped, metal case that houses the clutch.

belt A layer of material placed between the carcass and tread of a tire.

bench adjustment An adjustment performed while a part or assembly is removed from the vehicle.

beveled The tapered end of a part or tool.

bezel A retainer around a light or instrument.

bias belted tire A tire with bias plies and belts.

bias current A small amount of current applied to the emitter-base junction of a transistor to enable it to conduct.

bias ply tire A tire with plies that crisscross diagonally.

big end The end of the connecting rod that bolts around the crankpin.

billing The cost of servicing a vehicle.

binding force The force that attracts electrons and holds together atoms of matter.

black box A device that produces a certain output from a given input.

bleeder screw A variation of a bleeder valve.

bleeder valve A valve on a brake assembly that is opened when the brakes are bled.

bleeding A procedure that eliminates air from a hydraulic system.

blind rivet A small metal rivet inserted and fastened by expansion with a special tool.

blower A fan that pushes, or blows, air through a ventilation, heater, and/or air conditioning system.

blowgun A tool that controls and directs compressed air.

blown Ruptured through pressure.

blue To soften and turn blue, as steel, from excessive heat.

body The part of the automobile that holds and protects the occupants and cargo.

bolt A fastener with external threads, used with a nut.

bolt pattern The mounting configuration of bolts on various engine assemblies.

bonded Held in place with adhesive.

boost Manifold pressure created by a turbocharger.

booster valve A valve that increases mainline hydraulic pressure to clutches and bands in an automatic transmission.

bore The diameter of a machined hole.

boss A reinforced, protruding area around a hole.

bottom dead center (BDC) The position of the piston at the bottom of either the intake or power stroke.

bottom-feed injector In an electronic fuel injection system, an injector that sprays or feeds fuel into the system from the bottom.

Bourdon tube A springy, flexible metal tube that uncoils as pressure is applied; used in gauges and instruments.

box-end wrench A wrench with a closed head that encircles a fastener.

brake drum The part of a drum brake assembly that applies the pressure to slow or stop a vehicle.

brake dynamometer A dynamometer that applies a braking force to an engine to test its torque output.

brake horsepower (bhp) Output horsepower, as measured on a brake dynamometer.

brake lines Hoses and steel tubing that route hydraulic fluid from the master cylinder to the wheel cylinders.

brake lining The friction material that is riveted or bonded to a brake shoe and used to stop vehicle motion.

brake pad *See* brake shoe.

brakes Assemblies at each wheel that slow or stop an automobile.

brake shoe The part of a drum brake assembly that exerts pressure on the brake drum.

breakaway torque The point at which a driving wheel begins to rotate during a limited-slip clutch check.

breaker bar A long, hinged handle for a socket wrench.

breaker points A set of electrical contacts with a movable and a stationary contact. Used to interrupt the flow of current to the ignition primary coil.

break-in The initial wear period when an engine or other moving parts begin to conform to each other.

bridge rectifier circuit An assembly of diodes used to transform AC current into DC current.

British thermal unit (Btu) A customary unit of heat energy. The heat required to raise the temperature of 1 pound of water 1 degree F.

brush A carbon or copper graphite conductor that rides against a commutator to transfer electricity.

bushing A removable, hollow bearing surface for a hole to limit the size of the opening, resist abrasion, and/or serve as a guide.

bypass valve A valve used to redirect unfiltered oil through a clogged full-flow oil filter.

bypass wiring New wiring used to bypass inaccessible, defective wires.

C

cable linkage Clutch linkage operated by a cable.

caliper A U-shaped hydraulic gripping device that forces brake pads against a rotor to provide braking action on a rotor.

cam A projection on a rotating part that is used to cause another part to move.

cam follower A hollow lifter that fits over the valve stem, spring, and retainer; used in overhead camshaft engines.

cam grinder A machine that grinds a round piston into an oval cam shape.

cam grinding A machining process that produces an oval-shaped piston to counteract the effects of expansion.

camber The tilting of a wheel inward or outward from true vertical.

camshaft A metal shaft onto which cams have been cast or machined.

camshaft bearing A bearing used to support a camshaft. May be a fully cylindrical bushing, or a split, insert-type bearing.

capacitor An electronic device that can hold an electrical charge.

capscrew A screw with a six-sided head, which screws directly into a part.

carbon monoxide (CO) An odorless, colorless, toxic gas produced by incomplete burning of hydrocarbon fuel.

carbon tracking Formation of a line of carbonized dust within a distributor cap or on a rotor.

carburetion The enrichment of a gas (usually air) by combining it with a hydrocarbon fuel such as gasoline.

carburetor A vacuum-operated device that atomizes gasoline into a stream of air passing through the device.

carcass The inner part or body of a tire that holds air.

carriage bolt A bolt with a raised, square shoulder on the underside of the head.

castellated nut A nut held in position by slots and a cotter pin to prevent loosening.

caster The tilting of a steering knuckle forward or rearward from true horizontal.

casting A process of molding parts by liquifying material and pouring it into a shaped mold to solidify.

casting porosity A hole in a housing caused by a casting imperfection during manufacture.

catalyst A chemical that produces a chemical reaction between other substances, without being consumed itself.

center link *See* relay rod.

centrifugal advance mechanism A device, operated by centrifugal force, that advances ignition timing according to rpm.

centrifugal force An outward pull from the center of a rotating axis.

ceramic Material made of clay baked in a high-temperature oven.

cetane rating A rating that measures the ease with which diesel fuel will ignite under the heat and pressure of compression. Also, the fuel's anti-knock quality.

chafing Damage caused by friction and rubbing.

chain hoist A chain and pulley assembly attached to the ceiling or to a special structure. Used for lifting heavy parts, such as engines.

chamfer To produce an angled or beveled edge in a hole or around the end of a shaft.

charcoal canister A container that stores fuel vapors from the fuel tank and carburetor until they can be drawn into the engine by intake manifold vacuum and burned.

charge The presence of electricity; or, to induce electricity in a device.

charging Reversing the discharging process by connecting a source of DC current to a battery.

chassis The frame, unit body reinforcement, suspension, braking, and steering systems.

chassis dynamometer A dynamometer that tests the horsepower output of vehicles at the driving wheels.

check valve A valve that opens in the direction of flow and closes to prevent flow in the opposite direction.

chemical energy Energy produced or stored by chemical means.

chisel A wide, heavy tool used to separate or align parts.

choke plate A metal plate at the top of the carburetor air horn that can be closed to reduce air flow and provide a rich mixture for starting and warmup.

chuck The part of a drill that holds a drill bit.

circuit A completed pathway through which electricity can travel from a source to a load, and return.

circuit breaker A resettable device that interrupts a circuit to prevent excessive current flow.

circuit tester *See* continuity tester.

clearance The distance between adjacent parts.

clearance gauge A tool used to measure brake drum and brake shoe clearances.

C-lock axle assembly A driving axle held in place by a C-lock in the differential.

closed loop A feedback control system that uses output information to modify its own input information.

clouding The formation of tiny wax particles in diesel fuel, giving the fuel a cloudy appearance.

clutch An assembly that engages and disengages a manual transmission.

clutch chatter A shaking and shuddering from the clutch, felt everywhere in the automobile during clutch engagement.

clutch disc The part of the clutch that takes power from the flywheel and transfers it to the transmission.

clutch drag A condition in which the clutch disc and transmission input shaft do not come to a complete stop when the clutch is disengaged.

clutch fork A lever that pivots in the bellhousing and acts on the throwout bearing.

clutch-head screwdriver A screwdriver used to loosen or tighten clutch-head screws.

clutch linkage A series of parts that connects the clutch pedal and clutch fork.

clutch pedal pulsation A rapid up-and-down movement of the clutch pedal.

clutch plate *See* clutch disc.

clutch plate limited-slip A limited-slip differential system in which clutch discs lockup the driving axles.

clutch slippage A condition that causes the engine to run faster than the transmission.

clutch vibration A condition similar to clutch pedal pulsation but louder and not as rapid.

coefficient of drag (C_d) A factor that expresses the relative ease with which a shape passes through the air. The lower the coefficient of drag, the less air resistance.

coil spring A thick steel wire that is coiled to expand and compress under force.

coil wire An insulated cable that conducts secondary current from the coil to the central connection of a distributor cap.

coking Damage to lubricating oil consisting of gum, varnish, and carbon formation from excessive heat.

cold cranking amperes Current flow that can be maintained from a battery during a cold cranking test before voltage drops below 7.2. The test is performed at 0 degrees F (-32 degrees C).

cold tank A cleaning tank for aluminum and brass.

collapse To reduce in size/volume or a given distance.

collapsible steering column A steering column that collapses or bends upon impact.

collector A portion of a transistor to which electrons flow.

combination valve A valve in a disc/drum brake system that acts as a proportioning valve, metering valve, and failure warning switch.

combination wrench A wrench with one open end and one box end.

combustion chamber A space between the top of a piston at top dead center and the cylinder head, where combustion takes place.

combustion control computer (CCC) *See* electronic control unit (ECU).

combustion leak detector A diagnostic device used to detect combustion gases in the radiator.

commutator A part of an electrical motor through which electricity flows to armature windings.

compound Combinations of molecules that form a distinct type of matter.

compressed air Air under pressure.

compression Upward suspension movement.

compression ignition A way of igniting an air-fuel mixture through the heat of compression.

compression ratio A ratio comparing the volume above the piston at bottom dead center and the volume above the piston at top dead center.

compression ring One of the upper rings on a piston that prevents combustion gases from escaping past the piston.

compression test A diagnostic test to determine cylinder pressure and sealing.

compression tester A tool for measuring compression in an engine cylinder.

compressor A motorized pump that circulates refrigerant through an air conditioning system.

computer An electronic device that processes data into useful information.

computerized diagnostic tester A machine that sends electrical inputs to a computer, which compares engine performance against established criteria.

concave An inward-curved surface.

condenser Part of an air conditioning system that changes refrigerant from a gas to a liquid. *See also* capacitor.

conduction Heat transfer from a hotter object to a colder object upon contact.

conductor A material that allows the passage of electricity.

cone limited-slip A limited-slip differential system that uses friction-faced cones to lockup the driving axles.

configuration The shape, or layout, of an engine.

connecting rod A rod connected to the piston and to the crankshaft.

constant-mesh gears Multiple transmission gearsets on the transmission mainshaft and countershaft that are meshed at the same time.

constant-velocity joint Universal-like joints used on front-wheel drive axles which transmit power flow smoothly through sharp angles.

contact disk A heavy copper washer, part of a relay or solenoid, that forms a connection between terminals.

continuity tester A device used to determine if an electrical connection is broken.

continuous fuel injection Fuel injection that continues, in small amounts, without pause.

control arms Suspension parts that control coil spring action as a wheel is affected by road conditions.

control rods Solid steel rods that brace a torque-tube drive and control rear-end torque.

convection Upward heat transfer in a fluid-filled system.

conventional electrical theory The theory which states that electrical current flows from positive to negative.

conventional/metric conversion Tables or charts used to convert measurements from conventional to metric and from metric to conventional.

conventional system A system of measurement based on inches and pounds.

coolant A mixture of antifreeze and water that passes through the engine to help maintain proper engine temperatures.

coolant bypass A passage or hose that allows coolant to flow from the cylinder head to the water pump inlet when the thermostat is closed.

coolant hydrometer A testing device used to measure the freezing point of a coolant mixture.

cooling fins Thin metal projections cast or machined into parts to increase their surface area heat dissipation.

cooling system pressure tester A diagnostic device used to check whether cooling system components will hold pressure.

cord One of several layers of threads that run through the plies and belts of a tire.

core A solid plug used to form hollow areas in a casting mold.

cornering light A light that burns steadily, when turn signals operate, to light up the area in the direction of a turn.

corona effect A "glow" around faulty spark plug wires that can be seen in the dark. It is not necessarily harmful.

corrected horsepower Horsepower ratings that have been corrected for atmospheric conditions present during a dynamometer test.

corrosion inhibitor An oil additive that helps to prevent the formation of acids.

cotter pin A steel wire that is bent at the end to hold a part in place.

countergear The gear on the countershaft driven by the main drive gear.

countershaft The shaft, turned by engine power, that drives the gears on the mainshaft.

counterweight A weight formed on a crankshaft to balance the weight of the offset crankpins.

crane A portable lifting device used to lift engines or other heavy assemblies from an automobile.

crankcase The lower inside part of the cylinder block to which the crankshaft is bolted.

crankpin The machined, offset area of a crankshaft where the connecting rod journals are machined.

crankshaft A shaft with offset journals to which the connecting rods are attached at the bottom of the engine. It rotates, converting reciprocating motion to rotary motion, when the piston is pushed downward by combustion.

crankshaft end play The movement of the crankshaft forward or rearward in the block.

crimp The use of pressure to force a thin holding part to clamp to, or conform to the shape of, a held part.

crimp-type clamp A small, metal hose clamp band that is crimped to form a seal.

crossover pipe A metal pipe that connects two exhaust pipes on a V-type engine.

cubic centimeter (cc) The volume contained in a cube one centimeter, or 10 millimeters, on each side. (1,000 cc = 1 liter.)

cup plug A type of soft plug shaped like a cup whose flanges face outward.

curb idle speed The normal idling speed of an engine.

cushioning springs Clutch disc springs between the facings that absorb shock when the clutch is engaged.

customary unit A unit of measurement in feet, inches, ounces, or other conventional measurement unit.

cutting out A condition in which an engine dies momentarily at irregular intervals.

cylinder A round passageway inside the cylinder block, in which a piston travels.

cylinder block The largest part of an automotive engine which houses the major mechanical parts of the engine.

cylinder head The upper part of an engine that closes off the tops of the cylinders. Located within the head are ports, valves, and threaded openings for spark plugs.

cylinder liner A cylindrical sleeve used to form the actual cylinder wall within a block. Associated mainly with aluminum engine blocks, but also may be used in cast iron blocks with badly damaged cylinders.

cylinder ridge An unworn section at the top of a cylinder formed by the friction of piston rings acting on the rest of the cylinder.

cylinder ridge remover A tool used to machine the cylinder ridge from a cylinder so that the piston and connecting rod assembly can be removed.

cylinder taper The difference between a cylinder's diameter at the top and bottom.

D

damper pulley A pulley that usually is bolted to the torsional vibration damper. Used to help reduce irregular rotation.

dashpot A partially sealed, flexible diaphragm attached to a pushrod, which slows the closing action of the throttle when the accelerator is released.

data Uninterpreted raw facts and figures.

dead timing The process of correctly repositioning a distributor for engine starting purposes.

dealer An agent who sells new automobiles for an automobile manufacturer.

deceleration valve A valve that operates during deceleration to allow fresh air to bypass the carburetor to the intake manifold.

De Dion axle An independent-suspension driving axle system with open axles and U-joints.

defroster A distribution plenum system that directs heated air to the inside of the windshield.

dehumidified A condition in which moisture, or humidity, has been removed from air.

de-icer A chemical added to gasoline to prevent ice particles from forming and adhering to cold metal surfaces.

desiccant A special substance that absorbs moisture.

detergent-dispersant Chemical additive that keeps particles of carbon and other contaminants suspended in lubricating oil.

detonation Uncontrolled explosion of remaining air-fuel mixture from heat and pressure after spark ignition has occurred.

detonation sensor A device that senses sound waves produced by detonation and converts them into electrical signals.

diagnosis Investigation and analysis of an automotive service problem.

diagnostics Overall analysis of an automobile's operation.

diagonal cutting pliers Pliers with sharp-edge jaws, used to cut electrical and wire connections.

dial indicator An instrument used to measure runout and play.

diaphragm spring A single, thin sheet of convex-shaped metal in a pressure plate assembly, which "gives" when pressure is applied.

die A cutting tool used to form, or repair damaged, external threads.

dielectric compound An insulating material, often used to seal electrical connections against moisture and dirt.

diesel A hydrocarbon motor fuel, less refined and containing more heat energy than gasoline. Also, a condition in which a spark-ignition engine continues running after spark ignition ceases, may be called "run-on."

diesel engine A compression-ignition engine that operates on diesel fuel.

diesel injection pump A fuel injection pump, on a diesel engine, which creates high fuel pressures to force open a mechanical injection nozzle.

diesel knocking A condition similar to detonation in a gasoline engine. Fuel of too low a cetane rating ignites in an uncontrolled manner after an ignition lag.

differential An assembly of gears that drives the driving axles. The differential allows a difference in speed between the driving axles during turns.

differential bearing preload The load or pressure on bearings upon installation.

differential case A case that houses differential parts and is bolted to the ring gear. Sometimes called a carrier.

differential gearset A name given to the ring gear and drive pinion gear.

differential pinion gears Gears that help to direct more power to the outside wheel when the automobile is turning. They mesh with the side gears.

digital Information presented in the form of digits, or numbers.

dimple A tapered depression in sheet metal caused by an overtightened attachment device.

diode An electronic device that, in general, acts as a one-way valve and allows current to flow in only one direction.

diode trio Three diodes connected in parallel. Used in some alternators to supply rectified DC to the rotor coil, and to apply bias current to regulator transistors.

direct current (DC) Electricity, such as that produced by a battery, that flows in only one direction, from one terminal of an electrical source to the other.

direct ignition system (DIS) A high-performance ignition system with a double-ended coil for each two cylinderd instead of a distributor mechanism. Also known as a distributorless ignition system.

direct injection A diesel fuel injection system in which fuel is injected directly onto the top of the piston.

disc brake A brake assembly in which a caliper with shoes exerts pinching pressure on a disc to slow or stop the vehicle.

discharge nozzle A shaped opening for a main-metering or accelerator pump fuel discharge.

dispatch sheet A form used to keep track of time schedules and appointments in a service area.

displacement The volume of a cylinder through which a piston moves from bottom dead center to top dead center.

display An oscilloscope pattern that shows cylinder firings one after another, in firing order.

display device An instrument or warning light that alerts the driver to a condition of a vehicle system.

distribution plenum A chamber that accepts and distributes circulating air.

distributor A component of the ignition system that determines the firing order of the cylinders. It also directs electrical current to the correct spark plug at precisely the right time.

distributor cam A multi-lobed cam against which a rubbing block rides to open and close breaker points. It is driven by the distributor shaft.

distributorless ignition system See direct ignition system (DIS).

D-Jetronic The first electronic fuel injection system, developed by Robert Bosch Corporation, applied to production passenger vehicles. Airflow is measured by means of a manifold absolute pressure sensor.

doping The process of adding trace amounts of chemicals to semiconductors to produce positive (P) or negative (N) characteristics.

double-acting shock absorber A shock absorber that controls both upward and downward movements of a suspension.

double Cardan U-joint A constant-velocity joint consisting of two closely connected universal joints.

dowel A cylindrical part that fits through other parts to attach or position them.

dowel pin A cylindrically shaped fastener that aligns two parts.

downdraft carburetor A carburetor in which air flows through the barrel in a downward, vertical direction.

drag link See relay rod.

drain plug A pipe plug fitting used to drain lubricant.

drill bit A steel shaft that has a sharp, spiral groove and cuts into metal when turned.

drive belt A belt, driven by the engine, which operates equipment in the engine compartment.

drive belt tension guide A tool used to measure tightness of a drive belt.

drive link assembly A chain, or link, assembly that connects the turbine shaft to the input shaft in a front-wheel drive automatic transmission.

drive member A gear that drives, or provides power for, other gears in a planetary gearset.

drive pinion gear A pinion gear that takes power from the driveline and transfers it to the differential through the ring gear. It is part of the ring and pinion gearset.

drive size The diameter of the drive in a socket wrench handle, or the socket drive end.

driveability The degree to which a vehicle operates properly. Includes starting, running smoothly, accelerating, and delivering reasonable fuel mileage.

driveline The driveshaft and flexible joint assembly that transfers torque from the transmission to the differential in a rear-wheel drive automobile.

driven member The output gear during planetary gearset operation.

driveshaft A solid or hollow metal shaft forming part of the driveline.

drivetrain A series of components that transmit power from the engine to the driving wheels.

driving axle A shaft that accepts power from the differential and turns the driving wheel.

drop center The well in a wheel rim.

drum brake A brake that exerts pressure inside a drum to slow or stop the vehicle.

dry friction Friction that occurs when unlubricated objects come in direct contact, such as metal rubbing against metal.

dry liner A cylinder liner whose outer surface does not contact the coolant in a liquid-cooled engine.

dual bed A catalytic converter with two areas where catalytic reactions occur.

dual bed converter A catalytic converter that promotes the oxidation of HC and CO to CO_2 and H_2O. It also promotes the reduction of NO_X to N_2 and O.

dual overhead camshaft (DOHC) An engine design with two camshafts—one for intake valves, the other for exhaust valves—mounted above the cylinder head.

duty cycle The amount of time that a mixture-control solenoid is activated so that fuel cannot flow.

dwell The length of time that contact points remain closed during operation.

dwell angle The amount of rotation through which breaker points remain closed to energize the ignition primary coil.

dwell-tachometer An instrument used for measuring dwell and the speed of the engine.

dynamic balance The equal distribution of weight on each side of the centerline of a wheel.

dynamometer A device for measuring horsepower.

E

early fuel evaporation (EFE) A system on General Motors vehicles that increases intake charge vaporization. Uses a heat riser valve to heat the intake manifold with heat from the exhaust manifold.

eccentric An off-center circular lobe that rotates to move a part, similar to a cam.

electric impact wrench An electrically powered socket wrench.

electrical capacity The ability of a battery to produce current.

electrical connector plug A plastic, two-part, male-female plug with metal electrical terminals in each half. When the halves are pushed together, an electrical connection is made.

electrical manifold heater A device that uses information from computer sensors to increase the temperature of the air and surrounding metal in an intake manifold. The warmer air allows better fuel vaporization.

electrical schematic A "map" of electrical components that uses symbols and lines to indicate electrical circuits.

electrical solenoid A device that uses electricity to create a magnetic force to push, pull, or hold a mechanical linkage.

electrical system The system that provides electrical current to start and operate all the electrical components of the automobile.

electricity A form of energy that can be used to create light, heat, or force to cause motion.

electrochemical action A chemical reaction that produces the potential for electrical flow, as in a battery.

electrolysis A chemical and electrical decomposition process that can damage metals such as brass, copper, and aluminum in the cooling system.

electrolyte A material whose atoms become ionized, or electrically charged, in solution. Automobile battery electrolyte is a mixture of sulfuric acid and water.

electromagnet A magnet formed by electrical flow through a conductor.

electron electrical theory A theory that states that electrical current flows from negative to positive.

electronic control module (ECM) *See* electronic control unit (ECU).

electronic control unit (ECU) A small computer mounted on board the vehicle to control engine operation, including air-fuel ratio, spark timing, and other vehicle functions.

electronic fuel injection (EFI) A gasoline fuel injection system in which a microprocessor is used to energize solenoid-controlled injection nozzles. The nozzles spray finely atomized gasoline into the incoming air.

electronic ignition control unit A transistorized device used to switch current on and off to the ignition primary coil.

electronic mixture control Air-fuel mixture control that operates by moving a tapered or stepped needle within a carburetor jet. Movement is based on signals from an electronic control unit.

electronic muffler A muffler that uses "anti-noise" to cancel out naturally occurring noises. These types of mufflers are currently in developmental stages.

electronic power steering (EPS) A power steering system that uses computer controls to improve steering control and "feel." EPS increases the force applied to the steering linkage and reduces the effort required to turn the steering wheel.

electronics A branch of physics that deals with the behavior and effects of small amounts of electrons in semiconductor circuits.

electron An extremely small particle that orbits the nucleus of an atom.

element A specific, unique type of matter that cannot be reduced to a simpler form.

emission control Techniques and equipment that reduce the level of harmful gases in engine exhaust.

emitter A portion of a transistor from which electrons are emitted, or forced out.

emulsion tube A perforated tube immersed in a fuel. Air is passed through the tube to make a frothy air-fuel mixture.

end play The distance a shaft moves forward or rearward in a direction parallel to its length.

energy The capacity to do work. *See also* force, power, work.

engine A machine that converts energy produced by combustion into mechanical force.

engine calibration unit A separate computer module that contains the programs for a specific car model.

engine lift A crane used to remove or install an automobile engine.

engine support fixture A support used to hold an engine in place while a transaxle is being removed.

enrichment circuit A carburetor circuit that provides a richer air-fuel mixture during full-power operation.

epoxy cement An extremely strong plastic adhesive formed by mixing a base substance and an epoxy resin.

epoxy resin A thermosetting plastic material that hardens to form a strong adhesive bond when exposed to oxygen; the hardening agent in epoxy cement.

ethanol A form of alcohol used as a motor fuel and in liquor. Also known as grain alcohol.

evacuate The process of applying vacuum to a closed refrigeration system to remove air and moisture.

evaporation A procedure by which a liquid is turned into vapor.

evaporative emission controls Parts and systems designed to prevent fuel vapors from escaping directly into the atmosphere.

evaporator A radiator-like part of an air conditioner that absorbs heat as air passes over it.

exhaust gas analyzer An electronic diagnostic instrument used to determine the amounts of harmful gases in engine exhaust fumes.

exhaust gas recirculation (EGR) Reusing exhaust gases in the intake charge to reduce NO_X.

exhaust gas recirculation (EGR) valve A vacuum-operated valve that can open a passage between exhaust and intake manifold areas to recirculate exhaust gases.

exhaust manifold The hollow part connected to the cylinder head at the exhaust port that directs burned gases into the exhaust system.

exhaust pipe The metal pipe that connects the exhaust manifold to the rest of the exhaust system.

exhaust system A series of parts that direct heat and exhaust gases from the engine into the atmosphere.

exhaust valve A valve fitted into the exhaust port of a cylinder head.

expansion dome A raised portion of the upper fuel tank wall.

expansion plug A form of soft plug that expands to form a better seal as pressure is exerted against it.

extension A steel rod that adds distance between the drive and the socket on a socket wrench.

external combustion Term applied to an engine whose heat source is outside the area where force is applied to cause motion.

external threads Threads that are on the outside of a part.

external transmission cooler An automatic transmission cooler system that has a separate, external radiator.

F

face shield A clear, shatterproof bubble that protects a person's face while working.

failure warning switch A switch that activates a dashboard warning light when the front or rear brake system fails.

fast-idle cam A stepped or tapered part connected between the choke plate and the throttle plate; used to help the engine run faster during warmup.

fatigue Cracking or breaking damage that results from repeated flexing and bending of a part.

feedback carburetor A carburetor with a mixture control solenoid activated by a control unit based on information received from sensing units.

feedback control A method of combining the functions of using sensors, a computer, and servomechanisms. Data about the effects of the computer's output commands are fed back into the computer. The computer constantly compares the input data to a set goal, and adjusts its output accordingly.

feeler gauge A tool that contains a number of small, flat strips of metal or plastic that are used to measure the space between two surfaces.

ferrous Containing iron.

field coil A coil of wire on an alternator rotor or starter motor frame that produces a magnetic field when energized.

filament A metal element within a glass bulb that is heated to produce light.

file A steel tool with rows of teeth to remove, smooth, or polish metal.

fill plug A plug that is removed to check the oil level of, or add oil to, a manual transmission.

fillet A rounded area machined at the intersection of two flat sections that are at an angle to one another.

firing order A sequence in which ignition occurs in the cylinders of an engine.

firing position The crankshaft, piston, and valve position at which the spark occurs to ignite the air-fuel mixture in a cylinder.

fixed caliper A disc brake design that uses four pistons to force pads against a rotor.

flame front The edge of the burning air-fuel mixture during combustion.

flange A projecting lip, rim, or edge.

flare An expanded, shaped end on a metal tube or pipe.

flare-nut wrench A special wrench used on a tubular metal line fitting to prevent damage to the fitting.

flat rate An established length of time in which a repair should be completed. Also the estimated cost of a repair.

flat rate manual A book that lists flat rates for different automotive jobs, and provides an estimated repair cost for a customer.

flat spot *See* hesitation.

flat washer A wide washer used to prevent damage to a part surface.

fleet A group of vehicles operated by an organization.

fleet garage A shop that services fleet vehicles for an organization.

flexplate A smaller version of a flywheel, used with an automatic transmission.

float A lightweight part that rests, or floats, on the surface of a liquid.

float bowl A hollow fuel reservoir area within a carburetor from which fuel is drawn during carburetion.

floating caliper A disc brake design in which the caliper slides in and out against a rotor.

flooding A condition in which excess, unvaporized fuel in the intake manifold prevents the engine from starting.

floor jack A portable tool used to raise and lower an automobile.

fluid coupling A method of driving an automatic transmission by using a fluid to transfer rotation.

flushing gun A device that uses compressed air to force clean water and air through a cooling system for flushing purposes.

flywheel A heavy, circular wheel attached to a rotating part to increase its momentum when spinning.

foam inhibitor An additive that reduces the formation of foam in oil.

footprint The portion of the tire tread that makes contact with the road surface.

foot-pound (ft.-lb.) A customary unit of work, equivalent to the force necessary to move one pound a distance of one foot.

force The cause of motion. *See also* energy, power, work.

forge A process of hammering or pounding a heated metal piece into a desired shape. Forging produces structurally stronger assemblies than casting.

forward-biased A transistor to which a small current is applied to the emitter-base connection, which is turning on the transistor. A large current can flow through the emitter-collector path.

four-stroke cycle The four up-and-down movements, or strokes, of the piston within the cylinder that complete one firing cycle. The four strokes are intake, compression, power, and exhaust.

frame The section of the automobile that supports the body, engine, and drivetrain.

free-floating piston pin A piston pin that is free to turn in the piston and connecting rod small end.

free travel The distance a clutch pedal moves before it begins to take up slack in the clutch linkage.

friction The resistance between two rubbing or sliding materials.

friction bearing A bushing, sleeve, or insert bearing.

friction pads The gripping surface of a brake caliper that slows the rotor.

frictional horsepower (fhp) The horsepower necessary to overcome internal friction within an engine.

frontal area The area against which air resistance is present as a vehicle travels forward.

front-end geometry A term used to describe the angles between front suspension parts on an automobile.

fuel bars Hollow tubes above the throttle body assembly used to spray fuel.

fuel filter A device to trap and hold sediment, rust particles, and other contaminants in fuel.

fuel injection A fuel system that sprays a measured amount of fuel into the incoming air flow or into the combustion chamber.

fuel rail A line that supplies fuel to multi-point fuel injectors.

fuel-vapor separator A device that uses gravity to separate liquid fuel from vaporized fuel.

fuel wetting The accumulation of excess fuel on the walls and floor of the intake manifold and on the port.

full field To bypass the regulator and apply full battery current to the alternator field windings to produce maximum output.

full-floating axle A driveline axle that serves one purpose, to turn the wheel. These axles are used mainly on large, heavy-duty trucks.

full-flow oil filtering system A filtering system in which all of the oil passes through the filter before it can flow through the galleries and to the moving parts of the engine.

fuse A safety device that conducts a given current of electricity, then melts to open a circuit and prevent excessive current flow.

fuse block A plastic block with connectors for wires, fuses, circuit breakers, turn signal flashers, and relays.

fusible link A wire, smaller in diameter than the circuit it is to protect, that can burn out like a fuse.

G

garter spring A small, narrow coil spring that helps to apply pressure behind the lip of a seal against a rotating shaft.

gasket A soft material used to seal areas between nonmoving parts.

gasohol A mixture of alcohol (up to 10 percent) and gasoline.

gasoline The most common hydrocarbon motor fuel.

gassing A discharge of large amounts of hydrogen gas and loss of water during charging.

gear A toothed wheel that meshes with, or fits into, another toothed wheel.

gear clash A noise made when a manual transmission is shifted improperly from one gear to another.

gear drive A method of using gears of varying sizes to coordinate and turn shafts.

gear ratio A number indicating the relative number of turns and rotating speeds of a driving and a driven gear.

gear tooth contact pattern test A diagnostic procedure in which the differential ring gear and drive pinion gears are marked with a compound. The gears then are rotated to determine how they mesh.

general automobile technician A technician who is certified in all areas of automobile servicing.

general repair manual A service manual in condensed form, covering several models and years of automobiles.

generator *See* alternator.

glass bead blaster A cleaning machine that shoots glass beads at parts to clean them off.

glazed Polished by heat and friction to a slick, poorly gripping surface.

glow plug An electrical heating element used in a diesel engine to preheat the air and surrounding metal to a temperature sufficient to ignite the fuel.

governor pressure A pressure lower than mainline pressure in an automatic transmission that assists in upshifting.

grade markings Marks on fasteners that indicate strength.

graphic displays Drawings or pictures of a vehicle that can be illuminated to indicate problem areas.

graphited oil A thin lubricant containing graphite particles.

grease points Areas where grease or oil is added when lubricating.

greasy friction Friction that occurs when a thin film of grease or oil is present between touching surfaces.

grinder A tool that uses an electric motor to drive a grinder on one side and, usually, a wire wheel on the other.

gross horsepower A rating of an engine's horsepower without any driven accessories.

ground A common connection through which electricity can flow to complete a circuit.

group number A number code that indicates the physical dimensions and terminal types of a battery.

growler An electrical testing device that produces a strong magnetic field; used to test armatures.

H

hairspring A fine wire spring.

halogen light A small glass bulb that contains a filament surrounded by halogen gas. The small bulb fits within a larger reflector and lens element.

hammer A striking tool with a metal or wooden, head attached to a handle.

hand-held diagnostic computer A testing instrument that communicates with an automobile's computer to identify system malfunctions.

hand tool A tool that is operated by hand, without power.

hand vacuum pump A hand-operated device that produces and measures vacuum.

harmonic balancer A torsional vibration damper that has a central hub, an elastic intermediate sleeve, and an outer rim.

harness A bundled group of wires.

head The part of a wrench that grips a fastener.

head gasket The gasket between the cylinder block and the cylinder head.

header An aftermarket, high-performance combination of an exhaust manifold and a headpipe, custom-formed from steel tubing.

heat dam A small groove cut into the piston above the top ring groove or below the oil ring groove. It prevents excess combustion heat from being transferred.

heat energy Energy produced by burning.

heat exchanger A device that transfers heat from one fluid to another.

heat range The heat-transferring ability of a spark plug, indicated by a code on the side of the plug.

heat riser valve A flap or valve in an exhaust manifold that causes exhaust gases to flow out more slowly than in an unrestricted opening.

heat shield A device, usually made of pressed or perforated sheet metal, used to protect parts and people from excessive heat.

heat shrink A special type of plastic insulation for electrical wires that shrinks to fit tightly around parts when heated.

heat sink A device used to dissipate heat and protect parts.

heat stove A sheet metal enclosure around the exhaust manifold, through which air can flow and be heated by the manifold.

heater assembly A housing that holds the heater core and control doors for the distribution plenum chambers.

heater control system Switches and levers that are operated manually, by vacuum, and/or electrically to control heating system flow.

heater control valve A valve that controls coolant flow to the heater core inlet hose.

heater core A small radiator, or heat exchanger, used to heat air for the comfort of passengers.

heater hose A hose that connects the heater core to the engine cooling system.

heavy-duty cooling system cleaner Powdered phosphoric acid, usually packaged in a two-part container with a chemical neutralizer (baking soda).

height sensor An electronic sensor that signals a control module when the suspension is compressed by a heavy load.

hesitation A momentary lack of response as the accelerator is pressed down, such as when moving away from a stop.

hoist A lift that is stored in the floor of a work area and used to raise or lower an entire automobile.

hollow-head screw *See* Allen head screw.

horsepower A unit of power based on the ability of a horse to pull a load. One horsepower is equivalent to 550 foot-pounds of work per second, or 746 watts. In an engine, horsepower is the rate at which an engine produces torque.

hot-idle compensator A device that opens an air passage to lean the mixture slightly when the engine is hot.

hot tank A tank filled with corrosive solvent; used to remove rust, corrosion, and deposits from large ferrous metal parts.

Hotchkiss drive A suspension design that uses leaf springs and the frame to control rear-end torque.

hub nut A fastener that holds a transaxle driving-axle assembly to the driving wheel.

humidity The water vapor content in air.

hunting gearset A differential gearset in which one drive pinion gear tooth contacts every ring gear tooth after several rotations.

hydraulic lash adjuster A device similar to a hydraulic lifter; used in overhead-camshaft valve trains.

hydraulic lifter A lifter mechanism that uses oil pressure to transfer force to the valve train.

hydraulic linkage A clutch linkage that is operated by hydraulic fluid.

hydraulic press An assembly that uses the pressure of a hydraulic jacking unit or hydraulic cylinder to produce pushing force.

hydraulic tools Tools that operate on oil pressure.

hydrocarbon A chemical substance made up primarily of hydrogen and carbon atoms. Petroleum products are hydrocarbon compounds.

hypoid gears A type of spiral, beveled ring and pinion gearset in a differential. Hypoid gears mesh below the ring gear centerline.

I

idle discharge port A fuel discharge port positioned just below the edge of the throttle plate when it is closed for idling.

idle-speed solenoid An electrical solenoid used to position the throttle plate for proper idle speed.

idler arm A steering rod connected to the frame; helps to support the relay rod end opposite the Pitman arm.

idler gear A gear meshed between a driving gear and a driven gear.

idling Normal slow-speed engine operation without the accelerator pedal depressed.

ignition interlock A theft prevention device that locks the steering wheel if the key is not inserted and turned. Also used to prevent vehicles with automatic transmissions from starting in gear.

ignition timing The point at which spark occurs, as referenced by piston and crankshaft position.

ignition timing marks Marks on external engine and pulley or flywheel surfaces used to determine piston position in the No. 1 cylinder.

impeller A fan-like propeller that forces a fluid to flow.

impeller pump A pump with radiating blades that operates by drawing fuel in at its center and forcing it outward through centrifugal action.

inches of vacuum A measurement of engine vacuum represented by the height of a column of mercury, sometimes written as "inches of Hg."

inclinometer An instrument used to measure driveshaft inclination, or angle.

independent garage An automotive service and repair shop that deals with all parts of an automobile.

independent suspension A suspension system that allows wheels to move independently.

indicated horsepower (ihp) Theorectical horsepower calculated through pressure measurements within a cylinder.

indirect injection A diesel fuel injection system in which fuel is injected into a precombustion chamber for ignition.

induce To create another magnetic field and/or an electric current through the use of a magnetic field.

induction hardening A heat-treating process that hardens metals.

inductive ammeter An ammeter that reads the amount of current flowing through a wire by magnetic induction.

inductive pickup A device that can induce a signal current by being clamped around a conductor.

inductive timing light A timing light with an inductive pickup to trigger a flash.

information disclosure Protection for consumers against servicing frauds.

inline diesel injection pump A diesel injection pump using a camshaft similar to an engine camshaft to move plungers that pump diesel fuel.

inner tube A rubber tube that fits inside a tire and holds air.

input Data furnished for the use of a computer. Also, to supply data to a computer.

inside micrometer A tool used to measure hole diameters, or distances between parallel surfaces.

instrument voltage regulator A voltage regulator used to stabilize and limit voltage for accurate instrument operation.

insulator A material that does not conduct electricity.

intake charge The air-fuel mixture drawn from the carburetor through the intake manifold and into the cylinder on its intake stroke.

intake manifold A hollow part that directs air-fuel mixture from the carburetor into the intake port of a cylinder head.

intake port The port on a cylinder head connected to the intake manifold.

integral-carrier differential A design in which the differential carrier bearings are mounted directly in the rear axle housing.

integral power steering A power steering system that applies hydraulic pressure to a piston in the steering gearbox.

integral valve guide A hole machined in the metal of the cylinder head to align and support the valve stem.

integral valve seat A valve seat machined into the cylinder head.

integrated circuit (IC) A miniaturized series of electronic components connected together to perform specific functions. Also known as a "chip."

integrated control system An electronic control system that combines many functions into a single control unit.

intercooler Type of radiator heat exchanger mounted in the airstream to cool the intake air charge after compression in a turbocharger.

interference angle The difference in angle between two mating parts.

internal combustion Term applied to an engine whose heat source is enclosed within the area where force is applied to cause motion.

internal short A short circuit between positive and negative battery plates.

internal transmission cooler An automatic transmission cooler system that uses the engine radiator for cooling.

ion An element that has more or less electrons than the number needed to balance the number of protons in the nucleus of its atom.

iso-octane A hydrocarbon fuel assigned an octane rating of 100, to which other motor fuels are compared.

J

jack A portable tool that is used to raise an automobile off the floor.

jack pads Reinforced parts of a chassis where jacks can be placed to raise an automobile.

jackstand A support device to hold the automobile off the floor after it has been raised by a jack.

jet A precisely sized, calibrated hole in a hollow passage, through which fuel or air can pass.

jet spray booth A closed container in which large engine parts are cleaned with multiple high-pressure spray jets.

joule (J) A unit of energy or work, equivalent to one Newton-meter of force per second.

jounce Upward suspension movement.

journal The area of a rotating shaft that is in contact with a bearing.

jump starting Starting a vehicle with a weak battery by connecting its electrical system to a strong battery.

jumper cables Heavy cables with color-coded clamps at each end, used for jump starting.

jumper wire Wires used to make connections for test or replacement wiring purposes.

junction block A plastic base with metal terminals and connectors for common wiring connections.

K

key A small, narrow piece of metal used to fasten two rotating parts.

keyway A slot that holds a key fastener.

kickdown linkage An automatic transmission linkage system that allows downshifting by depressing the throttle pedal to the floor.

kilowatt (kw) An electrical measurement equal to 1,000 watts.

K-Jetronic A form of EFI developed by Robert Bosch Corporation in which fuel is sprayed constantly from all injectors. Air-fuel ratio is adjusted by varying fuel pressure.

knurl A raised ridge on a cylindrical part, produced by a machining operation. Also, to produce a knurl.

L

land The areas on a piston between the grooves.

lateral runout A measurement of the in-and-out movement of a tire and wheel.

lead dioxide (PbO_2) A porous, chocolate-brown, crystalline substance that forms the positive plate of a battery element.

lead sulfate ($PbSO_4$) A combination of lead, sulfur, and oxygen that forms on the plates of a battery. Sulfate prevents electrolyte from penetrating the plate and impairs electrochemical action.

leaded gasoline Gasoline that contains tetraethyl lead.

leaf spring A semi-elliptic suspension assembly that is made up of one or more long, thin strips of metal.

leak detector tool A butane-gas-burning tool used to locate refrigerant leaks in an air conditioning system.

leak-down tester A tester used to measure leakage from the combustion chamber through piston rings, intake and exhaust valves, and head gaskets.

lean mixture An air-fuel mixture that contains more than 14.7 parts of air to one part of gasoline, by weight.

lever and shaft assembly Part of a rod and lever clutch linkage that connects two levers to one shaft.

LH-Jetronic A variation of L-Jetronic electronic fuel injection that uses microprocessors to control the injector timing and duration.

lifter A part that changes the rotating motion of the cam lobe into a reciprocating motion.

limited-slip differential A design that reduces individual wheel spin through a series of clutches in the differential that lock the two driving axles together.

line technician—heavy work A technician whose work involves complicated service on an automobile.

line technician—light work A technician who performs minor service on an automobile.

linkage power steering A power steering system that applies hydraulic pressure to the steering linkage system.

lip-type seal An assembly consisting of a metal or plastic casing, a sealing element made of rubber, and a garter spring to help hold the seal against a turning shaft.

liquid cooling A method of cooling heated parts by circulating a liquid through hollow areas of the parts.

liquid cooling system cleaners A mild solution of phosphoric acid dissolved in water and mixed with detergent.

L-Jetronic A form of EFI developed by Robert Bosch Corporation in which airflow is measured by a pivoting flap in the airstream. Fuel is injected twice each crankshaft revolution.

LN-Jetronic A variation of L-Jetronic EFI that uses low-pressure bottom-feed fuel injectors and a hot-wire mass air flow sensor.

load A device or resistance that can cause work to be done by electrical energy.

load leveling system *See* automatic level control.

lobe An individual cam shape on a camshaft.

lock-to-lock The rotating distance of a steering wheel from extreme left to extreme right.

locking pliers Pliers with jaws that can be locked in position.

locking steering column A locking mechanism in the ignition switch. Prevents the steering wheel and shift lever from moving when the ignition switch is in the "lock" position.

locknut A nut used to keep a screw or bolt from loosening.

lockup system A friction assembly that locks the turbine and torque converter housing together in an automatic transmission.

lockwasher A washer with sharp edges that grip parts and keep them from moving.

longitudinal A mounting that is parallel to, or in line with, a front-to-back centerline.

longnose pliers Pliers with long, narrow jaws. Also called needlenose pliers.

lubricant Slippery materials, such as oil and grease, that reduce friction.

lubrication A substance, usually oil or grease, used to reduce friction between parts, thus reducing heat and wear.

lubrication system A number of parts, components, and passages, which function to keep engine parts lubricated.

lug nuts Nuts that hold a wheel on the wheel lugs.

lug pattern *See* bolt pattern.

M

machining A process of producing a desired size, shape, or finish on a part by using power machinery.

MacPherson strut A single-arm suspension system that combines a coil spring and shock absorber to act as an upper control arm.

magnetic clutch A clutch that rotates with a drive pulley and turns an air conditioning compressor on or off with a magnetic connection.

main body The central section of a carburetor which contains the float bowl and most passages, jets, and discharge ports.

main drive gear The gear, turned by the transmission input shaft, that drives the countergear.

main journal The central area of a crankshaft, where the main bearings support the shaft in the block.

main metering circuit A vacuum-operated fuel discharge system that draws fuel from a tube inserted into the float bowl. It discharges the fuel in the venturi area.

mainline pressure Pressure that is regulated in an automatic transmission hydraulic system.

mainshaft The shaft in the transmission that transfers torque to the driveline.

maintenance The care and upkeep of assemblies and individual parts of an automobile.

maintenance-free battery A battery that, in normal service, does not require the periodic addition of water to maintain an electrolyte level.

major thrust face The side of the piston to which most force is applied by side thrust.

malleable Able to be shaped.

mallet A small hammer.

manifold absolute pressure (MAP) A measure of the degree of vacuum or pressure within an intake manifold; used to measure air volume flow.

manifold gauge assembly A set of gauges and hoses, used on air conditioning systems, that determines high and low pressures, correct refrigerant charge, and helps to diagnose air conditioner problems.

manifold vacuum Vacuum present at the intake manifold.

manual bleeding A method of bleeding the brakes by operating the brake pedal.

manual control valve Directs fluid to switching and pressure-regulating valves in an automatic transmission.

manual steering A system in which the driver provides the steering effort.

manual transaxle A transaxle that is shifted by the driver.

manual transmission A transmission that is shifted by the driver.

manufacturer's service bulletin A one- or two-page bulletin describing a service change instituted by the manufacturer.

manufacturer's service manual A service manual, published by the manufacturer, covering one or more models for one year of an automobile.

map A three-dimensional graph.

mass air-flow sensor An EFI air intake sensor that measures the mass, not the volume, of the air flowing into the intake manifold.

master cylinder A hydraulic fluid reservoir for the brake system that also develops hydraulic pressure.

mate To fit together, as parts.

matter The material of which all things are made.

measurement inspection The use of measurement tools and specific measurement procedures to locate problem areas.

mechanical efficiency A comparison of how much horsepower an engine produces at the flywheel, compared to the theoretical horsepower the engine could produce from the pressures generated within the combustion chamber.

mechanical energy Energy stored or produced by mechanical means.

mechanical fuel injector A fuel injection nozzle assembly held closed by strong spring pressure and opened by fuel pressure.

mechanical fuel pump A pump driven by engine power, usually from the camshaft, through either an eccentric lobe or a lobe, pushrod, and rocker arm arrangement.

mechanical measuring tool A tool that is operated and adjusted by hand to obtain a measurement.

memory A computer feature that stores, or saves, information.

mercury switch A switch that completes a circuit by allowing liquid mercury metal to flow and conduct across two contacts.

mesh To fit together, as gear teeth.

meshing spring A spring on a starter motor driveshaft used to prevent excessive damage to gear teeth during engagement.

metal spray To deposit molten metal on a surface by spraying.

metering In carburetion, the addition of a measured amount of fuel into the airstream passing through a carburetor.

metering rod A small, tapered or stepped rod within a jet. The rod's up or down movement controls the effective cross-sectional area of the jet and the amount of fuel flowing.

metering valve A valve that delays hydraulic pressure to a front disc brake assembly when drum brakes are used at the rear.

methanol A poisonous form of alcohol used as a motor fuel, a carburetor cleaner solvent, and, formerly, as an antifreeze additive for liquid cooling systems. Also known as wood alcohol.

metric system A system of measurement based on meters and kilograms.

micron A very small metric dimensional measurement, equal to 39 millionths of an inch.

microprocessor A computer chip used to process raw data into usable information.

minor thrust face The side of a piston to which least force is applied by side thrust.

misfueling Adding leaded gasoline to vehicles with catalytic converters.

missing A lack of power in one or more cylinders.

modulate To control by degrees.

module A replaceable unit, such as a computer processing unit.

molecule A combination of atoms that form a given type of matter.

motion transfer Hydraulic pressure being transferred from one part to another.

Motronic A Robert Bosch electronic control system that combines ignition timing control with fuel injection control.

muffler A metal container with holes, baffles, and chambers that muffle, or soften, exhaust noise.

multiple-disc clutch A series of friction and steel discs that operate as an apply device in an automatic transmission.

multi-point fuel injection Fuel injection at each individual intake manifold runner.

multi-viscosity oil A chemically modified oil that has been tested for viscosity at cold and hot temperatures.

N

national coarse thread A bolt, screw, or stud with relatively wide spaces between threads.

national fine thread A bolt, screw, or stud with relatively narrow spaces between threads.

needle valve A variable restriction consisting of a tapered (convex) part that can move in or out of a matching (concave) opening. Also, a blunt, pointed valve that fits into a seat to modulate or stop fluid flow.

negative ground A way of connecting a single-wire vehicle system so that the negative battery terminal is connected to the metal of the vehicle to form a ground.

net horsepower A rating of an engine's horsepower with all normal drive accessories connected.

neutral position The position of a synchronizer between two gears when neither is being engaged to the mainshaft.

neutral safety switch An electrical series switch in the starting control circuit that will not close unless the transmission is in Park or Neutral.

neutron Part of the nucleus of an atom that is without charge.

nodular iron A metal, used in pressure plates, that contains graphite, which acts as a lubricating agent.

no-load rpm test A test made to check the free rotational speed of a starter motor.

nonhunting gearset A differential gearset in which one drive pinion gear tooth contacts only three ring gear teeth after several rotations.

non-sequential fuel injection Fuel injection that occurs without regard for the opening of individual intake valves.

nonthreaded fasteners Fasteners without threads.

normally aspirated A fuel system that draws air due to atmospheric pressure.

nose The high spot of a cam.

nucleus The center portion of an atom, consisting of parts called protons and neutrons.

nut A fastener with internal threads, used to tighten bolts.

O

observed horsepower Horsepower readings from a dynamometer test that have not been corrected to standardized units.

octane rating A measure of a motor fuel's ability to resist ignition from heat and pressure. Also, the anti-knock rating.

odometer A digital instrument that records how far a vehicle has traveled.

off-idle port A fuel-discharge port located just above the idle discharge port. It serves as an air bleed during idle and a discharge port at off-idle speeds.

offset Placed off center. Also, the measurement between the center of the rim and the point where a wheel's center is mounted.

oil control ring The lower grooved or perforated ring on a piston that scrapes excess oil from the cylinder walls.

oil filter A filtering element enclosed in a metal container, used to remove foreign particles from engine oil before it flows to moving engine parts.

oil galleries Small passageways that carry oil within the engine.

oil pan A storage area for oil, that is bolted to the bottom of the engine block.

oil pump A part of the lubrication system that moves oil, under pressure, to keep engine parts lubricated. Also, a pump that pressurizes the hydraulic system in an automatic transmission.

oil pumping A condition in a worn engine in which oil is drawn past the piston rings by intake vacuum.

oil spurt hole A small hole drilled in a connecting rod big end or bearing cap for external lubrication purposes.

open A break in electrical flow at some point in a circuit.

open-end wrench A wrench with two open ends.

open-loop A feedback control system operating in such a way that output information does not affect input information.

operating mode A specific way in which a vehicle might operate.

organic compound A material that contains carbon compounds.

organic brake linings Brake linings that contain asbestos.

orifice A precisely sized hole that controls fluid flow. Also, an opening.

oscilloscope A sophisticated electronic testing instrument that displays on a screen changing levels of voltage in an electrical system.

output The results of processing data in a computer.

output device A mechanism or indicating device that is controlled by computer output.

output shaft A driveshaft that is connected to a differential and to a driving axle on a front-wheel drive automatic transmission.

outside micrometer A measuring tool placed on the outside of a part to measure outside diameter.

overbore The dimension by which a machined hole is larger than the standard size.

overdrive The condition that occurs when a smaller driven gear turns faster than a larger driving gear.

overhead camshaft A camshaft mounted in a cylinder head, above the combustion chambers.

overinflation Excessive tire pressure that causes the center of a tire to wear quickly.

overrunning clutch A mechanism used to disengage a driven part from its driving member when another, connected part turns faster.

oversize Larger than the standard size.

oxidation The combination of a substance with oxygen to produce an oxygen-containing compound. Also, the chemical breakdown of a substance or compound caused by its combination with oxygen.

oxidation inhibitor An oil additive that helps to prevent oil oxidation

oxygen sensor A sensor that detects the amount of oxygen present in exhaust gases.

P

palladium A rare metal, used with platinum as an oxidizing catalyst.

parallel Passing simultaneously through two or more units.

parallelogram steering linkage A linkage system in which the Pitman arm, idler arm, and relay rod form three sides of a parallelogram.

parking brake A cable or linkage assembly that is operated by the driver to activate the rear brakes.

partial nonhunting gearset A differential gearset in which one drive pinion gear tooth contacts six differential ring gears after several rotations.

partial torque Tightening a fastener just enough to hold adjustment until final torque can be applied.

parts requisition A place on a repair order to log in parts and parts prices.

passive restraint A safety belt or air bag that operates automatically.

pedal reserve The distance between a brake pedal and the floor of an automobile when the brake pedal is depressed fully.

penetrating solvent A thin, oily solvent that flows between threads and helps to lubricate parts for removal.

pentroof A combustion chamber shaped like the roof of a house.

percolation Boiling of fuel within the carburetor.

phase A portion of a cyclical event, such as a given section of an AC sine wave.

Phillips screwdriver A screwdriver used to loosen or tighten Phillips head screws.

pickup coil assembly A small, permanent magnet and coil assembly that, in combination with a reluctor, forms an electronic position sensor.

pinion depth setting The in-and-out adjustment of the drive pinion gear.

pinion shaft A differential part that is mounted in the carrier and that holds the differential pinion gears.

piston A can-shaped part that moves up and down inside a cylinder during combustion.

piston clearance The distance between the piston and the cylinder wall.

piston pin A tubular steel pin that attaches the piston to the connecting rod.

piston ring compressor A tool used to compress piston rings around a piston so that it may be inserted into a cylinder bore.

piston ring expander A tool used to remove cast piston rings from a piston without breaking them.

piston side thrust The force developed against the cylinder wall by the tilting of the piston at the top or bottom of its strokes.

piston slap The noise made by the tilting of a piston against the cylinder wall on the power stroke. Caused by excessive piston clearance.

Pitman arm A steering rod that connects the steering gearbox to the relay rod.

pivot stud A stud that serves as a pivot point for a mechanism.

planetary gearset A set of gears in an automatic transmission that are in constant mesh and that can be shifted while they are spinning.

Plastigage Perfect Circle Plastigage, a thin thread of wax-like material that flattens when compressed between two parts. The amount of flattening indicates the clearance between the parts.

platinum A rare metal, used as an oxidizing catalyst in catalytic converters.

play Looseness in the connection of two parts.

plunger Part of a dial indicator that transfers play or contour to the indicator.

ply One of several layers of material that make up a tire carcass.

pneumatic tool A tool powered by compressed air.

pole piece An iron core that increases magnetic reluctance; used within field coils.

pole shoe *See* pole piece.

polisher A tool that shines and polishes an automobile's finish.

poppet valve A valve that moves up and down to open and close a port.

porosity Tiny holes in a casting caused by air bubbles.

port A hole or passage.

portable electric drill An electric-powered drill that can be carried.

ported vacuum A vacuum source on the carburetor, slightly above the throttle plate, which provides vacuum only after the throttle plate has opened slightly.

port-type injection Fuel injection within an intake port.

positive crankcase ventilation (PCV) system An emission control system that routes blowby gases and unburned oil fuel vapors to the intake manifold.

positive ground A way of connecting a single-wire vehicle system so that the positive battery terminal is connected to the metal of the vehicle to form a ground.

pour point depressant A chemical additive that helps cold oil to remain thin enough to flow to engine parts during starting.

power The time rate at which work is done or energy is emitted or transferred.

power brake booster unit An assist device that reduces driver effort on the brake pedal.

power steering A system that uses hydraulic power to assist the driver when turning.

power steering pump A part that creates hydraulic fluid flow to operate a power steering system.

power tools Tools that are operated by electricity, compressed air, or hydraulic pressure.

power train Engine and drivetrain.

prechamber A small opening within a cylinder head where fuel is ignited prior to ignition in the main combustion chamber.

precision insert bearings Friction bearings made in two halves and precisely formed and machined to proper shape and size.

preignition Ignition caused by a glowing deposit or metal part heated enough to begin ignition before a spark occurs at the spark plug.

press *See* stamp.

pressure The amount of force pushing on a surface.

pressure bleeding A method of bleeding a brake system by using air pressure at the master cylinder.

pressure plate A heavy, flat ring that presses against the clutch disc.

pressure plate assembly Parts of the clutch that hold the clutch disc tightly against the flywheel.

pressure regulator valve A valve in an automatic transmission that controls hydraulic pressure from the oil pump.

pressure relief valve A valve that opens under pressure to allow excessive pressure to escape.

pressure test A check of hydraulic pressures in an automatic transmission.

pressure transfer The ability of pressure to be equal in all areas of a hydraulic system.

primary brake shoe The smaller of two brake shoes, positioned toward the front of an automobile.

primary throttle plate The first, or main, throttle plate in a multiple-barrel carburetor.

printed circuit A thin sheet of nonconductive plastic material on which strips of conductive metal have been deposited.

process To change data into meaningful information by following program instructions.

program A set of instructions for a computer.

programmable read-only memory (PROM) A computer module containing information and programs for a specific car model. The information and programs may be used as source information but may not be changed.

programmed Provided with instructions, such as a computer.

propane enrichment Adding propane, a hydrocarbon fuel, to a lean air-fuel mixture to locate a vacuum leak.

propeller shaft *See* driveshaft.

proportioning valve A valve in a brake system that limits maximum pressures to rear drum brakes, preventing wheel lockup on hard braking.

proton Part of the nucleus of an atom that has a positive charge.

protractor An angle gauge.

prove-out sequence A display sequence that checks, or proves, that an indicating device's basic operation is correct.

pry bar A long, steel rod used to position or break free heavy parts.

pulse air valve A container with reed valves, used to provide small amounts of air for air injection.

pulse transformer A transformer that produces periodic surges of high voltage, such as an ignition coil.

pulse width The length of time in which energy is applied to open an EFI injector to spray fuel.

punch A thin, tapered metal tool used to remove rivets and pins.

Q

quench area The narrow, confined area farthest from the center of the combustion chamber where the flame front is extinguished.

R

rack and pinion A steering system that operates with a steering rack and a pinion gear.

radial ply tire A tire with plies that run straight across the width of the tire. Radial tires also have belts between the carcass and the tread.

radial runout A measurement that determines how much a tire and wheel are out of round.

radiation Heat transfer from one location to another through empty space.

radiator A heat exchanger that transfers heat from liquid coolant to air passing through a finned core.

raster An oscilloscope pattern that displays all firing order traces separately, in a stack.

reach The length of the threaded part of a spark plug.

reaction member A gear that is held from moving during planetary gearset operation.

reamer A tool that machines holes to exact sizes.

rear-end torque A reaction caused by torque transmitted to the differential that causes the differential and rear axle housings to rotate.

rebound Downward suspension movement.

recap tire *See* retreaded tire.

receiver/dryer A liquid refrigerant storage tank that removes moisture from refrigerant.

recess A shaped hollow space on a part.

reciprocating piston engine An engine in which pistons are driven back and forth in cylinders.

recirculating ball A steering gearbox design that includes a worm gear, sector gear, ball nut rack, and ball bearings.

reduction The chemical removal of oxygen from a compound containing oxygen.

reed valve A valve device with a flexible diaphragm to open or block an opening, similar in operation to a check valve.

refrigerant A liquid with a very low boiling point that is used to absorb heat in an air conditioning system.

refrigerant control A control that regulates the amount of refrigerant passing from the high-pressure side to the evaporator, or low-pressure side, of an air conditioning system.

refrigeration A process that maintains cold temperatures, or that cools.

regrind To machine to a smaller size with a grinding device.

relay An electromagnetically operated device that uses small amounts of electricity to operate a heavy-duty switch.

relay rod A steering rod that transfers steering movement from the Pitman arm toward both front wheels.

release levers Parts of a clutch pressure plate assembly that release the holding power of clutch springs when depressed by the throwout bearing.

reluctor A device that conducts lines of magnetic force easily; part of a switching device for electronic ignition.

remote starter A simple electrical switch with leads that can be connected to crank the engine with the starter motor.

remote starter switch A spring-operated electrical momentary contact switch and leads. Such a switch is used to activate the starter motor from a location away from the key switch.

removable-carrier differential A design in which the differential carrier bearings are mounted in a removable casting, separate from the rear-axle housing.

repair The replacement or fixing of automobile parts that wear out, break, or malfunction.

repair order A form that shows information about a customer, a car (service description, parts used), and billing.

replaceable valve guide A bushing pressed into a cylinder head to form a valve guide.

residual check valve A brake valve that applies residual pressure to a drum brake assembly to prevent leakage at wheel cylinders.

resistance The quality of a substance that resists electrical flow, measured in ohms.

resistor An electrical component that lowers the amount of voltage and current passing through an electrical circuit.

resonator A small, muffler-like device added to an exhaust system to absorb additional sound frequencies.

restraint anchor An attachment point where a seat and shoulder belt assembly is attached to the automobile.

retainer A holding device.

retainer-plate axle assembly A driving axle that is bolted to a retaining plate and to a wheel backing plate.

retaining ring A ring with spring action that snaps into grooves and holds gears onto shafts.

retard To make ignition spark occur later.

retractor A reel that rotates and locks a seat belt in position.

retreaded tire A used tire carcass with a new tread added.

returnability The ability of a steering wheel to return to a straight-ahead position after it has been turned.

reverse biased A transistor whose emitter-base junction receives a small current flow in a reverse direction. No current can flow through the emitter-collector path.

reverse flushing A procedure of forcing clean liquid backwards through the cooling system to loosen and carry away rust, scale, corrosion, and other contaminants.

reversed polarity A condition that occurs when the connections to a DC source are switched from one pole to another.

revolutions per minute (rpm) The number of times a device, shaft, wheel, gear, etc., makes one complete 360 degree turn.

rhodium A rare metal used with platinum as a reducing catalyst in catalytic converters.

rich mixture An air-gasoline mixture containing less than 14.7 parts of air to one part of gasoline, by weight.

rim diameter The size of the hole in a tire that fits over the wheel rim.

ring gap The space between the ends of a cylinder ring.

ring gear A large gear that redirects torque rotation in the differential; part of the ring and pinion gearset.

ring seal A ring-shaped seal made of rubber or hollow metal tubing; used to form a stationary or dynamic seal.

rivet A soft metal pin used to fasten materials together permanently.

rocker arm ratio The ratio between the distances moved by the pushrod end and the valve tip end of a rocker arm.

rocker follower A pivoting rocker arm used in an overhead-camshaft engine.

rocker shaft A shaft on which a rocker arm or rocker follower is mounted.

rod and lever linkage Clutch linkage that is operated by a series of rods and levers.

rod bearing knock A sharp knocking sound caused by excess connecting rod bearing clearance.

rod cap A heavy metal clamp, used to attach the connecting rod to the crankpin.

rod journal The offset crankpin area where a connecting rod is attached.

rolling resistance Friction between the tires and the road.

rollover valve A safety valve that automatically closes a fuel-vapor line to the front of the vehicle if the vehicle tilts excessively.

rope-type seal A seal made up of fibers braided like a rope.

rotary distributor type pump A diesel injection pump in which a rotating device is used to pump and distribute the diesel fuel.

rotary engine An engine whose internal movement is rotary rather than reciprocal.

rotary roller pump An EFI pump that uses rollers on a centrally mounted eccentric to pump fuel.

rotor The shaft, rotating coil, and field pole assembly of an alternator that produces a magnetic field. Also, a metal disc that is part of a disc brake system and that rotates with the axle.

rubbing block A plastic or resin-impregnated fiber block attached to a movable breaker point contact. It rests against the distributor cam.

runner A hollow, tubular area of a manifold.

runout The distance the surface or edge of a rotating part moves, in or out, during rotation.

RTV A room-temperature vulcanizing, or hardening, synthetic rubber sealant.

Rzeppa joint A constant-velocity joint with large-diameter ball bearings that run in the inner and outer parts of the joint.

S

SAE Society of Automotive Engineers.

SAE horsepower A customary, corrected horsepower rating.

safety glass Automobile glass that crumbles into small pieces to help prevent injuries.

safety glasses Goggles that are worn to protect eyes while working.

safety guards Shields positioned over moving parts to protect the worker from flying debris.

safety rim A wheel rim with raised sections that hold a tire on a wheel.

sampling probe A device, inserted in a tailpipe, that allows small amounts of exhaust gas to enter an exhaust gas analyzer.

sand cast Refers to materials cast in a mold made of sand; rough-surfaced.

sander A tool that smooths and removes metal.

saturate To fully develop the primary coil's magnetic lines of force.

scoring Parallel grooves worn by friction around the inside or outside circumference of a part.

screwdriver A tool that consists of a shaft of metal with a handle and a blade to loosen or tighten screws.

screw extractor A steel tool that removes broken bolts or studs.

scuff Momentary welding of two rubbing surfaces caused by excessive heat. When the weld is broken, a scuffed surface remains.

scuffing Damage caused by greasy friction between parts.

seal A part or assembly used to form a leak-proof seal around a rotating part.

sealed-beam A large, sealed, evacuated glass bulb reflector and lens housing that contains a filament.

seam A line of contact where two parts join.

seat To rub by friction until rubbing edges match each other almost perfectly.

secondary brake shoe The larger of two brake shoes, positioned at the rear of an automobile.

secondary circuit The high voltage side of an ignition system. Includes the ignition secondary coil, coil cable, rotor, distributor cap, spark plug cables, and spark plugs.

secondary progression system A system to open a secondary throttle plate in relation to the amount of opening of the primary throttle plate.

section width The distance from sidewall to sidewall on a tire.

sector gear A gear that has teeth in a semi-circle, or on a section of a circle.

segments Copper sections of a commutator that are connected to each end of an armature coil on a motor or generator.

seize To melt and stick, such as when an aluminum piston melts and sticks to the cylinder wall.

self-adjuster Part of a drum brake system that adjusts drum brake clearances automatically.

self-diagnostics The capability of a computer system to diagnose its own problems.

self-energizing A reaction that creates additional braking performance in drum brake systems.

self-locking nut A hump-shaped nut that locks to the bolt when tightened.

self-tapping screw A screw that cuts its own threads into metal.

semi-centrifugal clutch A clutch with weighted release levers, or fingers, that ease clutch pedal operation.

semiconductors Materials that partially conduct electricity.

semi-floating axle A driving axle that helps to support the weight of the automobile.

semi-floating piston pin A piston pin that is fixed to either the piston or the connecting rod.

semi-metallic brake linings Brake linings made of steel fibers, coarse sponge iron, graphite, friction modifiers, and a resin binder.

sending unit A device with variable resistance that operates a dash board gauge.

sensor A device that senses a condition and modifies an electrical signal in response to changes in the condition.

separator strip A porous plastic or paper insulator between battery plates.

sequential fuel injection (SFI) Fuel injection that occurs at an individual multi-point injector just before the opening of its intake valve.

series Passing consecutively through one unit after another.

series circuit A circuit in which each electrical load, in order, forms part of a circuit.

series-parallel A combination of series and parallel flow patterns. *See also* series and parallel.

service bay A working area, especially in a garage, where automobiles are serviced.

service manager Head of a service department.

service manual A book that explains how to service and repair an automobile.

service station A gas station with one or more service bays to service automobiles.

service writer A person who greets customers in a service department, makes a preliminary diagnosis of service requirements, and prepares a cost estimate.

servicing Maintenance and repair of automobiles.

servo A hydraulically operated part that controls a transmission band in an automatic transmission.

servomechanism A force amplifier used for positioning control.

SFI. *See* sequential fuel injection.

shell A spherical shape formed by spinning electrons that orbit the nucleus of an atom.

shift lever A lever that transfers motion from one part to another.

shift quadrant The letters that designate gear positions on an automatic transmission gearshift indicator.

shift valve A switching valve that selects the correct gear in an automatic transmission.

shifting arms Short levers attached to the shifting shafts on the side of a manual transmission.

shifting forks Semi-circular yokes that push or pull the synchronizers toward, or away from, a gear to be engaged to, or disengaged from, the mainshaft.

shifting shafts Metal shafts inside a manual transmission case that move the shifting forks.

shimmy Side-to-side wheel vibration.

shock absorber A hydraulic suspension part that dampens spring action and works with the springs to control movements of the body, wheel, and axle.

shop foreman A person in charge of several technicians, also provides quality control on repair and service operations.

shop layout The location of work areas and equipment in a shop.

short circuit A condition created when electricity accidentally flows incompletely through a circuit, bypassing a part of the intended circuit.

short finder An electrical testing device used to find the location of a shorted or broken wire.

shroud A hollow duct that helps to direct air toward the cooling system fan for better engine cooling.

shunt An electrical bypass.

siamese To form two adjacent intake or exhaust ports into a common port, perhaps separated by a thin center divider.

sidedraft carburetor A carburetor in which air flows through the barrel in a horizontal direction.

side gears Differential gears that are connected to the driving axles and mesh with the differential pinion gears.

side marker light A light visible from the side of a vehicle that increases vehicle visibility in poor lighting conditions.

side terminals An internally threaded connector to which all negative or positive battery plates are connected.

sidewall The side of a tire.

sight glass A glass-covered hole in an air conditioning system through which refrigerant level is checked.

sine wave A pictorial representation of an alternating current (AC) voltage.

single overhead camshaft (SOHC) An engine design with a single camshaft mounted in the cylinder head.

single-point fuel injection Fuel injection at a single point, rather than at each individual intake manifold runner.

single-viscosity oil An engine oil tested for viscosity at 212 degrees F (17.7 degrees C) or at 0 degrees F (-18 degrees C), but not both.

single-wire circuit A method of wiring a vehicle so that the metal of the vehicle forms a ground connection. One wire is used to connect the other pole of an electrical source to the load.

skirt The lower part of a piston, below the ring grooves.

slant Term describing an angle that is not perpendicular to the horizontal.

sleeve A thin cylinder that is inserted or pressed into the block as a replacement for a damaged or worn cylinder bore. Also, the part of a micrometer between the spindle and the thimble, with a scale that gives a measurement reading.

slide valve A device that slides back and forth to open or close an opening.

sliding caliper *See* floating caliper.

slip A condition caused when a driving part rotates faster than a driven part.

slip joint A tubular splined assembly that allows two rotating parts to remain connected as one or both move along a common horizontal axis.

slip-joint adjustable pliers Pliers whose jaws adjust to two different sizes; sometimes called combination pliers.

slip ring A circular, smooth ring on an alternator rotor through which electrical connections are made by brushes.

slip yoke A yoke attached to a slip joint.

slotted screwdriver A screwdriver with a flat blade.

slow charging Slow battery charging done at rates less than three amperes, sometimes called "trickle charging."

sludge A thick, soft, tarry substance formed in oil by water vapor and contaminant particles.

sluggishness A condition in which the engine will not deliver sufficient power under load or at high speed.

small hole gauge A tool for measuring holes too small for an inside micrometer.

snap ring *See* retaining ring.

sniffer A hose on a leak detector tool through which refrigerant can flow.

socket wrench A wrench with a handle and a drive that turns a socket.

soft plug A plug made of stamped sheet metal; used to seal holes, especially in cast metal parts.

solder An alloy, or mixture, of lead and tin that when heated can be used to join metals such as brass, copper, and iron. Also, to join by using solder.

solderless connector Metal devices used for joining or terminating wires. Stripped wires are placed in a tubular metal device which is crimped to hold wires securely.

solenoid A device that uses electromagnetism to exert a pulling or holding force.

solid lifter A lifter made from a single piece of solid metal.

solvent A chemical used to clean automotive parts.

solvent cleaner A tank used for cleaning small parts with cleaning solvent.

space frame A unitized body with additional body parts and braces built into it.

spaghetti tubing A hollow, straw-shaped flexible insulating material that can be slipped over wires.

spark ignition A way of igniting an air-fuel mixture by using a high-voltage electrical arc.

spark plug A removable part that screws into the cylinder head, extends into the combustion chamber, and ignites the air-fuel mixture. A spark plug consists of an outer metal shell with one electrode and a center ceramic insulator with another electrode. The electrodes are separated by a small air gap, across which a high-voltage current can arc.

specialty repair manual A manual published by someone other than the manufacturer. It covers in great depth a particular component or system of an automobile.

specialty shop A shop that specializes in one type of automotive service, such as transmissions, mufflers, or tune-ups.

specific gravity The relative density of a material when compared to the density of pure water.

specifications Measurements for tightening, fitting, and testing parts of an automobile.

speed handle A long cranking handle for a socket wrench.

speed nut A flat nut that is pressed onto a bolt or stud.

speedometer cable A cable that connects to the transmission and operates the speedometer.

spindle The movable measuring surface on a micrometer.

spindown time The time required for the clutch disc, input shaft, and transmission gears to come to a complete stop.

splash lubrication A method of distributing oil in which moving parts splash and throw oil to adjacent parts.

splay To spread or move outward from a central point.

splice To join. Electrical wires can be joined by soldering or by using crimped connectors.

spline A machined ridge on the inner or outer circumference of a round part, used to connect and drive another part.

split driveshaft A driveshaft consisting of multiple sections connected by universal joints.

split reservoir A master cylinder design that has two chambers to direct brake fluid to the front and rear brakes.

sponge lead (Pb) Porous lead that forms the negative plate of a battery element.

sponginess A condition in which the engine does not speed up as much as expected when the accelerator is depressed, especially during cruising.

spongy pedal A brake pedal action that feels springy.

spray painting A process that uses compressed air to spray paint on an automobile.

spring The part of the suspension that absorbs road shock and supports the frame, body, engine, and drivetrain above the wheels.

spring seat A saucer-like bracket that holds coil springs in position.

spring shackle A part that allows a leaf spring to change length as its leaves bend.

springs Coiled or leafed parts that absorb road shocks.

sprung weight The weight of all parts suspended by the springs.

stabilizer bar A long steel rod mounted longitudinally to the lower control arms and to the body to control body roll.

stagger To arrange parts in an alternating or zigzag pattern.

stalk A lever, usually attached to the steering column that contains switches for electrical devices.

stalling A condition in which the engine dies after starting.

stamp To form parts from sheet metal by pressing it between metal forms.

star wheel An adjusting screw that is used to adjust drum brake clearances.

starter drive The unit mounted on the armature shaft; it combines the overrunning clutch and the drive pinion gear into a single unit.

static balance The equal distribution of weight around a wheel.

stator A part between the impeller and the turbine in a torque converter that redirects and accelerates fluid flow. Also, a stationary coil within an alternator, in which current is induced.

steam cleaner A machine that produces steam and mixes it with a soap solution to clean parts of an automobile.

steering A turning system that moves the front wheels.

steering arm Linkage that connects a steering knuckle to a tie rod.

steering axis inclination An angle created when the steering knuckle slants inward.

steering column A housing that encloses the steering shaft and holds control devices, such as shift and turn signal levers.

steering gearbox A set of gears in a housing. These gears change rotating motion from the steering shaft to side-to-side motion at the steering linkage. Often referred to as a steering gear.

steering kickback A sharp, rapid movement of a steering wheel when an automobile strikes a bump or other obstruction.

steering knuckle A part of the wheel spindle assembly that connects to the ball joints. In many cases, a part that is forged into a unit with the wheel spindle.

steering linkage A system of rods that transfers steering motion from the steering gearbox to the front wheels.

steering ratio A gear ratio in the steering gearbox that determines the steering wheel rotations that are required to turn the wheels from lock to lock.

steering shaft A shaft that connects the steering wheel to the steering gearbox.

steering wheel surge A sudden jerk of the steering wheel.

Stellite An extremely hard alloy of cobalt, chromium, and tungsten.

step-down transformer A transformer in which the voltage produced in the secondary coil is less than that in the primary coil.

stepping motor An electrical motor that can move a plunger in or out a precise distance in response to electrical signals.

step-up transformer A transformer in which the voltage created in a secondary coil is greater than that in the primary, or first, coil.

stoichiometric A chemical term that indicates the proper or ideal mixture of chemicals for a particular reaction.

straightedge A tool shaped like a ruler; has straight edges to measure for warpage.

stratified charge An intake charge that is layered, with a richer mixture on top and a leaner mixture beneath it.

streamlining The practice of reducing the coefficient of drag of a vehicle's body.

stress Applied force that may cause a part to crack or break.

strip To remove insulation from a wire.

stroke The distance a piston moves in a cylinder, up or down.

strut A flat structural piece that controls expansion.

stub frame A short frame used to help support the weight of the engine and the transmission at the front of some automobiles.

stud A fastener with external threads at each end.

substrate A ceramic honeycomb grid structure coated with catalyst materials.

sulfation The formation of lead sulfate on battery plates.

sulfuric acid A powerfully corrosive substance formed from water and sulfur.

supercharger A compressor driven by mechanical engine power to force air into an intake manifold under pressure.

superimposed Placed one on top of the other, an oscilloscope pattern for comparison of all cylinder firings.

surface plate A steel plate, accurately machined until it is almost perfectly flat; used to check the flatness of other parts.

surging A condition in which the engine speeds up and/or slows down with the throttle held steady.

suspension Springs and shock absorbers connected to the frame or underbody to provide a smooth ride over uneven surfaces.

swing axle A driving axle that pivots from the differential and moves, or swings, the axle in an up-and-down motion.

swirl A circular or whirlpool-shaped flow of the intake charge.

switch A device that can complete (close) or interrupt (open) an electrical circuit.

synchronizer An assembly that can engage a gear to turn the mainshaft.

T

tampering Intentionally removing, modifying, disconnecting, or disabling emission control devices.

tang A projecting piece that fits into a matching opening.

tap A cutting tool used to form, or repair damaged, internal threads.

taper A difference in width or diameter between two ends of a component.

technician's stethoscope A device, similar to a doctor's stethoscope, used for listening to mechanical noises.

telescoping gauge A spring-loaded tool for measuring the inside size of a hole.

temperature grade A grading system that indicates how well a tire resists and dissipates heat.

terminal A location at which an electrical connection is made.

test drive A form of troubleshooting that requires specific driving procedures by a technician.

test lamp *See* continuity tester.

testing by substitution Substituting a known good part for a suspected defective part.

tetraethyl lead An extremely toxic chemical lead compound added to gasoline to increase its octane rating.

T-handle A socket wrench handle used in tight working areas and for greater working force.

thermal efficiency A comparison between the energy present in fuel and the energy output of an engine.

thermal gauge A gauge in which a heating coil and a bimetallic spring move an indicating needle.

thermodynamics The scientific study of the mechanical action or relations of heat.

thermostat A temperature regulating mechanism.

thermostatic coil A bimetallic coil, or spring, that responds to heat by curling or uncurling.

thermostatically operated air cleaner A device that mixes hot and cool air to promote fuel vaporization in the intake charge.

thimble The adjusting portion of a micrometer that moves the spindle and has a bevel scale to give a reading.

thread gauge A tool for measuring the number of threads per inch.

thread pitch The distance between each thread of a fastener.

thread sealant A sealing gel used over fasteners before installation.

threaded fasteners Fasteners that connect parts with grooves, or threads.

three-phase AC current Alternating current induced in three coils at slightly staggered times.

three-quarter floating axle A driveline axle, used on older passenger vehicles, that has two functions only: to turn and help attach the wheel.

three-way catalyst A catalytic converter that changes HC, CO, and NO_x into H_2O, CO_2, N, and O. Also called a three-way catalytic converter.

throttle body The lower section of a carburetor, which contains the throttle plates, idle mixture passages, and adjusting screws.

throttle body injection (TBI) Fuel injection with one or two injectors located in a throttle body.

throttle bore The opening in a carburetor through which air passes.

throttle plate A movable plate, or flap, in an opening at the bottom of a carburetor barrel.

throttle position sensor An electrical sending unit that relays information to a control system about how far the throttle plate of the carburetor is open.

throttle positioner An electrical device that mechanically holds a throttle plate open.

throttle pressure An increase in automatic transmission mainline pressure to help control shifting.

throttle valve A balancing valve that controls throttle pressure in an automatic transmission.

throttle valve cable *See* kickdown linkage.

throw An offset area on a crankshaft.

throw-off lubrication A system of lubrication in which oil is thrown from crankshaft main bearings onto cylinder walls.

throwout bearing A prelubricated ball bearing that is moved against the pressure plate release levers when the clutch is operated.

tie rod A steering rod connected between the relay rod and steering arm.

timing gear A gear that drives or is driven in a synchronized gear assembly.

timing light An instrument that produces a brief (0.0010 second), repeated flash of light that appears to "freeze" a moving pulley or flywheel.

tire plug A tire repair kit that contains a plug to fill the puncture and a head to seal the puncture.

tire pressure gauge A tool for checking air pressure in a tire.

tire rotation The switching of tire and wheel assemblies to different locations on an automobile to increase tire life.

tire valve A valve through which a tire is inflated or deflated.

toe-in A condition in which the front edges of the tires on the same axle are closer than the rear edges.

toe-out A condition in which the front edges of the tires on the same axle are farther apart than the rear edges of the tires.

toe-out on turns The difference in angles between the front wheels during turns.

top dead center (TDC) The position of the piston at the top of either the compression or exhaust stroke.

top-feed injector An EFI injector in which fuel is fed in from the top.

torque A turning or twisting force, such as that used to turn a shaft, measured in pounds-feet (lb.-ft.). *See also* power, work.

torque arm A long steel rod that is connected between the rear axle and the transmission to control rear-end torque.

torque converter An assembly that produces a fluid coupling between the engine and automatic transmission.

torque converter housing A case that encloses the parts of a torque converter.

torque specifications Required amount of tightening for fasteners.

torque steer A twisting axle movement in front-wheel drive automobiles that causes a pulling action under acceleration toward the side with the longer driving axle.

torque-tube drive A method of controlling rear-end torque by connecting a rigid tube between the differential and transmission or vehicle frame.

torque wrench A tool that measures the torque, or turning force, applied to a fastener.

torquing-over Engine twisting motion in a direction opposite to crankshaft rotation.

torsion bar A steel bar that is twisted to provide spring action in a suspension.

torsional coil springs Clutch disc springs in the hub that help smooth out rotation by absorbing torsional vibration.

Torx screwdriver A screwdriver used to loosen or tighten Torx head screws.

trace A characteristic oscilloscope pattern. Also, a very small amount.

tracer A contrasting colored stripe on plastic insulation, used to help trace a particular wire.

traction The ability of a tire to grip the surface on which it is riding.

traction grade A grading system that indicates a tire's ability to stop on wet pavement.

traction wheels Driving wheels.

trailing arm An arm on the rear suspension that extends rearward from the actual suspension mounting point.

tramp Wheel hop caused by static balance.

transaxle A driving system in which the transmission and differential are combined in one unit.

transfer port *See* off-idle port.

transformer A device with coils in which electricity can be produced through induction.

transistor A semiconductor device used to control or amplify current flow.

transitional port *See* off-idle port.

transmission A series of gears that take power from the engine and transfer it to more usable speeds while driving.

transmission band A friction-lined, steel strip that wraps around a clutch drum in an automatic transmission.

transmission input shaft The shaft through which engine power enters the transmission.

transverse Perpendicular, or at right angles, to a front-to-back centerline.

tread The part of the tire that touches the road.

tread wear indicator A series of bands, or bald spots, that appear on a tire when only 1/16 inch remains.

trim height A manufacturer's recommended height, from ground level, of certain suspension and body parts.

tripod joint A constant-velocity joint with a triangle-shaped center spider, needle bearings, and circular races in the outer part of the joint.

tripot joint *See* tripod joint.

trouble code Information in the form of numbers stored by an on-board computer to indicate specific problems.

trouble light A caged, insulated bulb light used when working on automobiles.

troubleshooting A step-by-step procedure for diagnosing an automotive problem.

troubleshooting chart A listing of problem conditions, possible sources of problems, and actions to be applied to resolve the problems.

trunnion One of the arms of the cross-shaped central part of a universal joint.

tubeless tire A tire that fits tightly against the wheel to keep air from escaping.

turbine A wheel with curved blades that can be used to turn a shaft or to pump fluids under pressure. Also, a fan-like part connected to the transmission input shaft and driven by a fluid coupling.

turbine-type pump A centrifugal EFI pump similar to a water pump or turbocharger compressor wheel.

turbo lag The time necessary for a turbocharger to begin compressing the intake air charge.

turbocharger A method of supercharging that uses the power of exhaust gases to turn a compressor. The compressor forces air, or an air-fuel mixture, into the intake manifold under pressure.

turbulence An irregular pattern of flow of an intake charge.

two-stroke cycle An engine operating cycle that combines the functions of intake and exhaust, and compression and ignition. This causes a power stroke to occur once each revolution of the crankshaft.

two-way catalyst A catalyst that promotes the oxidation of HC and CO exhaust emissions.

U

U-joint *See* universal joint.

underinflation Low tire pressure; causes the outside of a tire to wear quickly.

uneven firing Term used to characterize an engine whose crankshaft receives power pulses at uneven intervals.

Uniform Tire Quality Grading (UTQG) A federal government testing program that grades tires for traction wear and temperature resistance qualities.

unitized body A reinforced body structure that eliminates the need for a frame.

universal joint A joint that allows the driveshaft to transmit torque at different angles as the suspension moves up and down.

universal-joint adaptor A socket wrench adapter that allows the socket to be turned from an angle.

unpowered test light An electrical testing device that can be used to check for power and/or ground connections.

unsprung weight The weight of parts that are not suspended by the springs.

updraft carburetor A carburetor in which air flows through the barrel in an upward, vertical direction.

V

vacuum advance mechanism A mechanism that can advance ignition timing in response to engine load.

vacuum and pressure gauge A tool for measuring engine vacuum and fuel system pressure.

vacuum diaphragm tester A tool used to check for proper operation of and for leaks in vacuum systems.

vacuum modulator A diaphragm-operated part that controls throttle valve action in some automatic transmissions.

vacuum motor A vacuum diaphragm attached to a pull rod that can provide a mechanical pulling or pushing action.

vacuum relief valve A valve that opens to allow outside air into a sealed container to prevent the formation of a vacuum.

vacuum retard A mechanism that can retard ignition timing to decrease combustion pressures and lengthen combustion time.

valve A part that moves to open or close a port.

valve adjusting screw A threaded shaft located in a rocker arm or rocker follower; used to adjust valve clearance.

valve adjustment The process of adjusting valve clearance.

valve adjustment shim A thin, replaceable, precisely sized metal insert used to adjust valve clearance. Generally used with cam followers.

valve body A housing in an automatic transmission that contains a maze of passages and valves to direct hydraulic pressure to the proper locations.

valve clearance The distance between the tip of the valve stem and the rocker arm when the camshaft is positioned on its base circle. Also, the distance between a cam follower and the tip of the valve stem when the camshaft is positioned on its base circle.

valve float A condition in which the valves fail to follow the actions of the valve train, and remain open.

valve guide The area within the cylinder head that guides and supports the valve stem. May be integral or replaceable.

valve lash *See* valve clearance.

valve overlap A situation in which both the intake and exhaust valves are open at the same time.

valve rotator A mechanism that rotates a poppet valve while it is being opened and closed.

valve seat insert A circular, hardened metal ring pressed into a cylinder head to form a valve seat.

valve spring surge A valve spring vibration that causes erratic valve action.

valve train The parts that operate the valves, driven by the camshaft. May include lifters, pushrods, rocker arms, cam followers, and adjusting shims.

vane-type pump A pump with flat vanes, or blades, that move within an enclosure to trap and compress fluids.

vapor canister *See* charcoal canister.

vapor lock A collection of vapor in an area of a liquid fuel system, which prevents an engine from receiving fuel.

vapor recovery canister A container that stores fuel vapors for a vapor recovery system.

vapor recovery system A system that traps and holds fuel vapors until they can be drawn into an engine with an intake charge.

vapor/fuel separator A unit that separates fuel vapor from liquid fuel.

variable ratio steering A steering system within the steering gearbox that changes steering ratios as the steering wheel is turned.

varnish A dark, partially transparent hard coating caused by lubricant oxidation that can increase friction and reduce the size of oil passages.

ventilation system A passenger comfort system that circulates outside air through the passenger compartment.

venturi A narrow restriction in a carburetor bore that increases the velocity of air passing through the barrel.

vernier calipers A tool used to measure inside and outside diameters.

viscosity The thickness or thinness of a fluid and its ability to resist flowing.

viscosity index improver A chemical additive that helps oil to maintain an adequately thick lubricating film as the oil is heated.

viscous drive fan clutch A mechanism that uses fluid between an impeller and a stator to drive a fan.

viscous friction Friction between layers of liquid, such as oil.

voice simulator A device that produces sounds like human speech.

voice warning device A device that plays a recorded message or produces speech sounds to warn the driver of problems or conditions.

volatile Able to vaporize.

voltage drop Voltage lost by the passage of electrical current through resistance.

voltage drop test A test to measure the amount of voltage drop, or loss, across a conductor while electrical current is flowing. Excessive voltage indicates high resistance.

voltage regulator A device to control alternator current and voltage charging output to the battery.

voltage spike A pulse, or surge, of electricity that creates momentary high voltage. Usually caused by connecting or disconnecting a heavy electrical load in an electrical system.

voltmeter A measuring instrument used to check battery voltage level.

volumetric efficiency A comparison between how much fuel and air are actually drawn into an engine and how much could be drawn in.

vortex flow A swirling, twisting motion of fluid.

W

warpage Bending.

washer A thin spacer used between a screw or nut and the automotive part being tightened.

wastegate A pressure relief valve for a turbocharger.

water cooled An engine cooling system that operates with liquid.

water jacket Hollow spaces filled with coolant surrounding the cylinders in the block and combustion chamber areas in the cylinder head.

water pump A belt-operated pump that forces liquid coolant through an engine cooling system.

water sludge A soft, light-colored foamy substance formed from water whipped into engine oil by the action of the crankshaft.

watt A unit of power equivalent to an electrical current of one ampere at one volt. Also, the work done at the rate of one joule per second. Equals 1/746 horsepower.

wear indicator A round nipple mounting surface whose position on a ball joint surface indicates wear.

webbing Woven material, such as that in a seat or shoulder belt.

weep hole A ventilation hole located at the bottom of a water pump shaft housing.

wet compression test A compression test conducted after a small amount of engine oil has been added through the spark plug hole. Used to determine if piston rings are sealing properly.

wet liner A cylinder liner whose outer surface is in contact with the coolant in a liquid-cooled engine.

wheel alignment The positioning of steering and front suspension components and wheels.

wheel cylinder A cylinder in the brake assembly that transfers hydraulic pressure to the brake shoes.

wheel lugs Studs in a wheel hub on which a wheel is mounted.

wheel spindle A part that is connected through wheel bearings to a wheel.

wheel spindle assembly A suspension part that consists of a wheel spindle and a steering knuckle.

wheelbase The distance between the center of a front wheel and the center of a rear wheel.

whip Flexing of a shaft that occurs during rotation.

wide-open throttle (WOT) An operating mode in which the accelerator pedal is held to the floor for maximum engine power production.

winding The loops of wire that form a coil, as in an electric motor or generator.

wire gauge A feeler gauge that contains round, not flat, measuring strips.

wiring diagram A diagram that, like a road map, shows how wires are connected in circuits.

work The result of applying force to an object and causing it to move. *See also* energy, force, power.

work order Instructions to a technician on a repair order.

worm gear A gear with spiral threads.

worm-drive clamp A metal clamping band with a screw that engages slots on the band.

wrench A tool used to loosen or tighten a fastener.

wrist pin *See* piston pin.

Y

yoke A U-shaped part used to connect two other parts.

Z

zener diode A special form of diode that conducts in the opposite direction when sufficient voltage is applied.

Index

A

Acceleration, and carburetor function, 208
Accelerator pump, 208
Accelerator pump circuit, 208
Accessory circuits, function of, 14
Accidental ground, 280
Accumulator, 571, 729
Acid fog, 453
Acid rain, 452—453
Activated charcoal, 443
Active restraints, 745—746
Active suspension systems, 641—642
    malfunctions, 656
    parts, 641
Actuator, defined, 217
Additives, engine oil, 136
Adjustable open-end wrenches, 53
Adjusting sleeves, 659
Advance timing light, 372
Aeration, 198
Aerodynamics, defined, 22
After bottom dead center (ABDC), 130
After top dead center (ATDC), 130, 359
Age of automobiles, 3
Air bags, 746, 749
Air bleed, 459
Air chisels, 62
Air cleaner, thermostatically controlled, 444, 459—461
Air conditioning system, 727—730
    controls, 729—730
    evacuating, 739
    leakage detection, 737
    maintenance, 735—736
    parts, 727—728
    problem diagnosis, 733, 735
    recharging, 739—740
    safety precautions in servicing, 735, 736, 737, 739
    servicing, 737—740
Air cooling systems, 91
Air cushion restraint system, 746
    servicing, 749

Air filter, 180
    replacing, 196, 248
    in vapor recovery canister, 460
Air filter assembly, 180
Air gap, 351
Air horn, 210, 221
    replacement, 229
    tightening sequence, 221
Air horn gasket, 229
Air impact wrenches, 61—62
Air injection systems, 260, 450—451, 462—463
    types, 450—451
Air inlet, 725
Air nozzles, 36
    safety precautions, 37
Air pollution, 428—432
    health problems and, 429, 456
    incomplete combustion and, 429
    invisible, 429—430
    man-made, 428—429
    natural, 428
    sources, 430—432
Air pressure in tires, measuring, 69
Air pressure plate, 591
Air pressure test, 590—591
Air pump
    checking output, 462
    engine-driven, 451
Air ratchets, 62
Air resistance, 99
Air springs, 627
Air suspension system, 627
    malfunctions, 656
Air temperature, underhood, 444
Air-bleed jets, 205—206
Air-cleaner-to-carburetor gasket, 197
Air-cooled, defined, 12
Air-fuel mixture, 12, 180
    during acceleration, 208
    adjustment, 229, 411
    carburetor and, 85
    carburetor bores and, 211
    combustion of, 358—359
    compression of, 86
    at cruising speeds, 208

electronic control of, 361, 411
emissions and, 443—444
during full-power operation, 209
ignition of, 14
lean, 195
at low speeds, 208
metering, 204
stoichiometric ratio, 180, 411, 443—444

Air-hold adapter, 472

Airflow, 22

Airflow sensor, 239

Allen head screws, 45

Allen wrenches, 54

All-indirect gearing, 531, 536—537

Alloys, defined, 106

Alternating current (AC), 281
alternator and, 328
rectifying, 328, 330, 332—333
three-phase, 329—330

Alternator(s), 290
bearing noise, 340
bench tests, 346
bridge rectifier circuit, 332—333
diodes in, 330, 332
electrical output, 328—333
elements, 328
full-fielding, 342
function, 290, 328
grounding straps, 339
integrated circuits, 334—335
removing for off-car service, 346
undercharging and, 339
visual checks, 340
voltage regulators, 333—335
wiring connection checks, 339—340

Alternator drive belt, 339

Alternator output test, 342

Alternator ripple current, testing, 371

Altitude compensation valve, 215

Aluminum, 328

Ambient air, 444

American Petroleum Institute (API)
engine oil standards, 134, 135

Ammeter, 71—72, 281, 285, 389
as charging indicator, 335—336
inductive, 342

Amperage, testing, 371

Ampere, Andre M., 281, 282

Ampere (amp), 281

Ampere-hour rating, 295

Analog instruments, 388

Analog signals/displays, 236, 371, 388, 407

Anchor pin, 701

Aneroid bellows, 215

Anode, 286

Anti-friction ball and roller bearings, 106

Anti-lock braking system (ABS), 704—706
servicing, 721—722

Anti-seize compound, 48

Anti-skid braking system. *See* Anti-lock braking system (ABS)

Anti-theft system, 395—396, 744

Antifreeze, 71, 151—152

Antifreeze/water mixture. *See* Coolant

Anvil (micrometer), 67

API service ratings, 135

Apply devices, 571—572

Apprenticeship programs, 7

Approval of customer, 27—28

Arbor, 503

Arc-joint pliers, 55

Armatures, 308, 354
turning on lathe, 326

Asbestos in brake linings, 564

ASE/NIASE certifications, 5

Aspect ratio, 684

ATF. *See* Automatic transmission fluid

Atmospheric conditions, and horsepower output, 98

Atmospheric pressure, 157—158

Atomization, defined, 206

Atoms, 277, 278

Attitude, and safety, 33

Audible warning devices, 743

Audio systems, 395

Auto supply and accessory stores, 4

Automatic choke. *See* Choke, automatic

Automatic load leveling system, 628—629

Automatic temperature control system, 729—730

Automatic transaxle, 17, 578—583
ATF reservoir, 581
computer controlled, 579
design, 578—579, 581
hydraulic controls, 579
identification numbers, 589
operation, 579, 581—583
overhauling, 593
preventive maintenance, 587—589
problem diagnosis, 589—591
safety precautions in servicing, 586, 588, 591
servicing, 586—594
testing, 590—591

Automatic transaxle fluid, changing, 588

Automatic transmission, 17, 567—583
    computer controlled, 578—579
    computer tests, 593
    fluid coupling, 567—568
    gearshift lever, 574
    hydraulic control system, 574—578
    identification numbers, 589
    neutral safety switch and, 312
    oil pan removal, 588
    on-car adjustments, 591—593
    overhauling, 593
    preventive maintenance, 587—589
    problem diagnosis, 589—591
    safety precautions in servicing, 586, 588, 591
    servicing, 586—596
    solenoid valve tests, 593
    testing, 590—591
    torque converter, 568

Automatic transmission dipstick, 587

Automatic transmission fluid (ATF), 574
    adding, 587
    changing, 588—589
    checking, 589
    cooling of, 160, 578
    diagnosing, 587—588
    leakage sources, 593
    repeated low level, 592
    reservoir, 581
    screen, 588—589
    types, 574

Automobile, basic parts of, 10—22

Automobile body, 18—19, 22

Automotive industry career opportunities, 2—8

Automotive machinists, 485—486

Automotive service
    defined, 2
    education and training for, 7

Automotive service businesses, 3—4

Automotive specialty technicians, 6

Automotive technicians
    ASE certification areas, 5
    certification of, 4—5
    general categories of, 5—6

Axis, 526

Axles. *See* Driving axles

B

Babbitt metal, 106

Backing plate, 700, 720

Backlash, 618, 619
    adjustments for, 618—619

Backpressure, 186

Backpressure transducer, 448, 450

Baffles, 263

Balance pads, 115

Balance weights, 690

Balancing coil gauge, 389—390

Ball and needle roller bearings, 534

Ball joints, 630
    defined, 263
    inspecting, 647—648
    lubrication, 644—645
    replacing, 651—652
    wear indicators, 647

Ball nuts, 660

Ball return guides, 660

Ballast resistor, 353—354

Band adjustments, 592

Barrel of carburetor, 180

Base circle (of cam lobe), 128

Batteries, 39, 290—306
    ampere-hour rating, 302
    charging cycle, 294
    checking terminals, 316
    chemical reactions, 292, 294
    cleaning, 303
    connections, 280, 399
    construction, 291—292
    discharge cycle, 292
    electrical capacity, 295
    electrochemical action, 290
    electrolyte, 290—291
    function, 290—292
    grids, 291, 295
    group numbers, 296
    instrument testing, 300—302
    jump-starting procedures, 298—299
    life of, 294
    "magic" chemical rejuvenation, 305
    maintenance-free, 295—296, 300
    physical size, 296
    plates, 291
    posts, 292
    preventive maintenance, 299—304
    ratings of electrical capacity, 295
    removing, 302—303
    replacement, 295, 304
    safety precautions in servicing, 34, 70, 291, 294, 298, 299, 303, 304, 305, 317
    servicing, 298—306
    6-volt, 277
    terminal type, 296
    terminals, 276
    testing, 72, 300—302
    12-volt, 277, 295
    visual inspection, 299—300
    voltage checks, 399
    voltage while running, 336

watt-hour rating, 302
Battery cables, 310
    checking, 316—317
    cleaning, 303—304
    connections, 316—317
    corroded, 317
    repairing, 317
    replacing, 304, 317
    sizes, 316
Battery carriers, 303
Battery cells, 291—292
Battery chargers, 304
Battery connection pliers, 56
Battery hydrometers, 70
Battery protective spray, 303, 304
Battery testing, 72, 300—302
    with battery-starter tester, 301—302
    with hydrometer, 300
    with voltmeter, 301
Battery tray, cleaning, 303
Battery-starter tester, 301—302
BDC. *See* Bottom dead center
Beads, 682
Bearing(s), 83
    anti-friction ball and roller, 106
    camshaft, 125, 505
    clearance, 106, 108
    insert, 106
    main, 106
    metals used for, 106
    precision insert, 106
Bearing caps, 106
Bearing crush, 108
Bearing installation tool, 619
Bearing race, 550
Bearing retaining devices, 108—109
Bearing spread, defined, 108
Before bottom dead center (BBDC), 130
Before top dead center (BTDC), 130, 259
Bellhousing, 554
Belts, drive, 69
    alternator, 339
    cooling fan, 164—165
Belts (tires), 682
Bendix Corporation, 232
Bezel, 391
Bias current, 334
Bias-belted tires, 683
Bias-ply tires, 683
Billet, defined, 104
Billing, 28

Billing department, 28
Bimetallic spring/temperature sensor, 210, 384, 385, 444, 459
Binding force, 279
"Black box," 408
Bleeder screw, 715
Bleeder valve, 715
Bleeding, 714—715
Blowby gases, 116, 136, 431, 506
Blower, 725
Blowgun, 61
Body of automobile, 22
    unitized, 18—19
Body panels, 19
Body roll, 630—631
Bolt pattern, 694
Bolts
    defined, 45
    grade markings, 47
    head shapes, 45
    removing with solvent, 269
    size designations, 47
    tightening and loosening pattern, 147
Bonded, defined, 555
Boost pressure gauge, 391
Booster valve, 576
Booster venturis, 180
Bore, 83
    defined, 113
    engine displacement and, 94
Bosch, Robert, 239
Boss, 112
Bottom dead center (BDC), 86, 87
Bottom-feed injector, 237
Bound electrons, 280
Bourdon tube, 140, 391
Box-end wrenches, 53
Brake dynamometers, 98
Brake fade, 711
Brake fluid, 697, 700
    adding, 709
    checking, 709
Brake horsepower (bhp), 97
Brake lines, 696, 698
    inspection, 710
    routing systems, 698
Brake linings, 698—699
    thickness measurement, 712—713, 714
Brake pads, 700

Brake pedals
    checking action of, 710
    vibration, 711, 713

Brake shoes, 20, 700

Brake springs, 720

Brake system, 18, 696—722, 742. *See also* Disc brake system; Drum brake system; Power brake system
    bleeding, 714—716
    construction, 697—700
    electronic controls, 414
    hydraulic principles and, 696—697
    operation, 698
    preventive maintenance, 709—711
    problem diagnosis, 711—714
    safety precautions in servicing, 698, 711, 712, 714, 718
    servicing, 709—722
    testing, 711—712
    types, 20

Brake system warning light, 742

Brake-spring pliers, 56

Braking friction, 696

Breakaway torque, 610—611

Breaker bar (socket wrench), 54

Breaker point ignition systems, 352—354, 366
    cable, 357

Breaker points, 352, 353
    adjusting, 375—376
    replacing, 375—378
    visual inspection of, 369

Break-in procedures, 509

Bridge rectifier circuit, 332

British thermal unit (Btu), 99

Brushes, 308
    replacing, 326

Bucket lifter. *See* Cam follower

Bulbs, 393, 399

Bumpers, energy-absorbing, 745, 749—750
    cartridge replacement, 749—750

Bushings, 106, 648—649, 648
    replacing, 653

Business transactions, responsibilities in, 7

Bypass wiring, 403

C

Cable linkage, 557—558

CAFE standards, 435

California Air Resources Board (ARB), 432

Caliper, 20, 700

Cam/cam lobe, defined, 11, 84, 123, 128, 130

Cam follower, 123, 478

Cam grinders, 114, 116

Camber, 667, 677, 679
    MacPherson strut adjustment, 678

Camshaft, 84, 128—130. *See also* Overhead camshaft (OHC)
    bearings, 125, 505
    construction, 130
    drive mechanisms, 84
    function, 11
    lobes, 123, 482
    location, 122
    removing, 482, 500
    replacing, 482, 505
    servicing, 481—482

Camshaft lobe lift, measuring, 481—482

Camshaft bearing journals, regrinding, 503

Capacitor. *See* Condenser

Capscrew, 45

Carbon dioxide, 429

Carbon monoxide, 33, 411, 429, 430

Carbon tracking, 368

Carburetion
    defined, 204
    function, 180
    steps in, 204—206

Carburetor(s), 204—230
    checking for fuel in, 192—193
    air horn, 221
    air-fuel ratio adjustment, 229, 411
    altitude compensation valve, 215
    atomization in, 206
    checking for leakage, 221
    common problems, 221, 224—225
    control of air flow in, 204
    control of fuel flow in, 204—206
    feedback, 217
    filter replacement, 220
    float problems, 225
    fuel pump problems, 225
    fuel-bowl vent, 214—215
    function, 12, 85
    future of, 217
    idle speed adjustments, 228—229
    lack of fuel in, 224
    main body, 221
    mechanical fuel pumps and, 178
    multiple-venturi, 211—212
    overhaul procedures, 230
    preventive maintenance, 220—221
    servicing, 220—230
    solenoid adjustments, 228—229
    stuck automatic choke, 225
    temperature-control devices, 212—213
    throttle body, 221
    throttle-control dashpot, 215
    throttle-positioner solenoid, 215—217

vacuum operation, 206
vacuum leaks, 195
vacuum vents, 215
vaporization in, 206

Carburetor choke. *See* Choke *headings*

Carburetor circuits, 206—211
accelerator pump circuit, 208
choke circuit, 209—211
enrichment circuit, 208—209
float circuit, 206—207
idle circuit, 207—208
main metering circuit, 208

Carburetor spray cleaner, 195—196, 221

Carcass, 682, 690

Carcinogens, 431

Cardan joint, 516

Career opportunities, 2—8
training for, 7

Case porosity, 593

Cassette tape players, 395

Castellated nuts, 45

Caster, 668, 677, 679

Casting, 102
methods, 121

Catalysts, 452

Catalytic converter, 262, 443, 450, 451—453
bead replacement, 271—272
checking, 464
damaged, 271
dual bed, 262
fasteners, 464
function, 262, 451
leaded gasoline and, 453
problems, 271, 464
processes in, 452
three-way, 262

Cathode, 286

CCA. *See* Cold cranking amperes

Center punches, 56

Center support bearing, 516, 522

Centrifugal advance mechanism, 359, 360, 372

Centrifugal force, 557

Ceramics, 277

Cetane rating, 185

Chafing, 165

Chain hoist, 37

Chamfered, defined, 105

Chapman struts, 634

Charcoal, activated, 443

Charges, electrical, 278—279

Charging (batteries), 294
fast, 294, 304—305
slow, 305

Charging indicators, types, 335—336

Charging station, 737

Charging system, 14, 328—349. *See also* Alternator; Voltage regulator
alternator output test, 342
cautions in servicing, 342, 343
charging voltage test, 342
components, 328
electrical tests, 340, 342—343, 345—346
full-fielding, 342—343
function, 328
fusible links, 343, 345
oscilloscope trace test, 343
preventive maintenance, 338—340
problem indicators, 338
safety precautions in servicing, 338, 340, 346
servicing, 338—349
voltage drop test, 343

Charging voltage test, 342

Chassis, parts of, 18—19

Chassis dynamometer, 98

Check valve, 127—128, 443, 448

Chemical energy, 82

Cherry picker. *See* Crane

Chisels, 56
air-driven, 62

Chlorofluorocarbons, 727

Choke, automatic
checking, 221
fast-idle adjustment, 221
pull-off adjustment, 211, 221
richness adjustment, 221
stuck, 225

Choke break mechanism, 211

Choke circuit, 209—211
cold weather and, 143

Choke plate, 210—211
opening, 192
operation, 210—211

Choke protractor, 221

Chrysler fuel injection system, 236

Circuit(s), electrical. *See* Electrical circuit(s)

Circuit breakers, 384—385, 399, 744

Circuit resistance test, 343

Circuit testers, 71

Clamps
battery, 303
exhaust system, 263, 265
hose, 166

Clean Air Act, 432, 434, 435

Cleaning equipment. *See also* Solvents
    safe use of, 38
    types, 63—64
Clearance, defined, 106, 108
Clearance gauge, 714
Clearance volume, 95
C-lock axle assembly, 612
    reassembly, 620—621
    removing, 612
C-lock bearing, 619
Closed-loop systems, 411, 420—421
Clothing, 34
Clouding, defined, 199
Clutch, 14, 554—565
    bellhousing, 554
    defined, 16
    engagement/disengagement, 528, 556, 558
    free travel, 561
    function, 16
    inspecting, 564
    maintenance, 560—561
    with manual transaxles, 535
    noises, 563
    parts, 554—556
    problem diagnosis, 561—563
    removing, 563
    replacing, 564—565
    safety precautions in servicing, 563—564
    semi-centrifugal, 557
    servicing, 560—565
Clutch alignment tool, 565
Clutch chatter, 562
Clutch disc, 554—555
Clutch drag, 562
Clutch fork, 556, 556
Clutch linkage, 557—558
    adjusting, 560—561
    checking and adjusting, 542
    function, 557, 558
    lubricating, 560
    manual transaxle, 537
    removing, 545
    types, 557—558
Clutch pedal free travel, 561
Clutch pedal pulsation, 563
Clutch plate, 554-555
Clutch plate limited-slip differential, 601
Clutch slippage, 562
Clutch switch, 312
Clutch vibration, 563
Clutch-head screwdrivers, 55
Coefficient of drag, 99

Coil(s), 284—285, 308
    in ignition coil, 350—351
Coil cable, 357
Coil spring suspension system, 625, 649
    rear, 633—634
    servicing, 650—653
Coil springs, 514
Coil wire, 352
Coking, 187, 200
Cold cranking amperes (CCA), 295
Cold patch repair, 693
Cold stalling, 226
Cold tanks, 64
Cold weather, 209—210
    choke circuit and, 143
    exhaust gases and, 267—268
    oil dilution and, 143
    starting in, 294
Collapsible steering column, 666—667, 745
Collector, 287
Combination pliers, 55
Combination valve, 702
Combination wrenches, 53
Combustibles, storage and use of, 40
Combustion, 358—359
    incomplete, 429
    normal, 95
Combustion chambers, 11, 86—87
    cracked, 472
    pentroof, 445
    shape, 90
    temperatures, 150
    volume, and compression ratio, 95
Combustion control, and emissions, 445
Combustion control computer (CCC). *See* Electronic control unit
Combustion leak detector, 167
Comfort accessories, 394
Commutator, 308
    repairing, 326
Compressed air
    defined, 36
    pneumatic tools powered by, 61
    safety precautions in using, 36—37, 61
Compressed gas, cylinder storage, 40
Compression (of springs), 628
Compression ignition engine, 91
Compression pressure, and combustion, 359
Compression ratio, 95
    defined, 86, 95
    formula, 95

Compression rings, 116
    ring gaps, 116
Compression stroke, 86
Compression test, 470—472
    normal readings, 471
Compression tester, 70, 470
Compressional heat, and octane rating, 95
Compressors, 728
Computer chips. *See* Integrated circuits
Computer control system, 408—409, 411
Computer modules, 409
Computerized diagnostic testers, 73
Computers, 3, 407—408
    "on-board," 412
Concave, defined, 487
Condenser, 333, 353, 728
Conduction of heat, 150, 152
Conductors, 276, 277
Cone limited-slip differential, 601—602
Connecting rod(s), 83, 116—117
    big end, 116
    cleaning, 495
    enlarged end bore, 496
    function, 11
    motion of, 83
    offset, 117
    oil spurt holes, 139
    parts, 116
    removal preparations, 490—492
    removing, 492—493
    replacing, 506
    single throw shared by, 104
    throws and, 104
    twisted or bent, 495
Connecting rod bearing
    inspecting, 496
    servicing, 495—496
Connecting rod bearing(s)
    measuring clearance, 491—492
    replacing, 506—507
Connecting rod bearing caps, marking for removal, 491
Connecting rod side clearance, checking, 507
Connector plugs, electrical, 417
Constant-mesh gears, 528—529, 569
Constant-velocity joint, 517, 538
    boot replacement, 543
    inspecting, 543
    Rzeppa, 538
Contact disks, 310
Contact patches, 625, 681, 683
Contact points, 326

Continuity testers, 71
Continuous fuel injection, 181
Continuously variable transmission (CVT), 581
    control system, 582—583
    problem diagnosis, 594
    servicing, 594
Control arms, 514, 625, 629—630, 648
    replacing, 653
Control modules, 356
    diagnostic tools, 422
    replacing, 422
    testing, 422
Control rods, 638
Control systems, 408—409, 411. *See also* Electronic control unit
Control valves
    automatic transmission, 575—578
    power steering system, 663
Convection, 150, 152
Convenience lights, 392—393
Conventional electrical theory, 280
Conventional grade markings, 47
Conventional system of measurement, 43
    conversion to metric system, 43
    threaded fastener size designations, 46—47
    wrench sizes, 52
Conventional/metric conversion chart, 43
Coolant
    adding, 164
    boiling temperature, 152, 158
    checking, 163—164
    checking with coolant hydrometer, 71, 170
    defined, 12, 151
    excess water in, 170
    overheating of, 158
    thermostat, 155, 157
Coolant bypass, 157
Coolant hydrometer, 71, 170
Coolant pump. *See* Water pump
Coolant recovery system, 158, 159
Cooling fan, 153, 155
    belt-driven, 153
    blade types, 155
    electric, 153
Cooling fins, 91
Cooling system(s), 150—172
    air cooling, 91
    drive belt maintenance, 164—165
    flushing, 170—171
    function, 12
    gaskets, 166—168
    heat energy loss and, 99
    heat transfer in, 150—151

leakage detection and repair, 165—170
liquid cooling, 91
parts, 151
pressure checks, 165
routine maintenance, 163
safety precautions in servicing, 163, 166, 169, 170
163—172

Cooling system cleaners, 170—171

Cooling system hoses, checking, 165

Cooling system pressure tester, 165

Cooling system testers, 70

Copper, 276, 277

Cords, 682

Core, defined, 102

Core plugs, 102, 121

Cornering lights, 392

Corona effect, 367

Corporate average fuel economy (CAFE) standards, 435

Cotter pins, 45

Countergear, 528

Countershaft, 528

Counterweights, 104

Courtesy lights, 392—393

Crane, 37. *See also* Engine lift

Crankcase
defined, 102
emissions, 431—432

Cranking problems, 317—319, 323

Cranking vacuum test, 470

Crankpins, 104

Crankshaft(s), 83, 103—106
attachment to block, 106
balanced, 103
bearings, 106
checking, 506
cracked, 499
damper removal, 499
defined, 11
design of, 103—104
endplay, 498
forged, 104
front-to-back motion, prevention of, 109
inspecting, 499
manufacture of, 104—105
oil passageways, 108
removing, 499
safety precautions in servicing, 499
seals, replacing, 505
securing to engine block, 106
servicing, 498—499

Crankshaft journals
inspecting, 496
regrinding, 499, 503
replacing bearing inserts, 505

Crimped, defined, 168

Crossover pipe, 261

Cruise control, 394, 408, 422
diagnostic flowchart, 422

Cruising speeds, and carburetor function, 208

Cubic centimeter (cc), defined, 94

Cubic inch, defined, 94

Cup plug, 477

Curb idle speed, 216

Current Law, 283

Current, electrical. *See* Electrical current

Cushioning springs, 555

Cutters, 55

Cutting out, 228

CVT. *See* Continuously variable transmission

Cycle, defined, 86

Cycles per second (cps), 281

Cylinder(s)
banks of, 102
configurations, 87, 90
defined, 82
engine displacement and, 94
function, 10
honing, 503
measuring, 500
number of, 11, 88, 90, 104
numbering arrangements, 87, 360
pressure in, during four-stroke cycle, 97
reboring, 113, 503
volume measurements, 94

Cylinder block, 10, 83, 102, 104
aluminum, 102
aluminum, with silicon, 104
attaching cylinder head to, 121
casting, 102
cleaning, 500
coolant flow, 155
cores used in casting, 102
design, 102—103
gallery, 108
inline-type, 102
inspecting, 500
iron, 102
machining, 102
parts removal, 499—500
porosity, 501
preparing for reassembly, 505
reassembly procedures, 504—508
safety precautions in servicing, 489
servicing, 498—509
shapes, 102
soft plugs, 170
V-type, 102

Cylinder head, 83
    attaching to cylinder blocks, 121
    casting, 121
    checking for cracks, 484
    cleaning, 484
    construction, 121
    coolant passageways, 121—122
    counter flow design, 121
    cross-flow design, 121
    design, 120—123
    exhaust passageways, 121
    exhaust ports, 84
    function, 11
    hemispherical combustion chamber, 120, 123
    inspecting, 477
    installing, 487—488
    intake passageways, 121
    intake ports, 84, 446
    machining, 121, 485—486, 487
    materials, 120
    oil passageways, 121—122
    ports, 121
    reassembling, 487
    removing, 482—483
    repairing cracks in, 487
    replacing, 507—508
    safety precautions in servicing, 476, 477
    servicing, off-car, 482—487
    servicing, on-car, 476—482
    visual inspection, 484
    warpage, 485
    wedge combustion chamber, 120, 122—123

Cylinder head gasket leaks, 167

Cylinder honing tool, 503

Cylinder leakage test, 472—473

Cylinder leakage testers, 70

Cylinder liners, 102—103

Cylinder ridge remover, 492

Cylinder sleeves, 503

Cylinder taper, 500

D

Damper pulley, 87, 105

Damping, 640

Dashpot, 215

Data, defined, 407

Dead timing, 379

Deceleration valve, 445

De Dion axle assembly, 603

Defrosters, 725, 744

Dehumidification, 727

De-icers, 174—175

Delayed brake fade, 711

Detent valve, 577

Detonation, 360
    causes, 175
    defined, 175
    prevention of, 360—361

Detonation sensor, 360

Diagnostic equipment, 3
    computers, hand-held, 419, 656

Diagnostic testing, computerized, 73

Diagnostics, defined, 75

Diagonal cutting pliers, 55

Dial indicators, 69, 481

Diaphragm, defined, 139

Diaphragm pump, 234

Diaphragm spring, 556

Die stocks, 57

Dielectric compound, 417

Dies, function, 57

Diesel engines
    emissions, 434
    engine oil, 135
    fuel ignition in, 91
    oil cooler, 160
    pistons, 115
    pumping losses, 99
    thermal efficiency, 99
    timing, 362

Diesel fuel, 99, 185
    cetane rating, 185
    volatility rating, 185

Diesel knocking, 185

Diesel mechanical fuel injection system, 185, 241—243
    fuel filter assembly, 199
    fuel filter replacement, 255—256
    fuel pressures, 243
    injection pumps, 243
    safety precautions in servicing, 255, 256
    servicing, 254—256
    testing injection nozzles, 256
    water removal, 254—255

Dieseling, 177, 215
    diagnosis and repair, 227

Differential, 15, 597—602
    assemblies, 597—598
    cleaning, 616
    construction, 600
    defined, 18
    design, 597—600
    fill plug, 609
    for four-wheel drive, 604
    function, 18, 597—598
    identification code tag, 611
    inspecting, 609, 616—617

installation, 619
interaxle, 606
limited-slip, 600—603
movement of, 512—513
problem diagnosis, 610—611
removing, 613—616
safety precautions in servicing, 609, 611, 615, 616
servicing, 609—619

Differential bearing preload, 618—619

Differential case, 598

Differential drive pinion, replacing, 617

Differential gear ratios, 597
calculating, 599—600

Differential gearset, 611

Differential lubricant, 609—610

Differential pinion gears, 598, 611

Differential ring gear, 536, 537, 611
replacing, 617

Differential side gears, 598

Digital, defined, 371

Digital display, 407

Digital instruments, 388

Digital signals, 236

Dimples, 146

Diode(s), 286, 330, 332
function, 286

Diode trio, defined, 335

Dipstick, engine oil, 140

Direct (distributorless) ignition system (DIS), 362, 370

Direct current (DC), 281
alternator and, 328

Direct injection, 241

Disc and drum brake combination system, 702—703

Disc brake system, 20, 645, 699—700
function, 699—700
inspection, 712—714
pad replacement, 712
parts, 20, 700
reassembling, 713—714
rebuilding calipers, 717—718
rotor removal and replacement, 718—719
servicing, 716—719
types, 700

Discharge nozzle, 208

Discharge ports, 206

Dispatch sheets, 27

Displacement, 86, 94

Displays
analog, 407
graphic, 387
digital, 407
oscilloscope, 373

Display devices, 407

Distribution plenums, 725

Distributor, 351—352
components, 351
function, 12
removing, 379
replacing, 379

Distributor cam, 352, 353

Distributor cap, 357, 362
terminals, 368
visual inspection, 368

Distributor housing, 351

Distributor rotor, 351—352, 357, 362
positioning, 379
visual inspection, 368

Distributorless ignition system, 362, 370

Dive, 642

Diverter valve, 462—463

D-Jetronic electronic fuel injection system, 232, 239

DOHC. *See* Dual overhead camshafts

Doping, 286

Double A-arm suspension system, 631

Double Cardan universal joint, 516—517, 523

Double-acting, defined, 628

Dowel, defined, 109

Dowel pins, types, 49

Downshift valve, 577

Downshifting, 533

Drag racing, 599—600

Drill bits, 61

Drive axle ratio (DAR), 534
calculating, 599, 600

Drive belts
checking, 164
condition, 164
damaged, 339
defined, 69
replacing, 164—165
tension, 164
tightening, 339

Drive belt tension gauges, 69

Drive member, 569

Drive pinion
integral carrier, 617
reassembly, 617—618
removable carrier, 617—618
removing, 614

Drive pinion depth setting, 617

Drive pinion gear, 597

Drive sizes (socket wrench), 54
    adapters, 54

Driveability problems, 225—228
    fuel injection systems and, 249, 251

"Drive-by-wire," system, 414

Driveline, 15, 512—523
    balancing, 523
    cleaning, 519
    function, 17, 512, 513
    inspecting, 520
    maintenance, 519—521
    parts, 17—18, 512
    problem diagnosis, 521
    removal, 521—522, 544
    replacing, 523
    safety precautions in servicing, 519, 521—522
    servicing, 519—523
    test driving procedures, 75
    troubleshooting, 521

Driven member, 569

Driven yoke, 516

Driveshaft, 18, 512—513, 515—516
    function, 516
    with Hotchkiss type drive, 514
    runout, 521
    split, 515
    unbalanced, 516

Driveshaft tube, inspecting, 520

Drivetrain
    defined, 14
    parts, 512

Driving axles, 15, 602—603
    bearing installation, 619—620
    checking, 543
    dial indicator checks, 619
    front-wheel drive, 537—538
    function, 18
    rear, function of, 597
    removing, 544, 612—613
    runout, 619
    seal installation, 619
    servicing, 544, 619—621

Driving wheels. *See* **Wheels**

Driving yoke, 516

Drop center, 686

Droplights, 60

Drum brake system, 20, 645, 700—702
    combined with front disc brakes, 702—703
    function, 700—701
    inspection, 714
    parts, 20, 700
    resurfacing drum, 720
    self-adjusting mechanisms, 701—702
    self-energizing, 701—702
    servicing, 719—722

Dry friction, 97

Dry liners, 103

Dual bed converter, 262

Dual overhead camshafts (DOHC), 90, 123

Duty cycle, 217, 236

Dwell, 72, 371, 378

Dwell angle, 353, 378

Dwell meter, 419—421

Dwell time, 356

Dwell-tachometers, 72

Dye charge, 737

Dynamic balance, 690

Dynamometer, 98

E

Early fuel evaporation (EFE), 182, 261

Early fuel evaporation (EFE) valves, 270—271

Eccentric, defined, 130

Efficiency
    defined, 98
    of vehicle, overall, 99

EFI. *See* Electronic fuel injection system

EGR. *See* Exhaust gas recirculation

Electric current. *See* Electrical current

Electric drills, portable, 60—61

Electric impact wrenches, 60

Electric motors, 307—308

Electrical circuit(s), 276—277, 382, 384
    categories, 283
    common connections, 382
    common ground, 382
    defined, 206, 276
    protective mechanisms, 384—385, 387

Electrical circuit protection, 743—744

Electrical connector plugs, 417

Electrical cords, defective, 39

Electrical current, 61, 179, 281, 329—333
    alternator output, 328—329
    speed of, 281
    types, 281

Electrical current flow,
    direction of, 280
    measurement of, 71—72

Electrical devices, 382—404
    servicing, 398—404

Electrical fuel pump, replacing, 199

Electrical schematics, 277

Electrical solenoid, 215

Electrical solenoid valve, 439, 440

Electrical system
    basic checks, 399
    cautions in servicing, 399, 402
    ignition system as part of, 14
    inspecting, 400
    problem sources, 400
    replacing defective units, 404
    safety precautions in servicing, 398, 403, 404
    servicing, 398—405
    subsystems, 14
    test instruments for, 401—402
    troubleshooting, 399—402

Electrical testers
    connections, 280
    types, 71—72

Electrical theory, 280

Electrical tools, 60—61

Electricity, 276—286
    knowledge of, 400
    sources of, 276, 290

Electrochemical action, 290

Electrolysis, defined, 151

Electrolyte, 290—291
    fill level, 304
    testing with battery hydrometer, 70

Electromagnetic powder clutch, 581—582

Electromagnetism, 284, 307

Electromagnets, 285

Electromechanical regulators, 333—334

Electron electrical theory, 280

Electronic control module (ECM). *See* Electronic control unit

Electronic control unit (ECU), 182—183, 360—361, 440
    for EFI, 235, 254
    input information sensors, for EFI, 236
    for monitoring emissions, 217

Electronic devices, 3, 407—422
    cautions in servicing, 416, 422
    control systems, 408—409, 411, 414
    electrical connections, 417
    for emission control, 411
    for engine control, 411, 421
    examples, 408
    servicing, 416—422
    troubleshooting, 416—419

Electronic fuel injection (EFI) system, 181—182, 217, 232—257
    air and fuel filter replacement, 248
    air mass measurement, 239, 240—241
    air volume measurement, 240
    air-fuel mixture adjustments, 248
    air-measuring systems, 239—241
    basic parts, 233—236
    bleeding fuel lines, 246
    Bosch systems, 239
    bottom-feed injectors, 237
    checking attachments and mountings, 248
    checking for fuel in, 193
    Chrysler system, 236, 240
    continuous fuel injection, 181
    diesel, 185, 199
    D-Jetronic, 232, 239
    driveability problems and, 249, 251
    duty cycle, 236
    electrical connector repairs, 246—247
    electronic control unit, 235, 254
    fuel delivery check, 249, 251
    fuel injectors, 236
    fuel pressures, 233
    idle speed adjustments, 248
    injection grouping, 237
    injector replacement, 254
    input information sensors, 236
    instrument checks, 251
    K-Jetronic, 239
    L-Jetronic, 239
    leakage repair, 245—246
    LH-Jetronic, 239
    manufacturer's service procedures, 251
    Motronic, 239
    multi-point, 181—182, 237, 249
    multi-point injector balance test, 251
    non-sequential (pulsed), 182
    preventive maintenance, 245—248
    pumps, 179
    repairs, 251—256
    safety precautions in servicing, 245, 246, 247, 248, 249, 251, 254, 255
    sensor replacement, 251
    sequential fuel injection (SFI), 182, 237
    servicing, 245—256
    single-point, 181, 236, 249
    system identification, 251
    top-feed injectors, 237
    troubleshooting, 248—249
    troubleshooting charts, 251
    types, 236—237, 239—240
    vacuum hose repairs, 246

Electronic ignition control unit, 354—356
    cable, 357
    testing, 373

Electronic power steering (EPS) system, 663—665
    parts, 664
    servicing, 675

Electronic sensors, types, 354—356

Electronics, 286—287

Electrons, 278
    bound, 280
    charge of, 279
    flow of, 284
    shells, 279

Elements, defined, 277

Emergency brake linkage, removing, 545—546

Emergency flashers, 743

Emergency procedures, 40

Emergency telephone numbers, 34

Emission control, 3, 428—436
    defined, 122

Emission control systems, 378, 411, 438—466, 744
    carburetor and, 217
    checking, 457—464
    electronically operated, 440
    engine controls, 438—439
    hemispherical combustion chambers and, 123
    integrated control systems and, 457
    manufacturer's warranty, 456
    performance and, 455—456
    preventive maintenance, 456—458
    safety precautions in servicing, 455, 457, 461, 464
    sanctions for tampering with, 455
    servicing, 455—466
    subsystems, 440—453
    tampering with, 455—456
    tamper-proof adjustments, 444
    vacuum-operated, 439—440
    wedge combustion chambers and, 122—123

Emissions, 451—452
    from diesel engines, 434
    sources, 430—432
    vehicle operating mode and, 438
    vehicle parts and systems affecting, 438

Emitter, 287

Employees' responsibilities, 8

Employers' responsibilities, 7—8

End play, 498, 621

End play adjustment, 621

Energy
    defined, 81
    forms of, 81—82
    loss of, 99

Energy Policy and Conservation Act, 435

Engine
    break-in procedure, 509
    calculating maximum rpm, 600
    cleaning, 476—477
    compression tests, 470—472
    cylinder leakage tests, 472—473
    defined, 10, 82
    engine oil pressure test, 473—474
    function, 10
    longitudinally mounted, 153
    modifications for emissions control, 445—447
    noises, 474
    overhaul procedures, 400
    problem diagnosis, 469—474
    reinstallation procedures, 508—509
    removing, 490
    with rocker arms, 478, 479
    safety precautions in servicing, 469, 470, 472, 473, 474
    size, 94—95
    starting, 307
    stratified charge, 446
    transversely mounted, 153
    vacuum tests, 469—470

Engine analyzer, 371, 378

Engine bearings, 106, 108—109
    durability, 106
    lubrication, 106, 108

Engine block. *See* Cylinder block

Engine calibration unit, 422

Engine control devices, 411, 421
    closed-loop systems, 411
    emission controls and, 438—439
    open-loop systems, 411

Engine displacement, 86, 94

Engine efficiency, losses in, 98—99

Engine lift, 63

Engine load, 359—360

Engine measurements, 94—100

Engine oil. *See also* Oil *headings*
    acid formation, 143
    additives, 136
    API ratings, 135
    burning of, 136—137
    changing, 144—145
    checking level of, 140
    deterioration of, 143—144
    dilution of, 143—144
    function, 134, 140
    leakage sources, 136, 145
    multi-viscosity, 135
    natural, 136
    oxidation, 143
    pressure test, 473—474
    routine change intervals, 144
    SAE viscosity ratings, 134—135
    single-viscosity, 135
    sludge formation, 144
    synthetic, 136
    for turbocharged engines, 135
    water contamination, 165

Engine power, defined, 81

Engine size, 94—95

Engine speed, 378—379
    adjusting, 72
    engine vacuum and, 360
    ignition timing and, 359, 378—379
    specification of, 88
    testing, 371

Engine vacuum, 360

Enrichment circuit, 208—209

Environmental Protection Agency (EPA), 434
    gasoline pump stickers, 175
    new-vehicle car mileage ratings, 228

EPA. *See* Environmental Protection Agency
Epoxy cement, 500
Epoxy resin, 168, 194
Estimates, written, 27
Ethanol, 90
Ethylene glycol, 151
Evaporation, 727
Evaporative emission controls, 440—441
Evaporator tubes, 729
Even firing engines, 104
Exhaust cam lobes, 128—129
Exhaust emissions, 168—169, 430—431, 744
Exhaust gas analysis, 464
Exhaust gas analyzer, 72, 167—168, 229, 248, 458, 464
Exhaust gas pulse system, 451
Exhaust gas recirculation (EGR) system, 261
    checking, 463—464
    defined, 448
    temperature-controlled, 448
Exhaust gas recirculation (EGR) valve, 448
    checking, 463—464
    cleaning, 463
    controls, 448, 450
    replacing, 463
Exhaust gases, 86, 411. *See also* Emissions
    leakage, 267—268
    recirculation of, 448, 450
    safety precautions, 38
Exhaust manifold, 86, 260—261
    air injection, 260, 450—451
    function, 11
    removing, 483
    warped, 269—270
Exhaust manifold gasket, 260
Exhaust manifold valve, 182
Exhaust pipe, 261
    connections, 261
    replacing, 271
    seal, 261
Exhaust port, 84, 86
Exhaust stroke, 87
Exhaust system, 12, 86, 259—272
    backpressure, 263—264
    bolt and screw removal, 269
    catalytic converter repair/replacement, 271—272
    connecting pipes, 262—263
    detecting restrictions in, 268
    dual, 261
    EFE valve servicing, 270—271
    gas leaks, 267—268
    gaskets and seals, 262—263, 269
    hanger and clamp inspection and replacement, 269
    hangers, 269
    heat riser valve servicing, 270—271
    leakage detection, 268
    noise checks, 268
    parts, 259—265
    pipe connections, 262—263
    preventive maintenance, 268—271
    problems, 267—268
    safety precautions in servicing, 267, 269, 270, 271
    seal and gasket replacement, 269
    securing, 265
    servicing, 267—272
Exhaust valves, 86, 125, 445—446
    opening duration, 130
Exits, 34
Expanding rivets, 49
Expansion plugs, 477
Extension (socket wrench), 54
External combustion, defined, 82
External combustion engines, 82
External threads, 45
    cutting, 57
External transmission cooler, 578

F

Face shields, 33, 60
Fail-safe mode, 236
Failure warning switch, 702—703
Fan. *See* Cooling fan
Fast charging, 304—305
Fast-idle cam, 211
Fast-idle speed adjustment, 221
Fasteners, 44—50
    conventional size designations, 46—47
    function, 43
    grade markings, 47
    metric size designations, 47
    nonthreaded, 48—50
    threaded, 44—48
Feedback carburetor, 217, 236, 443—444
    mixture-control solenoid, 217
Feedback control, 409, 411
Feeler gauges, 66, 485
Ferrous materials, 284
Fiberglass leaf spring independent front suspension system, 633
Field coils, 308—309, 328
    series-wound, 312
    shunt-wound, 312

Field limiter relay, 334
Files, 56
Fill plug, 541
Fillet, 105
Filters
    air, 196—197
    ATF, 588—589
    fuel, 179, 196, 199, 235, 248, 255—256
    oil, 137, 138, 139, 144—145
Fire extinguishers, 34, 40
Fire prevention, 40
Firing order, 87—88, 360
    angle of throws and, 104
    defined, 367
    secondary cable connections and, 367
Firing patterns, problem analysis, 373
Firing position, defined, 478
First-aid kits, 34
Fixed caliper brake assembly, 700
Fixed restriction jet, 204
Flame front, 122
Flammable liquids, 62
Flanges, defined, 109, 292
Flare, double lap, 195
Flare-nut wrench, 53, 196
Flat rate manuals, function, 26
Flat washers, 46
Fleet, defined, 4
Fleet garages, 4
Flexible air chamber, 640
Flexplate, 568
Float, 207
Float bowl, 206
Float circuit, 206—207
Float level, adjustment, 229
Floating caliper brake assembly, 700
Flooding, 225
Floor jacks, 62
Floor of work area, 33
Fluid coupling, 567—568
Flushing gun, 170—171
Flywheel, 307, 554
    function, 85
    removing, 499
    replacing, 507
Flywheel ring gear, 307, 311, 324
Foot-pounds (ft-lb), defined, 96
Footprint (of tire). *See* Contact patches

Force
    application of, 81
    transfer, 573—574
    torque and, 96
Foremen. *See* Shop foremen
Forging, defined, 104
Forms, 27—28
Forward-biased transistors, 334
Four-stroke cycle, 86—87
    piston movement in, 11
    pressure in cylinder during, 97
    strokes, 86—87
Four-wheel drive (4WD), 15, 603—606
    design variations, 604
    differentials, 604
    drivelines, 18
    full-time, 606
    hub locking system, 604
    part-time, 603
    shifting into, 604
    transfer case, 18
    viscous coupling, 606
Four-wheel steering (4WS) system, 20
    mechanical, 665—666
    power-assisted, 666
Frame, 18
Franklin, Benjamin, 280
Free travel, defined, 561
Free-floating piston pins, 116
Freeze plugs. *See* Core plugs
Friction
    defined, 12, 96, 106, 696
    types, 97—98
Friction bearings. *See* Insert bearings
Friction facing, 554—555, 564
Friction material, 696, 698, 700
Friction pads, 20
Frictional horsepower (fhp), 97
Fringe benefits, 7
Front suspension system, 629—633
    parts, 629—631
    types, 631—632
Front-end geometry, 667—668
Front-end wheel alignment, 677—679
Front-wheel drive (FWD), 15
    driving axles, 537—538
    torque flow, 537534
    torque steer, reducing, 538
    transaxle design, 534, 536
Frontal area, defined, 99
Fuel(s), 174—177. *See also* Air-fuel mixture; Diesel fuel; Gasoline

alternatives, 90
economy requirements, 3
burning of, 82
contamination, 193
Fuel atomization system, 180—182
carburetor, 180
electronic fuel injection, 181—182
Fuel bars, 236
Fuel filler cap, defective, 461
Fuel filter, 179, 196
diesel fuel injection systems, 199, 255—256
for EFI systems, 235
replacing, 196, 248
types, 179
Fuel gauge, 407
Fuel gauge sender, 178
Fuel heater, 199
Fuel injection system, 85. *See also* Diesel mechanical fuel injection system; Electronic fuel injection (EFI) system
advantages, 232
fuel filter, 179
function, 12
mechanical, 232
Fuel injector, 236
Fuel injector connectors, 248
Fuel line, 178
bleeding, 246
checking for blockage, 198
EFI system, 233
ice in, 198
replacing, 194—195
Fuel line hoses, 458
Fuel mileage
emissions control technology and, 435—436
government regulation and, 435—436
Fuel pressure regulator, 234—235
Fuel pump, 178—179
checking inlet vacuum, 198
diaphragm-type, 197—198
for EFI systems, 233—234
electrical, 179, 199
mechanical, 178, 199
pressure output, 225
replacing, 198—199
testing, 197—198
Fuel pump eccentric, 130
Fuel rail, 237, 248
Fuel system, 85, 174—202
air filter replacement, 196—197
checking for fuel contamination, 193
checking for fuel in, 192
diesel, 185
driveability problems and, 226, 228

emissions control modifications, 443—445
function, 12
fuel filter replacement, 179, 196
gasket replacement, 195
leakage detection, 193—194
leakage repair, 194—196
parts, 174, 177
PCV filter replacement, 197
preventive maintenance, 193—197
safety precautions in servicing, 40, 192, 193, 194, 195, 196, 197, 198, 199, 200
servicing, 192—202
Fuel system emissions, 431
Fuel tank, 177—178, 441—442
cleaning, 177
domed, 442
EFI system, 233
repairing, 194
testing, 197—198
Fuel tank filler cap
nonvented, 441
replacing, 194
vented, 431
Fuel wetting, 237
Fuel-bowl vent, 214—215
Fuel-flow, defined, 139
Full-fielding, defined, 342
Full-power operation, carburetor function and, 208—209
Full-time four-wheel drive, 606
Fuse blocks, 384
Fuses, 280, 384, 399, 744
Fusible links, 343, 384, 399, 744
replacing, 345
FWD. *See* Front-wheel drive

G

Galleries, 108, 139
Garter springs, 110
Gas mileage, factors lowering, 228
Gas pressure shock absorbers, 628
Gases. *See* Emissions; Exhaust gases
Gaskets, 83. *See also* Head gasket
cooling system leaks and, 166—168
engine oil leakage and, 136
engine overhaul set, 109
fuel system, 195
function, 109
materials for, 109
replacement of, 109
Gasoline, 174—176
chemical stability, 177

contamination of, 177
de-icers, 175
leaded, 453
octane rating, 95, 175—176
resistance to normal combustion, 175
sulfur content, 177
volatility, 174

Gassing, 294

Gauges
analog, 388, 389
examples, 388
magnetic, 389
nonelectric, 390

Gear(s)
defined, 525
driven, 525
driving, 525
function, 16
jumping out of, 544
locked, 544
shifting, 16
torque on, 526—527

Gear clash, 544

Gear marking compound, 611

Gear ratios
defined, 525
determining, 525—526
for economy, 537
overdrive, 527

Gear tooth contact pattern test, 611—612, 619

Gearshift lever, automatic transmission, 574

Gearshift linkage, 531
adjustment, 591
automatic transmission, 574
lubrication, 589

General repair manuals, 25—26

G-forces, 639

Glass bead blasters, 64

Glow plug, 241—242

Goggles, 33, 36

Government regulation, 3
air pollution and, 432, 434—435
fuel mileage and, 435—436
on refrigerants, 727

Governor, 578

Governor drive gear system, servicing, 593

Grade markings, 47

Grams per mile (gpm), 432

Graphic displays, 387

Graphite, 308

Graphited oil, 270—271

Greasy friction, 97

Grids, lead-calcium, 295

Grinders, 61

Grinding wheels, 61

Gross horsepower, 98

Ground, 276—277
accidental, 280
common, 382
negative, 277
positive, 276

Grounding, 317

Group numbers (batteries), 296

Growlers, 325

H

Hair, 34

Hairspring, defined, 390

Hall-effect sensors, 354—355, 370

Halogen lights, 391

Hammers, 56

Hand tools, 52—57
defined, 34
proper choice of, 36, 52

Hand vacuum pump, 373

Hangers, exhaust system, 265

"Hard" ride, 639

Harmonic balancer, 105

Harnesses, 382, 384

Head gaskets, 83, 109
blown, 472
function, 109
leakage, 165

Header, 261

Headlights, 391—392, 742—743
height control and, 640

Heat
conduction of, 150, 152
convection of, 150
friction and, 696
radiation of, 150

Heat dam, 116

Heat energy, 82
converting to motion, 82
loss of, 99

Heat exchanger, 160

Heat ranges, spark plug, 357—358

Heat riser valve, 182

Heat shield, 265, 464

Heat sink, 333

Heat stove, 444

Heat transfer, 150—151

Heater, electrical manifold, 182—183
Heater control assembly, 726
Heater control valve, 169
Heater core, 160, 725
Heater hoses, 160
Heating system, 159—160, 725—726
    inspection, 732
    problem diagnosis, 733
    safety precautions in servicing, 733
    servicing, 733
Heat-shrink tubing, 317
Height control, 640
Height sensor, 628
Hertz (hz), 281
Hesitation, 226
High-pressure spray cleaners, 63
Hoists, 37, 63
Holding and drive devices, 571
Holes, enlarging, 57
Horn, 394
Horsepower (hp), 81
    brake, 97
    corrected, 98
    defined, 96
    equivalent in watts, 98
    frictional, 97
    gross, 98
    indicated, 97
    measuring, 98—99
    net, 98
    parasitic losses of, 98, 189, 216
    SAE, 98
    torque and, 96
Hose clamps, 166
Hoses. *See also* Cooling system hoses
    inspecting and replacing, 458
    for vapor recovery system, 443
Hot patch repair, 694
Hot stalling, 226
Hot tank, 64, 484
Hotchkiss drive, 514—515, 637—638
    with coil springs, 514
    with leaf springs, 514—515
Hot-idler compensator, 212, 214
Hot-wire sensor, 241
Hub (of wheel), 686
Hub locking system, 604
Hub nut, 544
Humidity, 727
Humidity control, 727
Hunting gearset, 611

Hydraulic actuators, 571
Hydraulic control system
    automatic transaxle, 579
    automatic transmission, 574—578
    computerized, 582—583
Hydraulic jack, 37
Hydraulic lash adjuster, 128
Hydraulic lifters, 127—128
Hydraulic linkage, 558
Hydraulic press, 63, 493
Hydraulic pressure, 572—573
    in automatic transmissions, 575—578
    in power brake system, 704
    in power steering system, 660
Hydraulic pressure tests, 590
Hydraulic tools, 62—63
Hydraulic valve lifter, 478
Hydraulic wheel cylinders, 698
Hydraulics, principles of, 572—574
Hydro-mechanical servo systems, power steering as, 660
Hydrocarbon (HC) molecules, 174
Hydrocarbon fuel, 143, 429
Hydrocarbons, 430
Hydrogen, 279
Hydrometers
    battery, 70, 300
    coolant, 71
Hypoid gears, 597

I

Ice, in fuel line, 198
Idle circuit, 207—208
Idle discharge port, 207
Idle mixture
    adjusting, 229
    control screws, 205
Idle positioner solenoid, 216
Idle speed, 207
    adjusting, 204
    increasing, 216
    setting, 215
Idle speed control (ISC) motor, 216
Idle speed solenoid, 216
Idler arm, 659
    idler gear, 531
Idler pulley, 164
Ignition advance, computer-control unit, 360, 370

Ignition coils, 350—351
    in distributor cap, 357
    primary, 351, 352, 353
    secondary, 351, 352, 353
Ignition interlock, 312
Ignition spark, 350, 351, 352, 353
    checking for, 365—366
Ignition stroke. *See* Power stroke
Ignition switch, 309
    positions, 309
Ignition system, 12, 14, 91, 350—379
    cables and wires, servicing, 367
    components, 350—358
    diagnostic instruments, 370—373
    disabling, 469
    distributorless, 362, 370
    electronic, 373
    firing pattern, 373
    primary circuit, 352—356, 367
    primary system switching mechanism, 369—370
    purpose, 350
    safety precautions in servicing, 365, 372, 374
    secondary circuit, 356—358
    servicing, 365—379
    troubleshooting, 366, 370—373
Ignition timing, 359—360, 411
    computer-controlled, 448
    emissions control and, 447—448
    engine load and, 359—360
    engine speed and, 359, 378—379
    retard, 360
    setting, 378—379
Ignition timing marks, 359
Impact tools, 60
Impact wrenches
    air-driven, 61—62
    electric, 60
Impeller, 567—568
    defined, 152
Impeller pump, 179
Inclinometer, 521
Independent garages, 4
Independent suspension systems, parts, 629—631
Indicating devices, 417, 743
Indicator light, for charging system, 335
Indirect injection, 241
Induction, 285—286
Induction hardened, defined, 125
Inductive ammeter, 342
Inductive pickup, defined, 371
Inductive timing light, 372
Information disclosure, 27—28
Injection grouping, 237

Injection timing, 237, 362
Inline diesel injection pump, 243
Inner liner, 682
Inner tubes, 681
Input, 408
Input information sensors, EFI, 236
Insert bearings, 83, 106
Inside micrometers, 68
Instrument voltage regulator, 388
Instruments, 388—391. *See also* Gauges
Insulation
    electrical, 384
    faulty, 367
Insulators, 277
Intake cam lobes, 128—129
Intake charge, defined, 86
Intake manifold, 85
    electrical heater, 182—183
    function, 11
    gasket leaks, 166
    heat in, 182, 185
    modifications for emissions control, 445—447
    redesigned, 446
    vacuum leaks, 195
    V-type, removing, 482
Intake manifold vacuum, and engine load, 359—360
Intake port, 84, 85
Intake stroke, 86
Intake valves, 85, 125, 445—446
    opening duration, 130
Integral power steering system, 660
Integral valve guides, 126
Integral valve seat, 125
Integral-carrier differential, 600, 613
    installation, 619
    reassembly, 618
    removing, 613—614
Integral-carrier drive pinion assembly, 617
Integrated circuits (IC), 287, 411
    in alternators, 334—335
Integrated control systems, 457
Interaxle differential, 606
Intercooler, 186
Interference angle, 125
Internal combustion, defined, 82
Internal combustion engines, 10, 82
Internal short, 292
Internal threads, cutting, 57
    internal transmission cooler, 578

International Standard of Units (SI), 94
International system of measurement. *See* Metric system
Inventory, 28
Inversion layer, 429
Ionization, 179
Ions, 179
Iso-octane, 175

J

Jacks, 37
Jackstands, 34, 37, 62
Jet, 204—206
    defined, 204
    types, 204—206
Jet spray booth, 484
Joule, James P., 96
Joule (J), defined, 96
Jounce, 628
Journal, defined, 103
Jump-starting procedures, 298—299
    safety precautions, 298, 299
Jumper cables, 298
Jumper wires, 401
Jumper wiring, 403, 404
Junction block, 400

K

Kettering, Charles F., 307, 350
Key fasteners, 49
Keyways, 49
Kickback, 664—665
Kickdown linkage system, 574, 591
Kickdown valve, 577
Kilopascal (kPa), 573, 681
Kilowatt (kw), 81, 98
Kirchhoff's Voltage Law, 283
K-Jetronic EFI system, 239
Knocking, 175, 360

L

Labor charges, 28
Lands, defined, 116
Lateral runout, 691
Lead dioxide, 291

Lead sulfate, 292
Leaded gasoline, 453
Leaf spring suspension system, 626—627
    rear, 635, 637
    transverse-mounted, 635, 637
Leaf springs, 514, 626—627
Leak detector tool, 737
Leakage
    automatic transmission fluid, 593
    coolant, 165—170
    of engine oil, 136
    exhaust gases, 267—268
    fuel system, 193—196
    oil drain plug, 144, 145
    refrigerant, 737
    thermostat gasket, 166
Leak-down tester, 472
Lean misfire, 195, 226
Lever and shaft assembly, 557
LH-Jetronic EFI system, 239
Lifters. *See* Valve lifter
Light-emitting diodes (LEDs), 355—356
Lights, 391—393, 399, 742—743
Limited-slip differential, 600—603
    clutch pack, 601, 616
    cone type, 601—602
    lubricants, 610
    problem diagnosis, 610
    reassembly, 618
    removing, 616
Limited-slip differential tool, 610
"Limp-in" mode, 236
Line boring, 503
Line technician—heavy work, 5—6
Line technician—light work, 5
Linings, 20
Linkage power steering system, 660, 663
Linkage systems, 591
Lip-type seals, 110
Liquid cooling system, 91
L-Jetronic EFI system, 239
LN-Jetronic system, 241
Load, electrical, 276
Load, engine, and ignition timing, 359—360
Lobes, defined, 123
Locking pliers, 55, 56
Locking steering column, 667
Locking steering wheel, 744
Locknuts (micrometer), 67
Locknuts (rocker assembly), 123

Lock-to-lock steering wheel movement, 659
Lockup system, 569
Lockwashers, 46
Longitudinal torsion bar suspension system, 632
Longitudinally, defined, 153
Longnose pliers, 55
"Low" gear, 16
Low-speed operation, and carburetor function, 207—208
Lubricants, for manual transmission, 528
Lubrication
    of ball joint, 644—645
    defined, 12
    of differential, 609—610
    of engine bearings, 106, 108
    of main bearings, 108
    of steering linkage system, 672
    of universal joints, 520—521
Lubrication system, 91, 134—148. *See also* Engine oil
    components, 137—141
    defined, 12
    friction and, 97—98
    function, 12, 134
    oil and filter change, 144—145
    pressurized, 91
    routine maintenance, 144—145
    safety precautions in servicing, 143, 144, 147, 148
    servicing, 143—148
    servicing with oil pan removed, 147—148
Lug nuts, 686
Lug pattern, 694

M

Machine screws, 45
Machining, 485
    defined, 102
    line boring, 503
    pistons, 495
Machinist, automotive, 485—486
MacPherson strut suspension system, 632, 649
    front-end alignment, 678
    problem diagnosis, 649
    rear, 634
    replacing, 653—656
    strut installation, 655—656
    strut reassembly, 655
Magnetic attraction and repulsion, 278—279
Magnetic clutch, 735
Magnetic fields, 284—285, 307
Magnetic force, and electric motors, 307—308
Magnetic gauges, 389
Magnetic induction, 285—286

Magnetic position sensors, 354
Magnetic probe timing adapter, 372
Magnetic pulse generator, 369
Magnetism, 283—286
Main bearing(s), 106
    clearance, 474, 499
    number of, 106
    replacing, 505
Main bearing bore, checking, 500, 502—503
Main drive gear, 528
Main journal, 103
Main metering circuit, 208
Mainline pressure, 576
Mainshaft, 528
Maintenance, defined, 2
Maintenance-free batteries, 295—296, 300
Major thrust face, 114, 115
Malleable, defined, 109
Mallets, 56
Manifold, defined, 11. *See also* Exhaust manifold; Intake manifold
Manifold absolute pressure (MAP), 239
Manifold gauge assembly, 737
Manifold heater, electrical, 182, 185
Manifold vacuum, 439, 443, 469
Manual bleeding, 715
Manual control valve, 576
Manual steering system, 660, 671
Manual transaxle, 534—538
    checking lubricant level, 541—542
    cleaning and inspecting, 550
    clutch linkages, 537
    components, 534
    disassembling, 549
    front-wheel drive, 535—536
    parts, 550
    problem diagnosis, 543—544
    reassembling, 550, 552
    removing, 546, 549
    replacing lubricant, 542—543
    servicing, 536
    shifting gears in, 16
    visual inspection, 543
Manual transaxle drivetrain, parts, 535—537
Manual transmission, 525—552
    checking lubricant level, 541
    cleaning and inspecting, 550
    clutch switch and, 312
    components, 528
    disassembling, 549
    five-speed, 534
    four-speed, 531, 533

gear selection, 530—531
gears, 528—529
gearshift linkage, 531
lubricants, 528
power flow, 531, 533
preparation for removing, 545—546
problem diagnosis, 543—544
reassembling, 550, 552
removing, 546, 549
replacing, 552, 553
replacing lubricant, 542—543
servicing, 541—553
shifting gears in, 16
visual inspection, 543

Manufacturer's service bulletins, 25
Manufacturer's service manuals, 25, 26—27, 33
troubleshooting charts, 249
Manufacturing careers and opportunities, 2
Map, defined, 360
Mass air-flow sensor, 240—241
Master Automotive Technicians, 5
Master cylinder, 696, 697—698
parts, 697—698
rebuilding, 716
Mating surfaces, defined, 509
Matter, 277
Measurement inspections, 78
Measurement systems
conventional system, 43
metric system, 43
Measurement tools, 66—73
for air and liquids, 69—71
diagnostic testers, 73
for electronics, 71—72
mechanical, 66—69
using for troubleshooting, 78
Mechanical efficiency, 98
Mechanical energy, 82
Mechanical fuel injection nozzle, diesel, 185
Mechanical fuel injection system, diesel, 241—243
Mechanical fuel pump, replacing, 199
Mechanical measuring tools, 66—69
Mechanical output devices, 422
Memory, 417
Mercury switch, 392—393
Metal sprayed, defined, 115
Meter, 371
Metering, 204
Metering rod, 208
Metering valve, 702
Methanol, 90
Metric grade markings, 47

Metric system of measurement, 43
conversion to conventional system, 43
threaded fastener size designations, 47
wrench sizes, 52
Micrometers, 67—68
Microprocessor, defined, 233
Minor thrust face, 114
Mirrors, electrically positioned, 394
Misfueling, defined, 453
Missing, defined, 226
Mixture-control solenoid, 217
Modem, 419
Modified strut suspension system, 632
Modulation, defined, 448
Modules, computer, 409. *See also* Control modules; Electronic control unit
Molecules, defined, 277
MON. *See* Motor Octane Number
Motion transfer, 573
Motor Octane Number (MON), 175
Motronic electronic fuel injection system, 239
Movable flap air measurement system, 239
Movable plate air measurement system, 239
Movable pole starter motor, 312
Moving car in shop, safety precautions, 40
Muffler, 86, 263—264
electronic, 264—265
function, 263
replacing, 271
Multi-leaf rear springs, 635, 637
Multi-point fuel injection system, 181, 237
checking fuel delivery, 249, 251
Multi-point injector balance test, 251
Multi-port fuel injection. *See* Multi-point fuel injection system
Multi-viscosity oils, 135
Multiple-cylinder engines, 87—88
total displacement in, 86
Multiple-disc clutch, 571
Multiple-venturi carburetors, 211—212

N

National course thread (NC), 46, 47
National fine thread (NF), 46, 47
National Institute for Automotive Service Excellence (ASE) certifications, 5
NC. *See* National course thread
Needle valve, 205

Needlenose pliers, 55
Negative plate (battery), 291
Net horsepower, 98
Neutral position, 530
Neutral safety switch, 312
Neutrons, 278
New-car dealerships, 3
Newton, Sir Isaac, 96
Newton-meters (Nm), defined, 96
NF. *See* National fine thread
Nitrogen, oxides of, 430, 445, 451
Noise generators, 264—265
Noises
    air conditioning system, 735
    alternator, 340
    chattering, 610
    clutch problems, 563
    detonation, 175
    diesel knocking, 185
    differential gear problems, 610
    engine, 474
    exhaust system, 264-265, 268—269
    gear clash, 544
    hissing, 473
    manual transmission/transaxle, 543—544
    piston slap, 113
    power steering pump, 674
    preignition, 176
    solid lifters, 128
    spark knock on acceleration, 463
    tire, 691
No-load RPM test, 324
Non-sequential fuel injection, 182
Nondetergent oils, 135
Nonhunting gearset, 611
Nonretainer-type drive pinion, 616, 617
Nonthreaded fasteners, 48—50
Normally aspirated engine, defined, 186
Nose (of camshaft lobe), 123
Nucleus, defined, 278
Nuts
    castellated, 45
    defined, 45
    grade markings, 47
    head shapes, 45

O

Occupant protection systems, 744—746
    problems, 748
    servicing, 748—749
Octane rating, 95

Odometer, 391
Odors
    fuel, in passenger compartment, 461
    of rotten eggs, 464
Off-idle port, 207—208
Offset, defined, 114, 694
Offset screwdrivers, 55
OHC. *See* Overhead camshaft
Ohm, George S., 282
Ohm, 282
Ohm's Law, 282
Ohmmeter, 371, 402
Oil. *See* Engine oil
Oil bypass valve, 139
Oil catch pan, 144
Oil control rings, 116
Oil cooler, replacing/repairing, 169
Oil distribution system, 139
Oil drain plug, 137
    leaks, 144, 145
    oil pan, 145
    opening, 144
    replacing, 145
Oil filter, 137, 138, 139
    changing, 144—145
    clogged, 139
    location, 137
Oil filter gasket, 144
Oil filtering systems, 139
Oil galleries, 12, 139
Oil level indicator, 140—141
    electronic, 141
Oil pan, 91, 137, 588—589
    replacing, 507
    sealant, 148
Oil pan gasket, 148
Oil passages, 104
Oil pressure indicator, 139—140
    dashboard warning lamp, 139—140
    electrical, 140
Oil pressure sender, replacing, 147
Oil pressure sensor, 473
Oil pressure test, 473—474
Oil pump, 12, 91, 575
    checking, with disassembled engine, 147—148
    gear operation, 137
    gears/rotors, 137
    housing, 137
    pickup assembly, 137
    pressure relief valve, 137
    replacing, 507

Oil spurt holes, 117, 139
Oil wedge, 98, 106
On-board computer system, 235
    trouble codes, 248—249
Open circuit, 280
Open-end wrenches, 52—53
Open-loop operation, 411, 419—420
Operating mode, 438
Optical sensors, 355—356
Ordering parts, 28
Organic brake linings, 698
Orifice, defined, 205, 459
Oscilloscope, 72, 343, 372
Oscilloscope trace test, 343
Otto, Nikolaus, 81
Output, 408
Output device, 408, 409, 422
Output volume and aeration test, 197—198
Outside micrometers, 67—68
Overall drivetrain gear ratio (ODR), 534
    calculating, 599
Overall transmission gear ratio (OTR), 533—534
    calculating, 599
Overall vehicle efficiency, and heat energy loss, 99
Overbore, 503
Overcharging, 338—339
Overdrive gear ratios, 527, 533
Overhead camshaft (OHC), 80, 84, 122
    bearing supports, 122
    cap attachment points, 122
    designs, 123
    drive mechanisms, 123—124
    engine size and, 125
    inspecting, 478
    speed and motion of, 123—124
    valve movement methods, 123
Overhead camshaft (OHC) engine, 478, 479
    valve clearance, 478
Overhead camshaft timing gear, 124
Overhead camshaft valve train, 123—125
Overheating, causes of, 171—172
Overinflation, 689
Overrunning clutch, 311, 572
Oxidation
    catalytic converter and, 452
    defined, 452
    of engine oil, 136, 143
Oxides of nitrogen, 430, 445, 451
Oxides of sulfur, 431

Oxygen sensor, 217, 411
    replacement, 421—422
Ozone, 431

P

Packing, defined, 110
Pads, 700
    replacement, 712
Painting equipment, safe use of, 38
Palladium, 452
Parallel circuits, 283, 382
Parallel flow, 155
Parallelism, 713
Parallelogram steering linkage, 658—659
Parasitic loss of horsepower, 98, 189, 216
Parking brake, 706
Partial nonhunting gearset, 611
Particulate matter, 429, 431
Parts, ordering, 28
Parts departments, 3, 6
Parts managers, 7
Parts per million (ppm), 429—430
Parts requisition, 28
Part-time four-wheel drive, 603
Passenger compartment, safety features, 745
Passive restraints, 746
Passive suspension systems, 639
PCV filter, 461
PCV system. See Positive crankcase ventilation
    (PCV) system
PCV valves, 443, 462
Pedal reserve, 710
Penetrating solvent, 269
Pentroof combustion chamber, 445
Percolation, 225
Performance, and emissions control, 435—436
Phillips screwdrivers, 55
Phillips screws, 45
Photochemical smog, 429
Physical energy, 82
Pickup coil assembly, 354, 369
Pickup coil output, testing, 371
Pilot, 487, 619
Pin punches, 56
Pinion bearing installation tool, 617

Pinion bearing preload adjustment, 618

Pinion gear, 307. *See also* Starter pinion gear

Pinion shaft, 598

Pinion depth tool, 617

Piston(s), 83, 112—116
    balance pads, 115
    cam grinding, 114
    cast, 115
    cleaning, 493—494
    clearance, 113
    construction, 115—116
    controlling expansion of, 113—114
    defined, 11, 82
    finishing, 115—116
    forged, 115
    heads, 112
    heat grooves, 116
    knurling, 495
    lands, 116
    machining, 115—116
    measuring, 494
    movement of, 11
    parts, 112
    pin bore wear, 494
    removal preparations, 490—492
    removing, 492—493
    ring groove wear, 494
    ring grooves, 115, 116
    seizing of, 113
    servicing, 493—495
    shapes, 112
    side thrust, 114
    skirt, 112
    thrust forces, 114

Piston clutch, 571

Piston engine. *See also* Engine *headings*
    basic systems of, 12, 14
    classification of, 90—91
    conversion of heat energy to motion by, 82
    external combustion, 82
    four-stroke cycle, 86—87
    fundamentals, 81—92
    internal combustion, 82
    major parts, 83—86
    rebuilding, 113

Piston motion
    in four-cylinder engines, 87—88
    in four-stroke cycle, 86—87

Piston pins, 112, 116
    free-floating, 116
    installing, 496
    offset, 115
    retainers, 116
    semi-floating, 116

Piston ring(s), 83, 115, 116
    fitting, 506
    installing, 496
    testing, 472
    types, 116

Piston ring compressor, 506

Piston ring expander, 493

Piston ring gaps, positioning, 506

Piston/connecting rod assembly, replacing, 506

Piston slap, 113, 115

Pitman arm, 658, 663

Pitman steering gearbox, 671

Pitman steering gearbox lubricant, 671

Pivot stud, defined, 128

Planet carrier, 569

Planet gears, 569

Planetary gearset(s), 569—571, 579
    components, 569
    compound, 570
    power flow, 570—571

Plastigage, 68, 491, 499

Platinum, 452

Play
    defined, 69
    measuring, 69

Pliers, 55—56

Plies, 682

Pneumatic tools, 61—62

Polarity
    defined, 280
    retaining, 280
    reversed, 281, 284

Pole shoes, 309

Polepieces, 309

Polishers, 61

Polishing discs, 61

Pollutant standards warning system, 434

Pollution, 727. *See also* Air pollution; Emissions
    defined, 428
    environmental, 456

Pop rivets, 49

Poppet valve, 125

Porosity, defined, 500

Port fuel injection. *See* Multi-point fuel injection system

Portable electric drills, 60—61
    drill sizes, 61

Ported vacuum, 215, 360, 440, 450
    canister purge circuit and, 443

Positive crankcase ventilation (PCV) system, 136—137, 431—432, 744
    filter replacement, 197, 248

function, 443
problems, 443
servicing, 461—462
Positive plate (battery), 291
Pounds per square inch (psi), 573
Pounds-feet (lb-ft), 96
Power
compression ratio and, 95
defined, 95, 282
Power brake booster unit, 703
Power brake system, 703—704
servicing, 721
Power circuit, 208—209
Power pulley. *See* Damper pulley
Power pulses, 88
angle of throws and, 104
Power seats, 394
Power steering fluid, 672
Power steering pump, 660—661
abnormal noise, 674
replacement, 675
rotors, 661
vane-type, 661
Power steering system, 660, 664—665
electronic, 663—665
gearbox, 661, 663
linkage system, 663
lubrication, 672
parts, 660
rack and pinion, 663
reservoir, 660
visual inspection, 672
Power stroke, 86—87. *See also* Power pulses
Power tools, 36, 60—64
safety precautions, 36, 60
Power train, defined, 14
Prechamber, 241
Precision insert bearings, 106
lubrication of, 106, 108
Precombustion chamber, 241
Preignition, 176—177
Pressure, 696
braking and, 696
defined, 82, 572
liquid under, 572—574
transfer of, 573
Pressure bleeding, 715—716
Pressure measurements, 70
Pressure plate
long style, 555
Pressure plate assembly, 555—556
Pressure regulator valve, 576

Pressure relief valve, 441—442
Pressure test, of fuel pump, 197
Pressurized lubrication systems, 91
Pride in work, 8
Primary brake shoe, 700
Primary circuit, 350, 352—356
Primary coil, cables, 367
Primary ignition system, checking wiring of, 367
Primary piston, 698
Primary throttle plate, 212
Printed circuits, 384
Processing of data, defined, 407
Program, defined, 407
Programmable read-only memory (PROM), 422
Programmed, defined, 236
PROM . *See* Programmable read-only memory, 422
Propane, 457—458
Propane adjustment procedure, 229
Propane enrichment, defined, 457
Propane enrichment tester, 457
Propeller shaft. *See* Driveshaft
Proportioning valve, 702
Protection systems. *See* Occupant protection systems
Protractor, 221
Prove-out sequence, 417
Pry bars, 57
Pulse air valve, 451
Pulse transformer, 350
Pulse width, 235—236
Pump seal, 593
Pumping losses, defined, 99
Punch, 56
Purge valves, 460
Pushrod, 128, 478
Pushrod engine, camshaft service, 481—482
Pushrod valve train, 80, 84, 125
camshafts, 130

Q

Quadrant adjustment, 591
Quench areas, 122, 329
Quick-splice connectors, 403

R

Race, 550

INDEX

Rack, 659

Rack and pinion steering system, 663
    gearbox lubrication, 671
    manual, 660
    parts, 659—660
    replacement, 675

Rack housing, 660

Radial runout, 691

Radial tires, 683

Radiation of heat, 150—151

Radiator, 152
    crossflow, 152
    downflow, 152
    drain, 152
    function, 12
    leak repairs, 168—169

Radiator pressure cap, 152
    cooling system leaks and, 166
    faulty, 164
    opening, 163
    operation of, 157—159

Radio, 395

Raster pattern, 373

Ratchets
    air-driven, 62
    defined, 53
    of micrometers, 67

Rate, defined, 81

Reach, spark plug, 357

Reaction member, 569

Reamers, 57

Rear axle, functions, 597

Rear suspension systems, 633—638

Rear-end torque, 513, 514, 633

Rear-end wheel alignment, 679

Rear-wheel drive (RWD)
    parts, 14—15
    self-energizing drum brakes and, 701

Rebound, 628

Recapped tires, 685

Receiver/dryer, 728

Recess, defined, 292

Reciprocating piston engine, 91. *See also* Engine; Piston engine
    basic parts, 10—11

Recirculating ball power steering gearbox, 660, 661
    adjustments, 674
    variable ratio steering and, 663

Rectify, defined, 328

Reduction
    catalytic converter and, 452
    defined, 452

Reed valve, 451

Refrigerant, 727
    checking, 736, 740
    control of, 728—729
    discharging, 738—739
    government regulations, 727
    leakage sources, 737
    recharging, 737

Regrinding, defined, 503

Relay(s), 310, 385, 387
    defined, 333
    field limiter, 334
    voltage limiter, 333—334

Relay rods, 658

Reluctors, 285, 286, 354

Remote starter, 375—376

Removable-carrier differential, 600, 614
    disassembling, 615—616
    installation, 619
    reassembly, 618
    removing, 614—616

Removable-carrier drive pinion, 617—618

Repair orders, 27—28
    parts requisition section, 28

Repairs, defined, 2

Replaceable valve guides, 126

Replacement of automobile, 3

Research Octane Number (RON), 175

Reserve capacity (battery), 295

Residual check valve, 703

Resistance, 282

Resistance Law, 283

Resistor, function, 140

Resonator, 264

Retainer, defined, 116

Retaining ring, 50

Retainer-plate type axle assembly, 612
    bearing and seal installation, 620
    end-play adjustment, 621
    reassembly, 621

Retainer-type drive pinion, 616
    reassembly, 618

Retard, 360

Retractors, 745—746

Retreaded tires, 685

Returnability, defined, 673

Reverse flushing, 170—171

Reverse gear, 531
Reverse-biased transistors, 334
Revolutions per minute (rpm), 88
Rhodium, 452
Ring gaps, 116
Ring gear, 307, 311, 324, 569
Ring groove, 494
Ring groove spacer, 494
Ring seals, 110
Ring side clearance, 506
Rings. *See* Piston ring(s)
Riveting tools, 49
Rivets, 49
Road draft tube, 431
Road test. *See* Test drive
Rocker arm, 123, 128, 478, 479
    adjustable, 128
    nonadjustable, 128
    removal of, 479
    roller, 128
Rocker arm ratio, 128
Rocker shaft, 123
Rod and lever linkage, 557
Rod bearing caps. *See* Rod caps
Rod bearing knock, 474
Rod caps, 116—117
Rod journals, 103
Roll, 642
Rolling radius, 600
Rolling resistance, defined, 99
RON. *See* Research Octane Number
Room temperature vulcanizing (RTV) sealant, 147
Rootes-type supercharger, 188
Rope-type seals, 505
Rotary distributor type pump, 243
Rotary engines, 91
Rotary roller pump, 234
Rotary valve, 640
Rotational speed, 16
Rotor, 20, 328, 330, 700
    checking, 713
    in distributor, 351—352
    inspection, 713
Rough idle, 104, 226
Rubbing block, 352
Rulers, 66
Runner, 474

Run-on. *See* Dieseling
Runout, 619, 691
    checking, 713
    defined, 69, 521
    measuring, 69
Rusted parts, freeing, 269
RWD. *See* Rear-wheel drive
Rzeppa joint, 538

S

Saddle (floor jack), 62
SAE viscosity ratings, 134—135
Safety, 33—40
Safety containers, 34
Safety glass, 745
Safety glasses, 33, 36, 60
Safety guards, 36, 39, 60
Safety inspections, 3
Safety practices, guide to, 33—34
Safety precautions. *See* subheading under names of
    systems serviced
Safety rim, 686
Safety rules, 40
Safety stands. *See* Jackstands
Safety switches, 744
Safety systems, 742—751
    servicing, 748—750
Sales careers and opportunities, 2
Sampling probe, 464
Sanders, 61
Sanding discs, 61
Saturate, defined, 353
Schematics, electrical, 277
Screw extractors, 57
Screwdrivers
    defined, 54
    sizes, 54
    slotted, 54
    special, 55
Screws
    defined, 45
    extracting, 57
    types, 45
Scuffing, 152
    defined, 116, 136
Sealants
    RTV, 147
    thread-locking, 48

Seals
- function, 109
- replacement of, 109
- types, 110

Seams, defined, 168

Seat belt assembly, 745—746
- automatic, 746
- checking, 748
- replacement, 748—749

Seating, defined, 509

Secondary brake shoe, 700

Secondary circuit, 350

Secondary coil, cables, 367

Secondary progression system, 212

Section width, 684

Sediment chamber, 292

Seizing, 152
- defined, 48, 113
- lubrication and, 134

Self-adjusting drum brakes, 701

Self-diagnostic codes, 656

Self-diagnostics, 412, 417, 419
- activating, 417, 419

Self-energizing drum brakes, 701—702

Self-locking nuts, 45

Self-starter, 307

Self-tapping screws, 45

Semi-active suspension systems, 639—640
- damping control, 640
- flexible air chambers, 640
- malfunctions, 656

Semi-centrifugal clutch, 557

Semi-conductors, 286

Semi-floating axles, 603

Semi-floating piston pins, 116

Semi-independent suspension system, rear, 634—635

Semi-metallic brake linings, 699

Sending units, 388—390

Sensors, 407, 408, 409
- anti-theft, 744
- defined, 407
- for electronic engine controls, 421
- ignition system, 350
- leakage detectors, 737
- replacement of, 251
- wheel speed, 705

Separator plate, 575

Separator strips, 291

Sequential fuel injection (SFI), 182, 237

Series circuits, 283, 382

Series flow of coolant, 155

Series-parallel circuits, 283, 382

Series-parallel flow of coolant, 155

Service advisors. *See* Service writers

Service bays, 4

Service managers, 6

Service manuals, 25—26. *See also* Manufacturer's service manuals
- troubleshooting charts, 75

Service stations, automotive services of, 4

Service writers, 6, 28

Servicing. *See* Automotive service

Servo, 571

Servomechanisms, 409

Shaft, defined, 11

Sheet metal screws, 45

Shell, 279

Shift forks, 529

Shift lever, 310

Shift linkage
- checking and adjusting, 542
- manual transaxle, 537
- removing, 545

Shift positions, automatic transmission, 574

Shift quadrant, 574

Shift rails, 529

Shifting problems, 544, 592

Shifting valves, 576

Shimmy, 689

Shock absorber fluid, 648

Shock absorbers, 628—629
- function, 19
- inspecting, 648
- replacing, 650—651

Shop equipment, defective, 39—40

Shop foremen, 6

Shop layout, 34

Short circuit, 280

Short finders, 403

Shoulder belts, 745—746, 748

Shunt field coils, 312

SI. *See* International Standard of Units

Siamesed, defined, 121

Side gears, 598

Side marker lights, 392

Side terminals, 292

Sidewalls, 682

Sight glass, 736
Silicon, 286
    cylinder block and, 103
Silicone sealers, 136
Sine waves, 328
    overlapping, 330
Single overhead camshaft (SOHC), 123
Single-point fuel injection system, 181, 236
    checking fuel delivery, 249
Single-viscosity oils, 135
Single-wire system, 276
Sleeves (micrometer), 67
Sleeves (cylinder liners), 103
Sleeving, 500, 503
Sliding caliper brake assembly, 700
    pad replacement, 712
Slip joint, 520
    parts, 513
Slip rings, 328
Slip yoke, 513
Slip-joint adjustable pliers, 55
Slippage (slip), 568
Slotted screwdrivers, 55
Slow charging, 305
Sluggishness, 226, 228
Small hole gauges, 68
Smog, 429
Smoke from tailpipe
    black, 429
    bluish-white, 429, 592
Smoking, 34, 40
Snap ring. *See* Retaining ring
Snap-ring pliers, 56
Sniffer, 737
Society of Automotive Engineers (SAE)
    engine oil standards, 134—135
    horsepower, 98
Socket wrenches, 53—54
Soft plugs, 477
    replacing, 170, 477, 487, 505
"Soft" ride, 639
SOHC. *See* Single overhead camshaft
Soldered, defined, 168
Soldering, electric wires, 403
Solderless connectors, 403
Solenoid, 285, 310
    computer-controlled, 450
Solenoid direct drive starter, 312

Solenoid gear reduction, 312
Solenoid starting system, 311
Solenoid valves, 459, 578, 579, 582, 666
    testing, 593
Solid lifters, 127, 128
Solvent cleaners, 63—64
Solvents
    defined, 38
    penetrating, 269
    safety precautions in using, 38
Sound warning devices, 387
Space frame, 19
Spaghetti tubing, 326
Spark advance, 360
Spark ignition, 91
"Spark map," 448
Spark plug(s), 351, 357—358
    color reading, 374
    firing patterns, 373
    function, 12, 14, 84
    heat ranges, 357—358
    in hemispherical-head cylinder head, 123
    installing, 375
    reach, 357
    removing, 374
    replacement intervals, 373
    replacing, 375
    visual inspection, 374
    in wedge-head cylinder head, 122
Spark plug cables, 357, 366
Spark plug gap, 378
Speakers, 395
Specialization, 6, 7
Specialty shops, 4, 230
Specialty technicians, 6
Specific gravity, 300
Specifications, defined, 25
Specifications tables, 27
Speed handle (socket wrench), 54
Speed nuts, 45
Speed of vehicle
    electronic power steering system and, 664
    height control and, 640
    maximum top speed, calculating, 599—600
Speedometer, 390
Speedometer cable, 545
Spider, 516
Spikes, 290
Spills, 34
Spindle, 630

Spindle (micrometer), 67
Spindown time, 562
Splash lubrication, 139
Splayed, defined, 104
Splicing, 403
Splines, 513
Split driveshaft, 515
Split-ball gauges, 68
Sponge lead, 291
Sponginess
    causes, 228
    defined, 226
Spongy pedal, 710
Spool valves, 582, 704
Spray cleaner booth, 484
Spray painting, 38
Spring damper, 126
Spring seats, 633
Spring shackle, 637
Springs
    compression of, 628
    function, 19, 625
    problem indicators, 648—649
    removing, 652—653
    types, 19, 625—627
Sprung weight, 625
Squat, 642
Stabilizer bar, 630—631, 648
    replacing, 651
Stalk, defined, 387
Stalling, causes of, 226
Starter drive, 310—311
Starter motor, 307—309
    brushes, 308
    checking wiring, 316
    circuit design, 312
    off-car checks, 324
    removing, 316, 323
    series field coils, 312
    servicing, 323—326
    shunt coils
    steps in rebuilding, 324—326
    testing, 324
    types, 311—312
    undercharging and, 339
    wiring, 312
Starter mounting bolts, 317
Starter mountings, 323
Starter pinion gear, 307, 310, 324
Starter relay, 310
    testing, 319, 323

Starter solenoid, 310
    testing, 319, 323
    wiring and connections, 319
Starter system, 307—326. *See also* Starter motor
    elements, 309—311
    function, 14
    on-car tests, 318—319, 323
    preventive maintenance, 316—317
    safety precautions in servicing, 316, 317, 319, 323, 326
    servicing, 316—326
    troubleshooting, 317—319
Starting, 307
Static balance, 690
Stator, 328, 569
    windings, 329—330
Steam cleaning, 38, 63
Steam engines, 82
Steering arms, 659
Steering axis, 668
Steering column, 666—667
    collapsible, 666—667, 745
    locking, 667
Steering effort, 664
    selectable, 665
Steering gear, 20, 675
Steering gearbox, 658
Steering kickback, 664—665
Steering knuckle and spindle assembly, 630
Steering linkage system
    lubrication, 672
    replacement, 674—675
Steering lubricant, 671
Steering ratio, 659—660
Steering shaft, 658
Steering system, 18, 658—667
    function, 20
    manual, 660
    power, 660, 664—665
    preventive maintenance, 671—672
    problem diagnosis, 672—674
    response, 664
    safety precautions in servicing, 671
    servicing, 674—675
    types, 20, 658
Steering wheel, 20
    locking, 744
Steering wheel kickback, 673
Steering wheel surge, 674
Stellite, 125
Step-down transformer, 286
Step-up transformer, 286

Stethoscope, technician's, 474
Stoichiometric air-fuel mixture, 180, 411, 443—444
Stop lights, 392
Stop pins, 555
Storage containers, 35
Straight control arm and strut rod suspension system, 632
Straightedge, 66, 485
Stratified charge engine, 446
Streamlining, 99
Stress, defined, 105
Stripping, 403
Stroke
    defined, 86
    determining, 94
    length of, 94
Strokes per cycle, 91
Strut rods, 648
Struts, 118
Stub frame, defined, 19
Studs, defined, 46
Substrate, defined, 452
Sulfation, 292, 293, 294
Sulfur, 177, 452, 453
Sulfur, oxides of, 431, 452, 464
Sulfuric acid, 177, 267, 452, 453
    in batteries, 290—291
Sun gear, 569, 570
Superchargers, 187—189
    Rootes-type, 188—189
    servicing, 201
Supercharging, 186, 187—189. *See also* Turbocharging
Superimposition, 373
Supervisors, 7
Surface area, 696
Surface plate, 481
"Surge tank," 333
Surging, 226
Suspension, 18, 19
Suspension system, 625—657
    front, 629—633
    functions, 625
    inspecting, 647
    preventive maintenance, 644—646
    problem diagnosis, 646—649
    rear, 633—638
    safety precautions in servicing, 644, 646, 647, 648, 649, 654
    servicing, 644—657

    springs, 625—627
    troubleshooting, 644—646
Sway bar. *See* Stabilizer bar
Swerving during acceleration, 610
Swirl, 175
Switches
    electronic, 354—356
    function, 276
    heavy-duty, 385
    mechanically operated, 352—354
    mercury, 392—393
    positions, 276
    on stalk, 387
    starter motor, 312
    transformers as, 287
Synchronizers, 529—531
    action of, 529—530
    neutral position, 530

T

Tachometer, 221, 324
Taillights, 392, 743
Tailpipe, 265
Tampering, 455—456
Tang, defined, 56, 108
Tap wrenches, 57
Taps, defined, 57
TDC. *See* Top dead center
Technician's stethoscope, 474
Telescoping gauges, 68
Temperature sensors, 210, 384, 385, 444, 459—460
Temperature warning systems, 157
Temperature-compensated accelerator pump, 214
Tension gauge, 164
Terminals, 276—277
Test drive
    following troubleshooting charts, 77
    procedures, 75, 77
    safety precautions, 40, 75, 77
Test instruments, for electrical system, 401—402
Test lamps, 71
Test light, 401
Testing by substitution, 373, 402
T-handle (socket wrench), 54
Thermal efficiency, 99
Thermal gauges, 390
Thermodynamics, 82, 99
Thermostat
    connected to choke plate, 210

cooling system, 155, 157
    replacing, 171
Thermostat gasket, leaks, 166
Thermostatic coil, 155
Thermostatically controlled air cleaner, 444, 459—460
Thermo-vacuum valves, 448
Thimble (micrometer), 67
Thread
    defined, 44
    external, 45
    internal, 45
    size designations, 46—47
Thread gauge, 47
Thread pitch, 47
Thread-locking compounds, 48
Threaded fasteners, 44—48
    chemical compounds for, 48
    diameters, 46—47
    size designations, 46—47
    thread pitch, 47
Three-phase AC current, 329—330
Three-quarter floating axle, 603
Three-way catalyst, 452
Three-way catalytic converter, 262
Throttle body, 181, 221
Throttle bore, 180
Throttle linkage, adjustment procedure, 422
Throttle plate, 204, 360, 438, 440, 445
    control of closing rate, 215
    control of position of, 215
    primary, 212
Throttle position sensor, 217
Throttle positioner, 445
Throttle pressure, 577
Throttle valve, 204, 577
Throttle valve cable, 574
Throttle-return dashpot, 215
Throw-off, 91
Throwout bearing, 556, 565
Throws
    angle of, and firing order, 104
    defined, 103
    number of, 106
Thrust bearings, 109
Tie rods, 20, 659
Timing chain, 508
Timing gear, 105
    overhead camshaft, 124
    replacing, 508

Timing light, 372
Timing marks, 359, 372, 379, 482
Tire plugs, 693
Tire pressure gauges, 69
Tire valves, 681—682
Tire-changing machine, 692
Tires, 18, 22, 681—685
    balancing, 690—691
    calculating rolling radius of, 600
    changing, 692
    checking pressure, 689
    coding marks, 683—684
    compact spare, 694
    construction, 681—685
    contact patches, 625
    function, 681
    materials, 682—683
    maximum air pressure, 681
    measuring inflation of, 69
    parts, 682
    preventive maintenance, 688—690
    pressure, recommended, 688—689
    problem diagnosis, 691
    proper inflation, 688—689
    puncture repair methods, 693—694
    puncture-sealing, 694
    retreaded, 685
    rotation, 690
    runout, 691
    safety precautions in servicing, 683, 685, 688
    safety systems, 742
    sizes, 22, 684—685
    speed rating, 683—684
    suspension system and, 625
    temporary use, 685
    tread design, 683
    types, 683
    unrepairable, 693
    visual inspection, 689
Toe-in, 668, 677—678, 679
Toe-out, 668, 677—678
Tools, proper choice and use of, 35, 36. *See also* Hand tools; Power tools; and names of specific tools
Top dead center (TDC), 86
Top-feed injector, 237
Torque
    defined, 10, 14, 81, 96
    driveline and, 512
    effects of, 513
    force and, 96
    formula for determining, 96
    on gears, 526
    horsepower in relation to, 96
    increasing/decreasing, 16
    manual transmission and, 528
    power and, 95

rear-end, 513, 514, 633
starter motor and, 308
starter pinion gear and, 307
transmission and, 525
transmission of, 17
units of measurement for, 96
universal joints and, 513

Torque arms, 633

Torque converter, 568—569, 579
draining ATF from, 588
function, 17
installing drainplugs in, 588
parts, 568

Torque converter housing, 568

Torque specifications, for fasteners, 47

Torque steer, 538

Torque wrenches, 54, 610

Torque-tube drive, 514

Torqued, defined, 44

Torquing-over, 536

Torsion bar suspension system, 627

Torsion bars, 627

Torsional coil springs, 555

Torx bolts, 45

Torx screwdrivers, 55

Torx wrenches, 54

Total engine displacement, defined, 94

Trace minerals, 286

Trace patterns, 373

Tracer, defined, 384

Traces, oscilloscope, 343

Traction, defined, 681

Traction wheels, 690

Trailing arm, 634—635

Tramp, 690

Transaxle. *See also* Automatic transaxle; Manual transaxle
functions, 15, 16—17
parts, 16

Transaxle case, 543

Transfer case, 604
defined, 18
servicing, 621—622

Transfer port. *See* Off-idle port

Transformers, 285—286, 350—351

Transistors, 286—287, 334

Transitional port. *See* Off-idle port

Transmission, 15. *See also* Automatic transmission; Manual transmission

function, 16—17, 525
safety precautions in servicing, 541, 543, 544, 546, 549, 552

Transmission bands, 571, 592

Transmission case, 543

Transmission coolers, 578

Transmission crossmember, 546

Transmission fluid. *See* Automatic transmission fluid

Transmission input shaft, 528

Transmission jack, 552

Transmission oil. *See* Automatic transmission fluid

Transmission plug, 522

Transmission torque multiplication factor (TMF), 534

Transverse leaf rear springs, 637

Transverse torsion bar suspension system, 632

Tread, 682, 683

Tread wear, 689—690

Tread wear indicator, 690

Trim height, 646, 648

Trip computer, 408, 412

Tripod (tripot) joint, 538

Trouble codes, 248—249, 417, 419

Troubleshooting, 75—78
charging system problems, 340
defined, 75
EFI systems, 248—249
electrical system, 399—402
ignition system, 366
pinpointing problem, 75, 77—78
rechecking after repairs, 78
steps in, 75

Troubleshooting charts, 75, 77, 249

Trunnions, 516

Truss rods, 514

Tubeless tires, 681

Tubing wrenches, 53

Tune-up
diagnosing need for, 370—373
emissions-legal, 456
services included in, 366

Turbine, 186, 567—568

Turbine-type pump, 234

Turbo lag, 186

Turbocharger, 186—187
exhaust connections, 261
lubrication problems, 200
oils for, 136
servicing, 200

Turbocharging, 186—187

Turbulence, 122, 123, 175
    engine speed and, 359
    improving, 445
Turn signal flashers, 384, 385, 392, 742
Two-stroke cycle, 91
Two-way catalyst, 452

U

U-joints. *See* Universal joints
Underbody, 19
Undercharging, 338—339
Underinflation, 689
Uneven firing engines, 104
Unified System (of measurement), 46
Uniform Tire Quality Grading (UTQG) system, 685
Unitized body, 18—19
Universal joints, 513, 514
    conventional, 516, 523
    defined, 18
    double Cardan, 516
    inspecting, 520
    lubricating, 520—521
    parts, 516
    removing/replacing, 521—523
    rusted, 520
    socket wrenches, 54
Unpowered light test, 401
Unsprung weight, 625
Upshifting, 533
Uses of engines, classification by, 92

V

Vacuum, ported, 215, 360
Vacuum advance mechanism, 360, 372, 447—448
Vacuum check valves, 448
Vacuum control systems, 459
Vacuum delay valves, 448, 458—459
Vacuum diaphragm, 360
Vacuum diaphragm testers, 70
Vacuum gauge, 70, 458
Vacuum hoses, 458
Vacuum leaks, 195—196, 457—458
Vacuum modulator, 577, 592—593
Vacuum motor, 444, 460
Vacuum power brake booster, 711
Vacuum power brake system, 703, 721

Vacuum pump, 458
Vacuum recovery system, canister replacement, 460
Vacuum relief valve, 441
Vacuum retard mechanism, 360, 447—448
Vacuum tester, 221
Vacuum tests, 469—470
Vacuum valves, 458, 459
Vacuum vent, 215
Vacuum-operated control system, for emission control, 439—440
Vacuum/pressure gauge, 70, 391
Valve(s), 125—126
    arrangement of, 90
    check, 443
    construction, 125
    defined, 11
    design, 125
    function, 84
    leaking, 472
    measuring, 485
    number of, 90
    opening, 84
    parts, 125
    poppet, 125
Valve adjusting screw, 123
Valve adjustment shim, 123
Valve body, 575
    oil pan filter/screen, 588—589
Valve clearance, 127
    adjustments, 478
Valve cover
    applying sealant to, 147
    bolt removal pattern, 477
    removing, 145—146
Valve cover gasket, replacing, 145—147
Valve float, 125
Valve guide(s), 121, 126—127, 485
    checking, 485
    reaming out, 485
Valve guide clearance, 481
Valve guide inserts, 485
Valve head, 125
Valve lash, 127
Valve lifter, 127—128, 478
    concave, 487
    hydraulic, 127—128
    removing, 483
    solid, 127
Valve locks, 126
Valve overlap, 130
Valve rotators, 126

Valve seat(s), 125
    integral, 125
    machining, 486
    width, 486
Valve seat inserts, 121, 125
Valve spring(s), 126
    checking, 481
    removing, 478—481
Valve spring assembly, parts, 126
Valve spring compressor, 480
Valve spring seat, 126
Valve spring surge, 126
Valve stem
    checking for wear, 485
    clearance around, 126—127
    seal, 126
Valve timing and duration
    determining, 130
    diagrams, 130
Valve train, 84
    function, 11
    overhead camshaft, 123
    pushrod, 125
    types, 90
    visual inspection, 477—478
Valve-in-head design, 90
Vane-type power steering pump, 462, 661
Vapor(s)
    dangerous, 33
    noticeable during hard cornering, 461
    at tailpipe, 267
Vapor/fuel separator, 443
Vapor lines, replacing, 194, 458
Vapor lock, 179
Vapor recovery canister, 443
Vapor recovery system, 214—215, 440—441
    components, 441
    filter replacement, 460
    problems, 460—461
Vaporization, 206
    improving, 445
Variable restriction jets, 205
Variable steering ratio, 663
Variable-width pulleys, 582
V-belts. *See* Drive belts
Ventilation system, 33, 38, 726
Venturi
    booster, 180
    defined, 180, 440
    multiple, 211—212
Venturi vacuum, 440
Vernier calipers, 69

Vibrations, 691
Viscosity, defined, 87, 574
Viscosity ratings, 134—135
Viscous coupling, 607
Viscous drive fan clutch, 155
Viscous friction, 97—98
Visual inspections, 77—78
Voice simulators, 407
Voice warning device, 407, 743
Volatility, of diesel fuel, 185
Volkswagen, 232
Volt, 282
Volta, Alessandro, 282
Voltage
    checking, 399
    measurement of, 71
    testing, 371
Voltage drop, defined, 283
Voltage drop test, 318—319, 343
Voltage Law, 283
Voltage limiter relay, 334
Voltage regulator
    electromechanical, 339
    for gauges, 388
    function, 14, 333
    overcharging due to, 339
    transistorized, 334, 339
    types, 333—335
    undercharging and, 339
    visual checks, 340
    wiring connection checks, 339—340
Voltage spikes, 342
Voltmeter, 71, 290, 301, 336, 402
Volumetric efficiency, 98
Vortex flow air measurement system, 240
V-type engines, exhaust connections, 261
Vulcanized, defined, 682

W

Wages, 7
Warning devices, 387, 407, 743
Warning lights, 387, 702—703, 742, 743
Warranty, defined, 3
Warranty repairs, defined, 3
Washers, 46
"Wasted spark," 362
Wastegate, 186
Water jacket, 151, 155

Water pump, 12, 152—153
  leakage checks/repairs, 169—170
  removing, 169—170
  replacing, 171
Water pump gasket, 166—167
Water sensors, 199
Water separators, 199
Water sludge, 144
Water-cooled, defined, 12
Watt, James, 96, 98, 282
Watt (w), 81, 98, 282, 283
Watt-hour rating, 295
Wear indicators, 647
Weather. *See also* Cold weather
  engine oil viscosity ratings and, 135
Webbing, 748
Weep hole, 169
Welding, defined, 263
Welding equipment, safe use of, 40
Wet compression tests, 472
Wet liners, 103
Wheel(s), 18, 22, 685—686
  inspecting, 689, 694
  removing, 692
  replacement, 694
  runout, 691
  sizes, 22
Wheel alignment, 667—668, 676—679
  equipment, 676
  front-end, 667—668, 677—679
  prealignment checks, 676
  rear-end, 679
Wheel balancing, 690—691
Wheel bearings, 645—646
  inspecting, 647
  problem indications, 647
  repacking, 645
Wheel calipers, 696
Wheel chocks, 34
Wheel cylinders, 20, 696, 720
Wheel hub, 686

Wheel lockup, 705
Wheel lugs, 686
Wheel speed sensors, 705, 721
Wheelbase, defined, 515
Whip, defined, 106
Wide-open throttle (WOT), 439
Windings, 308
Windshield washers, 393—394
Windshield wipers, 393—394, 744
Wire buffing wheels, 61
Wire gauges, 67
Wires/wiring, 382, 384
  AWG size, 404
  bundled, 382
  bypass, 400
  color coded, 384
  common ground, 382
  connections, 399
  connector plugs, 417
  inspecting, 400
  joints, 403
  loose or corroded, 339
  replacing, 402—403, 404
  soldering, 403
  splicing, 400
  terminals, 403
Wiring diagrams, 384, 400
Work, defined, 81, 96
Work areas, safety precautions, 33
Work orders. *See* Repair orders
Work schedules. *See* Dispatch sheets
Worm-drive clamp, 166
Wrenches, 52—54
  air impact, 61—62
  electric impact, 60
Wrist pins. *See* Piston pins

Y

Yokes, 516